S0-AJW-003

HANDBOOK OF HUMAN FACTORS AND ERGONOMICS IN HEALTH CARE AND PATIENT SAFETY

HUMAN FACTORS AND ERGONOMICS

Gavriel Salvendy, Series Editor

Aykin, N., (Ed.): *Usability and Internationalization of Information Technology.*

Bagnara, S. (Ed): *Theories and Practice in Interaction Design*

Carayon, P. (Ed): *Handbook of Human Factors and Ergonomics in Health Care and Patient Safety*

Hendrick, H., and Kleiner, B. (Eds.): *Macroergonomics: Theory, Methods and Applications.*

Hollnagel, E. (Ed.): *Handbook of Cognitive Task Design.*

Jacko, J. A., and Sears, A. (Eds.): *The Human-Computer Interaction Handbook: Fundamentals, Evolving Technologies and Emerging Applications.*

Karwowski, W., (Ed.): *Handbook of Standards and Guidelines in Ergonomics and Human Factors.*

Meister, D., (Au.): *Conceptual Foundations of Human Factors Measurement.*

Meister, D., and Enderwick, T. (Eds.): *Human Factors in System Design, Development, and Testing.*

Proctor, R., and Vu, K. (Eds.): *Handbook of Human Factors in Web Design.*

Schmorrow, D. (Ed.): *Foundations of Augmented Cognition.*

Stanney, K. (Ed.): *Handbook of Virtual Environments: Design, Implementation, and Applications.*

Stephanidis, C. (Ed.): *User Interfaces for all: Concepts, Methods, and Tools.*

Wogalter, M. (Eds.): *Handbook of Warnings.*

Ye, N. (Ed.): *The Handbook of Data Mining.*

Also in this Series

HCI International 1999 Proceedings 2 Volume Set
HCI International 2001 Proceedings 3 Volume Set
HCI International 2003 Proceedings 4 Volume Set
HCI International 2005 Proceedings 11 Volume CD Rom Set ISBN# 0–8058–5807–5

HANDBOOK OF HUMAN FACTORS AND ERGONOMICS IN HEALTH CARE AND PATIENT SAFETY

EDITED BY
PASCALE CARAYON
UNIVERSITY OF WISCONSIN, MADISON

2007
LAWRENCE ERLBAUM ASSOCIATES, PUBLISHERS
Mahwah, New Jersey London

Lawrence Erlbaum Associates, Inc., Publishers
10 Industrial Avenue
Mahwah, NJ 07430

Cover design by Tomai Maridou

Library of Congress Cataloging-in-Publication Data

Handbook of human factors and ergonomics in health care and patient safety / edited by Pascale Carayon.

p. cm.
Includes bibliographical references and index.
ISBN 0–8058–4885–1 (cloth : alk. paper)

1. Medical errors—Prevention—Handbooks, manuals, etc. 2. Human engineering—Handbooks, manuals, etc. 3. Patients—Safety measures—Handbooks, manuals, etc. 4. Health facilities—Design and construction—Handbooks, manuals, etc. I. Carayon, Pascale.

[DNLM: 1. Medical Errors—prevention & control. 2. Human Engineering. 3. Safety Management. WB 100 H2355 2007]

R729.8.H34 2007

610.28'9—dc22 2006020119

Books published by Lawrence Erlbaum Associates are printed on acid-free paper, and their bindings are chosen for strength and durability.

Printed in the United States of America
10 9 8 7 6 5 4 3 2 1

CONTENTS

SERIES FOREWORD

Gavriel Salvendy, Series Editor

Purdue University and Tshingua University, China

In the year 2000, the Institute of Medicine (IOM) published a report on "To Err is Human: Building a Safer Health System" that attracted much attention from health care institutions in the U.S. and all over the world. According to data reported by the IOM, between 44,000 and 98,000 Americans die each year as a result of medical errors. The report goes on to say that more people die every year as a result of medical errors than from motor vehicle accidents, breast cancer, or AIDS. This handbook presents state-of-the-art knowledge on the application of human factors and ergonomcs to medical errors, and to quality and safety of care.

With the rapid introduction of highly sophisticated computers, telecommunications, new ways of delivering patient care services, and innovative organizational approaches in the health care industry, a major shift has occurred in the way patients receive care and in the way health care staff use technology to deliver care. The objective of this book series on Human Factors and Ergonomics is to provide researchers and practitioners in health care a platform where important issues related to these changes can be discussed. Methods and recommendations are presented which will ensure that emerging technologies and work organizations will provide increased productivity, quality, satisfaction, safety, and health in the new workplace and in the Information Society.

The present volume is published at a very opportune time. Information Society Technologies are emerging as a dominant force, both in health care workplaces and in everyday life activities. In order for these new technologies to be truly effective and to be used safely, they must provide communication modes and interaction modalities that serve across different languages and cultures. The new technologies must also accommodate the diverse requirements of the user population at large, including disabled and elderly people. This will make the Information Society universally accessible, to the benefit of all mankind.

The present handbook provides a comprehensive account on how to manage and improve health care delivery in order to minimize medical errors and maximize patient satisfaction and well being in an economically responsive way. Hence, the effective use of this handbook may result in lower medical error which could reduce malpractice suits and thus reduce the cost of health care delivery. The handbook also provides invaluable information for reducing stress and workload and increasing the well being of health care personnel.

The 96 authors of this handbook include 72 from academia, 26 from the health care industry, and 7 from governmental agencies. These individuals are some of the very best and most respected in the field. The 131 tables, 159 figures, and 3,424 references provide an insightful presentation of the subject with pointers for further insight study. The handbook is especially invaluable to health care employees and managers, human factors specialists, hospital administrators, and health care policy makers.

FOREWORD

Peter J. Pronovost

Johns Hopkins University
Department of Anesthesiology and Critical Care Medicine.

In the six years since the Institute of Medicine released its landmark report, "To Err is Human," progress has been slow and arduous toward improving patient safety. Caregivers and researchers are struggling to advance the science of patient safety, understand its epidemiology, clarify priorities, implement scientifically sound yet feasible interventions, and develop measures to evaluate progress. As Robert Frost said "we have miles to go before we sleep."

Though errors have become more visible, our patients continue to suffer preventable harm and this makes patients, regulators, accreditators, and caregivers increasingly frustrated. While there is broad consensus that faulty systems, rather than faulty people, are the cause of most errors, the health care community struggles to find practical and scientifically sound ways to address and mitigate hazards. We are increasingly aware that health care information technology (HIT) and devices, believed by many to be a panacea for patient safety, are a double edged sword. These technologies, while defending against some mistakes, invariably introduce new mistakes and new harm. This reality does not impugn technology but rather impugns the methods by which we develop and introduce technology or other system changes into health care. Any system change may defend against some mistakes, but will invariably introduce new ones. The underlying problem with technology may lie, not in the operation, but in the design that is often invisible to the operator of the equipment.

Strategies that minimize these types of mistakes through the design, operation and monitoring of technologies, often called Human Factors Engineering (HFE), are well developed, though nascent, in their application to health care. This book seeks to accelerate the extent to which HFE are applied in health care, both conceptually and practically.

The *Handbook of Human Factors and Ergonomics in Health Care and Patient Safety*, edited by Dr. Pascale Carayon, provides a conceptual framework and practical strategies to more fully apply human factors principles in health care and improve patient safety. This book gives us the detail about human factors research, theories, and tools. It demonstrates, through examples, how the theories and tools have been applied. The authors are experienced researchers and practitioners of HFE in health care and their collective wisdom is an excellent 'field guide' for HFE. The book provides important information for health care safety and for quality leaders, researchers and practitioners. Perhaps, as the principles and tools discussed in this book become widely applied in health care, we will begin to realize significant improvements in patient safety. Indeed, we may proactively design technologies and a health care system that prevents most errors and defends or minimizes the harmful events that continue to occur.

PREFACE

As described by Dr. Pronovost in his foreword to this handbook, the discipline and profession of human factors and ergonomics (HFE) can help healthcare organizations, professionals and institutions to improve the design of healthcare systems and processes as well as the implementation of changes, and to prevent or mitigate medical errors. These efforts require active collaboration between HFE, on one hand, and healthcare providers, professionals and leaders, on the other hand. Both groups can benefit from reading this handbook by learning about HFE in health care and patient safety, HFE tools and intervention strategies, as well as specific applications of HFE in various healthcare settings.

The book is a multi-authored edited handbook. It contains 51 chapters which are organized in 9 sections: (1) Introduction, (2) Macroergonomics and Systems, (3) Job and Organizational Design, (4) Physical Ergonomics, (5) Technology, (6) Human Error, (7) Human Factors and Ergonomics Methodologies, (8) Human Factors and Ergonomics Interventions, and (9) Specific Applications. The nine sections cover theory, research, tools and methods, applications, and various perspectives on HFE in health care and patient safety.

The chapter authors are highly respected individuals in academia, health care, industry, and government. They come from Denmark, France, Germany, Japan, the Netherlands, the United Kingdom, and the USA. All of these authors contributed to the handbook at the invitation of the editor. Every chapter was reviewed by one or more reviewers (see below the list of reviewers) and the editor. An appreciative thank you is given to the reviewers for insightful comments and to the authors for producing excellent chapters.

—Pascale Carayon

Acknowledgments

This handbook would not exist if Mike Smith, of the University of Wisconsin-Madison, did not suggest the idea a few years ago. Not only did he contribute a chapter to this handbook, but he was also my Ph.D. advisor. He is a mentor, collaborator, and friend. A very special thank you goes to Mike Smith.

I also would like to thank my colleagues at the Center for Quality and Productivity Improvement and the Department of Industrial and Systems Engineering at the University of Wisconsin-Madison.

In particular, a very warm thank you goes to Carla Alvarado and Ann Schoofs Hundt for their support and friendship; I have learned a lot (and am still learning) from them regarding the specificity of health care and patient safety. Patti Brennan, Dave Gustafson, Bentzi Karsh, Mike Smith and David Zimmerman in the Department of Industrial and Systems Engineering have not only contributed to this handbook, but are also influential colleagues and close collaborators. A general thank you goes to my healthcare collaborators at various hospitals, healthcare systems, and schools of medicine, pharmacy and nursing at universities across the world.

I would also like to thank Gavriel Salvendy of Purdue University for his support during my academic career, as well as his encouragement from the very beginning of this project until the end. As the Human Factors/Ergonomics series editor at Lawrence Erlbaum and Associates, he was very instrumental in facilitating the creation of this handbook.

My greatest appreciation goes to the individuals associated with this project at Lawrence Erlbaum Associates, in particular Bill Webber who helped at the very beginning. Later, Anne Duffy took over Bill's position and was a marvelous person to work with. I would also like to thank Claudia Dukeshire, Margaret Rizzi, and Michelle White for their hard work, patience and help throughout this process.

Last but not least, a most sincere thank you goes to Peter, my parents, Laurence, Christian, Sophie, Camille, Clémence, Marie, Clara and Pierre-Dominique. Merci de tout coeur. Dank u wel.

List of reviewers

Carla Alvarado	Craig Harvey	Larry Murphy	Eric Williams
Claus Backhaus	Sue Hignett	Todd Molfenter	Yan Xiao
Sue Bogner	Ann S. Hundt	Audrey Nelson	Klaus Zink
Caroline Cao	Bentzi Karsh	Wilmar Schaufeli	
Frank Drews	Brian Kleiner	Bruce Thomadsen	
Nancy Elder	Kari Lindstrom	Eric Thomas	
John Gosbee	Ingo Marsolek	Bob Wears	

HANDBOOK OF HUMAN FACTORS AND ERGONOMICS IN HEALTH CARE AND PATIENT SAFETY

Section

I

INTRODUCTION

1

HUMAN FACTORS AND ERGONOMICS IN HEALTH CARE AND PATIENT SAFETY

Pascale Carayon
University of Wisconsin, Madison

Recently, much emphasis has been put on human factors approaches to patient safety (Bogner, 1994; Cook, Woods, & Miller, 1998; Leape, 1994; Wears & Perry, 2002). The health care field has embraced the various models and approaches to human error and safety to analyze and evaluate risk and improve the quality and safety of care provided to patients in various institutions and venues (Reason, 2000; Vincent, Taylor-Adams, & Stanhope, 1998). In Ergonomics in Design in 2004, Lucian Leape wrote: "Given the complexity of health care and the formidable obstacles it presents to change, to overcome those barriers and create a safe culture does indeed seem to be the ultimate challenge for those who specialize in human factors" (p.11). This handbook of human factors and ergonomics (HFE) in health care and patient safety addresses the challenge described by Leape by presenting and discussing a variety of HFE issues, concepts, and methods that can help understand, identify, mitigate, and remove the obstacles to safe health care. It includes nine sections with 51 chapters organized in sections:

1. Introduction.
2. Macroergonomics and Systems.
3. Job and Organizational Design.
4. Physical Ergonomics.
5. Technology.
6. Human Error.
7. Human Factors and Ergonomics Methodologies.
8. Human Factors and Ergonomics Interventions.
9. Specific Applications.

DEFINITION OF HFE

According to the International Ergonomics Association (IEA; 2000), ergonomics or human factors can be defined as follows:

Ergonomics (or human factors) is the scientific discipline concerned with the understanding of interactions among humans and other elements of a system, and the profession that applies theory, principles, data and methods to design in order to optimize human well-being and overall system performance.

In this definition, the *system* represents the physical, cognitive, and organizational artifacts that people interact with. The system can be a technology or device; a person, a team, or an organization; a procedure, policy, or guideline; or a physical environment. Interactions between people and the systems are tasks. Ergonomists are concerned with the *design* of systems to make them fit the needs, abilities, and limitations of people (IEA, 2000; Meister & Enderwick, 2001). As suggested by Leape and colleagues (1995) and many other experts and practitioners in health care and patient safety (Bogner, 1994; Carayon, 2005; Cook et al., 1998), the discipline of human factors can very much contribute to the safe *design* of health care systems by considering the various needs, abilities, and limitations of people involved in those systems. The quality and safety of care provided by health care systems are, of course, dependent on the patients' risk factors and the technical skills and knowledge of the health care staff but are also very much influenced by various characteristics of the system that can be manipulated, changed, and improved with HFE principles and methods (Vincent, Moorthy, Sarker, Chang, & Darzi, 2004). In health care, the "people involved" are diverse and include health care providers and workers, patients, and their families. People have varied needs, abilities, and limitations that change over time. It is important to recognize the diversity of the discipline of HFE in attempting to deal with those needs, abilities, and limitations.

Domains of HFE

The discipline of HFE covers three major domains: (a) physical ergonomics concerned with physical activity, (b) cognitive ergonomics concerned with mental processes, and (c) organizational ergonomics (also called macroergonomics) concerned with sociotechnical systems. Physical ergonomics focuses primarily on the physical characteristics of the person, cognitive ergonomics on the cognitive characteristics of the person, and organizational ergonomics on the psychosocial characteristics of the person. A number of topics relevant to each of the HFE domains, as well as examples of application in health care and patient safety, are presented in Table 1–1. Important physical HFE issues in health care and patient safety, such as the design of hospital facilities, the design of physical environment, patient handling, patient room design, noise, and alarms are addressed in the section Physical Ergonomics. Physical ergonomics issues are also relevant for specific applications in health care, such as intensive care environments and emergency rooms. The auditory environment in intensive care units and the impact of poor ergonomic design of alarms on health care providers' performance and stress are described in the chapter on noise and alarms in health care (see Bass & Friesdorf, chap. 22, this volume). Methods for assessing physical ergonomic factors in health care and for designing and implementing interventions aimed at reducing physical stress are addressed by several chapters in the Physical Ergonomics section and the Human Factors and Ergonomics Methodologies section.

A major cognitive issue in health care and patient safety that has received much attention is human error. The Human Error section includes chapters on complexity of patient care (Woods, Patterson, & Cook, chap. 29, this volume), taxonomies of medical failures (Thomadsen, chap. 30, this volume), human factors in the design of reporting systems (Johnson, chap. 31, this volume), and communication about unexpected errors and outcomes (Pichert, Hickson, & Vincent, chap. 32, this volume). Various approaches for human error reduction strategies are described in one chapter in the section on Human Factors and Ergonomics Interventions (Amalberti & Hourlier, chap. 32, this volume). There is increasing understanding among health care leaders and managers of the contribution of latent failures (i.e., those failures that lay dormant for a long time and that are removed in space and time from active failures; see Reason, 1990) and organizational factors to medical errors and patient safety. The HFE discipline has had a major impact in health care to help those health care leaders and managers understand the human mechanisms involved in medical errors and the influence of system characteristics on human behavior and human error (Bogner, 1994).

The design and implementation of technologies in health care have raised various cognitive ergonomics issues such as usability of medical devices and cognitive workload related to robotics and various types of information and communication technology. These issues are addressed by chapters in the Technology section of this handbook.

Various organizational ergonomics issues such as job stress and burnout of health care workers, organizational culture and learning, teamwork, and organizational design (e.g., transitions of care) are addressed in the Job and Organizational Design section. Multiple organizational ergonomics issues relevant to applications of HFE to specific health care contexts are addressed in the last section of the handbook. For instance, the challenges related to the family as a complex organization, such as roles and responsibilities for health information management, are highlighted in the chapter on HFE in home care (Zayas-Cabán & Brennan, chap. 49, this volume). Issues of staffing (e.g., understaffing and nursing shortage) and turnover and the impact of poorly designed working conditions on nursing home staff attitudes and behavior are discussed in the chapter on HFE in nursing home care (Zimmerman & Antonova, chap. 50, this volume).

TABLE 1–1. Domains of Human Factors and Ergonomics (HFE) in Health Care and Patient Safety

Domains of HFE	*Relevant Topics*	*Examples of Application to Health Care*	*Examples of Application to Patient Safety*
Physical ergonomics	Working postures Materials handling Repetitive movements Work related musculoskeletal disorders Workplace layout Safety and health	Reducing and preventing back injuries among nurses Designing workstations and work rooms (e.g., medication preparation room) for optimal human performance	Designing a patient room to facilitate and support safe patient care Designing medication labels so they are readable and understandable
Cognitive ergonomics	Mental workload Decision making Skilled performance Human–computer interaction Human reliability Work stress and training as these may Relate to human–system design	Evaluating the usability of technologies Designing training systems Designing usable interfaces for health information technologies	Designing an event reporting system Creating and implementing incident analysis processes
Organizational ergonomics	Communication Crew resource management Work design Design of working times Teamwork Participatory design Community ergonomics Cooperative work New work paradigms Virtual organizations Quality management	Designing healthcare jobs for reducing stress and burnout and improving satisfaction and retention Implementing improvement activities that consider HFE principles of teamwork and participation	Implementing crew resource management training in surgery teams Designing work schedules for reduced fatigue and enhanced performance

The diversity of clinical needs experienced by patients and the organizational design consequences such as difficulty in standardizing processes and the diversity of people, locations, and systems involved in the care of patients are described in the chapter on HFE in primary care (Beasley, Hamilton, & Karsh, chap. 51, this volume). Organizational ergonomics issues such as human factors risk management in medical products, assessment of safety culture and climate, and incident analysis are addressed by several chapters in the Human Factors and Ergonomics Methodologies section.

From Micro- to Macroergonomics

The diversity of topics addressed by HFE researchers and practitioners is clearly demonstrated in this handbook. Historically, the HFE discipline has developed to encompass an increasing number and type of interactions between people and systems. Hendrick (Hendrick, 1991, 1997; Hendrick & Kleiner, 2001) described five "human-system interface technologies" that have developed in the HFE discipline over time:

- Human–machine interface technology or hardware ergonomics.
- Human–environment interface technology or environmental ergonomics.
- Human–software interface technology or cognitive ergonomics.
- Human–job interface technology or work design ergonomics.
- Human–organization interface technology or macroergonomics.

In health care, hardware ergonomics is concerned with the design of controls (e.g., controls in a telemetry monitoring unit), displays (e.g., display of an anesthesia machine), workspaces (e.g., nursing station, patient room), and facilities (e.g., hospital, nursing home). Hardware ergonomics draws on knowledge, concepts, and methods of physical and cognitive ergonomics primarily. Environmental ergonomics issues in health care include noise, temperature, humidity, airflow, and vibration and draw mainly from the physical ergonomics domain. Human–software interface issues in health care are varied given the increasing number and diversity of devices, equipment, and technologies implemented and used. They include usability and various cognitive ergonomics issues (e.g., information overload, consistency of presentation). Work design ergonomics in health care include issues related to work schedule and work content. Macroergonomic issues in health care and patient safety are varied, as demonstrated by many of the chapters of this handbook, in particular in the Macroergonomics and Systems section and the Job and Organizational Design section. They include teamwork, organizational culture and learning, clinical microsystems, work systems, sociotechnical systems, high-reliability organizations, and collaborative initiatives.

HFE: DESIGN FOR PEOPLE

HFE professionals aim at understanding the interactions among humans and other elements of a system and use their knowledge to improve the design, implementation, and use of various systems. *Humans* in health care systems include the providers and workers either directly or indirectly involved in patient care as well as patients and their families and friends as highlighted in the first section of this chapter. Three categories of human characteristics need to be considered in any HFE effort: (a) physical, (b) cognitive, and (c) psychosocial characteristics.

Physical Characteristics of People

Working environments in health care can impose many physical stressors on the health care providers and workers as well as on patients and families. According to the Bureau of Labor Statistics (2000), nursing is an occupation at high risk for work-related musculoskeletal injuries. Designing the health care workplace for optimal human performance and using technologies such as lifting devices to minimize the need for human strength can help reduce the physical stressors to which nurses are exposed. Physical stressors experienced by health care workers, their causes, methods for measuring them, and approaches for either removing them or reducing their impact on workers are described in several chapters in the handbook.

To better design health care systems from a physical viewpoint, several physical human characteristics need to be understood: height, weight, reaching envelop, physical strength, physical movement, and sensory characteristics (e.g., vision, hearing).

For instance, a technology such as bar coding medication administration (BCMA) is often implemented on a personal digital assistant (PDA). Several physical characteristics of the BCMA technology users should be considered, such as the size of their hands for holding the PDA and their visual capacity in reading small font size on the PDA screen. Such technologies should be designed for the full range of physical human characteristics.

Physical human characteristics are also important for designing the physical layout of hospital units. The following ergonomic principles should be considered in health care workspace layout design: to minimize detection and perception time, to minimize decision time, to minimize manipulation time, and to optimize opportunity for movement (Carayon, Alvarado, & Hundt, 2003). These principles consider physical characteristics of people, such as reaching envelop in the design of a workstation, to allow opportunity for movement and reduce static physical loading. These principles also rely on knowledge of cognitive characteristics of humans, such as information processing and reaction time, to minimize perception time and decision time.

Cognitive Characteristics of People

Cognitive human characteristics of importance in health care and patient safety include information processing and decision making, knowledge and expertise, and human error. Information processing and decision-making models (Klein, Orasanu, & Calderwood, 1993; Rasmussen, Pejtersen, & Goodstein, 1994; Wickens, Lee, Liu, & Becker, 2004) provide useful information on cognitive human characteristics, for example, limitation of attention capacity, decision-making heuristics, and characteristics of short- and long-term memory. The usability of medical devices and associated instructions and training materials relies very much on knowledge regarding the cognitive characteristics of end users. Various usability methods that can be used to evaluate medical devices and other health care technologies are described in the chapter on usability evaluation (Gosbee & Gosbee, chap. 38, this volume).

The design of medical devices and technologies can be improved by considering the cognitive strengths and limitations of end users. Additional information on those aspects of HFE in health care and patient safety is provided in the chapters on HFE and the design of medical devices (Ward & Clarkson, chap. 23, this volume) and on human computer interaction in health care (Drews & Westenskow, chap. 27, this volume).

Cognitive human characteristics are also important to consider when designing training programs and materials. For instance, mental models of end users can provide valuable information on how end users perceive specific tasks. The introduction of a technology changes the way tasks are performed, therefore requiring end users to develop a new mental model for their work. This need for the development of a new mental model should be addressed in the training offered to end users regarding the new technology (Salas & Cannon-Bowers, 1997). See, for example, the chapter on teamwork training for patient safety (Salas, Wilson-Donnelly, Sims, Burke, & Priest, chap. 44, this volume).

Understanding why humans make errors has benefited very much from theories and models based on cognitive ergonomics (Reason, 1990; Wickens et al., 2004). The well-known taxonomy of slips-lapses and mistakes reflects two different cognitive mechanisms: one more routine sometimes automatic mechanism that is susceptible to distractions and breaks in the routine and another mechanism that involves higher-cognitive processing and decision-making processes (Rasmussen et al., 1994; Reason, 1990; Wickens et al., 2004). More recently, experts have also highlighted the role of organizational factors in creating the conditions (i.e., latent failures) for human errors and accidents (Reason, 1997). Several chapters in this handbook propose various system models of human error, for instance, "The Artichoke Systems Approach for Identifying the Why of Error" (Bogner, chap. 7, this volume).

Psychosocial Characteristics of People

In the previous section of this chapter, we discussed how the HFE discipline evolved over time from micro- to macroergonomics and has integrated organizational issues in system design. The role of psychosocial human characteristics in human factors-based effort is now widely recognized. Motivation and satisfaction are important to consider when implementing any organizational and technological change in health care institutions. A historical perspective and a look into the future regarding quality improvement in health care are provided by a chapter in the handbook (Molfenter & Gustafsen, chap. 42, this volume). Using human factors knowledge, concepts, and methods can lead to more effective and efficient health care quality improvement efforts in health care. For instance, a key HFE principle is user involvement. This principle has been clearly highlighted in the area of information technology design and implementation (Carayon & Karsh, 2000; Eason, 1988, 2001). Understanding how end users can be involved in an organizational or technological change to foster and support their motivation and satisfaction is critical for the success and sustainability of health care and patient safety improvement activities.

Important knowledge, concepts, and methods that consider social needs and characteristics in, for instance, the design of teamwork has been produced by HFE. A range of best practices and guidelines for implementing teamwork in health care is described in the chapter on teamwork training for patient safety (Salas, Wilson-Donnelly, Sims, Burke, & Priest, chap. 44, this volume). Given the collaborative nature of much health care and patient safety activity, it is important to understand the social fabric of work systems. For instance, the physical design of a hospital need to consider not only the physical needs and requirements of the end users (e.g., patients, nurses, physicians) but also their social needs such as need for communication and social interaction. HFE can provide the principles and tools for designing health care systems that support and encourage the social needs of health care providers, patients, and their families.

Job stress and burnout are becoming increasingly important in health care institutions. This is for instance demonstrated by the nursing shortage that has been linked to poor working conditions and stressful work environments (Wunderlich & Kohler, 2001). Many of the psychosocial issues of importance to providers in stressful working environments, such as workload and time pressure, are listed in the chapter on the nursing home care (Zimmerman & Antonova, chap. 50, this volume). Descriptions of psychosocial outcomes such as job stress and burnout and useful information on approaches for dealing with and reducing those negative outcomes that are increasingly experienced by health care providers are provided in other chapters in the handbook.

The HFE discipline has increasingly paid attention to the way HFE knowledge, concepts, and methods are being implemented. It is not sufficient to understand the physical and cognitive characteristics of people and their physical and cognitive interface with systems. Although such knowledge is important for understanding the positive and negative system interface characteristics from a human factors viewpoint, to successfully change systems by implementing HFE changes, knowledge about psychosocial human characteristics and methods such as participatory ergonomics are necessary. Participatory ergonomics is a method that has been developed for helping the design and introduction of HFE efforts (Noro & Imada, 1991) in which end users are involved in the identification and analysis of HFE risk factors in their work system as well as the design and implementation of ergonomic solutions (Noro & Imada, 1991; Wilson & Haines, 1997).

The actual implementation of participatory ergonomics varies according to dimensions such as permanency (is the participatory ergonomics effort a one-shot effort or part of an on-going organizational effort?), focus (e.g., analysis of a task, workstation, or job), and decision making (e.g., individual consultation vs. group delegation; Haines, Wilson, Vink, & Koningsveld, 2002). A range of methods has been proposed for implementing participatory ergonomics such as Design Decision Groups (Wilson, 1991), quality circles and other quality improvement methods (Noro & Imada, 1991; Zink, 1996), and representative ergonomic facilitators (Haims & Carayon, 1998).

Bohr, Evanoff, and Wolf (1997; Evanoff, Bohr, & Wolf, 1999) conducted several studies of participatory ergonomics in health care with the objective of dealing with physical stressors that contribute to work-related musculoskeletal disorders. A key challenge identified in these studies was the high workload and time pressure related to patient care. These conditions make the involvement of workers (e.g., ICU nurses) more difficult and challenging. Participatory ergonomics approaches need to be developed and tested to effectively and efficiently involve health care workers. Those innovative methods need to consider the psychosocial benefits of participation (e.g., motivation and satisfaction of health care workers), while at the same time not overburdening health care workers who are already experiencing high workload and time pressure and have primary commitment to patient care.

Diversity of People in Health Care

The definition of human factors (or ergonomics) by the IEA does not make any specific assumption about the "humans." Much HFE research and practice have targeted the workers, for example, nurses, physicians, pharmacists, and other health care providers and staff (e.g., technicians, orderlies, maintenance personnel, biomedical engineers). Many chapters of this handbook describe HFE issues for the workers in health care systems. On the other hand, HFE knowledge, concepts, and methods can equally apply to the design of systems with which patients and their families interact. For instance, the viewpoint of both the providers and the patients and their families is addressed by the chapter in this handbook on communicating about unexpected errors and outcomes (Pichert, Hickson, & Vincent, chap. 33, this volume). Several HFE issues of relevance to children as patients are listed in the chapter on the specific application of HFE to pediatrics (Scanlon, chap. 48, this volume). Health care workers, patients, and their families all have physical, cognitive, and psychosocial characteristics that need to be considered when studying and designing the systems with which they interact. Sometimes different groups with different characteristics and different objectives exist in the same environment or interact with the same technologies. For instance, an infusion pump is used by a nurse who sets the medicine and the IV. The same IV pump is also part of the system in which the patient's medical condition evolves. The IV pump may be attached to a pole that limits the mobility of the patient (e.g., going to the bathroom). Health care system design is therefore challenging because of the variety of end user groups and their different system interactions to achieve various objectives (i.e., various tasks).

HFE SYSTEMS APPROACHES

An important element of the IEA definition of HFE relates to system design. This handbook has an entire section dedicated to HFE systems approaches in health care and patient safety. These chapters propose systems approaches and frameworks on work system design (Carayon, Alvarado, & Hundt, chap. 4, this volume), sociotechnical system design (Kleiner, chap. 5, this volume), clinical microsystems (Mohr & Barach, chap. 6, this volume), a system model of the factors contributing to errors (the "Artichoke" model; Bogner, chap. 7, this volume), and organizational learning as a system characteristic (Hundt, chap. 8, this volume). Another chapter addresses the conflicts and possible compromises between the professional medical model and systems approaches (Smith & Bartell, chap. 9, this volume). In the context of increasing decentralization and distribution of care over time, organizations, and systems, another chapter describes collaborative initiatives in patient safety (Carayon, Kosseff, Borgsdorf, & Jacobsen, chap. 10, this volume).

System Boundaries, Objectives, and Interactions

A key issue in system design is the definition of the system, that is, the objectives of the system and the boundaries of the system. The boundaries of the system can be physical (e.g., a patient room as the system) or organizational (e.g., a hospital or a department as the system). Guidance for understanding the various elements and characteristics of systems is provided by many of the systems approaches and frameworks included in this handbook. For instance, the work system model developed by Smith and Carayon (Carayon & Smith, 2000; Smith & Carayon, 2000; Smith & Carayon-Sainfort, 1989) assumes that an *individual* (e.g., a health care worker, a patient) performs a variety of *tasks* using *tools and technologies* in a particular *physical environment* and location under *organizational conditions*. Specific facets of the five work system elements are described in several chapters of this handbook. The work system model is used to describe HFE issues related to specific applications (e.g., home care) in other chapters. Intervention and redesign approaches based on the work system model are proposed in a few chapters of this handbook, for instance, in the chapter on ergonomics programs and effective interventions (Smith, chap. 41, this volume).

Many HFE principles exist that can help design the systems, the system elements, and the interactions between system elements to produce high-quality safe care and to ensure the well-being, safety, and performance of end users. Increasingly, researchers and health care practitioners have begun to understand that interactions and interfaces between connected, dependent systems for patient care produce challenges for quality and safety of care. For example, medication reconciliation for patients being admitted to a hospital involves the admitting team of the hospital but may require communication with the primary care physician and the pharmacy where the patient gets his or her medications, both being located elsewhere. This example shows that the care provided to the patients depends on several systems (e.g., admitting team at hospital, primary care physician, pharmacy) that are connected to each other and depend on each other to produce high-quality safe care. Issues related to the interactions and interfaces between systems and organizations are addressed in a few chapters in the handbook. The chapter on collaborative initiatives for patient safety (Carayon, Kosseff, Borgsdorf, & Jacobsen, chap. 10, this volume) describes how health care institutions collaborate to solve patient safety problems. The chapter on Human Factors of Transition of Care (Harvey, Schuster, Durso, Matthews, & Surabattula, chap. 15, this volume) highlights many of the HFE issues of transitions of care, such as communication between providers during a shift change. How patient safety problems arise at the interface between systems and how effective interactions between systems can contribute to the discovery of errors therefore preventing patient harm are addressed by the chapters describing HFE in specific applications, such as the emergency care environment (Wears & Perry, chap. 47, this volume).

System Design

Three types of system design can be delineated (Carayon & Smith, 1993; Meister & Enderwick, 2001): (a) design of a new system (e.g., construction of a new hospital), (b) updating a system (e.g., technological change in medication administration such as BCMA), and (c) system redesign (e.g., effort aimed at redefining the objectives of a clinical microsystem, its systems, and processes). The design of a new system versus the updating or redesign of an existing system poses different kinds of challenges. In the design of a new system, there may not be any historical data or previous experience to build on and use to determine the system design characteristics. On the other hand, this may facilitate the creation and implementation of new structures and processes because of lack of resistance to change. Updating or redesigning a system should involve significant planning work to address not only the system design itself (i.e., content of the change) but also the implementation organization (i.e., process of the change). Many factors need to be considered at the stages of system implementation such as involvement and participation of end users, information about the change communicated to end users and other stakeholders, training and learning, feedback, and project management (Korunka & Carayon, 1999; Korunka, Weiss, & Karetta, 1993). The transition period during which the new system is being implemented can be a source of uncertainty and stress; therefore providing adequate support (e.g., "super users" or other experts available for consultation) and sufficient time for learning and adaptation are important.

We also need to emphasize that designing a new system, and updating or redesigning a system, may occur over a long period of time. In a study of the implementation of smart IV pump technology in an academic hospital, we described the continuous changes that occurred after the introduction of the new technology (Carayon, Wetterneck et al., 2005). This case highlights the need to understand technology implementation in a longitudinal manner. Once a technology is implemented, several changes can subsequently occur in the technology itself and/or in the tasks and processes associated with the technology. Therefore, the implementation of a technology in health care can have characteristics of both an episodic change and a continuous change (Weick & Quinn, 1999).

Levels of System Design

Much of the discussion on human factors and patient safety has focused on human error. There is increasing recognition in the human error literature of the different levels of factors that can contribute to human error and accidents (Rasmussen, 2000). If the various factors are aligned "appropriately" like "slices of Swiss cheese," accidents can occur (Reason, 1990). Table 1–2 summarizes the different approaches to the levels of factors contributing to human error. It is interesting to make a parallel between the different levels of factors contributing to human error and the levels identified to deal with quality and safety of care (Berwick, 2002; Institute of Medicine Committee on Quality of Health Care in America, 2001). The 2001 IOM report on Crossing the Quality Chasm defines four levels at which interventions are needed to improve the quality and safety of care in the United States: Level A, experience of patients and communities; Level B, microsystems of care, that is, the small units of work that actually give the care that the patient experiences; Level C, health care organizations; and Level D, health care environment. These levels are similar to the hierarchy of levels of factors contributing to human error (see Table 1–2).

Human error models and approaches provide much information on how to understand, analyze,

TABLE 1–2. Levels of System Factors Contributing to Human Error

Authors	*Factors Contributing to Human Error*
Rasmussen (2000): levels of a complex sociotechnical system	Work Staff Management Company Regulators/associations Government
Moray (1994): hierarchical systems approach that includes several layers	Physical device Physical ergonomics Individual behavior Team and group behavior Organizational and management behavior Legal and regulatory rules Societal and cultural pressures
Johnson (2002): four levels of causal factors that can contribute to human error in healthcare	Level 1: Factors that influence the behavior of individual clinicians (e.g., poor equipment design, poor ergonomics, technical complexity, multiple competing tasks) Level 2: Factors that affect team-based performance (e.g., problems of coordination and communication, acceptance of inappropriate norms, operation of different procedures for the same tasks) Level 3: Factors that relate to the management of health care applications (e.g., poor safety culture, inadequate resource allocation, inadequate staffing, inadequate risk assessment and clinical audit) Level 4: Factors that involve regulatory and government organizations (e.g., lack of national structures to support clinical information exchange and risk management)
For comparison, levels of factors contribution to quality and safety of patient care	
(Berwick, 2002; Institute of Medicine Committee on Quality of Health Care in America, 2001)	Level A: Experience of patients and communities Level B: Microsystems of care (i.e., the small units of work that actually give the care that the patient experiences) Level C: Health care organizations Level D: Health care environment

and evaluate near misses and accidents (Shojania, Wald, & Gross, 2002). However, there is another large body of literature in human factors that has been relatively ignored in the discussion on quality of care and patient safety. This body of literature provides much information on how to design and improve work systems (Hendrick, 1997; Hendrick & Kleiner, 2001; Salvendy, 1997). Human errors in health care and patient safety, as well as a range of other HFE issues that can help in describing, understanding, and designing high-quality safe health care systems, are addressed in chapters in this handbook.

HFE AS AN INNOVATION IN HEALTH CARE AND PATIENT SAFETY

The application of HFE in health care is not new. In the late 1950s, Al Chapanis, one of the founders of human factors, and colleagues at the Johns Hopkins University conducted a study of medication errors in hospitals (Chapanis & Safrin, 1960; Safren & Chapanis, 1960a, 1960b). Using the critical incident technique method, they collected data on 178 medication errors over a 7-month period (Safren & Chapanis, 1960a). The medication errors were classified in seven categories (e.g., wrong patient, wrong dose of medication, omitted medication) and 90% of the immediate causes of the medication errors fell in the following categories: failure to follow required checking procedures, misreading or misunderstanding written communication, transcription errors, medicine tickets misfiled in ticket box, and computational errors (Safren & Chapanis, 1960a). This research led to several recommendations for improving written communication (e.g., legibility of handwriting), medication procedures (e.g., double checking), and the working environment (e.g., design of the nurse station and the medication preparation area; Safren & Chapanis, 1960b). HFE has more recently been applied to improve the design of health care technologies such as patient controlled analgesia (PCA) pumps (Lin, Vicente, & Doyle, 2001) and infusion pumps (Zhang, Johnson, Patel, Paige, & Kubose, 2003). Several chapters in this handbook provide examples of the successful application of HFE in health care and patient safety, such as the design of health care facilities. However, much still needs to be learned about the applicability and application of HFE in health care (Carayon, 2005).

The reader of this chapter and the handbook will be able to appreciate the diversity of the HFE discipline and its various applications in health care. HFE can benefit many different functions within health care institutions to help solve many different kinds of problems, including patient safety (Carayon, 2005). For instance, HFE methods for analyzing the usability of technologies can be used by Information System staff in health care organizations that are involved in the design of computerized provider order entry, electronic medical record systems, and other information technologies. HFE has been used by a Canadian hospital in its decision-making process regarding infusion pumps (Ginsburg, 2005). Larsen, Parker, Cash, O'Connell, and Grant (2005) applied HFE principles to the redesign of pharmacy-generated medication labels. HFE knowledge on work system design for optimal worker well-being, health, and safety can be used by human resources staff in health care organizations that are dealing with worker turnover and retention. Lourisen, Houtman, Kompier, and Grundeman (1999) described the implementation of a work stress management program in a Dutch hospital. Various actions were implemented to deal with the diversity of problems experienced by different groups. For instance, the pharmacy staff experienced many physical ergonomic problems that were addressed by redesigning their workstation. Supervisors received training on effective performance reviews. New staff was hired to provide support and expertise in occupational safety and health and physical ergonomics.

To facilitate and support the use of HFE knowledge, concepts, and methods in health care, we can consider HFE as an innovation whose diffusion, dissemination, implementation, and sustainability needs to be understood and specified. Diffusion is the passive spread of innovations and changes, whereas dissemination involves active and planned efforts to convince target groups to adopt an innovation. The implementation of the innovation includes active and planned efforts to incorporate an innovation within an organization. Making an innovation routine is the goal of sustainability. In this section, we use the conceptual model of innovation developed by Greenhalgh, Robert, MacFarlane, Bate, and Kyriakidou (2004) to examine the potential challenges related to the use of HFE in health care (see Figure 1–1). This model is the outcome of a systematic literature review that focuses primarily but not exclusively on studies done in health care. The conceptual innovation model specifies several components that contribute to the diffusion of innovation in health care. Several components of this innovation model are discussed separately and their relevance for understanding the introduction of HFE in health care is described.

HFE as an Innovation

Greenhalgh et al. (2004) listed key attributes of innovations that influence their adoption: relative advantage, compatibility, low complexity, trialability, observability, potential for reinvention, fuzzy boundaries, risk, task issues, nature of knowledge required (tacit/explicit), and technical support. In this section, we discuss the following four attributes: relative advantage, compatibility, task issues, and nature of knowledge required.

HFE is more likely to be adopted by health care organizations if clear advantages in term of effectiveness or cost-effectiveness can be demonstrated. So far, we lack this kind of systematic evidence. Much knowledge has been developed on HFE variables that can negatively affect health care workers and organizations, such as poorly designed working conditions and low usability of health care technologies. However, there is a lack of evidence regarding the positive impact of HFE interventions on quality and safety of patient care. More research and knowledge needs to be produced to understand the *fundamental* HFE issues involved in health care and patient safety (Cook, 2003), as well as methods for *integrating* HFE in the organizational fabric and structure of health care organizations (Carayon, 2005). With regard to fundamental research, Cook and his colleagues have argued for a deeper understanding of the work of health care providers (Cook, 2003; Cook, 2004; Nemeth, Cook, & Woods, 2004). On the applied side, HFE professionals need to devise creative ways of applying HFE in health care. In a previous section of

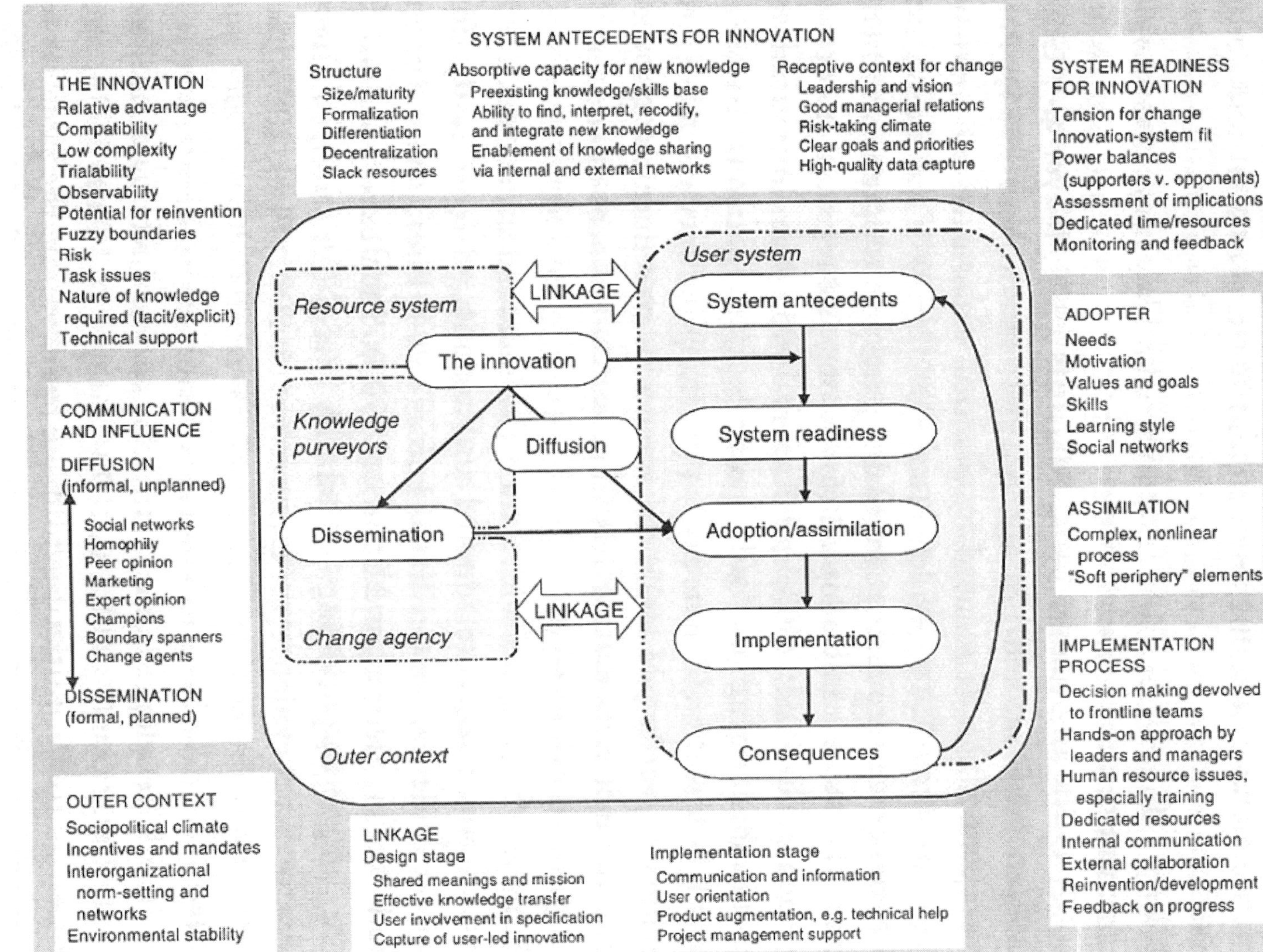

Figure 1–1. Model of innovation in health service delivery and organization (Greenhalgh et al., 2004).

this chapter, we discussed the challenges of applying participatory ergonomics in health care working environments in which the clinicians may not have time to effectively participate in improvement activities. If we are to involve end users in HFE improvement and design activities, we need to create new approaches that consider the specific characteristics of the health care context.

Innovations that are compatible with the adopter's values, norms, and needs are more likely to be adopted (Greenhalgh et al., 2004). The potential conflict between the physician's professional model and the systems approaches advocated by the discipline of HFE is addressed in the chapter on the relationship between physician professionalism and health care systems change (Smith & Bartell, chap. 9, this volume). A proposal to build on the cultural characteristics of health care organizations to foster and support system changes is presented in the chapter on tilting the culture in health care (Carroll & Quijada, chap. 45, this volume). Because of its emphasis on compatibility between the innovation and the adopter, this approach is more likely to contribute to the adoption of HFE innovations than an approach that lacks such compatibility.

The implementation and diffusion of HFE is likely to succeed because many HFE interventions produce conditions that are relevant to task performance. As specified in the IEA definition of ergonomics, the goal of HFE is to optimize human well-being and overall system performance. Therefore, HFE innovations can target specific areas of work with the aim of improving task performance. Such effort is more likely to be accepted by health care workers because they directly experience the benefits of such innovation. For instance, HFE innovations can help redesign work and processes of health care providers by removing performance obstacles and facilitating performance (Carayon, Gurses, Hundt, Ayoub, & Alvarado, 2005). Staff in outpatient surgery centers have reported the following performance obstacles: conflict among nurses and between nurses and physicians, lack of staffing, noise, and crowded environment (Carayon, Gurses, et al., 2005). HFE principles can be used, for instance, to redesign the physical layout of outpatient surgery centers to reduce noise and its propagation and to facilitate communication among the staff. This type of intervention can provide direct benefits to health care providers and are, therefore, more likely to be accepted by them.

The knowledge required to design, implement, and disseminate HFE interventions is very diverse. It relies on knowledge provided by basic scientific disciplines, such as physiology, sociology, and psychology, as well as by applied sciences, such as industrial engineering, business, and management. This diversity in the HFE knowledge base may hinder the dissemination of HFE in health care. One approach for dealing with this would be to use a phased approach in which HFE is applied in very specific applications, for instance when a particular technology is being implemented or when a renovation is being planned. Once these applications have demonstrated their effectiveness, HFE could be diffused and disseminated more broadly within the health care organization. The dissemination of HFE could also rely on a network of HFE-trained health care providers spread throughout the organization and available to work on specific projects (Vicente, 2003). This approach has been successfully used in the dissemination of office ergonomics knowledge within a public service organization (Haims & Carayon, 1998). This type of approach relies on intense involvement of HFE experts over a short period of time. In a "train-the-trainer" model, the HFE experts aimed at transmitting their knowledge and expertise to a group of employees. Over time, this group of employees gains HFE knowledge and experience, and the role of the HFE experts changes to a support role. The HFE-trained employees can be the seeds for the dissemination of HFE throughout the organization.

Adoption of HFE by Individuals

According to Rogers (1995), several aspects of adopters and the adoption process affect the innovation process (Greenhalgh et al., 2004): general psychological antecedents, meaning of the innovation for the adopter, adoption decision, concerns in preadoption stage, concerns during early use, and concerns in established users. In this discussion, we focus on the adoption process of HFE innovations. This adoption process was described by Rogers (1995) and Hall and Hord (1987). Rogers (1995) described the adoption process as having five stages: awareness, persuasion, decision, implementation, and confirmation. A core element of the Concerns-Based Adoption Model of Hall and Hord (1987) is change facilitators who have access to a resource system, and who are responsible for understanding the needs of the adopters. The model also highlights concerns occurring in various stages: preadoption stage (e.g., concerns for awareness and information), early use (e.g., concerns for information and training), and established use (e.g., concerns for feedback).

HFE is more likely to be adopted as an innovation if the adopters "are aware of the innovation; have sufficient information about what it does and how to use it; and are clear about how the innovation would affect them personally" (Greenhalgh et al., 2004). There is increasing recognition by health care providers and organizations of the discipline of HFE. However, health care leaders and top managers are somewhat unclear of the benefits that HFE efforts could bring to improve the quality and safety of care provided by their organizations (Carayon, 2005). As a discipline, we need to make a greater effort to reach out the "customers" of our knowledge, concepts, and methods. This issue is discussed in the chapter on human factors and patient safety: continuing challenges (Henriksen, chap. 2, this volume).

Assimilation of HFE by the System

Health care organizations, departments, and teams that adopt HFE innovations are likely to go through a "messy" process of assimilation, as opposed to a linear straightforward process (Greenhalgh et al., 2004; Van de Ven, Polley, Garud, & Venkataraman, 1999). Health care organizations that initiate, develop, and implement an HFE effort may experience setbacks and surprises. The *innovation journey* as described by Van de Ven and colleagues (1999) is open and dynamic and may take parallel, convergent, and/or divergent paths. Therefore, people and organizations involved in the implementation of HFE in health care should expect setbacks and surprises. These are part of a "normal" innovation journey in which HFE knowledge, concepts, and methods are assimilated by the entire health care organization. In the health care environment, this innovation journey is further complicated by the larger financial, legal, regulatory context.

Diffusion and Dissemination of HFE

The spread of innovation can be conceptualized as a continuum between pure diffusion and active dissemination. Diffusion is characterized by unplanned, informal, decentralized processes, whereas dissemination involves planning and formal processes. Various elements help spread innovations: network structure, homophily (similarity in terms of socioeconomic, educational, professional, and cultural backgrounds), opinion leaders, champions, boundary spanners, and formal dissemination programs (Greenhalgh et al., 2004). In this section, we discuss network structure and various individual roles (opinion leaders, champions, and boundary spanners) as contributing to the diffusion and dissemination of HFE innovations.

Strong social networks can support and influence the adoption of HFE innovations. The use of network structures to diffuse and disseminate patient safety innovations is described in the chapter on collaborative initiatives for patient safety (Carayon, Kosseff, Borgsdorf, & Jacobsen, chap. 10, this volume). Such network structures can also contribute to the adoption of HFE innovations for improving patient safety. In health care organizations, for example, physicians and nurses tend to operate in strong social networks. The presence of opinion leaders and champions in those networks can very much influence the adoption of HFE innovations. It is important to identify those opinion leaders and champions who are to support the implementation of HFE efforts. This process can be facilitated by training opinion leaders and champions in HFE and "giving away" HFE to the end users (Carayon et al., 2003). It is important to realize that opinion leaders can have either a positive or negative influence on the adoption of HFE innovations (Ferlie, Fitzgerald, Wood, & Hawkins, 2005). Identifying those negative influences early on in the implementation process is important to anticipate potential difficulties and setbacks.

System Antecedents for the Assimilation of HFE as an Innovation

Different organizations provide different environments for the assimilation of innovations. Organizations can be characterized on the following dimensions: structure, absorptive capacity for new knowledge, and receptive context for change (Greenhalgh et al., 2004). Many structural characteristics have been related to innovativeness. Organizations are more likely to assimilate innovations if they are large, mature, functionally differentiated and specialized, if resources are available for new projects, and if they have decentralized decision-making structures (Greenhalgh et al., 2004). However, it is important to recognize that these structural characteristics explain only a small percentage of organizational innovativeness.

HFE is more likely to be adopted by large health care organizations and systems. It may be more challenging for small health care organizations, such as rural hospitals, to have access to HFE expertise. In addition, those organizations may not have slack resources to invest in pilot studies of HFE. Health care organizations tend to be highly differentiated with many semiautonomous departments and units. Such an organizational structure can facilitate the adoption and assimilation of HFE knowledge. For instance, in the redesign of a hospital intensive care unit, HFE knowledge can be used to improve the physical environment for both health care providers and patients. Such a local effort can succeed because the HFE effort is focused on a single unit that has some autonomy regarding the redesign.

Health care organizations with a learning organization culture are more likely to adopt HFE innovations (Greenhalgh et al., 2004). Additional information on models and processes of organizational learning is provided in the chapter on organizational learning in health care (Hundt, chap. 8, this volume). Understanding how HFE knowledge can be "absorbed" by health care organizations deserves further attention. This can occur through direct participation in multidisciplinary research projects that involve HFE and health care disciplines (Gopher, 2004) or through practical applications involving HFE in which cycles of learning and feedback occur.

Strong leadership, strategic vision, and climate conducive to experimentation and risk are characteristics of organizations that are receptive to change (Greenhalgh et al., 2004). HFE innovations are more likely to be adopted by health care organizations that have those characteristics. This is particularly challenging in health care organizations that tend to have strong professional boundaries and in which professionals tend to function within unidisciplinary communities (Ferlie et al., 2005).

System Readiness for HFE as an Innovation

An organization is ready for HFE innovations if there is tension for change and HFE is seen as a promising solution to current problems. There is increasing pressure on health care organizations for improving the quality and safety of care; this creates an environment more receptive to change. Do health care leaders and top managers perceive HFE as having the potential to provide solutions for improving quality and safety of care? How can we help health care leaders and top managers understand the (potential) benefits of HFE? HFE professionals and their scientific organizations have a unique role to play to demonstrate how HFE can fit in and help in improving patient safety.

System readiness for innovation is also influenced by the innovation-system fit (does the HFE innovation fit with the organization's values, norms, strategies, goals, and ways of working?), assessment of implications (has the impact of the HFE effort been fully assessed and anticipated?), support and advocacy (is there support for HFE within the organization?), dedicated time and resources (have resources been allocated to the HFE effort?), and capacity to monitor and evaluate the innovation (is there a system in place for evaluating the actual and anticipated effects of the HFE effort?).

The Outer Context

Numerous external factors influence the adoption of innovations: informal interorganizational structures (see the chapter on collaborative initiatives for patient safety; Carayon, Kosseff, Borgsdorf, & Jacobsen, chap. 10, this volume), intentional spread strategies such as quality improvement collaboratives (see the chapter on quality improvement in health care: a historical overview and a look into the future; Molfenter & Gustafson, chap. 42, this volume), uncertainty in the wider environment, and political directives (Greenhalgh et al., 2004). Vicente (2003) conducted an interesting longitudinal case study analysis of how a manufacturer of a PCA pump was influenced by various external factors that finally led to increased attention paid by the manufacturer to patient safety and human factors engineering. This case study clearly demonstrates the larger environmental and sociotechnical factors that can influence the adoption of HFE by manufacturers of health care equipment. The influence of the larger environment on human performance and behavior is discussed in a number of chapters in this handbook. See, for instance, the chapter on the Artichoke Systems Approach for identifying the why of error (Bogner, chap. 7, this volume).

Implementation and Routinization of HFE

Once HFE has been identified as an innovation by a health care organization, the steps following the adoption include implementation and routinization. Factors that facilitate the implementation of HFE innovations are very similar to factors that facilitate the implementation of other organizational and technological changes. Various models and best practices for technology implementation are described in the chapter on new technology implementation in health care (Karsh & Holden, chap. 25, this volume) and the chapter on patient safety and technology: a two-edged sword (Battles, chap. 24, this volume). Elements for system readiness discussed in a previous section are also relevant for implementation. Additional elements relevant for implementation include: adaptive and flexible organizational structure, top management support and continued commitment, human resource issues, funding, intraorganizational communication, interorganizational networks, feedback, and adaptation/reinvention (Greenhalgh et al., 2004). We discuss human resource issues and feedback involved in the implementation and routinization of HFE innovations in health care in this section.

The implementation and continued use of HFE innovations by health care providers depend on their motivation, capacity, and competence. As recommended by the participatory ergonomics approach (Wilson & Haines, 1997), early involvement of individuals affected by the HFE innovations is critical for a successful implementation and for sustainability. If the HFE innovation changes the way work is done and tasks are performed, attention should be dedicated to training. Additional information on this important human resource issue related to innovation implementation is provided in the chapter on teamwork training for patient safety: best practices and guiding principles (Salas, Wilson-Donnelly, Sims, Burke, & Priest, chap. 44, this volume).

The implementation and routinization of HFE innovations can benefit from feedback on the impact of the innovation and the implementation process. Timely and accurate feedback can provide useful information for anticipating some of the potential difficulties and setbacks in the implementation process. It can also provide important information for the next innovation implementation, therefore creating a process of organizational learning. See the chapter on organizational learning in health care (Hundt, chap. 8, this volume) for additional information.

Linkage Among Components of the Innovation Model

We have described the different elements of the innovation model separately. However, it is important to understand the links among the elements of the model (Greenhalgh et al., 2004). External HFE experts can serve a vital role as change agents that create the links between the different components of the innovation process.

Research shows that external change agents are more effective if the adopters of the innovation perceive them as credible; if they exhibit social skills and can develop strong interpersonal relationships with the end users of the innovation; if they build bridges between the developer of the innovation and the end users of the innovation by, for instance, relaying end user needs and concerns to the developer of a particular technology; and if they help and empower the end users develop their own evaluation of the innovation (Greenhalgh et al., 2004). This research has a number of implications for external HFE experts hired by health care organizations as change agents. First, the HFE experts should be perceived as credible. This requires that HFE experts have extensive experience in the health care domain, have a strong capacity for learning and listening, and are flexible. Second, HFE experts should have strong interpersonal skills. This will facilitate communication between the HFE experts on one hand and the adopters and end users of the HFE innovation on the other hand. Third, the HFE experts should be knowledgeable about the HFE innovation itself, as well as the process for implementing the innovation. The training of HFE experts should allow for the development of skills and knowledge on how to be an effective change agent.

CONCLUSION

This handbook represents the diversity of the HFE discipline, and its application to health care and patient safety. Lucian Leape (2004) called for HFE researchers and practitioners to take on the challenges of designing high-quality safe health care systems. This handbook provides important information to HFE practitioners engaged in efforts to improve systems and processes in health care. It also clearly highlights the health care and patient safety gaps in HFE knowledge that HFE researchers need to consider. See the Appendix for additional Resources on HFE.

This handbook strongly emphasizes HFE systems approaches. The HFE discipline aims at improving the interactions between humans and various system elements (Wilson, 2000). Whether health care is a system has been debated; however, systems approaches are necessary to understand the multiple, complex interactions between humans and systems as well as between systems. This handbook can help practitioners in health care organizations become more familiar with HFE systems approaches, concepts, and methods. It provides many methods and tools as well as the underlying frameworks necessary to adequately apply HFE knowledge.

References

Berwick, D. M. (2002). A user's manual for the IOM's "Quality Chasm" report. *Health Affairs, 21*(3), 80–90.

Bogner, M. S. (Ed.). (1994). *Human error in medicine.* Hillsdale, NJ: Lawrence Erlbaum Associates, Inc.

Bohr, P. C., Evanoff, B. A., & Wolf, L. (1997). Implementing participatory ergonomics teams among health care workers. *American Journal of Industrial Medicine, 32*, 190–196.

Bureau of Labor Statistics. (2000). *Lost-worktime injuries and illnesses: Characteristics and resulting time away from work, 1998.* Washington, DC: U.S. Department of Labor, Bureau of Labor Statistics. Available from http://www.bls.gov/news.release/osh2.nr0.htm

Carayon, P. (2005). Top management's view on human factors and patient safety: Do they see it? In R. Tartaglia, S. Bagnara, T. Bellandi, & S. Albolino (Eds.), *Health care systems ergonomics and patient safety* (pp. 38–42). Florence, Italy: Taylor & Francis.

Carayon, P., Alvarado, C., & Hundt, A. S. (2003). *Reducing workload and increasing patient safety through work and workspace design.* Washington, DC: Institute of Medicine.

Carayon, P., Gurses, A. P., Hundt, A. S., Ayoub, P., & Alvarado, C. J. (2005). Performance obstacles and facilitators of health care providers. In C. Korunka & P. Hoffmann (Eds.), *Change and quality in human service work* (Vol. 4, pp. 257–276). Munchen, Germany: Hampp.

Carayon, P., & Karsh, B. (2000). Sociotechnical issues in the implementation of imaging technology. *Behaviour and Information Technology, 19*, 247–262.

Carayon, P,. & Smith, M. J. (1993). *The balance theory of job design and stress as a model for the management of technological change.* Paper presemted at the Fourth International Congress of Industrial Engineering, Marseille, France.

Carayon, P., & Smith, M. J. (2000). Work organization and ergonomics. *Applied Ergonomics, 31*, 649–662.

Carayon, P., Wetterneck, T. B., Hundt, A. S., Enloe, M., Love, T., Rough, S., et al. (2005, July). *Continuous technology implementation in health care: The case of advanced IV infusion pump technology.* Paper presented at the 11th International Conference on Human–Computer Interaction, Las Vegas, Nevada.

Chapanis, A., & Safrin, M. A. (1960). Of misses and medicines. *Journal of Chronic Diseases, 12*, 403–408.

Cook, R. (2004). *Observational and ethnographic studies: Insights on medical error and patient safety.* Paper presented at the 6th Annual NPSF Patient Safety Congress, Boston, MA.

NPSF: National Patient Safety Foundation

Cook, R. I. (2003). *Lessons from the war on cancer: The need for basic research on safety* (No. Testimony submitted for the AHRQ sponsored 2nd National Summit on Patient Safety Research, 7 Nov. 2003, Arlington, VA). Chicago, IL: University of Chicago, Cognitive Technologies Laboratory.

Cook, R. I., Woods, D. D., & Miller, C. (1998). *A tale of two stories: Contrasting views of patient safety.* Chicago, IL: National Patient Safety Foundation.

Eason, K. (1988). *Information technology and organizational change.* London: Taylor & Francis.

Eason, K. (2001). Changing perspectives on the organizational consequences of information technology. *Behaviour and Information Technology, 20*, 323–328.

Evanoff, V. A., Bohr, P. C., & Wolf, L. (1999). Effects of a participatory ergonomics team among hospital orderlies. *American Journal of Industrial Medicine, 35*, 358–365.

Ferlie, E., Fitzgerald, L., Wood, M., & Hawkins, C. (2005). The nonspread of innovations: The mediating role of professionals. *Academy of Management Journal, 48*, 117–134.

Ginsburg, G. (2005). Human factors engineering: A tool for medical device evaluation in hospital procurement decision-making. *Journal of Biomedical Informatics, 38*, 213–219.

Gopher, D. (2004). Why is it not sufficient to study errors and incidents: Human factors and safety in medical systems. *Biomedical Instrumentation & Technology, 38*, 387–409.

Greenhalgh, T., Robert, G., MacFarlane, F., Bate, P., & Kyriakidou, O. (2004). Diffusion of innovations in service organizations: Systematic review and recommendations. *The Milbank Quarterly, 82*, 581–629.

Haims, M. C., & Carayon, P. (1998). Theory and practice for the implementation of "in-house," continuous improvement participatory ergonomic programs. *Applied Ergonomics, 29*, 461–472.

Haines, H., Wilson, J. R., Vink, P., & Koningsveld, E. (2002). Validating a framework for participatory ergonomics (the PEF). *Ergonomics, 45*, 309–327.

Hall, G. E., & Hord, S. M. (1987). *Change in schools—Facilitating the process.* Albany: State University of New York Press.

Hendrick, H. W. (1991). Human factors in organizational design and management. *Ergonomics, 34*, 743–756.

Hendrick, H. W. (1997). Organizational design and macroergonomics. In G. Salvendy (Ed.), *Handbook of human factors and ergonomics* (pp. 594–636). New York: Wiley.

Hendrick, H. W., & Kleiner, B. M. (2001). *Macroergonomics—An introduction to work system design.* Santa Monica, CA: The Human Factors and Ergonomics Society.

Institute of Medicine Committee on Quality of Health Care in America. (2001). *Crossing the quality chasm: A new health system for the 21st. century.* Washington, DC: National Academy Press.

International Ergonomics Association. (2000). *The discipline of ergonomics.* Retrieved August 22, 2004, from www.iea.cc

Johnson, C. (2002). The causes of human error in medicine. *Cognition, Technology & Work, 4*, 65–70.

Klein, G. A., Orasanu, J., & Calderwood, R. (Eds.). (1993). *Decision making in action: Models and methods.* Norwood, NJ: Ablex.

Korunka, C., & Carayon, P. (1999). Continuous implementations of information technology: The development of an interview guide and a cross-national comparison of Austrian and American organizations. *The International Journal of Human Factors in Manufacturing, 9*, 165–183.

Korunka, C., Weiss, A., & Karetta, B. (1993). Effects of new technologies with special regard for the implementation process per se. *Journal of Organizational Behavior, 14*, 331–348.

Larsen, G. Y., Parker, H. B., Cash, J., O'Connell, M., & Grant, M. J. C. (2005). Standard drug concentrations and smart-pump technology reduce continuous-medication-infusion errors in pediatric patients. *Pediatrics, 116*, e21–e25.

Leape, L. (2004). Human factors meets health care: The ultimate challenge. *Ergonomics in Design, 12*(3), 6–12.

Leape, L. L. (1994). Error in medicine. *Journal of the American Medical Association, 272*(23), 1851–1857.

Leape, L. L., Bates, D. W., Cullen, D. J., Cooper, J., Demonaco, H. J., Gallivan, T., et al. (1995). Systems analysis of adverse drug events. *Journal of the American Medical Association, 274*, 35–43.

Lin, L., Vicente, K. J., & Doyle, D. J. (2001). Patient safety, potential adverse drug events, and medical device design: A human factors engineering approach. *Journal of Biomedical Informatics, 34*, 274–284.

Lourisen, E., Houtman, I., Kompier, M., & Grundeman, R. (1999). The Netherlands: A hospital "healthy working for health." In M. Kompier & C. Cooper (Eds.), *Preventing stress, improving productivity—European case studies in the workplace* (pp. 86–120). London: Routledge.

Meister, D., & Enderwick, T. P. (2001). *Human factors in system design, development, and testing.* Mahwah, NJ: Lawrence Erlbaum Associates, Inc.

Moray, N. (1994). Error reduction as a systems problem. In M. S. Bogner (Ed.), *Human error in medicine* (pp. 67–91). Hillsdale, NJ: Lawrence Erlbaum Associates, Inc.

Nemeth, C. P., Cook, R. I., & Woods, D. D. (2004). The messy details: Insights from the study of technical work in health care. *IEEE Transactions on Systems Man and Cybernetics Part A—Systems and Humans, 34*, 689–692.

Noro, K., & Imada, A. (1991). *Participatory ergonomics.* London: Taylor & Francis.

Rasmussen, J. (2000). Human factors in a dynamic information society: Where are we heading? *Ergonomics, 43*, 869–879.

Rasmussen, J., Pejtersen, A. M., & Goodstein, L. P. (1994). *Cognitive systems engineering.* New York: Wiley.

Reason, J. (1990). *Human error.* Cambridge, England: Cambridge University Press.

Reason, J. (1997). *Managing the risks of organizational accidents.* Burlington, VT: Ashgate.

Reason, J. (2000). Human error: Models and management. *British Medical Journal, 320*(7237), 768–770.

Rogers, E. M. (1995). *Diffusion of Innovations* (Fourth ed.). New York: Free Press.

Safren, M. A., & Chapanis, A. (1960a). A critical incident study of hospital medication errors—Part 2. *Hospitals, 34,* 53; 65–68.

Safren, M. A., & Chapanis, A. (1960b). A critical incident study of hospital medication errors—Part I. *Hospitals, 34*, 32–34;57–66.

Salas, E., & Cannon-Bowers, J. A. (1997). Methods, tools, and strategies for team training. In M. A. Quinones & A. Ehrenstein (Eds.), *Training for a Rapidly Changing Workforce: Applications of Psychological Research* (pp. 249–279). Washington, D.C.: American Psychological Association.

Salvendy, G. (Ed.). (1997). *Handbook of human factors and ergonomics* (2nd ed.). New York: Wiley.

Shojania, K. G., Wald, H., & Gross, R. (2002). Understanding medical error and improving patient safety in the inpatient setting. *Medical Clinics of North America, 86*, 847–867.

Smith, M. J., & Carayon, P. (2000). Balance Theory of job design. In W. Karwowski (Ed.), *International encyclopedia of ergonomics and human factors* (pp. 1181–1184). London: Taylor & Francis.

Smith, M. J., & Carayon-Sainfort, P. (1989). A balance theory of job design for stress reduction. *International Journal of Industrial Ergonomics, 4*, 67–79.

Van de Ven, A. H., Polley, D. E., Garud, R., & Venkataraman, S. (1999). *The innovation journey.* New York, NY: Oxford University Press.

Vicente, K. J. (2003). What does it take? A case study of radical change toward patient safety. *Joint Commission Journal on Quality and Safety, 29*, 598–609.

Vincent, C., Moorthy, K., Sarker, S. K., Chang, A., & Darzi, A. W. (2004). Systems approaches to surgical quality and safety—From concept to measurement. *Annals of Surgery, 239*, 475–482.

Vincent, C., Taylor-Adams, S., & Stanhope, N. (1998). Framework for analysing risk and safety in clinical medicine. *British Medical Journal, 316*(7138), 1154–1157.

Wears, R. L., & Perry, S. J. (2002). Human factors and ergonomics in the emergency department. *Annals of Emergency Medicine, 40*, 206–212.

Weick, K. E., & Quinn, R. E. (1999). Organizational change and development. *Annual Review of Psychology, 50*, 361–386.

Wickens, C. D., Lee, J. D., Liu, Y., & Becker, S. E. G. (2004). *An introduction to human factors engineering* (2nd ed.). Upper Saddle River, NJ: Prentice Hall.

Wilson, J. R. (1991). Participation—A framework and a foundation for ergonomics? *Journal of Occupational Psychology, 64*, 67–80.

Wilson, J. R. (2000). Fundamentals of ergonomics in theory and practice. *Applied Ergonomics, 31*, 557–567.

Wilson, J. R., & Haines, H. M. (1997). Participatory ergonomics. In G. Salvendy (Ed.), *Handbook of human factors and ergonomics* (pp. 490–513). New York: Wiley.

Wunderlich, G. S., & Kohler, P. O. (Eds.). (2001). *Improving the quality of long-term care.* Washington, DC: National Academy Press.

Zhang, J., Johnson, T. R., Patel, V. L., Paige, D. L., & Kubose, T. (2003). Using usability heuristics to evaluate patient safety of medical devices. *Journal of Biomedical Informatics, 36*, 23–30.

Zink, K. J. (1996). Continuous improvement through employee participation: Some experiences from a long-term study. In O. Brown Jr. & H. W. Hendrick (Eds.), *Human factors in organizational design and management-V* (pp. 155–160). Amsterdam: Elsevier.

APPENDIX

Bibliography on Human Factors and Ergonomics

Salvendy, G. (Ed.). (2006). *Handbook of human factors and ergonomics* (3rd ed.). New York: Wiley.
Stanton, N., Hedge, A., Brookhuis, K., Salas, E., & Hendrick, H. W. (Eds.). (2004). *Handbook of human factors and ergonomics methods*. Boca Raton, FL: CRC Press.
Wilson, J. R., & Corlett, N. (Eds.). (2005). *Evaluation of human work* (3rd ed.). Boca Raton, FL: CRC Press.

Key Journals in the Area of Human Factors and Ergonomics

- Human Factors
- Ergonomics
- Applied Ergonomics

Selected Journals in the Area of Health Care and Patient Safety

- *Journal of the American Medical Association* (JAMA)
- *Quality and Safety in Health Care*
- *Journal of Patient Safety*
- *Joint Commission Journal of Quality & Safety*

Key National and International Organizations in the Area of Health Care and Patient Safety

Agency for Healthcare and Research Quality (www.ahrq.gov)

- UK National Patient Safety Agency (www.npsa.nhs.uk)
- Vetrans Administration National Patient Safety Agency (http://www.patientsafety.gov/index.html)
- Institute of Medicine (http://www.iom.edu/)
- U.S. Food and Drug Administration (www.fda.gov)
- Joint Commission on Accreditation of Healthcare Organizations (www.jcaho.org)
- Institute for Healthcare Improvement (www.ihi.org)
- Institute for Safe Medication Practices (www.ismp.org)
- National Patient Safety Foundation (www.npsf.org)

2

HUMAN FACTORS AND PATIENT SAFETY: CONTINUING CHALLENGES

Kerm Henriksen
Agency for Healthcare Research and Quality
U.S. Department of Health and Human Services

Interest in human factors and ergonomic approaches as applied to patient safety and health care has increased considerably in the past 15 years. However, long before the current heightened interest in patient safety and quality of care, thoughtful analyses and work was rendering a clearer understanding of how human factors specialists would come to think about the garden variety term "human error" (e.g., Hollnagel, 1993; Norman, 1981, 1988; Perrow, 1984; Rasmussen, 1982, 1987; Reason, 1990; Sanders & Shaw, 1988; Senders & Moray, 1991). Uncritical usage of the term—viewing error as a cause rather than consequent and connoting blame—became less acceptable in serious investigations of accidents. More specifically in health care, pioneering work can be traced to critical incident studies of hospital medication errors (Safren & Chapanis, 1960a, 1960b) and anesthesia mishaps (Cooper, Newbower, Long, & McPeek, 1978). Early examples and calls for human factors applications and involvement in medicine and health care can be found in Rappaport (1970) and in an edited work by Pickett and Triggs (1975). By the late 1980s, a number of articles featuring human factors concepts and applications could be found in the literature, many of which dealt with anesthesia equipment (e.g., Allnutt, 1987; Gaba, Maxwell, & DeAnda, 1987; Thompson, 1987) and infusion pumps (LeCoq, 1987). Yet other articles during this period focused on medication errors in various settings, the effects of fatigue on resident and physician performance, perioperative deaths in surgical care, judgmental limitations in medical decision making, inadequate infection control, and detection problems in diagnostic radiology. Also noteworthy was an influential paper that appeared on how the culture of an organization contributes to its ability to be safe and highly reliable (Weick, 1987).

Throughout the 1990s, the unintended consequences of clumsy automation, task complexity, and excessive workloads on human performance in high-risk patient environments received serious focus (e.g., Cook & Woods, 1994; Gaba, Howard, & Fish, 1994; Weinger & Englund, 1990). Highly relevant work on teams, crew resource management, and the relationship between job design and ergonomics also was well underway during this period (e.g., Salas, Dickenson, Converse, & Tannenbaum, 1992; Smith & Carayon-Sainfort, 1989; Wiener, Kanki, & Helmreich, 1993). By 1994, the frontier of human factors applications to medical systems was sufficiently populated to warrant publication of a well received edited work on human error in medicine (Bogner, 1994). Research by Gopher and Donchin and their colleagues (Donchin et al., 1995; Gopher et al., 1989) underscored the role of communication problems in ICU mishaps. Other excellent works, underscoring

No official endorsement of this paper by the Agency for Healthcare Research and Quality or the Department of Health and Human Services is intended or should be inferred.

the role of systemic factors in relation to error, were prepared by Woods, Johannesen, Cook, and Sarter (1994); Leveson (1995); Reason (1997); and Vincent (2001).

Results from two studies on adverse events derived from large samples of hospital admissions—one from New York State using 1984 data (Brennan et al., 1991; Leape et al., 1991) and the other from Colorado and Utah using 1992 data (Thomas et al., 1999)—would come to figure quite prominently in gaining an understanding of the extent to which preventable adverse events serve as a leading cause of death and injury. The proportions of adverse events considered preventable were 58% and 53% in the New York and Colorado/Utah studies respectively. By the mid—1990s, a mainstream medical journal, the *Journal of the American Medical Association,* was publishing articles on the systemic components of error in medicine and adverse drug events (Leape, 1994; Leape et al., 1995). In 1996, health care experts, human factors professionals, and industrial accident specialists could be found working together and addressing patient safety issues at the first Annenberg Conference. In the same year, the National Patient Safety Foundation and the Veterans Health Administration's safety initiative were launched (Leape, 2004).

Without these human factors efforts, the research on vulnerabilities in high risk patient environments, and the studies of adverse events in hospitalized patients that were accumulating beneath the threshold of public awareness, it is doubtful that the frequently cited Institute of Medicine (IOM) report, *To Err Is Human* (Kohn, Corrigan, & Donaldson, 2000), could have been the tipping point that it came to be. As is well known, the media put the spotlight on the vast number of preventable deaths, the American public became alarmed, Congress appropriated funding, and the Agency for Healthcare Research and Quality (AHRQ) was designated as the lead agency for supporting research on patient safety.

Although the *To Err Is Human* (Kohn et al., 2000) report provided a service to the human factors and ergonomics (HFE) community by exposing a wide audience of health services researchers and practitioners to systems concepts and human factors principles, the extent to which these concepts and principles have been adopted by a broader health care audience and have had an impact on the care and lives of patients is open to question. These very promising beginnings stand in contrast to the considerable progress that is still needed to realize significant gains in patient safety.

This chapter focuses on a few fundamental issues that continue to thwart the full potential of human factors and ergonomic practices for enhancing patient safety. Specifically, the chapter addresses the following challenges, stated here in the form of questions: (a) What is the nature of acceptable evidence from human factors and ergonomic methodologies in relation to evidence-base medicine, (b) are we failing to take our own medicine in our reluctance to study not the components themselves but the interdependencies or alignments among system components, and (c) is there a role for human factors and ergonomic professionals to serve as leaders of transformational change in health care?

TWO CULTURES: EVIDENCE-BASED MEDICINE AND HUMAN FACTORS

The *To Err Is Human* (Kohn et al., 2000) report viewed quite favorably human factors contributions to system safety in other industries and urged health care "to apply the theory and approaches already used in other fields to reduce errors and improve reliability" (p. 60). At the same time, subsequent assessments of the evidence base for safe practices in health care appear to be at somewhat of a loss as to what to make of these contributions. Given the impetus for building a safer health care system, a certain amount of tension exists beneath the surface as to the nature of the evidence on which changes to processes and products will be based. To take advantage of contributions from outside of health care in affecting change, new ways of thinking about the evidence base are needed. A secondary purpose of what follows is to examine possible differences of scientific culture between health services researchers and human factors specialists and whether differences in the prevailing research paradigms of each serve as a determining factor in adoption of safety practices in health care settings.

What Constitutes Acceptable Evidence?

An underlying theme of evidence-based medicine is that medical decision making should be based on a solid foundation of scientific evidence. Medicine's past, like many fields, is replete with examples of authoritative misinformation and variation of practice. Millenson (1997) provides a very readable and informative account of how medicine has not always done the right thing or done the right thing right. One approach for assessing the soundness of the evidence-base of best practices for ensuring patient safety is to resort to a formal analysis—an assessment of the evidence base—for categorizing the strength of evidence regarding the impact and effectiveness of specific patient safety practices. Such an approach is found in a 668-page analysis sponsored by the AHRQ entitled *Evidence Report/Technology Assessment 43, Making Health Care Safer: A Critical Analysis of Patient Safety Practices* (Shojania, Duncan, McDonald, & Wachter, 2001). The report consists of 59 well-constructed, bite-size chapters that critically examine patient safety practices. Each safety practice's level of evidence for effectiveness is assessed in terms of a hierarchy of study designs, starting with the first level of randomized clinical trials (RCT) and ending at a fourth level of observational studies without controls. A hierarchy of outcome measures also is utilized, starting with the first level of clinical outcomes such as morbidity and mortality and ending with no outcomes relevant to decreasing medical errors and/or adverse events. Both level of study design and outcome measures along with other factors such as number of studies and number of patients in studies enter into a *study strength* rating of high, medium, or low.

The evidence report on patient safety practices was well received, by far exceeding the distribution of other evidence-based reports sponsored by AHRQ. However, many of the reviewed safety practices that have a basis in human factors engineering or have origins outside of health care received low ratings or the ratings were omitted because of the lack of sufficient evidence to demonstrate their efficacy. These included such practices as use of pre-anesthesia checklists, use of bar coding, aviation style crew resource management, napping strategies, use of human factors principles in evaluation of medical devices, and promoting a safety culture. Still other outside-of-health care practices were provided reviews to provide some background in the practice rather than because of their level of evidence. These included incident reporting, root cause analysis, computerized physician order entry and decision support, and simulators as training tools, among other practices.

Leape, Berwick, and Bates (2002) noted that the way in which the evidence-based assessment was framed (i.e., along the lines of a classic academic medical model), its authors were guided to focus on topics not of the greatest importance to patient safety, but instead on practices for which there were readily available data. Many simple and accepted medication practices endorsed by hospital associations were not included in the report because evidence of efficacy was lacking. Leape et al. (2002) further noted it is one thing to demonstrate the efficacy of a practice such as prophylactic anticoagulation for venous thromboembolism with randomized controlled trials and receive a high evidence base rating; it is another thing to ensure that systems problems are addressed such as making sure that anticoagulation gets to the right patients who need it at the right time and in the right dose. An unsettling irony is that *Evidence Report 43* (Shojania et al., 2001) does not receive a very high rating for focusing on significant system problems that the *To Err Is Human* (Kohn et al., 2000) report brought to light.

The editors and contributors to *Evidence Report 43* (Shojania et al., 2001) deserve credit for their effort to be inclusive of practices outside of health care, yet the treatment that these practices received in terms of actual assessment of their evidence base raises many questions (see chap. 2 of the report). On the one hand, the need to examine and reap the benefits of safety practices derived from other disciplines such as human factors and organizational theory was fully acknowledged. However, the research designs from these disciplines rate low in comparison to the *Report's* gold standard, the RCT. Given the acceptance differential that the rating hierarchies create and compared to the 11 patient safety practices that acquired the greatest strength of evidence regarding their impact and effectiveness (and that attracted considerable media attention), serious exploration of practices from the medium, lower, or lowest tables for impact and strength of evidence is more of a hard sell. It is fair to ask whether evidence hierarchies send an unintended message that thwart development of nontraditional approaches. On the other hand, engineering has had its share of embarrassing moments (Petroski, 1992), and many organizational and managerial innovations have not lived up to their advanced billing. To forsake the remarkable advances in health care brought about by evidence-based medicine would be foolhardy. As Shojania et al. (2001) commented, "we are left with our feet firmly planted in the middle of competing paradigms" (p. 26).

We could lament that what we need is more one-handed patient safety researchers (as Harry Truman did economists), but actually we should be rejoicing that we have two competing paradigms. Actually the paradigms are best conceived as complementary rather than competing. It is unlikely that one will subvert the other in a Kuhnian sense. Instead, we find that some research questions—those focusing on issues of clinical efficacy—are most likely best addressed by RCTs, whereas other questions such as understanding the barriers to technology adoption or improving the design of a troublesome interface are best addressed by qualitative techniques and test and evaluation procedures respectively. What constitutes useful evidence in patient safety research has very little to do with a yardstick of *study strength*, but depends on the nature of the questions being asked.

Understanding the Other Paradigm

Every school boy and girl knows, so we like to believe, that human factors engineering focuses on the interaction of people, technology and tools, and the physical and organizational environment in which tasks are performed to realize a specified goal. As a practice, HFE are guided by a different paradigm—one that has more of an engineering or process improvement framework that does not wait for unequivocal evidence to make changes.

Although most human factors specialists are academically trained in deriving hypotheses from scientific theory and putting them to test under controlled experimental conditions, human factors work in nonacademic settings typically is driven by systems development that, in turn, is driven by advancing technology. In terms of system design, the focus is on determination of objectives and performance specifications, incorporating human strengths and limitations in the design process, and then engaging in iterative test and evaluation. Usually a system is designed to improve on some already existing system. As adapted from Sanders and McCormick (1993), defining activities can be described as follows.

System Definition. This includes determining objectives, system specifications, and functions that need to be performed. Because functions need to fulfill system objectives and needs of intended users, interviews with representative users and field observations under operational conditions are conducted. Know thy user is the first commandment!

Basic Design. Here the principal human factors activities include allocating functions to human or hardware and software, determining human performance requirements, and conducting job/task analyses. A job/task analysis determines what tasks, performed in what manner, under what conditions, in response to what cues, to what standards of performance, make up the job. More recently, cognitive task analysis techniques are used for tasks that are predominantly knowledge-based and that require problem solving.

Detailed Design. Once the functions and tasks allocated to humans and machines have been delineated, the design of the human-device interface (be it displays, controls, consoles, computer navigation, and work stations) is undertaken. Working with engineers and designers, human factors specialists ensure that human considerations in terms of cognitive, perceptual, and behavioral capabilities are taken into account. The use of human factors design guidelines help to ensure that end users will be interacting with devices they can actually use and that improve system performance.

Test and Evaluation. Concurrent with detailed design is a considerable amount of iterative testing and evaluation of prototypes with representative users. Attribute evaluations assess design features (e.g., readability of displays) in terms of standards and guidelines. Given the increasingly opaque nature of electronic systems, it is reasonable to assess whether the user-system interface encourages the formation of appropriate mental models of system behavior and facilitates meaningful dialogue between device and user. Recovery from a malfunctioning device will be slim if the user does not have a workable mental model of its underlying functionality.

Postmarket Follow-Up Despite the iterative nature of the design process that leverages user feedback to improve subsequent design versions, it is not uncommon for less than ideal designs to reach market. Software code, for example, may be written in such a manner that it does not recognize the full range of data entry variations by human operators. Human-device interface inadequacies may reveal themselves only under unusual circumstances that unfortunately will eventually occur such as the infrequent but recurring malfunction of the Therac-25 linear accelerator that massively over-irradiated six people (Kaye, Henriksen, Morisseau, & Deye, 1993; Leveson, 1995). Hence postmarket follow-up is a necessary step to identify unanticipated threats to safety that can be removed by product recall and redesign.

The preceding sequence of design events suggests a linearity and order that does not exist in reality. Human factors specialists may be called on to work just specific aspects of these activities rather than having involvement through all phases of design. Design engineers draw on past experience and rely on intuitive judgment that can be faulty. Because of schedule pressures in the overall design cycle, human factors tests with mock-ups and quasiexperiments are not always conducted. Like many interdisciplinary exchanges, learning to work with another discipline is not always easy, but is necessary. Design engineers can benefit by being more receptive to guidelines and input from human factors experts; human factors personnel are of greatest value when they present cognitive and behavioral data in useful form. Fortunately, because of the iterative nature of the design process and the extensive test and feedback activity that occurs, serious design flaws are becoming fewer. Working in teams and developing effective interdisciplinary communication skills is as challenging in system design as it is in health care.

Differences and Similarities

Although the knowledge, skills, and aptitudes of human factors practitioners overlap to a certain extent with those of their health services research counterparts, there are differences in the prevailing paradigms to which each group has become acculturated. Table 2–1 summarizes some of these differences and similarities. Some of these differences parallel differences between research and evaluation (Isaac & Michael, 1981). Evidence-based medicine *qua* research seeks new knowledge, attempts to establish cause–effect relationships or associations, and formulates hypotheses from which study designs are derived. RCT, higher-order quasiexperimental designs, and epidemiological approaches serve as the most vaulted paradigms. Threats consist of factors that can undermine internal and external validity whereas remaining challenges involve incorporating the results into practice and having a favorable impact on the health and lives of patients. Although basic research is conducted by human factors professionals in university and government laboratory settings, the purpose of most human factors activity as a practice is to improve a process or product. Testing and evaluating new processes or designs to inform specific decisions represent different end values compared to testing hypotheses and reaching generalizable conclusions. As shown in the table, the predominant paradigm is an iterative specify–design–test and evaluate process using mock-ups, rapid prototyping techniques, as well as quasiexperimental designs when circumstances permit. Threatening elements consist of failing to understand user needs and inadequate definition of project scope (e.g., "We built a Lexus and all they wanted was a Volkswagen."). Presenting information in useful form to personnel not trained in human factors and demonstrating that recommended changes have a favorable impact on end users remain as challenges.

Since World War II, human factors engineering has come the way of technology in that changes typically are made in advance of a strong evidence base. With process improvement or product delivery as the goal, and working under tight timelines, establishing a RCT evidence base before making an improvement would be considered a luxury, if not misdirected. This does not mean that there is no evidence base or that changes are made in a vacuum. New prototype changes for an interface can be made and tested in simulated form on computer screens as was done with a patient controlled analgesia device (Lin, Vicente, & Doyle, 2001). Ideally, results can be fed back to the designers and manufacturers where changes can be made. Designers and manufacturers, however, frequently ignore recommendations from human factors and usability experts (Shneiderman, 2003). Vicente (2003) provided an interesting case study of a recalcitrant medical device manufacturer that eventually incorporated human factors input into the design process, but only after powerful macro- and megalevel sociotechnical forces converged to facilitate the change.

System Safety as an Emergent Process

Despite the nature of the evidence used in assessing the efficacy and effectiveness of patient safety practices, the prevalent response to managing medical errors continues to be to put out the fire, identify the individuals involved, determine their culpability, schedule them for retraining or disciplinary measures, introduce retro-fixes or new procedures,

TABLE 2–1. Comparison of Evidence-Based Medicine and Human Factors Paradigms

Property	*Evidence-Based Medicine*	*Human Factors Practice*
Purpose	New knowledge, truth	System improvement, product delivery
Conceptual basis	Cause-effect relationships, associations	Optimize human performance, minimize operator error
Impetus	Curiosity, lack of knowledge	Need for improvement
Cognitive pursuit	Hypothesis testing	Hypothesis testing, test and evaluation
End value	Generalizable conclusions	Specific decisions
Predominant paradigms	Randomize clinical trial	Systems design
	E R O X O C R O O	Specify – design – test and evaluate
	Quasi experimental designs	Quasi experimental designs
	Epidemiological approach	Mock-ups, rapid prototyping
Threats	Factors undermining internal and external validity	Users needs undetermined, inadequate definition of project scope
Challenges	Translating research results into practice, demonstrating patient safety impact	Presenting information in useful form, demonstrating patient safety impact

and issue proclamations for greater vigilance. Closing the barn door after the horse has bolted is a very human response; however, practitioners who ignore a fuller complement of system factors and their interdependencies will soon discover other open doors. Because many adverse events have a relatively low probability of occurrence, it is tempting but foolhardy to gain a false sense of security and infer the fix is working, especially in the face of external pressures to "fix the problem." With medicine emphasizing knowledge acquisition and nursing emphasizing rules, it is not surprising that errors are regarded as personal failings with corrective actions focusing on sharp-end personnel rather than system properties.

One of the lessons from other high risk industries outside of health care is that safety does not come about as the result of a collection of randomized controlled trials. A tongue-in-cheek *British Medical Journal* article noted that the effectiveness of parachutes has not been subjected to a rigorous evaluation of randomized controlled trials and that the most vigorous advocates of evidence-based medicine who distain observational studies may want to serve as subjects in a double blind, randomized controlled trial (Smith & Pell, 2003). Suffice it to say that impressive gains in aviation safety were not made on clear-cut evidence that certain practices reduced the frequency of crashes. Instead, these gains resulted from the widespread implementation of hundreds of small system changes in equipment, procedures, operational environment, training of personnel, and organizational practices that occurred not all at once but accrued over the years to yield an incredibly strong track record and culture for safety.

In health care, the current practice of anesthesia provides a similarly impressive example of how system changes accrue to yield a high level of safety. Process improvements in anesthesia over the past couple of decades have resulted in a 10-fold reduction of mortality—a level of safety to which the very best of other industries strive (Leape et al., 2002). Once again, this accomplishment came about not as a result of controlled studies, but by applying a broad array of process changes, human factors principles, new equipment (e.g., pulse oximeter), organizational commitment, and teamwork and training practices, which over time took their toll for the better. Many of these practices, taken singly, would probably be insufficient to result in noticeable improvement, but when combined in the aggregate across several system components, the improvement can be considerable. In examining the performance improvement of anesthesiologists through the use of simulation (which at the time also lacked a solid evidence base), Gaba (1992) asserted "no industry in which human lives depend on skilled performance of responsible operators has waited for unequivocal proof of the benefits of simulation before embracing it" (p. 494).

Safety is a characteristic of systems and not of their components (Cook, 1998; Kohn et al., 2000). Safety is an emergent property of systems. For this emergent property to flourish, small, incremental changes in the direction of simplification, standardization, improved team performance, and managing the unexpected need to be introduced. Rather than succumbing to a methodological straightjacket, our research designs and test and evaluation procedures need to be sufficiently adaptable to capture these changes.

INTERDEPENDENCE OF SYSTEM COMPONENTS: TAKING OUR OWN MEDICINE

Reason's (1990) distinction between latent and active errors, subsequently referred to as latent conditions and active failures (Reason, 1997), has had a tremendous impact on our thinking about the interdependent role that system components play in bringing about accidents. To use Reason's metaphor, *latent conditions* are the resident pathogens that lie dormant for some time, yet combine with other unsuspected pathogens to thwart the system's defenses and lead to error. Errors at the point of the provider–patient interface frequently result from the unique concatenation of several necessary but singly insufficient factors that are present in the system long before the occurrence of the adverse event. Latent conditions can include poor interface design, communication breakdown, gaps in supervision, incorrect equipment installation, faulty maintenance, gaps in information transfer, clumsy automation, hidden software bugs, fast paced production schedules, unworkable procedures, extended work hours, staffing problems, inadequate training, aloof management, absence of a safety culture, and poor organizational resilience, among other things. A number of these latent conditions were found operating in a human factors evaluation of the external beam radiation therapy environment for treating cancer patients (Henriksen, Kaye, Jones, Morisseau, & Serig, 1995; Henriksen, Kaye, & Morisseau, 1993). Just as health care in general has been slow compared to other hazardous industries in embracing human factors and ergonomic principles, patient safety researchers have been slow in embracing not the components themselves, but the interdependence among system components. The device people like to study devices, the reporting system people like to record near misses and adverse events, and the organizational people like to focus on organizational issues. But are we looking beyond our own silos and examining how our activities impact other components of the system? In brief, are we failing to take our own medicine and seriously examine the interdependent nature of complex and open systems?

One interesting facet of health care is that it is a target-rich environment for those who indeed like to study the interdependencies among components. There is no scarcity of interacting components to study. This is how the authors of the *To Err Is Human* (Kohn et al., 2000) report defined our health system:

> Health care is composed of a large set of interacting systems—paramedic, and emergency, ambulatory, inpatient care, and home health care; testing and imaging laboratories; pharmacies; and so forth—that are connected in loosely coupled but intricate networks of individuals, teams, procedures, regulations, communications, equipment and devices that function with diffused management in a variable and uncertain environment. (p. 158)

In fact, one may be taking unwarranted liberties by even referring to health care in the United States as a system, given the qualities of loose coupling, uncertainty, diffuse management, and lack of a common goal implied in the above definition. Health care in the United States is highly fragmented, ranging from thousands of small provider practices to large hospitals where decision making is highly decentralized and accountability is difficult to discern (Richardson & Corrigan, 2003). Efforts at improving or redesigning health care not only require a sound understanding of the processes of care in particular clinical specialty areas, but also an understanding of the interacting work system factors made up of people (disciplines) performing

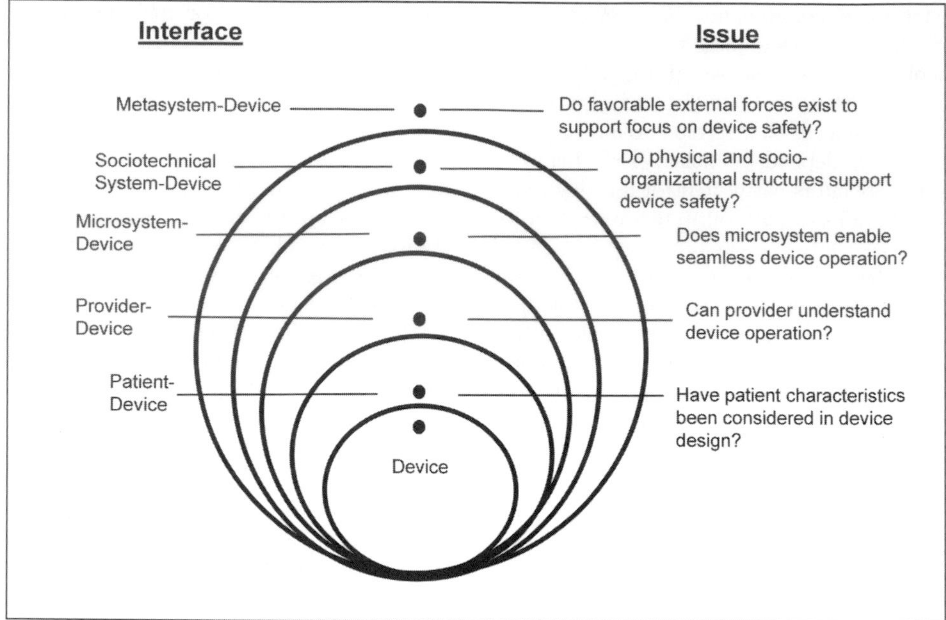

Fig. 2–1. An expanding progression of device interfaces.

tasks using various tools and technology within a physical environment in pursuit of organizational goals (Carayon & Smith, 2000). It further requires an understanding of the broadest of macroergonomic or metasystem forces—economic trends, legislation and regulation, medical and health care education, professional interest groups, information technology—that shape care delivery. With such a wealth of opportunity for improved understanding, a challenge facing the human factors researcher is determining what to focus on and organizing the focus in a systematic, programmatic framework.

A Progression of Interfaces

For the researcher interested in technology and medical devices, one approach is to recognize that medical device use involves an expanding progression of interfaces each with their own vulnerabilities and opportunities for confusion (Henriksen, 2003). A basic realization that is not always appreciated is that a device does not exist by itself. There are always attachments, couplings, tubing, cables, displays, and other peripheral items with which the device interfaces, as well as people, the immediate microsystems, a broader sociotechnical system, and even a metasystem. As shown in Figure 2–1, five expanding interface levels can be identified in health care settings. Starting at the very bottom and progressively working outward, we have a progression of interfaces: (a) patient-device, (b) provider-device, (c) microsystem-device, (d) sociotechnical system-device, and (e) metasystem-device. The basic idea is to identify interface incompatibilities at each of the expanding levels that are frequently neglected yet that have the potential to harm patients. By progressively working backward through an expanding series of interfaces—from sharp end to blunt end, from micro to macro, from active errors to latent conditions—a thorough, robust, and programmatic effort can be launched for realizing significant improvement in patient safety.

Patient-Device Interface. With respect to the patient-device interface, what physical, cognitive, and affective characteristics need to be taken into account in the design and use of the device? Does the device or accessory attachments need to be fitted or adapted to the patient? What sort of understanding does the patient need to have of device operation? Are there perceptual–motor skills that need to be acquired for device use? Does actual use of the device depend on remembering to use it? Given the increasing migration of sophisticated devices into the home and strong economic incentives to move patients out of hospitals as soon as possible, safe home-care device use becomes not only a human factors concern from the traditional perspective of patient-device interface incompatibilities, but these trends also pose a serious macroergonomic challenge, especially with patients who are aged, probably sicker as a result of shorter hospital stays, and where adequate assessment of the capability of home caregivers (also likely to be aged) and the suitability of the home environment for this purpose is lacking. In this one example, it is fairly easy to see that at least three levels require careful examination and alignment if significant improvements are to occur: patient limitations with respect to the device, the device in relation to the immediate environment (micro system) in which it is placed, and device use in relation to broader health socioeconomic (metasystem) factors.

Provider-Device Interface. A similar set of concerns exist for the provider (user)-device interface. Health care providers must be prepared not only to adapt to changes in the management and structure of health care organizations, but also to technological innovations such as new monitoring systems, smart infusion pumps, electronic medical records, bar coding, and computerized physician order entry, not to mention the seamless integration of these separately developed technologies. Can providers (e.g., nurses, physicians, anesthesiologists) confidently operate the controls and navigate the device, determine its current state, and maintain an accurate mental model that reflects the actual functioning of the device? Given the opaque nature of computer-controlled systems, it is not uncommon for users to have a poor understanding of the full functionality of the device or even what state the device is in. Flawed mental models exact their toll when the device malfunctions. With a flawed mental model, the malfunction remains a mystery, appropriate knowledge-based behavior cannot be acquired on short notice, and the probability of provider-device recovery is slim. The design challenge here is the creation of a user-device interface that promotes the formation of appropriate mental models and facilitates meaningful dialogue between user and device. Although human factors assessments traditionally have focused on the user-device interface and problems associated with underlying opaque functionality, relatively little attention has been directed at discerning the optimal and less-than-optimal conditions in the immediate environment for learning to use the device as intended. Variation among personnel in effective device use is especially likely to occur in those sociotechnical environments that fail to appreciate the "learning curve" associated with devices. These organizations fail to recognize that the manner in which provider personnel are introduced to new devices and systems is likely to have a considerable impact on user acceptability, the types of errors that occur, how quickly users become proficient with the device, and the ultimate costs associated with switching over to new systems. Many of these problems can be avoided with more sensible technology adoption and learning strategies. Without strong organizational leadership to support the intelligent introduction of new technology and systematic learning opportunities for sharp-end personnel, mastery in the operation and maintenance of relatively complex devices will continue to remain a haphazard art.

Microsystem-Device Interface. Medical devices in and of themselves are rarely stand-alone devices but instead are tied into and coupled with other components and accessories that collectively are intended to function as a seamless, integrated system even though the components may come from different vendors. Providers at the sharp end all too frequently experience situations that reinforce the notion that seamless integration cannot be taken for granted. Take, for example, the Sentinel Event Alert issued by the Joint Commission on Accreditation of Healthcare Organizations (JCAHO) regarding a series of deaths and injuries in patients on long term ventilation. Three contributing factors were identified: an alarm on the ventilator that malfunctioned or was misused, a tubing disconnect, and a dislodged airway tube. None of the cases were directly related to malfunctions of the ventilator itself. In the majority of cases, inadequate training, insufficient staffing levels, and communication breakdowns also were cited.

In hospital operating rooms and ICUs, there is a bewildering array of monitoring units from different manufacturers that interface with various devices that place an added "learning curve" burden on provider personnel. Greater involvement of the hospital's purchasing officer or chief financial officer in understanding the impact of purchasing decisions on provider performance, quality of care, and patient safety is a macroergonomic issue needing to receive more attention. Unsettling reports of medical gas mix ups leading to fatalities pinpoint another microsystem interface problem with wider ranging macroergonomic corrective actions. On several occasions, nitrogen and carbon dioxide have been mistakenly connected to the oxygen supply system as a result of someone over-riding the system's built-in safeguards by retrofitting adapters or connectors. The lesson learned here is that designing oxygen connectors so that they will only fit the oxygen container on the patient's gas supply system, in and of itself, may not be sufficient. Instances exist where maintenance personnel have changed this forcing function. Other components of the greater sociotechnical system need to be involved such as training, more frequent inspections, and an organizational safety culture that doesn't regard device safety as simply a device problem.

Sociotechnical-Device Interface. How well does the device fit into the surrounding physical environment and socioorganizational structure with which it will interact and be supported by for appropriate operation? This includes consideration of device requirements as well as patient and provider needs and characteristics with respect to facilities design and workspace layout, organizational policies and procedures regarding device use and quality assurance, ensuring that personnel are properly trained in the operation of the device, ensuring that the device is properly maintained and calibrated, ensuring that all entities (e.g., chief operating officers) that have something to do with device use—no matter how far upstream from the point of patient care—are collectively mindful of their role in enhancing patient safety.

Hendrick and Kleiner (2001) underscored the notion of mutual interdependence among four primary sociotechnical system components: the technological subsystem, personnel subsystem, external environment, and work system design. These four components encapsulate many of the latent conditions described by Reason (1997). When some aspect of one of the four components is changed, it has an impact on the other three. All too frequently, however, managers fail to recognize the interdependence among components, operating instead in "*stovepipes*" or "*silos*." Failing to recognize the interdependencies among system components, managers are caught off-guard when the ripple effects result in unique combinations of latent conditions, workplace factors, and provider actions that surely have their impact. Not all the impacts are benign, of course; too many manifest themselves as preventable harm to patients.

Metasystem-Device Interface. The metasystem-device interface represents the broadest or most macroergonomic of levels within which the device resides and interacts. What are the factors that operate at this level? For a start, they include economic trends, legislation and regulation, medical and health care education, medical malpractice and tort reform, new product concepts and designs, technological innovations, focusing events such as the Duke organ transplant mismatch, the national mood, and public opinion. When favorably aligned, these forces are sufficient to tip the usual inertia in altering the status quo and can actually usher in change. Vicente (2002) noted that a traditional human factors orientation with "knobs and dials" has limited the impact of human factors on improving patient safety. Many of the most significant threats to patient safety exist at the microsystem level and beyond. Drawing from risk management work by Rasmussen (1997), Vicente (2002) describes various levels—work, staff, management, company, regulators/associations, and government—of complex sociotechnical systems and acknowledged that although horizontal improvements can be made within levels, significant advances in patient safety and the health care system as a whole can only occur with appropriate vertical integration and alignment of the interdependencies across levels. These levels show a parallel progression of interdependencies to the progression of interfaces described here.

The undesirable consequences of clumsy medical devices are exacerbated when they add to the mental workload of providers who are already understaffed, when upper management remains aloof to what occurs at the microsystem level, where production goals and pressures outweigh safety protections, and where professional leadership and appropriate governance is missing. Participants at all levels need to focus not only on their own local concerns but also on how these concerns are shaped by interdependencies among other levels and interfaces of the entire metasystem.

Fragmentation Trumps Alignment

For the near term, health care in the United States is likely to remain fragmented and managed in a very loose and diffused fashion. Although health care administrators and providers are becoming more aware that they are part of a system that is not vertically aligned very well, understanding the functioning of macroergonomic factors has remained a somewhat elusive art form. Resources are rarely distributed evenly among a system's components; inequities create quality, reliability, and safety problems within the system. Decisions are made in a loose, diffuse, and disorderly fashion, usually under delayed, worsening conditions. The consequences accrue gradually over time and interact with other variables that are lying dormant and that are not fully recognized. Poor quality care is the consequence, yet contributing macrolevel factors are difficult to isolate and determine. Overworked providers at the sharp end are the last line of defense and the most vulnerable for it is they who inherit the less recognized sins of omission and commission of everyone else who has played a role in the design of the greater sociotechnical system (Reason, 1990).

Although no one set out purposefully to create such a daunting system, the complexity of health care has evolved immensely. Adding to the complex and open problem space, vested interests with strong backing and memberships are omnipresent. Information that is accessed and used serves to justify positions already taken. Rational attributions are offered to make sense out of what has already happened (Pfeffer & Salancik, 1977). Weick (2002) used the term *consensual neglect* to refer to the tendency of organizational decision makers to tacitly agree to ignore many of the unexpected events they encounter to achieve unity of purpose and act as a single entity. Disruptive and politically incorrect issues are ignored, overly simplified, or become homogenized into more acceptable terms. Yet if no one is willing to point out that the emperor is acting strangely and a bit too casual in his dress, the realization that fragmentation and misalignment are very much a part of the patient safety equation will go unrecognized.

In her analysis of the *Challenger* disaster, Vaughan (1996) described a more subtle form of organizational misalignment—those small, incremental erosions to safety and quality that over time become the norm—that she referred to as *normalization of deviance.* Disconfirming information is explained away and brought into the realm of acceptable risk, thereby reducing any cognitive dissonance and preserving the original belief that things are essentially okay. A similar normalization of deviance has happened in health care with the benign acceptance of shortages and adverse working conditions for nurses.

Rather than normalizing and accepting greater and greater risks, high reliability organizations operate much differently (Weick, 2002). In these organizations, there is a shared awareness at all levels of the possibility for failure and a heightened sensitivity to operations. There is an expectation that errors will be made and so the task is to create organizational resilience by anticipating, detecting, and recovering from error before harm occurs. To preclude adverse events, continued adjustments are made to variations, which if ignored could escalate and lead to disaster. Because of these continued adjustments, safety is viewed as a *dynamic nonevent* where things may appear smooth and routine on the surface, but there is plenty of corrective activity beneath. In brief, organizational members have a shared mental model—a collective mindfulness—of the potential for dynamic states to unfold and create new threats to safety.

Unfortunately, few high reliability organizations exist in health care. Although most patient safety research has focused on the downstream errors made by individual providers and work teams, the case is made here that it is time to start addressing the organizational, managerial, cultural, and external environmental issues that operate at the macroergonomic levels. In the drive to cut health care costs, there has been a steady drumbeat to do more with less, forcing overburdened personnel to the edge of the safety envelope where increased risks exist for patients. As production pressures increase, investment in safe practices declines. Clearly now is the time to start focusing on the interdependence of macroergonomic factors and to achieve the alignments that have been neglected far too long.

BRINGING ABOUT TRANSFORMATIONAL CHANGE: A ROLE FOR HUMAN FACTORS

Assume as a consultant you are hired by a hospital chief executive officer (CEO) to lay out a plan for transforming the hospital from a culture of blame to a culture of safety and to help lead the change. What would you do? It is probably safe to assume that human factors professionals typically do not perceive themselves as change agents or as planners of organizational transformations. After all, it is not something for which many of us have received any training. Relatively few professions train their members to be change agents. Yet there is a serious need for transformational change in health care and given that its traditional assumptions and paradigms have been called into question, there is plenty of opportunity for individuals to venture out of their accustomed silos, stretch their comfort zones a bit, and make a difference. In fact, during the past 20 years, a number of human factors pioneers (many of them contributing chapters to this handbook) have populated the frontiers of various clinical care settings, gaining a valued understanding of the demands, constraints, and uncertainties in these work environments so that their tools and skills could be applied appropriately. However, other knowledge, skills, and abilities are needed if we are to take advantage of favorable external forces (i.e., external jolts like the IOM report, increasing public awareness, changing political climate) that can facilitate the needed sociotechnical system alignments for bringing about a culture of safety. Knowledge and skill in implementing transformational change provides a potential vehicle for bringing about the needed alignments.

Transformational Change

Transformational change or leadership refers to the process of taking an organization or work unit that has probably performed well enough to maintain the status quo but now needs to "reinvent" or transform itself into an entity that is much better equipped to respond to the challenges of a rapidly changing external environment. It may involve taking fuller advantage of advanced technology, streamlining operational processes and procedures, redesigning work flow, cutting out redundancy and waste, significantly improving quality, shortening turn-around times, and putting the customer first. More specifically in health care, it may mean revamping serious system problems that undermine patient safety, redesigning systems of care that curtail deficiencies and disparities of quality, restructuring the work environment of provider personnel, addressing performance deficits resulting from staffing shortages, long work hours, and unsafe practices, or establishing a health information technology infrastructure that realizes a higher standard of care and culture of safety. Transformational initiatives also mean that employees are empowered to adapt successfully to change, that immediate self-interests are transcended, that higher-order values are the focal point, and that everybody feels they have a stake and are personally involved in organizational learning and in bringing about a higher standard of care.

Transformational Versus Transactional Leadership

In the management literature, transactional leadership—a form of leadership practice prevalent in many of our government agencies, private corporations, and health care facilities—serves as a convenient foil for transformational leadership. Under conditions of *transactional leadership,* followers are motivated by appealing to their self interests. The relationship between leaders and followers revolves around an equitable exchange or quid pro quo. As long as both parties hold up their end of the agreement, everything is fine. If one of the parties falls short and does not hold up its end of the agreement, the relationship sours and typically dissolves. Under conditions of transactional leadership, one's activities serve as a means to an end. The end goal is some form of extrinsic reward (e.g., remuneration, prestigious job).

Transformational leadership practices, on the other hand, strive to raise the consciousness of followers by appealing to higher motives and values rather than self-interest. Drawing heavily from the work of Burns (1978), the IOM report on *Keeping Patients Safe: Transforming the Work Environment of Nurses* (Page, 2004) noted that "transformational leadership is in essence a relationship of mutual stimulation and elevation that raises the level of human conduct as well as the aspirations of both the leader and those led, and has a transforming effect on both" (p. 110). Those who are energized and find new meaning in their work are more likely to assume leadership roles themselves. Transformational leaders are those who clearly articulate a compelling vision of the future that evokes commitment and who establish pathways that are realistic and credible for attaining the end goals. The joint attainment of higher goals is now the primary basis of the relationship between leaders and followers and not just quid pro quo. Rather than one's activities serving as a means to an end, one's activities now represent the end values. Realization of the end values serves as the reward. The reward is said to be intrinsic rather than extrinsic because activities leading up to it are inherently satisfying and worthwhile.

One has identified with the end values to such an extent that achieving them serves as its own reward.

A Study in What Can Go Wrong

Despite occasional efforts at shared governance and empowerment of provider personnel, traditional command and control modes of leadership continue to reinforce the prevailing authority gradients that exist in most health care facilities. This mode of leadership and management does not lend itself very well to notions of two-way communication, mutual trust and sharing of information, and to the common pursuit of a culture of safety. If patient safety and higher quality care are jointly held higher order goals, leaders and managers will need to become more involved in sharp-end activities and better understand the needs and frustrations of providers. Likewise, providers will need to learn that although their individual concerns and aims are important, there will be times that pursuit of the collective goal will preclude immediate satisfaction of individual interests.

There is relatively little in the literature that describes reengineering or major change efforts in health care organizations. One exception, however, is a study involving 14 hospitals that had undertaken restructuring initiatives and that varied in size, geographic location, and ownership (Walston & Kimberly, 1997). Intensive interviews were conducted with 60 executives, 121 midlevel managers, 31 physicians, 24 staff nurses, and 19 other staff members. The study identified a number of change management practices, many of them interrelated, that were inconsistently implemented in the hospitals undergoing redesign. Although all the redesign initiatives started with varied communication strategies (e.g., newsletters, meetings, forums) that were well thought out and intended, it was not long before communication became less frequent, fading to the point where employees lacked feedback on project status and fell back on older, accustomed routines. Related to the lack of communication was a lack of measurement and data tracking. As a consequence, failure to record and display reengineering progress resulted in the perception that outcomes were unsustainable and undermined efforts to sustain the process (Walston & Kimberly, 1997). In addition to low employee involvement in developing the change initiatives, unit nurse managers and staff nurses were frequently assigned managerial duties without the necessary training. Many employees described situations where the reengineering effort drifted and where attention to it was simply short lived. A survey of CEOs of U.S. general medical–surgical urban hospitals with more than a 100 beds, however, found that 40% of those responding had not codified the change process through written procedures manuals, guidelines, or budgets (Walston, Burns, & Kimberly, 2000). Although the risk of malicious compliance that can result from excessive rules needs to be avoided, sustaining major restructuring requires some formalizing of the changes in written form, prepared guidelines, and management tools. At the same time, Behal (2004) notes that solely making structural, system redesign, or policy changes will unlikely have a lasting impact for moving from a culture of blame to a culture of safety. As is noted in the following section, there are other aspects of organizational life that require attention.

Transformational Stages and Practices

Bennis and Nanus (1985) and Tichy and Devanna (1986) conducted two of the better known qualitative interview studies with CEOs in a variety of corporations that that were in need of revitalization to adapt to a changing marketplace. The interviews were analyzed to identify processes that occurred when leaders were able to successfully turn their organizations around.

Recognizing the Need for Change. An initial requisite process, serving as an impetus for subsequent change, was first recognizing the need for change. Recognizing the need for change is difficult when the downward changes are gradual and when mature, highly in-grained organizational cultures focus on past successes rather than needed changes for the future. Under these conditions, a major leadership challenge is to convince other key people in the organization as to the seriousness of the problem and the need to make major changes and not just incremental adjustments. For change to occur there has to be a clear sense of a problem. A clear picture of the costs and undesired consequences associated with not making changes helps foster a sense of urgency. Gaining a clear picture of a problem that is dispersed and that has lied dormant over the years sometimes occurs only after aggregating disparate data in a manner that has high crises-like impact as was evident in the *To Err* Is Human (Kohn et al., 2000) report. Because of the highly fragmented nature of health care in the United States where no one is in charge, the problem is not so much in recognizing a need for change as it is in recognizing the need to align the multiple competing interests more harmoniously under a common goal.

Developing a Vision. Successful leaders of transformations are able to energize their workforces with a vision of an attractive future that is sufficiently compelling to justify the costs and uncertainties of changing comfortable patterns of behavior (Bennis & Nanus, 1985; Tichy & Devanna, 1986). Kotter (1986) also underscored the importance of creating an attractive picture of the future as part of major organizational change. Berwick's Saving 100,000 Lives Campaign is an excellent example of an urgent and compelling vision. At the time of writing, over 3,000 hospitals had expressed a commiment to implement changes that would improve patient care and prevent avoidable deaths (http://www.ihi.org, May 30, 2006). Given a widely recognized need for change, a sense of urgency, and a picture of an attractive future, energies are then channeled in the direction of obtaining the vision. Clear and appealing visions have several desirable properties. They provide a new sense of direction, give new meaning to the work, and collectively raise consciousness levels in terms of being part of a worthwhile enterprise. With a commonly shared vision, everybody knows what is important and how to separate the trivial from the nontrivial. False starts and wasted initiatives become fewer, decision making is facilitated, and self-management of staff at all levels is improved.

Developing Commitment and Trust. A clear vision is not enough, however. Commitment to changes that the new vision will bring and trust in the organization's leadership are essential if the new values and assumptions are to be embodied in the organization's evolving culture. Commitment and trust are tied closely to consistency among the decisions, statements, and actions of the organization's leadership. Leaders communicate values, priorities, and what is important by simply what they attend to, what they respond to, and what gets them excited (Schein, 1992). Sometimes values and priorities compete with one another and the selected course of action says a lot about the values the organization considers of utmost importance. Reason's (1997) discussion of the universal production–protection space of organizations, where there is a trade-off or tension between the forces of production in an organization and those of protection (safety), is quite relevant here. Far too often, production trumps safety. However, when the actions of those in leadership positions consistently demonstrate that safety comes first and that quality standards are of prime importance, the greater the likelihood these values will become embedded in the culture of the workforce. Changes made to the design of information systems, to the design of organizational structures, and to the design of facilities need to be consistent with the values and objectives contained in the new vision. The new facilities, systems, and procedures that get developed say a lot about the value placed on patient safety as amply illustrated in the design and building of a new community hospital based on safety-driven design principles (Reiling et al., 2004).

It is not uncommon for personnel working in the sharp-end trenches and those in managerial positions to have perceptions that are far apart on the same issue. Earlier studies in communication patterns have shown considerable disparity between the perceptions of supervisors and subordinates with respect to understanding subordinates' problems and obtaining ideas from subordinates (e.g., Likert, 1961). A more recent survey of clinicians and managers in an academic medical center also found that management had a more exalted view of safety than those at the sharp end (Pronovost et al., 2003). Firth-Cozens (2003) noted that leaders "need to stay close to the action if their organizations are to be not just seen as safe, but actually to be safe" (p. 401).

Implementing Change. Transformational change has an element of venturing into the unknown, of going where no one else has gone before. There is no well crafted script. Organizations undergoing change that realize they are involved in a generative process, that are trying to learn how to implement the knowledge they generate for more productive change, and that are formulating new ways of perceiving and thinking about challenges to better shape them are known as learning organizations. In brief, learning organizations are those that take learning-to-learn seriously and that continually expand their capacity to generate their future (Carroll & Edmondson, 2002; Fiol & Lyles, 1985; Senge, 1990).

It is essential that members of the leadership team be prepared and committed to undertake implementation activities. Support from middle and lower levels of management will be needed if a broad coalition is to be realized. Ensuring that these pockets of support are in place helps to minimize downstream obstacles. Action-oriented change agents that have the energy, commitment, and that can communicate a contagious confidence in the realization of the vision, if not already present, need to be strategically placed in the key positions of targeted change. The change agents, in turn, oversee task forces or transition teams that guide and mold the implementation. As implementers of the change, task forces, by necessity, alter existing organizational structures. Introducing new information technology, for example, introduces changes in work procedures, work flow, communication networks, performance standards, subject matter expertise that is needed, relationships with vendors, training initiatives, personnel decisions, organizational subunits, and shifts in organizational hierarchies. As the task forces rely on an assortment of analyses for determining how the work is to be performed, the importance of involving the end user in the design process cannot be overstated. End users not only have expertise in the functional requirements of the work, but as early arrivers to the change process they are in a position to assuage fears and share relevant information with coworkers. They also can be called on to help monitor the progress of change and to detect unintended consequences. Given any venture into uncharted waters, unintended consequences are to be expected. Systematic monitoring efforts can help detect and collect data on new forms of vulnerability that are introduced unintentionally, allowing corrective actions to be taken before harm occurs.

Sustaining the Change. Drawing attention to the gains and making them visible to all the parties involved in the change is a crucial step. Making the gains visible provides early evidence that the sacrifices and disruption are worth it, shows that hard work is recognized, and demonstrates to late arrivers and reluctant supporters that the change momentum is irreversible and moving forward. Having used the early visible successes to allay fears, build confidence, and to establish positive expectations regarding subsequent change, the stage is set for sustaining the change. To ensure that the change process maintains a forward momentum, removing recalcitrant barriers to change and codifying the change in formal structures will be needed. This may entail altering organizational units, reassigning personnel, redesigning jobs, specifying new work flows, initiating training programs, incorporating change objectives in performance appraisal plans, and posting and making visible organizational outcome measures. Periodic reinvigoration may be needed. New people who see themselves as change agents, prudent risk takers, and who like to be on the leading edge of new initiatives need to be brought on board. Change itself becomes the norm. Leaders from the HFE community are needed who are willing to venture into the unknown, who can articulate a core set of values that guide change, who understand the interdependencies of sociotechnical systems, and who see themselves as primary change agents for higher quality and safer health care organizations.

CONCLUDING COMMENT

The preceding three challenges by no means exhaust the set of possible issues that could or should be addressed. The reader, no doubt, could add several to the list. Despite the disparate nature of the challenges discussed, they are linked together by the overarching concern that significant gains in patient safety will not be realized by working in our comfortable *silos* and going about human factors and ergonomic business as usual. Many of the fundamental issues, which have been consensually neglected over the years, remain with us as tomorrow's challenges. Three of these fundamental issues that require clear articulation and improved understanding have been discussed: (a) what constitutes an acceptable evidence base in the patient safety research community for human factors practices, (b) are we focusing sufficiently on the interdependencies and needed alignments among system components in complex health systems, and (c) are we willing to serve as agents of transformational change in clinical care environments needing to undergo reengineering efforts?

The take-home messages are straightforward. First, for health care to benefit from the safety gains made in other industries that employ human factors principles, a less rigid and more open approach to evaluating the evidence base—one that is based on the nature of the questions asked rather than a hierarchy of study designs—will be needed. Second, given an increasingly complex and open health care system, greater focus on the interdependencies and needed alignments both within and between the various sociotechnical levels needs to occur. And third, there currently exists a great opportunity (which will likely continue in the years ahead) for human factors researchers and practitioners to join with their clinical counterparts to serve in leadership roles and continually expand health care's capacity to generate safe and high quality patient care environments. As noted by Leape (2004), health care needs a human factors voice and human factors professionals must speak out and play active roles. Given that critical self-examination and open discussion are hallmarks of a mature profession, we should be able to identify, debate, and overcome the challenges that undermine the full realization of human factors and ergonomic contributions to safety and quality of health care.

References

Allnutt, M. F. (1987). Human factors in accidents. *British Journal of Anaesthesiology, 59,* 856–864.

Behal, R. (2004). An organization development framework for transformational change in patient safety: A guide for hospital senior leaders. In B. J. Youngberg & M. J. Hatlie (Eds.). *The patient safety handbook* (pp. 51–65). Sudbury, MA: Jones and Bartlett.

Bennis, W. G., & Nanus, B. (1985). *Leaders: The strategies for taking charge.* New York: Harper & Row.

Bogner, M. S. (Ed.). (1994). *Human error in medicine*. Hillsdale, NJ: Lawrence Erlbaum Associates, Inc.

Brennan, T. A., Leape, L. L., Laird, N. M., Hebert, L., Localio, A. R., Lawthers, A. G., et al. (1991). Incidence of adverse events and negligence in hospitalized patients: Results from the Harvard Medical Practice Study I. *New England Journal of Medicine, 324,* 370–376.

Burns, J. (1978). *Leadership*. New York: Harper & Row.

Carayon, P., & Smith, M. J. (2000). Work organization and ergonomics. *Applied Ergonomics, 31,* 649–662.

Carroll, J. S., & Edmondson, A. C. (2002). Leading organisational learning in health care. *Quality and Safety in Health Care, 11,* 51–56.

Cook, R. I. (1998, November). *Two years before the mast: Learning how to learn about patient safety.* Presented at Enhancing Patient Safety and Reducing Medical Errors in Health Care, Rancho Mirage, CA.

Cook, R. I., & Woods, D. D. (1994). Operating at the sharp end: The complexity of human error. In M.S. Bogner (Ed.), *Human error in medicine* (pp. 255–310). Hillsdale, NJ: Lawrence Erlbaum Associates, Inc.

Cooper, J. B., Newbower, R. S., Long, C. D., & McPeek, B. (1978). Preventable anesthesia mishaps: A study of human factors. *Anesthesiology, 49,* 399–406.

Donchin, Y., Gopher, D., Olin, M., Badihi, Y., Biesky, M., Sprung, C. L., et al. (1995). A look into the nature and causes of human errors in the intensive care unit. *Critical Care Medicine, 23,* 294–300.

Fiol, C. M., & Lyles, M. A. (1985). Organizational learning. *Academy of Management Review, 35,* 505–538

Firth-Cozens, J. (2003). Evaluating the culture of safety. *Quality and Safety in Health Care, 12,* 401.

Gaba, D. M. (1992). Improving anesthesiologists' performance by simulating reality. *Anesthesiology, 76,* 491–494.

Gaba, D., Howard, S., & Fish, K. (1994). *Crises management in anesthesiology.* New York: Churchill-Livingstone.

Gaba, D. M., Maxwell, M., & DeAnda, A. (1987). Anesthetic mishaps: Breaking the chain of accident evolution. *Anesthesiology, 66,* 670–676.

Gopher, D., Olin, M., Badihi, Y., Cohen, G., Donchin, Y., Sieski, M., et al. (1989). The nature and causes of human errors in the medical intensive care unit. *Proceedings of the Human Factors Society 33rd Annual Meeting* , 956–960.

Hendrick, H. W., & Kleiner, B. M. (2001). *Macroergonomics—An introduction to work system design.* Santa Monica, CA: Human Factors and Ergonomics Society.

Henriksen, K. (2003, August). Macroergonomic interdependence in patient safety research. *Proceedings of the XVth Triennial Congress of the International Ergonomics Association, and the 7th Joint Conference of Ergonomics Society of Korea/Japan Ergonomics Society, Seoul, Korea.*.

Henriksen, K., Kaye, R. D., Jones, R., Morisseau, D. S., & Serig, D. I. (1995). *Human factors evaluation of teletherapy, Vol. I: Identification of problems and alternative approaches* (Rep. NUREG/CR–6277). Washington, DC: U.S. Nuclear Regulatory Commission.

Henriksen, K., Kaye, R. D., & Morisseau, D. S. (1993). Industrial ergonomic factors in the radiation oncology therapy environment. In R. Nielsen & K. Jorgensen (Eds.), *Advances in industrial ergonomics and safety V* (pp. 267–274). London: Taylor & Frances.

Hollnagel, E. (1993). *Reliability of cognition: Foundations of human reliability analysis.* London: Academic.

Isaac, S., & Michael, W. B. (1981). *Handbook in research and evaluation.* San Diego, CA: EdITS.

Kaye, R. D., Henriksen, K., Morisseau, D. S., & Deye, J. A. (1993). Human factors in radiation oncology therapy: Some software control issues. *Computer-based medical systems. Proceedings of the Sixth Annual IEEE Symposium,* (pp. 258–265). Los Alamitos, CA: IEEE Computer Society Press..

Kohn, L. T., Corrigan, J. M., & Donaldson, M. S. (Eds.). (2000). *To error is human—Building a safer health system.* Washington, DC: National Academy Press.

Kotter, J. P. (1996). *Leading change:* Boston: Harvard Business School Press.

Leape, L. L. (1994). Error in medicine. *Journal of the American Medical Association, 272,* 1851–1857.

Leape, L. L. (2004, Summer). Human factors meets health care: The ultimate challenge. *Ergonomics in Design, 12*(3), 6–12.

Leape, L. L., Bates, D. W., Cullen, D. J., Demonaco, H. J., Gallivan, T., Hallisey, R., et al. (1995). Systems analysis of adverse drug events. *Journal of the American Medical Association, 274, 35–43.*

Leape, L. L., Berwick, D. M., & Bates, D. W. (2002). What practices will most improve safety? Evidence-base medicine meets patient safety. *Journal of the American Medical Association, 288,* 501–507.

Leape, L. L. Brennan, T. A., Laird, N. M., Lawthers, A. G., Localio, A. R., Barnes B.A., et al. (1991). The nature of adverse events in hospitalized patients: Results from the Harvard Medical Practice Study II. *New England Journal of Medicine, 324,* 377–384.

Le Coq, A. D. (1987). Application of human factors engineering in medical product design. *Journal of Clinical Engineering, 12*(4), 271–277.

Leveson, N. G. (1995). *Safeware: Systems safety and computers.* Reading, MA: Addison-Wesley.

Likert, R. (1961). *New patterns of management.* New York: McGraw-Hill.

Lin, L., Vicente, K. J., & Doyle, D. J. (2001). Patient safety, potential adverse drug events, and medical device design: A human factors engineering approach. *Journal of Biomedical Informatics, 34*, 274–284.

Millenson, M. L. (1997). *Demanding medical excellence: Doctors and accountability in the information age.* Chicago: University of Chicago Press.

Norman, D. A. (1981). Categorization of action slips. *Psychological Review, 88,* 1–15.

Norman, D. A. (1988). *The psychology of everyday things.* New York: Doubleday.

Page, A. (2004). (Ed.). *Keeping patients safe: Transforming the work environment of nurses.* Washington, DC: The National Academies Press, Institute of Medicine of the National Academies.

Perrow, C. (1984). *Normal accidents—Living with high-risk technologies.* New York: Basic Books.

Petroski, H. (1992). *To engineer is human—The role of failure in successful design.* New York: Vintage.

Pfeffer, J., & Salancik, G. R. (1977). Organizational design: The case for a coalition model of organizations. *Organizational Dynamics, 6,* 15.

Pickett, R. M., & Triggs, T. J. (Eds.). (1975). *Human factors in health care.* Lexington, MA: Health.

Pronovost, P. J., Weast, B., Holzmueller, C. G., Rosenstein, B. J., Kidell, R. P., Haller, K. B., et al. (2003). Evaluation of the culture of safety: Survey of clinicians and managers in an academic medical center. *Quality and Safety in Health Care, 12*, 405–410.

Rappaport, M. (1970). Human factors applications in medicine. *Human Factors, 12*, 25–35.

Rasmussen, J. (1982). Human errors: A taxonomy for describing human malfunction in industrial installations. *Journal of Occupational Accidents, 4,* 311–35.

Rasmussen, J. (1987). The definition of human error and a taxonomy for technical system design. In J. Rasmussen, K. Duncan, & J. Leplat (Eds.), *New technology and human error* (pp. 23–30). Chichester, UK: Wiley.

Rasmussen, J. (1997). Risk management in a dynamic society: A modeling problem. *Safety Science, 27,* 183–213.

Reason, J. (1990). *Human error.* New York: Cambridge University Press.

Reason, J. (1997). *Managing the risks of organizational accidents.* Brookfield, VT: Ashgate.

Reiling, J. G., Knutzen, B. L., Wallen, T. K., McCullough, S., Miller, R., & Chernos, S. (2004). Enhancing the traditional hospital design process: A focus on patient safety. *Joint Commission Journal on Quality and Safety, 30*(3), 115–124.

Richardson, W. C., & Corrigan, J. M. (2003). Provider responsibility and system redesign: Two sides of the same coin. *Health Affairs, 22*(2), 116–118.

Safren, M. A., & Chapanis, A. (1960a). Critical incident study of hospital medication errors: Part 1. *Hospitals, 34*(9), 32–34, 57–58, 60, 62, 64, 66.

Safren, M. A., & Chapanis, A. (1960b). Critical incident study of hospital medication errors: Part 2. *Hospitals, 34*(10), 53, 65–66, 68.

Salas, E., Dickerson, T. L., Converse, S. A., & Tannenbaum, S. I. (1992). Toward an understanding of team performance and training. In R. J. Swezey & E. Salas (Eds.), *Teams: Their training and performance.* (pp. 3–29). Norwood, NJ: Ablex.

Sanders, M. S., & McCormack, E. J. (1993). *Human factors in engineering and design.* New York: McGraw-Hill.

Sanders, M. S., & Shaw, B. (1988). *Research to determine the contribution of system factors in the occurrence of underground injury accidents.* Pittsburgh, PA: Bureau of Mines.

Schein, E. H. (1992). *Organizational culture and leadership* (2nd ed.). San Francisco: Jossey-Bass.

Senders, J. W., & Moray, N. P. (1991). *Human error: Cause, prediction and reduction.* Hillsdale, NJ: Lawrence Erlbaum Associates, Inc.

Senge, P. M. (1990). *The fifth discipline: The art & practice of the learning organization.* New York: Doubleday/Currency.

Shneiderman, B. (2003). *Leonardo's laptop: Human needs and the new computing technologies.* Cambridge, MA: MIT Press.

Shojania, K. G., Duncan, B. W., McDonald, K. M., & Wachter, R. M. (Eds.). (2001). *Making health care safer: A critical analysis of patient safety practices. Evidence Report/Technology Assessment No.43* (AHRQ Publication No. 01-E058). Rockville, MD: Agency for Healthcare Research and Quality.

Smith, G. C. S., & Pell, J. P. (2003). Parachute use to prevent death and major trauma related to gravitational challenge: Systematic review of randomised controlled trials. *British Medical Journal, 327,* 1459–1461.

Smith, M. J., & Carayon-Sainfort, P. (1989). A balance theory of job design for stress reduction. *International Journal of Industrial Ergonomics, 7*, 67–79.

Thomas, E. J., Studdert, D., Newhouse, J. P., Zbar, B. I. W., Howard, K. M., Williams, E. J., et al. (1999). Costs of medical injuries in Colorado and Utah in 1992. *Inquiry, 36,* 255–264.

Thompson, P. W. (1987). Safer design of anaesthesia equipment. *British Journal of Anaesthesia, 59,* 913–921.

Tichy, N. M., & Devanna, M. A. (1986). *The transformational leader.* New York: Wiley.

Vaughan, D. (1996). *The challenger launch decision: Risky technology, culture and deviance at NASA.* Chicago: Chicago University Press.

Vicente, K. J. (2002). From patients to politicians: A cognitive engineering view of patient safety. *Quality and Safety in Health Care, 11,* 302–304.

Vicente, K. J. (2003). What does it take? A case study of radical change toward patient safety. *Joint Commission Journal on Quality and Safety, 29*(11), 598–609.

Vincent, C. (2001). *Clinical risk management—Enhancing patient safety.* London: BMJ Books.

Walston, S., & Kimberly, J. (1997). Reengineering hospitals: Evidence from the field. *Hospital and Health Services Administration, 42*(2), 143–163.

Walston, S., Burns, J., & Kimberly, J. (2000). Does reengineering really work? An examination of the context and outcomes of hospital reengineering initiatives. *Health Services Research, 34*(6), 1363–1388.

Weick, K. E. (1987). Organizational culture as a source of high reliability. *California Management Review, 29*(2), 112–127.

Weick, K. (2002). The reduction of medical errors through mindful interdependence. In M. M. Rosenthal & K. M. Sutcliffe (Eds.), *Medical error: What do we know? What do we do?* San Francisco: Jossey-Bass.

Weinger, M. B., & Englund, C. E. (1990). Ergonomics and human factors affecting anesthetic vigilance and monitoring performance in the operating room environment. *Anesthesiology, 73*, 995–1021.

Wiener, E. L., Kanki, B. G., & Helmreich, R. L. (Eds.). (1993). *Cockpit resource management.* San Diego, CA: Academic.

Woods, D. D., Johannesen, L. J., Cook, R. I., & Sarter, N. B. (1994). *Behind human error: Cognitive systems, computers, and hindsight* (CSERIAC SOAR 94–01). Dayton: Wright-Patterson AFB OH.

Section

II

MACROERGONOMICS AND SYSTEMS

3

A HISTORICAL PERSPECTIVE AND OVERVIEW OF MACROERGONOMICS

Hal W. Hendrick
University of Southern California, Los Angeles

Ergonomics, or human factors as the discipline also is known, is considered to have had its formal inception in the late 1940s. Ergonomics initially focused on the design of human–machine interfaces and physical environment factors as they affect human performance. This emphasis on optimizing the interfaces between individual operators and their immediate work environment, or what herein will be referred to as *microergonomics,* characterized the ergonomics/human factors discipline for the first 3 decades of its formal existence.

In North America, this initial microergonomic focus was on the human factors (e.g., perception, response time, transfer of training, etc.) associated with the design of such sociotechical systems as aircraft cockpits, gun sites, and other military systems, and the field accordingly was labeled as *human factors.* In Europe, following World War II, the primary focus was on rebuilding war-torn industries, and the primary emphasis was on applying the science of work, or *ergonomics* (ergo = work; nomics = study of), to the design of factory workstations. Although the primary emphasis was somewhat different, the objective of both human factors and ergonomics was to apply scientific knowledge about human capabilities, limitations, and other characteristics to the design of operator controls, displays, tools, seating, workspace arrangements, and physical environments to enhance health, safety, comfort, and productivity and to minimize design induced human error (Chapanis, 1991, 1992; Hendrick, 2000). In more recent decades, not only has human–machine interface design been applied to a broader number of systems but also to a progressively increasing number of consumer products to make them safer and more usable, including medical devices.

Beginning with the development of the silicon chip, the rapid development of computers and automation occurred and along with this development came an entirely new set of human–system interface problems. As a result, a new subdiscipline of ergonomics emerged that centered on software design and became known as *cognitive ergonomics.* The emphasis of this new subdiscipline was on how humans think and process information and how to design software to dialog with humans in the same manner. As a direct result of the development of the cognitive ergonomics subdiscipline, the human factors/ergonomics discipline grew about 25% during the 1980s. However, the emphasis remained on individual human–system interfaces.

In the late 1970s, ergonomists began to realize that our traditional predominate focus on individual human–system interfaces, or *microergonomics*, was not achieving the kinds of improvements in overall total system functioning that both managers and ergonomists felt should be possible. Ergonomists

Adapted, updated, and modified from Hendrick, H. W., & Kleiner, B. M. (2001). Macroergonomics: An introduction to work system design. Santa Monica, CA: Human Factors and Ergonomics Society.

began to realize that they could effectively design human–machine, human–environment, and human–software interfaces and still have a poorly designed work *system*. Out of this realization developed the subdiscipline of *macroergonomics*.

THE BEGINNING: SELECT COMMITTEE ON HUMAN FACTORS FUTURES, 1980–2000

As with most new subdisciplines, one can identify a number of precursors within the larger discipline, or related disciplines, and this is the case with macroergonomics. However, the most direct link leading to the formal development of macroergonomics as a distinct subdiscipline can be traced back to the U.S. Human Factors Society's Select Committee on Human Factors Futures, 1980–2000. In the late 1970s, many dramatic changes were occurring in all aspects of industrialized societies and their built environments. Arnold Small, Professor Emeritus of the University of Southern California (USC) and former president of The Human Factors Society, noted these changes and believed that traditional human factors/ergonomics would not be adequate to effectively respond to these trends. At Professor Small's urging, in 1978 the Human Factors Society (now, the Human Factors and Ergonomics Society) formed a Select Committee on Human Factors Futures, 1980–2000 to study these trends and determine their implications for the human factors discipline. Arnold was appointed chair of this Select Committee. He, in turn, appointed me to that committee and specifically charged me to research trends related to the management and organization of work systems.

After several years of intensive study and analysis, the committee members reported on their findings in October 1980 at the Human Factors Society annual meeting in Los Angeles, California. Among other things, I noted the following six major trends as part of this report.

1. *Technology.* Recent breakthroughs in the development of new materials, microminiaturization of components, and the rapid development of new technology in the computer and telecommunications industries would fundamentally alter the nature of work in offices and factories during the 1980 to 2000 time frame. In general, we were entering a true information age of automation that would profoundly affect work organization and related human–machine interfaces.
2. *Demographic shifts.* The average age of the work populations in the industrialized countries of the world would increase by approximately 6 months for each passing year during the 1980s and most of the 1990s. Two major factors account for this "graying" of the workforce. First, the aging of the post-World War II "baby boom" demographic bulge that now has entered the workforce. The second factor was the lengthening of the average productive life span of workers because of better nutrition and health care. In short, during the next 2 decades, the workforce would become progressively more mature, experienced, and professionalized. As the organizational literature has shown (e.g., see Robbins, 1983), as the level of professionalism (i.e., education, training, and experience) increases, it becomes important for work systems to become less formalized (i.e., less controlled by standardized procedures, rules, and detailed job descriptions), tactical decision making to become decentralized (i.e., delegated to the lower-level supervisors and workers), and management systems to similarly accommodate. These requirements represent profound changes to traditional bureaucratic work systems and related human–system interfaces.
3. *Value changes.* Beginning in the mid 1960s and progressing into the 1970s, a fundamental shift occurred in the value systems of workforces in the United States and Western Europe. These value system changes and their implications for work systems design were noted by a number of prominent organizational behavior researchers and were summarized by Argyris (1971). In particular, Argyris noted that workers now both valued and expected to have greater control over the planning and pacing of their work, greater decision-making responsibility, and more broadly defined jobs that enabled a greater sense of both responsibility and accomplishment. Argyris further noted that, to the extent organizations and

work system designs did not accommodate these values, organizational efficiency and quality of performance would deteriorate. These value changes were further validated in the 1970s by Yankelovich (1979) based on extensive longitudinal studies of workforce attitudes and values in the United States. Yankelovich found these changes to be particularly dramatic and strong among those workers born after the World War II. In addition, Yankelovich particularly noted that these baby-boom workers expected to have meaningful tasks and the opportunity for quality social relationships on the job.

4. *Ergonomics-based litigation.* In the United States, litigation based on the lack of ergonomic safety design of both consumer products and the workplace is increasing, and awards of juries often have been high. The message from this litigation is clear: Managers *are* responsible for ensuring that adequate attention is given to the ergonomic design of both their products and their employees' work environments to ensure safety.

   One impact of this message, as well as from the competition issue noted previously, is that ergonomists are likely to find themselves functioning as true management consultants. A related implication of equal importance is that ergonomics education programs will need to provide academic courses in organizational theory, behavior, and management to prepare their students for this consultant role.

5. *World competition.* Progressively, U.S. industry is being forced to compete with high quality products from Europe and Japan; and products from other countries, such as Taiwan and Korea, soon will follow. Put simply, the post-World War II dominance by U.S. industry is gone. In light of this increasingly competitive world market, the future survival of most companies will depend on their efficiency of operation and production of state-of-the-art products of high quality. In the final analysis, the primary difference between successful and unsuccessful competitors will be the quality of the *ergonomic* design of their products and of their total work organization, and the two are likely to be interrelated.

6. *Failure of traditional (micro-) ergonomics.* Early attempts to incorporate ergonomics into the design of computer work stations and software have resulted in improvement but have been disappointing in terms of (a) reducing the work system productivity costs of white collar jobs, (b) improving intrinsic job satisfaction, and (c) reducing symptoms of high job stress.

As I noted several years later, we had begun to realize that it was entirely possible to do an outstanding job of ergonomically designing a system's components, modules, and subsystems yet fail to reach relevant systems effectiveness goals because of inattention to the *macro*ergonomic design of the over-all work *system* (Hendrick, 1984). Investigations by Meshkati (1986) and Meshkati and Robertson (1986) of failed technology transfer projects and by Meshkati (1990) of major system disasters (e.g., Three Mile Island and Chernobyl nuclear power plants and the Bhopal chemical plant) have all led to similar conclusions.

I concluded in my 1980 report that, "for the human factors/ergonomics profession to truly be effective, and responsive to the foreseeable requirements of the next 2 decades and beyond, there is a strong need to integrate organizational design and management (ODAM) factors into our research and practice" (p. 5). It is interesting to note that all of these predictions from 1980 have come to pass and are continuing. These needs, and the outstanding success of a number of macroergonomic interventions, would appear to account for the rapid growth and development of macroergonomics that since has occurred.

INTEGRATING ODAM WITH ERGONOMICS

As a direct response to my report, in 1984 an ODAM technical group was formed within the Human Factors Society. Similar groups were formed that year in both the Japan Ergonomics Research Society and the Hungarian society, and less formal interest groups were formed in other ergonomics societies internationally. In 1985, the International Ergonomics Association (IEA) formed a Science and Technology Committee comprised of eight technical committees. The first eight committees formed were based on the input from the various federated societies as to what areas could most benefit from an international-level technical committee. One of those first eight was an ODAM technical committee (TC). This TC consistently has been one of the IEA's most active. For example, the IEA ODAM TC helped organize the highly successful biennial IEA International Symposia on Human Factors in ODAM, with the proceedings of each being commercially published by either North-Holland or the IEA Press.

In 1988, in recognition of its importance to the ergonomics discipline, ODAM was made 1 of the 5 major themes of the 10th IEA Triennial Congress in Sidney, Australia. In recognition of both its importance and rapid growth, it was one of 12 themes for the 11th Triennial Congress in Paris, France in 1990. At the 12th Triennial Congress in Toronto, Canada in 1994, and again at the 13th Congress in Tempare, Finland in 1997, a major multisession symposium on Human Factors in ODAM was organized. For both of these Congresses, more papers were received on macroergonomics and ODAM than on any other topic. A similar symposium was held during the 14th Triennial Congress in the year 2000 where ODAM/macroergonomics was one of the three areas with the most presentations. A similar trend was evident at the 15th Triennial Congress in Seoul, Korea and is expected to continue in the foreseeable future.

By 1986, sufficient conceptualization of the ergonomics of work systems had been developed to identify it as a separate subdiscipline, which became formally identified as *macroergonomics* (Hendrick, 1986). In 1998, in response to the considerable methodology, research findings, and practice experience that had developed during the 80s and 90s, the Human Factors and Ergonomics Society's ODAM TG changed its name to the Macroergonomics Technical Group (METG).

Definition of Macroergonomics

Conceptually, *macroergonomics* can be defined as a top-down, sociotechnical systems approach to work system design, and the carry-through of that design to the design of jobs and related human–machine and human–software interfaces (Hendrick, 1997; Hendrick & Kleiner, 2001).

Although top-down conceptually, in practice macroergonomics is top-down, middle-out, and bottom-up. *Top down* an overall work system structure may be prescribed to match an organization's sociotechnical characteristics [e.g., see the Macroergonomics Analysis of Structure (MAS) method later in this chapter]. *Middle out* an analysis of subsystems and work processes can be assessed both up and down the organizational hierarchy from intermediate levels and changes made to ensure a harmonized work system design. *Bottom-up* most often involves identification of work system problems by employees and lower-level supervisors that result from higher-level work system structural or work processes. Most often, a true macroergonomic intervention involves all three strategies. The macroergonomics design process tends to be nonlinear and iterative and usually involves extensive employee participation at all organizational levels.

Purpose of Macroergonomics

The ultimate purpose of macroergonomics is to ensure that work systems are fully harmonized and compatible with their sociotechnical characteristics. In terms of systems theory, such a fully harmonized and compatible system can result in synergistic improvements in various organizational effectiveness criteria, including health, safety, comfort, and productivity.

THEORETICAL BASIS OF MACROERGONOMICS

Macroergonomics is soundly grounded in both empirically developed and validated sociotechnical systems theory and general systems theory.

Sociotechnical Systems Theory and the Tavistock Studies

The sociotechnical model of work systems was empirically developed in the late 1940s and 1950s by Trist and Bamforth (1951) and their colleagues at the Tavistock Institute of Human Relations in the United Kingdom. Follow-on research was carried out by Katz and Kahn (1966) at the Survey Research Center of the University of Michigan and many others. This follow-on research served to further confirm and refine the sociotechnical systems model.

The sociotechnical systems model views organizations as transformation agencies, transforming inputs into outputs. Sociotechnical systems bring three major elements to bear on this transformation process: a *technological subsystem, personnel subsystem, and work system design* comprised of the organization's structure and processes. These three elements interact with each other and a fourth element, the relevant aspects of the ex*ternal environment,* on which the work system is dependent for its survival and success. Insight into sociotechnical systems theory is provided by the classic studies of Welsh deep seam coal mining in the United Kingdom by the Tavistock Institute.

The Tavistock Studies

The origin of sociotechnical systems theory, and the coining of the term *sociotechnical systems,* can be traced to the studies of Trist and Bamforth relative to the effects of technological change in a deep seam Welsh coal mine (DeGreene, 1973). Prior to the introduction of new machine technology, the traditional system of mining coal in the Welsh mines was largely manual in nature. Small teams of miners who functioned relatively autonomously carried out the work. To a large extent, the group itself exercised control over work. Each miner could perform a variety of tasks, thus enabling most jobs to be interchangeable among the workers. The miners derived considerable satisfaction from being able to complete the entire "task." Further, through their close group interaction, workers could readily satisfy social needs on the job. As a result of these work system characteristics, the psychosocial and cultural characteristics of the workforce, the task requirements, and the work system's design were *congruent.*

The technological change consisted of replacing this more costly manual, or *shortwall,* method of mining with mechanical coal cutters. Miners no longer were restricted to working a short face of coal. Instead, they now could extract coal from a long wall. Unfortunately, this new and more technologically efficient *longwall* system resulted in a work system design that was *not* congruent with the psychosocial and cultural characteristics of the workforce. Instead of being able to work in small, close-knit groups, shifts of 10 to 20 workers were required. The new jobs were designed to include only a narrowly defined set of tasks. The new work system structure severely limited opportunities for social interaction, and job rotation no longer was possible. The revised work system required a high degree of interdependence among the tasks of the three shifts. Thus, problems from one shift carried over to the next, thereby holding up labor stages in the extraction process. This complex and rigid work system was highly sensitive to both productivity and social disruptions. Instead of achieving the expected improved productivity, low production, absenteeism, and intergroup rivalry became common (DeGreene, 1973).

As part of follow-on studies of other coal mines by the Tavistock Institute (Trist, Higgin, Murray, & Pollock, 1963), this conventional longwall method was compared with a new, *composite* longwall method. The composite work system design utilized a combination of the new technology and features of the old psychosocial work structure of the manual system. As compared with the conventional longwall system, the composite work system's design reduced the interdependence of the shifts, increased the variety of skills utilized by each worker, permitted self-selection by workers of their team members, and created opportunities for satisfying social needs on the job. As a result of this work system redesign, production became significantly higher than for either the conventional longwall or the old manual system; absenteeism and other measures of poor morale and dissatisfaction dropped dramatically (DeGreene, 1973).

Prior to the Tavistock studies there was a widely held belief in *technological determinism.* It incorporated the basic concept of Taylorism that there is "one best way" to organize work. It was a belief that the one best way for designing the work system is determined by the nature of the technology employed. Based on the Tavistock Institute studies, Emory and Trist (1960) concluded that *different organizational designs can utilize the same technology.* As demonstrated by the Tavistock studies, including the mining study, the key is to select a work system design that is compatible with the characteristics of (a) the people who will constitute the personnel portion of the system and (b) the relevant external environment, and then to (c) employ the available technology in a manner that achieves congruence.

Joint Causation and Subsystem Optimization

As noted previously, sociotechnical systems theory views organizations as open systems engaged in transforming inputs into desired outcomes (DeGreene, 1973). *Open* means that work systems have permeable boundaries exposed to the environments in which they exist. These environments thus permeate the organization along with the inputs to be transformed. There are several primary ways in which environmental changes enter the organization: through its marketing or sales function, through the people who work in it, and through its materials or other input functions (Davis, 1982).

As transformation agencies, organizations continually interact with their external environment. They receive inputs from their environment, transform these into desired outputs, and export these outputs back to their environment. In performing this transformation process, work systems bring two critical factors to bear on the transformation process: technology in the form of a *technological subsystem*, and people in the form of a *personnel subsystem*. The design of the technological subsystem tends to define the *tasks* to be performed including necessary tools and equipment, and so forth, whereas the design of the personnel subsystem primarily prescribes the *ways* in which tasks are performed. The two subsystems interact with each other at every human–machine and human–software interface. Thus, the technological and personnel subsystems are *mutually interdependent.* They both operate under *joint causation*, meaning that both subsystems are affected by causal events in the external environment.

Joint causation underlies the related sociotechnical system concept of *joint optimization*. The technological subsystem, once designed, is relatively stable and fixed. Accordingly, it falls to the personnel subsystem to adapt further to environmental change. Because the two subsystems respond jointly to causal events, optimizing one subsystem and fitting the second to it results in suboptimization of the joint work *system*. Consequently, maximizing overall work system effectiveness requires jointly optimizing *both* subsystems. Thus, to develop the best possible fit between the two, joint optimization requires the *joint design* of the technical and personnel subsystems given the objectives and requirements of each and of the overall work system (Davis, 1982). Inherent in this joint design is developing an optimal structure for the overall work system (see the MAS method later in this chapter).

Joint Optimization Versus Human-Centered Design

At first glance, the concept of joint optimization may appear to be at odds with human-centered interface design. It might seem that human-centered design would lead to maximizing the personnel subsystem at the expense of the technological subsystem and, thus, to suboptimization of the work system. In fact this has proven not to be the case. In human-centered design, the goal is to make optimal use of humans by the *appropriate* employment of technology so as to optimize jointly the capabilities of each and their interfaces. When a human-centered design approach is not taken invariably the capabilities of the technological subsystem are maximized at the expense of the personnel subsystem. In recent decades, this often has been exemplified by automating and giving the human the leftover functions to perform, which suboptimizes the personnel subsystem and, thus, the total work system. To achieve the appropriate balance, then, joint optimization is operationalized through (a) joint design, (b) a human-centered approach to function and task allocation and design, and (c) attending to the organization's sociotechnical characteristics.

General Systems Theory

General systems theory holds that all systems are more than the simple sum of their individual parts, that the parts are interrelated, and that changing one part will affect other parts. Further, when the individual parts are harmonized, the performance effects will be more than the simple sum of the parts would indicate. Thus, like biological and other complex systems, the whole of sociotechnical systems is more than the simple sum of its parts. When the parts are fully harmonized with each other and with the work system's sociotechnical characteristics, improvements in work system effectiveness can be much greater than the simple sum of the parts would indicate, such as will be illustrated in the case examples described later in the chapter.

IMPLEMENTING MACROERGONOMICS

A true macroergonomic effort is most feasible when a major work system change already is to take place—for example, when changing over to a new technology, replacing equipment, or moving to a new facility. Another opportunity is when there is a major change in the goals, scope, or direction of the organization. A third situation in which management is likely to be receptive is when the organization has a costly chronic problem that has *not* proven correctable with a purely microergonomic effort or via other intervention strategies. Recently, the desire to reduce lost time accidents and injuries, and related costs, has led senior managers in some organizations to support a true macroergonomics intervention. As will be illustrated with an actual case later, many of these efforts have achieved dramatic results.

Frequently, a true macroergonomic change to the work system is not possible initially. Rather, the ergonomist begins by making microergonomic improvements, which yield positive results within a relatively short period of time. This sometimes is referred to as "picking the low hanging fruit." When managers see these positive results, they become interested in and willing to support further ergonomic interventions. In this process, if the ergonomist has established a good working rapport with the key decision makers, the ergonomist serves to raise the consciousness level of the decision makers about the full scope of ergonomics and its potential value to the organization. Thus, over time, senior management comes to support progressively larger ergonomics projects—ones that actually change the nature of the work system as a whole. Based on my personal experience and observations, this process typically takes about 2 years from the time one has established the necessary working rapport and gained the initial confidence of the key decision maker(s).

Macroergonomics Versus Industrial and Organizational (I/O) Psychology

I/O psychology can be viewed conceptually as the flip side of the coin from ergonomics. Whereas ergonomics focuses on designing work systems to fit people, I/O psychology primarily is concerned with selecting people to fit work systems. This is especially true of classical industrial psychology as compared with microergonomics.

Organizational psychology can be viewed as the opposite side of the coin from macroergonomics, although there is greater overlap than in the case of industrial psychology and microergonomics. Both organizational psychology and macroergonomics are concerned with the design of organizational structures and processes, but the focus is somewhat different. In the case of organizational psychology, improving employee motivation and job satisfaction, developing effective incentive systems, enhancing leadership and organizational climate, and fostering teamwork are common objectives. Although these objectives are important and are considered in macroergonomics interventions, the primary focus of macroergonomics is to design work systems that are compatible with the organization's sociotechnical characteristics and then ensure that the microergonomic elements are designed to harmonize with the over-all work system structure and processes.

PITFALLS OF TRADITIONAL APPROACHES TO WORK SYSTEM DESIGN

Over a 20-year period, I was involved either by myself or with my graduate students in assessing more than 200 organizational units. Based on these assessments, I identified three highly interrelated work system design practices that frequently underlie dysfunctional work system development and modification efforts. These are *technology-centered design, a "left-over" approach to function and task allocation,* and *a failure to consider an organization's sociotechnical characteristics* and integrate them into its work system design (Hendrick, 1995).

Technology-Centered Design

In exploiting technology, designers incorporate it into some form of hardware or software to achieve some desired purpose. To the extent that those who must operate or maintain the hardware or software are considered, it usually is in terms of what skills, knowledge, and training will be required. Often, even this kind of consideration is not thought through ergonomically. As a result, the intrinsic motivational aspects of jobs, psychosocial characteristics of the workforce, and other related work-system design factors rarely are considered. Yet these are the very factors that can significantly improve work system effectiveness.

Typically, if ergonomic aspects of design are considered, it usually is after the equipment or software *already is designed.* Then, the ergonomist may be called in to modify some of the human–system interfaces to reduce the likelihood of human error, eliminate awkward postures, or improve comfort.

Often even this level of involvement does not occur until testing of the newly designed system reveals serious interface design problems. At this point in the design process, because of cost and schedule considerations, the ergonomist is severely limited in terms of making fundamental changes to improve the work system. Instead, the ergonomist is restricted to making a few "band aid" fixes of specific human–machine, human–environment, or human–software interfaces. Thus, the ultimate outcome is a suboptimal work system.

There is a well-known relationship between when professional ergonomics input occurs in the design process and the value of that input in terms of system performance: The earlier the input occurs in the design process, the greater and more economical is the impact on system effectiveness.

A related impact of a technology-centered approach to redesigning existing work systems is that employees are not actively involved throughout the planning and implementation process. As the organizational change literature frequently has shown, the result often is not only a poorly designed work system but also a lack of commitment and, not infrequently, either overt or passive–aggressive resistance to the changes. From the authors' personal observations, when a technology-centered approach is taken, if employees are brought into the process at all, it is only after the work system changes have been designed—the employee's role being to do informal usability testing of the system. As a result, when employees find serious problems with the changes (as often happens), cost and schedule considerations prevent any major redesign to eliminate or minimize the identified deficiencies.

Given that most of the so-called *reengineering* efforts of the early 90s used a technology-centered approach, it is not surprising that a large majority of them have been unsuccessful (Keidel, 1994). As Keidel noted, these efforts failed to address the "soft" (i.e., human) side of engineering and often ignored organizational effects.

"Left-Over" Approach to Function and Task Allocation

A technology-centered approach often leads to treating the persons who will operate and maintain the system as impersonal components. The focus is on assigning to the "machine" any functions or tasks that its technology enables it to perform. Then, what is left over is assigned to the persons who must operate or maintain it or be serviced by it. As a result, the function and task allocation process fails to consider the characteristics of the workforce and related external environmental factors. Most often the consequence is a poorly designed work system that fails to make effective use of its human resources.

As noted earlier from the sociotechnical systems literature, effective work system design requires *joint design* of the technical and personnel subsystems (DeGreene, 1973). Put in ergonomic terms, joint optimization requires a *human-centered* approach. In terms of function and task allocation, Bailey (1989) referred to it as a *humanized task* approach. He noted

> This concept essentially means that the ultimate concern is to design a job that justifies using a person, rather than a job that merely can be done by a human. With this approach, functions are allocated and the resulting tasks are designed to make full use of human skills and to compensate for human limitations. The nature of the work itself should lend itself to internal motivational influences. The left over functions are allocated to computers. (p. 190)

Failure to Consider the System's Sociotechnical Characteristics

As previously noted, from the sociotechnical systems of the literature, four major characteristics or elements of sociotechnical systems can be identified. These are the (a) *technological subsystem*, (b) *personnel subsystem*, (c) *external environment*, and (d) *organizational design* of the work system. These four elements interact with one another, so a change in any one affects the other three (and, if not planned for, often in dysfunctional or unanticipated ways). Because of these interrelationships, characteristics of each of the first three elements affect the fourth: The organizational design of the work system. Empirical models have been developed of these relationships that can be used to determine the optimal work system structure (i.e., the optimal degrees of vertical and horizontal differentiation, integration, formalization, and centralization to design into the work system) and have been integrated into the MAS method, described later.

Unfortunately, as first was documented by the Tavistock studies of coal mining more than 4 decades ago, a technology-centered approach to the organizational design of work systems does *not* adequately consider the key characteristics of the other three sociotechnical system elements. Consequently, the resulting work system design most often is *suboptimal* (Emory & Trist, 1960; Trist & Bamforth, 1951).

Criteria for an Effective Work System Design Approach

Based on the pitfalls cited previously, several criteria can be gleaned for selecting an effective work system design approach.

1. Joint design. The approach should be human-centered. Rather than designing the technological subsystem and requiring the personnel subsystem to conform to it, the approach should require design of the personnel subsystem jointly with the technological subsystem. Further, it should allow for extensive employee participation throughout the design process.
2. Humanized task approach. The function and task allocation process should first consider whether there is a need for a human to perform a given function or task before making the allocation to humans or machines. Implicit in this criterion is a systematic consideration of the professionalism requirements (i.e., education and training) and of the cultural and psychosocial characteristics of the personnel subsystem.
3. Consider the organization's sociotechnical characteristics. The approach should systematically evaluate the organization's sociotechnical system characteristics, and then integrate them into the work system's design.

One approach that meets all three of these criteria is macroergonomics. As was noted earlier, conceptually, macroergonomics is a top-down sociotechnical systems approach to work system design, and the carry-through of the over-all work system design to the design of human–job, human–machine, human–environment, and human–software interfaces. It is a top-down sociotechnical systems approach in that it begins with an analysis of the relevant sociotechnical system variables, and then systematically utilizes these data in designing the work system's structure and related processes. Macroergonomics is human-centered in that it systematically considers the worker's professional and psychosocial characteristics in designing the work system; and then carries the work system design through to the ergonomic design of specific jobs and related hardware and software interfaces. Integral to this human-centered design process is joint design of the technical and personnel subsystems using a humanized task approach in allocating functions and tasks. A primary methodology of macroergonomics is *participatory ergonomics* (Noro & Imada, 1991).

Macroergonomic Analysis of Structure (MAS)

Macroergonomics involves the development and application of human–organization interface technology. This technology is concerned with optimizing the organizational structure and related processes of work systems. Accordingly, an understanding of macroergonomics requires an understanding of the key dimensions of organizational structure.

The organizational structure of a work system can be conceptualized as having three core dimensions. These are *complexity, formalization, and centralization* (Bedeian & Zammuto, 1991; Robbins, 1983; Stevenson, 1993).

Complexity refers to the degree of *differentiation* and *integration* that exist within a work system. Differentiation refers to the extent to which the work system is segmented into parts; integration refers to the number of mechanisms that exist to integrate the segmented parts for the purposes of communication, coordination, and control.

Complexity: Differentiation

Work system structures employ three common types of differentiation: *vertical, horizontal,* and *spatial.* Increasing any one of these three increases a work system's complexity.

Vertical Differentiation. Vertical differentiation is measured in terms of the number of hierarchical levels separating the chief executive position from the jobs directly involved with the system's output. In general, as the size of an organization increases, the need for greater vertical differentiation also increases (Mileti, Gillespie, & Haas, 1977). For example, in one study size alone was found to account for 50% to 59% of the variance (Montanari, 1976). A major reason for this strong relationship is the practical limitations of span of control. Any one manager is limited in the number of subordinates that he or she can direct effectively (Robbins, 1983). Thus, as the number of first level employees increase, the number of first-line supervisors also must increase. This, in turn, requires more supervisors at each successively higher level and ultimately results in the creation of more hierarchical levels in the work system's structure.

Although span of control limitations underlie the size-vertical differentiation relationship, it is important to note that these limitations can vary considerably depending on a number of factors. Thus, for an organization of a given size, if large spans of control are appropriate, the number of hierarchical levels will be fewer than if small spans of control are required. A major factor affecting span of control is the *degree of professionalism* (education and skill requirements) designed into employee jobs. Generally, as the level of professionalism increases, employees are able to function more autonomously and thus need less supervision. Consequently, the manager can supervise effectively a larger number of employees. Other factors that affect span of control are the degree of formalization, type of technology, psychosocial variables, and environmental characteristics (see Hendrick, 1997, or Hendrick & Kleiner, 2001, 2002, for a more detailed discussion).

Horizontal Differentiation. Horizontal differentiation refers to the degree of departmentalization and specialization within a work system. Horizontal differentiation increases complexity because it requires more sophisticated and expensive methods of control. In spite of this drawback, specialization is common to most work systems because of the inherent efficiencies in the division of labor. Adam Smith (1876/1970) demonstrated this point over 200 years ago. Smith noted that 10 workers, each doing particular tasks (job specialization), could produce about 48,000 pins per day. On the other hand, if the 10 each worked separately and independently, performing all of the production tasks, they would be lucky to make 200.

Division of labor creates groups of specialists, or *departmentalization*. The most common ways of designing departments into work systems are on the basis of (a) function, (b) simple numbers, (c) product or services, (d) client or client class served, (e) geography, and (f) process. Most large corporations will use all six (Robbins, 1983).

Two of the most common ways to determine whether or not a work group should be divided into one or more departments are the degree of commonality of *goals* and of *time orientation.* To the extent that subgroups either differ in goals or have differing time orientations they should be structured as separate departments. For example, sales persons differ from research and development (R&D) employees on both of these dimensions. Not only do they have very different goals, but also the time orientation of sales personnel usually is *short* (1 year or less), whereas it usually is *long* (3 or more years) for R&D personnel. Thus, they clearly should be departmentalized separately and usually are (Robbins, 1983).

Spatial Dispersion. Spatial dispersion refers to the degree an organization's activities are performed in multiple locations. There are three common measures of spatial dispersion. These are (a) the number of geographic locations comprising the total work system, (b) the average distance of the separated locations from the organization's headquarters, and (c) the proportion of employees in these separated units in relation to the number in the headquarters (Hall, Haas, & Johnson, 1967). In general, complexity increases as any of these three measures increase.

Complexity: Integration

As noted previously, *integration* refers to the number of mechanisms designed into a work system for ensuring communication, coordination, and control among the differentiated elements. As the differentiation of a work system increases, the need for integrating mechanisms also increases. This occurs because greater differentiation increases the number of units, levels, and so forth that must communicate with one another, coordinate their separate activities, and be controlled for efficient operation. Some of the more common integrating mechanisms that can be designed into a work system are formal rules and procedures, committees, task teams, liaison positions, and system integration offices. Computerized information and decision support systems also can be designed to serve as integrating mechanisms. Vertical differentiation, in itself, is a primary form of integrating mechanism (i.e., a manager at one level serves to coordinate and control the activities of several lower-level groups).

Once the differentiation aspects of a work system's structure have been determined, a major task for the macroergonomics professional is to then determine the kinds and number of integrating mechanisms to design into the work system. Two few integrating mechanisms will result in inadequate coordination and control among the differentiated elements; too many integrating mechanisms stifle efficient and effective work system functioning and usually increase costs. A systematic analysis of the type of technology, personnel subsystem factors, and characteristics of the external environment all can be used to help determine the optimal number and types of integrating mechanism (see the MAS method, described later in this chapter, in the section "Sociotechnical System Considerations in Work System Design.").

Formalization

From an ergonomics design perspective, formalization can be defined as the degree to which jobs within the work system are standardized. Highly formalized designs allow for little employee discretion over what is to be done, when it is to be done, or how it is to be accomplished (Robbins, 1983). In highly formalized designs there are explicit job descriptions, extensive rules, and clearly defined procedures covering work processes. Ergonomists can contribute to formalization by designing jobs, machines, and software so as to standardize procedures and allow little opportunity for operator decision discretion. By the same token, human–job, human–machine, and human–software interfaces can be ergonomically designed to permit greater flexibility and scope to employee decision making (i.e., low formalization). When there is low formalization, employee behavior is relatively unprogrammed and the work system allows for considerably greater use of one's mental abilities. Thus, greater reliance is placed on the employee's professionalism and jobs tend to be more intrinsically motivating.

In general, the simpler and more repetitive the jobs to be designed into the work system, the higher should be the level of formalization. However, caution must be taken not to make the work system so highly formalized that jobs lack any intrinsic motivation, fail to effectively utilize employee skills, or degrade human dignity. Invariably, good macroergonomic design can avoid this extreme. The more nonroutine or unpredictable the work tasks and related decision making, the less amenable is the work system to high formalization. Accordingly, reliance has to be placed on designing into jobs a relatively high level of professionalism.

Centralization

Centralization refers to the degree to which formal decision making is concentrated in a relatively few individuals, groups, or levels, usually high in the organization. When the work system structure is highly centralized, lower-level supervisors and employees have only minimal input into the decisions affecting their jobs (Robbins, 1983). In highly decentralized work systems, decisions are delegated downward to the lowest level having the necessary expertise.

It is important to note that work systems carry out two basic forms of decision making, *strategic* and *tactical*, and that the degree of centralization often is quite different for each. Tactical decision making has to do with the day-to-day operation of the organization's business; strategic decision making concerns the long range planning for the organization. Under conditions of low formalization and high professionalism, tactical decision making may be highly decentralized, whereas strategic decision making may remain highly centralized. Under these conditions, it also is important to note that the information required for strategic decision making often is controlled and filtered by middle management or even lower-level personnel. Thus, to the extent that these persons can reduce, summarize, selectively omit, or embellish the information that gets fed to top management, the less is the actual degree of centralization of strategic decision making.

In general, centralization is desirable (a) when a comprehensive perspective is needed; (b) when it provides significant economies; (c) for financial, legal, or other decisions that clearly can be done more efficiently when centralized; (d) when operating in a highly stable and predictable external environment; and (e) where the decisions have little effect on employees' jobs or are of little employee interest. Decentralized decision making is desirable (a) when an organization needs to respond rapidly to changing or unpredictable conditions at the point where change is occurring; (b) when "grass roots" input to decisions is desirable; (c) to provide employees with greater intrinsic motivation, job satisfaction, and sense of self worth; (d) when it can reduce stress and related health problems by giving employees greater control over their work; (e) to more fully utilize the mental capabilities and job knowledge of employees; (f) to gain greater employee commitment to, and support for, decisions by involving them in the process; (g) when it can avoid overtaxing a given manager's capacity for human information processing and decision making; and (h) to provide greater training opportunity for lower-level managers.

SOCIOTECHNICAL SYSTEM CONSIDERATIONS IN WORK SYSTEM DESIGN

As noted in our coverage of sociotechnical systems theory, the design of a work system's structure (which includes how it is to be managed) involves consideration of the key elements of three major sociotechnical system components: (a) the technological subsystem, (b) the personnel subsystem, and (c) the relevant external environment. Each of these three major sociotechnical system components has been studied in relation to its effect on the fourth component, organizational structure, and empirical models have emerged that can be used to optimize a work system's organizational design. The models of each of these components that I have found most useful have been integrated into a macroergonomics method that I have labeled Macroergonomics Analysis of Structure (MAS).

Technology: Perrow's Knowledge-Based Model

Perhaps the most thoroughly validated and generalizable model of the technology-organization design relationship is that of Perrow (1967), which utilizes a *knowledge-based* definition of technology. In his classification scheme, Perrow began by defining technology by the action a person performs on an object to change that object. Perrow noted that this action always requires some form of technological knowledge. Accordingly, technology can be categorized by the required knowledge base. Using this approach, he identified two underlying dimensions of knowledge-based technology. The first of these is *task variability* or the number of exceptions encountered in one's work. For a given technology, these can range from routine tasks with few exceptions to highly variable tasks with many exceptions.

The second dimension has to do with the type of search procedures one has available for responding to task exceptions, or *task analyzability*. For a given technology, the search procedures can range from tasks being well defined and solvable by using logical and analytical reasoning to being ill defined with no readily available formal search procedures for dealing with task exceptions. In this latter case, problem solving must rely on experience, judgment, and intuition. The combination of these two dimensions, when dichotomized, yields a 2 × 2 matrix as shown in Table 3–1. Each of the four cells represents a different knowledge-based technology.

1. Routine technologies have few exceptions and well-defined problems. Mass production units most frequently fall into this category. Routine technologies are best accomplished through standardized procedures and are associated with high formalization and centralization.
2. Nonroutine technologies have many exceptions and difficult to analyze problems. Aerospace operations often fall into this category. Most critical to these technologies is flexibility. They thus lend themselves to decentralization and low formalization.
3. Engineering technologies have many exceptions, but they can be handled using well-defined rational–logical processes. They therefore lend themselves to centralization but require the flexibility that is achievable through low formalization.
4. Craft technologies typically involve relatively routine tasks, but problems rely heavily on experience, judgment, and intuition for decision. Problem solving thus needs to be done by those with the particular expertise. Consequently, decentralization and low formalization are required for effective functioning.

Personnel Subsystem

At least two major aspects of the personnel subsystem are important to organizational design.

TABLE 3–1. Perrow's Knowledge-Based Technology Classes.

	Task Variability	
Problem Analyzability	*Routine With few Exceptions*	*High Variety With Many Exceptions*
Well defined and analyzable	Routine	Engineering
Ill defined and unanalyzable	Craft	Nonroutine

These are the degree of professionalism and the psychosocial characteristics of the workforce.

Degree of Professionalism

Degree of professionalism refers to the education and training requirements of a given job and, presumably, possessed by the incumbent. Robbins (1983) noted that formalization can take place either on the job or off. When done on the job, formalization is *external* to the employee; rules, procedures, and the human–machine and human–software interfaces are designed to limit employee discretion. Accordingly, this tends to characterize unskilled and semiskilled positions. When done off the job, it is done through the professionalization of the employee. Professionalism creates formalization that is *internal* to the worker through a socialization process that is an integral part of formal professional education and training. Thus, values, norms, and expected behavior patterns are learned *before* the employee enters the organization.

From a macroergonomics design perspective, there is a trade-off between formalizing the organizational structure and professionalizing the jobs and related human–machine and human–software interfaces. As positions in the organization are designed to require persons with considerable education and training, they also should be designed to allow for considerable employee discretion. If low education and training requirements characterize the design of the positions, than the work system should be more highly formalized.

Psychosocial Characteristics

In addition to careful consideration of cultural differences (e.g., norms, values, mores, role expectations, etc.), I have found the most useful integrating model of psychosocial influences on organizational design to be that of *cognitive complexity*. Harvey, Hunt, and Schroder (1961) identified the higher-order structural personality dimension of concreteness–abstractness of thinking, or cognitive complexity, as underlying different conceptual systems for perceiving reality. We all start out in life relatively concrete in our conceptual functioning. As we gain experience we become more abstract or complex in our conceptualizing, and this changes our perceptions and interpretations of our world. In general, the degree to which a given culture or subculture (a) provides through education, communications, and transportation systems an opportunity for *exposure* to diversity and (b) encourages through its child-rearing and educational practices and *active* exposure to this diversity (i.e., an active openness to learning from exposure to new experiences), the more cognitively complex the persons of that particular group will become. An active exposure to diversity increases the number of conceptual categories that one develops for storing experiential information and number of "shades of gray" or partitions within conceptual categories. In short, one develops greater *differentiation* in ones conceptualizing. With an active exposure to diversity one also develops new rules and combinations of rules for *integrating* conceptual data and deriving more insightful conceptions of complex problems and solutions. Note from our earlier review that these same two dimensions of "differentiation" and "integration" also characterize *organizational* complexity. Relatively concrete adult functioning consistently has been found to be characterized by a relatively high need for structure and order and for stability and consistency in ones environment, closedness of beliefs, absolutism, paternalism, and ethnocentrism. Concrete functioning persons tend to see their views, values, norms, and institutional structures as relatively unambiguous, static, and unchanging. In contrast, cognitively complex persons tend to have a relatively low need for structure and order or stability and consistency and are open in their beliefs, relativistic in their thinking, and have a high capacity for empathy. They tend to be more people-oriented, flexible, and less authoritarian then their more concrete colleagues and to have a dynamic conception of their world: They *expect* their views, values, norms and institutional structures to change (Harvey, 1963; Harvey et al., 1961).

In light of this, it is not surprising that I have found evidence to suggest that relatively concrete managers and workers function best under moderately high centralization, vertical differentiation, and formalization. In contrast, cognitively complex workgroups and managers seem to function best under relatively low centralization, vertical differentiation, and formalization.

Although only weakly related to general intelligence, cognitive complexity is related to education. Thus, within a given culture, educational level can sometimes serve as a relative estimate of cognitive complexity.

Of particular importance is the fact that since the post-World War II baby boomers have entered the workforce in industrially developed societies, the general complexity level of educated or highly trained employees has become moderately *high*. This can be traced to the greater exposure to diversity and less authoritarian and absolutist child rearing practices these adults experienced while growing up as compared with those who experienced childhood prior to World War II (Harvey, 1963; Harvey et al, 1961). As a result, successful firms are likely to be those having work system designs that respond to the guidelines given herein for more cognitively complex workforces—particularly if their workforces are highly professionalized.

Environment

Critical to the success and indeed the very survival of an organization is its ability to adapt to its external environment. In open systems terms, organizations require monitoring and feedback mechanisms to follow and sense changes in their specific task environment and the capacity to make responsive adjustments. *Specific task environment* refers to that part of the organization's external environment that is made up of the firm's critical constituencies (i.e., those that can positively or negatively influence the organization's effectiveness). Neghandi (1977), based on field studies of 92 industrial organizations in five different countries, identified five external environments that significantly impact on organizational functioning. These are *socioeconomic* including the nature of competition and the availability of raw materials; *educational* including both the availability of educational programs and facilities and the aspirations of workers; *political* including governmental attitudes toward business, labor, and control over prices; *legal*; and *cultural* including the social class or caste system, values, and attitudes.

Of particular importance to us is the fact that specific task environments vary along two dimensions that strongly influence the effectiveness of a work system's design: These are the degrees of *environmental change* and *complexity*. The degree of change refers to the extent to which a specific task environment is dynamic or remains stable over time; the degree of complexity refers to whether the number of relevant specific task environments is few or many in number. As illustrated in Table 3–2, these two environmental dimensions in combination determine the *environmental uncertainty* of an organization. Of all the sociotechnical system factors that impact on the effectiveness of a work system's design, environmental uncertainty repeatedly has been shown to be the most important (Burns & Stalker, 1961; Duncan, 1972; Emory & Trist, 1965; Lawrence & Lorsch, 1969; Neghandi, 1977). With a high degree of uncertainty, a premium is placed on an organization's ability to be flexible and rapidly responsive to change. Thus, the greater the environmental uncertainty, the more important it is for the work system's structure to have relatively low vertical differentiation, decentralized tactical decision making, low formalization, and a high level of professionalism among its work groups. By contrast, highly certain environments are ideal for high vertical differentiation, formalization, and centralized decision making, such as found in classical bureaucratic structures. Of particular note is the fact that today most high technology corporations are operating in highly dynamic and complex environments. From my observations, although many of these corporations have increased the level of professionalism of their employees, they have not yet fully adapted their work system's design to their environments. I believe this is an issue to be thoroughly assessed in medical systems.

Integrating the Results of the Separate Assessments

The separate analyses of the key characteristics of a given organization's technological subsystem, personnel subsystem, and specific task environment each should have provided guidance about the structural design for the work system. Frequently, these results will show a natural convergence. At times, however, the outcome of the analysis of one sociotechnical system element may conflict with the outcomes of the other two. When this occurs, the macroergonomics specialist is faced with the issue of how to reconcile the differences. Based both on the suggestions from

TABLE 3–2. Environmental Uncertainty of Organizations.

	Degree of Change	
Degree of Complexity	*Stable*	*Dynamic*
Simple	Low uncertainty	Moderate high uncertainty
Complex	Moderate low uncertainty	High uncertainty

the literature and my personal experience in evaluating over 200 organizational units, the outcomes from the analyses can be integrated by weighting them approximately as follows: If the technological subsystem analysis is assigned a weight of 1, give the personnel subsystem analysis a weight of 2, and the specific task environment analysis a weight of 3. For example, let's assume that the technological subsystem falls into Perrow's "routine" category, the personnel subsystem's jobs call for a high level of professionalism and external environment has moderately high complexity. Weighting these three as suggested previously would indicate that the work system should have moderate formalization and centralization. Accordingly, the results would indicate that most jobs should be redesigned to require a somewhat lower level of professionalism, and attendant hardware and software interfaces should be designed/redesigned to be compatible.

It is important to note that the specific functional units of an organization may differ in the characteristics of their technology, personnel, and specific task environments—particularly within larger organizations. Therefore, the separate functional units may, themselves, need to be analyzed as though they were separate organizations, and the resultant work systems designed accordingly.

RELATION OF MACRO- TO MICROERGONOMIC DESIGN

Through a macroergonomic approach to determining the optimal design of a work system's structure and related processes, many of the characteristics of the jobs to be designed into the system, and of the related human–machine and human–software interfaces, already have been prescribed. Some examples are as follows (Hendrick, 1991).

1. Horizontal differentiation decisions prescribe how narrowly or broadly jobs must be designed and, often, how they should be departmentalized.
2. Decisions concerning the level of formalization and centralization will dictate (a) the degree of routinization and employee discretion to be ergonomically designed into the jobs and attendant human–machine and human–software interfaces; (b) the level of professionalism to be designed into each job; and (c) many of the design requirements for the information, communications, and decision support systems, including what kinds of information are required by whom, and networking requirements.
3. Vertical differentiation decisions, coupled with those concerning horizontal differentiation, spatial dispersion, centralization, and formalization will prescribe many of the design characteristics of the managerial positions including span of control, decision authority and nature of decisions to be made, information and decision support requirements, and qualitative and quantitative educational and experience requirements.

In summary, effective macroergonomic design drives much of the microergonomic design of the system and thus ensures *optimal ergonomic compatibility* of the components with the work system's overall structure. In sociotechnical system terms, this approach enables joint optimization of the technical and personnel subsystems from top to bottom throughout the organization. The result of this is greater assurance of optimal *system* functioning and effectiveness, including productivity, safety and health, comfort, intrinsic employee motivation, and quality of work life.

SYNERGISM AND WORK SYSTEM PERFORMANCE

From the previous discussion, it should be apparent that macroergonomics has the potential to improve the ergonomic design of organizations by ensuring that their respective work system's designs harmonize with the organizations' critical sociotechnical characteristics. Equally important, macroergonomics offers the means to ensure that the design of the entire work system, down to each individual job and workstation, harmonizes with the design of the over-all work system. As we noted earlier, a widely accepted view among general system theorists and researchers is that organizations are synergistic—that the whole is more than the simple sum of its parts. Because of this synergism, it is my experience that the following tend to occur in our complex organizations.

When Systems Have Incompatible Organizational Designs

When a work system's structures and related processes are grossly incompatible with their sociotechnical system characteristics, and/or jobs and human–system interfaces are incompatible with the work system's structure, the whole is *less than* the sum of its parts. Under these conditions, we can expect the following to be poor: (a) productivity, especially *quality* of production; (b) lost time accident and injury rates; and (c) motivation and related aspects of quality of work life (e.g., stress), and for human error incidents to be relatively high. Needless to say, these are literally life or death concerns for medical systems.

When Systems Have Effective Macroergonomic Designs

When a work system has been effectively designed from a macroergonomics perspective, and that effort is carried through to the microergonomic design of jobs and human–machine, human–environment, and human–software interfaces, then production, safety and health, and quality of work life will be much *greater* than the simple sum of the parts would indicate.

MACROERGONOMICS RESULTS

A major reason for the rapid spread and development of macroergonomics has been the dramatic results that have been achieved when it was applied to improving an organization's work system. Based on both sociotechnical and general systems theory, at ODAM-III in 1990, I predicted that 50% to 90% or greater improvements would be typical of macroergonomics interventions having strong management and labor support in comparison to the 10% to 20% gains expected from microergonomics. These results subsequently have been achieved around the world in a number of interventions. A good example is the work of my former University of Southern California (USC) colleague, Andy Imada. Working with a large petroleum distribution company in the southwestern United States, Andy achieved the results shown in Table 3–3.

Imada is convinced that taking a macroergonomic approach and using participatory ergonomics at all levels of the organization enabled not only the initial improvements but the integration of safety and ergonomics into the organizational culture, which accounts for the sustained results over time (see Imada, 2002, for a full description of this intervention).

In a macroergonomics project, which involved designing a new university college at the University of Denver, my colleagues and I achieved the results shown in Table 3–4 as compared with our prior similar organization at the USC. These results are particularly impressive in that the USC academic unit already was an efficient, well-functioning organization, refined by over 20 years of experience. Of particular note, a true humanized task approach was used in which we first determined what we needed humans to do, designed the jobs accordingly, and only then selected the equipment and designed or purchased the software to support the humans (see Hendrick & Kleiner, 2002, for a full description).

MACROERGONOMICS METHODOLOGY

Methods Developed for Macroergonomics Application

Since the first Human Factors in ODAM symposium in 1984, in addition to MAS described previously, a number of important new macroergonomic methodologies have been developed and successfully applied including the following (in addition to the references cited for some of these methods, see Part 2 of Stanton, Hedge, Hendrick, Brookhuis, & Salas, 2004, for more detailed descriptions).

MacroErgonomic Analysis and Design (MEAD)

MEAD is a 10-step process for systematically analyzing a work system's processes that was developed by

TABLE 3–3. Results of a Macroergonomic Intervention in a Petroleum Company.

	Reduction After	
	2 Years (%)	*9 Years (%)*
Motor vehicle accidents	51	63
Industrial accidents	54	70
Lost workdays	94	97
Off the job injuries	84	69
$60,000 savings in petroleum delivery costs per year.		

TABLE 3–4. Designing a new University College

Results (as compared with old University of Southern California academic unit)
27% savings in operating expenses
23% reduction in campus staffing requirements
20% reduction in off campus study center administrative time (30 centers)
67% reduction in average processing time for student registrations, grades, etc. from off campus locations

Brian Kleiner of Virginia Tech. Through MEAD's application, deficiencies in work system processes can be identified and corrected (see Hendrick & Kleiner, 2001 or 2002 for a detailed description).

HITOP

Developed by Ann Majchrzak of the USC and her colleagues, HITOP is a step-by-step manual procedure for implementing technological change (Marjchrizak, 1988). The procedure is designed to enable managers to be more aware of the organizational and human implications of their technology plans and thus better able to integrate technology within its organizational and human context. It has been successfully applied in a number of manufacturing organizations.

Top MODELER

Top MODELER is a decision-support system that also was developed by Ann Majchrzak of the USC and her colleagues. Its purpose is to help manufacturing organizations identify the organizational changes required when new process technologies are being considered (Marjchrizak & Gasser, 2000).

CIMOP

CIMOP is a knowledge-base system for evaluating computer integrated manufacturing, organization, and people system design. It was developed by J. Kantola and Waldemar Karwowski of the University of Louisville (Kantola & Karwowski, 1998; Karwowski, Kantola, Rodrick, & Salvendy, 2002). The intended users of CIMOP are companies designing, redesigning, or implementing a computer integrated manufacturing system (see Hendrick & Kleiner, 2002, for a more detailed description).

Systems Analysis Tool (SAT)

SAT is a method developed by Michelle Robertson of the Liberty Mutual Research Center for conducting systematic trade-off evaluations of work system intervention alternatives (Robertson, Kleiner, & O'Neill, 2002). It is an adaptation, elaboration, and extension of the basic steps of the scientific method. SAT has proven useful in enabling both ergonomists and managerial decision makers to determine the most appropriate strategy for making work system changes (see Hendrick & Kleiner, 2002 for a detailed description).

Anthropotechnology

Although not intentionally developed as a macroergonomics tool, anthropotechnology is, in effect, a macroergonomics methodology. It deals specifically with analysis and design modification of systems for effective technology transfer from one culture to another. It was developed by the distinguished French ergonomist, Alain Wisner. Wisner and others, such as Philippe Geslin of the Institut National de la Recherche Agronomique, have been highly successful in applying anthropotechnology to systems transferred from industrially developed to industrially developing countries. (See Wisner, 1995, for a detailed description).

Methods Adapted for Macroergonomics Application

In addition, most of the traditional research methods have been modified and adopted for macroergonomic application. These include Macroergonomic Organizational Questionnaire Surveys (MOQS)—an adaptation of the organizational questionnaire survey method by Pascale Carayon and Peter Hoonakker (2004) of the University of Wisconsin, laboratory experiment, field study, field experiment (perhaps the most widely used conventional method in macroergonomic studies—make work system changes to one part of the organization and if it improves organizational effectiveness, then implement throughout the work system), interview survey, and focus groups. Brian Kleiner has done a number of interesting laboratory studies in his macroergonomics laboratory at Virginia Tech to investigate sociotechnical variables as they relate to macroergonomics (see Hendrick & Kleiner, 2001, 2002, for some examples). Carayon and Hoonakker (2004) have successfully applied MOQS in a variety of interventions.

Several methods initially developed for microergonomics also have been adapted for macroergonomic application. Most notable are the following.

Kansai Ergonomics

This method initially was developed for consumer product design by Mitsuo Nagamachi and his colleagues at Hiroshima University but can be applied to evaluating worker affective responses to work system design changes (Nagamachi, 2002).

Cognitive Walkthrough Method (CWM)

The CWM is a usability inspection method that rests on the assumptions that evaluators are capable of taking the perspective of the user and can apply this user perspective to a task scenario to identify design problems. As applied to macroergonomics, evaluators can assess the usability of conceptual designs of work systems to identify the degree to which a new work system is harmonized or the extent to which workflow is integrated.

Participatory Ergonomics (PE)

PE is an adaptation of participatory management that was developed for both micro and macroergonomic interventions. When applied to evaluating the overall work system, the employees work with a professional ergonomist who serves as both the facilitator and resource person for the group. A major advantage of this approach is that the employees are in the best position to both know the problem symptoms and to identify the macroergonomic intervention approach that will be most acceptable to them. Equally important, having participated in the process, the employees are more likely to support the work system changes—even if their own preferred approach is not adopted. Finally, the participatory approach has proven to be particularly effective in establishing an ergonomics and safety culture, which sustains performance and safety improvements that initially result from the macroergonomic intervention. Participatory ergonomics is the most widely used method in successful macroergonomic interventions. It often is used in combination with other methods, such as those described previously.

MACROERGONOMICS AS A KEY TO IMPROVING HEALTH CARE AND PATIENT SAFETY

All sociotechnical systems are organizations that exist to accomplish some purpose, be it manufacturing, refining, product distribution, sales, or to provide some kind of service. Organizations accomplish their respective purposes by transforming some kind of input into output, that is, a product or service. As noted earlier, all sociotechnical systems are comprised of a personnel subsystem, technological subsystem, a work system design (structure and processes), and interact with relevant aspects of their external environment that permeate the organization and on which the system must be responsive to survive and be successful. It thus is not surprising that macroergonomics has successfully been applied to a wide spectrum of organizations. For example, during the first five IEA International Symposia on Human Factors in Organizational Design and Management (1984 to 1996), successful macroergonomic interventions were reported for 23 manufacturing, 8 construction, 2 community planning and development, 9 interorganizational networks, and 38 service organizations. The service organizations included various governmental organizations, architectural design firms, telecommunications companies, banks, credit card companies, and an advertising agency, printing business, insurance company, and industrial product supplier. They also included 4 hospitals.

As with the cases just mentioned, all types of health care organizations are complex sociotechnical systems. They share with these other organizations the same issues of employee health, safety, comfort, productivity, and the need to eliminate system-induced human error. Unlike most of the organizational types previously mentioned, the services health care systems provide to their customers may be truly life and death matters, as well as matters of enhancing customer quality of life. An employee medical error potentially can result in a patient's death. Poor or inefficient patient care can drastically degrade that customer's quality of life. In addition, inefficient health care systems increase health care costs. Accordingly, the potential of macroergonomics for enabling health care work systems to better meet their goals is of greater importance than for most other types of organizations. With real top management, and employee commitment at all organizational levels, macroergonomic interventions—including systematic carry-through to the microergonomic design/modification of jobs and related human–machine, human–environment, and human–software interfaces—can result in the same dramatic improvements in organizational effectiveness that have been experienced in many other types of sociotechnical systems.

References

Argyris, C. (1971). *Management and organizational development.* New York: McGraw-Hill.

Bailey, R. W. (1989). *Human performance engineering* (2nd ed). Englewood Cliffs, NJ: Prentice-Hall.

Bedeian, A. G., & Zammuto, R. F. (1991). *Organizations: Theory and design.* Chicago: Dryden.

Burns, T., & Stalker, G. M. (1961). *The management of innovation.* London: Tavistock.

Carayon, P., & Hoonakker, P. L. T. (2004). Macroergonomics organizational questionnaire survey (MOQS). In N. Stanton, A. Hedge, K. Brookhuis, E. Salas, & H. W. Hendrick (Eds.), *Handbook of human factors and ergonomics methods* (pp. 76-1–76-10). Boca Raton, FL: CRC Press.

Chapanis, A. (1991). To communicate the human factors message you have to know what the message is and how to communicate it. Keynote address to the HFAC/ACE Conference, Edmonton, Alberta, Canada (Part 1). *Human Factors Society Bulletin,* 34(11), 1–4.

Chapanis, A. (1992). To communicate the human factors message you have to know what the message is and how to communicate it. Keynote address to the HFAC/ACE Conference, Edmonton, Alberta, Canada (Part 2). *Human Factors Society Bulletin, 35*(1), 3–6.

Davis, L. E. (1982). Organizational design. In G. Salvendy (Ed), *Handbook of industrial engineering* (pp. 2.1.1–2.1.29). New York: Wiley.

DeGreene, K. (1973). *Sociotechnical systems.* Englewood Cliffs, NJ: Prentice-Hall.

Duncan, R. B.(1972). Characteristics of organizational environments and perceived environmental uncertainty. *Administrative Science Quarterly, 17,* 313–327.

Emery, F. E., & Trist, E. L. (1960). Sociotechnical systems. In C. W. Churchman & M. Verhulst (Eds.), *Management sciences: Models and techniques* (pp. 83–97). Oxford: Pergamon.

Emery, F. E., & Trist, E. L. (1965). The causal texture of organizational environments. *Human Relations, 18,* 21–32.

Hall, R. H., Haas, J. E., & Johnson, N. J. (1967). Organizational size, complexity, and formalization. *Administrative Science Quarterly, 12,* 303.

Harvey, O. J. (1963). System structure, flexibility and creativity. In O. J. Harvey (Ed.), *Experience, structure and adaptability.* New York: Springer.

Harvey, O. J., Hunt, D. E., & Schroder, H. M. (1961). *Conceptual systems and personality organization.* New York: Wiley.

Hendrick, H. W. (1980). *Human Factors in Management, 1980–2000.* Unpublished report, presented at the 1980 Annual Meeting of the Human Factors and Ergonomics Society, Los Angeles, CA.

Hendrick, H. W. (1984). Wagging the tail with the dog: Organizational design considerations in ergonomics. In *Proceedings of the Human Factors and Ergonomics Society 28th Annual Meeting* (pp. 899–903). Santa Monica, CA: Human Factors and Ergonomics Society.

Hendrick, H. W. (1991). Ergonomics in organizational design and management. *Ergonomics, 34,* 743–756.

Hendrick, H. W. (1986). Macroergonomics: A concept whose time has come. *Human Factors Society Bulletin, 30,* 1–3.

Hendrick, H. W. (1995). Future directions in macroergonomics. *Ergonomics, 38,* 1617–1624.

Hendrick, H. W. (1997). Organizational design and macroergonomics. In G. Salvendy (Ed.), *Handbook of human factors and ergonomics* (pp. 594–636). New York: Wiley.

Hendrick, H. W. (2000). Human factors and ergonomics converge—A long march. In I. Kourinka (Ed.), *History of the International Ergonomics Association: The first quarter of a century* (pp.125–142). Geneva: IEA Press.

Hendrick, H. W., & Kleiner, B. M. (2001). *Macroergonomics: An introduction to work system design.* Santa Monica, CA: Human Factors and Ergonomics Society.

Hendrick, H. W., & Kleiner, B. M. (Eds.). (2002). *Macroergonomics: Theory, methods and applications.* Mahwah, NJ: Lawrence Erlbaum Associates, Inc.

Imada, A. S. (2002). A macroergonomic approach to reducing work-related injuries. In H. W. Hendrick & B. M. Kleiner (Eds.), *Macroergonomics: Theory, methods and applications* (pp. 347–358). Mahwah, NJ: Lawrence Erlbaum Associates, Inc.

Kantola, J., & Karwowski, W. (1998). A fuzzy-logic based tool for the evaluation of compter-integrated manufacturing, organization and people system design. In W. Karwowski & R. Goonetilleke (Eds.), *Manufacturing agility and hybrid-automation II* (pp. 43–46). Hong Kong: IEA Press, HKUST.

Karwowski, W., Kantola, J., Rodrick, D., & Salvendy, G. (2002). Macroergonomic aspects of manufacturing. In H. W. Hendrick & B. M. Kleiner (Eds.), *Macroergonomics: Theory, methods, and applications* (pp. 223–258). Mahwah, NJ: Lawrence Erlbaum.

Katz, D., & Kahn, R. L. (1966). *The social psychology of organizations.* New York: Wiley.

Keidel, R. W. (1994). Rethinking organizational design. *Academy of Management Executive, 8*(4), 12–30.

Lawrence, P. R., & Lorsch, J. W. (1969). *Organization and environment.* Homewood, IL: Irwin.

Marjchrizak, A. (1988). *The human side of factory automation.* San Francisco, CA: Jossey-Bass.

Marjchrizak, A. (2005), TOP-Modeler, in N. Stanton, A. Hedge, K. Brookhauis, E. Salas, & H. W. Hendrick (Eds), *Handbook of human factors and ergonomics methods* (pp. 85–1 through 85–3). Boca Raton, FL: CRC Press.

Meshkati, N. (1986). Human factors considerations in technology transfer to industrially developing countries: An analysis and proposed model. In O. Brown, Jr. & H. W. Hendrick (Eds.), *Human factors in organizational design and management II* (pp. 351–368). Amsterdam: North-Holland.

Meshkati, N. (1990). Human factors in large-scale technological system's accidents: Three Mile Island, Bhopal, and Chernobyl. *Industrial Crisis Quarterly, 5,* 133–154.

Meshkati, N., & Robertson, M. M. (1986). The effects of human factors on the success of technology transfer projects to industrially developing countries: A review of representative case studies. In O. Brown, Jr. & H. W. Hendrick (Eds.), *Human factors in organizational design and management II* (pp. 343–350). Amsterdam: North-Holland.

Mileti, D. S., Gillespie, D. S., & Haas, J. E. (1977). Size and structure in complex organizations. *Social Forces, 56,* 208–217.

Montanari, J. R. (1975). *An expanded theory of structural determinism: An empirical Investigation of the Impact of Managerial Discression on Organizational Structure.* Unpublished doctorial dissertation, University of Colorado, Boulder.

Montanari, J. R. (1976). *An Expanded Theory of Structural Determinism: An Empirical Investigation of the Impact of Managerial Discretion on Organizational Structure.* Unpublished doctoral dissertation, University of Colorado, Boulder, CO.

Nagamachi, M. (2002), Relationships among job design, macroergonomics, and productivity. In H. W. Hendrick & B. M. Kleiner (Eds.). *Macroergonomics: Theory, methods and applications* (pp. 111–132). Mahwah, NJ: Lawrence Erlbaum.

Neghandi, A. R. (1977). A model for analyzing organization in cross-cultural settings: A conceptual scheme and some research findings. In A. R. Negandhi, G. W. England, & B. Wilpert (Eds.), *Modern organizational theory* (285–312) . Kent State, OH: University Press.

Noro, K., & Imada, A. S. (1991). *Participatory ergonomics.* London: Taylor & Francis.

Perrow, C. (1967). A framework for the comparative analysis of organizations. *American Sociological Review, 32,* 194–208.

Robertson, M. M., Kleiner, B. M., & O'Neill, M. J. (2002). Macroergonomic methods: Assessing work system processes. In H. W. Hendrick & B. M. Kleiner (Eds.). *Macroergonomics: Theory, methods and applications* (pp. 67–96). Mahwah, NJ: Lawrence Erlbaum.

Robbins, S. R. (1983). *Organizational theory: The structure and design of organizations.* Englewood Cliffs, NJ: Prentice-Hall.

Smith, A. (1970). *The wealth of nations.* London: Penguin. (Originally published in 1876)

Stanton, N., Hedge, A., Hendrick, H., Brookhuis, K., & Salas, E. (Eds.). (2004). *Handbook of human factors and ergonomics methods.* London: Taylor & Francis.

Stevenson, W. B. (1993). Organizational design. In R. T. Golembiewski (Ed.), *Handbook of organizational behavior* (pp. 141–168). New York: Marcel Dekker.

Trist, E. L., & Bamforth, K. W. (1951). Some social and psychological consequences of the longwall method of coal-getting. *Human Relations, 4,* 3–38.

Trist, E. L., Higgin, G. W., Murray, H., & Pollock, A. B. (1963). *Organizational choice.* London: Tavistock.

Wisner, A. (1995). Situated cognition and action: Implications for ergonomic work analysis and anthoroptechnology. *Ergonomics, 38*(8), 1542–1557.

Yankelovich, D. (1979). *Work, values and the new breed.* New York, NY: Van Norstrand Reinhold.

4

WORK SYSTEM DESIGN IN HEALTH CARE

Pascale Carayon, Carla J. Alvarado, and Ann Schoofs Hundt
University of Wisconsin, Madison

Designing the work of health care providers is an important issue for health care organizations, the providers themselves, patients and their families, and society at large. Poor work conditions can contribute to negative outcomes for health care organizations, such as high turnover and injuries (Cohen-Mansfield, 1997). Health care providers may experience a range of negative feelings and emotions and suffer from health problems and injuries when their work is not ergonomically designed (Linzer et al., 2000; Stubbs, Buckle, Hudson, Rivers, & Worringham, 1983). Patients and their families may be indirectly affected by poor design of the work environment in which care is provided to them (Agency for Healthcare Research and Quality, 2002; Institute of Medicine Committee on the Work Environment for Nurses and Patient Safety, 2004). Finally, society at large may be affected by poor design of health care working environments. For instance, nursing shortage has been linked to difficulty in attracting and keeping nurses because of poor working conditions (Institute of Medicine Committee on the Work Environment for Nurses and Patient Safety, 2004; Wunderlich & Kohler, 2001). Understanding the characteristics of "good" work and learning about the process of work system design provide the foundation for health care organizations to engage in work improvements that can ultimately lead to a range of positive outcomes for the organizations themselves (e.g., reduced turnover), the workers (e.g., increased job satisfaction), patients and their families (e.g., improved quality and safety of care), and society (e.g., decreased nursing shortage) (Carayon, Alvarado, Brennan et al., 2003; Carayon, Alvarado, & Hundt, 2003; Kovner, 2001; Sainfort, Karsh, Booske, & Smith, 2001).

Existing data show that the work of health care providers is in need of redesign. A national study found a dramatic aging of the registered nurse population along with a lack of nursing students in the pipeline to replace retiring nurses (Buerhaus, Staiger, & Auerbach, 2000). Over the past decade economic pressures within health care settings have led to efforts that have reduced hospital length of stays resulting in a higher level of patient illness and more activity in hospital and home care settings. Optimizing patient outcomes with a shrinking workforce while achieving financial stability is challenging to patient care managers, often leading to high voluntary turnover (Schryer, 2004). The change in the work environment has led to more stress and consequently less satisfaction on the part of staff nurses. The high-pressure environment along with declining resources has made the overall work environment less attractive, particularly for older nurses (Lasky & Wake, 2001). The work environment is in need of redesign that gives nurses the support and tools they need to spend as much time as possible caring for patients (Thrall, 2003). Improvements in work design are also necessary for many other health care job categories, as demonstrated for example by job dissatisfaction, stress, and burnout experienced by physicians (McMurray, Linzer, Douglas, & Konrad, 1999; Wetterneck et al., 2002).

WORK SYSTEM

The human error literature defines a variety of work-related factors that can contribute to accidents

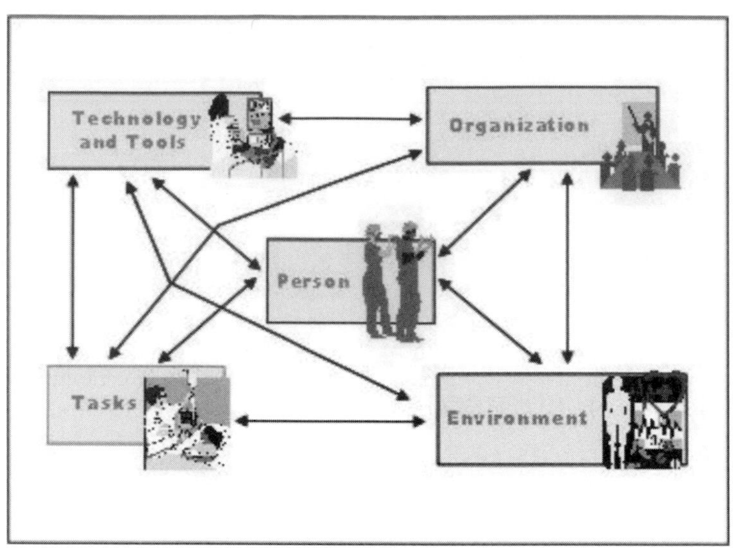

FIGURE 4–1. Model of the work system (Carayon & Smith, 2000; Smith & Carayon-Sainfort, 1989).

(Rasmussen, 1982; Reason, 1990). Whereas human error models describing the "chain of events" leading to accidents are interesting and useful for analyzing and evaluating specific accidents, other models and approaches need to be considered when redesigning work. Unlike the human error literature, these other models put little emphasis on understanding and dissecting the specific mechanisms that lead to human error and accidents. They provide information on *how* to change work and on *what* needs to be changed in the work system (Parker & Wall, 1998).

The concept of "system" in patient safety and quality of care has been largely emphasized (Berwick, 2002; Buckle et al., 2003; Donabedian, 1980; Institute of Medicine Committee on Quality of Health Care in America, 2001; Leape, 1997; Vincent, Moorthy, Sarker, Chang, & Darzi, 2004). In this chapter, the concept of system is used in a very specific sense: a *system* represents the various elements of work that a health care provider uses, encounters, and experiences to perform his or her job. Designing or redesigning work requires a systematic approach that takes into account the various elements of work. The work system model developed by Carayon and Smith is a very useful model to understand the many different elements of work (Carayon & Smith, 2000; Smith & Carayon-Sainfort, 1989). The work system is comprised of five elements: the *person* performing different *tasks* with various *tools and technologies* in a *physical environment* under certain *organizational conditions* (see Figure 4–1).

The person at the center of the work system has physical, cognitive, and psychosocial characteristics. Physical characteristics include, for instance, height, physical strength, and reach envelop. Cognitive characteristics include, for instance, information processing capacity, knowledge, and expertise. Psychosocial characteristics include, for instance, motivation, needs, and tolerance for ambiguity. In this chapter, the person is a health care provider such as a nurse, physician, pharmacist, or respiratory therapist. The work system model can also be applied to the patient as a person (Carayon, Alvarado, Hsieh, & Hundt, 2003).

The tasks performed by the individual represent the specific interactions between the person and his or her work to accomplish particular objectives. *Task* is a key concept in human factors and ergonomics (Kirwan & Ainsworth, 1992). It is important to distinguish between the "prescribed task" and the "activity" or the task actually carried out (Hackman, 1969; Leplat, 1989). The prescribed task is defined by the organization: it is what is expected of the worker, and what she or he is supposed to do. What the worker actually does may be different from the prescribed task. The activity or the task being carried out is what the worker actually does and can be characterized by some type of work analysis (e.g., direct observation, interview with the worker (Leplat, 1989). The task actually carried out is influenced by a redefinition process that considers the context and circumstances in which the work is being performed (Hackman, 1969). Various dimensions can be used to characterize the task: difficulty or challenge, content, variety, repetitiveness, skill utilization, autonomy and job control, clarity, uncertainty, demands, contact with others, and feedback (Carayon, Haims, & Yang, 2001). Several chapters in this handbook describe various aspects of the tasks of health care workers. See, for example, the chapter on job stress in health care workers (Murphy, chap. 11, this volume) and the chapter on burnout in health care (Schaufeli, chap.14, this volume).

The person performing a task uses various *tools and technologies*, which can be simple (e.g., paper and pencil), can be technologically advanced (e.g., bar coding medication administration technology, MRI or Magnetic Resonance Imaging), require the use of hands (e.g, hand tools such as surgical endoscopic tools), may involve remote sensing (e.g., robotic surgery), enhance the capacity of the operator (e.g., lifting aids, remote monitoring), and support other tools and technologies (e.g., nursing workstation where other equipment is located and used). The tools and technologies interact with the physical, cognitive, and psychosocial characteristics of the person and other elements of the work system. For instance, studies on health care information and communication technologies have identified various human factors problems: poor coordination between nurses and physicians when a bar coding medication administration system was implemented at VA hospitals (Patterson, Cook, & Render, 2002) and poor usability of a computerized provider order entry system (Koppel et al., 2005). Tools and technologies should be designed to support the work of the health care provider in performing the tasks necessary to accomplish quality and safety of care objectives.

The *physical environment* can be characterized on the following dimensions: noise, lighting, temperature–humidity–air flow, space and layout, and other characteristics (e.g., vibration, radiation). Much is known about how to design the various dimensions of the physical environment (Parsons, 2000). See, for example, the chapter on physical environment in health care (Alvarado, chap. 19, this volume) and the chapter on noise and alarms in health care (Buss & Friesdorf, chap. 22, this volume). It is important to recognize that the physical environment interacts with all of the characteristics of the person. Physical characteristics of the person need to be considered when designing, for instance, the height of a workstation and the placement of various pieces of equipment. The design of the physical environment can also affect the cognitive characteristics of the person, such as providing cues for attracting the attention of health care providers. Psychosocial characteristics of the person are also important to consider when designing a physical environment. For instance, the physical layout of a unit can facilitate communication between workers by creating a space where various providers can meet and discuss.

A range of *organizational conditions* govern and influence the way a person performs tasks using tools and technologies in a specific physical environment: work schedules, organizational support (e.g., social support from supervisors and managers, resources provided during a technological or organizational change), communication, collaboration, coordination, decision-making structure, role characteristics such as ambiguity and conflict, training, rewards, benefits, performance evaluation, teamwork, and organizational culture. A number of chapters in this handbook describe these organizational conditions, for instance, the chapter on safety culture in health care (Itoh, Andersen, & Madsen, chap. 13, this volume) and the chapter on organizational learning in health care (Hundt, chap. 8, this volume).

It is important to underline the system nature of work, that is, the work system is comprised of elements that interact with each other. Consider the example of the task of central venous cannulation in the intensive care unit (ICU). It is an essential part of patient management in a variety of clinical settings, for example, for critically ill patients requiring long-term venous access and patients with cancer requiring chemotherapy. Unfortunately, patient complications of central venous catheter cannulation are very common. Noise and other sensory disruptions abound in the modern ICU, creating stress, masking communication, patient responses, alarms, and so forth (characteristics of the environment and the task). The environment is crowded and messy and physically difficult to work in (characteristics of the physical environment that interact with task performance). Crowds of people are waiting to get a moment of the physician's and/or nurse's time, all making it more difficult for them to carry out their task without interruptions (organizational conditions that affect task performance).

TABLE 4–1. Possible Pathways Between Job Redesign and Performance (Kelly, 1992)

Job redesign leads to improved job perceptions and job satisfaction, which in turn may positively influence intrinsic motivation. Workers who are intrinsically motivated are more likely to perform more and better.
Job redesign may lead to increased extrinsic motivation because of pay raises or improved promotion prospects. Workers who are extrinsically motivated are more likely to perform more and better.
Performance may improve if workers perceive closer links between effort, performance, and valued rewards (expectancy theory).
Job redesign may be accompanied by goal setting: employees either set themselves new performance targets or management sets new performance targets. There is considerable evidence that goal setting is a powerful motivator, in particular under specific conditions (e.g., difficult goals).
It is possible that when redesigning a job one discovers structural inefficiencies such as inadequate access to equipment. These inefficiencies may be overcome by work methods improvements that are part of the job redesign.

Hospital supply stocking schedules and storage often make supplies difficult to locate or not available when needed (organizational condition and characteristic of the physical environment). The hospital administration may not budget for or approve new devices or technology for the tasks (organizational conditions and technology). To correct a perceived problem with task performance, the administration may introduce additional new tasks for a group of employees. With the introduction of nurses performing peripherally inserted central venous catheters, there is a possibility of physician workload reduction. Although an effective solution to balance the physician workload, it is important not to create negative aspects for the nursing workload. This example demonstrates how various elements of the work system interact with each other, and how elements of the work system of one worker (e.g., the physician) can affect elements of the work system of another worker (e.g., the nurse). Too often because of a lack of understanding of the entire structure and design of the overall work system and of how various components affect health care worker performance, a solution may be proposed that does not consider the potential for creating "imbalance" within the system work factors and, therefore, lead to increases in workload on the person.

In the context of increasing technological change in health care, we need to understand the interactions between tools and technologies and other elements of the work system. The following example describes the implementation of smart intravenous pumps in a hospital setting and their impact on the work system of different health care providers. Given the range of intravenous pump users and the variety of patient care environments and clinical circumstances surrounding their use, one might reasonably expect that organizations would consider this variation in "smart" IV pump use and in turn how it could influence implementation of smart IV pumps organization-wide. For example, features associated with smart pumps including alarm thresholds and drug libraries require careful consideration of user needs and expectations. Because more potent drugs are included in drug libraries, consideration must be given to the type and extent of training an ICU versus general medical/surgical versus pediatric nurse would receive. Use in the operating room by certified registered nurse anesthetists and anesthesiologists requires careful assessment of clinical practices to arrive at a consensus when determining upper and lower dose limits associated with unsafe levels of IV drug administration. Consensus in this case may be difficult to achieve.

WORK SYSTEM AND PATIENT SAFETY

The literature on work design has identified many different mechanisms on the relationship between work redesign and worker satisfaction, motivation, and performance (see Table 4–1; Kelly, 1992). These mechanisms relate to internal and external motivation of the workers (Steers, Mowday, & Shapiro, 2004), expectancy between effort and rewards (Donovan, 2002), goal setting (Ivancevich, 1976), and efficiency gains (Kelly, 1992).

We have identified four different pathways or mechanisms between work redesign and patient safety (see Table 4–2). First, redesigning work may directly target the causes or sources of patient safety problems. For instance, a number of studies have shown that excessive working hours and

TABLE 4–2. Pathways Between (Re)Design and Patient Safety

1. Work redesign may directly target the causes or sources of patient safety problems.
2. Work redesign may lead to improved efficiencies by removing performance obstacles, therefore freeing up time and reducing workload for nurses to provide better, safer patient care.
3. Work redesign may lead to the reexamination of who does what, i.e., the objectives of work, and indirectly improve quality and safety of care.
4. Work design can be considered as part of the "Structure" element of Donabedian's (1980) model of quality of care. Therefore, improving work can improve care processes and therefore patient outcomes, including patient safety.

poorly designed work schedules of nurses (Rogers, Hwang, Scott, Aiken, & Dinges, 2004; Smith, Colligan, Frockt, & Tasto, 1979) and physicians (Landrigan et al., 2004; Lockley et al., 2004) can lead to various patient safety problems. Several methods, such as FMEA and RCA, allow the identification of errors and specific work system factors that contribute to patient safety problems (Carayon, Alvarado, & Hundt, 2003). These methods can benefit from the use of the work system model to systematically address all elements of work that can contribute to an actual incident or event (e.g., Root Cause Analysis) or to anticipate the work factors that may contribute to vulnerabilities in a process (e.g., Failure Modes and Effects Analysis). Studies performed in manufacturing have shown relationships between ergonomic deficiencies (e.g., poor working postures) and poor quality of assembly (Axelsson, 2000; Eklund, 1995). In a similar manner, redesigning work in health care may remove the sources of possible error and improve quality of work, that is, patient safety. These sources of error can be contained in any of the work system elements or can be the product of interactions between elements of the work system.

Second, work redesign may lead to improved efficiencies. Inefficiencies can be considered as "performance obstacles" (Brown & Mitchell, 1991; Park & Han, 2002; Peters, Chassie, Lindholm, O'Connor, & Kline, 1982), that is, aspects of the work system that hinder health care workers' ability to provide good safe patient care (Carayon, Gurses, Hundt, Ayoub, & Alvarado, 2005). Removing inefficiencies may directly or indirectly improve patient safety. Inefficiencies may represent sources of error or conditions that enhance the likelihood of error. For example, reliance on paper copies of patients' medical records can result in delays and inefficiencies when there is a problem locating them. As a result unwise (and subsequently unsafe) clinical decisions may be made in the interest of averting delays when a provider determines sufficient (although not necessarily ideal) information is available. Sources of performance obstacles or inefficiencies can be categorized according to the work system model: task-related inefficiencies (e.g., lack of job-related information, interruptions), tools and technologies (e.g., poor functioning or inaccurate equipment, inadequate materials and supplies), organizational inefficiencies (e.g., inadequate staffing, lack of support, lack of time), and physical environment (e.g., noisy, crowded environment) (Carayon et al., 2005; Peters et al., 1982; Peters & O'Connor, 1980). Removing performance obstacles and inefficiencies may free up time and reduce workload (e.g., more efficient use of currently employed nurses), therefore reducing the likelihood of error. Preliminary research has been conducted to identify performance obstacles in health care, specifically among staff of outpatient surgery centers and intensive care nurses (Carayon et al., 2005). This research highlights the variety of work factors that can hinder performance of health care providers, that is, their capacity to provide high-quality safe care. For instance, among staff in several outpatient surgery centers, conflict between nurses and physicians, lack of staffing, lack of staff training, and poor work schedules were identified as major performance obstacles. Among ICU nurses, we found that the major performance obstacles were related to patient requirements and inadequate staffing (e.g., poor performance, inadequate skills, insufficient number of people) (Carayon et al., 2005).

Third, work redesign may lead to improved effectiveness. Whereas efficiency is the degree to which work is performed well, effectiveness relates to the content of work itself: Are we doing the right things? Work redesign may lead an organization to ask questions regarding task allocation (who is doing what?) and job content (what tasks should be performed that are not performed and what tasks are performed that should not be performed?). For instance, Hendrickson, Doddato, and Kovner (1990) conducted a study on how hospital nurses spend their time. Work sampling techniques were used to evaluate time spent in various activity categories (i.e., with patient, patient chart, preparation of therapies, shift change activities, professional interaction, miscellaneous activities, checking physician's orders, unit-oriented in-service, paperwork, phone communications, supplies, miscellaneous nonclinical). Results showed that within a typical 8-hr shift nurses spend an average of 31% of their time with patients. Such results may lead to the reexamination of the tasks and content of nursing jobs. Work redesign can contribute to increased efficiency, that is, the elimination of unnecessary activities may free up time for health care workers to provide better, safer patient care and may lead to a more efficient effective distribution of tasks.

Fourth, Donabedian's framework for assessing quality of care can fit the work system model and provide justification for another pathway between work design and patient safety (Donabedian, 1978, 1980, 1988). In Donabedian's model, the *structure* includes *the organizational structure* (work system model element 'organization), *the material resources* (work system model elements = environment, tools/technology), and *the human resources* (work system model elements = worker, tasks). Donabedian's two other means of assessing quality include evaluating the *process(es)* of care—how worker tasks and clinical processes are both organized and performed (e.g., Was the care provided in compliance with the clinical pathway? Was medically indicated care provided?), and evaluating the *outcome(s)* of care—assessing the clinical results and impacts of and patient satisfaction with the care provided. Donabedian (1980) proposed that a direct relationship exists between structure, process, and outcome.

The structure of an organization affects how safely care is provided (the process); the means of caring for and managing the patient (the process) affects how safe the patient is on discharge (outcome). We have developed a patient safety model that incorporates the work system model in Donabedian's framework (Carayon, Alvarado, Brennan et al., 2003). Overall, the *work system* in which care is provided affects both the work and clinical *processes*, which in turn influence the patient and organizational *outcomes* of care. Changes to any aspect of the work system will, depending on how the change or improvement is implemented, either negatively or positively affect the work and clinical processes and the consequent patient and organizational outcomes. Redesigning a work system requires careful planning and consideration to ensure that neither the quality of care nor patient or worker safety are compromised because of a lack of consideration to all of the elements of the work system. Likewise, subsequent review of the success of change in an organization cannot be fully accomplished without assessing the entire work system, the clinical and work processes, and both the clinical and organizational outcomes of patient care.

HUMAN FACTORS AND ERGONOMICS IN WORK SYSTEM DESIGN

A question raised by ergonomists is how to ensure that ergonomic criteria are considered in the early stage of work system design (Clegg, 1988; Dekker & Nyce, 2004; Luczak, 1995; Slappendel, 1994). Johnson and Wilson (1988) discussed two approaches for taking into account ergonomics in work system development: (a) provision of guidelines and (b) ergonomics input within collaborative design. To define work system guidelines, we need to better understand the specific work system elements (and their combinations) that affect the outcomes of quality and safety of health care. We have somewhat limited information on this. But a large body of literature on work system design principles is available (Salvendy, 1997). The other strategy proposed by Johnson and Wilson may be more rapidly implemented. This necessitates the close collaboration of ergonomists and health care professionals at various stages of work system design.

Work System Design Guidelines

Various principles for work system design can be derived from a large body of literature. The work design body of knowledge can be categorized into (a) ergonomics (or human factors), (b) job stress, and (c) job/organizational design. Tables 4–3, 4–4, and 4–5 provide a list of theories and models of ergonomics, job stress, and job/organizational design. Each of the theories or models provides information on one or several elements of the work system. This information helps in defining the positive and negative aspects of the work system and its elements, as well as identifying possible solutions for redesigning work and improving patient safety. For example, ergonomics knowledge can help in designing appropriate features of the physical work system, such as noise, workspace, and workstation. The Job

Strain model of Karasek and Theorell (1990) identified two key job stressors: workload and lack of control. According to this model, it is the combination of high workload and low job control that is the most stressful. When workload cannot be reduced (and this is often the case), this model recommends that workers be given more control over various aspects of their work. These are only examples of criteria for "good" work. Studies among nurses have examined aspects of work such as job characteristics (Tonges, Rothstein, & Carter, 1998; Woodcox, Isaacs, Underwood, & Chambers, 1994), role stress (Woodcox et al., 1994), patient handling tasks (Owen, Garg, & Jensen, 1992), and workload (Carayon & Gurses, 2005). These studies have examined various aspects of work among nurses but have not adopted the system approach to work design and patient safety that has been recommended by many experts (Bogner, 1994; Leape, 1997).

Work Design Process

Designing or redesigning work can represent a major organizational investment requiring the involvement of numerous people, substantial time to conduct evaluations, analyze data and design and implement solutions, adequate resources, and sufficient expertise and knowledge. Like any major organizational or technological change, work redesign needs to be "managed." A process needs to be implemented for coordinating all the personnel, activities and resources involved in the work redesign project.

The collaborative work redesign process involves the following pieces:

- A series of steps and activities logically and chronologically organized.
- A "toolbox" of tools and methods that one can use to evaluate the work system and design and implement solutions (Carayon, Alvarado, & Hundt, 2003).
- A set of overarching principles that can guide an organization when embarking into a work redesign project (e.g., participation and involvement of workers and various stakeholders, feedback).

Many work design processes have been identified and proposed (Wilson, 1995b). They take different forms, have different levels of specification, and use different terminologies. However, they all can fit into a sequence of analysis, synthesis, and evaluation. Wilson (1995b) proposed an ergonomics design process with 12 steps. Figure 4–2 describes the steps of the process. This ergonomics design process is very detailed with regard to analysis, but may lack details and specifications for synthesis and evaluation. The structured work redesign process proposed by Parker and Wall (1998) includes 8 phases and is displayed in Figure 4–3.

To implement the work design process, tools and methods need to be used for each of the different steps (Carayon, Alvarado, & Hundt, 2003). At the analysis stage, the following tools can be used: job analysis, FMEA, root cause analysis, process analysis, interdependence analysis/variance analysis, ergonomic analysis, and work sampling. Some of these same tools can be used to identify solutions for work redesign: process analysis, interdependence analysis/variance analysis, and ergonomic analysis. Other tools can be used to design solutions, such as task allocation methods and simulation.

WORK SYSTEM DEVELOPMENT

The work design process unfolds over time; three different stages of work system development can be distinguished (Clegg, 1988):

1 Design of work system.
2 Implementation of work system.
3 Operation of work system.

Phase of Work System Design

Ergonomic criteria should be considered as early as possible. Unfortunately, very little research has been conducted to examine how ergonomic criteria are considered in the design of new work systems. Wulff, Westgaard, and Rasmussen (1999a; 1999b) conducted a study of the implementation of ergonomics requirements in large-scale engineering projects of the design of off-shore installations. Exploratory case studies in two engineering design companies involved in two different design projects were conducted. Considerable resistance to using ergonomic requirements in their design was observed within the engineering teams. A reason for the resistance appeared to be the lack of familiarity with this new set of requirements in combination with high total workload. A solution to this problem may be to include an "active" ergonomics

TABLE 4–3. Ergonomics in the Work System

Fields in Ergonomics	*Work System Elements*					*For More Information*
	Individual	*Tasks*	*Tools and Technologies*	*Physical Environment*	*Organizational Conditions*	
Environmental design	Physical, cognitive, and psychosocial characteristics			Workspace design Noise, Vibration Lighting Climate		(Konz, 1983; Salvendy, 1997)
Tool design	Physical characteristics		Medical device design			(Sanders & McCormick, 1993)
Workspace and workstation design	Physical, cognitive and psychosocial characteristics		Workstation design	Workspace design		(Kroemer Kroemer, & Kroemer-Elbert, 2001; Sanders & McCormick, 1993)
Work schedules	Physical and psychosocial characteristics				Shiftwork system design	(Monk & Tepas, 1985)
Human–computer interaction	Experience, learning style, cognitive characteristics	Usability	Interface design			(Jacko & Sears, 2003)

TABLE 4–4. Job Stress in the Work System

Models and Theories of Job Stress	*Work System Elements*					*For More Information . . .*
	Individual	*Tasks*	*Tools and Technologies*	*Physical Environment*	*Organizational Conditions*	
Karasek's Job Strain model		Workload Job decision latitude			Social support	(Karasek & Theorell, 1990)
Cooper and Marshall's model	Level of anxiety Level of neuroticism Tolerance for ambiguity Type A behavioral pattern	Work overload and time pressure		Poor working conditions	Role ambiguity and conflict Responsibility Lack of job security Little or no participation Poor organizational climate	(Cooper & Marshall, 1976)
Role stress					Role ambiguity Role conflict Role overload	(Kahn, 1981)
Person–Environment Fit Theory	Misfit between the person's *needs*... Misfit between the person's *abilities*...	...and the work environment's *supplies*, e.g., boredom ...and the work environment's *demands*, e.g., overload				Caplan, et al., 1975; Edwards, 1988)

TABLE 4–5. Job/Organizational Design in the Work System

Models and Theories of Job/Organizational Design	*Work System Elements*					*For More Information . . .*
	Individual	*Tasks*	*Tools and Technologies*	*Physical Environment*	*Organizational Conditions*	
Scientific management		Simple, repetitive tasks	Adequate tools for the tasks			(Taylor, 1911)
Job rotation		Regular rotation between tasks				(Parker & Wall, 1998)
Job enlargement		Increasing the number of tasks				(Parker & Wall, 1998)
Job enrichment		Upgrading tasks to include extra "skilled tasks"			Increased employee responsibility such as making decisions about work scheduling and task allocation	Herzberg, 1974)
Job Characteristics Theory	Needs, such as growth need strength	Skill variety Task completeness Task significance AutonomyFeedback				(Hackman & Oldham, 1976)
Sociotechnical Systems Theory	Options to accommodate individual differences and circumstances	Challenge Ability to learn Scope of decision making			Social support Recognition Job future(Semi-) autonomous work groups	(Davis & Wacker, 1987; Emery & Trist, 1965; Trist, 1981)
Teamwork	Group design composition Knowledge and skills	Group design structure of task Level of effort			Organizational context (reward, education, information) Group synergy	(Hackman, 1987, 1989; Tannenbaum et al., 1996)
High involvement management					Power Knowledge Information Rewards	(Lawler, 1986)
Organizational climate and culture					Organizational climate and culture Safety climate and culture	(Helmreich & Merritt, 1998; Hofstede, 1997)

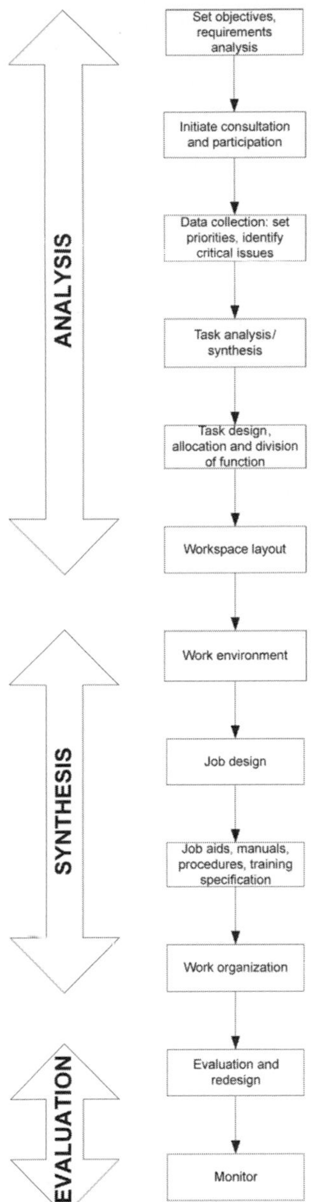

FIGURE 4–2. Ergonomics design process (Wilson, 1995b).

resource person in the design organization. When an ergonomist with high legitimacy was actively involved in the design process, ergonomics requirements were more likely to be used. In addition, organizational means can be used to ensure the implementation of ergonomic criteria in the design process. Examples of organizational means include emphasis on ergonomics in general company policy documents, high organizational status for ergonomics, and active support of senior management. How does this research translate to the design of work systems in health care to achieve high-quality safe patient care? This raises issues regarding the availability of human factors expertise

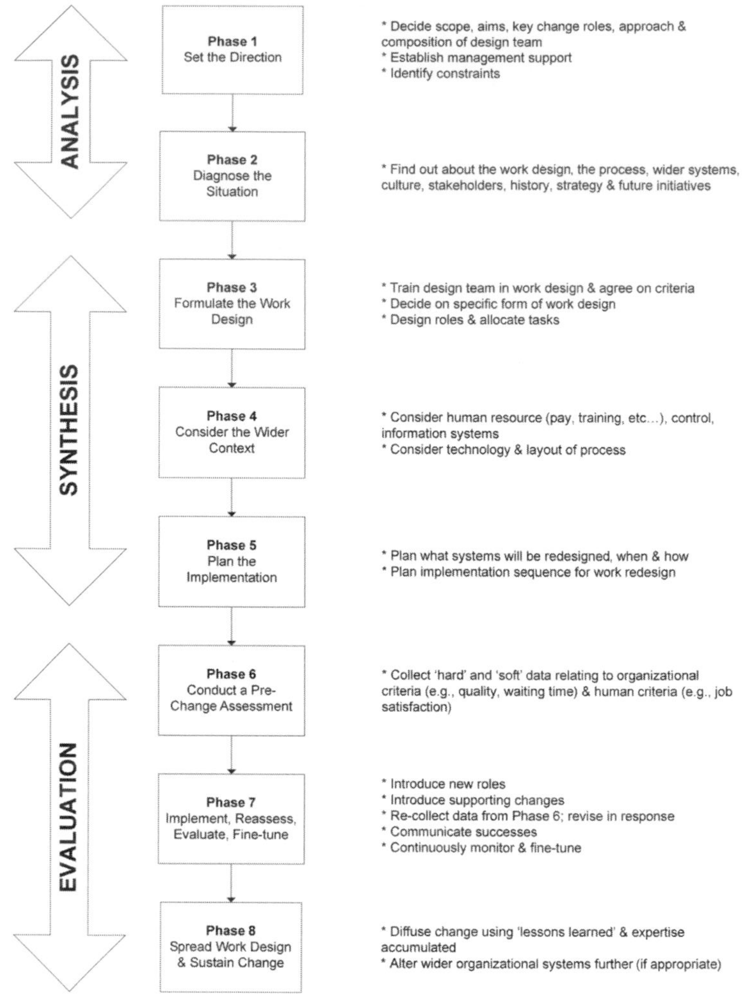

FIGURE 4–3. Structured work redesign (Parker & Wall, 1998).

within health care institutions, as well as the need to allocate adequate resources so that ergonomic criteria are considered at the work system design stage (Carayon, 2005). A variety of jobs and functions within health care organizations can benefit from knowledge, principles, and tools of human factors and ergonomics (Carayon, 2005). This human factors body of knowledge and expertise needs to be translated and transferred to the people involved in those jobs and functions.

Clegg (1988) argued that organizations have many choices when they design manufacturing systems. Decisions are made regarding the following factors:

- The type and level of technology. Different elements can impact decisions regarding the type and level of technology: resources available, expected return on investment, technology "push"
- The allocation of functions between humans and machines. In general, the human aspect is considered late in the design of manufacturing systems, therefore leaving the "leftover" tasks to people.
- The roles of humans in the system. Once allocation of function decisions have been made, the various tasks need to be

organized into job designs for the future operators of the system.

- The organizational structures to support workers. Companies that are introducing new manufacturing systems could usefully ask themselves what organizational structures are appropriate.
- The way in which people participate in their design. The type, extent, and timing of worker participation in the design of work systems are all important aspects to consider (Smith & Carayon, 1995).

Choices made regarding these different issues have important ergonomic implications. Similar choices are made by health care organizations when designing work systems and structures, patient care processes, as well as other processes. For example, usability testing, aside from evaluating aspects of learnability and memorability (Nielsen, 1993), can likewise be used when assessing technologies for selection and implementation by including worker input in the selection process. In turn this user feedback can be used to aid in the design of new user training modules. Once the technology is implemented, the organization must provide adequate support to monitor and sustain a successful implementation by being responsive to user needs, questions, and difficulties. Likewise it is important to monitor for unanticipated problems and/or patient safety concerns arising from the technology or provider interaction with it and, as appropriate, share findings with users.

How do we ensure that designers of health care work systems and technologies take into account human factors and ergonomics? One example of work system design is the construction of a new health care facility. Health care facility construction, whether a new building or an expansion of an existing medical center, can present a number of challenges and a number of opportunities, not the least of which is improving working conditions, quality of care, and patient safety. A far-sighted health care facility in West Bend, Wisconsin, is demonstrating that new construction projects actually present an opportunity to improve working conditions and patient safety. In April 2000, St. Joseph's Community Hospital of West Bend, Wisconsin, a member of SynergyHealth Inc., started focusing on how the design of a new facility could affect patient safety. A participatory learning laboratory developed recommendations that St. Joseph's could apply in the design process (Reiling & Chernos, 2004). St. Joseph's facility design process for patient safety is an interesting case study (for more information, see Reiling & Chernos, chap. 18, this volume).

Phase of Implementation

In the phase of work system implementation, the question arises as to the methods and processes to use to facilitate the change process and rapidly achieve the expected outcomes (i.e., improved quality and safety of care). The way change is implemented (i.e., process implementation) is central to the successful adaptation of organizations to changes. A "successful" work system implementation from the human factors viewpoint is defined by its "human" and organizational characteristics: reduced/limited negative impact on people (e.g., stress, dissatisfaction) and on the organization (e.g., delays, costs, medication errors) and increased positive impact on people (e.g., acceptance of change, job control, enhanced individual performance) and on the organization (e.g., efficient implementation process, safe patient care). Success also includes decreasing medical errors and improving quality of care. Several authors have recognized the importance of the process of implementation in achieving a successful organizational change (Korunka et al., 1993; Tannenbaum, Salas, & Cannon-Bowers, 1996).

Participatory ergonomics is a powerful method for implementing work system changes (Wilson, 1995a). Participation has been used as a key method for implementing various types of organizational changes, such as physical ergonomic programs (Haims & Carayon, 1996; Wilson & Haines, 1997), continuous improvement programs (K. J. Zink, 1996), and technological change (Carayon & Karsh, 2000; Eason, 1988). Noro and Imada (1991) defined participatory ergonomics as a method in which end-users of ergonomics (e.g., workers, nurses) take an active role in the identification and analysis of ergonomic risk factors as well as the design and implementation of ergonomic solutions. Evanoff and his colleagues have conducted studies on participatory ergonomics in health care (Bohr, Evanoff, & Wolf, 1997; Evanoff, Bohr, & Wolf, 1999). One study examined the implementation of participatory ergonomics teams in a medical center. Three groups participated in the study: a group of orderlies from the dispatch department, a group of ICU nurses, and a group of laboratory workers. Overall, the team members for the dispatch and the laboratory groups

TABLE 4–6. Questions to ask for Describing the Work System

1. What are the characteristics of the individual performing the work? Does the individual have the musculoskeletal, sensory, and cognitive abilities to do the required task? If not, can any of these be accommodated for the task?
2. What tasks are being performed and what are characteristics of the tasks that may contribute to unsafe patient care? What in the nature of the tasks allows the individual to perform them safely or assume risks in the process?
3. What in the physical environment can be sources of error or promotes safety? What in the physical environment ensures safe behavior or leaves room for unsafe behavior?
4. What tools and technologies are being used to perform the tasks, and do they increase or decrease the likelihood of untoward events?
5. What in the organization prevents or allows exposure to hazard? What in the organization promotes or hinders patient safety? What allows for assuming safe or unsafe behavior by the individual?

were satisfied with the participatory ergonomics process, and these perceptions seemed to improve over time. However, the ICU team members expressed more negative perceptions. The problems encountered by the ICU team seemed to be related to the lack of time and the time pressures due to the clinical demands.

A more in-depth evaluation of the participatory ergonomics program on orderlies showed substantial improvements in health and safety following the implementation of the participatory ergonomics program (Evanoff et al., 1999). The studies by Evanoff and colleagues demonstrate the feasibility of implementing participatory ergonomics in health care, but highlight the difficulty of the approach in a high-stress, high-pressure environment, such as an ICU, where patient needs are critical and patients need immediate or continuous attention. More research is needed to develop ergonomic methods for implementing work system changes that lead to improvements in human and organizational outcomes as well as improved quality and safety of care. This research should consider the high-pace, high-pressure work environment in health care organizations.

The implementation of any new work system always engenders problems and concerns. The process by which these problems and concerns are resolved is important from a human factors point of view but also from a quality of care and patient safety point of view. It is necessary to have the capability and tools to identify potential human factors and quality and safety of care problems in a timely manner.

Operational Phase

During the operational phase, the new work system is in place. What are the characteristics of a work system that lead to quality of care and patient safety? Much work is needed to specify the structural component of quality of care. For instance, what working conditions are related to quality of care? Much is known about the working conditions that affect stress, job satisfaction, and other human outcomes (Kalimo, Lindstrom, & Smith, 1997; Smith, 1987). However, we need to know more about the working conditions that affect quality and safety of care, and, more important, how to improve working conditions to improve both human outcomes (e.g., reduced stress, increased job satisfaction and reduced injuries) and patient outcomes (see Tables 4–3, 4–4, and 4–5 for a set of work design guidelines).

Workload is one working condition of particular importance in health care quality and safety. For instance, Tarnow-Mordi, Hau, Warden, and Shearer (2000) examined the relationship between mortality rates and the workload of hospital staff in one adult ICU in the United Kingdom. The measures of ICU workload most strongly associated with mortality were peak occupancy, average nursing requirement per occupied bed per shift, and the ratio of occupied to appropriately staffed beds. This study illustrates the current approach to workload and patient safety: Workload is typically measured at the unit level (e.g., an ICU). Such level of analysis does not reveal the system design characteristics that may contribute to workload and patient safety problems. In the study of Tarnow-Mordi et al. (2000), explanations for the association between high workload and mortality highlighted several work system characteristics, such as insufficient time for clinical procedures to be done appropriately, inadequate training or supervision, errors, overcrowding and consequently nosocomial infections, limited availability of equipment, and premature discharge from the ICU. Human factors conceptualization and assessment of workload would reveal the sources of workload and help identify ways of redesigning the work system to reduce workload and improve patient safety (Carayon & Gurses, 2005).

How does one ensure that continuous improvements in work system design and important outcomes (quality of care and patient safety, as well as human and organizational outcomes) are achieved? Various models and approaches to quality improvement and management have been proposed and implemented in health care (for example, Shortell et al., 1995). This research would benefit from an ergonomic point of view to simultaneously optimize work system design and improve quality and safety of care. Ergonomic approaches to quality management and improvement have emphasized the importance of job and organizational design and quality of working life (Carayon, Sainfort, & Smith, 1999), the link between ergonomic deficiencies and quality deficiencies (Axelsson, 2000; Eklund, 1995), and the importance of management approaches for improved safety and health (Zink, 2000). All of these ergonomic approaches have much to offer to design continuous improvement systems and processes in health care. The goal of the improvement systems and processes would be to improve human and organizational outcomes as well as quality of care and patient safety.

CONCLUSION

Describing the work system, its elements, and their interactions is only the first step in work design (see Table 4–6 for a set of questions to ask to describe the work system). We also need to evaluate the work system and its elements, that is, to define the positive and negative aspects of the work system elements. The evaluation of the positive and negative aspects relies on knowledge in the fields of ergonomics, job stress, and job/organizational design. These three fields of research have developed a number of theories, models, and principles that characterize "good" work (see Tables 4–3, 4–4, and 4–5). This information is important for not only evaluating the work system but also for designing solutions for improving the work system and subsequent outcomes, such as performance and patient safety. Once the positive and negative aspects of the work system and its elements have been defined we need to come up with solutions that will improve the work design and increase patient safety.

Solutions for improving work design and patient safety may involve the elimination of negative aspects of the work system. As an example, proper health care provider hand hygiene is the corner stone of any infection control effort. However, hand hygiene is performed less than half of the time it is required for patient care. Health care providers state negative aspects for this lack of hand hygiene, such as handwashing agents causing irritation and dryness, sinks inconveniently located or lack of sinks, lack of soap and paper towels, and too busy and insufficient time. Because eliminating these negative aspects may not be feasible, solutions for improving work design may involve other elements of the work system that compensate for or balance out the negative aspects. Placement of waterless alcohol-based hand hygiene agents in patient care areas or personal bottles of the agent for each health care provider eliminates the need for sinks, soap, and paper towels. The waterless alcohol hand rub is fast acting and can be used while performing other tasks such as communicating with patients, family members, and professional colleagues. This compensatory effect is the essence of the Balance Theory proposed by Carayon and Smith (Carayon & Smith, 2000; Smith & Carayon-Sainfort, 1989). In many situations, it may not be possible to eliminate negative aspects of the work system because of constraints due to physical design, the infrastructure, and costs, as well as constraints on recruitment. In those situations, it is important to come up with work design solutions that take into account those negative aspects and try to minimize their impact on workers and patient safety.

The five elements of work (i.e. the individual, tasks, tools and technologies, physical environment, and organizational conditions) represent a system: They influence each other and interact with each other. A change in any one element of the work system can have effects on other elements. This system concept has a number of consequences for work redesign:

- We cannot look at one work element in isolation.
- Whenever there is a change in work, we need to consider the effects on the entire work system.
- The activity of work redesign necessitates knowledge and expertise in a variety of domains, for example, environmental

design such as lighting and noise; job design such as autonomy and work demands; and organizational design such as safety culture, teamwork, and communication.

- Work redesign aims at removing the "negative" aspects of work, that is, the aspects contributing to poor performance and unsafe patient care. When this is not feasible, work redesign involves building or relying on other elements of work to compensate for or balance out the negative aspects of work.

References

Agency for Healthcare Research and Quality. (2002). *Impact of working conditions on patient safety* (Rep. No. 03-P003). Rockville, MD: Author.

Axelsson, J. R. C. (2000). *Quality and ergonomics—Towards successful integration.* Unpublished doctoral dissertation, Linkoping University, Linkoping, Sweden.

Berwick, D. M. (2002). A user's manual for the IOM's "Quality Chasm" report. *Health Affairs, 21*(3), 80–90.

Bogner, M. S. (Ed.). (1994). *Human error in medicine.* Hillsdale, NJ: Lawrence Erlbaum Associates, Inc.

Bohr, P. C., Evanoff, B. A., & Wolf, L. (1997). Implementing participatory ergonomics teams among health care workers. *American Journal of Industrial Medicine, 32*, 190–196.

Brown, K. Q., & Mitchell, T. R. (1991). A comparison of just-in-time and batch manufacturing: The role of performance obstacles. *Academy of Management Journal, 34*, 906–917.

Buckle, P., Clarkson, P. J., Coleman, R., Lane, R., Stubbs, D., Ward, J., et al. (2003). *Design for patient safety.* London: Department of Health Publications.

Buerhaus, P., Staiger, D., & Auerbach, D. (2000). Implications of an aging registered nurse workforce. *Journal of the American Medical Association, 283*, 2948–2954.

Caplan, R. D., Cobb, S., French, J. R. P., Harrison, R. V., & Pinneau, S. R. (1975). *Job demands and worker health.* Washington DC: U.S. Government Printing Office.

Carayon, P. (2005). Top management's view on human factors and patient safety: Do they see it? In R. Tartaglia, S. Bagnara, T. Bellandi, & S. Albolino (Eds.), *Healthcare systems ergonomics and patient safety* (pp. 38–42). Florence, Italy: Taylor & Francis.

Carayon, P., Alvarado, C. J., Brennan, P., Gurses, A., Hundt, A., Karsh, B.-T., et al. (2003). Work system and patient safety. In H. Luczak & K. J. Zink (Eds.), *Human factors in organizational design and management, VII* (pp. 583–589). Santa Monica, CA: IEA.

Carayon, P., Alvarado, C., Hsieh, Y., & Hundt, A. S. (2003, August). *A macroergonomic approach to patient process analysis: Application in outpatient surgery.* Paper presented at the 15th Triennial Congress of the International Ergonomics Association, Seoul, Korea.

Carayon, P., Alvarado, C., & Hundt, A. S. (2003). *Reducing workload and increasing patient safety through work and workspace design.* Washington, DC: Institute of Medicine.

Carayon, P., & Gurses, A. (2005). A human factors engineering conceptual framework of nursing workload and patient safety in intensive care units. *Intensive and Critical Care Nursing. 21*, 284–307.

Carayon, P., Gurses, A. P., Hundt, A. S., Ayoub, P., & Alvarado, C. J. (2005). Performance obstacles and facilitators of healthcare providers. In C. Korunka & P. Hoffmann (Eds.), *Change and quality in human service work* (Vol. 4, pp. 257–276). Munchen, Germany: Hampp.

Carayon, P., Haims, M. C., & Yang, C. L. (2001). Psychosocial work factors and work organization. In W. Karwowski (Ed.), *The international encyclopedia of ergonomics and human factors* (pp. 111–121). London: Taylor and Francis.

Carayon, P., & Karsh, B. (2000). Sociotechnical issues in the implementation of imaging technology. *Behaviour and Information Technology, 19*, 247–262.

Carayon, P., Sainfort, F., & Smith, M. J. (1999). Macroergonomics and total quality management: How to improve quality of working life? *International Journal of Occupational Safety and Ergonomics, 5*, 303–334.

Carayon, P., & Smith, M. J. (2000). Work organization and ergonomics. *Applied Ergonomics, 31*, 649–662.

Clegg, C. (1988). Appropriate technology for manufacturing: Some management issues. *Applied Ergonomics, 19,* 25–34.

Cohen-Mansfield, J. (1997). Turnover among nursing home staff. *Nursing Management, 28*(5), 59–64.

Cooper, C. L., & Marshall, J. (1976). Occupational sources of stress: A review of the literature relating to coronary heart disease and mental ill health. *Journal of Occupational Psychology, 49*, 11–25.

Davis, L. E., & Wacker, G. J. (1987). Job design. In G. Salvendy (Ed.), *Handbook of human factors* (pp. 431–452). New York: Wiley.

Dekker, S. W. A., & Nyce, J. M. (2004). How can ergonomics influence design? Moving from research findings to future systems. *Ergonomics, 47*, 1624–1639.

Donabedian, A. (1978). The quality of medical care. *Science, 200*, 856–864.

Donabedian, A. (1980). *The definition of quality and approaches to its assessment.* Ann Arbor, MI: Health Administration Press.

Donabedian, A. (1988). The quality of care. How can it be assessed? *Journal of the American Medical Association, 260*, 1743–1748.

Donovan, J. J. (2002). Work motivation. In N. Anderson & D. S. Ones (Eds.), *Handbook of industrial, work and organizational psychology, Vol. 2: Organizational Psychology* (pp. 53–76). Thousand Oaks, CA: Sage.

Eason, K. (1988). *Information technology and organizational change.* London: Taylor & Francis.

Edwards, J. R. (1988). The determinants and consequences of coping with stress. In C. L. Cooper & R. Payne (Eds.), *Causes, coping and consequences of stress at work* (pp. 233–266). Chichester, UK: Wiley.

Eklund, J. A. E. (1995). Relationships between ergonomics and quality in assembly work. *Applied Ergonomics, 26*, 15–20.

Emery, F., & Trist, E. (1965). The causal texture of organizational environments. *Human Relations, 18*, 21–31.

Evanoff, V. A., Bohr, P. C., & Wolf, L. (1999). Effects of a participatory ergonomics team among hospital orderlies. *American Journal of Industrial Medicine, 35*, 358–365.

Hackman, J. R. (1969). Toward understanding the role of tasks in behavioral research. *Acta Psychologica, 31*, 97–128.

Hackman, J. R. (1987). The design of work teams. In J. Lorsch (Ed.), *Handbook of organizational behavior* (pp. 315–342). Englewood Cliffs, NJ: Prentice Hall.

Hackman, J. R. (1989). *Groups that work (and those that don't): Creating conditions for effective teamwork.* San Francisco: Jossey Bass.

Hackman, J. R., & Oldham, G. R. (1976). Motivation through the design of work: Test of a theory. *Organizational Behavior and Human Performance, 16*, 250–279.

Haims, M. C., & Carayon, P. (1996). Implementation of an "in-house" participatory ergonomics program: A case study in a public service organization. In O. J. Brown & H. W. Hendrick (Eds.), *Human factors in organizational design and management, V* (pp. 175–180). Amsterdam: Elsevier.

Helmreich, R. L., & Merritt, A. C. (1998). *Culture at work in aviation and medicine.* Aldershot, England: Ashgate.

Hendrickson, G., Doddato, T. M., & Kovner, C. T. (1990). How do nurses use their time? *Journal of Nursing Administration, 20*(3), 31–37.

Herzberg, F. (1974, September/October). The wise old Turk. *Harvard Business Review, 52,* 70–80.

Hofstede, G. (1997). *Cultures and organizations—Software of the mind.* New York: McGraw Hill.

Institute of Medicine Committee on Quality of Health Care in America. (2001). *Crossing the quality chasm: A new health system for the 21st. century.* Washington, DC: National Academy Press.

Institute of Medicine Committee on the Work Environment for Nurses and Patient Safety. (2004). *Keeping patients safe: Transforming the work environment of nurses.* Washington, DC: The National Academies Press.

Ivancevich, J. M. (1976). Effects of goal setting on performance and job satisfaction. *Journal of Applied Psychology, 61*, 605–612.

Jacko, J. A., & Sears, A. (Eds.). (2003). *The human-computer interaction handbook.* Mahwah, NJ: Lawrence Erlbaum Associates, Inc.

Johnson, G. I., & Wilson, J. K. (1988). Future directions and research issues for ergonomics and advanced manufacturing technology (AMT). *Applied Ergonomics, 19,* 3–8.

Kahn, R. (1981). *Work and health.* New York: Wiley.

Kalimo, R., Lindstrom, K., & Smith, M. J. (1997). Psychosocial approach in occupational health. In G. Salvendy (Ed.), *Handbook of human factors and ergonomics* (pp. 1059–1084). New York: Wiley.

Karasek, R. A., & Theorell, T. (1990). *Healthy work: Stress, productivity and the reconstruction of working life.* New York: Basic Books.

Kelly, J. (1992). Does job re-design theory explain job re-design outcomes? *Human Relations, 45*, 753–774.

Kirwan, B., & Ainsworth, L. K. (1992). *A guide to task analysis.* London: Taylor & Francis.

Konz, S. A. (1983). *Work design: Industrial ergonomics* (2nd ed.). Columbus, OH: Grid.

Koppel, R., Metlay, J. P., Cohen, A., Abaluck, B., Localio, A. R., Kimmel, S. E., et al. (2005). Role of computerized physician order entry systems in facilitating medications errors. *Journal of the American Medical Association, 293*, 1197–1203.

Korunka, C., Weiss, A., & Karetta, B. (1993). Effects of new technologies with special regard for the implementation process per se. *Journal of Organizational Behavior, 14*, 331–348.

Kovner, C. (2001). The impact of staffing and the organization of work on patient outcomes and health care workers in health care organizations. *The Joint Commission Journal on Quality Improvement, 27*, 458–468.

Kroemer, K. H. E., Kroemer, H., & Kroemer-Elbert, K. (2001). *Ergonomics—How to design for ease and efficiency* (2nd ed.). Upper Saddle River, NJ: Prentice Hall.

Landrigan, C. P., Rothschild, J. M., Cronin, J. W., Kaushal, R., Burdick, E., Katz, J. T., et al. (2004). Effect of reducing interns' work hours on serious medical errors in intensive care units. *New England Journal of Medicine, 351*(18), 1838–1848.

Lasky, P., & Wake, M. (2001). *Will Wisconsin have a nursing workforce to meet future health care needs?* Retrieved July 10, 2005, from http://www.pophealth.wisc.edu/uwphi/publications/briefs/may01brief.htm

Lawler, E. E., III. (1986). *High involvement management: Participative strategies for improving organizational performance.* San Francisco: Jossey-Bass.

Leape, L. L. (1997). A systems analysis approach to medical error. *Journal of Evaluation in Clinical Practice, 3*(3), 213–222.

Leplat, J. (1989). Error analysis, instrument and object of task analysis. *Ergonomics, 32*, 813–822.

Linzer, M., Konrad, T. R., Douglas, J., McMurray, J. E., Pathman, D. E., Williams, E. S., et al. (2000). Managed care, time pressure, and physician job satisfaction: Results from the Physician Worklife study. *Journal of General Internal Medicine, 15*, 441–450.

Lockley, S. W., Cronin, J. W., Evans, E. E., Cade, B. E., Lee, C. J., Landrigan, C. P., et al. (2004). Effect of reducing interns' weekly work hours on sleep and attentional failures. *New England Journal of Medicine, 351*(18), 1829–1837.

Luczak, H. (1995). Macroergonomic anticipatory evaluation of work organization in production systems. *Ergonomics, 38*, 1571–1599.

McMurray, J. E., Linzer, M., Douglas, J., & Konrad, T. R. (1999, April–May). *Burnout in U.S. women physicians: Assessing remediable factors in worklife.* Paper presented at the 22nd Annual Meeting of the Society of General Internal Medicine, San Francisco, CA.

Monk, T. H., & Tepas, D. I. (1985). Shift work. In C. L. Cooper & M. J. Smith (Eds.), *Job stress and blue collar work* (pp. 65–84). New York: Wiley.

Nielsen, J. (1993). *Usability engineering.* Amsterdam: Morgan Kaufmann.

Noro, K., & Imada, A. (1991). *Participatory ergonomics.* London: Taylor & Francis.

Owen, B., Garg, A., & Jensen, R. C. (1992). Four methods for identification of most back-stressing tasks performed by nursing assistants in nursing homes. *International Journal of Industrial Ergonomics, 9*, 213–220.

Park, K. S., & Han, S. W. (2002). Performance obstacles in cellular manufacturing implementation—Empirical investigation. *International Journal of Human Factors and Ergonomics in Manufacturing, 12*, 17–29.

Parker, S., & Wall, T. (1998). *Job and work design.* Thousand Oaks, CA: Sage.

Parsons, K. C. (2000). Environmental ergonomics: A review of principles, methods and models. *Applied Ergonomics, 31*, 581–594.

Patterson, E. S., Cook, R. I., & Render, M. L. (2002). Improving patient safety by identifying side effects from introducing bar coding in medical administration. *Journal of the American Medical Informatics Association, 9*, 540–553.

Peters, L. H., Chassie, M. B., Lindholm, H. R., O'Connor, E. J., & Kline, C. R. (1982). The joint influence of situational constraints and goal setting on performance and affective outcomes. *Journal of Management, 8*, 7–20.

Peters, L. H., & O'Connor, E. J. (1980). Situational constraints and work outcomes: The influences of a frequently overlooked construct. *Academy of Management Review, 5*, 391–397.

Rasmussen, J. (1982). Human errors. A taxonomy for describing human malfunction in industrial installations. *Journal of Occupational Accidents, 4*, 311–333.

Reason, J. (1990). *Human error.* Cambridge: Cambridge University Press.

Reiling, J., & Chernos, S. (2004). Error reduction through facility design. In M. S. Bogner (Ed.), *Human error in healthcare: A handbook of issues and indications.* Mahwah, NJ: Lawrence Erlbaum Associates, Inc.

Rogers, A. E., Hwang, W.-T., Scott, L. D., Aiken, L. H., & Dinges, D. F. (2004). The working hours of hospital staff nurses and patient safety: Both errors and near errors are more likely to occur when the hospital staff nurses work twelve or more hours at a stretch. *Health Affairs, 23*, 202–212.

Sainfort, F., Karsh, B., Booske, B. C., & Smith, M. J. (2001). Applying quality improvement principles to achieve healthy work organizations. *Journal on Quality Improvement, 27*, 469–483.

Salvendy, G. (Ed.). (1997). *Handbook of human factors and ergonomics* (2nd ed.). New York: Wiley.

Sanders, M. S., & McCormick, E. J. (1993). *Human factors in engineering and design.* New York: McGraw-Hill.

Schryer, N. (2004). Implementing organizational redesign to support practice: The Tulane model. *Journal of Nursing Administration, 34*, 400–406.

Shortell, S. M., O'Brien, J. L., Carman, J. M., Foster, R. W., Hughes, E. F. X., Boerstler, H., et al. (1995). Assessing the impact of continuous quality improvement/total quality management: Concept versus implementation. *Health Services Research, 30*, 377–401.

Slappendel, C. (1994). Ergonomics capability in product design and development: An organizational analysis. *Applied Ergonomics, 25*, 266–274.

Smith, M. J. (1987). Occupational stress. In G. Salvendy (Ed.), *Handbook of human factors and ergonomics* (pp. 844–860). New York: Wiley.

Smith, M. J., & Carayon, P. (1995). New technology, automation, and work organization: Stress problems and improved technology implementation strategies. *The International Journal of Human Factors in Manufacturing, 5*, 99–116.

Smith, M. J., & Carayon-Sainfort, P. (1989). A balance theory of job design for stress reduction. *International Journal of Industrial Ergonomics, 4*, 67–79.

Smith, M. J., Colligan, M. J., Frockt, I. J., & Tasto, D. L. (1979). Occupational injury rates among nurses as a function of shift schedule. *Journal of Safety Research, 11*(4), 181–187.

Steers, R. M., Mowday, R. T., & Shapiro, D. L. (2004). The future of work motivation theory. *Academy of Management Review, 29*, 379–387.

Stubbs, D. A., Buckle, P., Hudson, M. P., Rivers, P. M., & Worringham, C. J. (1983). Back pain in the nursing profession, I: Epidemiology and pilot methodology. *Ergonomics, 26*, 755–765.

Tannenbaum, S. I., Salas, E., & Cannon-Bowers, J. A. (1996). Promoting team effectiveness. In M. A. West (Ed.), *Handbook of work group psychology* (pp. 503–530). New York: Wiley.

Tarnow-Mordi, W. O., Hau, C., Warden, A., & Shearer, A. J. (2000). Hospital mortality in relation to staff workload: A 4-year study in an adult intensive care unit. *Lancet, 356*, 185–189.

Taylor, F. (1911). *The principles of scientific management.* New York: Norton.

Thrall, T. H. (2003). Work redesign. *Hospitals & Health Networks, 77*(3), 34–38.

Tonges, M. C., Rothstein, H., & Carter, H. K. (1998). Sources of satisfaction in hospital nursing practice. A guide to effective job design. *Journal of Nursing Administration, 28*(5), 47–61.

Trist, E. (1981). *The evaluation of sociotechnical systems.* Toronto, Ontario, Canada: Quality of Working Life Center.

Vincent, C., Moorthy, K., Sarker, S. K., Chang, A., & Darzi, A. W. (2004). Systems approaches to surgical quality and safety—From concept to measurement. *Annals of Surgery, 239*, 475–482.

Wetterneck, T. B., Linzer, M., McMurray, J. E., Douglas, J., Schwartz, M. D., Bigby, J. A., et al. (2002). Worklife and satisfaction of general internists. *Archives of Internal Medicine, 169*, 649–656.

Wilson, J. R. (1995a). Ergonomics and participation. In J. R. Wilson & E. N. Corlett (Eds.), *Evaluation of human work—A practical ergonomics methodology* (2nd ed., pp. 1071–1096). London: Taylor & Francis.

Wilson, J. R. (1995b). A framework and a context for ergonomics methodology. In J. R. Wilson & E. N. Corlett (Eds.), *Evaluation of human work—A practical ergonomics methodology* (2nd ed., pp. 1–39). London: Taylor & Francis.

Wilson, J. R., & Haines, H. M. (1997). Participatory ergonomics. In G. Salvendy (Ed.), *Handbook of human factors and ergonomics* (pp. 490–513). New York: Wiley.

Woodcox, V., Isaacs, S., Underwood, J., & Chambers, L. W. (1994). Public health nurses' quality of worklife: Responses to organizational changes. *Canadian Journal of Public Health, 85*, 185–187.

Wulff, I. A., Westgaard, R. H., & Rasmussen, B. (1999a). Ergonomic criteria in large-scale engineering design, I: Management by documentation only? Formal organization vs. designers' perceptions. *Applied Ergonomics, 30*, 191–205.

Wulff, I. A., Westgaard, R. H., & Rasmussen, B. (1999b). Ergonomic criteria in large-scale engineering design, II: Evaluating and applying requirements in the real-world of design. *Applied Ergonomics, 30,* 207–221.

Wunderlich, G. S., & Kohler, P. O. (Eds.). (2001). *Improving the quality of long-term care.* Washington, DC: National Academy Press.

Zink, K. (2000). Ergonomics in the past and the future: From a German perspective to an international one. *Ergonomics, 43*, 920–930.

Zink, K. J. (1996). Continuous improvement through employee participation: Some experiences from a long-term study. In O. Brown Jr. & H. W. Hendrick (Eds.), *Human factors in organizational design and management, V* (pp. 155–160). Amsterdam: Elsevier.

5

SOCIOTECHNICAL SYSTEM DESIGN IN HEALTH CARE

Brian M. Kleiner
Virginia Polytechnic Institute and State University, Blacksburg

Increasingly, it is being recognized that health care systems need to be viewed as sociotechnical systems and that ergonomic interventions need to be holistic and systemic. Macroergonomics is the subdiscipline of ergonomics that focuses on this holistic work system analysis and design. Sociotechical systems theory is the recognized theoretical framework supporting macroergonomics. Through a macroergonomic approach, supported by sociotechnical systems theory, it is believed that 60% to 90% performance is possible. This is in contrast to the 10% to 20% improvements typically achieved through intervention. It is contended that health care systems could benefit from the larger error reduction/performance improvement scenario. This chapter begins with some background about sociotechnical systems. It then defines the health care system as a sociotechnical system. I then focus on the major contribution of this chapter, a major method for analyzing and designing work system processes, called Macroergonomic Analysis and Design (MEAD).

SOCIOTECHNICAL SYSTEMS DESIGN BACKGROUND

Health care systems are viewed as sociotechnical work systems. There are three work system design practices that often characterize dysfunctional work system development and modification efforts as causes of challenges to safety, health and performance. These are technology-centered design, a "left-over" approach to function and task allocation, and a failure to consider an organization's sociotechnical characteristics and integrate them into its work system design (Hendrick and Kleiner, 2001).

Technology-Centered Design

In employing technology in medical environments, designers incorporate automation into some form of hardware or software to achieve some intended purpose. To the extent that those who must operate or maintain the hardware or software are considered, it usually is in terms of the skills, knowledge, and training required. Often, even this kind of consideration is not considered at an ergonomic or interface level. As a result, the motivational aspects of jobs, psychosocial characteristics of the work force, and other related work system design factors rarely are considered properly. Yet, these are the very factors that guarantee implementation success and can significantly improve work system effectiveness.

If ergonomic aspects of design are considered, it is often after the equipment or software already is designed. Then, the ergonomist or the untrained professional may be secured to modify some of the human–system interfaces to reduce the likelihood of human error. Often, even this level of involvement does not occur until testing of the newly designed system reveals serious interface design problems. This has certainly been the case with the automation of many physicians' offices. Medical record keeping has been automated to the point where many physicians' have been relegated to data entry personnel. In many cases, this automation trend has been driven by the health maintenance organizations (HMOs) and other stakeholders' desire and perhaps need to drive costs down. However, at this point in the design process, because of cost and schedule considerations, the ergonomist is severely limited in terms of making fundamental changes to improve the work system. Instead, the ergonomist is restricted to making a few superficial improvements of specific human– machine, human–environment, or human–software interfaces. Thus, the ultimate outcome can be a suboptimized work system in which a given subsystem may be improved or "optimized" at the expense of one or more other subsystems. As observed in the pervasive automation of medical records, health care worker roles have significantly changed and have been defined by the capabilities and limitations of the software and have not necessarily been defined by a systematic function allocation strategy, whereby the tasks performed by human health care worker or machine have been systematically and scientifically determined. Although there have been positive outcomes reported, there are also difficulties associated with the human interface.

A related impact of a technology-centered approach to redesigning existing work systems is that employees typically do not participate in the planning and implementation process. As the organizational change literature frequently has demonstrated, the result often is not only a poorly designed work system but a lack of commitment and resistance to the proposed changes. When a technology-centered approach is taken, if employees are brought into the process at all, it is usually after the work system changes have been designed. The employee's role is delegated to usability testing of the system. As a result, when employees find serious problems with the changes (as often happens), cost and schedule considerations prevent any major redesign to eliminate or minimize the identified deficiencies.

"Left-Over" Approach to Function and Task Allocation

A technology-centered approach often leads to treating the persons who will operate and maintain the system as mechanistic components. That is, the motivational and other sociocultural aspects are ignored. The focus is on assigning to the "machine" any functions or tasks that the technology enables it to perform. Then, what is left over is assigned to humans. As a result, the function and task allocation process fails to consider the characteristics of the work force and related external environmental factors. Most often, the consequence is a poorly designed work system that fails to make effective use of its human resources. Returning to the medical records example, physicians report that their tasks in many cases are that which the computer system cannot perform. By observing the variability inherent in current automation of medical records systems, it is obvious that a sociotechnical systems approach to design has not been taken. For example, although one large HMO system has physicians performing the data entry tasks, another physician network has the nurses entering the data. Neither has undergone sufficient test and evaluation to determine the best work system design.

Joint optimization is a term used to describe the goal of making sure designers attend to both the technology and the personnel simultaneously. To achieve joint optimization, the designer jointly designs the technology and personnel (DeGreene, 1973). Put in ergonomic terms, joint optimization requires a "human-centered approach" (Hendrick & Kleiner, 2001), which means designing to support human capability and limitation. In terms of function and task allocation, Bailey (1989) referred to the process as a humanized task approach. He noted that

> this concept essentially means that the ultimate concern is to design a job that justifies using a person, rather than a job that merely can be done by a human. With this approach, functions are allocated and the resulting tasks are designed to make full use of human skills and to compensate for human limitations. The nature of the work itself should lend itself to internal motivational influences. The left over functions are allocated to computers. (p. 190)

With this definition, it is all too clear that medical record keeping can benefit from a more systematic analysis and design process.

Failure to Consider the System's Sociotechnical Characteristics

The primary structural and process characteristics of sociotechnical systems first were empirically identified by the Tavistock Institute in the United Kingdom, based on studies of longwall coal mining and other industries over 4 decades ago (DeGreene, 1973). Unfortunately, as first was documented by the Tavistock studies of coal mining 5 decades ago (Emery & Trist, 1965; Trist & Bamforth, 1951), a technology-centered or leftover approach to the organizational design of work systems does not adequately consider the key characteristics of the sociotechnical system elements. Consequently, the resulting work system design most often is suboptimal. In the example of automating medical records systems, the sociotechnical factors to consider include, but are not limited to, the sociocultural characteristics and preferences of the nurses and physicians, the capability of the software and hardware, the expectations of the insurance and other stakeholders for efficiency and other performance improvements.

Sociotechnical Systems Design Defined

A simplified descriptive model views the health care work system as being comprised of several important and related subsystems (see Figure 5–1.) Five major characteristics or elements of sociotechnical systems can be identified. These are the (a) technological subsystem, (b) personnel subsystem, (c) external environment, (d) internal environment, and (e) organizational design. The *technological subsystem* is concerned with the manner in which work is performed. This includes the technology (e.g. software, machinery, and equipment), tools, methods, and procedures. In a medical record keeping system, this includes the computer and specialized software adopted. It also includes the procedures used by the physician's office. The *personnel subsystem* concerns the sociocultural and socioeconomic characteristics of the people involved in health care, including their selection and training. The major roles under consideration in the medical records area are nursing and physicians. In addition to nurses and physicians, the major roles to consider with medical records also often include a Director of Medical Records and medical records clerks. The *external environment* is comprised of the political, economic, technological, educational, and cultural forces that affect the health care work system. The work system must successfully procure resources from its environment and efficiently and effectively produce products back to the environment to be successful. Key stakeholders in the medical records environment are the HMO, patients and their families, medical specialists, pharmaceutical companies, and insurance providers. The *internal environment* (e.g., physician's office) is also depicted. This is comprised of both the physical and cultural microenvironment of the office. *Interfaces* are the connections between components of a system. Of special importance are the interfaces among the various subsystems. This is important because it implies that a given component of a system (e.g., medical device) may not be as important in terms of potential errors as the connection between the medical device and the health care worker who uses the device. At the organizational level, this notion of interface can be applied to the interface between the organization and the external environment. Inadequate or inappropriate interfaces between a work system's organizational design and external environment can create an internal environment fraught with risk (Hendrick & Kleiner, 2001). An improper human–technology interface can create a hazard for the health care worker. An ineffective interface between a worker and his or her organization can encourage that short cuts be taken. Because these five elements interact with one another, a change in any one affects the other four.

Because of these interrelationships, characteristics of each of the first elements affect the last: The *organizational design* of the work system. Empirical models have been developed of these relationships that can be used to determine the optimal work system structure (i.e., the optimal degrees of vertical and horizontal differentiation, integration, formalization, and centralization to design into the work system; Burns & Stalker, 1961; Lawrence & Lorsch, 1969; Perrow, 1967; Woodward, 1965). *Vertical differentiation* is the segmentation of the organization in terms of hierarchical levels. *Horizontal differentiation* relates to departmentalization. *Formalization* has to do with standardization through policies, procedures, processes, and so forth. *Centralization* is the degree to which decision making is concentrated among a few individuals, typically at the apex of the organization.

HEALTH CARE AS A SOCIOTECHNICAL WORK SYSTEM

Work System Components and STS Factors

A fundamental assumption of this chapter is that health care systems, are *sociotechnical systems,* defined

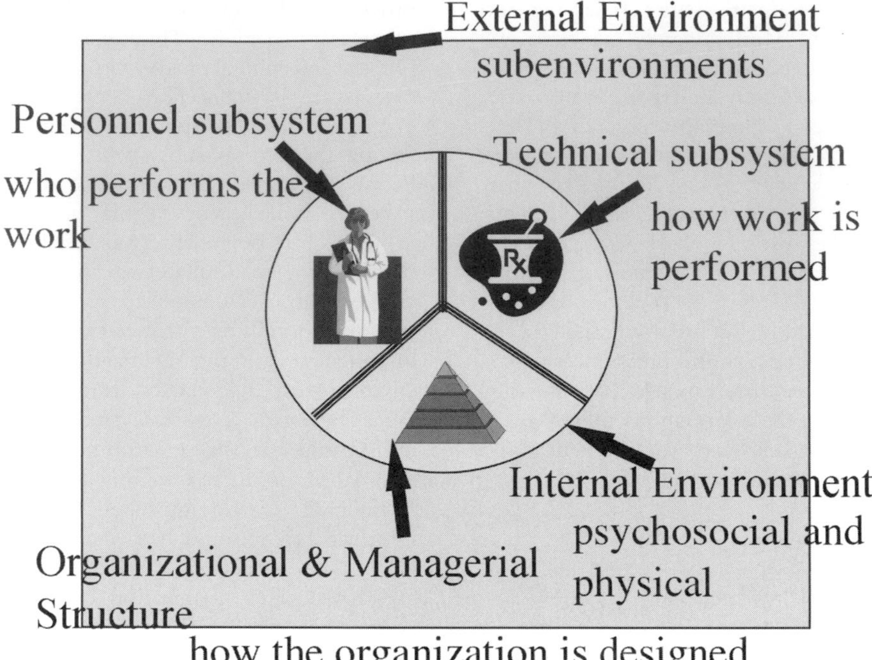

FIGURE 5–1. Simplified diagram of sociotechnical work system.

as two or more people (personnel subsystem) interacting with hardware or software (technological subsystem), an internal and external environment, and an organizational design (Figure 5–1; Hendrick & Kleiner, 2001).

Sociotechnical systems are complex systems that are highly dependent on the dynamic interactions between humans and other subsystems, including other humans. This complexity introduces needs and requirements for the design of support tools to facilitate work. Systems such as medical records keeping systems or medical event reporting systems (MERS) are technologies used by personnel within a work system. MERS technology consists of knowledge features and hardware and software features. For a MERS, the relevant work system components include personnel subsystem components such as the clinical effectiveness team, physician units, nursing units, hospital staff (pharmacy, radiology, etc.); the technical subsystem that includes the information system that stores and reports the medical events; the organizational and managerial structure that refers to the hospital organizational design in which the MERS will operate; the internal environment that includes the culture, its acceptance to new technology such as a MERS, and the external environment, including the insurers, patient families, hospital owner/operator, and forth.

MEASURING PERFORMANCE IN A HEALTH CARE SYSTEM

Regardless of other benefits that may be realized from macroergonomic improvements to the work system, organizations usually are not able to justify the intervention unless there is a clear economic benefit to be derived. A definition of macroergonomics can be found in chapter 1 and in the next section we describe MEAD, which is an example of macroergonomic intervention. Therefore, it is extremely important to clearly identify the costs and economic benefits that can be expected and outline how they will be measured. Kleiner's (1997) framework is often used to guide the characterization of performance metrics. In general, major benefits to be realized will be reduction of medical errors, improved job satisfaction, increased quality of care, and improved productivity. Major costs could include consulting costs and any hardware and/or software costs associated with the intervention.

SOCIOTECHNICAL SYSTEMS ANALYSIS AND DESIGN METHOD

The MEAD evaluation framework has been developed based on the work of Emery and Trist (1978), Clegg, Ravden, Corbertt, and Johnson (1989), and Kleiner's own experience with large scale physical organizational change in academia, industry, and government with specific emphasis on integrating sociotechnical systems (STS) theoretical propositions and prescriptions with microergonomics 1999; Kleiner, Drury, & Palepu, 1998).

The MEAD methodology is unique in its strong theoretical support and because it integrates ergonomics interface design, function allocation, and other macroergonomics tools. Specifically, Emery and Trist (1978) and Clegg et al. (1989) are used to perform STS analysis and nested function allocation. Function allocation strategies and methodologies have been under development since the original Fitts (1951) list was proposed. More recent function allocation methodologies assume the use of a group or team of analysts but do not specify work system design, decision making, or other requirements of such design teams. Clegg et al. (1989) provided requirements for function allocation methods. Included are the criteria that the allocation methods be (a) systematic (i.e., organized), (b) criterion-based (i.e., focused on results), (c) multidimensional (i.e., include technology and sociocultural issues), (d) capable of handling both large and small functions (i.e., scalable), (e) iterative (i.e., have an improvement loop), (f) linked to earlier design decisions (i.e., provide feeback), (g) face valid in an organizational context (i.e., appear to be useful and meaningful), and (h) promote participation (i.e., involve key stakeholders). Using this guidance, the methodology outlined herein is adapted and modified from the Emery and Trist (1978) sociotechnical analytical model. This methodology can be used to assess various health care environments. To help illustrate the methodology and its constituent phases and steps, the nursing home environment will be used as an example. Once all 10 phases have been introduced, we share MEAD applied to a MERS to illustrate the overall model.

Phase I. Environmental and Organizational Design Subsystems: Initial Scanning

In terms of sociotechnical systems theory, the open organizational system attempts to seek a steady state by adapting to forces or changes that pass through its borders. Because the external environment may be the most influential in terms of whether the sociotechnical system will be successful, achieving a valid organization–environment fit is quite important and, in fact, may be the strongest contributor to positive changes in the work system.

An example is found in the nursing care work system. In one observed example, the health care staff works 12 hr shifts formally. Informally, nurses and certified nursing assistants (CNAs) work an extra 1 to 3 hr/per shift to get the expected tasks accomplished. Physicians are connected to the work system only by a fax machine. Critical decisions are made on an hourly basis, usually by the nursing staff. They do their best to get physician approval and direction for actions, but the system does not allow for such external communication. Due to excessive physical and mental workloads risk of errors and accidents are commonplace. Environmental forces include the need to reduce costs, maintain a high quality restraint-free environment, and state regulatory agencies.

Step 1: Perform MVP Analysis. Mission, vision, and principles, or MVP, provide identity to the organization, for example, a nursing home. These often are called identity statements. These are the strategic drivers of the organization. Often, there are variances between what the organization professes as its defining characteristics and its actual identity as inferred from organizational behavior. For example, the nursing home may profess its focus on client care, but due to understaffing, in reality the staff:client ratio is not satisfactory. The formal company identity statements should be identified and evaluated with respect to their components. A mission statement describes the organization's purpose or goals and describes the current activity of the organization in terms of its products and/or services. The vision is forward looking—what the organization seeks to become in 5 to 10

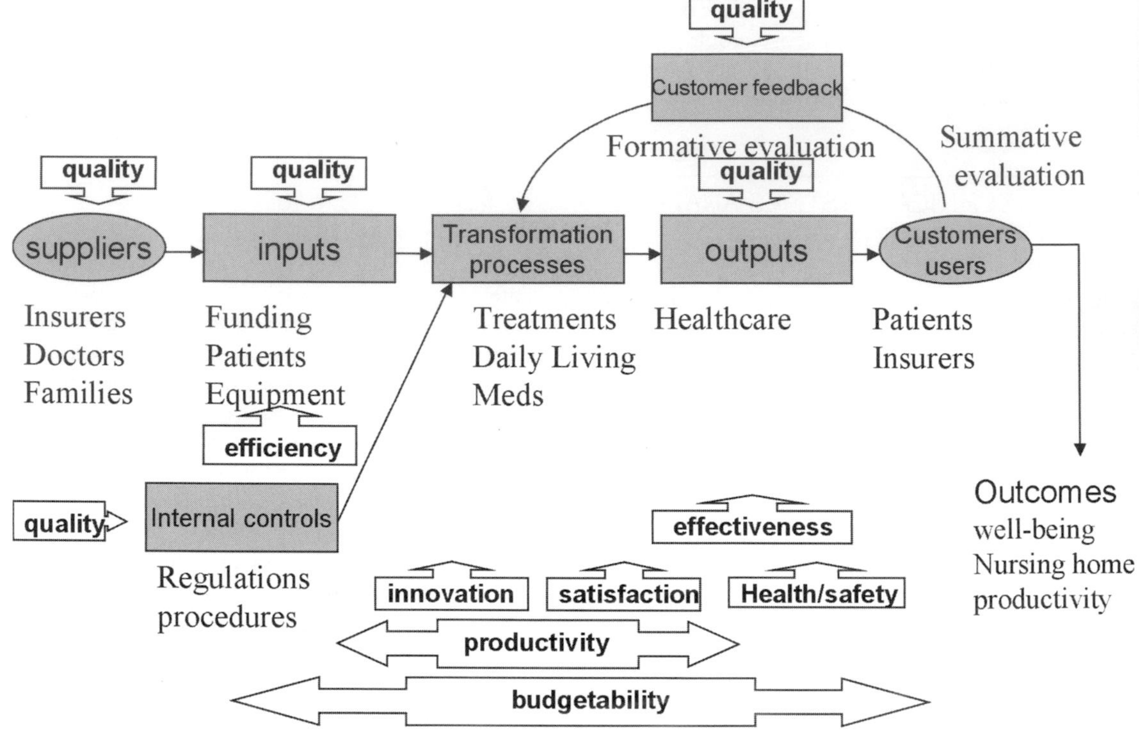

FIGURE 5–2. System model of nursing home work system.

years. The principles or values of the organization are the cultural elements that define the core attributes of the organization. Ultimately, these underlying attributes should drive behavior, which can be observed and measured (Hendrick & Kleiner, 2001).

In observing behavior, a gap may be perceived between inferred values (i.e., observed from behaviors and attitudes) and professed values of the organization, for example, the gap between intended care for clients and actual care for clients. This gap can be quantified by objective data collection or administering a survey, but rich qualitative information also can be gleaned from interviews. Has the vision been articulated in such a way that employees are genuinely excited about the organization's future? Finally, if there is a gap between how people behave and how the company expects people to behave, what are the underlying causes? Is management seen to "walk the talk"? Is the reward system misaligned with the organization's objectives? Are new or redesigned interfaces needed? These are some of the questions that can be used in interviews or converted to items in a Likert-type survey instrument for analysis.

Step 2: Perform System Scan. Work systems work within other work systems. For example, a radiology department operates within a hospital that operates within a network of hospitals. Scanning is the process of defining the workplace in systems terms including defining relevant boundaries around a particular work system. Several tools are available to assist with scanning. As illustrated in Figure 5–2, system outputs are the products or services provided by the organization.

System inputs are the resources transformed to produce products and services. Suppliers provide the inputs and consumers receive, pay for, and use the outputs. Other "customers," or stakeholders, may be beneficiaries (i.e., may have a vested interest) of the system but either do not pay for, or use directly, products and services. Because Deming 1982 was focused on statistical quality control, he emphasized internal control, or those mechanisms or processes that keep transformation processes

TABLE 5–1. Partial Environmental Scan

Subenvironment Stakeholder Groups	Expectations for Nursing Home From Subenvironment	Expectations From Nursing Home for Subenvironment
Families	Help my relative recover quickly	Medicine isn't perfect
Insurer	Run a tight ship	We need to consider patient safety and health before cost control
Physicians	We are professionals	Nursing home is a business

Note. this table displays some of the relevant stakeholders but is not all-inclusive

from spiraling out of control. Also, consumer feedback was seen as instrumental to the design and redesign of transformation processes. The system scan also establishes initial boundaries or domains of responsibility. As described by Emery and Trist (1978), there are throughput, territorial, social, and time boundaries to consider. Throughput boundaries represent the input owned by the system and transformed to output for distribution to consumers. Here, the internal physical environment is important as it affects work. Territorial boundaries have to do with the physical space used for product conversion. Increasingly, this space is distributed and therefore "virtual" or geographically dispersed. Social boundaries are the boundaries created by the formal organization chart's definition of jobs. Time boundaries will relate to an organization's mission—seasonality, number of shifts, and the like. In sociotechnical terms, it is instructive to identify variances passed into the production system from suppliers, consumers, and other customers. A list of design changes that could control these variances closer to the source (i.e., inside the organizational system) should be created.

Step 3: Perform Environmental Scan. The system scan helps to define the work system's boundaries. Everything outside of these boundaries becomes part of the external environment. In the environmental scan, the organization's subenvironments and the principle stakeholders within these subenvironments are identified (see Table 5–1). Their expectations for the organization are identified and evaluated. Conflicts and ambiguities are seen as opportunities for process or interface improvement. A realistic future scenario is created that predicts what would occur if the organization did nothing to change the conflicts and ambiguities. Then, an idealistic future scenario is created that defines a future state wherein stakeholders' expectations are fully realized. Variances are evaluated to determine design constraints and opportunities for change. The work system itself can be redesigned to align itself with external expectations or, conversely, the work system can attempt to change the expectations of the environment to be consistent with its internal plans and desires. In our experience, the gaps between work system and environmental expectations often are gaps of perception, and communication interfaces need to be developed between subenvironment personnel and the organization. Specifically, then, the macroergonomist designs or redesigns interfaces among the organizational system and relevant external subenvironments to improve communication and decision making. These interfaces are referred to as organization–or work system–environment interfaces.

Step 4: Specify Initial Organizational Design Dimensions. Environmental factors are determinants of organizational design, as are factors from the technological and personnel subsystems. Rather than wait to address organizational design until all subsystems have been analyzed, it is useful to develop organizational design hypotheses at this stage based on the environmental scan. By referring to the empirical models of the external environment (e.g., Burns & Stalker, 1961; Lawrence & Lorsch, 1969), we can hypothesize optimal levels of complexity (both differentiation and integration), centralization, and formalization.

Phase II. Technical Subsystem Analysis: Define Production System Type and Performance Expectations

Step 1: Define Production System Type. It is important to identify the work system's "production" type because the type of production system can help determine optimal levels of complexity, centralization, and formalization. The system scan performed in the previous phase should help in this regard, and the analyst should consult the production models discussed previously. Most nursing homes exhibit a high number of exceptions in the work system and the procedures available are somewhat formalized (Lawrence & Lorsch, 1969) .

Step 2: Define Performance Expectations. In this context, the key performance criteria related to the organization's purpose and technical processes are identified. First and foremost, this requires a determination of success factors for products and services but may also include performance measures at other points in the organization's system, especially if decision making is important to work process improvement. As described in Kleiner's (1997) framework adapted from Sink and Tuttle (1989), specific standardized performance criteria guide the selection of specific measures that relate to different parts of the work process. Performance measures are objective, or they can be subjective, as in the case of self-reports. In a nursing home environment, profitability, productivity, health, and safety appear to be key performance indicators.

Quality can be considered in terms of checkpoints or quality associated with the different components of the system. Quality Checkpoints 1, 3, and 5 are quality criteria popularized by Deming (1982) and the Total Quality Management (TQM) movement, still receiving attention in the health care community. Checkpoint 1 emphasizes the quality of suppliers, which has been operationalized within the quality movement in the form of supplier certification. Checkpoint 3, in-process control, relates to the use of statistical quality control charts to monitor and control processes. Checkpoint 5 refers to customer satisfaction, operationalized as the customer getting what is wanted and needed. Quality Checkpoints 2 and 4 correspond to traditional measures of quality control, traditionally assured through inspection of inputs and outputs respectively. Checkpoint 6 corresponds to total quality management or the management system by which the other criteria are managed. Because of the increasing need to manage and measure flexibility in systems, Kleiner (1997) added a flexibility criterion, which relates to each of these checkpoints as well. This is especially pertinent to the design of work systems as sociotechnical systems. The efficiency criterion focuses on input or resource utilization. Effectiveness relates to whether objectives are realized. Productivity is measured as the ratio of outputs to inputs. According to Sink and Tuttle (1989), quality of work life includes safety and health as a criterion. Innovation refers to creative changes to process or product that result in performance gains. Profitability is a standard business management criterion. For not-for-profit organizations, Sink and Tuttle (1989) introduced "budgetability" or expenditures relative to budget to replace the profitability criterion. A popular philosophy in modern organizational performance measurement is to achieve a balanced scorecard, which means managing performance multidimensionally across the system (Sink & Tuttle, 1989). In other organizations, it has been seen as useful to identify key performance indicators linked to a few vital strategic goals. In either case, the metrics should be linked to the organization's strategy.

Step 3: Specify Organizational Design Dimensions. Now that the type of production system has been identified and the empirical production models consulted, the organizational design hypotheses generated in the previous phase should be supported or modified until the personnel subsystem can be thoroughly analyzed as well.

Step 4: Define System Function Allocation Requirements. From the Clegg et al. (1989) function allocation methodology, this also is the appropriate time to specify system level objectives. Requirements specifications can be developed, including microergonomic requirements. Also included are system design preferences for complexity, centralization, and formalization. Clegg et al. (1989) also suggested the use of scenarios, which present alternative allocations and associated costs and benefits. Nursing home environments are also being inundated with computer technologies and in many instances, as previously discussed, these are technology-driven interventions (Al-Tarawneh, Stevens, & Arndt, 2004).

Phase III. Flowchart the Technical Work Process and Identify Unit Operations

Step 1: Identify Unit Operations. Unit operations are groupings of conversion steps that together form a complete piece of work and are bounded from other steps by territorial, technological, or temporal boundaries. Unit operations often can be identified by their own distinctive subproduct and typically employ 3 to 15 workers. They also can be identified by natural breaks in the process (i.e., boundaries) determined by state changes (transformation) or actual changes in the raw material's form, location (input), or storage. In the health care environment, typically, a department constitutes a unit operation. For example, on admission to a hospital, the patient may pass through several departments in which changes occur to the patient. In admissions, there is informational inputs, throughputs, and outputs. In radiology, images of the patient are the unique outputs from the sociotechnical radiological processing. In surgery, there are observable changes in the patient resulting from the process. For each unit operation or department, the purpose/objectives, inputs, transformations, and outputs are defined. If the technology is complex, additional departmentalization (horizontal differentiation) may be necessary. For example, some branches of medicine are quite specialized, such as oncology. Additional horizontal differentiation might include creating special units within oncology, perhaps on the basis of different types of cancers. If collocation is not possible or desirable, spatial differentiation (organizing geographically) and the use of integrating mechanisms may be needed. Integrating mechanisms are usually tools that coordinate and link people together. E-mail, voice recordings, and even medical charts can be considered integrating mechanisms. If the task exceeds the allotted schedule, then additional work groups or shifts may be needed. Ideally, resources for task performance should be contained within the unit, but in health care environments, interdependencies create additional challenges. Effective integrating mechanisms that facilitate communication and information transfer, training, and possibly job rotation are required. Service environments like a nursing home may have nonlinear work processes, but they can be documented nonetheless.

Step 2: Flowchart the Process. The current workflow of the transformation process (i.e., conversion of inputs to outputs) should be flow charted, including material flows, workstations, and physical as well as informal or imagined boundaries. In linear systems, such as most production systems, the output of one step is the input of the next. In nonlinear systems, such as many service or knowledge work environments, steps may occur in parallel or may be recursive. Unit operations identified during the previous stage are highlighted on the flowcharts. Also identified are the functions and subfunctions (i.e., tasks) of the system (Clegg et al., 1989). The purpose of this step is to assess improvement opportunities and coordination problems posed by technical design or the facility. Identifying the workflow before proceeding with a detailed task analysis can provide a meaningful context in which to analyze tasks. Once the current flow is charted, the macroergonomist or analyst can proceed with a task analysis for the work process functions and tasks.

Phase IV. Collect Variance Data

Step 1: Collect Variance Data. A variance is an unexpected or unwanted deviation from standard operating conditions or specifications or norms. It is assumed that controlling such variances is in the strategic interest of the organization. In statistical quality control terms, based on historical performance data, a system's upper and lower bounds (standard deviation) can be determined. This represents the range of expected performance for a system. When a data point occurs outside of this range, its cause is said to be "special" and the system is said to be "out of control." Data points that appear within the range are said to have "common" causes of variances. Special variances need to be tackled first to get the work process in control. Then, common system variation can be tackled for overall system improvement. For the ergonomist, identifying variances at the process level as well as the task level can add important contextual information for job and task redesign. By using the flowchart of the current process and the corresponding detailed task analysis, the macroergonomist or analyst can identify variances. Variances in a complex medical environment such as a nursing home include such diverse variations as staff skill levels, floor surfaces types, mental status of clients, medication labeling, and so forth.

Step 2: Differentiate Between Input and Throughput Variances. Deviations in raw material are called input variances; and deviations related to the process itself during normal operations are called throughput variances. These can both be identified at this stage. Differentiating

TABLE 5–2. Partial Variance Matrix for Nursing Home Work System.

	Partial Variance Matrix for Nursing Home									
Nursing	*Variance*	1	2	3	4	5	6	7	8	9
Home Unit Operation										
Input	1. Doctor skill[a]									
	2. Condition severity	X								
	3. CNA skill		X							
Throughput	4. Equipment reliabilitya	X	X							
	5. Proceduresa	X		X						
	6. Function allocation				X	X				
	7. Diagnosis	X	X	X	X					
	8. Hospitalization	X				X		X		
	9. Recovery		X			X			X	

[a] Key variance due to performance impact

between types of variances helps determine how to control the variances. Of the types just mentioned, all are input variances. Throughput variances would include procedures, treatments, administration of medications, and so forth.

Phase V. Construct Variance Matrix

Step 1: Identify Relationships Among Variances. Key variances are those that significantly impact performance criteria and/or may interact with other variances thereby having a compound effect. The purpose of this step is to display the interrelationships among variances in the transformation work process to determine which ones affect which others. The variances should be listed in the order in which they occur down the vertical Y-axis and across the horizontal X-axis. The unit operations (groupings) can be indicated and each column represents a single variance. The analyst can read down each column to see if this variance causes other variances. Each cell then represents the relationship between two variances. An empty cell implies the two variances are unrelated. The analyst or team also can estimate the severity of variances by using a Likert-type rating scale. Severity would be determined on the basis of whether a variance or combination of variances significantly affect performance. This should help identify key variances. See Table 5–2 for an initial, partial variance table.

Step 2: Identify Key Variances. A variance is considered key then if it significantly affects quantity of production, quality of production, operating costs (utilities, raw material, overtime, etc.), social costs (dissatisfaction, safety, etc.), or if it has numerous relationships with other variances (matrix). Typically, consistent with the Pareto Principle, only 10% to 20% of the variances are significant determinants of the quality, quantity, or cost of product.

Phase VI. Personnel Subsystem Analysis: Construct Key Variance Control Table and Role Network

Step 1: Construct Key Variance Control Table. The purpose of this step is to discover how existing variances are currently controlled and whether personnel responsible for variance control require additional support. The Key Variance Control Table includes the unit operation in which variance is controlled or corrected; who is responsible; what control activities are currently undertaken; what interfaces, tools, or technologies are needed to support control; and what communication, information, special skills, or knowledge are needed to support control.

Step 2: Construct Role Network. A job is defined by the formal job description, which serves as a contract or agreement between the individual and the organization. This is not the same as a work role, which is comprised of actual behaviors of a person occupying a position or job in relation to other people. These role behaviors result from actions and expectations of a number of people in a role set. A role set is comprised of people who are sending expectations and reinforcement to the role occupant. Role analysis addresses who interacts with whom, about what, and the effectiveness

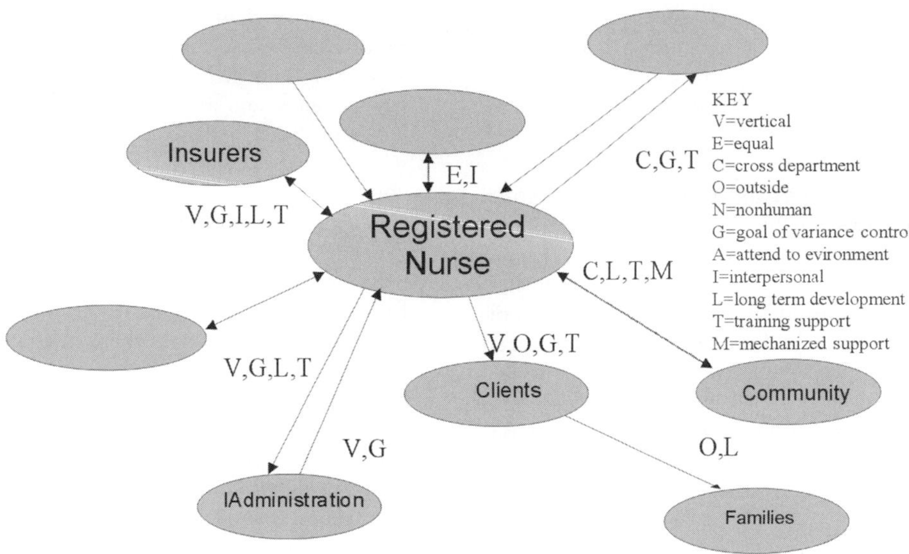

FIGURE 5–3. A role network map of relationships.

of these relationships. This relates to technical production and is important because it determines the level of work system flexibility. As illustrated in Figure 5–3, a role network is a map of relationships indicating who communicates with the focal role.

First, the role responsible for controlling key variances is identified. Although multiple roles may exist that satisfy this criterion, there is often a single role without which the system could not function, at least when the boundary is established within a small work system. For example, in radiology, the radiologist or perhaps the technician may serve this role. In a larger department, it is possible that multiple focal roles operate in which case the network can be illustrated accordingly.

With the focal role identified within a circle, other roles can be identified and placed on the diagram in relation to the focal role. Based on the frequency and importance of a given relationship or interaction, line length can be varied, where a shorter line represents more or closer interactions. Finally, arrows can be added to indicate the nature of the communication in the interaction. A one-way arrow indicates one-way communication and a two-way arrow suggests two-way interaction. Two one-way arrows in opposite directions indicate asynchronous (different time) communication patterns.

To show (a) the content of the interactions between the focal role and other roles and (b) an evaluation of the presence or absence of a set of functional relationships for functional requirements, GAIL and the following key are typically used:

G = Short term goal of controlling variances.

A = To meet goals, must adapt to short term fluctuations.

I = Must also integrate activities to manage internal conflicts and promote smooth interactions among people and tasks.

L = Must ensure long term development of knowledge, skills, and motivation in workers.

Also the presence or absence of particular relationships (VECON) is determined by describing the work process functions in terms of five types of relationships:

V = Vertical hierarchy.

E = Equals or peers.

C = Cross-boundary involving another unit or department.

O = Outside or external stakeholder.

N = Nonsocial component such as a computer.

Step 3: Evaluate Effectiveness. The relationships in the role network now can be evaluated by the analyst. A Likert-type survey instrument can be used to distinguish between effective and ineffective relationships or interactions. Internal and external customers of roles can be interviewed or surveyed for their perceptions of role effectiveness as well.

Step 4: Specify Organizational Design Dimensions. At this juncture the organizational design hypotheses can be tested against the detailed analysis of variance and variance control. The role analysis and variance control table may suggest, for example, a need to increase or decrease formalization or centralization. If procedures are recommended to help control variances, this increase in formalization must be evaluated against the more general organizational design preferences suggested by the environmental and production system analyses.

Phase VII. Function Allocation and Joint Design

Step 1: Perform Function Allocation. Having previously specified system objectives, requirements, and functions it now is time to systematically allocate functions and tasks to human(s) or machine(s), including computer(s). It is helpful to review the environmental scan data to check for any subenvironment constraints (e.g. political, financial, etc.) before making any mandatory allocations (Clegg et al., 1989). Next, provisional allocations can be made to the human(s), machine(s), both, or neither. In the latter case, a return to developing requirements specifications is required. Whether implementing a restraint system or an information system in the nursing home, the question that should be asked is, "How automated should a given system be?" See Kleiner (1998) for a review of macroergonomic directions and issues in function allocation.

Step 2: Design Changes to the Technological Subsystem. Technical changes are made to, at best, prevent, or, at worst, control key variances. Human-centered design of the following may be needed to support operators as they attempt to prevent or control key variances: interfaces, information systems to provide feedback, job aids, process control tools, more flexible technology, redesign work station or handling system, or integrating mechanisms.

Step 3: Design Changes to the Personnel Subsystem. After considering human-centered system design changes in the previous step, it is time to turn attention to supporting the person directly by addressing knowledge and/or skill requirements of key variances and any selection issues that may be apparent. In the variance control table, we identified who controls variances and the tasks performed to control these variances. At this stage, we suggest personnel system changes to prevent or control key variances. This may entail specific skill or knowledge sets that can be acquired through technical training, formal courses, workshops, or distance learning. Key variance control assumes there is strategic support for the intervention.

Also very important is to distinguish between communication and information requirements (e.g., Cano, Meredith, & Kleiner, 1998) and to identify specific communication and/or information requirements to perform variance control. This may involve support for synchronous (same time) or asynchronous (different time) communication to support information flow. The type of information may vary. For example, the information may be specific and targeted to a particular role or roles, or the information might be shared in a mass communication medium. Finally, it is important to consider the purpose of the information. Data are collected to be transformed by decision tools to information. Information can be shared for information's sake, or it can be the raw material for decision making, which ultimately should result in positive action in the organization.

Step 4: Prescribe Final Organizational Design. At this point in the process, organizational design hypotheses have been generated and adjusted iteratively as new analyses are performed. It now is time to take the specifications for organizational design levels of complexity, centralization, and formalization and produce specific structures. Depending on the level of work system process analysis, this may require design/redesign at the organizational level or at the group/team level or at both levels. The former level is exemplified by choices between functional structures, product structures, geographic structures, or matrix structures. The latter may specify a functional or cross-functional team, and a self-managed or high performance work team. Functional structures (organizing around technical functions such as customer service, accounting, risk management, medical) and functional teams can be somewhat tolerable in stable environments with workforces of less than 250 employees. Compared to other industries, large health care systems and medical practices are generally stable compared to other sectors such as manufacturing or retail business. Functional structures have the advantages of minimizing redundancies, professional identification, and professional advancement. They often have the disadvantages of communication, coordination, and internal competition problems. These problems often result from functional suboptimization in which a given functional department perceives its technical area as the purpose of the organization rather than the over-all organizational objectives such as producing the goods or services of the organization. Product structures and geographic structures organize around products/divisions and regions, respectively, and represent an attempt to solve the suboptimization problem seen with functional structures. Most health systems are organized in product configurations. Departments such as oncology, radiation, emergency, cardiology, and gastroenterology exemplify a product organization. However, functional redundancies can often result in such organizations. The matrix structure is used where considerable flexibility is required in highly dynamic environments. They attempt to exploit the advantages of both the product and functional structures but often result in conflicts due to the existence of dual lines of authority. Extra integrating mechanisms often are required with matrix designs. The cross-functional team typically is used as an integrating mechanism with a functional structure. As the name implies, personnel from different functions form an informal team to focus on issues (e.g. process improvement) that affect the members' functions. The decisions of cross-functional teams result in recommendations to a sponsor with formal organizational authority. Self-managed teams are not informal but formal in that they are actually part of the organization's overall structure. Rather than make recommendations, self-managed teams have the authority to make decisions about work scheduling, work design, selection/deselection, and so forth.

Phase VIII. Roles and Responsibilities Perceptions

Step 1: Evaluate Role and Responsibility Perceptions. It is important to identify how workers *perceive* the roles documented in the variance control table, especially if the table initially was constructed by those who do not occupy the roles identified. Through interviews, role occupants can participate in an analysis of their perceptions of their roles. Using the previously constructed table, expected roles, perceived roles, and any gaps can be identified. Gaps can be managed through training and selection as well as technological support. In the nursing home example, nurses are perceived by outsiders as the focal roles. However, nurses are frustrated by this distinction. The legal and professional responsibilities associated with being the focal role are inconsistent with the pay scale and formal role.

Step 2: Provide Training Support. Once a system is changed ergonomically, training support completes the intervention. In this step, there is particular concern about using training to close any gap between expected roles and actual roles. Clearly, in nursing homes, the formal role network is at odds with the informal role network. In terms of maximizing care and quality, it is very important that health care workers individually and collectively, perform the roles prescribed for them.

Phase IX. Design/Redesign Support Subsystems and Interfaces

Step 1: Design/redesign Support Subsystems. Now that the work process has been analyzed and jointly designed, other internal organizational support subsystems may require redesign. Of particular interest are the reward systems. Other subsystems, such as maintenance, also may require adjustment. In all cases, the goal is to determine: (a) the extent to which a given subsystem impacts the sociotechnical production system; (b) the nature of the variance; (c) the extent to which the variance is controlled; and (d) the extent to which tasks should be taken into account in redesign of operating roles in the supporting subsystem units. In the case of hourly pay systems for nursing home environments, in many cases, the hours expended beyond the formal shifts are not reimbursable.

Step 2: Design/Redesign Interfaces and Functions. According to the Clegg et al. (1989) method of function allocation, individual and cumulative allocations made on a provisional basis in Step 7 can be further evaluated against (a) requirements specifications (including the scenarios developed earlier), (b) resources available at the time of implementation (including human and financial), and (c) the sum total outcome. In addition

TABLE 5–3. Macroeconomic Analysis and Design (MEAD) Example Applied to Medical Event Reporting System (MERS)

Phase	*Description*
1. Initial scanning	Analysis of Clinical Effectiveness vision related to patient safety and organizational behavior; Describe mission of MERS in system terms; Identify needs, conflicts between relevant stakeholders (patients, user groups/healthcare providers); Develop broad design mock-ups of MERS
2. Define production system type and performance expectations	Identify the outcomes necessary for relevant work units (work systems); Identify key performance criteria related to patient safety; Modify previous MERS design; Identify requirements specifications for new MERS and computer-based design requirements
3. Flowchart the technical work process and identify unit operations.	Identify complete conversion steps to achieve unit operations Identify opportunities to use precursor data and points in which precursor data are lost
4. Collect and differentiate variance data	Identify deviations and incompatibilities in the MERS; Categorize deviations by type; Propose new MERS mock-up (paper-based)
5. Construct variance matrix	Select most critical deviations or incompatibilities and their relationships within the work system
6. Construct variance control table and role network	Identify most effective controls or system features to eliminate deviations or incompatibilities; Identify work roles necessary to support controls; Identify organizational design changes necessary to support controls; work roles, and system feedback to redesign work systems
7. Function allocation and joint design	Identify medical event tasks requiring computer-support (semiautomation), total automation, or user only; Identify recommendations for personnel changes and final work system design to support new allocations
8. Evaluate and support roles and responsibilities	Evaluate users' perceptions of roles and identify training needs and interface design features to support a shared mental model; Present new MERS mock-up (paper-based and computer-based)
9. Design/redesign	Redesign work system elements such as training systems and interfaces to support proactivity; Finalize design of paper-based system prototype; Finalize design of computer based system prototype
10. Implement, iterate, improve	Implement trial runs of new MERS (paper-based and computer-based); Analyze data and provide recommendations for further improvement

Note. Adapted from Smith-Jackson and Kleiner, 2003.

to a check of function allocation, interfaces among subsystems should be checked and redesigned at this juncture. For example, is the motivational system for nurses or other health care workers aligned with the technology implemented in the system? Are there appropriate interfaces between the personnel system and the external environment? For example, does the hospital follow up with patients to assess their level of satisfaction and service related to a recent visit? Do doctors "on the ground" have the proper informational flow that informs them of the expectations of the parent HMO?

Step 3: Design/Redesign the Internal Physical Environment. Especially at the team and individual levels of work, the internal physical environment should be ergonomically adjusted if necessary to promote effectiveness. Looking at the technical and personnel variance analyses, we can assess whether there are physical environmental changes that can be made to promote improvement. These might include changes to temperature, lighting, humidity, or noise (control/hearing protection, etc). Nursing home environments are ripe for ergonomic redesign. For example, slips and falls are commonplace. Ergonomics can help minimize their occurrence.

Phase X. Implement and Improve

Step 1: Implement. At this point, it is time to implement the work process changes prescribed. In most cases, the macroergonomics team will not carry the direct authority to implement the changes suggested by the analysis. Therefore, proposals with recommendations for change may have to be prepared for presentation within the formal organizational structure. Such proposals should be consistent with the macroergonomics guiding principles and include both technical and social objectives. They likely will include participatory ergonomics as a fundamental methodology. Finally, any proposal should state expected multidimensional performance improvement. Based on the feedback received from proposal presentations, modifications to the proposal may be necessitated that will require a return to the earlier step that represents a challenged assumption or design.

Step 2: Perform Evaluations. Once the proposal for change is accepted and implementation begins, regular reviews of progress are required. To compliment the weekly formative evaluations performed by the implementation team, semiannual summative evaluations should be performed by an objective outside party. This evaluation should be presented to the implementation team and a constructive dialogue about expectations and progress-to-date should be conducted.

Step 3: Iterate. In practice, the process described is iterative. For continuous improvement, evaluations may suggest a return to an earlier step in the process for renewed partial or full redesign.

SUMMARY AND CONCLUSIONS

As reported by the health care institutions themselves, medical errors need to be controlled. Many have stated that health care work systems are in need of redesign. This chapter has assumed that the traditional gains of 10% to 20% improvements are not acceptable for all medical systems. Thus, macroergonomics and in particular MEAD is potentially a useful approach to make major medical system improvements. As illustrated in Table 5–3, MEAD can be applied to specific potential improvements or interventions such as evaluating and designing a medical event reporting system or MERS.

Whether the application domain is a nursing home, medical event reporting system, or a medical record keeping system, the three pitfalls of design have been observed in many health care environments. Technology-centered design and a left-over approach to function and task design need to be addressed by adopting a sociotechnical systems approach to design. MEAD is a framework that attempts to address these pitfalls for improvement to health, safety, and performance by adopting a sociotechnical approach. As a result, major improvements to performance, including reduction of medical errors and improved quality of care, are possible.

References

Al-Tarawneh, I. S., Stevens, W. J., & Arndt, S. R. (2004). An analysis of home and hospital medical device incidents in the Maude database. *Proceedings of the Human Factors and Ergonomics Society 48th Annual Meeting* (pp. 1718–1722). Santa Monica, CA: The Human Factors and Ergonomics Society.

Bailey, R. W. (1989). *Human performance engineering: Using human factors/ergonomics to achieve computer system usability* (2nd ed.). Englewood Cliffs, NJ: Prentice Hall.

Burns, T., & Stalker, G. M. (1961). *The management of innovation.* London: Tavistock.

Cano, A., Meredith, J., & Kleiner, B. M. (1998). Distributed and collocated group communication vs. decision systems support. In O. Brown Jr. & H. W. Hendrick (Eds.) *Human factors in organizational design and management II* (pp. 501–506). Amsterdam: Elsevier.

Clegg, C., Ravden, S., Corbertt, M., & Johnson, S. (1989). Allocating functions in computer integrated manufacturing: A review and new method. *Behavior and Information Technology, 8*, 175–190.

DeGreene, K. (1973). *Socio-technical systems.* Englewood Cliffs, NJ: Prentice-Hall.

Deming, W. E. (1982) *Out of the crisis.* Cambridge, MA: MIT Center for Advanced Engineering Study.

Emery, F. E., & Trist, E. L. (1965). The causal texture of organizational environments. *Human Relations,* Vol. 18, pp. 21–32.

Emery, F. E., & Trist, E. L. (1978). Analytical model for socio-technical systems. In W. A. Pasmore & J. J. Sherwood (Eds.), *Socio-technical systems: A sourcebook* (pp. 120–133). LaJolla, CA: University Associates.

Fitts, P. M. (1951). Engineering psychology in equipment design. In S.S. Stevens (Ed.), *Handbook of experimental psychology* (pp. 365–379). New York: Wiley.

Hendrick, H. W., & Kleiner, B. M. (2001). *Macroergonomics: An introduction to work system analysis and design.* Santa Monica, CA: The Human Factors and Ergonomics Society.

Kleiner, B. M. (1997). An integrative framework for measuring and evaluating information management performance. *International Journal of Computers and Industrial Engineering, 32*(3), 545–555.

Kleiner, B. M. (1998). Macroergonomic directions in function allocation. In O. Brown Jr. & H. W. Hendrick (Eds.), *Human factors in organizational design and management VI* (pp. 635–640). Amsterdam: Elsevier.

Kleiner, B. M., & Drury, C. G. (1998). The use of verbal protocols to understand and design skill-based tasks. *Human Factors and Ergonomics in Manufacturing, 8*(1), 23–39.

Kleiner, B. M., & Drury, C. G. (1999). Large-scale regional economic development: Macroergonomics in theory and practice. *Human Factors and Ergonomics in Manufacturing, 9*(2), 151–163.

Kleiner, B. M., Drury, C. G., & Palepu, P. (1998). A computer-based productivity and quality management system for cellular manufacturing. *Computers and Industrial Engineering, 34*(1), 207–217.

Lawrence, P. R., & Lorsch, J. W. (1969). *Organization and environment.* Homewood, IL: Irwin.

Perrow, C. (1967). A framework for the comparative analysis of organizations. *American Sociological Review, 32,* 194–208.

Sink, D. S., & Tuttle, T. C. (1989). *Planning and measurement in your organization of the future,* Norcross, GA: Industrial Engineering and Management Press.

Smith-Jackson, T. L., & Kleiner, B. M. (2003). *Application of human factors methods to design a user-centered, proactive medical event reporting system.* Unpublished manuscript, Virginia Polytechnic Institute & State University, Blacksburg.

Trist, E. L., & Bamforth, K. W. (1951). Some social and psychological consequences of the longwall method of coal-getting. *Human Relations, 4,* 3–38.

Woodward, J. (1965). *Industrial organization: Theory and practice.* London: Oxford University Press.

6

CLINICAL MICROSYSTEMS IN HEALTH CARE: THE ROLE OF HUMAN FACTORS IN SHAPING THE MICROSYSTEM

Julie K. Johnson
University of Chicago

Paul Barach
Jackson Memorial Hospital and University of Miami

The clinical microsystem provides a conceptual and practical framework for thinking about the organization and delivery of care. The purpose of this chapter is to provide a detailed description of the role that human factors play in shaping microsystems and how they affect the microsystem's ability to provide high quality and safe care.

INTRODUCTION TO CLINICAL MICROSYSTEM CONCEPTS

A clinical microsystem is a group of clinicians and staff working together with a shared clinical purpose to provide care for a population of patients (Batalden et al., 1997; Mohr, 2000; Nelson, Batalden, Mohr & Plume, 1998). The clinical purpose and its setting defines the essential components of the microsystem. These include the clinicians and support staff, information and technology, the specific care processes, and the behaviors that are required to provide care to its patients. Microsystems evolve over time, responding to the needs of their patients, providers, and external pressures. They often coexist with other microsystems within a larger (macro) organization.

The conceptual theory of the clinical microsystem is based on ideas developed by Deming (1986), Senge (1990), Wheatley (1992), and others who have applied systems thinking to organizational development, leadership, and improvement. The seminal idea for the microsystem as an organizational unit stems from the work of James Brian Quinn (1992). Quinn's work is based on analyzing the world's best-of-best service organizations, such as FedEx, Mary Kay Cosmetics, McDonald's, and Nordstrom. He focused on determining what these extraordinary organizations were doing to achieve high-quality explosive growth, high margins, and wonderful consumer reputations. He found that these leading service organizations organized around, and continually engineered, the front-line relationships that connected the needs of customers with the organization's core competency. Quinn called this front-line activity that embedded the service delivery process the smallest replicable unit or the minimum replicable unit. *This smallest replicable unit, or the microsystem,* is the key to implementing effective strategy, information technology, and other key aspects of intelligent enterprise.

In the late 1990s, under the aegis of the Institute of Medicine (IOM) and with funding by the Robert

Wood Johnson Foundation, Mohr and Donaldson investigated high-performing clinical microsystems (Donaldson & Mohr, 2000; Mohr, 2000). This research was based on a national search for the highest-quality clinical microsystems. Forty-three clinical units were identified using theoretical sampling. Leaders of the units were interviewed using a semistructured interview protocol. The results of the interviews were analyzed to determine the characteristics that seemed to be most responsible for enabling these microsystems to be effective. The results suggested that eight dimensions were associated with high quality of care—integration of information, measurement, interdependence of the care team, supportiveness of the larger system, constancy of purpose, connection to community, investment in improvement, and alignment of role and training. These eight factors became a framework for evaluating health care microsystems. Each of the dimensions can be thought of on a continuum that represents the presence of the characteristic in the microsystem. Increased appreciation of the small front-line work unit known as a microsystem means recognizing the characteristics that contribute to their identity and being mindful of the reliability of these characteristics.

The Dartmouth study (funded by the Robert Wood Johnson Foundation to continue and build on the IOM study) was based on 20 case studies of high-performing clinical microsystems and included on-site interviews with each member of the microsystem and analysis of individual microsystem performance data (Nelson, Batalden, Huber, Mohr, Godfrey, Headrick & Wasson, 2002; Nelson; Batalden, Homa, Godfrey, Campbell, Headrick, Huber, Mohr & Wasson, 2003; Godfrey, Nelson, Wasson, Mohr & Batalden, 2003; Wasson, Godfrey, Nelson, Mohr & Batalden, 2003; Batalden, Nelson, Mohr, Godfrey, Huber, Kosnik, & Ashling, 2003; Mohr, Barach, Cravero, Blike, Godfrey, Batalden & Nelson; Kosnik & Espinosa, 2003; Huber, Godfrey, Nelson, Mohr, Campbell & Batalden, 2003; Batalden, Nelson, Edwards, Godfrey & Mohr, 2003). As a result of this work, the dimensions of high-performing microsystems were further refined and expanded to include two additional categories. Table 6–1 lists the dimensions of high-performing microsystems and provides an operational definition of each.

The Microsystem Assessment Tool (see Figure 6–1) first published in 2002 is based on these dimensions (Mohr & Batalden, 2002). As we continue to move beyond conceptual theory and research to application in clinical settings, the emerging fields of chaos theory, complexity science, and complex adaptive systems have influenced how these concepts have been applied to improving microsystems (Arrow, McGrath, & Berdahl, 2000; Peters 1987; Plsek & Greenhalgh, 2001; Plsek & Wilson, 2001). This is evident in the work to bring together microsystems from around the world to learn and share best practices (for updates on these efforts see http://clinicalmicrosystem.org.

IMPROVING SAFETY WITHIN THE CLINICAL MICROSYSTEM

The clinical microsystem—as a unit of research, analysis, and practice—is an important level at which to focus patient safety interventions. It is at this system level that patients and caregivers meet, and it is at this level that real changes in patient care occur. Human errors occur within the microsystem, and ultimately it is the functioning microsystem that can capture, attenuate, or mitigate these errors from causing patient harm. The case included in Box 6–1 is illustrative of a patient safety event. Throughout the story, as told from the perspective of a senior resident in pediatrics, there are many obvious system "failures." One method that we have found to be useful for systematically looking at patient safety events builds on William Haddon's framework for injury epidemiology (Haddon, 1972). As the first Director of the National Highway Safety Bureau (1966–1969), Haddon was interested in the broad issues of injury that result from the transfer of energy in such ways that inanimate or animate objects are damaged. The clinical microsystem offers a setting in which this injury can be studied. According to Haddon (1972), there are several strategies for reducing injuries. First, prevent the marshalling of the energy; second, reduce the amount of energy marshaled; third, prevent the release of the energy; fourth, modify the rate or spatial distribution of release of the energy, fifth; separate in time and space the energy being released and the susceptible structure; sixth, use a physical barrier to separate the energy and the susceptible structure; seventh, modify the contact surface or structure with which people can come in contact; eighth, strengthen the structure that might be damaged by the energy transfer; ninth, when injury does occur, rapidly detect it and counter its continuation and extension; tenth, when injury does occur, take all necessary reparative and rehabilitative steps. All these

Box 6–1. One Resident's Story

I had a patient who was very ill. We thought that an abdominal CT would be helpful and it needed to be infused. He was 12 years old and was completely healthy up until three months ago. Since then he has been in our hospital and two other hospitals, pretty much the entire time. He has been in respiratory failure, he's had mechanical ventilation – including oscillation, he's been in renal failure, he's had a number of mini strokes, and when I came on service he was having diarrhea – three to five liters per day – and we still didn't know what was going on with him. He was a very anxious child. Understandably, it's hard for the nurses, and for me, and for his mother to deal with. He thought of it as pain, but it was anxiety and it responded well to anxiolytics.

When I came in that morning, it hadn't been passed along to nursing that he was supposed to go to CT. I heard the charge nurse getting report from the night nurse. I said, "You know that he is supposed to go for a CT today." She was already upset because they were very short staffed. She heard me and then said that she was the charge nurse and also taking care of two patients, and one had to go to CT. She went off to the main unit to talk to someone. Then she paged me and said, "If you want this child to have a scan you have to go with him." I said, "OK." Nurses are the ones who usually go. But it didn't seem to be beyond my abilities . . . at the time.

So, I took the child for his CT and his Mom came with us. We gave him extra Ativan on the way there, because whenever he had a procedure he was extra anxious. When we got there, they weren't ready. We had lost our spot from the morning. My patient got more and more anxious and he was actually yelling at the techs "Hurry up!" We went in the room. He was about five or six hours late for his study and we had given him contrast enterally. The techs were concerned that he didn't have enough contrast anymore and they wanted to give him more through his G-tube. I said, "That sounds fine." And they mixed it up and gave it to me to give through his G-tube and I went to his side and unhooked --- not registering that it was his central line --- I unhooked his central line, not only taking off the cap but unhooking something and I pushed 70 cc's of gastrografin. As soon as I had finished the second syringe I realized I was using the wrong tube. I said, "Oh no!" Mom was right there and said, "What?" I said, "I put the stuff in the wrong tube. He looks OK. I'll be right back, I have to call somebody." I clamped him off and I called my attending and I called the radiologist. My attending said that he was on his way down. The radiologist was over by the time I had hung up the phone. My patient was stable the whole time. We figured out what was in the gastrografin that could potentially cause harm. We decided to cancel the study.... I sent the gastrografin – the extra stuff in the tubes – for a culture just in case he grew some kind of infection and then we would be able to treat it and match it with what I had pushed into the line. I filled out an incident report. I called my chiefs and told them.... They said, "It's OK. He's fine, right?" I said, "Yes." They came up later in the evening just to be supportive. They said, "It's OK. It's OK to make a mistake."

The attending that I had called when I made the mistake said, "I'm sorry that you were in that situation. You shouldn't have been put in that situation." Another attending the next day was telling people, "Well , you know what happened yesterday." As if it were the only thing going on for this patient. I thought it was embarrassing that he was just passing on this little tidbit of information as if it would explain everything that was going on. As opposed to saying, "Yes, an error was made, it is something that we are taking into account." And he told me to pay more attention to the patient. Yes, I made the mistake, but hands-down I still know and always did know that patient better than he did. I just thought that was mean and not fair. And the only other thing I thought (that) was not good was the next morning when I was pre-rounding some of the nurses were whispering and I just assumed that was what they were whispering about. I walked up to them and said, "I'm the one who did it. I made a mistake. How is he doing?" I tried to answer any questions they had and move on. The nurse that had sent me down with him told me "It's OK, don't worry about it." The others just listened politely and didn't say anything.

The next day, I felt really bad. I felt very incompetent. I was feeling very awkward about being the leader of this child's care – because I am still at a loss for his diagnosis. And after the event, when the Grandma found out – she was very angry. I apologized to the Mom and I thought it would be overdoing it to keep saying, "I am so sorry." So, the next day I went into the room and said to the Mom, "You need to have confidence in the person taking care of your son. If my mistake undermines that at all, you don't have to have me as your son's doctor and I can arrange it so that you can have whoever you want." She said, "No. No, it's fine. We want you as his doctor." Then we just moved on with the care plan. That felt good. And that felt appropriate. I couldn't just walk into the room and act like nothing had happened. I needed her to give me the power to be their doctor. So, I just went and asked for it.

TABLE 6–1. Microsystem Dimensions and Operational Definitions

Dimension	*Operational Definition*
1. Leadership	The role of leadership is to balance setting and reaching collective goals, empower individual autonomy as well as accountability through building knowledge, respectful action, and reviewing and reflecting
2. Organizational support	The larger organization looks for ways to connect to and facilitate the work of the microsystem and coordinate the hand-offs between microsystems
3. Staff focus	There is selective hiring of the right kind of people; the orientation process is designed to fully integrate new staff into culture and work roles; expectations of staff are high regarding performance, continuing education, professional growth, and networking
4. Education and training	All clinical microsystems, regardless of whether they are part of an academic medical center, have responsibility for the ongoing education and training of staff and for aligning daily work roles with training competencies; academic clinical microsystems have the additional responsibility of training students
5. Interdependence	The interaction of staff is characterized by trust, collaboration, willingness to help each other, appreciation of complementary roles, respect and recognition that all contribute individually to a shared purpose
6. Patient focus	The primary concern is to meet all patient needs—caring, listening, educating, and responding to special requests, innovating against needs, and ensuring smooth service flow
7. Community and market focus	The microsystem is a resource for the community and the community is a resource to the microsystem; there is a focus on establishing the relationship with the community
8. Performance results	Performance focuses on patient outcomes, avoidable costs, streamlining delivery, using data feedback, promoting positive competition, and frank discussions about performance
9. Process improvement	An atmosphere for learning and redesign is supported by the continuous monitoring of care, use of benchmarking, frequent tests of change, and a staff that has been empowered to innovate
10. Information and information technology	Information is THE connector—staff to patients, staff to staff, needs with actions to meet needs; technology can facilitate effective communication and multiple formal and informal channels are used to keep everyone informed all the time, listen to everyone's ideas, and ensure that everyone is connected on important topics

strategies have a logical sequence that is related to the three phases of human injury, that is: preinjury, injury, and postinjury.

The Haddon matrix is a 3×3 matrix that lists factors related to a car crash (human, vehicle, and environment) and phases of the event (preinjury, injury, and postinjury). Figure 6–2 is an example of the Haddon matrix that has been completed to analyze a car crash (Haddon, 1972). The use of the matrix focuses the analysis on the interrelationships between the three factors (human, vehicle, and environment) and the three phases of the crash (preevent, event, and postevent). A mix of countermeasures derived from Haddon's strategies previously outlined are necessary to minimize loss. Furthermore, the countermeasures can be designed for each phase—preevent, event, and postevent. This approach confirms what we know about adverse events in complex environments—it takes a series of counter strategies to prevent and/or mitigate harm and death. Understanding injury in its larger context helps us recognize the complexity of systems and the inherent "unsafe" nature of systems, while highlighting the important work of humans to mitigate the inherent hazards (Dekker, 2002). We can also use the Haddon matrix to think

Clinical Microsystem Assessment Tool

Instructions: Each of the "success" characteristics (e.g., leadership) is followed by a series of three descriptions. For each characteristic, ***please check*** the description that ***best describes*** your current microsystem and the care it delivers *OR* use a microsystem you are *MOST* familiar with.

	Characteristic and Definition	Descriptions			
Leadership	**1. Leadership:** The role of leaders is to balance setting and reaching collective goals, and to empower individual autonomy and accountability, through building knowledge, respectful action, reviewing and reflecting.	☐ Leaders often tell me how to do my job and leave little room for innovation and autonomy. Overall, they don't foster a positive culture.	☐ Leaders struggle to find the right balance between reaching performance goals and supporting and empowering the staff.	☐ Leaders maintain constancy of purpose, establish clear goals and expectations, and foster a respectful positive culture. Leaders take time to build knowledge, review and reflect, and take action about microsystems and the larger organization.	☐ Can't Rate
Leadership	**2. Organizational Support:** The larger organization looks for ways to support the work of the microsystem and coordinate the hand-offs between microsystems.	☐ The larger organization isn't supportive in a way that provides recognition, information, and resources to enhance my work.	☐ The larger organization is inconsistent and unpredictable in providing the recognition, information and resources needed to enhance my work.	☐ The larger organization provides recognition, information, and resources that enhance my work and makes it easier for me to meet the needs of patients.	☐ Can't Rate
Staff	**3. Staff Focus:** There is selective hiring of the right kind of people. The orientation process is designed to fully integrate new staff into culture and work roles. Expectations of staff are high regarding performance, continuing education, professional growth, and networking.	☐ I am not made to feel like a valued member of the microsystem. My orientation was incomplete. My continuing education and professional growth needs are not being met.	☐ I feel like I am a valued member of the microsystem, but I don't think the microsystem is doing all that it could to support education and training of staff, workload, and professional growth.	☐ I am a valued member of the microsystem and what I say matters. This is evident through staffing, education and training, workload, and professional growth.	☐ Can't Rate
Staff	**4. Education and Training:** All clinical microsystems have responsibility for the ongoing education and training of staff and for aligning daily work roles with training competencies. Academic clinical microsystems have the additional responsibility of training students.	☐ Training is accomplished in disciplinary silos, e.g., nurses train nurses, physicians train residents, etc. The educational efforts are not aligned with the flow of patient care, so that education becomes an "add-on" to what we do.	☐ We recognize that our training could be different to reflect the needs of our microsystem, but we haven't made many changes yet. Some continuing education is available to everyone.	☐ There is a team approach to training, whether we are are training staff, nurses or students. Education and patient care are integrated into the flow of work in a way that benefits both from the available resources. Continuing education for all staff is recognized as vital to our continued success.	☐ Can't Rate
Staff	**5. Interdependence:** The interaction of staff is characterized by trust, collaboration, willingness to help each other, appreciation of complementary roles, respect and recognition that all contribute individually to a shared purpose.	☐ I work independently and I am responsible for my own part of the work. There is a lack of collaboration and a lack of appreciation for the importance of complementary roles.	☐ The care approach is interdisciplinary, but we are not always able to work together as an effective team.	☐ Care is provided by a interdisciplinary team characterized by trust, collaboration, appreciation of complementary roles, and a recognition that all contribute individually to a shared purpose.	☐ Can't Rate
Patients	**6. Patient Focus:** The primary concern is to meet all patient needs — caring, listening, educating, and responding to special requests, innovating to meet patient needs, and smooth service flow.	☐ Most of us, including our patients, would agree that we do not always provide patient centered care. We are not always clear about what patients want and need.	☐ We are actively working to provide patient centered care and we are making progress toward more effectively and consistently learning about and meeting patient needs.	☐ We are effective in learning about and meeting patient needs — caring, listening, educating, and responding to special requests, and smooth service flow.	☐ Can't Rate

Side A

Please continue on Side B

Clinical Microsystem Assessment Tool

- CONTINUED -

	Characteristic and Definition		Descriptions			
Patients	**7. Community and Market Focus:** The microsystem is a resource for the community; the community is a resource to the microsystem; the microsystem establishes excellent and innovative relationships with the community.		☐ We focus on the patients who come to our unit. We haven't implemented any outreach programs in our community. Patients and their families often make their own connections to the community resources they need.	☐ We have tried a few outreach programs and have had some success, but it is not the norm for us to go out into the community or actively connect patients to the community resources that are available to them.	☐ We are doing everything we can to understand our community. We actively employ resources to help us work with the community. We add to the community and we draw on resources from the community to meet patient needs.	☐ Can't Rate
Performance	**8. Performance Results:** Performance focuses on patient outcomes, avoidable costs, streamlining delivery, using data feedback, promoting positive competition, and frank discussions about performance.		☐ We don't routinely collect data on the process or outcomes of the care we provide.	☐ We often collect data on the outcomes of the care we provide and on some processes of care.	☐ Outcomes (clinical, satisfaction, financial, technical, safety) are routinely measured, we feed data back to staff, and we make changes based on data.	☐ Can't Rate
Performance	**9. Process Improvement:** An atmosphere for learning and redesign is supported by the continuous monitoring of care, use of benchmarking, frequent tests of change, and a staff that has been empowered to innovate.		☐ The resources required (in the form of training, financial support, and time) are rarely available to support improvement work. Any improvement activities we do are in addition to our daily work.	☐ Some resources are available to support improvement work, but we don't use them as often as we could. Change ideas are implemented without much discipline.	☐ There are ample resources to support continual improvement work. Studying, measuring and improving care in a scientific way are essential parts of our daily work.	☐ Can't Rate
Information and Information Technology	**10. Information and Information Technology:** Information is THE connector - staff to patients, staff to staff, needs with actions to meet needs. Technology facilitates effective communication and multiple formal and informal channels are used to keep everyone informed all the time, listen to everyone's ideas, and ensure that everyone is connected on important topics. *Given the complexity of information and the use of technology in the microsystem, assess your microsystem on the following three characteristics: (1) integration of information with patients, (2) integration of information with providers and staff, and (3) integration of information with technology.*	A. Integration of Information with Patients	☐ Patients have access to some standard information that is available to all patients.	☐ Patients have access to standard information that is available to all patients. We've started to think about how to improve the information they are given to better meet their needs.	☐ Patients have a variety of ways to get the information they need and it can be customized to meet their individual learning styles. We routinely ask patients for feedback about how to improve the information we give them.	☐ Can't Rate
		B. Integration of Information with Providers and Staff	☐ I am always tracking down the information I need to do my work.	☐ Most of the time I have the information I need, but sometimes essential information is missing and I have to track it down.	☐ The information I need to do my work is available when I need it.	☐ Can't Rate
		C. Integration of Information with Technology	☐ The technology I need to facilitate and enhance my work is either not available to me or it is available but not effective. The technology we currently have does not make my job easier.	☐ I have access to technology that will enhance my work, but it is not easy to use and seems to be cumbersome and time consuming.	☐ Technology facilitates a smooth linkage between information and patient care by providing timely, effective access to a rich information environment. The information environment has been designed to support the work of the clinical unit.	☐ Can't Rate

Side B

Figure 6–1. Microsystem Assessment Tool.

Factors

Phases	Human	Vehicle	Environment
Pre-injury	Alcohol intoxication	Braking capacity of motor vehicles	Visibility of hazards
Injury	Resistance to energy insults	Sharp or pointed edges and surfaces	Flammable building materials
Post-injury	Hemorrhage	Rapidity of energy reduction	Emergency medical response

Figure 6–2. Haddon matrix applied to car crash analysis.

about analyzing patient safety adverse events and their prevention. We have adapted the Haddon matrix to include phases labeled "preevent, event, and postevent" instead of "preinjury, injury, and postinjury." Figure 6–3 shows that we have revised the factors to include "patient/family, health care professional, system and environment" instead of "human, vehicle, and environment." Note that we have added "system" to refer to the processes and systems that are in place for the microsystem. "Environment" refers to the context in which the microsystem exists. The addition of the system recognizes the significant contribution that systems and human factors make toward predisposing to patient harm in the microsystem. Figure 6–3 shows a completed matrix using the pediatric case presented in Box 6–1. The next step in learning from errors and adverse events is to develop and implement countermeasures to address the issues in each cell of the matrix.

Safety is a property of the clinical microsystem that can be achieved only through a systematic application of a broad array of process, equipment, organization, supervision, training, and teamwork changes. Table 6–2 builds on the research of high-performing microsystems by considering specific actions focused on improving safety within the organization linked to each of the dimensions of a high-performing microsystem. The list provides an organizing framework for applying patient safety concepts to microsystems.

THE ROLE OF HUMAN FACTORS IN CLINICAL MICROSYSTEMS

Human factors emerged as a recognized discipline during World War II. Its use improved military system performance by addressing problems in signal detection, workspace constraints, optimal task training, cockpit design, and teamwork (Association for the Advancement of Medical Instrumentation, 2001). Nearly half a century of research and hands-on experience have produced a substantial body of scientific knowledge about how people interact with each other and with technology. Human factors knowledge and techniques have been productively applied to enhance human performance in a wide range of domains, from fighter planes to kitchens, emergency rooms, and trauma units (Barach & Weinger, 2005).

Human factors research and application is a logical fit in clinical microsystems because it affords the opportunity to study human interactions with devices and technology and systems and processes, with the overarching goal of enhancing safety and quality.

Phases		*Factors*			
		Patient/Family	Health Care Professional	System	Environment
	Pre-event	Consent (process, timing) Anxiety (play therapy) Patient lines Mother's presence	Not familiar with procedure Lack of physician-nurse communication Focus on anxiety and not on procedure Assumed roles, made assumptions Arrogance/respect	Several lines in patient Silos	RN shortage Scheduling delays Manufacturing (performance shaping factors, human factors) Lack of process for risk analysis
	Event	Anxiety (patient & parent's) No shared expectations No active participation	Fatigue Aware of limitations Training	Work hours Protocols Standardization Double checking	Work hours for residents Rushed No other clinician
	Post-event	Lack of explanation Disclosure Who should talk to family?	Guilt Lack of confidence Loss of face	Lack of understanding of errors/systems Lack of supportive environment for resident Incidence report M&M Analysis of event	Regulatory

Figure 6–3. Completed patient safety matrix. (Case study is provided in Box 6–1.)

Furthermore, human factors research on team decision making in complex task environments is of relevance to clinical microsystems (Brannick, Salas, & Prince, 1997; Foushee & Helmreich, 1988; Helmreich & Schaefer, 1994; Huey & Wickens, 1993; Swezey & Salas, 1992). There are many "performance shaping factors" that are known to play a role in degraded human capabilities; these must be considered to understand how best to optimize clinical care (Weinger & Englund ,1990; Weinger & Smith, 1993).

TABLE 6–2. Linkage of the Microsystem Dimensions to Patient Safety

Microsystem Dimensions	*What This Means for Patient Safety*
1. Leadership	Define the safety vision of the organization Identify the existing constraints within the organization Allocate resources for plan development, implementation, and ongoing monitoring and evaluation Build microsystems participation and input to plan development Align organizational quality and safety goals Engage the Board of Trustees about the organizational progress toward achieving safety goals Create recognition program for prompt truth-telling about errors Certification of helpful changes to improve safety
2. Organizational support	Work with clinical microsystems to identify patient safety issues and make relevant local changes Put the necessary resources and tools into the hands of individuals
3. Staff focus	Assess current safety culture Identify the gaps the between current work culture Conduct periodic assessments of safety culture Celebrate examples of desired behavior, e.g., acknowledgment of an error
4. Education and training	Develop patient safety curriculum Develop a core of people with patient safety skills who can work across microsystems as a resource
5. Interdependence of the care team	Build PDSA[a] into debriefings Use daily huddles to debrief and to celebrate identifying errors
6. Patient focus	Establish patient and family partnerships Support disclosure and truth around medical error
7. Community and market focus	Analyze safety issues in community and partner with external groups to reduce risk to population
8. Performance results	Develop key safety measures Create feedback mechanisms to share results with microsystems
9. Process improvement	Identify patient safety priorities based on assessment of key safety measures Address the work that will be required at the microsystem level
10. Information and information technology	Enhance error reporting systems Build safety concepts into information flow (e.g., checklists, reminder systems)

[a]PDSA, Plan–Do–Study–Act.

The environment of the hospital greatly affects and shapes the outcomes of clinical microsystems. Factors that influence the team's effectiveness include the performance of individual team members, the equipment they use, the care environment (e.g., established care process and procedures), and the underlying organizational and cultural factors. For example, distracters such as information overload, noise, spectators, and physical obstacles can be a danger to both patient and health care professionals. Although there is insufficient space in this chapter to discuss all of the performance shaping factors of relevance to the clinical microsystems, a few of the more pertinent factors are described in more detail in the following sections, including sleep deprivation and fatigue, envirومental noise, and interpersonal communication.

Sleep Deprivation and Fatigue

There is extensive literature on the adverse effects of sleep deprivation and fatigue on individual clinician performance. (Grantcharov, Bardram, Funch-Jensen, & Rosenberg, 2001; Howard,

Rosekind, Katz, & Berry, 2002; N. Taffinder, McManus, Gul, et al., 1998; Taffinder, McManus, Gul, Russell, & Darzi, 1998; Veasey, Rosen, Barzansky, Rosen, & Owens, 2002). Most studies of recurrent partial sleep deprivation have suggested that sleeping only 5 to 6 hours a night can lead to performance impairment (Bonnet, 2000). Sleep loss is associated with reduced performance on tasks requiring vigilance, cognitive skills, verbal processing, and complex problem solving (Veasey et al., 2002; Weinger & Ancoli-Israel, 2002). Performance decrements begin with a lack of appreciation of the skills being degraded and accumulate with continued partial sleep deprivation. This may be seen in physicians working regularly recurring on-call or night shifts. In the early morning hours, after nearly 24 hours without sleep (e.g., at the end of difficult on-call shift), psychomotor performance can be impaired "to an extent equivalent to or greater than is currently acceptable for alcohol intoxication" (Dawson & Reid, 1997, p. 235). Two recent laboratory simulation studies, involving sleep-deprived surgeons, demonstrated significant impairment in surgical skill (both speed and accuracy) in a virtual reality simulation of laparoscopic surgery (Grantcharov et al., 2001; Taffinder et al., 1998). Although the impact of fatigue on *team performance* has thus far been sparsely studied, the results may be expected to be similar with trade-offs between the benefits of team compensation and redundancy on the one hand and impaired team communication on the other.

The effects of individual team member's sleep deprivation (or other performance detractors such as working when ill) on the overall microsystem's clinical performance will depend on many factors, including time of day (circadian effects), clinical experience, task demands, clinical workload, and other team members' level of functioning.

The message for organizational leaders is to acknowledge the potential effect of sleep deprivation and fatigue on individuals and on the microsystem and design work schedules to provide team members with adequate rest periods.

Environmental Noise

The environment of care contains a number of factors that influence microsystem performance including noise, lighting, temperature, the need for protective gear, clutter, disorganization, and impaired physical access to the patient or essential tools/equipment. In the interest of brevity, only the effects of noise are discussed in detail. More detailed discussions of environmental noise are included in the handbook in "Noise and Alarms in Health Care—A Dilemma" (see Buss & Friesdorf, chap. 22, this volume) and "Physical Environment in Health Care" (see Alvarado, chap. 19, this volume).

The noise level in acute care environments can be quite high. For example, continuous background noise in the operating room typically ranges from 75 dB to 90 dB, and can increase to almost 120 dB (e.g., during high-speed gas-turbine drill use; Weinger & Smith, 1993). Although apparently never measured, it is reasonable to assume that sound pressures in the typical trauma unit are similar or louder than those found in surgical suites. In the trauma unit, noise may be generated by multiple conversations, mechanical ventilation, suction, overhead pages, use of medical equipment, and uncoordinated alarms. High noise levels create a positive feedback situation, where noisy rooms require louder voices and louder alarms leading to increased noise levels, missed clinical events, and patient harm. There is a growing literature on how hospitals can incorporate sound insulation materials and avoid overhead paging to reduce noise levels.

High noise levels interfere with effective verbal communication, which is always important, but may be critical during certain events such as resuscitations during shift handoffs, and at other times when it is vital for team members to hear clearly other members of the team. High noise levels in trauma units can also detrimentally affect short-term memory tasks, mask task-related cues, impair auditory vigilance (e.g., the ability to detect and identify alarms), and cause distractions during critical periods (Weinger & Englund, 1990; Weinger & Smith, 1993). Exposure to loud noise activates the sympathetic nervous system affecting mood and performance. The resulting stress response has been suggested to interact with other performance-shaping factors resulting in impaired decision making during critical clinical incidents (Selye, 1976).

Interpersonal Communication

Both verbal and nonverbal communication are critical to the success of microsystem performance (Kanki, Lozito, & Foushee, 1989). Failures of team communication lead to medical errors and adverse outcomes (Donchin, et al., 1995). In nonmedical highly complex domains involving teamwork (e.g.,

TABLE 6–3. Problems and Pitfalls in the Teamwork ‡

Difficulties coordinating conflicting actions
Poor communication among team members
Failure of members to function as part of a team
Reluctance to question the leader or more senior team members
Failure to prioritize task demands
Conflicting occupational cultures
Failure to establish and maintain clear roles and goals
Absence of experienced team members
Inadequate number of dedicated members
Failure to establish and maintain consistent supportive organizational infrastructure
Leaders without the "right stuff"

Note. Modified from Schull, Ferris, et al. (2001)

aviation crews, submarines), the team has often been together a long time and is well practiced. Effective team communications involve unspoken expectations, body language, traditions, general assumptions about task distribution, command hierarchies, as well as individual emotional and behavioral components. Failures of adequate communication between teams of clinical care providers in the intensive care unit contributed to medical errors (Donchin et al., 1995). In this intensive care unit study more than one third of all patient care errors reported were associated with failures of verbal communication. These communication failures occurred not only between nurses and physicians but also between nurses. Similarly, analysis of videotaped trauma team performance showed that highly skilled teams communicated in a variety of ways, many of which were nonverbal and implicit (Xiao, Mackenzie, & Patey, 1998). Team coordination breakdowns were manifested by conflicting plans, inadequate support in crisis situations, failure to verbalize problems, and poor delegation of tasks.

Team performance can be adversely affected by dysfunctional interpersonal interactions among team members. Such "miscommunication" often stems from a lack of shared expectations, beliefs, or training (Barach, 2002). Table 6–3 outlines some of the potential problems in teamwork. This suggests that teams and microsystems can enhance their performance by spending more time together, not just during formal training, but also through joint conferences and social events (Baker, Salas, King, Battles, & Barach, 2005).

Team members must make special efforts to communicate clearly and unambiguously, especially when members of the team are new or less experienced. Effective team communication can be further compromised when some or all of the team are subjected to other stressors such as sleep deprivation and fatigue.

Simply bringing individuals together to perform a specified task does not automatically ensure that they will function as a team. Teamwork depends on a willingness of clinicians from diverse backgrounds to cooperate toward a shared goal, to communicate, to work together effectively, and to improve. Each team member must be able to: (a) anticipate the needs of the others, (b) adjust to each other's actions and to the changing environment, (c) monitor each other's activities and distribute workload dynamically, and (d) have a shared understanding of accepted processes and how events and actions should proceed.

CONCLUSION

The microsystem concepts have evolved from systems theory and primary research on characteristics of high-performing clinical units. Specific interventions can be made to embed quality and safety into the microsystem. We offer several suggestions related to each of the microsystem characteristics that might serve as a guiding framework to adapt to individual microsystems. Leaders should set the stage for making safety a priority for the organization, but they should allow individual microsystems to create innovative strategies for improvement. Table 6–4 provides examples of performance shaping factors that affect the microsystem.

Microsystems with clear goals and effective communication strategies can adjust to new information with speed and effectiveness to enhance real-time problem solving. Individual behaviors change more readily on a team because team identity is less threatened by change than are individuals. Behavioral attributes of effective teamwork including enhanced interpersonal skills, can extend to other clinical arenas.

TABLE 6–4. Examples of Performance Shaping Factors Affecting the Microsystem#

Individual Factors	Environment of Care
Clinician knowledge, skills, and abilities	Physical layout and design
Cognitive biases	Way finding
Risk preference	Noise
State of health	Ambient lighting
Fatigue (including sleep deprivation)	Temperature, humidity, and pressurization
Breaks and boredom	Motion and vibration
Substance use/abuse (e.g., alcohol hangover effects)	Physical constraints (e.g., crowding)
Stress	Distractions
Task Factors	**Equipment/Tools**
Task distribution	Device usability
Task demands	Alarms and warnings
Workload	Automation
Shiftwork	Maintenance and obsolescence
	Protective gear
Team Communication	**Organizational/Cultural**
Teamwork/team dynamics	Production pressures
Interpersonal communication (clinician–clinician, clinician–patient)	Worker empowerment
	Culture of safety
	Policies
Interpersonal influence	Procedures
Shared protocols	Staffing
	Management structure

Note. Modified from Barach, & Weinger (2005).

Turning a clinical unit into an effective microsystem requires substantial planning and practice. There is a natural resistance to move beyond individual roles and accountability to team mindset. One can facilitate this commitment by: (a) fostering a shared awareness of each member's tasks and role on the team through cross-training and other team training modalities; (b) training members in specific teamwork skills such as communication, situation awareness, leadership, follower-ship, resource allocation, and adaptability; (c) conducting team training in simulated scenarios with a focus on both team behaviors and technical skills; (d) training team leaders in the necessary leadership competencies to build and maintain effective teams; and (e) establishing and consistently utilizing reliable methods of team performance evaluation and rapid feedback.

References

Arrow, H., McGrath, J., & Berdahl, J. (2000). *Small groups as complex systems.* Thousand Oaks, CA: Sage.

Association for the Advancement of Medical Instrumentation (2001). *The human factors design process for medical devices.* Arlington, VA: Association for the Advancement of Medical Instrumentation.

Baker, D., Beaubien, J., et al. (in press). Medical team training: An initial assessment and future directions. *Joint Commission Journal on Quality and Safety.*

Baker, D. P., Salas, E., King, H., Battles, J., & Barach, P. (2005). The role of teamwork in the professional education of physicians: current status and assessment recommendations. *Joint Commission Journal on Quality and Patient Safety. 31*(4), 185-202.

Barach, P. (2002). Enhancing patient safety and reducing medical error: The role of human factors in improving trauma care. In E. Soriede & C. Grande (Eds.), *Prehopsital trauma care* (pp. 767–777). Baltimore, MD: International Trauma Anesthesia and Critical Care Society..

Barach, P., & Weinger , M. (2005). Trauma team performance. In W. Wilson, C. Grande, & D. Hoyt (Eds.), *Trauma: Resuscitation, anesthesia, & critical care.* New York: Marcel Dekker.

Batalden, P. B., Mohr, J. J., Nelson, E. C. Plume, S. K., Baker, G. R., Wasson, J. H., et al. (1997). Continually improving the health and value of health care for a population of patients: The panel management process. *Quality Management in Health Care* 5(3), 41–51.

Batalden P. B., Nelson E. C., Mohr J. J., Godfrey M. M., Huber T. P., Kosnik L, et al. (2003a). Microsystems in health care, Part 5: How leaders are leading. *The Joint Commission Journal on Quality Improvement, 29(6)*, 297–308.

Batalden, P., Nelson, E., Edwards, W., Godfrey, M., & Mohr, J. (2003b). Microsystems in health care, Part 9: Developing small clinical units to attain peak performance. *The Joint Commission Journal on Quality Improvement, 29*(11), 575–585.

Bonnet, M. (2000). Sleep deprivation. In M. Kryger, T. Roth, & W. Dement (Eds.), *Principles and practice of sleep medicine* (3rd ed., pp. 53–71). Philadelphia: Saunders.

Brannick, M., Salas, E., & Prince, C. (1997). *Team performance assessment and measurement.* Mahwah, NJ: Lawrence Erlbaum Associates, Inc.

Dawson, D., & Reid, K. (1997). Fatigue, alcohol and performance impairment. *Nature, 388,* 235.

Dekker, S. (2002). *The field guide to human error investigations.* Aldershot, England: Ashgate.

Deming, W. E. (1986). *Out of the crisis.* Cambridge: Massachusetts Institute of Technology, Center for Advanced Engineering Study.

Donaldson, M. S., & Mohr, J. J. (2000). *Improvement and innovation in health care microsystems. A technical report for the institute of medicine committee on the quality of health care in America.* Princeton, NJ: Robert Wood Johnson Foundation.

Donchin, Y., D. Gopher, et al. (1995). A look into the nature and causes of human errors in the intensive care unit. *Crit Care Med*(23): 294–300.

Foushee, H., & Helmreich, R. (1988). Group interaction and flight crew performance. In E. Wiener & D. Nagel (Eds.), *Human factors in aviation* (pp. 189–227). San Diego, CA: Academic.

Gaba, D., & Howard, S. (2002). Fatigue among clinicians and the safety of patients. *New England Journal of Medicine, 347,* 1249–1255.

Godfrey, M., Nelson, E., Wasson, J., Mohr, J., & Batalden, P. (2003). Microsystems in Health Care: Part 3: Planning Patient Centered Services. *Joint Commission Journal on Quality Improvement. 29,* 4.

Grantcharov, T., Bardram, L., Funch-Jensen, P., & Rosenberg, J. (2001). Laparoscopic performance after one night on call in a surgical department: prospective study. *British Medical Journal, 323,* 1222–1223.

Haddon, W. J. (1972). A logical framework for categorizing highway safety phenomena and activity. *Journal of Trauma, 12,* 197.

Helmreich, R., & Schaefer, H.-G. (1994). Team performance in the operating room. In M. Bogner (Ed.). *Human error in healthcare* (pp. 225–53). Hillsdale, NJ: Lawrence Erlbaum Associates, Inc.

Howard, S., M. Rosekind, et al. (2002). Fatigue in anesthesia. *Anesthesiology*(97): 1281–94.

Huber, T., M. Godfrey, et al. (2003). Microsystems in Health Care: Part 8. Developing People and Improving Worklife: What Front-Line Staff Told Us. *The Joint Commission Journal on Quality and Safety* 29(10): 512–522.

Huey, B., & Wickens, C. (1993). *Workload transition: Implications for individual and team performance.* Washington, DC: National Academy Press.

Kanki, B., Lozito, S., & Foushee, H. (1989). Communication indices of crew coordination. *Aviation, Space, and Environmental Medicine 60,* 56–60.

Kosnik, L., & Espinosa, J. (2003). Microsystems in health care: Part 7: The Microsystem as a platform for merging strategic planning and operations. *The Joint Commission Journal on Quality and Safety,* 29(9), 452–459.

Mohr, J. (2000). *Forming, operating, and improving microsystems of care.* Hanover, NH: Dartmouth College Center for the Evaluative Clinical Sciences.

Mohr, J., & Batalden, P. (2002). Improving safety at the front lines: The role of clinical microsystems. *Quality and Safety in Health Care, 11*(1), 45–50.

Mohr, J., Barach, P., Cravero, J., Blike, G., Godfrey, M., Batalden, P., et al. (2003). Microsystems in health care, Part 6: Designing patient safety into the microsystem. *The Joint Commission Journal on Quality and Safety, 29(8),* 401–408.

Nelson, E. C., Batalden, P. B., Mohr, J. J., & Plume, S. K. (1998). Building a quality future. *Frontiers of Health Services Management, 15*(1), 3–32.

Nelson, E., Batalden, P., Huber, T., Mohr, J., Godfrey, M., Headrick, L., et al. (2002). Microsystems in health care, Part 1: Learning from high-performing front-line clinical units. *The Joint Commission Journal on Quality and Safety,* 28(9), 472–493.

Nelson, E., Batalden, P., Homa, K., Godfrey, M., Campbell, C., Headrick, L., et al. (2003). Microsystems in health care, Part 2: Creating a rich information environment. *The Joint Commission Journal on Quality and Safety, 29*(1), 5–15.

Peters, T. (1987). *Thriving on chaos: Handbook for a management revolution.* New York: Harper & Row.

Plsek, P. E., & Greenhalgh, T. (2001). Complexity science: The challenge of complexity in health care. *British Medical Journal, 323*(7313), 625–628.

Plsek, P. E., & Wilson, T. (2001). Complexity, leadership, and management in healthcare organisations. *British Medical Journal, 323*(7315), 746–749.

Quinn, J. B. (1992). *The intelligent enterprise.* New York: Free Press.

Schull, M., Ferris, L., Tu, J. V., Hux, J. E., & Redelmeier, D. A. (2001). Problems for clinical judgement: 3. Thinking clearly in an emergency. *Canadian Medical Association Journal, 164,* 1170–1175.

Selye, H. (1976). *Stress in health and disease.* Boston: Butterworth.

Senge, P. (1990). *The fifth discipline.* New York: Doubleday.

Swezey, R., & Salas, E. (1992). *Teams: Their training and performance.* Norwood, NJ: Ablex.

Taffinder, N., McManus, I., Gul, Y., Russell, R. C., & Darzi, A. (1998). Effect of sleep deprivation on surgeon's dexterity on laparoscopic simulator. *Lancet, 352,* 1191.

Trustees of Dartmouth College (2004). Available at http://clinicalmicrosystem.org

Veasey, S., Rosen, R., Barzansky, B., Rosen, I., & Owens, J. (2002). Sleep loss and fatigue in residency training: A reappraisal. *Journal of the American Medical Association, 288,* 1116–1124.

Wasson, J. H., Godfrey, M. M., Nelson, E., Mohr, J., & Batalden, P. (2003). Microsystems in health care, Part 4: Planning patient-centered care. *The Joint Commission Journal on Quality and Safety, 29*(5), 227–237.

Weinger, M., & Ancoli-Israel, S. (2002). Sleep deprivation and clinical performance. *Journal of the American Medical Association, 287,* 955–7.

Weinger, M., & Englund, C. (1990). Ergonomic and human factors affecting anesthetic vigilance and monitoring performance in the operating room environment. *Anesthesiology, 73,* 995–1021.

Weinger, M., & Smith, N. (1993). Vigilance, alarms, and integrated monitoring systems. In J. Ehrenwerth & J. Eisenkraft (Eds.), *Anesthesia equipment: Principles and applications* (pp. 350–84). Malvern, PA: Mosby Year Book.

Wheatley, M. (1992). *Leadership and the new science: Learning about organization from an orderly universe.* San Francisco: Berrett-Koehler.

Xiao, Y., Hunter, W. A., Mackenzie, C. F., Jefferies, N. J., Horst, R., and the LOTAS Group (1996). Task complexity and emergency medical care and its implications for team co-ordination. *Human Factors, 38,* 636-4.

7

THE ARTICHOKE SYSTEMS APPROACH FOR IDENTIFYING THE WHY OF ERROR

Marilyn Sue Bogner
Institute for the Study of Human Error, LLC

Error in medicine or more generally error in health care strikes fear in the hearts and minds of every individual. That concern, which normally lies fallow, was whipped into a frenzy of public consternation and outrage by the media's repeated highlighting the revelation in the report on medical error by the Institute of Medicine (IOM) of the U.S. National Academy of Sciences that 44,000 to 98,000 people die annually because of medical error (Kohn, Corrigan, & Donaldson, 1999). That outrage was directed primarily at physicians and nurses because they commit the erroneous acts that result in adverse outcomes. The adverse outcome that received the most attention in the IOM report was death, but in common parlance, adverse outcomes include injuries and prolongation of treatment of the presenting health problem or the onset of a new health problem that requires extended treatment—problems related to the care provided rather than the condition of the patient.

Although this discussion addresses error as an act that results in an adverse outcome, it also pertains to near misses and conditions typically not considered yet a rich source of information—those of hazards, potential errors, accidents waiting to happen. (For the sake of completeness, however, it should be noted that an act that is considered an error can have an innocuous outcome or no outcome at all.) For a topic that has evoked so much emotion and been the focus of considerable research, there is a surprising lack of unanimity in definitions of health care error.

ERROR

For the purpose of the IOM report, an error was defined as "The failure of a planned action to be completed as intended (i.e., error of execution) or the use of a wrong plan to achieve an aim (i.e., error of planning)" (Kohn et al., 1999, p. 23). Errors have been differentiated into technical errors reflecting skill failures, judgmental errors that involve the selection of an incorrect strategy of treatment, and normative errors that occur when the larger social values embedded within medicine as a profession are violated (Bosk, 1979, p. 51). Errors also have been described in terms of the aspect of health care in which they occur, such as errors of missed diagnosis, mistakes during treatment, medication mistakes, inadequate postoperative care, and mistaken identity (Gibson & Singh, 2003, pp. 60–69).

Admittedly those definitions of health care error do not constitute an exhaustive list nor are they representative of the full range of nuances in definitions. The definitions nonetheless illustrate the pervasive attitude that is evident in a host of other definitions of error and in nearly all discussions of patient safety—that the care provider is the sole cause of the error, hence the cause of the attendant adverse outcome. This can have onerous connotations because the previous definitions of error do not state if the error and its outcome are expected or unexpected. If the error were presumed to be expected, a subtle message is conveyed that the adverse outcome was intentional so the error is indicative of negligence, carelessness, even malfeasance. This is not to say that such a situation cannot occur; several years ago the media reported that a male nurse "euthanized" elderly patients, but that is a rare exception. Although the innuendo exists that the preventable deaths from errors indicate carelessness, inattention and implicitly negligence, for the sake of unbiased scientific inquiry as well as fairness to the care provider, such should not be the initial assumption and will not be considered in the ensuing discussion.

Error Reporting

The IOM report (Kohn et al., 1999) advocates error reporting systems to address and reduce the likelihood of errors in health care by serving two functions: "They can hold providers accountable for performance, or alternatively, they can provide information that can lead to improved safety" (p. 74). The performance being addressed is health care, specifically providing care with errors that result in adverse outcomes. The sentence from the IOM report is interesting in that accountability seems to be an end unto itself indicating that the provider is the source of the problem and holding the provider accountable will enhance patient safety. The other function of error reporting is to "provide information that can lead to improved safety," however, one would hope that error reporting for accountability also would be to "provide information that can lead to improved safety" (p. 74). The reason for stating the two functions as alternatives, that is, *either* accountability *or* enhancing safety, is not clarified in the report. The significance of the different focus of each of those approaches in determining the information obtained by error reporting activities and the use of that information in activities to reduce error is illustrated by the discussion of two cases later in this chapter.

Error reporting to date primarily is directed toward the former function—accountability. This focus presumes that the health care provider whether a highly skilled professional, a lay person caring for a friend or a relative, or a lay person engaged in self care for transient or chronic health problems is the sole cause of an error and adverse outcome. Despite the variety of types of care providers, professional care providers in organizational settings have been the focus of error reporting endeavors; the range and dispersion of other care providers precludes their consideration at this time.

Error Reporting for Accountability

The numerous error reporting for accountability programs collect information that essentially addresses "Who did what." The number of errors reported is considered indicative of the existence of an active safety culture; the more errors reported, the more active the safety culture (Cohen et al., 2004). Although reported errors and safety culture may be related, the question can be raised if the information from the error reporting programs is effective in addressing error. If it were, then it could be expected that the incidence of errors would have decreased over the 5 years since the IOM report and subsequent implementation of error reporting programs although that implementation might have been only in the most recent 2 years. Admittedly there probably would be an initial increase in the number of errors reported reflecting an increased sensitivity to error; however, sufficient time has elapsed that a decrease in error might be expected. This apparently has not occurred.

A recent report (CNN, 2004) stated 195,000 deaths occur annually in hospitals—deaths caused by the care the patients receive. This is far greater than the 100,000 deaths extrapolated from the Harvard Medical Practice Study (HMPS) (Leape, 1994), or the 44,000 to 98,000 preventable deaths stated in the IOM report (Kohn et al., 1999). The increase in reported deaths due to errors might indicate increased sensitivity to error as mentioned previously, however this seems unlikely given the difference between the 195,000 in the CNN report and the highest estimate of preventable deaths—100,000 from the HMPS—is 95,000. That is far in excess of what would be expected as reflecting sensitivity to errors. The error reporting programs are varied and geographically diverse so there is little basis to attribute the absence of success to those considerations.

Another example that error reporting for accountability programs are not notably successful in addressing adverse outcomes is the estimate extrapolated from a 20% sample of all U.S. hospitals that adverse outcomes for the Agency for Healthcare Research and Quality patient safety indicators cost $9.3 billion in excess charges and 35,591 deaths annually (Zhan & Miller, 2003). Further lack of effectiveness of current error reporting programs is in the MedPAC (2004) report to Congress, which stated that patient safety adverse events in the Medicare population increased in the 3-year period from 2000 to 2002. The apparent lack of viability of provider accountability information in reducing the incidence of error may lead to activities that follow suggestions in the IOM report (Kohn et al., 1999) to modify error reporting programs to collect standardized information. No evidence, however, was cited indicating that standardized error reporting data have been more successful in addressing error than nonstandardized data.

The lack of effectiveness of the provider accountability approach to addressing error may indicate that it is not the appropriate focus. Indeed, reporting error for accountability could be considered as collecting facts:

> The simple collecting of facts is indispensable at certain stages of a science; it is a wholesome reaction against a philosophical and speculative building of theories. But it cannot give a satisfactory answer to questions about causes and conditions of events. ... Even from a practical point of view the mere gathering of facts has a very limited value. It cannot give an answer to the question that is most important for practical purposes—namely what must one do to obtain a desired effect in given concrete cases? (Lewin, 1936/1966, p. 4)

The committee that developed the IOM report appears to appreciate the *collection of facts* nature of the provider as the sole source of error approach:

> The common initial reaction when an error occurs is to find and blame someone. However, even apparently single events or errors are due most often to the convergence of multiple contributing factors. Blaming an individual does not change these factors and the same error is likely to recur. Preventing errors and improving safety for patients require a systems approach to modify the conditions that contribute to error. The problem is not the people; the problem is the system needs to be made safer. (Kohn et al., 1999, p. 42)

This suggests that the second function of error reporting previously quoted from the IOM report, "to provide information that can lead to improved safety", is the preferable approach to actually address errors. To obtain information for this function of error reporting, the IOM report states that "individuals be allowed to report errors they commit in a way that provides freedom of expression so that contributory factors can be identified" (p. 86). The ensuing discussion describes a means for acquiring such information and presents examples of how that information can be used to effectively enhance patient safety by reducing, indeed preventing, error.

Error Reporting of Safety Related Information

Error reporting for accountability addresses "Who did what." Error reporting to provide information that can lead to improved safety addresses why an error happened. To do that effectively, it is necessary to determine what will be reported, that is, operationally define error—discuss what an error is and describe an approach to ascertain the reasons why an error occurs. To demonstrate the value of information that addresses the why of error, examples of its application in two cases are presented. It should be noted that, although this discussion has been and will continue to be focused on error, the points made also apply to near misses or errors that almost happened and hazards—potential errors or errors that are waiting to happen.

The "Why" of Error

This discussion of why error occurs draws heavily on the work of Kurt Lewin who was one of the most creative psychological theorists of his time—some say of the 20th century. Although his work originally was published nearly three quarters of a century ago, there has been a recent resurgence of interest in it—his insights on behavior have few if any rivals. He made innumerable contributions to society by applying theory to real world issues. Although he was not one of those involved in human factors endeavors per se uring World War II, he conducted applied research in the spirit of the burgeoning discipline of human factors during that time and contributed significantly to the war effort. His work and his philosophy that "Nothing is as practical as a good theory" (Marrow, 1969, p. viii) are particularly relevant in the consideration of patient safety as is apparent in the following discussion.

To understand why an error occurs, it is necessary to understand the nature of error, how it comes into existence. The previously quoted definitions of error describe error in terms of generic procedures as well as aspects of health care in which an error occurs; however, there is no definition of error *qua error*—the key to understanding why error occurs. Interestingly enough, the definition is implicit in the consideration of error—error is an act performed by a person that for the purpose of this discussion has an adverse outcome. An act is behavior and behavior is a function of a person interacting with the environment (Lewin, 1936/1966). This centrality of environment in behavior is not unique to Lewin; it is evident throughout over 2 centuries of research and theory in psychology as well as the other social sciences and in the millennia of philosophical writings; the influence of environment also is considered in the research and theory of the biological and physical sciences.

ERROR AS BEHAVIOR

The simple statement, error is behavior, has profound implications for addressing error and for reducing and preventing its occurrence. Because behavior reflects the interaction of the individual and factors in the environment, the presumption that an individual care provider is the sole cause of an error is not valid. Admittedly a care provider performs the act that leads to an adverse outcome; however, that act is not solely of the care provider's volition. Factors in the context in which a person functions influence that person, hence the act is influenced by such factors. Because of this, the typical consideration of the care provider as the sole cause of error—error reporting for accountability—presents only one aspect of a complex phenomena, that of the person, and as such is incomplete and misleading.

A persuasive argument for considering context in adverse outcomes is stated regarding aviation accidents: "Without full information concerning the context and environment(s) in which accidents occur, it is not possible to understand their genesis and how to take rational steps to prevent future accidents" (Billings, 1997, p. 147). Thus, to understand why a care provider committed an error so that rational steps to prevent future adverse events may be taken, it is necessary to address factors in the context in which the error occurred. That might seem an impossible task given the complexity of the context in which health care is provided; however, the impossible task becomes possible by representing the myriad of factors in the environmental context as systems.

Environmental Context

Research on error conducted by the nuclear power, manufacturing, aviation, and other industries identified categories of environmental factors associated with error (Moray, 1994; Rasmussen, 1982; Senders & Moray, 1991). Those categories that are analogous across industries also are considered applicable to health care. The categories are comprised of factors that influence the care provider and might by impacting his or her task performance contribute to health care error. The categories and examples of factors in terms appropriate for health care are the following: *the patient* (the product of the industry) with personality, psychological factors such as anxiety, and physical characteristics that include weight and frailty in addition to the presenting health problem; *the means of providing care* (the means of performing a task) such as a medical device and its technological sophistication as well as cognitive workload; *the care provider* (the worker) with hard-wired human constraints that include limited perceptual acuity, cognitive functioning that is affected by stress and fatigue, limited capacity to retain and manipulate units of information, and anthropometric characteristics of height and handedness as well as knowledge, heuristics, memory, and anticipation. Those three categories are ensconced in a broader context comprised of five additional categories of factors identified by industry error research.

The additional 5 categories with examples of factors are *ambient conditions* of illumination, temperature, noise, altitude; *the physical environment* with placement of medical equipment, room size, clutter; *the social environment* of other care providers and personnel, family members, professional culture; *organizational factors* such as workload, hours worked, reports, policies for caring for uninsured persons, organizational culture; and *legal–regulatory–reimbursement–national culture factors* that include threat of litigation, regulatory constraints, reimbursement policies, and national cultural mores.

The factors in each of the categories interact. Because a complex of interacting elements defines a system (von Bertalanffy, 1968), each of the categories is a system. The categories of factors are

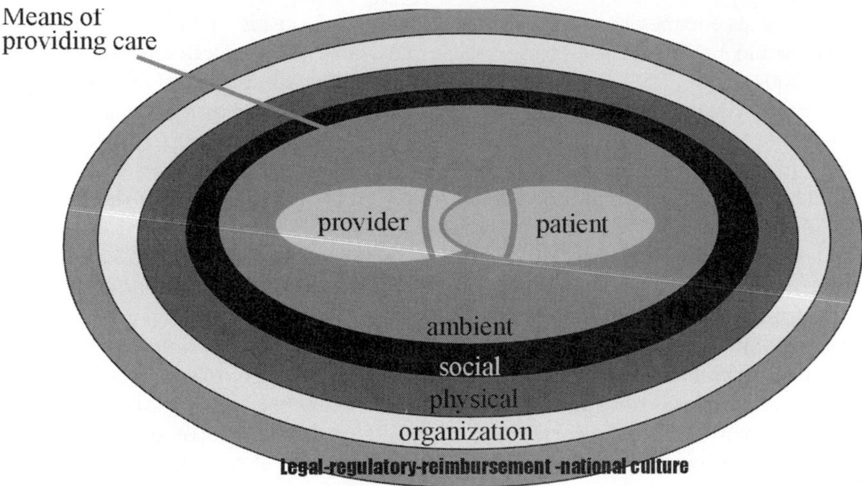

Figure 7–1. Systems of factors that influence behavior.

hierarchical nested systems represented by concentric circles (Figure 7–1). These systems constitute an overall context of care suprasystem of factors that affect the care provider. Changes in factors in one of the systems of categories of factors can affect factors in the other systems. The underlying concept of systems is applicable to all targets of analysis (von Bertalanffy, 1968), so this systems conceptualization of the environmental context of care is appropriate for addressing health care error

When determining why an incident occurred, the representation of systems encourages considering each of those systems for factors that affected the care provider. This is important in countering the Stop Rule, the tendency for people in seeking to explain why a situation occurred by backward chaining of possible causes to stop when a plausible, easily explained cause is identified (Rasmussen, 1990). In health care, the explanation that triggers the Stop Rule typically is the care provider who committed the error. This approach goes beyond the care provider to consider factors in other systems when determining why the act was committed. This is important because such a conceptualization affords an ordered approach to understanding complex phenomena—a systems approach.

SYSTEMS APPROACH

The previously described context of care suprasystem comprised of systems of factors that affect the care provider is manifest in the systems approach. This conceptualization differs from the systems and other systems approaches in the health care error literature. When considering error, the term system usually refers to error reporting systems or as in the IOM report, the health care system. The systems approaches in the literature typically refer to the health care system consisting of all aspects of health care however closely or loosely they are related as the source of factors that cause error. The study reported as "Systems Analysis to Adverse Drug Events" (Leape et al., 1995) found that errors were committed in physician ordering, transcription, pharmacy dispensing, and nurse administration—each of which can be interpreted as representing an aspect of the health care system. It is not clear how those findings, valid as they may be, identify targets for change to reduce error except perhaps to emphasize error reporting for accountability for persons in the identified groups. When considered with respect to the previously described systems approach of factors that can impinge on the person and provoke error, those findings are transformed into drug ordering, drug transcription, drug dispensing, and drug administration. That change is subtly powerful; the replacement of the care provider by a systems factor that affects the provider, the drug, reorients the focus and stretches the search for why the error occurred beyond the human to system factors. Thus, factors related to the drug(s) involved in the adverse events can be analyzed to identify possible error-provoking factors such as look alike names and nearly identical packaging that would be targets for change (U.S. Pharmacopeia, 2004).

Even in relatively constrained situations such as a hospital room, the context of care factors that are candidates for provoking error are legion and the prospect of identifying them is daunting. Lewin's very practical (and empirically supported) theory that behavior is a function of the person interacting with the environment (Lewin, 1946/1964a) provides a vehicle for transforming a challenging task into one that not only is manageable, but one that also has profound positive ramifications for patient safety. To determine why an error occurred so that actions can be taken to avoid its recurrence, it is necessary to identify factors in the context of care that impact the person and influence his or her behavior so they can be modified or removed to reduce or prevent the likelihood of recurrence of that error. The error-provoking factors a person experiences may or may not be readily observable. To appreciate the importance of such factors in understanding error, it is necessary to consider the life space of the person (Lewin, 1936/1966).

Life Space

The life space of a person consists of factors that influence the person at the point in time when he or she performs any task including one that is an error. Integrated in the life space concept is a consideration that few theorists and researchers have systematically approached—that of time (Lewin, 1943/1964b)—specifically the influence on present behavior of occurrences in the past and anticipation of future events. This is particularly important in addressing health care error because of the influence of past adverse outcomes and the concern about possible litigation. The influence of the past and anticipation of the future can be considered as driving the treatment of patients by care providers. The medical treatment of a given patient is influenced if not determined by the knowledge and experience of the effectiveness of certain regimens in treating comparable patients, so that the anticipated consequences of the regimen with the presenting patient would be favorable.

The influence on present behavior of occurrences in the past and anticipation of future events provides an example that underscores the power of a definition of error. The example involves a surgeon who developed a treatment plan for a procedure based on his or her experience and knowledge of what has been effective in addressing the presenting health problem of the patient. During that surgical procedure, the patient's condition unexpectedly and for no apparent reason deteriorated. Influenced by experience with a similar case, the surgeon deviated from the treatment plan to avoid the threatening adverse outcome. Under the altered treatment plan, the condition of the patient stabilized and the presenting health problem was addressed effectively—an adverse outcome was avoided. There was consensus among the surgical staff that the original treatment plan was appropriate and the deterioration of the patient could not have been anticipated. Despite the positive patient outcome, the surgeon could be accused of committing an error according to the definition of error in the IOM report: "An error is defined as the failure of a planned action to be completed as intended (i.e., an error of execution) or the use of a wrong plan to achieve an aim (i.e., error of planning)" (Kohn et al., 1999, p. 23).

As behavior at a given time can be influenced by factors in a person's life space from past and future occurrences as well as factors in the context that are not apparent to an observer, the person's behavior may not be influenced by seemingly relevant (to an observer) factors in the context of care. With respect to the latter point, it is not uncommon for a care provider to perform a task under conditions that an observer most likely would consider nonproblematic yet there are factors in those conditions that adversely affect the provider's abilities. For example, the conditions in which Dr. Brown performs a laparoscopic (keyhole surgery) procedure using the long instruments with complex handles designed for such surgery may appear routine to an observer; those conditions, however, set the stage for error.

Dr. Brown experiences severe shoulder muscle cramps because of the postures necessary to use the instruments to perform the procedure. The discomfort from those cramps causes him to be unusually distracted by an argument near the door to the operating room and exacerbates his concern about the news he received immediately before surgery that his application for research funding was rejected. Thus, the observed situation had little resemblance to what actually affected Dr. Brown; what is observable can be misleading. To actually identify error-provoking factors it is necessary to obtain information from a person's life space—information that essentially is available only to that

Figure 7–2. Life space artichoke.

person; an expert's sole interpretation of factors that caused an error is likely to be incomplete if not inaccurate.

It has been said that there are so many health care errors because the health care system in the United States is broken. That line of reasoning is clarified by Lewin's life space concept (1943/1964b) because a person, a care provider, is affected only by the influences that factors exert in his or her life space. Problems in a component of the health care system such as the emergency room, unless influential in the life space of a care provider at the time of performing a task such as delivering a child, do not exist for that person and as such do not affect his or her behavior. This is not to say that components of the health care system do not play a part in error; they play a part, but only via their factors in the systems of the context of care experienced by the care provider.

To understand why an error occurs, that which contributes to an error must be of the same quality and magnitude as the error. In other words, an erroneous act reflects the interaction of the person and factors in the context; addressing that act in terms of a component of the health care system or an organization is analogous to considering a mansion when a brick is needed—the component of the health care system is too much information at the same time it is not enough because it is not appropriate. Given that factors in the context of care induce error, the context should be explored to identify those factors—factors to factors, not an entire entity to a factor in context.

The systems approach in this discussion is unique because it refers to systems of factors experienced from a person's environment that influence the person to perform mental as well as physical behaviors, factors in that person's life space (Lewin, 1936/1966). The systems that influence behavior presented in Figure 7–1 are the systems of factors that comprise a person's life space. Admittedly, this can be a difficult concept to understand—it is after all, a paradigm change. To aid in this endeavor, the life space of a person as the suprasystem in Figure 7–1 is represented as an artichoke.

Life Space as an Artichoke

As representations of error as slices of Swiss cheese (Reason, 1990) and as an onion (Moray, 1994) have been helpful in conveying their concepts of error, the systems of factors that influence behavior (Figure 7–1) that comprise the life space of a person are presented as an "exploded" artichoke (Figure 7–2) with the person affected by those factors, the person whose life space is represented by the artichoke, as the center, the heart of the artichoke. The concentric circles of leaves of the artichoke represent the categories of systems of the environmental context.

A person's life space is unique to that individual. Consider care providers Nurse Green and Dr. Jones. When addressing factors that impact Nurse Green and lead to error, only the life space of Nurse Green can be considered—Nurse Green is at the

Figure 7–3. Systems affecting the provider.

heart of the Artichoke of her life space. If Dr. Jones were integral to the error behavior of Nurse Green, her life space would contain only the factors of Dr. Jones that influenced her behavior. To access the factors that affected Dr. Jones to perform the act that affected Nurse Green and led to her committing an error, it is necessary to consider Dr. Jones' life space, the Artichoke with Dr. Jones as the heart, separate and distinct from the life space of Nurse Green. The life space concept is applicable to only one individual at a time. To determine factors that affected several people involved an error, it is necessary to consider the life space of each of them. That information can be compared across people for patterns; however, it would not be surprising if the findings were similar to those in Rashomon (Akutagawa, 1952), that is, each life space consists of factors that create a different story.

The circles of leaves on the vertical artichoke in Figure 7–3 illustrate the influence of the systems of factors of an individual care provider's life space on that provider as the heart of the Artichoke. As that person may appear trapped by the Artichoke, so each of us is trapped by our experiences of our life space; that is, we act with respect to what we have experienced (Kelley, 1955). This representation underscores that the life space is unique to a person—only one person is at the center of the Artichoke.

The concept of life space, although insightful in discussion, must be assessable if its factors that affect error behavior are to be addressed. A tool has been developed, an Incident Worksheet, to elicit and document factors in a person's life space that he or she has experienced as contributing to error behavior. Once identified, those factors can be addressed to reduce or eliminate their error-inducing impact.

DOCUMENTATION TOOL

The suggestion in the IOM report that care providers "be allowed to report errors they commit in a way that provides freedom of expression so that contributory factors can be identified" (Kohn et al., 1999, p. 86) raises the issue of how best to report error to ensure freedom of expression. One option is that the provider may be asked questions about contributory factors in an interview or questionnaire. Questions, however, pose a problem eliciting the true subjective experiences of an individual—the well documented problem that questions can and do shape responses. Factors that seem to be error-inducing to the person developing the questions may bear little resemblance to the factors in the person's life space at the tine of an incident. Another option is focus groups that can be effective in eliciting opinions; however, groups exert social influence on members hence their responses. Because of this, focus groups do not provide the freedom of expression necessary for identifying error-provoking factors from the care provider's life space.

Lewin expressed the concept of life space in terms of mathematical topology (1936/1966). Topology has the decided advantage of providing an abstract framework for addressing influences of a person's behavior. That approach was criticized because it is questionable if the necessary conditions for topological representation could be met, so for that and other reasons it is not viable as a tool to identify life space error-provoking factors. Rather than pursuing some other mathematical representation of the life space as a tool as tempting as that may be, this discussion presents an empirically based simple tool for the care provider to express factors that he or she experienced as affecting his or her behavior at the time an incident occurred.

A lesson on providing the freedom of expression necessary for identifying error-provoking factors is offered by the aviation industry. The Aviation Safety Reporting System (ASRS), a free text written self-report, has been found to be a viable way to obtain comprehensive information about near misses and other safety related issues (Reynard, Billings, Cheaney, & Hardy, 1986). Free text has the advantage over a questionnaire or interview of not focusing the respondent's attention on factors that may be extraneous to that person's experience.

A totally free response, free text option, however, allows for the perpetuation of the tendency for people to belabor a given point that may or may not be relevant to the issue. For example, because of a belief that physician handwriting is the source of medication errors, almost filling a prescription for Cefotaxime with Ceftazidime will be attributed to the writing. Similarly because of a belief that technology is the ultimate problem-solver, it typically is presumed that computer entry for prescribing drugs will eliminate the errors resulting from handwriting. That, however, may not be what actually occurs. The use of computer entry for medications to prevent errors attributed to handwriting has not fully accomplished that purpose—errors occur with computer entry more than handwriting (Robeznieks, 2005). Indeed, the use of computer order entry can introduce new varieties of error that cause adverse outcomes. For example, a software programmer designated doxepin and doxycycline by the same mnemonic, DOX100. The pharmacist inadvertently selected the wrong drug, which resulted in an extended hospital stay for the elderly patient who received three doses of doxepin instead of doxycycline (U.S. Pharmacopoeia, 1995).
Another tendency that can be readily perpetuated by a totally free response option is the exercising of the previously discussed Stop Rule that would be manifest in the individual explaining an incident by attributing the cause to himself or herself or to another person rather than continuing the search for an explanation why the incident occurred (Rasmussen, 1990). The Stop Rule especially would be evident in a free text self-report by health care professionals because they are taught that assuming responsibility for error is their professional obligation even though the error involved a device or procedure that was nearly impossible to use.

The Stop Rule response and rambling text problems with the ASRS can be alleviated by providing prompts and limiting space in which the provider can respond. The tool for documenting the Artichoke Systems Approach to medical error is based on the belief that to understand a person's behavior it is necessary to determine how the person experienced the environment, the context in which that behavior occurred (Gold, 1999)—to obtain information from that person's life space. Because this information is subjective, experienced by the care provider, it is

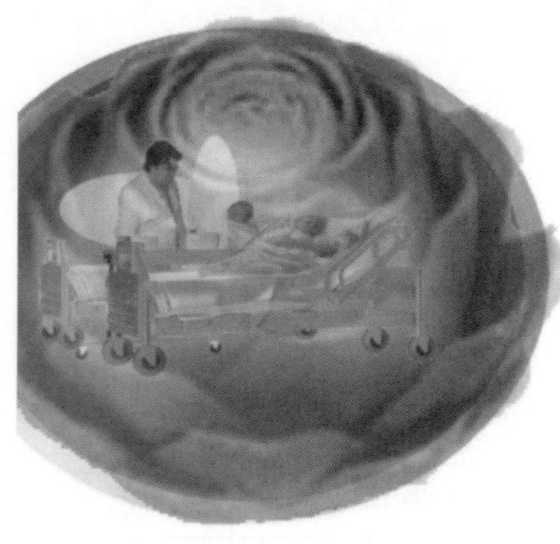

Figure 7–4. Conceptualization of tool.

necessary to elicit the information in such a way that it is not compromised or corrupted.

The documentation tool is in the spirit of the ASRS yet somewhat focused. It can be used to identify factors that not only affected behavior that lead to errors but also behaviors that became near misses and situations that are errors waiting to happen. It is a variation of the Critical Incident Technique (Flanagan, 1954) in which the person involved in a situation identifies factors that contributed to that situation. That technique, the CIT, has been modified and used in considering health care incidents (Short, O'Regan, Lew, & Oh, 1992).

The empirical basis for the tool is the results of the error research by industry represented as the systems of factors in a person's life space. More specifically, the conceptualization of the tool is the systems comprising the life space of the individual, that is, the nuclear care providing system—the care provider, the means of providing care and the patient—by which health care is provided in specific contexts be they technologically sophisticated health care facilities, homes without electricity or running water, all types of sports settings, battlefields, natural disaster sites, in space or under the polar ice cap—and the 5 systems that complete the definition of the context (Figure 7–4).

The format of the Artichoke Systems Approach documentation tool—the Incident Worksheet—is simple to allow the care provider maximum freedom of expression. The names of the eight life space systems of categories of factors appear along the left side of a sheet of paper or computer screen to serve as prompts for the care provider to consider each system for factors he or she experienced as contributing to the error. Each system name is followed by a line on which the provider can note the factors in that system that contributed to an error. Although there is a space to identify the incident, the place and time it occurred, the person completing the tool is not requested to identify himself or herself. The purpose of the tool is to identify factors that contribute to error, not the person whose act was an error. Completed worksheets from two examples of the application of the Artichoke Systems Approach appear as Figures 7–5 and 7–6. As illustrated by the worksheet in Figure 7–5, it is not necessary to fully complete the worksheet to obtain valuable information; however, additional information may be useful.

Although many would argue that obtaining information by observation is preferable to self-report, self-report is preferable for the Incident Worksheet for two reasons. First, as stated previously, factors in a care provider's life space are those experienced by that person and as such are not readily observable hence observation would not provide the desired information. Second, self-report has been demonstrated, albeit by a small number of physicians, to provide a more complete set of data elements than observation or review of medical records (Spies, Mokkink, Robbe, & Grol, 2004). Self-report also provides a modicum of confidentiality. The life space systems represented on the Incident Worksheet are general, so the worksheet can be used by all concerned with patient safety including those who have the most time to identify error-provoking factors in the context of their care—patients. Given that the

Incident Worksheet

Actual X Near Miss__ Potential __

Example of Initial Notes by Dr. Surgeon

Incident: Lacerated patient's liver.
Time: 7:30 a.m.
Location: OR 12.

Context Category	Factors
Patient:	5'2" 400 lb.
Means of providing care	Laparoscopic instrument use caused cramping.
Provider	Short, stood on stool.
Ambient:	
Physical	OR table too high, fell when pedal slipped off stool.
Social	
Organization	
Legal-regulatory-reimbursement-cultural	

Figure 7–5. Worksheet for a surgical incident.

emphasis in this approach is on factors that affect the person to commit errors, it may seem that accountability, which has been the thrust of most of the patient safety efforts, is ignored; it is not.

Accountability

Accountability in the Artichoke Systems Approach is similar to that of the ASRS—if a care provider is involved in an error or near miss or is aware of factors that can lead to error and does not identify and report them then he or she is accountable for the adverse events stemming from those factors. This accountability can be greater than for a person's own actions because a care provider by not reporting error-inducing factors that would be modified or removed contributes to the error behavior of the providers who encounter those factors.

Admittedly a problem exists in identifying those providers who although aware of error-provoking factors do not convey that information on an Incident Worksheet or otherwise notify others. That problem is not unique to the Artichoke Systems Approach; it is analogous to the lack of eagerness of care providers to report an incident to an error-reporting for accountability program. Indeed, the problem with reporting error-inducing system factors is considerably less because the onus of possible malpractice litigation is less for the care provider—the locus of causality lies in the identified systems factors. Brief descriptions of the two incidents reported on Incident Worksheets in Figures 7–5 and 7–6 illustrate accountability and the value of the Artichoke Systems Approach in effectively addressing health care error.

THE ARTICHOKE SYSTEMS APPROACH: TWO INCIDENTS

Laparoscopic Surgery Injury

Laparoscopic surgical procedures are performed using instruments and a tiny video camera to project images of the procedure on a monitor inserted into the body through small incisions. Such procedures incur less trauma to the body than the single often long incision necessary for open surgery.

Incident Worksheet

Actual X Near Miss __ Potential__

Example of Completed Report

Incident: Patient fall.

Time: 10 a.m.

Location: Bathroom, room 612.

Context Category	Factors
Patient:	Fred Smith, 2 days post-op.
Means of providing care	Assisted walking. Left pt on toilet.
Care provider:	Busy. Pt said fell reaching for toilet tissue.
Ambient:	Bright lights, comfortable temp.
Physical	Grab bars on bathroom wall.
Social	Nurse Rice needed help.
Organization	Understaffed.
Legal-regulatory-reimbursement-cultural	Fiscal constraints.

Figure 7–6. Worksheet for a patient fall.

Because of that, time in the hospital and for overall recovery for a laparoscopic procedure is markedly less than for open surgery. Although generally considered safe, incidents with adverse outcomes do happen in laparoscopic procedures. Dr. Surgeon during the procedure to remove Mrs. Patient's gall bladder lacerated her liver (Matern, 2003)—an incident that could be considered a case of negligence ripe for malpractice litigation. That possibility of litigation is the reason Dr. Surgeon did not report the incident to the error reporting for accountability program. Is this really a case of negligence?

Dr. Surgeon was concerned about the incident. Because it was not necessary for him to identify himself on an Incident Worksheet he jotted down a few factors although he knew anyone reading the worksheet could identify him. Dr. Surgeon did not indicate factors in each of the systems on the Incident Worksheet as contributing to lacerating Mrs. Patient's liver; nonetheless, the information he provided (Figure 7–5) is sufficient to identify factors that contributed to the incident. Mrs. Patient was morbidly obese; Dr. Surgeon is not tall. For Dr. Surgeon to have access to the areas of Mrs. Patient's abdomen for the insertion of the video camera and instruments, it was necessary that the operating table be lowered considerably more than for a procedure on a nonmorbidly obese person performed by a tall surgeon—that was not possible. The operating table because of its design could not be lowered to the appropriate height, so it was necessary for Dr. Surgeon to stand on a platform stepstool to attain something near the appropriate height to perform the surgery. A foot pedal that controls features of one of the laparoscopic surgical instruments was placed on the stepstool for Dr. Surgeon's use. As he moved his foot to press the pedal, it fell off the stepstool. This put Dr. Surgeon off balance causing his hand that held a surgical instrument to move in a way that lacerated Mrs. Patient's liver.

The Artichoke Systems Approach Information from the Incident Worksheet could be forwarded to a person who has the authority to initiate remedial actions for the identified factors: send a request to the biomedical engineering department to do whatever is possible to increase the capability of lowering the operating table, notify the manufacturer of the operating table of the problem emphasizing the necessity of retrofitting existing tables to be maximally lowered and designing new tables to have the capability of accommodating the characteristics of the patients including those who are morbidly obese, alert purchasing to the patient safety importance for any table purchased for any procedure including operating tables to have the capability to be maximally lowered to accommodate morbidly obese patients, and request maintenance to equip all platform stepstools with an attaching device to restrain foot pedals from falling. The effectiveness of these changes can be monitored and evaluated.

Results From Each of the Approaches The report error for accountability approach would identify Dr. Surgeon as the cause of the incident. He would be held accountable for it; the incident might be discussed in the weekly Morbidity and Mortality meeting causing him considerable embarrassment and distress. Dr. Surgeon would work under the cloud of possible litigation. Nothing involved in the incident would be changed—no lessons would be learned from the incident. Surgeons would continue to be at risk of committing errors from the inappropriate height at which they have to perform procedures and the risk of doing harm when the foot pedal falls from a stool would continue—risks that would inordinately apply to women who typically have a smaller stature than men. Patients would continue to be injured and surgeons sued.

The Artichoke Systems Approach identified factors that provoked the error—factors that when addressed can effectively reduce the likelihood of error for short as well as tall surgeons with morbidly obese patients undergoing not only laparoscopic surgery but also any procedure that uses a table. As for accountability, the error-inducing factors Dr. Surgeon identified point to entities that were accountable in part for the incident—the operating table manufacturer for not producing a table that meets the needs of surgeons by accommodating the range of patient and provider characteristics and the engineering department for not providing a means of securing the foot pedal on the stepstool. The error-provoking factors might not have been anticipated, so accountability strictures would be minimal; however, if the conditions were identified and not addressed accountability ramifications could be stringent.

Patient Fall

A major source of adverse outcomes in hospitals is patient falls. Consider the incident that Nurse Drew's patient Mr. Smith fell in his room. Typically that incident should be reported by Nurse Drew to the error reporting for accountability program. She did not do so for fear she would be considered negligent and that the incident would be recorded in her personnel file—although that would be the first incident for her, she knew she could lose her license if she were involved in three incidents and wanted a clean record. Mr. Smith was quite shaken and there was some bruising, but no bones were broken.

The hospital emphasized patient safety and sought to involve everyone, so an Incident Worksheet was available to Mr. Smith who believed he knew something about the incident because, after all, he was involved (Chapanis, 1996). He asked Nurse Drew to help him complete the Worksheet (Figure 7–6), which she did gladly because it raised an issue that was a long-standing concern for her. Mr. Smith's experience was that his fall was precipitated by reaching for toilet tissue. Because of the placement of the tissue holder, reaching for and acquiring tissue would be a difficult maneuver under any condition; it became nearly impossible with the awkwardness from his recent hip replacement surgery and somewhat compromised balance due to medications.

The Artichoke Systems Approach Worksheet information is submitted to a person who is responsible for addressing patient safety issues. That person notifies the appropriate person to change the location of tissue holder to be within easy reach of everyone not only in Mr. Smith's bathroom but also in all bathrooms in the facility. Even if a patient does not fall when reaching for the tissue, speedy and successful recovery from surgery such as a hip replacement can be threatened by the placement of a tissue holder that requires a patient to bend forward or down.

Results From Each of the Approaches Mr. Smith's fall when entered in the error reporting for accountability program would likely result in Nurse Drew being reprimanded and possibly sent to training, which would be a blow to her morale. Most important, no factors that contributed to the fall would be addressed because the concern about the incident was provider accountability—patients would continue to fall reaching for the tissue.

Information from the Artichoke Systems Approach for Mr. Smith's incident led not only to changes that reduced the likelihood of patient falls throughout the facility but also to avoiding compromising a patient's recovery—a safety enhancement that continues over time. The Artichoke Systems Approach to Mr. Smith's incident provides valuable lessons for enhanced patient safety at the current time and into the future.

RAMIFICATIONS

Error Prevention

The Artichoke Systems Approach advocates expanding the commonly expressed goal "if the health care industry is to improve patient safety, systems and processes must be designed to be more resistant to error occurrence and more accommodating to error consequence" (Spath, 2000, p. 202) to emphasize error prevention. Many efforts to address health care error are designed and implemented based on presumptions or experiences with other industries without systematically determining if those presumptions or experiences are valid for health care issues. Even more important, such activities are undertaken without identifying what it is that prompts behavior that leads to adverse outcomes. Because of this, energy and resources are expended with minimal if any diminution of the problem (HealthGrades Quality Study, 2004; MedPAC, 2004). This does not have to be.

Error reduction and ultimately prevention can be achieved by targeting efforts to error-provoking factors in the environmental context in which the error behavior occurred. Efforts are targeted to the factors that are the culprits in health care error and that are identified using the Incident Worksheet so the error-provoking qualities of those factors might be modified or removed. Thus, efforts are focused on identified rather than presumed causes of error. Recall the examples of Dr. Surgeon, who lacerated Mrs. Patient's liver, and Mr. Smith, Nurse Drew's patient who fell. The factors that contributed to those incidents were identified for correction—correction that would affect care providers and patients not only across similar environmental contexts but also into the future. It is to be noted that error prevention will at no time be universal; that is, error will occur not because people are innately prone to error as many propose often citing the first part of the literary quotation "To err is human, to forgive, divine" (Pope, 1711/2004) as support—such innate error proneness has yet to be empirically documented—but because factors in the context of care may not be in harmony with the characteristics, skills, and hard-wired capabilities of the care provider and as such induce error until identified and changed.

Expands Error-inducing Factors

Although reported in the error research literature of various industries (Moray, 1994; Rasmussen, 1982; Senders & Moray, 1991), findings that are the basis for the life space systems of the Artichoke System Approach, the overarching systems of factors represented by the outermost leaves of the Artichoke, the legal–regulatory–reimbursement– national culture factors, rarely are considered in discussions and research on health care error; consideration stops at the organization. The importance, indeed the critical importance, of going beyond the organization to the outermost system of leaves of the Artichoke is attested by experiences of the members of the Pittsburgh Regional Health Initiative (PRHI).

The PRHI is a patient safety collaboration that involves more than 40 hospitals in12 counties in southwestern Pennsylvania, hundreds of physicians, 4 major insurers covering 85% of the commercial market, 32 large and small business health care purchasers representing more than 200,000 workers and dependents, organized labor, and dozens of civic leaders. The PRHI has identified Barriers to Achieving Excellence. Those barriers are:

- Reimbursements that do not always reward the appropriate care for patients.
- Legislation such as Health Insurance Portability and Accountability Act.
- Multiple wasteful reporting systems that do not coordinate information or formulate it into useful insights.

The PRHI proposed that to achieve excellence it is necessary to remove barriers by "gathering information—real and perceived—to share with legislators, insurers, payers, and others in the maze of health care

TABLE 7–1 Each System Represents a Discipline(s), its Research, and Applications.

Patient – psychology, physiology
Means of providing care – human factors
Provider – psychology, physiology, human factors
Ambient – engineering
Physical – architecture, design
Social – sociology, anthropology, psychology
Organization – management, business
Legal, Regulatory, Reimbursement, National Culture – economics, cultural anthropology

requirements and ask for targeted improvements" (PRHI , 2004, p. 1). In other words, to achieve excellence it is necessary to address the factors of the outermost leaves of the Artichoke, the legal–regulatory–reimbursement–national culture factors. Thus, the Artichoke Systems Approach addresses the full range of real world error related factors.

Integrates Approaches to Health Care Error

This Artichoke Systems Approach is unlike other systems approaches because it addresses only factors that affect behavior by influencing the person through the systems of the Artichoke of that person's life space. The provider is the heart of the Artichoke as illustrated in Figure 7–2, so the consideration of error-inducing factors is the view from the heart of the Artichoke. This approach by addressing the life space of a care provider incorporates all approaches to health care error such as crisis resource management, just culture, and team training by addressing factors of those approaches that affect a provider's life space. This approach synthesizes the work of the other approaches to error at the point where an error occurs—the behavior of the care provider. The act of error is the source of adverse outcomes and as such must be the target of efforts to address such outcomes—the only way to accomplish that is by targeting factors that affect the person to perform such an act (see Bogner, 1994/2003a, 2003b 2000, 2002, for further discussion of this systems approach and its application to health care and the understanding of human error).

An Interdisciplinary Approach

The Artichoke Systems Approach also points to disciplines that not only can enrich the study of error, they also can contribute to effectively reducing the likelihood of error. Each of the categories of factors of the systems of the Artichoke has at least one field of study with a literature of research findings and methodologies as well as individuals with knowledge appropriate to address error-inducing factors identified in that specific system. Many people involved in heath care error study the literature of disciplines other than their own because that literature is relevant to their interest. People established in their careers such as health care providers cannot become as knowledgeable in other fields as people with careers in those fields. Rather than working to become an expert in another field, collaboration with those who are experts in the field can provide the synergy to effectively address error and enhance patient safety (Gopher, 2004). The designation of the disciplines to the specific systems in Table 7–1 is for the purpose of illustration; clearly some disciplines are appropriate to address error-inducing factors in several systems.

Examples of opportunities for interdisciplinary collaboration noted in Table 7–1 are provided by further consideration of laparoscopic surgical procedures as discussed earlier in this chapter as the case of Dr. Surgeon. Laparoscopic cholecystectomies have become the gold standard for gall bladder removal; however, error is not unknown in that procedure. Contributions to the investigation of an incident as well as studies of how to avoid error in performing that procedure could be enhanced by collaboration with a *physiologist* who could address the affect of the *physiology of the patient,* whether morbidly obese or extremely frail, on the position of the surgeon that is necessary to perform the procedure hence *physiology of the shoulders of the surgeon.* The positioning of the surgeon's shoulders (see Figure 7–7) causes muscle cramping that affects his or her manual dexterity and leads to error.

To explore how the means of providing care could be addressed to avoid the muscle cramping experienced by the surgeon, collaboration with a *human factors specialist* might identify ways to adjust *the means of providing care*, the devices involved in the surgical procedure, for short term error avoidance and recommend design modifications to the

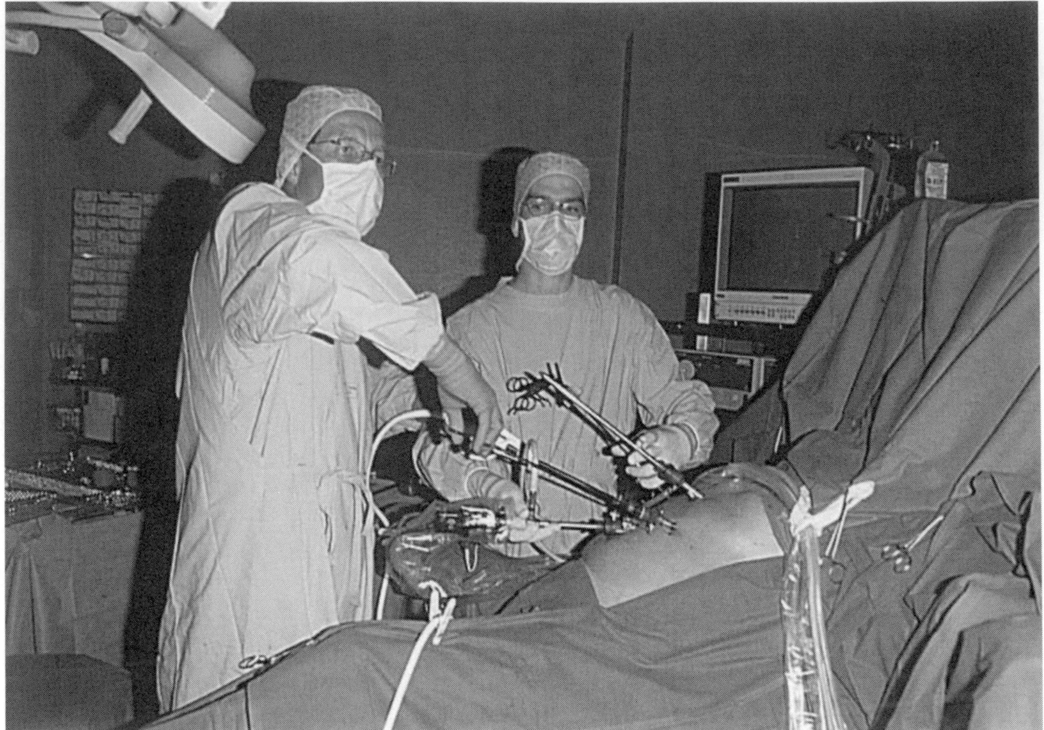

Figure 7–7. Laparoscopic surgery.

manufacturer to reduce the likelihood of error over time. Surgeons removing a gall bladder laparscopically have erroneously yet purposefully cut the common bile duct; *psychologists* can study that error as an indication of *perception of contiguous structures in two dimensional space that belie a three dimensional orientation of the structures.* The results of such a study could be incorporated in training for the short term and a target for modification of the video projection by industry to address that perceptual issue.

An *engineer* might be consulted to ascertain if ambient conditions such as *illumination* were sufficient so the surgeon could clearly see the image of the surgical field on the monitor. An *architect, engineer,* or *human factors specialist* could arrange the *placement of the monitor so it would be in line with the actual surgical field* rather that at an angle that makes it necessary for the surgeon to construe his or her view of the surgical field from the monitor to correspond with the actual surgical site (see Figure 7–7). The social issue of *lack of communication* among the care providers involved in the procedure could be addressed by a *sociologist, psychologist,* or *anthropologist.* The *time constraints* that drive the surgeon to complete the procedure quickly that could lead to an error although a seemingly management or business consideration of the organization reflects *reimbursement policies* that necessitate maximizing the number of procedures performed.

A PARADIGM FOR PATIENT SAFETY

The Artichoke Systems Approach by addressing factors in the life space of an individual involved in an error changes the paradigm from one that considers the person involved in an error as the sole cause of that error with remedial activities directed to that individual to the consideration of error as indicating the influence of contextual factors on health care providers that results in behavior with an adverse outcome. The paradigm changes from addressing "who" and "what" to "why." The change is from attributing the cause of error to a person with little if any consideration of what actually affects that individual, from interpreting, indeed projecting the causes of behavior onto a person to acknowledging that the individual involved in an error is the one whose life space was influenced by the factors, the person who experienced the factors affecting his or her behavior, and as such is the legitimate source of information about those factors. To effectively enhance patient safety, it is time, indeed past time, to change the paradigm for studying health care error to one that identifies why an error occurs, a paradigm that "provides information that can lead to improved safety" as represented by the Artichoke Systems Approach.

References

Akutagawa, R. (1952). *Rashomon.* Singapore: Berkeley.

Billings, C. E. (1997). *Aviation automation: The search for a human-centered approach.* Mahwah, NJ: Lawrence Erlbaum Associates, Inc.

Bogner, M. S. (1994). *Human error in medicine.* Mahwah, NJ: Lawrence Erlbaum Associates, Inc.

Bogner, M. S. (2000). A systems approach to medical error. In C. Vincent & B. DeMol (Eds.), *Safety in medicine* (pp. 83–100). Amsterdam: Pergamon.

Bogner, M. S. (2002). Stretching the search for the "why" of error: The systems approach. *Journal of Clinical Engineering, 27*,110–115.

Bogner, M. S. (2003a). *Misadventures in health care: Inside stories.* Mahwah, NJ: Lawrence Erlbaum Associates, Inc.

Bogner, M. S. (2003b). Understanding human error. In M. S. Bogner (Ed.), *Misadventures in health care: Inside stories* (pp. 41–58). Mahwah, NJ: Lawrence Erlbaum Associates, Inc.

Bosk, C. (1979). *Forgive and remember: Managing medical failures.* Chicago: University of Chicago Press.

Chapanis, A. (1996). Perspective: Musings from a hospital bed. *Ergonomics in Design, 6,* 35–36.

CNN.com. (2004, July 28). Study: Hospital errors cause 195,000 deaths. Retrieved June 1, 2006, from http://www.CNN.com.

Cohen, M. M., Kimmel, N. L., Benage, M. K., Hoag, C. C., Burroughs, T. B., & Roth, C. A.(2004 August). Implementing a hospital-wide safety program for cultural change. *Joint Commission Journal of Quality and Safety, 30,* 424–431.

Flanagan, J. C. (1954). The critical incident technique. Psy*chological Bulletin, 51*, 327–358.

Gibson, R., & Singh, J. P. (2003). *Wall of silence: The untold story of the medical mistakes that kill and injure millions of Americans.* Washington, DC: Lifeline.

Gold, M. (1999). Cassirer's philosophy of science and the social sciences. In M. Gold (Ed.), *The complete social scientist* (pp. 23–37). Washington, DC: American Psychological Association.

Gopher, D. (2004). Why it is not sufficient to study errors and incidents: Human factors and safety in medical systems. *Biomedical instrumentation & technology, 38*(5), 387–409.

HealthGrades Quality Study. (2004). *Patient safety in American hospitals.* Retrieved June 1, 2006, from http://www.healthgrades.com.

Kelly, G. A. (1955). *A theory of personality.* New York: Norton.

Kohn, L. T., Corrigan, J. M., & Donaldson, M.S. (Eds.). (1999). *To err is human: Building a safer health system.* Washington, DC: National Academy Press.

Leape, L. L. (1994). The preventability of medical injury. In M. S. Bogner (Ed.), *Human error in medicine* (pp. 13–26). Hillsdale, NJ: Lawrence Erlbaum Associates, Inc.

Leape, L.L., et al. (1995). Systems analysis of adverse drug events. *Journal of the American Medical Association, 274,* 35-43.

Lewin, K. (1964a). Behavior and development as a function of the total situation. In D. Cartwright (Ed.), *Field theory in social science* (238–303). New York: Harper & Row. (Original work published 1946)

Lewin, K. (1964b). Defining the "field at a given time." In D. Cartwright (Ed.), *Field theory in social science*(pp. 43–59). New York: Harper & Row. (Original work published 1943)

Lewin, K. (1966). *Principles of topological psychology.* New York: McGraw-Hill. (Original work published 1936)

Marrow, A. J. (1969). *The practical theorist: The life and work of Kurt Lewin.* New York: Basic Books.

Matern, U. (2003). The laparoscopic surgeon's posture. In M. S. Bogner (Ed.), *Misadventures in health care: Inside stories* (pp. 41–58). Mahwah, NJ: Lawrence Erlbaum Associates, Inc.

MedPAC. (2004, March). *Quality of care for Medicare beneficiaries. Report to the Congress: Medicare Payment Policy. Retrieved June 1, 2006, from http://www.medpac.gov.*

Moray, N. (1994). Error reduction as a systems problem. In M. S. Bogner (Ed.), *Human error in medicine* (pp. 67–92). Hillsdale, NJ: Lawrence Erlbaum Associates, Inc.

Pittsburgh Regional Healthcare Initiative. (2004, June). *PRHI Executive Summary.* Retrieved June 1, 2006; from http://www.prhi.org

Pope, A. (2004). *An essay on criticism.* Whitefish, MT: Kessinger. (Original work published 1711)

Rasmussen, J. (1982). Human errors: A taxonomy for describing human malfunction in industrial installations. *Journal of Occupational Accidents, 4,* 311–333.

Rasmussen, J. (1990). Human error and the problem of causality in analysis of accidents. *Philosophical Transactions of the Royal Society of London, 337*, 449–462.

Reason, J. (1990). *Human error.* New York: Cambridge University Press.

Reynard, W. D., Billings, C. E., Cheaney, E. S., & Hardy, R. (1986). *The development of the NASA Aviation Safety Reporting System* (NASA Ref. Publ. 1114). Washington, DC: National Aeronautics and Space Administration Scientific and Technical Information Branch.

Robeznieks, A. (2005, January 24). Data entry is a top cause of medication errors. *AMNews* June 1, 2006, from http://www.amednews.com

Senders, J. W., & Moray, N. P. (1991). *Human error: Cause, prediction, and reduction.* Mahwah, NJ: Lawrence Erlbaum Associates, Inc.

Short, T. G., O'Regan, A., Lew, J., & Oh, T. E. (1992). Critical incident reporting in an anaesthetic department quality assurance program. *Anesthesia, 47*, 3–7.

Spath, P. L. (2000). Reducing errors through work system improvements. In P. L. Spath (Ed.), *Error reduction in health care: A systems approach to improving patient safety* (pp. 199–204). Chicago: Health Forum.

Spies, T. H., Mokkink, H. G. A., Robbe, P. F. DeV., & Grol, R. T. (2004). Which data source in clinical performance assessment?

A pilot study comparing self-recording with patient records and observation. *International Journal for Quality in Health Care, 16*(1), pp. 65–72.

U.S. Pharmacopeia. (1995). Computers: Errors in—errors out. *U.S. Pharmacopeia Quality Review, 48*. Retrieved June 1, 2006, from http://www.usp.org/patientsafety/Newsletters/USP Quality Review/Archives/No.48

U.S. Pharmacopeia. (2004). Too much Similarity. *U.S. Pharmacopeia Quality Review, 78*. Retrieved June 1, 2006, from http://www.usp.org/patientsafety/Newsletters/USPQuality Review/Archives/No.78

von Bertalanffy, L. (1968). *General system theory: Foundations, development, applications*. New York: George Brazilier.

Zhan, C., & Miller, M. R. (2003). Excess length of stay, changes, and mortality attributable to injuries during hospitalization. *Journal of the American Medical Association, 290*(14), 1868–1874.

8

ORGANIZATIONAL LEARNING IN HEALTH CARE

Ann Schoofs Hundt
University of Wisconsin, Madison

Learning is a notion not at all foreign to health care. Individual practitioners, and practitioners as a whole through their respective specialties, continually attempt to improve the means of providing care to patients albeit technically, technologically, and/or interpersonally. **Organizational** learning, however, poses challenges that require a willingness to change when and where change is needed. The "organizational" aspect also requires developing shared mental models, a well-articulated mission, and a willingness to surrender personal goals in lieu of the organization's goals. This must continuously occur between all providers and managers working within an organization despite the fact that these individuals may not be accustomed to working together. These groups may, in fact, represent various cultures within the organization as suggested by Carroll and Quijada (see chap. 45, this volume).

Organizational learning requires that those working within health care organizations understand that they must function as a system comprised of numerous, sometimes complicated, interdependent parts. And to the extent possible, they must understand this system. Those providing care, because of the type and extent of training they receive as well as their status in society, frequently practice in a fashion that makes communication and teamwork and other aspects of effectively functioning systems a challenge. The varying philosophical underpinnings that promote individual accountability as suggested by Smith and Bartell (chap. 9, this volume) can also present barriers to understanding organizations as systems and, ultimately, organizational learning.

Tremendous progress has been made over the years in providing better health care as well as more options to care. Providers have, in turn, generally been receptive to adopting proven alternatives that result in better patient outcomes. This poses a common indisputable goal for all working in health care, regardless of their role, be they housekeepers, technicians, surgeons, administrators, or nurses (Bohmer & Edmondson, 2001): Every patient should be provided the best possible care. From this, we suggest that the commitment to individual learning, and in some instances group learning, that already exists in health care can serve as a platform from which management should espouse and promote organizational learning.

This is especially important today as health care undergoes greater scrutiny than ever by individuals receiving care, those paying for care, agencies accrediting health care organizations, and government legislating health care. *Clinical microsystems* in health care organizations that espouse learning have been shown to (a) produce better outcomes through ongoing measurement and process improvement, (b) exhibit greater efficiency (with less waste), (c) utilize varying levels of information technology to promote ongoing communication, (d) function as a team to manage their patients, (e) maintain a focus on both staff and the patient, (f) have considerable top management support and autonomy, and (g) display a strong culture (Nelson, Batalden, Huber, Mohr, Godfrey, et al., 2002). A valid question to raise here is if learning can accomplish such significant results at a lower (microsystem) level, how can it *not* accomplish equally—or more—significant results at an organizational level? What must be done to achieve learning

TABLE 8–1

Researchers	*Tenets*
Cyert & March, 1963	"behavioral theory of a firm"; organizations are adaptive; learning occurs from experience; organizations have short-term focus
Cangelosi & Dill, 1965	"adaptation" (from Cyert & March) results from stress; learning is sporadic & stepwise; failure leads to change; need to consider individual learning
Argyris & Schon, 1978	"theory of action"; learning must be communicated; introduce "single- & double-loop learning"
Duncan & Weiss, 1979	focus on knowledge knowledge must be communicated
Shrivastava, 1983	pointed out fragmented and multidisciplinary approach to organizational learning to date; proposed two-dimensional typology based on origin and the "how" of learning
Fiol & Lyles, 1985	change does not imply learning; learning results in improved performance; need balance between stability & change; four contextual factors to organizational learning: culture, strategy, structure, environment
Leavitt & March, 1988	learning is: based on experience, history-dependent, attempts to meet targets
Senge, 1990	learning organization; leadership is key
Simon, 1991	individual learning in organizations is a "social not solitary phenomenon"
Crossan, et al, 1999	"4I"s – intuiting, interpreting, integrating and institutionalizing – progression of learning in organizations

at a higher level? In other words, how do we achieve organizational learning in health care?

In this chapter, we will review the concept of organizational learning and aspects of it that have been studied in various areas of research. We review a selection of papers that are considered seminal works, frequently cited or especially pertinent to the topics we develop. A summary of the discussion is found in Table 8–1. We will also present an overview on organizational learning in health care as well as a health care case study that exemplifies various aspects of organizational learning. More thorough reviews of the full body of research on organizational learning exist elsewhere (Crossan & Guatto, 1996; Crossan, Lane, White, & Djurfeldt 1995; Huber, 1991; Leavitt & March, 1988; Popper & Lipshitz, 1998; Shrivastava, 1983; Wang & Ahmed, 2003).

ORGANIZATIONAL LEARNING RESEARCH—PAST AND PRESENT

One challenge to organizational learning, be it in health care or any other industry, is determining both how to define it and how to characterize it. Considerable work has been published that presents both conflicting reviews of work in the area as well as complementary discussions of views presented. Cyert and March (1963) are generally credited, through their presentation of the *behavioral theory of the firm*, as being among the first to introduce the notion of organizational learning. Since then, others have presented varying definitions and conceptual frameworks that describe their perspective on organizational learning. Some of these follow.

Some Early Research

In their early work, Cyert and March (1963) presented organizational learning as one of the four major underpinnings of business decision making (p. 116). They characterize organizations as being adaptive (much like individuals) that respond to changes, both internal and external, to their environment. They posit that learning results from an organization's experience and that, given the short-term focus in most organizations, current and pressing issues are in turn most frequently and easily addressed. Preference is given to routine actions, standard operating procedures, and similar existing structures within the organization when considering change.

Shortly afterward, Cangelosi and Dill (1965) presented an early experimental case study based on premises of Cyert and March (1963) in which graduate students were given a management simulation project that required them to develop an organizational structure for their assigned detergent company along hierarchical functionally specialized lines. Findings suggested that (a) adaptation results from stress (frequently due to an unanticipated negative outcome), (b) learning is "sporadic and stepwise rather than continuous and gradual" (Cangelosi & Dill, 1965, p. 203), and (c) determining what to learn requires an understanding of how to learn. They also found in their experiment that failure invariably led to change. In conjunction with the concept of a system, Cangelosi and Dill suggested that there are three types of stress: those stimulating subsystem (e.g., individual and/or group) learning, those stimulating total-system learning, and those stimulating both types of learning. As a result of their work they proposed the need to (a) evaluate organizational learning in light of individual learning (or an interaction of the two), (b) identify facets of an organization—primarily its environment—that define the necessary learning tasks and the ability required to achieve them, and (c) determine when learning actually occurs as well as what is learned.

Argyris and Schon's (1978) seminal work presented organizational learning through a *theory of action* framework. Explaining their theory in light of systems terminology, individual workers or groups of workers receive *inputs* that may or may not coincide with their understanding of the system's premises, structure, or strategic plans. These inputs are then *processed* by the workers that in turn may cause them to respond to the input by either changing or maintaining the way they do things to conform to the system's requirements. As a result, any response, or *output* (either affirming or negative), must be communicated and then stored in the organization's memory to ensure that the organization learned from the experience. If this does not occur, Argyris and Schon pointed out that "the individual will have learned but the organization will not have done so" (p. 19). This notion is later described by Shrivastava (1983) as *assumption sharing*. Like many following them, Argyris and Schon (1978) placed great emphasis on the relationship between individual and organizational learning: "Just as individuals are the agents of organizational action, so they are the agents for organizational learning" (p. 19). We will continue to discuss this concept of individual and organizational learning throughout this chapter.

Another concept Argyris and Schon (1978) presented in their work is that of single-loop and double-loop learning. *Single-loop learning* occurs within the constraints of the existing organizational structure, policies, and procedures. The issue (in this case the learning opportunity) in question undergoes minimal in-depth scrutiny despite the fact that the best solution may require a significant change to the way in which the organization or unit operates. This type of learning is generally less dramatic and frequently does not result in organizational learning. It coincides with what March (1991) later termed *exploitation*. Conversely the inherent nature of *double-loop learning,* which requires deep scrutiny of organizational norms, operations, and strategies prior to making a decision almost universally results in organizational learning. Conflict frequently occurs in double-loop learning due to the scrutiny involved. March referred to double-loop learning as *exploration* due to the creativity and risk taking involved in solving problems and, ultimately, learning.

Other early work conducted by Duncan and Weiss (1979) recognized the power that organizational learning offered in helping researchers understand how organizations and the people in them function over time. They defined organizational learning as "the process within the organization by which **knowledge** about action– outcome relationships and the effect of the environment on these relationships is developed" (p. 84). They pointed out, however, that ideally this knowledge should produce a change in behavior if a change is appropriate or necessary and that access to and use of this knowledge is central to organizational learning. Like Argyris and Schon (1978), Duncan and Weiss (1979) stated that regardless of its acceptance and change (or not) this knowledge must be communicated to achieve a *system of learning* (p. 89).

Synthesis and Expansion of Organizational Learning Concepts

The fact that the research was "fragmented and multidisciplinary" was noted by Shrivastava (1983, p. 9) in his review of past organizational learning research. On review of the work already summarized here, he found the concepts to primarily be complementary but in want of empirical research. This led him to propose a two-dimensional typology on organizational learning. One dimension spanned the *origin* of learning: individual–organizational (resulting in three components accounted for by a "middle" element); the other spanned the *how* of learning: evolutionary–design (based on how intentional vs. spontaneous the learning is). As a result, this model offered six categories of organizational learning. Interestingly the origin concept continues to undergo considerable discussion whereas the how facets have received little attention.

When distinguishing between learning and adaptation Fiol and Lyles (1985) suggested that change alone does not imply learning. They stated that no consensus existed surrounding the definition of organizational learning, at least in part because various divergent bodies of research assessed different aspects of the concept. They pointed out that general agreement existed around the notion that learning resulted in improved performance and in turn proposed a definition for organizational learning incorporating their belief: "the process of improving actions through better knowledge and understanding" (p. 803). They (like others such as Argyris & Schon, 1978; Duncan & Weiss, 1979) emphasized that knowledge, when communicated properly, resulted in organizational learning that is more than the sum of individuals' (within the organization) learning. They also asserted that organizations require a balance between the change versus the stability associated with learning.

Fiol and Lyles' (1985) synthesis of past work resulted in proposing that four contextual factors affect the likelihood that (organizational) learning occurs. This included a *culture* that represented shared beliefs, ideologies, and norms that affect an organization's actions; *strategy* that drives organizational goals and objectives by providing "bounds and context" for decision making as well as "momentum" to learn (p. 805); *structure* within which decisions are made that may vary by or within the unit (i.e., limited vs. extensive) depending on the context of the decision; and *environments*, both internal and external to the organization having varying degrees of complexity and change. This coincides with Senge (1990a, 1990b) and other learning organization researchers whom we will discuss later in this chapter.

Leavitt and March (1988) reviewed the organizational learning literature and summarized it by building on behavioral studies of organizations. Using this framework, they stated that behavior (i.e., learning) in organizations is based on *experience*, either direct or interpretive, and sometimes results in drawing a conclusion from a limited number of experiences (referred to as *superstitious learning*); is *history-dependent* and therefore draws from memory both internal and external to the organization; and is geared to meeting *targets* that requires a form of intelligence meaning that organizations can learn from their own limitations and/or mistakes. They also build on the diffusion of innovation literature (Rogers, 2003) to suggest how the transfer of knowledge from one organization to the next occurs (Leavitt & March, 1988).

Organization Science devoted its first issue of 1991 to the topic of organizational learning. In this issue, March, Sproull, and Tamuz presented their seminal work on how organizations can learn from samples of one or fewer. Both error and high reliability literature also build on this notion where catastrophic incidents are relatively rare (e.g., aviation, nuclear power, military, etc.), yet when an event of grave consequence occurs the entire industry responds. In addition, they point out the value of and need to review historical events to promote organizational learning. Similar events should be recognized as data points rather than detailed stories. This work supports the careful review of near misses and hypothetical consequences of actions as identified through prospective analysis (Krouwer, 2004; Solomon & Petosa, 2001) relatively recently introduced to health care. (Interestingly, this paper was reprinted 12 years later; March, Sproull, & Tamuz, 2003.) Numerous other philosophical discussions are included in this issue of *Organization Science* and include March's (1991) discussion of exploration and exploitation as presented here previously and Weick's (1991) presentation of strategies for conducting organizational learning research by either retaining its traditional definition based primarily on the concept of *stimulus–response* and thus focusing on behavior or replacing it with a definition based more on knowledge acquisition that does not require a change in behavior.

Learning Organizations

The concept of learning organizations as introduced and most popularly discussed by Peter Senge (1990a, 1990b) brought the notion of organizational learning to a more functional level for managers. Learning organizations offer a way of managing an organization by espousing and implementing rather clear, although not simple, tenets. The distinction between learning organizations and organizational learning stems from the cognitive basis for adaptation and improvement that exists in learning organization research as explained through the requirement that shared mental models are developed much like the "stated and known shared paradigm" proposed by

Duncan and Weiss (1979) and that organizations must be looked at as systems (Edmondson & Moingeon, 1998). Learning organizations offer yet another perspective on organizational learning.

Senge posited that **leadership** is the distinguishing factor affecting whether or not a learning organization can be achieved. The leader is a *designer* who clearly and consistently states the organization's purpose and priorities, a *teacher* to help all in the organization (leader included) develop clear views of "current reality" that result in shared mental models (Senge, 1990b), and *steward* of those they lead and those they serve as defined by the organization's mission. When performing these roles the leader helps build shared vision; identify, challenge, and test mental models (much like double-loop learning as defined by Argyris & Schon, 1978); and promote systems thinking by motivating everyone in the organization to recognize and understand the big picture and their respective contribution to it. The leader is responsible for developing an environment that promotes and facilitates learning, and then ensures that idea (information) exchange occurs within and between all levels of the organization (Garvin, 1993).

Given these premises we suggest the leaders serve as the conduits between top-level managers, individuals functioning within the organization, and all others. Considerable time and effort is required to achieve effective communication between everyone functioning at varying levels within the organization, yet the gates must be open in all directions for learning to occur. Looking at learning in this manner we believe learning can be both—and at other times either—individual and/or organizational. Regardless, leaders must recognize that "whichever direction ideas flow through the organization, it is clear that nothing will happen unless they do flow" (Simon, 1991, p. 131).

Individual and Organizational Learning

Building on the tenets of learning organizations, Stata (1989) provided a useful distinction between individual and organizational learning. He suggested that unlike individual learning, organizational learning requires shared insights, knowledge, and mental models among all major decision makers but cautions that learning occurs at the rate at which the "slowest link learns" (p. 64). He also emphasized that learning builds on previous knowledge and experience that together constitute organizational memory. He saw systems thinking as the mechanism by which both individual and organizational learning occur. This is further developed by Kim (1993) who offered two components to his definition of learning: an operational and a conceptual aspect—the "know-how" and "know-why" of learning, respectively (p. 38). He stated that a balance between the two must exist for organizational learning to occur and that the **transfer** of learning "is at the heart of organizational learning" (p. 37).

Simon (1991) went so far as to suggest that **individual** learning in organizations does not occur at the one-person level. He sees individual learning in organizations as a "social, not a solitary, phenomenon" (p. 125). The transfer of information from one individual (or group) to another individual (or group) is critical to internal learning and once internal learning is part of the organization's memory it becomes organizational learning. These ideas coincide with his view of organizations as being systems comprised of interrelated roles.

More recent work by Crossan and her group (Crossan, Lane, & White, 1999; Vera & Crossan, 2004) offered a means of learning in organizations that progresses through a "4I" cycle comprised of four processes: intuiting, interpreting, integrating, and institutionalizing, similar to the framework proposed by Huber (1991). This research offers a link between the individual, the group or team, and the organization levels and explains the manner in which learning occurs, either (a) through a feedback mechanism (much like March's, 1991, exploitation), which promotes learning based on current organizational procedures, structures, and norms, or (b) a feed forward mechanism (exploration) in which individual and/or group learning affects the organizational level in a manner that may require scrutiny and subsequent changes within the organization, sometimes at the highest levels. Others build on this work by introducing the notion of politics in organizations (Lawrence, Mauws, Dyke, & Kleysen., 2005).

ORGANIZATIONAL LEARNING RESEARCH APPLIED TO HEALTH CARE

Interestingly, chapter 1, paragraph 1 of Argyris and Schon's (1978) seminal work reads:

> There has probably never been a time in our history when members, managers, and students of organizations were so united on the importance of organizational learning. Costs of **health care** … have risen precipitously and we urge agencies concerned with

> these services to learn to increase their productivity and efficiency. (p. 8)

Little did they know that costs were to be one of many issues (that now include quality and safety, among others) that now draws even more public attention and scrutiny to health care than existed in the 1970s (Kohn, Corrigan, & Donaldson, 2000). Unfortunately, relatively little has been accomplished in health care to promote or achieve organizational learning. Leaders, known to be key to the change process (Senge, 1990b), must express a willingness to change and lead it; this must be more than lip service. One way to accomplish this, that builds on values all in health care espouse, is to promote organizational learning as a way of both improving the system in which providers function (in Donabedian's, 1978, terms both the structure and processes) and then achieving the outcomes they so passionately desire. In other words, we need to mine the "know what" and "know how" and communicate findings throughout the organization. Simon's (1991) "flow of ideas" must occur.

Organizational Structure Considerations and Levels of Learning

"Human beings are naturally programmed to learn, organizations are not" (Carroll & Edmondson, 2002). As we know health care organizations are collections of individuals who sometimes work alone and other times work in groups or teams that are generally aligned by their department or the service provided. There is a fluid nature to the work of those in health care that requires constant adaptation. In addition, some of those practicing in organizations (i.e., the physicians) are generally not employees of it. This poses even greater challenges to building administrative structures and strategic management that promotes open dialog between mutually exclusive entities. Given these realities, how can an organization be managed to promote change and learning while achieving better health outcomes for patients and staff alike?

As stated previously, a major strength of all who work in health care is the fact that they are in it for the patient. The current scientific environment affords patients and providers alike continually changing and continuously improving health services. At the research level, the notion of improvement is generally clearly defined, and once achieved there are many ways of communicating newly discovered or developed practices. Professional societies, peers, and manufacturers offer three avenues for communicating new information to their audiences. It is then up to the respective individual or groups to introduce new practices to their peers and/or organization. Of course, the notion of change or learning of this sort must first be encouraged in an organization.

An example of this is discussed in various papers by Edmondson and others (Edmondson, 2003; Edmondson, Bohmer, & Pisano, 2001; Pisano, Bohmer, & Edmondson, 2001) who studied the introduction of minimally invasive cardiac surgery in sixteen hospitals. This technology required considerable changes to the routines of those working on the cardiac surgery team within the operating room. Technically the hospitals studied were similar (all were among top-ranked cardiac surgery programs) but they experienced widely varying levels of success when implementing the new technology. From this study, the researchers found that the extent of success depended heavily on factors related to individual and organizational learning, including: (a) the manner in which team members were selected, their roles defined, and learning planned; (b) the extent of practice the team underwent and the willingness of its members to adapt to changes required in their respective practices; (c) an openness by the team leader to request and obtain input and accept and promote changes to standard operating procedures; and (d) a willingness of the team to review data and the team's perceived performance and make improvements when deemed appropriate.

In the successful implementations, the research team saw double-looped learning within the team due to the fact that individuals and the group as a whole were both willing to overlook past standard operating procedures and create an environment for new practice and learning. Likewise there was ongoing multidirectional **communication** within the team and an ongoing desire to **improve** and gain more **knowledge**. This example presents a success story beyond local learning (Carroll & Edmondson, 2002). It goes beyond March's (1991) exploitation to his exploration and is similar to the performance and improvements noted in clinical microsystems (Mohr & Batalden, 2002; Nelson et al., 2002).

At a different level of analysis, Tucker reported on cases of individual nurses solving problems—opportunities for learning—they incurred in their daily practice (Tucker, 2004; Tucker & Edmondson,

2002; Tucker, Edmondson, & Spear, 2002) In nearly every case observed, the respective nurse dealt with his job obstacle as a unique occurrence, never saw it in light of the system in which he was practicing, and rarely informed his manager of the event nor any pattern of events he experienced or observed. An opportunity for organizational learning was lost in nearly every one of the cases. Why? Here we do not see individuals recognizing that they work within a system comprised of highly interdependent parts. In addition, there does not appear to be a culture that promotes individual input in generating opportunities for system improvements.

Learning (or Not) From Failure

Here we introduce the concept of learning from failure. Although Cangelosi and Dill (1965) found failure to consistently result in learning in their case study, this was not supported in other research conducted in health care (Edmondson, 1996; Tucker & Edmondson, 2003). In one study (Edmondson, 1996), the detection of drug errors, subsequent reporting of them, and the likelihood that learning would occur from errors varied both between and within two hospitals. The primary factor that most affected the identification of errors was the respective unit's willingness to openly discuss mistakes. Although the researchers could only speculate that more open units with accessible nurse managers were likely to perform better (i.e., less medication errors), they were able to prove that detection and reporting was a function of the unit's leadership.

In the Tucker and Edmondson (2003) study of 26 nurses at nine hospitals of varying size and patient population, 196 "failures" were observed. Of these, the vast majority (86%) of the failures were deemed problems rather than errors. Focusing on the problems, the research team investigated the type of problem solving that occurred in each case. Coined *first-order problem solving*, the team found that most frequently (93% of the time) the nurses dealt with the problems on their own by simply obtaining the missing information or supply and then proceeding with their work. This form of problem solving is considered counterproductive due to the fact that little if any communication occurs because the worker fails to report the event that could otherwise result in a system improvement or an opportunity for learning. In isolated problem solving instances *second-order problem solving* occurred when, after the worker addressed the problem, she then communicated it to the individual or department responsible and shared the problem with her manager who occupied a role that could affect analysis and possible change.

The more common first-order problem solving inhibited an opportunity for organizational learning to occur. Kim (1993) referred to this as situational learning that some might not even consider learning. Tucker and Edmondson (2003) suggested three reasons why first-order problem solving occurred: (a) there is an emphasis on individual vigilance in health care—each provider is personally responsible for solving his or her own problems; (b) concerns of unit efficiency interfere with encouraging workers to perform outside of the realm of what is perceived as their job—taking care of patients; and, finally, (c) more often today, health care workers have been empowered to solve problems because front line managers frequently are removed from the local aspect of their job and therefore are inaccessible to their staff. They are dealing with issues at a different level and have lost touch with the issues that individuals who report to them incur. As demonstrated here opportunities for organizational learning that arise from failure are lost.

Recently Chuang and others (Berta & Baker, 2004; Chuang, Soberman-Ginsburg, & Berta, 2004) posited that varying factors affect the extent of responsiveness to patient safety issues within an organization, which in turn affect the extent of organizational learning that occurs. They suggested event-specific, group, organizational, and interorganizational factors that influence an organization's willingness and ability to respond to adverse events and then, possibly, learn from them. Although not yet tested, it poses an interesting framework when assessing both failure and learning in organizations.

So, how can we learn from failure? How can opportunities for organizational learning be recognized and fostered? Much of the current literature supports the overwhelmingly influential role of leadership. We briefly discuss this in the next section.

Leadership

"Leadership must be distributed broadly if organizations are to increase their capacity for learning and change" (Carroll & Edmondson, 2002, p. 54). Much of the focus of patient safety and organizational learning in current health care literature points to the need for leadership to promote and instill learning within the organization at all levels and to then communicate the learning that occurred throughout the organization (Batalden & Splaine, 2002; Batalden, Stevens, & Kizer, 2002; Mohr, Abelson, & Barach, 2002). Care must be taken, however, because simultaneous learning does not require interdependence (Leavitt & March, 1988). Therefore learning must be managed (Bohmer & Edmondson, 2001). Likewise, learning must coincide with and complement an organization's mission. Change, when deemed appropriate and necessary, must be allowed to occur, yet there must be a subtle balance between the need for improvement and change versus the need for routine and constancy.

One aspect of the continuous improvement movement that was also emphasized by the Institute of Medicine in its well-known *To Err Is Human* (Kohn et al., 2000) is the notion of health care as a system. In many instances, errors occur due to the nature of the system and must be addressed, resolved, learned from, and then communicated as such—system-wide. Top down (feedback) and bottom up (feed forward) learning must occur to achieve organizational learning (Crossan et al., 1999; Vera & Crossan, 2004). The politics of organizations must be recognized but must not interfere with the organization's ability to learn (Lawrence et al., 2005). Nor can politics interfere with the need for all within the organization to develop shared vision and mental models.

Likewise, we must learn from the science of training (Baker, Salas, Barach, Battles, & King, chap. 17, this volume; Salas, Wilson-Donnelly, Sims, Burke, & Priest, chap. 44, this volume) to capitalize on opportunities for learning and promote learning through proven training methods. All too often learning occurs but the communication and ongoing monitoring of what changed as a result of learning does not. In these instances the knowledge gained never becomes a part of the organization's memory. Varying competencies and priorities obviously exist by those working in health care organizations (ranging from managerial to technical to interpersonal to clerical) , but the ultimate beneficiary of services—the patient and his or her family—must not be lost sight of.

The aforementioned can be summarized as the combination of structure, values, and skills that Carroll and Edmondson (2002) suggested enhance organizational learning. Amid the countless priorities of the future of health care one cannot overlook the influence, value, and power organizational learning offer health care organizations, those providing care and services, and, most important, those receiving the services.

CASE STUDY

We conclude this chapter by presenting a case study that offers a health care organization that fully espouses organizational learning. We do not intend to oversimplify the challenges associated with organizational learning yet suggest that by simplifying and clearly stating the organizations' priorities and missions, communicating effectively between all levels and individuals within the organization, sharing mental models which guide decision making throughout the organization, recognizing the value of everyone in organization and their importance in the system, promoting ongoing learning (and communication of its results) at all levels, and recognizing when change is necessary and appropriate will all make a difference in an organization's ability to achieve organizational learning.

Valley Hospital is a 350-bed midwestern hospital formally affiliated with a large tertiary care facility—Lincoln Hospital—located less than 100 miles away from Valley. The current hospital leadership has been in place for the past 6 years and inherited an organization having a long-standing reputation for providing high quality care to area residents. Individuals who choose to not receive care at Valley Hospital (local population: approximately 47,000) generally go to Lincoln Hospital because the nature of their condition can be better managed there where oncologic, endocrine, orthopedic, and transplant services are all ranked nationally. Valley physicians freely refer these patients and any patients requesting or requiring a second opinion to Lincoln Hospital medical staff.

The mission of Valley has been clearly expressed since its inception 75 years ago when the hospital was founded by a group of visionary managers with the support of 25 well-respected community physicians. Offspring of some of these physicians still practice in the community. "Patients first and foremost" serves as the basis for every management decision made. Although the organization espouses innovation, it does not consider itself an early innovator (Rogers, 2003) because management believes that waiting a few years to introduce new technology not only results in more highly refined equipment and service but also allows the organization to choose from a wider array of suppliers.

The management team at Valley is well positioned. The hospital president is well respected by everyone in the organization. He makes a concerted effort to walk the halls of the hospital at least once a week on each shift. His undergraduate degree is in nursing, which he practiced for 3 years prior to attending graduate school and receiving a masters degree in health administration. Subsequently he served as an assistant administrator at a university hospital on the east coast and was promoted within the organization on a number of occasions. At the age of 41 he received a doctorate in business with an emphasis in organizational behavior. The following year he was offered his current role at Valley. The chief operating officer is a visionary who enjoys challenging the opinions of the rest of top management. His role, whether intentional or not, is to question every major decision the organization makes, regardless of who is involved. The chief information officer (CIO) has formal education and experience in both IT and finance. She is well liked by everyone she works with. The organization's recent implementation of computerized provider order entry (CPOE) occurred over a 12-month period and was deemed a success due to the commitment to preparing everyone affected by the technology, training them, and monitoring their use. The chief financial officer (CFO) complements the CIO and president. He is a reserved individual and a creative but deliberate thinker. Managers in the hospital are afforded a significant degree of autonomy by top management but are also offered support both by those they report to as well as their peers on an ongoing basis. Leadership training is emphasized through ongoing education and staff development activities offered on site as well as off-site coursework and conferences.

Valley has survived the health care reimbursement environment fairly well. The indigent population in Valley is less than 5% of the area population. As members of the medical community retire, many of them volunteer 3 days a month at a free clinic that offers numerous ongoing wellness and health screening programs to needy area residents. This clinic is administratively supported by the hospital.

Valley offers strong cardiac services, both medical and surgical. Their cardiac patients are known to achieve high levels of quality of life after completing their cardiac program at Valley. Other general surgery, medicine, obstetric, and pediatric services provide high quality care and, as documented through ongoing peer review, achieve good outcomes as well.

Those who work at Valley, regardless of their role, tend to have significant longevity with the organization. Many attribute this to the numerous employee-focused wellness and family-centered programs and activities Valley offers, as well as a stable community in which they live. Likewise optimal working conditions and quality of working life issues are regularly addressed and continuously improved. Community employers also provide a healthy mix of jobs.

Valley Hospital is unconventional in many ways. One is that its management structure promotes multidirectional communication. Considerable resources are invested in staff regardless of their role in the organization. Three monthly meetings are held (one for each shift) that serve to update staff on impending changes, events, and opportunities for learning and also to obtain feedback from staff. Recognizing that attendance for many is difficult, the meeting is recorded and also played live on a hospital employee cable TV network. Televisions are readily accessible in staff break rooms throughout the building. Live internet access is also available.

Likewise, there is considerable investment made in staff development and employee health programs. This serves as a strong recruitment tool. When individuals interview at the organization, they are introduced to current staff who fill similar roles as the person being recruited to answer questions of any nature. Various employee health programs are offered at the beginning and end of work shifts as well as during the day. Short relaxation programs are designed to fit breaks.

Two years ago a decision was made to replace the infusion pumps used on the patient care units as well as those used in the operating room (OR; at the time a different type of pump was used in the OR from that used on the units). Both types of old pumps were becoming obsolete, repairs were occurring more frequently, and the pumps did not offer many of the significant safety features currently available on the market. A team of stakeholders convened to determine what the most important factors were that would lead to product selection. Anesthesia personnel, although satisfied with how the pumps worked in the operating room, identified their greatest patient safety concern: when patients came to the operating room from other units already on a pump or when they left the OR to go to a patient room (intensive care unite [ICU] or general care) and were on a pump, they had to switch the pump being used. They were concerned that an error having grave consequences was waiting to happen. Nurses agreed with this concern but were also acutely aware of the fact that the way in which they, like all floor nurses, used the pump was different than in the OR. Anesthesiologists and certified registered nurse anesthesiologists (CRNAs) rarely leave an operating room for any significant length of time, unlike floor and even ICU nurses who nearly always were responsible for two or more patients at a time. Nurses felt that features ensuring correct programming of the pumps were a necessity. New "smart" pumps were on the market due to the increased attention being paid to decimal and order-of-magnitude errors in programming. Pediatric nurses agreed with these concerns as well, but also pointed out the need to have pumps that offered syringe pump features to address the small volumes of most pediatric infusions.

Learnability and memorability (Nielsen, 1993) were important to all nurses on the team. This was especially true for nurses who realized the patients they cared for were less likely to require high risk medications and therefore less frequently require the specialized features of a smart pump. Anesthesia personnel displayed considerable confidence in their tinkering skills and were more focused on reliability and efficiency based on the risks their patients pose when administering intraoperative medications. Pharmacy representatives on the team recognized the decrease in infusion errors that would most likely occur as a result of using the smart intravenous technology. Their position was to develop a business case for the pumps by showing how a decrease in medication errors would likewise decrease the hospital's liability.

Once a selection matrix was developed, representatives from the top three selected manufacturers were invited to give a presentation to the team members. Subsequently, one representative from each use area on the team (i.e., Nursing, Anesthesiology, Pharmacy) made day-long site visits to observe and talk with their peers in three organizations, each using one of the three different pumps under consideration. On their return, these individuals presented their findings (and personal preference) to the team. Because the recommendations of the various users did not unanimously support one pump, discussion, as facilitated by the chief operating officer, resulted in lively debate between team members. The alarm feature of the pumps generated the most discussion. Three months after the team first met, the members came to a consensus and recommended that Valley purchase and implement the Nifty® pump in 6 months.

The CFO was able to work out a deal with the manufacturer of Nifty® so 25 sets of pumps (including programming and pumping modules) would be delivered six weeks prior to implementation. This afforded an HFMEA© (Healthcare Failure Mode and Effects Analysis) team (DeRosier, Stalhandske, Bagian, & Nudell, 2002) six weeks to conduct a prospective risk assessment, get sufficient feedback from those affected by implementation of new pumps (including biomedical engineers, those in the hospital reprocessing and supply area, unit clerks and nurse aides, pharmacists, nurses, and anesthesiologists and CRNAS), and identify areas of concern that needed to be incorporated in the training program required of all pump users.

One volunteer from each unit was given the status of super-user. She was given significant amounts of training and preparation on all features associated with the pump, however special emphasis was offered to aspects of the pump her patient mix was most likely to require. The alarm and other subtle issues identified from the site visits as well as the on-site super-user training were then incorporated in small-group user training. One week prior to house-wide implementation, pumps were delivered to each of the care areas for formal training and tinkering. Feedback from previous technology implementations emphasized the need to ensure that all users were confident with their ability to appropriately use the new technology on the first day of the implementation. The role of the super-users was deemed critical and each super-user was relieved of any patient care responsibilities during the preimplementation week as well as the first week in which the pumps were in use. A pump coordinator had been hired as well and began her job when the HFMEA team first met. Her relationship with the super-users was critical to the success of the implementation as well as the learning associated with familiarity and use of the pumps. A follow-up team was identified to address, resolve, monitor, and communicate any problems that arose.

Although relatively smooth, the implementation did incur unanticipated problems. The alarm issue, as noted by the selection team, caused the most consternation on the part of users and patients alike. Dealing with the annoyance of an alarm was paramount in the minds of most users. What the follow-up team recommended and successfully accomplished was to work with the manufacturer to make minor software changes. Because the alarms initially represented equal levels of urgency to the programmer, and because Valley was not the only user of the Nifty® pump to complain to the manufacturer about the alarm, software changes were made to the pumps that modified the most common alarms. These alarms were deemed as annoyances by the users and occurred when there was a 20-sec delay in programming the pump. When the delay alarm was extended to 1 min, nearly all complaints associated with it were eliminated. Other problems identified by users were first reported to their respective super-user. Logs were kept on the types of problems reported and what follow up occurred. The pump coordinator then gave twice-a-month reports to all groups affected by the pump and communicated the problems identified and offered updated training to all users on an on-going basis. Administration also included this report in its monthly hospital-wide meetings. At these meetings, the learning that occurred was shared and opportunities to continue learning more about the pumps were identified.

Discussion

So, what makes this case an example of organizational learning? First and foremost, the leaders of the organization see their role as one of ensuring and promoting the communication of learning from and to all levels within the organization, reinforcing the mission, and developing common mental models for those who work at Valley. This is well-received through the monthly staff meetings as well as the fact that top management is recognized and known by the workers. The leaders understand the countless opportunities through which the organization can learn: whether through major technology implementations (such as the infusion pump) or through equally significant feedback from providers, clerks, technicians, and others who practice and work at the organization. Likewise customer feedback is regularly collected, monitored, and, as necessary, acted on.

All learning is shared and communicated to offer continuous improvements to the services the organization provides. When failures, large or small, occur within the organization, they are captured, addressed, and, in most instances, rectified. In turn this information is shared with others in the organization who are made aware of instances when learning from failure can occur. If necessary, ongoing or further training is offered to ensure understanding and acceptance of change.

Employees are regularly reminded of their role in the system in which they work and receive feedback that recognizes the value of what they do caring for patients, interacting with families, coordinating activities with coworkers, and participating on teams. When teams are formed, the systems aspect of the team is expressed via multidisciplinary representation of all process owners. Aside from ensuring that all necessary viewpoints are expressed, others participating on the team also learn to appreciate the viewpoints of others from different disciplines.

Finally, the organization's mission is succinct, clearly indisputable, and facilitates development of a common mental model and direction for all in the organization to strive toward. Emphasis is placed on the patient, but staff received considerable attention through the many employee development and health programs. Clearly a healthy work organization (Sainfort, Karsh, Booske, & Smith, 2001) exists at Valley where organizational learning is prized.

CONCLUSION

Briefly we have reviewed seminal organizational learning research and presented it in light of health care. Given the incredible amount of scrutiny it receives, never has the need to change health care organizations and the way in which care is provided been greater than it is today. Espousing organizational learning offers a powerful means of responding to the scrutiny and improving the care we provide as well as the skills and work environment of those working in the health care industry.

References

Argyris, C., & Schon, D. A. (1978). *Organizational learning: A theory of action perspective.* Reading, MA: Addison-Wesley.

Batalden, P. B., & Splaine, M. (2002). What will it take to lead the continual improvement and innovation of health care in the twenty-first century? *Quality Management in Health Care, 11*(1), 45–54.

Batalden, P. B., Stevens, D. P., & Kizer (2002). Knowledge for improvement: Who will lead the learning? *Quality Management in Health Care, 10*(3), 3–9.

Berta, W. B., & Baker, R. (2004). Factors that impact the transfer and retention of best practices for reducing error in hospitals. *Health Care Management Review, 29*(2), 90–97.

Bohmer, R. M. J., & Edmondson, A. C. (2001). Organizational learning in health care. *Health Forum Journal, 44*(2), 32–35.

Cangelosi, V. E., & Dill, W. R. (1965). Organizational learning: Observations toward a theory. *Administrative Science Quarterly, 10*(2), 175–203.

Carroll, J. S., & Edmondson, A. (2002). Leading organisational learning in health care. *Quality and Safety in Health Care, 11*(1), 51–56.

Chuang, Y.-T., Soberman-Ginsburg, L., & Berta (2004, August). *Why others do, but you don't? A multi-level model of responsiveness to adverse events.* Academy of Management Annual Meeting, New Orleans, LA.

Crossan, M., & Guatto, T. (1996). Organizational learning research profile. *Journal of Organizational Change Management, 9*(1), 107.

Crossan, M. M., Lane, H. W., White, R. E., & Djurfeldt, L. (1995). Organizational learning: Dimensions for a theory. *The International Journal of Organizational Analysis, 3*(4), 337–360.

Crossan, M. M., Lane, H. W., & White (1999). An organizational learning framework: From intuition to institution. *Academy of Management Review, 24*(3), 522–537.

Cyert, R. M., & March, J. G. (1963). *A behavioral theory of the firm.* Englewood Cliffs, NJ: Prentice-Hall.

DeRosier, J., Stalhandske, E., Bagian, J. P., & Nudell, T. (2002). Using health care failure mode and effect analysis: The VA National Center for Patient Safety's prospective risk analysis system. *Joint Commission Journal on Quality Improvement, 28*(5), 209, 248–267.

Donabedian, A. (1978). The quality of medical care: Methods for assessing and monitoring the quality of care for research and for quality assurance programs. *Science, 200,* 856–864.

Duncan, R., & Weiss, A. (1979). Organizational learning: Implications for organizational learning. *Research in Organizational Behavior, 1,* 75–123.

Edmondson, A. C. (1996). Learning from mistakes is easier said than done: Group and organizational influences on the detection and correction of human error. *The Journal of Applied Behavioral Science, 32*(1), 5–28.

Edmondson, A. C. (2003). Speaking up in the operating room: How team leaders promote learning in interdisciplinary action teams. *Journal of Management Studies, 40*(6), 1419–1452.

Edmondson, A. C., Bohmer, R. M., & Pisano, G. P. (2001). Disrupted routines: Team learning and new technology implementation in hospitals. *Administrative Science Quarterly, 46,* 685–716.

Edmondson, A., & Moingeon, B. (1998). From organizational learning to the learning organization. *Management Learning, 29*(1), 5–20.

Fiol, C. M., & Lyles, M. A. (1985). Organizational learning. *Academy of Management Review, 10*(4), 803–813.

Garvin, D. A. (1993). Building a learning organization. *Harvard Business Review, 71,* 78–91.

Huber, G. (1991). Organizational learning: The contributing processes and the literatures. *Organization Science, 2*(1), 88–115.

Kim, D. H. (1993). The link between individual and organizational learning. *Sloan Management Review, 35*(1), 37–50.

Kohn, L. T., Corrigan, J. M., & Donaldson. (2000). *To err is human: Building a safer health system.* Washington, DC: National Academy Press.

Krouwer, J. S. (2004). An improved failure mode effects analysis for hospitals. *Archives of Pathology and Laboratory Medicine, 128*(6), 663–667.

Lawrence, T. B., Mauws, M. K., Dyck, B., Kleysen, R. F. (2005). The politics of organizational learning: Integrating power into the 4I framework. *Academy of Management Review, 30*(1), 180–191.

Leavitt, B., & March, J. G. (1988). Organizational learning. *Annual Review of Sociology, 14,* 319–340.

March, J. G. (1991). "Exploration and exploitation in organizational learning." *Organization Science 2*(1), 71–87.

March, J. G., Sproull, L. S., & Tamuz, M. (1991). Learning from samples of one or fewer. *Organization Science, 2*(1), 1–13.

March, J. G., Sproull, L. S., & Tamuz, M. (2003). Learning from samples of one or fewer. *Quality and Safety in Health Care, 12,* 465–472.

Mohr, J. J., Abelson, H. T., & Barach, P. (2002). Creating effective leadership for improving patient safety. *Quality Management in Health Care, 11*(1), 69–78.

Mohr, J. J., & Batalden, P. B. (2002). Improving safety on the front lines: The role of clinical microsystems. *Quality and Safety in Health Care, 11*(1), 45–50.

Nelson, E. C., Batalden, P. B., Huber, T. P., Mohr, J. J., Godfrey, M. M., Headrick, L. A., et al. (2002). Microsystems in health care, Part 1: Learning from high-performing front-line clinical units. *The Joint Commission Journal on Quality Improvement, 28*(9), 472–493.

Nielsen, J. (1993). *Usability engineering.* San Diego, CA: Academic.

Pisano, G. P., Bohmer, R. M. J., & Edmondson, A. C. (2001). Organizational differences in rates of learning: Evidence from the adoption of minimally invasive cardiac surgery. *Management Science, 47*(6), 752–768.

Popper, M., & Lipshitz, R. (1998). Organizational learning mechanisms: A structural and cultural approach to organizational learning. *Journal of Applied Behavioral Science, 34*(2), 161–179.

Rogers, E. M. (2003). *Diffusion of innovations.* New York: Free Press.

Sainfort, F., Karsh, B., Booske, B. C., & Smith, M. J. (2001). Applying quality improvement principles to achieve healthy work organizations. *Journal on Quality Improvement, 27*(9), 469–483.

Senge, P. M. (1990a). *The fifth discipline: The art and practice of the learning organization.* New York: Currency Doubleday.

Senge, P. M. (1990b). The leader's new work: Building learning organizations. *Sloan Management Review, 32,* 7–23.

Shrivastava, P. (1983). A typology of organizational learning systems. *Journal of Management Studies, 20*(1), 7–28.

Simon, H. A. (1991). Bounded rationality and organizational learning. *Organization Science, 2*(1), 125–134.

Solomon, R., & Petosa, L. (2001). Appendix F: Quality improvement and proactive hazard analysis models: Deciphering a new Tower of Babel. *Patient Safety: Achieving a New Standard of Care*, Institute of Medicine.

Stata, R. (1989). Organizational learning: The key to management innovation. *Sloan Management Review, 30*(3), 63–74.

Tucker, A. L. (2004). The impact of operational failures on hospital nurses and their patients. *Journal of Operations Management, 22,* 151–169.

Tucker, A. L., & Edmondson, A. C. (2002). Managing routine exceptions: A model of nurse problem solving behavior. *Advances in Health Care Management, 3,* 87–113.

Tucker, A. L., & Edmondson, A. C. (2003). Why hospitals don't learn from failures: Organizational and psychological dynamics that inhibit system change. *California Management Review, 45*(2), 1–18.

Tucker, A. L., Edmondson, A. C., & Spear, S. (2002). When problem solving prevents organizational learning. *Journal of Organizational Change Management, 15*(2), 122–137.

Vera, D., & Crossan, M. (2004). Strategic leadership and organizational learning. *Academy of Management Review, 29*(2), 222–240.

Wang, C. L., & Ahmed, P. K. (2003). Organizational learning: A critical review. *The Learning Organization, 10*(1), 8–17.

Weick, K. E. (1991). The nontraditional quality of organizational learning. *Organization Science, 2*(1), 116–124.

9

THE RELATIONSHIP BETWEEN PHYSICIAN PROFESSIONALISM AND HEALTH CARE SYSTEMS CHANGE

Maureen A. Smith and Jessica M. Bartell
University of Wisconsin, Madison

The idea of physician professionalism dates back to the Hippocratic oath, which instructed physicians to practice medicine for "the benefit of patients and abstain from whatever is deleterious and mischievous" (Shortell, Waters, Clarke, & Budetti, 1998). Patient trust in the physician–patient relationship is based on the idea that physicians have responsibility and control over medical decision making and that physicians prioritize the needs of patients over all other considerations (Mechanic & Schlesinger, 1996). In recent years, the issue of professionalism among physicians has been the central focus of numerous national and international medical conferences, medical journal publications, and public initiatives by professional organizations (American Board of Internal Medicine Foundation, 2003). Concern by these groups over the effects of a changing U.S. health care delivery system on physicians' primary dedication to patients has spawned a call for renewed commitment to the principles of the primacy of patient welfare, patient autonomy, and social welfare (ABIM Foundation, ACP-ASIM Foundation, & European Federation of Internal Medicine, 2002). In part, this professionalism "movement" has been motivated by perceived negative effects of the efforts by health care organizations to improve quality and decrease costs on the ability of physicians to serve as advocates for their patients (Shortell et al., 1998). In the current era in which competition between health care organizations is increasingly based on reducing inappropriate variations in care (Wennberg, 1998) and improving quality of care and patient satisfaction (Enthoven & Vorhaus, 1997), the perceived conflict between the goals of physicians as professionals and organizational management efforts may help to explain why so many organizational attempts to change physician behavior meet with failure (Grimshaw et al., 2001).

There has been increasing recognition that quality medical care is a property of systems not just of individuals. In 2001 the Institute of Medicine issued its report on health care quality, *Crossing the Quality Chasm*, that stressed the importance of system design in creating health care environments that are both safe and that produce quality health-related outcomes (Institute of Medicine, 2001). Human factors engineering, a concept adapted from the field of industrial engineering that promotes system design as a method to improve the interactions between the worker and the work environment, is increasingly used to address issues of work efficiency and safety (Helander, 1997). Systems design has also been identified as an important component of quality improvement and patient safety. Research has shown system improvements that decrease reliance on individual memory and attention, decrease error rates, and improve the quality of care (Institute of Medicine, 2001). This systems approach to changing physician behavior may not only be a more effective approach but may be better accepted by physicians than traditional organizational incentives to decrease costs and improve quality. Furthermore, this approach may be more reconcilable with the concept of physician professionalism because of its more indirect effects on physician behavior.

In this chapter, we examine the tension between physician professionalism and the financial and nonfinancial incentives that currently dominate the relationships between physicians and organizations. We examine the roles of system redesign in health care quality improvement and discuss the implications of a systems approach to health care organizational management for physician job satisfaction, professionalism, accountability, and the quality of patient care.

PHYSICIAN PROFESSIONALISM

The concept of professionalism implies that professionals are bound by the ethics of their professions to serve in their clients' best interests. Professionals are drawn to their professions because of an altruistic desire to serve, and this altruism restrains them from responding primarily to their own economic self-interest. They are also embedded in a community of peers with similar altruistic intentions and values that serve to further restrain their self-interest (Sharma, 1997). Historically, the physician was seen as the central force from which healing power originated and was controlled. With this conceptualization of physician professionalism came great responsibility and great respect. However, it has been noted that physician professionalism also fosters a "culture of blame" when things go wrong because, if physicians are responsible for the entirety of the medical process, then they are also exclusively to blame for poor quality care (Leape, 1994). It is this desire to maintain autonomy and responsibility related to professional status that may, in part, explain physician resistance to incentives connected with health care organizational management.

ORGANIZATIONAL INCENTIVES AND AGENCY THEORY

Multiple and sometimes conflicting organizational incentives have created a sense among physicians that their role as professionals may be threatened (Edwards, Kornacki, & Silversin, 2002). These organizational incentives include financial as well as nonfinancial incentives to manage utilization, decrease inappropriate variations in care, encourage evidence-based practice of medicine, and increase productivity. Examples of these incentives include salary withholds and bonuses, utilization review, practice guidelines, treatment reminders and prompts, and peer comparisons (Flynn, Smith, & Davis, 2002). These incentives are increasing in prevalence and in influence: A study of the arrangement that physicians make with managed care organizations demonstrated that extensive utilization review occurred in 62% of physician practices, 63% of physicians used practice guidelines, and 68% received personalized profiles of their practice patterns (Gold, Hurley, Lake, Ensor, & Berenson, 1995). These incentives, which are often viewed as a form of bureaucratic control, may be resented by practicing physicians who see control and authority over their practice decisions as their professional responsibility. As a result, these incentives have often been less than fully successful in motivating actual changes in physician behavior. Even with additional training and incentives, changing physician behavior is recognized as a very difficult process (Grimshaw et al., 2001).

The role and power of these incentives—as well as physicians' resistance to them—can be explained by agency theory. Agency theory was developed in the context of corporate interactions to describe the relationship between those who contract for services (principals) and those who perform such services (agents). This theory has been applied in the health care setting to describe the interactions between health care organizations (i.e., principals) and physicians (i.e., agents; Flood & Fennell, 1995). Agency theory says that financial and nonfinancial incentives are offered by principals to induce behaviors in agents that are consistent with the goals of the principal (Sappington, 1991). These incentives cause physicians to balance the interests of their patients (physicians' primary concern as professionals) with another set of interests, namely the interests of the organization with which the physicians are associated. This concept has been coined "double agency" and refers to the tension between these different influences and loyalties (Shortell et al., 1998). Because physician professionalism suggests that physicians should act as "perfect agents" for their patients by holding patient goals paramount (Shortell et al., 1998), the uncomfortable balance of having to act as agent for both patients and organizations may explain some of the physician resistance to organizational incentives.

There is growing distress among physicians, which appears largely related to the incentives present in medical practice associated with managed care. This distress may be contributing to the growing crisis of confidence in physicians among patients (Mechanic & Schlesinger, 1996) as well as to increased career dissatisfaction among physicians involved in managed care (Linzer et al., 2000). Furthermore, the inherent conflicts between organizational incentives and physician professionalism may undermine effectiveness of health care organizational quality improvement initiatives. For these reasons, a new approach to managing medical practice and physician behavior is needed. A systems perspective, which focuses on the structure of health care, health care processes, and the interface between workers and the work environment, may be more compatible with physician professionalism than are direct incentives. It may also be a more effective method of quality improvement and may help to improve physician acceptance of organizational change efforts.

THE SYSTEMS PERSPECTIVE—AN ENGINEERING APPROACH

The Institute of Medicine (2001), in its report *Crossing the Quality Chasm*, called on health care organizations to take a systematic approach to quality improvement. The report pointed to evidence that systems changes can effectively change practice and improve health care quality by reducing medical errors and decreasing inappropriate variations in care. Although the safety and efficiency arguments are convincing, it is becoming increasingly apparent that system redesign may also help physicians to improve their worklives and to better meet their patients' needs while functioning within the required constraints of health care delivery systems (Morrison & Smith, 2000).

The balance theory of human factors engineering, which focuses on system redesign to optimize the interactions between workers and the system they work in, stresses the interactions between the individual, the task, technology, environment, and organizational factors (Carayon & Smith, 2000; Smith & Sainfort, 1989). The balance model places the individual in the center of the system with his or her work affected by the other aspects of the work system (i.e., the task, technology, the work environment, and organizational factors). The individual then makes decisions that are also affected by stress and other modulating factors such as individual competence, age, motivation, and gender (Helander, 1997). Finally, the individual manifests a response output. The interaction between this action and the work system produces an outcome. In designing systems for better efficiency and safety, human factors engineers concentrate on redesigning environmental and workplace factors rather than selection and training of individuals. Training is utilized to ease organizational changes, especially with the implementation of new technology.

Human factors engineering has been increasingly employed to change the way that health care is delivered (Gosbee, 1999). For example, human factors engineering has been employed in anesthesiology in the development of engineered safety devices such as a system of gas connectors that does not allow a gas hose or cylinder to fit the wrong site. This type of advance, combined with new technologies, standards, and guidelines, and an emphasis on a "culture of safety," has helped to dramatically decrease the rate of death due to anesthesia (Gaba, 2000). Another example of system change to improve safety is the institution of computerized physician order entry, which has been shown to reduce physician and nursing transcription errors (Mekhjian et al., 2002). Both of these examples illustrate a new approach to health care: the redesign of the environment and important tools to optimize the performance of the operator (i.e., the physician).

Efficiency in health care has similarly been helped by systems improvements. The enormous expansion of knowledge and technology in medicine in recent years has made it impossible for physicians to apply all available preventive screening, convey all known information about relevant medical conditions, and provide all available treatments for each patient in the increasingly limited time available for an individual office appointment. This situation, in which physicians find themselves on a hypothetical treadmill and "running faster just to stand still," has been referred to as "hamster health care" (Morrison & Smith, 2000, p. 1541). Systems redesign, such as computerized screening protocols and electronic communications, may relieve the physician of some of the less demanding functions of health care and therefore save time for more meaningful patient care.

Quality improvement efforts that incorporate system redesign may also be more successful in changing physician behavior, maintaining physician job satisfaction, and preserving physician professionalism than traditional incentive approaches. System redesign, which is based on the idea of

TABLE 9–1 Structure-Process–Outcome Model versus Human Factor Engineering Perspective

Domain	*Structure-Process–Outcome*	*Human Factors Engineering*
Structure/environment	1. Material resources responsibilities, training, coworkers 2. Human resources 3. Organizational structure motivation, gender, stress 4. Technology features	1. Organization communication, 2. Ambient noise, climate, illumination 3. Operator competence, age,
Process/tasks	1. Physician technical quality and feedback 2. Physician interpersonal relationships	1. Task composition, allocation, 2. Operator perception, decision making, and response
Outcomes	1. Effects of health care on health status of individuals and populations	1. Negative outcomes: errors, accidents, injuries, physiological stress 2. Positive outcomes: productivity, time, quality

making it "easy to do things right and hard to do things wrong" may be a more indirect, easier approach to changing physician behavior (Leape, 1994). Well-designed systems may be less resented than direct financial and nonfinancial incentives set forth by health care organizations. Often, systems changes are quite transparent, so they are not in conflict with physician autonomy and control. However, involvement of physicians in the process of system design is important to both success of the system and to physician acceptance (Leape, 1994). Maintenance of a sense of job control and participation in the process of organizational change may improve physician acceptance of and adherence to organizational quality improvement programs (Edwards et al., 2002).

A MEDICAL APPROACH TO SYSTEMS CHANGE

Avedis Donabedian (1988) was instrumental in developing a theory and a language for talking about quality improvement in health care. His model of quality assessment—the Structure–Process–Outcomes (SPO) Model—foreshadowed many of the ideas of human factors engineering. First, the SPO model acknowledges that the structural factors such as material resources, human resources, and organizational structure likely influence the quality of care. Second, amenities of care, which include convenience, comfort, quiet, and privacy (all characteristics of the environment surrounding the health care interaction) are emphasized as important components of quality. Third, the interactions between the provider and patient are prioritized.

These components of the SPO model parallel the components of human factors engineering (HFE) models of quality assessment (see Table 9–1). Where the structure/environment is represented by material, organizational, and human resources in the SPO model, it is represented by organizational factors, work environment, individual characteristics, and technology features in HFE. In the SPO model, process/tasks is represented by physician technical quality and interpersonal relationship; in the HFE model it is represented by aspects of the task and the process surrounding the individual's decision making. Finally, outcomes are defined as effects of health care on health status of individuals and populations according to the SPO model and as both negative and positive effects such as errors, accidents, injuries, stress, productivity, time, and quality in the HFE model.

There are, however, a number of differences between the SPO model and a HFE approach. These differences stem largely from the fact that physician professionalism, which influences the medical approach to quality assessment, is highly reflected in the SPO model. First, the primary emphasis in the SPO model is on the performance of practitioners. Donabedian (1988) wrote "Our power, our responsibility and our vulnerability all flow from the fact that we are the foundation for that ladder, the focal point for that family of concentric circles" (p. 1743) that define quality health care. Because of the centrality of the physician role, the SPO model places the responsibility for other elements of quality within the health care system (the physician–patient interaction, the "amenities of care," and the structural elements) mostly on physicians and minimally on patients or health care systems. In Donabedian's view, structural elements should not carry much weight in quality assessment because prior evidence has demonstrated a weak link between the structure and processes of health care (Palmer & Reilly, 1979) and because structural characteristics are a "rather blunt instrument" in quality assessment (Donabedian, 1988, p. 1745–6). Because of this lack of emphasis on structure as a determinant of quality health care, the responsibility for quality of care remains the physician's.

Another difference between the SPO model and the HFE model is that researchers who use the SPO model have become quite sophisticated in understanding the relationships of process to patients' health-related outcomes and in measuring these outcomes (Brook, McGlynn, & Cleary, 1996; Clancy & Eisenberg, 1998). HFE, on the other hand, focuses primarily on the interaction of the individual and his or her work surroundings, including structure and process, and individual and system outcomes; but HFE has only recently started to focus on patient outcomes.

A NEW APPROACH

Any attempts to apply a systems perspective to health care must take into account the current medical environment, which has been shaped by physician professionalism and the tension between physician professionalism and the current organizational incentives structure. Because of the historical centrality of the physician in the design and conceptualization of health care systems in the past, there has been a lack of focus within quality improvement efforts on the structural aspects of health care. There has instead been a focus on clinical processes of care carried out by the physician and other practitioners. If health care organizations are going to both improve the quality of health care and acknowledge the difficulties that physicians have with the change process due to their belief in physician professionalism, a new approach to quality improvement is needed, one that incorporates more of a systems perspective and emphasizes the structure of health care delivery and reengineers the processes of care to relieve physicians of exclusive responsibility for the quality of health care provided.

There have been a number of developments in the way health care is provided that focus on incorporating a systems approach to quality improvement. These approaches to health care have largely been developed by organizations looking for ways to provide quality health care while balancing the needs of physicians, patients, and health care organizations (Berwick, 1996). These developments include disease management programs and collaborative multidisciplinary patient care teams. These approaches, although still in the early stages of development and evaluation, show great potential for incorporating a systems approach to health care while preserving physician professionalism.

Disease Management Programs

Disease management is a concept that first appeared in the U.S. health care field approximately 15 years ago. It is a system for managing chronic conditions at a health care system level to improve the effectiveness and efficiency of care across the continuum of care. High prevalence, high cost conditions are selected. Disease maps and care maps are established to determine how the disease is managed within the health care organization and what drives the costs of care. The most expensive and highest risk patients are identified and interventions to improve care for these patients are evaluated. The programs are monitored and adjusted as necessary using a continuous quality improvement framework. Some examples of these successful disease management programs include focused discharge planning and postdischarge care for patients with congestive heart failure (Phillips et al., 2004); intensive monitoring, education, and follow-up for patients with diabetes mellitus by nurses and nurse case managers (Sidorov et al., 2002); and tertiary prevention of coronary heart disease risk factors (DeBusk et al., 1994).

Disease management programs take advantage of system-level analyses to identify the need for interventions and implement changes in the structure and processes of care around certain high-prevalence, high-cost diseases. Interventions may include physician incentives such as education, bonuses, and profiles, but these incentives are established as only a part of multiple interventions contributing to system redesign (Weingarten et al., 2002). The responsibility for management of the condition is therefore spread more equally across the many participants in the delivery of health care—physicians, other practitioners, employees more indirectly involved in patient care (e.g., receptionists, administrators, other clinic personnel), and patients (Bodenheimer, 1999). If implemented well, disease management programs should improve physician worklife through better coordination and support of care (Bodenheimer, Wagner, & Grumbach, 2002).

Multidisciplinary Patient Care Teams

The development of multidisciplinary patient care teams within health care organizations takes advantage of system-level work redesign to provide effective health care. These teams have been especially effective for managing the health care of patients with chronic diseases (Wagner, 2000). By taking advantage of team members' individual skill sets, patient care teams can create a system-level health care delivery system in which individuals' roles and functions are optimized and effectiveness of care is maximized. This type of quality improvement intervention has the potential to improve care through population management, protocol-based regulation of medication, self-management support, and intensive follow up (Wagner, 2000). Similar to other system-level approaches, this type of quality improvement effort does not take away from the professional role of physicians or create pressure on physicians to change the way they practice medicine with financial incentives or negative consequences for noncompliance. Instead, these teams create a collaborative, positive environment to supplement physician expertise in the day-to-day management of chronic illness.

ACCOUNTABILITY

One of the major issues left to be resolved in this new approach to health care systems change is that of accountability. Accountability is a crucial component of any health care system because, ultimately, someone or something must be responsible for the activities, actions, and outcomes of health care. Under the professional model of health care, accountability is firmly established: Physicians are responsible to their professional colleagues and organizations and to their individual patients. In addition, with the current structure of U.S. health care in which health care organizations operate as economic entities within a larger marketplace, accountability occurs when patient–consumers show disapproval of a physician or health plan through "exit" (i.e., switching providers or plans; Emanuel & Emanuel, 1996). These different accountabilities are not entirely compatible, but at least they are well defined. Under the systems model, however, accountability is much less clear. The systems model emphasizes the multifactorial nature of error and a nonpunitive approach (Wears et al., 2000). Accountability under this model occurs through a method of root cause analysis, which evaluates the many contributors to error. Although this more diffuse structure of accountability has many advantages, health system change that employs a systems approach needs to carefully define which entities are responsible for which sets of activities and who is responsible for explaining or answering for which actions (Emanuel & Emanuel, 1996).

CONCLUSIONS

Incorporation of a systems perspective into quality improvement efforts may help to improve physician acceptance of quality management activities and to decrease the tension between these activities and physician professionalism. Formation of disease management programs and multidisciplinary teams are examples of system-level interventions that may be effective in improving quality while preserving physician professionalism. More research is needed to determine which system-level interventions are most effective and to what extent these interventions truly have positive effects on physician job satisfaction, physician attitudes toward quality improvement efforts, and the ability of physicians to maintain a sense of professionalism. HFE has much to add to the medical approach to health care quality improvement, but the optimal approach and methodology that combines a systems approach takes into account physician professionalism and clearly establishes accountability has yet to be defined.

References

ABIM Foundation, ACP-ASIM Foundation, & European Federation of Internal Medicine. (2002). Medical professionalism in the new millennium: A physician charter. *Annals of Internal Medicine, 136*(3), 243–246.

American Board of Internal Medicine Foundation. (2003). *The Medical Professionalism Project and the Physician Charter.* Available from http://www.abimfoundation.org/professional.html

Berwick, D. M. (1996). A primer on leading the improvement of systems. *British Medical Journal, 312*(7031), 619–622.

Bodenheimer, T. (1999). Disease management—Promises and pitfalls. *New England Journal of Medicine, 340*(15), 1202–1205.

Bodenheimer, T., Wagner, E. H., & Grumbach, K. (2002). Improving primary care for patients with chronic illness: The chronic care model, Part 2. *Journal of the American Medical Association, 288*(15), 1909–1914.

Brook, R. H., McGlynn, E. A., & Cleary, P. D. (1996). Quality of health care. Part 2: Measuring quality of care. *New England Journal of Medicine, 335*(13), 966–970.

Carayon, P., & Smith, M. J. (2000). Work organization and ergonomics. *Applied Ergonomics, 31*, 649–662.

Clancy, C. M., & Eisenberg, J. M. (1998). Outcomes research: Measuring the end results of health care. *Science, 282*(5387), 245–246.

DeBusk, R. F., Miller, N. H., Superko, H. R., Dennis, C. A., Thomas, R. J., Lew, H. T., et al. (1994). A case-management system for coronary risk factor modification after acute myocardial infarction. *Annals of Internal Medicine, 120*(9), 721–729.

Donabedian, A. (1988). The quality of care. How can it be assessed? *Journal of the American Medical Association, 260*(12), 1743–1748.

Edwards, N., Kornacki, M. J., & Silversin, J. (2002). Unhappy doctors: What are the causes and what can be done? *British Medical Journal, 324*(7341), 835–838.

Emanuel, E. J., & Emanuel, L. L. (1996). What is accountability in health care? *Annals of Internal Medicine, 124*(2), 229–239.

Enthoven, A. C., & Vorhaus, C. B. (1997). A vision of quality in health care delivery. *Health Affairs, 16*(3), 44–57.

Flood, A. B., & Fennell, M. L. (1995). Through the lenses of organizational sociology: The role of organizational theory and research in conceptualizing and examining our health care system. *Journal of Health and Social Behavior, 35* (Special Issue), 154–169.

Flynn, K. E., Smith, M. A., & Davis, M. K. (2002). From physician to consumer: The effectiveness of strategies to manage health care utilization. *Medical Care Research & Review, 59*(4), 455–481.

Gaba, D. M. (2000). Anaesthesiology as a model for patient safety in health care. *British Medical Journal, 320*(7237), 785–788.

Gold, M. R., Hurley, R., Lake, T., Ensor, T., & Berenson, R. (1995). A national survey of the arrangements managed-care plans make with physicians. *New England Journal of Medicine, 333*(25), 1678–1683.

Gosbee, J., & Lin, L. (1999). The role of human factors engineering in medical device and medical system errors. In: C. Vincent (Ed.), *Clinical Risk Management: Enhancing Patient Safety* (pp. 309–330). London: BMJ Publishing.

Grimshaw, J. M., Shirran, L., Thomas, R., Mowatt, G., Fraser, C., Bero, L., et al. (2001). Changing provider behavior: An overview of systematic reviews of interventions. *Medical Care, 39*(8 Suppl 2), II2–45.

Helander, M. G. (1997). The human factors profession. In G. Salvendy (Ed.), *Handbook of human factors and ergonomics.* New York: Wiley. (pp. 3–16).

Institute of Medicine. (2001). *Crossing the quality chasm: A new health system for the 21st century.* Washington, DC: National Academy Press.

Leape, L. L. (1994). Error in medicine. *Journal of the American Medical Association, 272*(23), 1851–1857.

Linzer, M., Konrad, T. R., Douglas, J., McMurray, J. E., Pathman, D. E., Williams, E. S., et al. (2000). Managed care, time pressure, and physician job satisfaction: Results from the physician worklife study. Jurnal of *General Internal Medicine, 15*(7), 441–450.

Mechanic, D., & Schlesinger, M. (1996). The impact of managed care on patients' trust in medical care and their physicians. *Journal of the American Medical Association, 275*(21), 1693–1697.

Mekhjian, H. S., Kumar, R. R., Kuehn, L., Bentley, T. D., Teater, P., Thomas, A., et al. (2002). Immediate benefits realized following implementation of physician order entry at an academic medical center. *Journal of the American Medical Informatics Association, 9*(5), 529–539.

Morrison, I., & Smith, R. (2000). Hamster health care. *British Medical Journal, 321*(7276), 1541–1542.

Palmer, R. H., & Reilly, M. C. (1979). Individual and institutional variables which may serve as indicators of quality of medical care. *Medical Care, 17*(7), 693–717.

Phillips, C. O., Wright, S. M., Kern, D. E., Singa, R. M., Shepperd, S., & Rubin, H. R. (2004). Comprehensive discharge planning with postdischarge support for older patients with congestive heart failure: A meta-analysis. *Journal of the American Medical Association, 291*(11), 1358–1367.

Sappington, D. E. M. (1991). Incentives in principal–agent relationships. *The Journal of Economic Perspectives, 5*(2), 45–66.

Sharma, A. (1997). Professional as agent: Knowledge asymmetry in agency exchange. *Academy of Management Review, 22*(3), 758–798.

Shortell, S. M., Waters, T. M., Clarke, K. W., & Budetti, P. P. (1998). Physicians as double agents: Maintaining trust in an era of multiple accountabilities. *Journal of the American Medical Association, 280*(12), 1102–1108.

Sidorov, J., Shull, R., Tomcavage, J., Girolami, S., Lawton, N., & Harris, R. (2002). Does diabetes disease management save money and improve outcomes? A report of simultaneous short-term savings and quality improvement associated with a health maintenance organization-sponsored disease management program among patients fulfilling health employer data and information set criteria. *Diabetes Care, 25*(4), 684–689.

Smith, M. J., & Sainfort, P. C. (1989). A balance theory of job design for stress reduction. *International Journal of Industrial Ergonomics, 4*, 67–79.

Wagner, E. H. (2000). The role of patient care teams in chronic disease management. *British Medical Journal, 320*, 569–572.

Wears, R. L., Janiak, B., Moorhead, J. C., Kellermann, A. L., Yeh, C. S., Rice, M. M., et al. (2000). Human error in medicine: Promise and pitfalls, part 1. *Annals of Emergency Medicine, 36*(1), 58–60.

Weingarten, S. R., Henning, J. M., Badamgarav, E., Knight, K., Hasselblad, V., Gano, A., Jr., et al. (2002). Interventions used in disease management programmes for patients with chronic illness—Which ones work? Meta-analysis of published reports. *British Medical Journal, 325*(7370), 925.

Wennberg, D. E. (1998). Variation in the delivery of health care: the stakes are high. *Annals of Internal Medicine, 128*(10), 866–868.

10

COLLABORATIVE INITIATIVES FOR PATIENT SAFETY

Pascale Carayon
University of Wisconsin, Madison

Andrew Kosseff
St. Mary's Hospital

Amanda Borgsdorf
formerly of the Madison Patient Safety Collaborative

Kendra Jacobsen
Madison Patient Safety Collaborative

Fostering collaboration has been emphasized by many as a key strategy for improving patient safety (Goode, Clancy, Kimball, Meyer, & Eisenberg, 2002; Institute of Medicine Committee on Quality of Health Care in America, 2001). Collaboration within health care organizations has been mostly emphasized. For instance, team work and crew resource management (CRM) are seen as methods for improving communication and fostering collaboration between health care providers for increased quality and safety of care (Helmreich & Schaefer, 1994; Morey et al., 2002; Thomas, Sherwood, & Helmreich, 2003). Collaboration across health care organizations should also be encouraged to improve patient safety. First, collaboration between organizations that experience similar patient safety problems may contribute to sharing of experience, improve the solutions to the patient safety problems, and facilitate the implementation of those solutions. Second, many patient safety problems occur at the interface between organizations. For example, the creation and maintenance of accurate medication lists require both inpatient and outpatient organizations to work together (Cornish et al., 2005; van Walraven, Mamdani, Fang, & Austin, 2004). The problems occurring at the interface between multiple organizations cannot be solved by a single organization but need to be addressed collaboratively by the various organizations involved in the problems. Collaborative initiatives are one mechanism for encouraging collaboration across organizations to improve patient safety. Collaborative initiatives can also contribute to increasing the pace at which research is put into practice (Fraser, Lanier, Hellinger, & Eisenberg, 2002).

WHAT ARE COLLABORATIVE INITIATIVES FOR PATIENT SAFETY?

In 1999, the report issued by the Institute of Medicine (IOM), *To Err Is Human*, heightened attention of health care institutions regarding medical errors and patient safety (Kohn, Corrigan, & Donaldson, 1999). Since then, the number of professional, government, and business organizations involved in patient safety has increased dramatically. The number of recommendations, guidelines, and standards has become overwhelming for any health care institution. Health care institutions are starting to engage in various efforts to improve their systems and processes, while at the same time facing tremendous pressure from purchasers, consumer groups, and the public to increase the speed at which changes are implemented. A number of strategies can be used by health care organizations to improve patient safety; patient safety collaboratives represent only one of those strategies (Leatherman, 2002).

Collaborative initiatives for patient safety (CIPS) are multiorganizational structured networks whose aim is to improve patient safety. CIPS involve multiple organizations working together either on a specific patient safety issue over a specific period of time or working together on a range of patient safety issues over a long-term basis. Often these organizations are located within a specific geographical region (e.g., a city, a state). Sometimes these organizations are geographically spread and come to work together because they share a specific patient safety concern or because they are part of a larger health care system interested in safety improvement. CIPS can be categorized by different levels, such as system-, city-, county- or state-wide collaboratives. CIPS can be categorized by type of patient safety problem. For example, Institute for Healthcare Improvement (IHI) collaboratives have addressed a wide range of specific patient safety problems, such as reducing adverse drug events (Leape et al., 2000) and outcomes in intensive care (Kilo, 1999). Sometimes, CIPS are organized across a large health care system. For instance, in 1998, SSM Health Care launched a series of clinical collaboratives aimed at improving quality and safety of care (Kosseff & Niemeier, 2001).

In the area of health care quality, a group of researchers has identified key characteristics of quality collaboratives that use quality improvement methods (Ovretveit et al., 2002):

- Participation of a number of multiprofessional teams that are committed to improving services within a specific area and that are willing to share their experience with others. The teams come from various organizations that support those objectives.
- A focused clinical or administrative subject.
- Existence of variations in care or of gaps between best and current practice.
- Participants learn from experts about the research evidence, change concepts and methods, and quality improvement methods.
- Participants use an improvement/change method such as Plan-Do-Check-Act (PDCA) or rapid cycle improvement.
- Teams have measurable goals and track their progress toward accomplishment of those goals.
- Participants meet, report on their progress, and share and disseminate their results.
- Collaborative organizers can provide support to participants.

A number of collaborative improvement projects in the United States and abroad have been derived from the IHI Breakthrough Series (BTS) model (Wilson, Berwick, & Cleary, 2003). The implementation of a quality improvement program in critical care units led to significant improvement in various clinical outcomes, such as antibiotic use, adult respiratory distress syndrome survival, laboratory use, and appropriate use of sedation (Clemmer, Francioli, Bille, Oniki, & Horn, 1999). This intervention was based on principles of quality improvement coupled with building multidisciplinary collaborative relations and a safe environment that uses the knowledge and skills of the primary care providers to improve practice. A multidisciplinary collaborative quality improvement team in a group of 10 self-selected neonatal intensive care units (ICUs) led to improvements in clinical outcomes when compared to a control group of 66 other neonatal ICUs (Horbar et al., 2001), and to cost savings (Rogowski et al., 2001). The IHI BTS collaborative improvement model attempts to (a) identify, describe, and diffuse best practices to the collaborative's organizations; (b) improve outcomes in participating organizations; (c) develop expertise in health care quality improvement for each specific area; and (d) disseminate knowledge gained by the collaborative effort (Kilo, 1999). Recently, IHI has embarked on a large-scale collaborative effort, the 100K Lives Campaign (http: // www .ihi. org / IHI /Programs/ Campaign/), that targets six patient safety interventions: (a) deployment of rapid response teams; (b) delivery of reliable, evidence-based care for acute myocardial infarction; (c) prevention of adverse drug events (ADEs); (d) prevention of central line infections; (e) prevention of surgical site infections; and (f) prevention of ventilator-associated pneumonia.

WHAT IS THE NEED FOR CIPS?

Most research studies and actual patient safety initiatives occur within a single institution, rarely across institutions. Much benefit can occur, however, if collaborative initiatives are implemented to solve patient safety problems that occur either within organizations or across organizations. Various regional initiatives have been shown to be successful at improving quality of care (e.g., IHI BTS; Leape et al., 2000). These projects have addressed limited-scope patient safety problems (e.g., reducing adverse drug events) and have been limited in time (i.e., they exist for the duration of the specific project). Other initiatives have been broader in scope and time (see the section "Examples" later in this chap.). It is interesting to make a parallel between CIPS and the formation and sustained development of networks that have been successful for industrial development (e.g., Silicon Valley; Porter, 1998) and for solving community problems (e.g., Quality Improvement (QI) efforts in the city of Madison; Box, Joiner, Rohan, & Sensenbrenner, 1989). This approach to network development and implementation has not been widely used to improve patient safety in a community.

There are a number of reasons for the creation of patient safety collaborative initiatives:

1. Improving patient safety entails a number of complex issues. Collaborative initiatives allow health care providers confronted with similar complex patient safety problems (e.g., elimination of dangerous abbreviations, reducing patient falls) to share knowledge, expertise, and experience. The interactions between the collaborative members allow for a larger knowledge base to contribute to solving the safety problems. This broad knowledge base can foster the identification of innovative solutions for those complex patient safety problems.
2. While participating in patient safety collaborative initiatives, health care providers have numerous learning opportunities and can share their knowledge and skills. Learning and sharing of "common knowledge" (i.e., the knowledge one gains from doing the tasks; Dixon, 2000) can facilitate, improve, and accelerate the identification of patient safety problems, the creation of solutions, and the development of effective and efficient implementation plans. The collaborative environment facilitates and speeds up the spread of best practices.
3. Many of the patient safety problems occur at the interface between organizations, therefore, cannot be fully resolved by single organizations and necessitate the collaboration of these interfacing organizations. For instance, medication reconciliation across the continuum of care is a problem that involves multiple transitions of information across multiple organizations and entities, such as community pharmacy and hospital (Cornish et al., 2005). The collaboration of multiple organizations allow for the breakdown of organizational barriers, even in competing health care organizations. The breakdown of those barriers allows the safety problem to be solved more effectively and facilitates the spread and implementation of the solution.
4. Change in health care is becoming increasingly fast and systemic. Collaborative initiatives that allow learning and sharing may allow health care providers to be better prepared and equipped to deal with changes and to accelerate their safety improvement efforts.
5. Health care organizations that participate in patient safety collaborative initiatives are able to highlight the importance of safety as a strategic priority to their stakeholders and employees. Solutions developed in the context of collaborative initiatives may benefit from the peer

pressure generated by the interactions within the collaboratives. New knowledge from outside an organization can be a stimulus for change and improvement (Inkpen & Tsang, 2005).

6. Patient safety collaborative initiatives may allow the building of linkages between research and practice. The involvement of researchers within CIPS may allow "putting practice into research" (Fraser et al., 2002, p. 1). On the other hand, new knowledge stemming from research may be more easily implemented through collaborative initiatives.
7. Patient safety collaborative initiatives allow the creation of networks of patient safety experts. Collaborative members are therefore able to respond to new issues in safety improvement rapidly. The experts can also help in analyzing the safety problems and the proposed solutions and comment on the appropriateness of the analysis.

In a discussion of competition in health care, Porter and Teisberg (2004) stated:

> An ideal health care system would encourage close working relationships between local providers (for most routine and emergency services and follow-up care) and a wide array of leading providers (for definitive diagnoses, treatment strategies, and complex procedures in certain areas). These relationships would speed up the diffusion of state-of-the-art clinical care and would help to increase quality and efficiency throughout the system—but they are often resisted. (p. 70)

CIPS represent one approach for building relationships between local providers, as advocated by Porter and Teisberg (2004).

LEARNING FROM COLLABORATIVE INITIATIVES OUTSIDE OF HEALTH CARE

The role of organizational networks has been largely emphasized in the business and management literature. In practice, a number of network arrangements have emerged, such as clusters (Porter, 1998), industrial districts (Paniccia, 1998), and interorganizational networks for community development (Chisholm, 1998). A network form of organization has been defined as "any collection of actors (N≥2) that pursue repeated, enduring exchange relations with one another and, at the same time, lack a legitimate organizational authority to arbitrate and resolve disputes that may arise during the exchange" (Podolny & Page, 1998, p. 59). This definition emphasizes that the creation, development, and sustainability of collaborative initiatives are very much dependent on conditions facilitating cooperation, such as learning, trust, norms, equity, and context (Brass, Galaskiewicz, Greve, & Tsai, 2004). Members of a network are loosely coupled and voluntarily participate in network-related activities (Chisholm, 1998).

Interorganizational networks tend to be created by people who are "boundary spanners" in their own organizations. Boundary spanners are individuals whose job is to integrate the work of others and to build linkages between various jobs, functions, and organizations (Lawrence & Lorsch, 1967). Within health care organizations, boundary spanners are, for example, case managers who are responsible for coordinating the care of patients. Boundary spanners can contribute to quality of care via improvement in coordination and relations among providers (Gittell, 2002). Boundary spanners also operate at the environment–organization interface. Boundary spanners, such as top managers, can help create linkages with external organizations, and therefore contribute to the creation, development, and sustainability of CIPS.

Organizational learning plays an important role in building ties between the members of the network. Four levels of network functioning have been found to contribute to learning: (a) information sharing, (b) knowledge sharing, (c) knowledge combination when multiple knowledge bases (i.e., knowledge coming from multiple organizations) are combined into knowledge that goes beyond the original knowledge bases, and (d) self-design when new knowledge leads to new practices (Mohrman, Tenkasi, & Mohrman, 2003). A number of conditions can facilitate (or hinder) the transfer of information and knowledge. Inkpen and Tsang (2005) have categorized those conditions into (a) structural conditions, that is, network ties, network configuration, and network stability; (b) cognitive conditions, that is, shared goals and shared culture; and (c) relational conditions, that is, trust among network members. For instance, spatial proximity can be beneficial because participants in the collaborative initiative are located physically close, therefore facilitating interpersonal interactions and exchange of information and knowledge. When members of an interorganizational network do not share goals or have unclear goals, frustration and dissatisfaction may occur. Shared goals can help orient the work and activities of the network (Chisholm, 1998). Trust among members of the collaborative initiative is critical for facilitating information transfer and knowledge exchange. When suspicion replaces trust in the relationships between members of the network, knowledge sharing may be sacrificed.

Interorganizational networks can contribute to the creation of *communities of practice* (Wenger, McDermott, & Snyder, 2002). The purpose of communities of practice is "to create, expand and exchange knowledge, and to develop individual capabilities" (Wenger et al., 2002, p. 43). Communities of practice involve people and organizations on a self-selection basis: Volunteer people and organizations tend to have some expertise in a specific area of interest and/or are eager to discuss a particular issue. Wenger et al. have identified the following factors as contributing to the creation and development of communities of practice: passion, commitment, and identification with the group and its expertise. Communities of practice will attract people and organizations that are passionate about the particular issue being addressed. Members of the community of practice are committed to the community because they want to explore the practice domain and to share the relevant knowledge. The community of practice can provide an arena for solving problems experienced by the community members, provide access to expertise, allow for more perspectives on problems, and represent a forum for expanding skills and expertise. CIPS may facilitate the creation of communities of practice across health care organization boundaries.

Wenger et al. (2002) have identified seven principles for designing communities of practice. Those principles can be applied to CIPS.

1. Design for evolution. CIPS are dynamic in nature. New members may join the collaborative, bring new interests and pull the focus of the collaborative in different directions. Changes inside the member-organizations of the collaborative may have a similar impact. Physical, social, and organizational structures can facilitate the evolution of communities of practice. For instance, the creation of a website, hiring of a coordinator, and strategic planning sessions may contribute to the evolution of the collaborative.
2. Open a dialogue between inside and outside perspectives. Information from the outside should be brought inside the communities of practice. CIPS need to bridge the gap between members involved directly in the specific patient safety improvements and other people and organizations who are directly or indirectly affected by the issue addressed. For instance, CIPS may need to address how to communicate with the public at large.
3. Invite different levels of participation: core group, active participation, peripheral participation, and outsider. The core group of a collaborative initiative for patient safety involves the people who lead the initiative and actively participate in the creation and development of the collaborative. Over time, these people take a leadership role. The active participants may be deeply involved for some time, for instance in a particular patient safety improvement project, and then disengage once the project is completed. Peripheral members are rarely directly engaged in the patient safety improvement projects led by the collaborative. They are external observers who may have some interest in the changes fostered by the collaborative. Outsiders may be people who do not belong to the region covered by the collaborative but are interested in learning about the collaborative.
4. Develop both public and private community spaces. The CIPS communicate about their patient safety improvement project through public mechanisms (e.g., web site, meeting, conference). These public forums are used, for instance, to exchange and share information and knowledge. CIPS perform other activities privately, for example, within working groups.
5. Focus on value. Communities of practice exist because they deliver value to the participant organizations. CIPS are created to improve patient safety in a particular

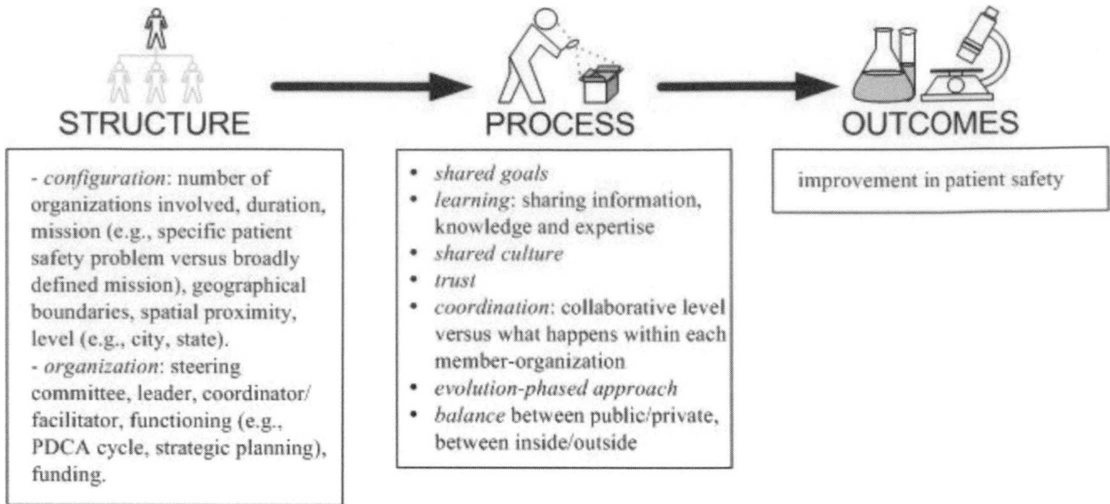

STRUCTURE
PROCESS
OUTCOMES
- configuration: number of organizations involved, duration, mission (e.g., specific patient safety problem versus broadly defined mission), geographical boundaries, spatial proximity, level (e.g., city, state).
- organization: steering committee, leader, coordinator/ facilitator, functioning (e.g., PDCA cycle, strategic planning), funding.
• shared goals
• learning: sharing information, knowledge and expertise
• shared culture
• trust
• coordination: collaborative level versus what happens within each member-organization
• evolution-phased approach
• balance between public/private, between inside/outside
improvement in patient safety

Figure 10–1. Model of the effectiveness and impact of collaborative initiatives for patient safety.

domain, in a particular geographical region, and/or within a specific period of time. Commitment to the collaborative is heightened if the participants see "value" in the collaborative. The value of CIPS is described in the section "What Is the Need for CIPS" earlier in this chapter. Ultimately, CIPS will survive and strive if improvement in patient safety is achieved.

6. Combine familiarity and excitement. On one hand, routine activities allow the development of relationships and connections among members of the communities. On the other hand, "exciting events provide a sense of common adventure" (Wenger et al., 2002, p. 62). Successful CIPS need to strike a balance between familiarity and excitement. Familiarity occurs via regular meetings and projects and may allow for the development of trust. Trust is important for allowing sharing of information and experience, in particular when dealing with issues as sensitive as medical errors. Special events can add the excitement necessary to keep the community "alive" and "lively." For instance, the discussion of a controversial patient safety solution can spark the interest of the collaborative members.
7. Create a rhythm for the community. CIPS may create their own tempo of interactions. For instance, special milestones (e.g., learning sessions around a particular patient safety topic) can be incorporated in the tempo of interactions. Yearly strategic planning sessions can also contribute to the creation of a rhythm for the collaborative initiative.

EFFECTIVENESS AND IMPACT OF CIPS

Using the Structure–Process–Outcome model of Donabedian (1988) as a framework and existing literature reviewed earlier, we can characterize CIPS on each of the three dimensions (see Fig.10–1). The structure of a CIPS can be characterized on two dimensions: (a) configuration (e.g., city-wide CIPS such as the Madison Patient Safety Collaborative, see next section) and (b) organization (e.g., leaders of the collaborative, working groups). As discussed in previous sections, various process characteristics can influence the effectiveness and impact of the collaborative. For instance, trust between the collaborative members contribute largely to sharing of knowledge, as well as the creation of new knowledge (Dixon, 2000). The major outcome of a CIPS is improvement in patient safety. The collaborative initiative may have other outcomes, such as increased participation of member-organizations in patient safety improvement and development of new patient safety knowledge.

EXAMPLES

In this section, we describe several examples of CIPS. The first CIPS is located within the city of Madison, Wisconsin. The second CIPS is located within a large health care system, SSM. In the third section, we briefly describe other CIPS, such as the Pittsburgh Regional Healthcare Initiative and state initiatives.

Madison Patient Safety Collaborative (www.madisonpatientsafety.org)

The Madison Patient Safety Collaborative (MPSC) in Madison, Wisconsin, was created in September 2000 after the Institute of Medicine report *To Err Is Human* (Kohn et al., 1999) stimulated health care providers to conduct thorough reviews of current safety practices and to identify opportunities for improvement. Administrators and physicians from several Madison hospitals and medical groups met to conduct a group analysis of the current state of patient safety, shared progress that had been made on patient safety topics, and discussed areas in which a collective effort might enable progress on a greater scale. As a result of these conversations, the Madison hospitals and medical groups decided to create a formal structure through which area health care providers could work together to develop, share, and implement patient safety solutions and the MPSC, a community-wide effort to reduce and prevent medical errors, was established (Carayon, Borgsdorf, Hundt, & Alvarado, 2003).

MPSC was founded on a "rising-tide" philosophy and the tenets that patient safety is a common goal rather than a competitive issue and that, through cooperation, Madison providers will be able to implement safety improvements that go well beyond the scope of any single organizational effort. Although medical errors occur in health care settings (Institute of Medicine Committee on Quality of Health Care in America, 2001), little information exists on the extent of the problem and effective strategies to improve patient safety, in particular outside acute care. Hospital patients clearly represent only a portion of those at risk of being party to a medical error. Distinctively, the MPSC, comprised of both hospitals and medical groups, seeks to improve patient safety community wide by implementing solutions that affect care provided both in the local hospitals and in area clinics.

Members of the MPSC include Dean Health System, Group Health Cooperative of Southcentral Wisconsin, Meriter Hospital, St. Mary's Hospital Medical Center, University of Wisconsin Hospital and Clinics, University of Wisconsin Medical Foundation, and the William. S. Middleton Memorial Veterans Affairs Hospital. Each member of the collaborative has made an organizational commitment to improving patient safety by dedicating infrastructure and resources to the MPSC's projects and by encouraging staff participation in the MPSC initiatives.

The core of the MPSC is comprised of key administrative and quality improvement leadership from each of the member organizations. This council provides strategic direction and serves as the decision-making body for the MPSC. In addition to directing the efforts of the collaborative, council members serve as champions for the community-wide effort and the implementation of patient safety solutions at their respective organizations. The larger council meets each month and more frequently when necessary to sustain momentum. Smaller, focused workgroups, comprised of staff from each provider and multiple disciplines, meet to design and implement each of the MPSC's specific initiatives.

With funding from Dean Health System, Meriter Hospital, St. Mary's Hospital Medical Center, University of Wisconsin Hospital and Clinics, and University of Wisconsin Medical Foundation, the MPSC established a community-based center in September 2000 that coordinates and communicates patient safety initiatives. The center houses an administrator that coordinates efforts among member organizations, shares "learnings" and progress, secures external support for MPSC activities, and maintains key partnerships with local and national organizations concerned with patient safety.

In these first years of formation, MPSC operations have been largely project based. The MPSC council identifies an area for collective work based on frequency, severity, and control of the related adverse event and potential for collaboration. For example, based on incident reports and literature, the most frequent type of harm experienced by patients is medication errors. The topic was then narrowed down to a manageable focus area in which improvement strategies could be employed and measurable results could be obtained on a timely basis to sustain enthusiasm and momentum for collaborative activities. For its initial effort related to medication errors, MPSC focused on eliminating the use of error-prone abbreviations in medication orders. Literature suggests that a selected number of commonly used abbreviations contribute to medication errors. The most commonly misinterpreted abbreviations cited are u for units, QD for daily, and inappropriate use of leading and trailing zeros when a decimal point is used. In December 2000, the MPSC generated a common policy on abbreviation use and shared baseline information about the current use of these targeted abbreviations at each hospital. Utilizing accelerated improvement cycles, each member enacted interventions thought to be most appropriate and effective for their respective organization and, after several months, data were again collected and analyzed. Members then determined whether interventions found to be effective at one organization could be adopted by others to yield greater results in the next improvement cycle. Prevailing interventions were staff education activities including articles in organization publications, mandatory web-based modules on error-prone abbreviations, and medication errors and communications from influential physician leaders. Tactics also included structural changes within the process such as the creation of a separate medication order form with a grid designed to increase legibility; separate columns for medication, dose, route, frequency, and indications; and a section highlighting unacceptable abbreviations and incorrect use of zeros with a decimal point. Audits show that the MPSC initiative has been successful in reducing the presence of these abbreviations by greater than 90%. Members have now added additional error-prone abbreviations that are not to be used, and intensive work continues in an effort to completely eliminate the targeted abbreviations and sustain results.

After medication errors, falls are often cited as the second most common untoward event experienced by patients during hospitalization. In June 2001, the MPSC established a workgroup charged with reducing the occurrence of and harm from patient falls. The initiative convened nurses, physicians, pharmacists, physical therapists, and managers. These content experts then developed and implemented numerous processes to prevent falls and harm including strengthened patient assessment and identification of those at risk for falls, environmental interventions like "Safe Room Set-up," and benzodiazepine and sleeping aid reduction protocols to address the role of medications. In the pilot of interventions, falls were down an average of 51% on target units.

The MPSC has used similar processes to identify and begin comprehensive projects, implement improvement strategies, and achieve results in a number of safety areas, including improving provider adherence to recommended hand hygiene practices, teaching clinicians how to communicate with patients and families about disappointing outcomes and medical errors, preventing surgical infections, and facilitating access to information that plays a key role in medication safety and medication reconciliation by creating and distributing an educational brochure that includes a tear-out wallet card on which key medical information can be recorded and carried by community members.

The formal, community-wide effort of the MPSC has many strengths. Safety practices and programs can be instituted that reach far beyond the scope of any single organizational effort and help to create a community climate of safety. The MPSC allows Madison providers to share scarce fiscal, supportive, and structural resources to realize a common goal. Knowledge, expertise, and experiences, often kept within the four walls of an organization, are shared and used to improve patient care community wide. The group effort allows the safety message and interventions to reach a larger population of both health care professionals and patients. Efforts that span the health care community also provide the opportunity for reduction in variance in practice, policies, and procedures. Ultimately, the MPSC acts as a catalyst and helps to sustain Madison providers' attention and energy devoted to making patient safety improvements.

Challenges to collective work do exist, however. Although there has been no resistance or hesitation among the provider organizations in this community to share data and solutions, a significant barrier to collaborative work is time pressure. Provider organizations and individuals at health care organizations face great workloads and increasing demands for additional activity, particularly from external bodies, such as regulators, accreditation bodies, and purchasers. To help provider organizations and physicians and staff manage the need for ever increasing activity in the patient safety area, the MPSC has developed several other exchange mechanisms that provide information and tools on a faster basis than a comprehensive project.

In "Sharing Sessions," health care professionals from the Madison hospitals and medical groups are convened in person to share best practices, strategies, policies and processes, "lessons learned," and ask questions of their peers on patient safety topics. Topics addressed in this fashion include surgical site identification (marking) and surgical verification (use of "time-outs"), patient identification, safe use of "sitters" (now called "patient safety attendants"), and communicating test results. Where there is a great interest on the part of the participants and the organizations, some Sharing Session topics may become comprehensive projects, such as with surgical infection prevention.

In "Just-In-Time Sharing," an individual from a member organization working on a particular patient safety topic contacts the MPSC. On inquiry, other MPSC members are queried as to practice and resources in the particular area of patient safety. Information and resources are then shared, typically via e-mail and within 1 to 3 business days, to speed improvements in patient safety as members work internally. Safety matters addressed through this mechanism include prevention of retained surgical items, informed consent, and safety in psychiatry departments.

Having well established the MPSC's infrastructure and produced a number of achievements through collaboration, the MPSC looks forward to accelerated implementation of patient safety solutions across the community. The MPSC is currently developing plans to expand the role of the coordination center to that of a Patient Safety Education Center. The center would serve as a community resource for study and education on best practices in patient safety and how multidisciplinary groups of health care professionals can best respond as a team in critical safety situations. By working together, the MPSC aims to make improvements in health care safety at a level not possible at each individual organization alone and serve as a model that will encourage other communities to adopt such a cooperative approach in the best interests of all health care consumers. The MPSC initiative is unique because of its broad agenda and community-wide focus.

Clinical Collaboratives at SSM Health Care

SSM Health Care is a large Catholic health care system with 20 hospitals, 3 nursing homes, and a home health care agency. It covers 4 states in the Midwest of the United States. A total of 5,000 physicians and 23,000 employees care for patients. In 1998 SSM launched a series of clinical collaborative initiatives (Kosseff & Niemeier, 2001). Topics addressed by the clinical collaboratives include improving the secondary prevention of ischemic heart disease, improving prescribing practices, using patient information to improve care, enhancing patient safety through safe systems, improving the treatment of congestive heart failure, achieving exceptional safety in health care, improving the care of community acquired pneumonia patients, and improving critical care.

Clinical collaboratives develop over time in three phases: (a) prework phase, (b) active phase, and (c) continuous improvement phase. During the prework phase, SSM leaders send out invitations to all SSM entities for participation in the particular clinical collaborative. Several characteristics of the prework phase contribute to the success of the collaboratives: top management support as demonstrated by the invitation letter signed by SSM leaders, clearly defined goals, relevant background information, and specification of baseline data to be collected. All SSM entities are asked to join the collaborative at the very beginning. During the prework phase, directors for the collaborative are selected. Careful attention is given to the selection of those leaders: They know the SSM system and have a particular interest in the topic addressed by the collaborative. Before the active phase begins, a series of two or three conference calls are organized for the participants to get acquainted with each other and learn more about the baseline data collection (and get help if necessary). The goals of the collaborative are reiterated during the initial conference calls to ensure that the collaborative team members understand and share the same goals.

The active phase lasts 6 months to 3 years. During this phase, the collaborative is expected to accomplish the project goals and to demonstrate measurable improvements. The active phase involves three major activities: (a) a 6-hr learning session led by the SSM chief executive officer (CEO) and the senior vice president (VP) for strategic development, (b) project work and completion, and (c) a second 6-hr learning session in which members of the team present their results to an audience that includes SSM senior management. Actual project work is managed by the collaborative director and facilitators and involves regular (monthly) conference calls, listserv discussions, and individual team coaching.

The final phase, that is, continuous improvement phase, aims at maintaining the gains achieved during the active phase. In addition, new goals related to the particular topic may come up and are then addressed. This phase involves regular data collection (e.g., every 3 months) and bimonthly conference calls.

Factors that contribute to the success of the SSM clinical collaboratives include (Kosseff & Niemeier, 2001):

- Careful selection of topics. The topic should be important: achieving the goals should lead to improvement in patient care. There should be scientific literature addressing the topic and providing support for the proposed changes. Data collected during the prework phase should demonstrate the existence of a gap in patient care and outcome therefore demonstrating the need for change.
- Leadership support. SSM leaders and top managers are very much involved at the inception of the teams as well as during the active phase. SSM leaders actively support the continuous improvement phase of the various collaboratives.
- Clear aims and measurable outcomes. Data are collected at the beginning of the project to provide baseline information on the current status. Data are regularly collected to monitor progress toward the goals.
- Participation of process users before implementation of changes. Health care providers and staff involved in the particular process targeted by the collaborative provide input before any change is implemented.

Implementation of changes is facilitated by a range of factors. For instance, teams attempt to identify changes that are easier to perform as compared to current care process. Pilot projects are used to test the proposed change. The local context is very important in "shaping" the particular change to be implemented in a specific health care organization: "local reinvention" is expected. The continuous improvement phase is critical to sustaining change over time and to maintain gains. This phase allows for the incorporation of new knowledge in care processes.

Other CIPS

A number of other patient safety collaboratives have emerged recently. In the Delaware Valley (Philadelphia and surrounding area) of Pennsylvania, a consortium of health care providers in collaboration with the Health Care Improvement Foundation, ECRI, and the Institute for Safe Medication Practices (ISMP) has been created to reduce medication errors at the regional level (McCarter, Centafont, Daly, Kokoricha, & Leander Po, 2003).

The Pittsburgh Regional Healthcare Initiative (PRHI) is a regional initiative aimed at improving quality of care and patient safety within the Pittsburgh area (Sirio et al., 2003). PRHI's vision is "to achieve perfect patient outcomes by identifying and solving system problems at the point of patient care" (Sirio et al., 2003, p.158). The characteristics of PRHI include:

1. Region-wide data collection and reporting systems. For instance, PRHI has endorsed the use of USP's (United States Pharmacopeia) Medmarx to evaluate medication errors.
2. Linking patient outcomes to processes of care. The 13 major cardiac surgery groups in southwestern Pennsylvania are working together to improve cardiac care outcomes. They are working on establishing a common cardiac registry for measuring patient outcomes related to CABG (Coronary Artery Bypass Graft) surgery.
3. Sharing information to improve performance. Based on analysis of the medication error data reported via the USP Medmarx system, a series of safe medication practice guidelines have been created on issues such as preventing the use of unsafe abbreviations.
4. Changing care processes to improve outcomes. For instance, a number of programs have been implemented to reduce infections.

PRHI's "region-wide shared learning" requires changes in the patient safety environment, culture, and infrastructure (Sirio, Keyser, Norman, Weber, & Muto, 2005). First, leaders of individual health care organizations must be committed to patient safety improvement in their own organizations. Second, health care workers in the various institutions members of the network need to be able to identify patient safety problems and find solutions. In addition, those lessons learned need to be shared widely. Third, changes in system design and clinical processes need to be linked to patient outcomes therefore demonstrating important improvements in patient care.

A number of states have developed patient safety centers. A review of six state patient safety centers by the National Academy for State Health Policy showed that they share the following common objectives: fostering a culture of safety, educating about patient safety, and potentially serving as a data repository (Rosenthal & Booth, 2004). These state centers have a range of roles, including educating providers about safety practices and promoting health information technology to improve patient safety. However, these state patient safety centers have only an indirect impact on patient safety improvement activities. The state centers are not directly involved in the design or redesign of systems of care. This feature distinguishes state patient safety centers from the other collaboratives described earlier. Those collaboratives have as a primary objective to implement changes in the systems and processes of care to improve patient safety.

CONCLUSION

"Trying to solve the shortcomings of the present care system at only one level or another—only at the level of aims and accountabilities, or microsystems, or organizations, or the environment—will not work" (Berwick, 2005, p. 334). CIPS represent an important mechanism for bridging various levels of the care system. They allow various health care providers to work together to improve patient safety. These collaboratives can lead to efficient and effective improvements in patient safety within the participating organizations. More important, they allow for the discovery of interactions between health care organizations. Understanding those interactions and their impact on patient safety can ultimately lead to solutions of patient safety problems that occur in between levels of care.

References

Berwick, D. M. (2005). The John Eisenberg lecture: Health services research as a citizen in improvement. *Health Services Research, 40*, 317–335.

Box, G. E. P., Joiner, L. W., Rohan, S., & Sensenbrenner, F. J. (1989). *Quality in the community: One city's experience*. Madison: Center for Quality and Productivity Improvement of the University of Wisconsin–Madison.

Brass, D. J., Galaskiewicz, J., Greve, H. R., & Tsai, W. (2004). Taking stock of networks and organizations: A multilevel perspective. *The Academy of Management Journal, 47*, 795–817.

Carayon, P., Borgsdorf, A., Hundt, A. S., & Alvarado, C. J. (2003, August 24–29). *Making a community safer for patients: The development of the Madison Patient Safety Collaborative*. Paper presented at the International Ergonomics Association Conference, Seoul, Korea.

Chisholm, R. F. (1998). *Developing network organizations: Learning from practice and theory*. Reading, MA: Addison-Wesley.

Clemmer, T. P., Francioli, P., Bille, J., Oniki, T. A., & Horn, S. D. (1999). Results of a collaborative quality improvement program on outcomes and costs in a tertiary critical care unit. *Critical Care Medicine, 27*, 1768–1775.

Cornish, P. L., Knowles, S. R., Marchesano, R., Tam, V., Shadowitz, S., Juurlink, D. N., et al. (2005). Unintended medication discrepancies at the time of hospital admission. *Archives of Internal Medicine, 165*, 424–429.

Dixon, N. M. (2000). *Common knowledge—How companies thrive by sharing what they know*. Boston, MA: Harvard Business School Press.

Donabedian, A. (1988). The quality of care. How can it be assessed? *Journal of the American Medical Association, 260*, 1743–1748.

Fraser, I., Lanier, D., Hellinger, F., & Eisenberg, J. M. (2002). Putting practice into research. *Health Services Research, 37*, 1–14.

Gittell, J. F. (2002). Coordinating mechanisms in care provider groups: Relational coordination as a mediator and input uncertainty as a moderator of performance effects. *Management Science, 48*, 1408–1426.

Goode, L. D., Clancy, C. M., Kimball, H. R., Meyer, G., & Eisenberg, J. M. (2002). When is "good enough"? The role and responsibility of physicians to improve patient safety. *Academic Medicine, 77*, 947–952.

Helmreich, R. L., & Schaefer, H. (1994). Team performance in the operating room. In M. S. Bogner (Ed.), *Human error in medicine* (pp. 225–253). Mahwah, NJ: Lawrence Erlbaum Associates, Inc.

Horbar, J. D., Rogowski, J., Plsek, P. E., Delmore, P., Edwards, W. H., Hocker, J., et al. (2001). Collaborative quality improvement for neonatal intensive care. *Pediatrics, 107*(1), 14–22.

Inkpen, A. C., & Tsang, E. W. K. (2005). Social capital, networks, and knowledge transfer. *Academy of Management Review, 30*(1), 146–165.

Institute of Medicine Committee on Quality of Health Care in America. (2001). *Crossing the quality chasm: A new health system for the 21st. century* Washington, DC: National Academy Press.

Kilo, C. M. (1999). Improving care through collaboration. *Pediatrics, 103*, 384–393.

Kohn, L. T., Corrigan, J. M., & Donaldson, M. S. (Eds.). (1999). *To err is human: Building a safer health system*. Washington, DC: National Academy Press.

Kosseff, A. L., & Niemeier, S. (2001). SSM health care clinical collaboratives: Improving the value of patient care in a health care system. *The Joint Commission Journal on Quality Improvement, 27*, 5–19.

Lawrence, P. R., & Lorsch, J. W. (1967). *Organization and environment.* Cambridge, MA: Harvard Graduate School of Business Administration.

Leape, L. L., Kabcenell, A. I., Gandhi, T. K., Carver, P., Nolan, T. W., & Berwick, D. M. (2000). Reducing adverse drug events: Lessons from a breakthrough series collaborative. *Joint Commission Journal on Quality Improvement, 26*, 321–331.

Leatherman, S. (2002). Optimizing quality collaboratives. *Quality & Safety in Health Care, 11*, 307.

McCarter, T. G., Centafont, R., Daly, F. N., Kokoricha, T., & Leander Po, J. Z. (2003). Reducing medication errors. A regional approach for hospitals. *Drug Safety, 26*, 937–950.

Mohrman, S. A., Tenkasi, R. V., & Mohrman, A. M. J. (2003). The role of networks in fundamental organizational change. *The Journal of Applied Behavioral Science, 39*, 301–323.

Morey, J. C., Simon, R., Jay, G. D., Wears, R. L., Salisbury, M., Dukes, K. A., et al. (2002). Error reduction and performance improvement in the emergency department through formal teamwork training: Evaluation results of the MedTeams project. *Health Service Research, 37*, 1553–1581.

Ovretveit, J., Bate, P., Cleary, P., Cretin, S., Gustafson, D., McInnes, K., et al. (2002). Quality collaboratives: Lessons from research. *Quality & Safety in Health Care, 11*, 345–351.

Paniccia, I. (1998). One, a hundred, thousands of industrial districts. Organizational variety in local networks of small and medium-sized enterprises. *Organization Studies, 19*, 667–699.

Podolny, J. M., & Page, K. L. (1998). Network forms of organization. *Annual Review of Sociology, 24*, 57–76.

Porter, M. E. (1998). Clusters and the new economics of competition. *Harvard Business Review, 76*(6), 77–90.

Porter, M. E., & Teisberg, E. O. (2004). Redefining competition in health care. *Harvard Business Review, 82*(6), 65–76.

Rogowski, J., Horbar, J. D., Plsek, P., Baker, L. S., Deterding, J., Edwards, W. H., et al. (2001). Economic implications of neonatal intensive care unit collaborative quality improvement. *Pediatrics, 107*, 23–29.

Rosenthal, J., & Booth, M. (2004). *State patient safety centers: A new approach to promote patient safety.* Portland, ME: National Academy for State Health Policy.

Sirio, C. A., Keyser, D. J., Norman, H., Weber, R. J., & Muto, C. A. (2005). Shared learning and the drive to improve patient safety: Lessons learned from the Pittsburgh Regional Healthcare Initiative. In *Advances in Patient Safety: From Research to Implementation* (AHRQ Publication Nos. 050021, 1–4 ed., vol. 3, Implementation Issues, pp. 153–165). Rockville, MD: Agency for Health care Research and Quality.

Sirio, C. A., Segel, K. T., Keyser, D. J., Harrison, E. I., Lloyd, J. C., Weber, R. J., et al. (2003). Pittsburgh Regional Healthcare Initiative: A systems approach for achieving perfect patient care. *Health Affairs, 22*(5), 157–165.

Thomas, E. J., Sherwood, G. D., & Helmreich, R. L. (2003). Lessons from aviation: Teamwork to improve patient safety. *Nursing Economic$, 21*, 241–244.

van Walraven, C., Mamdani, M., Fang, J., & Austin, P. C. (2004). Continuity of care and patient outcomes after hospital discharge. *Journal of General Internal Medicine, 19*, 624–631.

Wenger, E., McDermott, R., & Snyder, W. M. (2002). *Cultivating communities of practice.* Boston, MA: Harvard Business School Press.

Wilson, T., Berwick, D. M., & Cleary, P. D. (2003). What do collaborative improvement projects do? Experience from seven countries. *The Joint Commission Journal on Quality Improvement, 29*, 85–93.

Section III

JOB AND ORGANIZATIONAL DESIGN

11

JOB STRESS IN HEALTH CARE WORKERS

Lawrence R. Murphy
National Institute for Occupational Safety and Health (NIOSH)

Health care professions have traditionally been identified as "high stress" occupations in the literature (e.g., Payne & Firth-Cozens, 1987, 1999) as a result of exposure to psychosocial factors and infectious, hazardous, and physical agents (Rogers, 1997). Early studies identified health care workers as having more occupational illnesses and injuries than workers employed in other service industries (Bureau of Labor Statistics, 1995; Calhoun, 1980), higher rates of mental health admissions (Colligan, Smith, & Hurrell, 1977), more absenteeism, and higher staff turnover (Firth and Britton, 1989). For example, Colligan and colleagues (1977) examined 8,450 cases from mental health center records and found a disproportionate incidence of mental health disorders among workers in health care occupations, which comprised seven of the top-ranked occupations. Later studies confirmed excess mental health problems and stress-related disorders among health care workers (e.g., Gunderson & Colcord, 1982; Hoiberg, 1982).

The problem of job stress in health care workers is not merely an academic issue but one that is associated with high costs across a range of measures. For instance, each additional patient over four in a registered nurse's workload increases the risk of the death by 7% for surgical patients. Patients in hospitals with the lowest nurse staffing levels (eight patients per nurse) have a 31% greater risk of dying than those in hospitals with four patients per nurse. On a national scale, staffing differences of this magnitude could result in as many as 20,000 unnecessary deaths annually (Aiken, Clarke, Sloane, Sochalski, & Silber, 2002). That same study also reported a 23% increase in burnout and a 15% increase in job dissatisfaction among nurses for each additional patient beyond four. Because burnout has a major influence on nurse turnover, and the costs of turnover have been estimated at $40,000 to $60,000 per registered nurse, the costs become an issue quickly. For instance, in a hospital with 500 nurses and 15% turnover (the U.S. national turnover rate in 2003), the total costs for replacing 75 nurses would be $4.5 million.

In this chapter, the review of the literature is organized around job risk factors that have consistently emerged as important in studies of health care workers. The grouping of studies according to risk factors is somewhat arbitrary, because many studies included measures of multiple risk factors and multiple health outcomes. However, the grouping provides a convenient structure for categorizing the many studies that have been published in the literature. Before examining risk factors, a brief summary will be presented of the main mental health and organizational outcomes that have been examined in health care workers.

JOB-RELATED STRESS IN HEALTH CARE WORKERS

Mental Health Outcomes

Most studies of stress and health among health care workers have focused on three types of mental health indicators: burnout, perceived stress, and job satisfaction. The most common mental health problem associated with health care occupations has been burnout, especially emotional exhaustion (Cherniss, 1980; Duxbury, Armstrong, Drew, & Henly, 1984; Fruedenberger, 1974; Golembiewski, Munzenrider, & Stevenson, 1986; Maslach, 1982; Maslach & Jackson, 1982; Numerof & Abrams, 1984). The finding of elevated burnout in health care workers has appeared consistently in the literature over the years and continues to be a problem in the modern workplace (e.g., Robinson, Clements, & Land 2003).

Job stress in health care workers has been examined almost as often as burnout. Record studies noted earlier (Colligan et al., 1977; Gunderson & Colcord, 1982; Hoiberg, 1982) pointed to excess mental health problems in health care occupations. Caplan (1994) reported that approximately half of general practitioners scored over 5 on the General Health Questionnaire (GHQ), which was nearly double the rate for professional and managerial occupations. More recently, in a study of more than 7,000 workers in 13 occupations, Sparks and Cooper (1999) found the highest levels of mental stress among nurses and physicians compared with other occupations. Arafa, Nazel, Ibrahim, and Attia (2003) found that nearly 22% of nurses from five health organizations in Alexandria, Egypt, reported moderate to severe psychological symptoms on the GHQ. Weibel, Gabrion, Aussedat, and Kreutz (2003) found that emergency medical dispatchers showed significantly increased cortisol levels during their work time compared with levels found in controls.

Job satisfaction in health care workers is often found to be low relative to other occupations. Mottaz (1988) compared nurses with workers in other occupations and found that job satisfaction scores among nurses were more similar to those obtained from factory managers and factory workers than those obtained from groups such as university faculty, elementary school teachers, and administrative personnel.

The relationship between job stress and burnout is worth examining, because they are often associated in the research literature and appear to be measuring similar, if not the same, outcome. It is common to find a relationship between job stress and burnout, such as that recently reported by Stordeur, D'Hoore, and Vandenberghe (2001) in their study of 625 hospital nurses. Schaufeli and van Dierendonck (1993) demonstrated that the emotional exhaustion component of burnout shares approximately 30% of the variance with generic measures of mental and physical symptoms of job stress. Schaufeli and Buunk (2003) suggested that burnout is a type of job stress, characterized by its chronic nature and its tendency to be experienced by people with high goals and expectations at work. It is noteworthy that in its original description, burnout was characterized as situation specific (Fruedenberger, 1974), whereas job stress was often seen as more pervasive. In any case, for the purposes of this chapter, the concepts of burnout and job stress will be considered related but separate. A similar relationship appears to exist with respect to job stress and job satisfaction in that they co-vary in many studies. Thus, Arafa and colleagues (2003) reported a link between GHQ symptoms and job dissatisfaction.

Organizational Outcomes

In addition to health indicators, some studies have assessed organizational outcomes such as turnover and absenteeism. Gauci Borda and Norman (1997a) reviewed the literature on job satisfaction, turnover, and absenteeism and reported that intent to stay in current employment was the best indicator of actual turnover, and job satisfaction had the greatest influence on intent to stay. Less consistent results were found for the relationships between job satisfaction and absenteeism. The authors developed a model relating job satisfaction, intent to stay, and turnover, which they tested in a second study (Gauci Borda & Norman, 1997b). Tests of the model were confirmed for some subgroups (males but not females) and, in contrast to the conclusions from the literature review, job satisfaction was clearly linked to both absenteeism and intent to stay. Lu, Lin, Wu, Hsieh, and Chang (2002) collected questionnaire data from a random sample of 2,197 nurses in southern Taiwan and confirmed the positive association between job satisfaction and intent to turnover.

Job Risk Factors

On the risk factor side of the equation, a wide variety of job characteristics and organizational factors associated with worker well-being has been assessed, and a handful of these risk factors have emerged as the most significant.

Shift Work. Substantial evidence has accumulated over the years that indicates that the temporal scheduling of work can have a significant effect on psychological, behavioral, social, and physical well-being. Rotating shifts and permanent night work, in particular, have been linked to a variety of such disturbances (Jamal, 1981; Johnson, Tepas, Colquohoun, & Colligan 1981; Monk & Tepas, 1985; Rutenfranz, Colquhoun, Knauth, & Ghata, 1977). For instance, in a questionnaire and records study of 1,200 nurses, Tasto, Colligan, Skjei, and Polly (1978) found that nurses on rotating shifts took more sick days and gave more serious reasons for sick leave than workers on other shifts. Nurses on rotating shifts had more accidents at work. Questionnaire results identified more sleep problems, gastrointestinal complaints, and fatigue among rotating and night shift workers than day or afternoon shift workers. With respect to the deleterious effects of shift work on psychological and physiological functioning, workers on rotating shifts fared the worst, followed by night shift workers (Tasto et al., 1978). These shift-related complications have been attributed to a disruption of physiologic circadian rhythms and to social interactions associated with a work schedule that is at odds with the normal diurnal activity cycle (Aschoff 1981).

Role Stressors. Role stressors (ambiguity, conflict, and overload) constitute one of the earliest and most researched sources of stress at work among health care workers (Beehr & Newman, 1978; Kahn, Wolfe, Quinn, & Snoek, 1964). Most commonly, role ambiguity and role conflict have been assessed and found to be associated with outcomes such as job dissatisfaction (Jackson & Schuler, 1985) and symptoms of ill-health (Cooper & Marshall, 1976). Within health care occupations, especially nursing, role ambiguity and conflict have been consistently identified as major sources of stress and have occupied a central place in intervention efforts. For instance, Gray-Toft and Anderson (1981) recommended role conflict and ambiguity as key targets for improvement in efforts to lower nurse stress. An intervention study by Jackson (1983) sought to improve participation in decision making among workers in an outpatient hospital facility to reduce role conflict and ambiguity and to lower worker distress, absenteeism, and turnover intention. Revicki and May (1989) and Revicki and Whitley (1995) identified role ambiguity as a major source of stress for nurses, and Ellis and Miller (1994) found a relationship between role ambiguity and emotional exhaustion among nurses in an acute care hospital. More recently, Tovey and Adams (1999) reported that the main sources of job dissatisfaction were role conflict, lack of job security, inadequate resources, and a perceived lowering of standards of patient care.

Job Demands and Worker Control. A number of studies have examined the role of high demands and low control among health care workers, either separately or in combination, based on the Karasek model of job strain (Karasek, 1979; Karasek, Schwartz, & Theorell, 1982). Landsbergis (1988) found that burnout, job dissatisfaction, and psychosomatic symptoms were significantly higher among health care workers in jobs with high demands and low control. Fox, Dwyer, and Ganster (1993) also found support for the demand–control model in a study of nearly 200 hospital nurses that linked high job demands and low control with various physiological indicators of stress, including elevated blood pressure and cortisol. Munro, Rodwell, and Harding (1998) found lower satisfaction and higher GHQ scores among nurses with high job demands and low control, and content analyses of semistructured interviews with 35 nurses revealed that job dissatisfaction was most highly associated with feeling overloaded (McNeese-Smith, 1999).

In a study from the Netherlands, van Vegchel, deJonge, Meijer, and Hamers (2001) surveyed health care workers in nursing homes and found more psychosomatic complaints among workers who reported high effort and low rewards. Tummers, Janssen, Landeweed, and Houkes (2001) identified high workload as a major contributor to emotional exhaustion, and Yip (2001) found an association between work stress, manual lifting, and low back pain prevalence, suggesting a need for continued focus on ergonomics in the health care workplace.

Social Support. Many of the early studies pointed to a key relationship between low supervisory support and burnout and stress among health care workers (Duxbury et al., 1984; Golembiewski et al., 1986; Gray-Toft & Anderson, 1981, 1985; Maslach, 1982; Maslach & Jackson, 1982; Numerof & Abrams, 1984); more recent studies have confirmed this association. For instance, Landsbergis (1988) noted the contributing role of social support to burnout and job strain among 771 hospital and nursing home employees. Mottaz (1988) found that supervisory assistance was a major determinant of job satisfaction among 312 nurses from four hospitals. Likewise, Ellis and Miller (1994) found that role ambiguity and lack of social support were related to emotional exhaustion among 492 nurses in a large U.S. acute care hospital. Moreover, social support was clearly linked to employee commitment and turnover intention.

Revicki and May (1989) developed and tested a model of stress and mental health among health care workers. Using questionnaire data obtained from 300 hospital nurses, the authors found that a positive organizational climate was strongly related to supervisory behavior, which in turn was related to role clarity, better work group relations, and higher job satisfaction. A later study (Revicki & Whitley, 1995) reported increased job stress among emergency medicine residents who had low role clarity and low support at work. Based on cluster analyses of survey responses from 260 hospital nurses, Hillhouse and Adler (1997) identified social support, workload, and role conflict as the major factors causing nurse stress. A longitudinal study by Baldwin (1999) followed newly graduated British nurses and discovered that support and communication from senior staff were directly related to lower GHQ scores each year over a 3-year follow-up period.

More recent studies suggest that stress and burnout continue to be a problem for health care workers. Tummers and colleagues (2001) examined job characteristics and health outcomes and reported that high workload and low social support were linked to emotional exhaustion, a primary characteristic of burnout. Aiken, Clarke, and Sloane (2002) conducted a cross-national study of adult acute care hospitals in the United States, Canada, England, and Scotland involving more than 10,000 nurses working in 303 hospitals. The authors found that organizational and supervisory support for nursing was associated with dissatisfaction, burnout, and quality of care. Nurse reports of low quality care were three times as likely in hospitals with low staffing and low support for nurses as in hospitals with high staffing and high support for nurses.

Visser, Smets, Oort, and de Haes (2003) found strong links between job stress, satisfaction, and burnout among 1,500 Dutch medical specialists. Job satisfaction was most highly related to social support at work, and levels of stress were related to work overload, low autonomy, and job insecurity. McGrath, Reid, and Boore (2003) surveyed 171 nurses in Northern Ireland and found the most commonly cited stressors were too little time to perform their duties and rationing of services or resources.

In a recent review of the literature on job stress in nursing, the main sources of stress identified among nurses were workload, leadership or management style, professional conflict, and the emotional cost of caring (McVicar, 2003), which agrees with conclusions from another recent review of the literature (Lundstrom, Pugliese, Bartley, Cox, & Guither, 2002). Arafa and colleagues (2003) found that nearly 22% of nurses reported moderate to severe psychological symptoms on the GHQ, and these symptoms were related to lack of social support and job dissatisfaction. van Yperen and Hagedoorn (2003) suggested that social support increases intrinsic motivation, even under conditions of high job demands and low control. Munro and colleagues (1998) found lower satisfaction and higher GHQ scores among nurses with high job demands and low control and noted the importance of social support at work as a key element contributing to nurse well-being. In a similar vein, McGowan (2001) found that stress at work resulted from a perceived lack of organizational support and involvement among health care workers. As noted earlier, social support was one of the factors identified by Seo, Ko, and Price (2004) related to nurse job satisfaction.

Staffing. In a review of studies of health care workers in Finnish municipal health care centers, Lindstrom (1992) identified a wide range of potential job stressors, including time pressure (related to high workload and insufficient staff), responsibility for the health of others, lack of autonomy, and shift work. These stressors traditionally have been associated with burnout among health care workers. Inadequate staffing was one of the most common sources of stress reported by British mental health nursing students in an open-ended survey item about anticipated sources of stress (Kipping, 2000). Recent focus group data revealed that the three most important etiological factors among nurses were high workload, physical work environment, and inadequate staffing (Shamian, O'Brien-Pallas, Thomson, Alksnis, & Kerr, 2003). The factors most highly linked to perceived stress were workload, feelings of emotional exhaustion, and lack of social support. Shamian and colleagues (2003) provided evidence that psychosocial factors at work contribute to musculoskeletal injury, including reduced staffing levels and heightened intensity of work that resulted from downsizing and reorganization. Such injuries lead to higher rates of absenteeism and disability (Shamian et al., 2003).

Aiken, Clarke, Sloane, and colleagues (2002) found higher levels of burnout and job dissatisfaction among nurses who worked in hospitals with high patient-to-nurse ratios. This study is significant owing to the large sample size (more than 10,000 nurses) and inclusion of a large number of general hospitals in Pennsylvania. The authors also found that more than 40% of nurses who reported high burnout and job dissatisfaction also reported high turnover intent within the next 12 months, compared with 11% of satisfied, nurses who were not burned out. Finally, in a study of more than 800 nurses in 100 British hospitals, Adams and Bond (2003) noted that adequate staffing was linked to better interdisciplinary collaboration, an increased ability of nurses to cope with workload, and job satisfaction.

Fear of Violence. A noteworthy stressor identified by Lindstrom (1992) in his review was fear of violence at work, which only recently has come under intense research scrutiny (Lipscomb, 1999; National Institute for Occupational Safety and Health [NIOSH], 1996; Nolan, Dallender, Soares, Thomsen, & Arnetz, 1999; Simonowitz, 1996). This source of stress is common among health care workers who have direct patient contact and has been especially prevalent among nurses (Fottrell, 1980; Martin 1984).

Responding to an open-ended survey item about anticipated sources of stress, recently graduated mental health nursing students (n = 556) from colleges of nursing in England rated violence on the job, poor relations with other staff, and inadequate staffing as the most common sources of stress (Kipping, 2000). In a cross-national study of Swedish and English mental health nurses, Nolan, Soares, Dallender, Thomsen, and Arnetz (2001) reported that more than 70% of English nurses and nearly 60% Swedish nurses reported experiencing violence at work in the prior 12 months. Approximately 60% of both groups reported multiple incidents of violence during the 12-month period. The most common type of violence reported were verbal threats not involving physical contact.

Infectious Diseases. Health care workers are exposed to a variety of health-endangering factors at work, including physical agents, chemical agents, and infectious diseases (Gillmore, 1990; Rogers, 1997). For instance, in the 1980s, HIV/AIDS represented a new class of exposures for health care workers and a great deal of fear and worry. In the mid-1980s, the U.S. Centers for Disease Control and Prevention (CDC) published recommendations that encouraged hospitals to voluntarily adopt an infection control policy of universal precautions, in which health care workers are encouraged to assume that all patients are potentially infectious for HIV and other blood-borne pathogens (Centers for Disease Control and Prevention, 1987).

Despite these precautions, health care workers continued to experience heightened stress associated with caring for HIV/AIDS patients (Henderson 1989) and additional stress associated with fear of exposure to HIV/AIDS (Murphy, Gershon, & DeJoy, 1996). Regarding the latter, Gerbert, Maguire, Badner, Altman, and Stone (1989) reviewed research studies that explored the various reasons why HIV/AIDS generates stress among health care workers. These authors identified three primary reasons for the continuing fear among health care professionals: (a) the risk of occupational transmission of HIV/AIDS is real and the consequences are serious; (b) infection control procedures cannot prevent all occupational exposures, such as accidental needle sticks; and (c) communication barriers exist between medical experts and health care providers. Stress generated from fear of contracting HIV/AIDS at work has been evidenced by worker reported symptoms like nightmares, anxiety, constant worry and overall distress (Gerbert et al., 1989). Recent cases of severe acute respiratory syndrome (SARS) add yet another potential exposure for health care workers (World Health Organization, 2003).

Medical Errors. A final area of recent concern among health care workers is patient safety. Reports by the Institute of Medicine (Aspden, Corrigan, Wolcott, & Erickson, 2004; Kohn, Corrigan, & Donaldson, 1999) laid out the problem of medical errors and suggested the need for a new health care delivery system, one that would prevent errors from occurring and that would learn from the systems that do occur. The Kohn and colleagues report called for every health care organization in the United States to establish a "comprehensive patient safety system" that allows for immediate access to patient information and documents both adverse events and "near misses" in which patients were not harmed but could have been. Yet another IOM report (Page, 2004) provided guidance on ways to change the work environment of nurses to improve patient safety. One of the recommendations was the need to create and sustain a culture of safety that would include adequate staffing and organizational support for continued learning.

Another article provided a link between burnout and risk for workplace accidents. Clarke, Sloane, and Aiken (2002) found that nurses on units with lower staffing and higher levels of emotional exhaustion were twice as likely to report needle stick risks, such as carelessness and frequent recapping of needles. The authors noted that needle stick injuries are not randomly distributed but rather tend to occur more often in those nursing units characterized by lower staffing and poor organizational climates.

Summary

The literature provides strong evidence that health care workers experience elevated levels of burnout, stress, and job dissatisfaction, and these outcomes are closely related to turnover intention and absenteeism. Job characteristics and organizational factors linked to mental health outcomes include heavy workload, insufficient staffing, lack of supervisory support, role ambiguity, lack of autonomy, and shift work. Many of these have become exacerbated in the 1990s because of layoffs, restructuring, and job redesign (Armstrong-Stassen, Horsburgh, & Cameron, 1994; Pindus & Greiner, 1997; Short, 1997). To this list can be added more recently documented stressors such as job insecurity, work and family interference, and being threatened at work.

This chapter examines traditional and emergent sources of job stress among health care workers and the relationship of these sources to mental health and organizational outcomes using data obtained from a nationally representative sample of U.S. adults in 2002.

METHOD

General Social Survey

The General Social Survey (GSS) is a biannual, U.S. national survey conducted by the National Opinion Research Center. In addition to a core module that evaluates a variety of demographics and social attitudes, each year, the GSS contains a number of specialized modules. In the 2002 GSS, NIOSH added a module on the quality of work life (QWL). The purpose of the QWL module was to provide national estimates of how the experience of work is changing and how these changes might influence worker health and safety and to shed light on the types of preventive interventions that are needed to address the changing nature of work.

The QWL module was developed by NIOSH with input and advice from a panel of experts in organizational behavior, occupational safety and health, sociology, and human resource management (see Acknowledgements for list of panel members). A small group process was used to establish criteria for assembling the QWL survey module. The starting point for group discussions was the Quality of Employment Surveys (QES) conducted by the University of Michigan in 1969, 1972, and 1977. The QES were sponsored by the Department of Labor and contained more than 300 questions covering a wide range of topics dealing with the nature of job tasks, working hours, job security, and job satisfaction. The goals of these surveys were to establish baseline and normative data on working conditions and worker mental health, to examine trends in these factors, and to assess the impact of working conditions on worker well-being (Quinn & Shepard, 1974; Quinn & Staines, 1979; Survey of Working Conditions, 1971).

A first task for the panel was to assess whether topic areas and individual items from the QES were still relevant in 2002. Where QES topic areas were found lacking in relevancy for contemporary work issues, the panel made recommendations for updating the topic area. Members of the panel completed a premeeting exercise in which they rated the importance of each QES topic area as high, medium, or low with respect to its inclusion on the 2002 QWL module. Because the QES used a large number of items and scales, and the proposed QWL module would be relatively small, criteria were needed to narrow topic areas for inclusion in the module. Proposed criteria included the timeliness and relevance of the topic in modern work organizations, a balance of job content, work organization, outcome measures, and sensitivity of the item or scale to labor force trends. The panel decided that the QWL module should contain as many indicators as possible from the older QES to allow for assessments of trends, but that the QWL module should also reflect

TABLE 11–1. Frequency of Healthcare Occupations in the 2002 General Social Survey Dataset.

Occupational Title	N	%
Physicians	4	4%
Dentists	1	1%
Registered nurses	32	29%
Pharmacists	2	2%
Dietitians	1	1%
Inhalation therapists	1	1%
Occupational therapists	5	5%
Physical therapists	2	2%
Clinical laboratory technologists and technicians	4	4%
Radiologic technicians	3	3%
Health technologists and technicians, not elsewhere classified.	9	8%
Dental assistants	2	2%
Health aides, except nursing	11	10%
Nursing aides, orderlies, and attendants	32	29%
Dental laboratory and medical appliance technicians	1	1%
Total healthcare =	110	100%

a balance of measures of job content, work environment, organizational climate/culture, and health and performance measures.

Once a set of key quality-of-work life constructs was identified, individual items measuring each construct were selected. This process produced a rather large, 120-item survey that was nearly twice as long as desired. The next step was either to reduce the number of constructs being measured or to reduce the number of items used to measure each construct. The latter option was selected, and the number of items used per scale was decreased. A 90-item QWL module was created and pilot tested by NORC. Pilot testing revealed that the module was still too large for the 20-min interview time, and an additional 12 items were eventually dropped. The final module contained 76-items and assessed a broad range of work environment and health and safety factors (NIOSH, 2002).

Sample

The target population for the GSS in 2002 was the U.S. adult, noninstitutionalized, English-speaking population, 18 years or older (See Davis & Smith, 2002, for additional details on the sample, interviewing specifications, and data collection procedures). A total of 2,765 people responded to the survey, representing a response rate of 70%. For this chapter, the sample was restricted to those who were currently employed, worked at least 20 hr per week, and were not self-employed. The final sample size was 1,459 workers. All statistical analyses reported in this chapter were based on unweighted data.

Fifteen detailed health care occupations were contained in the GSS 2002 dataset, and the unweighted frequency of workers in each occupation is shown in Table 11–1. These health care occupations were grouped into four categories for statistical analyses: registered nurses (n = 32); nurse and health aides (n = 43); health technicians and therapists (n = 27); and physicians, dentists, and other (n = 8). The GSS sample was similar to national estimates of workers in the four occupational categories, according to the Bureau of Labor Statistics (BLS): registered nurses (GSS = 1.6%, BLS = 1.8%); nurse and health aides (GSS = 2.4%, BLS = 2.0%); health technicians and therapists (GSS = 1.8%; BLS = 1.8%); physicians, dentists, and other (GSS = 1.8%, BLS = 1.8%). Overall, health care workers were 7.2% of the GSS sample compared with 7.4% of the BLS occupation estimates.

Measures

Based on the job characteristics and organizational factors identified in the literature, fifteen dimensions of the work environment were examined, as shown in Table 11–2: adequacy of staffing (1 item), work/family interference (2 items), role clarity (1 item), skill utilization (1 item), work overload (1 item), autonomy (1 item), supervisory support (2 items), job security (1 item), pay equity (1 item), teamwork (1 item), ease of finding a simar job (1 item), threatened on the job (1 item), safe work behavior (2 items), safety and health conditions (1 item), and repeated lifting (1 item). Four well-being outcomes were included: job stress (1 item), burnout (1 item), self-rated health (1 item), and job satisfaction (1 item). Four measures of organizational outcomes were included: employee commitment (1 item), organizational effectiveness (2 items), accidents (1 item), and turnover intention (1 item).

Analysis Strategy

The analyses were conducted in several phases. First, analyses of covariance (ANCOVA) were used to compare health care workers as a group with non–health care workers on measures of job characteristics, health, and organizational outcomes using demographic variables as covariates. ANCOVA analyses were used because they provide mean scores for each group after adjustment for differences in demographic factors. Then, correlations among the measures were examined separately for health care and non–health care groups. Next, stepwise multiple regressions were used to explore relationships among job characteristics and measures of health and organizational outcomes for health care and non–health care groups separately. Finally, subgroups of health care workers were formed, and ANCOVAs were performed on the job, health, and organizational outcome variables to determine whether the detailed health care occupations were homogenous with respect to their scores.

RESULTS

Table 11–3 shows demographic information for health care and non–health care groups. The health care group members tended to be slightly older, more racially diverse, and female dominated compared with the non–health care group members. As a group, health care workers had slightly higher job tenure, education, and income, were more often paid by the hour, and worked other than day shifts compared with non–health care workers. ANCOVA revealed statistically significant differences between health care and non–health care groups on gender, race, job tenure, work shift, and pay schedule (p = .01).

Table 11–4 shows least square adjusted mean scores on job characteristics, health, and organizational outcomes for health care and non–health care groups. The adjusted mean scores were obtained from ANCOVA in which demographics, job tenure, and shift schedule were used as covariates. On many measures, health care workers differed from non–health care workers; the differences were statistically significant for approximately half of the measures. On the positive side, health care workers reported more clarity about what is expected of them on the job, higher utilization of skills and abilities, more opportunities to find a similar job, and more ability to work on teams than non–health care workers. On the negative side, health care workers reported significantly more repeated lifting on the job, more occasions of being threatened at work, lower ratings of pay equity, and lower scores on safety climate. Moreover, health care workers tended to report less supervisory support and more occasions of having too few staff members to get the work done (*ps* ≤ .10). For measures of health and organizational outcomes, health care workers tended to score higher on measures of job stress, burnout, and accidents at work but only the differences in job stress and accidents approached statistical significance (ps = .10).

Relationships among job characteristics and health and organizational outcomes were examined for health care and non–health care groups separately. Table 11–5 shows zero-order correlations among variables for health care and non–health care groups. For both health care and non–health care workers, none of the correlations among measures of job characteristics were greater than .40, except for the correlation between skill use and role clarity (*r* = .54) among health care workers and the correlations among the three safety items (*rs* = .59 to .80). Measures of outcomes were more highly interrelated in both health care and non–health care groups; pride at work was correlated with organizational effectiveness (*r* = .61 and .64, respectively), job satisfaction correlated with pride at work (*r* = .54 and .54, respectively), organizational effectiveness (*r* = .57 and .51, respectively), and turnover intent (*r* = –.25 and –.40, respectively), and stress correlated with burnout (*r* = .47 and .51, respectively).

The size of the correlations among variables in health care and non–health care groups were similar for the most part, but exceptions are noteworthy. For instance, among health care workers, the correlation between repeated lifting at work with job stress and burnout were significant (*r* = .38 and .49, respectively), whereas among non–health care workers, the correlations were small and in the opposite direction. Among health care workers, being understaffed was associated with job stress, burnout, and job satisfaction (*rs* = .46, .37, and

TABLE 11–2. Questionnaire Items and Scales From the National Institute for Occupational Safety and Health Quality of Work Life Module

JOB CHARACTERISTICS
Staffing (1 item)
*How often are there not enough people or staff to get all the work done? OF
Work/family (2 items, $\alpha = .63$)
How often do the demands of your family interfere with your work on the job? OF
How often do the demands of your job interfere with your family life? OF
Role clarity (1 item)
*On my job, I know exactly what is expected of me. AG
Skill utililization (1 item)
*My job lets me use my skills and abilities. AG
Workload (1 item)
*I have too much work to do everything well. AG
Autonomy (1 item)
*I am given a lot of freedom to decide how to do my own work. VT
Supervisor support (2 items, $\alpha = .74$)
*My supervisor is concerned about the welfare of those under him or her. VT
*My supervisor is helpful to me in getting the job done. VT
Job security (1 item)
*The job security is good. VT
Pay equity (1 item)
How fair is what you earn on your job in comparison to others doing the same type of work you do? ML
Teamwork (1 item)
In your job, do you normally work as part of a team, or do you work mostly on your own? YN
Ease of finding new job (1 item)
How easy would it be for you to find a job with another employer with approximately the same income and fringe benefits as you have now? VE
Threatened at work (1 item)
In the last 12 months, were you threatened or harassed in any way by anyone while you were on the job? YN
Safety and health conditions (1 item)
The safety and health conditions where I work are good. AG
Safe work behavior (2 items)
There are no significant compromises or shortcuts taken when worker safety is at stake. AG
Where I work, employees and management work together to insure the safest possible working conditions. AG
Repeated lifting (1 item)
Does your job require you to do repeated lifting, pushing, pulling or bending? YN

HEALTH AND ORGANIZATIONAL OUTCOMES
Job stress (1 item)
How often do you find your work stressful? OF3
Burnout (1 item)
How often during the past month have you felt used up at the end of the day? OF2
Job satisfaction (1 item)
All in all, how satisfied would you say you are with your job? VS
Self-rated health (1 item)
Would you say that in general your health is Excellent, Very good, Good, Fair, or Poor?
Intent to leave (1 item)
*Taking everything into consideration, how likely is it you will make a genuine effort to find a new job with another employer within the next year? VL
Employee commitment (1 item)
I am proud to be working for my employer. AG
Organizational effectiveness (2 items, $\alpha = .71$)
Conditions on my job allow me to be about as productive as I could be. AG
The place where I work is run in a smooth and effective manner. AG

*Item taken directly from the 1977 Quality of Employment Survey (Quinn and Staines, 1979).

Note. AG = Response scale is four point (Strongly agree to strongly disagree); ML = Response scale is four point (Much more than deserve to much less than deserve); OF = Response scale is four point (Often to never); OF2 = Response scale is five point (Very often to never); OF3 = Response scale is five point (Always to never); VE = Response scale is three point (Very easy to not at all easy); VT = Response scale is four point (Very true to not at all true); VL = Response scale is 3 point (Very likely, somewhat likely, not at all likely); VS = Response scale is four point (Very satisfied to not at all satisfied); YN = Yes, no.

TABLE 11–3. Demographic Information for Health Care and Non–Health Care Groups.

Occupational Group	*Health Care* (N = 110)	*Non–Health Care* (N = 1337)
Age (in years)	41.1	39.8
S.D.	12.2	12.2
Sex** (male)	0.15	0.49
S.D.	0.36	0.5
Race**		
White	55%	75%
Black	30%	15%
Hispanic	13%	8%
Other	3%	3%
Education		
Less than high school	8%	10%
High school/Junior college	60%	61%
College	18%	15%
Graduate	14%	13%
Marital status		
Married	41%	47%
Divorced/sep/wid	29%	23%
Never married	30%	31%
Work shift**		
Day Shift	61%	74%
Afternoon/nite shift	23%	15%
Rotate	17%	11%
Pay schedule*		
Salary	23%	38%
Hourly	75%	56%
Other	2%	5%
Job tenure (years)	6.8	6.5
S.D.	7.0	7.9
Income	15.4	15.2
S.D.	4.5	5.2
Second job	0.19	0.17
S.D.	0.40	0.38

Note. *p = .05, **p = .01

–.31, respectively) to a larger degree than non–health care workers (*rs* = .25, .21, and –.12, respectively). Being understaffed also correlated more highly with work–family interference among health care (r = .34) than non–health care (r = .18) workers.

Multiple Regression Analyses

The next set of analyses involved stepwise multiple regressions for each outcome measure, using job characteristics and demographics as predictors. These analyses were done for health care and non–health care groups separately and were performed to indicate whether the job characteristics that predict health outcomes would be the same for the two groups. Tables 6 and 7 show the results of the regression analyses for health care and non–health care groups, respectively.

For most of the health and organizational outcome measures, demographic variables as a group were not good predictors, and the regression models explained little of the variance ($R^2 \leq .04$). The exceptions were measures of job stress among non–health care workers (Black race; $R^2 = .11$), and the measure of intent to look for a new job for both health care (age; $R^2 = .14$) and non–health care groups (age, tenure, Black race; $R^2 = .09$). For the latter, job tenure was the single most significant factor in the models for intent to look for a new job among health care and non–health care groups ($R^2 = .09$ and .10, respectively).

The best predictor of job stress among non–health care workers was work–family interference (B = .22), followed by overwork (.18), being understaffed (.12), job security (–.11), and working together for safety (–.11). The overall model R^2 was .24. For health care workers, the best predictors of job stress were repeated lifting at work (.36), being understaffed (.34),

TABLE 11–4. Least–Square Adjusted Means for Healthcare and Non–Healthcare Groups.

	Healthcare (N = 110)	*Nonhealthcare* (N = 1337)
Job Characteristics		
Work/family interference	2.10	2.20
Supervisory support	3.10*	3.30
Insufficient staff	3.00*	2.80
Role clarity	3.50***	3.30
Skill utilization	3.60***	3.20
Workload	2.30	2.20
Autonomy	3.40	3.30
Job security	3.40	3.30
Pay equity	2.20***	2.50
Teamwork	0.70**	0.60
Repeated lifting	0.63***	0.44
Find new job	2.20***	1.80
Threatened at work	0.19**	0.11
Occupational safety & health conditions	3.23	3.23
No safety shortcuts	3.09**	3.26
Work together for safety	3.04 *	3.21
Health and safety outcomes		
Job stress	3.30*	3.10
Burnout	3.50	3.30
Self–rated health	3.80	3.70
Job satisfaction	3.40	3.30
Work accidents	0.31*	0.21
Organizational outcomes		
Organizational effectiveness	3.00	3.00
Employee commitment	3.30	3.20
Intent to leave	0.41	0.41

Note: *p £ 10 **p = 0.05 ***p = .01

and being White (–.25). The overall adjusted model R^2 was .35. The models for burnout also differed between health care and non–health care groups. For health care workers, the model resembled that for stress, with repeated lifting (.35) and insufficient staff (.31) although for non–health care workers, the model included work–family interference (.18), workload (.18), and insufficient staff (.12).

Regression models for job satisfaction were more similar for health care and non–health care groups. In both cases, supervisory support was the strongest predictor (B = .37 and .22, respectively), followed by skill utilization (.40 and .20, respectively), and age (.23 and .09, respectively). For non–health care workers only, job security, pay equity, and autonomy were also linked to job satisfaction.

Models for the three organizational outcomes showed a higher degree of similarity for health care and non–health care workers than the health measures. For instance, the models for organizational effectiveness for health care and non–health care groups contained supervisory support (B = .47 and .15, respectively), skill utilization (.40 and .20, respectively), and safety and health conditions (B = .33 and .15, respectively). For pride at work, the best predictors for health care and non–health care groups were working together to insure safety (B = .45 and .28, respectively), supervisory support (.31 and .18, respectively) and skill utilization (.24 and .20, respectively). For the remaining organizational outcome, intent to look for a new job, the main predictor for health care workers was supervisory support (B = –.30) although for non–health care workers, the predictors were low job security (–.19), job tenure (–.17), supervisory support (–.14), and the ease of finding a similar job (.12).

Health Care Subgroups

The next series of analyses examined scores among four separate health care subgroups (nurses, aides, technicians, and other). ANCOVAs were performed, and least square adjusted means scores

TABLE 11–5. Correlations Among Study Variablesa

	1	2	3	4	5	6	7	8	9	10	11	12	13	14
1, Age		–.12	–.06	.12	–.16	.09	–.10	.02	.05	–.03	.11	.16	–.10	–.05
2. Sex	.00		.31	.24	.07	–.10	.03	–.15	–.18	–.05	.03	.01	.03	.06
3. Education	.07	–.05		.49	.18	–.14	.15	–.20	.00	.07	.01	.06	.00	.15
4. Income	.22	.21	.27		.15	–.01	.04	–.07	.05	.10	.10	.12	.16	.25
5. Work/family interference	–.08	–.03	.13	.13		–.11	.34	.04	–.01	.18	–.07	.11	.08	.04
6. Supervisory support	.03	–.04	.04	–.01	–.10		–.23	.14	.18	–.24	.23	.38	.29	.08
7. Insufficient staff	.02	.02	.05	.08	.18	–.16		.06	–.04	.33	–.21	–.01	–.24	.17
8. Role clarity	.05	–.05	–.14	.01	–.06	.18	–.08		.54	.05	–.05	.13	–.02	.17
9. Skill utilization	.11	.02	.10	.14	.00	.30	.00	.34		.01	–.04	.31	.16	.04
10. Workload	.01	–.05	.03	.02	.26	–.16	.28	–.11	–.07		–.15	–.01	–.14	.08
11. Autonomy	.08	.02	.11	.11	–.07	.31	–.07	.10	.33	–.13		.10	.08	–.16
12. Job security	.01	–.01	.01	.04	–.06	.36	–.10	.11	.18	–.12	.26		.25	.10
13. Pay equity	.02	.03	.08	.08	–.11	.24	–.14	–.01	.08	–.13	.16	.14		.10
14. Teamwork	–.06	.02	–.04	.01	–.02	.16	.02	.04	.04	.03	–.04	.07	.06	
15. Repeated lifting	–.09	.17	–.31	–.21	–.03	–.06	.05	.07	–.12	.00	–.13	.02	–.07	.11
16. Find new job	–.15	.08	.04	–.07	–.01	.04	–.02	.01	.01	–.05	.03	.06	–.07	.00
17. Threatened at work	–.03	–.02	.02	.04	.15	–.15	.11	–.01	.01	.10	–.06	–.04	–.06	.01
18. Occupational Safety & Health conditions	.04	.02	.05	.09	–.06	.36	–.09	.21	.33	–.13	.26	.23	.17	.11
19. No safety shortcuts	.05	.01	.03	.07	–.07	.35	–.11	.20	.33	–.13	.23	.18	.18	.08
20. Work together for safety	.04	–.02	.01	.06	–.05	.42	–.10	.24	.36	–.11	.26	.23	.16	.13
21. Job stress	–.05	–.03	.07	.10	.34	–.20	.25	–.14	–.03	.32	–.15	–.17	–.14	–.01
22. Burnout	–.10	–.08	–.02	.00	.27	–.16	.21	–.07	–.05	.27	–.14	–.07	–.14	.00
23. Self–rated health	–.06	.02	.15	.09	–.05	.16	–.02	.05	.12	–.05	.13	.14	.11	.03
24. Job satisfaction	.16	.00	.11	.14	–.11	.45	–.12	.19	.40	–.19	.35	.37	.30	.10
25. Work accidents	–.09	.08	–.05	–.03	.09	–.06	.11	–.03	–.06	.01	–.12	–.03	–.07	.03
26. Organizational. effectiveness	.10	–.03	.01	.05	–.17	.44	–.19	.33	.44	–.18	.32	.28	.20	.07
27. Employee commitment	.11	–.02	.10	.08	–.09	.49	–.12	.23	.46	–.15	.33	.33	.20	.13
28. Intent to leave	–.26	–.02	–.04	–.22	.04	–.25	.04	–.11	–.21	.10	–.18	–.29	–.20	–.06

[a]Health care workers shown above the diagonal; non-health care below the diagonal

Health care workers (above diagonal): $r = .19, p \leq .05$; $r = .24, p \leq .01$; $r = .30, p \leq .001$

Non-health care workers (below diagonal): $r = .05, p \leq = .05$; $r = .07, p \leq = .01$; $r = .09, p \leq .001$

TABLE 11–5. Correlations Among Study Variablesa

		15	16	17	18	19	20	21	22	23	24	25	26	27	28
1	Age	–.20	–.14	–.03	.04	.14	.13	–.07	–.21	.02	.17	.03	.14	.18	–.31
2.	Sex	–.24	.09	.05	–.20	–.18	–.22	.04	–.10	.05	.03	.00	–.07	–.15	.12
3.	Education	–.24	.08	.17	–.05	–.14	–.21	.01	–.05	.17	–.05	–.11	–.14	–.12	–.07
4.	Income	–.20	.03	.27	–.08	–.11	–.09	.09	–.07	.18	.03	.05	–.14	–.01	–.13
5.	Work/family interference	.26	.03	.26	–.19	–.20	–.18	.39	.41	.02	–.09	.21	–.26	–.09	–.02
6.	Supervisory support	–.07	–.03	–.10	.22	.25	.44	–.24	–.15	.16	.47	–.20	.54	.52	–.28
7.	Insufficient staff	.29	–.04	.11	–.04	–.08	–.20	.46	.37	–.09	–.31	–.01	–.27	–.19	.13
8.	Role clarity	–.18	.05	–.11	.19	.25	.17	–.11	.08	.18	.22	–.10	.26	.33	–.07
9.	Skill utilization	–.21	.07	–.05	.22	.13	.10	–.14	–.01	.32	.42	–.03	.36	.50	–.32
10.	Workload	.19	.02	.11	–.16	–.20	–.23	.21	.33	.01	–.15	.04	–.22	–.12	.18
11.	Autonomy	–.10	.04	–.04	.16	.05	.19	.05	–.06	–.02	.12	–.04	.19	.18	–.07
12.	Job security	–.10	.16	.12	.20	.18	.16	–.01	.03	.14	.35	.13	.40	.43	–.26
13.	Pay equity	–.16	.15	.06	.10	.10	.19	–.14	–.06	.07	.24	–.07	.23	.27	–.07
14.	Teamwork	.02	.03	.01	.04	.09	.12	.05	.10	.04	–.07	.04	–.08	.06	.02
15.	Repeated lifting		–.05	.10	–.16	–.17	–.16	.38	.49	–.28	–.20	.14	–.25	–.19	.20
16.	Find new job	.04		.00	.06	.07	–.01	.04	.09	.28	.07	.00	–.03	.02	.18
17.	Threatened at work	.06	–.02		–.31	–.30	–.24	.18	.27	.08	.04	.35	–.22	–.18	–.02
18.	Occupational Safety & Health conditions	–.07	.05	–.11		.69	.75	–.14	–.21	.07	.28	–.23	.49	.57	–.05
19.	No safety shortcuts	–.06	.01	–.12	.59		.80	–.18	–.09	.09	.19	–.25	.42	.46	–.05
20.	Work together for safety	–.05	.01	–.13	.71	.68		–.19	–.16	.05	.29	–.18	.49	.59	–.10
21.	Job stress	–.02	.01	.18	–.15	–.13	–.16		.47	–.08	–.33	.21	–.37	–.19	.16
22.	Burnout	.06	–.01	.17	–.11	–.10	–.12	.51		–.16	–.20	.22	–.20	–.15	.17
23.	Self–rated health	–.10	.06	–.07	.14	.12	.12	–.10	–.19		.34	–.07	.20	.28	–.15
24.	Job satisfaction	–.12	–.07	–.12	.34	.30	.34	–.27	–.22	.20		.08	.57	.54	–.25
25.	Work accidents	.17	.00	.11	–.10	–.12	–.11	.07	.08	–.04	–.13		–.15	–.12	–.04
26.	Organizational. effectiveness	–.06	.01	–.15	.50	.44	.52	–.25	–.20	.18	.51	–.14		.61	–.26
27.	Employee commitment	–.11	.00	–.09	.56	.50	.60	–.18	–.16	.16	.54	–.09	.64	–	.22
28.	Intent to leave	.09	.17	.04	–.17	–.14	–.16	.09	.08	–.05	–.41	.06	–.27	–.32	

[a]Health care workers shown above the diagonal; non-health care below the diagonal

Health care workers (above diagonal): $r = .19, p \leq .05$; $r = .24, p \leq .01$; $r = .30, p \leq .001$

Non-health care workers (below diagonal): $r = .05, p \leq = .05$; $r = .07, p \leq = .01$; $r = .09, p \leq .001$

TABLE 11–6. Results of Stepwise Multiple Regression Analyses Among Health Care Workers[a]

	Job Stress	*Burnout*	*Self-Rated Health*	*Job Satisfaction*	*Work Accidents*	*Organizational Effectiveness*	*Employee Commitment*	*Intent to Leave*
Age				0.23				
Sex								
Job tenure			−0.24					−0.21
Income						−0.19		
Afternoon/night shift								
Rotating shifts								
Hourly pay								
Other pay								
Black	−0.25					0.21		
Hispanic								
Other race								
Work/family interference								
Supervisory support				0.37		0.47	0.31	−0.30
Insufficient staff	0.34	0.31						
Role clarity								
Skill utilization			0.35	0.40		0.24	0.41	
Workload								
Autonomy								
Job security								
Pay equity								
Teamwork								
Repeated lifting	0.36	0.35						
Find new job								
Threatened at work					0.36			
Occupational Safety & Health conditions						0.33		
No safety shortcuts								
Work together for safety							0.45	
Model Adjusted R^2 =	0.35	0.24	0.14	0.34	0.08	0.53	0.59	0.15

[a]Cell Entries are Standardized Regression Coefficients.

were calculated for job characteristics, health, and organizational outcome measures. The overall F ratio for each ANCOVA provides an indication of whether any of the groups differ on the measure. If the F ratio was significant, then post hoc tests were constructed to examine group differences, controlling for experimentwise error. It is noteworthy that the sample sizes for health care subgroups were small, especially for the doctor, dentist, other group, which reduces the power of statistical tests to detect significant differences among groups. Nevertheless, the ANCOVA results are reported here to provide a sense of how the detailed health care occupations compared on various measures.

Table 11–8 shows demographic factors and least square adjusted mean scores on job characteristics, health, and organizational performance indicators among the four occupational subgroups within health care. The health care occupations differed on some demographic factors, with registered nurses and health aides being predominately female although the doctor and dentist group was predominately male. Registered nurses and doctors and dentists tended to be older and have higher incomes than technicians and health aides. Scores on most measures were similar among the health care subgroups, but significant F ratios ($p \leq .05$) were obtained on four of the job characteristics (supervisory support, job security, teamwork, ease of finding a new job) and one of the outcome measures (job satisfaction). However, none of the post hoc tests was statistically significant after adjusting for demographics and controlling experimentwise error.

DISCUSSION

This study was the first nationally representative, personal interview survey of the quality of working life of U.S. adults conducted by the federal government since 1977. Unlike studies conducted within a single organization, this study has high external

TABLE 11–7. Stepwise Multiple Regression Analyses Among Non–Healthcare Workers[a]

	Job Stress	*Burnout*	*Self-Rated Health*	*Job Satisfaction*	*Work Accidentsic*	*Organizational Effectiveness*	*Employee Commitment*	*Intent to Leave*
Age	−0.07	−0.11	−0.09	0.09			0.07	−0.10
Sex		−0.08			0.06			
Job tenure	0.07	0.07						−0.17
Income			0.11					−0.10
Afternoon/night shift						0.05		
Rotate shifts								0.07
Hourly pay	−0.07				0.08			
Other pay								
Black					−0.09			0.10
Hispanic								
Other race								
Work/family interference	0.22	0.18				−0.09		
Supervisory support			0.13	0.22		0.15	0.18	−0.14
Insufficient staff	0.12	0.12			0.09	−0.08		
Role clarity	−0.07					0.10		
Skill utilization				0.20		0.20	0.20	−0.08
Workload	0.18	0.16		−0.09		0.07		
Autonomy		−0.09		0.10		0.07		
Job security	−0.11		0.09	0.16		0.06	0.13	−0.19
Pay equity				0.16				−0.10
Teamwork								
Repeated lifting					0.18			
Find new job								0.12
Threatened at work	0.09	0.10			0.09			
Occupational Safety & Health conditions			0.08	0.08		0.15	0.20	
No safety shortcuts					−0.10			
Work together for safety	−0.10					0.22	0.28	
Model Adjusted R^2 =	0.24	0.16	0.06	0.37	0.09	0.46	0.51	0.26

[a]Cell Entries are Standardized Regressions Coefficients.

validity and allows conclusions to be drawn about the entire U.S. workforce.

The results revealed that supervisory support was as an important predictor of job satisfaction, organizational effectiveness, employee commitment, and turnover intention among health care workers. In addition, skill utilization was an important correlate of job satisfaction, organizational effectiveness and employee commitment, as well as self-rated health. Repeated lifting at work was the best predictor of burnout among health care workers and was significant in the regression models for job stress and self-rated health. Being threatened on the job was the single best predictor of accidents and was present in the regression model for burnout among health care workers. Confirming both scientific and anecdotal reports, insufficient staffing was found to be linked closely to perceptions of job stress and burnout among health care workers.

GUIDELINES

These findings suggest some specific targets for intervention to reduce stress and improve performance in health care settings. One conceptual model that appears particularly useful in the current context is that of healthy work organizations, defined as organizations that foster both employee health and organizational performance (Cox and Leiter, 1992; Murphy, 2000; Murphy, 2002; Rosen, 1991; Sauter, Lim, & Murphy, 1996). The term "healthy work organizations" refers to the notion that worker well-being and organizational effectiveness can be fostered by a common set of job and organizational characteristics. This concept represents a significant departure from traditional work organization models that seek to improve either worker health (e.g., health promotion) or organizational effectiveness (e.g., total quality management) but rarely both.

TABLE 11–8. Comparisons of Least–Square Adjusted Mean Scores Among Detailed Health Care Occupations.

	Healthcare Total (N = 110)	*Registered Nurses* (N = 2)	*Nursing/ Health Aides* (N = 43)	*Technicians/ Therapists* (N = 27)	MD/DDS/ *Other* (N = 8)
Demographics					
Age	41.1	43.8	41.0	37.3	43.0
Sex	0.15	0.03	0.02	0.37	0.63
Education	14.2	15.3	12.5	15.0	16.1
Years on job	6.8	6.1	6.2	8.3	7.4
Income	15.4	18.3	12.0	16.2	19.7
Job factors					
Work/family interference	2.1	2.2	1.8	2.3	2.3
Supervisory support	3.1	3.3	3.1	3.3	2.4
Insufficient staff	3.0	3.0	3.0	3.0	3.1
Role clarity	3.5	3.5	3.5	3.7	3.7
Skill utilization	3.6	3.5	3.4	3.7	3.9
Workload	2.3	2.2	2.2	2.3	2.6
Autonomy	3.4	3.6	3.2	3.1	3.2
Job security	3.4	3.6	3.2	3.6	2.7
Pay equity	2.2	2.4	2.0	2.4	2.2
Teamwork	0.7	0.7	0.5	0.8	1.0
Repeated lifting	0.6	0.7	0.7	0.6	0.2
Find new job	2.2	2.5	2.0	2.2	1.8
Threatened at work	0.2	0.3	0.2	0.1	0.1
Occupational Safety & Health conditions	3.2	3.2	3.2	3.3	3.1
No safety shortcuts	3.1	3.0	2.9	3.2	3.0
Work together for safety	3.0	3.1	3.0	3.3	3.0
Health outcomes					
Job stress	3.3	3.7	3.3	2.9	2.9
Burnout	3.5	3.7	3.5	3.6	3.2
Self–rated health	3.8	3.9	3.3	4.0	3.8
Organizational outcomes					
Job satisfaction	3.4	3.4	3.2	3.7	3.0
Work accidents	0.3	0.3	0.3	0.2	0.2
Organizational effectiveness	3.0	3.0	3.0	3.1	2.6
Employee commitment	3.3	3.3	3.2	3.5	3.1
Intent to leave	0.4	0.4	0.5	0.5	0.2

Not coincidentally, many of the characteristics noted here are also prominent in descriptions of magnet hospitals (Kramer & Schmalberg, 1988; Scott, Sochalski, & Aiken, 1999; Sochalski, Aiken, & Fagin, 1997). Magnet hospitals are characterized by high nurse participation in decision making, staff development, nurse self-scheduling, and unit-based staffing. These organizational characteristics suggest a high level of supervisory support, high staff autonomy, more control over resources, and better relationships between nurses and physicians than nonmagnet hospitals (Aiken & Sloane, 1997).

In the this study, the finding that supervisory support was the strongest predictor of both job satisfaction and organizational effectiveness suggests that efforts directed toward improving support will reap benefits in terms of both worker satisfaction and overall effectiveness. Indeed, a recent study by Stamper and Johlke (2003) demonstrated that perceived organizational support acts as a buffer for role stress, thereby lowering overall stress levels and improving job satisfaction. Likewise, Heaney, Price, and Rafferty (1995) developed a caregiver support program to raise awareness of the importance of support at work, provide education in participatory problem-solving approaches to work-related problems, and equip workers with skills for implementing such approaches in work settings. The intervention was successful in improving levels of social support, as well as improving job satisfaction and worker mental health.

The research on innovative coping (Bunce & West, 1994, 1996; West, 1989) is relevant in discussions of the value of social support. Innovative coping refers to new strategies or tactics devised and applied by workers as a means of reducing excessive demands at work. West (1989) found that innovation among health care workers tended to occur when workload and supervisory support were high. Most of the innovations dealt with changes in working methods, relationships, and new skill development. Later studies confirmed the value of innovative coping by health care workers as a stress reduction strategy (Bunce & West, 1994, 1996).

Finally, these results suggest that although efforts to improve staffing ratios will have strong effects on lowering job stress and exhaustion, smaller but significant effects will be seen on job satisfaction and organizational effectiveness. In each case, changes in work organization can be expected to produce positive effects on both health and performance indicators.

Job redesign interventions can be successful if worker involvement is high and there is a strong management commitment to the intervention (e.g., Abts, Hofer & Leafgreen, 1994; Molleman & van Knippenberg, 1995; Murphy, 1999; Murphy, Pearlman, Rea & Papazian-Boyce, 1994). In the same way, large-scale restructuring efforts in health care settings, exemplified by the introduction of a new, patient-centered care model, can be successful in improving worker satisfaction and meaningfulness of work (Parsons & Murdaugh, 1994).

Overall, the present results provide clear directions and targets for improving the quality of work life for health care workers through attention to factors at multiple levels: job/task (staffing, repeated lifting, and skill utilization), supervision (supervisory support), and the organizational culture (safety climate).

ACKNOWLEDGEMENTS

The author is indebted to members of an expert panel who worked closely with NIOSH to design the QWL module. Panel members included Dave DeJoy, Jerry Halamaj, Michael Hodgson, James Jackson, Robert Kahn, Arne Kalleberg, Gwendolyn Keita, Ann Kempski, David LeGrande, Peter Marsden, Stacey Moran, David Moriarty, Michael Pergamit, Richard Price, and Tom Smith. Thanks also are due to the NIOSH members of the panel: Michael Colligan, Lynn Jenkins, Robert Peters, Roger Rosa, Steve Sauter, John Sestito, and Naomi Swanson. Don Dusenbury, Sheila Pederson, Paul Young and Yvonne Lewis facilitated the expert group meeting from the NOVA research company.

References

Abts, D., Hofer, M. & Leafgreen, P. K. (1994). Redefining care delivery: A modular system. *Nursing Management, 25,* 40–46.

Adams, A., & Bond, S. (2003). Staffing in acute hospital wards: Part 1. The relationship between number of nurses and ward organizational environment. *Journal of Nursing Management, 11,* 287–292.

Aiken, L. H., Clarke, S. P. & Sloane, D. M. (2002). Hospital staffing, organization, and quality of care: Cross-national findings. *International Journal for Quality in Health Care, 14,* 5–13.

Aiken, L. H., Clarke, S. P., Sloane, D., Sochalski, J., & Silber, J. H. (2002). Hospital nurse staffing and patient mortality, nurse burnout, and job dissatisfaction. *Journal of the American Medical Association, 288,* 1987–1993.

Aiken, L. H., & Sloane, D. M. (1997). The effects of specialization and client differentiation on the status of nurses: The case of AIDS. *Journal of Health and Social Behavior,* 38, 203–221.

Arafa, M., Nazel, M. W. A., Ibrahim, N. K., & Attia, A. (2003). Predictors of psychological well-being of nurses in Alexandria, Egypt. *International Journal of Nursing Practice, 9,* 313–320.

Armstrong-Stassen, M., Horsburgh, M. E., & Cameron, S. J. (1994). The reactions of full-time and part-time nurses to restructuring in the Canadian health care system. *Best Papers Proceedings, The Academy of Management, USA,* 96–100.

Aschoff, J. (1981). Circadian rhythms: Interference with and dependence on work-rest schedules. In L.C. Johnson, D. I. Tepas, W. P. Colquohoun, & M. J. Colligan (Eds.), *The Twenty-Four Hour Workday: Proceedings of a Symposium on Variation in Work-Sleep Schedules* (pp. 163–166). Washington, DC: U.S. Government Printing Office). DHHS Publication No. 81-127.

Aspden, P., Corrigan, J. M., Wolcott, J., & Erickson, S. M. Eds. (2003). *Patient Safety: Achieving a New Standard for Care.* Washington, DC: National Academies Press. Available from Institute of Medicine Web site, http://www.iom.edu/report.asp?id'16173

Baldwin, P. J. (1999). Nursing. In R. Payne & J. Firth-Cozens, *Stress in Health Professionals* (2nd ed.) (pp. 93–104). Chichester, UK: Wiley.

Beehr, T. A., & Newman, J. (1978). Job stress. Employee health and organizational effectiveness: A facet analyses, model and literature review. *Personnel Psychology, 31,* 665–699.

Bunce, D., & West, M. A. (1994). Changing work environments: Innovative responses to occupational stress. *Work & Stress, 8,* 319–341.

Bunce, D., & West, M. A. (1996). Stress management and innovation interventions at work. *Human Relations, 49,* 209–232.

Bureau of Labor Statistics. (1995). *Work Injuries and Illnesses by Selected Characteristics, 1993* (Summary 95-142).Washington, DC: U.S. Department of Labor.

Bureau of Labor Statistics. (2006). List of SOC Occupations. Retrieved March 15, 2006, from http://stats.bls.gov/oes/2002/oes_stru.htm#00-0000

Calhoun, G. (1980). Hospitals are high stress employers. *Hospitals, 54,* 171–176.

Caplan, R. P. (1994). Stress, anxiety and depression in hospital consultants, general practitioners, and senior health managers. *British Medical Journal, 309,* 1261–1263.

Centers for Disease Control and Prevention. (1987). Update: Human immunodeficiency virus infection in health-care workers exposed to blood of infected patients. *Morbidity and Mortality Weekly Reports, 36,* 285–289.

Cherniss, C. (1980). *Professional burnout in human service organizations.* New York: Praeger.

Clarke, S. P., Sloane, D. M., & Aiken, L. H. (2002). Effects of hospital staffing and organizational climate on needle stick injuries to nurses. *American Journal of Public Health, 92,* 1115–1119.

Colligan, M. J., Smith, M. J., & Hurrell, J. J. (1977). Occupational incidence rates of mental health disorders. *Journal of Human Stress, 3,* 34–39.

Cooper, C. L., & Marshall, J. (1976). Occupational sources of stress: A review of the literature relating to coronary heart disease and mental ill health. *Journal of Occupational Psychology, 49,* 11–28.

Cox, T., & Leiter, M. (1992). The health of health care organizations. *Work & Stress, 3,* 219–227.

Davis, J. A., & Smith, T. W. (2002). *General Social Surveys, 1972–2002* [machine-readable data file]. Principal Investigator, J. A. Davis; Director and Co-Principal Investigator, T. W. Smith; Co-Principal Investigator, P. V. Marsden, NORC ed. Chicago: National Opinion Research Center, producer, 2002. Storrs, CT: The Roper Center for Public Opinion Research, University of Connecticut, distributor.

Duxbury, M. L., Armstrong, G. D., Drew, D. J., & Henly, S. J. (1984). Head nurses' leadership style with staff burnout and job satisfaction. *Nursing Research, 33,* 47–101.

Ellis, B. H., & Miller, K. I. (1994). Supportive communication among nurses: Effects of commitment, burnout, and retention. *Health Communication, 6,* 77–96.

Firth, H., & Britton, P. (1989). Burnout, absenteeism, and turnover amongst British nursing staff. *Journal of Occupational Psychology, 62,* 55–59.

Fottrell, E. (1980). A study of violent behaviour among patients in psychiatric hospitals. *British Journal of Psychiatry, 136,* 216–221.

Fox, M., Dwyer, D., & Ganster, D. (1993). Effects of stressful demands and control on physiological and attitudinal outcomes in a hospital setting. *Academy of Management Journal, 36,* 289–318.

Fruedenberger, H. J. (1974). Staff burn-out. *Journal of Social Issues, 30,* 159–165.

Gauci Borda, R. & Norman, I. J. (1997a). Factors influencing turnover and absence of nurses: A research review. *International Journal of Nursing Studies, 34,* 385–394.

Gauci Borda, R., & Norman, I. J. (1997b). Testing a model of absence and intent to stay in employment: A study of registered nurses in Malta. *International Journal of Nursing Studies, 34,* 375–384.

Gerbert, B. Maguire, B. Badner, V. Altman, D., & Stone, G. (1989). Fear of AIDS: Issues for health professional education. *AIDS Education and Prevention, 1,* 39.

Gillmore, V. L. (1990). Occupational health risks for hospital nurses: Significant concerns for administrators. *The Health Care Manager, 9,* 57–63.

Golembiewski, R. T., Munzenrider, R. F., & Stevenson, J. G. (1986). *Stress in Organizations: Toward a Phase Model of Burnout.* New York: Praeger.

Gray-Toft, P. A., & Anderson, J. G. (1981). Stress among hospital nursing staff: Its causes and effects. *Social Science and Medicine, 15,* 639–647.

Gray-Toft, P. A., & Anderson, J. G. (1985). Occupation stress in the hospital: Development of a model for diagnosis and prediction. *Health Services Research, 19,* 753–774.

Gunderson, E. K. E., & Colcord, C. (1982). *Health Risks in Naval Operations: An Overview* (Report No. 82-1). San Diego, CA: Naval Health Research Center.

Heaney, C. A., Price, R. H., & Rafferty, J. (1995). The caregiver support program: An intervention to increase employee coping resources and enhance mental health. In L. R. Murphy, J. J. Hurrell, S. Sauter, & G. Keita (Eds.), *Job Stress Interventions* (pp. 93–108). Washington, DC: American Psychological Association.

Henderson, D. K. (1989). Perspectives on the risk for occupational transmission of HIV-1 in the health-care workplace. In C. E. Becker (Ed.), *Occupational Medicine State of the Art Reviews: Occupational HIV Infection: Risk and Risk Reduction* (pp. 000–000). Philadelphia, PA: Hanley and Belfus, Inc.

Hillhouse, J. J., & Adler, C. M. (1997). Investigating stress effect patterns in hospital staff: Results of a cluster analysis. *Social Science and Medicine, 45,* 1781–1788.

Hoiberg, M. S. (1982). Occupational stress and illness incidence. *Journal of Occupational Medicine, 24,* 445–451.

Jackson, S. E. (1983). Participation in decision making as a strategy for reducing job related strain. *Journal of Applied Psychology, 68,* 3–19.

Jackson, S. E. & Schuler, R. (1985). A meta-analysis and conceptual critique of research on role ambiguity and role conflict in work settings. *Organizational Behavior and Human Decisions, 36,* 16–28.

Jamal, M. (1981). Shift work related to job attitudes, social participation and withdrawal behavior: A study of nurses and industrial workers. *Personnel Psychology, 34,* 535–547.

Johnson, L. C., Tepas, D. I., Colquohoun, W. P., & Colligan, M. J. (1981). *The Twenty-Four Hour Workday: Proceedings of a Symposium on Variations in Work-Sleep Schedules* (DHHS Publication No. 81-127). Washington, DC: U.S. Government Printing Office.

Kahn, R. L., Wolfe, D. M., Quinn, R. P., & Snoek, J. D. (1964). *Organizational Stress: Studies in Role Conflict and Ambiguity.* New York: Wiley.

Karasek, R. A. (1979). Job demands, job decision latitude, and mental strain: Implications for job redesign. *Administrative Science Quarterly, 24,* 285–308.

Karasek, R. A., Schwartz, J., & Theorell, T. (1982). Job characteristics, occupation, and coronary heart disease. Cincinnati, OH: U.S. Government Printing Office.

Kipping, C. J. (2000). Stress in mental health nursing. *International Journal of Nursing Studies, 37,* 207–218.

Kohn, L. T., Corrigan, J. M., & Donaldson, M. S. (Eds.). (1999). *To Err Is Human: Building a Safer Health System.* Washington, D.C.: National Academies Press. Available from Institute of Medicine Web site, http://www.iom.edu/report.asp?id' 16173

Kramer, M., & Schmalenberg, C. (1988). Magnet hospitals: Part I. Institutions of excellence. *Journal of Nursing Administration, 18,* 1, 13–24.

Landsbergis, P. A. (1988). Occupational stress among health care workers: A test of the job demands-control model. *Journal of Organizational Behavior, 9,* 217–239.

Lindstrom, K. (1992). Work organization and well-being of Finnish health care personnel. *Scandinavian Journal of Work, Environment and Health, 18*(Suppl. 2), 90–93.

Lipscomb J. (1999). Violence in the workplace: A growing crisis among health care workers. In W. Charney & G. Fragala. (Eds.). *The epidemic of health care worker injury.* Boca Raton, FL: CRC Press.

Lu, K. Y., Lin, P. L., Wu, C. M., Hsieh, Y. L. & Chang, Y. Y. (2002). The Relationships among turnover intentions, professional commitment, and job satisfaction of hospital nurses. *Journal of Professional Nursing, 18,* 214–219.

Lundstrom, T., Pugliese, G., Bartley, J., Cox, J., & Guither, C. (2002). Organizational and environmental factors that affect worker health and safety and patient outcomes. *American Journal of Infection Control, 30,* 93–106.

Martin, J. P. (1984). *Hospitals in Trouble.* Oxford, UK: Basil Blackwell.

Maslach, C. (1982). *Burnout: The cost of caring.* Englewood Cliffs, NJ: Prentice Hall.

Maslach, C., & Jackson S. E. (1982). Burnout in health professions: A social psychological analysis. In **G. S. Sanders & J. Suls** (Eds.), *Social Psychology of Health and Illness.* Hillsdale, NJ.: Lawrence Erlbaum Associates, Inc.

McGowan, B. (2001). Self-reported stress and its effects on nurses. *Nursing Standard, 15,* 33–38.

McGrath, A., Reid, N., & Boore, J. (2003). Occupational stress in nursing. *International Journal of Nursing Studies, 40,* 555–565.

McNeese-Smith, D. (1999). A content analysis of staff nurse descriptions of job satisfaction and dissatisfaction. *Journal of Advanced Nursing, 29,* 1332–1341.

McVicar, A. (2003). Workplace stress in nursing: A literature review. *Journal of Advanced Nursing, 44,* 633–642.

Molleman, E., & Van Knippenberg, A. (1995). Work redesign and the balance of control within the nursing context. *Human Relations, 48,* 795–814.

Monk, T., & Tepas, D. A. (1985). Shiftwork. In C. L. Cooper & M. I. Smith (Eds.), *Job stress and Blue Collar Work.* Chichester, UK: Wiley.

Mottaz, C. J. (1988). Work satisfaction among nurses. *Journal of Health Care Management, 33,* 57–74.

Munro, L., Rodwell, J., & Harding, L. (1998). Assessing occupational stress in psychiatric nurses using the full job strain model: The value of social support to nurses. *International Journal of Nursing Studies 35,* 339–345.

Murphy, L. R. (1999). Organizational interventions to reduce stress in health care professionals. In R. Payne and J. Firth-Cozens (Eds.), *Stress in Health Professionals* (2nd ed., pp. 149–162). Chichester, UK: Wiley.

Murphy, L. R. (2000). Models of healthy work organizations. In L. R. Murphy & C. L. Cooper (Eds), *Healthy and Productive Work: An International Perspective.* London: Taylor-Francis.

Murphy, L. R. (2002). Job stress research at NIOSH: 1972–2002. In P. Perrewe & D. Ganster (Eds.), *Research in Occupational Stress and Well-Being* (pp. 1–55). Oxford, UK: JAI.

Murphy, L. R., Gershon, R. M., & DeJoy, D. (1996). Stress and occupational exposure to HIV/AIDS. In C. L. Cooper (Ed.), *Handbook of Stress Medicine* (pp. 176–190). Boca Raton, FL: CRC Press.

Murphy, R., Pearlman, F., Rea, C., & Papazian-Boyce, L. (1994). Work redesign: A return to the basics. *Nursing Management, 25,* 37–39.

National Institute for Occupational Safety and Health (NIOSH). (1996). *Current Intelligence Bulletin 57: Violence in the Workplace; Risk Factors and Prevention Strategies.* Cincinnati, OH: U.S. Department of Health and Human Services, Public Health Service, Centers for Disease Control and Prevention, National Institute for Occupational Safety and Health (Publication No. 96-100). Available from the Centers for Disease Control and Prevention Web site, http://www.cdc.gov/niosh/pdfs/2002-101.pdf

National Institute for Occupational Safety and Health (NIOSH). (2002). *Quality of Worklife Questionnaire.* Retrieved March 15, 2006 from http://www.cdc.gov/niosh/qwlquest.html

National Opinion Research Center. Retrieved May 23, 2006, from http://www.norc.uchicago.edu/projects/gensoc.asp

Nolan, P., Dallender, J., Soares, J., Thomsen, S., & Arnetz, B. (1999). Violence in mental health care: The experiences of mental health nurses and psychiatrists. *Journal of Advanced Nursing, 30,* 112–120.

Nolan, P., Soares, J., Dallender, J., Thomsen, S., & Arnetz, B. (2001). A comparative study of the experiences of violence of English and Swedish mental health nurses. *International Journal of Nursing Studies, 38,* 419–426.

Numerof, R. E., & Abrams, M. N. (1984). Sources of stress among nurses. *Journal of Human Stress, 10,* 88–B100.

Page, A. (2004). *Keeping patients safe: Transforming the work environment of nurses.* Washington, D.C.: National Academy Press.

Parsons, M. L., & Murdaugh, C. L. (1994). *Patient-Centered Care: A Model for Restructuring.* Gaithersburg, MD: Aspen.

Payne, R., & Firth-Cozens, J. (1987). *Stress in Health Care Professionals.* Chichester, UK: Wiley.

Payne, R., & Firth-Cozens, J. (1999). *Stress in health professionals* (2nd ed.). Chichester, UK: Wiley.

Pindus, N. M., & Greiner, A. (1997). *The Effects of Health Care Industry Changes on Health Care Workers and Quality of Patient Care: Summary of Literature and Research* (Contract Report DOL-J-9-MB5-0048, #17 prepared by The Urban Institute). Washington, DC: U.S. Department of Labor.

Quinn, R. P., & Shepard, L. J. (1974). *The 1972–73 Quality of Employment Survey.* Ann Arbor, MI: Institute for Social Research, 1977.

Quinn, R. P., & Staines, G. L. (1979). *The 1977 Quality of Employment Survey: Descriptive Statistics, With Comparison Data From the 1969–70 and the 1972–1973 Surveys.* Ann Arbor, MI: The University of Michigan.

Revicki, D. A., & May, H. J. (1989). Organizational characteristics, occupational stress, and mental health in nurses. *Behavioral Medicine, 15,* 30–36.

Revicki, D. A., & Whitley, T. W. (1995). Work-related stress and depression in emergency medicine residents. In S. L. Sauter & L. R. Murphy (Eds.), *Organizational Risk Factors for Job Stress.* Washington, DC: American Psychological Association.

Robinson J. R., Clements K., & Land C. (2003). Workplace stress among psychiatric nurses: Prevalence, distribution, correlates, and predictors. *Journal of Psychosocial Nursing, 41,* 33–41.

Rogers, B. (1997). Health hazards in nursing and health care: An overview. *American Journal of Infection Control, 25,* 248–261.

Rosen, R. (1991). *The Healthy Company.* Los Angeles: Jeremy P. Tarcher, Inc.

Rutenfranz, J., Colquhoun, W. P., Knauth, P., & Ghata, J. N. (1977). Biomedical and psychosocial aspects of shift work. *Scandinavian Journal of Work Environment Health, 3,* 165–182.

Sauter, S. L., Lim, S. Y., & Murphy, L. R. (1996). Organizational health: A new paradigm for occupational stress research at NIOSH. *Japanese Journal of Occupational Mental Health, 4,* 248–254.

Schaufeli, W. B., & Buunk, B. P. (2003). Burnout: An overview of 25 years of research and theorizing. In C. L. Cooper, M. Schabrac, & J. A. M. Winnebust (Eds.), *Handbook of Work and Health* (2nd ed., pp. 311–348). Chichester, UK: Wiley.

Schaufeli, W. B., & van Dierendonck, D. (1993). The construct validity of two burnout measures. *Journal of Occupational Behavior, 14,* 631–647.

Scott, J., Sochalski, J., & Aiken, L. (1999). Review of magnet hospital research: Findings and implications for professional nursing practice. *Journal of Nursing Administration, 29,* 9–19.

Seo, Y., Ko, Y., & Price, J. L. (2004). The determinants of job satisfaction among hospital nurses: A model estimation in Korea. *International Journal of Nursing Studies, 41,* 437–446.

Shamian, J., O'Brien-Pallas, L. Thomson, D. Alksnis, C., & Kerr, M. S. (2003). Nurse absenteeism, stress and workplace injury: What are the contributing factors and What can/should be done about it? *International Journal of Sociology and Social Policy, 23,* 81–103.

Short, J. (1997). Psychological effects of stress from restructuring and re-organization. *AAOHN Journal, 45,* 597–604.

Simonowitz, J. A. (1996). Health care workers and workplace violence. *Occupational Medicine: State of the Art Reviews, 11,* 277–291.

Sochalski, J., Aiken, L. H., & Fagin, C. M. (1997). Hospital restructuring in the United States, Canada, and Western Europe: An outcomes research agenda. *Medical Care, 35*(10)(Special Suppl.), OS13–OS25.

Sparks, K., & Cooper, C. L. (1999). Occupational differences in the work-strain relationship: Towards the use of situation-specific models. *Journal of Occupational and Organizational Psychology, 72,* 219–229.

Stamper, C., & Johlke, M. C. (2003). The impact of perceived organizational support on the relationship between boundary spanner role stress and work outcomes. *Journal of Management, 29,* 569–588.

Stordeur, S., D'Hoore, W., & Vandenberghe, C. (2001). Leadership, organizational stress, and emotional exhaustion among hospital nursing staff. *Journal of Advanced Nursing, 35,* 533–542.

Survey of Working Conditions (1971). Ann Arbor, MI: Institute for Social Research.

Tasto, D. L., Colligan, M. L., Skjei, E. W., & Polly, S. J. (1978). *Health Consequences of Shift Work* (DHHS [NIOSH] Publication No. 78-154). Washington, DC: U.S. Government Printing Office.

Tovey, E., & Adams, A. E. (1999). The changing nature of nurses' job satisfaction: An exploration of sources of satisfaction in the 1990s. *Journal of Advanced Nursing, 30,* 150–158.

Tummers, G., Janssen, P. M., Landeweed, A. B., & Houkes, I. (2001). A comparative study of work characteristics and reactions between general and mental health nurses. *Journal of Advanced Nursing, 36,* 151–162.

van Vegchel, N., deJonge, J., Meijer, T., & Hamers, J. (2001). Different effort constructs and effort-reward imbalance: Effects on employee well-being in ancillary health care workers. *Journal of Advanced Nursing, 34,* 128–136.

van Yperen, N. W., & Hagedoorn, M. (2003). Do high job demands increase intrinsic motivation or fatigue or both? The role of job control and job social support. *Academy of Management Journal, 46*, 339–348.

Visser, M., Smets, E., Oort, F. J., & de Haes, H. (2003). Stress, satisfaction and burnout among Dutch medical specialists. *Canadian Medical Association Journal, 168,* 271–275.

Weibel, L., Gabrion, I., Aussedat, M., & Kreutz, G. (2003). Work-related stress in an emergency medical dispatch center. *Annals of Emergency Medicine, 41,* 500–506.

West, M. A. (1989). Innovation amongst health care workers. *Social Behavior, 4,* 173–189.

World Health Organization. (2003). WHO issues a global alert about cases of atypical pneumonia: Cases of severe respiratory illness may spread to hospital staff. Retrieved March 15, 2006 from http://www.who.int/csr/sars/archive/2003_03_12/en/

Yip, Y. (2001). A study of work stress, patient handling activities and the risk of low back pain among nurses in Hong Kong. *Journal of Advanced Nursing, 36,* 794–804.

12

EFFECT OF WORKPLACE STRESS ON PATIENT OUTCOMES

Eric S. Williams
University of Alabama, Tuscaloosa

Julia McMurray and Linda Baier-Manwell
University of Wisconsin, Madison

Mark.D. Schwartz
New York University

Mark Linzer
University of Wisconsin, Madison

*Memo Investigators:**

Today's health care workers are under increased stress in their daily work lives as a result of changes in contemporary medical care that include disparities in access and quality, inequities in remuneration, and increased work demands with less control over multiple aspects of daily work life. These changes are compounded by a perceived loss of public trust in a setting increasingly bureaucratized and technologically dominated, with multiple financial and legal complexities. Put simply, the massive changes in the past two decades have made a stressful job even more stressful.

---

*Members of the MEMO investigative team include; Perry G. An, New York University School of Medicine; Elizabeth Arce, Division of General Medicine & Primary Care, Cook County Hospital; Stewart Babbott, M.D., Baystate Medical Center; JudyAnn Bigby, M.D., Brigham & Women's Hospital; James Bobula, Ph.D., University of Wisconsin, Department of Family Medicine; Karla Felix, Department of Medicine, VA New York Harbor Health Care System; Deborah Dowell, M.D., Division of Primary Care, New York University School of Medicine; John Frey, M.D., Department of Family Medicine, University of Wisconsin; Jessica Grettie, Department of Family Medicine, University of Wisconsin; Barbara Horner-Ibler, M.D., Department of Internal Medicine, Milwaukee Clinical Campus, University of Wisconsin; Robert Konrad, Ph.D., Sheps Center for Health Services Research, University of North Carolina; Peggy Leatt, Ph.D. Department of Health Policy & Administration, University of North Carolina; Ann Maguire, M.D., M.P.H., Division of General Internal Medicine, University of Wisconsin - Milwaukee; Bernice Man, M.D., Division of General Medicine & Primary Care, Cook County Hospital; Marlon Mundt, M.A., M.S., Department of Family Medicine, University of Wisconsin; Laura Paluch, B.A. Center for Urban Population Health, Aurora Sinai Medical Center; Mary Beth Plane, Ph.D., Department of Family Medicine, University of Wisconsin; Joseph Rabatin, M.D., Bellevue Hospital; Elianne Riska, Ph.D., Department of Sociology, Abo Akademi University, Finland; William Scheckler, M.D., Department of Family Medicine, University of Wisconsin; Jessica Sherrieb, Section of General Internal Medicine, University of Wisconsin; Anita Varkey, M.D., Division of General Medicine & Primary Care, Cook County Hospital.

For the individual worker, excessive long-term workplace stress can create illness, burnout, and dissatisfaction. One study of academic physicians (Ramirez et al., 1995) found burnout rates of between 37% and 47%, whereas others (Deckard, Hicks, & Hamory, 1992; Whippen & Canellos, 1991) reported an alarming 55%–67% burnout rate among physicians in private practice. Most ominously, Ramirez and colleagues (1995) found that younger physicians have nearly twice the burnout incidence compared with older physicians. Research on nurses and dentists also documents low satisfaction, with a burnout rate assessed at upwards of 27% (Astrom, Nilsson, & Winblad, 1990). For the organization, excessive worker stress is associated with turnover, absenteeism, poor morale, reduced efficiency, and poor performance (Ivancevich & Matteson, 1980). However, there is a far more subtle and pernicious aspect of stress: its impact on the quality of patient care. That is, are stressed workers able to deliver safe, high quality patient care? Sadly, the little available research suggests not. Several studies have found that dissatisfaction among physicians (Haas et al., 2000; Linn et al., 1985) and nurses (Weisman & Nathanson, 1985) is associated with patient dissatisfaction. Dissatisfied patients are less likely to adhere to medical treatments (DiMatteo et al., 1993; Weisman & Nathanson, 1985) and more likely to leave their physician (Keating et al., 2002). At the organizational level, Jones and colleagues (1988) found a linkage between organizational stress and the frequency of being sued for malpractice. Similarly, Keijsers, Schaufeli, LeBlanc, Zwerts, and Miranda (1995) found linkages between all three burnout dimensions with both objective and subjective measures of unit performance. Poor patient outcomes are suggested by several studies (Becker, Stolley, Lasagna, McEvilla, & Sloane, 1971; Grol et al., 1985; Melville, 1980) that link physician dissatisfaction with poor prescribing behaviors.

Several other studies directly focus on the linkages between satisfaction, stress, burnout, and patient safety. Bond and Raehl (1992) found that pharmacists' estimates of their risk of dispensing errors were correlated with career satisfaction. DeVoe, Fryer, Hargraves, Phillips, and Green (2002) found that physicians dissatisfied with their careers reported more difficulties caring for patients, including feeling that they had insufficient clinical autonomy, less than adequate time with patients, and little ability to provide quality care. In a study of an internal medicine residency program, Shanafelt, Bradley, Wipf, and Back (2002) found that three-quarters of the respondents exhibited symptoms of burnout. These physicians also reported more suboptimal patient care practices. However, of the three components of burnout, only the depersonalization scale was associated with suboptimal patient care. A report prepared for the Agency for Healthcare Quality and Research (Hickam et al., 2003) on the connection between working conditions (including job stress, dissatisfaction, and burnout) and patient safety cautioned that although these results are suggestive, the evidence available is not sufficient to firmly conclude that stress, dissatisfaction, or burnout cause higher rates of medical errors.

In addition to the limited research on associations between health care worker stress and patient outcomes, there is also a lack of any common theoretical model. In this work, we present a conceptual model and explore the empirical evidence supporting the intuitive idea that worker stress is a driver of poor quality care. This notion has been the focus of extensive research in other literatures (e.g., marketing), but its application is fairly novel to health care. The remainder of this chapter will discuss this conceptual model and its implications for research and practice.

THEORETICAL BASIS

Our conceptual model (Fig. 12–1) connecting health care worker job stress with patient outcomes is based on the job stress models of Lazarus and Folkman (1984) and Ivancevich and Matteson (1980). Lazarus and Folkman (1984) contributed a general theoretical framework and a cognitive process view of the stress process to this discussion; the value that Ivancevich and Matteson (1980) bring is a focus on the organizational contributors to stress.

At the heart of the Lazarus and Folkman model (1984) is the definition of stress as a troubled relationship between the person and the environment in which stressors tax or exceed an individual's coping resources. Step one of the model contains the causal antecedents. These personal (e.g., values, sense of control) and environmental (e.g., demands, constraints, potential for harm) stressors may then prompt cognitive processing (the second step) as they demand attention from the individual. This processing is dynamic and occurs across time and encounters with the stressors. This cycle starts with primary appraisal, proceeds to secondary appraisal, and finishes with reappraisal. Primary

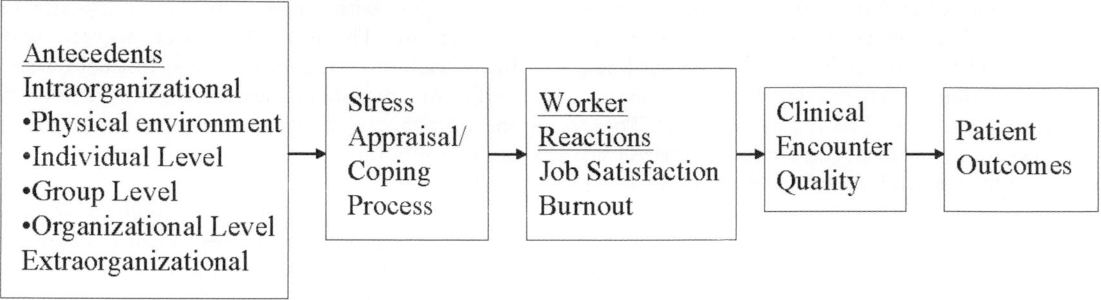

Figure 12–1. Conceptual model.

appraisal evaluates an encounter as to its "threat potential." If individual well-being is threatened (e.g., equilibrium is disturbed), then secondary appraisal takes place. It involves determining an appropriate *coping* mechanism for managing the threatening encounter. Reappraisal takes place soon there after and is the further assessment of the encounter on the basis of new information. For example, an intake receptionist at a busy ambulatory practice experiences an increasing sense of overwork as the practice takes on more patients than it can handle. Initially, he or she appraises the rise in the workload as "no big deal." As the volume of work increases, he or she appraises this event as a real threat, and his or her secondary appraisal results in the receptionist's asking for help. His or her reappraisal suggests that the request for help reduced the workload to a manageable level.

The third step of the stress process takes place when the environmental demands tax or exceed the coping resources of the individual. Lazarus and Folkman (1984) suggest that three sets of reactions occur. One reaction is the activation of what Selye (1956) termed the general adaptation syndrome (GAS) or stress response in which certain physiological changes associated with the "flight or fight" response are initiated. This occurs when the individual is "tensed up" and ready for action. Another aspect is the positive or negative feelings associated with the event. A wedding, for example, may be experienced as stressful because of the large, but positive, environmental demands. Going back to the receptionist, if the workload continues to increase and his or her coping mechanisms no longer work, then he or she is likely to feel "frazzled," emotionally drained, and increasingly dissatisfied with the job. The final aspect is the level of performance in the presence of excessive environmental demands. In the receptionist example, this is the question of his or her ability to continue to effectively serve incoming patients despite the large workload.

Stressors can trigger the coping mechanism in two different manners. One scenario involves a sudden and immediate environmental demand. The receptionist may have a patient who is late to an appointment and, on hearing that it will be at least a 1 h wait, bursts into tears or berates the receptionist. The environmental stressor (the tearful or yelling patient) is a sudden stressor and is likely to produce both physiological and emotional reactions. The other scenario is more subtle and chronic in nature. That is, the environmental demands build up over time and overwhelm coping mechanisms through relentless pressure. For example, the receptionist has done well in coping with an increasing amount of work. However, as the heavy workload has continued and his or her coping (e.g., asking for help; reorganizing the office) has yielded diminishing returns, he or she feels more and more tense and less and less satisfied with the job. This is the subtle effect of chronic stressors. Although many sudden stressors occur in health care environments (e.g., emergencies), most of the stressors are more chronic (e.g., workload).

The fourth step of Lazarus and Folkman (1984) concerns the long-term effects of the chronic stressors. As these stressors continue to affect the individual, she or he feels increasingly tense and may experience diminishing job satisfaction. Over a long period of time, this tension may result in poorer health (e.g., tension headaches; heart disease), lower employee morale (e.g., increased turnover), and decreased social functioning (e.g., social withdrawal).

As previously discussed, the model of Ivancevich and Matteson (1980) focuses on stress in organizations but follows the stressor–stress processing–outcomes organization of Lazarus and Folkman (1984).

Ivancevich and Matteson (1980) identified five sets of antecedents, four of which are intraorganizational (physical environment, individual level, group level, and organizational level) and one that is extraorganizational (e.g., family and economic pressures). These stressors cause job, career, and life stress as perceived by the individual and measured physiologically or psychologically. The outcomes described in Ivancevich and Matteson (1980) are separated into short-term and long-term effects. The short-term outcomes are physiological (e.g., blood pressure, serum catecholamines) and behavioral (e.g., satisfaction, performance, turnover) in nature. Long-term, sustained stress is thought to cause diseases of adaptation, most notably coronary heart disease, depression, apathy, burnout, and nervous exhaustion.

WORK STRESS AND THE QUALITY OF THE MEDICAL ENCOUNTER

The point of departure for our model from those just discussed is the consideration of specific effects of job stress and burnout on the quality of medical encounters and, ultimately, on patient outcomes. That patient outcomes should be considered important in a model of job stress appears intuitive, but the models reviewed here and those discussed in reviews of the stress and burnout literatures (Edwards & Burnard, 2003; Kahn & Byosiere, 1992) focus on outcomes to the organization (e.g., turnover) and worker (e.g., physical health) rather than outcomes to the patient. This is a bit perplexing given that the genesis of the burnout literature came from research in the helping professions, most notably nursing.

We know from the literature that the quality of patient interactions with health care workers can have profound effects on patient satisfaction (Haas et al., 2000), trust (Keating et al., 2002), continuity of care (Williams et al., 2001), adherence to medical treatments (DiMatteo et al., 1993), and switching to another physician (Keating et al., 2002). Although investigators have tentatively supported the links described here, the underlying mechanisms by which quality of health care worker–patient interactions lead to certain patient outcomes have generally been unexplored. We hypothesize that patient encounters with health care workers are the arena in which worker dissatisfaction, stress, burnout, and mental health issues brought on by stressful working conditions can affect patient outcomes. We contend that stressed workers are less able to fully engage with patients than their less-stressed counterparts. This lower level of engagement manifests itself in a variety of ways, from the overly "efficient" receptionist to the distracted nurse to the doctor who focuses on the patient's medical condition in preference to psychosocial matters.

The mechanism linking worker distress to patient outcomes begins with a consideration of the process model of worker burnout (Leiter & Maslach, 1988). It proposes a relationship between the three dimensions of burnout: emotional exhaustion, depersonalization, and lack of personal achievement. As difficult working conditions tax or exceed a person's coping resources, they cause that person to feel stressed. Over time, sustained stress begins to drain the emotional resources of the person, resulting in the first dimension of burnout, emotional exhaustion. This is accompanied by lowered job satisfaction, diminished organizational and professional commitment, and decrements in mental health. As emotional resources are sapped by sustained stress, the process model contends that the health care worker will attempt to cope by depersonalizing patient interactions. Essentially, this means workers will increase their emotional distance from others to conserve their own diminishing emotional resources. Depersonalization is accompanied by negative attitudes toward patients. As care providers continue to distance themselves from their patients, the process model of burnout suggests that they will feel a lack of personal accomplishment caused by their lack of attentiveness to their patients' needs.

Of course, depersonalization, negative attitudes toward patients, and feeling a lack of personal accomplishment do not occur in isolation. Rather, they take place in the context of a health care worker–patient interaction in which both the worker and patient react to the actions of the other. Bakker, Schaufeli, Sixma, Bosveld, and Dierendonck (2000) investigated the effects of depersonalization in a longitudinal study that provided a good theoretical frame for understanding the impact depersonalization has on the health care worker–patient relationship. Bakker's theoretical frame begins with equity theory (Adams, 1965), which suggests that all people have an innate tendency to pursue reciprocity in relationships. Inequity in this relationship will create motivation to return the relationship to a state of balance. The greater the perceived inequity is, the greater the motivation is to restore equity to the relationship. Bakker and colleagues argued further that this reciprocity extends even to the inherently unequal relationship between physicians and patients. Drawing on social exchange theory (Blau, 1964), Bakker and colleagues theorized that the balance of this relationship is maintained by the provision of professional services on the physician side and the gratitude, compliance with medical advice, and symptom alleviation on the patient side. Applying exit-voice-loyalty theory (Hirschman, 1970), patients can respond to a "distant" (depersonalizing) clinician in three ways: (a) increase their efforts to "get what they need" from their clinician (voice), (b) take what they can get from their relationship and remain loyal (but unhappy), or (c) leave the practice or hospital (exit). From the worker view, none of these responses is particularly good. Patients making (as they see it) reasonable demands of health care workers may be seen as overly "demanding," which adds to workers' level of stress and emotional exhaustion (Bakker et al., 2000). Loyal patients may not appear to be a problem, but a more distant, less satisfying relationship with their clinician is likely to result in lower patient trust (Keating et al., 2002), decreased treatment compliance (DiMatteo et al., 1993), and a higher probability of continuing illness. The problem with patients' exercising the third option (leaving) lies in its economic impact and disrupted continuity of care.

In their longitudinal study, Bakker and colleagues (2000) investigated the effect of physician depersonalization on subsequent patient demands and physician burnout. They found that physician depersonalization predicted the perceived level of patient demands *5 years later*. They also found that physician perceptions of excessive patient demand were associated with perceptions of lack of reciprocity from their patients and all three elements of burnout. This, of course, suggests a vicious cycle in which workers use distancing behaviors (depersonalization) in their patient encounters. Some patients, when confronted with apparently indifferent or unconcerned caregivers, become more demanding of their caregiver. This higher level of demand is perceived as a lack of reciprocity on the "ungrateful" patient's part, and the entire vicious cycle repeats itself.

A MODEL OF THE EFFECTS OF HEALTH CARE WORKER STRESS ON PATIENT OUTCOMES

Having explored the mechanisms of how worker stress affects patient outcomes, we now turn to an explanation of the model beginning with the stressors, continuing with the coping process, and concluding with the quality of medical encounters and ensuing patient outcomes.

Intra- and Extraorganizational Stressors

Following the model of Ivancevich and Matteson (1980), our model suggests five sets of antecedents, four intraorganizational and one extraorganizational. The first set of intraorganizational stressors concerns the physical work environment. Here the stressors can be related to light, noise, temperature, vibration, and motion (see the chapter on "Noise and Alarms in Healthcare" in this handbook). Despite their popular conception as clean places of quasi military order, hospitals can be rather noisy, cluttered, and busy places. Topf and Dillon (1988) investigated the effect of noise on critical care nurses on six intensive care units and found that stress caused by noise was positively correlated with burnout. In a more recent work, Topf (2000) proposed a model that linked ambient stressors (e.g., noise) to both subjective and physiological indicators of stress.

The second set of stressors, those at the individual level, appears to be the most thoroughly explored, largely because stress is thought of as primarily an individual phenomenon. Although many individual stressors have been set forth (Ivancevich & Matteson, 1980) only workload, role stressors, and patient demands will be considered here briefly. Workload in medical settings consists of more than total number of work hours; it also involves the time in direct contact with patients and the intensity of that contact. In a review of the correlates of burnout among psychiatric nurses, Melchior, Bours, Schmitz, and Wittich (1997) noted that "work pressures" had the highest correlation with nursing burnout. Likewise, a review by McVicar (2003) noted a similar relationship between workload and nursing stress. Using a large dataset from the Physician Worklife Study, Linzer et al. (2002) found that measures of time pressure during office visits and overall work hours were directly related to both stress and job satisfaction. Biaggi, Peter, and Ulich (2003) examined stressors, emotional exhaustion (a component of burnout), and aversion to patients among Swiss resident physicians. They found that quantitative overload, unmet needs to reduce workload, experienced work intensity, inadequate time off, and restricted autonomy to manage time were related to both emotional exhaustion and aversion to patients.

Role theory (Katz & Kahn, 1964) has played an important part in the development of the literature through stimulating research into the concepts of role ambiguity, role conflict, and role overload and their relationships to such variables as job satisfaction, job stress, and burnout. Role ambiguity can be defined as the level of information an individual has about his or her role in an organization although role conflict is the level of inconsistency among expectations inherent among different roles (Rizzo, House, & Lirtzman, 1970). Role overload is simply having too many roles for one person to balance (Kahn, Wolfe, Quinn, Snoek, & Rosenthal, 1964). In their review of the predictors of psychiatric nursing burnout, Melchior and colleagues (1997) noted that "role confusion" and workload had the highest correlations with burnout. In a review of the role of burnout in social workers, Lloyd (2002) reported that role ambiguity was related to both job satisfaction and job stress. Lloyd cited Balloch, Pahl, and McLean (1998) who reported that among 1276 social workers, elements of role ambiguity (e.g., being exposed to conflicting demands, being expected to do non-job tasks, and being unclear about what is expected) were the most frequently mentioned sources of dissatisfaction. Regression results from a large-scale study of organizational stress and emotional exhaustion among hospital nurses found that both lack of role clarity and presence of role conflict were significantly related to emotional exhaustion (Stordeur, D'hoore, & Vandenberghe, 2001).

Because health care is the quintessential helping profession, it is not surprising that many health care professionals report patient demands as a job stressor. McVicar (2003) reported in his review of the nursing burnout literature that the *emotional demands of caring* is one of the most consistently identified sources of stress. Likewise, Lloyd, King, and Chenoweth (2002) discussed the tension between the values of social work and the demands of daily work lives as a major stressor. Johnson and colleagues (1995) examined the effects of patient demand on job satisfaction and job stress in 581 Johns Hopkins Medical School graduates. Level of patient demand was significantly related to job stress but not to job satisfaction. Using a UK sample of general practitioners, Cooper, Route, and Faragher (1989) found that a measure combining job and patient demand was the most significant predictor of job dissatisfaction and a significant predictor of poor mental health.

The third set of stressors is those stressors that occur at the group level. These happen in interactions between small groups or teams in what has been referred to as the microsystem (Berwick, 2002) (see the chapter on "Clinical Microsystems in Healthcare: How Human Factors Shape the Microsystem" in this handbook). Stressors at this level include lack of teamwork, team cohesiveness, conflict between groups, status differentials, and group dissatisfaction. Stordeur and colleagues (2001) found that after heavy workloads, conflict between nurses and nurses with doctors had the highest impact on nursing stress. Similarly, a survey of Swiss physicians (Biaggi et al., 2003) found that good working relationships with colleagues were the most valued work characteristic. Members of community mental health teams reported high levels of satisfaction when they had a positive sense of team belonging and a clear picture of its role (Onyett, Pillinger, & Mujen, 1997). The same project also reported that contact with team colleagues and multidisciplinary interactions were the most rewarding parts of their jobs (Onyett, Pillinger, & Mujen, 1995).

Organizational stressors make up the fourth set of stressors. This set of variables includes organizational culture and climate, technology, management styles, and organizational design. Nystrom (1993) investigated the association between two cultural elements—task norms and pragmatic values—and job satisfaction and organizational commitment of health care managers and clerical staff. He found that both having strong task norms and valuing pragmatism were positively associated with satisfaction and commitment. Further, he found that organizations with stronger cultures (e.g., stronger norms and values) also pursued consistent organizational strategies. Bradley and Sutherland (1995) reported that certain elements of organizational structure and climate (e.g., a low morale work climate, lack of performance feedback, and perceived lack of communication and consultation on important matters) were associated with higher levels of stress among social workers. Stordeur and colleagues (2001) looked at the effect of different leadership styles on organizational stress and emotional exhaustion among hospital nursing staff. They found that nurse supervisors who continuously monitored performance to anticipate mistakes generated increased emotional exhaustion among the staff. After controlling for demographic and organizational stressors, no other leadership style was related to emotional exhaustion. Williams and colleagues (2002) found that when physicians lacked control over workplace and administrative issues, they were experienced higher stress levels.

The fifth and final set of stressors consists of those external to the organization and often reflects the struggle to balance work and life concerns. Majomi, Brown, and Crawford (2003) reported on a qualitative study of work–life balance among community mental health nurses. Not surprisingly, they found substantial impact on job stress via the competing role demands of home and work lives. However, they set forth a theory of "punctuated equilibria" that described the build-up of tension on both work and family lives that often lead to changes in work or family roles. Similarly, Dumelow, Littlejohns, and Griffiths (2000) examined work–life balance issues among U.K. hospital consultants. They described three types of work–life balance: career dominant, segregated, and accommodating. Although most doctors segregated their work and family lives, female doctors were more likely to have career dominant or accommodating styles. They also found considerable dissatisfaction among both groups with work–life balance. Visser, Smets, Oort, and de Haes (2003) found in their sample of 1308 Dutch medical specialists that work–home interference predicts job stress but not job satisfaction. These studies raise the possibility that workplaces with work–home balance policies could lower worker stress levels and increase quality of care.

Perceptions of Job Stress and Coping Strategies

All five of the antecedent sets can have an impact on the individual. However, Lazarus and Folkman (1984) assert that the processes of appraisal and coping mediate the relationship between these antecedents and their impact. They define coping as "constantly changing cognitive and behavioral efforts to manage specific external or internal demands that are appraised as taxing or exceeding the resources of the person" (p. 141). This process-oriented definition emphasizes the cyclic interplay of cognitive appraisal and the behavior–cognitive coping that occurs within and across situations. Stress is perceived when coping resources are "taxed or exceeded." In this conception, coping behaviors assume central importance in the recognition and the reactions to stress, especially for recurring encounters with patients.

Lazarus and Folkman (1984) grouped coping behaviors into problem-focused coping and emotion-focused coping categories. Problem-focused coping strategies use analytic processes aimed at problem definition, solution generation, and solution evaluation for resolving both environmental and internal issues. Emotion-focused coping strategies, which include "avoidance, minimization, distancing, selective attention, positive comparisons and wresting positive value from negative events" (p. 141), focus on lessening the emotional distress. One key point made by Lazarus and Folkman (1984) is that problem-focused and emotion-focused coping strategies can both facilitate and impede each other in the coping process.

The literature documents a rich array of coping strategies. For example, Firth-Cozens and Morrison (1989) used a qualitative approach to document coping strategies among a group of 173 first-year physicians in the United Kingdom (junior house officers). Five coping strategies were identified: tackled the problem (29%), asked for help (28%), rationalized (14%), failed to cope (12%), and dismissed the event (6%). The most important statistic in this list may be the 12% who failed to cope with their stresses and were thus "at risk" for adverse outcomes. In a similar vein, Linn, Yager, Cope, and Leake (1986) documented the health habits and coping behaviors of practicing physicians. They factor analyzed a list of behaviors "in which physicians engaged in the past 6 months to make themselves feel better" (p. 486). The six factors identified included socializing, exercising, attempting to become better organized, restructuring medical practice, arguing or withdrawing, and discussing feelings.

Nevertheless, what is the effect of stress, dissatisfaction, and poor mental health on the use of these strategies? Firth-Cozens and Morrison (1989) examined the use of these coping strategies among junior doctors who were qualified as a "case" under the General Health Questionnaire coding system (e.g., experiencing a clinically significant mental health issue). These highly stressed people manifested the same use of coping strategies except that they were much more likely to dismiss the event

($p << .05$). Interestingly, they also reported significantly higher concern for personally making mistakes ($p << .05$) than the nonstressed group. Linn and colleagues (1986) correlated their six coping strategies with a number of psychological outcomes. Arguing or withdrawing was positively correlated with anxiety ($r = .259$, $p << .001$), depression ($r = .166$, $p << .05$), job stress ($r = .156$, $p << .05$), work conflict ($r = .313$, $p << .001$), and life satisfaction ($r = -.234$, $p << .001$). Job satisfaction was positively associated with attempting to become better organized ($r = .203$, $p << .01$) and restructuring medical practice ($r = .192$, $p << .05$).

According to Lazarus and Folkman (1984), if these coping strategies work, then the effect of the antecedents on perceptions of stress and subsequent job satisfaction, burnout, and mental health will be muted. If the coping mechanisms are taxed or ineffective, then the individual will feel "stressed" and, as discussed earlier, will manifest lower job satisfaction, greater burnout, and poorer mental health.

Impact of Stress, Burnout, and Job Dissatisfaction on Patient Outcomes

The mechanism linking workplace stress to poor quality medical encounters was introduced earlier in this chapter. Now we turn our attention to a practical consideration of the dynamics of the worker–patient encounter. Nowhere is this dynamic better illustrated than in the physician–patient communication literature. Hall, Roter, and Katz (1988) published a meta-analysis of the correlates of provider behavior in medical encounters that include information giving, question asking by providers, and partnership building. Patient satisfaction, compliance, and recall and understanding were the patient outcome variables examined. Patient satisfaction was strongly associated with amount of information. All other correlates except for question asking were associated with patient satisfaction. Patient compliance was more weakly related with provider behaviors overall but was associated with giving more information, asking fewer overall questions, asking more questions about compliance, and engaging in more positive talk and less negative talk. Patient recall and understanding were predicted by more information giving, less question asking, more partnership building, and more positive talk. They concluded that the concept of reciprocity is a mechanism for the association between provider and patient behaviors. They asserted that providers' task or socioemotional behaviors are reciprocated by patients' task or socioemotional behaviors. Thus, to the extent that providers engage in more information giving, more partnership building, more positive talk, and less negative talk, patients will be more satisfied, more compliant, and have higher recall and greater understanding. If stress attenuates the time or energy providers have for information sharing, then the quality of the physician–patient relationship, and thus the quality of care, may suffer.

Stewart (1995) reviewed the literature on the effect of physician–patient communication on patient health outcomes. Among the 21 studies examined, 16 reported positive associations between quality of communication and patient health outcomes. They found that question asking (especially about feelings, patient understanding of the problem, and its impact on functioning) during history taking was related to reduced patient distress and anxiety and increased symptom relief. Another set of studies about discussions of the medical management plan revealed that when a patient is encouraged to ask more questions, is successful at obtaining information, is provided with information programs, and is given clear information, the patient responds with less anxiety and distress and perceives less physical and role limitations. Drawing on these two studies, one can infer that a provider using depersonalization as a coping strategy would be likely to give less information, ask more (interruptive) questions, and engage in less partnership building. These behaviors would be likely to result in less patient satisfaction, less compliance, and lower recall and could potentially lead to increased patient distress and less symptom relief.

This assertion was tested to a limited extent by Levinson, Roter, Mullooly, Dull, & Frankel (1997). They examined malpractice claims records among primary care physicians (PCPs) and surgeons and compared the communication patterns of no-claim versus two or more claim PCPs. They found that no-claim PCPs used more orientation statements (educated patients on expectations), laughed and used humor more, and tended to use more facilitation (checked understanding and encouraged patient talk) than PCPs who had been sued for malpractice. Further, visit times for non-sued PCPs were significantly longer than for sued PCPs. However, they found no such relationship for surgeons. They concluded that both process and emotional affect of routine office encounters predicted malpractice risk. In considering whether to pursue legal action over an adverse outcome, patients are likely to base their decision on their liking of the clinician and the extent to which their PCP appeared open to their concerns and educated them. Thus, if stress diminishes the quality of the physician–patient relationship, an unfortunate consequence could be a risk in malpractice claims—a situation that could result in yet another cycle of stress.

Additional support for this position can be found in the relationships between provider satisfaction and a number of patient outcomes. Although this literature is fairly sparse, a range of patient outcomes have been explored, and findings indicate that provider dissatisfaction influences patient outcomes in negative ways. The most obvious and well researched of these relationships is between provider satisfaction and patient satisfaction. Linn and colleagues (1985) investigated this association using data from 16 university-based primary care practices. The authors found strong correlations between patient satisfaction and the job satisfaction of faculty and house staff. Haas and colleagues (2000) replicated this work with a sample of general internists. This investigation included 11 general internal medicine practices comprising 2620 patients who had at least one visit to their physician (n = 166) in the preceding year. The patients of physicians who rated themselves to be very or extremely satisfied with their work had higher scores for overall satisfaction with their health care and with their most recent physician visit controlling for both patient and physician characteristics.

Weisman and Nathanson (1985) examined the professional satisfaction of 344 community nurses in 77 family planning clinics as it related to patient satisfaction and subsequent contraceptive compliance. The strongest determinant of aggregate satisfaction level of patients was nursing staff satisfaction. Leiter, Harvie, and Frizzell (1998) looked at the effect of burnout among nurses and its correspondence with patient satisfaction. Sixteen inpatient units were studied, which included 605 patients and 711 nurses. Aggregated to the unit level, they found that patients were less satisfied with their care in units where nursing staff felt more emotionally exhausted and more frequently expressed the intention to quit.

Two studies have investigated the relationship between provider satisfaction and adherence to medical treatments. Controlling for baseline adherence DiMatteo and colleagues (1993) found a positive correlation between physicians' global job satisfaction and patients' adherence at 2 years. Weisman and Nathanson (1985) found that patient compliance with contraceptive advice was directly linked to patient satisfaction and indirectly linked to nurse satisfaction.

Three studies investigated the relationship between physician satisfaction and prescribing behaviors. Becker and colleagues (1971) examined the characteristics associated with the prescribing of chloramphenicol, an antibiotic that can result in aplastic anemia and, thus, is only recommended for a few serious conditions. In a sample of 29 PCPs, the authors found that higher levels of prescribing were associated with lower satisfaction with the community and with patients. Using a larger sample of 124 U.K. general practitioners, Melville (1980) examined the relationship between job satisfaction scores and prescribing of three drugs deemed inappropriate by medical consensus. Using state supplied prescribing data, it was found that physicians who were high prescribers of practolol ($p << .05$), monamine oxidase inhibitors ($p << .05$), and anti-infective agents acting locally on the intestinal tract ($p << .05$) were less satisfied than low or nonprescribers. Another important set of results involved the prescribing of low doses of major tranquilizers to elderly patients, which may cause adverse reactions. The least satisfied physicians prescribed the most often and were more likely to allow ancillary staff to write prescriptions for major tranquilizers. Within this group of physicians, a further correlation was found between the level of staff prescribing and job satisfaction. Thus, the greatest level of prescribing overall and staff prescribing was found with the least satisfied physicians. Lastly, Grol and colleagues (1985) investigated prescribing behaviors with a satisfaction measure involving positive and negative feelings about work. Within their sample of 57 Dutch general practitioners, the authors found a positive correlation between negative feelings about work and volume of prescriptions written; there was no such correlation for positive feelings.

Hickam and colleagues (2003) prepared a report for the Agency for Healthcare Quality and

Research (AHRQ) on the effects of working conditions on patient safety. Among the variables considered were job stress, job dissatisfaction, and burnout. They reported two studies looking at the relationship between stress and patient safety. In a particularly impressive manuscript, Jones and colleagues (1988) examined the critical relationship between stress and medical malpractice. In the first study, they compared 91 hospital departments for malpractice risk. They found that those departments at highest risk (e.g., sued within the last year) manifested significantly higher levels of job stress across three of four measures used. Using a different set of 61 hospitals, the second study investigated the malpractice frequency and the four stress measures. They found strong correlations ($r = .39$ to $.56$) with three of the four measures. When the number of hospital beds was controlled, two of the four correlations remained significant. Using a sample of nurses, Dugan and colleagues (1996) examined the effect of job stress on turnover, absenteeism, injuries, and patient incidents. Zero order correlations were found between a perceived stress index and patient incidents overall ($r = .43$), medication errors ($r = .40$), patient falls ($r = .33$), but not IV errors ($r = .15$). However, a second stress measure composed of reported stress symptoms found more modest, nonsignificant correlations.

Hickam and colleagues (2003) also looked at the relationship between job satisfaction and burnout with patient safety. They identified six studies, two of which will be discussed here. Bond and Raehl (1992) examined the self-reported risk of dispensing errors and pharmacist workplace characteristics including workload, workflow, and job satisfaction. They found that pharmacists' estimates of risk of dispensing errors were correlated with career satisfaction. The strongest predictors, however, of estimated risk of dispensing errors was the amount of time available for dispensing each prescription. DeVoe and colleagues (2002) examined the relationship between career satisfaction and ability to provide high quality patient care. Using data from a nationally representative sample (from the Center for Studying Health System Change's Community Tracking study) of family physicians, they found that doctors dissatisfied with their careers were more likely to report difficulties caring for patients including feeling that they had sufficient clinical autonomy, adequate time with patients, and ability to provide quality care. The report for AHRQ (Hickam et al., 2003) concluded that although there was anecdotal evidence that stress, dissatisfaction, and burnout were related to patient safety, there was insufficient evidence to make a definitive conclusion.

Implications for Future Research

The model discussed in this chapter contains a number of potential directions for future research as well as one area in which future research efforts may yield diminishing returns. That area is the identification of predictors of job stress. Any search of Medline or ABI/Inform will yield thousands of articles on stress, with the majority of them documenting stress levels among some group or studying stress predictors. Although documenting the stress levels across time is useful for tracking purposes, the continued focus on studying predictors of stress risks becoming redundant. We argue that more fertile ground would be found by looking at job stress as an independent variable. As our work links stress with patient outcomes, we will confine our remaining discussion to four promising directions for future research suggested by our model.

The first direction involves better elucidation of the specific mechanisms linking stress, coping behaviors, emotional exhaustion, and depersonalization. In the previous section, we provided some insight by linking the work of Lazarus and Folkman (1984) on appraisal and coping with the process burnout model of Leiter and Maslach (1988). What is not entirely clear, either theoretically or empirically, is the process by which ineffective coping causes the exhaustion of emotional resources. Lazarus and Folkman (1984) indicate that long-term social functioning in a stressful environment is dependent on coping effectiveness (p. 190). Drawing on their dynamic interpretation of the interplay between appraisal and coping across encounters and time, it is possible that emotional exhaustion is part of the interplay when ineffective coping saps emotional resources to the point of emotional exhaustion. Depersonalization enters into this equation during the later stages of emotional exhaustion as a coping mechanism. That is, as a person fights a losing battle against stress and becomes more emotionally exhausted, he or she will turn to depersonalization as a way of conserving emotional resources. Future research will need to examine this mechanism and determine the veracity of this hypothesis.

A second specific arena for future research lies in how health care worker stress affects patient outcomes. Our model suggests that the quality of the medical encounter provides an important link. We draw on the physician–patient communication literature to support this contention. The weakness in this argument is that the connection between depersonalization and the physician communication behaviors associated with patient outcomes (e.g., patient satisfaction) is entirely theoretical. To be sure, it is based on a common sense notion, but no empirical studies have investigated this entire mechanism. Roter's landmark work on physician–patient communication could serve as a model (Roter, Hall, & Katz, 1987). Levels of stress, burnout, and depersonalization would need to be collected on the health care worker before patient encounters, and patient outcomes would need to be collected several times after the encounter. In addition, the data collection on both parties and of the worker–patient encounter would need to be performed across a large number of workers and patients over several encounters. Such a project would be necessarily intrusive and logistically challenging. However, such a study would reveal which worker behaviors are related to depersonalization, stress, and burnout and what patient outcomes can be associated with those worker behaviors.

The third arena for future research involves the linkage between health care worker stress and medical errors. We know that the relationship between stress and performance follows the Yerkes–Dodson inverted U in which the highest level of performance is at moderate levels of stress and the lowest levels of performance are associated with low and high levels of stress (Jex, 1998). Related research has also shown, for example, that role overload is negatively correlated with nurse supervisory ratings of nurse performance, job motivation, and quality of patient care (Jamal, 1984). Research on physicians has shown that excessive workload is negatively related to performance on tasks requiring monitoring or constant attention (Spurgeon & Harrison, 1989). According to Kjellberg (1977), sleep deprivation leads to poor performance because it impairs the ability to sustain attention for long periods of time. What has not been thoroughly investigated is the question of the role that stress plays in medical errors. That is, do stressed workers commit more medical errors than nonstressed workers? Jex and colleagues (1991) examined the impact of workload on a behavior composite of job performance, including "making mistakes." They found that workload was negatively correlated with this behavior composite. Firth-Cozens and Morrison (1989) found that among their cohort of former students, those who were stressed as students and as practitioners were more likely to report making mistakes. As the AHRQ report discussed earlier suggests, the existing work suggests that job stress may play some role in medical errors, but the evidence is not definitive.

A fourth area for future research lies at the organizational level. Most of the research into stress (and stress interventions, as discussed later) has occurred at the individual level, leaving an opportunity to investigate job stress and related variables at the unit or organizational levels. Keijsers and colleagues (1995) explored the connection between burnout and unit performance in 20 Dutch intensive care units. Performance was measured subjectively through the perception of unit nurses and objectively (APACHE III Standard Mortality Ratio). Interestingly, both emotional exhaustion and depersonalization were moderately negatively correlated with perceived unit performance and perceived personal performance but were slightly positively correlated with objective unit performance. A structural equation analysis reversed the causal chain suggested in this chapter. Among four competing models, a model suggesting that perceived unit performance was indirectly related to emotional exhaustion and depersonalization through its effect on perceived personal performance had the best fit. Another exemplar of this work is that of Jones and colleagues (1988), which is discussed in detail in other parts of this chapter.

Another stream of scholarship at the organizational level worthy of extension comes from the "healthy organizations" movement. Murphy (1995) offered four directions for future research on occupational stress management, three of which merit comment here. The first suggests that stress interventions must target not only the individual level but also the job and organizational characteristics. Such an approach requires substantial knowledge of job redesign, as well as organizational change. The second discusses the improvement of the measurement of stressors. Just as all firms have accounting systems to measure financial health, likewise there should be a system to assess "organizational health." The third involves making sure that workers are key participants in these organizational change efforts. That is, organizational level stress management much proceed from a grassroots base rather than being dictated from the top. Similarly, Karasek and Theorell (1990) titled their book Healthy Work: Stress, Productivity, and the Reconstruction of Work Life. Although their book begins with four chapters summarizing their work, the

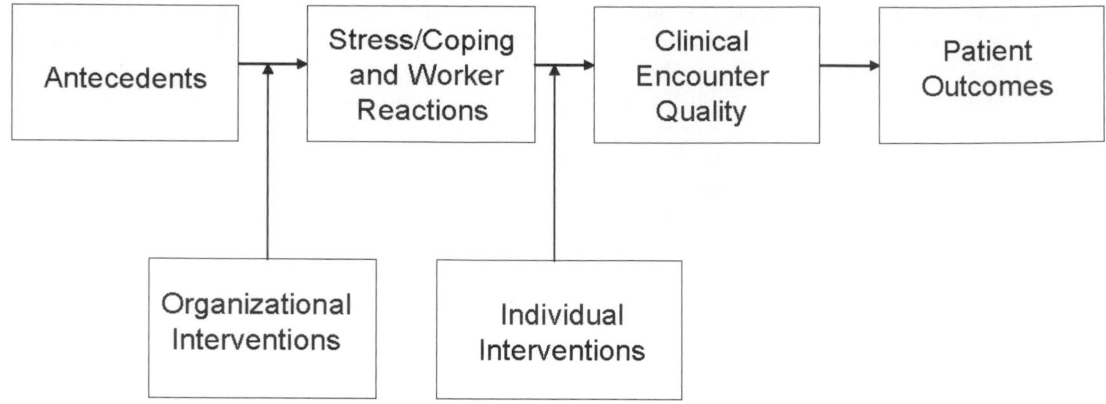

Figure 12–2. Intervention model.

bulk of the book focuses on the redesign of work and organizations to be healthier.

Implications for Stress Management

The implications for stress management mainly lie in the various interventions that can be used to manage stress. Not surprisingly, the literature on individual stress management is much more developed than for organizational interventions (Murphy, 1995). Individual interventions are substantially easier to implement and do not require substantial investment of organizational resources. Figure 12–2 illustrates the areas in which both individual and organizational interventions have an impact within our conceptual model (Fig. 12–1). The remainder of this section will discuss some individual level stress management techniques and then turn to a discussion of the importance and implications of further research into organizational level interventions.

Ways for managing stress are as varied as the stressors themselves. Some stress management techniques focus on the physical body, including exercise and physical relaxation techniques such as deep breathing exercises, yoga, massage, and progressive muscular relaxation. Other stress management techniques focus on the mind and involve meditation, self-hypnosis, and embracing solitude. A third set of stress management techniques involve time management (e.g., organizing, prioritizing), developing better communication skills, talking with others, finding a hobby, minimizing competition, and writing approximately stressors.

Most stress management techniques for physicians target the individual. McCue and Sachs (1991) tested the efficacy of a stress management workshop with 43 medicine and pediatric residents with a control group of 21 residents. The workshop taught interpersonal skills, prioritization, and positive outlook skills, as well as techniques to increase stamina and meet self-care needs. A stress measure and the Maslach Burnout Inventory were administered to all subjects 2 weeks before and 6 weeks after the workshop. Results showed that the overall level of stress declined for the intervention group, whereas the control group scores increased. The intervention group also experienced decreased emotional exhaustion and depersonalization than the control group.

Some interventions target specific populations of physicians. For example, female physicians face unique stressors. They are 60% more likely than male physicians to report signs or symptoms of burnout (McMurray, Linzer, Konrad, Dougas, Shugerman, & Nelson, 2000). Although this is due in part to the difficulty of balancing domestic responsibilities with career demands (Spickard, Gabbe, & Christensen, 2002), there are also patient expectations of female physicians to spend more time listening which leads to increase time pressure during office visits (McMurray et al., 2000). Adjusting panel sizes (patient lists) by patient gender is one step that can be taken to mitigate the work stress of female physicians.

Garside (1993) organized a mutual aid group as a stress management tool for physicians providing primary health care to people with AIDS. The themes identified and discussed by group members included anger, helplessness, lack of job satisfaction, the pain of seeing young patients die, and the pain of repeatedly delivering crushing messages to patients. Garside concluded that, "the commitment of the physicians to the group over a prolonged period and their own testimony indicated that with appropriate facilitation, physicians under stress from providing care to people with AIDS can benefit from the experience of mutual aid that has been so beneficial to other caregivers" (p. 42).

Similarly, much of the nursing literature is dedicated to common stress management techniques such as exercise, development of coping skills, and relaxation techniques. Nursing, however, is also moving toward alternative therapies such as philophonetics, reflexivity, and personality hardiness. For example, Sherwood and Tager (2000), applied philophonetics counseling to nurses who scored high on a self-reported burnout scale. Philophonetics involves the development of individually tailored, self-controlling emotional and cognitive strategies that extended conversational counseling and psychotherapy to include body awareness, movements and gestures, visualization, and sound. These strategies addressed feelings of victimization, disorientation, loss of decision-making power, lack of interpersonal boundaries, and disconnection from one's inner being and internal resources. Participating nurses reported significant reductions on all of the burnout scale items.

Encouraging reflexivity in professionals has long been recognized as a positive educational exercise. Reflection on the action taken as the results of an event allows the practitioner to review the situation and to gain insights for future practice (Hewitt, 2003). Reflexivity also can be used as a stress management technique to explore personal values and beliefs around an issue. For example, Peerson and Yong (2003) focused on four issues relative to nursing clinical practices: seeking technological solutions to health and ill health, moving from the nurse–patient relationship to the patient–healer relationship, using critical pathways, and supporting evidence-based nursing. Based on their experience, Peerson and Yong (2003) urged nurses to "engage in reflexivity and not to lose sight of their selves (knowledge, expertise and skills) ... in their contribution to health care" (p. 42)

Enhancing personality hardiness is another technique used to manage stress in the nursing milieu. Hardiness is defined as the interrelatedness of three factors controlled by the individual through lifestyle: control of the environment, commitment to self-fulfilling goals, and reasonable levels of challenge in daily life. Thomsens, Arnetz, Nolan, Soares, and Dallander (1999) found that these traits serve as buffers and appear to protect people from the psychological repercussions of stress. Nikou (2003) designed a study to investigate the relationships among hardiness, stress, and health-promoting behaviors in students attending a nursing student conference. The results indicated that hardiness is inversely related to stress and positively related to health-promoting behaviors.

Brake, Gorter, Hoogstraten, & Eijkman (2001) studied a group of Dutch dentists to determine the long-term effects of a burnout-intervention program. Using the Maslach Burnout Inventory, they identified a group of high-risk subjects who received feedback on their scores and were invited to participate in an intervention program. Two surveys were administered postintervention—several weeks afterward the intervention and a year later. In the first survey, the 92 dentists responding indicated an improvement in all elements of the Maslach Burnout Inventory. The second survey indicated substantial relapse. Interestingly, a group of controls showed no changes at the first survey, but a subgroup that had specifically pursued some strategy to reduce burnout demonstrated reduction in burnout symptoms.

The importance of organizational interventions to reduce stress has taken on a new importance given the emphasis on "systems thinking" in the Institute of Medicine report *Crossing the Quality Chasm* (2001). However, few organizational interventions are reported in the health care literature (van der Klink, Blonk, Schene, & van Dijk, 2001) even though many good recommendations have been made (Ivancevich & Matteson, 1987; Murphy, 1995). Of this small literature, four studies of note will be discussed.

Karasek, Baker, Marxer, Ahlbom, & Theorell's model (1981) shows that control over the workplace is the major moderator of the link between extreme job demands and stress.

Employee participation in decision making is one of the primary ways to enhance perceptions of employee control. The work by Jackson (1983) proposed a conceptual model in which participation in decision making decreases role ambiguity and conflict. Decreases in role stressors in turn were hypothesized to increase job satisfaction and decrease emotional distress, absenteeism, and turnover intentions. Using an experimental design (Solomon four-group design) with a sample of 126 nurses, nursing assistants, and unit clerks, Jackson found strong support for the effectiveness of increasing participation in decision making. Manipulations of participation decreased role conflict and ambiguity and increased perceived influence, which in turn resulted in greater job satisfaction and lower levels of stress, absence frequency, and turnover intentions.

Landsbergis and Vivona-Vaughan (1995) drew on work by both Karasek and colleagues (1981) and Jackson (1983) in their development of a conceptual model and employee participation intervention. Specifically, they proposed the creation of employee problem-solving committees, which would (a) reduce stressful working conditions and (b) enhance the functioning of the departments. The committees identified significant (but different) organizational stressors (e.g., communication, workloads), and a number of specific proposals were implemented. Unfortunately, one department showed little change and another showed mixed results. The author attributed the lack of results to substantial problems in the implementation of the interventions.

Similarly, Heaney, Price, and Rafferty (1995) evaluated the effectiveness of a multifaceted program called the Caregiver Support Program. The two goals of the program were to train workers to (a) develop and tap into workplace support systems and (b) develop and use participatory problem-solving techniques in work team meetings. They used a simple experimental design with a sample of 1375 direct care staff from group homes for mentally ill adults. Results showed that those who had participated in the program reported better team climate, more supervisory support, better coping abilities, and reduced depressive symptoms.

As part of their impressive four-study manuscript, Jones and colleagues (1988) presented the results of two organizationwide stress management programs. The first of these two used a sample of 700 hospital employees and looked at the impact of a stress management program on the reduction of medical errors. Responses to an employee survey and a series of discussions about how to manage stress resulted in a series of policy changes and a stress management training program. Results showed a significant reduction in medication errors. The second study presented examined the impact of an organizationwide stress management program on malpractice frequency. A set of 22 hospitals participated in the intervention and was matched with a control group of 22 similar hospitals. The results of this quasi experiment showed that the experimental group experienced a statistically significant decline in malpractice frequency.

Organizational interventions to reduce stress can have an effect, but they are very costly. More importantly, the management of stress in organizations appears to be an activity that occurs on an infrequent ad hoc basis. For organizational stress management to have a real impact, it must be incorporated into the everyday activities of the organization. Fortunately, the concept of strategic human resource management has gained currency in recent years (Ulrich, 1997). Quite simply, it argues that an organization's human resources practices and policies must be aligned with the company's strategy for the company to receive maximum benefit from its costliest resource—its people. Managing employee attitudes, including stress, is one of the more important aspects of human resource management. Both strategic human resource management and the "system thinking" espoused in the Institute of Medicine reports are based on the idea that only by changing organizational systems and strategies will a safer and less stressful healthcare workplace be achieved.

CONCLUSION

Given the massive changes in our health care system, it is imperative that we understand the impact that these changes have on the health care worker and on their most important product—care to the patient. Continuing to ignore the impact on health care workers of changes in the health care system inevitably will exacerbate the shortages felt in many health care professions as many qualified professionals leave for less stressful environs. The result will be a continuation of the current shortage of health care workers and a more subtle, and all the more pernicious, decline in the quality of care in spite of all the energy spent on engineering "system" solutions. Until policy makers realize that the health care worker is the center of the health care "system," any long-term improvement of patient safety is doomed to failure.

ACKNOWLEDGMENTS

Supported by a grant from the Agency for HealthCare Research and Quality (AHRQ), RO-1 #HS11955-01.

References

Adams, J. S. (1965). Inequity in social exchange. In L. Berkowitz (Ed.), *Advances in Experimental Social Psychology* (Vol. 2, pp. 267–299). New York: Academic.

Astrom, S., Nilsson, M., & Winblad, B. (1990). Empathy, experience of burnout and attitudes toward demented patients among nursing staff in geriatric care. *Journal of Advanced Nursing, 15,* 1236–1244.

Bakker, A., Schaufeli, W., Sixma, H., Bosveld, W., & Dierendonck, D. (2000). Patient demands, lack of reciprocity, and burnout: A five-year longitudinal study among general practitioners. *Journal of Organizational Behavior, 21,* 425–441.

Balloch, S., Pahl, J., & McLean, J. (1998). Working in the social services: Job satisfaction, stress, and violence. *British Journal of Social Work, 28,* 329–350.

Becker, M. H., Stolley, P. D., Lasagna, L., McEvilla, J. D., & Sloane, L. M. (1971). Characteristics and attitudes of physicians associated with the prescribing of chloramphenicol. *HSMHA Health Reports, 86*(11), 993–1003.

Berwick, D. M. (2002). A user's manual for the IOM's "Quality Chasm" Report. *Health Affairs, 21*(3), 80–90.

Biaggi, P., Peter, S., & Ulich, E. (2003). Stressors, emotional exhaustion, and aversion to patients in residents and chief residents. *Swiss Medical Weekly, 133,* 339–346.

Blau, P. M. (1964). *Exchange and power in social life.* New York: Wiley.

Bond, C. & Raehl, C. (2001). Pharmacists' assessment of dispensing errors: Risk factors, practice sites, professional functions, and satisfaction. *Pharmacotherapy 21*(5): 614-626.

Bradley, J., & Sutherland, V. J. (1995). Occupational stress in social services: A comparison of social workers and home help staff. *British Journal of Social Work, 25,* 313–331.

Brake, H. T., Gorter, H., Hoogstraten, J., & Eijkman, M. (2001). Burnout intervention among Dutch dentists: Long-term effects. *European Journal of Oral Science, 109,* 380–387.

Cooper, C., Route, U., & Faragher, B. (1989). Mental health, job satisfaction, and job stress among general practitioners. *British Medical Journal, 298,* 366–370.

Deckard, G. J., Hicks, L. L., & Hamory, B. H. (1992). The occurrence and distribution of burnout among infectious disease physicians. *Journal of Infectious Diseases, 16,* 224–228.

DeVoe, J., Fryer, G. E., Hargraves, J. L., Phillips, R. L., & Green, L. A. (2002). Does career satisfaction affect the ability of family physicians to delivery high-quality patient care? *Journal of Family Practice, 51*(3), 223–228.

DiMatteo, M. R., Sherbourne, C. D., Hays, R. D., Ordway, L., Kravitz, R. L., McGlynn, E. A., et al. (1993). Physicians' characteristics influence patients' adherence to medical treatment: results from the Medical Outcomes Study. *Health Psychology, 12*(2), 93–102.

Dugan, J., Lauer, E., Bouquot, Z., Dutro, B. K., Smith, M., & Widmeyer, G. (1996). Stressful nurses: The effect on patient outcomes. *Journal of Nursing Care Quality, 10*(3), 46–58.

Dumelow, C., Littlejohns, P., & Griffiths, S. (2000). Relation between a career and family life for English hospital consultants: Qualitative, semistructured interview study. *British Medical Journal, 320,* 1437–1440.

Edwards, D., & Burnard, P. (2003). A systematic review of stress and stress management interventions for mental health nurses. *Journal of Advanced Nursing, 42*(2), 169–200.

Firth-Cozens, J., & Morrison, L. (1989). Sources of stress and ways of coping in junior house officers. *Stress Medicine, 5,* 121–126.

Garside, B. (1993). Mutual aid group: A response to AIDS-related burnout. *Health & Social Work, 18,* 259–267.

Grol, R., Mokkink, H., Smits, A., van Eijk, J., Beek, M., Mesker, P., et al. (1985). Work satisfaction of general practitioners and the quality of patient care. *Family Practice, 2*(3), 128–135.

Haas, J., Cook, E. F., Puopolo, A., Burstin, H., Cleary, P., & Brennan, T. (2000). Is the professional satisfaction of general internists associated with patient satisfaction? *Journal of General Internal Medicine, 15,* 122–128.

Hall, J., Roter, D., & Katz, N. (1988). Meta-analysis of correlates of provider behavior in medical encounters. *Medical Care, 26*(7), 657–675.

Heaney, C. A., Price, R. H., & Rafferty, J. (1995). Increasing coping resources at work: A field experiment to increase social support, improve work team functioning, and enhance employee mental health. *Journal of Organizational Behavior, 16*(4), 335–352.

Hewitt, B. E. (2003). The challenge of providing family-centered care during air transport: an example of reflection on action in nursing practice. *Contemporary Nurse, 15,* 118.

Hickam, D., Severance, S., Feldstein, A., Ray, L., Gorman, P., Schuldheis, S., et al. (2003). *The effect of health care working conditions on patient safety* (No. 03-E031). Washington, DC: Agency for Healthcare Quality and Research.

Hirschman, A. O. (1970). *Exit, Voice, Loyalty: Responses to Decline in Firms, Organizations, and States.* Cambridge, MA: Harvard University Press.

Institute of Medicine. (2001). *Crossing the Quality Chasm: A New Health System for the 21st Century.* Washington, DC: National Academy Press.

Ivancevich, J. M., & Matteson, M. T. (1980). *Stress and Work: A Managerial Perspective.* Glenview, IL: Scott Foresman.

Ivancevich, J. M., & Matteson, M. T. (1987). Organizational level stress management interventions: A review and recommendations. *Journal of Organizational Behavior Management, 8*(2), 229–248.

Jackson, S. E. (1983). Participation in decision making as a strategy for reducing job-related strain. *Journal of Applied Psychology, 68*(1), 3–19.

Jamal, M. (1984). Job stress and job performance controversy: An empirical assessment. *Organizational Behavior & Human Decision Processes, 33,* 1–21.

Jex, S. M. (1998). *Stress and Job Performance: Theory, Research, and Implications.* Thousand Oaks, CA: Sage.

Jex, S. M., Hughes, P., Storr, C., Baldwin, D. C., Jr., Conrad, S., & Sheehan, D. V. (1991). Behavior consequences of job-related stress among resident physicians: The mediating role of psychological strain. *Psychological Reports, 69,* 339–349.

Johnson, J. J., Hall, E. M., Ford, D. E., Mead, L. A., Levine, D. M., Wang, N., et al. (1995). The psychological work environment of physicians. *Journal of Occupational & Environmental Medicine, 37*(9), 1151–1159.

Jones, J., Barge, B., Steffy, B., Fay, L., Kunz, L., & Wuebker, L. (1988). Stress and medical malpractice: Organizational risk assessment and intervention. *Journal of Applied Psychology, 73*(4), 727–735.

Kahn, R. L., & Byosiere, P. (1992). Stress in organizations. In M. D. Dunnette & L. M. Hough (Eds.), *Handbook of Industrial and Organizational Psychology, Second Edition* (2 ed., Vol. 3, pp. 571–650). Palo Alto, CA: Consulting Psychologists Press.

Kahn, R. L., Wolfe, D., Quinn, R., Snoek, J., & Rosenthal, R. (1964). *Organizational Stress: Studies in Role Conflict and Ambiguity.* New York: Wiley.

Karasek, R., Baker, D., Marxer, F., Ahlbom, A., & Theorell, T. (1981). Job decision latitude, job demands and cardiovascular disease: a prospective study of Swedish men. *American Journal of Public Health, 71*(1), 694–705.

Karasek, R., & Theorell, T. (1990). *Healthy Work: Stress, Productivity, and the Reconstruction of Working Life.* Chicago, IL: Basic Books.

Katz, D., & Kahn, R. L. (1964). *The Social Psychology of Organizations.* New York: Wiley.

Keating, N. L., Green, D. C., Kao, A. C., Gazmararian, J. A., Wu, V. Y., & Cleaary, P. D. (2002). How are patients' specific ambulatory care experiences related to trust, satisfaction, and considering changing physicians? *Journal of General Internal Medicine, 17,* 29–39.

Keijsers, G. J., Schaufeli, W. B., LeBlanc, P. M., Zwerts, C., & Miranda, D. R. (1995). Performance and burnout in intensive care units. *Work and Stress, 9*(4), 513–527.

Kjellberg, A. (1977). Sleep deprivation and some aspects of performance: II. Lapses and other attentional effects. *Waking and Sleeping, 1,* 145–148.

Landsbergis, P. A., & Vivona-Vaughan, E. (1995). Evaluation of an occupational stress intervention in a public agency. *Journal of Organizational Behavior, 16*(1), 29–48.

Lazarus, R. S., & Folkman, S. (1984). *Stress, Appraisal, and Coping.* New York: Springer.

Leiter, M., Harvie, P., & Frizzell, C. (1998). The correspondence of patient satisfaction and nurse burnout. *Social Science & Medicine, 47*(10), 1611–1617.

Leiter, M., & Maslach, C. (1988). The impact of interpersonal environment on burnout and organizational commitment. *Journal of Organizational Behavior, 9,* 297–308.

Levinson, W., Roter, D., Mullooly, J., Dull, V., & Frankel, R. (1997). Physician-patient communication: The relationship with malpractice claims among primary care physicians and surgeons. *Journal of the American Medical Association, 277*(7), 553–559.

Linn, L., Yager, J., Cope, D., & Leake, B. (1986). Health habits and coping behaviors among practicing physicians. *Western Journal of Medicine, 144,* 484–489.

Linn, L. S., Brook, R. H., Clark, V. A., Davies, A. R., Fink, A., & Kosecoff, J. (1985). Physician and patient satisfaction as factors related to the organization of internal medicine group practices. *Medical Care, 23*(10), 1171–1178.

Linzer, M., Gerrity, M., Douglas, J., McMurray, J., Williams, E. S., & Konrad, T. R. (2002). Physician stress: Results from the physician worklife study. *Stress and Health, 18,* 37–42.

Lloyd, C., King, R., & Chenoweth, L. (2002). Social work, stress, and burnout: A review. *Journal of Mental Health, 11*(3), 255–265.

Majomi, P., Brown, B., & Crawford, P. (2003). Sacrificing the personal to the professional: Community mental health nurses. *Journal of Advanced Nursing, 42*(5), 527–538.

McCue, J. D., & Sachs, C. L. (1991). A stress management workshop improves residents' coping skills. *Archives of Internal Medicine, 151,* 2273–2277.

McMurray, J., Linzer, M., Konrad, T., Dougas, J., Shugerman, R., & Nelson, K. (2000). The work lives of women physicians results from the physician work life study. The SGIM Career Satisfaction Study Group. *Journal of General Internal Medicine, 15*(6), 372–380.

McVicar, A. (2003). Workplace stress in nursing: A literature review. *Journal of Advanced Nursing, 44*(6), 633–642.

Melchior, M. E., Bours, G. J., Schmitz, P., & Wittich, Y. (1997). Burnout in psychiatric nursing: A meta-analysis of related variables. *Journal of Psychiatric and Mental Health Nursing, 4,* 193–201.

Melville, A. (1980). Job satisfaction in general practice: Implications for prescribing. *Social Science and Medicine, 14A*(6), 495–499.

Murphy, L. R. (1995). Occupational stress management: Current status and future directions. In C. L. Cooper & D. M. Rousseau (Eds.), *Trends in Organizational Behavior* (Vol. 2, pp. 1–14). New York: Wiley.

Nikou, V. R. (2003). *The Relationships of Hardiness, Stress, and Health-Promoting Behaviors in Undergraduate Female Nursing Students.* Paper presented at the Promoting Students' Success, 14th International Nursing Research Congress, Sigma Theta Tau International, St. Thomas, U.S. Virgin Islands.

Nystrom, P. C. (1993). Organizational cultures, strategies, and commitments in health care organizations. *Health Care Management Review, 18*(1), 43–49.

Onyett, S., Pillinger, T., & Mujen, M. (1995). *Making CMHT's work: CMHT's and the people who work in them.* London: Sainsbury Centre for Mental Health.

Onyett, S., Pillinger, T., & Mujen, M. (1997). Job satisfaction and burnout among members of community mental health teams. *Journal of Mental Health, 6,* 55–66.

Peerson, A., & Yong, V. (2003). Reflexivity in nursing: Where is the patient? Where is the nurse? *Australian Journal of Holistic Nursing, 10,* 30–45.

Ramirez, A. J., Graham, J., Richard, M. A., Cull, A., Gregory, W. M., Leaning, M. S., et al. (1995). Burnout and psychiatric disorder among cancer clinicians. *British Journal of Cancer, 71,* 1263–1269.

Rizzo, J., House, R., & Lirtzman, S. (1970). Role conflict and ambiguity in complex organizations. *Administrative Science Quarterly,* 150–163.

Roter, D., Hall, J., & Katz, N. (1987). Relations between physicians' behaviors and analogue patients' satisfaction, recall, and impressions. *Medical Care, 25*(5), 437–451.

Selye, H. (1956). *The Stress of Life.* New York: McGraw-Hill.

Shanafelt, T. D., Bradley, K. A., Wipf, J. D., & Back, A. L. (2002). Burnout and self reported patient care in an internal medicine residency program. *Annals of Internal Medicine, 136,* 358–367.

Sherwood, P., & Tagar, Y. (2000). Experience awareness tools for preventing burnout in nurses. *Australian Journal of Holistic Nursing, 7,* 15–20.

Spickard, A., Jr., Gabbe, S. G., & Christensen, J. F. (2002). Mid-career burnout in generalist and specialist physicians. *Journal of the American Medical Association, 288,* 1447–1450.

Spurgeon, P., & Harrison, J. M. (1989). Work performance and the health of junior hospital doctors: A review of the literature. *Work and Stress, 3,* 117–128.

Stewart, M. (1995). Effective physician-patient communication and health outcomes: A review. *Canadian Medical Association Journal, 152*(9), 1423–1433.

Stordeur, S., D'hoore, W., & Vandenberghe, C. (2001). Leadership, organizational stress, and emotional exhaustion among hospital nursing staff. *Journal of Advanced Nursing, 35*(4), 533–542.

Thomsens, S., Arnetz, B., Nolan, P., Soares, J., & Dallander, J. (1999). Individual and organizational well-being in psychiatric nursing. *Journal of Advanced Nursing, 30,* 749–757.

Topf, M. (2000). Hospital noise pollution: An environmental stress model to guide research and clinical intervention. *Journal of Advanced Nursing, 31*(3), 520–528.

Topf, M., & Dillon, E. (1988). Noise-induced stress as a predictor of burnout in critical care nurses. *Heart and Lung, 17,* 567–574.

Ulrich, D. (1997). *Human Resource Champions: The next agenda for adding value and delivering results.* Boston, MA: Harvard Business School Press.

van der Klink, J. J. L., Blonk, R. W. B., Schene, A. H., & van Dijk, F. J. H. (2001). The benefits of interventions for work-related stress. *American Journal of Public Health, 91*(2), 270–276.

Visser, M., Smets, E., Oort, F. J., & de Haes, H. (2003). Stress, satisfaction, and burnout among Dutch medical specialists. *Canadian Medical Association Journal, 168*(3), 271–275.

Weisman, C. S., & Nathanson, C. A. (1985). Professional satisfaction and client outcomes. A comparative organizational analysis. *Medical Care, 23*(10), 1179–1192.

Whippen, D. A., & Canellos, G. P. (1991). Burnout syndrome in the practice of oncology: Results of a random survey of 1,000 oncologists. *Journal of Clinical Oncology, 9*(10), 1916–1920.

Williams, E. S., Konrad, T. R., Linzer, M., McMurray, J., Pathman, D. E., Gerrity, M., et al. (2002). Physician, practice, and patient characteristics related to primary care physician physical and mental health: Results from the physician worklife survey. *Health Services Research, 37*(1), 121–143.

Williams, E. S., Konrad, T. R., Scheckler, W. E., Pathman, D. E., Linzer, M., McMurray, J. E., et al. (2001). Understanding Physicians' Intentions to Withdraw from Practice: The Role of Job Satisfaction, Job Stress, Mental and Physical Health. *Health Care Management Review, 26*(1), 9–21.

13

SAFETY CULTURE IN HEALTH CARE

Kenji Itoh
Tokyo Institute of Technology, Japan

Henning Boje Andersen and Marlene Dryløv Madsen
Risø National Laboratory, Denmark

SAFETY CULTURE AND ITS LINKS TO PATIENT SAFETY

It is widely recognized that human factors play a crucial role in safety in modern workplaces, particularly in safety-critical domains such as aviation, ship handling, and operations in nuclear power plants. Thus, human errors and human factors–related failures lie behind the majority of accidents in aviation (e.g., Amalberti, 1998), at sea (e.g., Bryant, 1991) and in other high-tech industries (e.g., Hollnagel, 1998). By contrast, it is only fairly recently that it has become generally realized that health care shares many of the characteristics of both high-tech and low-tech human–machine system operations found in industry and that it is equally vulnerable to human failures (Kohn, Corrigan, & Donaldson, 1999). For example, a number of health care activities are performed by teams that are vulnerable to the same error mechanisms that may jeopardize the performance of cockpit crews or control room teams in process industry. Safe and efficient teamwork require that communication, coordination, and task allocation be conducted so that shared situation awareness may be maintained both in traditional human–machine systems (e.g., Glendon & McKenna, 1995) and in health care (e.g., Helmreich, 2000b). Finally, human error has now been recognized as the most important cause of adverse events in health care (Kohn et al., 1999). Therefore, appears useful to adapt—and not necessarily copy—some of the modes of analysis and survey and assessment methods that have been developed for the mainly high-tech human–machine system domains to investigate human factors aspects of patient safety in the medical domain (Helmreich, 2000b). This is not to suggest that health care is just like production industry or the transport sector. Health care differs from other safety critical domains, as it has been observed in the U.S. Institute of Medicine report (Kohn et al., 1999), " ... mostly because of huge variability in patients and circumstances, the need to adapt processes quickly, and the rapidly changing knowledge base ...." However, methods and techniques to coordinate team performance in high-hazard human–machine environments, such as air traffic control, process industry, aircraft carrier operations, may be successfully adapted to health care. Indeed, anesthesiology has long been leading other health care fields in adapting human performance management techniques, such as the crew resource management approach pioneered by the aviation industry (Gaba, Howard, Fish, Smith, & Sowb, 2001).

The recognition that operational safety (and, in health care, patient safety) depends on our abilities to control human error does not mean that efforts should be directed exclusively to the psychological mechanisms underlying human error. Rather, effective safety management should be directed at factors that are conducive to human failure (Rasmussen, 1986)—in particular factors that are within the direct control of the organization. Thus, *organizational factors* have long been acknowledged to be of critical importance to safety in human–machine system operations (Griffiths, 1985; Reason, 1993). Reason's thesis indicates that organizational problems are frequently *latent causal factors* that contribute or even lead to the occurrence of human error made by frontline personnel and has become part of the industry standard in this field (Reason, 1997). Indeed, the majority of contributing causes to major accidents may be attributed to the organizations themselves (Reason, 1997). For example, it has been reported that 40% of incidents in the Dutch steel industry were caused by organizational failures (van Vuuren, 2000). Similarly, based on studies in aviation and maritime operations, quality and safety, by which the operators accomplish their tasks, are affected not only by their professional and technical competence and skills but also by their attitudes to and perceptions of their job roles, their organization, and management (Helmreich & Merritt, 1998).

The term *safety culture* was introduced in the late 1980s when, in the aftermath of the Chernobyl nuclear power accident in 1986, the International Atomic Energy Agency (IAEA) issued several reports analyzing and describing the causes of the disaster, arguing forcefully that the root causes of the accident were to be found in the safety culture that existed in the organization running the power plants (IAEA, 1991). Thus, the concept of safety culture was invoked to explain an organizational mindset that tolerated gross violations and individual risk taking behaviors (INSAG 1991). The INSAG (International Nuclear Safety Advisory Group) defined the notion as follows: "Safety culture is that assembly of characteristics and attitudes in organizations and people which establishes that, as an overriding priority, nuclear plant safety issues receive the attention warranted by their significance" (INSAG, 1991, p. 4).

Since the publication of the INSAG reports, a large number of studies have developed models and measures of safety culture. Although we return to the definitions of safety culture in the subsequent section, it should be noted that, in recent years, a number of studies of safety culture have appeared in health care (see chapter 39 on assessing safety culture and safety climate in health care in this handbook), building on methods and techniques in other safety critical domains, e.g., of health care (Gershon et al., 2000; Helmreich & Merritt, 1998; Singer et al., 2003).

This chapter describes, first, how the notions of safety culture and safety climate are defined and used in the human factors literature about nonmedical domains as well as in the patient safety literature. Then, based on our experiences of safety culture studies, not only in health care but also in maritime, railway, process industry and other industrial domains, we illustrate some main themes of safety culture by reviewing results of recent questionnaire surveys. Finally, we discuss some issues in developing and maintaining a good safety culture.

SAFETY CULTURE AND SAFETY CLIMATE

Since the 1930s (Krause, Seymour, & Sloat, 1999), and thus long before the term safety culture was introduced, it has been acknowledged that differences in organizational factors have an effect on the rate of occupational injuries. Thus, organizational factors such as psychosocial stressors, work team atmosphere, and perceptions of leadership have been studied for decades in relation to occupational accidents, usually under the heading of "organizational climate" or "safety climate."

The distinction between safety culture and safety climate has been discussed by a number of authors. Safety culture is most commonly conceptualized as a three-layer structure, following Schein (1992) work on organizational culture, and summarized with small variations in literally hundreds of papers: (a) an inner layer of basic assumptions (i.e., a core of largely tacit underlying assumptions taken for granted by the entire organization), (b) a middle layer of "espoused" values (i.e., values and norms that are embraced and adopted), and (c) an outer layer of "artifacts" that include tangible and overt items and acts such as procedures, inspections, and checklists. Safety culture is seen as part of the overall culture of the organization that affects members' attitudes and perceptions related to hazards and risk control (Cooper, 2000). By contrast, the distinct but closely related notion of *safety climate* is viewed as governed by safety culture *and* contextual and possibly local issues, so climate refers to employees' context-dependent attitudes and perceptions about safety related issues (Flin, Mearns, O'Connor, & Bryden, 2000; Glendon & Stanton, 2000).

Because the term safety culture is much more commonly used than safety climate, we shall, having no need for distinguishing between underlying and overt attitudes and perceptions, mostly use the term safety culture in its inclusive sense. Still, we will sometimes use the term climate to emphasize the need for referring to local, changeable, and explicit attitudes and perceptions.

Definition of Safety Culture

Numerous definitions have been proposed of the concepts of safety culture and climate (e.g., Flin et al., 2000; Pidgeon & O'Leary, 1994; Zohar, 1980). The most widely accepted and most often quoted safety culture definition is the one put forward by the Advisory Committee on the Safety of Nuclear Installations (in the United Kingdom):

> The safety culture of an organization is the product of individual and group values, attitudes, perceptions, competencies and patterns of behavior that determine the commitment to, and the style and proficiency of, an organization health and safety management. Organizations with a positive safety culture are characterized by communications founded on mutual trust, by shared perceptions of the importance of safety and by confidence in the efficacy of preventive measures. (Advisory Committee on the Safety of Nuclear Installations [ACSNI], 1993, p. 23)

According to this well-known definition, an organization's safety culture involves the shared *values, attitudes, perceptions, competencies and patterns of behavior* of its members. When an organization has a positive safety culture, there is a mutual and high level of trust among employees, and employees share the belief that safety is important and can be controlled.

However, a number of other, though not very dissimilar, definitions have been offered. For instance, Hale defines safety culture as "the attitudes, beliefs and perceptions shared by natural groups as defining norms and values, which determine how they act and react to risks and risk control systems" (Hale, 2000, p. 7). The Confederation of British Industry (CBI, 1990) defined safety culture as "the ideas and beliefs that all members of the organization share about risk, accidents and ill health."

In a comprehensive review, Guldenmund (2000) cited and discussed 18 definitions of safety culture and climate appearing in literature from 1980 to 1997. This proliferation of definitions has led to some difficulty in interpretation. It is not surprising, therefore, that Reason has noted in his discussion of safety culture and organizational culture that the latter notion "has as much definitional precision as a cloud" (Reason, 1997, p. 192).

Still, the most often cited definitions are similar in that they refer to normative beliefs and *shared* values and attitudes about safety related issues by members of an organization. Nearly all analysts agree that safety culture is a relatively stable, multidimensional, holistic construct that is shared by organizational members (e.g., Guldenmund, 2000). At the same time, safety culture is regarded as being relatively *stable* over time course; for instance, De Cock et al. (1986; cited in Guldenmund, 2000) found no significant change of *organizational* culture over a 5-year interval. The content of safety culture—the norms and assumptions it is directed at—is held to consist in underlying factors or dimensions, such as perceptions of commitment, leadership involvement, willingness to learn from incidents (see below). Moreover, safety culture has a *holistic nature* and is something shared by people, members, or groups of members within an organization. Indeed, the holistic or shared aspect is stressed in most definitions, involving terms such as "molar" (Zohar, 1980), "shared" (Cox & Cox, 1991), "group" (Brown & Holms, 1986), "set" (Pidgeon, 1991), and "assembly" (IAEA, 1991). When members successfully *share* their attitudes or perceptions, "the whole is more than the sum of the parts" (ACSNI, 1993, p. 23). Finally, safety culture is *functional* in the sense that it guides members to take adaptive actions for their tasks, as Schein (1992) regards organizational culture as "the way we do things around here."

Overlapping Types of Safety Culture

Viewing safety culture from a different perspective, we may distinguish between four overlapping spheres of safety culture, each of which may distinguish the perceptions and attitudes of different organizational units: national, domain specific, professional, and organizational culture. Thus, the safety culture shared by a given team of doctors and nurses will be shaped not only by their organization (hospital) but also by their different national cultures (see Helmreich & Merritt, 1998, for a review of a number of studies of especially aviation personnel). Again, when comparing different safety critical domains—healthcare, aviation, process industry, nuclear industry, and the like—we sometimes find considerable differences in attitudes and perceptions of safety relevant aspects (Gaba, Singer, Sinaiko, Bowen, & Ciavarelli, 2003). Finally, within the same domain, we should expect that there might well be possibly-task-related-differences between professions, such as between doctors and nurses or pilots and cabin staff (Helmreich & Merritt, 1998). These different and sometimes overlapping layers of cultures are intertwined to shape the safety culture of any particular organizational unit, and each of them can have both positive and negative effects on patient safety (Helmreich, Wilhelm, Klinect, & Merritt, 2001).

Regarding the national culture, Hofstede (1991) built a famous model based on responses about values and norms across different countries involving four dimensions: power distance, individualism–collectivism, uncertainty avoidance, and masculinity–femininity. Hofstede carried out a number of questionnaire surveys with 20,000 IBM employees working in more than 50 countries, mapping the different national cultures in terms of these four dimensions. Adapting parts of this paradigm, Helmreich and his collaborators (e.g., Helmreich et al., 1998; 2001) have conducted comparative studies on national cultures in aviation, finding national differences similar to Hofstede results.

Shifting our attention to cultures within an organization, we may sometimes observe *local* variations of culture within an organization (we may even suppose, for simplicity, that the organization is characterized by the same national, domain-level, and professional culture). For instance, Itoh, Andersen, Seki, and Hoshino (2001) identified local differences in safety culture characterizing each of the different contractors of a railway track maintenance company. The study found a relationship between the cultural types and accident and incident rates. Such variations of culture within a single organization are often called local cultures or subcultures. The variations in subcultures may depend on both task- and domain-related characteristics (e.g., differences in work load, department, and specialties; exposure to risks; and work shift patterns) and on demographic factors (e.g., age, experience, gender, seniority, and positions in the organization); they may also, it is natural to speculate, be influenced by individual factors related to local leadership and the team atmosphere defined by the most dominating and charismatic team members.

Links to Safety Management and Performance Shaping Factors

The rationale behind studying the safety culture of a given organization and groups of members is, briefly put, that by measuring and assessing safety culture we may be able to identify weak points in the attitudes, norms, and practices of the target groups and organizations; in turn, knowledge of weak points may be used to guide the planning and implementation of intervention programs directed at enabling the target groups and organizations to develop improved patient safety practices and safety management mechanisms.

As can be seen from the definitions mentioned previously, safety culture is coupled not only to employees' beliefs and attitudes about safety-related issues (e.g., motivation, morale, risk perceptions, and attitudes toward management) and factors that affect safety such as fatigue, risk taking, and violations of procedures, all of which can be partially controlled or are influenced by efforts of safety management, but also to safety management issues such as management's commitment to safety, its communication style, and the overt rules for reporting errors, etc. (Andersen, 2002). In addition, as will be further discussed in the section on dimensions of safety culture, an organization's safety culture reflects its policies about error management and sanctions against employees who commit errors or violations, the openness of communications between management and operators, and the level of trust between staff members and senior management (Helmreich, 2000a).

It is well known that the probability of human error is influenced by various work environment factors such as training, task frequency, human–machine interfaces, quality of procedures, supervision and management quality, work load, production and time pressure, and fatigue. Such error-inducing factors are called performance shaping factors (PSFs), a term name, now widely used, that was introduced in the early 1980s by human reliability assessment (HRA) analysts who seek to structure and, ultimately, quantify the probability of human failure in safety critical domains and in particular nuclear power production (Hollnagel, 1998; Swain & Guttmann, 1983). On the HRA approach, different types of relevant PSFs are identified and analyzed for a given set of tasks. In fact, the traditionally cited PSFs are tightly related to safety culture insofar as that a "good" safety culture will reflect a shared understanding of the importance and the means to control the factors that have an impact on human reliability.

To put it briefly, although effective safety management consists of identifying the factors that affect the safety of operations and having the means available to implement control mechanisms, safety culture comprises the mutual awareness of and supportive attitudes directed at controlling risks (Duijm, Andersen, Hale, Goossens, & Hourtolou, 2004). There is, therefore, a growing awareness that applications of human reliability assessment may contribute to a focus for the strengthening of safety culture when developing practices and strategies that help health care staff to control the potentially detrimental effects of PSFs (Department of Health, 2000; Kohn et al., 1999).

DIMENSIONS OF SAFETY CULTURE: WHAT, WHERE, AND HOW WE SHOULD MEASURE?

Dimensions of Safety Culture

A number of dimensions (components or aspects) of safety culture have been proposed, ranging from psychosocial aspects (e.g., motivation, morale, team atmosphere) to behavioral and attitudinal factors regarding management, job, incident reporting, and others. Many of these dimensions have been elicited by applying multivariate analysis techniques such as factor analysis and principal component analysis to questionnaire data. Some dimensions proposed by different researchers are quite similar, although the terms may differ, whereas others differ significantly. The number of dimensions also differs among studies, ranging from 2 to 16 (Guldenmund, 2000). This variation in safety culture dimensions has several sources: first, researchers have employed different questionnaires (so they have different questions to respondents); second, surveys have been made of quite different fields or domains with quite different hazard levels, recruitment criteria, training requirements, regulation regimes (e.g., nuclear power production, aircraft carriers, airline piloting, construction and building industry, offshore oil production platforms, shipping, railways, health care organizations and units); third, when aggregating a group of question items, the choice of a label is essentially a subjective interpretation. However, although the labels of dimensions will often vary, it is clear that similar dimensions (according to the meaning of the labels) are found across the different sets of dimensions proposed by different research groups. At this stage of safety culture research, however, it does not appear possible to infer a limited number of generalized, core dimensions of safety culture, which can satisfactorily match any purpose of application to any kind of profession or domain, from the diverse dimensions that have appeared in the literature.

In his seminal study of Israeli manufacturing workers Zohar (1980) proposed an eight-dimensional model that consisted of (a) the importance of safety training, (b) management attitudes toward safety, (c) effects of safe conduct on promotion, (d) level of risk at the workplace, (e) effects of required work pace on safety, (f) status of safety officer, (g) effects of safe conduct on social status, and (h) status of the safety committee. The eight-dimensional model was aggregated by Zohar into a two-dimensional model: (a) perceived relevance of safety to job behavior and (b) perceived management attitude toward safety. Cox and Flin (1998) identified the following five dimensions, which they called "emergent" factors: (a) management commitment to safety, (b) personal responsibility, (c) attitudes to hazards, (d) compliance with rules, and (e) workplace conditions. Pidgeon and O'Leary (1994) proposed four dimensions to capture good safety culture: (a) senior management commitment to safety; (b) realistic and flexible customs and practices for handling both well-defined and ill-defined hazards; (c) continuous organizational learning through practices such as feedback system, monitoring, and analyzing; and (d) care and concern for hazards, which is shared across the workforce. These four dimensions appear to relate to the dimensions proposed both by Zohar (1980) and by Cox and Flin (1998), and they have broader aspects of safety-related organizational issues such as a concept of organizational learning, which was not overtly included in Zohar's (1980) original factors.

In their study of safety culture in health care, Itoh, Abe, and Andersen (2002) derived nine indices of safety culture: (a) power distance, (b) recognition of communication, (c) recognition of teamwork, (d) recognition of own performance under high stress, (e) stress management for team members, (f) morale and motivation, (g) satisfaction with management, (h) recognition of human error, and (i) awareness of own competence. In their study of health care staff perceptions, Gershon and colleagues (2000) extracted six dimensions from questionnaire responses with a specific focus on blood borne pathogen risk management: (a) senior management support for safety programs, (b) absence of workplace barriers to safe work practices, (c) cleanliness and orderliness of the work site, (d) minimal conflict and good communication among staff members, (e) frequent safety-related feedback and training by supervisors, and (f) availability of personal protective equipment and engineering controls.

These examples illustrate the variation in factors we have described. In general, the criterion for including a putative dimension as a safety culture factor is that there is evidence (or reason to believe) that it correlates with safety performance. A given presumed safety culture dimension should be included only if a more positive score on this dimension is correlated with a higher safety record. If two comparable organizations or organizational units have different safety outcome rates (e.g., accident and incident rates), we should expect that the high-risk unit has a lower score on the dimension in question. However, using this criterion also means that safety culture dimensions may not directly relate to safety (e.g., Andersen, Hermann, Madsen, Østergaard, & Schiøler, 2004a; Itoh, Abe & Andersen, 2005). In fact, it would be wrong (and naïve) to expect the dimensions of safety culture that have the strongest link with safety outcome to be ones that most directly are about safety. On the contrary, there is evidence that the strongest indicators of safety performance are those that relate to dimensions that have an oblique or indirect link to safety, such as motivation, morale, and work team atmosphere (Andersen, et al., 2004a; Itoh et al., 2005).

Methods of Study and Assessment

Several approaches have been applied to investigating the safety culture of particular organizations: safety audits, peer reviews, performance indicator measures, structured interviews with management and employees, behavioral observations and field studies, and questionnaire surveys. Among these approaches, safety audits use methods that combine qualitative interviews and factual observations of (missing) documentation, (faulty) protections and barriers, actual behaviors, and the like to identify hazards existing in a workplace. Several safety audit tools have been developed consisting of safety performance indicators to determine whether a given safety management delivery system (say, the mechanism to ensure that staff members are trained to operate their equipment safely or the mechanism to ensure that learning lessons are derived from incidents) is fully implemented, or only partially so, or possibly only at a rudimentary level (e.g., Duijm et al., 2004). Audits are particularly useful for estimating the extent to which the organization policies and procedures have been defined and are being followed and, to some extent, how they might be improved. On the behavioral observation (or field study) approach, employees' behaviors and activities are observed during their task performance to identify potentially risky behaviors. However, the most frequent and widespread method for safety culture studies is the questionnaire-based survey technique. This is a useful and, depending on the quality of the questionnaire tool, a reliable method for collecting response data from larger groups, and it is especially time-efficient for large respondents.

In the last couple of decades, a large number of questionnaires have been developed and an even larger number of studies have been carried out to measure perceptions and attitudes of employees in a number of nonmedical domains (e.g., Cox & Cheyne, 2000; Williamson, Feyer, Cairns, & Biancotti, 1997; Zohar, 1980; compare reviews in Davies, Spencer, & Dooley, 1999, and Guldenmund, 2000). Similarly, a number of tools and surveys have been created in health care (e.g., Nieva & Sorra, 2003). Scott, Mannion, Davies, and Marshall (2003) reviewed instruments and tools available to measure organizational culture in health care.

As mentioned earlier, attempts have been made to compare scores on safety culture dimensions with safety-related outcome data. Several methods of measuring safety performance have been suggested. The most intuitive and strongest measure of safety outcome involves the *accident* or *incident rate* of an organization. This type of data can be evaluated repeatedly and regularly for an entire organization and its work units, such as clinical departments and wards, and it may also allow a comparison between work units or organizations in the same domain, for example, hospital wards. At the same time, there are many reasons why we should be wary of using accident data, even within one and the same domain and the same type of tasks. First, such data may be essentially dependent on external factors and may not reflect internal processes. For instance, university hospital clinics may be more likely to admit patients who are more ill and may therefore experience a greater rate of adverse events (Baker et al., 2004). Second, accident data may be of dubious accuracy because of underreporting by some organizations and overreporting by others (Glendon & McKenna, 1995).

Reporting of near misses and incidents may be a useful measure of safety performance, although its overriding goal is not to derive reliable statistics regarding rates of different types of incidents but to enable organizations to learn from such experiences (Barach & Small, 2000). Thus, Helmreich stresses that one should not seek to derive rates from incident reporting but rather focus on the valuable lessons they contain (Helmreich, 2000b). However, the wide variation observed across, for instance, hospitals and individual departments may have much more to do with local incentives and local reporting culture than with actual patient risks (Cullen et al., 1995). Therefore, when performing comparative studies between organizations or between work units within a single organization, we should be careful in interpreting the incident rate. As noted earlier, different patient profiles may entail that incident rates may not be directly comparable. In addition, there is another source of even greater uncertainty in the interpretation of rates. Thus, the incident rate may be interpreted either as a measure of risk (the greater the rate of reported incidents of a given type, the greater is the likelihood that a patient injury may take place) *or* as an index of the inverse of risk (i.e., as an index of safety), that is, the more staff members are demonstrating willingness to report, the greater is their sensitivity to errors and learning potential, and so, the greater is the safety in their department (Edmondson, 1996, 2004).

Outside health care, however, there are occasionally opportunities for acquiring safety performance data that may be less susceptible to local and random vagaries. Still, most studies that have compared safety culture measures and safety performance have referred to self-reported accidents and incidents. For instance, Diaz and Cabrera (1997), using questionnaire responses of three comparable companies, argued that rank orders of employees' attitudes toward safety coincided with those of employees' perceived (self-reported) safety level. Until recently, a strictly limited number of studies had examined the relation between safety culture measures and accident risk or ratio based on *actual* accident data. Among the small number of studies addressing actual accident and incident rates, Sheehy and Chapman (1987), however, could not find evidence of a correlation between employee attitudes and accident risk based on objective data on accidents. Itoh, Andersen, and Seki (2004), focusing on operators of track maintenance trains, identified a correlation between the train operators' morale and motivation with actual incident rates for each of the five branches belonging to a single contractor of a Japanese high-speed railway. A branch that employed train operators having higher morale and motivation exhibited a lower incident rate. Itoh and colleagues (2004) also found the very same correlation for company-based responses collected from all track maintenance companies working for the high-speed railway. Andersen and colleagues (2004) found that two daughter companies, both of which were involved in the same procedures and regulations, had different lost-time incident rates as well as self-reported incident rates and that the high-rate company showed more negative attitudes and perceptions on 56 out of 57 items on which they showed significantly different responses.

In summary, several methods are available for investigating safety culture in an individual organization or work unit: safety audits, peer reviews, performance indicator measures, structured interviews, behavioral observations, questionnaire surveys, and so forth. Each method has its strengths and weaknesses, its highly useful and less useful applications. Among these methods, some tools involving questionnaire-based surveys can produce safety culture index values that have been shown to correlate with safety performance. It is important to select appropriate methods and tools that match the purposes of safety culture assessment as well as the characteristics of an organization under study. One of the greatest advantages of using the methods mentioned here is their applicability to diagnosing safety status *proactively*, that is, safety culture indices can be measured *before* an actual accident takes place. Moreover, it is well known that only subset of all incidents is reported in health care (Antonow, Smith, & Silver, 2000). Therefore, the methods may serve as an independent means of assessing the risk level of specific organizations and units.

STAFF ATTITUDES TOWARD SAFETY-RELATED ISSUES: HOW EMPLOYEES PERCEIVE THEIR JOBS, MANAGEMENT AND SO FORTH

Safety Culture Perceived by Health Care Staff

In this section, we illustrate characteristics of safety culture in health care based on the results of our questionnaire surveys of Japanese hospitals (Itoh,

TABLE 13–1. Percentage Agreements and Disagreements for Indexes of Safety Culture[a]

Safety Culture Indexes	*Doctors*	*Nurses*	*Pharmacists*	*Total*
Power distance	30%	22%	28%	23%
	60%	60%	59%	60%
Communication	88%	86%	90%	86%
	5%	4%	3%	4%
Team work	58%	65%	55%	64%
	26%	16%	25%	18%
Own performance under high stress	49%	41%	43%	42%
	38%	36%	33%	36%
Stress management for team member	70%	69%	67%	69%
	20%	16%	22%	17%
Morale and motivation	73%	66%	66%	67%
	16%	15%	19%	15%
Satisfaction with management	46%	51%	52%	51%
	40%	29%	32%	30%
Recognition of human error	61%	61%	55%	60%
	26%	21%	29%	22%
Awareness of own competence	58%	45%	40%	46%
	27%	25%	31%	26%

Note. Upper row = percentage agreement; Lower row = percentage disagreement

[a]Adapted from Itoh et al., 2002

Abe, & Andersen, 2002, 2003). We describe health care staff responses and summarize results at the level of organizations (hospitals), specialties and wards, and job ranks or positions. Using multinational data from Denmark and Japan, we also conducted a comparison of health care staff with ship officers (Andersen, Garay, & Itoh, 1999; Itoh & Andersen, 1999). The comparison results are briefly described here.

Most of the safety culture scales contained in the questionnaire used in the Japanese survey were adapted from Helmreich's Operating Room Team Resource Management Survey (Helmreichs & Merritt, 1998), which contains 57 five-point Likert-type items about perceptions of and attitudes to jobs, teamwork, communication, hospital management, and other safety-related issues. As a framework of safety culture analysis, we employed the nine-dimensional model mentioned earlier. Each safety culture index included several items in the questionnaire. We illustrate this by taking, as an example, the index of power distance that contains 12 items. Power distance (derived from Hofstede, 1991) refers to the psychological distance between superiors and subordinate members—a small distance means that leaders and their subordinates have open communication initiated not only by leaders but also by juniors. Representative items of this index are as follows: "The senior person should take over and make all the decisions in life-threatening emergencies"; "senior staff deserves extra benefits and privileges"; and "doctors who encourage suggestions from team members are weak leaders."

A summary of the responses of approximately 600 health care staff (doctors, nurses, and pharmacists) from five Japanese hospitals is shown in Table 13–1 (the mean response rate of this sample was 91%). In the table, the percentage [dis]agreement is defined as the following rate: the nominator represents "strongly agree" and "slightly agree" ["strongly disagree" and "slightly disagree"] responses, excluding a neutral response option from the five-point scale, and hence, the sum of percentage agreement and disagreement is lower than 100% for each index; and the denominator represents the total number of responses for the specific items of each safety culture index. As can be seen in this table, Japanese health care providers have relatively high *morale and motivation*, and they exhibit good awareness of *communication* among team members and within their organization. Their satisfaction with *teamwork* is also relatively high, with nurses' perception of the value of teamwork being the highest. By contrast, respondents' *satisfaction with management* is rather low, and doctors' satisfaction is significantly lower than that of the other two professional groups.

An unexpected result is that all three groups show a relatively small power distance (higher agreement of this index indicates large power distance). Previous studies (e.g., Spector, Cooper, & Sparks,

2001) have shown that the Japanese are around the "upper middle" when compared with other nations in terms of power distance—so, although not at the extreme high end (e.g., Arab countries and Malaysia), the Japanese are not at the extreme low end either (e.g., Denmark and Ireland). The sample results also indicate that a large part of health care staff has realistic recognition of human error (high agreement indicates realistic recognition). That is, all three hospital groups recognize that "human error is inevitable," and they do not agree with the item "errors are a sign of incompetence." By contrast to these items, however, there was a large difference in an error reporting item "I am encouraged by my leaders and co-workers to report any incidents that I may observe" between the three groups. These results appear to point toward a distinction between attitudes that are relatively independent of and those that are influenced by local culture. Views about human fallibility and human limitations appear to be influenced by education, national culture, and possibly professional culture, whereas willingness to report errors and incidents is influenced by respondents' perceptions of their own specific safety management in an organization. In this respect, it may be noted that it is widely recognized that the incident reporting system is better managed for the nursing staff than for doctors and other staff groups.

Most hospital members recognize the need for monitoring colleagues' levels of stress and workload. For instance, more than 90% of respondents agreed that team members should be monitored for signs of stress and fatigue during tasks. By contrast, respondents do not exhibit any great awareness of the effects of stress on their own performance. More than half of doctors and one third of the nurses disagreed with the item "I am more likely to make errors or mistakes in tense or hostile situations." Similarly, only 5% of doctors agreed that their performance is reduced in a stressed or fatigued situation.

Differences Between Clinical Specialties Between Wards

Table 13–2 indicates percentage agreements and disagreements based both on specialties of the doctors (e.g., internal medicine, surgery) and on the types of wards of nurses. Nurses are classified into eight ward groups: internal medicine, surgery, intensive care unit (ICU), outpatient, pediatrics, mixed ward, and operating room (OR). Although the statistical analysis involved rank-based tests (Kruskal–Wallis, Mann–Whitney) to each question item, the aggregated percentage scores for each factor (agreement vs. disagreement) were not tested. When two groups differ significantly (and in the same direction) on all or most items within a given index, they are said to differ significantly on this index. There appears to be only small or nonsignificant differences between doctors' specialties for most safety culture indexes except for power distance and perception of communication. There is an overall trend of differences across specialties in that surgeons have slightly more positive attitudes and perceptions with regard to safety culture; that is, they demonstrate a bit higher morale and motivation, stronger satisfaction with management, better awareness of teamwork, more realistic recognition of human error, and so on. This appears to indicate that doctors belonging to different specialties may develop different types of safety cultures and even local subcultures.

Responses of nurses across the eight wards also appear to show differences in several of the safety culture indexes: communication, stress management for team member, morale and motivation, and recognition of human error. Among the eight ward groups, two stood out as remarkable in terms of responses to these indices. One type consists of nurses working in the operating room and pediatrics. Compared with the other ward groups, these nurses expressed greater agreement with the importance of communication, and they showed a higher level of realistic acknowledgment of their own performance limitations under stress conditions as well as more realistic recognition of human error, and, finally, a relatively lower level of morale and motivation. Nurses working in the internal medicine ward and with outpatients made up the other ward type. By contrast to the first type, these nurses had the highest morale and motivation and expressed greater agreements with stress management for team members but a lower level of appreciation of their own performance limits under stress condition. These results might, in part, reflect differences in tasks and work conditions, for example, more technical work (operating room) versus more clerical work (outpatients) and work with babies and infants (pediatrics) versus work with many elderly patients (internal medicine). Such variations in task characteristics, in turn, may give rise to ward-based subcultures in a hospital.

TABLE 13–2. Specialty- and Ward-Based Comparisons for Safety Culture Indexes[a]

	Speciality-Based Doctors			Ward-Based Nurses						
Safety Culture Indexes	*Physician*	*Surgeon*	*Others*	*Internal medicine*	*Surgery*	ICU	*Outpatient*	*Pediatrics*	*Mixed ward*	OR
Power distance	31%	33%	28%	20%	23%	27%	21%	24%	20%	21%
	57%	60%	61%	62%	59%	57%	61	65%	60%	63%
Communication	86%	90%	88%	87%	84%	83%	88%	88%	85%	91%
	8%	3%	5%	2%	6%	2%	5%	3%	4%	3%
Team work	52%	67%	54%	66%	64%	61%	66%	65%	67%	68%
	27%	20%	31%	14%	16%	14%	20%	22%	14%	15%
Own performance	54%	43%	51%	40%	41%	43%	36%	48%	42%	50%
under high stress	36%	43%	36%	36%	37%	31%	42%	35%	30%	33%
Stress management	66%	70%	72%	72%	69%	71%	71%	70%	68%	66%
for team member	21%	19%	20%	14%	17%	11%	15%	15%	16%	20%
Morale & motivation	68%	77%	73%	70%	62%	62%	77%	63%	60%	62%
	2%	13%	16%	13%	18%	15%	11%	26%	14%	18%
Satisfaction with	37%	53%	46%	55%	50%	47%	57%	53%	50 %	48%
management	51%	34%	36%	27%	28%	33%	32%	34%	27%	34%
Recognition of	57%	62%	62%	65%	59%	53%	61%	61%	61%	68%
human error	30%	23%	26%	20%	22%	24%	23%	26%	18%	21%
Awareness of	56%	66%	53%	49%	43 %	46%	48%	42%	40%	42%
own competence	34%	20%	28%	23%	26%	20%	26%	32%	20%	31%

Note. Upper row = percentage agreement; lower row = percentage disagreement.

[a] Itoh et al., 2003a

Hospital-Specific Cultures

Table 13–3 summarizes hospital-based responses of nurses in terms of mean, maximum, minimum, and range of percentage agreement and disagreement for each index over the five hospitals surveyed. In the table, maximum and minimum values were picked up across the five hospitals for each index, and range of percentage agreement/disagreement was indicated as subtracted value of the minimum from the maximum. Mean values were shown in this table as averaged percentage agreements and disagreements over the five hospitals. Differences across the five hospitals are larger than across wards (the latter reviewed in the previous subsections) for most indices. Results from a recent, comparable survey of approximately 6,000 health care staff responses from 22 Japanese hospitals show similar results: the observed difference across the 22 hospitals is even greater for every safety culture index than in the former five-hospital survey (Itoh et al., 2005). In fact, for most of the safety culture indices, the ranges of percentage agreement and disagreement across the hospitals are more than three times greater than across doctors specialties and nurses wards.

From these survey results, we conclude that there are large variations in safety culture in Japanese hospitals and somewhat smaller variations across specialties and wards. One may speculate that similar findings might be found in other settings. In addition, one may speculate approximately the reasons why hospital differences are greater than ward and specialty differences. At the very least, the latter result appears to indicate that the organizational culture of the individual hospitals plays a greater role than the variation in work and task conditions across specialties and wards.

Multinational and Multiprofessional Comparisons

In this subsection, we review some cultural differences and similarities across professions and across nationalities, comparing hospital safety culture with that of ship officers elicited from similar questionnaire based surveys (Andersen et al., 1999; Itoh & Andersen, 1999). Approximately 2,600 seafarer responses were collected from two Japanese and five Scandinavian ship companies, using an earlier, derivative version of the hospital questionnaire, the Ship Management Attitudes Questionnaire (SMAQ). The SMAQ sample included responses not only from ship officers but also from nonofficer seafarers. In this comparison of identical items from the two different domains, we used only the

TABLE 13–3. Variations of Safety Culture Indexes Across Five Hospitals[a]

Safety Culture Indexes	*Mean*	*Maximum*	*Minimum*	*Range*
Power distance	22%	24%	19%	5%
	60%	63%	59%	4%
Communication	86%	88%	84%	4%.
	4%	4%	3%	1%
Team work	65%	68%	58%	10%
	16%	20%	13%	7%
Own performance under high stress	41%	48%	35%	13%
	36%	42%	31%	11%
Stress management for team member	70%	72%	66%	6%
	16%	17%	14%	3%
Morale & motivation	66%	74%	60%	14%
	15%	21%	11%	10%
Satisfaction with Management	51%	59%	41%	18%
	29%	34%	24%	10%
Recognition of human error	61%	67%	54%	13%
	21%	24%	20%	4%
Awareness of own competence	45%	48%	42%	6%
	25%	31%	20%	11%

Note. Upper row = percentage agreement; lower row = percentage disagreement.

[a]Adapted from Itoh et al., 2003a

subsample of ship officers (approximately 1,500 responses), because this group may be regarded to have a somewhat higher social status comparable to that of health care staff. Table 13–4 shows percentage agreement and disagreement on safety culture indices for ship officer groups as well as for three health care groups using responses to the identical items in the two questionnaires. The entire sets of items in these questionnaires were not identical; the overlapped part was a subset of each questionnaire, and, therefore the percentage agreements and disagreements for health care staff groups, which were calculated from the same response sample, did not coincide with those in Table 13–1. The ship officer sample includes responses from Danish officers in Scandinavian companies and from Japanese and non-Japanese Asian officers.

Before discussing culture differences between health care staff and ship officers, we briefly describe national culture comparisons using ship officer's data. Table 13–4 shows that there are only minor cultural differences between Danish and Japanese officers for most indexes except for morale and motivation. An interesting finding is that the relatively larger differences were observed between Japanese and Asian officers, many of them are Pilipino and Indonesian, for most indexes, although the latter group is also working for Japanese companies, which may follow a Japanese management system. This suggests that a company nationality does not affect a professional culture as much as the nationality of its employees.

With regard to characteristics of professional culture, a comparison of the two Japanese samples shows noticeable differences between health care staff and ship officers in most indexes with a few exceptions. There are only small differences between the two professions for power distance (small distance is perceived by doctors, nurses and officers), for morale and motivation (where doctors and officers are alike), and for acknowledgement of human error (where nurses and officers are alike). For the other safety culture indexes, ship officers have more positive attitudes and perceptions than health care staff. They assign a greater importance to communication during task performance and to stress management for team members than do both doctors and nurses. Moreover, ship officers attitudes about human error are more realistic than that of doctors and are identical to the attitudes of nurses.

Integrating these cross-professional comparisons, we concluded that the safety culture among ship officers appears to be characterized by somewhat greater safety awareness than that of hospital staff. Helmreich and his associates (Sexton, Thomas, & Helmreich, 2000) reached similar conclusions when comparing responses from medical staff and airline pilots, although they found that surgeons and ICU staff were more like pilots than other specialties in their responses about teamwork and error recognitions.

TABLE 13–4. Comparison of Health Care Staff with Ship Officers on Safety Culture Indexes[a]

	Healthcare Staff			Ship Officers		
Safety Culture Indexes	*Doctor*	*Nurse*	*Pharma*	*Japanese*	*Asian*	*Danish*
Power distance	6%	9%	5%	8%	17%	8%
	89%	79%	88%	81%	63%	82%
Communication	86%	85%	86%	99%	80%	96%
	6%	4%	7%	1%	6%	1%
Own performance under high stress	49%	43%	44%	38%	24%	43%
	38%	35%	32%	43%	54%	38%
Stress management for team member	72%	67%	68%	92%	68%	82%
	19%	18%	20%	3%	6%	4%
Morale and motivation	81%	74%	72%	82%	81%	66%
	11%	10%	13%	8%	4%	12%
Recognition of human error	36%	53%	39%	51%	63%	51%
	48%	34%	45%	36%	18%	29%

Note. Upper row = percentage agreement; lower row = percentage disagreement.

[a]Adapted From Itoh at al., 2003a

INCIDENT AND ERROR REPORTING ATTITUDES: HEALTH CARE STAFF ACTIONS AFTER MISTAKES

Learning From Adverse Event Experiences

One of the most important issues in safety culture concerns the ability of an organization or a group to learn from situations when things go wrong (Department of Health, 2000). Therefore, staff attitudes to incident reporting and discussing their own and colleagues' adverse events and errors are vital. It is widely recognized that health care staff members being willing to report errors and incidents contributes to and is an expression of a healthy and good safety culture. Conversely, when health care staff members keep errors to themselves, it is a sign of a poor safety culture (Department of Health, 2000). Only by bringing up errors and other incidents will health care staff have a chance to learn from experience; and conversely, if such experience is kept suppressed, the risk of repeating this type of incident is greater.

Another key to successful learning from adverse events is management's commitment to incident reporting, approach to feedback to "reporters," and analysis of significant events (Department of Health, 2000). Such a commitment includes support to risk managers who have responsibilities for giving personal feedback to the reporting staff members and for sharing information and knowledge about the reported case within an organization.

Staff Attitudes to Error Reporting

In this subsection, we describe results of surveying health care staff attitudes and views about error reporting and related actions, primarily based on results of questionnaire survey in Japanese hospitals (Itoh et al., 2002). To uncover health care providers' error-reporting attitudes, we used a section of the questionnaire used in the Survey of Staff Attitudes to Reporting (Andersen, Madsen, Hermann, Schiøer, & Østergaard, 2002) and collected 550 responses from Japanese doctors and nurses. The data sample is the same as the one mentioned earlier in the chapter, excluding responses from pharmacists (the mean response rate was 91%). The questionnaire includes question items that refer to two fictitious adverse event cases, one in which the patient suffers a relatively severe outcome (resulting in long-term heart problem and little possibility of returning to work), and the other involving a relatively minor injury (no permanent impairment but a 1 week extension of hospitalization). Respondents are asked to imagine that they themselves were the acting doctor or nurse and to indicate the likelihood of bringing up the event with their leader and colleagues (5 questions) and informing the affected patient (6 questions). The questions asked included whether they would "keep secret about the error," "discuss it with colleagues," "report the event to the local hospital system," "inform the patient about the adverse event," or "express regrets to the patient." Respondents indicated their answers on a 5-point Likert scale ranging from definitely yes to definitely no.

Willingness to Report. Japanese doctors' and nurses' attitudes to error reporting for the two

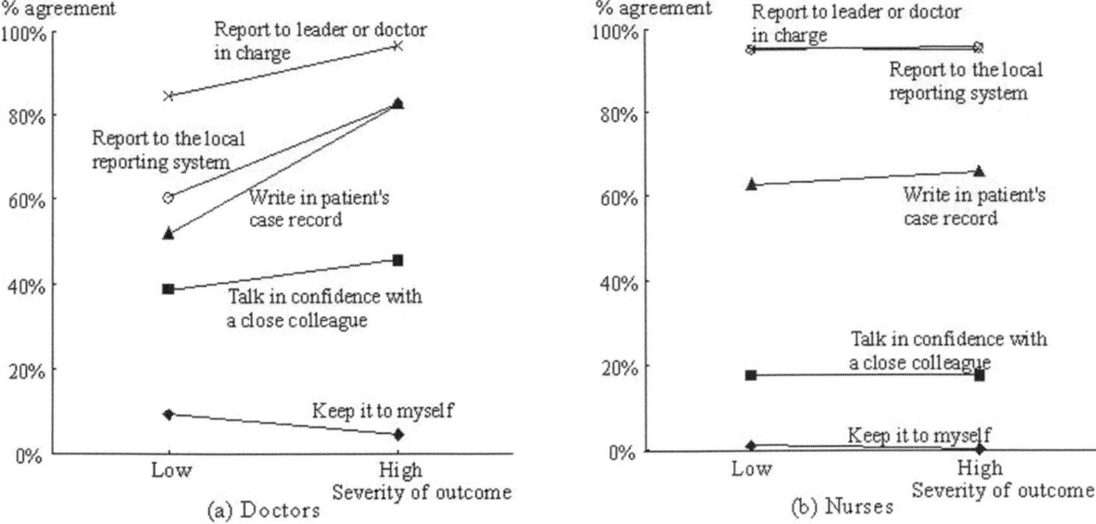

Figure 13–1. Willingness to report incident for low and high severity cases (Itoh et al., 2002).

fictitious cases are depicted in Figure 13–1. As an overall trend, doctors indicated that they would take significantly more positive actions on reporting for the severe than for the mild outcome case. The item, "talking in confidence with a close colleague to get support," is the only one that does not show a significant difference between the two levels of severity. The differences in nurse responses between the minor and the major injury case are minute, however (but because of the large sample size, some of them are nevertheless significant).

As can be seen in Figure 13–1, the attitudes of the nurses toward error reporting are significantly more positive than those of the doctors for almost all items. Nurses' willingness to report errors in the mild case is even higher for most items than those of doctors in the severe case. For the severe outcome case, there is no significant difference between doctors and nurses in entering the event into the patient case record, but doctors are more willing to do so. For the rest of the items, (e.g., "keep secret about the event" and "report the event to the local reporting system"), the nurses' responses are, in the severe case, much more positive than those of the doctors. For the mild outcome case, responses to all the items on incident/error reporting are significantly different between the two health care professional groups: Nurses showed much more positive attitudes to error reporting than doctors.

Reason proposed (Reason, 1997; Reason & Hobbs, 2003) that to be efficient in its pursuit of safety, a safety culture must rely on three interrelated component cultures: (a) a reporting culture—an organizational climate in which people are prepared to report their errors and near-misses, (b) a just culture—an atmosphere of trust in which people are encouraged, even rewarded, for providing essential safety-related information, and (c) a learning culture—a willingness and the competence needed to draw right conclusions from a safety information system such as the incident reporting system and they will to implement major reforms when their need is indicated. From the comparison results as well as the results described earlier, we may conclude that the reporting culture surrounding nurses is more mature than that of doctors. Assuming no differences in the just culture and the learning culture between doctors and nurses, the professional culture of the nurses currently may be *safer* than that of the doctors. In addition, considering the other two components of safety culture are interlocked with the reporting culture, the just culture and the learning culture in the nurses' sections might, we may speculate, be better than in the doctors sections.

Interaction With Patient Responses to interactions with the patient showed a similar trend to error-reporting attitudes across the two cases, as

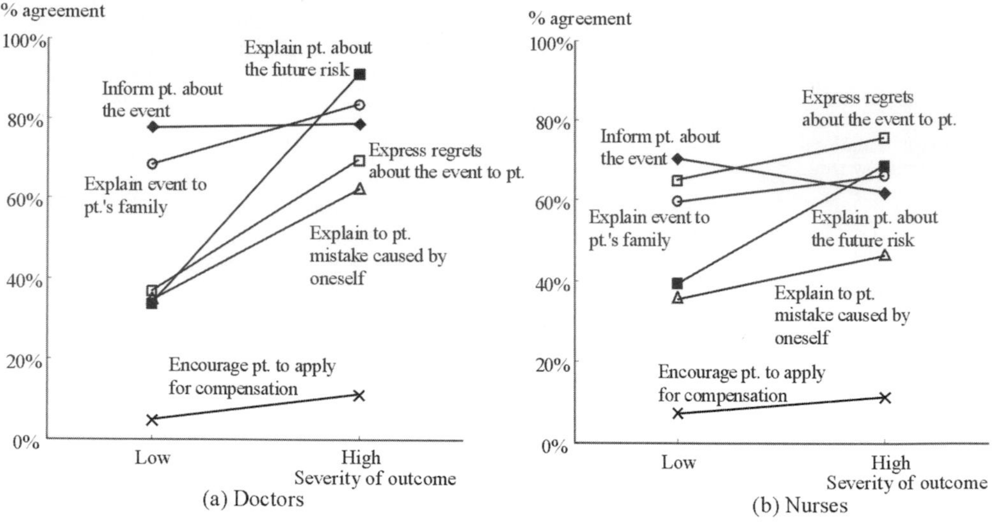

Figure 13–2. Interactions with respect to patient for low and high severity cases (Itoh et al., 2002).

shown in Figure 13–2. For almost all the proposed actions about interaction with the patient, both doctors and nurses provide more positive responses for the severe case than for the mild outcome case. The more severe the outcome of an error, the more likely it is that the consequence will be explained to the patient, that the patient will be told that the event was caused by the doctor's or nurse's mistake, that the event will be explained to the patient family, and that the doctor or nurse will express regret to the patient about the event.

Nurse responses show that they are more willing to apologize to the patient about the event than doctors. For the other actions in relation to the patient, doctors responded more positively than nurses, no doubt because doctors have the primary responsibility for carrying out these acts vis-à-vis the patient when such events occur.

Multinational Comparisons of Reporting Attitudes

Using the same adverse event cases and questions, a similar survey was conducted in Denmark, and we compared health care staff attitudes to error reporting and interaction with patients between Denmark and Japan (Andersen, Itoh, & Peraudeau, 2003). Comparative results of staff attitudes for the mild case between these two countries for some of the questions are shown in Figure 13–3. The overall trend in differences in staff attitudes to error reporting between doctors and nurses in Danish data is qualitatively similar to that of the Japanese results mentioned in the previous subsection but with some slight variations. Like in the Japanese hospitals, the Danish nurses' attitudes to error reporting are more positive than those of Danish doctors. In addition, Danish nurses are more willing to report to a leader or a doctor in charge and to enter the event into the patient case record. By contrast, as for interaction with patients, Danish doctors are a bit more positive than nurses, but only slightly so, just as is the case in Japan. The doctor is more likely to explain about the event and future risk, to explain the event was caused by him or herself, and to express personal regrets to the patient than the nurse. Some of the differences may be because doctors, and not nurses, will be critically involved when an adverse event is to be disclosed to a patient.

With regard to comparisons between the two countries, large differences can be seen both for doctors and nurses in Figure 13–3. Both Japanese doctors and nurses are more willing to submit incident reports to the hospital reporting system, compared with the Danish health care staff who, at the time of the survey, had only a few such systems available (therefore, most respondents indicated that this item is not available). However, for the other actions regarding error reporting and interaction with patients, the likelihood of any given action (except keeping the event secret) by Danish staff far exceeds that of their Japanese colleagues: Danish doctors and nurses are more likely to write

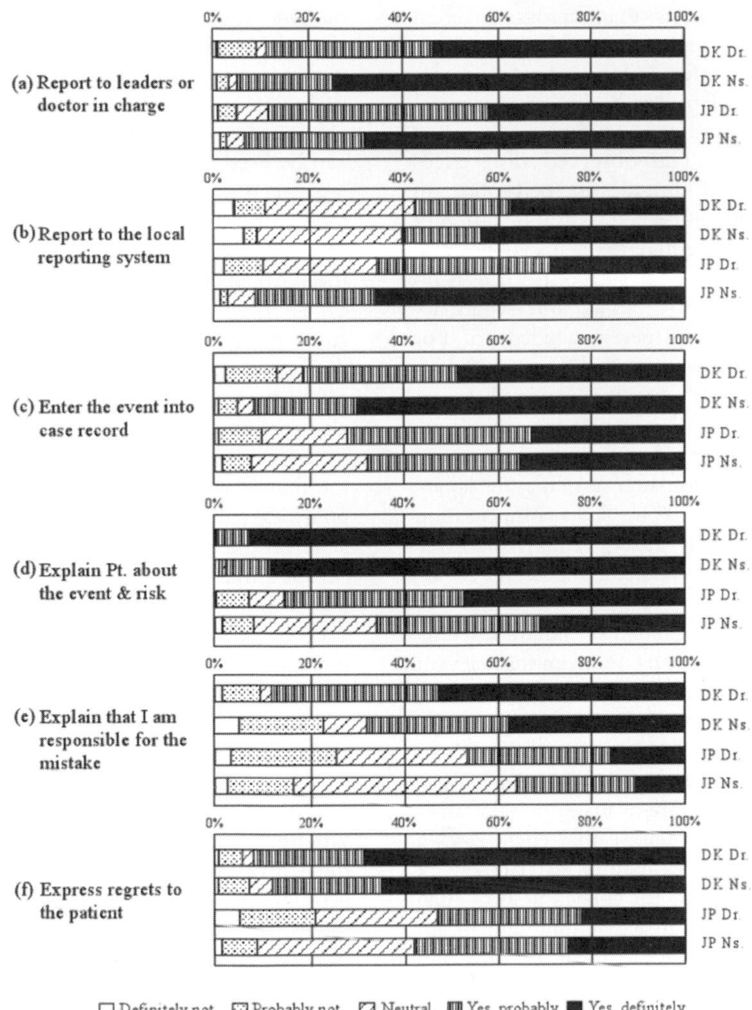

Figure 13–3. Comparison of responses on error reporting between Danish and Japanese health care staff (in mild case) (adapted from Andersen et al., 2003).

the event in the patient case record and more willing to explain to the patient about the event and future risk, to express their regrets, and to acknowledge their mistake than Japanese health care staff.

In summary, the comparative results about health care staff attitudes showed large differences between Denmark and Japan in terms of error and incident reporting and interactions with the patient who suffered an adverse event. Sources of such variations may be caused by a range of very dissimilar factors: for example, to cultural aspects of the specific ward, specialty and organization (e.g., organizational differences in safety culture issues), to differences in national culture, and to differences in safety management structures (e.g., availability of proper reporting system).

CONCLUSION: PERSPECTIVES FOR IMPROVING PATIENT SAFETY

In this chapter, we reviewed the concepts of safety culture and safety climate and some widely quoted definitions and dimensions of safety culture. In addition, we illustrated how some recent survey results of safety culture factors may be correlated to other indicators of safety outcomes. We described that much larger safety culture differences exist across different organizations or hospitals than across different departments, wards, or specialties within any single hospital. Comparing safety culture related attitudes between Danish and Japanese health care organizations, we also observed large differences between these two countries. The observed differences between health care staff attitudes and perceptions in the two countries should be tied to at least different national, and possibly different organizational cultures. Another determinant of such differences is no doubt tied to institutional, legal, and regulatory aspects, which we have merely alluded to. For instance, an important source of difference in staff attitudes to interactions with the patient between Denmark and Japan may be in part the complaints and compensation systems for medical accidents and incidents. There are no such systems in Japan, although Denmark has a no-fault compensation scheme and has recently introduced a nationwide confidential reporting system (Andersen, et al., 2004a).

The ability to derive and disseminate learning from adverse events depends on acquiring relevant information through incident reporting (Itoh, Seki, & Andersen, 2003). As described in earlier in the chapter, an organization (or a group) will be able to derive learning from negative experiences only if there are both an appropriate institutional framework and a strong local culture that encourages a trustful atmosphere. The institutional framework must provide mechanisms for collecting, providing feedback, analyzing, and implementing and monitoring the possible revisions to procedures and guidelines. Recent initiatives to establish national reporting systems (e.g., the National Patient Safety Reporting System for England and Wales introduced in February 2004, and the Danish national confidential reporting system, introduced in January 2004) have been inspired by experience with confidential reporting in other safety critical domains (e.g., Orlady & Orlady, 1999).

One of the greatest advantages of safety culture assessment is its potential for supporting *proactive* patient safety activities. Results of the survey should be used prospectively and, in combination with a proactive regime, support the identification of points at which a specific local safety culture may need to be strengthened. Equally, survey tools can be used to measure changes over time, including effects of intervention programs within departments or hospitals.

Finally, we would like to stress the importance of the participation and involvement of all members of an organization in safety activities, including frontline staff, technicians, orderlies, leaders in departments, and senior managers. Commitment to safety is required not only by management but by all the members within an organization—as illustrated in the stress on shared norms described in the first section of this chapter. Through such hospitalwide participation, it is possible to tackle safety activities continuously; to build on the experience of such initiatives, their outcomes, the knowledge and techniques involved, and the repeated applications to new cases; and, finally, to help in reinforcing safety awareness of health care practices with frontline staff and management.

ACKNOWLEDGMENTS

We are heavily indebted to Robert L. Helmreich, the University of Texas at Austin, who provided not only the original Flight Management Attitudes Questionnaire but also invaluable suggestions for our patient safety project. We would like to acknowledge Takako Aoki, Tomioka General Hospital, as well as Toshiko Abe and Naomi Kitazawa, Tokyo Medical and Dental University for their cooperation in this project. We are also grateful to members of the Danish patient safety survey project group, particularly Niels Hermann, Doris Østergaard and Thomas Schiøler for their contributions to the elaboration of survey instruments and the carrying out of questionnaire surveys.

References

Advisory Committee on the Safety of Nuclear Installations (ACSNI) (1993). *Human Factors Study Group Third Report: Organization for Safety.* Sheffield, UK: HSE Books.

Amalberti, R. (1998). Automation in aviation: A human factors perspective. In D. Garland, J. Wise, & D. Hopkin (Eds.), *Aviation Human Factors* (pp. 173–192). Hillsdale, NJ: Lawrence Erlbaum Associates, Inc.

Andersen, H. B. (2002). *Assessing Safety Culture* (Technical Report R-1459). Roskilde, Denmark: Risø National Laboratory.

Andersen, H. B., Garay, G., & Itoh, K. (1999). *Survey Data on Mariners: Attitudes to Safety Issues* (Technical Report I-1388). Roskilde, Denmark: Systems Analysis Department, Risø National Laboratory.

Andersen, H. B., Hermann, N., Madsen, M. D., Østergaard, D., & Schiøler, T. (2004a). Hospital staff attitudes to models of reporting adverse events: Implications for legislation. In C. Spitzer, U. Schmocker, & V. N. Dang (Eds.), *Proceedings of the 7th International Conference on Probabilistic Safety Assessment and Management* (pp. 2720–2725.). London: Springer-Verlag.

Andersen, H. B., Itoh, K., & Peraudeau, A. (2003). *Comparative Results of Danish-Japanese Surveys of the Attitudes of Doctors and Nurses to Reporting Adverse Events* (Technical Report I-2040 (EN)). Roskilde, Denmark: Risø National Laboratory.

Andersen, H. B., Madsen, M. D., Hermann, N., Schiøler, T., & Østergaard, D. (2002, July). Reporting adverse events in hospitals: A survey of the views of doctors and nurses on reporting practices and models of reporting. *Proceedings of the Workshop on the Investigation and Reporting of Incidents and Accidents* (pp. 127–136), Glasgow, UK.

Andersen, H. B., Nielsen, K. J., Carstensen, O., Dyreborg, J., Guldenmund, F., Hansen, O. N., et al. (2004b, May). Identifying safety culture factors in process industry. *Proceedings of Loss Prevention 2004 11th International Symposium.* Loss Prevention and Safety Promotion in the Process Industries, Prague, Czech Republic.

Antonow, J. A., Smith, A. B., & Silver, M. P. (2000). Medication error reporting: A survey of nursing staff. *Journal of Nursing Care Quality, 15*(1), 42–48.

Baker, G. R., Norton, P. G., Flintoft, V., Blais, R., Brown, A., Cox, J., et al. (2004). The Canadian adverse events study: The incidence of adverse events among hospital patients in Canada. *Canadian Medical Association Journal, 170*(11), 1678–1686.

Barach, P., & Small, S. (2000). Reporting and preventing medical mishaps: Lessons from non-medical near miss reporting systems. *British Medical Journal, 320,* 759–763.

Brown, R. L., & Holmes, H. (1986). The use of a factor-analytic procedure for assessing the validity of an employee safety climate model. *Accident Analysis and Prevention, 18*(6), 455–470.

Bryant, D. T. (1991). *The human element in shipping causalities.* London: HMSO.

Confederation of British Industry (CBI) (1990). *Developing a Safety Culture: Business for Safety.* London: Confederation of British Industry.

Cooper, M. D. (2000). Towards a model of safety culture. *Safety Science, 36,* 111–136.

Cox, S. J., & Cheyne, A. J. T. (2000). Assessing safety culture in offshore environments. *Safety Science, 34*(1–3), 111–129.

Cox, S. J., & Cox, T. (1991). The structure of employee attitudes to safety: An European example. *Work and Stress, 5*(2), 93–106.

Cox, S. J., & Flin, R. (1998). Safety culture: Philosopher's stone or man of straw? *Work and Stress, 12*(3), 189–201.

Cullen, D., Bates, D., Small, S., Cooper, J., Nemeskal, A., & Leape, L. (1995). The incident reporting system does not detect adverse events: A problem for quality improvement. *Journal of Quality Improvement, 21*(10), 541–548.

Davies, F., Spencer, R., & Dooley, K. (1999). *Summary Guide to Safety Climate Tools* (Offshore Technology Report 1999/063). London: HSE Books.

Department of Health (2000). *An Organisation with a Memory. Report of an Expert Group on Learning from Adverse Events in the NHS.* London: The Stationery Office.

Diaz, R. I., & Cabrera, D. D. (1997). Safety climate and attitude as evaluation measures of organizational safety. *Accident Analysis and Prevention, 29*(5), 643–650.

Duijm, N. J., Andersen, H. B., Hale, A., Goossens, L., & Hourtolou, D. (2004). Evaluating and managing safety barriers in major hazard plants. In C. Spitzer, U. Schmocker, & V. N. Dang (Eds.), *Proceedings of the 7th International Conference on Probabilistic Safety Assessment and Management* (pp. 110–115). London: Springer-Verlag.

Edmondson, A. C. (1996). Learning from mistakes is easier said than done: Group and organizational influences on the detection and correction of human error. *Journal of Applied Behavioral Science, 32*(1), 5–8.

Edmondson A. C. (2004). Learning from failure in health care: Frequent opportunities, pervasive barriers. *Quality & Safety in Health Care, 13*(Suppl. 2), ii3–ii9.

Flin, R., Mearns, K., O'Connor, P. and Bryden, R. (2000). Measuring safety climate: Identifying the common features. *Safety Science, 34*(1–3), 177–192.

Gaba, D. M., Howard, S. K., Fish, K. J., Smith, B. E., & Sowb, Y. A. (2001). Simulation-based training in Anesthesia Crisis Resource Management (ACRM): A decade of experience. *Simulation and Gaming, 32*(2), 175–193.

Gaba, D. M., Singer, S. J., Sinaiko, A. D., Bowen, J. D., & Ciavarelli, A. P. (2003). Differences in safety climate between hospital personnel and naval aviators. *Human Factors, 45*(2), 173–185.

Gershon, R. R. M., Karkashian, C. D., Grosch, J. W., Murphy, L. R., Escamilla-Cejudo, A., Flanagan, P. A., et al. (2000). Hospital safety climate and its relationship with safe work practices and workplace exposure incidents. *American Journal of Infection Control, 28*(3), 211–221.

Glendon, A. I., & McKenna, E. F. (1995). *Human Safety and Risk Management.* London: Chapman & Hall.

Glendon, A. I., & Stanton, N. A. (2000). Perspectives on safety culture. *Safety Science, 34,* 193–214.

Griffiths, D. K. (1985). Safety attitudes of management. *Ergonomics, 28,* 61–67.

Guldenmund, F. W. (2000). The nature of safety culture: A review of theory and research. *Safety Science, 34,* 215–257.

Hale, A. R. (2000). Culture's confusions. Editorial for the special issue on safety culture's and safety climate. *Safety Science, 34,* 1–4.

Helmreich, R. L. (2000a). Culture and error in space: Implications from analog environments. *Aviation, Space, and Environmental Medicine, 71*(9), A133–A139.

Helmreich, R. L. (2000b). On error management: Lessons from aviation. *British Medical Journal, 320,* 781–785.

Helmreich, R. L., & Merritt, A. C. (1998). *Culture at Work in Aviation and Medicine: National, Organizational and Professional Influences.* Aldershot, UK: Ashgate.

Helmreich, R. L., Wilhelm, J. A., Klinect, J. R., & Merritt, A. C. (2001). Culture, error, and crew resource management. In E. Salas, C. A. Bowers, & E. Edens (Eds.), *Improving Teamwork in Organizations* (pp. 305–331). Hillsdale, NJ: Lawrence Erlbaum Associates, Inc.

Hofstede, G. (1991). *Cultures and Organizations: Software of the Mind.* London: McGraw-Hill.

Hollnagel, E. (1998). *Cognitive Reliability and Error Analysis Method (CREAM).* London: Elsevier.

International Atomic Energy Agency (1991). *Safety Culture* (Safety Series No. 75-INSAG). Vienna: International Atomic Energy Agency.

International Nuclear Safety Advisory Group (1991). *Safety Culture* (Safety Series No. 75-INSAG).Vienna: International Atomic Energy Agency.

Itoh, K., Abe, T., & Andersen, H. B. (2002, July). A survey of safety culture in hospitals including staff attitudes about incident reporting. *Proceedings of the Workshop on the Investigation and Reporting of Incidents and Accidents* (pp. 144–153), Glasgow, UK.

Itoh, K., Abe, T., & Andersen, H. B. (2003, September). Health care staff attitudes towards management, job, teamwork and leadership in Japanese hospitals. *Proceedings of the 9th European Conference on Cognitive Science Approaches to Process Control* (pp. 67–74), Amsterdam, The Netherlands.

Itoh, K., Abe, T., & Andersen, H. B. (2005, March–April). A questionnaire-based survey on health care safety culture from six thousand Japanese hospital staff: Organisational, professional and specialty/ward differences. *Proceedings of the International Conference on Health Care Systems Ergonomics and Patient Safety.* Florence, Italy.

Itoh, K., & Andersen, H. B. (1999, May). Motivation and morale of night train drivers correlated with accident rates. *Proceedings of the International Conference on Computer-Aided Ergonomics and Safety.* Barcelona, Spain [CD ROM].

Itoh, K., Andersen, H. B., & Seki, M. (2004). Track maintenance train operators' attitudes to job, organisation and management and their correlation with accident/incident rate. *Cognition, Technology & Work, 6*(2), 63–78.

Itoh, K., Andersen, H. B, Seki, M., & Hoshino, H. (2001, June). Safety culture of track maintenance organisations and its correlation with accident/incident statistics. *Proceedings of the 20th European Annual Conference on Human Decision Making and Manual Control* (pp. 139–148), Copenhagen, Denmark.

Itoh, K., Seki, M., & Andersen, H. B. (2003). Approaches to transportation safety: Methods and case studies applying to track maintenance train operations. In E. Hollnagel (Ed.), *Handbook of Cognitive Task Design* (pp. 603–632). Mahwah, NJ: Lawrence Erlbaum Associates, Inc.

Kohn, L. T., Corrigan, J. M., & Donaldson, M. S. (Eds.) (1999). *To Err is Human: Building a Safer Health System.* Washington DC: National Academy Press.

Krause, T. R., Seymour, K. J., & Sloat, K. C. M. (1999). Long-term evaluation of a behavior-based method for improving safety performance: A meta-analysis of 73 interrupted time-series replications, *Safety Science, 32,* 1–8.

Nieva, V. F., & Sorra, J. (2003). Safety culture assessment: A tool for improving patient safety in health care organizations. *Quality & Safety in Health Care, 12*(Suppl. 2), ii17–ii23.

Orlady, H. W., & Orlady, L. M. (1999). *Human Factors in Multi-Crew Flight Operations.* Aldershot, UK: Ashgate.

Pidgeon, N. F. (1991). Safety culture and risk management in organizations. *Journal of Cross-Cultural Psychology, 22*(1), 129–140.

Pidgeon, N. F., & O'Learry, M. (1994). Organizational safety culture: Implications for aviation practice. In N. A. Johnston, N. McDonald, & R. Fuller (Eds.), *Aviation Psychology in Practice* (pp. 21–43), Aldershot, UK: Avebury Technical Press.

Rasmussen, J. (1986). *Information Processing and Human-Machine Interaction: An Approach to Cognitive Engineering.* New York: Elsevier.

Reason, J. (1993). Managing the management risk: New approaches to organizational safety. In B. Wilpert & T. Qvale (Eds.), *Reliability and Safety in Hazardous Work Systems* (pp. 7–22). Hove, UK: Lawrence Erlbaum Associates, Inc.

Reason, J. (1997). *Managing the Risk of Organizational Accidents.* Aldershot, UK: Ashgate.

Reason, J., & Hobbs, A. (2003). *Managing Maintenance Error: A Practical Guide.* Aldershot, UK: Ashgate.

Schein, E. H. (1992). *Organizational Culture and Leadership* (2nd ed.). San Francisco: Jossey-Bass.

Scott, T., Mannion, R., Davies, H., & Marshall, M. (2003).The quantitative measurement of organizational culture in health care: A review of the available instruments. *Health Services Research, 38*(3), 923–945.

Sexton, J. B., Thomas, E. J., & Helmreich, R. L. (2000). Error, stress, and teamwork in medicine and aviation: Cross sectional surveys. *British Medical Journal, 320*, 745–749.

Sheehy, P. N., & Chapman, A. J. (1987). Industrial accidents. In C. L. Cooper & I. T. Robertson (Eds.), *International Review of Industrial and Organizational Psychology* (pp. 201–227). New York: Wiley.

Singer, S. J., Gaba, D. M., Geppert, J. J., Sinaiko, A. D., Howard, S. K., & Park, K. C. (2003). The culture of safety in California hospitals. *Quality and Safety in Health Care, 12*(2), 112–118.

Spector, P. E., Cooper, C. L., & Sparks, K. (2001). An international study of the psychometric properties of the Hofstede Values Survey Module 1994: A comparison of individual and country/province level results. *Applied Psychology An International Review, 50*(2), 269–281.

Swain, A. D., & Guttmann, H. E. (1983). *Handbook of Human Reliability Analysis with Emphasis on Nuclear Power Plant Applications* (NUREG 278). Washington, DC: U.S. Nuclear Regulatory Commission.

van Vuuren, W. (2000). Cultural influences on risks and risk management: Six case studies. *Safety Science, 34* (1–3), 31–45.

Williamson, A. N., Feyer, A.-M., Cairns, D., & Biancotti, D. (1997). The development of a measure of safety climate: The role of safety perceptions and attitudes. *Safety Science, 25*(1–3), 15–27.

Zohar, D. (1980). Safety climate in industrial organizations: Theoretical and applied implications. *Journal of Applied Psychology, 65,* 96–101.

14

BURNOUT IN HEALTH CARE

Wilmar B. Schaufeli
Utrecht University, the Netherlands

Burnout is the index of the dislocation between what people are and what they have to do. It represents an erosion in values, dignity, spirit, and will—an erosion of the human soul. With these words, Maslach and Leiter (1997, p. 17) describe a particular negative psychological state that is also observed among health care professionals. Burnout is a metaphor that is commonly used to describe a state or process of mental exhaustion, similar to the smothering of a fire, the extinguishing of a candle, or the draining of a battery that cannot be recharged anymore. In the dictionary "to burn out" is described as follows:

> *to fail, wear out, or become exhausted by making excessive demands on energy, strength, or resources.*

Burnout was first described in greater detail by Herbert Freudenberger in the mid-1970s (Freudenberger, 1974). As an unpaid psychiatrist in a treatment center for drug addicts, he observed that young and idealistically motivated volunteers experienced a gradual energy depletion and loss of motivation and commitment, which was accompanied by a wide array of other mental and physical symptoms. Freudenberger labeled this particular state of exhaustion "*burnout*," a colloquial term used to refer to the devastating effects of chronic drug abuse. Independently, and at about the same time, Christina Maslach—a social psychological researcher—stumbled across that very term in California. She studied the ways in which health care professionals cope with emotional arousal at work and observed that many of them were emotionally exhausted, had developed negative perceptions about their patients, and experienced a crisis in their professional competence (Maslach & Schaufeli, 1993). In a way, the almost simultaneous discovery of burnout by the clinician Freudenberger and by the researcher Maslach marks the beginning of two different traditions that approach burnout from a practical and from a scientific point of view, respectively. So far, both traditions have developed relatively independent from each other. Initially the clinical approach prevailed, which was supplemented by a more scientific approach from the early 1980s, particularly after the introduction of a brief self-report questionnaire: the Maslach Burnout Inventory. By the turn of the century, approximately 6,000 scientific articles and books about burnout had appeared, and every year approximately 200 publications are added to this list. Most empirical studies were carried out in health care (34%) and in teaching (27%), with nurses (17%) as the most frequently studied health profession, followed by psychologists and psychotherapists (4%) and physicians (2%) (Schaufeli & Enzmann, 1998; p. 69–73).

Three conclusions can be drawn from the history of burnout: (a) burnout emerged as a social problem and *not* as a scholarly construct; (b) from the outset, burnout was strongly associated with "people work" in the human services, including health care; and (c) a clinical approach and a research approach to burnout have developed and coexist more or less independently.

The purpose of this chapter is to present a brief overview of burnout in health care. In the following sections, attention is paid to (a) symptoms and assessment; (b) related concepts such as job stress and depression; (c) prevalence; (d) correlates, causes, and consequences; (e) psychological explanations; and (f) interventions.

SYMPTOMS AND ASSESSMENT

Initially when the clinical approach to burnout prevailed, more than 130 possible symptoms were identified (Schaufeli & Enzmann, 1998. pp. 20–31), ranging from *a* (anxiety) to *z* (lack of zeal). These symptoms can be grouped in five clusters: (a) *affective* (e.g., depressed mood, emotional exhaustion), (b) *cognitive* (e.g., poor concentration, forgetfulness), (c) *physical* (e.g., headaches, sleep disturbances), (d) *behavioral* (e.g., poor work performance, absenteeism), and (e) *motivational* (e.g., loss of idealism, disillusionment). Remarkably—except for motivational symptoms—these clusters appear to match perfectly with the usual categorization of stress symptoms. Typically, burnout is not restricted to symptoms at the individual level; in addition, interpersonal symptoms in relation to colleagues and recipients of one's care or services are observed (e.g., irritability, dehumanization, indifference) as well as symptoms that are related to organizational outcomes (e.g., job dissatisfaction, job turnover, low morale). Hence, burnout not only constitutes a problem for the individual health care worker but also for his or her colleagues and one's recipients and for the organization at large. The significance of burnout is particularly evident because of this multifaceted nature.

However, a laundry-list approach that merely sums up all possible symptoms is inappropriate to define the syndrome, because it denies that burnout is a process and confuses symptoms with precursors, correlates, and consequences. Instead, Schaufeli and Enzmann (1998, p. 36), after reviewing a dozen or so definitions of burnout, proposed the following description:

> *Burnout is a persistent, negative, work-related state of mind in "normal" individuals that is primarily characterized by exhaustion, which is accompanied by distress, a sense of reduced effectiveness, decreased motivation, and the development of dysfunctional attitudes and behaviors at work. This psychological condition develops gradually but may remain unnoticed for a long time for the individual involved. It results from a misfit between intentions and reality at the job. Often burnout is self-perpetuating because of inadequate coping strategies that are associated with the syndrome.*

This definition of burnout specifies its symptoms, its preconditions, and its domain. More specifically, it narrows down more than 100 burnout symptoms to one core indicator (exhaustion) and four accompanying, symptoms: (a) distress (affective, cognitive, physical, and behavioral), (b) a sense of reduced effectiveness, (c) decreased motivation, and (d) dysfunctional attitudes and work behaviors. Furthermore, frustrated intentions and inadequate coping strategies play a role as preconditions in the development of burnout, and the process is considered to be self-perpetuating, despite the fact that it may not be recognized initially. Finally, the domain is specified: The symptoms are work related and burnout occurs in "normal" people who do not suffer from psychopathology.

This description is somewhat more comprehensive than the probably most often cited definition of burnout by Maslach, Jackson, and Leiter (1986, p. 1):

> *'Burnout is a syndrome of emotional exhaustion, depersonalization, and reduced personal accomplishment that can occur among individuals who do > people work' of some kind"*

The Maslach Burnout Inventory (MBI) is the most widely used self-report questionnaire that consists of the three dimensions included in the definition. Emotional exhaustion refers to the depletion or draining of emotional resources. Depersonalization points to the development of negative, callous, indifferent, and cynical attitudes toward the recipients of one's care or services. The term depersonalization may cause some confusion, because in psychiatry it denotes a person's extreme alienation from self and the world. By contrast, in the definition of Maslach and colleagues, depersonalization refers to an impersonal and dehumanized perception of one's recipients. Finally, lack of personal accomplishment is the tendency to evaluate one's work with one's recipients negatively. It is believed that the objectives are not achieved, which is accompanied by feelings of insufficiency and poor professional self-esteem.

Because the MBI is *the* instrument to measure burnout, the definition of burnout of Maslach and Jackson (1986) has gradually become equivalent with the way it is measured—in other words, burnout is what the MBI measures. This tautology hampers burnout research, because the concept of burnout is broader and more comprehensive than the MBI assumes. At least with regard to one issue, the original definition of burnout by Maslach and her colleagues was broadened. Initially, Maslach and her co-workers claimed that burnout occurs exclusively among professionals who do "people work"; that is, those who deal directly with patients, students, or clients, as is the case in health care, education, and social work, respectively. Hence, in their original view, burnout was restricted to these human services professions. However, meanwhile, the concept of burnout is broadened and defined as a crisis in one's relationship with work in general and not necessarily as a crisis in one's relationship with people at work (Maslach, Schaufeli, & Leiter, 2001). As a consequence, the three original burnout dimensions have been slightly redefined, and a new general version of the MBI (Schaufeli, Leiter, Maslach, & Jackson, 1996) has appeared that can been used among those who do not deal with patients or clients and the like, such as administrative hospital staff or maintenance and catering staff working in health care.

Generally speaking, the psychometric quality of the MBI is encouraging: The three scales are internally consistent, and the three-factor structure has been confirmed in various studies (Schaufeli and Enzmann, 1998, p. 51–53). The core symptom of burnout—emotional exhaustion—is the most robust scale of the MBI that is strongly related to other burnout measures (convergent validity). Paradoxically, as we will see, it is also the least specific scale that is related to measures of other concepts such as depression (discriminant validity). Although the MBI originates from the United States, similar positive psychometric results have been obtained with the French, German, Dutch, and Swedish versions of the MBI. Moreover, its cross-national validity has been demonstrated across German, French, and Dutch health care samples (Enzmann, Schaufeli, & Girault, 1994).

The MBI is designed as a research instrument to be used at the group level. Except for The Netherlands, there are no clinically validated cutoff points available for the MBI that allow to discriminate between burnout "cases" and "noncases" (Schaufeli, Bakker, Hoogduin, Schaap, & Kaldler, 2001). The MBI-test-manual presents only numerical cutoff points based on arbitrary statistical norms (Maslach, Jackson & Leiter, 1996). Although the test authors—correctly—warn that these should not be used for diagnostic purposes, there is a strong temptation to do so, especially for practitioners. Clearly, this is wrong, not only because the cutoff points are based on arbitrary statistical norms but also because they are computed from a composite, nonrepresentative, convenience U.S. sample. Only nation-specific and clinically validated cutoff points should be employed for the purpose of individual assessment.

BURNOUT, JOB STRESS, AND DEPRESSION

Burnout has been equated with a myriad of terms; most of them are plagued by the same sort of definitional ambiguity. The most prominent examples are job stress and depression. Can burnout be distinguished from these concepts?

Job stress is a generic term that refers to the temporary adaptation process at work that is accompanied by mental and physical symptoms. By contrast, burnout can be considered as a final stage in a breakdown in adaptation that results from the long-term imbalance of demands and resources. In other words, burnout results from prolonged job stress. Moreover, burnout includes the development of dysfunctional attitudes and behaviors toward the recipients of one's care or services (depersonalization), one's job, and the organization one is working for. By contrast, job stress is not necessarily accompanied by such negative attitudes and behaviors. This assertion is empirically supported by Schaufeli and Van Dierendonck (1993), who showed in a sample of nurses that burnout can be distinguished from generic job-related mental and physical distress, albeit that emotional exhaustion shares approximately 30% of its variance with distress. Finally, it has been claimed that everyone can experience stress, whereas burnout can only be experienced by those who entered their careers enthusiastically with high goals and expectations. For example, Pines (1993) argued that people who expect to derive a sense of significance from their work are susceptible to burnout, whereas those without such expectations would experience job stress instead of burnout.

Clearly, burnout and depression are characterized by similar dysphoric symptoms. Nevertheless, clinical practice suggests that the syndromes differ (Liu & Van Liew, 2003): Depressive patients are generally overwhelmed by listlessness and lethargy and hold steadfastly to their ideas of guilt, whereas burnout victims present their complaints much more vigorously—they feel disappointed and aggrieved. Furthermore, burnout includes specific dysfunctional attitudes and behaviors that are not typically found in depression. Finally, burnout tends to be job related and situation specific rather than pervasive, affecting other spheres of life as well. For instance, burnout is related to lack of reciprocity between investments and outcomes at work, whereas depression is related to lack of reciprocity between investments and outcomes in private life (Bakker, Schaufeli, Demerouti, Janssen, Van der Hulst, & Brouwer, 2000). So it is not surprising that after reviewing nearly 20 studies on burnout and depression, Glass and McKnight (1996, p. 33) wrote:

> *Burnout and depressive symptomatology are not simply two terms for the same dysphoric state. They do, indeed, share appreciable variance, especially when the emotional exhaustion component is involved, but the results do not indicate complete isomorphism. We conclude, therefore, that burnout and depressive symptomatology are not redundant concepts.*

Thus, it appears that burnout can be distinguished conceptually as well as empirically from job stress and from depression. Nevertheless, emotional exhaustion shows some overlap with both concepts. The fact that depersonalization and reduced personal accomplishment are less substantively related to both other concepts implies that burnout is a unique, multidimensional, chronic stress reaction that goes beyond the experience of mere exhaustion.

THE PREVALENCE OF BURNOUT IN HEALTH CARE

How often does burnout occur in health care? Principally, the MBI can be used to answer this question because, at least in The Netherlands, this instrument can discriminate between cases and noncases. Based on the clinically validated MBI cutoff points, 4% of the Dutch working population was identified as burnout cases; that is, they exhibit burnout levels similar to those who receive specialized psychotherapeutic treatment (Bakker, Schaufeli & Van Dierendonck, 2000). More specifically, the prevalence of this so-called clinical burnout was particularly high among occupational physicians (11%), psychiatrists (9%), teachers (9%), general practitioners (8%), community nurses (8%), midwives (7%), and social workers (7%). With the exception of teachers and social workers, these are all health care professionals. Moreover, it appears that particularly those who work in the community such as occupational physicians, general practitioners, community nurses, and midwifes, suffer from severe burnout. This agrees with a British study that showed working in the community is more stressful than working in inpatient services (Prosser et al., 1999). Conversely, in The Netherlands, relatively low levels of burnout were found among police officers (1%), hospice workers (2%), intensive care unit (ICU) nurses (2%), oncology nurses (2%), staff working with the mentally retarded (2%), and correctional officers (3%). Probably the low level of burnout in law enforcement is caused by a selection effect, because police officers and correctional officers are screened psychologically (i.e., those who score high on neuroticism are excluded and drop out, and neuroticism is positively related to burnout). Finally and quite remarkably, highly specialized nurses working in ICUs or in oncology wards do not appear to suffer much from burnout.

Schaufeli and Enzmann (1998, p. 61) analyzed mean burnout scores of various professions, including nurses, physicians, and mental health professionals using more than 70 U.S. studies published between 1979 and 1988. They found that, compared with physicians, nurses experience slightly less emotional exhaustion but *much* less depersonalization and personal accomplishment. Although gender bias cannot be ruled out—nurses are predominantly women, and women tend to report less depersonalization than males—it is likely that differences in depersonalization reflect different professional roles and attitudes: the nurturing and caring role of the nurse versus the more distant curing role of the doctor. This observation agrees with a recently longitudinal study that showed that high levels of depersonalization protect doctors from future stress (McManus, Winder, & Gordon, 2002). In other words, developing a distant attitude toward their patients appears to be adaptive for doctors. According to Schaufeli and Enzmann (1998, p. 61), in mental health care, nursing staff members experience more emotional exhaustion and less accomplishment than psychologists and counselors. Probably this reflects the more stressful nature of the jobs of the former that are characterized by poor decision latitude (probably leading to emotional exhaustion), as well as less extensive training (probably leading to a sense of reduced personal accomplishment).

Finally, it appears that compared with those who are not involved in direct care, burnout levels are

TABLE 14–1. Correlates, Causes, and Consequences of Burnout in Health Care

Correlates	*Causes*	*Consequences*
Demographic	*Job demands*	*Individual health*
• Young age	• Work overload	• Psychosomatic complaints
• Male (depersonalization)	• Time pressure	• Depression
Personality characteristics	• Emotional demands	• Cardiovascular disease
• Nonhardy personality	• Role problems	*Work-related attitudes*
• Poor self-esteem	*Poor job resources*	• Job dissatisfaction
• Avoiding coping style	• Social support	• Low organizational commiment
• High level of neuroticism	• Job control	• Organizational behavior
• "Feeling type"	• Feedback	• Turnover
	• Participation in decision-making	• Absenteeism
		• Poor performance

higher in health care workers who deal intensively with patients on a daily basis. For instance, a large study among more than 200 Japanese health care facilities showed that burnout scores were significantly higher among direct care staff members compared with facility directors, middle managers, and other types of staff personnel (Ito, Kurita, & Shiiya, 1999).

THE CORRELATES, CAUSES, AND CONSEQUENCES OF BURNOUT IN HEALTH CARE

Despite the impressive quantity of empirical publications on burnout, their quality is often questionable. For instance, the vast majority of studies are cross-sectional in nature so that only correlates of burnout can be identified, because to establish cause–effect relationships, longitudinal designs are necessary. In the next several paragraphs, the major variables associated with burnout in health care are discussed (see also Table 14–1), whereby special attention is paid to longitudinal studies. For a more detailed overview, see Cordes and Dougherty (1993), Lee and Ashforth (1996), Schaufeli and Enzmann (1998, pp. 69–99), and Schaufeli and Buunk (2002).

Demographic Characteristics.

Burnout is observed more often among younger health care professionals than among those aged older than 30 or 40 years. This is in line with the observation that burnout is negatively related to work experience. The greater incidence of burnout among the younger and less experienced health care workers may be caused by "reality shock" or by an identity crisis caused by unsuccessful occupational socialization. However, a cautionary note should be made because survival bias cannot be ruled out: Those who burn out early in their careers are likely to quit their jobs leaving behind the survivors, who exhibit lower levels of burnout. Finally, it is consistently found that men report higher depersonalization scores than women, a finding that is in line with other gender differences such as higher prevalence of aggression among men and higher interest in the nurturing role among women.

Personality.

Burnout is *less* common among those with a "hardy" personality who are characterized by involvement in daily activities, a sense of control over events, and openness to change (Maddi, 1999). By contrast, burnout is more common among those with an external locus of control who attribute events and achievements to powerful others or to chance compared with those with an internal locus, who ascribe events and achievements to their own ability, effort, or willingness to risk. Moreover, burnout is related to poor self-esteem and an avoidant, nonconfronting coping style.

Burnout appears to be particularly related to neuroticism, which is characterized by low levels of emotional stability, as well as anxiety, sadness, and irritability. Typically, emotional exhaustion shares approximately 30% of its variance with neuroticism, and 10%–15% with the two remaining burnout dimensions. Depersonalization and reduced personal accomplishment share approximately 5%–10% of their variance with personality factors, with the latter being somewhat stronger related to an avoiding coping style (15%) (Schaufeli & Enzmann, 1998, pp. 77–80). Because neurotic people are emotionally unstable and prone to psychological distress, neuroticism may act as a vulnerability factor that predisposes professionals to experience burnout.

Compared with "thinking types," "feeling types" are more prone to burnout, especially to depersonalization (Garden, 1991). The former are more hard-boiled and achievement oriented and tend to neglect others, whereas the latter are more tender minded and are characterized by concern and awareness for people. According to Garden (1991), feeling types are overrepresentated in health care and thinking types are more often found in business, which might explain the relatively high prevalence of burnout in this sector.

The question remains how important personality characteristics are relative to actual work experiences in explaining burnout. To answer this question, Burisch (2002) followed a sample of German female nurses on seven occasions during the first 3 years of their careers. He found that emotional exhaustion was particularly predicted by the nurses' concurrent work experiences, whereas depersonalization and reduced personal accomplishment were particularly predicted by dispositional variables.

Personality variables not only have a direct main effect on burnout levels, they might also play a more complex moderating role. For instance, nurses who experienced high job demands and high job control, *and* who have an active coping style experienced lower levels of burnout compared with those who behaved rather passively (De Rijk, Le Blanc, Schaufeli, & De Jonge, 1998). Hence, it appears that the interaction hypothesis of the job demands control model, stating that job control buffers the negative effects of high job demands, is only supported for active coping nurses; for those who have a passive coping style, job control in fact increases burnout.

General Job Stressors.

Workload and time pressure explain approximately 25% to 50% of variance of burnout, especially of emotional exhaustion (Lee & Ashforth, 1996). Relationships are much weaker with the other two MBI dimensions. The high correlation with workload must be qualified, however, because this stressor is often operationalized in terms of experienced strain so that considerable overlap in item content exists with emotional exhaustion. This agrees with the results of a longitudinal study (Hillhouse, Adler, & Walters, 2000) among U.S. residents that showed that future burnout was ***not*** predicted by quantitative workload (i.e., the number of working hours and hours on call) but rather by qualitative workload (i.e., experienced job stress). In addition, Krausz, Sagie, and Bidermann (2000) found that instead of work schedules per se, the extent to which work schedules were *preferred* (i.e., matched with personal preferences) predicted nurses' burnout levels. Conversely, use of technology—*objectively* assessed by the number of ICU patients who were mechanically ventilated—was substantively related to nurses' burnout levels (Schaufeli, Keijsers, & Reis-Miranda, 1995). In a somewhat similar vein, unfavorable patient-to-nurse ratios were shown to be positively related to burnout in more than 200 Pennsylvania hospitals (Aiken, Clarke, Sloane, Sochalski, & Silber, 2002). This study showed that an increase of one patient per nurse to a hospital's staffing level increased nurse burnout by 23% and patient mortality by 7% (after controlling for patient and hospital characteristics).

Role conflict (i.e., conflicting demands at the job have to be met) and role ambiguity (i.e., no adequate information is available to do the job well) are moderately to highly correlated with burnout. Role conflict shares approximately 24% of variance with emotional exhaustion, 13% with depersonalization, and only 2% with personal accomplishment; the percentages for role ambiguity are 14%, 8%, and 10%, respectively (Schaufeli & Enzmann, 1998, pp. 82–83). A recent longitudinal study among Spanish primary health care professionals showed that role conflict and role ambiguity predicted future exhaustion as well as depersonalization, whereas personal accomplishment was predicted only by role ambiguity (Pieró, Gozalez-Romá, Tordera, & Mañas, 2001).

Clear evidence exists for a positive relationship between lack of social support and burnout. In particular, lack of social support from supervisors is related to burnout. On the average, support from supervisors explains 14% of the variance of emotional exhaustion, 6% of depersonalization, and 2% of personal of personal accomplishment; for coworkers the amounts of variance are 5%, 5%, and 2%, respectively (see Lee & Ashforth, 1996). The longitudinal study of Leiter and Durup (1996) among health care professionals showed that emotional exhaustion predicted work overload and supervisor support, instead of the other way round, suggesting a cyclical process rather than straight causation. Finally, three factors that determine self-regulation of work activities are related to burnout: lack of feedback, poor participation in decision making, and lack of autonomy (e.g., Landsbergis, 1988).

Recently, two studies among health care professionals have successfully tested the so-called job-demands resources model that states that job demands are particularly related to exhaustion, whereas lacking job resources are particularly related to depersonalization (Bakker, Demerouti, Taris, Schaufeli, & Schreurs, 2003; Demerouti, Bakker, Nachreiner, & Schaufeli, 2000). More specifically, the model assumes two underlying processes: (a) energy depletion is driven by high job demands (e.g., time pressure, emotional demands, shift work, cognitive demands) and is associated with exhaustion, and (b) erosion of motivation is driven by lack of job resources (e.g., supervisor support, feedback, control, task variety, financial rewards) and is associated with disengagement (depersonalization).

Specific Job Stressors.

A review of 16 studies revealed that, overall, and contrary to expectations, common job-related stressors such as workload, time pressure, or role conflicts correlate higher with burnout than patient-related stressors or emotional demands such as interaction with difficult patients, problems in interacting with patients, frequency of contact with chronically or terminally ill patients, or confrontation with death and dying (Schaufeli & Enzmann, 1998, pp. 84–85)

For instance, Mallett, Price, Jurs, and Slenker (1991) found only weak correlations between death of patients and emotional exhaustion and depersonalization in a sample of nurses. Instead, lack of staffing and insufficiently qualified staff were considered the most stressful aspects of their work. Obviously, confrontation with death and dying of patients is not the most disturbing part of the nurses' job. It is likely that nurses have developed adaptive mechanisms that prevent negative long-term effects such as burnout.

Individual Health.

Significant correlations with self-report measures of depression and psychosomatic distress are often reported (Schaufeli & Enzmann, 1998, pp. 86–89). As noted earlier, burnout cannot be reduced to mere depressed mood or distress; it is related to both conditions. As far as self-reported frequency of various illnesses is concerned, Corrigan, Holmes, and Luchins (1995) reported a shared variance with emotional exhaustion plus depersonalization of 18% among psychiatric hospital staff. In a similar vein, Landsbergis (1988) found a significant positive relationship between nurses' self-reported symptoms of coronary heart disease and emotional exhaustion (3% shared variance) and depersonalization (4%); the relationship with reduced personal accomplishment was not significant (2 %).

Work-Related Attitudes.

Although high and unrealistic expectations are related to burnout, this association is not as strong and unequivocal as might be expected. This is probably because different concepts are used, such as omnipotence, irrational beliefs, idealism, unmet expectations, disillusionment, and outcome expectations. Furthermore, it is not always clear whether expectations refer to the organization, to patient's progress, or to personal effectiveness. Job dissatisfaction, poor organizational commitment, and intention to quit—all indicators of psychological withdrawal—share considerable amounts of variance with burnout: 5%–25%, depending on the dimension involved (Schaufeli & Enzmann, 1988, p. 80). The strongest associations are found with emotional exhaustion and depersonalization.

A longitudinal study among Israeli nurses showed that those most prone to leaving the hospital perceived their work as offering little challenge, autonomy or opportunity to express abilities or skills; in addition they also experienced high burnout levels (Krausz, Koslowski, Shalom, & Elkyakim, 1995).

Organizational Behavior.

Despite the popular assumption that burnout causes absenteeism, the effect of burnout on absenteeism is rather small and is best confirmed with respect to emotional exhaustion and next depersonalization. On the average, only approximately 2% of variance is shared with absenteeism as registered in the organization's records; relations with reduced personal accomplishment are marginal but significant (less than 1% of shared variance) (Schaufeli & Enzmann, 1998, pp. 91–92).

Generally speaking, intention to quit is positively associated with burnout, whereas the evidence for a relation with actual turnover is mixed. For instance, an older study revealed that levels of depersonalization predict nurses' actual job turnover within 2 years (Firth & Britton, 1989), but a more recent study among HIV health care professionals showed that burnout scores at baseline are unrelated to job retention over 4 years (Brown et al., 2002). The fact that the relationship of turnover intention to burnout is much stronger than with actual turnover suggests that a large percentage of burned out professionals stay in their jobs involuntarily.

It is important to distinguish between self-ratings of performance and objective measures or ratings by others such as co-workers or supervisors. Self-rated performance correlates weakly with burnout; approximately 5% of variance is shared with all three MBI dimensions against less than 1% for supervisor-rated or objectively assessed performance (e.g., Parker & Kulik, 1995). Most health care professionals perform their jobs in teams. So the question becomes this: Is team performance negatively affected by burnout levels of its members? Empirical results are mixed. On the one hand, as expected, team-level emotional exhaustion and reduced accomplishment of psychiatric rehabilitation teams appeared to be negatively related with patient satisfaction ratings (Garman, Corrigan, & Morris, 2002). On the other hand, positive correlations with burnout are found as well. For instance, Keijsers, Schaufeli, Le Blanc, Zwerts, and Reis-Miranda (1995) obtained an objective measure of ICU performance by calculating for each unit a standard mortality ratio—the ratio of actual versus predicted death rates adjusted for several patient characteristics such as diagnosis and severity of illness. Contrary to expectations, they found a small but significant *positive* correlation of objective ICU performance with emotional exhaustion (explained variance 2%) and no relationship with depersonalization or personal accomplishment. It appeared that the nurses who felt especially exhausted were those who were employed in objectively (and subjectively) well-performing ICUs but who scored low on self-reported personal accomplishment. A possible explanation is that nurses in well-performing ICUs exert themselves more and, as a consequence, they feel more exhausted. An alternative explanation is that nurses in well-performing units have a higher standard of comparison and thus feel that they accomplish less. At any rate, it appears that—in contrast to the prevailing view—burnout is not necessarily linked to low levels of *actual* performance.

In summary, various correlates of burnout in health care have been identified. The most consistent and strong relationships—particularly with emotional exhaustion—are found with general job stressors, such as workload, role problems, and lack of social support. Relation with specific job stressors pertaining to interactions with patients, as well as with personality factors, and with negative outcomes such as individual health, withdrawal from the organization, and poor work performance are somewhat less strong. Strictly speaking, to date, few causes or consequences of burnout can de identified, probably because the considerable stability of burnout across time—approximately 40%–45% of the variance of burnout is explained by the level of burnout 1 year before (Schaufeli & Enzmann, 1998, pp. 96–97)—which leaves little room for other causal factors. Also, mean levels of burnout among health care professionals do not change significantly over time (e.g. Prosser et al., 1999).

PSYCHOLOGICAL EXPLANATIONS FOR BURNOUT

Many different psychological explanations exist for burnout. These explanations emphasize the importance of individual, interpersonal, organizational, and societal factors, respectively (for an overview, see Schaufeli & Enzmann, 1998, pp. 100–142). This section describes three interpersonal approaches that are assumed to be of special importance for explaining burnout in health care, because they emphasize the role of emotionally demanding relationships with patients and the role of work teams in transmitting burnout symptoms.

Burnout as Emotional Overload.

According to Maslach (1993), interpersonal demands resulting from the helping relationship are considered to be the root cause of burnout. She argues that patient contacts are emotionally charged by their very nature, because health care professionals deal with troubled people who are in need. To deal with emotional demands and perform efficient and well, professionals may adopt techniques of detachment. When patients are treated in a more remote, objective way, it becomes easier to do one's job without suffering strong psychological discomfort. A functional way to do this is to develop an attitude of detached concern—the medical profession's ideal blending of compassion with emotional distance. A dysfunctional way to do this is depersonalization (i.e., to develop a persistent callous, indifferent, and cynical perception of patients). Ironically, the structure of the helping relationships in health care is such that it promotes this dysfunctional strategy. For instance, the focus is on the patients' problems rather than on their positive aspects, and there is a lack of positive feedback from patients because they only return when things go wrong. As a result of depersonalization, quality of care is likely to deteriorate because the major vehicle for the success—compassion with and concern for others—has been destroyed in an attempt to protect psychological integrity. Because success is more and more lacking, the professional's sense of personal accomplishment erodes and feelings of insufficiency and self-doubt develop.

Based on the theoretical approach of Maslach, Leiter (1993) conducted a series of studies among health care workers in which he distinguished quantitative job demands (e.g., work overload, hassles), qualitative job demands (e.g., interpersonal conflict), and lack of resources (e.g., lack of social support, poor patient co-operation, lack of autonomy, and poor participation in decision making). Both types of demands were expected to be related to emotional exhaustion, whereas resources were expected to be related to depersonalization and lack of personal accomplishment. Indeed, these hypothesized relationships were observed (Leiter, 1993). Furthermore, as expected, emotional exhaustion leads to depersonalization, but contrary to expectations, personal accomplishment appears to develop rather independently from both of the other burnout dimensions. This agrees with Lee and Ashforth (1996), who concluded that the results of their meta-analysis are largely consistent with Leiter's (1993) mixed sequential and parallel development model of burnout.

Thus, burnout appears to be a multidimensional construct whose dimensions are differentially related to job demands and job resources. Emotional exhaustion develops in reaction to job demands, including interpersonal demands, and appears to lead to depersonalization. Personal accomplishment is positively influenced by the presence of resources, but it largely develops in parallel to the other two burnout dimensions.

Burnout as Lack of Reciprocity.

By definition, the professionals' relationship with patients is complementary, which is semantically well-illustrated by the terms "caregiver" and "recipient of care"; the former is supposed to give care, assistance, advise, support, and so on, whereas the latter is supposed to receive care. Nevertheless, health care professionals look for some rewards in return for their efforts; for example, they expect their patients to show gratitude, to improve, or to at least make a real effort to get well. Because in practice these expectations are seldom fulfilled, it is likely that, over time, lack of reciprocity develops: Professionals feel that they continuously put much more in relationships with their patients than they receive back in return. As Buunk and Schaufeli (1993) have pointed out, lack of reciprocity—an unbalanced helping relationship—drains professionals' emotional resources and eventually leads to emotional exhaustion. This is typically dealt with by decreasing one's investments in the relationships with patients; that is, responding to patients in a depersonalized way instead of expressing genuine empathic concern. Accordingly, depersonalization can be regarded as a way of restoring reciprocity by withdrawing psychologically from patients. However, this way of coping with an unbalanced interpersonal relationship is dysfunctional because it deteriorates the helping relationship, increases failures, and thus fosters a sense of diminished personal accomplishment.

Indeed, positive relationships were found between lack of reciprocity at the interpersonal level and all three dimensions of burnout among various health professionals such as student nurses (Schaufeli, Van Dierendonck, & Van Gorp, 1996), general hospital nurses (Schaufeli & Janczur, 1994), medical specialists (Smets, Visser, Oort, Schaufeli, & De Haes, in press), and general practitioners (Van Dierendonck, Schaufeli, & Sixma, 1994). Although these studies are cross-sectional, there is some longitudinal evidence for a curvilinear relationship between lack of reciprocity and emotional exhaustion: feeling more deprived as well as feeling more advantaged results in higher exhaustion levels (Van Dierendonck, Schaufeli, & Buunk, 2001). Another longitudinal study showed that depersonalizing patients at Time 1 increases the likelihood of feeling harassed by them 5 years later, which in its turn fostered a lack of reciprocity, eventually leading to burnout (Bakker, Schaufeli, Sixma, Bosveld, & Van Dierendonck, 2000). Thus, a lack of reciprocity in the caregiver–recipient relationship appears to play an important role in the development of burnout in conjunction with the impairment of the quality of the doctor–patient relationship.

Furthermore, the relationship between lack of reciprocity and burnout appears to be moderated by personality factors. For instance, VanYperen, Buunk, and Schaufeli (1992) found that nurses who felt they invested highly in the relationships with patients showed elevated levels of burnout only when they were low in communal orientation, a personality characteristic that refers to a general responsiveness to the needs of others.

Similar social exchange processes that are observed in interpersonal relationships govern the relationship of the professional with the organization and with the team they work on. Each health care worker has a so-called psychological contract (Roussau, 1995) with the organization, which entails expectations about the nature of the exchange with that organization. Expectations concern concrete issues such as workload and career prospectives, as well as less tangible matters such as esteem and dignity at work and support from supervisors and colleagues. In other words, the psychological contract reflects the employees' subjective notion of reciprocity: (S)he expects gains or outcomes from the organization that are proportional to his or her investments or inputs. When the psychological contract is violated because experience does not match expectancies, reciprocity is corroded. Schaufeli, Van Dierendonck & Van Gorp (1996) showed in two samples of student nurses that in addition to withdrawal from the organization (i.e., reduced organizational commitment), violation of the psychological contract may also lead to burnout. Moreover, studies among therapists from a forensic psychiatric clinic and staff working with the mentally disabled confirmed that burnout is related to perceptions of inequity at the organizational level (Van Dierendonck, Schaufeli, & Buunk, 1996).

In addition to the individual level and the organizational level, social exchange processes play a role among colleagues in work teams. For instance, Buunk and Hoorens (1992) found some evidence that nurses keep a "support bookkeeping" that is based on the balance between giving and receiving support from others in their team. Given the centrality of the relationships with colleagues for work related outcomes, it appears plausible to expect that lack of reciprocity in the exchange relationship with one's colleagues is an important determinant of burnout as well. Indeed, Smets et al. (in press) found among medical specialists that in addition to lack of reciprocity at the interpersonal level, which was associated with emotional exhaustion and depersonalization, and to lack of reciprocity at the organizational level, which was related to emotional exhaustion, lack of reciprocity at the team was related to emotional exhaustion as well. Hence, it appears that team members who experience an imbalance between their investments in and their outcomes from the work are likely to feel emotionally exhausted.

Burnout as an Emotional Contagion.

It has been suggested that colleagues may act as models whose symptoms are imitated through a process of emotional contagion (Buunk & Schaufeli, 1993). That is, people under stress may perceive symptoms of burnout in their colleagues and automatically take on these symptoms. In addition to this nonconscious emotional contagion, there is an alternative way in which people may catch emotions of others. Contagion may also occur through a conscious cognitive process by "tuning in" to the emotions of others. This will be the case when an individual tries to imagine how (s)he would feel in the position of another, and, as a consequence, experiences the same feelings. The professional attitude of health care workers that is often characterized by empathic concern is likely to foster such a process of consciously tuning in to emotions of others.

The contagious nature of burnout is exemplified by the observation that burnout tends to concentrate in particular teams, whereas it is virtually not observed in comparable other groups as was shown by Bakker, Le Blanc, and Schaufeli (2004) in a sample of almost 80 European ICUs. Of course, this concentration of burnout in particular ICUs also may be explained by higher workloads, which would contradict a symptom contagion explanation. However, this alternative hypothesis was rejected, it appeared that—after controlling for job autonomy, subjective workload and objectively assessed workload (i.e., complexity of nursing tasks)—nurses' levels of experienced burnout remained higher in some units compared with other units. Moreover, nurses from these units observed more burnout complaints among their colleagues then their fellows did in the other units. These intriguing results support the contagion hypothesis of burnout. In a similar vein, Miller, Birkholt, Scott, and Stage (1995) found among professionals who work with the homeless that emotional contagion was directly as well as indirectly—through communicative responsiveness—related to burnout. Recently, a study with general practitioners (Bakker, Schaufeli, Sixma, & Bosveld, 2001) showed that those who perceived burnout complaints among their colleagues reported higher levels of emotional exhaustion and subsequent negative attitudes (depersonalization and reduced personal accomplishment) than those who did not perceive such complaints. In addition, individual susceptibility to emotional contagion was positively related to burnout, particularly in combination with the perception of burnout symptoms in their colleagues. That is, doctors who perceived burnout complaints among colleagues *and* who were susceptible to emotional contagion reported the highest exhaustion scores.

In summary, in a way, the three psychological explanations that are outlined here are complementary. The first approach assumes that burnout results from emotionally charged relationships between caregivers and recipients. It stipulates a dynamic process in which depersonalization is considered to be a dysfunctional attempt to deal with feelings of emotional exhaustion. However, it remains unclear *why* the relationship between caregiver and recipient is so demanding. This is where the second approach links in by emphasizing that this helping relationship is often characterized by a lack of reciprocity from the part of the caregiver. In addition, it is this lack of reciprocity, not only in interpersonal relationships but also in the relationship with the team and with organization that lies at the core of the burnout syndrome. Instead of working too long, too hard with too difficult recipients, it appears that the *balance* between give and take is crucial for the development of burnout. Finally, once burnout has occurred among individual team members, a group-based process of emotional contagion appears to play a role in spreading it among other team members.

INTERVENTIONS

Because individual characteristics as well as workplace characteristics are involved in the etiology of burnout, interventions may focus on the person or on the job. Although it is generally acknowledged that a *combination* of both approaches would be most effective, burnout interventions have been conducted predominantly on the individual level. This final section briefly reviews individual and workplace interventions to reduce burnout in health care. For a more extensive, general review of burnout interventions see Schaufeli and Enzmann (1998; pp. 144–183).

Individual Interventions.

Individual approaches to prevent or reduce burnout include cognitive–behavioral techniques such as stress inoculation training, rational emotive therapy, cognitive restructuring and behavioral rehearsal. A cognitively oriented approach is relevant because burnout often involves "wrong" cognitions such as unrealistic expectations. In addition, relaxation techniques are often used to reduce burnout. A recent meta-analysis of nearly 50 (quasi) experimental studies revealed the effectiveness of these individual interventions in terms of symptom reduction, including burnout (Van der Klink, Blonk, Schene, & Van Dijk, 2001). It appeared that individual stress management interventions are effective, and, more specifically, that cognitive–behavioral interventions alone or in combination with relaxation are more effective than relaxation training. In addition, other measures have been recommended to combat burnout, including time management, balancing work and private life, physical training, dieting, and increasing one's social skills—particularly assertiveness. To counteract the reality shock experienced by many beginning health care professionals, preparatory training programs may provide them with more realistic images of their profession, instead of fostering wrong expectations. Unfortunately, the effects of such measures on reducing burnout are largely unknown.

By contrast, the effectiveness of so-called burnout workshops has been studied in greater detail. Basically, these workshops rest on two pillars: (a) increasing the participants awareness of work-related problems and (b) augmenting their coping resources by cognitive and behavioral skills training and by establishing support networks. More specifically, workshops may include self-assessment, didactic stress management, relaxation, cognitive and behavioral techniques, time management, peer support, and the promotion of a healthy life style and a more realistic image of the job (Schaufeli & Enzmann, 1998, pp. 179–182). In other words, burnout workshops combine many rather general strategies for one specific purpose: preventing and combating burnout.

But what is known about their effectiveness? Pines and Aronson (1983) evaluated a 1-day burnout workshop for employees of two social services that combined several individual approaches (e.g., relaxation techniques, cognitive stress management, time management, social skills training, didactical stress management, and attitude change). The participants' level of exhaustion decreased slightly but not significantly. However, compared with the control group that did not participate in the workshop, satisfaction with co-workers went up significantly in the experimental group. Schaufeli (1995) evaluated a somewhat similar burnout workshop for community nurses but found that only the symptom levels (i.e., emotional exhaustion, psychological strain, and somatic complaints) of the participating nurses decreased significantly. However, no significant changes were observed in levels of depersonalization and reduced personal accomplishment. Van Dierendonck, Schaufeli, and Buunk (1998) evaluated a 3-day burnout workshop for staff working in direct care with mentally disabled. The workshop was cognitive–behaviorally oriented and included aspects such as cognitive restructuring, didactic stress management, and relaxation, as well as career counseling. Results showed that emotional exhaustion dropped significantly for the experimental group compared with the control group at each follow-up after 6 and 12 months, but again no effects were observed for depersonalization and personal accomplishment. Registered absenteeism significantly decreased in the experimental group, but it increased in the control group. Freedy and Hobfoll (1994) used stress inoculation training among nurses to enhance their social support and individual mastery resources. Participants experienced significant enhancements in social support and mastery compared to the no-intervention control group. Particularly, nurses with low initial levels on both resources showed significant reductions in emotional exhaustion and depression. Similar positive results were obtained by West, Horan, and Games (1984), who used didactic stress management, training coping skills (i.e., relaxation, assertiveness, cognitive restructuring, and time management), and exposure via role playing. A 4-month follow-up showed that burnout (i.e., emotional exhaustion and reduced personal accomplishment) decreased significantly, as did anxiety and systolic blood pressure. More detailed analysis revealed that coping skills were the main ingredient of the program.

Workplace Interventions.

Only occasionally, workplace interventions are explicitly carried out to reduce stress or burnout. Generally, other purposes are targeted such as increased productivity and efficiency, cost-effectiveness, smooth communication, or organizational flexibility. Nevertheless, there is an increasing awareness that preventing burnout is important because of the high direct and indirect costs associated with it. Workplace interventions to reduce burnout are more rare than individual approaches, and their effectiveness is seldom evaluated.

Recently, Spickard, Gabbe, and Christensen (2002) listed several suggestions for health care organizations to prevent physician burnout: establishing a mentor program, providing confidential support groups, establishing a physician health committee, providing an annual well-being retreat, institutionalizing a sabbatical program, providing membership in a fitness center, offering periodic continuing medical education programs, providing flexible time-scheduling, and reducing paperwork. In addition, from the outset work redesign (i.e., job enlargement, job rotation, and job enrichment) has been considered as a major tool to decrease quantitative and qualitative workload and thus to prevent burnout (Pines & Maslach, 1980). Because many burnout candidates feel "locked in" their careers, career development programs and career counseling are other organizational approaches that can prevent burnout. Two-way communication between management and employees, adequate procedures for conflict management, and participative decision making also have been proposed as antidotes to burnout. Moreover, it is suggested that social support from colleagues and superiors should be institutionalized in the form of regular consultations and team meetings.

What about the effectiveness of such workplace interventions? As noted earlier, few studies exist. The introduction of a system of planned nursing care in a Swedish psychogeriatric clinic in which each patient was assigned a particular nurse who was responsible for all nursing tasks, led to a reduction of burnout at the 1-year follow-up compared with the traditional ward system in which one patient was nursed by many different nurses (Berg, Welander-Hansson, & Hallberg, 1994). Unfortunately, a similar job redesign project among psychiatric nurses in The Netherlands failed to confirm this positive result (Melchior, Philipsen, Abu-Saad, Halfens, Van den Berg, & Gassman, 1996). The effectiveness of a training program in emotion-oriented care for cognitively impaired elderly persons was demonstrated (Schrijnemaekers et al., 2003): Compared with the control nursing homes, caregivers' levels of burnout in the experimental homes had decreased significantly at follow-up after 6 and 12 months.

All three of these workplace interventions had been designed beforehand and were then implemented. However, a few case examples exist of participatory action research projects in which specific workplace interventions were introduced *during* the project in which employees, researchers, and consultants closely collaborated. For instance, Van Gorp and Schaufeli (1996) carried out an organizational development program to reduce burnout and sickness absenteeism in four large Dutch community mental health centers. Specific measures were introduced in each center to tackle the three common problems: work overload, poor team leadership, and poor collaboration between professional staff and support staff. At the 1-year follow-up, levels of burnout had not markedly decreased, but workers were more satisfied in all centers and registered absenteeism had decreased significantly in one of them. Le Blanc and Peeters (2003) described a participatory training program for functional teams of oncology care providers (nurses, physicians, and radiotherapy assistants). Based on a thorough assessment, specific team problems were identified for each team, and the teams were subsequently coached by a team- counselor to solve these issues during the next 6 months. At the 1-year follow-up, levels of emotional exhaustion and depersonalization of the control teams had increased, whereas *no* change was observed for the experimental teams. Therefore, it appears that the team-based training was effective to prevent burnout from increasing. Finally, the effectiveness was evaluated of a large-scale participatory job stress reduction program that was implemented in 81 Dutch domiciliary care institutions including over 26,000 workers (Taris et al., 2003). Each of these institutions had implemented a specific set of measures that might include any combination of person-directed interventions (e.g., job mobility programs, opportunity to visit conferences) and work-directed interventions (e.g., introducing rules for lifting patients, employee participation in planning tasks and shifts). It was concluded at the 2½-year follow-up that (a) institutions with employees who reported high levels of exhaustion implemented many interventions of various kinds, and (b) employees working in institutions that had implemented many (more than four) interventions reported lower levels of exhaustion than did employees of other institutions. Thus, the more measures that were taken by the home-care institution, the more levels of burnout of its members had decreased.

In summary, it appears that individual burnout interventions work. At the least, the core symptom of burnout—emotional exhaustion—can be reduced by training health care professionals to use particular coping skills, most notably cognitive restructuring and relaxation techniques. Conversely, levels of personal accomplishment and depersonalization appear rather resistant to change. This is not surprising because most techniques focus on reducing arousal rather than on changing attitudes (depersonalization) or on enhancing specific professional resources (personal accomplishment). Compared with individual interventions, the effects of workplace intervention programs to reduce burnout are somewhat disappointing. This is at least partly caused by methodological reasons because workplace programs are often participatory in nature so that a pretest posttest (quasi) experimental design is not feasible, and effectiveness cannot be demonstrated. Nevertheless, some examples in health care exist of participatory intervention programs to reduce burnout.

CONCLUSION

Based on estimates of its prevalence, severe burnout appears to be a problem in health care, particularly for those working in the community. In the past decades, burnout received quite a lot of attention, both from researchers and practitioners, because it is not only detrimental to the individual involved but also for the organization. Burnout appears to be related to job stress and depression but can nevertheless be distinguished from these alternative conditions on conceptual and empirical grounds. On the empirical level, burnout—particularly emotional exhaustion—is related to personality factors, job stressors, and individual and organizational outcomes. Theoretical approaches that emphasize the social nature of burnout by taking into account the emotional overload resulting from patient contacts, the disturbed balance between give and take, and emotional contagion on work teams appear to offer a promising route for explaining burnout in health care settings. Particularly individual-based interventions appear to be effective in reducing burnout (emotional exhaustion), whereas the effectiveness of organization-based interventions still stands out.

References

Aiken, L. H., Clarke, S. P., Sloane, D. M., Sochalski, J., & Silber, J. H. (2002). Hospital nurse staffing and patient mortality, nurse burnout, and job satisfaction. *Journal of the American Medical Association, 288,* 1987–1993.

Bakker, A. B., Demerouti, E., Taris, T., Schaufeli, W. B., & Schreurs, P. J. G. (2003) A Multi-group analysis of the Job Demands-Resources Model in four home-care organizations. *International Journal of Stress Management, 10,* 16–38.

Bakker, A. B., Le Blanc, P. M., & Schaufeli, W. B. (2004) *Burnout contagion among intensive care nurses.* Manuscript submitted for publication.

Bakker, A. B., Schaufeli, W. B., Demerouti, E., Janssen, P. M. P., Van der Hulst, R., & Brouwer, J. (2000). Using equity theory to examine the difference between burnout and depression. *Anxiety, Stress, and Coping, 13,* 247–268.

Bakker, A. B., Schaufeli, W. B., Sixma, H., & Bosveld, W. (2001). Burnout contagion among general practitioners. *Journal of Social and Clinical Psychology, 20,* 82–98.

Bakker, A. B., Schaufeli, W. B., Sixma, H. J., Bosveld, W., & Van Dierendonck, D. (2000). Patient demands, lack of reciprocity, and burnout: A five-year longitudinal study among general practitioners. *Journal of Organizational Behaviour, 21,* 425–441.

Bakker, A., Schaufeli, W. B., & Van Dierendonck, D. (2000). Burnout: Prevalentie, risicogroepen en risicofactoren [Burnout: Prevalence, groups at risk and risk factors]. In I. L. Houtman, W. B. Schaufeli, & T. Taris (Eds.), *Psychische vermoeidheid en werk: Cijfers, trends en analyses* (pp. 65–82). Alphen a/d Rijn, The Netherlands: Samsom.

Berg, A., Welander-Hansson, U., & Hallberg, I. R. (1994). Nurses' creativity, tedium and burnout during 1 year of clinical supervision and implementation of individually planned nursing care: Comparisons between a ward for severely demented patients and a similar control ward. *Journal of Advance Nursing, 20,* 742–749.

Brown, L. K., Schultz, J. R., Forsberg, A. D., King, G., Locik, S. M., & Butler, R. B. (2002). The predictors of retention among HIV/hemophilia health care professionals. *General Hospital Psychiatry, 24,* 48–54.

Burisch, M. (2002). A longitudinal study of burnout: The relative importance of dispositions and experiences. *Work & Stress, 16,* 1–17.

Buunk, B. P., & Hoorens, V. (1992). Social support and stress: The role of social comparison and social exchange processes. *British Journal of Clinical Psychology, 31,* 445–457.

Buunk, B. P. & Schaufeli, W. B. (1993). Burnout: A perspective from social comparison theory. In W. B. Schaufeli, C. Maslach & T. Marek (Eds.), *Professional Burnout: Recent Developments in Theory and Research* (pp. 53–69). Washington, DC: Taylor & Francis.

Cordes, C. L., & Dougherty, T. W. (1993). A review and an integration of research on job burnout. *Academy of Management Review, 18,* 621–656.

Corrigan, P. W., Holmes, E. P., & Luchins, D. (1995). Burnout and collegial support in state psychiatric hospital staff. *Journal of Clinical Psychology, 51,* 703–710.

Demerouti, E., Bakker, A. B., Nachreiner, F., & Schaufeli, W. B. (2000). A model of burnout and life satisfaction among nurses. *Journal of Advanced Nursing, 32,* 454–464

De Rijk, A., Le Blanc, P. M., Schaufeli, W. B., & De Jonge, J. (1998). Active coping and need for control as moderators of the job demand-control model: Effects on burnout. *Journal of Occupational and Organizational Psychology, 71,* 1–18.

Enzmann, D., Schaufeli, W. B., & Girault, N. (1994). The validity of the Maslach Burnout Inventory in three national samples. In L. Bennett, D. Miller, M. Ross (Eds.), *Health Workers and AIDS: Research, Interventions and Current Issues* (pp. 131–150). London: Harwood.

Firth, H., & Britton, P. G. (1989). "Burnout," absence and turnover amongst British nursing staff. *Journal of Occupational Psychology, 62,* 55–59.

Freedy, J. R., & Hobfoll, S. E. (1994) Stress inoculation for reduction of burnout: A conservation of resources approach. *Anxiety, Stress and Coping, 6,* 311–325.

Freudenberger, H. J. (1974). Staff burn-out. *Journal of Social Issues, 30,* 159–165.

Garman, A. N, Corrigan, P. W., & Morris, S. (2002). Staff burnout and patient satisfaction: Evidence of relationships at the care unit level. *Journal of Occupational Health Psychology, 7,* 235–241.

Garden, A. M. (1991). The purpose of burnout: A Jungian interpretation. *Journal of Social Behavior and Personality, 6,* 73–93.

Glass, D. C., & McKnight, J. D. (1996). Perceived control, depressive symptomatology, and professional burnout: A review of the evidence. *Psychology and Health, 11,* 23–48.

Hillhouse, J. J., Adler, C. M., & Walters, D. N. (2000). A simple model of stress, burnout and symptomatology in medical residents: A longitudinal study. *Psychology, Health & Medicine, 5,* 63–73.

Ito, H., Kurita, H., & Shiiya, J. (1999). Burnout among direct-care staff members of facilities for persons with mental retardation in Japan. *Mental Retardation, 37,* 447–481.

Keijsers, G. J., Schaufeli, W. B., Le Blanc, P. M., Zwerts, C., & Reis-Miranda, D. (1995). Performance and burnout in intensive care units. *Work & Stress, 9,* 513–527.

Krausz, K., Koslowski, M., Shalom, N., & Elkyakim, N. (1995). Predictors of intentions to leave the ward, the hospital, and the nursing profession: A longitudinal study. *Journal of Organizational Behavior, 16,* 277–288.

Krausz, M., Sagie, A., & Bidermann, Y. (2000). Actual and preferred work schedules and scheduling control as determinants of job-related attitudes. *Journal of Vocational Behavior, 56,* 1–11.

Landsbergis, P. A. (1988). Occupational stress among health care workers: A test of the job demands-control model. *Journal of Organizational Behavior, 9,* 217–239.

Le Blanc, P. M. & Peeters, M. C. W. (2003). Towards a group and organization oriented approach to stress management in health care. In J. Hellgren, K. Näswall, M. Sverke, & M. Söderfeldt (Eds.). *New Organizational Challenges for Human Service Work* (pp. 207–226). München, Germany: Rainer Hampp Verlag.

Lee, R. T., & Ashforth, B. E. (1996). A meta-analytic examination of the correlates of the three dimensions of job burnout. *Journal of Applied Psychology, 81,* 123–133.

Leiter, M. P. (1993). Burnout as a developmental process: Consideration of models. In W. B. Schaufeli, C. Maslach & T. Marek (Eds.), *Professional Burnout: Recent Developments in*

Theory and Research (pp. 237–250). Washington, DC: Taylor & Francis.

Leiter, M. P., & Durup, M. J. (1996). Work, home, and in-between: A longitudinal study of spillover. *Journal of Applied Behavioral Science, 32,* 29–47.

Liu, P. M., & Van Liew, D. A. (2003). Depression and burnout. In J. Kahn & A. M. Langleib (Eds.), *Mental Health and Productivity in the Workplace* (pp. 433–457). San Francisco: Jossey-Bass.

Maddi, S. R. (1999). The personality construct of hardiness: Effects on experiencing, coping, and strain. *Consulting Psychology Journal, 51,* 83–95.

Mallett, K. L., Price, J. H., Jurs, S. G., & Slenker, S. (1991). Relationships among burnout, death anxiety, and social support in hospice and critical care nurses. *Psychological Reports, 68,* 1347–1359.

Maslach, C. (1993). Burnout: A multidimensional perspective. In W. B. Schaufeli, C. Maslach, & T. Marek (Eds.), *Professional Burnout: Recent Developments in Theory and Research* (pp. 19–32). Washington, DC: Taylor & Francis.

Maslach, Jackson, & Leiter. (1986).

Maslach, C., Jackson, S. E., & Leiter, M. (1996). *Maslach Burnout Inventory. Manual* (3rd ed.). Palo Alto, CA: Consulting Psychologists Press.

Maslach, C., Jackson, S. E., & Leiter, M. P. (1986). *Maslach Burnout Inventory: Manual* (2nd ed.). Palo Alto, CA: Consulting Psychologists Press.

Maslach, C., & Leiter, M. P. (1997) *The Truth About Burnout: How Organizations Cause Personal Stress and What to Do About It.* San Francisco, CA: Jossey-Bass.

Maslach, C., & Schaufeli, W. B. (1993). Historical and conceptual development of burnout. In W. B. Schaufeli, C. Maslach & T. Marek (Eds.), *Professional Burnout: Recent Developments in Theory and Research* (pp. 1–16). Washington, DC: Taylor & Francis.

Maslach, C., Schaufeli, W. B. & Leiter, M. P. (2001). Job burnout. *Annual Review of Psychology, 52,* 397–422.

McManus, I. C., Winder, B. C., & Gordon, D. (2002). The causal links between stress and burnout in a longitudinal study of UK doctors. *The Lancet, 359,* 2089–2090.

Melchior, M. E. W., Philipsen, H., Abu-Saad, H. H., Halfens, R., Van den Berg, A. A., & Gassman, P. (1996). The effectiveness of primary nursing on burnout among psychiatric nurses in long-stay settings. *Journal of Advanced Nursing, 24,* 694–702.

Miller, K., Birkholt, M., Scott, C., & Stage, C. (1995). Empathy and burnout in human service work: An extension of a communication model. *Communication Research, 22,* 123–147.

Parker, P. A., & Kulik, J. A. (1995). Burnout, self- and supervisor-rated job performance, and absenteeism among nurses. *Journal of Behavioral Medicine, 18,* 581–599.

Pieró, J.-M., Gozalez-Romá, V., Tordera, N., & Mañas, M-A (2001). Does role stress predict burnout over time among health care professionals. *Psychology & Health, 16,* 511–526.

Pines, A. M. (1993). Burnout: An existential perspective. In W. B. Schaufeli, C. Maslach, & T. Marek (Eds.), *Professional Burnout: Recent Developments in Theory and Research* (pp. 33–51). Washington, DC: Taylor & Francis.

Pines, A., & Aronson, E. (1983). Combating burnout. *Children and Youth Services Review, 5,* 263–275.

Pines, A., & Maslach, C. (1980). Combating staff burnout in a day care setting: A case study. *Child Quarterly, 9,* 5–16.

Prosser, D., Johnson, S., Kuipers, E., Dunn, G., Szmukler, G., Reid, Y., et al. (1999). Mental health, "burnout," and job satisfaction in a longitudinal study of mental health staff. *Social Psychiatry and Psychiatric Epidemiology, 34,* 295–300.

Rousseau, D. M. (1995). *Psychological Contracts in Organizations: Understanding Written and Unwritten Agreements.* Thousand Oaks, CA: Sage.

Schaufeli, W. B. (1995) The evaluation of a burnout workshop for community nurses. *Journal of Health and Human Services Administration, 18,* 11–31.

Schaufeli, W. B., & Buunk, B. P. (2002) Burnout: An overview of 25 years of research and theorizing. In M. J. Schabracq, J. A. M. Winnubst, & C. L. Cooper, *Handbook of Work and Health Psychology* (pp. 383–425). Chichester, UK: Wiley.

Schaufeli, W. B., & Janczur, B. (1994). Burnout among nurses. A Polish-Dutch comparison. *Journal of Cross-Cultural Psychology, 25,* 95–113.

Schaufeli, W. B., & Enzmann, D. (1998). *The Burnout Companion to Study and Practice: A Critical Analysis.* London: Taylor & Francis.

Schaufeli, W. B., Keijsers, G. J., & Reis-Miranda, D. (1995). Burnout, technology use, and ICU-performance. In S. L. Sauter & L. R. Murphy (Eds.), *Organizational Risk Factors for Job Stress* (pp. 259–272). Washington, DC: American Psychological Association.

Schaufeli, W. B., Leiter, M. P., Maslach, C., & Jackson, S. E. (1996). The MBI-General Survey. In C. Maslach, S. E. Jackson & M. Leiter (Eds.), *Maslach Burnout Inventory* (3rd ed., pp. 19–26). Palo Alto, CA: Consulting Psychologists Press.

Schaufeli, W. B., & Van Dierendonck, D. (1993). The construct validity of two burnout measures. *Journal of Organizational Behavior, 14,* 631–647.

Schaufeli, W. B., Van Dierendonck, D., & Van Gorp, K. (1996). Burnout and reciprocity: Towards a dual-level social exchange model. *Work & Stress, 3,* 225–237.

Schaufeli, W. B., Bakker, A., Hoogduin, C. A. L., Schaap, C., & Kladler, A. (2001). Burnout in an out-patient sample: The clinical validity of the Maslach Burnout Inventory and the Burnout Measure. *Psychology and Health, 16,* 565–582.

Schrijnemaekers, V. J. J., van Rossum, E., Candel, M. J. J., Frederiks, C. M. A., Drix, M. M. A., Sielholst, H., et al. (2003). Effects of emotion-oriented care on work-related outcomes of professional caregivers in homes of elderly persons. *Journal of Gerontology, 28(B),* 50–57.

Smets, E. M. A., Visser, M. R. M., Oort, F., Schaufeli, W. B., & De Haes, H. C. J. M. (in press). Perceived inequity: Does it explain burnout among medical specialists? *Journal of Applied Social Psychology.*

Spickard, A., Gabbe, S. G., & Christrensen, J. F. (2002). Mid-career burnout in generalist and specialis physicians. *Journal of the American Medical Association, 288,* 1447–2450.

Taris, T. W., Kompier, A. J., Geurts, S. A. R. E., Schreurs, P. J. G., Schaufeli, W. B., De Boer, E., et al. (2003). Stress management interventions in the Dutch domiciliary care sector: Findings from 81 organizations. *International Journal of Stress Management, 10,* 297–325.

Van der Klink, J. J. L., Blonk, R. W. B., Schene, A. H., & Van Dijk, F. J. H. (2001). The benefits of interventions for work related stress. *American Journal of Public Health, 91,* 270–276.

Van Dierendonck, D., Schaufeli, W. B., & Buunk, B. P. (1996). Inequality among human service professionals: Measurement and relation to burnout. *Basic and Applied Social Psychology, 18,* 429–451.

Van Dierendonck, D., Schaufeli, W. B., & Buunk, B. P. (1998). The evaluation of an individual burnout program. The role of inequity and social support. *Journal of Applied Psychology, 83,* 392–407.

Van Dierendonck, D., Schaufeli, W. B., & Buunk, B. P. (2001). Burnout and inequity among human service professionals: A longitudinal study. *Journal of Occupational Health Psychology, 6,* 43–52.

Van Dierendonck, D., Schaufeli, W. B., & Sixma, H. J. (1994). Burnout among general practitioners: A perspective from equity theory. *Journal of Social and Clinical Psychology, 13,* 86–100.

Van Gorp, K., & Schaufeli, W. B. (1996). *Een gezonde geest in een gezonde organisatie: Een aanzet tot burnout-interventie in de ambulante GGZ.* [A healthy mind in a healthy organization: Burnout intervention in community mental health centers] The Hague, The Netherlands: VUGA.

VanYperen, N. W., Buunk, A. P., & Schaufeli, W. B. (1992). Communal orientation and the burnout syndrome among nurses. *Journal of Applied Social Psychology, 22,* 173–189.

West, D. J., Horan, J. J., & Games, P.A. (1984) Component analysis of occupational stress inoculation applied to registered nurses in an acute care hospital setting. *Journal of Consulting Psychology, 31,* 209–218.

15

HUMAN FACTORS OF TRANSITION OF CARE

Craig M. Harvey
Louisiana State University

Richard J. Schuster
Wright State University

Francis T. Durso
Wright State University

Amy L. Matthews
United State Air Force, Ohio

Deepti Surabattula
Louisiana State University

Continuity of care—what exactly does it mean? What do human factors have to do with achieving continuity of care? Let us review a case, refer to figure 15–1, from a 2002 New York Times article discussing continuity of care in a New York hospital that implemented 8- to 12-hr shifts for physicians with regard to sleep error issues (Zuger, 2002). This case raises several issues related to continuity of care.

Communications among care providers, responsibility for the patient, shift-change handovers, coordination among providers and different areas of the hospital, and physician–nurse interaction are just some of the problems evident in the case. The landmark 1999 Institute of Medicine (IOM) report, *To Err is Human: Building a Safer Health System,* reported that 44,000–98,000 yearly deaths are caused by medical care errors (Kohn, Corrigan, & Donaldson, 2000) highlighting a critical problem clinicians face in the increasingly technical realm of modern health care. Its next report, *Crossing the Quality Chasm,* IOM outlined 10 key recommendations for improving health care quality, with the final recommendation emphasizing collaboration and communication between physicians as critical to "ensure an appropriate exchange of information and coordination of care" (Committee on Quality of Health Care in America, Institute of Medicine, 2001, p. 62). This chapter discusses the human factors and system elements that affect the continuity of care for patients and provide a framework for human factors experts to work with the medical community to improve medical quality.

DEFINING CONTINUITY OF CARE IN HEALTH CARE

According to Shortell (1976, p. 377), continuity of medical care can be defined as "the extent to which medical care services are required as a coordinated and uninterrupted succession of events consistent with the medical care needs of the patients."

At 4 p.m. a middle-aged woman was brought to the emergency room by ambulance. She told paramedics she had just been released from prison upstate and had arrived in the city without her asthma medications. She was calm, coherent and wheezing loudly.

In the E.R. her first resident (Resident 1) looked through the manila envelope of medical records she had with her and confirmed a diagnosis of labile asthma requiring many medications to keep in check.

By evening she was getting the usual treatments for asthma, but her wheezing grew worse. She began to cough. Resident 1 had gone home. Resident 2 re-evaluated her. He had been told she had asthma. But here she was coughing up a storm. Could she also have an underlying infection, like tuberculosis? He ordered a chest X-ray and sputum collection.

Hospital protocol mandates that any patient suspected of tuberculosis must be placed in isolation, to prevent the microbe from infecting others. Into isolation the patient went, and the door closed behind her. Arrangements were made to admit her to the hospital.

At 3 a.m. Residents 3 and 4 who were in charge of the wards upstairs, arrived in the E.R. to evaluate her. She was not wheezing much, they found, but seemed unwilling to answer their questions. They looked through the folder. Sure enough, she had a psychiatric history that could account for her odd behavior. They decided she had mild asthma, possible tuberculosis and psychosis. They started her on antibiotics in case she had bronchitis, too, and went on to other duties.

At 7 a.m. a paper mask was put over her nose and mouth and she was moved to an isolation room upstairs. Her nurse found her to be agitated, tearing off her clothes and refusing to be touched. Perhaps she needed sedation?

The nurse paged Resident 5, who was just beginning his shift. "I don't know the patient," he said. He would formally accept her case at 9 a.m. rounds. A conscientious resident, he took a look at her anyway. She was naked, disheveled and would not let him near her. He shrugged, and went to rounds.

On rounds Resident 3 presented her case: a woman with a history of asthma, possible tuberculosis and a serious psychiatric illness who was brought in off the street by paramedics.

At 10:30 a.m., Resident 5 and the attending doctor legally responsible for the patient's care went to check on her together.

They found her crouched on her bed on all fours, hallucinating wildly, gasping for air, in the very last stages of asthma so severe that she was no longer wheezing or coughing because no air was getting into her lungs, and no oxygen was getting to her brain. She was minutes from cardiac arrest. She was resuscitated and brought to the intensive care unit, where she stayed for a week, slowly getting well. (Zuger, 2002)

Figure 15–1. Transition of care case example.

According to Anderson and Helms (1995), continuity of care is a series of interconnected patient care events within a health care institution and among multiple settings that requires coordination across time, settings, and providers of health care. Lamb and Stempel (1994) state that a recent strategy for promoting continuity of care is to follow the progress of a patient through all the health care settings and to arrange for services based on the changing needs and goals of the patient (Sparbel & Anderson, 2000). Two core elements that distinguish continuity from other health care situations are care of an individual patient and care delivered over time. Though patients' experiences can be categorized to different group levels such as doctors' practices, hospital wards, and health care organizations, the unit of measurement of continuity is the individual.

Continuity is how the patient experiences integration and coordination of services (Haggerty Reid, Freeman, Starfield, & Adair, 2003). In a review of various disciplines of medicine (e.g., family, mental health, nurses, disease management), Haggerty and colleagues found three types of continuity of care: (a) informational, (b) management, and (c) relational. Informational continuity is defined as the common thread that links two health care providers. Documentation about the patient's history and treatment preferences forms the basis of this type of care. In the case discussed, it is easy to see that information was not handled appropriately in the transition of care between the residents. In fact, shift changes between physicians are one of the least studied areas and yet one of the most risky to patient care (Wears et al., 2003a). To achieve continuity requires extensive collaboration and coordination on the part of medical providers. Given the importance of communication in the physician's daily activities, it is interesting to note that medical schools rarely include courses in interphysician communication (Meyers & Miller, 1997). This is not too surprising given the earlier case description. The second type of continuity—management continuity—occurs when multiple specialists are working on a complex medical problem. Although each specialist has the patient's best interest in mind, the collective group of specialists may ultimately work at odds with one another, albeit unknowingly, unless a plan is in place to facilitate the coordination of the medical providers. Finally, relational continuity entails supplying the patient a core staff of providers such that the patient can receive a sense of predictability and coherence (Haggerty et al., 2003).

Continuity of care is viewed as one of the primary dimensions of the quality of patient health care (Greenberg, Rosenheck, & Fontana, 2003). According to Greenberg and colleagues, continuity of care can be defined by three concepts. These concepts begin to define areas in which problems in patient transitions can occur:

1. Regularity of care, which is indicated by the evenness of the health services being provided without a gap in the health care
2. Continuity of treatment across organizational boundaries (e.g., consistency and continuity in the transition from inpatient to outpatient health care services).
3. Provider consistency, meaning that the patient is involved with a few consistently available health care providers

The complexity involved in the continuity of care among health care providers can lead to a number of problems. Some of these factors are technological (e.g., electronic medical record standards), whereas although others are social (e.g., the culture in the medical community that sometimes drives patient care). In both cases, there are opportunities for human factors experts to provide expertise. Some problems include the following:

- Existence of a diverse array of health care providers and settings
- Lack of standardized communication processes
- Varying levels of knowledge and expertise among health care providers in the management of conditions related to continuity of care
- Lack of agreement and inconsistency in the diagnosis and treatment of the patient's condition
- Improper planning and communication for the ensuing continuity of care (Hadjistavropoulas Pierce, Biem, & Franko, 2005)
- Lack of understanding through the narrative and contextual information exchange (Weirs et al., 2003a)

HANDOFFS FOR TRANSITIONING PATIENTS

Shortell (1976) stated that continuity of care involves a coordinated, uninterrupted succession of events. In essence, the patient goes through a series of transitions. In many cases, physicians and nurses handoff a patient to another caregiver. Once they handoff that patient, providers give up responsibility for the patient. The goal of this handoff is to maintain continual care for the patient, "continuity of care." However, from our perspective one needs to understand the transitions taken between medical providers so as to identify the issues that ultimately affect the ability to achieve the goal of continuous care. Thus, we will many times use the

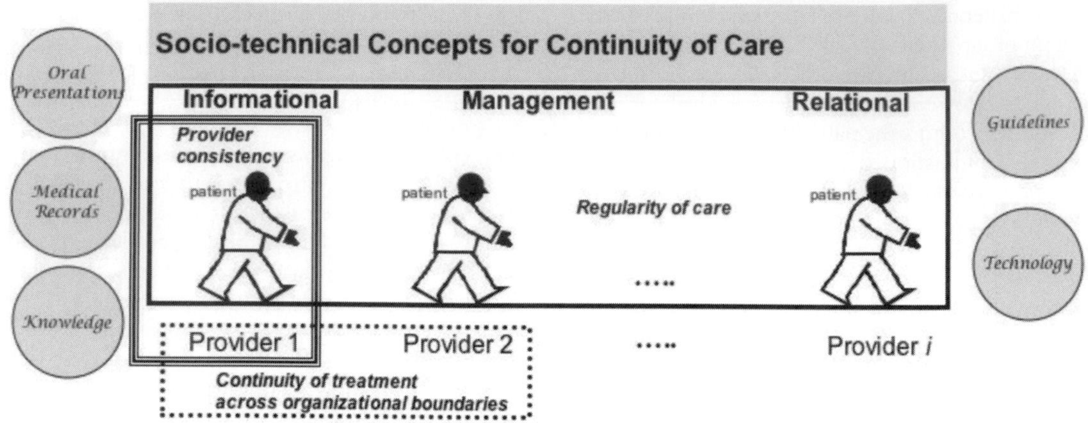

Figure 15–2. Continuity of care transitions.

terms transition or handoff to refer to the process of maintaining continuity of care.

Achieving continuity of care sometimes requires that a patient transition from one health provider to another, as illustrated in Figure 15–2. Pictured are several elements (e.g., oral presentation, medical records), which will be discussed in the chapter, that can ultimately affect the success of patient transitions. It is interesting to note that the characteristics of patient transitions or handoffs have seldom been studied. Roughton and Severs (1996) found that doctors believe their handoff practices need improvement and go on to state, "The lack of advice and guidance on the structure of the handoff has impeded good practice, and a standard of professional practice needs to be set" (Landucci & Gipe, 1999, p. 4).

One may like to think of medical practitioners as members of a team. However, because continuity of care requires patients to transition from one provider or team of providers to another, continuity of care is more akin to shift changes that occur in other industries as opposed to traditional team settings. This should not be misinterpreted to imply medical providers do not work in teams. One can find many settings in which providers do work in teams (e.g., surgical teams, emergency medicine trauma centers). What is evident though is that in many cases a patient is transitioned between either individual practitioners or medical teams. In the case of the patient discussed earlier, several people were responsible for her care, including the five residents, attending physicians (albeit not discussed in the case), nurses, and medical assistants. Although each of these people has specific duties and responsibilities, they are not necessarily working as a team. Instead, as this case illustrates, health care professionals many times are working solely as a group of people. Thus, the authors view this process as more of a handoff. Durso, Crutchfield, and Harvey (2006) outlined general characteristics of shift changes that have the same features as a patient handoff:

- Transfer of responsibility—when a patient is transferred from one entity (e.g., physician, medical department) to another, there is an implied transfer of responsibility or obligation to care for the patient
- Transfer of authority—when a patient is transferred from one entity (e.g., physician, medical department) to another, there is an implied transfer of authority or the power to ensure a patient is cared for
- Minimal co-presence—when a patient is transferred, generally there is little time during which both the receiving and transferring entity are together; in many cases, this co-presence is achieved outside the presence of the patient through either a telephone call or through written documentation
- No division of labor—once responsibility for a patient is transferred, the party

relieved of the patient is generally no longer called on to provide services for that patient; however, there are cases in which a provider may be responsible for delivering some services on a temporary basis (e.g., after surgery nonsurgical orders when patient is transferred to surgery)

- Same environment—typically the environment (e.g., hospital) remains the same in the transition or at least the capabilities provided by the environment are similar
- Fundamental minimal skill set—it is generally assumed that both entities involved in the transition share a fundamental set of skills and knowledge, although their tools, expectations, and assumptions may differ
- Routineness of the transfer—transfers are considered routine because the medical providers have experienced these same types of situations in the past (This element may be fundamentally why handoffs are not taught—they are routine!)

By understanding the elements that begin to define the transitions in continuity of care, one can begin to evaluate where potential errors can occur. Transitions can occur between physicians, medical facilities, and medical teams within a medical facility or medical department to medical department (e.g., the Emergency Department to the Laboratory). Let us take a look at the many human factors issues related to the communication methods during transitions that can affect the quality of patient care as presented in Figure 15–2.

COMMUNICATION AND COORDINATION MECHANISMS USED IN HEALTH CARE TRANSITIONS

The reliability of health care systems depends on people—doctors, administrators, nurses, technicians, aids, pharmacists, accountants, engineers, and maintenance technicians. The health care system is people centered and people driven (Van-Cott, 1994). The interactions and communication between physicians during handoffs are critical to patient care continuity. Each handoff stage has different information-gathering requirements, and communications between physicians may vary. Grusenmeyer (1995) found three communication characteristics of the shift change that would also apply to the medical community and inpatient admission handoffs:

- Communication is based on verbal exchanges, with the purpose of cooperation and coordination to maintain current production or service by exchanging information, instructions, diagnoses, and the like.
- Communication requires representations of the current state by both the incoming and outgoing operators and involves a sharing of expectations of likely future paths that the patient might take.
- Communication requires operators to maintain a shared knowledge and representation of the current system or patient in this case.

Coiera, Jayasuriya, Hardy, and Bannan (2002) found that 90% of information exchange takes place through interpersonal discussion rather than interaction with information sources (e.g., medical records). It was also found that poor communication wastes time, threatens poor patient care, and is the main culprit in preventable adverse events in clinical practice.

Physicians spend a great deal of time interacting with other physicians—seeking advice, referring patients for continuing treatment, and providing follow-up information. Dr. Go (Director of Emergency Medicine Student Education, Assistant Professor of Emergency Medicine, University of Missouri-Kansas City) and Dr. Watson (Clinical Professor in the Division of Pharmacy Practice, University of Missouri-Kansas City) claimed that although physicians and other medical health care providers frequently communicate with each other, professional communication skills are generally neither taught in depth in medical school nor covered in clinical textbooks (Go & Watson, 1997). This problem is not new. Rudd, Siegler, & Byyny (1978) found that in approximately one-third of the physician consultations, either the requesting physician failed to ask specific questions or the consulting physician failed to answer specific questions. Kunkle (1964) found that in more than one-half of the cases, the referring physician gave little or no clinical information.

Despite efforts to outline the characteristics of interphysician communication, no valid research exists as to what constitutes an effective interphysician or internurse communication. Patterson, Roth,

Woods, Chow, & Gomes (2004) found 21 handoff strategies workers used in high-risk environments (e.g., nuclear power plants, ambulance dispatch, NASA mission control) that differed depending on the setting. Wears and colleagues (2004a) only found 9 of Patterson and colleague's strategies used in emergency department (ED) handoffs. Still unanswered in the literature is the importance of specific components of this type of communication. A few of the questions that arise include the following:

- Which is more important—defining a specific question or providing relevant clinical information?
- Does better/improved physician communication result in better patient care? (Go & Watson, 1997)
- Does the structure of interaction (e.g., nature of participants, number of patients involved, and probability of future interaction) of the handoff affect it effectiveness? (Wears et al., 2004b)

In answering these questions, other factors in addition to individual communication might affect the transition of a patient. In fact, one must look at the overall sociotechnical aspects of the work environment. McNeese (2000, p. 164) states "sociocognitive factors help team members make sense of a situation, converge multiple perspectives toward a solution, and transfer knowledge from one context to another." As described earlier, sharing of information and outlining a plan of care for the patient are perceived to be the critical goals of the handoff (or other physician communication). Much of sociotechnical research states "the social and technical subsystems of an organization must be optimized jointly for the greatest overall performance results" (Hammond, Koubek, & Harvey, 2001, p. 37; see the Hendrick chapter in this handbook and the Kleiner chapter in this handbook for further discussion). The "interactions used by teams, including criticisms, opinions, clarifications, and summaries, are key to optimizing decisions" (Hammond et al., 2001, p. 38). Hammond et al. go on to state that modifications to the relations of the social subsystem can have a direct impact on the information that is shared between the people. These possible restrictions (if the communication is inhibited) or opportunities (if communication is enhanced) will directly influence the success or failure of the team. Most of the handoffs that are conducted in the hospitals today take place in a distributed manner; communication typically takes place over the phone rather than face-to-face. Thus, technology also can have a profound impact on the success of teams in meeting their goals (Hammond et al., 2001; Rognin, 2000).

In reality, various methods of communication exist in health care. As is illustrated by the example in Figure 15–3, communication is achieved through many methods and among many people (Matthews, Harvey, Schuster, & Durso, 2002). A sociotechnical perspective suggests that one should evaluate each of these mechanisms to understand how best to facilitate continuity of care, because each method of communication can affect the amount of information transferred or retrieved. The various methods and mechanisms of communication used in transition of care and further discussed here are the following:

- Oral presentation
- Medical records (e.g., physical, electronic records)
- Use of guidelines
- Communication technologies (e.g., phone, video)

Oral Presentations

Oral presentation skills are vital to any transition of care. Communication between physician and patient has received increased interest, and numerous studies have been conducted on this interaction. Less research attention has been paid to communication among health care providers. Haber and Lingard (2001, p. 308) stated, "oral presentation of patient cases provides a vehicle for the collaborative conduct of medical work, the teaching and evaluation of clinical competence, negotiation of professional relationships and the reproduction of professional values." Nursing shift changes and communication strategies have been thoroughly studied (Ekman & Segesten, 1995; Hardey, Payne, & Coleman, 2000; Lamond, 1999; Payne, Hardey, & Coleman, 2000). These studies found communication was incomplete; more information regarding the patient's visit and medical history was contained in the physical chart than was communicated during the oral shift report (Lamond, 1999). Lamond argued that failing to provide accurate and complete patient information during the shift report is a critical

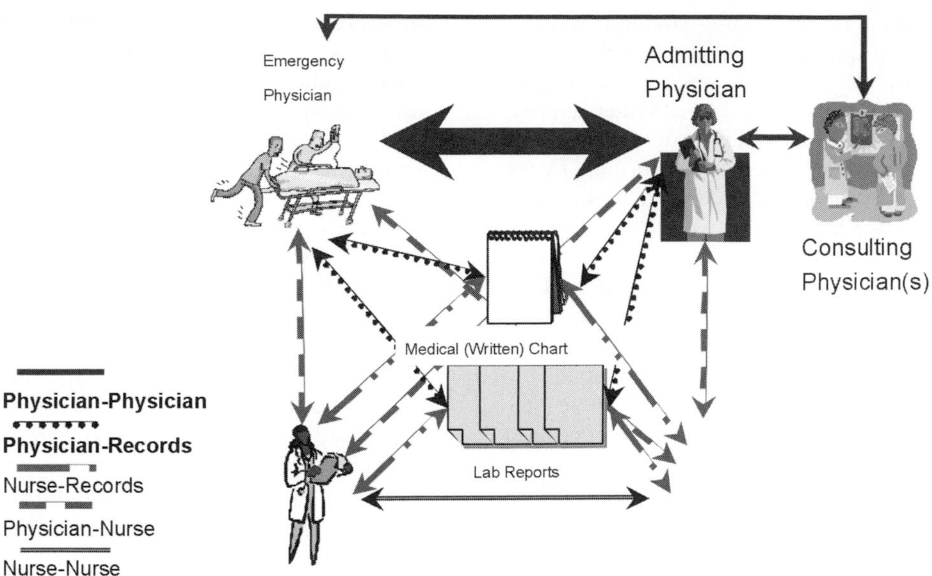

Figure 15–3. Example of communication patterns in an emergency department (Matthews et al., 2002).

communication failure. Payne et al. (2000) found information shared during nursing shift change reports relied more on a nurses' personal notes than on formal patient charts. The ability to transfer this "unofficial" (undocumented) information from one clinician to another should allow for more comprehensive handoffs. However, one should also consider a method in which to capture this information, because it ultimately contributes to a patient's continuity of care.

Haber and Lingard's (2001) study of the perception of oral presentation of third-year medical students and their teachers found that the students described and conducted their presentation according to rigid rules and structure. Their presentation did not change much from the written records. They presented the case in the same manner to the residents and to the attending physicians although the information requirement for both the audiences was different. By contrast, the teachers' presentations were like a conversation between two physicians during which the case was described as a story and the diagnostic was described as an argument. The more expert presenters (interns and residents) changed their presentation according to the purpose of the case presentation. A resident modified the same case for three different contexts: a telephone request for a specialist doctor, an acute care presentation to the Intensive Care Unit (ICU), and a presentation to a medical team faculty physician at attending rounds. The interns followed the strategies of the residents as to what should be included in their postcall rounds presentation. They also learned that in busy postcall context, offering less and letting the resident choose was better than offering more.

Wears and colleagues (2003a) performed naturalistic observations of caregivers' transitions in the ED as part of an ongoing study on safety in emergency care. They cited two cases of transitions—failure and recovery. As an example of transition of failure, the case of a patient who was brought in to the ED for an apparent drug overdose in a suicide attempt, was described. A drug screen of the patient showed a high level of acetaminophen, a drug that produces fatal liver damage 2–3 days after ingestion. Although an antidote, (N-acetylcysteine or NAC), exists, it must be started within 8 hr of ingestion and follows a strict regimen of doses every 4 hr for a total of 17 doses. The ED physician ordered a loading dose of NAC; however, he did not order continuing doses. Thus, the admitting team, as is local practice, was left with completing the task. During shift change, the physician reported to the oncoming physician that the ICU team had been consulted for admission and that

"He (the patient) *is getting* NAC" (emphasis added)." The nurses on shift made a similar report. Because the patient became stable, he was admitted instead to the ward. Orders for NAC were written approximately 7 hr after the loading dose was given. Before the orders were performed in the ED, the patient was transmitted to the ward, and the status of NAC dosing was not mentioned during this transition. The ward nurse did not get to the NAC order until near the end of the night shift and the second dose of NAC was finally given approximately 15 hrs after the loading dose. Thus, the second and third doses of NAC were omitted, and treatment resumed at approximately the time the fourth dose was due. The patient developed liver failure and died.

Wears and colleagues also provided a case in which the transition resulted in a recovery of an error. An elderly woman came to the ED complaining of difficulty walking and back pain. Her initial evaluation showed slight weakness in the left leg, and a computed tomography scan of her brain was normal. She was admitted to the neurology service at approximately 6:30 a.m. with a diagnosis of acute stroke. Because the symptoms were mild and appeared to be resolving, she was not given thrombolytic (blood clot dissolving) therapy. The ED shift change occurred at 0700, and because the patient had not yet been transported to an inpatient bed, her care in the ED was handed over to the oncoming day shift physicians. When the case was discussed at handover, the association of back pain and transient difficulty moving one leg led the oncoming physician to question the diagnosis of stroke and suggest another possibility, aortic dissection (which requires different medical management and is a greater life threat). Further studies were ordered, and this evaluation revealed that aortic dissection was present, leading to a profound change in the patient's management (surgical repair of the damaged aorta instead of medical treatment of stroke). In spite of the recovered error and the change to a more appropriate treatment, the dissection was quite extensive by the time it was discovered. Unfortunately the patient died.

As a follow-on study of ED handoffs, Wears and colleagues (2004b) completed ethnographic observations of caregivers' transitions in five EDs in United States and Canada. They found that at all the institutions the nurse and physician turnovers were performed separately and followed the four phases outlined by Matthews, et al. Both physicians and nurses claimed that the discussion content was different during the two handoffs, although some observers of both handoffs disagreed. Incoming caregivers generally took notes, whereas outgoing caregivers rarely referred to their notes and patients' charts were seldom used. Observed turnovers were interactional rather than transactional. The authors claimed that the efforts to improve the quality of handoffs in emergency care have dealt with their superficial mechanisms rather than the abstract structure underlying them.

In all cases, oral handoffs have positive and negative aspects. During oral representation, any questions about the patient care can be answered immediately, and information that is difficult to understand can be repeated or explained. Nevertheless, sometimes during an oral presentation the speaker and the audience tend to drift away from the main purpose of the interaction and end up talking about personal or unrelated information. In addition, they are sometimes conducted under less-than-ideal circumstances. For example, a shift handoff may require all the shift nurses to be present during the handoff meeting. During the handoff, some of the nurses may be called away because of an emergency and they may not be updated on the care plans. Nurses agree that they perform handoffs from memory, and the information passed on is generally never written or documented. Sometimes important information may be entirely overlooked, or the receiver may neglect it, which results in a serious error (Kennedy, 1999). In most cases, the information handed over is retrospective and states what nurses plan to do rather than what they have done. Feely (1973) stated the shift handoff should concentrate on patients rather than tasks and procedures. In some cases, shift coordinators, who have limited knowledge about patient care, do the handoffs. This can lead to serious misunderstandings of the information.

Another important factor affecting the continuity of care is the communication gap among health care providers. Physicians and nurses have been exchanging data and information about patients both verbally and in writing since the beginning of the profession (Landucci & Gipe, 1998). Organizational messages may exhibit considerable variation in the language, audience, diffusion methods, and purpose as opposed to interpersonal messages (Anderson & Helms, 2000). When clinicians communicate regarding patient information, there is a strong probability that essential information is omitted or incorrect (Landucci & Gipe, 1999). The human tendency not to listen (or to listen poorly) and lack of organization of data are a few of the major factors affecting communication of patient information. Magnitude of fatigue experienced by the health care providers, their anxiety level during communication, quality and complexity of the information being transferred, degree of similarity between cases, and extent of trust and respect also affect the information being transferred (Landucci & Gipe, 1998). Because of excessive workload, prioritization of information takes place. Only the relevant information is given, and large amounts of data are omitted. Of course, there is a chance that important information is being omitted in this process (Anderson & Helms, 2000).

In a study on nonverbal handoffs, Kennedy (1999) found that bedside handoff was more effective than oral presentation. The nurses could identify the patient from the records, they had to document and read all the information, and no information was overlooked or missed. The time that had been spent on the verbal handoff was now used documenting and discussing patient problems. Another advantage was that even if a nurse was off for a couple of days he or she could look through the documentation and be updated quickly. Kennedy concluded that nonverbal handoffs allow better time management, improve continuity of care, and save nursing hours.

Oral presentations are not the only method of communication, although they are one of the primary forms. Sometimes, multiple organizations provide health care to a single patient. Communication and coordination of information in such situations is relatively more complex. Interorganizational communication among multiple providers is one of the core functions of coordination and is vital to patient care (Anderson & Helms, 2000). Written records are one of the primary means organizations use to communicate on a patient (e.g., referral letters, laboratory analyses).

Medical Records

Medical records are a form used for communication among health care providers to provide complete and accurate information. Medical records provide a bridge between the care given to the patient and the evaluation of the care. Medical records help providers maintain continuity of high quality care by keeping the health care providers aware of the patients' status. Documentation of nursing care is the foremost source of information for nurses as well as other health care providers. One of the challenges faced by practicing nurses is the documentation of health care provided to patients. This documentation is the evidence of the care provided to patients and is a prerequisite for safe care. It is estimated that 15% to 20% of the nurses' working time is spent on documentation (Martin, Hinds, & Felix, 1999). Researchers found that the time spent on documentation was one of the main causes of overtime.

The four main purposes of documentation are as follows:

- Communicate client health information
- Facilitate quality assurance
- Facilitate research
- Demonstrate nurse accountability

The documentation of the patient care should be such that the patient's status, care provided, and other clinical data should be represented in a concise and comprehensive manner (Martin et al., 1999). The volume of communication also is affected by the mode of transmitting data among health care providers. Formal structured communication channels increase the amount and richness of the information transferred. The timing of communication is also related to how information is transmitted. One standard of evaluation is how promptly the receiving organization receives the information. Patient records in nonhospital settings reveal the extent of personal communication between health care providers (Anderson & Helms, 2000). In most cases, new health care providers begin treatment without complete information or with some wrong information. There is no continuity of care in this case, which could put the patient's health at risk and increase the already high health care costs (Tessier & Waegemann, 2003). According to Sullivan, Hoare, and Gilmour (1992), interorganizational communication plays an important role in continuity of care. Difference in expectation of referral outcome, confusion about the reason for referral, and vague referral data contribute to problems in continuity (Sparbel & Anderson, 2000).

An advance in technology has given some providers the use of electronic medical records (EMRs) to replace traditional paper documentation and charts. With the advent of the Internet and increasing use of EMRs, communication among health care providers has advanced a great deal and has the potential to improve further. EMRs now make it easier to electronically exchange patient information. EMRs also can play a major role in the patient–physician communication. EMRs store patient information in a database that can be updated as a patient's history changes. EMRs make it easier for a doctor to go through the patient's history, thus increasing the accuracy of the diagnosis and accuracy in prescribing the correct medication. Sending the patient details to other health care providers for consultation is also easier and faster. The introduction of EMRs has had, on the whole, a positive impact on the speed of communication among health care providers (Makoul et al., 2001).

Studies done in Europe showed that physicians used computers 8.5% to 91% of time during their consultations (Richards, Sullivan, Mitchell, & Ross, 1998; Waring, 2000; Watkins, Harvey, Langley, Faulkner, & Gray, 1999). Close study of videotape evidence of physician consultations found that although physicians claimed they used a computer during consultation, only 51% of the physicians actually did so. The Watkins' study also showed that although 91% of physicians claimed they use a computer during a consultation, 75% of them use it to record the problem list and only 36% use it on every visit. Studies have shown that the interface designs that did not involve the end-user physicians during the design have not been accepted completely. O'Dell, Tape, & Campbell (1991) found that the installation of computers in the consultation room resulted in poor acceptance, because of poor typing skills of the physicians and poor workflow integration.

A solution to this problem may be the use of handwriting and pen computing in EMRs. Arvary (2002) found that adding handwriting to a medical record via a pen computer with digital ink increased user satisfaction and usability. When the physicians used a computer to type the details of the case, they were forced to enter the computer environment and completely isolate themselves from the patients. This was not the case when using a wireless pen-based system. The physicians were able to conduct a more fluid conversation.

There are a few negative aspects of EMRs. Patients generally feel comfortable when the physician has his or her full concentration on them. Even a small additional point of focus like a chart may be distracting. Patients also fear that EMRs reduce the confidentiality of their records and can be accessed easily if it is entered in the computer database. Studies also show that the use of EMRs increased the length of the physician consultation yet patient-initiated conversation decreased in comparison with the physician-initiated conversation. In addition, in many cases, a physician's workplace is not designed to accommodate computer hardware. A computer must be placed such that the physician can look at the computer and the patient without shifting positions. An even better design would be to arrange the computer such that the patient can also see the monitor. This gives the patient a sense of security about his medical records. Most physicians are reluctant to use EMRs because of their low efficiency in typing and hence the increase in consultation time (Arvary, 2002; Makoul et al., 2001). Although current standards for EMRs currently do not exist, many companies and agencies are attempting to develop standard means for interacting. In addition, hand-held computing devices may reduce the problems identified with EMRs and patient interaction (Arvary, 2002).

Electronic systems also potentially can negatively affect medical collaboration. Wears and colleagues (2003b) compared two different types of status boards for managing critical information in the ED handoffs in four hospitals. Two of the EDs used electronic tracking systems, and two used manual tracking systems. The manual systems appeared to be a source of interaction among the caregivers especially at shift change and handoffs. Staff members at all levels were seen starting their shift by looking at the manual, and during the shift, they were seen scanning the information on the manual system and discussing the condition of a patient with other co-workers. No such interaction was observed in the electronic tracking system. The electronic system tended to move the work away from a collaborative effort and toward individual work environment. More information could be stored in the electronic system, but it was found that many blanks were left unfilled by the caregivers. It was found that the manual systems were constantly undated by the caregivers, and important notes about critically ill patients were posted.

Acceptance of computers in consultation can be found in nations practicing socialized medicine with strong government advocacy. In countries like the United States, health care is mostly privatized. The use of EMRs is less common because of high cost, lack of standardization, and poor workflow design. In market-based health care systems, physicians' income depends on their productivity. Reengineering the workflow and introduction of desktop computers has reduced physician productivity (Arvary, 2002).

Use of Guidelines in Medical Transition

The U.S. Agency for Health Research in Quality (AHRQ) is the leader in the development of clinical guidelines. This agency defines clinical guidelines

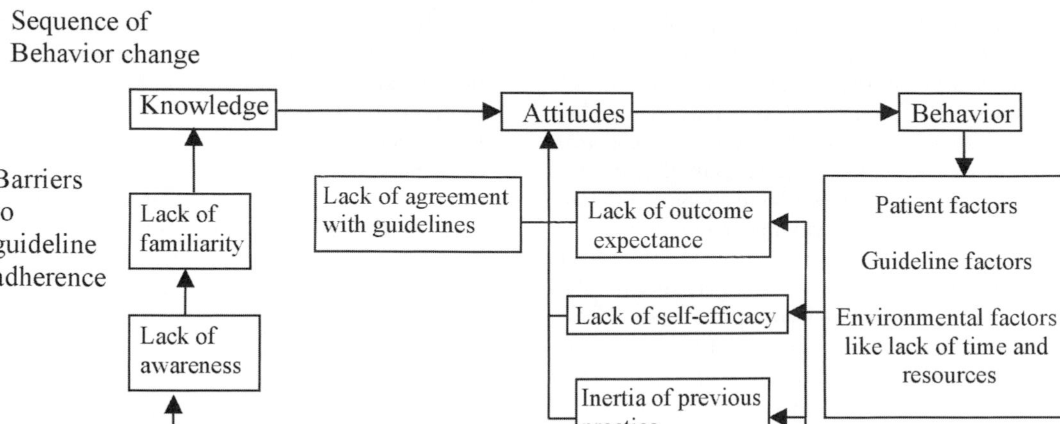

Figure 15–4. Barriers to physician adherence to guidelines in relation to behavior change (Adapted from Cabana et al., 1999).

as "systematically developed statements to assist practitioners' and patients' decisions about health care in specific clinical circumstances" (AHRQ, 2000, p. 1). Keiser and Howard (1998) state that critical pathways (e.g., standards of care and mechanism for multidisciplinary documentation) provide the most efficient and effective method of treatment protocols. Pathways outline the role that all the necessary and appropriate members of the health team provide in managing a patient. As the resources in the hospitals vary, each hospital must develop its own pathway. The three main goals of critical pathways are as follows:

- To improve quality
- To improve efficiency
- To improve costs

Numerous medical organizations and entities have created admission guidelines to be used by physicians as tools to aid in the decision-making process for admission of patient with certain diseases. These guidelines have tremendous potential to be used as a mechanism for communication between physicians, allowing them to address known admission criteria early in the process. Admission guidelines enable the emergency providers to conduct the preplanning and provide the admitting physician with evidence of the need for admission, if required. Based on the foundations of evidence-based medicine, multiple organizations have created disease treatment guidelines, an "integration of the best available clinical evidence from systematic research" (Mulrow & Lohr, 2001, p. 249). These guidelines have the potential both to increase the amount of information shared and to improve the evaluation of the handoff. These guidelines are for guidance, and they need not be followed to the letter. However, they assist in effective information transfer and retrieval and help health care providers focus on the patient's diagnosis. Goldszer and colleagues (2004) found that outcomes for patients with pancreatitis were significantly better when treatment followed pancreatic pathways than when traditional methods were followed.

Although many guidelines were developed to assist physicians in patient handoffs, physicians seldom abide by them. In addition, little information is available as to how the physicians change their practice methods when they become aware of new guidelines. Many barriers hinder physicians' adherence to guidelines. Figure 15–4 displays barriers to physician adherence to guidelines, including lack of familiarity, agreement, self-efficacy, and inertia of previous practice (Cabana et al., 1999)

Communication Technologies

The technology explosion means that communication among health care providers can be achieved via many different modes of communication—video, tape recorder, Internet, and the like. The first recorded telephone call for medical help occurred when Alexander Graham Bell called after he spilled sulfuric acid on himself. Since the invention of the telephone in 1876, it has been used as a tool for delivering health care. Speed, improved access, convenience to patients, and cost savings are the principal advantages of consultations by telephone. Although most patients are happy with this option, as it is time and cost-effective, physicians have mixed feelings about telephone consultation. Physicians prefer to physically see and examine the patient before they diagnose their problems and prescribe medication. Use of the telephone reduces time and effort, yet it has its limitations (Car & Sheikh, 2003).

Although the telephone was really the first technology available to allow health care providers and patients to interact at a distance from one another, other types of technologies are being used today to support continuity of care. Prouse (1995) stated that regardless of the method of communication, a handoff should provide adequate information about patient care to help the other health provider to continue patient care safely and effectively. Tape-recorded handoffs were studied at Trinity Hospice in Clamphan, England. It was found that hospital handoffs, involving the staff of the early and the late shift, cost the organization an average of 12.75 nursing hours at an estimated cost of £80 (approximately $140 U.S.) per report. To reduce costs and the loss of nursing hours, the tape-recorded handoff was introduced. The reporting nurses collected information from their colleagues 2 hr before the shift handoff and started recording one-half hour later. The reporting nurses, registered nurses, took information from the health care assistants, added their own information, and prepared a comprehensive report. The report identified the patients by location, name, age, diagnosis, and reason for admission. Initially, the staff felt that they needed time for questions, for which 10 minutes were allotted. Later this proved unnecessary. The introduction of tape-recoded handoffs proved to be very efficient in reducing the handoff time. Nurses who gave long reports with unnecessary details were now able to present information clearly and concisely. These nurses also reported that the information was clear and concise and that they no longer needed to search for their colleagues long after the shift began to ask questions. It also was observed that 50% more nursing time was available during the handoff periods. Tape-recorded shift handoff proved to be an effective method of communication in this hospital (Prouse, 1995).

With the advent of the Internet, the term "telemedicine" has been coined. Information can be passed among health care providers via the Internet. In the late 1950s, Drs. K. Bird and T. Fitzpatrick of Massachusetts General Hospital, in cooperation with Raytheon Corporation, set up a microwave and closed-circuit television link between the hospital and a health care center at Boston's Logan Airport. At that time, radiographic and dermatologic images and physical diagnosis could be transmitted between the two sites. Drs. Bird and Fitzpatrick were leaders in setting forth a vision of telemedicine. Today the Internet has evolved from an academic and scientific instrument to an instrument for social interaction and business. Information about a patient's history can be sent to specialists via the Internet. This information can be stored centrally or in a messaging database for further archiving. Physicians can use a centralized document storage designed around a common client interface to send and store images from a central location. Teleconferences and videoconferences also enabled consultation between physicians at different locations (including different continents) and allow for the exchange of data or images. The fact that a patient can receive medical care while not residing in the same location as the provider has brought a new era to transition of care that raises many new questions with regard to continuity of care (Crump & Tessen, 1997; Nester, 2001).

Explicit and Tacit Knowledge Differences

Information-sharing methods are critical in the accuracy and completeness of transitions between health care providers. However, it is also important to understand what is being shared between them. The type of shared information will differ depending on whether they are dealing with explicit or tacit knowledge. Patel, Arocha, and Kaufman (1999, p. 75) define *explicit knowledge* as the application of the scientific dimension, "correlating or applying principles in an axiomatic or deductive fashion to a patient's symptoms, yielding a precise diagnostic solution." They go on to describe *tacit knowledge* as using "intuition, experience, and holistic perceptions" (p. 75) to diagnose the patient and proceed with clinical treatments.

Explicit knowledge is the typical curriculum at medical schools in which medical students apply their "book knowledge" to the patients and

	Outgoing Provider			
Responsibility	Yes	Yes	Yes	No
Goals	Job + Patient Synchronization	Job + Patient Synchronization	Job + Transfer Information	
Hand-off Tasks	Identifies pertinent information Retrieves information Organizes information Simplify situation	Identifies pertinent information Retrieves information Organizes information Simplify situation	Communicate information Simplify situation	Possibly Assist Transition
Mental Workload	High: Prospective memory Divided attention	High: Prospective memory Divided attention	Very High: Divided Attention Mutual understanding Communication	Low
Supporting technology	Medical records/EMRs Tape recorder		Telephone Video conference	
Etc				
Grusenmeyer Phases	**End of Shift**	**Arrival**	**Meeting**	**Taking Post**
Responsibility	No	No	No	Yes
Goals	At most, Patient Synchronization	Patient Synchronization + Understanding	Understanding	Job + Understanding
Hand-off Tasks	Information gathering	Information gathering	Encode Information Ask questions	Take over job task Continue information gathering
Mental Workload	Low	Medium: Develop situation awareness (SA)	Very High: Develop SA Divided Attention Mutual understanding Communication	High: Develop SA
Supporting technology		Medical records/EMRs	Telephone Video conference	Medical records/EMRs Tape recorder
Etc				
	Incoming Provider			

Figure 15–5. Issues to be considered in patient transitions (Adapted from Durso et al., 2006).

symptoms they see (Cimino, 1999). Tacit knowledge is obtained through experience and cannot be taught overtly (Patel et al., 1999). Cimino (1999) maintains that physicians acquire much of their tacit knowledge during their residency years. In general, as tacit knowledge increases and supplements explicit knowledge, physicians acquire an expertise that couples medical facts with the gestalt of patient disease management.

The difference between explicit and tacit knowledge, and the expertise that comes with each, greatly affects the sharing of information. As we have seen, explicit knowledge contains direct findings and symptoms, whereas tacit knowledge uses nonanalytic reasoning to consider the patient as a whole rather than the individual symptoms. Cimino (1999) argued that more expert (experienced) physicians have a better understanding of what patient information is critical and what is not. In some cases, the inclusion of concrete physiological findings and test results may take precedence over the physician's impression of the patient, but, if the physician has a number of years of experience, the overall impression of the patient may be more important than the finding (or absence) of abnormal signs. Likewise, information required by physicians may differ on the basis of their level of expertise and the importance they place on the explicit versus tacit information.

With the integration of computers into the workplace, oral shift handoffs, in some cases, have been replaced by computerized documentation that enhances the explicit factual knowledge while potentially hindering the development of tacit knowledge. Hence, one must consider the redesign of the working structures to allow all the health care providers to reflect on their experiences as an individual as well as an organization (Herbig, Bussing, & Ewert, 2001).

IMPROVING CONTINUITY OF CARE TRANSITIONS

Sanders and McCormick (1993) defined a human factors engineer as one who discovers and applies information about human behavior, abilities, limitations, and other characteristics to the design of tools, machines, systems, tasks, jobs, and environments for productive, safe, comfortable, and effective human use. As one can see from this discussion, the transition of patients between health care providers provides a complex environment in which the abilities and limitations of humans are stretched. In addition, the tools and technologies in place do not always meet the needs of the system. Thus, there is still much human factors work to be done.

When one considers the elements that come into play in making these transitions, one might reflect back on Gusenmeyer's phases of an handover. Figure 15–5 was adapted from Durso and colleagues to show some elements that could be studied in evaluating transitions in heath care by human factors experts. The list provided is in no way complete; however, it does begin to provide a framework for how one can consider the role different elements of human abilities and technology may support the handoff process. In addition, it begins to define the task phase characteristics that provider's experience. Figure 15–5 breaks down the handoff process into four phases according to Gusenmeyer's previous work outside the medical domain: end of shift, arrival, meeting, and taking afterward. Mathews et al. (2002) and Wears et al. (2004b) both found support for these phases in the medical domain. In looking at each phase, several dimensions may require exploration, including (a) the tasks being done during the handoff, (b) the mental workload demands during the handoff, and (c) the technologies and potential technologies that might support a particular phase of the handoff. Likewise, one can look at the handoff from the perspective of the outgoing provider and the incoming provider. Both people and teams of people may have different demands or needs. Figure 15–5 begins to define a framework of the work elements and cognitive demands that must be considered in the handoff process. Exploring these phases in detail will help illuminate places that can be improved to ensure continuity of care.

Dealing with transitions in health care will take a combined effort between medical personnel and human factors experts. As health care costs continue to rise at double percentage rates each year, it is important that we as human factors begin to explore more areas in medicine to find solutions that may ultimately help curb skyrocketing costs. In addition, it is our belief that the Institute of Medicine has given us a mandate by outlining the need to "ensure an appropriate exchange of information and coordination of care."

References

Agency for Healthcare Research and Quality (2000). The National Guideline Clearinghouse™ Fact sheet. AHRQ Publication No. 00–0047, July 2000. Agency for the Healthcare Research and Quality, Rockville, MD. http://www.ahrq.gov/clinic/ngcfact.htm.

Anderson, M. A., & Helms, L. B. (2000). Talking about patients: Communication and continuity of care. *The Journal of Cardiovascular Nursing, 14*(3), 15–28.

Arvary, G. (2002). A primary care physician perspective survey on the limited use of handwriting and pen computing in the electronic medical record. *Informatics in Primary Care, 10,* 161–172.

Cabana, M. D., Rand, C. S., Powe, N. R., Wu, A. W., Wilson, M. H., Abboud, P. C., et al. (1999). Why don't physicians follow clinical practice guidelines? *Journal of American Medical Association, 282*(15), 1458–1465.

Car, J., & Sheikh, A. (2003). Information in practice: Telephone consultations. *British Medical Journal, 326,* 966–969.

Cimino, J. J. (1999). Development of expertise in medical practice. In R. J. Sternberg & J. A. Horvath (Eds.), *Tacit Knowledge in Professional Practice: Researcher and Practitioner Perspectives* (pp. 101–120). Mahwah, NJ: Lawrence Erlbaum Associates, Inc.

Coiera, E. W., Jayasuriya, R. A., Hardy, J., & Bannan, A. (2002). Communication loads on the clinical staff in the emergency department. *Medical Journal of Australia, 176*(9), 415–418.

Committee on Quality of Health Care in America, Institute of Medicine. *Crossing the Quality Chasm: A New Health System for the 21st Century.* (2001). Washington, DC: National Academy Press.

Crump, W. J., & Tessen, R. J. (1997). Communication in integrated practice networks: Using interactive video technology to build the medical office without walls. *Texas Medicine, 93*(3), 70–75.

Durso, F. T., Crutchfield, J., & Harvey, C. M. (2006). ATC position relief briefings: Example of the cooperative shift change. *Theoretical Issues in Ergonomics.* Manuscript submitted for publication.

Ekman, I., & Segesten, K. (1995). Deputed power of medical control: The hidden message in the ritual of oral shift reports. *Journal of Advanced Nursing, 22,* 1006–1011.

Feeley, E. M. (1973). The challenge of inter-shift reporting. *Supervisor Nurse, 4*(2), 43–46.

Go, S., & Watson, W. A. (1997). Enhancing physician-physician communication skills. *Academic Medicine, 72*(11), 934.

Goldszer, R. C., Rutherford, A., Banks, P., Zou, K. H., Curley, M., Rossi, B. R., et al. (2004). Implementing clinical pathways for patients admitted to a medical service: Lessons learned. *Critical Pathways in Cardiology, 3*(1), 35–41.

Greenberg, G. A., Rosenheck, R. A., & Fontana, A. F. (2003). Continuity of care and clinical effectiveness: Treatment of posttraumatic stress disorder in the department of veterans affairs. *The Journal of Behavioral Health Services and Research, 30*(2), 202–214.

Grusenmeyer, C. (1995). Shared functional representation in cooperative tasks—The example of shift changeover. *The International Journal of Human Factors in Manufacturing, 5,* 163–176.

Haber, R. J., & Lingard, L. A. (2001). Learning oral presentation skills: A rhetorical analysis with pedagogical and professional implications. *Journal of General Internal Medicine, 16*(5), 308–314.

Hadjistavropoulos, H., Pierce, T. D., Biem, H. J., & Franko, J. (2003). What are the initial perceptions of multidisciplinary personnel new to integrated care pathway development? *Health Care Management Forum, 16*(2), 14–18.

Haggerty, J. L., Reid, R. J., Freeman, G. K., Starfield, B. H., & Adair, C. E. (2003). Education and debate: Continuity of care: A multidisciplinary review. *British Medical Journal, 327,* 1219–1221.

Hammond, J., Koubek, R. J., & Harvey, C. M. (2001). Distributed collaboration for engineering design: A review and reappraisal. *Human Factors and Ergonomics in Manufacturing, 11,* 35–52.

Hardey, M., Payne, S., & Coleman, P. (2000). "Scraps": Hidden nursing information and its influence on the delivery of care. *Journal of Advanced Nursing, 32,* 208–214.

Herbig, B., Bussing, A., & Ewert, T. (2001). The role of tacit knowledge in the work context of nursing. *Journal of Advanced Nursing, 34*(5), 687–695.

Keiser, J. F., & Howard, B. J. (1998). Critical pathways: Design, implementation, and evaluation. *Clinical Laboratory Management Review, 12*(5), 317–332.

Kennedy, J. (1999). An evaluation of non-verbal handover. *Professional Nurse, 14*(6), 391–394.

Kohn, L. T., Corrigan, J. M., & Donaldson, M. S. (Eds.). (2000). *To err is human: Building a safer health system.* Washington, DC: National Academy Press.

Kunkle, E. C. (1964). Communication breakdown in referral of the patient. *Journal of American Medical Association, 187,* 63.

Lamb, G. S. & Stempel, J. E. (1994). Nurse case management from the client's view: Growing as insider-expert, *Nursing Outlook, 42* (1), 7–13.

Lamond, D. (1999). The information content of the nurse change of shift report: A comparative study. *Journal of Advanced Nursing, 31,* 794–804.

Landucci, D., & Gipe, B. T. (1998). The art and science of the handoff: How hospitalists share data. *Cost and Quality Quarterly Journal, 4*(3), 8–10.

Landucci, D. & Gipe, B. T. (1999). The art and science of the handoff: How hospitalists share data. *Hospitalist, 3* (1), p.4.

Makoul, G., Curry, R. H., & Tang, P. (2001). The use of electronic medical records communication patterns in outpatient encounters, *Journal of the American Medical Informatics Association, 8,* p 610–615.

Martin, A., Hinds, C., & Felix, M. (1999). Documentation practices of nursing in long term care. *Journal of Clinical Nursing, 8,* 345–352.

Matthews, A. L., Harvey, C. M., Schuster, R. J., & Durso, F. T. (2002). Emergency Physician to Admitting Physician Handovers: An Exploratory Study. *Proceedings of Human Factors and Ergonomics Society 46th Annual Meeting,* 1511–1515.

McNeese, M. D. (2000). Socio-cognitive factors in the acquisition and transfer of knowledge. *Cognition, Technology and Work, 2,* 164–177.

Mulrow, C. D., & Lohr, K. N. (2001). Proof and policy from medical research evidence. *Journal of Health Politics, Policy, and Law, 26,* 249–266.

Nestor, M. L. (2001). Telemedicine and the Internet: A collaboration and history of communication technologies. *Dermatologic Clinics, 19*(2), 379–385.

O'Dell, D. V., Tape, T. G., & Campbell, J. R. (1991). Increasing physician acceptance and use of the computerized ambulatory medical record. *Proceedings of the Annual Symposium on Computer Applications in Medical Care,* 848–852.

Patel, V. L., Arocha, J. F., & Kaufman, D. R. (1999). Expertise and Tacit Knowledge in Medicine. In R. J. Sternberg & J. A. Horvath (Eds.), *Tacit Knowledge in Professional Practice: Researcher and Practitioner Perspectives* (pp. 75–99). Mahwah, NJ: Lawrence Erlbaum Associates, Inc.

Patterson, E. S., Roth, E. M., Woods, D. D., Chow, R., & Gomes, J. O. (2004). Handoff strategies in settings with high consequences for failure: Lessons for health care operations. *International Journal for Quality in Health Care, 16*(2), 125–132.

Payne, S., Hardey, M., & Coleman, P. (2000). Interactions between nurses during handovers of elderly care. *Journal of Advanced Nursing, 32,* 277–285.

Prouse, M. (1995). A study of the use of tape-recorded handovers. *Nursing Times, 91*(49), 40–41.

Richards, H. M., Sullivan, F. M., Mitchell, E. D., & Ross, S. (1998). Computer use by general practioners in Scotland. *British Journal of General Practice, 48,* 1473–1476.

Rognin, L. (2000). Cooperation, reliability of socio-technical systems and allocation of function. *International Journal of Human-Computer Studies, 52,* 357–379.

Roughton, V. J., & Severs, M. P. (1996). The junior doctor handover: Current practices and future expectations. *Journal of the Royal College of Physicians of London, 30,* 213–214.

Rudd, P., Siegler, M., & Byyny, R. L. (1978). Perioperative diabetic consultation: A plea for improved training. *Journal of Medical Education, 53,* 590–596.

Sanders, M. S., & McCormick, E. J. (1993). *Human Factors in Engineering and Design.* New York: McGraw-Hill.

Shortell, S. M. (1976). Continuity of medical care: Conceptualization and measurement. *Medical Care, 14*(5), 377–391.

Sparbel, K. J., & Anderson, M. A. (2000). Integrated literature review of continuity of care: Part 1, Conceptual issues. *Journal of Nursing Scholarship, 32*(1), 17–24.

Sullivan, F. M., Hoare, T., Gilmour, H. (1992). Outpatient clinic referrals and their outcome. *British Journal of General Practice, 42* (356), 111–115.

Tessier, C., & Waegemann, C. P. (2003). The continuity of care record. Organizations unite to create a portable, electronic record for use at transfer points in patients' care. *Health care Informatics: The Business Magazine for Information and Communication Systems, 20*(10), 54–56.

Van Cott, H. (1994). Human errors: Their causes and reduction. In M. S. Bogner (Ed.), *Human Error in Medicine* (pp. 55–67). Hillsdale, NJ: Lawrence Erlbaum Associates, Inc.

Waring, N. (2000). To what extent are practices "paperless" and what are the constraints to them becoming more so? *British Journal of General Practice, 50,* 46–47.

Watkins, C., Harvey, I., Langley, C., Faulkner, A., & Gray, S. (1999). General practitioners' use of computers during consultation. *British Journal of General Practice, 45,* 381–383.

Wears, R. L., Perry, S. J., Eisenberg, E., Murphy, L., Shapiro, M., Beach, C., et al. (2004a). Conceptual framework for studying shift changes and other transitions in care. *Proceedings of the Human Factors and Ergonomics Society 48th Annual Meeting, New Orleans, LA,* 1615–1619.

Wears, R. L., Perry, S. J., Eisenberg, E., Murphy, L., Shapiro, M., Beach, C., et al. (2004b). Transitions in care: Signovers in the emergency department. *Proceedings of the Human Factors and Ergonomics Society 48th Annual Meeting, New Orleans, LA,* 1625–1628.

Wears, R. L., Perry, S. J., Shapiro, M., Beach, C., Croskerry, P., & Behara, R. (2003a). Shift changes among emergency physicians: Best of times, worst of times. *Proceedings of the*

Human Factors and Ergonomics Society 47th Annual Meeting, Denver, CO, 1420–1423.

Wears, R. L., Perry, S. J., Shapiro, M., Beach, C., Croskerry, P., & Behara, R. (2003b). A comparison of manual and electronic status boards in the emergency department: What's gained and what's lost? *Proceedings of the Human Factors and Ergonomics Society 47th Annual Meeting, Denver, CO*, 1415–1419.

Zuger, A. (2002). Sleep won may come at a price. *New York Times on the Web.* Retrieved March 23, 2006 from http://query.nytimes.com/gst/fullpage.html?sec=health&res=9A01E0DB1E3CF93BA25755C0A9649C8B63

16

RELIABILITY ENHANCEMENT AND DEMISE AT BACK BAY MEDICAL CENTER'S CHILDREN'S HOSPITAL

Karlene H. Roberts, Vinit Desai, and Peter Madsen
University of California, Berkeley

High reliability organizations (HROs) are organizations in which errors can have catastrophic consequences, but in which errors are avoided most of the time. In these organizations, error avoidance and safety are as much a part of the bottom line as is productivity. Massive manmade catastrophes are impossible for individuals to produce, because individuals lack the resources needed to do so. Thus, manmade catastrophe requires organizations. Increasing technological innovation also contributes to the possibility of catastrophe because our technologies are sufficiently complex that mere humans can easily mismanage them.

The purpose of this chapter is to define HRO processes and then detail the story of their successful application and ultimate defeat in one healthcare organization. Over the course of doing research designed to uncover what it takes to be an HRO and observing the application of HRO processes in a number of organizations, it has become quite apparent that building and keeping an HRO is an enormously difficult process. It requires continual attention to making certain HRO processes do not disappear.

HRO PROCESSES

Research on HROs has been conducted since the mid-1980s when a group of scholars at the University of California at Berkeley coined the term "high reliability organizations" and headed off to study three organizations in which errors are unacceptable. Those three organizations were a commercial nuclear power plant, the U.S. Federal Aviation Administration's operation of its air traffic control system, and the U.S. Navy's aircraft carrier aviation program. Other scholars in other places have joined this research effort, and yet other scholars do similar research under slightly different rubrics. By this time, the concepts derived from the work are many, and it is difficult to categorize them. This is possibly because the scholars in the area have drawn concepts from both micro and macro organizational theory and have not been very clear about the crossover.

Most recently some of them (see for example, Grabowski &Roberts, 1999; Roberts, Madsen, & Desai, 2005; Tadmor & Roberts, 2006) have moved away from focusing on single organizations, finding that errors often occur across boundaries between organizations or in other organizations that influence and cause errors in a focal organization. By focusing narrowly on a single organization, the error and its consequences may be missed. For example, patients are regularly transferred in hospitals from emergency to acute and less acute ca and from hospitals to long or short-term care ar then to home care. Each of these junctures offe numerous possibilities for error.

Another example of this is a hypothetical petroleum company identified as an HRO. Its product provides fuel for a major aviation disaster. The aviation system (consisting of air traffic control personnel, airline flight crews, maintenance crews, and others) is likely not an HRO in this situation. Is the petroleum company? The errors occurred elsewhere, and the petroleum company likely bore no direct or even partial technical responsibility for the accident. It does bear a social relationship with the aviation system, and its fuel now blankets an accident site. By focusing on the innocent petroleum company, we miss the larger picture of unreliability. The system level work is, as yet, too underdeveloped to discuss with any confidence and was unavailable at the time the organization analyzed here was in the process of enhancing its reliability.

The rest of the research on HROs focuses on either organizational processes or cognitive processes as they foster and define reliability enhancement. A number of studies examined the organizational structural processes that contribute to reliability enhancement (e.g., Eisenhardt, 1993; Bigley & Roberts, 2001; Reason, 1997; Roberts, 1990).

Eisenhardt (1993), for example, found that there is simultaneous centralization–decentralization in organizations in what she calls "high velocity environments." It is in the form of a pattern she calls "consensus with qualification," which is a two-step decision process in which everyone in the organization affected by a decision tries to reach consensus on it but when that cannot occur, the decision is made centrally. Roberts (1990) finds that HROs move back and forth between centralization and decentralization. When nothing important is transpiring, they can afford hierarchy but as the tempo heats up and complex interactions are required, centralization breaks down. Reason, too, (1997) discussed flexibility and shifting structures.

Bigley and Roberts (2001) focused on three structural mechanisms that give the organization flexibility. These are structuring mechanisms to support flexibility, organizational support for constrained improvisation on the part of lower level people, and cognition management methods that, in combination, lead to exceptional organizational reliability under volatile environmental conditions. Structuring mechanisms include elaborating the structure, switching roles within it, migrating authority, and resetting the system as external conditions change. Cognition management has to do with the way organizational members are able to build and maintain viable understandings of the activity system to which they belong.

The Bigley and Roberts work is closely associated with Weick's notions of mindfulness and sense making in organizations (Weick, 1995). Weick and Roberts (1993) developed the concept of collective mind to explain how organizations such as nuclear-powered aircraft carriers produce reliable performance. Collective mind is conceptualized as a capacity for heedful interrelating. Variation in this capacity affects the ability of a system to maintain a working similarity between itself and a complex dangerous environment. The more developed the capacity for mindful performance is, the greater the comprehension is and the fewer errors that result. Other writers talk about distributed cognition and how cognitions are melded together in organizations to complete tasks and reach goals. Often these writers take a decision-making rather than a structural perspective, as Weick and Roberts do (See for example, Hutchins, 1995; Klein, 1989).

Consistent with this intraorganizational focus is the work on crew resource management coming from the airline industry (e.g., Helmreich & Foushee, 1993). Crew resource management takes a variety of forms and is essentially a mechanism for reducing the authority gradient and improving communication and decision making in flight crews.

In addition to these intraorganizational processes, some human resource management strategies that contribute to reliability enhancement were identified. In her Ph.D. thesis, based on data collected in banks, Carolyn Libuser (1993) identified five human resource processes that contribute to reliability. To identify these processes, she compared failing and successful banks. These processes are process auditing, using appropriate reinforcement (Kerr, 1975), trying to be the gold standard in the industry, perceiving risk appropriately, and command and control. Elements of command and control include migrating decisions to people with appropriate information regardless of rank, providing a redundancy of people or hardware, developing situational awareness or "the big picture," adhering to formal rules and procedures, and providing constant training.

In summary, research on HROs is now done at a variety of universities. It generally focuses on cognitive or organizational processes that contribute to reliability enhancement. Suggestions from this work are slowly turning researcher attention to the larger systems in which these processes occur, but this work is under developed. We will examine a health care organization that adopted some of these organizational processes to enhance reliability.

BACK BAY CHILDREN'S HOSPITAL[1]

At the time the organization adopted high reliability processes, Back Bay Children's Hospital (BBCH), with 250 beds, was the tertiary children's hospital for a geographic area more than three times the size of Vermont. The population was 2.5 million people with 500,000 under the age of 15. The catchment area included urban, rural, and wilderness areas, with a large number of desert and mountain communities. BBCH is attached to a large university medical center that is the teaching facility for the Schools of Medicine, Nursing, and Allied Health Professions.

In 1988 and 1989, the co directors of BBCH's Pediatric Intensive Care Unit (PICU) joined the hospital staff and began instituting HRO principals. The PICU had 25 beds, with an average daily census of 21, 9 on ventilators. A total of 105 registered nurses were assigned to the PICU, with 14 on duty at any one time. Twenty respiratory care practitioners (RCPs) were assigned, with four working at any one time. Four resident physicians (residents) rotated through the PICU for 1 month at a time, one from emergency medicine and three from pediatrics. Pediatric residents received 2 months of PICU experience, emergency medicine residents received 1 month. Occasionally an anesthesia resident spent 2 consecutive months in the PICU. Fellows from pediatric emergency medicine or pediatric anesthesia spent time in the PICU supervising resident physicians.

Five attending physicians, specialists in pediatric critical care (pediatric intensivists), rotated through the unit for 1 week at a time. One intensivist was the PICU's medical director.

Medical care was provided by the intensivist who saw each patient on twice-daily rounds. Morning rounds were for teaching, and the group consisted of all residents, the fellow if on service, lead respiratory therapist, charge nurse, pharmacist, social worker, and the patient's bedside nurse and respiratory therapist. Students also accompanied the team. Students came from nursing, respiratory care, medicine, social work, and pediatrics. Afternoon rounds were work rounds to review the day's progress and response to therapy and discuss the plan for the night. Generally, only the attending physician and the two on-call residents made these rounds.

During rounds, the patient was presented to the group for discussion of the diagnosis, general plan of treatment, potential problems, and the family's response to the situation. All participants had an opportunity to present their perceptions and ideas, and questions were solicited. As a general rule, the team did not move on until all caregivers felt comfortable with the plan. Rounds lasted 3 to 4 hr in the morning. Afternoon rounds lasted approximately 1 hr.

The PICU had 1704 admissions in 1996, making it one the largest PICUs in the United States both in number of beds and admissions. PICUs of 4 to 6 beds made up 40% of PICUs, those with more than 18 beds made up fewer than 6% at the time. The latter averaged 1277 ± 63 patient admissions per year (Pollack, Cuerdon, & Getson, 1993). Pollack and colleagues (1993) found mortality rates of 7.8 ± 0.8% for PICUs with more than 18 beds. The PICU at BBCH had a 5.2% mortality rate in 1996.

Approximately half of the admissions came through BBCH's pediatric critical care transport system, now one of the larger transport services in the country (McCloskey & Johnston, 1990). Most (75%) pediatric specialized transport systems transport fewer than 400 patients per year, with the largest transporting 720 patients. The year of the McCloskey study, BBCH pediatric critical care transport brought 599 children to the PICU. This did not include neonates transported to the Neonatal Intensive Care Unit. In 1996, BBCH transported 871 children.

Before HRO efforts, transported children had approximately twice the mortality rate as those admitted from within the institution (via the Emergency Department, Operating Room, or Acute Care Ward). McNab (1991) found posttransport mortality to be 11.5% when he reviewed 130 charts of children with decreased level of consciousness. In the first 6 months of 1990, three

[1]Back Bay Children's Hospital is not the name of the organization studied. Data for this study were drawn from archives, written documents, participant observation, and interviews with key personnel.

BBCH children went into cardiac arrest and died during transport. In 1991, there was one deterioration caused by cardiac arrest, which did not result in death.

In the spring of 1993, BBCH experienced three difficulties with endotracheal intubation (placing a breathing tube into the windpipe) while on transport. These difficulties had the potential for catastrophic effects on the child. There were no complications or physiologic deteriorations during transport in 1996 (871 transports).

Between 1989 and 1996, the PICU became highly reliable. It grew in total admissions, average daily census, and average daily ventilators. As admissions and acuity (average daily census of ventilators) increased, the mortality rate remained between 3.3% and 5.3%.

During this time, accidental extubations also decreased. (This is the dislodging of the tube that enters the windpipe and connects to the ventilator tubing.) Such events can lead to cardiac arrest and consequent death or severe brain injury. BBCH's PICU rate compared favorably with published data (Little, Koenig, & Newth, 1990). Consequential events dropped to zero in 1995. (Consequential events refer to events that lead to an increased level or amount of care, neurologic injury, or death.)

Pharmacists joined the medical team rounds in 1991. Drug errors were then lower in the PICU than on the general ward (Furr & Van Stralen, 1995). Even though the rate for each is less than 1%, the potential for harm is far greater in the PICU. A critically ill child frequently receives drugs that are harmful, if not lethal, when given as the wrong dose or wrong drug.

RCPs had been an integral part of rounds since one of the PICU directors came in 1989. Their participation grew into a system of protocols and specialized, weekly rounds. The goal of these rounds was to manage the mechanical ventilation of patients. As a result of ventilator rounds, patient time on the ventilator decreased from 1993 to 1994 (Scott et al., 1997). The longer a patient is on a ventilator the greater the likelihood of a complication. Complications include lung damage, pneumonia, and accidental extubation. For these reasons, intensivists adjust the ventilators to the mildest settings that will support the child and wean as soon as, and as quickly as, safety allows.

Respiratory therapists were assigned from a core group of experienced people. If the needs exceeded the resources of this group, therapists were drawn from the Neonatal Intensive Care Unit core group and lastly from the adult ICUs. Therapists were generally selected from those with the most experience and skill that was age specific for the patient population. Premium was also placed on their ability to work with the greater degree of decision-making responsibility provided in the PICU. Turnover for the Department of Respiratory Care was approximately 15%; for the PICU it was also 15%. However, the turnover was most commonly for job advancement, with PICU experience increasing academic opportunities (e.g., medical, dental, or nursing schools) or professional advancement at other hospitals. This turnover occurred in a small portion of therapists, as almost 80% were with the institution for more than 5 years.

Nurses began formal presentation of the patient on rounds in 1994. This process brought the nurse and the nurse's concerns and impressions into the medical plan. Nurses became aware of the medical risks to the patient and the risks and plans of the therapies to be used. A clinical dietitian joined rounds on a regular basis in 1994 to review the nutritional status and care of the critically ill or injured child.

The PICU had more long-term employees (greater than 5 years) than other ICUs at the medical center. The turn over was approximately 5%, which tended to be lower than adult ICUs. Institutional stress was lower, although there tended to be greater stress from the nature of the patients cared for in the PICU. When nurses left, they tended to leave for personal reasons outside of the job or for academic or professional advancement. In recent years, there was a significant increase in PICU nurses obtaining baccalaureate or master's degrees.

HOW THEY DID IT

The overall philosophy of the PICU was to support the bedside caregiver, particularly the nurse in an environment that has numerous social and psychological hazards. Teamwork and team formation were fostered. Shaming, naming, and blaming, particularly after a bad outcome, were not accepted. Objectives were developed for each patient and problem. Freedom to safely try interventions to identify what worked for a particular patient was encouraged. Attending physician support was always available, either immediately by phone or within 20 min in person. Deficiencies in care were used as teaching opportunities because every caregiver was valued as a long-term member of the team. There were many ways to approach the care in the PICU, and various services were able to pursue their own unique philosophies. All of this is consistent with HRO principles.

By adopting this philosophy, the management of BBCH's PICU fostered a setting in which there was simultaneous centralization and decentralization. This setting provided a foundation for structural mechanisms that support flexibility, constrained improvisation, and developed shared cognitions about the meaning of unfolding events. A reason to use teamwork and to avoid shaming, naming, and blaming is to foster the development of collective mind. Only when the probability of people engaging in defensive behaviors is reduced and possibilities for rich cognitive representations of ongoing events are opened up will heedfulness and collective mind result.

Although HRO processes may well play themselves out differently in different situations, BBCH's staff directly applied the Libuser principles (1993). The rich philosophical tapestry laid down in the organization allowed for the application of more specific HRO processes. Described here is an accounting of how each of Libuser's components was addressed at BBCH.

Process Auditing

Process auditing refers to systematic checks and formal audits to inspect for problems in the "process." At BBMC's PICU, the process was providing critical care medicine in an environment of physiologic uncertainty and instability. The whole staff constantly entertained the thought they had missed something. They encouraged questioning and the presentation of data that supported or refuted working hypotheses. Any team member could question care at any time. Answering questions educated everyone. This often led to immediate changes in patient care. Process auditing also occurred when the team systematically reviewed all cardiac arrests, deaths, accidental extubations, medication errors, and incident reports.

Appropriate Rewards

Appropriate rewards were made to encourage participation in patient care. All disciplines participated in seeking to lower accidents and stress and improve morale of caregivers. Often rewards were intrinsic. As team members demonstrated knowledge, insight, and discretion in patient care, they tended to play a greater role in tactical and strategic management. Their opinions were more frequently sought and incorporated into care plans. Attending physicians listened to nurses and listened to them in public.

Quality Review

Quality review was performed to ensure the PICU had the lowest rate of potentially preventable mortality and morbidity. Quality improvement reviews were a part of process auditing in formal, standing committees of the institution. When an event produced consequential injury, the administration (medical center and medical staff) became involved. This was after the fact but with the intent of using such a situation as a marker of deficiency that needed review. Quality referent levels were adopted from nationally accepted norms and the medical, respiratory care, and nursing literatures.

Risk Awareness

Risk awareness was a more significant problem early in the growth of the PICU, because caretakers had to become accustomed to looking for risk. It remained a problem later as the PICU introduced new, high-risk, therapies. In the first few years, as the acuity of patients increased, the staff (residents, nurses, and respiratory therapists) often missed covert signs of compensated physiologic dysfunction. Although unstable physiology was at first frequently unrecognized, when the situation was explained, the staff immediately recognized its severity.

The directors of the PICU used this experience for bedside teaching. They began a program of in-service lectures specific to the various disciplines (nursing, respiratory care, resident physicians) and developed two regularly scheduled conferences, one directed to emergency medical service providers, and the other directed to nurses in emergency departments and ICUs. It then became rare for a patient to unexpectedly deteriorate in the PICU. The team generally worked to remember what went on when things went wrong. This is contrary to caring for a child by only remembering when care was done correctly.

Command and Control

Command and control played a role in providing BBCH's PICU its most successes. This concept includes decision migration that overcomes an authority gradient, redundancy, situational awareness, rules and procedures, and training.

PICU management prescribed *decision migration* to the best-qualified team member, given the particular situation. Particularly during time of emergency, the most qualified person to make or guide decisions is the bedside caregiver. Frequently, the team could not predict what would work in a specific situation. The member present could observe for a response to therapy. Quick decisions can bring stability to a rapidly changing situation. To allow this freedom to adapt to the evolving situation, the attending physician or charge nurse often engaged in decision making of a more strategic nature. This involved developing situational awareness and the ability to see into the near future for planning.

Decisions particularly migrated during busy events such as multiple resuscitations. In these cases, a small group of staff were given a few objectives to reach and several alternative therapies to use in reaching the objective. This allowed the attending physician to psychologically back away from the immediate multiple case situation to focus on the more critically ill child and simultaneously retain situational awareness.

The authority gradient that often occurs between physician or surgeon and other team members can lead to tragedy. Often when an authority gradient comes into play, the person further down the gradient responds with anger or indifference or fails to respond. This is typical in nurse–doctor situations. To counteract this problem, the nursing staff used a form for professional interactions. These forms moved up the chain of command from the nurse to administration. They then moved downward to the physician involved through his or her chain of command. This insulated the lower level caregiver from reprisal.

The team engaged in a variety of decision strategies borrowed from public safety and military decision making, in addition to using the medical decision-making model. For long-term care, they adopted the medical model of gathering information, processing it, and coming to a conclusion. With the correct diagnosis, the medical care given will result in a good outcome. This is a deterministic system, in which A determines B. Often, however, necessary information is not available or is hidden in the noise of the situation; this apparently random variation characterizes the situation. Such situations are called severely stochastic, and in these situations, the team used an emergency decision-making model borrowed from the military and public safety. The objective is first clarified. Then decisions are made to bring some control, to reaching the objectives. The caregiver learns what works by action. This closed loop process was borrowed from Boyd's OODA loop (observe, orient, decide, act; Mason, 2003). When the situation is uncertain actions are directed at identifying the structure of the problem. The OODA loop teaches what works through actions.

Redundancy

Redundancy ensures thoroughness in evaluating the patient and in choosing a therapy. Many signs the team monitored were measured by two methods, and during resuscitations, several team members monitored the same vital sign. Therapies were also monitored for success or failure using several methods. Redundancy is able to detect subtle, early heralds of change and to ensure that small, random changes in patient condition do not lead to major changes in therapy. A subtle sign of deterioration (e.g., respiratory dysfunction) generally presents as a pattern that can be monitored. This reduced the incidence of hypervigilance; that is, responding strongly to subtle changes that occur not because of the disease but that are random. Redundancy may cause a narrowing of attention to the wrong problem.

Patient alertness, respiratory rate and rhythm, oxygen saturation, work of breathing, and air entry were used to monitor the respiratory system. Monitors include nurses' recording vital signs, respiratory care practitioners' recording respiratory findings and blood gas measurements, and physicians' conducting a physical examination and data interpretation. Any change in a finding will alert caregivers to look further, either with closer inspection or calling another individual (e.g., a nurse calling a respiratory care practitioner to evaluate the chest). Data begin to accumulate that indicate a random change and, therefore, no emergency, or the accumulation of data begins to paint the picture of respiratory dysfunction. Further data will develop into a pattern of impending respiratory failure, upper airway obstruction, or lower airway obstruction. This redundancy helps filter noise and allows identification of early heralds of deterioration. The latter permits earlier intervention when therapies are safest and most effective.

Situational Awareness

Situational awareness came both with experience in the PICU and experience as a supervisor. The absence of situational awareness occurs when caregivers narrow their attention to their own patients and, frequently, to some other alarm with greater signal strength. What is perceived as greater signal strength is idiosyncratic to the individual. It could be newer staff such as an inexperienced nurse who focuses on fever rather than signs of septic shock or an inexperienced respiratory therapist who focuses on carbon dioxide level in the blood rather than compensation to a normal pH. Members of the PICU team had a sense of the function of the whole unit. In numerous instances when a child began deteriorating, the right number of staff arrived. They came from several patient pods away to assist. This almost uniformly occurred without a call for assistance or a Code Blue.

The existence of **rules and procedures** allowed respiratory therapists and nurses to influence the medical care to a great degree and engage in quick responses to changes. Therapist-driven protocols permitted respiratory therapists to take into consideration the patient's past responses and expected responses while deciding therapy. Many times, the bedside therapist developed a feel for a given patient that the resident did not have. Consequential events from accidental extubations also decreased as the nurses used the protocols for sedation and paralysis to give the patient more comfort and safety.

Training

Because they were a part of a teaching institution, and one that develops new therapies, PICU staff members had the goal of always considering themselves in **training**. Consequently, they observed each other's performance and gave assistance through mutual teaching and learning. Formal training was also a part of the system, as discussed earlier.

Summary

From 1989 to 1999, the PICU outcome data all went in the direction hoped for in a highly reliable organization. Admissions, daily census, ventilator use, and transports went up, and mortality, consequential events, and calls refused went down. Things were clearly moving in the right direction!

A TURN AROUND AT BACK BAY CHILDREN'S HOSPITAL

Between 1999 and 2000, both proponents of reliability enhancement at BBCH's PICU left the unit. Other pediatric intensivists were put in charge. They strictly adhered to the medical model of patient care. To expand on our earlier mention of this model in both a legal and a billing sense, the physician is fully responsible and fully accountable for patient care. The management of care is determined by the physician, and nurses cannot make diagnoses. If a nurse generates an order and it is incorrect, he or she can lose his or her license. The things nurses and RCPs can decide are very circumscribed. Even verbal orders must be signed off on by the physician. Adherents to this model (almost all physicians) focus on the things they do correctly and almost never mention failure. Failure is seen by physicians as a character flaw and not a hint that something is inherently wrong with the system in which they operate. Failures are brushed under the rug and not discussed. The physician is the central and final authority. This model is a part of the medical culture complete with its extreme authority gradients. Physicians are within their legal duties to perform in this manner.

But what is the situation in the PICU? The PICU is like a number of organizations that try to operate as reliably as possibly, because failure to do so has catastrophic outcomes. Ninety-five percent of the time, the situation can be characterized as boring; the other 5% of the time, it is sheer terror. During 95% of the time, the medical model of decision making works pretty well. Situations are reasonably linear, and the medical model of gathering information, processing it and coming to a conclusion, then adopting a plan, further refining diagnosis and further refining the plan, works well. It does not work at all well for those fluid, indefinable, and confusing situations that characterize the other 5% of the situations. Other high reliability organizations have found that when nothing is happening, a hierarchical organizations works well and is perhaps even the most efficient and effective structuring strategy. But when activity increases and uncertainty abounds, this kind of hierarchy breaks the organization (see, e.g., Eisenhardt, 1993; Roberts, 1990; Weick, 1987).

At BBCH'S PICU immediately after the two protagonists of high reliability operations, left the new intensivists did away with the high reliability strategy because it is diametrically opposite of the medical model, and possibly they failed to understand it. At first, nurses and RCPs continued on with processes to enhance reliability, but after awhile this, too, went by the wayside. Turnover increased, resulting in the loss of organizational memory, and management did not reward reliability-enhancing behaviors among caregivers.

First, the structural mechanisms that contribute to flexibility, such as elaborating the structure, changing roles, and migrating authority were eliminated. The purpose of the medical model is to rely on hierarchy with the physician at the top. Discarding structural flexibility contributes to the removal of at least two other processes, migrating decision making or "consensus with qualification" and cognitive efforts to build mindfulness and sensemaking. Teamwork, such as that required by the airlines and taught through crew resource management programs, was also eliminated. The mere fact of making such changes removes the possibility of developing rich cognitive representations of what is going on.

But what happened to the human resource processes so carefully put into place? It is doubtful that unit carefully **audited its processes**. One piece of evidence that it did not is that in 1999 there were three deaths on the unit but in 2000 there were six, in 2001 eight, 2002 eight, and 2003 ten. One would think that if the unit was aware of this steady change it would have addressed the issue of why. Asking questions was not encouraged. In fact, people were at least implicitly punished for doing so.

The **reward** system was dramatically changed. Under the new system, people are rewarded for keeping quiet, not for questioning activities on the unit. In fact, one RCP was criticized for suggesting a different way to change a ventilator.

It does not appear the unit is looking either to itself or outside itself for a **high quality standard**. A reason to admit a child to the PICU was that the child's doctor requested it. Physicians report that they are unable to get their patients into the unit. This policy must reflect on the standard of care provided (or not provided) to sick children. In addition, 15 to 20% of the children released to a lower level of care returned within 24 to 48 hr. This, too, reflects a standard of care provided in the unit.

Risk perception does not appear to have been high. In the first place, the mortality rate steadily increased in a 3-year period. Notice of this trend should have been taken in the first year, with ameliorative actions immediately taken. When a patient is deteriorating, there is no national standard for treating the patient. Today, in this PICU, physicians often do not come in for resuscitations. Yet there is some window in the resuscitation process in which a patient can often be turned around.

What about **command and control**? We have already seen that decisions no longer *migrate* around the unit until they are made by the people with the best knowledge about what is occurring (often the bedside caregiver). There is currently high turnover of RCPs on the unit, suggesting that the positive values of *redundancy* may be undermined. The deteriorating teamwork also reduces redundancy. People in the unit complain of high stress, which may well be related to the high turnover.

After the high reliability protagonists left the unit, the hospital began to hire respiratory care therapists with little training. Combined with a reward system that positively rewards lack of questioning automatically leads to a situation in which people are unable to develop *situational awareness or* see *the big picture*. In addition to using respiratory care therapists with less training, no one was trained in the fluid decision-making processes characterized by the military and public safety. Finally, there cannot be constant on-the-job training if people are encouraged not to think about or talk about errors.

It appears that by failing to implement reliability-enhancing processes, BBCH's PICU also put itself in harms way vis à vis patient outcomes. Admissions, daily census, transports, and children on ventilators went down while mortality, consequential events, and calls refused went up Finally, BBCH's PICU is an example of what we have seen in numerous other organizations. It takes a considerably long time to build an HRO, and HROs require constant attention (1989–1999). It takes a relatively short time to dismantle an HRO (1999–2003).[2]

CONCLUSIONS

Healthcare relies on "evidenced-based medicine" to determine forms of care. Caretakers are encouraged to follow policies that are often very rigid.

[2]Examples of HROs that have fallen into disrepute include offshore oil platforms, military aircraft squadrons, oil refineries, and banks.

This probably works much of the time, and is probably why it is embraced in the field of medicine. However, healthcare providers are often faced with unusual, unknown situations in which trained intuition and previous experience with cognitive practice in developing situational awareness and sensemaking can guide future experiences.

Health care professionals want the best outcomes for their patients. Nevertheless, the model of care they learn reduces their effectiveness in terms of patient safety. First, we encourage healthcare managers to examine their total situations for processes that impede safe care. Coming back to our initial example of organizations that are socially interdependent, we can ask if hospital policies or policies developed in other agencies are detrimental to or supportive of safe care.

Then we can examine those processes encouraged in our own care centers. Are we structured such that care is given by the person or persons best trained and able to give it? Does the structure allow migrating decision making? Can we change organizational arrangements or structuring in such a way as to maximize the quality of care? Is teamwork a part of our care strategy?

We also need to examine the human resource processes initially used at BBCH to determine whether they are useful to other health care situations. Is process auditing in place? Are we rewarding behavior A while hoping for behavior B (Kerr, 1975)? Do we know what the quality standards are or should be and are they in place? Do we perceive that risk is inherent in our situations, and if we perceive that it is, what are we doing about it? Finally, are the command and control processes in place?

We do not know which, if any, of these processes can be overlooked in an organization that is still highly reliable. We do know that organizations that include them are more likely to maintain reliability than those that do not.

Possibly the most important lesson for practitioners from this study is the fact that HROs require constant nurturance and are highly susceptible to dismantling. Without constant efforts directed at structuring, process auditing, using appropriate rewards, distributed decision making, and numerous other processes, organizations will lose their resilience and ability to respond to changing and uncertain events. All these efforts require resources, and managers need to think about the potential consequences to reliability of cutbacks, mergers, and other strategies designed to save money. Business models and appropriate health care models are probably incompatible with one another.

With regard to enhancing reliability in health care there is still much researchers need to investigate. The BBCH experience suggests that more attention needs to be directed to the larger system in which the PICU operates. For example, how influential was the larger system on the PICU's return to the authority gradient? How interactively effective were the transport system and the PICU? The HRO processes identified in the PICU are similar to those identified by Libuser (1993) in her original research. These processes probably were differently articulated in Libuser's works than in the PICU. Are some articulations more acceptable and credible in health care than in other settings? Do other HRO processes exist that should be added to the list? Should some be eliminated? Although we provide a number of general conclusions and specific prescriptions for managers, HRO research has a long way to go before completing the picture of how risks are mitigated in health care organizations.

ACKNOWLEDGMENTS

This work was partially supported by National Science Foundation grant SESB0105402.

References

Bigley, G. A., & Roberts, K. H. (2001). Structuring temporary systems for high reliability. *Academy of Management Journal, 44* 1281–1300.

Eisenhardt, K. (1993). High reliability organizations meet high velocity environments: Common dilemmas in nuclear power plants. In K. H. Roberts (Ed.), *New Challenges to Understanding Organizations* 33–54. New York: Macmillan.

Furr Y. J., & van Stralen D. (1995, October) Routine clinical pharmacist interactions decrease significant interventions in the pediatric intensive care unit. Paper presented at the California Society of Hospital Pharmacists Conference, Palm Springs, CA.

Grabowski, M., & Roberts, K. H. (1999). Risk mitigation in virtual organizations. *Organization Science, 10,* 704–721.

Helmreich, R. L., & Foushee, H. C. (1993). Why crew resource management?: The history and status of human factors training programs in aviation. In I. E. Weiner, B. Kanki, & R. L. Helmreich (Eds.), *Cockpit Resource Management.* San Diego, CA: Academic Press.

Hutchins, E. (1995). *Cognition in the Wild.* Cambridge, MA: MIT Press.

Kerr, S. (1975). On the folly of rewarding A while hoping for B. *Academy of Management Journal, 18,* 769–783.

Klein, G. (1989). Recognition-primed decision *Advances in Man-Machine Systems Research, 5,* 47–92.

Libuser, C. (1993). Organizational structure and risk mitigation. Unpublished doctoral dissertation, University of California, Los Angeles.

Little L. A., Koenig, J. C., & Newth, C. J. L. (1990). Factors affecting accidental extubations in neonatal and pediatric intensive care patients. *Critical Care Medicine 18*(2),163–165.

Mason, S. (2003, July) John Boyd and strategic naval air power. *U.S. Naval Institute Proceedings, 129*:76–78.

McCloskey, K. A., & Johnston, C. (1990). Critical care interhospital transports: Predictability of the need for a pediatrician. *Pediatric Emergency Care, 6,* 89–92.

MacNab, A. J. (1991). Optimal escort for interhospital transport of pediatric emergencies. *Journal of Trauma, 31*(2), 205–9.

Pollack M. M., Cuerdon, T. C., & Getson, P. R. (1993). Pediatric intensive care units: Results of a national survey. *Critical Care Medicine, 21*(4), 607–14.

Roberts, K. H. (1990). Some characteristics of one type of high reliability organization. *Organization Science, 1,* 160–176.

Reason, J. (1997). Managing the risks of organizational accidents. Aldershot, UK: Ashgate.

Roberts, K. H., Madsen, P. M., & Desai, V. M. (2005). The space between in space transportation: A relational analysis of the failure of STS 107. In M. Farjoun & W. Starbuck (Eds.), *Organization at the Limit: NASA and the Columbia Disaster.* Oxford, UK: Blackwell.

Scott, R., Perkin, R. M., Elliott, B., Rogers, M., Malinowski, T., & Langga, L. (1997). Testing the team. *RT: The Journal for Respiratory Care Practitioners. 10,* 75–78.

Tadmor, C., & Roberts, K. H. (2006). Structural Failures and the Development of an Organizational Breakdown: The Tragedy of the USS Greeneville. Manuscript submitted for publication.

Weick, K. E. (1987). Organizational culture as a source of high reliability. *California Management Review, 29,* 112–127.

Weick, K. E. (1995) *Sensemaking in Organizations.* Thousand Oaks, CA: Sage.

Weick, K. E., & Roberts, K. H. (1993) Collective mind and organizational reliability: The case of flight operations on an aircraft carrier deck. *Administrative Science Quarterly*, 38, 357–381.

17

THE RELATION BETWEEN TEAMWORK AND PATIENT SAFETY

David P. Baker
American Institutes for Research

Eduardo Salas
University of Central Florida

Paul Barach
University of Miami

James Battles
Agency for Healthcare Research and Quality

Heidi King
Department of Defense, TRICARE Management Activity

This chapter reviews the empirical evidence concerning the relation between teamwork and patient safety. The available evidence suggests that training teams of health care providers constitutes a pragmatic, effective strategy for enhancing patient safety by reducing medical errors.

BACKGROUND

In 1999, the Institute of Medicine (IOM) published *To Err is Human: Building a Safer Health System,* a frightening indictment of the inadequate safety that the United States medical establishment too often provides its patients (Kohn, Corrigan, & Donaldson, 1999). Extrapolating from data gathered through the Harvard Medical Practice Study (HMPS) and the Utah-Colorado Medical Practice Study (UCMPS; Studdert, Brennan, & Thomas, 2002), the IOM report concluded that medical errors cause between 44,000 and 98,000 deaths annually (Kohn et al., 1999). The report also noted that medical errors are financially costly. The IOM estimated that, among U.S. hospital inpatients, medication errors alone cost approximately $2 billion annually. Besides their direct costs, errors result in opportunities lost, given that funds spent in correcting mistakes cannot be used for other purposes, as well as in higher insurance premiums and co-payments. In addition, because of their effect on diminished employee productivity, decreased school attendance, and a lower state of public health, such errors exact a price from the society at large. Specifically, the IOM estimated that the total indirect cost of medical errors that result in patient harm lies between $17 and $29 billion annually. Finally, medical errors undermine patients' and health professionals' confidence in the health care system itself.

Key to this chapter's orientation toward teamwork-related research, the IOM noted that the majority of medical errors result from health care system failures, rather than from individual providers' substandard performance. Thus, in conjunction with its drive to implement organizational safety systems by delivering safe practices (Tier 4), the IOM recommended establishing interdisciplinary team-training programs (Kohn et al., 1999).

The primary responsibility for conducting and supporting research to address the IOM's recommendations currently rests with the Agency for Healthcare Research and Quality (AHRQ). This responsibility encompasses three broad areas: (a) identifying the causes of errors and injuries in health care delivery; (b) developing, demonstrating, and evaluating error-reduction and patient-protection strategies; and (c) distributing effective strategies throughout the U.S. health care community (AHRQ, 2000). One of AHRQ's initial efforts to address the patient safety crisis was to commission a review of the existing evidence base for different safe patient practices. Evidence Report 43 titled, *Making Health Care Safer: A Critical Analysis of Patient Safety Practices,* presents existing data on practices viewed as having the potential to improve patient safety. Within this report, Pizzi, Goldfarb, and Nash (2001) identified crew resource management (CRM)—a team training approach—as a strategy that has tremendous potential to improve patient safety based on its success in aviation.

STRUCTURE OF THIS CHAPTER

The next sections of this chapter review the evidence concerning the extent to which training medical personnel as teams is likely to improve patient safety outcomes. This chapter extends the work of Pizzi and colleagues (2001) by providing a more comprehensive review of the evidence base that supports the importance of teamwork in high-risk industries and presents a compelling argument for teamwork's relation to patient safety. In the next section, we define the key characteristics of a team and discuss the principles that underlie successful teamwork. Next, we describe and evaluate research concerning the relation between teamwork and safety in real world, high-risk settings. Third, we introduce current trends and issues in medical team training. Finally, we offer a set of conclusions and recommendations for future research.

TEAMWORK

Definitional Issues Concerning Teams and Teamwork

Teams and teamwork have received an increasing amount of attention over the last 20 years (Driskell & Salas, 1992; Dyer, 1984; Foushee, 1984; Salas, Bowers, & Cannon-Bowers, 1995; Stone, 2000). Numerous chapters (Dyer, 1984; Salas et al., 1995) and books (Brannick, Salas, & Prince, 1997; Guzzo & Salas, 1995; Salas, Bowers, & Edens, 2001a; Wiener, Kanki, & Helmreich, 1993) have specifically addressed critical issues related to team performance. In fact, organizations that do not rely on teams, at least to some extent, are scarce.

Given the prevalence of teams in the workplace, the literature reflects substantial agreement regarding their defining characteristics. Any inconsistencies among definitions are due, at least in part, to the reality that teams reflect a variety of purposes (e.g., learning, producing a product, solving problems, gaining acceptance), forms (e.g., virtual, co-located), sizes, and longevity (e.g., ad hoc, long term) (Cohen & Bailey, 1997).

What Is a "Team"? To identify the key features of a team, we reviewed several often-cited definitions (Dyer, 1984; Guzzo & Shea, 1992; Mohrman, Cohen, & Mohrman, 1995; Salas, Dickinson, Converse, & Tannenbaum, 1992), as well as other relevant literature. The definition we adopted includes the following four characteristics: (a) teams consist of a minimum of two or more people; (b) team members have specific roles, perform specific tasks, and interact or coordinate to achieve a common goal or outcome; (c) team members possess specialized knowledge and skills and often work under conditions of high workload; and (d) teams embody the coordination that results from task interdependency, which distinguishes teams from small groups.

The Nature of Effective Teamwork. Teamwork has traditionally been described in terms of classical systems theory, which posits that team inputs, team processes, and team outputs are arrayed over time. In particular, team inputs include the characteristics of the task to be performed, the elements of the context in which work occurs, and the attitudes team members bring to a team situation. Team process constitutes the interaction and coordination required among team members if the team is to achieve its specific goals. Team *outputs* consist of the products that result from team performance (Hackman, 1987; Ilgen, 1999; McGrath, 1984). Thus, teamwork per se occurs in the process phase, during which team members interact and work together to produce team outputs. Finally, teamwork does not require team members to work together permanently; it is sustained by a shared set of teamwork skills, not by permanent assignments that carry over from day to day.

However, simply installing a team structure in an organization does not automatically result in effective teamwork. Effective team performance requires team members' willingness to cooperate for a shared goal. Moreover, effective teamwork depends on effective within-team communication and adequate organizational resources and support. In short, teamwork requires team members to develop a shared awareness of one another's roles and abilities. Without this awareness, serious but avoidable adverse outcomes may result from a series of apparently trivial errors that effective teamwork would have prevented.

Extensive research has yielded numerous models of effective teamwork (Campion, Medsker, & Higgs, 1993; Fleishman & Zaccaro, 1992; Hambrick, Cho, & Chen, 1996; Stevens & Campion, 1994; West & Anderson, 1996). Historically, this literature has sought to identify generic teamwork skills associated with most teams. However, the focus has more recently shifted toward identifying the specific *competency requirements* that team members exhibit (Cannon-Bowers, Tannenbaum, Salas, & Volpe, 1995; O'Neil, Chung, & Brown, 1997; Stevens & Champion, 1994). Although the term *competency* signifies a variety of meanings, it is generally used to denote the qualities needed by a jobholder (Boyatzis, 1982).[1] Specifically, Parry (1998) defined the term "competencies" as a cluster of related knowledge, skills, and attitudes that affects a major part of one's job (e.g., represent one or more key roles or responsibilities; correlate with successful job performance; can be measured against well-accepted standards; and can be improved through training and development).

Generally speaking, team competencies are the attributes team members need to engage successfully in teamwork: As has been suggested, " ... It is essential to understand the nature of competencies required to function in a team as a means to define selection criteria, design and conduct training, and assess team performance" (Cannon-Bowers et al., 1995, p. 334). To explicate this understanding, Cannon-Bowers and colleagues identified three types of competencies that are critical for effective teamwork: (a) teamwork-related knowledge (K), (b) teamwork-related skills (S), and (c) teamwork-related attitudes (A).

In summary, teams know things, do things, and feel things; moreover, they know, do, and feel within specific environments. Thus, their effective performance depends on integrating a host of interrelated personal and situational characteristics.

Training Teams

Team training can be defined as applying a set of instructional strategies, that rely on well-tested tools (e.g., simulators, lectures, videos), to specific team competencies (Salas & Cannon-Bowers, 2000; Salas, Rhodenizer, & Bowers, 2000; Salas, Rozell, Mullen, & Driskell, 1999). Effective team training reflects general principles of learning theory, presents information about requisite team behaviors, affords team members the opportunity to practice the skills they are learning, and provides remedial feedback.

A great deal of research has been devoted to the most effective strategies and techniques for training specific team knowledge, skill, and attitude (KSA) competencies. A comprehensive review of this research has presented an extensive collection of principles and guidelines concerning the design and delivery of team training. For example, guidelines exist for assertiveness training (Smith-Jentsch,

[1]Boyatzis (1982, p. 10), in his seminal work on competencies, defined a job competency as "an underlying characteristic of a person, which results in effective or superior performance in a job."

KNOWLEDGE COMPETENCIES	
Competency	**Definition**
Shared Task Models/Situation Assessment	A Shared understanding of the situation and appropriate strategies for coping with task demands
Teammate Characteristics Familiarity	Knowing the Task-related competencies, preferences, tendencies, strengths, and weaknesses of teammates.
SKILL COMPETENCIES	
Mutual Performance Monitoring	Tracking fellow team members' performance to ensure that the work is running as expected and that proper procedures are followed
Flexibility/Adaptability	Ability to recognize deviations from expected course of events to readjust one's own actions accordingly
Supporting/Back-Up Behavior	Providing feedback and coaching to improve performance or when a lapse is detected; assisting teammate in performing a task; and completing a task for the team member when an overload is detected.
Team Leadership	Ability to direct / coordinate team members, assess team performance, allocate tasks, motivate subordinates, plan/organize, and maintain a positive team environment.
Closed-Loop Communication/ Information Exchange	The initiation of a message by the sender, the receipt and acknowledgement of the message by the receiver, and the verification of the message by the initial sender
ATTITUDE COMPETENCIES	
Team Cohesion	The total field of forces that influence members to remain in a group; an attraction to the team as a means of task accomplishment
Mutual Trust	A positive attitude held by team members regarding the aura, mood, or climate of the team's internal environment
Collective Orientation	The belief that a team approach is better than an individual one
Importance of Teamwork	The positive attitude that team members exhibit toward working as a team

Exhibit 17–1. Essential Team Knowledge, Skill, and Attitude (KSA) Competencies—adapted from (Salas et al., 2001a)

Salas, & Baker, 1996), cross-training (Volpe, Cannon-Bowers, Salas, & Spector, 1996), stress management training (Driskell & Johnston, 1998), and team self-correction (Smith-Jentsch, Zeisig, Acton, & McPherson, 1998).

In addition to the available team training research and practical guidance, the team competencies presented in Exhibit 17–1 provide an excellent resource for designing team training programs. Cannon-Bowers and colleagues contended that team KSA competencies should serve as the starting point for conducting training needs analyses. After establishing a team's specific competency requirements, trainers must specify appropriate training strategies.

To meet this requirement, Cannon-Bowers and colleagues offered detailed information on the nature of training required for developing particular team competencies and the strategies that are likely to be successful (Cannon-Bowers et al., 1995).

Finally, a successful team training program constitutes more than developing team members' KSAs. For example, because organizational factors outside the training program itself affect the program's success, conducting a needs analysis before designing a training intervention is essential to determining the best delivery method or instructional strategy. In addition, training developers should take advantage of the increased practice opportunities provided by certain training tools, such as advance organizers (e.g., outlines, diagrams, graphic organizers), preparatory information, prepractice briefs, attentional advice, goal orientation, and meta-cognitive strategies (Cannon-Bowers & Salas, 1998).

SUMMARY

The preceding section discussed the elements that typify effective teamwork and effective team training. High-performing teams exhibit a sense of collective efficacy, are dependent on each other, and believe that they can solve complex problems. Moreover, effective teams are dynamic: they optimize their resources, engage in self-correction, compensate for each other by providing back-up behaviors, and adapt as necessary. Because they can often coordinate without communicating overtly, effective teams can respond efficiently in high-stress, time-restricted environments.

Designing training that will improve teamwork skills on the job is a challenge. In virtually any field, team training requires a comprehensive, sustained strategy that targets many aspects of teamwork. Teams operate in complex environments. Yet team training is charged with improving trainee KSAs and facilitating desirable performance outcomes (e.g., safety, timely and accurate responding, patient welfare) under these conditions. Therefore, effective training programs must (a) systematically represent sound theory and a thorough needs analysis; (b) provide trainees with information, demonstrations, guided practice and timely diagnostic feedback; and (c) reflect organizational cultures that encourage the transfer of the trained competencies to the task environment.

TEAMWORK IN HIGH-RISK CONTEXTS

Commercial Aviation

Because aviation constitutes a field in which mistakes can cause an unacceptable loss of life and property, the flight industry has been on the forefront of developing teamwork training to reduce risk. Among the team training programs that have evolved within aviation, the best known is crew resource management (CRM) training.

Recent research suggests that CRM training results in heightened safety-related attitudes; improved communication, coordination, and decision-making behaviors; and enhanced error-management skills (Helmreich & Merritt, 1998; Wiener et al., 1993). CRM training also has demonstrated consistently positive results across a wide range of team structures, including pilot crews, maintenance crews, dispatch crews, and air traffic control teams (Helmreich & Foushee, 1993; Oser, Salas, Merket, & Bowers, 2001; Smith-Jentsch, Baker, Salas, & Cannon-Bowers, 2001).

CRM has grown from focusing solely on awareness and attitude change, to incorporating behavioral skills training, to integrating training in teamwork with training in technical flying skills, as is the case with the U.S. Federal Aviation Administration's new Advanced Qualification Program (AQP). Recent reviews suggest that CRM training results in positive reactions to teamwork concepts, increased knowledge of teamwork principles, and improved teamwork performance in the simulator (Salas, Burke, Bowers, & Wilson, 2001b).

CRM's effect on the ultimate criterion—a reduction in the number of accidents—has yet to be empirically established (Salas et al., 2001b). However, accidents represent a poor criterion methodologically, because they exhibit an extremely low base rate (Helmreich & Foushee, 1993). Thus, researchers have relied on surrogate measures—like improvements in teamwork-related knowledge and skills, behavioral demonstrations of CRM skills on simulated flights, instructor evaluations of trained versus untrained crews, and changes in an organization's safety culture—to demonstrate the effectiveness of CRM training.

Viewed in isolation, each piece of evidence concerning the effectiveness of CRM training can be disputed; nevertheless, the pattern of results suggests that CRM training does improve the margin of aviation safety. In short, the reasonable inference is that if trainees demonstrate improvement in critical teamwork-related competencies and if they apply these skills in the posttraining environment, the ultimate outcome will be safe flights.

As practiced currently, CRM training meets the criteria we mentioned in our discussion of team training. CRM focuses on trainable, measurable skills that are crucial to successful performance. As such, CRM's methodologies could theoretically apply to virtually any medical domain in which effective teamwork minimizes error and enhances patient safety.

Military

The other high-risk industry in which team training has grown to prominence is the military. The watershed for this research was USS Vincennes' accidentally shooting down an Iranian Airbus in the Persian Gulf in 1988. In response to this incident, the Navy began a multiyear, multimillion-dollar research program to formally study teamwork and team training interventions. The program, called Tactical Decision Making Under Stress (TADMUS), began in 1990 and led to breakthrough advances in the science and practice of team training. Results of the program have highlighted new approaches, such as interpositional knowledge (cross-) training, mental-model training, and team self-correction training.

Following the Navy's lead, the U.S. Air Force and the U.S. Army also supported research on team performance and team training (Keesling, Ford, & Harrison, 1994; Siebold, 1994). In fact, as Salas and colleagues (1995, p. 71) pointed out, "Much [had] been accomplished since Dyer's (1984) seminal review." These research and practice results were incorporated into what has become the current state of military team training.

Some type of formal team training is now a major component of training in most branches of the U.S. Armed Forces. For example, all branches of the U.S. Armed Forces give pilots and other aircrew CRM (Spiker, Silverman, Tourville, & Nullmeyer, 1998). Current military aviation team training is again adopting the best practices of civil aviation, such as AQP, and combining these practices with traditional military training and cutting-edge technology.

In addition to pilots and other aircraft crewmembers, many sailors, soldiers, airmen, and Marines receive team training. For example, the U.S. Navy has recently adopted an approach called Team Dimensional Training (TDT), which resulted from the TADMUS program (Cannon-Bowers & Sales, 1998). TDT addresses team-related KSAs, provides practice in briefing and debriefing, and trains trainers and team leaders to evaluate and critique team skills (Tannenbaum, Smith-Jentsch, & Behson, 1998). It has been tested in teams as different from one another as submarine attack center teams, seamanship and shipboard damage control teams, naval aircrews, and surface warfare teams. In short, TDT trains teams to correct themselves, in addition to training team leaders to guide their members through the self-correction process (Smith-Jentsch et al., 1998).

Summary

This section briefly examined the empirical evidence concerning team performance and team training in high-risk contexts. Given that serious misfortunes resulting from human error are relatively rare in aviation and in military forces not involved in warfare, empirically linking team performance to the "ultimate criterion" of reducing these errors is difficult. Nevertheless, taken as a whole, research on teams has demonstrated a strong relation between effective team performance and desirable "proxy" criteria, such as flexibility, adaptability, ability to reallocate resources, and resistance to stress. In addition, teams yield valuable process-oriented benefits, such as cohesion, retention, and morale (Gully, Devine, & Whitney, 1995; Gully, Incalcaterra, Joshi, & Beaubien, 2002).

Given the pervasiveness of these findings, inferring that successful teamwork might substantially reduce severe life-threatening medical errors is not unreasonable. Thus, we view the relations documented in this section as relevant to medical team training. Now we turn to a discussion of existing medical team training programs.

TEAMWORK AND HEALTH CARE

The Case for Teamwork

Throughout the health care community, small groups of people work together in intensive care units (ICU), operating rooms, labor and delivery wards, and family medicine practices. To make safe and efficient patient care a priority, physicians, nurses, pharmacists, technicians, and other health professionals must coordinate their activities. However, even though a myriad of the conditions addressed by health care professionals require interdisciplinary teams, members of these teams are rarely trained together; furthermore, they often come from separate disciplines and diverse educational programs.

Given the interdisciplinary nature of the work, teamwork is critical to ensuring patient safety. Teams make fewer mistakes than do individuals, especially when each team member knows his or her responsibilities, as well as those of other team members. Consistent with health care's increased emphasis on teamwork, AHRQ advocates a cultural shift from denying the presence of medical errors to discussing, learning about, and preventing them. CRM training—which falls under the domain of "team training"—is one means for helping to bring about this cultural shift.

Moreover, AHRQ is not alone in recognizing the importance of teamwork for improving the performance of medical professionals. In concert with AHRQ's goals, the Accreditation Council for Graduate Medical Education (ACGME) recently identified several teamwork-related competencies that residents must master. These competencies include communication with patients and significant others, patient counseling and education, working with other health care professionals, and facilitating the learning of students and other health care professionals (Dunnington & Williams, 2003).

Similarly, the Association of American Medical Colleges (AAMC) funded a "critical incident" analysis to investigate the behaviors that result in successful and unsuccessful performance during medical school and residency. The results revealed the importance of a number of teamwork-related competencies, including interpersonal skills and professionalism, interactions with patients and family, ability to foster a team environment, and ability to mentor or educate other students (Adams, Goodwin, Searcy, Norris, & Oppler, 2001).

Medical team training programs derived from CRM began with the introduction of anesthesia crisis resource management (ACRM) training at Stanford University School of Medicine and at the Anesthesiology Service at the Palo Alto Veteran Affairs Medical Center (Howard, Gaba, Fish, Yang, & Sarnquist, 1992). AHRQ's 2001 review of in-place patient safety practices critiqued the ACRM model, citing it as high in impact but low in evidence supporting its effectiveness (Pizzi et al., 2001). More recently, the Department of Defense (DOD) funded several other CRM-derived team training initiatives. Specifically, MedTeams™ (Morey, Simon, Jay, & Rice, 2003) has been implemented in a number of U.S. Army and U.S. Navy hospitals.

ACRM Program

Developed by David Gaba and his colleagues at Stanford University and the Palo Alto Veteran Affairs (VA) Medical Center, ACRM is designed to help anesthesiologists effectively manage crises by working in multidisciplinary teams that include physicians, nurses, technicians, and other medical professionals (Gaba, 1998; Gaba, Howard, Fish, Smith, & Sowb, 2001a; Howard et al., 1992). To facilitate this goal, ACRM training provides trainees with critical incident case studies to review (Davies, 2001). In addition, ACRM provides training in technical skill and in team KSAs. Training in the selected teamwork skills is intended to enable trainees to learn from adverse clinical occurrences and to work more effectively with different leadership, followership, and communication styles (Gaba, Howard, Fish, Smith, & Sowb, 2001b).

ACRM training takes place in a simulated operating room (OR), after completing the reading assignments that precede each module. The simulated OR includes actual monitoring equipment, a full-patient simulator, a video station for recording the team's performance, and a debriefing room that is equipped with a variety of audiovisual equipment. The full-patient simulator incorporates a series of complex mathematical models and pneumatic devices to simulate a patient's breathing, pulses, heart and lung sounds, exhaled CO_2, thumb twitches, and other physiological reactions (Gaba, Howard, Fish, Smith, & Sowb, 2001a; Murray & Schneider, 1997).

The ACRM curriculum comprises 3 full separate days of simulation training, over 3 years of anesthesiology training. Day 1 provides an introduction to ACRM principles and skills. Day 2 provides a refresher on these skills and analyzes clinical events from the perspective of the clinician's technical and teamwork skills and from the perspective of the organization as a larger system. Day 3 emphasizes leadership training, debriefing skills, and adherence to the procedures established to deal with adverse clinical events. Each training module consists of a similar structure: preassigned readings, course introduction and review of materials, familiarization with the simulator, case study analysis and videotape reviews, and 6 hr of participating in simulator scenarios, followed by an instructor-led debriefing and a postcourse data collection. Each scripted training scenario is approximately 45 min long; each debriefing session lasts approximately 40 min (Gaba et al., 2001a).

Several instructors are required to run the ACRM training scenarios. They might include a retired OR nurse who role-plays the circulating nurse and an anesthesiologist instructor who role-plays the operating surgeon. In addition, a director monitors and records the simulation from another room, communicating with the instructors via two-way radios. Throughout the simulation, trainees rotate through various roles, such as "first responder," "scrub technician," and "observer"(Gaba et al., 2001a).

ACRM training, complete with yearly refresher training, is currently used at several major teaching institutions in the United States and around the world (Australia, Israel, Denmark). At some centers, ACRM training is offered for experienced practitioners as well as for trainees. Moreover, some malpractice insurers (i.e., Harvard Risk Management Foundation) have lowered their rate structure for ACRM-trained anesthesiologists (Gaba et al., 2001a).

An ACRM evaluation typically assesses a variety of process-oriented criteria. *Teamwork performance* is typically assessed using behavioral markers of the 10 teamwork skills specified in the previous section (Gaba et al., 1998). One measure of these teamwork behaviors consists of a checklist, which, as noted previously, is analogous to the Line/LOS Checklist used in CRM programs (Helmreich, Butler, Taggart, & Wilhelm, 1995). Using a 5-point rating scale, trained raters evaluate team performance on each dimension (Gaba et al., 1998). Measures of interrater agreement exhibited r_{wg} values (James, Demaree, & Wolf, 1984) ranging between .60 and .93 (Gaba et al., 1998); an r_{wg} of .70 is considered sufficiently high to reflect a satisfactory degree of agreement among raters.

Most of the thousands of participants who have undergone ACRM training evaluate it favorably, even the "death scenario," which is specifically designed to assess how trainees handle losing a patient; these positive responses generally last for up to 6 months after training (Gaba et al., 2001b). Furthermore, recent research suggests that participation in ACRM training also increases trainees' self-efficacy and decreases their self-reported anxiety (Tays, 2000).

Despite these positive assessments, to our knowledge, no studies have taken the next logical step of directly investigating the link between team process and patient-safety criteria. In fact, virtually no research has tested the effect of any aspect of ACRM training on actual performance outcomes. With respect to *individual* (i.e., technical) performance, this lack of outcome-related validity derives, at least in part, from the difficulties associated with quantifying the performance of anesthesiologists (Gaba et al., 1998).

However, with respect to assessing the effects of *team process*, the lack of outcome-related validity cannot be explained so easily because programmed outcomes are embedded into the ACRM training scenarios (e.g., the "death scenario"). Thus, we believe that developing measures to assess the effectiveness of teamwork in facilitating positive outcomes and in successfully managing, if not avoiding, negative outcomes would constitute a constructive focus for future research. Furthermore, given the current state of simulation, devising training scenarios for which the outcome is contingent on the level of trainees' demonstrated teamwork skills might be worthwhile.

MedTeams' Purpose and Strategy

The primary purpose of MedTeams is to reduce medical errors through interdisciplinary teamwork. MedTeams was initially developed for emergency departments (EDs) on the premise that most errors result from breakdowns in systems-level defenses that occur over time (Simon et al., 2000). According to the MedTeams' ED curriculum, each team member has a vested interest in maintaining patient safety and is expected to take an assertive role in breaking the error chain. MedTeams defines a core ED team as a group of 3–10 (average = 6) medical personnel who work interdependently during a shift and who have been trained to use specific teamwork behaviors to coordinate their clinical interactions. Each core team includes at least one physician and one nurse. A coordinating team that assigns new patients to the core teams and provides additional resources as necessary manages several core teams.

MedTeams' training was developed from an evaluation-driven course design. Based on needs-analysis data, five critical dimensions were identified. Then, 48 specific, observable behaviors were linked to these dimensions and Behaviorally Anchored Rating Scales (BARS; Smith & Kendall, 1963) were constructed. Finally, to establish its content validity, the MedTeams curriculum was reviewed and refined during three 5-day expert panel sessions that included ED physicians and nurses from 12 hospitals of various sizes (Simon et al., 1998). Expert panel review and modification of the curriculum were used to create labor and delivery and OR versions of MedTeams.

MedTeams uses a train-the-trainer approach to implement the training. Trainers, designated by their facility, receive comprehensive training on how to teach MedTeams and are certified as MedTeams' instructors. The course consists of an 8-hr block of classroom instruction that contains an introduction module, five learning modules, and an integration unit. After completing the classroom training, each team member participates in a 4-hr practicum that involves practicing teamwork behaviors and receiving feedback from a trained instructor. Coaching, mentoring, and review sessions also are provided during regular work shifts (Simon et al., 1998).

MedTeams' training has been evaluated using a quasi-experimental research design (Morey et al., 2002, 2003) in which a variety of process factors (e.g., quantity of teamwork behaviors) and enabling factors (e.g., attitudes toward teamwork, staff burnout) were measured over a 1-year period. An analysis of these data indicated a positive effect of training on *outcome criteria* (e.g., medical errors, patient satisfaction; Morey et al., 2002). However, this study suffered two significant limitations; participating hospitals self-selected into either the experimental or control groups, and observers were not blind to the experimental conditions. To address this limitation, a subsequent evaluation of MedTeams in labor and delivery units is in progress, using a randomized clinical trial design (Goldman, Shapiro, Mann, Risser, & Greenberg, 2002).

Summary

This chapter has summarized the general state of medical team training. We concentrated our discussion on ACRM and MedTeams, because these are the most thoroughly documented medical team training programs. These programs have made progress in improving patient safety; nevertheless, despite the encouraging nature of the extant data, the degree to which medical, CRM-inspired training will enhance patient safety remains in question. Thus, to provide a strategy for further investigation, the final section of this chapter integrates our findings into conclusions and recommendations relevant to medical team training.

CONCLUSIONS AND RECOMMENDATIONS

Conclusion 1: The Medical Field Lacks a Theoretical Model of Team Performance.

To date, research has not developed a comprehensive model of team performance in medical settings; consequently, medical team training programs have not been grounded in a scientific understanding of effective teamwork in the medical community. Given this gap in knowledge, the first research effort we advocate is to develop a theoretical medical team performance model that hypothesizes (a) the relations among predictors of performance and (b) the relations between predictors and outcome criteria. Nevertheless, despite the absence of a team performance model that focuses on medical teams per se, previous research has provided considerable relevant knowledge; the availability of this knowledge underlies several of the remaining conclusions.

Conclusion 2: The Science of Team Performance and Training can Help the Medical Community Improve Patient Safety.

As discussed in this chapter, a more general science of team performance and training has evolved and matured over the last 20 years. This science has produced a number of principles, lessons learned, tools, and guidelines that will serve the patient safety movement. Our recommendations are as follows: (a) that the medical community continue to inform itself of the progress of this science through a variety of venues (e.g., specialized workshops, books) and (b) that the medical community enlist the help of team training experts to apply to patient safety the principles, guidelines, and learning afforded by previous research.

Conclusion 3: Research Has Already Identified Many of the Competencies Necessary for Effective Teamwork in Medical Environments.

Previous investigations have identified the competencies required for effective team functioning in a number of complex settings. Many of these competencies apply to the medical community. However, as Cannon-Bowers and colleagues (1995) noted, the team skills literature is confusing, contradictory, and plagued with inconsistent labels and definitions. For example, across studies, different labels are used to refer to the same teamwork skills, and the same labels are used to refer to different skills. Thus, we recommend using a two-step process to develop a taxonomy with standard nomenclature; this taxonomy would name and define teamwork-related KSAs that constitute the core *competencies* related to successful teamwork in the medical domain.

The first step in developing such a taxonomy is to determine an appropriate level of explanation; the constructs included in the taxonomy must be conceptualized broadly enough to span the medical field yet be specific enough to facilitate valid measurement. Further, although this list of core competencies should reflect all relevant aspects of team performance, it must be concise enough to generate teamwork and team training research and to facilitate team training needs analyses in organizations.

The second step, determining relevant core competencies, encompasses two activities. One task is to establish which of the many competencies manifested in previous research are relevant to virtually all medical teams; another task is to identify core medical team competencies that have not emerged from team research in other domains. We believe that using task analytic techniques (e.g., survey questionnaires, structured interviews, and unobtrusive observations) will yield the most valid information. We also emphasize the importance of large-scale, stratified data collections, because the goal is to identify generic competency requirements with which the medical community at-large concurs.

Conclusion 4: A Number of Proven Instructional Strategies are Available for Promoting Effective Teamwork.

The science of team performance and training has developed and validated numerous training strategies that can provide requisite competencies to teams who perform in complex environments. Through a variety of formats and objectives, these strategies extend beyond CRM training. We recommend (a) that the medical community use these strategies wherever possible, given that some are relatively easy to design and deliver and (b) that the community explore strategies other than CRM to improve patient safety.

Conclusion 5: Team-Training Strategies Must Be Further Adapted to Medical Needs.

We are convinced that no single model of team training can be applied across all medical practices and contexts. For the purposes of this discussion, we define a "practice" as a medical specialty or subspecialty, such as emergency medicine, general or family medicine, intensive care, surgical medicine, or obstetrics. Medical practices differ dramatically across a variety of criteria: size, purpose, duration, redundancy of expertise, decision time, and consequence of error, to name but a few.

Moreover, a particular practice may operate in a number of diverse contexts. As an example, emergency medicine providers function in hospital EDs, in emergency-response mobile units, and on battlefields. Similarly, to mention several obvious distinctions, urban and rural general providers operate in independent or multipractitioner offices, as well as in community walk-in clinics. Neither the competencies that impel successful teamwork nor an optimal team-training strategy can be expected to generalize across all these contexts. And, of course, not all members within the same team will necessarily need the same KSAs.

Therefore, in addition to the core-competency taxonomy, we also recommend developing practice-specific taxonomies. These putative taxonomies would not be redundant with the generic, core-competency taxonomy. Rather, a practice-specific taxonomy would denote the specific KSA requirements that are central to teamwork in a given practice. The medical content and procedures that define this practice would drive the identification of relevant team competencies.

Virtually no previous research has addressed the manner in which differences within and between medical practices should be reflected in practice-specific taxonomies. Yet we find this issue sufficiently compelling to warrant further investigation. Because these taxonomies are derived from the medical characteristics of specific practices (and contexts within them), subject-matter experts who represent each practice might be invaluable in identifying practice-specific team competencies that are not redundant with the generic core-competency taxonomy. Nevertheless, we also suggest that researchers avail themselves of survey questionnaires, structured interviews, and unobtrusive observations.

Conclusion 6: The Medical Community Has Made Considerable Progress in Designing and Implementing Team Training Across a Number of Settings.

Our review of team training programs clearly shows that the medical community is striving to implement CRM training across a number of medical domains. We recommend that this trend be continued. However, the extent to which these programs are being implemented with the help of what we know from the science of learning, of team performance, and of training is less clear. Thus, we recommend strengthening the link between scientific knowledge and medical team training. Furthermore, as noted previously, the medical community should explore other strategies that can be effectively applied to medical team training. Specifically, we first recommend that medical team training be *developed* to reflect the established instructional principles that underlie team-training research. Second, we recommend that the quality of these programs be *evaluated* on the basis of confirmed scientific criteria (e.g., assessing the degree to which training transfers to the actual work environment).

Conclusion 7: The Institutionalization of Medical-Team Training Across Different Medical Settings Has Not Been Addressed.

Our final conclusion focuses on what we consider the imperative need to embed medical team training in professional development. By "embedding" we mean implementing and regulating medical team training throughout a health care provider's career. As noted earlier, the ACGME identified several teamwork-related competencies that residents must master. Similarly, AAMC funded a "critical incident" analysis to investigate the behaviors that result in successful and unsuccessful performance during medical school and residency. Although not originally targeted toward team performance, the results revealed the importance of a number of teamwork-related competencies.

Simply stated, for medical team training to deliver the impact that it can potentially exert on patient safety, it must be instantiated at every stage of a provider's working life. For example, certain medical school assignments might require students to prepare team projects. Interns and residents might observe, participate in, and evaluate practicing teams in hospitals. The larger challenge, however, occurs after providers have completed their formal training.

We believe that the structure of health care, as currently conceptualized, does offer appropriate junctures where teamwork skills could be evaluated. For example, like the examinations that are constructed for board certification in medical specialties, it might ultimately be useful to develop a board certification test for teamwork. Such an exam might combine a written test of knowledge and situational judgment with performance in a simulated scenario. Because the board examinations are practice specific, their teamwork component could assess practice-specific teamwork competencies. In addition, the Joint Commission on Accreditation of Healthcare Organizations currently evaluates hospitals on criteria that range from medical practices to managerial systems to facilities maintenance. At some point in the future, folding generic competency criteria into the Joint Commission evaluation might focus providers' attention on the importance of teamwork in medical settings, as well as yielding valuable research data.

ACKNOWLEDGMENTS

This research was supported by a grant from the AHRQ and the DOD, TriCare Management Activity. The views expressed herein are those of the authors.

References

Adams, K. A., Goodwin, G. F., Searcy, C. A., Norris, D. G., & Oppler, S. H. (2001). *Development of a Performance Model of the Medical Education Process. Technical Report Commissioned by the Association of American Medical Colleges.* Washington, DC: American Institutes for Research.

Agency for Health Care Research and Quality (AHRQ). (2000). *Doing What Counts for Patient Safety: Federal Actions to Reduce Medical Errors and Their Impact.* Rockville, MD: Author.

Boyatzis, R. E. (1982). *The Competent Manager.* New York: Wiley.

Brannick, M. T., Salas, E., & Prince, C. (1997). *Team Performance Assessment and Measurement.* Mahwah, NJ: Lawrence Erlbaum Associates, Inc.

Campion, M. A., Medsker, G. J., & Higgs, A. C. (1993). Relations between work group characteristics and effectiveness: Implications for designing effective work groups. *Personnel Psychology, 46,* 823–850.

Cannon-Bowers, J. A., & Salas, E. (1998). *Making Decisions under Stress: Implications for Individual and Team Training.* Washington, DC: American Psychological Association.

Cannon-Bowers, J. A., Tannenbaum, S. I., Salas, E., & Volpe, C. E. (1995). Defining competencies and establishing team training requirements. In R. A. Guzzo, E. Salas, & Associates (Eds.), *Team Effectiveness and Decision-Making in Organizations* (pp. 333–380). San Francisco: Jossey-Bass.

Cohen, S. G., & Bailey, D. E. (1997). What makes teams work: Group effectiveness research from the shop floor to the executive suite. *Journal of Management, 23,* 239–290.

Davies, J. M. (2001). Medical applications of crew resource management. In E. Salas, C. A. Bowers, & E. Edens (Eds.), *Improving Teamwork in Organizations: Applications of Resource Management Training* (pp. 265–281). Mahwah, NJ: Lawrence Erlbaum Associates, Inc.

Driskell, J. E., & Johnston, J. H. (1998). Stress exposure training. In J. A. Cannon-Bowers & E. Salas (Eds.), *Making Decisions Under Stress—Implications for Individual and Team Training* (pp. 191–217). Washington, DC: American Psychological Association.

Driskell, J. E., & Salas, E. (1992). Collective behavior and team performance. *Human Factors, 34,* 277–288.

Dunnington, G. L., & Williams, R. G. (2003). Addressing the new competencies for residents' surgical training. *Academic Medicine, 78,* 14–21.

Dyer, J. L. (1984). Team research and training: A state of the art review. In F. A. Muckler (Ed.), *Human Factors Review* (pp. 285–323). Santa Monica, CA: Human Factors and Ergonomics Society.

Fleishman, E. A., & Zaccaro, S. J. (1992). Toward a taxonomy of team performance functions. In R. W. Swezey, & E. Salas (Eds.), *Teams: Their Training and Performance* (pp. 31–56). Norwood, NJ: Ablex.

Foushee, H. C. (1984). Dyads and triads at 35,000 feet: Factors affecting group processes and aircrew performance. *American Psychologist, 39,* 885–893.

Gaba, D. M. (1998). Research techniques in human performance using realistic simulation. In L. C. Henson & A. H. Lee (Eds.), *Simulators in Anesthesiology Education* (pp. 93–101). New York: Plenum.

Gaba, D. M., Howard, S. K., Fish, K. J., Smith, B. E., & Sowb, Y. A. (2001a). Simulation-based training in anesthesia crisis resource management (ACRM): A decade of experience. *Simulation & Gaming, 32,* 175–193.

Gaba, D. M., Howard, S. K., Fish, K. J., Smith, B. E., & Sowb, Y. A. (2001b). Simulation-based training in anesthesia crisis resource management (ACRM): A decade of experience. *Simulation & Gaming, 32,* 175–193.

Gaba, D. M., Howard, S. K., Flanagan, B., Smith, B. E., Fish, K. J., & Botney, R. (1998). Assessment of clinical performance during simulated crises using both technical and behavioral ratings. *Anesthesiology, 89,* 8–18.

Goldman, M. B., Shapiro, D. E., Mann, S., Risser, D. T., & Greenberg, P. (2002). *Study design considerations for the MedTeams Evaluation* (unpublished technical report). Andover, MA: Dynamics Research Corporation.

Gully, S. M., Devine, D. J., & Whitney, D. J. (1995). A meta-analysis of cohesion and performance: Effects of level of analysis and task interdependence. *Small Group Research, 25,* 497–520.

Gully, S. M., Incalcaterra, K. A., Joshi, A., & Beaubien, J. M. (2002). A meta-analysis of team efficacy, potency, and performance: Interdependence and level of analysis as moderators of observed relationships. *Journal of Applied Psychology, 87,* 819–832.

Guzzo, R. A., & Salas, E. (1995). *Team Effectiveness and Decision Making in Organizations.* San Francisco: Jossey-Bass.

Guzzo, R. A., & Shea, G. P. (1992). Group performance and inter-group relations in organizations. In M. D. Dunnette & L. M. Hough (Eds.), *Handbook of Industrial and Organizational Psychology* (2nd ed., pp. 269–313). Palo Alto, CA: Consulting Psychologists Press.

Hackman, J. R. (1987). The design of work teams. In J. W. Lorsch (Ed.), *Handbook of Organizational Behavior* (pp. 315–342). Englewood Cliffs, NJ: Prentice Hall.

Hambrick, D. C., Cho, T. S., & Chen, M. (1996). The influence of top management team heterogeneity on firms' competitive moves. *Administrative Science Quarterly, 41,* 659–684.

Helmreich, R. L., Butler, R. E., Taggart, W. R., & Wilhelm, J. A. (1995). *Behavioral Markers in Accidents and Incidents. Report Commissioned By the FAA* (NASA/UT/FAA Technical Report 95–1). Austin, TX: The University of Texas.

Helmreich, R. L., & Foushee, H. C. (1993). Why crew resource management? Empirical and theoretical bases of human factors training in aviation. In E. L. Weiner, B. G. Kanki, & R. L. Helmreich (Eds.), *Cockpit Resource Management* (pp. 3–45). San Diego, CA: Academic.

Helmreich, R. L., & Merritt, A. C. (1998). *Culture at Work in Aviation and Medicine: National, Organizational, and Professional Influences.* Brookfield, VT: Ashgate.

Howard, S. K., Gaba, D. M., Fish, K. J., Yang, G., & Sarnquist, F. H. (1992). Anesthesia crisis resource management training: Teaching anesthesiologists to handle critical incidents. *Aviation, Space, and Environmental Medicine, 63,* 763–770.

Ilgen, D. R. (1999). Teams embedded in organizations: Some implications. *American Psychologist, 54,* 129–139.

James, L. R., Demaree, R. G., & Wolf, G. (1984). Estimating within group interrater reliability with and without response bias. *Journal of Applied Psychology, 69,* 85–98.

Keesling, W., Ford, P., & Harrison, K. (1994). Application of the principles of training in armor and mechanized infantry units. In R. F. Holz, J. H. Hiller, & H. H. McFann (Eds.), *Determinants of Effective Unit Performance: Research on Measuring and Managing Unit Training Readiness* (pp. 137–178). Alexandria, VA: U.S. Army Research Institute for the Behavioral & Social Sciences.

Kohn, L. T., Corrigan, J. M., & Donaldson, M. S. (Eds.). (1999). *To Err Is Human: Building a Safer Health System.* Washington, D.C.: National Academy Press. Available from Institute of Medicine Web site, http://www.iom.edu/report.asp?id' 16173

McGrath, J. E. (1984). *Groups: Interaction and Performance.* Englewood Cliffs, NJ: Prentice Hall.

Mohrman, S. A., Cohen, S. G., & Mohrman, A. M. (1995). *Designing Team-Based Organizations: New Forms for Knowledge Work.* San Francisco: Jossey-Bass.

Morey, J. C., Simon, R., Jay, G. D., & Rice, M. M. (2003). A transition from aviation crew resource management to hospital emergency departments: The MedTeams story. In R. S. Jensen (Ed.), *Proceedings of the 12th International Symposium on*

Aviation Psychology (pp. 1–7). Dayton, OH: Wright State University Press.

Morey, J. C., Simon, R., Jay, G. D., Wears, R., Salisbury, M., Dukes, K. A., et al. (2002). Error reduction and performance improvement in the emergency department through formal teamwork training: Evaluation results of the MedTeams project. *Health Services Research, 37,* 1553–1581.

Murray, W. B., & Schneider, A. J. L. (1997). Using simulators for education and training in anesthesiology. *American Society of Anesthesiologists Newsletter, 61,* 1–2.

O'Neil, H. F., Chung, G. K. W. K., & Brown, R. S. (1997). Use of network simulations as a context to measure team competencies. In H. F. O'Neil, Jr. (Ed.), *Workforce Readiness: Competencies and Assessment.* Mahwah, NJ: Lawrence Erlbaum Associates, Inc.

Oser, R. L., Salas, E., Merket, D. C., & Bowers, C. A. (2001). Applying resource management training in naval aviation: A methodology and lessons learned. In E. Salas, C. A. Bowers, & E. Edens (Eds.), *Improving teamwork in organizations: Applications of resource management training* (pp. 283–301). Mahwah, NJ: Lawrence Erlbaum Associates, Inc.

Parry, S. B. (1998). Just what is a competency? (And why should we care?). *Training, 35,* 58–64.

Pizzi, L., Goldfarb, N. I., & Nash, D. B. (2001). Crew resource management and its applications in medicine. In K. G. Shojana, B. W. Duncan, K. M. McDonald, & R. M. Wachter (Eds.), *Making Health Care Safer: a Critical Analysis of Patient Safety Practices* (pp. 501–510). Rockville, MD: AHRQ.

Salas, E., Bowers, C. A., & Cannon-Bowers, J. A. (1995). Military team research: 10 years of progress. *Military Psychology, 7,* 55–75.

Salas, E., Bowers, C. A., & Edens, E. (2001a). *Improving Teamwork in Organizations: Applications of Resource Management Training.* Mahwah, NJ: Lawrence Erlbaum Associates, Inc.

Salas, E., Burke, S. C., Bowers, C. A., & Wilson, K. A. (2001b). Team training in the skies: Does crew resource management (CRM) training work? *Human Factors, 43,* 641–674.

Salas, E., & Cannon-Bowers, J. A. (2000). The anatomy of team training. In N. S. Tobias & D. Fletcher (Eds.), *Training and Re-Training: A Handbook for Business, Industry and Military* (pp. 312–335). Farmington Hills, MI: Macmillan.

Salas, E., Dickinson, T. L., Converse, S. A., & Tannenbaum, T. S. (1992). Toward an understanding of team performance and training. In R. W. Swezey & E. Salas (Eds.), *Teams: Their Training and Performance* (pp. 3–29). Norwood, NJ: Ablex.

Salas, E., Rhodenizer, L., & Bowers, C. A. (2000). The design and delivery of crew resource management training: Exploiting available resources. *Human Factors, 42,* 490–511.

Salas, E., Rozell, D., Mullen, B., & Driskell, J. E. (1999). The effect of team building on performance: An integration. *Small Group Research, 30,* 309–329.

Siebold, J. B. (1994). The relation between team motivation, leadership, and small unit performance. In H. F. O'Neil, Jr., & M. Drillings (Eds.), *Motivation: Theory and Research* (pp. 171–190). Mahwah, NJ: Lawrence Erlbaum Associates, Inc.

Simon, R., Langford, V., Locke, A., Morey, J. C., Risser, D., & Salisbury, M. (2000). A successful transfer of lessons learned in aviation psychology and flight safety to health care: The MedTeams system. In *Patient Safety Initiative 2000-Spotlighting Strategies, Sharing Solutions* (pp. 45–49). Chicago: National Patient Safety Foundation.

Simon, R., Morey, J. C., Rice, M. M., Rogers, L., Jay, G. D., Salisbury, M. et al. (1998). Reducing errors in emergency medicine through team performance: The MedTeams project. In A. L. Scheffler & L. Zipperer (Eds.), *Enhancing Patient Safety and Reducing Errors in Health Care* (pp. 142–146). Chicago: National Patient Safety Foundation.

Smith, P. C., & Kendall, L. M. (1963). Retranslation of expectations: An approach to the construction of unambiguous anchors for rating scales. *Journal of Applied Psychology, 47,* 149–155.

Smith-Jentsch, K. A., Baker, D. P., Salas, E., & Cannon-Bowers, J. A. (2001). Uncovering differences in team competency requirements: The case of air traffic control teams. In E. Salas, C. A. Bowers, & E. Edens (Eds.), *Improving Teamwork in Organizations: Applications of Resource Management Training* (pp. 31–54). Mahwah, NJ: Lawrence Erlbaum Associates, Inc.

Smith-Jentsch, K. A., Salas, E., & Baker, D. P. (1996). Training team performance-related assertiveness. *Personnel Psychology, 49,* 909–936.

Smith-Jentsch, K. A., Zeisig, R. L., Acton, B., & McPherson, J. A. (1998). Team dimensional training. In J. A. Cannon-Bowers & E. Salas (Eds.), *Making Decisions Under Stress: Implications for Individual and Team Training* Washington, DC: American Psychological Association.

Spiker, V. A., Silverman, D. R., Tourville, S. J., & Nullmeyer, R. T. (1998). *Tactical Team Resource Management Effects on Combat Mission Training Performance* (USAF AMRL Technical Report AL-HR-TR–1997–0137). Brooks Air Force Base, San Antonio, TX: U.S. Air Force Systems/Materiel Command.

Stevens, M. J., & Campion, M. A. (1994). The knowledge, skill, and ability requirements for teamwork: Implications for human resource management. *Journal of Management, 20,* 503–530.

Stone, F. P. (2000). *Medical Team Management: Improving Patient Safety Through Human Factors Training* (HCR Reference Number: 00080). Eglin Air Force Base, FL: Military Health System Health Care.

Studdert, D. M., Brennan, T. A., & Thomas, E. J. (2002). What have we learned from the Harvard Medical Practice Study? In M. M. Rosenthal & K. M. Sutcliffe (Eds.), *Medical Error: What Do We Know? What Do We Do?* (pp. 3–33). San Francisco: Jossey-Bass.

Tannenbaum, S. I., Smith-Jentsch, K. A., & Behson, S. J. (1998). Training team leaders to facilitate team learning and performance. In J. A. Cannon-Bowers & E. Salas (Eds.), *Making Decisions Under Stress: Implications for Individual and Team Training* (pp. 247–270). Washington, DC: American Psychological Association.

Tays, T. M. (2000). *Effect of Anesthesia Crisis Resource Management Training on Perceived Self-Efficacy.* Unpublished doctoral dissertation, Pacific Graduate School of Psychology, Palo Alto, CA.

Volpe, C. E., Cannon-Bowers, J. A., Salas, E., & Spector, P. E. (1996). The impact of cross training on team functioning: An empirical investigation. *Human Factors, 38,* 87–100.

West, M. A., & Anderson, N. R. (1996). Innovation in top management teams. *Journal of Applied Psychology, 81,* 680–693.

Wiener, E. L., Kanki, B. G., & Helmreich, R. L. (1993). *Cockpit Resource Management.* San Diego: Academic.

Section IV

ERGONOMICS

18

HUMAN FACTORS IN HOSPITAL SAFETY DESIGN

John Reiling and Sonja Chernos
St. Joseph's Community Hospital of West Bend

The physical environment has a significant effect on health and safety (AIA, 2001; Ulrich, Quan, Zimring, Joseph, & Choudhary, 2004); however, hospitals have not been designed with the explicit goal of enhancing patient safety through facility design. Designing and constructing new hospitals or remodeling existing hospitals create latent conditions and active failures affecting human error, current processes, new processes, a patient safety culture, and overall patient safety. Although no significant research has been conducted on the effect of facility design on patient safety, there is some research on human factors, the effect of human error on patient safety, and facility effect on behavior.

Human factors analysis is defined as "the study of the interrelationships between humans, the tools they use, and the environment in which they live and work" (Weinger, Pantiskas, Wiklund, & Carstensen, 1998, p. 1484). In a health care environment, humans interact with the facility, and its fixed and moveable equipment and technology. As a result, the design of the facility, with its fixed and moveable components, can have a significant impact on human performance.

The work of Reason (1990), Norman (1988), and Leape and colleagues (1995) tells us that human error is attributed to human cognition and the limitations of memory and thought processes. Most daily activities are routine and completed with little or no higher level thought processes. In these types of activities, errors known as slips or mistakes (lapses) can occur for multiple reasons, including distractions, interruptions, multitasking, or any deviation from the routine activity. Other behaviors require a conscious, knowledge-based, or rule-based thought process that often borrows from past experiences. Errors in these behaviors are referred to as mistakes and can result from a lack of knowledge, experience, communication, or even misjudgment.

According to the Institute of Medicine (IOM, 2004), tens if not hundreds of thousands of errors occur every day in the U.S. health care system. Numerous studies have documented the impact of human error on patient safety. In the 1991 Harvard Medical Practice Study, researchers reported that 69% of injuries suffered by patients hospitalized in New York State in 1984 were the result of errors (Leape, 1994). Nearly 14% of these injuries were fatal. In another study conducted by the United States Pharmacopeia's (USP) Center for the Advancement of Patient Safety (2001), 2.4% of errors reported in 2001 (2,539/105,603) resulted in harm. Patients involved in these harmful errors received intensive patient care resulting in prolonged hospital stays, additional testing and monitoring, and increased drug therapy—ultimately increasing the use of hospital resources and costs.

Facility impact on behavior is a documented and researched relationship (Kleeman, 1983; Norman, 1988; Moray, 1994; Ulrich et al., 2004).In the book *Challenge of Interior Design,* Kleeman (1983) stated, "There are those who assert that essentially the design of an interior space and its location not only can communicate with those who enter it but also controls their behavior." In *The Psychology of Everyday Things,* Donald A. Norman (1988) reports that humans do not always behave clumsily, and humans do not always err, but they are much more likely to err when things they use are badly conceived and designed. Finally, Moray (1994, p. 67–91) sums it up well by saying "people of good intentions, skilled and experienced may none the less be forced to commit errors by the way in which the design of their environment calls forth their behavior."

The importance of creating a patient-safe culture and its impact on patient safety is also well documented (IOM, 2004; Kizer, 1998; Reason, 1997). The facility design process "engineered properly" will enhance or create a patient-safe culture.

Finally, the impact of facility design on systems is self-evident. The movement of materials, the movement of patients, the way people and patients see each other, and clinical processes are all heavily influenced by facilities and the fixed and moveable equipment and technology contained within.

SynergyHealth, St. Joseph's Hospital, with the help of leaders in patient safety, quality improvement, and human factors developed facility design principles aimed at creating a safe hospital facility. The application of the facility design principles discussed in this chapter and the use of a design process to enhance or create a patient safety culture is the key initiative of St. Joseph's to reduce human error and improve the quality of care.

LEARNING LAB

St. Joseph's decided to relocate an 80-bed acute care hospital located in West Bend, Wisconsin, near Milwaukee. Its affiliate, SynergyHealth, West Bend Clinic is a multispecialty group of more than 50 physicians serving patients in five locations. In the unique position of building a new hospital, St. Joseph's recognized the opportunity to increase patient safety and promote a patient-safe culture by improving the traditional hospital facility design process. With billions of dollars spent annually on health care construction, St. Joseph's identified the need to develop a set of safety-driven design principles that could be used by all health care organizations, whether they were building a new facility, remodeling, or expanding an existing facility.

Inspired by the Institute of Medicine (IOM) report, *To Err is Human: Building a Safer Health System* (1999), internal discussion at St. Joseph's focused on how design of a new facility could affect patient safety. St. Joseph's contacted leaders in patient safety, quality improvement, and human factors to seek their advice. The belief was that there was an opportunity to learn collectively about how a facility could be designed to improve patient safety. In April 2002, leaders in systems-engineering, health care administration, health services research, human behavior research, hospital quality improvement and accreditation, hospital architecture, medical education, pharmacy, nursing, and medicine participated in a conference titled "Charting the Course for Patient Safety—A Learning Lab." The conference was sponsored in part by a grant from the University of Minnesota, Carlson School Program in Health Administration (Carlson School of Management, 2002). Derived from the Learning Lab was the collective belief that safe hospitals could be designed (Reiling et al., 2004)

- Using a process that supports the anticipation, identification, and avoidance of failure
- By designing against latent conditions and active failures, which compromise physical and organizational defenses
- By creating an organizational culture of safety

Participants in the Learning Lab were instructed to consider human error, why it occurs, and the factors that undermine human performance. A great deal of information has been published on dealing with human error from a systems approach (Moray, 1994; Nolan, 2000; Reason, 2000). In a systems approach, error reduction is achieved by strategically building defenses, barriers, and safeguards into the facility, equipment, and processes that make up the system. To achieve error reduction, an understanding of the types of errors present is necessary. Reason classified errors found in complex systems such as health care as either active failures or latent conditions (Reason, 1990, 2000). Examples of latent conditions present in a health care facility include lack of standardization of equipment and procedures, poor visibility, high noise levels, and excessive movement of patients. Unlike active failures that are difficult to predict, latent conditions can be identified and remedied with safety barriers before they can contribute to an adverse effect. Safety barriers act to prevent a health care provider from committing an active failure or by mitigating the effect of an active failure (Ternov, 2000).

SAFETY DESIGN PRINCIPLES

Participants of the Learning Lab were instrumental in developing a set of safety-driven design principles

TABLE 18–1. Latent Conditions

Visibility of patients to staff
Standardization
Automate where possible
Scalability, adaptability
Immediate accessibility of information, close to the point of service
Patients involved with care
Noise reduction
Minimize fatigue

that would guide the design process. These design principles are aimed at minimizing latent conditions, reducing active failures, and creating a facility design process focused on patient safety.

Latent Conditions

When designing a new health care facility or remodeling an existing facility, we are concerned with identifying and preventing both active errors and latent conditions. Table 18–1 lists the latent conditions identified in the Learning Lab proceedings as areas of concern to patient safety. These areas were specifically addressed in the facility design process.

Visibility of Patients to Staff. There is little debate that health care organizations trail their peers in service innovation. In the 19th century, it was said that form follows function. In the 21st century, it is becoming clear that form shapes function. A well-chosen form helps providers deliver services more efficiently and inexpensively (Drake, 2001). A pod structure allowing close proximity to their patients allows nurses to deliver improved quality by enabling them to quickly respond to patient needs and more effectively monitor patient progress.

Unit designs must allow caregivers to be visual in proximity to the patients under their care as well as accommodate the more traditional orientations of broader-based patient responsibility. This can be accomplished by designing multiple mini-nursing stations throughout the unit, offering alcoves for charting and dictation and allowing for wall desks either in corridors or in the patient rooms. Visibility allows the staff to observe changes in patient's skin color or trouble with breathing or to be "nudged" into remembering medications or to change an IVs. Decreasing the risk of falls when patients attempt to transfer into and out of bed without assistance is also an important target safety goal (Agostini, Baker, & Bogardus, 2002).

Lighting is an important consideration in improving the visibility of patients to staff. The light source chosen can change the appearance of patients, causing unnecessary concerns about potential changes in the patient's medical condition. Proper lighting is also necessary to conduct an accurate assessment of a patient.

Natural light versus other lighting sources is an important consideration. A study conducted at Montefiore University Hospital in Pittsburgh, Pennsylvania showed that patients in rooms with greater natural light took less pain medication, resulting in a 21% reduction of drug costs compared with equally ill patients assigned to darker rooms (Walsh, Rabin, Day, Williams, Choi, & Kang, 2004).

Additional design features to enhance visibility of patients include windows in alcove doors, which allow caregivers to conveniently check on patients without disturbing them; patient care areas wired for cameras; and a family area in every patient room to encourage family members to stay with patients for a longer period of time.

Standardization. Care standardization could substantially affect the basic consequences of organizational factors, enough to reduce medical errors and improve quality. Standardization has been documented as an important strategy in human factors design (IOM, 1999; Norman, 1988). Standardization reduces reliance on short-term memory and allows those unfamiliar with a given process or design to use it safely (IOM, 1999). Much of the work in human factors focuses on improving the human-system interface by designing better systems and processes. This includes standardization of patient rooms, treatment areas, equipment, and procedures. The standardization of the facility and room design—including the location of the outlets, the bed controls, the cupboard the latex gloves are stored in, the charting process, even switches on light fixtures, down to the most minute detail—all have an impact on behavior.

In *The Psychology of Everyday Things,* Norman (1988) discussed design novelty, stating, "Users don't want each new design to use a different method for a task. Users need standardization." Specific examples to consider in facility standardization are as follows:

- Providing truly standardized patient rooms
- Standardizing all emergency exam rooms, postrecovery rooms, ambulatory and

diagnostic exam rooms, and admission and observation rooms

- Standardizing the locations of all gases throughout the facility
- "Migrating" toward standardized IVs, beds, monitors, and all equipment
- Standardizing medication system and other care systems

Automate Where Possible. In its 2001 report, Crossing the Quality Chasm: A New Health System for the 21st Century, the IOM identifies information technology solutions as a necessary component to improving patient safety. Many errors occur because clinicians do not have access to complete and accurate patient information and because they are often responsible for remembering large amounts of knowledge such as drug-to-drug interactions for a large number of medications (IOM, 2004). When designing a new health care facility, technology planning should begin on Day 1 of the design process. If possible, on-going assessment of the type and frequency of errors within the existing institution will assist in determining specific technological needs and in setting priorities. To facilitate the decision process, it is beneficial to create a list of those information technology systems needed or desired and to characterize those systems according to their relationship to one another, impact on patient safety, financial investment, implementation time, and impact on facility design and operation. The financial investment should reflect the total cost of ownership, including costs for purchase, implementation, maintenance, and staff support. Safe hospitals will be highly digitized, minimizing human touches of data and providing robust decision support, with complete and accurate information available at the point of service to multiple providers simultaneously.

Design features include proper wiring for current and future needs, with wired and wireless technologies strategically located throughout the facility. Some examples include computers, radiology screens, monitors, uninterruptible power systems, and information infrastructure. What and how technologies are chosen and used (e.g. nurse call systems, security, materials, bar coding, Radio Frequency Identification Device) could have significant impact on room design and adjacencies.

Scalability, Adaptability. Scalability is the ability to expand or remodel easily so that latent errors are not designed into the building expansion. If radiology does not have adjacent space to expand or evolve to, major diagnostic technologies could be physically separated causing additional handoffs and potential errors. Adaptability is the ability to adapt space for different or evolving services so that latent errors are not created. Patient rooms in the future could be the location where more procedures are performed (surgical and diagnostic), minimizing transfers and handoffs.

Many design and construction concepts can be applied to achieve a scalable or adaptable health care facility—everything from open spaces with modular systems to infrastructure requirements to expansion zones that support scalable and adaptable buildings. Specific examples that St. Joseph's has incorporated into the facility design include ceiling heights (floor to floor) to allow for expansion or changes and wiring or wireless technology that will allow future technology to be easily implemented. Key services located on outside walls allow for expansion sizing of patient rooms that provide greater adaptability.

Immediate Accessibility of Information Close to the Point of Service. Lack of knowledge and information can lead to errors (Reason, 1990). As stated earlier, two significant causes of adverse drug events and potential adverse drug events are lack of drug knowledge and inadequate availability of patient information (Leape et al., 1995).

Technologies have been developed that can assist physicians and other caregivers with complex cognitive tasks such as diagnosis and treatment by providing "real-time" medical information. Examples of these technologies include the Internet, computer-based patient records, and clinical decision support systems. In addition to technological solutions, other examples for improving the immediate accessibility of information close to the point of service include providing charting alcoves directly adjacent to patient rooms, providing caregivers with easy access to patient charts without disrupting the patient, having surgery operating rooms equipped with large boards or monitors that display patient information and scheduled procedures, and verifying physician instructions and the patient's chart and medical records.

Noise Reduction. A report by the World Health Organization (Berglund, Lindwall, & Schwela, 1999) indicated that noise interferes with communication, creates distractions, affects cognitive performance and concentration, causes annoyance, and contributes to stress and fatigue. Mental activities involving a demand on working memory are particularly sensitive to noise and can result in degradation of performance. In a study by Murphy and colleagues (as cited in Morrison, Haas, Shaffner, Garrett, & Fackler, 2002), anesthesia residents exhibited reduced mental efficiency and poorer short-term memory under the noisy conditions of an operating room averaging 77dB(A).

In addition to safety considerations, noise affects the quality of the healing environment for patients. It may elevate blood pressure, increase pain, alter quality of sleep, and reduce overall perceived patient satisfaction (Onen, Alloui, Gross, Eschallier, & Dubray, 2001; Berens, 1999; Topf, 2000). Studies in pediatric and intensive care units (ICUs) have shown that noise routinely disrupts sleep necessary for patient comfort or recovery (Berens, 1999).

The nature of sound and the reverberation rate, or how long the sound remains, has a direct effect on the noise level. When the reverberation rate is long, there is greater opportunity for sounds to blend together, increasing the noise level. With speech communication, a longer reverberation time combined with background noise makes speech perception increasingly difficult (Berglund & Lindwall, 1995).

St. Joseph's has many design features to minimize noise. Examples include no overhead paging; quiet floor coverings (carpet, rubber); "quiet" heating, ventilating, and air conditioning systems; private rooms, standardized with insulation between rooms; more absorbent ceiling tiles; and "quiet" equipment and technology.

Patients Involved With Care. Being informed gives patients the opportunity to participate in shared decision making with physicians and may help patients better articulate their individual views and preferences to physicians and nurses (Deyo, 2000; Wensing, 1998). During a hospital stay, keeping patients and family members informed also can potentially reduce error. Patients or their family members can receive a daily schedule of prescribed medication and treatments. They should be encouraged to verify this information with the caregivers administering the medication or treatment. They can assure that all caregivers have washed their hands. Every patient room has a designed "family" area distinct from the "caregiver" area, which encourages family members to stay with the patient. This family area includes a couch that folds out to be a bed, a desk with appropriate access to the Internet and other power sources, and closets for storage for patient and family members belongings. A large window creates a well-lit and inviting space. In addition, a portable computer on a movable stand, stand will be available for access to patients and families, along with caregivers, in the room, so patients and families can be fully informed of medication and treatment plans. A sink with soap dispenser and alcohol dispenser will be in direct view of the patients, helping them assure compliance with this critical infection control effort.

Minimize Fatigue. Research has identified fatigue as possibly contributing to human error (Ternov, 2000; Jha, Duncan, & Bates, 2002). Although research has not proven the effects of fatigue on patient safety, studies have shown that fatigue has a negative effect on alertness, mood, and psychomotor and cognitive performance, all of which can have an effect on patient safety (Gaba & Howard, 2002; Warltier, Howard, Rosekind, Katz, & Berry, 2002; Jha et al., 2002).

Shift work and, in certain circumstances, long hours and increased workloads are inevitable in patient care. As a result, minimizing fatigue is a complex issue in hospitals that requires a comprehensive approach. In facility design, this could mean minimizing the distances staff must travel between patient rooms, nurses' stations, and treatment areas. This step could affect not only the number of patient rooms per floor but also vertical and horizontal adjacencies of departments. The use of technology can increase the efficiency of workload and reduce the reliance on short-term memory and thought processes. One example is the computerized tube transport system that can either eliminate or significantly reduce the need for staff to hand deliver laboratory specimens, blood products, or medical supplies and can increase the efficiency of workloads (Pevco, 2002). Noise reduction, as previously recommended, is another method of minimizing fatigue in both hospital staff and patients.

Several design features can be implemented to minimize noise for both patients and staff. Building standardized, private patient rooms in which head walls are not shared and walls are insulated reduces the transfer of noise between rooms. Triple-glazed windows reduce outside noise. Noise from inside the hospital can be reduced by eliminating overhead paging, installing absorbent floor coverings

TABLE 18–2. Active Failures/Precarious Events
Operative/postoperative complications/infections
Events relating to medication errors
Deaths of patients in restraints
Inpatient suicides
Transfusion related events
Correct tube—correct connector—correct hole
Patient falls
Deaths related to surgery at wrong site
MRI hazards

(carpet, rubber) and ceiling tile, and by selecting a variety of "quiet" equipment and technology from heating, ventilation, and air conditioning systems to paper towel dispensers.

Active Failures

The active failures that a health care institution focuses on should be based on the institution's own database of sentinel events, adverse events, medical errors and near misses and then on national databases such as those from the Joint Commission on Accreditation of Hospital Organizations (JCAHO). An additional source could be the report *Serious Reportable Events in Health Care,* issued by the National Quality Forum (2002). The facility design process at St. Joseph's focused on preventing the occurrence of nine precarious events, shown in Table 18–2, and their root causes identified from a review of the database of JCAHO and input from the Veterans Administration National Center for Patient Safety. These events will be described in the following paragraphs.

Operative and Postoperative Complications and Infections. The most harmful and costly active failures are operative or postoperative complications and infections. Postoperative infections, surgical wounds accidentally opening, and other often-preventable complications led to more than 32,000 U.S. hospital deaths and more than $9 billion in extra costs annually (Associated Press, 2003). The Center for Disease Control and Prevention estimated that nearly 2 million patients in U.S. hospitals each year develop an infection (JCAHO, 2003a). Approximately 90,000 of these patients die as a result of their infection.

Facility design considerations include placing sinks in every patient care area, allowing providers to wash their hands in front of the patient; using high efficiency particulate air (HEPA) filters areas and ultraviolet lights in all public areas and in key patient care areas to reduce airborne pathogens, eliminating blinds on the windows to reduce condensation, designing single-patient rooms, using air flow in patient rooms (modified "laminar flow"), and standardizing the location of prominently visible sanitizer dispensers. At St. Joseph's, a separate air system, with HEPA and ultraviolet filters, was installed for the Emergency Room (ER) waiting area. In addition, the number of times the air is removed and returned to rooms has been increased from our current location, increasing the probability of filtering out bacteria and viruses.

Events Relating to Medication Errors. Medication-related errors are one of the most common types of errors occurring in hospitals. The IOM report *To Err is Human: Building a Safer Health System* (1999) cited studies showing that in 1993, approximately 7,000 deaths were attributable to medication errors and that one of every 854 inpatient hospital deaths resulted from a medication error. The IOM estimated that increased hospital costs resulting from preventable adverse drug events affecting inpatients are approximately $2 billion for the nation as a whole.

Studies have shown that technology can have a significant impact on patient safety and quality of care (Evans, Pestotnik, & Classen, 1998; Hunt et al., 1998). Facility designs should make certain that proper wiring and cabling is included in all "non-traditional" areas where medication may be dispensed or delivered. In addition, pneumatic tube system wiring should be installed in all areas where medication will be dispensed or ordered with wireless and wired capability for computers and other technologies such as bar coding. Technology plans should include an integrated system consisting of electronic medical records, decision support, CPOE (Computerized Physician Order Entry), bar coding, and an automated pharmacy system. Design plans to reduce medications errors include single-patient rooms where medications will be stored in locked containers and verified using bar-coding technology when medications are delivered.

Deaths of Patients in Restraints. Despite efforts to reduce the use of restraints in medicine, preventing restraint deaths continues to be an area of concern in health care. In 2001, approximately 8% of all sentinel events reported to the

JCAHO (n.d.) were restraint deaths. Facility design considerations to reduce the risk of restraint deaths include providing a comfortable space for family members to stay with patients to provide patients with additional support and comfort. In addition, increasing the visibility of patients so that staff can provide increased or continuous observation of patients would minimize the needs for restraints. Visibility could be increased by using cameras, windows in alcove doors, or space in the caregiver or family areas for someone to remain with the patient for longer periods of time.

Inpatient Suicides. Inpatient suicides cause a great deal of distress to both relatives and caregivers. Research indicates there are approximately 30,000 suicides in the United States per year, 5% or 6% of which occur in hospitals (Busch, Fawcett, & Jacobs, 2003).

Because of the challenges faced in predicting suicide risk, facility design considerations are an important component in preventing inpatient suicides. Patient rooms should be designed for maximum visibility through cameras or glass in alcove doors and by providing a comfortable space for family members to remain with the patient for longer periods of time. Patient rooms should be equipped with break-away curtain rods and "suicide proof" shower heads or other comparable equipment. Special attention should be given to ensuring that sharp objects or other harmful items are not stored in patient rooms.

Transfusion-Related Events. The implementation of an integrated technology system for pharmacy also will reduce potential errors and adverse outcomes in the blood transfusion process. Tube system, electronic medical records, bar coding, physician order entry decision support, and automated lab systems are examples of technology that will reduce the probability of transfusion-related events. Proper wiring and location of computers are some facility features necessary to implement the technology.

Correct Tube, Correct Connector, and Correct Hole. Most types of correct tube, correct connector, and correct hole are inadvertent mix-ups of gases or tubes being attached to the wrong locations. Facility design elements to consider include providing common storage of gases, using color coding, standardizing head wall location of gases throughout the hospital, and using different size connectors for all gases and all tube connections. Other "forcing functions" should be developed to catch inadvertent mix-up of gases or tube connection errors.

Patient Falls. Despite considerable research on identifying risk factors and developing fall prevention programs, patient falls remain a common adverse event. Risk factors have been identified to aid health care providers in identifying patients who have a greater likelihood of falling. These risk factors include altered mental status, decreased mobility, history of falls, toileting needs such as needing assistance or incontinence, medications, age, and other factors such as dizziness or length of hospitalization (Evans, Hodgkinson, Lambert, & Wood, 2001).

Falls often occur at night as patients attempt to get out of bed because they are disoriented or need to use the restroom. One innovative suggestion is to develop technology to "catch" high-risk patients before they leave their beds, thus eliminating potential falls. In a JCAHO review of 22 cases, 13 occurred in general hospitals, with one occurring in a psychiatric unit (2000b). Most of the patients were older than 80 years of age. Approximately one-third of the cases involved falling from a bed. Other falls occurred while walking; in the bathroom; or from a commode, gurney, or chair (JCAHO, 2000b).

Facility design considerations that can assist in reducing patient falls include enhancing visibility of patients to staff. Hospitals need to create or use existing technology to "catch" high-risk patients before they leave their beds, thus eliminating potential falls. Facility considerations include bathrooms located near the head of the bed so patients always have a handrail to hold on to and do not have open space to cross. Designing patient rooms with alcoves with windows so medical and surgical patients are more visible to staff would help reduce falls. Beds that have the ability to drop down near the floor minimize the potential of a fall. Flooring choices that are softer (e.g. carpet, rubber) would reduce injuries from falls. "Infrared" technologies that "catch" patients leaving the bed can turn on lights and create an emergency call to the nurse call system. Wiring for digital cameras that could observe patients with a high risk of falling may be useful.

Wrong-Site Surgery. The problem of wrong-site surgery has received widespread media attention in recent years. Although efforts to address this problem have increased, it still remains a significant concern across the nation. Facility design considerations that complement the Universal Protocol and other process recommendations include standardizing

Figure 18–1. Design of patient room.

operating room suites, installing proper lighting, and installing cable for access to digital images and photographs of surgery sites along with x-rays so that "reverse" readings do not occur, and this information about the patient is ever present.

Magnetic Resonance Imaging (MRI) Hazards. Magnetic metal objects are still brought into the MRI with some frequency, causing death and harm to patients. One source is the patient's own metals, such as a pacemaker, or bringing in magnetic metals. Design features that can reduce MRI hazards include USING a three-room process to enter, locating gases in rooms, color-coding all stretchers, wheelchairs and such nonmagnetic, and using a hand-held metal detector to check staff before entering.

Patient Rooms

Based on the safety design principles, including designing around precarious events, our patient rooms evolved from a "traditional" patient room. Each patient room in the new facility will be a private room allowing more space for staff to provide care and for family members who want to stay close to the patient. A small charting alcove adjacent to the room will allow nurses to observe without disturbing the patient's rest, creating a better healing environment.

All rooms will be truly standardized in layout and placement of equipment and supplies. A cabinet or "nurse server" in each alcove will hold the patient's bar-coded medication, kept in a locked box, and all other supplies needed for patient care, allowing the nurse to remain in the room with the patient, reducing fatigue and increasing time spent with the patient. Noise reduction will be achieved through the use of special noise-absorbing ceiling tiles, enhanced steel strength to reduce vibrations, the elimination of overhead paging, and improved insulations between rooms.

For added safety, patient rooms will be wired for the use of cameras to assist with monitoring of high-risk patients. The cameras will connect directly to the nurses' station and will be used only with the consent of the patient. Additional technologies such as the use of automatic lights that go on when the patient attempts to get out of bed are being considered to reduce the potential for patient falls. Bedside computers are planned to allow nurses or other staff to double check medication or other scheduled treatment before administration. Patients will have access to their scheduled medication or other treatments prescribed, encouraging them to become more involved with their own care.

TABLE 18–3. Process Recommendations

Matrix (Post Learning Lab)
Failure Mode and Effects Analysis (FMEA) at each stage of design
Patients/families involved in design process
Equipment planning Day 1
Mock-ups Day 1
Design for the vulnerable patient
Articulate a set of principles for measurement
Establish a checklist for current and future design

FACILITY DESIGN PROCESS CHANGES TO ACHIEVE SAFETY

The traditional hospital design process requires that architects be given program objectives (role and program), room requirements (functional space program), and constraints such as the need to locate next to other departments (block diagrams). Once this preliminary information has been provided, a detailed room-by-room layout is completed and eventually developed into construction documents (design development). Typically, no issues about patient safety are raised, which creates an opportunity to repeat latent conditions or active failures existing in current hospital design that contribute to errors.

St. Joseph's challenge was to change the traditional hospital design process to incorporate the safety-driven design recommendations gleaned from the Learning Lab. Rather than just directing architects to design the new facility based on existing models, the new development process followed specific process recommendations, shown in Table 18–3, to ensure a constant focus on designing for the safety of patients.

A Team Approach to Design

Initially, the design and construction team, consisting of architects, contractors, mechanical, electrical and plumbing architects, the equipment planner, and the owner's representative, met with physicians, hospital executives, and staff members to discuss the facility design principles and to brainstorm ideas about safe design. A matrix was developed as a result of these sessions that identified potential safe design features and their costs.

Hospital leadership recognized that a cross-departmental team approach would be needed and formed an 11-member facility design advisory council. Members of the council represented various departments within the hospital and included management, staff members, and physicians. The council was led by the chief operating officer, and it was responsible for overseeing the design process and providing updated design information to hospital employees and administration.

In addition to the facility design advisory council, design teams were formed for each department within the hospital. The department design teams, ranging from 3 to 10 members, were put together by the department managers. As required by JCAHO, each design team had multidisciplinary representation, including physicians from clinical areas. Each team, with the aid of the architects, was responsible for ensuring that the safety considerations for each facility design principle were met within their department. They were required to complete the guiding principles checklist for their area and report their proposed design recommendations to hospital leadership and the facility design advisory council.

Throughout the design process, hospital employees were encouraged to share their opinions about existing safety concerns and give suggestions for improvements through e-mail, voicemail, a suggestion box, or by completing a staff survey. All suggestions were reviewed by either the department design teams or the Facility Design Advisory Council.

Equipment and Technology Planning

In keeping with the facility design recommendations, equipment planning began early in the design process. An on-site technology fair was held at which staff members were given the opportunity to evaluate information systems and other technologies and generate ideas for application. A long-range technology plan was developed to determine which systems could be implemented immediately or at the completion of the new facility and which would be acquired in the future. One goal is to implement an integrated system that would provide complete patient information at any point of care. Initial technology plans included centralized scheduling; a nurse call system; pneumatic tube transport; and automated systems for pharmacy, and management of materials. Bar coding, electronic medical records, CPOE (Computerized Physican Order Entry) and emergency room systems are scheduled to be implemented in the future.

Application of Failure Modes and Effects Analysis

The management team, facility advisory committee, and department design teams met routinely with architects during the design process. Experts were brought in to advise on critical considerations, such as noise reduction and lighting, and to educate the design teams on the use of failure mode and effects analysis (FMEA). FMEA is a tool commonly used in other industries, such as aviation and manufacturing, to identify and prevent problems associated with product and process design (General Motors, 2001; See the chapter on "Human factors risk management in medical products" for additional information on FMEA.) In hospital design, as in other industries, it is easier to fix potential failures during the planning stages than after construction has begun.

FMEA was conducted at each stage of the design process such as with block diagrams (adjacencies). Although the use of FMEA is very time-consuming and often labor intensive, it is very beneficial in identifying potential failures and developing innovative solutions associated with design considerations. One example in which FMEA was used was in assessing the movement of a very ill patient from the ER to Radiology to ICU. The initial design had ICU on a different floor from the ER and Radiology. Using FMEA, we identified that the more frequently a patient must be moved, especially between floors, the greater likelihood that errors could occur or equipment could fail, resulting in harm to a patient. In addition, ill patients will need nurses to accompany them, therefore leaving the department short of staff and creating the potential for errors or adverse events. For example, as a result of the application of FMEA to the design, the ICU is now adjacent to the ER.

Mock-ups

Mock-up designs of patient rooms began immediately for the ICU, Medical/Surgical floor, and the New Life Center. Many different types of mock-ups exist, from two-dimensional to computer generated to physical construction. At St. Joseph's, two mock-up rooms were constructed: one on the Medical/Surgical floor and the other in the New Life Center. Physicians, nurses, staff, patients, and family members were invited to view and evaluate the rooms. Suggestion forms were placed in each room to encourage feedback. As a result of suggestions received from staff, the alcove storage was redesigned, desk heights were changed, and the configuration of the bathroom evolved.

The mock-up rooms can serve two functions in addition to designing a safer environment: simulation on systems and future education and orientations. With a mock-up, simulations on redesigned or current processes can occur. They can be routine functions, medication delivery in a patient room, or complex circumstances such as an emergency code. Ongoing training until the completion of the remodeled or new space can be quite helpful in minimizing transition errors.

Evaluation

Evaluation has been an integral part of the design process. Ongoing evaluation of the existing facility and processes has been critical to the design process. Identifying where errors occur and what latent conditions contribute to those errors has assisted in identifying needed technology and improved processes and creating safer, more efficient space in the new facility. Retreats have been periodically conducted to review our commitment to, and results of, designing around patient safety. The ongoing discussion is changing our culture to be more focused on patient safety. We are redesigning many of our processes to address the safety design principles, including designing around precarious events (excerpt from Reiling et al., 2004).

CULTURE OF SAFETY

When the facility process began, St. Joseph's did not recognize that patient safety principles, applied to facilities, would affect patient safety. In addition, recognition of the facility impact on precarious events and their prevalence within St. Joseph's did not exist. "The State of Wisconsin does not have errors like the rest of the country and clearly St. Joseph's does not have them," was our initial belief. Our efforts to establish a culture of safety led to the implementation of an anonymous reporting system, education of employees and physicians, use of checklists, a focus on capital to improve safety, and involving patients in the design process.

CONCLUSION

The safety design principles and design process recommendations discussed in this chapter apply not only to facility design but also to system redesign. Standardization, noise reduction, the use of checklists, and FMEA are examples of facility design principles that can be beneficial in creating patient-safe systems. Human factors applied to system redesign to reduce latent conditions and active failures are an important application of the national Learning Lab findings.

The facility design process at St. Joseph's involved a continuous review of human factors. Efforts focused on reducing reliance on short-term memory and thought processes; improving communication and dissemination of information between providers; and maximizing standardization, automation, and all the design principles. The focus on improving safety influenced every aspect of the facility design process from department adjacencies to building materials. The focus on improving safety has also changed our systems. The new facility has literally been designed "from the inside out" with functional relationships between spaces inside driving the exterior shape of the building. The architects on this project have labeled the innovative design "The Synergy Model" to describe the process of shaping the entire building, its spaces, and systems to work efficiently as a whole for the care and safety of patients.

References

Agostini, J. V., Baker, D. I., Bogardus, Jr., S. T. (2002). In K. G. Shojania, B. W. Duncan, K. M. McDonald, & R. M. Wachter (Eds.), *Evidence Report/Technology Assessment, No. 43. Making Health Care Safer: A Critical Analysis of Patient Safety Practices.* Rockville, MD: Agency for Health Care Research and Quality. Retrieved December 17, 2002, from www.ahcpr. gov/clinic/ptsafety.

American Institute of Architects (AIA). *Guidelines for Design and Construction of Hospital and Health Care Facilities.* Washington, DC: AIA Press, 2001.

Associated Press. (2003). Hospitals take a huge toll. *MSNBC.* Retrieved March 28, 2004, from http://www.msnbc.com/news/977139.asp

Berens, R. J. (1999). Noise in the pediatric intensive care unit. *Journal of Intensive Care Medicine. 14,* 118–129.

Berglund, B., & Lindwall, T. (Eds.). (1995). Community noise. Prepared for the World Health Organization. *Archives of the Center for Sensory Research, 2.* Retrieved January 6, 2003, from http://www.sensoryresearch.org/AbsBBTL.asp

Busch, K. A., Fawcett, M. D., & Jacobs, D. G. (2003). Clinical correlates of inpatient suicides. *Journal of Clinical Psychiatry, 64,* 14–19.

Carlson School of Management. (2002). *Designing a Safe Hospital.* Publication 1 Series. Minneapolis, MN: University of Minnesota.

Deyo, R. A. (2000). Tell it like it is. *Journal of General Internal Medicine, 15,* 752.

Drake, R. (2001, August 31). Hospital design can help make the bottom line better. *Washington Business Journal, 20,* 35. Retrieved December 17, 2002, from http://washington.bizjournals.com/washington/stories/2001/09/03/focus4.html

Evans, D., Hodgkinson, B., Lambert, L., & Wood, J. (2001). Falls risk factors in the hospital setting: A systematic review. *International Journal of Nursing Practice, 7,* 38–45.

Evans, R. S., Pestotnik, S. L., Classen, D. C., et al. (1998). A computer-assisted management program for antibiotic and other antiinfective agents. *New England Journal of Medicine, 338,* 232–238.

Gaba. D. M., & Howard, S. K. (2002). Fatigue among clinicians and the safety of patients. *New England Journal of Medicine, 347,* 1249–1254.

General Motors Corporation. (2001). *Potential Failure Mode and Effects Analysis (FMEA): Reference Manual* (3rd ed.). Detroit, MI: Author.

Gosbee J., & DeRosier, J. (2002). *Oxygen (Compressed Gas) Cylinder Hazard Summary.* Washington DC: VA National Center for Patient Safety.

Hunt, D. L., Haynes, R. B., Hanna, S. E., Smith, K., et al. (1998). Effects of computer-based clinical decision support systems on physician performance and patient outcomes. *JAMA, 280,* 1339–1345.

Institute of Medicine. (1999). *To Err Is Human: Building a Safer Health System.* Washington DC: National Academy Press.

Institute of Medicine. (2001). *Crossing the Quality Chasm: a New Health System for the 21st Century.* Washington DC: National Academy Press.

Institute of Medicine. (2004). *Patient Safety: Achieving a New Standard for Care.* Washington DC: National Academy Press.

Jha, A. K., Duncan, B. W., & Bates, D. W. (2002). Fatigue, sleepiness, and medical errors. In K. G. Shojania, B. W. Duncan, K. M. McDonald, & R. M. Wachter (Eds.), *Evidence Report/Technology Assessment, No. 43. Making Health Care Safer: A Critical Analysis of Patient Safety Practices.* Rockville, MD: Agency for Health Care Research and Quality. Retrieved January 6, 2003, at http://www.ahcpr.gov/clinic/ptsafety/chap46a.htm

JCAHO. (2000b). Fatal falls: Lesson for the future. *Sentinel Event Alert,14.* Retrieved March 28, 2004, from http://www.jointcommission.org/SentinelEvents/SentinelEventAlert/sea_14.htm

JCAHO (2001). A Follow up review of wrong site surgery *Sentinel Event Alert 24.* Retrieved March 28, 2004 from http://www.jointcommission.org/SentinelEvents/SentinelEventAlert/sea_24.htm

JCAHO. (2003a). Infection control related sentinel events. *Sentinel Event Alert, 28.* Retrieved March 28, 2004, from http: // www. jcaho. org / about+us / news + letters /sentinel + event+alert/print/sea_28.htm.

Kizer, K. W. (1998). Large system change and a culture of safety. *Proceedings of Enhancing Patient Safety & Reducing Errors in Health care.* Rancho Mirage, CA.

Kleeman, W. B., Jr. (1983) *The challenge of interior design.* New York: Van Nostrand Reinhold.

Leape, L. L. (1994) Error in Medicine, *Journal of the American Medical Association 272*(23), 1851–1857

Leape L. L, Bates D. W., Cullen D. J., Cooper, J., Demonanco, H. J., Gallivan, T., Hallisey, R. et al. (1995). Systems analysis of adverse drug events. *JAMA, 288,* 501–513.

Moray, N. (1994). Error reduction as a systems problem. In M. S. Bogner (Ed.), *Human Error in Medicine* (pp. 67–91). Hillsdale, NJ: Lawrence Erlbaum Associates, Inc.

Morrison, (2002). Noise, Stress and Annoyance in a Pediatric Intensive Care Unit, *Critical Care Medicine, 30.*

National Quality Forum. (2002). *Serious Reportable Events in Health Care.* Washington, DC: Author.

Nolan, T. W. (2000). System changes to improve patient safety. *British Medical Journal, 320(7237) 771-773.*

Norman, D. A. (1988). *The Psychology of Everyday Things.* _______: Basic Books.

Onen, S. H., Alloui, A., Gross, A., Eschallier, A., & Dubray, C. (2001). The effects of total sleep deprivation, selective sleep interruption and sleep recovery on pain tolerance thresholds in healthy subjects. *Journal of Sleep Research, 10,* 35–42.

PEVCO Systems International. (2002). *PEVCO Smart System: Health Care's Most Advanced Computerized Pneumatic Tube Transport System.* Baltimore, MD: Author.

Reason, J. (1990). *Human Error.* New York: Cambridge University Press.

Reason, J. (1997). *Managing the Risks of Organizational Accidents.* _______, United Kingdom: Ashgate Publishing Limited.

Reason, J. (2000). Human error: Models and management. *BMJ, 7237,* 768–770.

Reiling, J. G. Knutzen, B. L., Wallen, T. K. McCullough, S., Miller R., & Chernos S. (2004) Enhancing the traditional hospital design process: a focus on patient safety. *Joint Commission Journal on Quality & Safety,* 30, 115–124.

Ternov, S. (2000). The human side of mistakes. In P. L. Spath (Ed.), *Error Reduction in Health Care* (pp. 97–*138).* San Francisco, CA: Jossey-Bass.

Topf, M. (2000). Hospital noise pollution: An environmental stress model to guide research and clinical interventions. *Journal of Advanced Nursing, 31,* 520–528.

Ulrich, R., Quan, X., Zimring, C., Joseph, A., & Choudhary, R. (2004). *The Role of the Physical Environment in the Hospital of the 21st Century: A Once-in-a-Lifetime Opportunity.* The Center for Health Design. Retrieved December 29, 2004, from http://www.healthdesign.org/research/reports/

U.S. Pharmacopeia (USP). (2002). *USP Releases New Study on Medication Errors at U.S. Hospitals.* Rockville, MD: Author.

Walsh, J. M., Rabin, B. S., Day, R., Williams, J. N., Choi, K., & Kang, J.D. (2004). The effect of sunlight on post-operative analgesic medication usage: *Psychosomatic Medicine, 67*(1), 156–163.

Warltier, D. C., Howard, S. K., Rosekind, M. R., Katz, J. D., Berry, & A. J. (2002). Fatigue in anesthesia. *Anesthesiology, 97,* 1281–1294.

Weinger, M. B., Pantiskas, C., Wiklund, M., & Carstensen, P. (1998). Incorporating human factors into the design of medical devices. *JAMA, 280,* 1484.

Wensing, M., & Grol, R. (1998). What can patients do to improve health care? *Health Expectations, 1,* 37–49.

World Health Organization (2001). *Fact sheet: Occupational and community noise, 258.* Retrieved January 6, 2003, from http://www.who.int/mediacentre/factsheets/fs258/en/index.html

19

THE PHYSICAL ENVIRONMENT IN HEALTH CARE

Carla J. Alvarado
University of Wisconsin, Madison

> *"... it cannot be denied that the most unhealthy hospitals are those situated within the vast circuit of the metropolis ... Facts such as these (and it is not the first time that they have been placed before the public) have sometimes raised grave doubts as to the advantage to be derived from hospitals at all, and have led many an one to think that in all probability a poor sufferer would have a much better chance of recovery if treated at home,*
>
> *It may appear a strange principle to enunciate as the very first requirement in a hospital that it should do the sick no harm."*
>
> Florence Nightingale
> London, England 1863

Florence Nightingale is mostly known for her radical innovations in nursing care. Besides being a nurse, reformer, statistician, epidemiologist, and humanitarian, Nightingale was a hospital designer. Nightingale's *Notes on Hospitals* (Nightingale, 1863) provided detailed recommendations on the proper physical environment for civilian health care institutions. Some of these recommendations remain today. Nightingale felt patients needed fresh air circulation for infection prevention and sunlight and windows with outside views to raise patient spirits and to quicken healing. Just as infection control and patient safety and the patient experience were at the forefront of Nightingale's work in the design of the hospital environment, so it continues in the design of modern health care facilities.

The goal of this chapter is to identify overall the human factors components and specific key environmental elements (e.g., light, sound, climate, arrangement of space) that directly affect the patient experience in the health care system. These human factors components and elements not only influence patients' satisfaction, privacy, confidentiality, safety, stress but also their health care providers' safety, quality of care, and quality of working life.

This chapter discusses facilities common to communities in the United States. Facilities with unique services, such as freestanding ambulatory surgical centers, long-term care or assisted living, hospice care, home care, mobile emergency care, psychiatric hospitals, and special care facilities such as Alzheimer's and other dementia units, may require special environmental considerations. However, sections of the chapter will be useful for all care settings, and the human factors principles presented are applicable in most settings.

The care environment discussed in this chapter consists of those features in a built health care entity that are created, structured, and maintained to support quality health care [American Institute of Architects (AIA) & Facility Guidelines Institute (FGI) 2001]. As more patient care moves out of the traditional health care settings and into the patient's home, patients and their families have become more involved in the course of care. Medical devices continue to increase the technology of care both in the built environment, the mobile environment, and

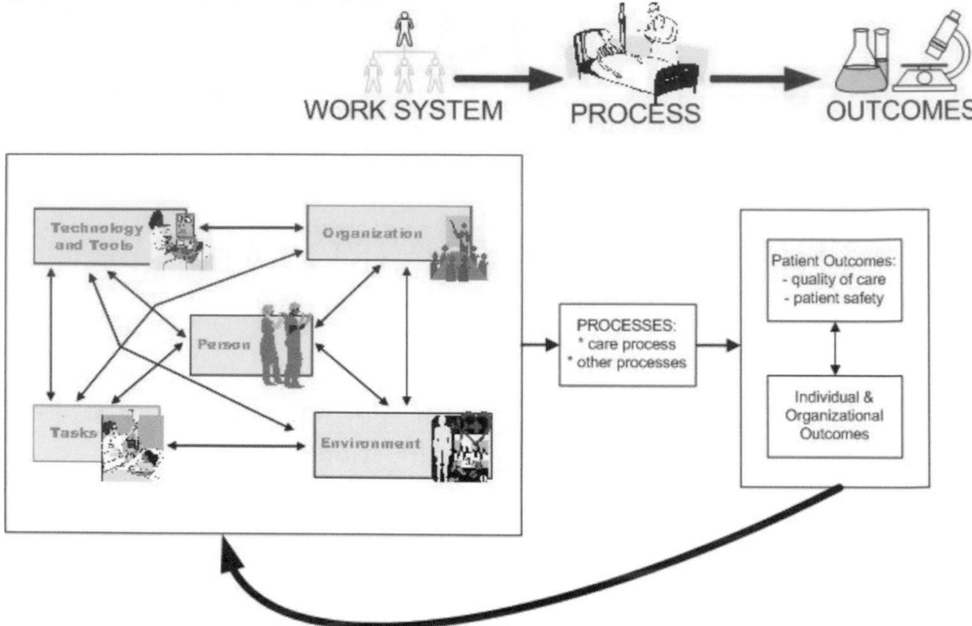

Figure 19–1. Systems Engineering Initiative for Patient Safety (SEIPS) model of work system and patient safety.

now the home care setting. Facilities need to respond to the changing physical environment requirements for accommodations for both the people and the technologies, considering human factors principles in their response to the surrounding needs for physical space.

The physical environment does not constitute the total environment of care for patient safety. Social factors and organizational factors are also highly important and contribute to the human factors/systems approach to patient safety. It is important to ensure that the physical environment enhances the dignity of the patient and complies with federal and state regulations through environmental features that permit patient privacy and confidentiality in all care settings. Other chapters in this handbook address the social, organizational, and systems factors related to health care and patient safety.

"Human factors" is generally defined as the study of human beings and their interaction with products, environments, and equipment in performing tasks and activities (Czaja, 1997). These interactions take place within a physical environment. The late Dr. Avidis Donabedian's classic model of quality of care, which developed into the now-standard method for evaluating the best hospitals in the country, incorporates three major components of care: structure, process, and outcome (Donabedian 1988). Within Donabedian's well-known structure–process–outcome model, Carayon and colleagues (Carayon 2003; Carayon, Alvarado et al. 2003) embeded a work system/ human factors model (Figure 19–1) developed by Carayon and Smith (Smith & Carayon-Sainfort, 1989; Carayon & Smith 2000; Smith & Carayon 2000).

The Systems Engineering Initiative for Patient Safety (SEIPS) is funded by the Agency for Healthcare Research and Quality. The SEIPS model explicitly recognizes the interdependent nature of the five major aspects of a work system: a *care provider* performing various *tasks* using *tools and technology* in a given *environment* within an established *organization*. Thus, the model provides necessary specification to the structural elements posed by Donabedian. The physical environment affects the processes of care or the task, the people performing or receiving this care, and the tools and technology being used in the specific care organization. Research has shown the effect of the physical or built environment on clinical outcomes. Some elements of the physical environment affect staff

TABLE 19–1. Human Factors Texts

Title	Authors	Publisher
Handbook of Human Factors and Ergonomics, 2nd Edition (1997)	Salvendy, G., editor	John Wiley & Sons
An Introduction to Human Factors Engineering, 2nd Edition (2004)	Wickens, C. D., Lee, J. D., Liu Y., and Gordon Becker, S. E.	Pearson Prentice Hall
Human Factors in Engineering and Design, 7th Edition (1993)	Sanders, M. S. and McCormick, E. J.	McGraw-Hill, Inc.
Kodak's Ergonomic Design for People at Work, 2nd Edition (2004)	Chengalur, S. N., Rodgers, S. J., and Bernard, T. E.	John Wiley & Sons

performance more directly. Nosocomial infections and physiologic responses to light, noise, temperature, and the like are the most often studied (Topf 1985; Topf 1988; Topf & Dillon 1988; Topf 1992; Buemi, Allegra et al. 1995; Cabrera & Lee 2000; Topf 2000; Topf & Thompson 2001; Morrison, Haas et al. 2003; Sehulster & Chinn 2003). Potential patient safety hazards in the environment are often localized in the medical devices used by the worker; therefore, much of patient safety analysis in the literature focuses on hazards associated with care devices and procedures. However, as shown in Figure 19–1, patient safety hazards are created by a combined interaction of the task, equipment, organizational, and environmental conditions with the patient or the health care provider. The physical environment can either support or impede what the patients and staff want and need to do in a health care facility. Thus, the physical environment can add to caregiver or patient stress and can be a major detriment to the course of a patient's care. The physical environment should be designed to reduce patient, family, and staff stress wherever possible.

Patient and employee satisfaction is another outcome that may be influenced by the physical environment, but this outcome is less studied. Although elements of the physical environment may support or detract from patients' health care experience and staff members' quality of working life, they will not be addressed in this chapter. Research- and evidence-based materials are available on the subject of patient satisfaction and staff members' quality of working life to support these environmental design goals (Rubin, Owens et al., 1998). In addition, medical device controls and displays, workstation design, mechanical hazards (strains, sharps injury, slips, and falls), electrical hazards (shock, fire), toxic substances (medical gases, therapy agents, radiation) and fire hazards are all associated with the physical environment of patient care. Well-established standards and guidelines exist for these areas and are referenced throughout in the chapter but will not be expanded on.

ENVIRONMENT OF CARE: STANDARDS, GUIDELINES, RULES, AND REFERENCE TEXTS

Several excellent human factors texts are available for a detailed discussion of the physical environment. Much of the discussion and recommendations made in this chapter reference these texts (see Table 19–1)

In addition to these human factors texts, many professional society guidelines; accreditation standards; codes; and federal, state, and local rules that address the environment of care. But be aware; some manufacturers' recommendations and societal guidelines or standards are based on research documentation in reviewed scientific literature and some are not. Specific contents of many standards and guidelines are based on older technologies and may not meet current needs; others simply are not in agreement in content or recommendations, and the extent of their disagreement can be quite large. Applying external recommendations in isolation without considering the timeliness, science, and interactions of other related documents may result in less-than-optimum outcomes. In addition, the interactions of the various components of the SEIPS model (Figure 19–1) should always be taken into consideration in the design of the built environment. Based on the model's elemental interactions, the physical environment and its impact on quality care can be ever changing.

This chapter will not generate yet another set of detailed recommendations for environmental design nor will it go into great detail regarding the standards,

TABLE 19–2. Originators of U.S. Codes, Guidelines, Standards and References Applicable to the Health Care Environment

Organization	*Web Address*
Agency for Healthcare Research and Quality (AHRQ)	http://www.ahrq.gov/
American Society of Heating, Refrigerating and Air–Conditioning Engineers (ASHRAE)	http://www.ashrae.org
American National Standards Institute	www.ansi.org
American Society of Mechanical Engineers (ASME)	http://asme.org
American Society for Testing Materials (ASTM)	http://www.astm.org
Americans With Disabilities Act (ADA)	http://www.usdoj.gov/diabilities.htm
Association for the Advancement of Medical Instrumentation (AAMI)	http://www.aami.org
Association for Professionals in Infection Control and Epidemiology (APIC)	http://apic.org
Centers for Disease Control and Prevention (CDC)—Hospital Infection Control Practices (HICPAC)	http://www.cdc.gov
Department of Defense (DOD)	http://astimage.daps.dla.mil/quicksearch
ECRI	http://ecri.org
Food and Drug Administration (FDA)	http://www.fda.gov
Human Factors and Ergonomics Society	http://www.hfes.org
Illuminating Engineering Society of North America (IESNA)	http://www.iesna.org
International Code Council (ICC)—International Building Code	http://www.iccsafe.org
Joint Commission on Accreditation for Healthcare Organizations (JCAHO)	http://jcaho.org
National Council on Radiation Protection and Measurements (NCRP)	http://www.ncrp.com
National Fire Protection Association (all fire codes are adopted for use by Centers for Medicare and Medicaid Services (CMS))	http://www.nfpa.org/codes/index/html
Nuclear Regulatory Commission (NCR)	http://www.ncr.gov
Occupational Safety and Health Administration	http://www.osha.gov

rules etc. on the various environmental elements. Therefore, Table 19–2 presents a summary of current U.S. standards, guidelines, and rules to consider when addressing the human factors elements and design of any health care environment.

In addition to the entities in Table 19–2, the AIA and the FGI revised (anticipated July 2006) the 2001 edition of the *Guidelines for Design and Construction of Hospitals and Health Care Facilities* (AIA & FGI 2001). At present, more than 40 states and the Joint Commission on Accreditation for Healthcare Organizations (JCAHO) reference the *Guidelines* for licensure or accreditation of health care facilities. In addition, the U.S. federal government continues to reference the *Guidelines* in regulating Department of Health and Human Services medical facilities. Applicable environmental aspects of this document include the therapeutic environments discussed in this chapter, and the *Guidelines* should be consulted. Perhaps the most controversial issue the AIA revised document addresses is the need for single versus semiprivate room occupancy. Currently, this revision is under discussion, because in addition to fiscal issues, there are patient safety pros and cons to single versus semiprivate rooms. Consider for example the human factors elements in this proposed recommendation. Sources of stress for patients are perceived lack of privacy and control, noise, and crowding (Schmaker & Pequegnat, 1989); a single room may address these issues. Nosocomial infections may be passed between patients occupying the same room via the hands of their health care providers or shared equipment. However, most falls occur in patient rooms, among elderly patients who are attempting to go to the bathroom when they are alone (Kerzman, Chetrit, et al., 2004). Multioccupancy patient rooms may provide more constant supervision and therefore may be more appropriate for patient safety in this scenario. In new construction, the maximum number of beds per room shall be one unless the functional program demonstrates the neccessity of a two-bed arrangement. Approval of two-bed arrangement shall be obtained from the licencing authority.

Environment Design for the Disabled

The Americans with Disabilities Act (ADA) became law in 1990. Under Titles II and III of the

ADA, public, private, and public service hospitals and other health care facilities environment must comply with the accessibility and usability guidelines (U.S. Department of Justice [DOJ], 1990,1992,1994). It should be recognized, however, that the users of health care facilities often have very different environmental accessibility needs from the typical adult individual with disabilities addressed by the ADA model standards and rule. Hospital patients, and especially long-term care nursing facility residents, because of their stature, reach, and strength characteristics, typically require the assistance of caregivers during transfer maneuvers or activities of daily living. Many prescriptive requirements of model accessibility standards place both older persons and caregivers at greater risk of injury. Flexibility may be permitted for the use of assistive devices or configurations that provide safety considerations for this assistance (DOJ, 1990,1992,1994), and these should be considered in the human factors and ergonomics assessment of the physical environment.

Environmental Design Provisions for Disasters

In health care settings where there is recognized potential for hurricanes, tornadoes, flooding, earthquakes, or other regional disasters, environment design should consider the need to protect the lives and safety of all health care facility patients and workers. In addition, the environment must address the very significant need for continuing care services after such disasters. Acute care facilities with emergency and intensive care unit (ICU) services can serve as receiving, triage, and treatment centers in the event of nuclear, biological, or chemical exposures. Human factors principles are applicable to the designation and design of specific areas for these functions and the threat of terrorism initiatives (Institute of Medicine [IOM], 2002). The U.S. Department of Defense (DOD) determined mass smallpox vaccinations can be conducted safely with very low rates of serious adverse events. Program implementation emphasized human factors: careful staff training, contraindication screening, recipient education, and attention to bandaging. Their experience suggests that broad vaccination programs may be implemented with fewer serious adverse events than previously believed (Grabenstein & Winkenwerder, 2003). In addition, health care data security is an essential requirement in all antibioterrorism efforts. Developers of medical information systems should use the existing security development and human factors evaluation methods to foresee as many as possible of the environmental and technical factors that may endanger data security and apply appropriate precautions against this threat (Niinimaki, Savolainen, & Forsstrom, 1998).

INDIVIDUAL BUILT ENVIRONMENT COMPONENTS

Space and Physical Constraints

The goal of human factors is to design systems that reduce human error, increase productivity, and enhance safety and comfort (Wickens, Lee et al., 2004). Workspace design and component placement (e.g., medical equipment, furniture, workstations) is one of the chief fields that improve the fit between humans, medical devices, and the patient care physical environment. The health care environment presents many challenging venues in space and physical constraints. Important issues include workspace, equipment and furnishings, placement, reach dimensions, clearance dimensions, and visual dimensions in the design of the ambient environment. Ideally, we would like to place each component in an optimum location for providing safe, high-quality care. This optimum location would be predicated on the human capabilities and characteristics, including sensory capabilities, anthropometric (data on body dimensions of various populations) and biomechanical (mechanics of muscular activity) characteristics, and the interindividual differences of the users (Sanders & McCormick, 1993). Unfortunately, in the crowded environment of a health care facility, this goal is not always possible. Only so many medical device displays can fit in the optimum viewing area; only so many patient rooms can be close to the nursing station; only so many controls can be placed in an area for rapid response. An additional factor that must be considered is that in the same people, these differences change over time. The average age of a nurse in the United States is now in the mid to late 40s. People's sensory and musculoskeletal systems change over time, and the physical environment must be accommodated as the human body ages.

Sanders and McCormick (Sanders & McCormick, 1993) provide specific principles that should be applied when determining the arrangements of components within a physical space:

1. Importance Principle. This principle states that important components should be placed in convenient locations. "Importance" refers to the degree to which the component is vital to the achievement of the objectives of the system; for example, the scanning device used in a medication bar coding system would need to be located in a convenient location for the nursing staff use, because it is essential to the bar-coding scan.
2. Frequency-of-Use Principle. This principle states that frequently used components should be placed in convenient locations; for example, medication carts should be accessible and convenient to nurses and pharmacy personnel alike, and computer order entry equipment should be convenient to patient care venues, not hidden in a report room.
3. Function Principle. The function principle of arrangement provides for the grouping of components according to their function. Thus, temperature indicators and temperature controls might well be grouped; central venous catheter placement equipment and sterile field supplies might be stored together on a designated "line-placement cart."
4. Sequence-of-Use Principle. In the use of items, sequences or patterns of relationships frequently occur in performing some task. In applying this principle, the items are arranged so as to take advantage of such patterns; for example, patient resuscitation equipment is located at the head of the patient bed, and commercial packs for urinary drainage placement start with sterile gloves on the top layer of the pack to facilitate removal and for donning before catheter insertion.

Arrangement of Components

Arranging components in a work space—be they the computer workstations and monitors in a nursing station or the controls and displays at an ICU bed—requires the availability of the relevant data and the use of certain methods in applying that data. The types of data relevant for arranging components in the environment generally will fall into three broad categories (Sanders & McCormick, 1993):

1. Basic Data About Human Beings. These are anthropometric and biomechanical data and data on sensory, cognitive and psychomotor skills. Such data generally comes from research and are published in various tables, source books and papers (Panero & Zelnick, 1979; Gordon, Churchill et al. 1989; Kroemer 1989; (ISO) 1996; Kroemer, Kroemer et al. 2001).
2. Task Analysis Data. These are data about the work activities of the people who are involved in the specific system or work environment. Task analysis is at the heart of the human factors contributions to arranging components in a workspace. Often care providers have experience performing some level of task analysis as they participate in Root Cause Analysis (RCA), failure modes and effects analysis or FMEA, or usability testing exercises in their health care facilities.
3. Environmental Data. This category covers any relevant environmental features of the patient care situation such as lighting, noise, temperature, and the like. A discussion of these topics follows in the chapter.

When modifying an existing environment, data relating to this environment is appropriate to collect. The data can be obtained in various methods. Observations, videotaping, and interviews with both new and experienced personnel are all useful methods of obtaining the necessary data.

In the case of new systems, facilities information about the activities to be performed can be based on the same methods and can be inferred from tentative drawings, plans, procedures, or concepts available (see the chapter, Human Factors in Hospital Safety Design, in this handbook for a discussion of these topics).

Relationships between components, whether they are care providers, patients, or things, are called links. Links fall generally into three classes:

- Communication links:
  - Visual (person to person or person to equipment)

  - Auditory or voice (person to person, person to equipment, or equipment to person)
  - Touch (person to person, person to equipment, or equipment to person)
- Control links:
  - Equipment controls (person to equipment)
- Movement links from one location to another:
  - Eye movements
  - Manual or foot movements or both
  - Body movements

The kind of information collected about links includes how often the components are linked and in what sequence the links occur. For example, how often does the nurse leave the patient bedside to procure care supplies? Where do these needed supplies reside? How often is a patient status monitor viewed? Is information communicated to another care provider immediately before or after the patient status monitor viewed?

Probably the most common method of arranging components by using link data is through trial and error. The designer physically arranges scale drawings of the components, trying to keep the most frequently used components in the most advantageous locations. In recent literature on designing and building a patient-safe hospital, environmental FMEA or (E)FMEA was performed (see the Chapter: Human Factors in Hospital Safety Design, in this handbook). Regardless of what technique is used, it is important that the resulting arrangement be validated by using a mock-up or simulator with actual health care providers carrying out actual or simulated tasks (Sanders & McCormick, 1997). No matter how thorough the design data, end-users' "hands on" input is vitally important.

In addition, hospitals and clinics are forever adding patient care equipment and technology. Unfortunately, onsite storage of this equipment is most often not considered at the time of purchase. The environmental assessment needs to consider equipment storage requirements and how the equipment interacts with the available space and other critical equipment in that space. Devices designed for clinical environments are often difficult to modify for use in patient homes or mobile care units such as ambulances and helicopters, where use and storage space may be far more limited.

The design of the inanimate environment has influence on the care providers' personal ease, with important components for patient care being placed at convenient locations for frequent retrieval, line of sight and control, and consistency with other layouts within the system. Human factors engineering as it applies to this user–environment–control strategy can effectively reduce the patient risk within these components.

The following physical space issues should be considered in all health care venues:

- Clearance problems are among the most often encountered health care space issues. They are most important with medical care devices. The space needed for the medical equipment, as well as between and around, it is critical for the operators' proper and safe use and the patients' safety. The transport care environment, such as ambulances and helicopters, are extremely limited in space and often pose a clearance challenge. Home care may pose an extreme challenge, where the use and storage of the medical equipment must compete with the family's personal belongings, and no governmental safety standards or fire codes are applicable.
- Reach requirements of the medical devices and environment should not exceed the arms of the smallest user as they operate a hand-operated device or activate a foot pedal (Wickens, Lee et al. 2004).
- In addition to reach requirements, critical controls or devices must be accessible and must not be placed in tandem with other critical controls or devices that may inadvertently be activated or deactivated. Although link analysis may suggest minimizing distances between components, such as the sequential links between controls, there is physical space required in the operation of various controls to avoid touching other controls. One must consider the combination of anthropometric factors (such as larger fingers and hands or longer and shorter arms) and the psychomotor movements made in the use of the device controls.
- A well-designed environment should also consider the cleaning and maintenance requirements of the medical equipment

and environmental surfaces. What space is required for the reprocessing or maintenance? What physical hazards (e.g., moisture, electrical cords, brushes, tools) will be added to the physical space for this work? What cleaning requirements would pose hazards to patients and care providers in daily environmental hygiene?

- Medical equipment placement should ensure that the visual displays can be easily seen and read by the care providers in the inanimate environment. This requires proper positions and clear line of sight with respect to the medical equipment design and other equipment in the area. The normal line of sight is usually considered to be approximately 15° below the horizon (Sanders & McCormick, 1993). The area for most convenient visual displays has generally been considered to be defined by a circle roughly 10°–15° in radius around the normal line of sight. However when arranging the care environment, one must be mindful of patient privacy and patient-sensitive information being displayed inappropriately.
- Falls resulting in injury are relatively common in health care for both patients and workers (Herwaldt & Pottinger, 2003; Tinetti 2003; Kerzman, Chetrit et al., 2004). Fall hazards may be created by a combination of medical equipment placement and environmental conditions. People can fall and sustain injuries in a number of ways associated with the placement of objects, including furniture placement; proximity of bathroom or portable commode to patient bed; wet surfaces created by the patient, their care equipment, medications, or placement of patients' drinking water; equipment cords or tubing associated with the equipment; or the equipment placement in the care environment (Edelberg, 2001; Morse, 2002; Kerzman, Chetrit et al., 2004).

Environment Components and Nosocomial Infection

The inanimate health care facility environment as it directly relates to medical devices and patient care procedures is rarely implicated in disease transmission, except among patients who are immunocompromised (Maki, Alvarado et al., 1982; Sehulster & Chinn, 2003). Nonetheless, inadvertent exposures to environmental pathogens via (a) medical devices (e.g., endoscopes, respiratory therapy equipment; Rutala 1996); (b) improper heating ventilation, and air conditioning (HVAC) systems and airborne pathogens (e.g., *Mycobacterium tuberculosis* and *varicella-zoster* virus); (c) contaminated environmental surfaces contact and lack of provider hand hygiene (e.g., multidrug resistant organisms); (d) device-related chemical exposure (e.g., heat intolerant medical devices requiring lower temperature liquid sterilants or ethylene oxide (ETO) sterilization and aeration); (e) or medical devices creating potentially infective aerosols in the environment can result in adverse patient outcomes and cause illness among patients and health care providers (Centers for Disease Control and Prevention [CDC], 1994; Rutala, 1996; Sehulster & Chinn, 2003; Tablan, Anderson et al., 2003).

The incidence of health care environment–associated adverse outcomes, including patient infections can be minimized as follows:

- Accurate and clearly written manufacturers' instructions on medical device care, use, and storage in infection prevention
- Appropriate use of cleaners and disinfectants with medical devices and in the inanimate environment
- Appropriate maintenance of the HVAC system and water systems
- Adherence to water-quality standards for equipment using water from main lines (e.g., water systems for hemodialysis, ice machines, hydrotherapy equipment, dental unit water lines, and automated endoscope reprocessors);
- Ambient temperature and ventilation standards for specialized care environments (e.g., airborne infection isolation rooms, protective environments, or operating rooms)
- Following general human factors guidelines in designing individual patient room and workplace accessibility to personal protective and isolation gear such as gloves, masks, gowns, and, especially, hand hygiene opportunities

THERMAL ENVIRONMENTS

Six main factors should be considered to assess the health care thermal environment:

- Air temperature
- Radiant temperature
- Air velocity and quality
- Humidity
- The activity of the occupants—both patients and their care providers;
- The clothing worn by the patients (or lack there of) and their care providers

When the body becomes "too hot or too cold," it reacts in a way that is consistent with maintaining core temperature at a relatively constant level. The initial reaction to heat is vasodilatation, in which peripheral blood vessels dilate and transfer blood, and hence heat, to the surface of the body where it is lost to the surrounding environment and sweating occurs. When the body is exposed to cold, the initial reaction is vasoconstriction, in which the peripheral blood vessels constrict and reduce the flow of blood to the body surface to reduce heat loss (Parsons, 2000).

There have been numerous studies of the effects of thermal environments on human performance (Parsons, 1993). A number of general conclusions can be made on these effects. Important individual considerations are as follows: subjective and psychological parameters, degree of acclimatization, and factors that contribute to individual differences (e.g., body composition, gender). As heat stress increases, there will be effects on physical and mental performance but decrements in performance occur not only at high environmental temperatures. Performance at vigilance tasks, which is frequent in health care, can be lowest in slightly warm environments that can have soporific effects (Parsons, 2000). For example, tasks caregivers perform comfortably in street clothes or hospital uniforms become "warmth discomfort" when carried out in patient isolation grab (e.g., moisture-resistant gowns, masks, gloves) over clothes or uniforms. Tasks such as monitor surveillance may be compromised as the technician becomes drowsy in a warm environment.

The effects of cold on health care providers' performance are often ignored and can be very significant. Although there are few effects on mental performance, cold can act as a "secondary task" (i.e., shivering or physical movement to "keep warm"), hence increasing workload and possibly decreasing mental and physical performance of the given task. The effects of cold on manual performance are attributed to physiological reactions to cold. Slowing of movements because of stiffening of joints and slow muscle reaction, numbness, and a loss of strength are all associated with vasoconstriction caused by cold. These reactions cause a deterioration in manual dexterity and hence performance at many manual tasks (Parsons, 1993). Tactile sensitivity is related to skin temperature. In general, cold stress is less of an occupational hazard than heat stress. However, providing emergency care in the out of doors often exposes the care provider to the cold and elements such as snow and ice. Under these extreme conditions, a person may not even be able to determine what they are touching without visual reference. In addition, because of the requirements of the Occupational Safety and Health Administration Bloodborne Pathogens Rule (DOL, 2001) for moisture resistance and the use of personal protective gowns during surgical procedures, many operating rooms have lowered ambient room temperature to accommodate the medical staff wearing this extra protective gear. Other operating room health care providers who are not directly performing the surgical procedure are often cold in this setting, causing shivering and reduced tactile sensitivity. All patients undergoing surgical procedures, no matter how minor, are at risk, to varying degrees, of developing hypothermia.

The heat exchange process is very much affected by four environmental conditions: air temperature, humidity, airflow, and the temperature of surrounding surfaces (e.g., walls, ceilings glass windows, heat-producing equipment; Sanders & McCormick, 1993). The interaction of these conditions is quite complex and cannot be covered in detail here. However, some familiar examples in health care environments are the extremes in temperatures associated with reprocessing of medical devices and instruments; instrument washers and steam sterilizers; laundry facilities; kitchen equipment such as stovetops, ovens, and automated dishwashers and sanitizers. As mentioned earlier, modern health care environments rarely exhibit extremes of cold, but health care providers and patients can suffer reactions to cold in air conditioning and in medical transport. Table 19–3 shows recommended

TABLE 19–3. Recommended Temperature and Humidity Settings[a]

Functional Space	*Temperature* °F (°C)	% *Relative Humidity*
Operating rooms	68–75 (20–23.9)	30–60
Delivery rooms	68–75 (20–23.9)	30–60
Recovery rooms	68–75 (20–23.9)	30–60
Intensive care units	70–75 (21–23.9)	30–60
Nursery	75–80 (23.9–26.7)	30–60
Emergency DEPARTMENT	70–75 (21–23.9)	30–60
Radiology	70–75 (21–23.9)	30–60
Clinic rooms	70–75 (21–23.9)	30–60
Patient rooms (general)	70–75 (21–23.9)	30–60
X-ray (surgery/cardiac catheterization lab)	72–78 (22.2–25.6)	30–60
Physical therapy	72–78 (22.2–25.6)	30–60
Central supply	72–78 (22.2–25.6)	30–60
Laboratory (clinical and pathology)	70–75 (21–23.9)	30–60
Endoscopy suites	70–75 (21–23.9)	30–60

[a] Modified from *American Society of Heating, Refrigerating and Air-conditioning Engineers, Inc. (ASHRAE)*, HVAC Design Manual for Hospital and Clinics, *SP–91, 2003.*

temperature and humidity settings in various hospital built environments (Robert, 2004), and these are applicable in most out of the hospital settings such as clinics, extended care facilities, and specialty care settings.

Care provider and patient-safety related to extremes in temperature become more problematic in prehospital, emergency care, and transportation environments (Rossman, 1992; Helm, Castner et al., 2003). For example, drugs used in prehospital emergency medical service, in principle, are subject to the same storage restrictions as hospital-based medications. The prehospital emergency environment, however, often exceeds these storage recommendations. Main stress factors are sunlight, vibration, and extreme temperature, which may lead to alteration in chemical and physical stability of stored pharmaceuticals, as well as microbiological contamination and concentration enhancement of pharmacological inserts (Helm, Castner et al., 2003).

Effects of Clothing on Heat Exchange

The clothes care providers wear can have a profound effect on the heat exchange process. The insulating effect of clothing reduces heat loss to the environment. When the environment is cold, such as out of doors or in extreme air conditioning, reduced heat loss is beneficial; however when the environment is hot, clothing interferes with the heat loss and can be harmful (Sanders & McCormick, 1993). The insulation value of most clothing materials is associated with the amount of trapped air within the weave and fibers and the permeability of the material to moisture. In hot environments, evaporation of sweat is vital to maintaining thermal equilibrium, and materials that interfere with this process can result in heat stress or even heat stroke. In a cold environment, if evaporation of sweat is impeded, a garment can become soaked with perspiration, thus reducing its insulating capacity and warmth. Consider the example of the operating room and surgical instrument decontamination and reprocessing. The surgical reprocessing technician is required to be outfitted according to bloodborne pathogens precautions, because exposure to human blood and other infectious materials is highly anticipated in the job duties. The technician will most likely wear undergarments, socks, shoes and shoe covers, a tight weave cotton scrub suit or uniform, a moisture proof cover gown or apron, protective gloves, head cover, a splash or moisture resistant mask, and a face shield or safety glasses. The work environment contains sources of high heat—hot water, steam, and hot stainless steel surgical instrument trays. Under prolonged exposures to this scenario, heat stress could result, affecting quality of working life and perhaps quality of patient care. Although most modern health care facilities are closely climate controlled, specific areas of high heat and cold may remain an environmental hazard to patient and worker safety.

In addition, in cold environments, the amount of insulation capacity needed may interfere with the performance of patient care tasks; heavy gloves reduce manual dexterity, heavy coats or jackets may reduce range of motion, hats or hoods may reduce hearing and vision, and so on. When health care providers are in cold or hot environments, efforts should be made to optimize their comfort, protect their health, and maximize their safe performance.

AIR QUALITY

In addition to the effects of heat and cold, there are other aspects of thermal comfort associated with the air. Draft can be a serious problem in ventilated and air-conditioned buildings. Draft has been identified as one of the most annoying environmental factors in workplaces. People are most sensitive to air movement at the head region (i.e., head, neck, and shoulders; Sanders & McCormick, 1993). Draft in the health care environment may be associated with HVAC systems requiring rapid air exchange such as in operating rooms or using high-efficiency particulate air (HEPA) filtration for patient isolation techniques.

Humidity is the amount of moisture in the air. The indoor humidity, to a large extent, tracks the prevailing outdoor humidity. In cool, dry, winter weather, heating the air lowers the relative humidity; in summer, because most air-conditioning systems do not humidify, dry conditions cannot be avoided. Exposure to low humidity can occur in situations such as out of doors (e.g., emergency medical technicians, first aid stations) care in most desert regions during summer and patient care in air-conditioned buildings. Similar dry conditions can occur during winter months in heated buildings. Low humidity can result in dryness in noses and throats, dry skin, and chapped lips (Sanders & McCormick, 1993). A perceivable level of eye irritation is experienced by both contact lenses wearers and those who do not wear contact lenses when the relative humidity is at or below 30%; the effect becomes pronounced after 4 hr (Rohles & Konz, 1987). Although no peer-reviewed scientific publications associate adverse patient safety events and humidity-associated dry eyes, theoretically, there could be caregiver vision irritation and impairment caused by low environmental humidity.

Popular belief holds that weather influences moods. The Weather Channel© discovered that the most frequent expressions used to describe the effects of heat and humidity are lazy and tired, slow or unproductive, sick or uncomfortable, and irritable and angry. However, little support in the scientific literature for this belief has been provided by the few studies assessing the effect of humidity and temperature in the workplace. With regard to associated injury illness, humans are more tolerant of high humidity than high temperatures (Sanders & McCormick, 1993). Most high humidity effects on thermal comfort are conditions associated with the high temperature of the environment. If a person cannot efficiently evaporate heat-induced sweat away from their body because of the high moisture content in the surrounding air, their thermal comfort declines.

Humid conditions associated with patient safety can occur and may lead to error when care providers do not consider that medical devices are now used in many environments of care. For example, in home use, changing temperatures or high humidity (such as a bathroom or shower) may affect medical device performance. Areas of high humidity associated with patient care may not take into account medical device storage requirements and possible degradation of device components, such as adhesives and chemical reagents and relative humidity requirements for optimal use. Elevated humidity above the desired range could result in condensation or moist supplies, damaged equipment, and microbial build up in the environment. There is no prescribed specific humidity level for health care settings. However, the recommendations made in Table 19–2 provide a range of acceptable humidity levels and are more reasonable than requiring an absolute of 50% humidity, as there will always be some variation, depending on temperature fluctuation.

Air-Handling Systems in Health Care Facilities

When designing an air-handling system for use in health care facilities, use AIA construction and environmental guidelines (AIA & FGI, 2001) and American Society of Heating, Refrigerating and Air-Conditioning Engineers (ASHRAE) standards *ASHRAE HVAC Design Manual for Hospitals and Clinics* (ASHRAE, 2003) as minimum standards where state or local regulations are not in place for design and construction of health care facility ventilation systems. People performing the environmental human factors analysis should be familiar with these guidelines and consider all the potential interactions between the people, medical devices, and the air-handling systems in various care settings. JCAHO environment of care standards require hospitals to install and maintain appropriate pressure relationships, air exchange rates, and filtration efficiencies for ventilation systems serving areas designed to control airborne contaminants, such as biological agents, gases, fumes, and dust (JCAHO, 2005). Care technology, medical devices or equipment, should not generate any airborne contaminants that exceed these prescribed containment standards and guidelines. More importantly, medical devices used in this environment must not create an exhaust flow sufficient to modify the direction of the room's pressure (i.e., that an infectious disease isolation room would remain negative pressure to the general unit hallway, or an operating room positive pressure to corridor air).

Hospital Environments With Special Air Handling

In addition to general hospital ventilation, four areas of patient care have additional or special air-handling requirements; therefore, these requirements need to be considered when conducting a human factors and environmental assessment or design for these areas (CDC, 1994; Sehulster & Chinn, 2003; Tablan, Anderson et al., 2003).

1. Airborne infection isolation (AII) refers to the isolation of patients infected with organisms spread via airborne droplet nuclei <<5 μm in diameter. This isolation area receives numerous air changes per hour (ACH) (>>12 ACH for new construction as of 2001; >>6 ACH for construction before 2001) and is under negative pressure, such that the direction of the airflow is from the outside adjacent space (e.g., the corridor) into the room. The air in an AII room is preferably exhausted to the outside, but it may be recirculated, provided that the return air is filtered through a HEPA filter. The use of personal respiratory protection is also indicated for persons entering these rooms when caring for tuberculosis or smallpox patients and for staff members who lack immunity to airborne viral diseases (e.g., measles or varicella-zoster virus infection). If the SEIPS model is applied to this environment, it is apparent that medical equipment should be accessible and usable when the operator is in full personal protective equipment: mask, isolation gown, gloves, and protective eyewear. Medical equipment used in this environment must not create an exhaust flow sufficient to modify the direction of the isolation room's negative pressure. In addition, if not disposable, the medical device often stays in the room environment; hence, power supplies, device care, and maintenance must be considered in the human factors analysis and design of the environment for patient safety, because the device often does not come out of the room until the patient leaves.
2. Protective environment is a specialized patient care area, usually in a hospital, with a positive airflow relative to the corridor (i.e., airflows from the room to the outside adjacent space). The combination of HEPA filtration, high numbers of ACH, and minimal leakage of air into the room creates an environment that can safely accommodate the immunocompromised patient. Immunocompromised patients are those patients whose immune mechanisms are deficient because of immunologic disorders (e.g., HIV infection or congenital immune deficiency syndrome), chronic diseases (e.g., diabetes, cancer, emphysema, cardiac failure), or immunosuppressive therapy (e.g., radiation, cytotoxic chemotherapy, antirejection medication, steroids). These patients, who are identified as high-risk patients, have the greatest risk of infection caused by airborne or waterborne microorganisms. The caregiver will often be gowned, gloved, and masked to protect the patient from microbes from outside the room. The medical device in this environment must facilitate use under such protective gear and conditions. Medical devices used in this care environment should not generate or collect dust (possible reservoir Aspergillus spp.), not create a wet environment or reservoirs of water, (e.g., Pseudomonas spp., atypical mycobacteria associated with contaminated device reservoirs), and not produce unfiltered exhaust. If the medical equipment generates exhaust, HEPA filtration of the exhaust must be incorporated into the environment or the equipment design.

3. Ventilation Requirements for Operating Rooms maintain a positive-pressure ventilation with respect to corridors and adjacent areas. People and medical equipment for the operating room use should not block the room ventilation or create an unfiltered exhaust in this environment. The equipment must be easily used by care personnel in full surgical gowns, gloves, hair coverings, and protective eyewear. Do not use ultraviolet lights in the equipment design, because they pose a risk to both patient and operator and are without efficacy in preventing air associated surgical site infections. Do not design an environment that requires operating room doors to remain open for extended periods of time. Operating room doors should be closed except for the passage of equipment and entry of essential personnel and patients.
4. Procedures for Infectious Tuberculosis Patients who also require emergency surgery. All medical devices in this setting must be adaptable for use with the operator wearing an N95 respirator approved by the National Institute for Occupational Safety and Health without exhalation valves or powered air-purifying respirators. Because of these personal protective equipment (PPE) requirements, other environmental components such as hearing, sight, range of motion, and the like may be compromised for patient safety and should be addressed.

Other potentially infectious hazards associated with the air in the health care setting include medical lasers. In addition to the direct hazards to the eye and skin from the laser beam itself, it is important to address other hazards associated with the use of lasers. These nonbeam hazards, in some cases, can be life threatening (e.g., electrocution, fire, and asphyxiation). Because of the diversity of these hazards, the human factors environmental analysis may wish to include laser safety or industrial hygiene personnel to effect the hazard evaluations. Surgical lasers and dental lasers should be designed for operator use while wearing appropriate personal protective equipment (PPE), including N95 or N100 respirators, to minimize exposure to laser plumes and splatter. Most likely, the operator of the laser and the assistants will be gowned, masked, gloved, and wearing laser protective eyewear. The physical environment cannot interfere with any of this in the use of the patient care equipment or protective gear; in turn, the protective gear should not compromise hearing, sight, range of prescribed motion, or the like.

NOISE

Before the era of technology, our health care environmental noise consisted of human voices and activities of daily living, such as housekeeping, meal preparation and distribution, or the sounds of nature coming from the outdoors. The hospital atmosphere of the 1940s and 1950s was still one of austere silence, as in a library reading room. Hallways displayed a ubiquitous picture of a uniformed nurse, finger to the lips, sometimes accompanied by the words, "Quiet please." Signs on the street read, "Hospital Zone—Quiet." The occasional overhead page for a physician signaled a true emergency. That subdued setting has gradually been replaced by one of turbulence and frenzied activity. People now dart about in a race against time; telephones ring loudly; intercom systems blare out abrupt, high-decibel messages that startle the unsuspecting listener. These sounds are superimposed on a collection of beeps and whines from an assortment of electronic gadgets—pocket pagers, call buttons, telemetric monitoring systems, electronic intravenous machines, ventilator alarms, patient-activity monitors, and computer printers. The hospital, designed as a place of healing and tranquility for patients and of scholarly exchanges among physicians, has become a place of beeping, buzzing, banging, clanging, and shouting (Grumet, 1993). Noise has become such a pervasive aspect of health care settings and daily life that we refer to it as "noise pollution," as we might reference a chemical spill or infectious agent introduced into the environment. Noise is a source of stress for patients (Schmaker & Pequegnat, 1989). Excess noise can lead to increased anxiety and pain perception, loss of sleep, and prolonged convalescence (Baker, 1984; Williams, 1988; Baker, Garvin et al., 1993; Baker, 1993). Noise can mask the sound of critical patient monitor alarms, interfere with communication, and cause misinterpretation of care measures, such as breath or chest sounds or blood pressure. The impact of sound in the ICU environment affects staff members as well as patients and their families and may contribute to elevated heart rate, stress, and annoyance levels independent of the stress produced by caring for critically ill patients. These stressful conditions may contribute to "burnout" and may negatively affect staff retention (Oates & Oates, 1995, 1996; Walsh-Sukys, Reitenbach et al., 2001; Morrison, Haas et al., 2003).

Nevertheless, noise in certain environmental circumstances may actually be helpful. For example, low levels of continuous noise (the hum of an HVAC system fan) can mask the more disruptive effects of distracting noise (the tick of the clock or the conversation at the nursing station). Soft background music in the environment may even be soothing and stress reducing to patients and care staff (Cabrera & Lee, 2000).

The stimulus for hearing is sound—a vibration (actually compression and rarefaction) of the air molecules. The acoustic stimulus can therefore be presented as a sine wave, with amplitude and frequency. The frequency of the stimulus more or less corresponds to its pitch, and the amplitude corresponds to its loudness. When describing the effects of sound on hearing, the amplitude is typically expressed as a ratio of sound pressure, which is measured in decibels (dB). In addition to amplitude and frequency, two critical dimensions of the sound are its temporal characteristics, sometimes referred to as the envelope in which a sound occurs, and its location (Wickens, Lee et al., 2004). For example, temporal characteristics are what distinguish the wailing of an ambulance siren from a steady horn blast of someone's car alarm, and the location (relative to the listener) is of course what distinguishes the ambulance siren pulling up behind from that of the ambulance about to cross the intersection to the front (Wickens, Lee et al., 2004).

A detailed description of the ear as a sensory organ is found in many texts (Sanders & McCormick, 1993; Wickens, Lee et al., 2004) and will not be discussed here. However, it should be mentioned that a physical factor responsible for loss in sound transmission is the potential loss of hearing of the listener. Simple human aging is responsible for a large portion of hearing loss, prebycusis, particularly sound in the high-frequency regions. Degeneration of the hair cells of the organ of corti in the cochlea of ear results in aging nerve deafness. Once nerve degeneration has occurred, it rarely can be remedied, only accommodated with hearing devices. As the care provider work force continues to age, hearing critical alarms and performing tasks that rely on hearing may become compromised. In addition, our aging patient population becomes an important consideration in alarm choice and design for warnings, especially in nursing homes.

Noise and Performance

The effects of noise on human physical and mental performance can be divided into effects on nonauditory task performance and auditory task performance (e.g., interference with speech communication). The effects of noise on nonauditory task performance in the literature have been inconclusive; different studies indicate that noise has no effect on or even increases task performance (Sanders & McCormick, 1993; Parsons, 2000).

Noise can interfere with auditory communication of information (e.g., speech, warning signals) and therefore decrease task performance and possibly safety. Humans can detect signals within a background noise. It is important to know the "efficiency" of this detection within a specific type of background noise to be able to assess the effects of background noise on communication or to design a warning signal for that environment. Without going into great detail, the detection threshold of a signal within a background noise can be represented as the signal-to-noise ratio over noise frequency; criteria exist and indexes have been established defining an articulation index (Parsons, 2000). As with other environmental stresses, noise can add to the workload of a care task and can potentially affect safe performance in this way. In addition, the onset of a loud noise will cause a startle response, often characterized by muscle contractions, blink, and head or hand jerk movement (Sanders & McCormick, 1993). These responses are relatively transient and settle back to normal or near normal levels very quickly; but, for example, consider the safety outcomes of a startle response in delicate, precise surgery or invasive bedside procedures such as central venous catheter placement. The startle response can disrupt perceptual motor function.

A variety of health care environment noise sources have been measured and guidelines established for the ambient environment and auditory warning and alarm signals (see Chapter 22). As previously discussed, masking occurs when one aspect of the sound environment reduces the sensitivity of the ear to another component of the sound environment. The basic idea is that if an auditory signal occurs in the presence of noise, the threshold of detection of that signal is raised, and this elevated threshold should be exceeded by the signal if it is to be detected (Sanders & McCormick, 1993). This may lead to what could be referred to as "auditory signal inflation" as the environment gets louder, so must the alarm. Perhaps the better solution is to reduce the background noise and, in turn, the signal. This of course may not be possible in medical transport where the competing noise is generated by the vehicle engine itself, helicopter rotors, ambulance sirens, and the like. Studies found that blood pressure values obtained by medical personnel in a quiet environment were significantly more accurate than those obtained in an ambulance (Prasad, Brown et al., 1994) and that although emergency care providers could correctly identify 96% of breath sounds in a quiet environment, only 54% could in an ambulance (Brown, Gough et al., 1997).

The World Health Organization and the U.S. Environmental Protection Agency (EPA) recommend that hospital noise levels not exceed 40–45 dB(A) during the day and 35 dB(A) at night. Sound levels above 50 dB(A) are sufficient to cause sleep disturbance, and sustained levels above 85 dB(A) can damage hearing (Morrison, Haas et al., 2003). Although the levels of noise in hospitals do not approach industrial time-weighted averages of noise (manufacturing plant lines can yield continuous noise at 80 to 95 dB(A), requiring PPE hearing protection), substantial levels of noise are encountered in hospitals. The average equivalent decibel level found in an operating room was in the range of 60 to 65 dB(A), but the sound-reflective tile surfaces and stainless-steel equipment tended to enhance harsh and reverberant noises. A simple task like popping open an envelope of sterile surgical gloves yields a sound of 86 dB(A), dropping a stainless-steel bowl can generate an abrupt or explosive "impulse" sound at 108 dB(A), and the sound of escaping gas as anesthesia tanks are changed reaches 103 dB(A). When an impulse sound exceeds the background level by approximately 30 dB, a startle reaction can occur in the listener. Patients in the recovery room are generally exposed to 50 to 70 dB(A), but shouts between staff members and the shifting of bedrail positions expose them to short-term noise at 90 dB(A). Similarly, hospital laboratories generally have background hums in the 60–68-dB(A) range, whereas the noisy automated machines in laboratory workplaces are in the 65–74-dB(A) range. Computer rooms at peak activity reach 85 dB(A), and a pneumatic-tube carrier arrives with an 88-dB(A) thud (Grumet, 1993).

Sleep loss is a major contributor to "ICU psychosis," especially within the surgical ICU, which is generally noisier than its medical counterpart (Falk & Woods, 1973). A patient in the ICU on a life-support system may be attached to10 different audible warning devices (Pownall, 1987), and hospitalized patients—who often feel estranged and fearful to begin with—experience hospital noises as far more sinister than staff members realize (Nolen, 1973): they often falsely construe the various alarm systems as indicating life-threatening events.

Alarms

In a study of alarm systems used by anesthesiologists during surgical procedures, 75% of the alarms were found to be spurious, caused by factors such as the patient's movement or simple mechanical events; only 3% of the sounds indicated an actual risk to the patient (Kestin, Miller et al., 1988). A survey of Canadian anesthesiologists found that 67% had deliberately disabled audible signals to cope with the many false alarms and "sonic overkill" during surgery (McIntyre, 1985). Although human factors and alarms are discussed at length in Chapter 22, a brief overview of effective critical alarms strategies is provided here with broad-based recommendations based on the human factors literature.

In 2002, JCAHO set a 2003 national patient safety goal to improve the effectiveness of clinical alarm systems. This goal takes the form of two recommendations for hospitals (JCAHO, 2002):

- Implement regular preventive maintenance and testing of alarm systems
- Assure that alarms are activated with appropriate settings and are sufficiently audible with respect to distances and competing noise within the [care] unit.

As in other areas of human factors environment and task analysis, most guidelines and principles need to be accepted with a few grains of salt, because specific circumstances, trade-off values, and the like may argue for their violation. With such a caveat in mind, a few guidelines for the use of auditory displays and alarms are as follows (Sanders & McCormick, 1993; Salvendy, 1997; ECRI, 2002; Wickens, Lee et al., 2004):

- Principles of Auditory Display
- Compatibility—signal should exploit learned or natural relationships

- Wailing signals with emergency (tornados)
- Approximation—Two-stage signals
- Attention-demanding signal to attract attention and identify a general category of information
- Designation signal to follow with the exact information identified with the first signal (e.g., an auditory paging system "fire alarm" first the fire alarm sounds, and then the operator announces the exact location of the fire in the facility)
- Dissociability—signals should be discernible from any ongoing audio input
- Parsimony—input signals to operator should not provide more information than is necessary
- Invariance—the same signal always equals the same information at all times
- Principles of Presentation
- Avoid extreme dimensions—high intensity can cause a startle
- Use an intensity relative to ambient noise—signal should not be masked by ambient noise
- Use interrupted or variable signals—avoid steady-state signals, vary the frequency
- Do not overload the auditory channel—too many signals overload the operator (At Three Mile Island, more than 60 different warning signals were activated during the disaster.)

Both JCAHO and ECRI believe that hospitals should not determine audibility by measuring sound levels of alarms. Instead, users should examine whether an alarm is sufficiently audible in the environment in which it is being used (ECRI, 2002). Various factors of the SEIPS model need to be included in this examination, such as organizational staffing levels, arrangement of components in the built environment, tasks that may compete with the alarm, background noise in the environment, and frequency of the alarm signal. Hearing, when coupled with the other senses, can offer the brain an overwhelming array of information. Be mindful that each sensory modality appears to have particular strengths and weaknesses and must not be considered independent of the others.

VIBRATION

Vibration may be distinguished in terms of whether it is specific to a particular limb, such as the vibration produced by a surgical drill or saw, or whether if influences the whole body, such as that from a helicopter or ambulance (Wickens, Lee et al., 2004). Vibration can significantly affect the comfort and performance of health care providers in emergency vehicles (Bruckart, Licina et al., 1993). It is unlikely that the patients and care providers would be exposed to such vibration levels in hospitals. However, vibration to the hands of surgeons, dentists, and other care providers from vibrating dental or surgical tools or devices such as drills, saws, and the like may be significant and could cause physical damage. Poorly designed medical devices not only jeopardize performance and productivity but are a major cause of cumulative trauma disorders such as tendonitis, neuritis, Raynaud's phenomenon, trigger finger, epicondylitis (tennis elbow, golfers elbow), and carpal tunnel syndrome (Armstrong, Buckle et al., 1993; Chaffin, Andersson et al., 1999).

Vibration appears to principally affect visual and motor performance. Visual performance is generally impaired most by vibration frequencies in the range of 10 Hz to 25 Hz. The degradation in performance is probably caused by the movement of the image on the retina, which causes the image to appear blurred. Impacts of vibration include medical device display vibration, which can reduce the ability to see fine detail in displays, and impaired performance using isometric controls (Griffin, 1997). Shape and orientation of controls may affect performance in vibration to the extent that they alter vibration breakthrough or proprioceptive feedback to the operator, but impacts also may be related to task difficulty and the axis of vibration (Griffin, 1997). Its presence in an emergency vehicle, for example, makes medical devices with touch screens extremely unreliable as control input devices (Wickens, Lee et al., 2004). In summary, considerable research has demonstrated the effects of vibration on performance. The effects of vibration are dependent on the difficulty of the task, type of display, and type of device used and physical attributes of the user (Chaffin, Andersson et al., 1999).

Low frequency vibration—such as the regular sea swells on a ship, the rocking of a light aircraft, or the environment of a closed cab in a moving ambulance—can lead to motion sickness. Quite simply, the discomfort of the sickness is sufficiently intrusive that it is hard to concentrate on anything else, including the care tasks at hand (Thornton & Vyrnwy-Jones, 1984; Sanders & McCormick, 1993). The results of motion sickness, including emesis and anxiety, not only are uncomfortable but also could adversely affect the patient's care and outcome (Fleischhackl, Dorner et al., 2003; Weichenthal & Soliz, 2003). The incidence of patient motion sickness during ambulance transport is reported to be between 20% and 33%, this is often associated with nausea and vomiting and may, therefore, be a risk factor for aspiration (Fleischhackl, Dorner et al., 2003).

ILLUMINATION

The design of artificial illumination systems does have an effect on the performance and comfort of those using the environment as well as on the effective responses of the people to the environment. It is not the intention of this chapter to make anyone an expert on illumination—illumination engineering is both an art and a science (Sanders & McCormick, 1993). The Illuminating Engineering Society of North America (IESNA) Recommended Practice (IESNA, 1995) provides guidelines for good lighting in those areas unique to health care facilities and is intended for both lighting designers and health care professionals. The aim of this section is to familiarize the reader with some basic characteristic of illumination—illuminance, luminance, contrast, glare, and shadow—and to illustrate the importance of proper illumination in the patient care environment.

Essentially all visual stimuli that humans can perceive may be described as a wave of electromagnetic energy. The wave is represented as a point along the visual spectrum and referred to as its wavelength. Although we can measure or specify the hue of a stimulus reaching the eye by its wavelength, the measurement of brightness is more complex, because there are several different meanings of light intensity. The source of light may be characterized by its luminous intensity or luminous flux, the actual light energy of the source, the lumen. However, the actual amount of this energy that strikes the surface of an object to be seen is described as the illuminance and is measured in units of 1 lumen/ft^2 (a footcandle) or 1 lumen/m^2 (a lux). Hence, the term illumination characterizes the lighting quality of a given working environment. How much illuminance an object receives depends on the distance of the object from the light source. Luminance of an object is the amount of light reflected off the object to be detected, discriminated, and recognized by the observer when these objects are not themselves the source of light (Wickens, Lee et al., 2004). Luminance is different than illuminance because of differences in the amount of light that surfaces either reflect or absorb. Glare is produced by brightness within the field of vision that is sufficiently greater than the luminance to which the eyes are adapted. Although we may be concerned about the illumination of light sources in direct viewing (i.e., looking directly into the operating room lights or the sun, the direct glare produced by the light source), reflected glare also can cause annoyance, discomfort, or loss in visual performance (Sanders & McCormick, 1993).

Poorly designed lighting environment, such as too little or too much light, veiling reflections, and glare, can contribute to visual discomfort and eyestrain (Boyce, 1981). The effects of lighting on performance are somewhat complex, and limits of the visual system influence the nature of the visual information that arrives at the brain for the more elaborate perceptual interpretation (Wickens, Lee et al., 2004). Defects occur in the person's visual system, color defects ("color blind"), myopia, and so on that may place additional constraints and limits on this visual information. Table 19–4 suggests the environment minimum average illuminance for environments in a general hospital (Sanders & McCormick, 1993; IESNA 1995; AIA & FGI, 2001).

Besides the measurement of the environmental illuminances recommended in Table 19–4, there are basic concepts of illumination that should be considered for any human factors environment and task analysis in a patient care setting:

- Sources of illumination in health care settings
- Use of color
- Visibility
- Distribution of light
- Glare
- Reflectance
- Effects of the lighting on performance
- Use of lighting and elderly workers
- Lighting and video display terminal use

TABLE 19–4. Health Care Environment Illuminance Recommendations

Environment	Ambient LightLux/Footcandles	Task Lighting Lux/Footcandles
Patient rooms	300/30	750/75
Nursing station (day)	300/30	500/50
Nursing station (night)	100/10	500/50
Medication preparation	300/30	1000/100
Exam room	300/30	1000/100
Operating rooms	300–500/30–50	10,000–20,000/1000–2000
Hallways (active shifts)	300/30	
Hallways (sleeping)	100/10	

Furthermore, the human factors environmental analysis must consider that excessive exposure to light can cause direct effects on health. Medical devices such as ultraviolet lights and medical lasers can cause damage to the eye. The sun's radiation can damage and cause cancer to the unprotected skin of care providers in the outdoors. In addition to these direct effects on health, eyestrain can be caused by inadequate lighting conditions. Too little or too much light, veiling reflections, disability, and discomfort glare and flicker can cause eyestrain (Boyce, 1981). This can cause irritation in the eyes, a breakdown of vision, and possible patient safety risks. There are nonvisual effects of light on the body (e.g., seasonal affective disorder, influence of various glands in the body); however, not much is known about their effects quantitatively (Parsons, 2000) and certainly not as they relate to patient care and safety issues. The circadian rhythms of night shift workers do not usually adjust to their unusual work and sleep schedules, reducing their quality of life and producing potentially dangerous health and safety problems. An illumination study area is now focusing on the patient care environment, but no environmental recommendations can be made at this time (Eastman & Martin, 1999).

Light can cause discomfort as well as positive sensations in patients and care providers. Lighting conditions that produce definite discomfort can usually be identified, and criteria for the concepts of illumination previously mentioned are readily available for application and remedy (IESNA, 1995). The conditions that create emotional responses or pleasant environments are not as well understood, and designing for these conditions remains both an art and a science (Parsons, 2000).

Visual performance is the only performance outcome that changing the lighting conditions can affect directly. Lighting itself cannot produce work output. What it can do is make details easier to see and colors easier to discriminate without producing discomfort or distraction; the greater the contribution of vision to the task, the greater the effect of lighting is. Recommendations for improving the visual performance include the following:

- Changes in the task
- Increase the size of detail; make the patient care procedures or medication instructions or device monitors in a larger font.
- Increase luminance contrast of the detail in task; use color contrast or light and shadow to define boundaries or highlight important areas.
- In a cluttered visual field, make the object to be detected clearly different from the surrounding objects (e.g., size, color, shape, contrast).
- Changes in environment
- Increase luminance.
- Change to a lamp with better color properties.
- Provide lighting that is free from disability glare and veiling reflections.
- Add to or change lighting to increase apparent size or luminance contrast of the object.

Lastly, special attention should be given to the good visual performance for the elderly care provider or elderly patient. Individual visual performance differences accompany aging. Presbyopia (farsightedness), a decrease in visual acuity, contrast sensitivity, color discrimination, an increase in time to adapt to sudden changes in luminance, and sensitivity to glare are common attributes associated with age. Possible environmental solutions may be increasing illuminance, taking note that glare and veiling reflections associated with illuminance also can increase, and creating transition zones in the envionment (e.g., going from a movie theater to the theater lobby to the outdoors), which give the person more time to acclimate to the brighter conditions (Sanders & McCormick, 1993). Finally, most visual physical disability is not one of total blindness but rather one of partial sight. The most common causes for visual disability involving partial sight are cataracts, macular degeneration, glaucoma, and diabetes—a disease currently at epidemic levels in the United States, affecting both patients and care provider populations. Several illumination strategies are suggested for improving visual performance under these conditions (Trace Center Engineering et al., 2005):

- Cataracts: limit glare
- Macular degeneration: provide more light and magnification
- Glaucoma: limit patterns, use high contrast color, spatially separate items
- Diabetes: increase light in the early stages of retinal change and magnification.

Just as our patients' acuity increases as patients' age increases, the economics and worker shortages are keeping older patient care providers in the workforce, and they are retiring later. The environment must be designed to accommodate the interaction of conditions associated with this aging population.

CONCLUSION

Aspects of the health care environment that were not addressed in this chapter and that can have direct influence on patient and care provider safety include safety and security, energy and cost-effectiveness, disaster planning, infection control, inforation technology, and health care technology and communications. For an extended understanding of the health care facility environment, the reader is encouraged to consult current reference literature on these subjects.

Although built or physical environments of care are usually assessed in terms of the effects of their separate component parts, patients and care staff members are exposed to whole, integrated environments (Parsons, 2000). This built environment is one part of the system, and it constantly changes in the interactions of other systems components (Smith & Carayon, 2000; Carayon, 2003; Carayon, Alvarado et al., 2003). This last point brings me back to reemphasize one final human factors/ patient safety issue that this chapter has touched on repeatedly: the importance of a thorough task analysis (Wickens, Lee et al., 2004). The full impact of the physical environment on human performance and patient safety can never be adequately predicted without a clear understanding of the parts of the SEIPS model that are present and that relate with the other; which people will interact with them; which people must interact with them; and the cost to patients, care providers, and the organization if task performance is degraded.

References

American Institute of Architects (AIA) & Facility Guidelines Institute (FGI). (2001). *Guidelines for Design and Construction of Hospital and Health Care Facilities.* Washington, DC: The American Institute of Architects.

American Society of Heating, Refrigerating and Air-Conditioning Engineers (ASHRAE). (2003). *ASHRAE—HVAC Design Manual for Hospitals and Clinics.* Atlanta, GA: ASHRAE.

Americans With Disabilities Act of 1990. Pub. L. No. 101-336, § 2, 104 Stat. 328 (1991).

Armstrong, T. J., Buckle, P., Fine, L. J., Hagberg, M., Jonsson, B., Kilbom, A., et al. (1993). A conceptual model for work-related neck and upper-limb musculosceletal disorders. *Scandinavian Journal of Work and Environmental Health,* 19, 73–84.

Baker, C. F. (1984). Sensory overload and noise in the ICU: Sources of environmental stress. *CCQ Critical Care Quarterly, 6*(4), 66–80.

Baker, C. F. (1993). Annoyance to ICU noise: A model of patient discomfort. *Critical Care Nursing Quarterly, 16*(2), 83–90.

Baker, C. F., Garvin, B., & Polivka, B. (1993). The effect of environmental sound and communication on CCU patients' heart rate and blood pressure. *Research in Nursing & Health, 16*(6), 415–421.

Boyce, P. R. (1981). *Human Factors in Lighting.* New York: Macmillan.

Brown, L. H., Gough, J. E., Bryan-Berg, D. M., & Hunt, R. C. (1997). Assessment of breath sounds during ambulance transport. *Annals of Emergency Medicine, 29*(2), 228–231.

Bruckart, J. E., Licina, J. R., & Quattlebaum, M.. (1993). Laboratory and flight tests of medical equipment for use in U.S. Army Medevac helicopters. *Air Medical Journal, 1*(3), 51–56.

Buemi, M., Allegra, A., Grasso, F., & Mondio, G.. (1995). Noise pollution in an intensive care unit for nephrology and dialysis. *Nephrology, Dialysis, Transplantation 10*, 2235–2239.

Cabrera, I. N., & Lee, M. H. (2000). Reducing noise pollution in the hospital setting by establishing a department of sound: A survey of recent research on the effects of noise and music in health care. *Preventive Medicine, 30*(4), 339–345.

Carayon, P. (2003). Macroergonomics in quality of care and patient safety. In H. Luczak & K. J. Zink (Eds.), *Human*

Factors in Organizational Design and Management (pp. 21–35). Santa Monica, CA: IEA Press.

Carayon, P., Alvarado, C., Hsieh, Y., & Hundt, A. S. (2003, August 24-29). *A macroergonomic approach to patient process analysis: Application in outpatient surgery.* Paper presented at XVth Triennial Congress of the International Ergonomics Association, Seoul, Korea.

Carayon, P., & Smith, M. J. (2000). Work organization and ergonomics. *Applied Ergonomics, 31*, 649–662.

Centers for Disease Control and Prevention. (CDC) (1994). Guidelines for preventing the transmission of mycobacterium tuberculosis in health-care facilities. *MMWR. Recommendations and Reports: Morbidity and Mortality Weekly Report, 43*(RR-13), 1–132.

Chaffin, D. B., Andersson, G. B. J., & Martin, B. J. (1999). *Occupational Biomechanics.* New York, NY: John Wiley and Sons, Inc.

Czaja, S. J. (1997). Systems Design and Evaluation. In G. Salvendy (Ed.), *Handbook of Human Factors and Ergonomics* (pp. 17–40). New York, NY: John Wiley and Sons, Inc.

Donabedian, A. (1988). The quality of care. How can it be assessed? *Journal of the American Medical Association, 260*(12), 1743–1748.

Eastman, C., and Martin, C. (1999). How to use light and dark to produce circadian adaptation to night shift work. *Annals of medicine 31*(2), 87–98.

ECRI (2002). Critical Alarms and Patient Safety. *Health Devices, 31*(11), 397–417.

Edelberg, H. K. (2001, March). How to prevent falls and injuries in patients with impaired mobility. *Geriatrics 56*, 41–45.

Falk, S. A., & Woods, N. F. (1973). Hospital noise—levels and potential health hazards. *New England Journal of Medicine, 289*(15), 774–81.

Fleischhackl, R., Dorner, C., Scheck, T., Fleischhackl, S., Hafez, J., Kober, A., et al. (2003). Reduction of motion sickness in prehospital trauma care." *Anaesthesia, 58*(4), 373–377.

Gordon, C. C., Churchill, T., Clauser, C. E., Bradmiller, B., McConville, J. T., Tebbitts, I., et al. (1989). *1988 Anthropometric Survey of U.S. Army Personnel: Summary Statistics Interim Report.* (No. Technical Report NATICK/TR-89-027). Natick, MA: U.S. Army Natick Research, Development and Engineering Center.

Grabenstein, J. D., & Winkenwerder, W. (2003). U.S. Military Smallpox Vaccination Program experience. *JAMA, 289*, 3278–3282.

Griffin, M. J. (1997). Vibration and motion. In G. Salvendy (Ed.), *Handbook of Human Factors and Ergonomics* (pp. 828–857). New York, NY: John Wiley and Sons.

Grumet, G. (1993). Pandemonium in the modern hospital. *New England Journal of Medicine, 328*(6), 433–437.

Helm, M., Castner, T., & Lampl, L. (2003). Environmental temperature stress on drugs in prehospital emergency medical service. *Acta Anaesthesiologica Scandinavica, 47*(4), 425–429.

Herwaldt, L., & Pottinger, J. (2003). Preventing falls in the elderly. *Journal of the American Geriatrics Society, 51*(8), 1175–1177.

Illuminating Engineering Society of North America (IESNA). (1995). *Lighting for Hospitals and Health Care Facilities ANSI Approved.* (Vol. IESNA Publication RP-29-95). New York, NY: Illuminating Engineering Society of North America.

Institute of Medicine (IOM) (2002). *Countering Bioterrorism: The Role of Science and Technology.* Washington, DC: The National Academy Press.

International Standards Organization (1996a). *Basic human body measurements for technological design.*

International Standards Organization (1996b). *ISO 7250.*

Joint Commission on Accreditation of Healthcare Organizations (2003). 2003 JCAHO National Patient Safety Goals. *Joint Commission Perspectives on Patient Safety, 3*(1), retrieved June 2006 from http://www.jcrinc.com/subscribers/patientsafety.asp?durki=3746#goal3746.

Joint Commission on Accreditation of Healthcare Organizations (JCAHO). (2005). *Environment of Care® Essentials for Health Care.* (5th Edition ed.). Chicago: Joint Commission Resources.

Kerzman, H., Chetrit, A., Brin, L., & Toren, O. (2004). Characteristics of falls in hospitalized patients. *Journal of Advanced Nursing, 47*(2), 223–229.

Kestin, I., Miller, B., & Lockhart, C. (1988). Auditory alarms during anesthesia monitoring. *Anesthesiology, 69*, 106–109.

Kroemer, K. H. E. (1989). Engineering anthropometry. *Ergonomics, 32*, 767–784.

Kroemer, K. H. E., Kroemer, H., & Kroemer-Elbert, K. (2001). *Ergonomics—How to Design for Ease and Efficiency.* Upper Saddle River, NJ: Prentice Hall.

Maki, D., Alvarado, C., Hassemer, C., & MA, Z. (1982). Relationship of the inanimate environment to endemic nosocomial infection. *New England Journal of Medicine, 307*: 1562–1566.

McIntyre, J. (1985). Ergonomics: anaesthetists' use of auditory alarms in the operating room. *International Journal of Clinical Monitoring and Computing, 2*, 47–55.

Morrison, W. E., Haas, E. C., Shaffner, D. H., Garrett, E. S., & Fackler, J. C. (2003). Noise, stress, and annoyance in a pediatric intensive care unit. *Critical Care Medicine, 31*(1), 113–119.

Morse, J. M. (2002). Enhancing the safety of hospitalization by reducing patient falls. *American Journal of Infection Control, 30*(6), 376–380.

Niinimaki, J., Savolainen, M., & Forsstrom, J. J. (1998). Methodology for security development of an electronic prescription system. *Proceedings AMIA Symposium:* 245–249.

Nightingale, F. (1863). *Notes on Hospitals.* London: Longman, Green, Longman, Roberts, and Green.

Nolen, W. (1973). Can you hear me? *New England Journal of Medicine, 289*, 803–804.

Oates, P. R., & Oates, R. K. (1996). Stress and work relationships in the neonatal intensive care unit: Are they worse than in the wards. *Journal of Paediatrics and Child Health, 32*, 57–59.

Oates, R. K., & Oates, P. (1995). Stress and mental health in neonatal intensive care units. *Archives of Disease in Childhood, 72*, F107–F110.

Occupational Safety & Health Administration (2001). *Occupational exposure to bloodborne pathogens.* 29 Code of Federal Regulations 1910.

Panero, J. and Zelnick, M. (1979). *Human Dimensions and Interior Space.* New York: Whitney Library of Design/Watson-Guptill Publications, Inc.

Parsons, K. C. (1993). *Human thermal environments.* London: Taylor & Francis.

Parsons, K. C. (2000). Environmental ergonomics: A review of principles, methods and models. *Applied Ergonomics, 31*, 581–594.

Pownall, M. (1987). Medical alarms: ringing the changes. *Nursing Times, 83*, 18–19.

Prasad, N. H., Brown, L. H., Ausband, S. C., Cooper-Spruill, O., Carroll, R. G., & Whitley, T. W. (1994). Prehospital blood pressures: Inaccuracies caused by ambulance noise? *American Journal of Emergency Medicine, 12*(6), 617–20.

Robert, C. (2004). Effective Design of Heating, Ventilation and Air-conditioning Systems for Healthcare Facilities. *Hospital Engineering and Facilities Management.*

Rohles, F. H., & Konz, S. A. (1987). *Climate.* New York: Wiley.

Rossman, L. (1992). Protecting yourself. *Journal of Emergency Medical Services, 17*(11), 48–49.

Rubin, H. R., Owens, A. J., et al. (1998). *An Investigation to Determine Whether the Built Environment Affects Patients' Medical Outcomes.* Martinez, CA: The Center for Health Design, Inc.

Rutala, W. A. (1996). APIC guideline for selection and use of disinfectants. *American Journal of Infection Control, 24,* 313–342.

Salvendy, G. (Ed.). (1997). *Handbook of Human Factors and Ergonomics.* New York, NY: John Wiley & Sons.

Sanders, M. S., & McCormick, E. J. (1993). *Human Factors in Engineering and Design.* New York: McGraw-Hill.

Schmaker, S. A., & Pequegnat, W. (1989). Hospital design, health providers, and the delivery of effective health care. In E. H. Zube and G. T. Moore (Eds.), *Advances in Environment, behavior, and design* (p. 355). New York, NY: Plenum Press.

Sehulster, L., & Chinn, R. Y. W. (2003). Guidelines for environmental infection control health-care facilities. *CDC, 52,* 1–42.

Smith, M. J., & Carayon, P. (2000). Balance theory of job design. In W. Karwowski (Ed.) *International Encyclopedia of Ergonomics and Human Factors* (pp. 1181–1184). London: Taylor & Francis.

Smith, M. J., & Carayon-Sainfort, P. (1989). A balance theory of job design for stress reduction. *International Journal of Industrial Ergonomics, 4,* 67–79.

Tablan, O. C., Anderson, L. J., Besser, R., Bridges, C., & Hajjeh, R. (2003). Guidelines for preventing health-care—Associated Pneumonia, 2003. *CDC, 53,* 1–36.

Thornton, R., & Vyrnwy-Jones, P. (1984). Environmental factors in helicopter operations. *Journal of the Royal Army Medical Corps, 130*(3), 157–61.

Tinetti, M. (2003). Clinical practice. Preventing falls in elderly persons. *New England Journal of Medicine, 348*(1), 42–49.

Topf, M. (1985). Noise-induced stress in hospital patients: Coping and nonauditory health outcomes. *Journal of Human Stress, 11*(3), 125 34.

Topf, M. (1988). Noise-induced occupational stress and health in critical care nurses. *Hospital Topics, 66*(1), 30–4.

Topf, M. (1992). Stress effects of personal control over hospital noise. *Behavioral Medicine 18*(2), 84–94.

Topf, M. (2000). Hospital noise pollution: an environmental stress model to guide research and clinical interventions. *Journal of Advanced Nursing, 31*(3), 520–528.

Topf, M., & Dillon, E. (1988). Noise-induced stress as a predictor of burnout in critical care nurses. *Heart & Lung: Journal of Acute & Critical Care, 17*(5), 567–574.

Topf, M., & Thompson, S. (2001). Interactive relationships between hospital patients' noise-induced stress and other stress with sleep. *Heart & Lung, 30*(4), 237–243.

Trace Center, C. o. Engineering, et al. (2005). Madison. 2005.

U.S. Department of Labor (DOL). (2001). Revised Occupational Safety and Health Administration Bloodborne Pathogens Act.

U.S. Department of Justice (DOJ). (1990). Americans with Disabilities Act (ADA).

U.S. Department of Justice (DOJ). (1992). ADA Regulation for Title II.

U.S. Department of Justice (DOJ). (1994). ADA Regulation for Title III.

Vanderheiden, G. C., & Vanderheiden, K. R. (1992). *Guidelines for the design of consumer products to increase their accessibility to persons with disabilities or who are aging.* Retrieved June 22, 2006.

Walsh-Sukys, M., Reitenbach, A., Hudson-Barr, D., & DePompei, P. (2001). Reducing light and sound in the neonatal intensive care unit: An evaluation of patient safety, staff satisfaction and costs. *Journal of Perinatology, 21*: 230–235.

Weichenthal, L., & Soliz, T. (2003). The incidence and treatment of prehospital motion sickness. *Prehospital Emergency Care,* 7(4), 474–476.

Weiss, S. J., Ellis, R., Ernst, A. A., Land, R. F., & Garza, A. (2001). A comparison of rural and urban ambulance crashes. *American Journal of Emergency Medicine, 19*(1), 52–56.

Wickens, C. D., Lee, J. D., Liu, Y., & Becker, S. E. G. (2004). *An Introduction to Human Factors Engineering.* Upper Saddle River, NJ: Prentice Hall.

Wickens, C. D., Lee, J. D., Liu, Y., & Gordon Becker, S. E. (2004). *An Introduction to Human Factors Engineering.* Upper Saddle River, NJ: Pearson Prentice Hall.

Williams, M. A. (1988). The physical environment and patient care. *Annual Review of Nursing Research, 6,* 61–84.

20

PHYSICAL ERGONOMICS IN HEALTH CARE

Sue Hignett
Loughborough University

There are increasing numbers of publications, in particular conference proceedings, describing projects looking at physical ergonomic and musculoskeletal problems in all areas of hospital and health care work. This chapter starts by presenting a general review of previously researched topics that found the largest proportion of projects were addressing musculoskeletal problems in nursing staff but also included a small and growing number of research studies looking at other staff groups.

Aspects of physical ergonomics in health care relating to the design of workplaces, equipment, and systems, including participatory projects, will be covered to give examples for a range of clinical and support areas: patient handling, operating theatres, dental work, midwifery, radiography (ultrasonography and mammography), cytology, and ambulance work. Some of the more popular postural analysis tools, which have been used to assess physical workload as part of a full ergonomic workplace assessment, are also described. This chapter does not provide a full overview of the extent of the literature and is not based on a systematic review protocol.

PHYSICAL ERGONOMICS

A review was carried out to look at the scope of healthcare ergonomics (Hignett, 2002) by searching conference publications (*Ergonomics Society,* [UK], *Human Factors and Ergonomics Society* [USA], *International Ergonomics Association*) and peer-reviewed journals (*Applied Ergonomics, Ergonomics, and Human Factors*). In just over 10 years of ergonomics research (1989 to 2000), 348 papers were found representing approximately 3% of the total papers. The topics covered ranged from musculoskeletal research (36%) through to accident investigation (8%) as shown in Figure 20–1. It is likely that a review in 2010 would find a different distribution of research work in health care ergonomics due to the increased funding available for patient safety initiatives from about the beginning of the twenty-first century.

Physical ergonomics projects usually focus on staff (caregiver) well being and so it is interesting to look at the distribution of projects by staff groups (Figure 20–2). It can be seen that nurses (registered and unregistered) account for the largest group (41%), followed by doctors at 24%. The focus for this chapter is musculoskeletal issues for health care workers. It will include dynamic and static activities and starts with an overview of some of the more commonly used postural analysis tools.

Postural Analysis

Several postural analysis tools have been used to assess physical stress in healthcare tasks. Some are used to diagnose problems and evaluate changes these include OWAS, NIOSH, RULA, and REBA. Other more general assessment tools might be used to review compliance with legislation or professional recommendations (e.g., MAC, www.hse.gov.uk/msd/mac/index.htm).

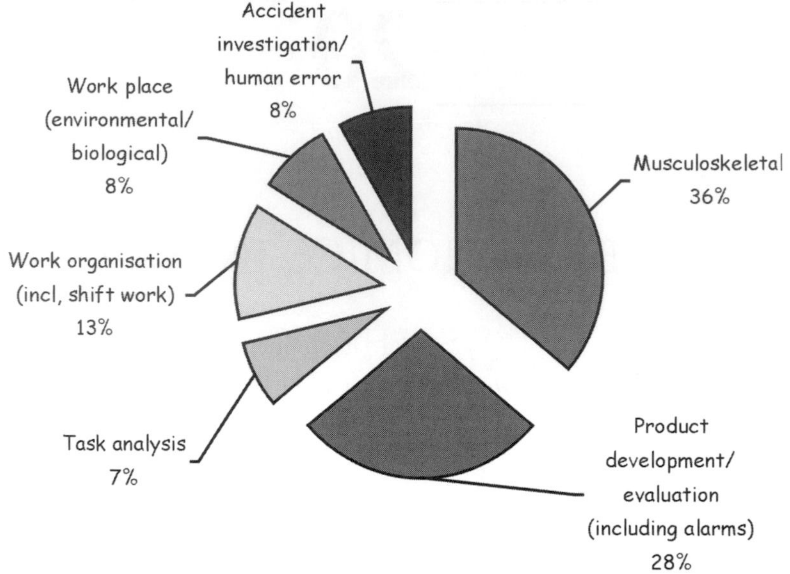

Figure 20–1. Research on Healthcare Ergonomics (1989–2000)

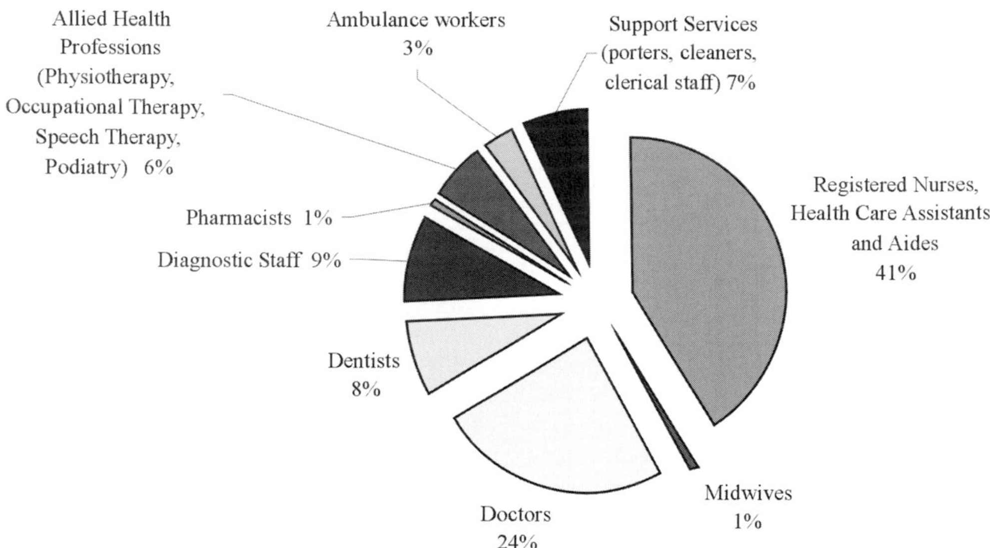

Figure 20–2. Staff Groups in Healthcare Ergonomics Projects (1989–2000)

OWAS

The Ovako Working posture Analysis System (OWAS) was developed as an observational technique for evaluating work postures, with a set of criteria for redesigning working methods and places based on the evaluation of experienced workers and ergonomics experts (Karhu, Kansi, & Kuorinka, 1977). Although OWAS was developed in the steel industry, it has been widely used in other industrial sectors from forestry to health care.

OWAS has 84 basic posture types (four back, three arm, seven legs, and three weight/force), which are combined to give a total possible 252 posture types. Data are collected by split-second observations (snap shots) of the workers' posture. Equal interval observations (time sampling) are recommended (Hignett & McAtamney, 2005), and a minimum of 80 observations is recommended for accuracy (Suurnäkki et al., 1988), although a 10% error limit is reported for 100 observations (Institute of Occupational Health, 1992).

The postures are analyzed to produce an action category (AC) rating, which gives an indication of the level of severity of the postural load (Louhevaara & Suurnäkki, 1992) starting from AC1, where no action is required, through to AC4, where action is required immediately.

NIOSH

The National Institute for Occupational Safety and Health (NIOSH) first developed a lifting equation in 1981 to assist safety and health practitioners evaluate lifting demands in the sagittal plane based on biomechanical demands of weight, frequency of lift, and horizontal and vertical displacement (Walters, Putz-Anderson, Garg, & Fine, 1993). This equation was revised and expanded in 1991 to include asymmetrical lifting tasks and complex coupling situations. Although it is not intended for use in the evaluation of lifting people, it has been used to evaluate non-patient handling tasks.

Steinbrecher (1994) used NIOSH to assess the task of loading linen sacks into a disposal chute. She found that the recommended weight limit (RWL) of 23 kg was exceeded when the linen sack was being placed in the chute. The recommendations included the provision of linen carts to facilitate the movement of the sack.

RULA

Rapid Upper Limb Assessment (RULA) was developed by Lynn McAtamney and Nigel Corlett at the Institute for Occupational Ergonomics, University of Nottingham (McAtamney & Corlett, 1992). It is a quick survey method that can be used as part of a ergonomic workplace assessment, where musculoskeletal disorders are reported, and it was validated on computer operators and sewing machinists. RULA assesses biomechanical and postural loading on the neck, shoulders, and upper limbs and was designed to assess predominantly sedentary work. It allocates scores based on the position of groups of body parts with additional scores for force/load and muscle activity. The final RULA score is a relative rather than an absolute score, and it gives an indication of the risk level on a four-point AC scale from AC1, where the posture is acceptable, through to AC4, where investigation and changes are needed immediately.

REBA

Rapid Entire Body Assessment (REBA) is a whole-body assessment tool, which was initially designed to provide a pen-and-paper postural analysis tool to be used in the field by direct observations or with photographic stills/video (Hignett & McAtamney (2000). It is currently being further developed for use on a pocket personal computer (Janik, Münzbergen, & Schultz, 2002).

REBA was developed to assess the type of unpredictable working postures found in health care and other service industries and was validated using examples from the electricity, health care, and manufacturing industries. Data are collected about the body posture, forces used, type of movement or action, repetition, and coupling. A final REBA score is generated giving an indication the level of risk and urgency with which action should be taken on a five-point AC category scale of 0 to 4, from no action required through to action necessary now.

PHYSICAL ERGONOMICS: RESEARCH AND PRACTICAL APPLICATIONS

The research and case study examples that are presented in the following sections are not exhaustive and just seek to give examples of the type of physical ergonomic and musculoskeletal problems encountered in the health care industry.

Patient Handling

There is a separate chapter on patient handling, so just a few studies are included in this section to show how working postures have been assessed using a variety of ergonomic, physiological, biomechanical, and observational tools. The following abstracts illustrate some of the measures and tools, with Stubbs, Buckle, Hudson, and Rivers (1983) using intra-abdominal pressure (IAP); Marras, Davis, Kirking, and Bertsche (1999) and Schibye and Skotte (2000) using electromyography and the lumbar motion monitor; and Owen, Garg, and Jensen (1992) using a 3-D static biomechanical model and rated perceived exertion.

Stubbs and Osborne (1979) evaluated the loads of activities used on a nursing shift. Data were collected using an observation study, weighing of loads (direct or by comparative assessment), and IAP. They rated three commonly used lifts and recommended the shoulder lift because the musculoskeletal loading was lower than the orthodox and three-person lifts. Stubbs et al. (1983) continued this work to look at four techniques: shoulder (Australian) lift; orthodox (traditional) lift; through arm lift; and underarm drag. The techniques were evaluated using IAP and a five-point comfort rating scale. They again found that the shoulder lift produced significantly lower pressures than the other three lifts, but there was no significant difference between the other three lifts.

Winkelmolen, Landeweerd, and Drost (1994) looked at the differences in working postures by calculating back compression and collecting data on rated perceived exertion (RPE) for six techniques: (a) lifting up in bed with two caregivers; (b) shoulder lift; (c) orthodox lift; (d) barrow lift, by grasping hands under the patient waist, with one other hand under thigh and one other hand under head; (e) through arm lift; (f) under arm lift. They recorded the working posture in terms of flexion, lateral flexion, and rotation; the biomechanical stress as the compressive force relative to the maximum permissive load; and the subjective stress as RPE and body part discomfort. The results found that the barrow lift caused significantly more stress than the shoulder or orthodox lifts and that the orthodox and barrow lifts produced more subjective stress than the underarm and through arm lifts. The shoulder lift was the least stressful.

Marras et al. (1999) aimed to identify the nature and range of low-back spinal forces and risk of low-back disorder associated with the various patient-handling tasks and techniques. Outcome measures included an electromyography (EMG)-assisted biomechanical model; lumbar motion monitor (LMM); and bipolar electrodes at the ten major trunk muscle sites. They found that the single caregiver hug method (pivot) resulted in about 10% higher risk than any of the two-caregiver transferring methods, with lifting from the hospital chair having the highest risk and compression values.

Schibye and Skotte (2000) looked at the working postures and mechanical load on the low back during the most common patient handling tasks and then compared the variation in the load between the different tasks: (a) turning in bed from supine to left-side lying (toward caregiver); (b) repositioning from the middle of the bed to the nearest bedside; (c) turning from supine to right-side lying (away from caregiver); (d) elevating from supine in bed to sitting on the edge of the bed; (e) lifting the patient from sitting on the edge of the bed to standing; (f) moving from sitting to a supine position in bed; (g) transferring from sitting on the bed to sitting in a wheelchair; (h) positioning backward in the seat of the wheelchair; (i) repositioning higher upward in bed. The working postures and mechanical load were measured using a 3-D biomechanical dynamic multi-segment model for L4/5 with two force plates to stand on at the edge of the bed; EMG of Erector Spinae; and LMM to measure the kinematics of the low back. The highest moments were found during the two tasks typically involving a lifting procedure of the whole body weight, lifting from sitting on the edge of the bed to standing and repositioning backward in a wheelchair, and also for two repositioning tasks, repositioning from the middle of the bed to the nearest side to the caregiver and repositioning higher upward in the bed.

Owen et al. (1992) described four methods to identify (comparison and ranking) the most back-stressing tasks performed by nursing assistants in a nursing home. They identified 153 tasks, which were grouped into sixteen task categories. The outcome measures included: RPE; comparative ranking for stressfulness; postural measurement for the compression force on L5/S1; and 3-D static biomechanical model to estimate the tensile strength on the Erector Spinae muscles. They found that the most stressful tasks involved transferring patients from one destination to another: toileting (toilet-chair); bed transfers (bed-chair); and bathing.

All the studies were able to evaluate the tasks and equipment and make recommendations to assist staff in deciding between the different technique and equipment options.

Surgery (Operating Theaters)

The ergonomic design of the operating room (OR) working environment has been the subject of a number of research projects. The complexity of the environment is outlined in Kumar (1998) in Figure 20–3 by listing the areas that all support different tasks and activities.

The environmental design and layout of the OR will vary depending on the clinical specialty (e.g.,

• Anaesthetic room • Scrub up area • Sterile supply area • Dirty utility area • Sterilizing room • Unsterile stock and heavy equipment area	• Plaster room • Medical, nursing and technicians' lounge room • Dark rooms for processing x-rays • Staff changing rooms • Cleaners room

Figure 20–3. The Rooms and Areas in an Operating Theater (Kumar, 1998)

orthopedic, burns, obstetric), but all will have specific requirements for lighting, ventilation, storage, communication, and so forth. The equipment used will also vary with specialist operating tables and attachments for plastic surgery (e.g., hand tables), laparoscopies (monitor display), and orthopedics (e.g., shoulder tables). There are a number of key tasks that contribute to musculoskeletal injury risks; these include patient positioning (e.g., for lateral thoracotomy), static loading (e.g., scrub nurse in long procedures), and awkward working postures (e.g., vaginal hysterectomy). The following abstracts give examples for these issues where the problems were investigated using OWAS (Kant, de Jong, van Rijssen-Moll, & Borm, 1992), RULA (Cowdry & Graves, 1997), and WATBAK (Jozwiak & Gadzincki, 1993).

Kant et al. (1992) looked at the postures of operating room staff (surgeons, assistant anesthetists, instrumentation nurses, and circulating nurses) using OWAS as an occupational health survey and research tool. They found that the postures adopted in this environment were predominantly static with instrumentation nurses and surgeons showing a high level of poor postures. Jozwiak and Gadzincki (1993) used WATBAK to analyze postures from video recordings for operating room staff. They concluded that the workload for all the professional groups was harmful, and this mainly depended on long-duration static loading for surgeons and on lifting excessive weights for nurses.

Cowdry and Graves (1997) used hierarchical task analysis and postural analysis (RULA) to look at surgical access to pediatric patients. They found that many surgeons were using adult tables with the result that the tables were much wider than the patients. This contributed to the awkward working postures observed, particularly for seated surgeons. Pediatric operating tables and chairs were recommended, as well as consideration of alternative patient positioning.

Laparoscopic surgery has specific problems, some related to the use of the monitor (Erfanian, Luks, Kurkchubasche, Wesselhoeft, & Tracy, 2003), where an in-line image projection might be preferable to viewing the image on a monitor at or above eye level across the table. Other problems relate to an increase in static postures for neck and trunk in comparison with open surgery (Nguyen et al., 2001). Upper-limb discomfort is also increased with the awkward positions required to manipulate the longer laparoscopic instruments. Matern, Eichenlaub, Waller, and Ruckauer (1999) measured the forces required to operate laparoscopic instruments and found that four to six times more force was required to complete the same task compared with the instruments used for open surgery.

Dental Work

The working posture of dentists has been assessed in a number of studies including Finsen, Christensen, and Bakke (1998), Kawamoto, Yoshihara, Inoue, and Taniguchi (2003), Ozawa, Tahata, Nakano, Shiozawa, and Arai (2003), and Turner, Rohrer, and James (2002). Finsen et al. (1998) used a questionnaire to identify musculoskeletal risk factors (65% neck/shoulder and 59% back) and EMG to look at common work activities, finding prolonged neck flexion and upper limb abduction. Ozawa et al. (2003) carried out a detailed analysis of the upper-limb postures in relation to the tooth position of the patient. They used EMG to analyze wrist postures and found that the most extreme wrist and forearm postures occurred when the dentists were working on the rear molar teeth. Kawamoto et al. (2003) examined the change in the working postures of dentists in Japan over the last 50 years through a review of educational documents. They found that until about 1970 a dentist was most likely to adopt a standing posture, with the patient sitting, from the late 1980s this was changed to the patient in a reclined position and the dentist sitting. Turner et al. (2002) describe a practical study aiming to identify an appropriate chair for use in a dental clinical. The chairs were evaluated over 8 months, and discomfort ratings were collected for the head, neck, lower back, and shoulders. The dentists preferred a saddle-shaped seat and arm rests but did not use the reverse (chest) support option that had been provided to address the forward- leaning working posture typically found in dental work.

Midwifery (Obstetric Nursing)

Midwifery is a very reactive area of health care with respect to both childbirth and breast feeding. The establishment of breast feeding is strongly encouraged in the UK (Department of Health, 1993), and major problem is that assisting breast feeding can be a lengthy (up to 1 hour per feed) and frequent task with midwives reporting that they can spend most of a night shift (8 hours) going from one mother to the next to assist with breast feeding (Hignett, 1996). There have been two studies looking at breast feeding. Steele and Stubbs (2002) used several methods to investigate midwifery involvement in breast feeding: focus groups, postural analysis, and body mapping. The physical discomfort was measured using the body mapping and the Quick Exposure Checklist (Li and Buckle, 1999). The QEC results suggested that midwives were frequently flexed (trunk) between 20° and 60° whereas the focus group data provided detailed information about task and posture rationale and possible alternatives.

Case Study: Breast Feeding

Hignett (1996) explored practical options to decrease the musculoskeletal stress on midwives by looking at the interactions between the mother, baby, and midwife. Mothers may choose to feed their babies sitting in an armchair, sitting on the bed, or lying on the bed. The midwives explained that they need to be able to see the baby's mouth/nipple interface so will position themselves with this as the key requirement, often resulting in sustained, awkward postures. Two design recommendations were proposed and tested. The first was a reverse chair that supports the midwife in a forward-leaning posture. Although this was found to be successful for some staff, unfortunately, the midwives were still wearing a dress as their uniform, which resulted in an unacceptable combination of the dress and chair. Since this work was undertaken, the uniform code has changed to include trousers, so the reverse chair may now provide a feasible design solution. The second recommendation was based on the observation that breast feeding seemed to become easier in a squashy armchair when the mother returned home. A foam armchair back support was sourced and tested by mothers, and their feedback was positive. This back support enabled the mother to adopt a supported sitting posture on both the bed and chair and resulted in a reduced need for assistance from the midwife.

Radiography

Radiography or x-ray technology is an essential service in health care. It is one of the primary methods of diagnosing pathology and trauma and can be found in most hospitals, clinics, and trauma centers with varying numbers of staff and shift patterns (some providing 24-hour cover). This section outlines two studies that look at general issues in radiography and then goes on to give a more detailed review for two specialist areas: ultrasonography and mammography.

Kumar, Moro, and Narayan (2003) reported a study looking at the activities of seven x-ray technologists working in different areas, and they analyzed the biomechanical stress for 16 tasks: (a) wearing a lead apron; (b) loading a small x-ray cassette (10N); (c) loading a large x-ray cassette (20N); (d) pushing/pulling an x-ray tube; (f) pushing a mobile x-ray unit on the floor; (g) pulling a mobile x-ray unit on the floor; (h) pushing a patient stretcher in the hallway; (i) pushing a wheelchair with a patient in the hallway; (j) repositioning a patient horizontally in bed; (k) repositioning a patient on side in bed; (l) repositioning a patient to an upright seated position in bed; (m) repositioning a cassette under a patient; (n) slider board transfer

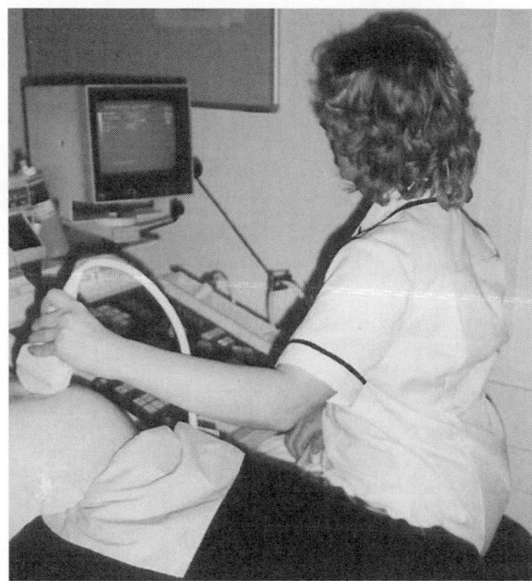

Figure 20–4. Obstetric Sonography (Hignett, 2005)

of a patient; (o) spine board transfer of a patient; (p) pulling the spine board; and (q) lifting a patient from a wheelchair. The patient handling tasks were found to produce the highest levels of biomechanical stress, but the more radiographic-specific (a–f, l) also contributed to the biomechanical load due to repetition (b–c), with the task of repositioning a cassette under a patient (l) exceeding the maximum NIOSH 1981 limit.

Rothmore (2003) also looked at the effect of wearing a lead apron to compare three different designs (two-piece, one-piece, and one-piece with a waist belt apron), using visual analogue scales for fatigue and ease of movement and a body map to record discomfort. The one-piece apron was found to be significantly more uncomfortable (neck/shoulder and low back) than the other two options.

Ultrasonography

Ultrasonography is used for general investigations of soft tissues, as well as specific diagnostic tests in a range of clinical specialties, for example, gynecology, obstetrics, cardiology, vascular, and pediatrics. Sonography has been available as a diagnostic tool since 1942 and was recognized as a separate profession in 1974 (Ransom, 2002). Sonographers use a hand-held transducer, which is applied to the area needing investigation, linked to a scanning machine, which relays the images collected by the probe. It typically involves a static posture to maintain the arm in a fixed position (unsupported abduction) while pressing the transducer against the patient (Figure 20–4).

Russo, Murphy, Lessoway, and Berkowitz (2002) reported on a survey in British Columbia, finding that the majority of sonographers (91%) had work-related musculoskeletal problems. They collected data about the work environment and corporate culture, schedule, tasks, and equipment to look for associations with physical symptoms. Factors included the number of hours scanning, static and awkward postures, and psychosocial factors, including social support (coworker, supervisor, and senior management) and decision-making with respect to planning the workload and taking breaks.

Crawford, McHugo, and Vaughan (2002) used RULA to assess the musculoskeletal risk to obstetric sonographers. This highest scores were recorded when performing the scan. This was suggested to relate to both the complexity of the sonographer-transducer-patient interface and the lack of flexibility in design of the equipment and workstation.

Practical examples of operational changes include job design (to encourage breaks, reduce repetition, and introduce job rotation), replacement equipment and furniture, as well as training and rehabilitation programs (Kilbourn, 2004; Anderson & Denison, 2003; Redden, 2003).

Case Study: Cardiac Ultrasonography

A problem was identified by ultrasonographers working in cardiology who were performing cardiac scans on patients throughout their working day. The patient has to lie on his left-hand side, and the sonographer reaches over the patient's chest to place the probe on the left-hand side of the chest. Initially, the plinth was not designed to accommodate the sonographer, so they either stood or perched on the edge of it in an awkward twisted position (Figure 20–5). As the work required both upper limbs to carry out fine motor activities, a high level of concentration was required, so a slumped posture was often observed as the sonographer concentrated on the (a) visual display, (b) machine controls, and (c) probe placement. On initial assessment using RULA, a score of 7 was recorded, which placed this posture in AC 4, indicating that immediate investigation and change was required. A modified plinth was designed in collaboration with the manufacturer with a cut-out seat section. Post-implementation, a reduced RULA

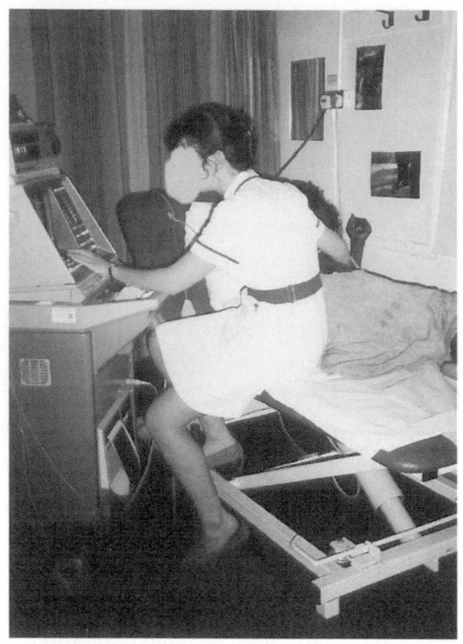

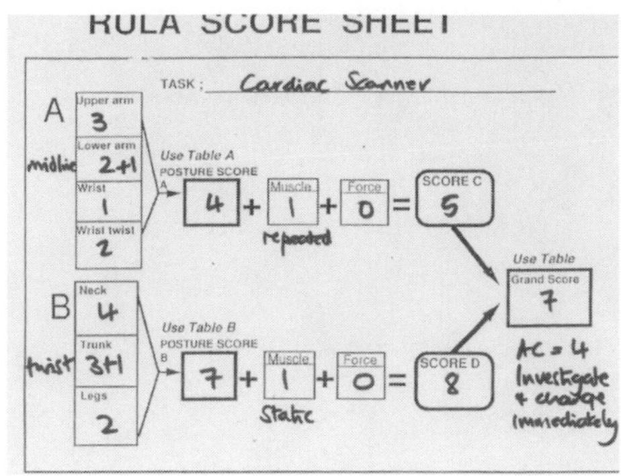

Figure 20–5. Cardiac Ultrasonography (Hignett, 2005)
RULA Score = 7, Action Category = 4 (Investigate and Change Immediately)

score of 6 (AC 3) was recorded. However, it was felt that there was scope for further improvement (and reduction in the RULA score), so further discussions were held with the manufacturer to improve both the design of the seat and the plinth ultrasound machine interface.

Mammography

Two studies were found that looked at the musculoskeletal problems in mammography (x-ray for breast screening). Hearn and Reeves (2003) sent a questionnaire to 81 breast-screening centers in the UK with an 80% response rate. They found a high rate of work-related upper-limb symptoms, mostly in the neck, shoulder, and thumb and suggested that motorized mammography machines caused less strain on the upper limbs.

May and Gale (1997) carried out an assessment of mammography work using hierarchical task analysis, observation, hourly body mapping, and workplace measurements. They found that the radiographers experienced difficulty when placing the film cassette in the bucky (holder) and that they experienced increasing discomfort in the lower back throughout the working day.

Both studies showed how the design of the equipment was resulting in awkward and uncomfortable working postures. Recommendations were made to improve the working position and equipment design.

Cytology

The difficulties of working with microscopes have been identified for some years, with headaches and neck discomfort and visual strain listed as the main three problems (Emanuel & Glonek, 1975). A number of studies have been carried out to determine the optimal working conditions, focusing on the neck angle (Lee, Waikar, Aghazadeh, and Tandon 1986) or standing rather than sitting (Sillanpää, Nyberg, and Laippala 2003), but more recently the focus has been on the design of the microscope itself.

Kreczy, Kofler and Gschwendtner (1999) reported that 80% of professionals working daily for a long time with a light microscope complain of eye strain, back pain, fibromyalgia, or tension headaches. They developed a work station design that permitted a more neutral working posture with respect to neck and back posture. This work station was evaluated by 12 individuals and compared with a control work station. They found a significant reduction in muscle activity for the neck, upper limb, and back for the new design and recommended that fully adjustable ergonomic microscopes should be used to minimize physical discomfort from prolonged microscope use.

Hopper, May and Gale (1997) looked at the entire workstation (microscope, computer, work bench, and chair), as well as the environmental

1. Adjustment of eyepiece to achieve a comfortable position 2. Comfort of the right arm and hand 3. Comfort of the left arm and hand 4. Overall body comfort 5. Adjustment of microscope body length to comfortable length	6. Position of controls – distance 7. Position of controls – height 8. Usability of controls 9. Overall Usability 10. General impression (shape, colour etc.)

Figure 20–6. Microscope Questionnaire

conditions and lighting. They recommended minimum ergonomic guidelines for cytology, which were incorporated into national guidelines (NHS Cancer Screening Programme, 2003).

Case Study: Microscope Design

Nottingham City Hospital is one of the largest cervical screening centers in the UK, handling over 65,000 cervical smears each year. This results in cytology staff having to spend long periods of time at the microscope, which was causing neck, shoulder, and back problems for some staff. A task analysis was undertaken as a precursor to a comparative product evaluation to recommend replacement microscopes to the department (Nikon, 1999). This resulted in a 10-point questionnaire (Figure 20–6).

The existing microscopes were found to have minimum adjustment options, and one of the key dimensions (eye–elbow distance, 300 mm) fit less that 1 in 100 of the British female adult working population. A better design solution was to provide a spacer giving an adjustment in the eye–elbow height to accommodate a range of users (e.g., 95th percentile British male adult = 615 mm, 5th percentile British female adult = 455 mm).

Ambulance Work

Ambulance ergonomics previously has been an under-researched area of health care; however, there have recently been a number of studies in Canada, Australia, and the UK that have addressed the musculoskeletal problems encountered by ambulance workers (Hignett et al., 2003). The following two case studies look at the difficulties encountered by paramedics in lifting from the floor and when working on patients in the ambulance.

Case Study: Paramedics Lifting From the Floor

A project was conducted in 2002 to comparatively evaluate three methods of raising a patient from the floor (Hignett & Rankin, 2003) as shown in Figure 20–7: (a) manual technique, (b) air cushion, and (c) hoist.

The manual lift was found to be the fastest lift, but it produced the highest postural risk scores as shown in Figure 20–8.

It was recommended that the air cushion should be used on a regular basis but that a hoist should also be available if required. The ambulance service has since deployed specialist vehicles to provide a secondary response service by bringing additional moving and handling equipment to the scene. As the population of bariatric people increases, ambulance workers are finding that problems that they already faced (e.g., stretcher loading, ergonomically challenging equipment, difficult working environments) are being exacerbated (Hays, Hignett, and Grimshaw, 2004), and the specialist vehicles are also being used to ensure that appropriate equipment and expertise are available

Case Study: Layout of Ambulance Equipment

A detailed link analysis was undertaken to look at the working layout of the interior of a UK ambulance as shown in Figure 20–9 (Ferreira & Hignett, in

(a) Manual

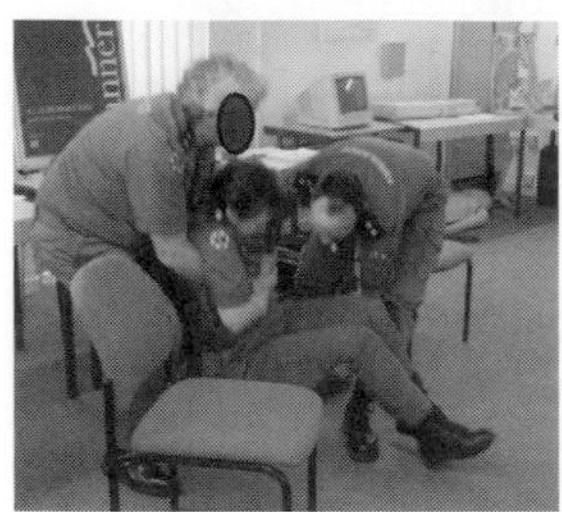

(b) Air Cushion

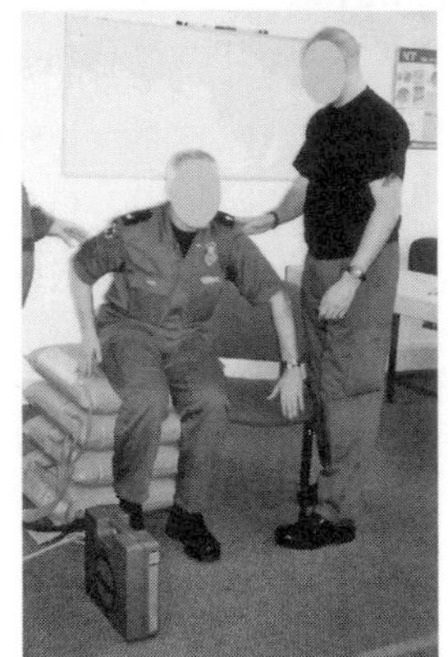

(c) Hoist

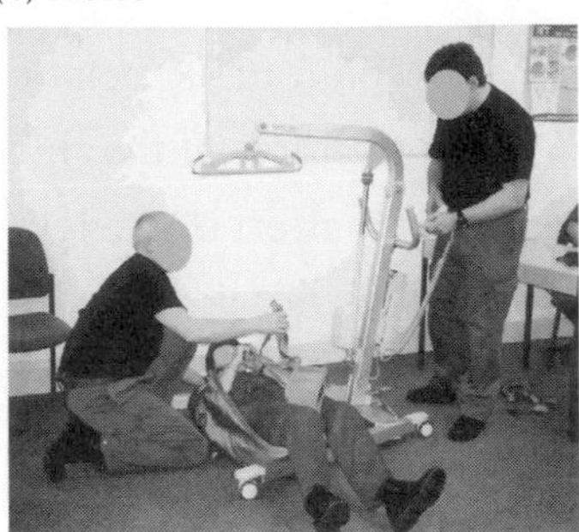

Figure 20–7. Lifting from the Floor in Ambulance Work (Hignett & Rankin, 2003)

(a) Time to complete task

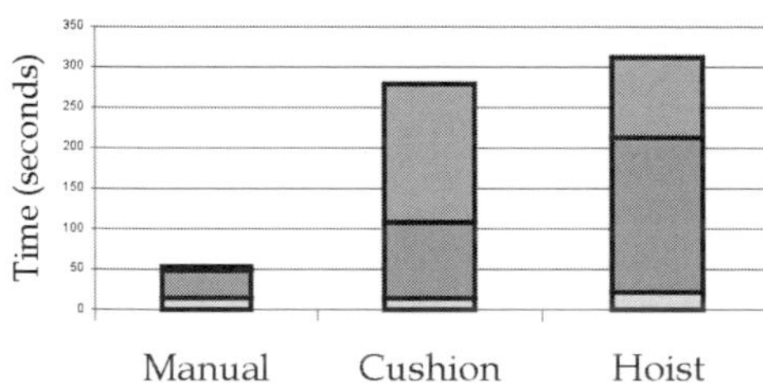

(b) REBA score for set and lift phases

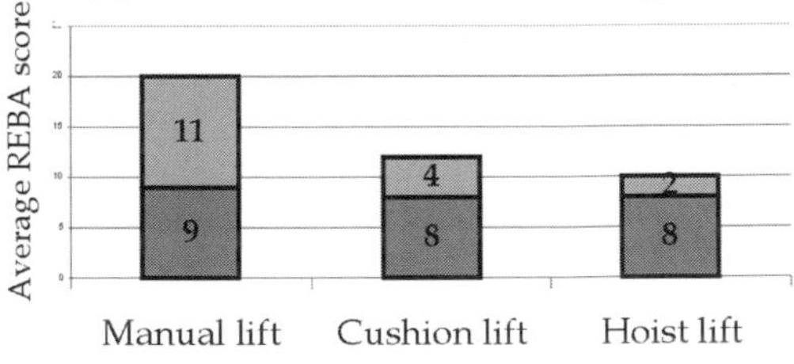

Figure 20–8. Lifting from the Floor: Time and Postural Analysis

press). From this analysis it was possible to identify that the attendant seat (seat A, Figure 20–9) was rarely used, with the side passenger seat being the preferred seat (seat B, Figure 20–9).

It was found that the preferred working positions resulted in the paramedic having to perch on the edge of seat B in order to reach the patient. So the paramedics were working without seat belt protection and not as the manufacturer had designed the task for this vehicle, producing safety risks for both the paramedic and the patient. Further research work is currently ongoing to look at the issue in more detail and to consider alternative layouts.

Participatory Ergonomic Projects for Musculoskeletal Problems

The definition of participatory approaches includes interventions at macro (organizational, systems) levels, as well as micro (individual), where workers are given the opportunity and power to use their knowledge to address ergonomic problems relating to their own working activities. Participatory ergonomics (PE) has been used as an approach to tackle musculoskeletal problems (Vink & Wilson, 2003). The projects using PE have historically tended to be at a micro level, with mixed results; however, participatory interventions at macro levels also have much to offer (Hignett, Wilson, and Morris, 2005) as shown in the following abstracts.

Evanoff, Bohr, and Wolf (1999) reported a PE project carried out with hospital orderlies to see if direct worker participation in problem solving would improve job satisfaction, injury rates, lost time, and musculoskeletal symptoms. The intervention was evaluated using the OSHA 200 log, workers compensation insurance records and self-administered surveys of workers at 1, 7, and 15 months. They found a decrease in risks of work injury, with a reduction in the relative risk of 50%, for both OSHA 200 log and injury rate, as well as a reduction in total days lost. The survey found a large and statistically significant reduction in the

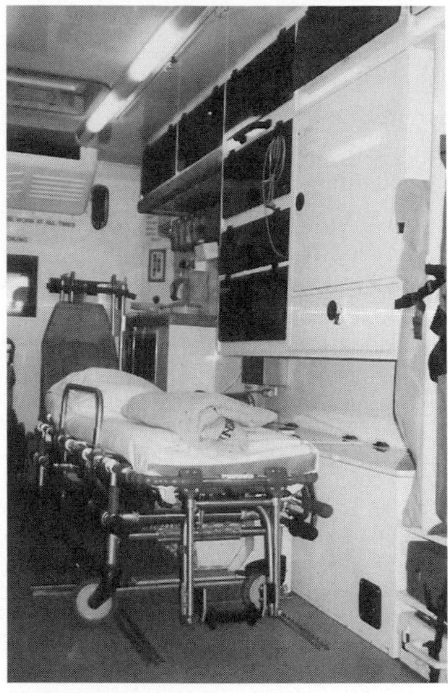

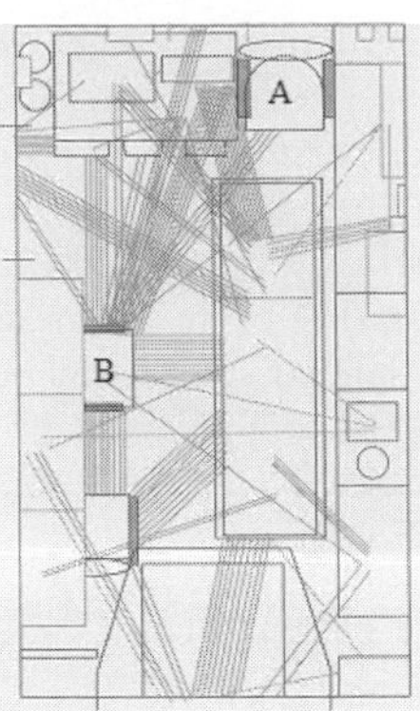

Figure 20–9. Interior Layout of Ambulance and Link Analysis Diagram (Ferreira & Hignett, 2005)

proportion of employees with musculoskeletal symptoms.

Estryn-Behar et al. (2000) used a PE methodology for the conception of a new hospital laboratory. A number of different analyses (activity analysis of seven professional groups, space analysis, noise map, and lighting map) were conducted to provide information for discussion at three workshops. The workers were able to plan a more practical new space layout. This was tested with a 3-D scale model to give every worker the opportunity to modify the model. The final layout produced an improved functional distribution of the available space.

In many clinical areas staff are required to work in sterile conditions, which may involve working in a flow cabinet. Lee, Lochang, Engst, and Robinson (2002) carried out an initial assessment of the task of working at a laminar flow hood using injury data, staff questionnaires, task analysis, and RULA. The results were used to develop design criteria for three prototypes, which were subsequently evaluated by staff in a participatory ergonomics project. The resultant recommendations supported a design offering variability in working height and depth. Other laboratory work often involves the use of pipettes in the preparation and testing of samples. Mattie, Watzke, Raschke, and Birch (2002) evaluated the design of pipetting workstations using questionnaires, observational data, biomechanical analysis, and simulation testing. They found that few ergonomic principles with respect to working height, adjustability, lighting, or workstation comfort had been incorporated into existing workstations.

CONCLUSION

This chapter has explored a range of research areas looking at musculoskeletal issues for health care workers. Previous research was found to have looked predominantly at nursing and medical workers. A number of postural analysis tools have been used to assess health care activities, and brief descriptions for four of these (field) tools were given.

Dynamic and static postural loading was explored with examples for patient handling (see also chap. 22 of this vol.), operating theaters, dentistry, midwifery, radiography, ultrasonography, mammography, cytology screening, and ambulance work. The chapter closes with a short section on participatory ergonomics as the most promising approach to achieve change in the health care sector. It has been suggested that there is a major difference in working in health care ergonomics with respect to implementing change, whereby 80% of the effort will be needed to progress a project and only 20% on understanding the problem. The reverse is probably more usual for ergonomics projects in other industries (Hignett, 2003).

There is an increasing body of research looking at human error and safety issues. It will be interesting to look at the two bodies of literature to see whether links can be drawn. It would seem likely that working in a posture with a high level of physical stress might be a contributory factor to patient safety incidents due to poor sight lines or difficulties in manipulating equipment, consumables, and controls.

References

Anderson, K. H., & Denison, A. R. (2003, August). Musculoskeletal injury in venous mapping. *Synergy*, 14–16.

Cowdry, I. M., & Graves, R. J. (1997). Ergonomic issues arising from access to patients in paediatric surgery. In S. Robertson (Ed.), *Contemporary ergonomics* (pp. 32–38). London: Taylor & Francis.

Crawford, J. O., McHugo, J., & Vaughan, R. (2002). Diagnostic ultrasound: the impact of its use on sonographers in obstetrics and gynaecology. In P. T. McCabe (Ed.), *Contemporary ergonomics* (pp. 21–26). London: Taylor & Francis.

Department of Health. (1993). *Changing childbirth*. Report of the Expert Maternity Group. London: HMSO.

Emanuel, J. T., & Glonek, R. J. (1976). Ergonomic approach to productivity improvement for microscope work. Paper presented at the AIIE Systems Engineering Conference, Institute of Industrial Engineering, Norcross, GA.

Erfanian, K., Luks, F. I., Kurkchubasche, A. G., Wesselhoeft, C. W., Jr., & Tracy, T. F., Jr. (2003). In-line image projection accelerates task performance in laparoscopic appendectomy. *Journal of Pediatric Surgery, 38*(7), 1059–1062.

Estryn-Behar, M., Wilanini, G., Scialom, V., Rebouche, A., Fiette, H., & Artigou, A. (2000). New conception of an hospital laboratory with a participatory ergonomics methodology. In *Proceedings of the 14th triennial congress of the International Ergonomics Association and the 44th annual meeting of the Human Factors and Ergonomics Society* (San Diego, California, July 29–August 4, 2000). Santa Monica: Human Factors and Ergonomics Society.

Evanoff, B. A, Bohr, P. C., & Wolf, L. D. (1999). Effects of a participatory ergonomics team among hospital orderlies. *American Journal of Industrial Medicine, 35,* 358–365.

Ferreira, J., & Hignett, S. (2005). Reviewing ambulance design for clinical efficiency and paramedic safety. *Applied Ergonomics. 36,* 97–105.

Finsen, L., Christensen, H., & Bakke, M. (1998). Musculoskeletal disorders among dentists and variation in dental work. *Applied Ergonomics, 29*(2), 119–125.

Hays, R., Hignett, S., & Grimshaw, M. (2004, September). Transporting very heavy (bariatric) ambulance patients. *Ambulance Today,* 19–21.

Hearn, T., & Reeves, P. (2003, September). RSI in mammography, *Synergy*, 16–19.

Hignett, S. (2005). Ergonomics in health and social care. In J. Smith (Ed.), *The guide to the handling of patients* (5th ed., pp. 41–48). Teddington, Middlesex: Back Care/Royal College of Nursing.

Hignett, S. (2003). Hospital ergonomics: a qualitative study to explore the organisational and cultural factors. *Ergonomics, 46*(9), 882–903.

Hignett, S. (2002). Hospital ergonomics. Keynote lecture. Paper presented at the occupational health conference, Porto do Varzim, Portugal, October 31, 2002.

Hignett S. (1996). Manual handling risks in midwifery: identification of risks factors. *British Journal of Midwifery, 4*(11), 590–596.

Hignett, S., Crumpton, E., Alexander, P., Ruszala, S., Fray, M., & Fletcher, B. (2003). *Evidence-based patient handling: tasks, equipment and interventions.* London: Routledge.

Hignett, S., & McAtamney, L. (2000). Rapid Entire Body Assessment (REBA). *Applied Ergonomics,* 31, 201–205.

Hignett, S., & McAtamney, L. (2006). REBA and RULA: Whole body and upper limb rapid assessment tools. In W. Marras & W. Karwowski (Eds.), *Occupational ergonomics handbook* (2nd ed.). Boca Raton, FL: CRC Press. 42–1 through 42–12.

Hignett, S., & Rankin, S. (2003, December). Use your head when lifting patients. *Ambulance Today,* 18–19.

Hignett, S., Wilson, J. R., & Morris, W. (2005). Finding ergonomic solutions: participatory approaches. *Occupational Medicine. 55,* 200–207.

Hopper, J. A., May, J., & Gale, A. (1997). Screening for cervical cancer: the role of ergonomics. In S. Robertson (Ed.), *Contemporary ergonomics* (pp. 38–43). London: Taylor & Francis.

Institute of Occupational Health. (1992). *OWAS: a method for the evaluation of postural load during work* (training publication no. 11). Helsinki, Finland: Institute of Occupational Health, Centre for Occupational Safety.

Janik, H., Münzbergen, E., & Schultz, K. (2002). *REBA-verfahren (rapid entire body assessment) auf einem pocket computer.* Paper presented at the 42 Jahrestagung der Deutschen Gesellschaf für Arbeitsmedizin un Umweltmedizin (DGAUM), München, April, 10–13, 2002.

Jozwiak, Z. W., & Gadzicki, E. (1993). Biomechanical analysis of work postures of operating room staff. In W. S. Marras, W.Karwowski, J. L. Smith, & L. Pacholski (Eds.), *The ergonomics of manual work* (p. 241–244). [Proceedings of the 12th triennial congress of the International Ergonomics Association, Poland, June 14–17.] London: Taylor & Francis.

Kant, I., de Jong, L. C. G., van Rijssen-Moll, M., & Borm, P. J. A. (1992). A survey of static and dynamic work postures of operating room staff. *International Archives of Occupational and Environmental Health, 63*, 423–428.

Karhu, O., Kansi, P., & Kuorinka, I. (1977). Correcting working posture in industry: a practical method for analysis. *Applied Ergonomics, 8*(4), 199–201.

Kawamoto, M., Yoshihara, M., Inoue, M., & Taniguchi, T. (2003). Principle for education of operating posture during dental practice. Paper presented at the XVth triennial congress of the International Ergonomics Association and the 7th joint conference of Ergonomics Society of Korea/Japanese Ergonomics Society, "ergonomics in the digital age," Seoul, Korea, August, 24–29 2003.

Kilbourn, P. (2004). Reducing the health risks to sonographers, *Synergy,* March. 13–18.

Kumar, B. (1998). *Working in the operating department* (2nd ed., 132–143). London: Pearson Professionals Ltd.

Kumar, S., Moro, L., & Narayan, Y. (2003). A biomechanical analysis of loads on x-ray technologists: a field study. *Ergonomics, 46*(5), 502–517.

Kreczy, A., Kofler, A., & Gschwendtner, A. (1999). Under-estimated health hazard: proposal for an ergonomic microscope workstation. *Lancet, 354*(13), 1701–1702.

Lee, E. J., Lochang, J., Engst, C., & Robinson, D. (2002). Applying an ergonomic design process to injury prevention in a hospital pharmacy. Paper presented at the 33th annual conference of the Association of Canadian Ergonomists, Banff, Canada, October, 2003.

Lee, K. S., Waikar, A. M., Aghazadeh, F., & Tandon, S. (1986). An electromyographic investigation of neck angles for microscopists. In *Proceedings of the Human Factors Society 30th annual meeting,* 548–551.

Li, G. and Buckle, P. (1999). Evaluating change in exposure to Risk for musculoskeletal disorders—a practical tool. http://www.hse.gov.uk/Research/CRR-pdf/1999/CRR99251.pdf.

Ljundberg, A-S., Kilbom, A., & Hägg, G. M. (1989). Occupational lifting by nursing aides and warehouse workers. *Ergonomics, 32*(1), 59–78.

Louhevaara, V., & Suurnäkki, T. (1992) *OWAS-a method for the evaluation of postural analysis during work.* Helsinki, Finland: Institute of Occupational Health, Centre for Occupational Safety.

Marras, W. S., Davis, K. G., Kirking, B. C., & Bertsche, P. K. (1999). A comprehensive analysis of low-back disorder risk and spinal loading during the transferring and repositioning of patients using different techniques. *Ergonomics, 42*(7), 904–926.

Matern, U., Eichenlaub, M., Waller, P., & Ruckauer, K. (1999). MIS instruments. an experimental comparison of various ergonomic handles and their design. *Surgical Endoscopy,* 13, 756–762.

Mattie, J., Watzke, J., Raschke, S., & Birch, G. (2002). Proposed experimental design for evaluating best practices amongst laboratory technologists involved in pipetting work. Paper presented at the proceedings of the 33th annual conference of the Association of Canadian Ergonomists, Banff, Canada, October, 2003.

May, J., & Gale, A. G. (1997). Understanding musculoskeletal discomfort in mammography. In S. Robertson, (Ed.), *Contemporary ergonomics* (pp. 26–31). London: Taylor & Francis.

McAtamney, L., & Corlett, E. N. (1993). RULA: a survey method for the investigation of WRULD. *Applied Ergonomics, 24*(2), 91–99.

NHS Cancer Screening Programmes (2003). Ergonomics Working Standards for Personnel Engaged in the Preparation, Scanning and Reporting of Cervical Screening Slides. NHSCSP Publication No. 17. Sheffield: NHS Cancer Screening Programmes. http://www.cancerscreening.nhs.uk/cervical/publications/nhscsp17. pdf.

Nguyen, N. T., Hung, M. D., Ho, M. D., Smith, W. D., Philipps, C., Lewis, C., De Vera, R. M., & Berguer, R. (2001). An ergonomic evaluation of surgeons' axial skeletal and upper extremity movements during laparoscopic and open surgery. *The American Journal of Surgery, 182*, 720–724.

Nikon Company. (1999). Ergonomic assessment at the Nottingham City Hospital: cytology microscopes. Retrieved from www.nikon.co.uk/Inst/downloads/microworldletters/micro_issues.15.pdf.

Owen, B., Garg, A., & Jensen, R. C. (1992). Four methods for identification of most back-stressing tasks performed by nursing assistants in nursing homes. *International Journal of Industrial Ergonomics, 9*, 213–220.

Ozawa, T., Tahata, K., Nakano, M., Shiozawa, K., & Arai, T. (2003). The effects of tooth positions on dentists hand muscle EMG activities and forearm movements during root canal instrumentation. Paper presented at the XVth triennial congress of the International Ergonomics Association and the 7th joint conference of Ergonomics Society of Korea/Japanese Ergonomics Society, "ergonomics in the digital age," Seoul, Korea, August 24–29, 2003.

Ransom, E. (2002). *The causes of musculoskeletal injury among sonographers in the UK.* Available from Society of Radiographers, 207 Providence Square, Mill Street, London SE1 2EW.

Redden, D.M. (2003), Musculoskeletal discomfort amongst radiotherapy staff. Paper presented at the 34th annual conference of the association of Canadian Ergonomists, "fit, form, and function," Ontario, Canada, October, 15–18, 2003.

Rothmore, P. (2003). Lead aprons, radiographers and discomfort. Paper presented at the XVth triennial congress of the International Ergonomics Association and the 7th joint conference of Ergonomics Society of Korea/Japanese Ergonomics Society, "ergonomics in the digital age," Seoul, Korea, August 24–29, 2003.

Russo, A., Murphy, C., Lessoway, V., & Berkowitz, J. (2002). The prevalence of musculoskeletal symptoms among British Columbia sonographers. *Applied Ergonomics, 33*, 385–393.

Schibye, B., & Skotte, J. (2000). The mechanical loads on the low back during different patient handling tasks. In *Proceedings of the IEA2000/HFES 2000 Congress. The Human Factors and Ergonomics Society. Santa Monica, California, 5,* 785–788.

Sillanpää, J., Nyberg, M., & Laippala, P. (2003). A new table for work with a microscope, a solution to ergonomics problems. *Applied Ergonomics, 34,* 621–628.

Steele, D., & Stubbs, D. (2002). Measuring working postures of midwives in the health care setting. In P. McCabe (Ed.), *Contemporary ergonomics* (39–44). London: Taylor & Francis.

Steinbrecher, S. (1994). The revised NIOSH lifting guidelines. *AAOHN Journal, 42*(2), 62–66.

Stubbs, D. A., & Osborne, C. (1979). How to save your back. a comparison between the nursing profession and the construction industry. *Nursing, 3,* 116–124.

Stubbs, D. A., Buckle, P., Hudson, M. P., & Rivers, P. M. (1983). Back pain in the nursing profession. the effectiveness of training. *Ergonomics, 26*(8), 767–779.

Suurnäkki, T., Louhevaara, V., Karhu, O., Kuorinka, I., Kansi, P., & Peuraniemi, A. (1988). Standardised observation method of assessment of working postures: the OWAS method. In A. S. Adams, R. R. Hall, B. McPhee, & M. S.Oxenburgh, (Eds.), *Ergonomics International 88. proceedings of the 10th triennial congress of the International Ergonomics Association* (281–283).

Turner, C., Rohrer, B., & James, T. (2002). Use of ergonomic chairs in a dental clinic. In P. T. McCabe (Ed.), *Contemporary ergonomics* (pp. 27–32). London: Taylor & Francis.

Vink, P., & Wilson, J. R. (2003). Participatory ergonomics. Paper presented at the XVth triennial congress of the International Ergonomics Association and the 7th joint conference of Ergonomics Society of Korea/Japanese Ergonomics Society, "ergonomics in the digital age," Seoul, Korea, August 24–29, 2003.

Walters, T. R., Putz-Anderson, V., Garg, A., & Fine, L. J. (1993). Revised NIOSH equation for the design an evaluation of manual lifting tasks. *Ergonomics, 36*(7), 749–776.

Winkelmolen, G. H. M., Landeweerd, J., & Drost, M. R. (1994). An evaluation of patient lifting techniques. *Ergonomics,* 37(5), 921–932.

21

EVIDENCE-BASED INTERVENTIONS FOR PATIENT CARE ERGONOMICS

Audrey Nelson, Andrea S. Baptiste, and Mary Matz
James A. Haley Veterans Affairs Medical Center, Tampa, Florida

Guy Fragala
Environmental Health & Engineering, Newton, Massachusetts

Many patient handling recommendations are based on tradition and personal experience rather than scientific evidence. The purpose of this paper is to summarize current evidence for interventions designed to reduce work-related musculoskeletal disorders (WMSDs) associated with patient handling. Many commonly used interventions, such as manual patient lifting, classes in body mechanics, training in safe lifting techniques, and back belts have strong evidence that they are NOT effective in reducing caregiver injuries. There is significant evidence to support use of patient handling technologies. There is growing operational support for use of patient care ergonomic assessment protocols, no lift policies, and lift teams, but more research is needed in these areas. Promising new interventions, which are still being tested, include use of unit-based peer leaders and clinical tools, such as algorithms and patient assessment protocols. Given the complexity of this high-risk, high-volume, high-cost problem, multifaceted programs are more likely to be effective than any single intervention. The authors conclude with several recommendations to promote safe patient handling and movement across clinical practice settings. A call for action includes a new curriculum for schools of nursing, legislative initiatives, research translation tools, and a research agenda for patient care ergonomics.

STATEMENT OF THE PROBLEM

Many patient handling tasks are considered to be high-risk, due to the magnitude of weight lifted, awkwardness and unpredictable nature of the load lifted (patient), and sustained awkward positions used to provide nursing care. The cumulative weight lifted by a nurse in a typical 8-hour workday is equivalent to 1.8 tons (Tuohy-Main, 1997). Many patient lifts are accomplished in awkward positions, such as bending over beds or chairs while the nurse's back is flexed (Blue, 1996; Videman et al., 1984). In one study, nurses spent an average of 1.6 hours per shift in a stooped posture (Stubbs et al., 1984). Inadequate space and poorly designed work environments contribute to these awkward positions. Unfortunately, nurses accept back pain as part of their job, with 52–63% of nurses reporting musculoskeletal pain that lasts for more than 14 days; in 67% of cases, pain was a problem for at least 6 months (Vasiliadou, Karvountzis, Somalis, Roumeliotis, & Theodosopulou, 1995).

Nursing injuries associated with patient handling tasks have far-reaching consequences for the injured nurse, the patient, and the organization. The injured nurse faces guilt and may be blamed for the injury. Other personal concerns include a fear of re-injury (Dillman, 1993), adverse career prospects (Wicker, 2000), impaired quality of life, and chronic pain or disability. The injury can result in the need to transfer to another nursing position or may necessitate an unwanted career change.

There are significant clinical consequences of nurse injuries as well, including a negative impact on quality of care, patient safety, and patient comfort (Wicker, 2000). In many cases injured nurses are not replaced when they are away from work or on restricted duty, creating staff shortages, which detract from patient care quality. Further, nurses with significant musculoskeletal pain or those on modified duty may not be physically able to get patients out of bed as frequently or perform other physically demanding tasks, decreasing patient services. When an injured nurse attempts to perform patient handling tasks with limited strength or musculoskeletal pain, the results can include resident discomfort (U.S. Dept. of Labor, 2002) or adverse events associated with dropping patients, dragging patients across surfaces, or frightening patients with jerky movements during transfer tasks. Patients can respond to these practices with fear, pain, damage to shoulder, hip fractures, bruises, loss of dignity during lifting procedure, increased dependency, skin tears, and pressure area damage (Tuohy-Main, 1997).

Work-related musculoskeletal injuries in nursing have serious consequences to the organization. A loss of productivity results from absenteeism (U.S. Dept. of Labor, 2002), lost or modified workdays (Dillman, 1993), and the difficulties imposed with scheduling and coverage on affected nursing units. The cumulative effect of nursing injuries can lead to staff shortages (Dillman, 1993) and permanent loss of experienced nursing staff to disability or decision to leave the job (Corlett, Lloyd, Tarling, Troup, & Wright, 1993). Recruitment and retention of nurses is a serious problem, exacerbated by nurses with job-related injuries. In one survey of injured nurses, 12% of respondents indicated they were considering making an employment transfer, and another 12% said they were thinking of leaving the nursing profession due to back pain (Owen, 1989). A study in England found that 12% of nurses who intended to leave nursing permanently cited back pain as a main or contributing factor (Stubbs, 1986). A third study in 1992 queried 99,955 RNs who left the profession; 18.3% said it was because of concern for safety in the health care environment (Moses, 1992).

Another organizational concern is the significant costs associated with injuries to care providers (Dillman, 1993; U.S. Dept. of Labor, 2002). In addition to the direct costs of medical treatment and compensation, the organization faces hidden factors that inflate the average claim loss by a factor of five, including temporary hires for replacement personnel, overtime to absorb the duties of injured worker, legal fees; time loss costs for claim processing, witnesses; decreased output following traumatic event; training temporary and/or replacement personnel (Charney, Zimmerman, & Walara, 1991; U.S. Dept. of Labor, 2002). Organizations may be concerned about the risk of liability (Wicker, 2000; Corlett et al., 1993), as well as regulatory deficiencies associated with hazardous work environments. Lastly, the organization can face the effects of diminished staff morale and job satisfaction (U.S. Dept. of Labor, 2002), exacerbating the serious problems associated with recruitment and retention of nurses.

SIGNIFICANCE OF THE PROBLEM

In 2001, the incidence rates of work-related injury or illness or lost work days were 8.8 per 100 hospital workers and 13.5 per 100 among nursing home workers in 2001 (Bureau of Labor Statistics [BLS], 2002). The magnitude of work related injuries in health care may be better understood if these injury rates were compared to other industry. In 2001, the construction industry's annual rates of work-related injuries were 4.0 per 100 workers in mining, 7.9 per 100 workers for construction ,and 8.1 per 100 workers for manufacturing. In 2002, the BLS reported the incidence rate for occupational injury in nursing as 12.6 per 100 full-time workers (BLS, 2002). This number is considered to be a low estimate, since underreporting of injuries in nursing is common (U.S. Dept. of Health and Human Services, 1999). A national survey in Scotland (McGuire & Dewar, 1995) was sent to 5,184 randomly selected nurses, with a 73% response rate; one area examined was reporting of injuries. While 33.4% of subjects had sustained an injury related to patient handling, only 52% reported the injury. Data from over 80 studies across a number of countries indicated that back injury in nurses have a worldwide point prevalence of approximately 17%, an annual prevalence of 40–50%, and a lifetime prevalence of 35–80% (Hignett, 1996). There has been a steady decline in occupational injuries starting in 1992, but when work-related injuries for patient care providers are examined, no such improvement is noted (Fragala & Bailey, 2003).

Given the high incidence and prevalence of WMSDs in nursing, it is not surprising their associated costs are also significant. In 1990, the estimated cost of back pain ranged from $50 to $100 billion annually in the United States (Frymoyer & Cats-Baril, 1991). Employee turnover, reduced production, and medical cost reimbursement are estimated at an additional $30 billion annually. Back pain is second to only the common cold as the most frequent cause for sick leave (Klein, Jensen, & Sanderson, 1984) and is the most common reason for filing of workers' compensation claims.

EVIDENCE RELATED TO HIGH-RISK PATIENT HANDLING TASKS

High-risk patient handling tasks are defined as duties that impose significant biomechanical and postural stressors on the care provider (Nelson, Lloyd, Menzel, & Gross, 2003b). Factors such as the patient's weight, transfer distance, confined workspace, unpredictable patient behavior, and awkward positions such as stooping, bending, and reaching are threats to the nurse's safety. The frequency and duration of these tasks, as well as the physical environment in which care is delivered, varies from one clinical setting to another. Few would argue that one of the highest risk tasks associated with patient handling is a manual patient transfer. Patient transfers can start with the patient in a sitting position (vertical transfer) or when the patient is supine (lateral transfer) (Nelson & Fragala, 2004). However, not all high-risk tasks involve patient transfers. Other high-risk patient handling tasks include repositioning a patient in bed or chair and transporting a patient in a bed or stretcher. Further, risk for injury extends beyond tasks involving patient movement. Patient-handling tasks can be designated as high-risk if the tasks are performed in a forwardly bent position with the torso twisted, such as feeding, bathing, or dressing a patient. It is the combination of frequency and duration of these high-risk tasks that predisposes a caregiver to musculoskeletal injuries and make some clinical practice settings more dangerous than others (Nelson et al., 2004).

EVIDENCE FOR MANUAL PATIENT-LIFTING TECHNIQUES

There is no safe way to manually lift a physically dependent adult patient without risk to the caregiver. Simply adding additional care providers for lifting patients is not a safe alternative. Marras (2004) found that when additional caregivers are involved in manual patient lifting, only 10% of the full burden is achieved for each additional person. This lack of efficiency is explained by the odd distribution of patient weight—most of weight is in the middle, so care providers at legs or head have a lighter load (Corlett et al., 1993). It becomes increasing impossible to safely lift or move a patient with a large abdominal girth, not only due to weight distribution, but because of the inability of staff to reach around and securely grasp the patient.

Hignett and associates (2003) have systematically reviewed research related to manual lifting techniques used in health care. In the United States the most common manual lifting technique is the drag lift (also known as the under arm, axilla, auxiliary, through arm, or hook and toss) (Owen, Keene, Olson, & Garg, 1995). Using this manual lifting technique, a patient is lifted by their armpits on the crooks of the caregivers' elbows. Performed by two caregivers, this method is used to lift a patient up in bed, transfer from bed to chair, lift off the floor, assist to standing from seated position, and support during walking (Hignett et al., 2003). This lift has been banned in the United Kingdom (UK) since 1981 (Hignett et al., 2003). This technique imposes significant risk to the patient, placing considerable pressure against the patient's brachial plexus, which can cause nerve damage in the patients' neck, shoulder, arm, or hands and can result in shoulder subluxation (Owen, 1999).

Owen (1985) discussed the controversies about proper techniques for manual lifting and then compared biomechanics of four lifting techniques on a sample of injured and non-injured nurses. The four techniques were (a) kinetic, (b) straight back/flexed knees, (c) bent over, and (d) variation of b and c. The kinetic lift is a manual lifting technique that includes a gentle rocking motion to gain momentum for the lift. Owen found significant differences in how injured nurses lifted compared with non-injured nurses, with regard to the distance feet were apart and distance of load from body at knee level. None of the subjects maintained a completely straight back—and she concluded that this approach, while taught for 50 years in the United States, may not work in nursing.

EVIDENCE FOR BODY MECHANICS CLASSES AND TRAINING IN LIFTING TECHNIQUES

The traditional approach to reducing back injuries in the health care industry was to develop and implement a training program in safe lifting techniques and strategies for performing lifts properly. Numerous studies have been conducted over the past 30 years regarding the effectiveness of such training in reducing the impact of occupational back injuries as a whole and, in specific, to the health care industry. These studies concluded that lifting instruction had little or no effect in injury prevention (Anderson, 1980; Brown, 1972; Buckle, 1981; Daltroy et al., 1997; Daws, 1981; Dehlin, Hedenrud, & Horal, 1976; Harber et al., 1994; Lagerstrom & Hagberg, 1997; Owen & Garg, 1991; Snook, Campanelli, & Hart, 1978; Stubbs, Buckle, Hudson, Rivers, & Worringham, 1983b). There is some evidence to suggest that training programs are ineffective if they constitute only education without work modification (Feldstein, Valanis, Vollmer, Stevens, & Overton, 1993; Hignett, 1996). Further studies in agreement with the ineffectiveness of education alone are Hack and Potter (1996), Hollingdale and Warin (1997) and Stubbs and colleagues (1983b).

The most common approach taken is to develop and implement classes in body mechanics and/or training in lifting techniques. This approach is not new, and many organizations over the years have delivered many training programs aimed at teaching proper lifting techniques. With all this activity, the incidence of these back injuries in nursing personnel are not declining, and the question arises, "how effective have these training programs been in reducing the impact of this huge problem?" Suggested reasons why this approach has been ineffective include:

1. Behavior modification is difficult to achieve, and even if it is achieved, new behaviors are often short lived. Most times workers will revert to old behavior styles.
2. Optimum theoretical principles are taught in a classroom setting. When workers move into the patient care environment, it is often very difficult to apply these theoretical principles.
3. There may not be optimum principles, which universally apply to all workers because of differences among people. With the wide variety of possible situations, it is often difficult to prescribe a one best way technique.
4. Even if there were a best way to conduct a manual lift, due to the loads involved, there is no safe way for a worker to manually lift a dependent patient.

Daws' findings indicate that following the introduction of a training program for nursing within a health authority in southeast England, there was a pronounced decrease in the number of reported back injuries. However, after this impressive start, the results turned disappointing. He states that these studies raised the question as to whether existing training for the nursing profession with respect to manual handling is appropriate (Daws, 1981). Buckle (1981) published a case study of patients with severe low back pain attending a rehabilitation unit, finding that there was no difference related to history of previous training in lifting techniques, indicating that amount of training received was not a factor separating those who had experienced low back pain problems. Stubbs and associates (1983b) found no relationship between time spent training and subsequent prevalence of back pain. The current emphasis on training was questioned and the need for controlled perspective trials stressed. An approach requiring the development of intrinsically safe systems at work, with particular emphasis on the contribution of ergonomics, was recommended (Stubbs et al., 1983b).

In 2003, Hignett and colleagues summarized research to create an evidence-base for patient handling (Hignett et al., 2003). Based on 12 studies conducted in six countries between 1987 and 2001, they concluded that training interventions had no impact in improving work practices or decreasing injury rates (Billin, 1998; Engkvist et al., 2001; Fanello et al., 1999, Harber et al., 1994; Lagerstrom & Hagberg, 1997; Nussbaum & Torres 2001; Scopa 1993; St. Vincent, Tellier, & Lortie, 1989; Stubbs et al., 1983b; Troup & Raudala, 1987; Wachs & Parker, 1987; Wood 1987). While conceding that training initiatives may have positive short-term effects on knowledge and skills, these outcomes do not translate into reduced injury rates or decreased musculoskeletal pain (Best 1997; Feldstein et al., 1993; Scholey 1983). Harber et al. (1994) indicated that training at nursing school or on-the-job did not have a protective effect in preventing back injuries among new nursing graduates. Further implementation of engineering job redesign was suggested. Larese and Fiorito (1994) demonstrated that training courses are often useless when work organization and the number of nurses involved in patient care do not change. The effectiveness of a three-year education and training program was studied, where 90% of the subjects questioned were positive about participating in the program, and 88% expected that participation would lead to decrease in musculoskeletal disorders; however, no decrease in the prevalence of musculoskeletal symptoms was reported during the study time period (Lagerstrom & Hagberg, 1997).

Why have these traditional education and training approaches been ineffective? The most important reason is that no amount of training can negate the biomechanical impact of high-risk patient handling tasks. Biomechanical evaluations have found that forces exerted on the musculoskeletal system when nurses perform patient handling tasks are beyond the reasonable limits and capabilities, regardless of the technique used to perform the task manually (Nelson et al., 2003b; Owen & Garg, 1991), and unacceptably high levels of stress on the low back occurs during patient transfer tasks (Gagnon, Sicard, & Sirois, 1986; Garg & Owen, 1992; Ulin, Chaggin, Patellos, & Blitz, 1997; Winkelmolen, Landeweerd, & Drost, 1994). Another key reason education and training has not been effective is that in a classroom setting, participants are taught theoretical principles under optimum conditions. However, when they get out in the real work environment, because of the design of patient care areas and the equipment in use, it is often difficult to apply these optimum theoretical principles to real life situations.

While traditional education and training has not been effective, a new educational approach shows promise. Unit-based peer leaders include credible coworkers who receive specialized training and return to the unit to coach other coworkers to promote safe patient handling. In the Veterans Administration, these peer leaders are called "Back Injury Resource Nurses" (BIRNs), and they are a prime example of how to garner nursing support for patient care ergonomics programs (Nelson et al., in press [b]).

BIRNs foster team building and provide an outlet and conduit for staff safety concerns and suggestions for improvement. Key BIRN activities included:

1. Introducing new technology and/or practices related to safe patient handling;
2. Conducting ongoing hazard evaluations on the unit;
3. Orienting new staff to patient handling technology and practices;
4. Conducting annual competency assessments of staff in safe use of patient handling equipment;
5. Maintaining/sustaining patient care ergonomics program over time (Nelson, 2003; Nelson et al., 2003b, 2003c).

BIRNS were selected by their nurse manager based on their informal leadership abilities, respect of coworkers, enthusiasm, time management skills, and interest. Ergonomic experience was not required. All levels of nursing staff were given the opportunity to become a BIRN.

Another peer leader found in the health care industry, the Ergo Ranger, is utilized as an integral part of BJC Health Care (Matz & Wolf, 2004). As BIRNS, they are motivated and enthusiastic and no previous ergonomic experience is required. Different from BIRNS, Ergo Rangers are facility-wide rather than unit-based, and they provide support to all hospital employees on global ergonomic issues. As with BIRNS, providing education and support to patient care staff in proper patient handling techniques and use of equipment is key to their role in patient care environments. Generally, their responsibilities include:

- ergonomic and equipment training;
- implementation of new equipment;
- input in new construction design;
- analysis of facility injury data to identify trends;
- response to all facility safety concerns and issues; and
- response to employee request for ergonomic assessments.

These case studies provide baseline support for these educational initiatives, but further studies are needed to evaluate the effectiveness of peer leaders in reducing the incidence and severity of work-related injuries associated with patient care.

EVIDENCE FOR USE OF BACK BELTS TO PREVENT INJURIES

Back belts are used in a variety of industries including health care in hopes of reducing the risk from lifting. They are made of a lightweight breathable material, usually with double-sided pulls that allow for adjusting the degree of tightness and pressure. Lifting produces a variety of forces within the body, which contributes to pressure on the spine. This is termed "spinal loading." Back belts are touted as being protective of spinal loading because they are said to decrease internal forces of the spine, increase intra-abdominal pressure to counter the spinal forces, and stiffen the spine to decrease forces on the spine. Other postulated protective qualities include their restriction of bending motions and that they remind the wearer to lift properly (Nelson et al., 2003b).

There is strong evidence that back belts are NOT effective as a strategy to prevent job-related injuries in nursing (National Institute for Occupational Safety and Health [NIOSH], 2002). NIOSH completed an extensive review of the literature and later conducted a large epidemiological study and two laboratory evaluations of back belts. The two laboratory studies evaluated the physiological and human motion effects of the same belt used in the epidemiological study, but they did not study the association of back injury and pain with the use of back belts. Physiologically, during use, there was a significant reduction in mean oxygen consumption but not heart rate, blood pressure, or breathing rate (Bobick, Beland, Hsiao, & Wassell, 2001). Giorcelli, Hughes, Wassell and Hsiao (2001) found that during box lifting tasks, the belts affected spine bending. The epidemiological study examined the effects of back belt use on back injuries among manual material handlers in a retail setting. This study found that elastic support back belt use was not associated with reduced incidence of back injury clams or low back pain (Wassell, Gardner, Landsittel, Johnston, & Johnston, 2000).

EVIDENCE FOR ERGONOMIC ASSESSMENT PROTOCOLS

Ergonomic assessment protocols outline key ergonomic assessment factors for patient care environments, such as the presence of confined spaces (particularly around bed or wheelchair); building and room design; selection and design of furniture, fittings, and equipment; communication; definition of workload; working capacity of each nurse; and the assessment of specific hazardous tasks (Corlett et al., 1993). Recommendations from ergonomic evaluations might include administrative solutions, such as development of repair and maintenance protocols; designation of a room for storage purposes; or implementation of the use of standardized criteria for matching the patient handling tasks, patient, and equipment as in the assessment, care plan, and algorithms for safe patient handling and movement (Nelson, 2003 [chap. 5]).

Facilities that have developed and implemented ergonomic-based injury prevention programs using effective engineering controls have achieved considerable success in reducing work-related injuries and costs. Studies show that ergonomic approaches have reduced staff injuries, workers compensation costs, and lost time due to injuries (Bruening, 1996; Bureau of National Affairs, 1996; Fragala, 1993; Fragala, 1995; Fragala, 1996; Fragala & Santamaria, 1997; Logan, 1996; Perrault, 1995; Sacrificial lamb stance, 1999; Villeneuve, 1998). Other studies have shown that an ergonomic approach to decreasing risk in nurses has been successful in the laboratory, as well as in long-term care and hospital settings (Owen & Garg, 1993; Owen & Fragala, 1999).

While the NIOSH and the Occupational Safety and Health Administration (OSHA) have long endorsed an ergonomic approach, and many studies report the successful use of ergonomics to alleviate stress from patient handling, no one prior to Nelson had translated the operational steps identified for industry and manufacturing to the health care industry in the United States. Nelson et al. (2004) developed a risk-assessment protocol to conduct ergonomic assessments of patient care environments as a part of their Veterans' Health Administration study. They stress that this systematic and comprehensive approach is a necessary prerequisite to ensure required infrastructure is in place prior to instituting a no lift policy. Nelson's nine-step protocol is detailed in chapter 3 of the *Patient care ergonomics resource guide* (Nelson, 2003). Further research is needed to refine this intervention and evaluate the outcomes of its use in a variety of practice settings.

1. Collect baseline data. This provides data for use in identifying patient handling tasks and causes of injuries that will direct

seletion of appropriate patient handling equipment. This information can also be used when monitoring the success of the patient care ergonomics program.

2. Identify high-risk units. Implementation of a patient care ergonomic program in high-risk units provides risk interventions for staff in an environment that will in all probability benefit the most from such interventions. This also provides a cost effective way to demonstrate the impact of such a program. Successful data from the high-risk unit/s can be used to support dissemination to other areas.
3. Obtain pre-site visit data. A critical element of an ergonomic protocol is a physical site visit of the unit under consideration by a team of experts in ergonomics. Prior to this visit, the team must have an understanding of characteristics of the unit. Pre-site visit data provides the team with information on the physical characteristics of the unit, storage conditions, repair and maintenance protocols, patient population characteristics, staff demographics, and equipment availability and use.
4. Identify high-risk tasks. High-risk tasks vary from unit to unit based on patient populations. During this activity, staff are asked to rank high-risk tasks from high to low risk for injury, which provides the site visit team staff perception of high-risk tasks.
5. Conduct site visit. During the site visit, the team will hold pre- and post-conference with staff and management to get a better understanding of all of the information already discussed. They will walk through the unit and observe physical layout and barriers, use of equipment, staff attitudes, and other conditions that might influence staff safety.
6. Perform risk analysis. Risk analysis summarizes and analyzes all of the information gleaned from the previous steps.
7. Formulate recommendations. Results from the risk analysis will be used to determine appropriate intervention strategies to reduce risk from patient handling.
8. Implement recommendations. Intervention and program implementation requires support and buy-in from all. Understanding barriers and facilitators specific to staff, patients, and management and developing an action plan for implementation, makes this process easier.
9. Monitor results on an ongoing basis to continuously improve safety.

Others have used ergonomic approaches but not a systematic process. Lynch and Freund (2000) employed an ergonomic consultant using observational analysis and the NIOSH lifting equation to estimate back injury risk and identify areas where equipment was needed and work practice training was appropriate. Equipment was purchased to lower the risk in these areas. Owen, Keene and Olson (2002) suggested that an ergonomic approach requires the assessment of stressful tasks, the development of alternative methods to decrease physical stress to the body, and application of these methods to the job. Through this process, the physical demands of the job are changed. Their ergonomic program included the institution of patient handling equipment deemed appropriate through laboratory analyses to control for tasks staff perceived to be the most stressful. They concluded that perceived physical stresses could be reduced through the implementation of an ergonomic program and that the ergonomic program was sustainable over time. One of the main messages of Hignett et al.'s (2003) systematic review of intervention strategies to reduce musculoskeletal injuries associated with patient handling activities was that multifaceted interventions that based on risk assessment programs are most likely to be successful in reducing risk factors associated with patient handling activities. She suggested that risk assessment programs might include feedback to staff and supervisors and the discussion of goals with clients. She relayed the suggestion that risk assessment, in the context of interventions to reduce risks associated with patient handling, provides the framework that is needed for an intervention to be embedded within an organization's structure and culture (Hignett, 2001).

The OSHA *Ergonomic guidelines for nursing homes* advocates use of ergonomics to successfully resolve the complicated patient handling problem in health care environments. They set the ergonomic process as the foundation for safer nursing home environments. The guideline recommends that employers develop a process for systematically addressing ergonomics issues in their facilities and incorporate this process into an overall program to recognize and prevent occupational safety and health hazards (U.S. Dept. of Labor, 2002). The National Research Council (NRC)/Institute of Medicine (IOM) (Panel on Musculoskeletal Disorders, 2001) report on musculoskeletal disorders and the workplace concluded that evidence supports the introduction of interventions to reduce the risk of musculoskeletal disorders of the low back and upper extremities, including the application of ergonomic principles to reduce physical as well as psychosocial stressors. The report noted that the application of ergonomic principles forms the basis for much of the successful intervention literature. Owen and Garg [1991] emphasized the need for unit-based ergonomic assessments. Without such a systematic analysis of the job and the environment, it will be difficult to solve many of the occupationally related problems.

EVIDENCE FOR PATIENT-HANDLING EQUIPMENT AND DEVICES

Evidence supporting the use of mechanical aids, lifting devices, and other technology in the health care industry has been mixed. Some reasons for this may be that educational programs, training, and use of technology as an interventional approach has been implemented simultaneously; and due to the confounding variables, it is difficult to ascertain which intervention was most effective. Other explanations may be that studies relied on retrospective data or that the intended result was not shown (Aird, Nyran, & Roberts, 1988). This stresses the important need for carefully controlled clinical trials of musculoskeletal injury-prevention programs in the health care environment. While there is adequate literature to support the use of technology in health care, there are also studies that explain some of the barriers to using this equipment. The following evidence demonstrates the use of technology for various categories of equipment:

Adjustable Beds: A high-risk task for home care nursing is providing care in non-adjustable beds (Knibbe & Friele, 1996). There is moderate evidence to substantiate that electric beds reduce strain relating to patient handling tasks (Hampton, 1998) and that height adjustable beds should be used (De Looze et al., 1994) for patient handling tasks such as bathing (Knibbe & Knibbe, 1995).

Mobile Mechanical Lifting Devices: Mobile mechanical assistive devices have been found to be effective in reducing discomfort, fatigue, and physical demands (Yassi et al., 2001). Several studies have demonstrated that mobile lifts significantly reduce work-related injuries in nursing staff (Daynard et al., 2001; Evanoff, Wolf, Alton, Canos, & Collins, 2003; Garg, Owen, Beller & Banaag, 1991a, 1991b).

Ceiling Mounted Patient Lifts: Ceiling lifts are used in patient transfers by placing a sling under the patient and suspending this sling onto an overhead track. There is strong evidence that this technology is well-accepted by nursing staff and patients, takes less time that mobile lifts, and is cost-effective (Holliday, Fernie, & Plowman, 1994; Ronald et al., 2002;Villeneuve, 1998). However, injuries associated with repositioning are not changed (Ronald et al., 2002; Villeneuve, 1998).

Friction Reducing Devices: The lateral transfer of a patient is a task performed frequently in nursing that places caregivers at high risk of injury (Nelson et al., 2003c). This task places high compressive forces on the low back due to the horizontal reach exerted by the caregiver when leaning over the stretcher to hold onto to the draw sheet of the bed prior to moving the patient. Studies have shown that the use of friction reducing devices has proven to serve as a beneficial technological solution in lateral patient transfers (Baptiste, Boda, Nelson, Lloyd & Lee, 2006; Bohannon 1999; Bohannon & Grevelding, 2001; Lloyd & Baptiste, 2006; Zelenka, Floren, & Jordan, 1996) and in turning patients in bed (Gagnon, Akre, Chehade, Kemp, & Lortie, 1987a; Gagnon, Chehade, Kemp, & Lortie, 1987b; Gagnon, Roy, Lortie, & Roy, 1988).

Gait Belts: Gait belts are used to assist caregivers in walking and in sit to stand transfers. These devices typically go around the patients' waist. Gait belts with handles are preferable because the caregiver can grasp the belt more comfortably with a reduced risk of a hand or wrist injury. There is moderate evidence to show that gait belts should not be used with one caregiver for vertical transfers of weight bearing patients (Gagnon et al., 1987a, 1987b, 1988).

Despite the positive aspects of patient handling equipment/devices, some improvements are warranted. Moderate evidence indicates that there needs to be enhancements with the design of both overhead and mobile mechanical lifts, including the design of sling support, position and design of handles and brakes, raising and lowering mechanisms, as well as training and proper use of lifts (Olsson & Brandt, 1992). Other problems noted in lift design have been identified: maneuverability (Bell, 1984); providing teaching packages for ongoing competency training (Bell, 1984); improve patient dignity and stability of lifts (McGuire, Moody, Hanson, & Tigar, 1996; LeBon & Forrester, 1997); match physical environment and other equipment (Love, 1996); spreader bar and sling attachment (Olsson & Brandt, 1992; McGuire, Moody, & Hanson, 1997); instructions on selecting sling type and size (Olsson & Brandt, 1992); and improved brakes (McGuire et al., 1997). Another major issue is that sling use and comfort are compromised by difficulties in selecting the correct size, design of leg pieces, and method of application (Medical Devices Agency [MDA] 1994; Norton, 2000).

Patient handling equipment is deemed to be essential for safe patient handling (Daynard et al., 2001; Garg et al., 1991a, 1991b; Smedley, Egger, Cooper, & Coggon, 1995); however, there are complex reasons why these devices are not fully utilized, including: (a) the patient does not like the equipment (Bell, 1987; Evanoff et al., 2003; McGuire et al., 1996; Retsas & Pinikahana, 2000); (b) the equipment is unstable or hard to operate; (c) the equipment is inconveniently located/storage deficits (McGuire & Dewar, 1995; Nelson et al., 2003a; Retsas & Pinikahana, 2000; Yassi et al., 2001); (d) the equipment is poorly maintained (Garg, et al., 1991a; Green, 1996; McGuire & Dewar, 1995; McGuire et al., 1997; Nelson, 2003; (e) lack of time (Bell, 1987; Daynard et al., 2001; Evanoff et al., 2003; Green, 1996; Laflin & Aja, 1995; McGuire et al., 1996; Nelson et al., 2003b; Owen & Garg, 1991; Retsas & Pinikahana, 2000; Takala & Kukkonen, 1987); (f) insufficient number of lifts available (Bell, 1987; Bewick & Gardner, 2000; Green, 1996; McGuire & Dewar, 1995; Moody, McGuire, Hanson, & Tigar, 1996; Nelson et al., 2003b; Newman & Callaghan, 1993); (g) lack of training (Bell, 1987; Bewick & Gardner, 2000; McGuire & Dewar, 1995; McGuire et al., 1997; Meyer, 1995; Moody et al., 1996; Retsas & Pinikahana, 2000), especially on units with high levels of turnover; (h) lack of space to operate equipment (Bell, 1987; Green, 1996; McGuire & Dewar, 1995); (i) unsuitable equipment purchased (Bewick & Gardner, 2000; McGuire & Dewar, 1995; Nelson, 1993) and weight limitations (Meyer, 1995).

EVIDENCE FOR CLINICAL TOOLS TO SUPPORT PATIENT ASSESSMENT AND DECISIONS ABOUT SAFE PATIENT HANDLING

There is significant variation in practices related to high-risk patient handling and movement tasks. Health care staff members have become accustomed to using whatever limited lifting aids are available, if they are available, rather than carefully matching equipment to specific patient characteristics. (Nelson, 2003 [chap. 5]). Use of patient assessment protocols and algorithms can provide a standardized way to assess patients and make appropriate decisions about how to safely perform high-risk tasks (Nelson et al., 2003b; Nelson et al., 2004; U.S. Dept. of Labor, 2002). Such a system also provides a clear communication channel between nurses on a unit (Nelson et al., 2003c).

In Britain, Hayne developed a Hazard Movement Code (Hayne, personal correspondence, in Collins, 1990); in Canada, a system was developed by the Health Care Occupational Health and Safety Association in Ontario (Health Care Occupational Health and Safety Association (1986); and in Australia, a system using logos was developed (Health Care Occupational Health and Safety Association, 1986). In the United States, Nelson and colleagues (Nelson, 2003; Nelson et al., 2003b) developed a patient assessment and series of algorithms for safe patient handling. Each of these tools assist nurses in selecting the safest equipment, technique, and number of staff needed to perform safe patient handling tasks based on specific patient characteristics (Nelson et al., 2003c). Using the assessment and algorithms ensure patients receive assistance appropriate for their functional level, improving safety for patients as well as staff. Nelson's assessment and algorithms were included in OSHA's *Ergonomic guidelines for nursing homes* (U.S. Dept. of Labor, 2002). They can also be found in the Veterans' Health Administration's *Patient care ergonomics resource guide* (Nelson, 2003).

Individual Patient Assessment

Owen et al. (2002) ensured proper patient movement technique by having nursing staff determine the appropriate method of transfer for each patient, then placing the explicit directions for these techniques in the patients' charts and at the patients' bedsides as a part of their study on the effectiveness of an ergonomic program in reducing risk of injury in nurses. They concluded that a good patient assessment of mobility needs is important to facilitate the use of the proper patient-handling technique. They also noted the need for a good communication system, so all staff were aware of the correct method to use.

The patient assessment developed by Nelson and associates (Nelson, 2003; Nelson et al., 2004) offers health care staff a means by which to document patient handling data for each patient that will affect decisions for determining appropriate equipment, technique, and number of staff. The assessment collects information on the following.

- Ability of the patient to provide assistance;
- Ability of the patient to bear weigh;
- Upper extremity strength of the patient;
- Ability of the patient to cooperate and follow instructions;
- Patient height and weight;
- Special circumstances likely to affect transfer or repositioning tasks, such as abdominal wounds, contractures, or presence of tubes; and
- Specific physician orders or physical therapy recommendations that relate to transferring or repositioning patients (e.g., a patient with a knee or hip replacement may need a specific order or recommendation to maintain the correct angle of hip or knee flexion during transfer).

Prior to implementation of the assessment, care plan, and algorithms, logistical factors must be considered and integrated into the assessment and care planning process.

- Who completes the assessment?
- How often is the assessment is completed?
- What is the communication plan?
- How will the plan be updated/revised as needed?

Algorithms for Decisions on Techniques, Equipment, and Number of Staff

After the assessment is completed, the information is used to direct recommendations in regard to the proper technique, equipment, and number of staff required when performing high-risk patient handling tasks. Algorithms for safe patient handling utilize the assessment information for this purpose. This ensures the recommendations will be individualized for each patient, based on their specific characteristics and needs. The algorithms were developed by a group of nursing experts and tested with different patient populations in a variety of settings. They should be used for every patient who needs help moving. But it is important to note that the algorithms provide general direction. Caregivers must use their professional judgment in applying the algorithms. Nelson et al. (in press [b]) found that the way patient assessment criteria and algorithms were used varied within and between the facilities. The units that were most successful incorporated them into routine practice by prominently displaying them in locations easily accessible to staff (Nelson et al., in press [b]).

Nelson and colleagues (Nelson, 2003) developed algorithms for the following high-risk tasks:

- Transfer to and from: bed to chair, chair to toilet, chair to chair, or car to chair;
- Lateral transfer to and from: bed to stretcher, trolley;
- Transfer to and from: chair to stretcher, chair to chair, or chair to exam table;
- Reposition in bed: side to side, up in bed;
- Reposition in chair: wheelchair or dependency chair;
- Transfer a patient up from the floor;
- Bariatric transfer to and from: bed to chair, chair to toilet, or chair to chair;
- Bariatric lateral transfer to and from: bed to stretcher, trolley;
- Bariatric reposition in bed: side to side, up in bed;
- Bariatric reposition in chair: wheelchair, chair, dependency chair;
- Patient handling tasks requiring sustained holding of a limb/access; and
- Bariatric transporting (stretcher, wheelchair, walker).

Further research is needed to evaluate the process and outcomes associated with use of these clinical tools in various clinical settings (acute care, specialty units, long-term care, and home care).

EVIDENCE FOR NO-LIFT POLICIES

Internationally, health care institutions individually, or as part of national legislation, have developed and implemented policies to address the great risk found in health care environments related to patient handling and movement. Such policies are termed "No Lift," "Zero Lift," "Minimal Lift," "Lift-free," or "Safe Patient Handling and Movement" because they call for staff to avoid manual handling in virtually all patient care situations. Manual handling is broadly defined as the transporting or supporting of a load by hand or bodily force and includes any pushing, pulling, carrying, holding, and supporting of loads (Health and Safety Executive, 1998). The concept of a no lift policy has been implemented internationally, at the facility level, state, level, or as a national policy. There are many variations ranging from national mandates to suggested guidelines, from being restrictive and prescriptive to allowing great latitude in making patient handing decisions.

No-Lift Policy in Australia

In 1998, the Australian Nurses Federation implemented a No Lift Policy based on the policy writ ten by the Royal College of Nursing (UK). The key elements include the following (Australia Nurses Foundation, 1998; Retsas & Pinikahana, 2000):

- The manual lifting of patients is to be eliminated in all but exceptional or life threatening situations.
- Patients are encouraged to assist in their own transfers and handling aids must be used whenever they can help to reduce risk, if this is not contrary to a patient's needs.
- Manual lifting may only be continued if it does not involve lifting most or all of a patient's weight.

No-Lift Policy in Canada

In Canada, a No Lift Policy was released in March of 2003 (Interior Health Authority, 2004). The following is a breakdown of important elements of this policy:

- The distinction is made that the no-lift policy does not mean health care providers will never transfer or reposition any resident manually. However, the criteria for who can be transferred or repositioned manually is defined and is based on patients' physical and cognitive status and medical conditions.
- Proper infrastructure must be in place before a No Lift Policy is enforced. Infrastructure is defined as management commitment and support, equipment, equipment maintenance, employee training, advanced training for resources, and culture of safety approach. The culture of safety approach includes collective attitude of employees at all levels taking a shared responsibility for safety in the work environment and by doing so providing safe environment from themselves and the patients.
- The policy bans the one-person low pivot manual transfer and two person side-by-side transfer.
- High-risk tasks must be assessed in advance to determine the safest way to perform the task; the assessment and solution must be communicated to all staff.
- Specific responsibilities are identified for managers/supervisors, peer leaders, employees, and safety/risk management staff.

No-Lift Policy in United Kingdom

Although the concept of a lift free hospital was first proposed by the Royal College of Nursing in 1993, some say it is still not a reality because no amount of policy or legislation will miraculously change provider behavior and prevent all inappropriate patient handling. The issue is a personal one, and the new approaches must become a way of life, which then will lead to support for policies and procedures. A culture of safety must be developed first to assure these policies are not just complied with but, rather, are supported and endorsed by front line care providers (Wicker, 2000). Despite the national policy, patient handling practices have been slow to change and identifies both training and resources to provide needed equipment as factors (Kneafsey, 2000). But, contradictory to this, the National Audit Office (NAO) report shows that reportable manual handling incidents have decreased over time. In 1996–97 there were 3751 manual handling incidents reported by NHS Trusts. This figure fell to 2705 for the year 2001–02 (NAO, 2003).

No-Lift Policy in United States

In the United States, the effectiveness of No Lift Policies has been documented through research in diverse settings in private and government-owned hospitals and long-term care facilities (Collins, Wolf, Bell, & Evanoff, 2004; Garg, 1999; Nelson et al., 2004). Nelson et al. (2004) successfully implemented a No Lift Policy as an integral part of a comprehensive safe patient handling and movement programs in acute care hospitals and long-term care facilities. Similar findings were obtained in studies conducted by the NIOSH and BJC Health Care in long-term care facilities (Collins et al., 2004) and in seven nursing homes and one hospital by researchers at the University of Wisconsin-Milwaukee (Garg, 1999).

In 2003, OSHA released ergonomics guidelines for nursing homes (U.S. Dept. of Labor, 2002) to reduce the number and severity of WMSDs. They serve as advisory recommendations rather than an enforceable standard. Specific recommendations include: (a) Manual lifting of residents should be minimized in all cases and eliminated where feasible; and (b) Employers should implement an effective ergonomics process that provides management support, involves employees, identifies problems, implements solutions, addresses reports of injuries, provides training, and evaluates ergonomics efforts (U.S. Dept. of Labor, 2002).

Facility-Based No-Lift Policies

Facility based no-lift policies are being used more frequently in the United States, and further research is needed related to process and outcome measures. In one study, Nelson et al. (in press [b]) implemented a no-lift policy as part of a comprehensive safe patient handling and movement program. Details of this policy and a template can be found at in the Patient Care Ergonomics Guide, Chapter 6, at www.visn8.med.va.gov/patientsafetycenter (Nelson, 2003). The policy includes the following key elements:

- The focus is on creating a safe workplace for caregivers. The policy is non-punitive and establishes methods to evaluate the transfer needs of patients, procedures to avoid hazardous patient handling, and movement tasks. Non-compliance indicates a need for retraining.
- Avoid hazardous patient handling and movement tasks whenever possible. If unavoidable, assess them carefully prior to completion.
- Use mechanical lifting devices and other approved patient handling aids for high-risk patient handling and movement tasks except when absolutely necessary, such as in a medical emergency.
- Use mechanical lifting devices and other approved patient handling aids in accordance with instructions and training
- Staff complete and document program and equipment training initially and as required to correct improper use/understanding of safe patient handling and movement.
- Equipment must be provided for safe working practice for staff and caregivers.
- The responsibilities of all medical, maintenance, and administrative staff with regard to the safe handling and movement of patients are communicated.
- The responsibilities of frontline caregivers, nurse managers, therapists, maintenance personnel, supervisors, and administrators are delineated.
- Required infrastructure must be in place before the policy is implemented. This includes:
    1 Adequate number and variety of patient-handling aids and mechanical-lifting equipment on the patient care unit;
    2 Sufficient numbers of staff trained and competent in the use of patient handling aids and equipment;
    3 Staff trained and skilled in applying safe patient handling and movement algorithms; and
    4 Administrators and supervisors who support the comprehensive approach.

In a second study, Evanoff et al. (2003) evaluated the effectiveness of lifting devices in long-term and acute care institutions and found greater reductions in injuries in the long-term care facilities where they proposed that the presence of a policy of mandatory lift usage made the difference. And, when evaluating best practices for back injury prevention in nursing homes, Collins et al. (2004) included a written Zero (No) Lift policy as a best practice. Best practices included in this study were determined through evaluation of the scientific literature on safe resident handling and movement and through public peer review of the research protocol by experts in patient handling. Their written policy established procedures for safe handling and movement of residents and for the evaluation of the transfer needs of residents. Results from this study demonstrate that compliance with a policy that requires the use of lifting equipment is possible.

The American Nurses Association (ANA) recently issued a position statement supporting actions and policies that result in the elimination of manual patient lifting to promote a safe environment of care for nurses and patients. This statement can be accessed at the American Nurses Association's Web site (ANA, 2003).

EVIDENCE TO SUPPORT USE OF LIFTING TEAMS

The lifting-team approach has demonstrated the ability to be effective in the reduction of injuries, lost days, restricted workdays, and compensable injury costs (Caska, Patnode, & Clickner, 1998; Charney, 1992; Charney, 1997, 2000; Charney, Zimmerman, & Walara, 1991; Davis, 2001; Donaldson, 2000; Meittunen, Matzke, McCormack, & Sobczak, 1999). A lifting team is defined as two or more physically fit people, competent in lifting techniques, working together to accomplish high-risk patient transfers (Meittunen et al., 1999). High-risk lifting tasks are concentrated on a few well-trained staff, rather than assigning these tasks to nursing staff who provide other patient-handling tasks. Members of the lifting team are carefully selected based on their physical attributes and no previous history of musculoskeletal injury. Lift-team members receive specialized training and mechanical lifting devices with which to perform their job safely. Charney and associates (1991) state that lift teams can eliminate seven critical risk factors that lead to back injuries for nursing personnel: (1) uncoordinated lifts, (2) unprotected personnel who lift, (3) lifting pairs with height/weight disparities, (4) nurses lifting with fatigue, (5) nurses lifting with injured partner, (6) failure to use mechanical lifting devices, and (7) under-trained lifters.

The logistics of setting up a lift team poses significant challenges. First, selecting and hiring the lift-team members can be difficult in an era of nursing staff shortages. Pre-requisites for becoming a lift-team member often include no history of back injury, physical exam, and a radiograph of spine. Members of lift teams have been male orderlies (Charney, 1992, 1997; Charney et al., 1991), or nursing staff (Caska & Patnode, 2000; Caska et al., 1998). Finding nursing assistants or orderlies experienced in health care with no previous musculoskeletal injury can be difficult (Caska & Patnode, 2000) and may be discriminatory in the United States. The efficiency of the lift team comes into question if there are staff shortages in one area, and lifts are not being performed. One way to address this problem may be to identify the high-risk units and allocate numbers of scheduled versus unscheduled lifts. One lift team was designed at a 220-bed acute care facility using an existing transport team of 20 persons as a lift team and adding mechanical patient lifting equipment (Charney, 2000).

Second, while lift teams have been effective in small community hospitals where the number of lifts needed is relatively small, the logistics of staffing a lift team of sufficient numbers where the daily number of lifts is very high (e.g., geriatric units or extended care) may be cost prohibitive. While Charney and associates (1991) found that a team of orderlies was able to meet 95% of day-shift lifts in a large hospital, completing 30 lifts per day and resulting in a significant decrease in lost time accidents (Charney, 1992), one would question how general these findings would be, because most facilities have a need for far more than 30 patient lifts per day. No study has carefully examined whether these positive effects are the result of the lift team or simply the addition of more staff (orderlies and nursing assistants). While one study of lift teams specified that the team would come from within existing staff, only process measures were addressed, and the incidence and severity of injuries were not reported (Caska,Patnode, and Clickner 1998).

Third, the lift teams are designed to address "scheduled lifts" and are often unable to accommodate unscheduled lifts and other high-risk tasks such as bed repositioning. Some facilities have addresses the need to accommodate unplanned lifts using a pager system, but this requires a larger lift team. Many high-risk nursing tasks are not lifts (e.g., toileting, dressing a patient, making an occupied bed) and are not therefore addressed in this intervention.

Fourth, to be effective in reducing risk to nursing personnel, the lift team should be available 24 hours a day, seven days a week. Many of the studies implementing lift teams provide this service only on the day shift, which does nothing to minimize exposure on evening and night shifts. A few studies have expanded lift teams across all shifts and days of the week, but costs and recruitment challenges associated with this expansion have not been adequately addressed. Extended time taken for a lift team to arrive may be a concern, especially if a patient has fallen to the floor or if a patient has to get to a scheduled appointment. However, one of the most important restrictions of why a lift team will not be successful revolves around the infrastructure. A facility must have the proper equipment available, programmatic support at all levels, set policy and procedures, adequate training of lift-team staff, a culture of safety and good communication among all parties.

Lift teams hold promise as an effective intervention in facilities where the volume of lifts per day can be effectively balanced with sufficient numbers of lift team members to effectively meet the demands for scheduled and unscheduled lifts. Further study is needed on the cost effectiveness of these teams in areas such as extended care, where the volume of lifts per day exceed a manageable lift team size.

EVIDENCE TO SUPPORT SINGLE VERSUS MULTIFACETED PROGRAMS

Both single and multifaceted interventions have been implemented to decrease risk of injury from patient-handling tasks. Single factor interventions have included those that use only training and education, or physical conditioning, or equipment interventions. Multifactor interventions have been found to incorporate different as well as common program elements, such as training and equipment (Collins et al., 2004; Hignett et al., 2003, Nelson et al., 2004; Collins, 1990 [p. 45]). The preponderance of evidence points to multifactor rather than single factor interventions as most effective in reducing risk for musculoskeletal injuries from patient handling tasks.

According to the Panel on Musculoskeletal Disorders and the Workplace (2001), there is no single intervention that has been found to be universally effective. This report found that no studies showed that personal modifiers were successful by themselves and isolated approaches such as use of back belts, training, relaxation, health education, and worker selection either have not been proven effective or have been proven ineffective. Hignett et al.'s (2003) systematic review of intervention strategies noted the lack of evidence in the literature for interventions using training alone. Lynch and Freund (2000) suggest that knowledge of risk factors alone is insufficient to alter behavior and prevent musculoskeletal injuries. They concluded that increases in knowledge might have a limited impact on behaviors over which individuals have control and less impact when sufficient organizational supports are absent, such as lifting equipment. Nelson (Nelson, Fragala, & Menzel, 2003a; Nelson et al., 2003c, 2004; in press [a]) continually repeats that previous research has clearly demonstrated the traditional education models, such as body mechanics classes and training in lifting techniques are not effective (Anderson, 1980; Brown, 1972; Buckle, 1987; Dehlin et al., 1976; Owen & Garg, 1991; Hayne, 1984; Shaw, 1981; Snook et al., 1978; Stubbs, Buckle, Hudson, & Rivers, 1983a; Stubbs et al., 1983b; Venning, 1988; Wood, 1987). Studies have shown that the use of lifts decrease the amount of perceived physical stress and reduce lifting forces required for patient transfers (Evanoff et al., 2003; Garg & Owen, 1992; Holliday et al., 1994; Laflin & Aja, 1995; Owen & Garg, 1991). Hignett et al. (2003) found that such single factor interventions based on the provision of equipment and those using the lifting-team approach provided moderate levels of evidence of effectiveness. But Nelson et al. (2004) resolutely advises that the proper infrastructure, that is, other program elements, are required for compliance in equipment use and for successful program implementation and continuation. Evanoff et al. (2003) relayed the findings of Galka (1991) and Gundewall, Lijequist, and Hansson (1993) that physical conditioning, alone or in combination with other changes, may be effective in reducing back injuries, but they also noted the difficulty in implementing such programs due to logistics and employee participation issues.

The Panel on Musculoskeletal Disorders and the Workplace (2001) Report on Musculoskeletal Disorders and the Workplace concluded from their scientific review that multiple intervention strategies are often reported as a means used to control the risk of low back pain in industry, and this is consistent with the overall logic influencing musculoskeletal risk. The report stated that to be effective, intervention programs should include the development of integrated programs that address equipment design, work procedures, and organizational characteristics. The IOM report offers a summary of best practices for control of upper-body extremity WMSDs using programs from 27 successful industry representatives and from six high-quality reviews of the literature related to low-back interventions. Both summarized similar findings. Collectively, data in the six high-quality reviews of the literature related to low-back interventions indicate that certain engineering controls (e.g., ergonomic workplace redesign), administrative controls (specifically, adjusting organizational culture), addressing modifying individual factors (specifically, employee exercise), and the inclusion of a combination of interventions are the only strategies that have been shown to be positively associated with the reduction of work-related low back pain.

One of the main messages of Hignett et al.'s (2003) systematic review of intervention strategies to reduce musculoskeletal injuries associated with patient handling activities was that multifaceted interventions based on risk assessment programs are the most likely to be successful in reducing risk factors associated with patient handling activities. Additional findings include that multifactor intervention studies not based on risk assessment can show improvements but cannot support the statement that they are successful. She suggests seven strategies to include in a generic intervention program: equipment provision/purchase, education and training (e.g., risk assessment, use of equipment, patient assessment), risk assessment, policies and procedures, patient assessment system, work environment redesign, and work organization/practice change.

By integrating clinical and safety improvement components in the VA SPHM program, more effective, workable and timely interventions to prevent and mitigate work-related injuries in nursing were instituted. The VA SPHM program successfully integrated evidence-based practice, technology, and safety improvement and included these elements: Ergonomic Assessment Protocol, Care Plan, Assessment, and Clinical Decision Algorithms for Safe Patient Handling, Back Injury Resource Nurses (BIRNs), State-of-the-art Equipment/Equipment Resource Guide, After Action Review Process, and No-Lift Policy. The VA SPHM program was also designed to facilitate provider acceptance as well as knowledge transfer throughout the VHA and health care industry (Nelson et al., in press [b]).

Collins (1990) states that in recognizing the multifaceted nature of the problem of occupational back pain in nursing, the need is apparent to develop multifaceted prevention programs capable of addressing the full range of problems involved. WorkSafe's original draft code of practice recommended a multifaceted approach that included problem identification, job task redesign, training and education, and rehabilitation. Others support multi-faceted interventions (Collins et al., 2004; Lynch & Freud, 2000; U.S. Dept. of Labor, 2002).

Considerations for multifaceted studies include the need for phasing in of program elements. Often there is a natural progression from one program element to the next. For instance, during the VA SPHM study, the BIRNS were put in place first to implement remaining program elements. Then the After Action Review process was implemented to foster openness and increase awareness for the program. The ergonomic evaluation was then completed to know what equipment was necessary. After phasing in a variety of patient handling equipment, the Care Plan, Assessment, and Algorithms for Safe Patient Handling could be instituted. Finally, the No Lift, or Safe Patient Handling Policy was adopted to tie all program elements together. Also, the many variables in multifaceted intervention programs create difficulty in sorting out the most effective interventions (Collins, 1990; Nelson et al., 2004). And, as with any multi-site intervention study, the dose of the intervention varies by site and over time.

Confirming one of the main messages of Hignett et al.'s (2003) systematic review of intervention strategies to reduce musculoskeletal injuries associated with patient handling activities, many recent intervention studies and reports purport that for success, patient-care ergonomic programs must approach the multifaceted nature of the problem by developing multifactor prevention programs capable of addressing the full range of problems involved. The programs should include supporting elements as well as equipment. Each reference mentioned here incorporated different as well as common program elements. (Collins, 1990; Collins et al., 2004; Hignett et al., 2003; Nelson et al., 2004). The NRC/IOM (Panel on Musculoskeletal Disorders, 2001) panel concluded that the evidence suggests that the most effective strategies involve a combined approach that takes into account the complex interplay between physical stressors and the policies and procedures of industries.

CONCLUSION

The purpose of this paper was to summarize current evidence for interventions designed to reduce WMSDs. Work-related musculoskeletal injuries associated with patient care have been a problem for decades. Despite strong evidence, published internationally over three decades, most clinical settings have used significant resources to implement strategies that are NOT evidence-based and have been found to be ineffective. There is a growing body of evidence to support interventions that are effective or show promise in reducing musculoskeletal pain and injuries in care providers. Key interventions, based on the level of evidence available today, is summarized below according to three key headings:

1. Over-used interventions that are NOT evidence-based;
2. Under-utilized patient care ergonomics interventions that ARE evidence-based; and
3. Patient care ergonomics interventions that show promise.

Over-Used Interventions that are NOT Evidence-Based

Four interventions are commonly used in the United States, even though there is strong evidence that they are NOT effective in reducing work-related injuries associated with patient care. First, the old, unsafe manual-handling techniques have been found to be unsafe for the nurse and the patient (Corlett et al., 1993). Sadly, an evaluation of current practices reveal that 98% of nurses are using a manual patient-lifting technique known as the drag lift (hook and toss) method (Owen, 1999), which has been deemed unsafe since 1981. This technique is currently taught by 83% of nurse educators in schools of nursing in the United States (Owen, 1999). It was described in the literature as "deplorable … inefficient, dangerous to the nurses, and often painful and brutal to the patient" (Hardicre, 1992; Owen, 1999). Old habits are difficult to break.

Despite over three decades of research that denounces the effectiveness of classes in body mechanics and training lifting techniques, these interventions are widely used in facility-based and academic-based curricula (Anderson, 1980; Billin, 1998; Brown, 1972; Buckle, 1987; Daltroy et al., 1997; Daws, 1981; Dehlin et al., 1976; Engkvist et al., 2001; Fanello et al., 1999; Harber et al., 1994; Hayne, 1984; Lagerstrom & Hagberg, 1997; Nussbaum & Torres, 2001; Owen & Garg, 1991; Scopa, 1993; Shaw, 1981; Snook, Campanelli, & Hart, 1978; St. Vincent et al., 1989; Stubbs et al., 1983a, 1983b; Troup & Raudala, 1987; Venning, 1988; Wachs & Parker, 1987; Wood, 1987).

Lastly, back belts were widely used in the early 1990's, but their use is diminishing, as evidence to support their utility in health care was lacking. A series of studies conducted by the NIOSH determined that they were not protective of health care workers performing patient-handling tasks (NIOSH, 2002).

Under-Utilized Interventions that ARE Evidence-Based

There are four interventions with evidence to support use in health care setting to reduce WMSDs associated with patient care. These interventions include Ergonomic Assessments of Patient Care Settings, Patient Care Equipment/Devices, No-Lift Policies, and Lifting Teams. There is a strong trend that multifaceted programs are more effective than any single intervention (Panel on Musculoskeletal Disorders, 2001).

New Interventions that Show Promise

While traditional education and training programs have not been effective, the use of unit-based peer leaders shows promise as a new education strategy. Called "Back Injury Resource Nurses (BIRNs)" or "Ergo Rangers," these nurse educators coach coworkers in safe patient handling. Another new intervention is the use of clinical tools, such as patient care assessment protocols and algorithms for standardizing patient specific decisions about safe patient handling and movement. Patient care assessment forms may assist in selecting the type of equipment and number of people needed for patient handling in such a way that maximizes patient function and independence. Algorithms show promise in standardizing decisions related to type of equipment and number of people needed to perform a task safely. Further research is needed for these new interventions to assess their level of effectiveness.

RECOMMENDATIONS

There is a growing body of evidence in patient-care ergonomics to support sweeping changes in the way patient care is provided across diverse clinical practice settings. However, "it is a serious mistake to believe that ergonomic improvements are easily made and that the results of ergonomic improvements will be substantial and immediate" (Kourinka & Forcier, 1995, p. 271). Significant restructuring of patient care settings as well as schools of nursing is needed to improve the health and safety of nursing staff. With the growing nursing shortage, we can no longer afford the "human sacrifice" approach to patient handling, defined as replacing the steady stream of injured nurses with newly recruited nurses. Recommendations to support evidence-based practice related to Patient Care Ergonomics in the United States will be organized in five categories:

1. Call for Action to Change Clinical Practice Settings
2. Call for Action to Update Curricula Related to Safe Patient Handling and Movement
3. Call for Political Action Towards No Lift Mandates
4. Call for Implementation Toolkits to Support Research Translation in Patient Care Ergonomics
5. Call for Implementation Toolkits to Support Research Translation in Patient Care Ergonomics
6. Research Agenda for Patient Care Ergonomics

Call for Action to Change Clinical Practice Settings

It is time to overturn traditionally used approaches to nurse safety when the evidence does not support their utility—back belts, biomechanics classes, training in lifting techniques, and manual patient lifting. This will involve a restructuring of patient-care settings to improve the health and safety of nursing staff. To be effective, interventions will need to be multifaceted and include employee involvement, employer commitment, and the development of integrated programs that address equipment design, work procedures, and organization characteristics (Panel on Musculoskeletal Disorders, 2001). Safe organizations that have fewer occupational injuries and lost workdays have the following characteristics (Panel on Musculoskeletal Disorders, 2001).

- Maintaining safe equipment (Hunt & Habeck, 1993);
- Investigating risks and accidents promptly (Hunt & Habeck, 1993);
- Emphasizing safety in all aspects of operation (culture of safety) (Hunt & Habeck, 1993);
- Active involvement of injured worker and supervisor in return to work process (Hunt & Habeck, 1993);
- Creative strategies to accommodate injured workers (Hunt & Habeck, 1993);
- Greater experience of workforce (Shannon et al., 1996);
- Greater involvement of workers in decision-making (Shannon et al., 1996);
- Interventions must mediate physical stressors, largely through application of principles of ergonomics (Panel on Musculoskeletal Disorders, 2001);
- Employee involvement is essential to successful implementation (Panel on Musculoskeltal Disorders, 2001); and
- Employer commitment to occupational safety (Panel on Musculoskeletal Disorders, 2001).

Nurse executives need to re-evaluate traditional approaches, favoring evidence-based approaches likely to reduce work-related injuries resulting from patient care. Given the high prevalence rates of nursing injuries (ANA, 2003; Hignett, 1996), nurses providing direct patient care need to consider the safety of their work environment when selecting places of employment.

Call for Action to Update Curricula Related to Safe Patient Handling and Movement

Significant restructuring of schools of nursing is needed to improve the health and safety of nursing staff in the United States. Curricula need to include evidence-based approaches to safe patient handling and movement, rather than relying on outdated manual-lifting techniques that are taught by 83% of nurse educators in schools of nursing (Owen 1999).

Call for Implementation Toolkits to Support Research Translation in Patient Care Ergonomics

Resolving safety issues is never an event but, rather, the outcome of a process. For the process to be effective, appropriate policies, procedures, and structures have to be established (Collins, 1990). Implementation toolkits would facilitate implementation (Collins, 1990). Examples of tools that would be useful for implementing evidence-based practices and best practices include:

- Performance criteria for patient handling equipment and guidelines for selection of equipment and furniture by clinical setting and/or patient population;
- Web-based education programs to increase nurse awareness of evidence-based approached for safe patient handling and movement;
- Curriculum for facility-based ergonomic training that includes evidence-based content as well as standardized processes for the dose and timing of the program;
- Curriculum for academic-based ergonomic training, targeting schools of nursing;
- Valid and reliable patient assessment tools;
- Valid and reliable clinical decision algorithms;
- Technology resource guides that reflect new and emerging patient-handling technologies;
- Clinical protocols for sling selection to optimize patient safety and comfort;
- Guidelines for patient care ergonomic program implementation;
- Templates for facility-based no lift policy;
- Clinical protocols for safe patient handling and movement for bariatric patients, including "just-in-time" training protocols; and
- Web-based database and mechanism for communicating "best practices".

Research Agenda for Patient Care Ergonomics

Many changes have been suggested for reducing WMSDs associated with patient care tasks, and many of these make sense from a physiological, psychological, or biomechanical viewpoint. Unfortunately, much work is needed to evaluate their effectiveness across diverse clinical practice settings. We need to build a coherent research agenda on safe patient handling applicable to multiple settings. Key research priorities include:

1. Building a business case for patient-care ergonomics, to facilitate implementation of evidence-based practice related to patient-care ergonomics (NIOSH, 2001).
2. Impact of aging workforce on WMSDs (NIOSH, 2001), particularly the changes in risk factor tolerance in aging nurses, recovery from injuries, and changes needed in workplace to accommodate older nurses.
3. Delineation of specific high-risk patient handling tasks in diverse clinical practice settings.
4. Examine reasons nurses underreport musculoskeletal injuries and the impact of reporting on outcomes
5. Promote translation of patient-care ergonomics research to practice, which includes significant changes in practice settings and a curriculum change in schools of nursing.
6. Design technologies to address high-risk, high-volume patient-handling tasks for which there are no viable technology solutions, such as repositioning a patient in bed and transporting patients.
7. Develop credible performance criteria for patient-handling equipment and test these devices in a laboratory and diverse clinical settings, comparing brands and features.

8. Need to integrate technologies (vertical and lateral lifts) and locate in each patient room, preferably ceiling out of the way.
9. Further research is needed to validate patient assessment tools and algorithms, in a variety of patient-care settings.
10. Evaluate strategies to effectively implement multifaceted programs and sustain them over time (Westgaard, 2000). In particular, evidence related to work organization factors (NIOSH 2001), such as participatory approaches, management commitment, financing, work interaction, job content and scheduling, behavioral intervention, safety and fitness effects, accommodation and functional capacity, development of participative approaches that include empowerment of nursing staff, and commitment and support at all levels, including management and the union (Collins, 1990).
11. Intervention research that addresses (NIOSH 2001):
    i) Alternative design of work tools and processes;
    ii) Optimizing work demands (e.g., posture) and temporal patterns of exposure;
    iii) Manual handling alternatives; and
    iv) Content, dose, and timing of effective ergonomic training and education.
12. While the ergonomics community is relatively confident as to the nature of risk factors, the exposure-effect relationship has not yet been worked out, particularly in patient care (Westgaard, 2000).
13. Need more research related to the predictive role and nature of relationships (singular, additive, multiplicative) of direct and indirect patient care activities in injury cessation (Retsas & Pinikahana, 2000).

References

Aird, J. W., Nyran, P., & Roberts, G. (1988). Comprehensive bank injury program: an ergonomics approach for controlling back injuries in health care facilities. In F. Aghandeh (Ed.), *Trends in ergonomics/human factors V* (pp. 705–712). Amsterdam: Elsevier.

American Nurses Association (ANA). (2003). *Position statement on elimination of manual patient handling to prevent work related musculoskeletal disorders.* Retrieved Feb. 2, 2005, from www.nursingworld.org/readroom/position/workplac/pathand.pdf.

Anderson, J. (1980). Back pain and occupation. In M. I. V. Jayson (Ed.), *The lumbar spine and back pain,* 2nd ed. (pp. 57–82). London: Pitman Medical.

Australian Nurses Foundation (Victoria Branch). (1998). *No lifting policy.* Melbourne, Australia: Author.

Baptisten, A., Boda, S. V., Nelson, A. L., Lloyd, V. J., & Lee, W. E. 3rd (2006). Friction-reducing devices for Lateral Patient Transfers: Clinical evaluation. *American Association of Occupational Health Nurses Journal, 54* (4), 173–180.

Bell, F. (1984). *Patient lifting devices in hospitals.* London: Croom Helm.

Bell, F. (1987). Ergonomic aspects of equipment ... patient lifting devices. *International Journal of Nursing Studies, 24*(4), 331–337.

Best, M. (1997). An evaluation of manutention training in preventing back strain and resultant injuries in nurses. *Safety Science, 25* (1–3), 207–222.

Bewick, N., & Gardner, D. (2000). Manual handling injuries in health care workers. *International Journal of Occupational Safety and Ergonomics, 6*(2), 209–221.

Billin, S. L. (1998). Moving and handling practice in neuro-disability nursing. *British Journal of Nursing, 7*(10), 571–578.

Blue, C. L. (1996). Preventing back injury among nurses. *Orthopaedic Nursing, 15*(6), 9–21.

Bobick, T. G., Beland, J. L., Hsiao, H., & Wassell, J. T. (2001). Physiological effects of back belt wearing during asymmetric lifting. *Applied Ergonomics, 32*(6), 541–547.

Bohannon, R. (1999). Horizontal transfer between adjacent surfaces: forces required using different methods. *Archive Physical Medicine Rehabilitation, 80,* 851–853.

Bohannon, R., & Grevelding, P. (2001) Reduced push forces accompany device use during transfers of seated subjects. *Journal of Rehabilitation Research and Development, 38,* 135–139.

Brown, J. (1972). *Manual lifting and related fields: an annotated bibliography.* Toronto, Ontario, Canada: Labour Safety Council of Ontario, Toronto.

Bruening, J. (1996). Keeping health care workers healthy. *Ergonomics News, Mar/Apr,* 20–21.

Buckle, P. (1981). A multidisciplinary investigation of factors associated with low back pain. Ph.D. thesis, Bedfordshire, UK: Cranfield Institute of Technology.

Buckle, P. (1987) Epidemiological aspects of back pain within the nursing profession. *International Journal of Nursing Studies, 24*(4), 319–324.

Bureau of National Affairs. (1996). *Empowering workers helps nursing home find answers to injury problem, cut costs.* BNA Workers' Compensation Report. Washington, DC: Bureau of National Affairs (BNA).

Caska, B. A., & Patnode, R. E. (2000, Sept. 26). Reducing lower back injuries in VAMC nursing personnel. Research report #94-136 to the Veterans Health Administration. Washington, DC: Department of Veterans Affairs (VHA).

Caska, B. A., Patnode, R. E., & Clickner, D. (1998). Feasibility of nurse staffed lift team. *AAOHN Journal, 46*(6), 283–288.

Charney, W. (1992). The lifting team: second year data reported (News). *AAOHN Journal, 40*(10), 503.

Charney W. (1997). The lifting team method for reducing back injuries: a 10 hospital study. *AAOHN Journal 45*(6), 300–304.

Charney, W. (2000). Reducing back injury in nursing: a case study using mechanical equipment and a hospital transport team as a lift team. *Journal of Health care Safety, Compliance, and Infection Control, 4*(3), 117–120.

Charney, W., Zimmerman, K., & Walara, E. (1991). The lifting team: a design method to reduce lost time back injury in nursing. *AAOHN Journal, 39*(5), 231–234.

Collins, M. (1990). Occupational back pain in nursing: development, implementation and evaluation of a comprehensive prevention program. Australia: Worksafe Australia, National Occupational Health and Safety Commission.

Collins, J. W., Wolf, L., Bell, J., & Evanoff, B. (2004). An evaluation of a "best practices" musculoskeletal injury prevention program in nursing homes. *Injury Prevention, 10*, 206–211.

Corlett, E. N., Lloyd, P. V., Tarling, C., Troup, J. D. G., & Wright, B. (1993). *The guide to handling patients,* 3rd ed. London: National Back Pain Association and the Royal College of Nursing.

Daltroy, L. H., Iversen, M. D., Larson, M. G., Lew, R., Wright, E., Ryan, J., Zwerling, C., Fossel, A. H., & Liang, M. H. (1997). A controlled trial of an educational program to prevent low back injuries. *The New England Journal of Medicine,* 337(5), 322–328.

Davis, A. (2001). Birth of a lift team: experience and statistical analysis. *Journal of Health care Safety, Compliance and Infection Control, 5*(1), 15–18.

Daws J. (1981). Lifting and moving patients, 3, a revision training programme. *Nursing Times. 77*(48), 2067–2069.

Daynard, D., Yassi, A., Cooper, J. E., Tate, R., Norman, R., & Wells, R. (2001). Biomechanical analysis of peak and cumulative spinal loads during patient handling activities: a substudy of a randomized controlled trial to prevent lift and transfer injury health care workers. *Applied Ergonomics, 32*, 199–214.

De Looze, M. P., Zinzen, E., Caboor, D., Heyblom, P., Van Bree, E., Van Roy, P., Toussaint, H. M., & Clarijs, J. P. (1994). Effect of individually chosen bed-height adjustments on the low-back stress of nurses. *Scandinavian Journal of Work Environment & Health, 20*, 427–434.

Dehlin, O., Hedenrud, B., & Horal, J. (1976) Back symptoms in nursing aides in a geriatric hospital. an interview study with special reference to the incidence of low-back symptoms. *Scandinavian Journal of Rehabilitation Medicine, 8*(2), 47–53.

Dillman, S. (1993). An ergonomic approach to the prevention of back injuries in the health care industry. 1993 AHEHP conference presentation by Guy Fragala. *Journal of Hospital Occupational Health*, 13–15.

Donaldson, A. W. (2000). Lift team intervention: a six year picture. *Journal of Health care Safety, Compliance and Infection Control, 4*(2), 65–68.

Engkvist, I.-L., Kjellberg, A., Wigaeus, H. E., Hagberg, M., Menckel, E., & Ekenvall, L. (2001). Back injuries among nursing personnel—identification of work conditions with cluster analysis. *Safety Science, 37*, 1–18.

Evanoff, B., Wolf, L., Aton, E., Canos, J., & Collins, J. (2003). Reduction in injury rates in nursing personnel through introduction of mechanical lifts in the workplace. *American Journal of Industrial Medicine, 44*, 451–457.

Fanello, S., Frampas-Chotard, V., Roquelaure, Y., Jousset, N., Delbos, V., Jarmy, J. & Penneau-Fontbonne, D. (1999). Evaluation of an educational low back pain prevention program for hospital employee. *Revue du Rhumatisme (Eng. Edn.), 66*(12), 711–716.

Feldstein, A, Valanis, B., Vollmer, W., Stevens, N., & Overton, C. (1993). The back injury prevention project pilot study: assessing the effectiveness of back attack, an injury prevention program among nurses, aides, and orderlies. *Journal of Occupational Medicine, 35*, 114–120.

Fragala, G. (1993) Injuries cut with lift use in ergonomics demonstration project. *Provider, October*, 39–40.

Fragala, G. (1995). Ergonomics: the essential element for effective back injury prevention for health care workers. *American Society of Safety Engineers, Mar*, 23–25.

Fragala, G. (1996). Ergonomics: how to contain on-the-job injuries in health care. *Joint Commission on Accreditation of Health care Organizations* (ch. 6, pp. 55–57). Oakbrook Terrace, IL: Joint Commission on Accreditation of Healthcare Organizations (JCAHO).

Fragala, G., & Bailey, L. P. (2003). Addressing occupational strains and sprains: musculoskeletal injuries in hospitals. *AAOHN, 51*(6), 252–259.

Fragala, G., & Santamaria, D. (1997). Heavy duties? *Health Facilities Management, May*, 22–27.

Frymoyer, J. W., & Cats-Baril, W. L. (1991). An overview of the incidences and costs of low back pain. *Orthopaedic Physical Therapy Clinics of North America, 22*(2), 263–271.

Gagnon, M., Sicard, C., & Sirois, J. P. (1986) Evaluation of forces on the lumbo-sacral joint and assessment of work and energy transfers in nursing aids lifting patients. *Ergonomics, 29*(3), 407–421.

Gagnon, M., Chehade, A., Kemp, F., & Lortie, M. (1987). Lumbo-sacral loads and selected muscle activity while turning patients in bed. *Ergonomics, 30*(7), 1013–1032.

Gagnon, M., Roy, D., Lortie, M., & Roy, R. (1988) Evolution of the execution parameters on a patient handling task. *La Travail Humain, 51*(3), 193–210.

Gagnon, M., Akre, F., Chehade, A., Kemp, F., & Lortie, M. (1987). Mechanical work and energy transfers while turning patients in bed. *Ergonomics, 30,* 1515–1530.

Galka, M. L. (1991). Back injury prevention program on a spinal cord injury unit. *SCI Nursing, 8*, 48–51.

Garg, A. (1999). *Long-term effectiveness of "zero-lift program" in seven nursing homes and one hospital* (contract no. U60/ CCU51208902, 1999). Milwaukee: University of Wisconsin-Milwaukee.

Garg, A., & Owen, B. (1992). Reducing back stress in nursing personnel: an ergonomic intervention in a nursing home. *Ergonomics, 35*(11), 1353–1375.

Garg, A., Owen, B., Beller, A., & Banaag, J. (1991a). A biomechanical and ergonomic evaluation of patient transferring tasks: bed to wheelchair and wheelchair to bed. *Ergonomics 34*(3), 289–312.

Garg, A., Owen, B., Beller, D., & Banaag, J. (1991b). A biomechanical and ergonomic evaluation of patient transferring tasks: wheelchair to shower chair and shower chair to wheelchair. *Ergonomics, 34*(4), 407–419.

Giorcelli, R. J., Hughes, R. E., Wassell, J. T., & Hsiao, H. (2001). The effect of wearing a black belt on spine kinematics during asymmetric lifting of large and small boxes. *Spine, 26*(16), 1794–1798.

Green, C. (1996) Study of the moving and handling practices on two medical wards. *British Journal of Nursing, 5*(5), 303–311.

Gundewall, B., Lijequist, M., & Hansson, T. (1993) Primary prevention of back symptoms, and absence from work: a prospective randomised study among hospital employees, *Spine 18*(5), 587–594.

Hack, L., & Potter, S. (1996). Lifting and handling: strategies for altering practices. *Nursing Times, 92*(15), 29–30.

Hampton, S. (1998) Can electric beds aid pressure sore prevention in hospitals? *British Journal of Nursing, 7*(17), 1010–1017.

Harber, P., Pena, L., Hsu., P., Billet, E., Greer, D., & Kim, K. (1994). Personal history, training and worksite as predictors of back pain of nurses. *American Journal of Industrial Medicine, 25*(4), 519–526.

Hardicre, J. (1992). Lifting techniques: put your back out of danger. *Nursing Standard, 7*, 54.

Hayne, C. (1984). Safe patient movement—an alternative approach. *Medical Education (International) Ltd*, 931–965.

Health Care Occupational Health and Safety Association. (1986). *Transfer and lifts for caregivers.* Ontario, Canada: HCOSHA Publications.

Health and Safety Executive (1998). *Manual handling operations—guidance on regulations* (p. 6). Sudbury England: HSE Books.

Hignett, S. (1996). Work-related back pain in nurses. *Journal of Advanced Nursing, 23*(6), 1238–1246.

Hignett, S. (2001) Embedding ergonomics in hospital culture: top-down and bottom-up strategies. *Applied Ergonomics, 31*, 61–69.

Hignett, S., Crumpton, E., Ruszala, S., Alexander, P., Fray, M., & Fletcher, B. (2003). Evidence-based patient handling: rasks, equipment and interventions. New York: Routledge.

Holliday, P. J., Fernie, G. R., & Plowman, S. (1994). The impact of new lifting technology in long term care. *AAOHN Journal, 42*(12), 582–599.

Hollingdale, R., & Warin, J. (1997) Back pain in nursing and associated factors: a study. *Nursing Standard 11*(39), 35–38.

Hunt, H. A., & Habeck, R. V. (1993). *The Michigan disability prevention study: research highlights.* Kalamazoo, MI: Uphohn Institute for Employment Research.

Interior Health Authority. (2004, October). *MSIP: a practical guide to resident handling* (section 2.0, pp. 1–7). British Columbia: Author. Retrieved February 2, 2002, from www.interiorhealth.ca/NR/rdonlyres/4AF034BF-78B6-48BF-A18D-2AA6B39F99FE/1908/FullGuideforMSIPAPracticalGuidetoResidentHandling.pdf.

Klein B. P., Jensen R. C., & Sanderson, L. M. (1984). Assessment of workers' compensation claims for back strains/sprains. *Journal of Occupational Medicine. 26*(6), 443–448.

Kneafsey, R. (2000). The effect of occupational socialization on nurses' patient handling practices. *Journal of Clinical Nursing, 9*, 585–593.

Knibbe, J. J., & Friele, R. D. (1996). Prevalence of back pain and characteristics of the physical workload of community nurses. *Ergonomics, 39*(2), 186–198.

Knibbe, N., & Knibbe, J. J. (1995). *Postural load of nurses during bathing and showering of patients.* Internal Report, Locomotion Health Consultancy, The Netherlands.

Kourinka, I., & Forcier, L. (1995). *Work-related musculoskeletal disorders: a reference book for prevention.* London: Taylor & Francis.

Laflin, K., & Aja, D. (1995). Health care concerns related to lifting: an inside look at intervention strategies. *The American Journal of Occupational Therapy, 49*(1), 63–72.

Lagerstrom, M., & Hagberg, M. (1997). Evaluation of a 3-year education and training program for nursing personnel at a Swedish hospital. *AAOHN Journal, 45*(2), 83–92.

Larese, F., & Fiorito, A. (1994). Musculoskeletal disorders in hospital nurses: a comparison between two hospitals. *Ergonomics, 37*(7), 1205–1211.

LeBon, C., & Forrester, C. (1997). An ergonomic evaluation of a patient handling device: the elevate and transfer vehicle. *Applied Ergonomics, 28*(5–6), 365–374.

Lloyd, J., & Baptiste, A. (2006). Friction-reducing devices for patient transfers: a biomechanical evaluation. *American Association of Occupational Health Nurses Journal, 54*(3), 113–119.

Logan, P. (1996). Moving and handling. *Community Nurse, Apr*, 22–24.

Love, C. (1996). Injury caused by lifting: a study of the nurse's viewpoint. *Nursing Standard, 10*(46), 34–39.

Lynch, R. M., & Freund, A. (2000) Short-term efficacy of back injury intervention project for patient care providers at one hospital. *American Industrial Hygiene Association Journal, 61*(2), 290–294.

Marras, W. S. (2004) State-of-the-art research perspectives on musculoskeletal disorder causation and control: the need for an intergraded understanding of risk. *Journal of Electromyography and Kinesiology, 14*(1), 1–5.

Matz, M., & Wolf, L. (2004, March). *Peer leaders as a new educational strategy.* Fourth annual safe patient handling and movement conference, Lake Buena Vista, FL.

McGuire, T., & Dewar, J. (1995). An assessment of moving and handling practices among Scottish nurses. *Nursing Standard, 9*(40), 35–39.

McGuire, T., Moody, J., & Hanson, M. (1997). Managers attitudes towards mechanical aids. *Nursing Standard, 11*(31), 33–38.

McGuire, T., Moody, J., Hanson, M., & Tigar, F. (1996). A study into clients attitudes towards mechanical aids. *Nursing Standard 11*(5), 35–38.

Medical Devices Agency (MDA). (1994). *Slings to accompany mobile domestic hoists.* London: Her Majesty's Stationery Office (HMSO).

Meittunen, E. J., Matzke, K., McCormack, H., & Sobczak, S. C. (1999). The effect of focusing ergonomic risk factors on a patient transfer team to reduce incidents among nurses associated with patient care. *Journal of Healthcare Safety, Compliance and Infection Control, 2*(7), 306–12.

Meyer, E. (1995). Patient lifter in a practical test: a spine-saving aid or bulk in the storage room? *Pflege Aktuell, 49*(9), 597–600.

Moody, J., McGuire, T., Hanson, M. & Tigar, F. (1996) A study of nurses attitudes towards mechanical aids. *Nursing Standard 11*(4), 37–42.

Moses, E. B. (Ed.). (1992). *The registered nurse population: findings from the national sample survey of registered nurses.* Washington, DC: U.S. Department of Human Services, U.S. Public Health Service, Division of Nursing.

National Audit Office (NAO). (2003). *A safer place to work—improving the management of risks to staff in NHS trusts, appendix 3(i).* London: Stationery Office.

National Institute for Occupational Safety and Health (NIOSH). (2001). *National occupational research agenda for musculoskeletal disorders.* Washington, DC: U.S. Department of Health and Human Services (Public Health Service, CDC, NIOSH).

National Institute for Occupational Safety and Health (NIOSH). (2002, March). *Summary of NIOSH back belt studies.* Retrieved February 3, 2005, from www.cdc.gov/niosh/topics/ergonomics/.

Nelson (Ed.). (2003). *Patient care ergonomics resource guide.* Tampa, FL: U.S. Dept. of Veterans Affairs, Patient Safety Research Center. Retrieved February 1, 2005, from www.patientsafetycenter.com/Safe%20Pt%20Handling%20Div.htm.

Nelson, A. L., & Fragala, G. (2004). Equipment for safe patient handling and movement. In W. Charney & A. Hudson (Eds.), *Back injury among health care workers* (pp. 121–135). Washington, DC: CRC Press (Lewis Publishers).

Nelson, A. L., Fragala, G., & Menzel, N. (2003a). Myths and facts about back injuries in nursing. *American Journal of Nursing, 103*(2), 32–40.

Nelson, A. L., Lloyd, J., Menzel, N. & Gross, C. (2003b). Preventing nursing back injuries: redesigning patient handling tasks. *AAOHN Journal, 51*(3), 126–134.

Nelson, A. L., Collins, J., Knibbe, H., Cookson, K., de Castro, A. B., & Whipple, K. L. (in press [a]). Myths & facts about no lift policies. *Nursing Management.*

Nelson, A.L., Matz, M., Chen, F., Siddharthan, K., Lloyd, J., & Fragala, G. (in press [b]). A multifaceted ergonomics program to prevent injuries associated with patient handling tasks. *International Journal of Nursing.*

Nelson, A. L., Powell-Cope, G., Gavin-Dreschnack, D., Quigley, P., Bulat T., Baptiste, A., Applegarth, S. & Friedman, Y. (2004). Technology to promote safe mobility in elderly. *Nursing Clinics of North America, 39*(3), 649–671.

Nelson, A. L., Owen, B., Lloyd, J., Fragala, G., Matz, M., Amato, M., Bowers, J., Moss-Cureton, S., Ramsey, G., & Lentz, K. (2003c). Safe patient handling & movement. *American Journal of Nursing, 103*(3), 32–43. (See also www.patientsafetycenter.com.)

Newman, S., & Callaghan, C. (1993). Work-related back pain. *Occupational Health (London), 45*, 201–205.

Norton, L. (2000). *An ergonomic evaluation into fabric slings used during the hoisting of patients.* Unpublished M.Sc. dissertation. Nottingham, UK: University of Nottingham.

Nussbaum, M. A.,& Torres, N. (2001) Effects of training in modifying working methods during common patient-handling activities. *International Journal of Industrial Ergonomics 27*, 33–41.

Olsson, G.,& Brandt, A. (1992). An investigation of the use of ceiling mounted hoists for disabled people. Denmark: Danish Centre for Technical Aids for Rehabilitation and Education.

Owen, B. (1985). The lifting process and back injury in hospital nursing personnel. *Western Journal of Nursing Research, 7*(4), 445–459.

Owen, B. D. (1989). The magnitude of low-back problem in nursing. *Western Journal of Nursing Research, 11*(2), 234–242.

Owen, B. D. (1999). Decreasing the back injury problem in nursing personnel. *Surgical Services Management, 5*(7), 15–21.

Owen, B., & Fragala, G. (1999). Reducing perceived physical stress while transferring residents. *AAOHN Journal, 47*(7), 316–323.

Owen, B., & Garg, A. (1991). Reducing risk for back pain in nursing personnel. *AAOHN Journal, 39*(1), 24- 33.

Owen, B. D., & Garg, A. (1993) Back stress isn't part of the job. *American Journal of Nursing, 93*, 48–51.

Owen, B., Keene, K., & Olson, S. (2002). An ergonomic approach to reducing back/shoulder stress in hospital nursing personnel: a five year follow up. *International Journal of Nursing Studies, 39*(3), 295–302.

Panel on Musculoskeletal Disorders and the Workplace; Commission on Behavioral and Social Sciences and Education; National Research Council; and Institute of Medicine. (2001). *Musculoskeletal disorders and the workplace: low back and upper extremities.* Washington, DC: National Academy Press.

Perrault, M. (1995). Investing in ergonomics. *OH&S Canada, Sep/Oct*, 39–45.

Retsas, A., & Pinikahana, J. (2000). Manual handling activities and injuries among nurses: an Australian hospital study. *Journal of Advanced Nursing, 31*(4), 875–883.

Ronald, L. A., Yassi, A., Spiegel, J., Tate, R. B., Tait, D., & Mozel, M. R. (2002). Effectiveness of installing overhead ceiling lifts. *AAOHN Journal, 50*(3), 120–126.

Sacrificial lamb stance is killing healthy backs. (1999). *Hospital Employee Health*, 29–33.

Scholey, M. (1983) Back stress; the effects of training nurses to life patients in a clinical situation. *International Journal of Nursing Studies, 20,* 1–13.

Scopa, M. (1993). Comparison of classroom instruction and independent study in body mechanics. *The Journal of Continuing Education in Nursing, 24*(4), 170–173.

Shannon, H. S., Walters, V., Lewchuk, J., Richardson, L. A., Moran, L. A., Haines, T., & Verma, D. (1996). Workplace organizational correlates of lost time accident rates in manufacturing. *American Journal of Industrial Medicine, 29*, 258–268.

Shaw, R. (1981). Creating back care awareness. *Dimensions of Health Service, 58*(3), 32–33.

Smedley, J., Egger, P., Cooper, C., & Coggon, D. (1995). Manual handling activities and risk of low back pain in nurses. *Occupational and Environmental Medicine, 52*, 160–165.

Snook, S., Campanelli, R., & Hart, J. (1978). A study of three preventative approaches to low back injury. *Journal of Occupational Medicine. 20*(7), 478–481.

St. Vincent M., Tellier, C, & Lortie, M. (1989). Training in handling: an evaluative study. *Ergonomics, 32*(2), 191–210.

Stubbs, D. (1986). Backing out: nurse wastage associated with back pain. *International Journal of Nursing Studies, 23*(4), 325–336.

Stubbs D., Buckle, P., Hudson, M., & Rivers, P. (1983a). Back pain in the nursing profession: II. The effectiveness of training. *Ergonomics, 26*(8), 767–779.

Stubbs, D., Buckle, P., Hudson, M., Rivers, P., & Worringham, C. J. (1983b). Back pain in the nursing profession: I. Epidemiology and pilot methodology. *Ergonomics, 26*(8), 755–756.

Stubbs, D. A., Baty, D., Buckle, P. W., Fernandez, A. F., Hudson, M. P., & Rivers, P. M. (1984). *Patient handling and back pain in nurses.* Interim report. DHSS report no. JR 125/120, section 2. Washington, DC: US Department of Health & Human Services (DHHS).

Switzer, S., & Porter, J. M. (1993). The lifting behavior of nurses—in their own words. In F. Darby & P. Turner (Eds.), *Proceedings of the seventh conference of the New Zealand Ergonomics Society*, 2–3 August 1996 (pp. 33–43). Wellington: New Zealand Ergonomics Society.

Takala, E. P., & Kukkonen, R. (1987). The handling of patients on geriatric wards. *Ergonomics, 18*(1), 17–22.

Troup, J. D., & Raudala, H. H. (1987). Ergonomics and training. *International Journal of Nursing Studies, 24*(4), 325–330.

Tuohy-Main, K. (1997). Why manual handling should be eliminated for resident and carer safety. *Geriatrician, 15*, 10–14.

Ulin, S., Chaffin, D. B., Patellos, C., & Blitz, S. (1997) A biomechanical analysis of methods used for transferring totally dependent patients. *Scientific Nursing, 14*(1), 19–27.

U. S. Department of Health and Human Services (HHS). (1999). *Federal Register*, Part II, Dept. of Labor, Occupational Safety and Health Administration, 29 ACR Part 1910; Ergonomics Program: Proposal Rule (November 23, 1999).

U.S. Department of Labor, Bureau of Labor Statistics (BLS). (2002). Survey of occupational inquiries and illnesses, 2001. December 19, 2002.

U. S. Department of Labor, Occupational Safety and Health Administration (OSHA). (2002). *Ergonomics guidelines for nurs-*

ing homes. Retrieved February 1, 2005, from www.osha.gov/ergonomics/guidelines/nursinghome/index.html.

Vasiliadou, A., Karvountzis, G. G., Soumilas, A., Roumeliotis, D., & Theodosopoulou, E. (1995). Occupational low-back pain in nursing staff in a Greek hospital. *Journal of Advanced Nursing, 21(1),* 125–130.

Venning, P. J. (1988) Back injury prevention. Instructional design features for program planning, *AAOHN Journal, 36*(8), 336–341.

Videman, T., Nurminen, T., Tolas, S., Kuorinka, I., Vanharanta, H., & Troup, J. (1984). Low back pain in nurses and some loading factors of work. *Spine, 9*(4), 400–404.

Villeneuve, J. (1998). The ceiling lift: an efficient way to prevent injuries to nursing staff. *Journal of Health care Safety, Compliance & Infection Control, Jan,* 19–23.

Wachs, J. E., & Parker, J. E. (1987). Registered nurses' lifting behaviour in the hospital setting. In S. Asfour (Ed.), *Trends in Ergonomics/Human Factors IV* (pp. 883–890). Holland: Elsevier Science.

Wassell, J .T., Gardner, L. I., Landsittel, D. P., Johnston, J. J., & Johnston, J. M. (2000). A prospective study of back belts for prevention of back pain and injury. *JAMA, 284*(21), 2727–2732.

Westgaard, R. H. (2000, October 22). Work-related musculoskeletal complaints: some ergonomics challenges upon the start of a new century. *Applied Ergonomics, 31*(6), 569–580.

Wicker, P. (2000). Manual handling in the perioperative environment. *British Journal of Perioperative Nursing, 10*(5), 255–259.

Winkelmolen, G. H. M., Landeweerd, J., & Drost, M. R. (1994). An evaluation of patient lifting techniques. *Ergonomics, 37*(5), 921–932.

Wood, D. (1987). Design and evaluation of a back injury prevention program within a geriatric hospital. *Spine 12(2*), 77–82.

Yassi, A., Cooper, J. E., Tate, R. B., Gerlach, S., Muir, M., Trottier, J., & Massey. K. (2001). A randomized controlled trial to prevent patient lift and transfer injuries of health care workers. *Spine, 26*(16), 1739–1746.

Zelenka, J. P., Floren, A. E., & Jordan, J. J. (1996). Minimal forces to move patients. *The American Journal of Occupational Therapy, 50*(5), 354–361.

·22·

ERGONOMICS; NOISE AND ALARMS IN HEALTH CARE—AN ERGONOMIC DILEMMA

Beate Buß and Wolfgang Friesdorf
Technical University of Berlin, Germany

Today's medical High Dependency Work Environments, such as Emergency Care Units (ECU), Operating Rooms (OR), and Intensive Care Units (ICU), involve a massive use of technology. Technology can overtake several tasks, such as the monitoring of a seriously invalid patient and his physiological functions. The usage of technology, especially information technology, represents huge progress in this area of health care work. Medical devices are supposed to provide for safe alerting of critical events. It turns out that many devices produce an information overload with a large number of "false-positive" alerts. Up to now, technical support of the health care worker has not been fully satisfactory. Technically generated acoustic alarms have become a noise source instead of informing or alerting the worker in specific situations (Biot, Carry, Perdrix, Eberhard, and Baconnier 2000; Block, Nuutinen, and Ballast, 1999; Chambrin et al., 1999; see also Friesdorf, Buß, and Göbel 1999; Tsien & Fackler, 1997; Solsona et al., 2001; Xiao, Mackenzie, Seagull, and Jaberi, 2001). The U.S. Joint Commission on Accreditation of Health Care Organizations (JCAHO) has raised the level of awareness for this issue in U.S. hospitals by making alarm safety one of its National Patient Safety Goals for 2003.

Medical High Dependency Environments (HDE), and so work fields with a large influence of technology, are characterized by their high degree of complexity:

- individual variation of patients and often uncertainty regarding the patient's status,
- a multi-factorial setting of goals (positive medical outcome, patient satisfaction, economic efficiency, staff satisfaction, etc.),
- the time pressure and the irreversibility of acting,
- a high degree of dynamic where there may be delayed reactions and effects,
- changing content of care activities and often lack of capacity to plan for specific care activities,
- interaction and need for cooperation of experts with different qualifications and knowledge in an often local distributed environment (Friesdorf & Göbel, 2003; see also Carayon & Friesdorf, 2006).

Thus, in order to design ergonomic effective alarm systems, it is necessary to adopt a system ergonomic approach. This system approach should consider the signal information, as well as the recipient who perceive the information in his or her actual work context.

In this chapter the relation between acoustic alarms and disturbing sound is discussed. Different approaches toward avoiding the existing dilemma between necessary acoustic warning and the generation of noise are shown.

NOISE—PERMANENTLY CONNECTED TO MEDICAL HIGH DEPENDENCY ENVIRONMENTS

Current Situation in Medical High Dependency Environments

In the medical field, noise pollution has been discussed for a number of years (e.g., Woods & Falk, 1974; Hilton, 1985; Berg, Stuttmann, and Doehn, 1997; Topf, 2000; Biot et al., 2001). Hilton defined noise as "sound at levels above those recommended for hospitals and perceived by patients as undesirable" (Hilton 1985, p. 283). A number of established standards exist. The standard issued by the U.S. Environmental Protection Agency (1974) recommends a sound pressure level of less than 35 dB(A) during the night (periods of sleep) and 45 dB(A) during the day. The German workplace ordinance §15 (Verein Deutscher Ingenieure, 1999) allows a maximum sound pressure level of 55 dB(A). Even though a number of standards and guidelines exist, the acoustic situation in medical HDE is still suboptimal.

Hilton (1985) examined sound pressure levels in different Intensive Care Units (ICU). She registered a mean equivalent continuous sound pressure level (mean Leq) of 32.5 dB(A) up to 57 dB(A) mean Leq for smaller units. This corresponds to the typical sound pressure level of a quiet conversation or radio music turned down to low volume. However, in major hospitals, sound levels of over 50 dB(A) Leq, and sometimes 68.5 dB(A) Leq, over a 24-hour period have been measured in the recovery room, as well as in the ICU (Hilton, 1985). This level of noise is comparable to street noise. In Hilton's study (1985), the specific sources of sound were identified, and measures of sound were performed. Sound produced by the staff ("talking" but also patient-related noise such as "making the bed" and "adjusting and adding medications to the intravenous drip") and sound produced by the use of work-related, non-medical equipment (e.g., telephone, printer, bed rails) were identified as major sources of sound. A message tube produced sound levels between 90 dB(A) and 100 dB(A). As a reminder, it is important to know that exposure of more than 85 dB(A) for 8 hours can lead to permanent damage of the human hearing (Crocker, 1997). Steady noise can come from equipment such oxygen going through a mask, oxygen going through nasal prongs, and ventilator machines (between 49 and 70 dB [A] mean Leq). Results of this study led to the recommendation for smaller rooms with fewer beds and the need for staff to be trained to be very sensitive concerning noise. The staff was trained to close the doors of patient rooms, especially at night, in order to reduce the impact of noise on patients.

In a manner similar to aviation (Noyes, Starr, Frankish, and Rankin, 1995) and control rooms in nuclear power stations (Yuki, 2002), the number of safety-relevant information signals in health care work has increased. The use of more sensitive sensor technology and the increase in information requirement by the medical staff as consequence of growing medical-diagnostic knowledge has caused this change (McKinnon, 1983; Xiao, Mackenzie, Seagull, and Jaberi, 2000). Recent studies have found even higher sound levels caused by the technologies. For instance, Moore et al. (1998) measured an average sound level of 60 dB(A) in an ICU and an average sound level of 62.33 dB(A) in an Acute Care Unit (ACU). They examined the effectiveness of a "door closing" intervention. Results showed a slight increase in the noise exposure for patients in the ICU: the noise was mainly produced by the medical devices in patient rooms. In contrast to the ICU, the ACU showed a reduction of 6 dB (A) caused by the "door closing" approach. In the ACU, noise was generated more often by hospital personnel than by the equipment nearby the patients. Another study by Kahn et al. (1998) found noise with high peak levels of 74.8 dB(A) up to 84.6 dB(A) in a medical ICU and a respiratory ICU. This study was able to identify 12 different causes of noise. The most common cause of sound peak were talking, usually involving conversations between the staff of the unit, television, patients, and monitor alarms. The contribution of technically generated acoustic alarms to noise found in clinical work systems has also been shown by Balogh, Kittinger, Benzer, and Hackl (1993). An average number of 2.1 ± 0.8 alarms per hour per patient was found. An instable patient with hemodialysis experienced on average 42 alarms per hour. Sound pressure levels of 60 until 70 dB(A) and sometimes more than 80 dB(A) were measured.

Besides the objective component of sound, it is important to recognize the impact of the subjective aspect of sound. A specific level of sound, its frequency, the absolute durance, and its slew rate can influence the perception of sound. Moreover, personal constitution and attitude toward sound play an important role. Individual variables (such as intrinsic sensitivity to a specific ambient stressor, restricted capacity, personal and cultural preferences, age, and past experience with similar sounds) also influence the assessment of sound as noise (Hilton, 1985; Topf, 2000).

Noise and Its Effect on Clinicians

As discussed previously, noise has a variable effect on medical staff. If a high sound level is perceived as negative, it can produce stress. If it is perceived as "necessary," its effect is reduced. Topf (2000) points out the different effects of noise on stress. With the help of an expanded environmental stress model, she defines the relation between environmental stressors, enhancement of person–environment compatibility and health. According to her model, sound becomes a state of subjective and/or physiological arousal that occurs when one is unable to exert personal control over the sound stressor. In a survey of 100 critical care nurses, Topf (1989) found that lower commitment to work was significantly related to greater perceived hospital noise disturbance. If an increased sound level is perceived as undesirable, it becomes an ambient stressor and can cause stress with all its negative effects, i.e.: strains on the cardiovascular system and low immunity. "Ambient stressors are chronic, negatively valued aversive aspects of physical environment" (Campbell, 1983; quoted in Topf, 2000, p. 521). Topf and Dillon (1988) surveyed nurses in a Critical Care Unit (CCU) and reported a significant correlation between self-reported, noise-induced subjective stress and emotional exhaustion, which is a component of burnout.

When sound level increases, health care workers showed physiological effects in addition to psychological effects. In a group of 11 nurse volunteers in a pediatric ICU, Morrison, Haas, Shafner, Garrett, and Fackler (2003) found an increase of tachycardia with a higher average sound level. The negative psychological effects could be compensated partly by the level of experience of the nurses. The relationship between environmental stressors and human performance has been examined in several studies (Murthy, Malhotra, Bala, and Raghunathan 1995; Jones & Broadbent, 1991; Parsons, 2000; Furnham & Strbac, 2002; Belojevic, Jakovljevic, and Slepcevic 2003). Human performance depends on many factors, including complexity of the tasks, consequences if these tasks are not accomplished, and predictability and controllability of ambient sounds. The exact effect of noise on human efficiency cannot be predicted by a simple linear model. Gawron (1982) reported that 7 of 58 studies showed that noise improved performance because of a focus in attention and increased arousal. Twenty-nine of the 58 studies reported reduced performance due to noise (see also Crocker, 1997 and Parsons, 2000). A field study by Flynn et al. (1996) showed that the frequency of dispensing errors committed by pharmacists depends on the sound level of environmental noise. The loudness (mean Leq for every half hour) that was determined in that experiment showed an effect on the dispensing error rate that was not entirely expected. The dispensing error rate rose according to increasing loudness (maximum mean Leq of 74.0 dB[A] per half hour) but fell again as soon as a certain value was reached. The reduction in the error rate after a certain sound level could be explained as a coping mechanism and adaptation of environment by pharmacists.

Noise and Its Effect on Patients

Sleeplessness in patients has been connected to hospital noise. Increased alertness and a change in sleeping patterns, such as reduced REM sleep, and awakening in the middle of sleep have been shown to result from noise (Topf, 1992; Krachman, D'Alonzo, and Criner 1995; Kroon & West, 2000). Noise can also have a negative effect on the healing process (e.g., see McCarthy, Ouimet, and Daun 1991) and even cause so-called ICU psychosis in the patient. The ICU psychosis syndrome manifests itself in patients withdrawing from interpersonal relationships; being suspicious, irritable, and delusional; and hallucinating (Dracup, 1988). Very high noise levels over a long period can even cause permanent hearing damage (Topf, 1983). A study in the neonatal ICU found a relationship between high noise and a rise in the heart and respiratory rate of neonates (Bremmer, Byers, and Kiehl 2003).

Both hospital staff and patients can evaluate sound in a very subjective way. A study performed by Allaouchiche, Duflo, Debon, Bergeret, and

Chassar (2002) in a Postanesthesia Care Unit (PACU) found that a raised mean Leq of 67.1 dB(A) did not have a significant correlation to patient feelings of comfort. This subjectivity dimension of noise implies that different noises will have different effects and sound levels will be perceived differently by patients. Whalen (1992) argues for making patients who are awake familiar with the acoustic environment in the ICU. Prepare them for certain alarms and noises and explain to them the meaning of the various alarms and noises. Such information can help patients better deal with the acoustical situation and perceive sound as less disturbing. This again reduces all influences that have a bad impact on the healing process. So the effect of "learned helplessness," which has a proved impact on the noise exposure (Hatfield et al., 2002) can be reduced. The patients get the feeling of control.

Noise Reduction Approaches

The high-technology health care work in medical HDE seems to be inseparable from noise. In this area of work, noise can be defined as a multi-faceted problem. Objectively measurable sound levels do not always fit with subjective perceptions of sound. Different sources of noise, such as the hospital staff itself, equipment, and acoustic alarms, have been identified. The important question is then how do we decrease the very high noise levels?

Noise can be reduced by hospital staff through Behavior Modification Training. New organizational approaches, like holding team meetings in rooms away from patients, can lead to a reduction in noise (Kahn et al., 1998; Moore et al., 1998; Hilton, 1985). Various survey instruments addressing issues of staff training and adaptation of the work environment to reduce noise have been developed. Johnson (2004) developed a survey instrument for the Neonatal Intensive Care Unit (NICU) environment; Cmiel, Karr, Gasser, Oliphant, and Neveau (2004), for Thoracic Surgery Units; and Walder, Franciolo, Meyer, Lancon, and Romand (2000), for a Surgical ICU. Besides reducing sources of noise by team training, music can be used to 'hide' sources of noise that are hard to remove (Perlman, 2003; Shertzer & Keck, 2001).

For medical devices it is possible to use new noise-reducing procedures and materials. For instance, Berg et al. (1997) implemented specific modifications to a device, that is, modification of T-piece used with intubated patients, and they achieved a reduction in noise of 15dB(A).

The three-level ergonomic approach of noise reduction (Compes, 1970) can be used to ascertain the feasibility of various noise-reduction strategies. This approach defines the following steps to fight noise: (a) primary noise abatement (prevention of noise at the place of origin: staff training, improved design of medical technology and acoustic alarms, and wearing noise-reducing clothes), (b) secondary noise abatement (device insulation and sound attenuation: architectural changes like smaller rooms, noise-absorbing materials such as wallpapers), and (c) tertiary noise abatement (personal protection: ear protectors, music). The three-level ergonomic approach has to be completed by the development and training of various coping strategies for dealing with increasing sound levels by hospital staff, as well as patients.

The question remains of how one can prevent noise that is caused by acoustic alarms. Are high sound-pressure level and high fundamental frequencies important determinants of the effectiveness of acoustic alarms? Noise reduction may seem rather paradoxical in this context. Woods (1995) talks about an "alarm problem" for comparable working systems. Alarms should help to identify abnormal conditions. But, today, nuisance alarms occur; poorly designed alarms distract and disrupt other tasks and hinder the clinical staff in efficient fault management. The following sections describe deficits in alarm systems, the problems of integrating alarms into clinical work processes, the safe use of alarms, and new approaches to alarm design.

ALARM SYSTEM DESIGN

Fundamental Characteristics of Alarms

In every day language, the term 'alarm' is used in various ways. One can find nine different definitions in the Colin's Dictionary (1986; quoted from Stanton, 1994, p. 2):

- to fill with apprehension, anxiety, or fear;
- to warn about danger: alert;
- fear or terror aroused by awareness of danger: fright;
- a noise, signal, etc., warning of danger;
- any device that transmits such a warning: a burglar alarm;

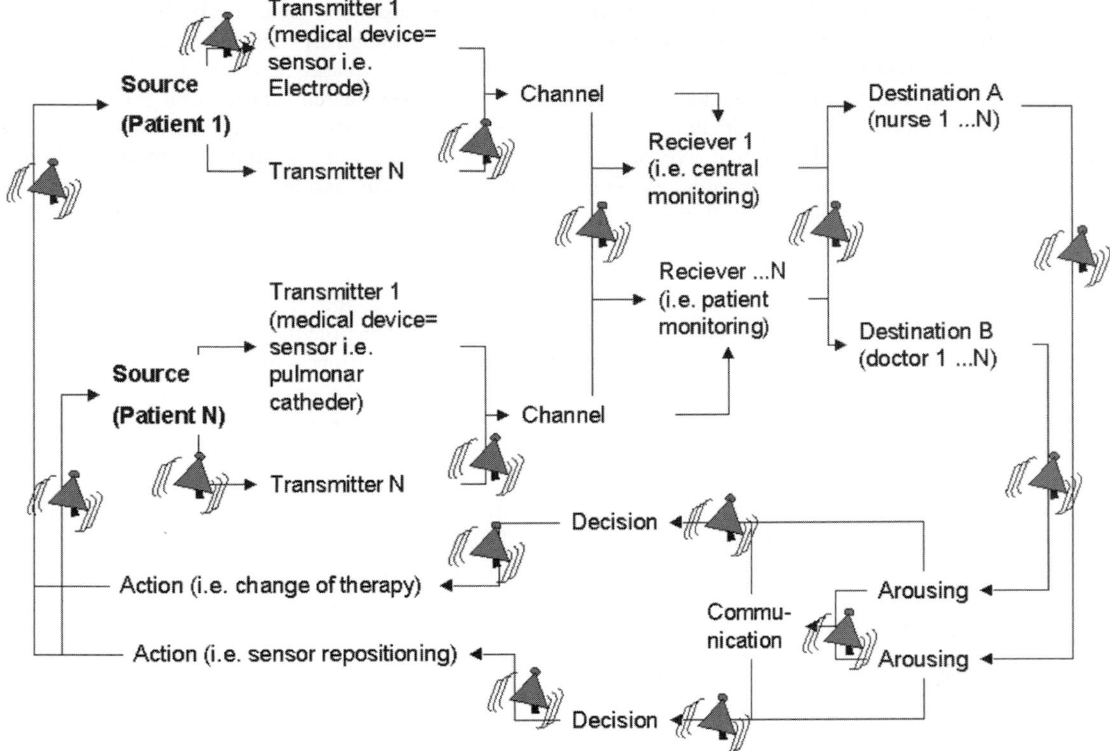

Figure 22–1. Risks in the Information Flow Between Patient–Machine(s)–Team of Clinicians (bells symbolize potential risk points in the information flow)

- the device in an alarm clock that triggers off the bell or buzzer;
- a call to arms;
- a warning or challenge.

The problem of defining an alarm or an alarm design concept derives from the quite unclear orientation of the term "alarm." On one hand, the term of alarm describes the stimulus. On the other hand, alarm means the reaction to a specific event. Stanton (1994, pp. 2–3) recommends to make a distinction between a *stimulus-based* and a *response-based model*. According to the stimulus-based model, a specific alarm has a certain effect on the individual. This assumes that every alarm is clearly describable in the way it affects the perceiving individual: the alarm is predictable and affects every individual the same way (engineering approach). According to the response-based model, an alarming situation causes a state of alarm in the individual. Thus, every individual shows a different reaction to an alarm depending on his/her experience, attitudes, and personal aims. Different individuals perceive different stimuli in a differently alarming way (psychological approach).

For the complex work field of health care, we propose that a combination of both models is necessary. While the *stimulus-based model* provides a starting point for selecting relevant events that should be linked to alarms and the creation of acoustic alarms, the response-based model emphasizes the description of the individual perceiving the signals and his or her background (e.g., qualification, specific training).

In health care HDEs many different sensor-technology alarms function as a special kind of information and may contradict the continuous information flow between human and machines (e.g., staff and patient monitor, staff and respiratory monitoring, etc.). "Formally, information is defined as the reduction of uncertainty." (Wickens, 1992, p. 50). In the current health care work environment, with its information overload, it is a problem to distinguish a signal from noise and, thus, to detect information as relevant or alarming.

Figure 22–1 shows potential risks of the information flow between patient, machine, and a team of specialized staff (nurses and doctors) that have to be signalized. There are not only common physiological or technological deviations, which need to have alarms. Risks caused by wrong detection of critical constellations by staff or errors in the communication process, especially between representatives of different disciplines, are relevant, too. The model is related to the mathematical theory of communication by Shannon and Weaver (1949). The bells show potential risks and breakdowns in the information flow. Risks such as ignoring or misunderstanding of signals are system immanent and can be caused by disconnected technical devices and result in lacks of information. Misinterpretation of signals are due to both technical failures, as well as human failures caused by less knowledge and experience of the staff.

In order to perceive appropriate information and to use it, the staff has to follow the steps of detection, discrimination, identification, recognition, and comprehension (for a more detailed description of the steps, see Easterby, 1984, p. 21). In every step lies the potential of misinterpretation or forgetting.

Current alarms support only a small part of potential breakdowns in the information flow, or they may provoke new breakdowns. Through the situation awareness process (Endsley, 2000), every specialized staff acting in a team develops his or her own mental model of the specific situation at hand. Individual Situation Awareness has to be brought in line with the team in order to prevent misunderstanding while dealing with critical situations (Orasanu & Salas, 1991; Buß & Friesdorf, 2000).

The complex health care work environment allows only a brief analysis of the information flow between people and machines. Thus, real critical physiological changes in the patients should be recognized in a timely manner. They should be classified appropriately according to their meaning or their risk.

Kerr (1985) describes two critical types of alarms that are present in the medical field. The so-called *immediate alarms* produce a signal only if a potentially damaging state has been reached in the patient. A typical immediate alarm would be the signalling of brain oxygen concentration. There are also *anticipatory alarms* that become active before deleterious physiological changes occur. Anticipatory alarms occur when the decrease in pressure in the oxygen supply to an anesthetic machine or ventilator, for instance. Anticipatory alarms should definitely be preferred to immediate alarms because they help the health care provider recognize potentially critical situations and thus allow an adequate reaction by the provider. Nevertheless, the current clinical work environment is characterized mainly by immediate alarms that provide alerts "at the last minute" (Smith, Mort, Goodwin, and Pope, 2003).

Common acoustic alarms of medical devices are prioritized based on the urgency of the required operator response. A combination of the speed of onset of potential harm (immediate, prompt, delayed) and the potential result of the alarm condition (death or irreversible injury, reversible injury, minor injury, or discomfort) provides for the distinction of four states of alarm signal: the information signal, and alarms with low, medium, and high priority. According to the standard of the International Electrotechnical Commision (IEC) [IEC/FDIS 60601-1-8 2003 medical electrical equipment, part 1-8—general requirements for safety; collateral standard: alarm systems; requirements, test and guidelines; general requirements and guidelines for alarm systems in medical electrical equipment and in medical electrical systems], all types of acoustic alarm signals can be defined by specific sound parameters, such as pitch range, the use of harmonics, and the character of the melody formed by three pulses sounding in succession (bursts). A pulse forms a basic unit of a signal as recommended by Patterson (1982). A pulse is defined as a sound with an amplitude envelope, which has an onset, an offset, and a specific duration.

The Tone Defines the Mental Picture—the Psycho-Acoustic Perspective

An essential problem of alarms used today is their acoustic design. They are too loud, which results in confusion, and are sometimes difficult to localize (Meredith & Edworthy, 1994; Momtahan, McIntyre, & Hogan, 1993). The "*better safe than sorry principle*" used by manufacturers and the absence of standardized prioritization of different devices using a systemic approach cause the noisy situation in health care environments.

Detection of Alarms

In contrast to sight, hearing is omni-directional: every auditory signal can be considered as an input from any direction and does not need any kind of scanning. First of all, there is a pre-attentive storage of the signal. The content of this pre-attentive short- term auditory store can get into the center of attention as soon as key signals cause a switch. Loudness or intensity of an acoustic signal will almost always grab the attention of the operator (Wickens, 1992). If the acoustic signal has any semantic relation to the topic that is the current focus of attention, the signal will grab the attention of the operator, too (Wickens, 1992). There are problems of filtering acoustic alarms out of a noise. The danger of confusion between different auditory messages or masking of auditory message, that is, an overlap of frequencies that hide the actual alarm, exist (Meredith & Edworthy, 1994).

Masking and Problems of Localization Alarms

The masking of sounds by environmental noise and the problems of localizing acoustic signals in the large open work spaces are additional difficulties in the design of acoustic alarms in HDE. Problems are usually caused by frequencies that are too high and not affected by the temporal effect—a difference in binaural hearing (Loeb, 1986). Masking can also be caused by unclear sound characteristics of the devices. In this context, Block (1996) discusses the importance of non-medical products, such as paging devices, radios, and portable computers, whose sound design is not oriented in any standardization corresponding to the clinical work system or work place.

Perceived Urgency

The perceived urgency of an acoustic signal is defined by various characteristic parameters. Buß, Neth, and Friesdorf (1998) conducted a study of 19 experienced ICU nurses. The nurses were asked about their expectations of a typical alarm signal. A recognizable alarm was characterized as high, clear, and harsh sounding and was formed by a rising and fast melody with a mid-length. Results of various laboratory experiments have demonstrated the same characteristics of 'good' alarms. Haas and Casali (1995) conducted several experiments that demonstrated that shorter inter-pulse intervals were associated with greater perceived urgency, as well as a raised pulse level. Hellier et al. (1993, quoted by Edworthy 1994, pp. 26 ff.) assessed four factors influencing perceived urgency using Stevens Power Law (a description of the relationship between objective and subjective variables; Stevens, 1957):

- the speed of a melody,
- the fundamentally frequency,
- number of units repetition,
- and the degree of inharmonicity.

The largest effect of subjective ratings of urgency as consequence of a change in the physical parameter produce the "Speed" parameter, before the "number of repeating units." The data show the large influence of rhythm to the subjective feeling of urgency (see also Hellier & Edworthy, 1999).

Sound Confusion

Meredith and Edworthy (1994) found out that simple, continuous sounds can very often lead to confusion although they are characterized by strongly differentiated pitches. This effect can be found especially if the sound is heard for longer periods. Alarms with a regular temporal pattern also lead to confusion, although their actual pulse speed diverges strongly from each other. To prevent confusion, several sound characteristics such as pitch, pulse speed, and harmonic should be varied simultaneously.

In a laboratory test Meredith found that acoustic qualities of the sound are not always the source of confusion (Meredith, 1992, quoted from Meredith & Edworthy, 1994, p. 217). Terms such as "high pitched," "continuous tone," or "complex tone," which people use to recognize specific alarms, can also lead to confusion. These terms are too vague and do not describe a defining characteristic of the alarm sound. Thus, confusion due to false memories even with large acoustic differences can be expected.

Another aspect with large influence in sound identification is the artificial versus natural character of a sound. Natural sounds, sets of acoustic stimuli, such as pitch speed and associations with existing contents of memory, can be recalled more easily and safer than synthesized sounds (Block, 1992; Edworthy, Hellier, and Hards 1995). The ecological approach to sound perception (e.g. Gaver, 1993) explains this effect: in everyday listening we try to reconstruct the event that generates the sound. This is connected to the fact that pitch, loudness, the rhythm, and other sound characteristics are tools to estimate the meaning and place of events and, thus, are not explicitly perceived.

Context plays an important role in the recognition of alarm and its evaluation. In today's health care work environment different studies, find out a gap between the perceived (psycho-acoustic) urgency and the real urgency of the situation (e.g., Edworthy, Loxley, and Dennis 1991; Mondor & Finley, 2003).

To summarize the section above, the main focus in the design of clinical alarm sounds from a psycho-acoustic point of view depends on the relation between the physical characteristic of the alarm and the subjective feeling or impression. Aspects of the physiological determined perception of sounds, such as localization, should be considered, as well as the mechanisms of human memory, that is, sound confusion by wrong coding. Emotional aspects should take into consideration, as well.

The Goal and the Task Direct Attention—the System–Ergonomic Approach

Besides the psycho-acoustic effect of an audio signal on the receiver (the team), the task to be completed and its work context play an important role in the recognition and clear identification of an alarm.

Bliss and Dunn (2000) showed that with increasing task workload and a simultaneously increasing alarm workload, the reaction to alarms decreases. Alarms are being ignored and/or the reaction time to alarms decreases. The dominance of the task in relation to acoustic stimuli was confirmed by Flynn et al. (1996). Flynn et al. (1996) showed that pharmacists were able to perform their tasks despite an increased level of sound. Noise should not be equated with alarms, which should have their own information content. But, there is a smooth transition between noise and alarm, especially in today's medical HDE.

Cry-Wolf Effect and Silencing

In today's clinical practice, most of the time false-positive alarms occur: alarms go off unnecessarily when there is no need for action (Biot et al., 2000; Hedley-Whyte & Sheridan, 2003). The large number of false-positive alarms is usually caused by the specific alarm limits chosen, which are normally too close and are not adapted to the patient's characteristics (Block et al., 1999; Solsona et al., 2001). Specialized medical staff (physicians and nurses) often use default alarm settings prescribed by the manufacturer. Those alarm settings do not adapt to the specific characteristics of the patient being monitored. The better safe than sorry-principle applies most of the time and often alarms are activated too early (Block & Schaaf, 1996).

False-positive alarms do not only create noise, but they also lead to reduced safety because of staff behavior related to the *cry-wolf effect* (Breznitz, 1983). Because of their limited reliability, the alarms lose their attention-creating purpose for specialized staff. The cry-wolf effect is work domain independent (e.g., Kerstholt & Passenier, 2000). Clinical staff get used to the alarms, then start to ignore them, and finally silence them. Xiao, Mackenzie, Jaberi, Harper, and LOTAS Group (1990) emphasized the potential for danger caused by silenced alarms, because of the resulting lack of presentation of really relevant alarms in critical situations.

Bliss, Gilson, and Deaton (1995) examined the cry-wolf effect and its relation to situational and alarm-dependent factors. A laboratory experiment was conducted with students using the dual-task method. Bliss and colleagues were able to prove the cry-wolf effect but not in the expected manner. In the experiment, the cry-wolf effect could be reduced up to a certain degree. In spite of the acquired knowledge of falls alarms, the number of alarm reactions level out. "The particular form the cry-wolf effect takes is dependent upon situational or alarm-dependent factors" (Bliss et al., 1995, p. 2307). Alarms with a high urgency want to be observed any time even if a high rate of false alarms exist. The general human predisposition to respond to alarms leads to a certain tolerance of false alarms: "...alarms can be as low as 90% accurate, with no loss of operator responding" (Bliss et al., 1995, p. 2310). Results show that humans overmatched, that is, their response frequencies were greater than stated alarm reliability levels. A specific training could reduce the cry-wolf effect in the experiment in the same way.

Despite this fact, falls alarms, especially falls positive alarms, represent an unnecessary strain on clinical staff and can lead to a reduction in patient safety.

Task-Related Use

There are still a number of deficiencies in the design of technical sensors, especially according to the connection of single signals to useful units.

Moreover, we need more information regarding medical-diagnostic development in order to understand more clearly the connection between physiological parameters, therapeutic treatment, and their effects on the patient's outcome. The relation between work tasks and work processes that produce specific demands on humans and technology and the characteristics of available alarms is important to consider from the perspective of Human Factors Research.

As described above, acoustic alarms in health care environments help the operator focus attention on a specific place connected to a critical event. Moreover, they can reveal possible causes related to the critical event, thus, leading to relevant decisions. On the basis of the given work structure and its resulting work load (i.e., in OR comparably different than in ICU), one can determine the requirements expected by the clinical staff for a specific alarm. Seagull and Sanderson (2001), for instance, published a field experiment that measured the reaction time of anesthetists on audible alarms in an OR depending on the specific operation phase. In this study, four types of surgical procedures (laparoscopy, arthroscopy, cardiac, and intracranial) and three phases of procedures (induction, maintenance, and emergence) were distinguished. The acoustic alarms were classified as: (a) requiring a corrective response, (b) being the intended result of a decision, (c) being ignored as a nuisance alarm or (d) functioning as a reminder. It was found that, especially in the introduction and emergence phase, a great number of acoustic alarms occurred, but most of them were ignored. In the maintenance phase, correcting performance occurred (i.e., changing the patient's configuration and changing the equipment setting). In this phase, the alarms were relevant to the anesthetists. One result of this study was a desirably larger sensitivity of acoustic alarms within the context of work. "...many alarms do not require action, such as new alarms during intubations" (Seagull and Sanderson, 2001, p. 76).

Seagull,Xiao, Mackenzie, and Wickens (2000) developed a Taxonomy of Purposes of Monitoring. According to the taxonomy, monitoring is "not a single unified task, but rather is composed of a number of discrete tasks: Routine monitoring, goal-state verification, problem solving, and interpretation of historical data. Each of these tasks contains different information demands" (Seagull et al., 2000, p. 1–224). Based on the large number of studies, Seagull et al. (2000) conclude that today's auditory alarm signals support Routine Monitoring. Its task is to track unexpected changes while the patient reaches a stable state (i.e., anesthesia). According to Seagull et al. (2000), no satisfying design solution has been proposed for the other three types of information demands. Acoustic alarms should support audio displays but also inevitably support these monitoring subtasks as well. According to Xiao and Seagull (1999), acoustic alarms should not only function as a simple status display. They have much more potential to introduce interruptions (e.g., as a reminder to do or to check something), to improve human limitations (e.g., attention limits, short-term memory limits), and to be source of problem-solving (e.g., to give a clue that something is not going right and that there is a need of diagnosis in the technical field).

WAY(S) FOR TASK- AND CONTEXT-ORIENTED ALARM SYSTEM DESIGN TO TAKE CARE OF SAFETY AND NOISE REDUCTION

> "A critical criterion for the design of fault management systems is how they support practitioner attention focusing, attention switching and dynamic prioritisation"
>
> (Woods 1995, p. 2375.)

Alarms should be the first indicator for the development of critical situations, and thereby they are a very important part of a monitoring system and thus a fault management system. The quote chosen by Woods (1995) reveals the great variety of design requirements to acoustic alarm systems: attention focusing, support attention switching, and dynamic prioritization.

The role of clinical staff has changed by increasing use of technology. Tasks will be chaired more and more between technology and men. A main task is now to monitor patient status and the effect of treatment. On the other side, the clinical expert must be able to detect technical defects, too. Both aspects have to be bring together in a practicable way. A lot of different and unstructured information that is given by patient monitors must be integrated and must be formed in a correct model of the current situation in staffs' minds. To reach this challenge, large cognitive performances have to be realize by the clinical staff, under conditions of high work load. Acoustic alarms have to fit the expectations of the receiver in the most optimal way. The operators' adaptation to the increasing number of acoustic alarms has to be considered, as well as the generally laws and principles of perception psychology. Acoustic conditions, such as the fundamental frequency or the time characteristic of an alarm, play an important role as does existing noise on a parallel level to the alarm system.

Most essentially, acoustic alarms have to be regenerative with actual work tasks and course of treatment. It is important to consider both: the actual and future state of the patient and the performance of the receiver (physician or nurse) at the time of the alarm. Different approaches are presented below that reflect historical changes in perspectives regarding alarm problem and its solution.

The Technological Way

The main field of work dealing with the optimization of acoustic alarms is in the development of better and more exact sensors and especially in the definition of algorithm to connect single signals in a most meaningful way. Today, integrated monitoring is considered as State of the Art for alarm design (Block & Schaaf, 1996). But, if one considers the previously quoted results of false alarms, it becomes obvious that there is a need for further research in this field. Co-operation between physicians who can define an adverse event from a physiological point of view and technicians or information scientists who can set up systems to process this knowledge is needed (Bates et al., 2003). Knowledge of adverse events and knowledge of the physiological system can support the design of valid alarms. Developments with a stronger orientation toward technical capacities on actual tasks, the need for information that has to be satisfied on the side of human actors, is revealed by the theoretical model of *Compound organism subsystems* (Arbeitsgruppe Medizintechnik, 1986). These information units describing the patient's state mark the beginning of the development of the so-called *smart monitors* and *intelligent alarms.* With smart monitors and intelligent alarms the integration of different physiological and technical parameters, oriented in the physiological and therapy model of a patient, is possible. Adverse events could be signalized that are oriented in the human organism against technological determined coherences. Specific algorithms allow to detect important changes within a physiological time series and their integration to therapy-relevant patterns. The use of intelligent alarms and smart monitors helps to reduce the number of false-positive alarms and so to reduce the noise at HDE. Faster and more exact diagnosis of critical events will be possible (Bates & Gawande, 2003; Fried, Gather, and Imhoff 2001; Tsien, Kohane, and McIntosh 2000). Different prototypes of smart monitors exist (Steimann & Adlassnig, 1994; Orr & Westenskow, 1994; Becker et al., 1997; Ireland, James, Howes, and Wilson, 1997; Krol & Reich, 2000; Broscheit, Lorenz, Baumeister, Tannert, and Greim 2002). Different companies (e.g., Philips) develop new monitor concepts with the potential for event detection. The manufacturer realizes a default setting, and the clinical staff is incumbent on definition of the attributes of an event, as a retrospective approach.

A really new perspective gives the approach of VISICU. VISICU's smart alarms module uses the possibilities of a totally interlinked ICU, called eICU. The module receives data from multiple bedside monitors and devices and processes these data through a series of algorithms. Physiologic thresholds and trends are considered, as well as laboratory and medication information (Bates & Gawande, 2003; Breslow, et al., 2004).

The Way of Human Resource Management and Alarm Design

Related to the available staff in the medical HDE and the recognition of critical events, two solutions for the *alarm problem* have been identified. On the one hand, there is a need of employment of a sufficient number of qualified staff. The new Institute of Medicine (U.S.) report (IOM et al., 2004) shows the difficult situation concerning the qualification of nursing staff and the "care key" per patient. The report emphasizes that an increased number of nurses would raise patient safety in critical care environments. Critical events can be recognized much faster, almost as soon as they begin, and the clinical staff is not always obliged to share its attention. Besides the capability to recognize and to be able to deal with critical events, the staff should be qualified in understanding the monitoring technology. Several studies showed that a specific adjustment of the alarm limits toward the patient specifications decreases the rate of false-positive alarms with out a decrease of patient safety (e.g., Asbury & Rolly, 1999; Cust, Donovan, and Colditz, 1999; Rheineck-Leyssius, & Kalkman, 1997, 1998; Solsona et al., 2001; Biot et al., 2001). Thus, the noise on the units can be reduced dramatically.

Another aspect of a better fit of technical generated alarms to human performances is the psycho-acoustic character of signals. The psycho-acoustic potential of the respective active alarm should be better designed. Block (1992) and Buß et al. (1998) conducted experiments examining attempts to provide alarms of medical–technical instruments with further semantics because of their specific sound characteristics. Block investigated (1992) in a laboratory setting a system of six musical alarm tones. It was designed with musical themes from popular songs used for different organ systemic aspects: oxygenation, ventilation, cardiovascular monitoring, temperature monitoring, artificial perfusion, and drug administration systems. A group of anesthesiologists and others took part. They had to remember the correct combination of a melody learned and the compatible organ system. Forty-two of 79 (53%) respondents got all six answers correct on the second testing. Buß et al. (1998) analyzed the ability to coding medical devices by typical alarm sounds, using the CEN EN 745 standard (CEN EN 475 1995 medical devices: electrically generated alarm signals [foreign standard]) as a framework for alarm design. Seventy-two different alarm bursts were developed and played randomly to 20 ICU nurses. Without training, the nurses had to assign each burst to one of five devices (monitor, ventilator, infusion pump, syringe pump, hemodialysis) and to assess the suitability. The bursts consisted of three pulses. The frequency of the first pulse was 440 Hz. The second and third pulses showed a higher, lower, or constant height. For three devices typical characteristic melodies could be found, with a maximum rate of correspondence of 45% by clinical experts.

In both experiments previous experiences and attitudes of clinical experts were used to coding typical alarm sounds. This way, starting with the recipient and not with the stimulus or transmitter, offers an increase in the ability to detect and recognize clinical alarms and lead to an increase in patient safety.

Solution by Standardization

The standardization of acoustic alarms in the medical field follows a long tradition. Compared to earlier standards, there is an upward trend toward a larger context orientation and an increased emphasis on the adaptation of acoustic alarms in accordance to the relevant work system (e.g., IEC/FDIS 60601-1-8, 2003). In order to prevent confusion, masking, and other problems, the development of standards in complex work system is considered as important. However, such standards cannot consider the heterogeneity of the personnel, spatial, and technical equipment of the single units and the heterogeneity of the patients. The scope of the standards may lead to a transfer of responsibility from the manufacturer to the staff, therefore leading to a wrong application of acoustic alarms (see previous section). Standards should provide a framework for alarm design but should not be that restrictive. Customisable alarms can help medical and nursery staff to cope with the complexity of the work system. Watson, Sanderson, and Russell (2004) found considerable benefits of anesthesia alarms that are tailorable to the needs of the clinical staff.

The Alliance of a Macro-Ergonomic and Micro-Ergonomic Approach

As the previous analysis has shown, the problem of alerting for critical events and the support of problem-solving underlies a great number of influential variables. The acoustic design itself represents a small part of the overall noise/alarm problem. Organizational conditions such as work system integration and existing work tasks, as well as the current patient treatment, have great relevance, too. The macro-ergonomic approach of Hendrick and Kleiner (2001) can provide the conceptual framework to deal with such complexity. In contrast to hardware ergonomics, cognitive ergonomics and work design ergonomics (see Hendrik, 1997, or the chapter by Hendrick in this handbook for a description of the levels of human factors or ergonomics), macro-ergonomics proposes an integral approach and considers technical as well as human components, the organization structures, and their respective dynamic changes. The relevance of a macro-ergonomic approach with regard to the introduction of new technologies in complex work systems is stated by Carayon: "Whenever implementing a technology, one should examine the potential positive AND negative influences of the technology on the other work system elements" (Carayon, 2003). This approach should be supplemented by a micro-ergonomic perspective, which focuses on the alarm system itself. Using the alliance of a macro-ergonomic approach with a micro-ergonomic approach can help to solve the *alarm problem* (see Backhaus & Friesdorf, in this volume for a description of the benefits of combining both ergonomic approaches).

FUTURE PERSPECTIVES

Our previous description of approaches to reduce noise in medical HDE and the simultaneous attempt to exploit the potential of acoustic alarming has demonstrated that the technical solutions have to be related to solutions for a suitable Human Resource Management through a systemic approach. What are the important issues in the field of critical care in the future and how will they affect the design of acoustic alarm systems?

Buß and Friesdorf (2003) conducted an international survey of experts on future environmental conditions for alarm systems in surgical ICUs. The survey focused on following aspects: staff qualification, technological development, and organizational changes. Using the method of networked thinking (Probst & Gomez, 1991), 29 thesis were formulated by an interdisciplinary team of physicians, nursery staff, human factors specialists, and engineers, a so-called monitoring group. Three main topics were analyzed, the future in:

- the individual expert and its nearly surrounding work field,
- the clinical team, and its organization and co-operation supported by technology,
- and the surgical ICU within the hospital organization and development of the Health Care market.

For every theses probable time of realization and a voting of the technological, organizational, or educational developments required for progress were asked. The study was realized as a three-round Delphi and was accomplished in the end of 2002. Figure 22–2 represents the survey design.

The study was based on the results of a Delphi Study by Friesdorf, Classen, Konichezky, and Schwilk (1997). A total of 51 heads of surgical ICUs and manufacturers participated in the first round. The rate of participation in the third round of the survey was 7.3%. The study revealed the following results.

The relevant infrastructure for the technical alarm solution, as well as the integration of various devices and the possibility of a "plug and play" approach, is only considered realistic in the remote future (between 2010 and 2015). Only little agreement was reached as assessed by the Inter-Decile Range (IDR). The IDR stands for the range of the middle 80% of all data. It is a measure of the uncertainty of the group of experts concerning the response in term of voted time of realization of these items. The belief in technical progress as shown in results of earlier studies (e.g., see Bender, Strack, Ebright, and von Haunalter, 1969 and Friesdorf et al., 1997) seems to be relative to the work field in focus. Solutions that seem futuristic at this point, that might be relevant for a safe and less noisy alarm, like for the broad application of portable devices were also considered to be realizable only in the far future (mid-time realization between 2010 and 2015 with an IDR of 2.9). If these prognoses are fulfilled, integrated alarm systems may not be realizable yet, even if they allow, besides the connection to all given patient-relevant data, a well-aimed alarm that is related to the respective tasks by a physician (i.e., via portable devices).

CONCLUSION

The ergonomic dilemma of creating a stress-free and quiet work environment on the one hand, and of exploiting the great potential of acoustic alarms in a visually demanding field of work to achieve a safe recognition and evaluation of critical events on the other hand, should be resolved in the next decade. It can only be resolved by a combination of a macro-ergonomic and a micro-ergonomic perspective. Alarms have to be understood within the system. The relation of the alarms with other noises and the work tasks people have to cope with must be considered in particular. Therefore, the various, mostly parallel approaches for designing electrically produced acoustic alarms should be combined in an interdisciplinary approach. Physicians, computer scientists, engineers, psychologists, and other human factors specialists should develop new systems of alarms that fit with the requirements of the work system. Alarms should be exposed and should be understood in their own function according to an arythmia alarm or pulse oximeter. The possibility to reduce the sound pressure of acoustic alarms of different devices offered by manufacturers can only help to reduce the noise, but does not solve the problem of detecting critical situations very early and safe. Alarms cannot be the

Thesis	My own expertise on this thesis				Thesis is important for		Time of probable realization						This requires ...			
	high	me-di-um	low	none	the design of new safety concepts Yes	No	never	today until 2003	until 2005	until 2010	until 2015	> 2015	techno-logical progress	new organi-zational concepts	new qualifi-cation concepts	other progress
Statement (Please tick in this row!)	**Only one cross, please!**			X		X	**Only one cross, please!**					X	**You can chose more than one alternatives!**		X	X
Results of the second Delphi round	51%	31%	15%	0%	86%	10%	10%	25%		Md / 75%			31%	76%	55%	6%

Figure 22–2. Structure of the Delphi Survey, Third-Voting Round

solution for safety problems: they are more the result of an unpredictable and very complex work system. Monitoring of the patient in a more treatment-oriented manner can help to reduce acoustic alarms automatically and give the personnel a better sense of situation awareness (Friesdorf, Konichezky, Groß-Alltag, Fattroth, and Schwilk 1994; Letho & Salvendy, 1995).

References

Arbeitsgruppe Medizintechnik (1986). Compound organism subsystems. *Ausarbeitung der Ergebnisse einer Arbeitstagung vom Zentrum für Anästhesiologie der Universität Ulm*, 02.Maibis 04.Mai in Haldensee/Tirol.

Allaouchiche, B., Duflo, F., Debon, R., Bergeret, A., & Chassard, D. (2002). Noise in the postanaesthesia care unit. *British Journal of Anaesthesia, 88*(3), 369–373.

Asbury A. J., & Rolly, G. (1999). Theatre monitor alarm settings: a pilot survey in Scotland and Belgium. *Anaesthesia, 54*(2), 176–180.

Balogh, D., Kittinger, E., Benzer, A., & Hackl, J.M. (1993). Noise in the ICU. *Intensive Care Meicine., 19*(6), 343–346.

Bates, D. W., & Gawande, A. A. (2003). Improving safety with information technology. *New England Journal of Medicine, 348*, 2526–2534.

Bates, D. W., Evans, R. S., Murff, H., Stetson, P. D., Pizziferri, L., & Hripcsak, G. (2003). Detecting adverse events using information technology. *Journal of the American Medical Informatics Association, 10*(2), 115–128.

Becker, K., Thull, B., Kasmacher-Leidinger, H., Stemmer, J., Rau, G., Kalff, G., & Zimmermann, H. J. (1997). Design and validation of an intelligent patient monitoring and alarm system based on fuzzy logic process model. *Artificial Intelligence in Medicine, 11*(1), 33–35.

Belojevic, G., Jakovljevic B., & Slepcevic, V. (2003). Noise and mental performance: personality attributes and noise sensitivity. *Noise Health, 6*(21), 67–89.

Bender, A. D., Strack, A. E., Ebright, G. W., & von Haunalter, G. (1969). Delphi study examines developments in medicine. *Future, 1*, 289–303.

Berg, P. W., Stuttmann, R., & Doehn, M. (1997). Lärm auf Intensivstationen; Ein Beitrag zur Lärmreduktion durch Modifikation der Atemgasbefeuchtung. *Der Anaesthetist, 46*(10), 856–859.

Biot, L., Carry, P. Y., Perdrix, J. P., Eberhard, A., & Baconnier, P. (2000). Évaluation clinique de la pertinence des alarmes en reanimation. *Annales Francaises de Anesthesie et de Réanimation, 19*, 459–466.

Biot, L., Holzapfel, L., Becq, G., Melot, C., Bouletreau, P., & Baconnier, P. (2001). Guidelines for selective activation of alarms for blood pressure monitoring: effect on noise pollution in ICU. *Annales Francaises de Anesthesie et de Réanimation*, 20(8), 677–685.

Bliss, J. P., & Dunn, M. C. (2000). Behavioral implications of alarm mistrust as a function of task workload. *Ergonomics, 43*(9), 1283–1300.

Bliss, J. P., Gilson, R. D., & Deaton, J. E. (1995). Human probability matching behaviour in response to alarms of varying reliability. *Ergonomics, 38*(11), 2300–2312.

Block, F. E., Jr. (1992). Evaluation of users' abilities to recognize musical alarm tones. *Journal of Clinical Monitoring, 8*, 285–290.

Block, F. E., Jr. (1996). Non-localizable alarm. *International Journal of Clinical Monitoring and Computing, 13,* 57–58.

Block, F. E., Jr., Nuutinen, L., & Ballast, B. (1999). Optimization of alarms: a study on alarm limits, alarm sounds, and false alarms, intended to reduce annoyance. *Journal of Clinical Monitoring and Computing, 15*(2), 75–83.

Block, F. E., Jr. & Schaaf, C. (1996). Auditory alarms during anesthesia monitoring with an integrated monitoring system. *International Journal of Clinical Monitoring and Computing, 13,* 81–84.

Bremmer, P., Byers, J. F., & Kiehl, E. J (2003). Noise and the premature infant: physiological effects and practice implications. *Journal of Obstetric, Gynecologic, and Neonatal Nursing, 32*(4), 447–454.

Breslow, M. J., Rosenfeld, B. A., Doerfler, M., Burke, G., Yates, G., Stone, D. J., Tomaszewicz, P., Hochman, R., & Plocher, D. W. (2004). Effect of a multiple-site intensive care unit telemedicine program on clinical and economic outcomes: an alternative paradigm for intensivist staffing. *Critical Care Medicine, 32*(1), 31–38.

Breznitz, S. (1983). *Cry-wolf: the psychology of false alarms* (pp. 1–100). Hillsdale: Lawrence Erlbaum Associates.

Broscheit, J., Lorenz, K. W., Baumeister, J., Tannert, A., & Greim, C. A. (2002). Determinantes of alarm-rate and the potential of intelligent monitoring systems in routine anaesthesiological use. *Journal of Clinical Monitoring and Computing, 17*, (7–8), 476.

Buß, B., & Friesdorf, W. (2003). Networked thinking and analysis of the future context of use-tools for the design of context determined auditory alarm signals (CDAAS). Paper presented at the 14th annual meeting of the ESCTAIC, Berlin, Oct. 1–4, 2003.

Buß, B., & Friesdorf, W. (2000). Situation awareness—the precondition for decision making in high dependency environments. In G. Trillo (Ed.), *ESCTAIC—11th annual meeting book of abstracts* (pp. 29–30), Trieste: European Society for Computing and Technology in Anaesthesia and Intensive Care.

Buß, B., Neth, K.-U., & Friesdorf, W. (1998). Auditory alarm signals in high dependency environments. In *ESCTAIC—9th annual meeting book of Abstracts* (p .9). Austria: European Society for Computing and Technology in Anaesthesia and Intensive Care.

Carayon, P. (2003). Macroergonomics in quality of care and patient safety. In H. Luczak & K. J. Zink (Eds.), *Human factors in organizational design and management*—VII (pp. 21–34). Santa Monica, CA: IEA Press.

Carayon, P., & Friesdorf, W. (2006). Human factors and ergonomics in medicine. In G. Salvendy (Ed.), *Handbook of human factors and ergonomics*, 3rd ed. (pp. 1517–1537). New York, NY: John Wiley.

Chambrin, M. C., Ravaux, P., Calvelo-Aros, D., Jaborska, A., Chopin, C., & Boniface. (1999). Multicentric study of monitoring alarms in the adult intensive care unit (ICU): a descriptive analysis. *Intensive Care Medicine 25*(12), 1360–1366.

Cmiel, C., Karr, D., Gasser, D., Oliphant, L., & Neveau, A. (2004). Noise control: a nursing team's approach to sleep promotion. *American Journal of Nursing, 104* (2), 40–48.

Compes, P. (1970). *Sicherheitstechnisches Gestalten—Gedanken zur Methodik und Systematik im projektiven und konstruktiven Maschinenschutz.* Habilitationsschrift, RWTH Aachen.

Crocker, M. J. (1997). Noise. In G. Salvendy (Ed.), *Handbook of human factors and ergonomics,* 2nd ed. (pp. 790–827). New York: John Wiley.

Cust, A. E., Donovan, T. J., & Colditz, P. B.(1999). Alarm settings for the Marquette 8000 pulse oximeter to prevent hyperoxic and hypoxic episodes. *Journal of Paediatrics and Child Health, 35*(2), 159–162.

Dracup, K. (1988). Are critical care units hazardous to health? *Applied Nursing Research, 1,* 14–21.

Easterby, R. (1984). Tasks, processes and display design. In R. Easterby & H. Zwaga (Eds.), *Information design* (pp. 19–36). Chichester: John Wiley.

Edworthy, J. (1994). Urgency mapping in auditory warning signals. In N. Stanton (Ed.), *Human factors in alarm design* (pp. 15–44). London: Taylor & Francis.

Edworthy, J., Hellier, E., & Hards, R. (1995). The semantic associations of acoustic parameters commonly used in the design of auditory information and warning signals. *Ergonomics, 38*(11), 2341–2361.

Edworthy, J., Loxley, S., & Dennis, I. (1991). Improving auditory warning design: relationship between warning sound parameters and perceived urgency. *Human Factors, 33,* 205–232.

Endsley, M. (2000). Theoretical underpinnings of situation awareness: a critical review. In M. R. Endsley & D. Garland (Hrsg.), *Situation awareness analysis and measurement* (pp. 3–32). London: Lawrence Erlbaum Associates.

Flynn, E. A., Barker, K. N., Gibson, J. T., Pearson, R. E., Smith, L. A., & Berger, B. A. (1996). Relationships between ambient sounds and the accuracy of pharmacists' prescription-filling performance. *Human Factors, 38*(4), 614–622.

Fried, R., Gather, U., & Imhoff, M. (2001). Online pattern recognition in intensive care medicine. In S. Bakken (Ed.) *Visions of the Future and Lessons From the Past. Proceedings of the 2001 American Informatics Association (AMIA) Annual Symposium* (pp. 184–188). Philadelphia, PA: Hanley & Belfus.

Friesdorf, W., & Göbel, M. (2003). Safety and reliability of clinical work processes. In H. Strasser, K. Kluth, H. Rausch, & H. Bubb (Hrsg.), Quality of work and products in enterprises of the future. *Proceedings of the annual spring conference of the GfA on the occasion of the 50th anniversary of the foundation of Gesellschaft für Arbeitswissenschaft* (pp. 669–672). Stuttgart: Ergonomie Verlag.

Friesdorf, W., Buß, B., & Göbel, M. (1999). Monitoring alarms—the key to patients safety in the ICU? *Intensive Care Medicine, 25,* 1350–1352.

Friesdorf, W., Classen, B., Konichezky, S., & Schwilk, B. (1997). Events which will influence intensive care units in future. *Technology and Health Care, 5,* 319–330.

Friesdorf, W., Konichezky, S., Groß-Alltag, F., Fattroth, A. & Schwilk, B. (1994). Data quality of bedside monitoring in an intensive care unit. *International Journal of Clinical Monitoring and Computing, 11,* 123–128.

Furnham, A., & Strbac, L. (2002). Music is as distracting as noise: the differential distraction of background music and noise on the cognitive test performance of introverts and extraverts. *Ergonomics, 45*(3), 203–217.

Gaver, W. W. (1993). What in the world do we hear? an ecological approach to auditory source perception. *Ecological Psychology, 5,* 1–25.

Gawron, V. J. (1982). Performance effects of noise intensity, psychological set, and task type and complexity. *Human Factors, 24,* 225–243.

Haas, E. C., & Casali, J. G. (1995). Percieved urgency of and response time to multi-tone and frequency-modulated warning signals in broadband noise. *Ergonomics, 38*(11), 2313–2326.

Hatfield, J., Job, R. F., Hede A. J., Carter, N. L., Peploe, P., Taylor, R., & Morell, S. (2002). Human response to environmental noise: the role of perceived control. *International Journal of Behavioral Medicine,* 9(4), 341–359.

Hedley-Whyte, J., & Sheridan, D. S. (2003). Helping to ameliorate a health care crisis- General requirements and guidelines for the application of alarms in medicine. *ISO Bulletin,* 23–25.

Hellier, E., & Edworthy, J. (1999). On using psychophysical techniques to achieve urgency mapping in auditory warnings. *Applied Ergonomics, 2,* 167–171.

Hendrick, H. W. (1997). Organizational design and macroergonomics. In G. Salvendy (Ed.), *Handbook of human factors and ergonomics* (pp. 594–636). New York: John Wiley.

Hendrick, H. W., & Kleiner, B. M. (2001). Macroergonomics—an introduction to work system design. *HFES issues in human factors and ergonomics book series,* vol. 2. Santa Monica, CA: Human Factors and Ergonomics Society.

Hilton, B. A. (1985). Noise in acute patient care areas. *Research in Nursing and Health, 8*(3), 283–291.

Ireland, R. H., James, H. V., Howes, M., & Wilson, A. J. (1997). Design of a summary screen for an ICU patient data management system. *Medical and Biological Engineering and Computing, 35,* 397–403.

Institute of Medicine (IOM), (U.S.) Board on Health Care Services Committee, Committee on Work Environment for Nurses. (2004). *Keeping patients safe: transforming the work environment of nurses.* Washington, DC: National Academy Press.

Johnson, A. N.(2004). Adapting the neonatal intensive care environment to decrease noise. *Annals of Emergency Medicine, 43*(1), 71–76.

Jones, D. M., & Broadbent, D. E. (1991). Human performance and noise. In C. M. Harris (Ed), *Handbook of acoustical measurements and noise control,* 3rd ed.. (pp. 24.1–24.24). New York: McGraw-Hill.

Kahn, D. M., Cook, T. E., Carlisle, C. C., Nelson, D. L., Kramer, N. R., & Millman, R. P. (1998). Identification and modification of environmental noise in an ICU setting. *CHEST, 114,* 535–540.

Kerr, J. H. (1985). Warning devices. *British Journal of Anesthesia, 57,* 696–708.

Kerstholt, J. H., & Passenier, P. O. (2000). Fault management in supervisory control: the effect of false alarms and support. *Ergonomics, 43*(9), 1371–1389.

Krachman, S. L., D'Alonzo, G. E., & Criner, G. J. (1995). Sleep in the intensive care unit. *CHEST, 107,* 1713–1720.

Krol, M., & Reich, D. L. (2000). Development of a decision support system to assist anaesthesiologists in operating room. *Journal of Medical Systems, 24*(3), 141–146.

Kroon, K., & West, S. (2000). 'Appears to have slept well': assessing sleep in an acute care setting. *Contemporary Nurse, 9*(3–4), 284–294.

Letho, M., & Salvendy, G. (1995). Warnings: a supplement not a substitute for other approaches to safety. *Ergonomics, 38*(11), 2155–2163.

Loeb, M. (1986). *Noise and human efficiency.* Chichester: John Wiley.

McCarthy, D. O., Ouimet, M. E., & Daun, J. M. (1991). Shades of Florence Nightingale: potential impact of noise stress on wound healing. *Holist Nursing Practice, 5*, 39–48.

McKinnon, S. (1983). Maximizing your ICU patient's sensory and perceptual environment. *Canadian Nurse, 79*(5), 41–45.

Meredith, Ch., & Edworthy, J. (1994). Sources of confusion in intensive therapy unit alarms. In N. Stanton, (Ed.), *Human factors in alarm design* (pp. 207–219). London: Taylor & Francis.

Moore, M., Nguyen, D., Nolan, S., Robinson, S., Ryals, B., Imbrie, J., & Spotnitz, W. (1998). Interventions to reduce decibel levels on patient care units. *American Surgeon, 64*(9), 894–899.

Momtahan, K. L., McIntyre, J. W. R, & Hogan, J. T. (1993). Audibility and identification of auditory alarms in the operating room and intensive care unit. *Ergonomics, 36*, 1159–1176.

Mondor, T. A., & Finley, G. A. (2003). The perceived urgency of auditory warning alarms used in the hospital operating room is inappropriate. *Canadian Journal of Anaesthesia 50*(3), 209–214.

Morrison, W. E., Haas, E. C., Shaffner, D. H., Garrett, E. S., & Fackler, J. C. (2003). Noise, stress, and annoyance in a pediatric intensive care unit. *Critical Care Medicine, 31*(1), 113–119.

Murthy, V. S., Malhotra, S. K., Bala, I., & Raghunathan, M. (1995). Detrimental effects of noise on anaesthetists. *Canadian Journal of Anaesthesia, 42*, 608–611.

Noyes, J. M., Starr, A. F., Frankish, C. R., & Rankin, J. A. (1995). Aircraft warning systems: application of model-based reasoning techniques. *Ergonomics, 38*(11), 2432–2445.

Orasanu, J., & Salas, E. (1991). Team decision making in complex environments. In G. Klein, J. Orasanu, R. Calderwood, & C. Zsambock (Eds.), *Decision making in action* (pp. 327–345). Norwood, NJ: Ablex Publishing.

Orr, J. A., & Westenskow, D. R. (1994). A breathing circuit alarm system based on neural networks. *Journal of Clinical Monitoring, 10*, 101–109.

Parsons, K. C. (2000). Environmental ergonomics: a review of principles, methods and models. *Applied Ergonomics, 31*, 581–594.

Patterson, R. D. (1982). Guidelines for auditory warning systems on civil aircraft. *Civil Aviation Authority Paper 82017.* Civil Aviation Authority Airworthiness Division, Cheltenham, England.

Perlman, J. M. (2003). The genesis of cognitive and behavioural deficits in premature graduates of intensive care. *Minerva Pediatrica, 55*(2), 89–101.

Probst, G. J. B., & Gomez, P. (1991). Die Methodik des vernetzten Denkens zur Lösung komplexer Probleme, In G. J. B. Probst & P. Gomez (Hrsg.), *Vernetztes Denken, Ganzheitliches Führen in der Praxis* (S. 4–20). Auflage, Wiesbaden: Gabler.

Rheineck-Leyssius, A. T., & Kalkman, C. J. (1997). Influence of pulse oximeter lower alarm limit on the incidence of hypoxeamia in the recovery room. *British Journal of Anaesthesia, 79*, 460–464.

Rheineck-Leyssius, A. T., & Kalkman, C. J. (1998). Influence of pulse oximeter settings on the frequency of alarms and detection of hypoxemia: theoretical effects of artifact rejection, alarm delay, averaging, median filtering or a lower setting of the alarm limit. *Journal of Clinical Monitoring and Computing, 14*(3), 151–156.

Seagull, F. J., & Sanderson, P. M. (2001). Anaesthesia alarms in context: an observational study. *Human Factors, 43*(1), 66–78.

Seagull, F. J., Xiao, Y., Mackenzie, C. F., & Wickens, C. D. (2000). Auditory alarms: from alerting to informing. *Proceedings of the IEA 2000/HFES 2000 Congress*, 1223–1226.

Shannon, C. E., & Weaver, W. (1949). *Mathematical theory of communication.* University of Illinois Press.

Shertzer, K. E., & Keck, J. F. (2001). Music and the PACU environment. *Journal of Perianesthia Nursing 16*(2), 90–102.

Solsona, J. F., Altana, C., Maull, E., Rodriguez, L., Bosque, C., & Mulero, A. (2001). Are auditory warnings in the intensive care unit properly adjusted? *Journal of Advanced Nursing, 35*(3), 402–406.

Smith, A. F., Mort, M., Goodwin, D., & Pope, C. (2003). Making monitoring "work": human-machine interaction and patient safety in anaesthesia. *Anaesthesia, 58*, 1070–1078.

Stanton, N. (1994). A human factors approach. In N. Stanton (Ed.), *Human factors in alarm design* (pp. 1–14). London: Taylor & Francis.

Steimann, F., & Adlassnig, K. P. (1994). Clinical monitoring with fuzzy automata. *Fuzzy Sets and Systems, 61*, 37–42.

Stevens, S. S. (1957). On the psychophysical law. *Psychological Review, 64*, 153–181.

Topf, M. (1992). Effects of personal control over hospital noise on sleep. *Research in Nursing & Health, 15*(1), 19–28.

Topf, M. (2000). Hospital noise pollution: an environmental stress model to guide research and clinical interventions. *Journal of Advanced Nursing, 31*(3), 520–528.

Topf, M. (1983). Noise pollution in the hospital. *New England Journal of Medicine, 309*, 53–54.

Topf, M. (1989). Sensitivity to noise, personality hardiness, and noise-induced stress in critical care nurses. *Environment and Behaviour, 21*, 717–733.

Topf, M., & Dillon, E. (1988). Noise-induced stress as a predictor of burnout in critical care nurses. *Heart & Lung, 17*, 567–574.

Tsien, C. L., & Fackler, J. C. (1997). Poor prognosis for existing monitors in the intensive care unit. *Critical Care Medicine, 25*, 614–619.

Tsien, C. L., Kohane, I. S., & McIntosh, N. (2000). Multiple signal integration by decision tree induction to detect artefacts in the neonatal intensive care unit. *Artificial Intelligence in Medicine, 19*(3), 189–202.

U.S. Environmental Protection Agency. (1974). *Information on levels of environmental noise requisite to protect health and welfare with an adequate margin of safety.* Washington, DC: EPA.

Verein Deutscher Ingenieure. (1999). Assessment of noise in the working area with regard to specific operations, *VDI 2058 Blatt 3/Part 3 Ausg. deutsch/engl.*

Walder, B., Franciolo, D., Meyer, J. J., Lancon, M., & Romand, J. A. (2000). Effects of guidelines implementation in a surgical intensive care unit to control night-time light and noise levels. *Critical Care Medicine, 28*(7), 2242–2247.

Watson, M., Sanderson, P., & Russell, W. J. (2004). Tailoring reveals information requirements: the case of anaesthesia alarms. *Interacting with Computers, 16*(2), 271–293.

Whalen, L. (1992). Noise in the intensive care setting. *Canadian Critical Care Nursing Journal, 89*(41), 9–10.

Wickens, C. D. (1992). *Engineering psychology and human performance,* 2nd ed. New York: Harper Collins.

Woods, D. D. (1995). The alarm problem and directed attention in dynamic fault management. *Ergonomics, 38*(11), 2371–2393.

Woods, N. F., & Falk, S. A. (1974). Noise stimuli in the acute care area. *Nursing Research, 23*, 144–150.

Xiao, Y., & Seagull, F. J. (1999). An analysis of problems with auditory alarms: defining the roles of alarms in process monitoring tasks. *Proceedings of the Human Factors and*

Ergonomics Society 43rd annual meeting, Santa Monica, CA: Human Factors and Ergonomics Society, 256–260.

Xiao, Y., Mackenzie, C. F., Seagull, J., & Jaberi, M. (2000). Managing the monitors: an analysis of alarm silencing activities during an anaesthetic procedure. *Proceedings of the Joint Meeting of the Human Factors and Ergonomics Society and the International Ergonomics Association (IEA 2000/ HFES 2000).* Santa Monica, CA: Human Factors and Ergonomics Society, 4, 250–253.

Xiao, Y., Mackenzie, C. F., Jaberi, M., Harper, B., & LOTAS Group. (1990). Alarms: silenced, ignored, and missed. *Anaesthesiology*, *73*, 995–1021.

Xioa, Y., Seagull, F. J., Mackenzie, C., Wickens, Ch., & Via, D. (2001, August). Auditory warning signals in critical care settings. *Final report prepared for National Patient Safety Foundation.*

Yuki, Y. (2002). Alarm system optimization for increasing operations productivity. *ISA Transactions, 41*(3), 383–387.

Section

TECHNOLOGY

·23·

HUMAN FACTORS ENGINEERING AND THE DESIGN OF MEDICAL DEVICES

James Ward and John Clarkson
University of Cambridge

Human Factors Engineering (HFE) involves the application of knowledge about human capabilities to the design of products. Through good design practice, HFE aims to assist users to accomplish their tasks efficiently, effectively, and safely as they interact with the product. The design of medical devices requires particular attention to good HFE practice, due to the complexity and variability of their environments of use; the need to gain regulatory approval on HFE practice before the device can be used; and the high-risk role that devices play in diagnosing and treating patients who may be suffering from severe medical conditions. Indeed, it is likely that failures to employ good HFE practice result in a significant number of cases of patient harm when using medical devices each year.

This chapter begins by highlighting various sources of difficulty that are faced when designing medical devices. Demonstrating the degree of these challenges, the scale, nature, and causes of "device-related" medical errors are then described. Subsequently, the chapter identifies the various regulations that relate to the use of HFE in the medical device design process and discusses the degree to which such regulations are currently followed by device design companies.

The second half of the chapter describes a "good practice" approach to HFE design, which follows a systems-approach based on a study for the United Kingdom's Department of Health (Engineering Design Centre [EDC], 2004). The chapter concludes by outlining challenges and opportunities for the use of good HFE design practice in the future.

BACKGROUND

> Simply defined, medical devices are health care products, excluding drugs, which are used for human beings for the purpose of prevention, diagnosis, monitoring, treatment or alleviation of an illness or injury. A formal definition of a medical device can be found through the web page of the Medicines and Health Care Products Regulatory Agency (MHRA) (formerly known as the Medical Devices Agency [MDA] in the United Kingdom (UK:http://www.mhra.gov.uk.

They are ubiquitous in the community (including use by general practitioners in surgery and in the home, in nursing homes, and in hospitals) and are used by an estimated 38 million people each day in the UK alone (MDA, 2001b). They employ a great diversity of technologies (Geddes, 1998) and range from the simple (e.g., tongue depressors and scalpels) to the complex (e.g., MRI scanners and patient monitors). In addition, medical devices share a combination of other characteristics that put particular pressure on the design process:

Complexity and variability of health care. Health care is perhaps the most complex of all industries (Leape, 2004), given the products used, the needs of those who use these products, the environments of use, and the tasks that may be performed. The complexity of this picture has continued to increase over the years (Geddes, 1998; Giuntini, 2000; Kaye, 2000; Meldrum, 1998; Swain, 2004; Wiklund, 2004), creating further challenges for safe and efficient operation. The contexts in which such devices are used are also growing in complexity (e.g., the range of health care treatments available and the number that are carried out), as is the information associated with such health care. Other sources of complexity include the social and cultural characteristics of health care (e.g., there can often be difficulties in lines of communication between health care professionals in different job roles, such as nurses and consultants), which can also have a strong effect on patient safety. Furthermore, great variability occurs in the health care setting, in terms of intrapersonal variation (e.g., the level of attention a device operator is able to provide toward device use may vary considerably during the course of their shift of work) and interpersonal variability (e.g., patients with seemingly similar medical conditions may react in a diverse way to the same type of treatment), and in terms of practices and procedures, hospitals and wards and, of course, devices.

Regulations. The design and manufacture of these products is a regulated process, which is governed in the UK by the MHRA and in the United States by the Center for Devices and Radiological Health (CDRH), a division of the Food and Drug Administration (FDA). In Europe and the United States, medical devices are grouped into different classes, depending on the risk that they present when put into service (Global Harmonization Task Force [GHTF], 1999). The stringency of the regulations increases correspondingly with the device class.

High risk. The risks of using the numerous different types of medical device can vary considerably from one device to another. However, many medical devices can be regarded as high-risk products (e.g., heart valves, balloon catheters, and infusion pumps), as illustrated by the number of device-related injuries and deaths that occur each year. Furthermore, high workload, inexperience, under-staffing and haste (e.g., in emergency situations) all contribute to the risk of medical device use. Work-related distractions, for example, have been found in a U.S.-based study of over 100,000 records of medication errors to be the most frequently reported contributing factor, occurring in nearly half of all cases (United States Pharmacopeial Convention [USPC], 2002).

Other characteristics. While medical devices may have been designed for use by health care professionals, more and more devices are being used in the home setting (MDA, 2000, 2001a), creating the potential for additional problems because devices may be operated by users whose ability to use them may be compromised by their medical condition or inexperience (National Reporting Council [NRC], 1996; Swain, 2004).

Device-Related Errors and HFE

Considering the challenges that device designers face, device-related problems are, indeed, not uncommon, and they often result in medical errors of various types. In some cases it is likely that poor HFE practice leads to these errors.

Medical errors occur in a medical setting and are caused by departures from correct practice, which result in near-misses or actual harm to patients or other individuals. In the last few years, much attention has been paid to this subject in the medical and academic literature and in the press. Research suggests that around 10% of patients admitted to UK hospitals are harmed through errors that occur during medical practice, equating to some 850, 000 episodes of harm each year, and an annual cost to the National Health Service (NHS) of $3.7bn, not to mention the wider human and economic costs (The Stationery Office [TSO], 2000). In the United States it has been calculated that medical errors may result in as many as 100,000 deaths per year (Kohn, Corrigan, & Donaldson, 2000).

Medical device-related errors (a sub-category of medical errors, occurring when a medical device is involved) have received far less published attention. Such errors tend to be caused by the device manufacturer or by the behavior of the user, as shown in Figure 23–1.

Manufacturer-related errors may occur as a result of inappropriate manufacture or inappropriate device design and, hence, may in some cases be the product of poor HFE practice. User errors are reported to be one of the major causes of errors involving medical devices. However, confusion exists over their definition, because different reports use this term in a variety of ways, yet often do not explain what they mean. Due caution is therefore required when interpreting the results of such reports because it is often impossible to determine whether these "user errors" are genuinely the result of errors made

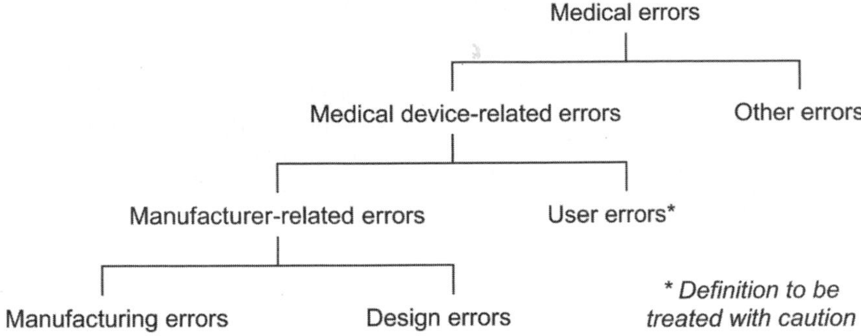

Figure 23–1. Classifying Causes of Device-Related Medical Errors

exclusively by users (i.e., due to negligence or sabotage) or whether, in fact, other factors (such as poor design or user fatigue due to under-staffing) have played a significant role in contributing to the error. Determining root causes can indeed be difficult, and it is perhaps due to this reason that the CDRH refers to errors that occur when using medical devices as use errors. This is a more general term than user errors and allows other causal factors (e.g., poor device design) to be taken into consideration. The International Electrotechnical Commission has now also adopted this term in its standard, IEC 60601-1-6 (International Electrotechnical Commission [IEC], 2004).

Behind the causes depicted in Figure 23–1 lie other factors that can contribute to medical errors, such as the cultural and social practices that take place in health care systems. It should also be noted that medical errors rarely result from a single cause but are often the product of a complex web of factors. Figure 23–1 should therefore be seen as an elementary illustration of the causes of device-related errors.

This chapter is concerned in particular with device design errors. These may manifest themselves in the form of device failures due, for example, to unreliable designs. However, of particular relevance are HFE-related design errors, such as counter-intuitive or over-complex operating sequences, which may lead to use errors.

A very limited literature exists that describes the scale, nature, and causes of such device-related errors, as described below.

Figures on Device-Related Errors

In the mid-1980s the FDA conducted an investigation into the causes of 1,664 quality problems (manufacturer-related errors) that led to recalls of medical devices in the United States. The resulting report (Food and Drug Administration [FDA], 1990) found that one-third were due to design-related errors, including poor device design, poor software design, and the selection of unsuitable components or materials. The FDA concludes that in 1991 as many as 59 fatalities and 929 serious injuries in the United States were due to device design errors, even without taking into account design errors in lower risk devices (FDA, 1996a).

The MHRA annually publishes a document that provides statistics on the number and causes of errors in the UK over the previous year. In 2003 nearly 9,000 reports were filed regarding device-related problems (Medicines and Health Care Products Regulatory Agency [MHRA], 2004). These resulted in 420 device recalls and field corrections, suggesting a highly significant level of device-related problems. Perhaps even more telling are the figures for manufacturers' promises to undertake design, manufacture, or quality system improvements: in 2003, nearly 1,000 such promises were made. The MHRA has stated that "many adverse incidents are the result of user error" (MDA, 2001a) and defines user errors as those where the device was not at fault. However, in contrast to the strict definition in this chapter, the MHRA includes the possibility that user errors may occur as a result of other underlying causes, such as poor maintenance or training. Between 2001 and 2003, 15–20% of investigations concluded that a user error had occurred (MHRA, 2004).

Evidence from the MHRA's counterpart in the United States, the CDRH, suggests that use error resulting from poor design is by no means uncommon:

> Use error caused by designs that are either overly complex or contrary to users' intuitive expectations for operation is one of the most persistent and critical problems encountered by FDA. (FDA, 2001b, page 18)

The FDA has reported that more than one-third of the incident reports each year involve use error (FDA, 2000a). Furthermore, senior individuals from the FDA have even suggested that human factors issues are "inherent in virtually every [device-related error]" (FDA, 2001a, page 9). Other, albeit considerably older, studies, specifically in the area of anesthesia equipment, suggest that use errors occur in approximately 70–90% of anesthesia equipment-related errors (Cooper, Newbower, & Kitz, 1984; Cooper, Newbower, Long, & McPeek, 1978; Dripps, Lamont, & Eckenhoff, 1961).

Although the proportion of use errors may be similar, the absolute number of errors may be far higher than the statistics above show because under-reporting is rife, with estimates of the scale of this problem range from 80% of true value (FDA, 1996a) to as little as 10% (Senders, 1994).

Given the different contexts of treatment (e.g., anesthesia or general medicine), the lack of clarity in definitions of "user" error, the age of some of the reports, the different locations of reporting (e.g., UK vs. United States), and the scale of under-reporting, it is difficult to identify a precise figure as to what extent poor HFE practice currently leads to medical errors involving medical devices. However, it is indeed likely that poor design—and poor use of HFE practice—has led to a significant number of medical errors, resulting in tragic consequences for a great many people. Applying good HFE in medical device design is therefore a critical activity.

Device Design Regulations and HFE Guidance

Regulations for medical device design are extensive. Therefore, the following sections outline only key areas that relate to the use of HFE during device design. Many of the documents described can be downloaded for free from the Web, and links are provided in the references section at the end of this chapter.

Design Controls

Design controls are a set of practices and procedures that should be applied during the design and development process, and they are detailed in a document called the "Quality System Regulation (QSR)" in the United States (FDA, 1996a), and "ISO standards" in the UK and European Union (British Standards Institution [BSI], 2000a, 2001). The aim of design controls is to assist designers to correct problems before they are embedded into the device design, to improve the chance of running an efficient design process and to design a device that is fit for purpose. Under design control, attention is paid to various design-related elements, including planning, identifying requirements, control of changes, the compilation of documentation, verification, and validation. Of the 59 fatalities and 929 serious injuries in the FDA's report (FDA, 1996a), it was estimated that three quarters could be preventable if design controls were introduced, with a further economic saving of around $370million due to anticipated improvements in patient safety (FDA, 1996a).

Although not all devices require the use of design controls, the spirit behind them is relevant to all device design projects. For example, HF-related data should be collected, no-matter the inherent risk of the device. HFE is not stated explicitly in the QSR, but it is nevertheless an expectation: "design requirements [must be] appropriate and address the intended use of the device, including the needs of the user and patient" and manufacturers must "ensure that devices conform to defined user needs and intended uses and shall include testing of production units under actual or simulated use conditions" (FDA, 1996a, page 52657). If user needs are to be met, HFE principles will need to be applied during design.

Another element of design control is design validation, which "means establishing by objective evidence that device specifications conform with user needs and intended use(s)" (FDA, 1996a, page 52656). Clearly, HFE principles will need to be followed to some extent if a device is to be suitably validated. Indeed, the CDRH states that HFE is a "critical component" of device design (FDA, 2001c).

The FDA has also produced a document entitled *Design control guidance for medical device manufacturers* (FDA, 1997), which is intended as an explanatory supplement to the QSR. This describes in more detail how design controls may be applied and highlights connections between elements of design control and the practice of HFE. For example, validation of the device "may have significant human factors implications" (FDA, 1997, page 34). Although the link between design control and HFE is expressed only rarely in this document, it does serve as an excellent general guide on the use of design controls and is consequently a most helpful guide for general device design.

Risk Analysis

Risk management is critical for effective device design. However, it was only in 2000 that a comprehensive standard was published; namely *BS EN ISO 14971:2000 Medical devices—application of risk management to medical devices* (BSI, 2000b). It includes the requirement to identify hazards associated with device use throughout its lifecycle, and provides lists of possible hazard sources and a variety of questions that designers may need to answer in order to facilitate designing their device in an appropriate manner. *Medical device use-safety: incorporating human factors engineering into risk management* also provides assistance on following a risk-based approach in terms of characterizing use-related problems, controlling these through good design, and demonstrating that these have been controlled (FDA, 2000c). For additional information, see the chapter on "Human factors risk management in medical products" in this handbook.

Standards

Standards have a particularly important role in defining expected design practice, and since the late 1990s, the collection of HFE-related standards for medical devices has been growing. Key standards are:

- ISO 13407:1999 Human-centered design processes for interactive systems (International Organization for Standardization [ISO], 1999) details the requirements for implementing HFE-related activities during the design of computer-based systems, although many of the principles in this document are applicable to a wider range of medical devices.
- ANSI/AAMI HE48-1993 Human factors engineering guidelines and preferred practices for the design of medical devices (American National Standards Institute/ Association for the Advancement of Medical Instrumentation [ANSI/AAMI], 1993) is an older standard, which incorporates ergonomic and device design data and also presents an overview of HFE-related design and evaluation activities.
- ANSI/AAMI HE74-2001 Human factors design process for medical devices (ANSI/AAMI, 2001) has been introduced much more recently than its sister standard and is, unusually, a process-based standard that focuses on the procedures and practices that should take place during the design process and when they might occur. It replaces parts of HE48, above, but does not fully supercede it.
- IEC 60601-1-6 Medical electrical equipment— part 1-6: general requirements for safety—collateral standard: usability (IEC, 2004) is very similar in content to ANSI/AAMI HE74, but also includes helpful discussion on other issues associated with HFE and the design of medical devices, such as what constitutes reasonably foreseeable misuse and other factors that should be considered during the design process.

Guidance Documents

Guidance helps to encourage the adoption of good design practice. While not mandatory, guidance often helps in the interpretation and implementation of standards. The CDRH operates a Web page on HFE (www.fda.gov/cdrh/humanfactors/ index.html), which is a gateway to various documents that provide guidance to designers on this subject. One such document is *Do it by design: an introduction to human factors in medical devices* (FDA, 1996b), which is an excellent starting point to the use of HFE in device design. Other documents include:

- *Make sure the medical device you choose is designed for you* (FDA, 2000b) is a checklist to help health care professionals to match the right medical device with user needs, and it also serves as a useful guide for designers.

- *Integrating human factors engineering into medical device design and development* (FDA, 2001c) is a powerful video-based guide that provides graphic insights into various environments of use and user-related problems. It also provides an understanding of the FDA's requirements from a regulatory perspective and can be purchased through the U.S. National Technical Information Service.
- *Safe, comfortable, attractive, and easy to use: improving the usability of home medical devices* (NRC, 1996) is perhaps the only comprehensive guide that provides insights into the needs of various stakeholders in the home-use medical device market.

Other device-specific guidance on HFE may, depending upon the device in question, be found through the CDRH's list of guidance documents, which can be searched by the name of the device (see www.fda.gov/cdrh/guidance.html).

Other Documents

The Medical Devices Directives apply to the design of devices for use in the European Union. Three directives have been published, and the relevant publication must be selected according to the type of device to be developed:

1. The active implantable medical devices directive, 90/385/EEC (1990).
2. The in-vitro diagnostic medical devices directive, 98/79/EEC (1998).
3. The medical devices directive, 93/42/EEC (1993).

The active implantable medical devices directive applies to devices that are introduced into the human body, have a source of power other than that supplied by the human body or gravity, and are intended to remain in the body after the procedure to introduce them. The in-vitro diagnostic medical devices directive applies to devices intended for in-vitro analysis of specimens derived from the human body, and the medical devices directive applies to the medical devices not covered by the other directives. Each of the directives is similar in scope, covering requirements for manufacturers of devices, such as the use of standards, company registration with the regulatory authorities, and the procedures for regulatory approval.

The directives also stipulate general design requirements for medical devices, but little mention is made of the use of HFE . For example, the Medical Devices Directive states simply that the benefits of the device must outweigh the risks and that it must provide an appropriate level of performance.

In addition to the directives, other policies and procedures, specific to a particular company, may also provide guidance on the use of HFE. These may vary greatly in scope and detail.

The Effect of Regulations on Good Device Design

It is difficult to assess the impact of the increasing availability of HFE guidance on medical device design. However, the following provides a brief insight into how well device designers currently use HFE and how effective this use is.

Matthew Weinger, a professor of anesthesiology in the University of California, San Diego, and Director of the San Diego Center for Patient Safety, stated in 2000: "It is troubling that many medical device user interfaces are poorly designed, fail to adequately support the clinical task for which they are intended, and frequently contribute to medical error … There are many examples—primarily in hospital-based acute-care medicine—of how medical device design flaws have contributed to user error, resulted in clinical inefficiency, or caused adverse patient events. In most circumstances, if manufacturers had employed a greater use of HFE, and followed more closely a user-centered approach to the design process, a superior device performance would have resulted" (Weinger, 2000). Peter Carstensen, Human Factors Team Leader at the CDRH approaches this area from a regulatory perspective and has said that "A lot of companies … still don't understand what's needed" (Medical Device & Diagnostic Industry [MDDI], 2001).

More recently, Michael Wiklund, another medical device related expert on HFE and author of *Medical device and equipment design: usability engineering and ergonomics* (Wiklund, 1995) suggests that the days when poor device design is blamed on the user are very much in decline. However, there is still scope for improvement: "Generally, manufacturers still do not

"Reports received throughout the Medical Device Reporting (MDR) system, recall data, and other postmarket information indicate that device design and related use errors are often implicated in adverse events" (FDA, 1996b).

"An emerging concern of great importance to the Agency is the implementation of good human factors practices in the design of medical devices" (FDA, 1996b).

"...better human factors design... It's the FDA's most critical safety problem" (Sawyer, D., CDRH Human Factors Team Member, personal co mmunication, April 30, 2001).

"Use error caused by designs that are either overly complex or contrary to users' intuitive expectations for operation is one of the most persistent and critical problems encountered by FDA" (FDA, 2001b).

"I think we [the FDA] still see plenty of evidence that companies aren't doing as good a job as they should... We also see key indications that there are a lot of companies out there that still don't understand what's needed" (Peter Carstensen, CDRH Human Factors Team Leader, in (MDDI, 2001)).

"Designers need a more complete and accurate understanding of device use... Relatively few user actions that can cause the device to fail other than the most apparent (e.g., fire or explosion), or well-known instances of use problems are considered by designers. This limitation during device design increases the likelihood of unexpected use scenarios... and use-related hazards for users and patients" (FDA, 2000c).

Figure 23–2. Uptake of HFE in the Medical Device Industry

regard human factors engineering as a fundamental need, equivalent to mechanical engineering or software programming," states Wiklund (2004). Commenting in a report by Swain, Wiklund adds: "[Not] enough device companies are doing a very good job in addressing human factors [design]" (Swain, 2004). Further quotations on the use of HFE in the design of medical devices are provided in Figure 23–2.

Speaking from a more general design-related perspective, the author of a public inquiry, which investigated a series of problems in a UK hospital, states that although there is a considerable pool of knowledge among the various medical agencies and professionals on the effect of design-led solutions on improving safety, it is "well recognised that the adoption of an approach to solving or addressing specific hazards by designing equipment differently is under-explored." (TSO, 2001, page 372). Concern has also been voiced over the introduction of new technologies into the health care system, without fully appreciating their effects. For example, in the community, mix-ups are not uncommon between medicines containers of similar appearance. Suggestions have been made in the past to improve differentiation between the containers by color-coding each and every medicine. However, this "solution" brings with it a number of dangers, including the danger of mix-ups between medicines due to differences in individuals' color perception, differences in appearance of color under different lighting conditions, and the limited range of available colors for print (Medicines Control Agency [MCA], 2001).

In 2003, research was conducted by the Universities of Cambridge and Surrey and the Royal College of Art in the UK, to investigate such systems-related problems. The study explored the use of design by the UK's National Health Service (NHS) to improve patient safety (EDC, 2004). Among other research methods, focus groups were conducted with staff from various settings in the NHS (midwifery, accident and emergency, cancer and palliative care). These revealed a number of problems related to the use of medical devices, including:

- problems with the large range of devices used on a single ward and the confusion that results;
- difficulties due to the lack of compatibility between different components from different devices (e.g., connectors on cardiology devices and giving sets on infusion devices);
- a lack of training of staff and under-staffing, meaning that safe and effective use can be seriously reduced;
- a lack of understanding of device user needs when users other than health care professionals operate devices;
- a lack of involvement of actual device users in purchasing decisions, meaning that their requirements were not necessarily considered; and
- purchasing decisions that were often made simply on lowest price, rather than fitness for purpose.

The report concluded that there was a particular need for improved device design, involving end-users in this process, and a reduction in the variety of similar equipment seen within departments and hospitals.

Summary

There remains an opportunity for considerable improvement in the uptake of HFE in the area of medical device design. While much of the above evidence is anecdotal, bringing together the information from many different sources suggests that a conclusion can be made. The following sections outline key principles related to HFE that should be followed when designing medical devices.

A FRAMEWORK FOR GOOD HFE DESIGN PRACTICE

As highlighted earlier in this chapter, much guidance exists on the use of HFE for medical device design. Such guidance generally states that HFE must be used throughout the design process. Early in the design process, for example, due consideration must be given to the needs of users—in the form of device design requirements—to drive the design of the device. In addition, from a regulatory and liability perspective, designers must be able to produce documented evidence of such practice. HFE may also be used iteratively throughout the design process, leading to repetition of HFE-related activities, such as task analysis and user testing.

It is not the purpose of this chapter to reproduce such guidance in detail, particularly since much of it is readily accessible on the Web. However, the following sections provide a framework for good design practice from an HFE perspective and are intended to complement the current HFE-specific guidance.

HFE—A Systems Engineering View

Design is the process by which something is created, whether it is a product, a protocol, or a service. It is helpful to consider what design is in the context of systems development because this will shed light on the role of design in improving patient safety. Systems engineering is defined by the International Council on Systems Engineering (INCOSE) as an approach that aims to enable the realization of successful systems (i.e., a collection of objects that may, for example, be the components in a device or the totality of a device, user, task, and environment and may include contextual factors such as social or cultural conditions). Starting with due consideration of customer needs, the wider system design problem is then considered throughout the design process.

Successful device designs require a systematic understanding of what happens in health care situations. This applies as much to specific tasks as it does to higher levels of system interaction. Without this knowledge, it is not possible to effectively undertake or evaluate designs with regard to patient safety. The Universities of Cambridge and Surrey and the Royal College of Art in the UK have proposed a model to help foster such understanding and to explain and support the use of systems engineering during health care design. An adapted form of this diagram is illustrated in Figure 23–3.

The model is based upon a convergence of views from the fields of ergonomics, engineering design, and user-centered design. In essence it can be read from top left, to the middle, and to top right, following a V-shape. Beginning with establishing a "knowledge base," through a series of activities, enclosed by the outer box in the figure, "safe medical care" can be provided. This is the fundamental aim of the health care design process and can occur as long as the other elements of the

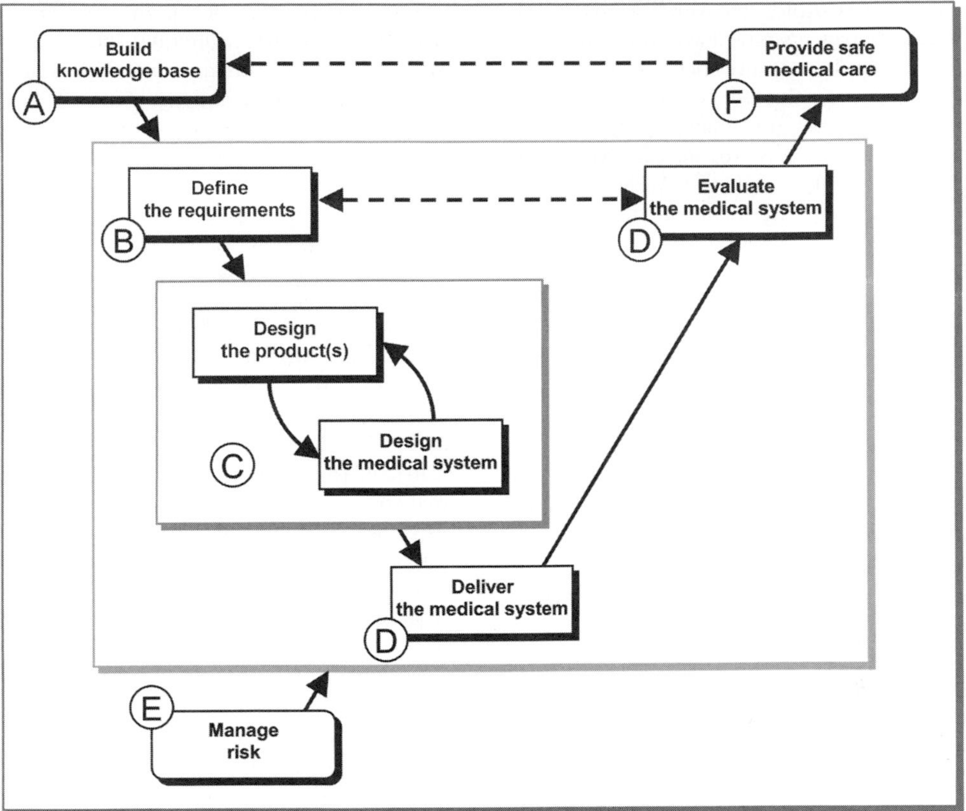

Figure 23–3. A Systems-Based, User-Centered Approach to Health Care Design, adapted from EDC (2004)

diagram are put into place and are carried out in the right order and in a suitable manner. The following sections describe this approach, within the context of HFE, following the activities in order. A much more detailed explanation about the origin and details of the model can be found in the document *Design for patient safety: a scoping study to identify how the effective use of design could help to reduce medical accidents* (EDC, 2004). This document describes a user-centered systems engineering view for the design of any products or services in the health care system.

Build a Knowledge Base

Knowledge is the essential foundation upon which evidence-based decisions can be made. Such knowledge is required so that patient safety problems can be successfully and systematically identified, prioritized and acted upon. The first stage of the design process is to develop a good understanding of the problem since without this, design solutions are unlikely to be as effective. Starting with a knowledge base, more effective design requirements can then be developed. The knowledge base can be built up through gaining an understanding of components within organizations, the associated environments of use, the clinical tasks to be performed, and the needs of the users and supports all the activities within the middle box in the figure.

Organizational/Environment Issues

An understanding needs to be gained of the interactions within and between organizational components of the health care system because these may impact on device design. There are many different parts of this system, from home health care to the ambulance, from pathology services to the operating theatre.

Consider, for example, "recovery" wards, which provide care for patients who have just undergone surgery. These wards often use specialist monitoring equipment. However, due to bed shortages, such wards can become full, and consequently patients may be moved to general wards instead. Nurses on general wards may be far less familiar with the equipment attached to the patient and, hence, device designs may need to take into account the needs of these users, as well as those for which they were directly intended. Even if, as in this case, the device is designed for use by specialists, designers armed with knowledge of the variability of actual-use conditions are at least in the position to make better informed decisions regarding their device design.

Environmental conditions can also have a great impact upon the ability of a user to operate a device safely and effectively. Consider, for example, the difference in demands placed on users between operating a patient monitoring device in a distraction-free and well-lit environment, compared with operation in a road-side accident and emergency situation, where sunlight falls onto the display making reading difficult and where tasks may change rapidly and must be performed quickly. The demands can be very different, and designs intended for use in both conditions must take into account these environmental factors.

Task Analysis

Complex chains of actions are required to deliver health care to individual patients, among the different parts of the health care system. At each stage, safety-critical tasks are carried out by health care professionals who are often caring for many other patients at the same time (with very different health conditions and care requirements). While the goal might appear straightforward (e.g., ensuring a patient is provided with the appropriate medicine in the right quantity), the tasks required to ensure this happens may be complex, involving checking and careful attention to detail. Task analysis enables goals to be defined, steps that are used to achieve these goals to be accurately assessed, and interactions between the person and the system (including use of equipment) to be identified. It should be kept in mind that organizational and regional variations exist in undertaking routine tasks, and actual-use of devices may differ from that which is expected (EDC, 2004). Critical tasks, shortcuts, and false assumptions about the state of the system may all occur and may be identified using task analysis. Techniques such as task analysis can therefore be instrumental in developing safe and effective device designs.

User Needs

An understanding of the requirements of device users must also be gained. Ergonomics literature has published much about human capabilities, although this has been developed primarily from study into working situations in offices and industry and for military purposes. More recently such data has been extended to cover older adults and children. However, much of the information is not regularly consulted by designers and decision-makers, either because they are not aware of its existence or because the formats in which it is presented are not appropriate or designer-friendly (EDC, 2004). The requirement for accurate data on user behavior and characteristics is heightened in health care due to the variety of situations in which equipment and information have to function and the range of potential users. Designers must therefore appreciate in depth the conditions and conventions of use and the range of abilities of the various users who will use the device, ensuring that the demands made during device operation match the sensory, cognitive, and motion capabilities of the intended users and the degree of risk involved.

An appreciation of these conditions and interactions requires engagement with various stakeholders in the health care system, and various research techniques may be used such as interviews, observations, and focus groups. Pathway mapping can also be used to help understand the flows of components within the health care system, such as medicines, materials, devices, information, and patients. Furthermore, records of device-related adverse incidents can be valuable in highlighting poorly designed device features and also the types of scenarios of use that can result in problems. These are published by the FDA (http://www.fda.gov/medwatch/safety.htm) and through MHRA's web page, under "Safety Warnings" (http://devices.mhra.gov.uk/).

Define the Requirements

The production of device design requirements is strongly linked to the previous activity of developing a knowledge base, and there is often a considerable overlap in these activities. However, as opposed to establishing a knowledge base, requirements are much more specific and form the starting point for device design (which is represented by the inner box in Figure 23–3). They are produced by elaborating, expanding and developing needs into engineering terms. This can be done by considering the business (e.g., cost and scheduling of tasks), technical (i.e., what functions the device must perform throughout its lifecycle of use and how well it should do them), and regulatory (e.g., laws, standards, and guidance) needs for the device (Shefelbine, Clarkson, Farmer, & Eason, 2002).

Device requirements are recorded in a document called a requirements specification, which provides a means of communication between various stakeholders in and around the project team. It defines the design challenge (i.e., what is to be achieved) but not how it is to be achieved. Developing suitable device requirements is a critical process because medical equipment, medications, packaging, and information have to function across a wide range of situations and for users with very different capabilities. If these are not all taken into account and documented, patient safety may well be compromised.

Design requirements, user needs, and other relevant factors must be explicitly stated so that design solutions can be developed against the background of well-understood contexts in which they will function. This will also enable the solutions to be more easily evaluated in terms of their effectiveness in improving patient safety, supporting care-givers and health care professionals in carrying out their work and enhancing patients' experiences (Shefelbine et al., 2002; Ward, Shefelbine, & Clarkson, 2003).

A helpful approach to device requirements capture is to consider the "who, what, where, when, and why of device use" (Shefelbine et al., 2002). This requires an appreciation of a wide range of influences, as illustrated in Figure 23–4. Input from relevant stakeholders (often including patients) may be particularly helpful in fleshing-out and developing the details of these influencing factors.

The lower half of the figure provides examples of the types of questions that a design company might ask when considering the who, what, where, when and why questions. When designing a syringe injector, for instance, nurses, doctors, and patients are key users who will be required to deliver a specific drug dose through venous delivery with a single injection. The device must be low in cost, must produce accurate readings, and must be transportable, as it will be used in a variety of locations. Following the citation of these outline requirements, further work is necessary to develop the details that can be used as a precise yardstick for success. Designers would need to consider, for example, what constitutes an "accurate" reading. This process is described further in Shefelbine et al. (2002).

Design the Product/Design the System

At the heart of this model (see Figure 23–3) are the activities of designing the product and designing the system around it. Successful product development cannot be achieved in isolation of the system or environment into which it will be introduced and, as such, this process will be unique to a particular product. It should be informed by all the relevant stakeholders and should be actively managed to minimize technical and commercial risk. While in this chapter the product considered is a medical device, the model applies equally to the design of other elements of health care, including procedures and protocols, information and environments.

The purpose of this part of the diagram is to show that whenever a product is designed, the system surrounding it must also be considered. The following example illustrates the importance of this "big picture" thinking when designing medical devices:

In the UK in 2001, a nine-year-old boy died from asphyxiation during routine surgery (British Broadcasting Corporation [BBC], 2001). It was found that the death had occurred because the tube delivering oxygen had become blocked. Sabotage was suspected, and a criminal investigation was launched. The investigation subsequently revealed that the blockage was caused by a most unexpected series of events. The delivery components were routinely stored in a drawer, where plastic caps used for intravenous administration sets had been discarded. It was found that the action of opening and closing the drawer had led to a cap becoming lodged inside the gas delivery tube. The blockage had not been identified by health care staff, and the delivery set had subsequently been used. A number of additional cases of patient harm were found to have occurred under similar circumstances across the UK over the previous few years.

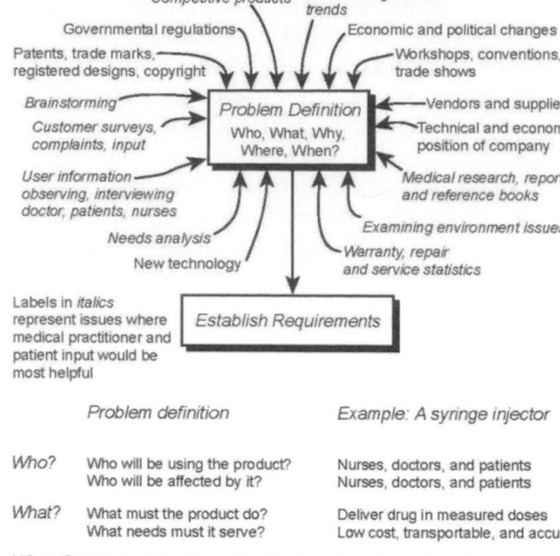

	Problem definition	Example: A syringe injector
Who?	Who will be using the product? Who will be affected by it?	Nurses, doctors, and patients Nurses, doctors, and patients
What?	What must the product do? What needs must it serve?	Deliver drug in measured doses Low cost, transportable, and accurate
Where?	Where will the product be used?	In the home, doctor's surgery, or hospital
When?	When will the product be used?	When venous delivery of a drug is necessary
Why?	Why will the product be used?	Delivery of a drug through single injection

Figure 23–4. Requirements Capture for Medical Device Design (Shefelbine et al., 2002)

Given the number of operations in UK hospitals each year, this incident was rare. However, had a better appreciation of the risks been gained, in particular of human and device interactions, the tragic consequences could perhaps have been avoided. To appreciate these risks, suitable device requirements would have been needed, based on a knowledge base being formed, followed by a comprehensive risk management program, as will be discussed below in the section on managing risk. As part of the product/systems design process, a number of "good practice" HFE-related design principles can be used, as follows.

Standardization

In the case of medical devices, a lack of standardization is known to cause problems for users. In the case of infusion devices, for example, wards that make frequent use of such devices (e.g., oncology) may use many different types (not least as a legacy of sporadic purchasing practices over the years), leading to a varied ward-stock of devices. In the UK an audit of the range of infusion devices in six NHS hospitals revealed that, on average, each hospital used 31 different types of infusion devices, where a much lower number would suffice (National Patient Safety Agency [NPSA], 2004). The result is confusion for the operators. A greater degree of standardization in control layouts may help to reduce this problem (Hertz, 2000).

Simplification and Intuition

If the incidence of such errors is to be reduced, device design that imposes minimal load on the user—design that is simple and intuitive to use—is of high importance. In the design of electronic device displays, for example, consideration should be given to a balance between presenting too much information, which can cause confusion, and too little, which can lead to oversights by the operator (Engelke & Olivier, 2002). A further, international challenge is that of the significant numbers of non-native health care practitioners being introduced into health care systems. In the UK in 2002–2003, for instance, over one-third of nurses added to the UKCC/NMC register were trained outside the UK. Although most people might be familiar with the English language, there may be significant cultural differences that can result in problems with patient safety (Nursing and Midwifery Council [NMC], 2004).

Prevention Rather Than Warning

It is better to trap errors by good design before they occur than to warn users of the possibility of them occurring. For example, devices such as cardiac monitoring equipment may be connected together using male and female plugs. Providing shrouding around the pins on the male-type connectors can prevent accidental insertion into a power point because the shrouding provides a physical barrier to such insertion. Providing a warning sign may be less effective because it may not be noticed. Designs should also be error-tolerant (Engelke & Olivier, 2002), which again requires a good understanding of the types of errors that users may make. For example, effective inhaler use requires coordination between the acts of producing a dose and breathing inward. Patients who find this difficult can be provided with a "spacer" device, which holds the medication particles in suspension for sufficient time to allow them to be inhaled relatively normally. For such patients, proper spacer use can vastly improve the effectiveness of the medication.

Deliver/Evaluate the Medical System

A key part of successful device delivery is to evaluate it in its setting of use, either under simulated or actual-use conditions. Evaluation ensures that evidence of satisfactory performance is available and can take the form of verification or validation. Verification occurs during the design process and involves gathering evidence that the device design, as it develops, is meeting its requirements. Essentially, it helps check that the device design is on-track. As such, verification can show that a requirement has been met, but—crucially—it tends not to demonstrate the appropriateness of such requirements. Validation is a complementary technique to verification and is a more comprehensive form of evaluation that involves gathering evidence that the device design is truly fit for purpose. This is usually done in the context of the device's expected use.

For example, a design requirement for a surgical lamp might be the intensity of light it can produce at a certain distance from its source over a designated area. Verification might involve measuring this light intensity using a prototype lamp and a light meter. Validation, however, might seek to determine whether the light provides sufficient illumination to carry out surgery. Ideally, it would therefore involve testing the design with surgeons in an operating theatre.

Effective evaluation requires suitable timing in the design process and a method that provides results of an appropriate validity. Evaluation that occurs late in the design process may risk a serious level of rework should a problem be discovered. Early evaluation may require repetition later in the design process because the design of the product may change in the interim, and extra evaluation costs may consequently be accrued. The method for evaluation must also be of a suitable validity—creating a balance between evaluation costs and the risk of a false-positive or a false-negative. Such methods may range from the "quick and cheap" (which may be appropriate for a "ball-park" result) to extremely extensive evaluation, where tens of thousands of units are tested as they leave a production line (Ward, Clarkson, Bishop, & Fox, 2002).

Evaluation methods do not necessarily have to be elaborate early in the design process. In the example above, measuring light intensity from a prototype surgical lamp may be a straightforward process, requiring little in the way of expensive equipment. It may also be possible to judge whether the device appears "on track" by evaluating it with a very limited number of health care practitioners. More comprehensive evaluation may take place when the design is judged to being close to completion.

Evaluation—even with users—will often require minimal investment for the potential to highlight areas for significant improvements in device quality. Electronic user interfaces may be evaluated to investigate the presence of various well-known HFE problems, such as poor feedback about device state and behavior, complex and arbitrary sequences of operation, different operating modes intended for different contexts, and ambiguous alarms (Obradovich & Woods, 1996), although good device design should minimize the presence of these in the first place.

In addition, an understanding of substandard use conditions (such as use in poor lighting or under awkward viewing or operating angles) can be important for developing evaluation methods, particularly later in the design process. Output from earlier activities, such as task analysis, can help to drive this process. For example, arranging a device evaluation with nursing staff who have just finished a night shift could ensure that the users are suffering from sufficient fatigue to be likely to uncover examples of poor device design, such as confusing operating sequences.

Much of the decision-making regarding timing and validity of evaluation can be made by balancing risks, as discussed in the following section. Further guidance on device evaluation can be found in the document *Good design practice for medical devices and equipment—design verification* (Ward, Clarkson, Bishop, & Fox, 2002).

Managing Risk

Risk is the product of the likelihood of an adverse outcome and its impact should it occur. The process of using risk to drive the device design—risk management—is an extremely powerful decision-making technique, which should be employed throughout the design process. This process supports all of the activities within the middle box in Figure 23–3. To employ risk management, the hazards (the potential sources of harm) need to be identified: from the perspective of overt device failures (e.g., due to poor reliability) and—from a HF perspective—in terms of failures of interactions between device user and the device itself. The knowledge base and device requirements stages clearly have a major role to play in contributing to successful hazard identification.

Following the identification of hazards, risks need to be estimated. Techniques such as Failure Mode and Effects Analysis (FMEA) and Fault Tree Analysis (FTA) can be most helpful in this regard. For additional information on FMEA and FTA, see the chapter on "Human Factors Risk Management in Medical Products" in this handbook. Next, risk mitigation strategies (such as device redesign or, less preferably, warning labelling) must be put into place to reduce any unacceptably high risks to a suitable level. Monitoring of the risks should be conducted. Again, in the case of HF, user feedback may be required. ISO 14971 provides much additional guidance on the risk management process (ISO, 1999).

Provide Safe Medical Care

Although comprehensive evaluation will have occurred during the design of the product, it cannot be certain that the new design is entirely fit for purpose. To provide safe medical care, further monitoring of the design's longer-term impact upon the health care system is therefore required. Feedback to the knowledge base should be provided to facilitate the continuous improvement of further designs.

CONCLUSIONS AND HFE IN THE FUTURE

The scale of poor HFE design for medical devices is still somewhat unclear, although it appears that this is by no means an insignificant problem. The effects, however, are clear—and range from inconvenience to users to patient morbidity and mortality. The demand for well-designed devices continues to rise (Engelke & Olivier, 2002), as do the penalties for designs that do not take good HFE practice into account. As a consequence, device designers are facing increasing incentives to follow good HFE practice.

For example, in many health care systems, purchasing has historically been based on identifying the lowest price for a product or service. Keselman et al., for example, investigated device purchasing practice in a large U.S. hospital and found that a number of groups and committees were involved in the process (Keselman, Patel, Johnson, & Zhang, 2003). However, despite the consideration of human factors requirements by one of these groups, restricted communication between the groups and committees led to under-representation of usability-related needs when the final purchasing decisions were made. Keselman et al. suggest that such practice is likely to be typical of the device purchasing process in other health care establishments.

However, purchasing strategies are becoming more mature, taking into account life-cycle costs, as well as the effects of good design on improving patient safety (EDC, 2004). As this maturity develops and spreads, through the adoption of guidance such as that produced by the MHRA (MDA, 2001a), purchasers will be interested in the longer-term costs of use, such as training, life-expectancy, and maintenance. This provides an opportunity for device designers who conduct good HFE design to gain competitive advantage over other companies that see this is as a lower priority, even if the devices of their competitors retail at a lower price. The NHS' Purchasing and Supply Agency (PASA) is already starting to buy products (such as medicines with easily readable and identifiable packaging) that have been demonstrated to be safe for patients, to the exclusion of products that have not.

Development of feedback loops, from device users to manufacturers, within the health care system will help to ensure that proper device evaluation is undertaken and also to reward companies that provide appropriate evidence of the fitness for purpose of their device. As the benefits of improved safety and ease of use are reaped in the health care setting, additional cost-savings will result, meaning that health care institutions will have new resources to pay for the costs of these necessary changes. Good HFE practice, therefore, is a crucial part of medical device design and can only grow in importance in the future.

References

ANSI/AAMI. (1993). *Human factors engineering guidelines and preferred practices for the design of medical devices* (no. ANSI/AAMI HE48-1993). American National Standards Institute, Association for the Advancement of Medical Instrumentation.

ANSI/AAMI. (2001). *Human factors design process for medical devices* (no. ANSI/AAMI HE74-2001). American National Standards Institute, Association for the Advancement of Medical Instrumentation.

BBC. (2001). *Fault clue to oxygen blockages*. Retrieved January 7, 2005, from http://news.bbc.co.uk/hi/english/health/newsid_1665000/1665040.stm.

BSI. (2000a). *BS EN ISO 9001:2000 Quality management systems—requirements*. London: British Standards Institution.

BSI. (2000b). *BS EN ISO 14971:2000 Medical devices—application of risk management to medical devices*. London: British Standards Institution.

BSI. (2001). *BS EN ISO 13485:2001 Quality systems—medical devices—particular requirements for the application of EN ISO 9001*. London: British Standards Institution.

Cooper, J. B., Newbower, R. S., & Kitz, R. J. (1984). An analysis of major errors and equipment failures in anesthesia management: considerations for prevention and detection. *Anesthesiology, 60*, 34–42.

Cooper, J. B., Newbower, R. S., Long, C. D., & McPeek, B. (1978). Preventable anesthesia mishaps: a study of human factors. *Anesthesiology, 49*, 399–406.

Dripps, R. D., Lamont, A., & Eckenhoff, J. E. (1961). The role of anesthesia in surgical mortality. *Journal of the American Medical Association, 178*, 261–266.

EDC. (2004). *Design for patient safety: a scoping study to identify how the effective use of design could help to reduce medical accidents*. Cambridge: Engineering Design Centre, University of Cambridge, UK.

Engelke, C., & Olivier, D. (2002, July). Putting human factors engineering into practice. *Medical Device and Diagnostic Industry*. Retrieved January 7, 2005, from www.devicelink.com/mddi/archive/02/07/003.html.

FDA. (1990). *Device recalls: a study of quality problems* (no. FDA/CDRH-90/26). Rockville, MD: Center for Devices and Radiological Health, Food and Drug Administration.

FDA. (1996a). *21 CFR Parts 808, 812, and 820. Medical devices; Current Good Manufacturing Practice (CGMP); final rule*. Federal Register. Vol. 61, No. 195. Office of the Federal Register, U. S. Government.

FDA. (1996b). *Do it by design: an introduction to human factors in medical devices*. Rockville, MD: Center for Devices and Radiological Health, Food and Drug Administration. Retrieved January 7, 2005, from www.fda.gov/cdrh/humfac/doit.html.

FDA. (1997). *Design control guidance for medical device manufacturers*. Rockville, MD: Center for Devices and Radiological Health, Food and Drug Administration. Retrieved January 7, 2005, from www.fda.gov/cdrh/comp/designgd.html.

FDA. (2000a). *IEC develops standard for medical device human factors design—ensuring a safe device-user interface*. Paper presented at The Human Factors and Ergonomics Society meeting, San Diego, CA.

FDA. (2000b). *Make sure the medical device you choose is designed for you* (no. (301) 443-2436). Rockville, MD: Center for Devices and Radiological Health, Food and Drug Administration. Retrieved January 7, 2005, from www.fda.gov/cdrh/humfac/you_choose_checklist.html.

FDA. (2000c). *Medical device use-safety: incorporating human factors engineering into risk management*. Rockville, MD: Center for Devices and Radiological Health, Food and Drug Administration. Retrieved January 7, 2005, from www.fda.gov/cdrh/humfac/1497.html.

FDA. (2001a). *Gaps in your postmarket safety net for medical devices* (Presentation by Larry Kessler to the Drug Information Association). Rockville, MD: Center for Devices and Radiological Health, Food and Drug Administration, Retrieved January 7, 2005, from http://www.fda.gov/cdrh/present/dia–jan-2000/dia-jan-2001.html.

FDA. (2001b). *General principles of software validation*. Rockville, MD: Center for Devices and Radiological Health, Food and Drug Administration. Retrieved January 7, 2005, from www.fda.gov/cdrh/comp/guidance/938.pdf.

FDA. (2001c). *Integrating human factors engineering into medical device design and development*, video no. AVA20885VNB1. Springfield, VA: National Technical Information Service.

Geddes, L. A. (1998). *Medical device accidents* (1st ed.). Boca Raton, FL: CRC Press.

GHTF. (1999). *Recommendation on medical devices classification*. Global Harmonization Task Force.

Giuntini, R. E. (2000, October). Developing safe, reliable medical devices. *Medical Device and Diagnostic Industry*. Retrieved January 7, 2005, from www.devicelink.com/mddi/archive/00/10/009.html.

Hertz, E. (2000). Focus on medical errors provides clinical engineering opportunities. *Biomedical Instrumentation & Technology, 34*(2), 131–132.

IEC. (2004). *Medical electrical equipment—part 1-6: general requirements for safety—collateral standard: usability* (no. IEC 60601-1-6). Geneva: International Electrotechnical Commission.

ISO. (1999). *ISO 13407:1999 Human-centred design processes for interactive systems*. Geneva: International Organization for Standardization.

Kaye, R. (2000). Human factors in medical device use safety: how to meet the new challenges (Presentation). Retrieved January 7, 2005, from www.fda.gov/cdrh/humfac/sandiego2.pdf.

Keselman, A., Patel, V. L., Johnson, T. R., & Zhang, J. (2003). Institutional decision-making to select patient care devices: identifying venues to promote patient safety. *Biomedical Informatics, 36*, 31–44.

Kohn, L., Corrigan, J., & Donaldson, M. (2000). *To err is human*. Washington, DC: National Academy Press.

Leape, L. (2004). Human factors meets health care: the ultimate challenge. *Ergonomics in Design, 12*(3), 6–12.

MCA. (2001). *Report to the committee on safety of medicines from the working group on labelling and packaging of medicines*. London: Medicines Control Agency.

MDA. (2000). *Equipped to care: the safe use of medical devices in the 21st century*. London: Medical Devices Agency.

MDA. (2001a). *Devices in practice—a guide for health and social care professionals*. London: Medical Devices Agency.

MDA. (2001b). *Minutes of the committee on safety of devices meeting 19 July 2001*. London: Medical Devices Agency.

MDDI. (2001). Human factors roundtable part I: the regulatory perspective. *Medical Device and Diagnostic Industry*. Retrieved

January 7, 2005, from www.devicelink.com/mddi/archive/01/01/006.html.

Meldrum, S. J. (1998). Where do I look for good ideas for my new device? In A. Murray (Ed.), *Medical equipment industry—potential for growth* (pp. 17–20). London: Institution of Electrical Engineers.

MHRA. (2004). *Adverse incident reports 2003* (no. MHRA DB 2004(01)). London: Medical Devices Agency.

NPSA. (2004). *Improving infusion device safety* (no. NPSA safer practice notice 01). London: National Patient Safety Agency. Retrieved January 7, 2005, from http://81.144.177.110/site/media/documents/526_npsa_saferpractice_01.pdf.

NMC. (2004). *Statistical analysis of the register, 1 April 2002 to 31 March 2003.* London: Nursing and Midwifery Council.

NRC. (1996). *Safe, comfortable, attractive, and easy to use: improving the usability of home medical devices.* Washington, DC: National Research Council.

Obradovich, J. H., & Woods, D. D. (1996). Users as designers: How people cope with poor HCI design in computer-based medical devices. *Human Factors, 38*(4), 574–592.

Senders, J. W. (1994). Medical devices, medical errors, and medical accidents. In M. S. Bogner (Ed.), *Human error in medicine* (pp. 159–177). Hillsdale, NJ: Lawrence Erlbaum Associates.

Shefelbine, S., Clarkson, P. J., Farmer, R., & Eason, S. (2002). *Good design practice for medical devices and equipment—requirements capture.* Cambridge: University of Cambridge Engineering Design Centre/University of Cambridge Institute for Manufacturing.

Swain, E. (2004, October). Nursing shortages and device design: a hidden connection. *Medical Device & Diagnostic Industry.* Retrieved January 7, 2005, from www.devicelink.com/mddi/archive/04/10/022.html.

TSO. (2000). *An organisation with a memory: report of an expert group on learning from adverse events in the NHS chaired by the chief medical officer.* London, UK: The Stationery Office.

TSO. (2001). *Learning from Bristol: the report of the public inquiry into children's heart surgery at the Bristol Royal Infirmary 1984–1995.* London, UK: The Stationery Office.

USPC. (2002). *Summary of information submitted to MedMARx in the year 2001.* Rockville, MD: United States Pharmacopeial Convention.

Ward, J. R., Clarkson, P. J., Bishop, D., & Fox, S. (2002). *Good design practice for medical devices and equipment—design verification.* Cambridge: University of Cambridge Engineering Design Centre/University of Cambridge Institute for Manufacturing.

Ward, J. R., Shefelbine, S. J., & Clarkson, P. J. (2003, August 19–21). *Requirements capture for medical device design.* Paper presented at the International Conference on Engineering Design, Stockholm.

Weinger, M. B. (2000). User-centered design: a clinician's perspective. *Medical Device and Diagnostic Industry,* from http://www/devicelink.com/mddi/archive/00/01/007.html. *22*(1).

Wiklund, M. E. (1995). *Medical device and equipment design: usability engineering and ergonomics.* Buffalo Grove, IL: Interpharm Press.

Wiklund, M. E. (2004, August). Human factors: moving in the right direction. *Medical Device & Diagnostic Industry.* Retrieved January 7, 2005, from www.device link.com/mddi/archive/04/08/016.html.

24

PATIENT SAFETY AND TECHNOLOGY, A TWO-EDGED SWORD

James B. Battles
Center for Quality Improvement and Patient Safety, Rockville, MD

The opinions and assertions contained herein are the private views of the author and are not to be construed as official or as reflecting the views of the Agency for Healthcare Research and Quality, the United States Department of Heath and Human Services.

The report of the Institute of Medicine (IOM) *To Err is Human* (Kohn, Corrigan, & Donalson, 2000) stressed the importance of automating repetitive, time-consuming, and error-prone tasks in the delivery of health care through the use of technology. While automation holds substantial promise for improved patient safety, safety science experts caution that all technology introduces the potential for new and different risk and hazards. It is critical that any new automated system in health care be tested in actual operational settings to determine what, if any, unanticipated failures exist. Field-based research is essential in the emerging field of patient safety to create the evidence as to what technologies actually improve patient safety and those that may well increase the potential for harm.

Patient safety has become a major concern throughout the world. In the United States (Quality Interagency Coordination Task Force, 2000) the United Kingdom (Department of Health, United Kingdom [DOH/UK], 2000), and in a number of other countries, major patient safety program activities have been launched in response to startling figures on the number of individuals that are harmed from health care associated injuries. Health care associated injuries are those associated with the process or structure of care rather than to a patient's underlying or physiological, environmental, or disease-related antecedent conditions (Battles & Lilford, 2003).

The goal of patient safety is to reduce the risk of injury or harm to patients from the structure and process of care. This can be accomplished by eliminating or minimizing unintended risks and hazards associated with the structure and process of care. Given this focus on health care associated injury contained within patient safety events, we can have as a vision for patient safety zero health care associated injuries or harm (Battles & Lilford, 2003).

How can we achieve this vision of zero heath care associated injuries or harm? Eisenberg (2000) presented an analogy of patient safety having the characteristics of an epidemic of worldwide portions. He suggested that a three-stage attack plan representing a continuum using the epidemic metaphor as an organizing principle for a patient safety initiative.

> Stage 1—identify the risks and hazards that cause or have the potential to cause health care associated injury or harm.
>
> Stage 2—design, implement, and evaluate patient safety practices that eliminate known hazards, reduce the risk of injury to patients, and create a positive safety culture.
>
> Stage 3—maintain vigilance to ensure that a safe environment continues and patient safety cultures remain in place (Battles & Lilford, 2003).

While many of the activities in patient safety have focused on Stage 1 of the Eisenberg epidemic model, major efforts are beginning to be devoted to Stage 2, involving designing and implementing activities that are designed to eliminate the risks and hazards from health care associated injury or harm. "Risk" is the possibility/probability of occurrence or recurrence of an event multiplied by the severity of the event (Department of Health, United Kingdom [DOH/UK], 2000). "Hazard" is anything that can cause harm (Department of Health, United Kingdom [DOH/UK], 2000). Both risks and hazards reveal themselves nested within an event. An event has been described by Freitag and Hale (1997) as deviations in activities or technologies that lead toward unwanted negative consequences. In the case of patient safety, the negative consequences are health care associated injuries/harm or the potential to cause such injuries/harm.

Adverse/harm events are those where actual harm and or injury has taken place with the harm representing some degree of severity from death to minor injury. There are, of course, events that occur that involve human failure and an interaction with latent conditions, but the outcome does not result in actual harm to the patient. These no harm events represent potential rather than actual harm with warning levels of potential severity as well. The near miss event on the other hand does not manifest itself in actual harm to a patient because there was intervention and recovery. Again, there is potential for harm with a similar level of potential severity (Battles & Shea, 2001). Zapt and Reason (1994) point out that detection or identification is the first step in error (patient safety) management. From an organizational point of view, it is important that detection or identification rates be high because unintended risks and hazards that are not detected can have disastrous consequences. For the overall detection process to be effective, one needs to identify not only adverse/harm events, but also no harm and near miss events as well. Adverse events caused by unintended risks and hazards are much less common than no harm and near miss events. Rather than wait for the adverse event, it is essential to detect and correct the unintended risks and hazards. Kaplan, Battles, Van der Schaaf, Shea, and Mercer(1998) state that one of the goals of patient safety management is to increase detection and reporting rates or detection sensitively level (DSL) of patient safety events in order to decrease risk of harm to patients resulting in a decrease in the event severity level (ESL).

But just identifying risks and hazards are insufficient to make health care safe. This requires moving to Stage 2 of Eisenberg's plan for patient safety. At the intervention stage, one can begin by considering safety by design. In designing interventions there has been a call for increased applications of technology to be applied to the patient safety problem. The application of technology was supported by recommendations in two of the IOM reports, *To Error is Human* and *Patient Safety: Achieving a New Standard of Care* (Aspden, Corrigan, Wolcott, & Erickson, 2004) These recommendations were based in part on the experience in high-risk industries, such as nuclear power and aviation, where automation has been introduced as a means of improving human performance.

A STRUCTURE AND PROCESS MODEL FOR PATIENT SAFETY

Clearly patient safety design and development need to focus on the sources of risks and hazards that can lead to health care associated injury. Donabedian's (1980) process-structure-outcome model is a helpful concept for describing the location and type of risks and hazards within the health care structure and process. The aim of safety interventions must be to improve outcome, that is, minimize harm to patient through failures in the process or structure of care. Improvements in safety will come from changes (improvements) in these areas. Of course, not all adverse outcomes, including death, are due to problems with either the process or structure of care. Some patients enter the health care system with antecedent conditions that are incompatible with life. Nothing that is done for these patients changes an adverse outcome—death or serious disability. In reality, process of care occurs within the structure of the health care system. One criticism of examining only the outcomes of care is that it does not take into account the patient's condition prior to entry the health care system. Coyle and Battles (1999) modified the Donabedian approach emphasizing the need to account for a patient's antecedent conditions or the patient's underlying physiological or behavioral status prior to entering the health care systems.

Although antecedent condition confounds comparisons between institutions, complex patients with co-morbid conditions do test the system's resilience. Thus, when it comes to the investigation

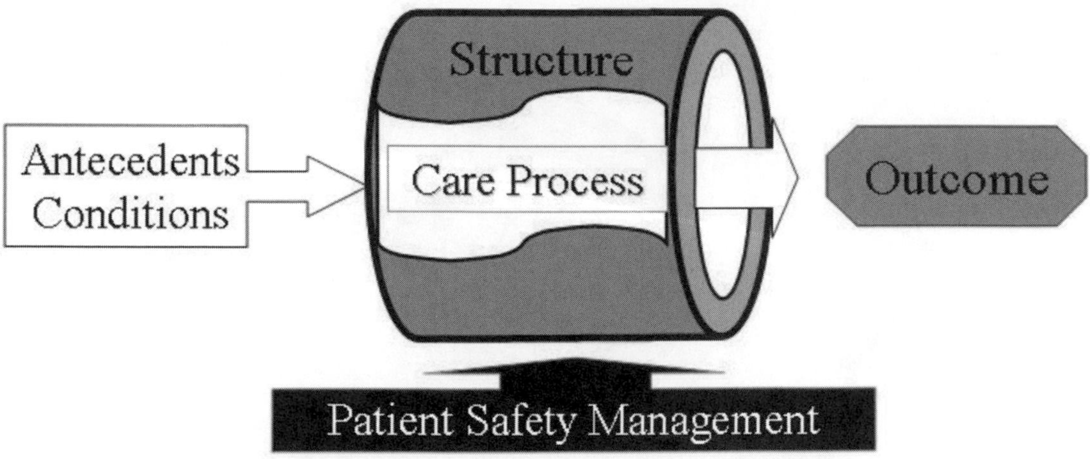

Adjust structure and process to eliminate or minimize risks and hazards of health care associated injuries before they have an adverse impact on the outcomes of care

Figure 24–1. A Structure and Process Model for Patient Safety Based on Donabedian

of particular events, careful examination of what went wrong in a complex case may identify weakness in a system, which could put healthier patients at increased risk. In this sense, such complex cases are like the canary in the mine shaft. For example, administration of an incompatible unit of blood in an emergency might lead to the identification of weak systems, which could eventually place a patient requiring elective transfusion at risk. Such weakness could lie either in the structure or the process of care. Figure 24–1 is an illustration of the patient safety version of the Donabedian model showing structure within process. Rather than the usual boxes aligned in linear fashion, the illustration presents the process as occurring within the structure of health care. Structure represents an outer boundary for the process of care, which occurs within the cylindrical containment of the structure of the health care delivery. Structure represents items including physical facilities, equipment (hardware), and personnel staffing and resources (budget) that are relative fixed. Process represents the procedures and protocols used to deliver the actual care and include such items as software, standard operating procedures (SOP), clinical guidelines, and standards of care. To optimize the outcomes of care, one would need to adjust the structure and process of care to accommodate the antecedent physiological conditions of the patient to minimize the variations in the outcome that are the consequences of the interaction of the care process within structure.

Types of Risks and Hazards

What are the sources of these risks and hazards that are embedded in the process and structure of care? Reason (1990, 1997) has classified two categories of failure (risks and hazards) based on who initiated them and how long it takes to have an adverse effect. Active failures are those committed by individuals in direct contact with the human system interface. These active failures are often referred to as human error. Errors can be considered actions or inactions that lead to deviations from intentions or expectations. Errors are most often manifested as risks that either lead to or have potential to lead to harm. Latent failures are created by the delayed consequences of technical and

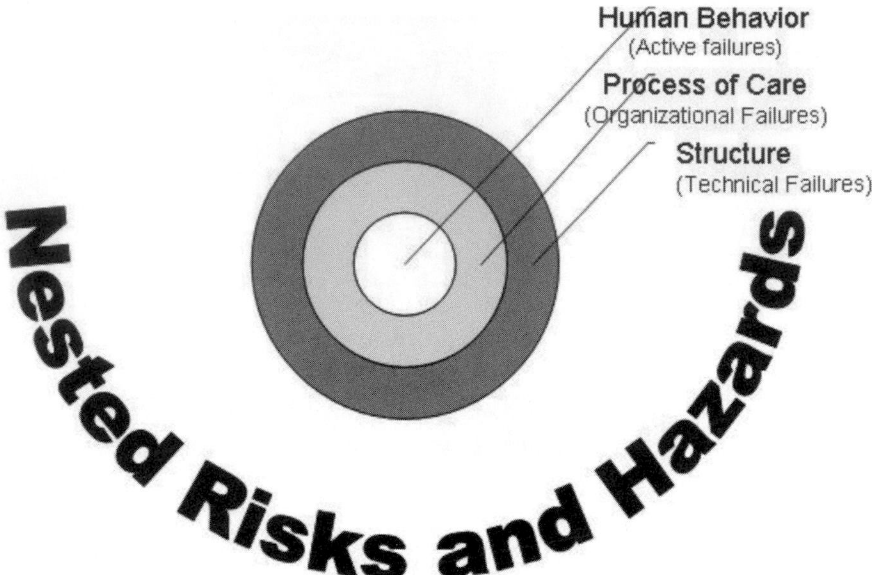

Figure 24–2. A Model for the Nested Relationship of Risks and Hazard in Patient Safety

organizational actions and decisions. Latent failures or conditions are also sometimes referred to as system errors; however; they might be more accurately considered to be hazards embedded within the structure and process of the health care system rather than risks associated with human behavior.

Rasmussen (1976) looked at the cognitive basis for human behavior underlying failures or error and identified three types—skill-based, rule-based, or knowledge-based failures—which again are considered specific types of risk. Knowledge-based behavior involves the conscious application of existing knowledge to manage novel situations. Rule-based decision-making involves the application of existing rules or schemes to manage familiar situations. Prolonged, active processing is not required, simply the selection and application of the appropriate rule. Mistakes are rule-based failures of planned actions/rules to be completed as intended or selecting the wrong rule to achieve an aim. Skill-based behaviors refer to "automatic" tasks requiring little or no conscious attention during execution. Slips are inadvertent skill-based failures of commission (doing the wrong thing). While lapses are skill-based failures of knowing what to do but failing to do it, omitting a step, or losing one's place in a process, namely, failures of omission. Fumbles are skill-based failure of whole body movement, such as dropping something. Van der Schaaf (1992) has organized the active failures and latent conditions into human, organizational, and technical failures. Relating this classification schema to the Donabedian model, we would have process within structure and human behavior nested within process. Active or human failures represent risks to process, with latent conditions representing hazards embedded in the process and structure (organizational or technical) of health care. As Reason (1990, 1997) has pointed out, it is a combination of active failures interacting with latent conditions that result in adverse events. Figure 24–2 is a representation of this nested concept within the framework of the Donabedian model. In this figure we have a nested set of interacting cylinders. The outer cylinder represents the structure that can be a source of embedded latent failures or hazards. It is within structure where technical failures and some organizational failures lurk waiting to interact with process hazards in the next cylinder, process of care. Process of care is where organizational hazards are contained. These are the relatively fixed policies and procedures that govern the delivery of care by the health professionals interacting with the patient. Within these organizational elements are embedded hazards dealing with organizational culture, team interaction and communication, and the

SOPs. The inner most cylinders represent the risk of individual and group behavior, or what Reason calls "active failures."

Risk Informed Design

It is the ability to know where in the structure and process of care the actual or potential risk and hazards are located. Complicating the aspects of health care is the fact that patient care occurs at both a micro and macro level. Thus, there are both clinical micro systems that have elements of structure process and behaviors, as well as the larger macro system of care, which links individual clinical micro systems. The search for these embedded risks and hazards must include looking both within the clinical micro systems being address, as well as those related micro systems both up stream and down of the system in question. These micro systems are, of course, contained within the larger macro systems.

Traditional approaches to assessing risk in health care have been retrospective in nature. There is a growing awareness that a more proactive approach to risk assessment in health care is required, especially if one is beginning to design interventions to eliminate risks and hazards (Battles & Lilford, 2003).

Process mapping has been a powerful technique used in a variety of industries to identify potential failure points and system breakdowns that represent significant risks and hazards. The application of process mapping approaches are gaining a greater degree of acceptance in various clinical areas as a valid approach to identifying potential hazards and risks associated with various clinical processes. Failure mode effect analysis or Failure Mode Analysis (FMEA) combines the attributes of process mapping with failure identification possibilities (Senders, 2004). Another approach that builds upon both process mapping and FMEA is that of probabilistic risk assessment (PRA; Spitzer, Schmocker, & Dang, 2004; Wreathall & Nemeth, 2004). Designers of nuclear power plants, aircraft, and spacecraft have been using this technique for decades. However, many applications of PRA have targeted mechanical systems rather than work processes that have a high degree of human interaction, such as exists in health care. Socio-technical probabilistic risk assessment (ST-PRA) combines the best of rigorous and well-tested engineering methodologies, with the science of human factors to provide a new methodology for modeling human systems (Marx & Slonim, 2003; Battles & Kanki, 2004; Hale, Slonim, Allen, Marx, & Kirland, 2004).

The application of prospective risk assessment as an essential part of the design process is leading to the acceptance of risk informed design. Without an adequate understanding of the real or potential risk, a patient safety technology based intervention may not achieve the designed result and only impose increased risk without addressing the real target risk that was the hazard in the first place.

SAFETY AND TECHNOLOGY

Despite the promise of improved safety through technology, a number of safety experts note that there are limits to technology. Billings (1996) stresses that there is an important human-centered aspect related to the ways in which individuals interact with automated systems. Cook (1998) emphasizes that future failures of automated systems cannot be forestalled by providing simply another layer of defense against failure. "Rather, safe equipment design and use depends on a chain of involvement and commitment that begins with the manufacturer and continues with careful attention to the vulnerabilities of a new device or system" in a real-world environment (Cook, 1998, p. 36). The IOM, in *To Error is Human*, issued a cautionary note relative to the application of technology and automation for patient safety.

Health care organizations should anticipate that any new technology can introduce additional sources of error and should, therefore, cautiously adopt automation, alert to the potential for accidental harm. Despite designers' best intentions, the IOM report committee emphasized that "All technology introduces new errors, even when its sole purpose is to prevent errors." Consequently, as change occurs, health systems should anticipate and prevent trouble (Kohn, Corrigan, & Donalson, 2000, p. 175)

Thus the application of any new technology has the potential of creating a latent structural and or process hazard the moment it is introduced in the health care delivery process.

Any study of a new technology in medicine should be completed within the context of what others have identified as useful criteria and issues relative to automation and safety. These criteria and issues highlight three general areas: design requirements, human–machine interface, and reliability.

Design Requirements

Kukla, Clemens, & Morse (1992) lists four design criteria or system requirements for an automated system. It must be: (a) technically efficient (i.e., reduce costs, increase ease of operation, or increase productivity of the process); (b) easy to use (i.e., users must be able to focus on their work rather than on the technology; (c) a better way for operators to do their job (or at least as good as current methods); and (d) adaptable to shifting constraints and priorities of changing business conditions. Safety concerns are focused primarily on the second and third criteria: ease of use and a new or better way of performing.

Human–Machine Interface

Reason (1990) indicates that in automated systems, the human operator is separated from direct contact with the process by a minimum of two systems: (a) task-interactive systems, including the task specific controllers, valves, and detectors that constitute the process of testing and recording, and (b) the human–machine interface. Because the operator has much less, if any, direct manipulation of the tested materials, the issue of ease of use mostly revolves around the design of the interface.

Corbett (1992) has indicated that as many as 75% of companies have achieved less than their expected benefit from advanced manufacturing technology primarily because of unanticipated problems with human-machine interaction. He identified five points that define the design issues relating to the interface: (a) allocation of functions between humans and machines, (b) configuration of the architecture of the system, (c) control characteristics of the human–machine interface, (d) informational characteristics of the human–machine interface, and (e) allocation of responsibilities among users such as operating and support personnel.

Reliability

Weiner (1995) has addressed the issue of reliability of automation indicating that as systems become more and more reliable, overconfidence may develop on the part of both operators and managers in the system's infallibility creating the situation that, "nothing recedes like success." He addresses three issues relating to the influences of automation: (a) situation awareness, (b) a definition of the optimum level of automation, and (c) an appropriate level of trust in the system's reliability. To the degree that automation may cause the operator to be out of the loop, there may be a loss of awareness of the automated system's immediate (present) state and less capability of the operator to intervene when needed. The challenge for system designers is to sufficiently engage the operator without sacrificing the benefits of automation. A correlate of this concern relates to trust. Because no system operates with absolute reliability, too high a level of trust may be problematic. Two major constituents define the reliability of an automated system, accuracy/precision and availability. Both are critical in the successful operation of a system. However, paradoxically, if reliability of the system is very high, operator vigilance is likely to decrease, and early signs of system change are less likely to be detected. Additionally, operator and systems manager's confidence in the availability of the technology may decrease the probability of system redundancy and backup. Thus, lack of back up in equipment or system components represents a potential hazard at the organizational and structural level. At the process and individual human behavior level, there are potential risks. Individual skills for backup or crisis mode operation are less likely to be well-practiced, and the risk of an active human failure becomes greater when unplanned, unpracticed changes are performed under pressure.

Operational complacency remains problematic unless process control and an external quality assurance function help adjust the degree of trust to more appropriate levels. A lesson can be learned from Xerox when they moved from technical reliability to perceived reliability (Rheinfrank, 1992, p. 39). Instead of focusing exclusively on failure proof systems, Xerox designed for ease of use and provided learning experiences for the operators—"when users learn how to recover from paper jams effortlessly, then paper jams become much less of a usability problem." Awareness is another important aspect of this concept. When users or operators of automated systems are aware of potential failures and understand recovery, they are more mindful and alert.

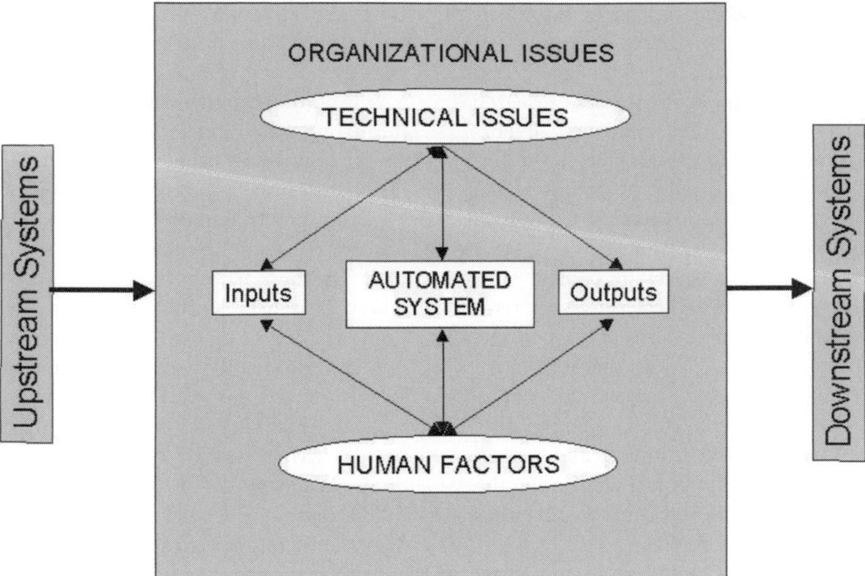

Figure 24–3. The Nested Contextualization Case Study Model

TECHNOLOGY AS PART OF THE SYSTEM

The important lessons learned from aviation and other industries as indicated by both Billings (1996) and Reason (1997) are that when new automated systems are introduced into an existing process, important interaction occurs between human behavior, organizational procedures, policies, and culture. It is within this nested context and under actual operational conditions that a new technology needs to be studied to examine actual and potential failure points. It is not only the equipment itself (technical design) that must be studied, but also the organizational and human factor aspects of its actual working environment as discussed in relationship to Figures 24–1 and 2. A complete understanding of system operations, potential and actual failures associated with the entire work flow process are necessary for the design of any new componenet technology.

Figure 24–3 is a further conceptualization of this nested context for understanding and technology or automated equipment of process. The automated system or technology is in the center surrounded by all of the latent organizational factors that may impact its use. Technical issues related to the device, such as design and installation, interact with human factors associated with the operators of the technology. These factors also interact with immediate inputs of the micro system, as well as outputs of the micro system under study. However, the context should not be limited just to the immediate proximity of the technology within the linear process of care. There is a need to examine clinical micro systems that are both up stream as well as down stream of the use of the device or technology. Often actions or inactions that occur outside of the immediate micro system up stream from the actual use of the technology itself as well as downstream may influence how safe the technology really is. Without exploring the impact of the technology outside the immediate micro system environment, the evaluation of the new technology is incomplete and lacking a more macro system view of the health care process.

Evaluating Impact

With any new system that has undergone extensive design and prototype testing, it is not until a system has been operating in the actual environment that some potential failures and weak points can be identified. In the early stages of implementation, it is critical that any new automated system be studied in actual operational settings to determine what any potential failures might be, whether they are technical, organizational, or human failures. Much can be learned from studying both actual events reported and observations of potential safety concerns and near miss events by those closely associated with system operations.

The case study approach as outlined by Yin (1994) is considered the most appropriate research method for this study to determine how events occur and why they occur in relationship to the setting when a new system is fully operational. Yin (1999) points out that the value of the case study in health services research is its intense focus on a single phenomenon within its real life context.

Thus, one must consider the relationship of risks and hazards associated with the structure, process, and human behavior associated with the technology that is being applied at both the design and implementation stage of the patient safety continuum. It is also necessary to continue to monitor and remain vigilant that the technology is in fact eliminating the targeted risks and hazards that was intended, rather than becoming a hazard itself.

CONCLUSION

The emerging field of patient safety will benefit from the creation of research evidence that documents which applications of technology lead to improved patient safety. Such evidence should be derived from field-based research approaches that examine given technological applications in a real-word context. Studies should not be limited to just the device or piece of equipment and its efficacy but, rather, should focus on how and in what context the technology exists. It is possible for an individual device to operate just exactly as it was designed and still represent a major patient safety hazard in the manner in which it is used within a system of health care delivery

Technology, if used appropriately, is a tool for improved safety. However, technology is a two-edged sword. We need to identify hazards to patient safety within our system and eliminate them. We have to be vigilant of things that set individuals at the sharp end for failure. We are obliged to incorporate redundancy and back-up capabilities once we automate. We must automate with caution and always expect the unexpected. Field-based research is essential in the emerging field of patient safety to create the evidence as to what technologies actually improve patient safety and those that may well increase the potential for harm.

References

Aspden, P., Corrigan, J. M., Wolcott, J., & Erickson, S. M. (Eds.) (2004). *Patient safety: achieving a new standard of care.* Washington, DC: National Academy Press.

Battles, J. B., & Kanki, B. G. (2004). The use of socio-technical probabilistic risk assessment at AHRQ and NASA. In C. Spitzer, U. Schmocker, & V. N. Dang (Eds.), *Probabilistic safety assessment and management 2004* (vol. 4, pp. 2212–2217). Berlin: Springer.

Battles, J. B., & Lilford, R. J. (2003). Organizing patient safety research to identify risks and hazards. *Quality and Safety in Health Care, 12*(Suppl II), ii2–ii7.

Battles, J. B., & Shea, C. E. (2001). A system of analyzing medical errors to improve GME curricula and programs. *Academic Medicine, 76*(2), 125–133.

Billings, C. E. (1996). *Aviation automation: the search for a human centered approach (human factors in transportation).* Hillsdale, NJ: Lawrence Erlbaum Associates.

Cook, R. I. (1998). Two years before the mast: learning how to learn about patient safety. In *Enhancing patient safety and reducing errors in health care.* Washington, DC, National Science Foundation, p. 36.

Corbett, J. M. (1992). Work at the interface: advanced technology and job design. In P. Adler, T. Winograd (Eds.), *Usability: turning technology into tools.* (pp. 133–166). New York, NY: Oxford University Press.

Coyle, Y. M., & Battles, J. B. (1999). Using antecedents of medical care to develop valid quality of care measures. *International Journal for Quality Health Care, 11*(1), 5–12.

Department of Health, United Kingdom (DOH/UK). (2000). *An organization with a memory: a report of an expert group on learning from adverse events in the NHS.* London: National Health Service.

Donabedian, A. (1980). *Explorations in quality assessment and monitoring: the definition of quality and approaches to its assessment, vol. I.* Ann Arbor, MI: Health Administration Press.

Eisenberg, J. M. (2001, September) *Medical errors as an epidemic.* Paper presented at the national summit on medical errors and patient safety research. Washington, DC: The Quality Interagency Coordination Task Force (QuIC).

Freitag, M., & Hale, A. (1997). Structure of event analysis. In A. Hale, B. Wilpert, & M. Freitag (Eds.), *After the event: from accident to organization learning* (pp. 11–22). New York: Elsevier.

Hale, M., Slonim, A., Allen, B., Marx, D., & Kirland, J. (2004). Socio-technical probabilistic risk assessment: its application to patient safety. In C. Spitzer, U. Schmocker, & V. N. Dang (Eds.), *Probabilistic safety assessment and management 2004* (pp. vol. 4 2218–2223). Berlin: Springer.

Kaplan, H. S, Battles, J. B., Van der Schaaf, T. W., Shea, C. E., & Mercer, S. Q. (1998). Identification and classification of the causes of events in transfusion medicine. *Transfusion, 38,* 1071–1081.

Kohn, L. T., Corrigan, J. M., & Donalson, M. S. (Eds.) (2000). *To err is human: building a safer health system*. Washington, DC: National Academy Press.

Kukla, C. D., Clemens, E. A., & Morse, R. S. (1992). Designing effective systems: a tool approach. In P. Adler & T. Winograd (Eds.), *Usability: turning technology into tools* (p. 144). New York, NY: Oxford University Press.

Marx, D. A., & Slonim, A. D. (2003). Assesing patient safety risk before the injury occurs: an introduction to socio-technical probabilistic risk modelling in health care. *Quality and Safety in Health Care, 12*(Suppl II), ii33–ii37.

Quality Interagency Coordination Task Force. (2000). *Doing what counts for patient safety: federal actions to reduce medical errors and their impact*. Washington, DC, Department of Health and Human Services.

Rasmussen, J. (1976). Outlines of a hybrid model of the process operator. In T. B. Sheridan, & G. Johannsen (Eds.), *Monitoring behavior and supervisory control*. (pp. 136–147) New York, NY: Plenum Press.

Reason, J. (1990). *Human error*. New York: Cambridge University Press.

Reason, J. (1997). *The organizational accident*. New York: Ashgate.

Rheinfrank, J. J. (1992). Design for usability: crafting a strategy for the design of a new generation of Xerox copiers. In P. Adler & T. Winograd (Eds.), *Usability: turning technology into tools* (p. 39). New York: Oxford University Press.

Senders, J. W. (2004). FMEA and RCA: the mantras of modern risk management. *Quality and Safety in Health Care, 13*(4), 249–250.

Spitzer, C., Schmocker, U., & Dang, V. N. (Eds.) (2004). *Probabilistic safety assessment and management 2004*. Berlin: Springer.

Van der Schaaf, T. W. (1992). *Near miss reporting in the chemical process industry*. Ph.D. thesis, Eindhoven University of Technology Eindhoven, NL.

Weiner, J. (1995). *Research techniques in human engineering*. Upper Saddle River, NJ: Prentice Hall.

Wreathall, J., & Nemeth, C. (2004). Assessing risk: the role of probabilistic risk assessment (PRA) in patient safety improvement. *Quality and Safety in Health Care, 13*, 206–212.

Yin, R. K. (1994). *Case study research: design and methods* (2nd ed.). Thousand Oaks, CA: Sage.

Yin, R. K. (1999). Enhancing the quality of case studies in health services research. *Health Services Research,* 34(5), 1209–1223.

Zapt, D., & Reason, J. T. (1994). Introduction to error handling. *Applied Psychology*, 43, 427–432.

·25·

NEW TECHNOLOGY IMPLEMENTATION IN HEALTH CARE

Ben-Tzion Karsh and Richard J. Holden
University of Wisconsin-Madison

Research shows that a variety of new technologies are able to improve the quality and safety of care delivered by health care facilities such as hospitals, primary care clinics, nursing homes, and surgi-centers. That evidence has lead to public and private pressure for health care delivery organizations to adopt the latest technologies, which is leading many such organizations into a state of continuous technological change. This situation may lead to more harm than good unless organizations design technology implementation strategies that follow science-based guidelines. This chapter reviews the science of technology implementation and then translates the evidence into practical design guidelines, which if followed, should help to promote successful technology change and adoption.

New technologies are increasingly becoming a common sight in health care delivery organizations, especially hospitals. These new technologies have the potential to improve all aspects of health care delivery, from diagnosis and treatment to administration and billing. Diagnostics have improved with the introduction of higher resolution functional magnetic resonance imaging (fMRI), positron emission tomography (PET), and computed tomography (CT) scans, not to mention advances in laboratory medicine technology for superior analysis of blood, urine, and cultures. Technologies used for treatment span the gamut from new infusion devices, such as smart IV pumps, to surgical technologies, such as endoscopic surgical tools, improved lasers, and even surgery-assisting robots like the da Vinci™. There are also other types of technologies used in the delivery of care, such as automated medication-dispensing devices and computerized decision-support systems that can help reduce the likelihood of medical errors. The pace of new technology implementation in health care delivery has been accelerating over the years, and there is good reason to believe that this will not change any time soon.

There are a wide variety of pressures on health care delivery organizations to continually adopt new technologies because of pressures from the government, purchasing groups, and consumers. Major government-sponsored reports from the Institute of Medicine and the Agency for Healthcare Research and Quality (AHRQ) (Institute of Medicine, 2001; Kohn, Corrigan, & Donaldson, 2000; Shojania, Duncan, McDonald, & Wachter, 2001) have concluded that advanced technologies have the ability to help improve quality of care and reduce medical errors. The Leapfrog Group, which represents the employers of nearly 35 million health care consumers throughout the United States (Leapfrog Group, 2004), has made the implementation of computerized provider order entry (CPOE) one of the three practices that the group is pushing hospitals to implement. This pressure seems to be having an effect. Recent estimates suggest that up to 40% of U.S. hospitals are planning on implementing electronic order entry within the next five years (Merli & Stickler, 2004). Similarly, a 2002 ISMP survey found that 50% of the responding hospitals were considering implementing bar coding technology. With the passage of the Food and Drug Administration (FDA) "Bar Code Label Requirement for Human Drug

Products and Biological Products" in February of 2004 (Food and Drug Administration, 2004), it is quite likely that the number of hospitals that will implement bar coding systems for medication dispensing and administration will grow even more rapidly. It is clear, then, that the rapid pace of new technology introductions into health care delivery organizations will continue into the foreseeable future.

Perhaps the most talked about type of new health care technologies are those designed to improve the capture, organization, analysis, tracking, access, and dissemination of information—information technologies (IT). These include the aforementioned computerized provider order entry systems and bar coded medication administration systems, but they also include computer decision-support systems (CDSS), electronic medical records (EMR), automated laboratory information management systems, pharmacy information management systems, and all other electronic systems that are used for the capture, organization, analysis, tracking, access, and dissemination of information. Because of the emphasis on IT for improving patient safety, the remainder of this chapter will focus on IT. However, the theories and principles that will be discussed are relevant to other types of new technologies, such as advanced surgical tools.

The new technologies being implemented in primary, secondary, and tertiary care facilities have the potential to greatly improve the quality of health care delivery and patient safety. Some research suggests that the potential benefits, specifically patient safety benefits, can, in fact, be realized. Research on CPOE has shown the technology can reduce serious medication errors, preventable adverse drug events (ADEs), inappropriate doses, and inappropriate frequency, as well as improve the rate of corollary orders, medication turnaround times, radiology completion times, and laboratory result reporting times (Kaushal & Bates, 2001; Kaushal, Shojania, & Bates, 2003; Mekhjian et al., 2002). The patient safety benefits of CPOE have been found in pediatric hospitals also (King, Paice, Rangrej, Forestell, & Swartz, 2003; Potts, Barr, Gregory, Wright, & Patel, 2004).

Bar coding systems have also been shown to improve patient safety. Bar codes can be used to improve patient identification, specimen handling and tracking, medication dispensing and administration, and medical record tracking (Wald & Shojania, 2001). The most direct impact on patient safety comes from its use for medication dispensing and administration, as well as patient identification, and the little evidence available suggests reductions in medication error are possible (Kaushal, Barker, & Bates, 2001; Puckett, 1995; Wald & Shojania, 2001). But the evidence for the impact of bar coding on patient safety is very scant, and additional research is needed. Patient safety benefits have also been shown for other IT (Shojania et al., 2001), as well as for new technologies that are not information technologies, such as endoscopes designed with disposable sheaths to protect working surfaces (Rothstein & Littenberg, 1995). It is thus clear that new technologies, especially new information technologies, have great potential for improving patient safety. Great care has to be taken, though, to achieve potential benefits.

Bates warned that "the net effect (of information technologies) is ... not entirely predictable, and it is vital to study the impact of these technologies" (Bates, 2000, p. 789). Others have pointed out the need for research to study the barriers to successful adoption of IT in health care, one of the three main focuses for IT research at AHRQ (Ortiz, Meyer, & Burstin, 2001). Even CPOE, which has been shown to improve patient safety, has failed in a number of large health care delivery systems (Prabhu, 2003), and there is also literature explaining how daunting of a task it is to implement such a technology (Miller, 2000a, 2000b, 2000c) due to issues related to training, the amount of time dedicated to system design, implementation and maintenance, provider buy-in, and similar issues.

Decades of research in manufacturing and other service industries have shown that information technologies alone rarely solve problems by themselves. That is, solutions entail much more than simply turning on the switch. Systems in health care are highly complex and are "characterized by multiple actors, multiple choices, multiple hand-offs ... no ownership, no natural team, and no one with hospital-wide authority to make changes and insure quality" (Leape, 1997), which makes changing such a system overwhelming. Researchers in cognitive psychology and human factors have shown that most safety problems result from the systems in which individuals work. These errors tend to result from either failures in the design of a process, a task, or equipment or failures within an organization or the environment (Sainfort, Karsh, Booske, & Smith, 2001; Smith & Carayon, 1995). The implication is that one cannot simply throw a technological solution at a problem that has multiple interacting causes because ITs create system-wide changes in that they affect workflow, working

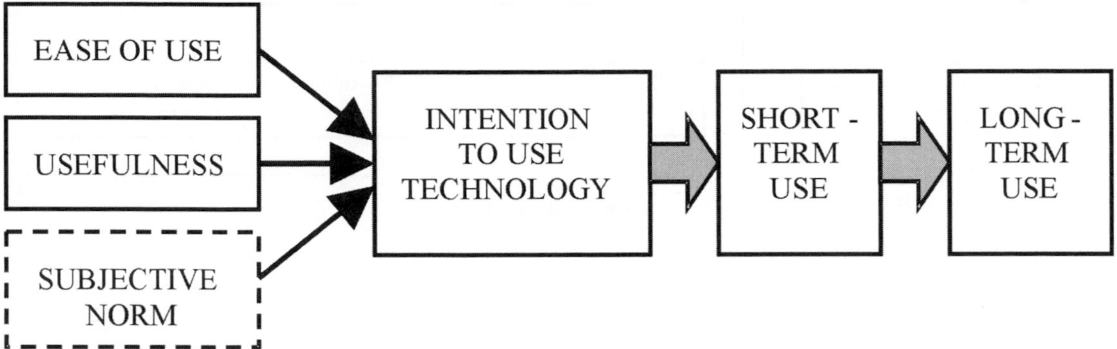

Figure 25–1. Technology Acceptance Model (Davis, Bagozzi, and Warshaw 1989); additions based on TAM2 (Venkatesh & Davis, 2000) in dashed boxes

conditions, communication networks, job security, training needs, and many other system factors (Carayon & Karsh, 2000; Eason, 1997; Eason, 2001; Eason, 1982; Eason, 1990; Karsh, 1997; Nolan, 1998, 2000; Reason, 2000; Smith & Carayon, 1995). Without studying these system changes, new and unforeseen problems are likely to arise (Majchrzak, 1992; Nolan, 2000; Smith & Sainfort, 1989) and thwart patient safety improvements.

If the total system implications of new technologies are not dealt with, not only might patient safety efforts suffer, but so too might overall quality of care, clinician performance, communication, decision making, and, ultimately, return on investment. For these reasons, the objectives of this chapter are to review the theories explaining how to correctly design IT implementation processes and then present specific design principles for effective implementation design. It is our hope that this chapter will convince the interested reader that technologies that are designed to improve patient safety, even if designed well, may not produce desired results unless the implementation — that is, the carrying out of all processes related to achieving successful use (e.g., training, financial support, maintenance, simulations, end user participation, etc.) — is designed appropriately as well.

NEW TECHNOLOGY IMPLEMENTATION THEORY

In general, models depicting the technology implementation process posit a number of system factors that jointly and/or independently predict the success of the implementation. Figures 25–1 and 25–2 illustrate two such models: the Technology Acceptance Model (TAM) and Innovation Diffusion Theory (IDT), respectively. How success is defined varies somewhat from model to model. Individual users' and organizations' behavioral intention to use and/or actual usage have been used as measures of IT implementation success in TAM (Davis, 1989), IDT (Moore & Benbasat, 1991; Rogers, 1995), the Model of PC Utilization (Thompson, Higgins, & Howell, 1991), and technology-based versions of the Theory of Planned Behavior (TPB; Harrison, Mykytyn, & Riemenschneider, 1997; Taylor & Todd, 1995), and Social Cognitive Theory (SCT; Compeau & Higgins, 1995; see Venkatesh, Morris, Davis, & Davis, 2003 for a review and comparison of "acceptance" models). Alternatively, in their review of dependent measures of IT implementation success DeLone & McLean (1992), describe five outcomes in addition to IT usage: system quality; information quality; user satisfaction; individual impact; and organizational impact. Others have used different combinations, such as usage, satisfaction, and attitudes (e.g., Al-Gahtani & King, 1999), and satisfaction alone (e.g., Adamson & Shine, 2003). Yet other outcomes of interest have included task-technology fit (Goodhue, 1995; Goodhue & Thompson, 1995) and implementation success at the organizational level (Leonard-Barton & Deschamps, 1988).

Although the various outcomes of implementation differ from one another, there is some tendency in the literature to treat them as determinants of the same construct—some desired result following the introduction of technology. To facilitate the current synthesis of the implementation literature,

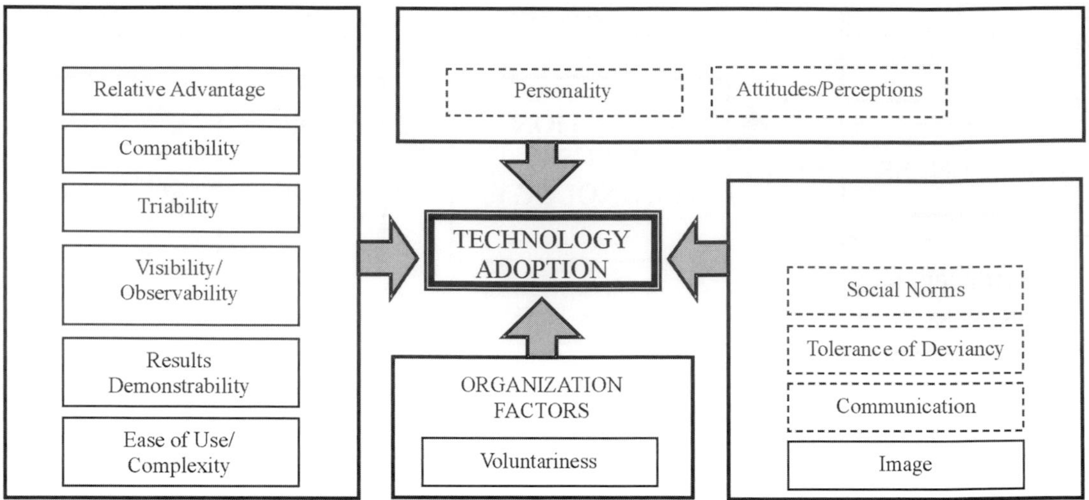

Figure 25–2. Diffusion of Innovation Theory, based on Moore and Benbasat's (1991) adaptation of Rogers (1983); dashed boxes denote additions based on Rogers (1995).

we refer to this end result as implementation *success*. Conceptually, success encompasses the entirety of outcomes studied in existing implementation literature (DeLone & McLean, 1992), irrespective of how these constituent outcomes interact. In discussing research findings we make sure to report the actual dependent measures used to assess implementation outcomes.

Few approaches have modeled implementation success by considering the system in its entirety (Klein & Ralls, 1995; Kukafka, Johnson, Linfante, & Allegrante 2003; but see Al-Gahtani & King (1999) and Klein & Sorra (1996)). More frequently, success has been investigated as a function of individual factors in isolation and sometimes as a function of a few factors combined. Kukafka et al.'s (2003) review of the IT implementation literature failed to find a single whole-system approach to implementation success, despite stating that in implementation "the picture that emerges is a complex nexus of contributing factors, as well as a rich context for the application of multi-level intervention strategies to promote IT use" (p. 219). Thus, narrow models (i.e., those that focus on only one or two kinds of predictors of success) are oversimplifying the true system into which IT is introduced (Liker, Haddad, & Karlin, 1999). From a practical standpoint, organizations investing in new IT will be more likely to succeed by considering the entirety of system factors in their implementation plan. For theoretical and pragmatic reasons, it would be crucial to understand how numerous factors previously studied interact to predict success. In the absence of supporting research, this chapter cannot provide such complete understanding. Instead, the chapter presents the implementation literature in four *a priori* categories that jointly describe a complex system: technological factors; individual/person factors; organizational factors; and social/cultural factors (see Lorenzi, Riley, Blyth, Southon, & Dixon, 1997, and Smith & Sainfort, 1989, for similar taxonomies). Whenever possible, we present evidence from the literature of how some of these factors interact.

TECHNOLOGY FACTORS

Quite predictably, a large portion of the implementation literature addresses how factors concerning the IT itself impact implementation success (Table 1). The parsimonious Technology Acceptance Model (Davis, 1989; Davis, Bagozzi, & Warshaw, 1989) posits two technology-specific predictors that have been repeatedly found to account for a large proportion of variance in individuals' intention to use new IT (see Venkatesh, 1999 for a review). TAM identifies these two predictors as perceived usefulness and perceived ease of use (see Figure 25–1). Perceived usefulness refers to potential users' perceptions about positive changes to job methods and performance as a result of using the new IT.

TABLE 25–1. Summary of Technology Factors

Factor	*Example references*
Ease of use	Davis, 1989, 1993; Venkatesh & Davis, 2000; Venkatesh et al., 2003
Usefulness	Davis, 1989, 1993; Venkatesh & Davis, 2000; Venkatesh et al., 2003
Compatibility	Chau & Hu, 2001
Impact	Chau & Hu, 2001
Complexity	Venkatesh et al., 2003
Enjoyment	Compeau & Higgins, 1995; Venkatesh et al., 2002

Similar constructs include extrinsic motivation in the Motivational Model (Davis, Bagozzi, & Warshaw, 1992), job-fit in the Model of PC Utilization (Thompson et al., 1991), relative advantage and results demonstrability in Innovation Diffusion Theory (Moore & Benbasat, 1991, 1996), and outcome expectations in Compeau and Higgins' (1995; Compeau, Higgins, & Huff, 1999) adaptation of Social Cognitive Theory.

The second major predictor in TAM, perceived ease of use, is defined as the expected ease with which potential users could learn to use and then actually utilize new IT. Thompson et al. (1991) similarly posit the construct of complexity: how complicated and difficult the IT is to learn and use (see also ease of use in IDT). Tests of Venkatesh et al.'s (2003) Unified Theory of Acceptance and Use of Technology (UTAUT) demonstrated that effort expectancy factors like ease of use predict intention to use new IT (see also Agarwal & Prasad, 1997, 1998; Karahanna, Straub, & Chervany, 1999). Additionally, perceived ease of use is thought to affect perceived usefulness (e.g., Venkatesh & Davis, 2000; but see Chau & Hu, 2001; Chau & Hu, 2002b; Hu, Chau, Sheng, & Tam, 1999, who fail to replicate both the ease of use–usefulness and ease of use–intention to use relationships with a sample of physicians). Current conceptions of TAM show that perceptions of usefulness and ease of use directly predict intention to use IT. In the original TAM, however, these perceptions were thought to indirectly influence intention to use by first affecting one's attitude toward technology. Although subsequent versions of TAM have excluded the attitude construct for the sake of parsimony, other work based on the theory of reasoned action (TRA) and the original TAM still contends that attitude is a useful predictor and that it may be especially important in certain situations, for example, when IT use is mandatory (Brown, Massey, Montoya-Weiss, & Burkman, 2002; see also the meta-analysis of implementation success predictors including attitude by Mahmood, Burn, Gemoets, & Jacquez, 2000).

Other technology characteristics predictive of implementation success include compatibility and impact. Compatibility is the degree to which implemented IT is consistent with work practices, needs, and values (Chau & Hu, 2001; Moore & Benbasat, 1991). Chau and Hu (2001) describe this construct using a health care example:

> Over time, a physician becomes increasingly accustomed to and deeply entrenched in a particular practice style and, thus, is unlikely to accept a technology that is incompatible (or perceived to be incompatible) with his or her work practices. (p. 704)

Accordingly, Chau and Hu (2002a; 2001) found that compatibility very strongly predicted perceived usefulness, which in turn strongly predicted physicians' self-reported use of telemedicine IT. Elsewhere, compatibility was predictive of both perceived usefulness (or, relative advantage) and perceived ease of use (Al-Gahtani & King, 1999; see also Venkatesh and Davis, 2000). In addition, Chau and Hu (2001) suggested, but were not able to formally test, that the impact of IT (i.e., the amount of change associated with new IT) may in part determine implementation success: physicians will be less likely to perceive IT as easy to use if it brings about major changes (see also Scholtz & Weidenbeck, 1990). Although the relationship between complexity of new IT and implementation success has not been formally modeled independently from measures of perceived ease of use, numerous models described above have been supported in respect to both complex and simple technologies (Venkatesh et al., 2003).

Another technology factor that has been studied is the enjoyment associated with new IT. Although enjoyment is a somewhat elusive factor that is difficult to manipulate, some have used the constructs of enjoyment (Davis et al., 1992) and fun (Igbaria, Schiffman, & Wieckowski, 1994) to predict implementation success. Enjoyment measures are very

TABLE 25–2. Summary of Individual/Person Factors

Factor	Example references
Age	Morris & Venkatesh, 2000; Venkatesh et al., 2003
Gender	Venkatesh & Morris, 2000; Venkatesh et al., 2003
Innovativeness	Agarwal and Prasad 1998
Learning	Agarwal and Prasad 1998
Experience	Agarwal & Prasad, 1999; Taylor & Todd, 1995
Self-efficacy	Agarwal et al., 2000; D.-R. Compeau & C.-A. Higgins, 1995
Anxiety	Compeau & Higgins, 1995

similar to those used to assess other constructs, such as intrinsic motivation (Venkatesh & Speier, 1999), affect (Compeau & Higgins, 1995; Thompson et al., 1991), and even attitude in some models (Agarwal & Prasad, 1999). Enjoyment is a perceived quality of the IT that directly predicts implementation success in some models (e.g., Compeau & Higgins, 1995; Venkatesh, Speier, & Morris, 2002) or predicts it indirectly in others by first influencing users' attitudes toward IT (e.g., Al-Gahtani & King, 1999). Finally, models based on Theory of Planned Behavior include the predictor perceived behavioral control. According to Venkatesh et al. (2003), this construct "reflects perceptions of internal and external constraints on behavior and encompasses self-efficacy, resource facilitating conditions, and technology facilitating conditions" (p. 454). Measures of perceived behavioral control have been found to predict intention to use (e.g., Brown et al., 2002; Taylor & Todd, 1995).

We transition to our next category of success-predicting factors by noting that many of the above technology factors were presented in terms of end-user perceptions. Certainly, these perceptions are not independent of actual system factors and are, in fact, based on the actual state of affairs (e.g., perceived ease of use reflects the actual amount of effort required to use the IT). One appeal to predicting success through perceptions is that perceptions are far easier to measure than are intangible system characteristics like enjoyment, for example. Additionally, such measures may capture unfair assessments of technology that do not reflect its actual characteristics yet still influence individuals' use of, satisfaction with, and attitudes toward the technology. Moreover, perception measures are uniquely capable of assessing the crucial interaction between technology characteristics and the individuals who are to use the IT. Indeed, what the user brings to the IT may be just as important in determining implementation outcomes as what the IT brings to the user (Agarwal & Prasad, 1999). We now go on to discuss the individual differences and person factors affecting implementation success.

INDIVIDUAL/PERSON FACTORS

A number of theoretical positions and findings suggest that individual differences are important determinants of implementation success (see reviews by Agarwal & Prasad, 1999; Kukafka et al., 2003; and Mahmood et al., 2000; see Table 2). Both demographic factors (e.g., age, gender) and factors more instrumental to human-technology interaction (e.g., computer experience and self-efficacy, level of education) directly and/or indirectly affect attitudes, usage, and satisfaction (Agarwal & Prasad, 1999). We first address demographic and trait factors.

A recent study by Venkatesh et al. (2003) demonstrated that individual differences in age and gender moderated a number of relationships between various predictors and reported intention to use new IT. Performance expectancy was a more important predictor of intention to use a new IT for younger men; while effort expectancy and social influence (which we will discuss later) were more important predictors of intentions to use a new IT for older women. Venkatesh and Morris (2000) also found that only women's initial intentions to use were affected by social influence (see also Venkatesh, Morris, & Ackerman, 2000) and perceived ease of use, whereas men's intentions were more influenced by usefulness (and by attitude toward technology, putatively a proxy for usefulness, in Venkatesh & Morris, 2000). Morris and Venkatesh (2000) reported that attitude toward technology was more salient for predicting behavioral intentions for younger workers, perceived behavioral control was more salient for older workers, and social influence was most salient to older women. We refer the interested reader to the original works for further descriptions and discussions of the complicated interactions involving age, gender, and a variety of other predictors and moderators. A great deal of past work has also addressed the effect of individual differences in personality and cognitive styles on implementation success (see Zmud, 1979, and more recently Lu, Yu, & Lu, 2001), though no modern research has explicitly examined cognitive factors like general fluid intelligence and executive (cognitive) control (for a review of these constructs, see Kane & Engle, 2002). Recently, Agarwal and Prasad (1998) proposed that a new stable personality trait indirectly affects perceived ease of use: personal innovativeness with IT. This trait, defined as one's willingness to try out new IT, is perhaps a reflection of technology-related risk taking (see also Lewis, Agarwal, & Sambamurthy, 2003). Similarly, one could consider attitude toward technology, which we have discussed as a technology factor, to reflect an affective response determined by one's experience and personality.

Other individual differences found to be predictive of implementation success are situational; that is, they are learned and, therefore, modifiable. Agarwal and Prasad (1999), for example, theorized that TAM factors are influenced by individual differences in factors related to learning and experience (see also Bostrom, Olfman, & Sein, 1990; Levine, 2001; and Pisano, Bohmer, & Edmondson, 2001, who discuss the importance of learning in determining the success of implementation). In support of their model, Agarwal and Prasad (1999) found that individual differences in level of education and in amount of prior, relevant experience positively influenced perceived ease of use. These findings are consistent with other studies showing that prior, relevant experience (e.g., prior computer experience is relevant to IT use) and skills related to the new IT directly and indirectly predict implementation success. User experience and skill, often measured together, have been modeled as determinants of perceived ease of use and perceived usefulness, self-efficacy, attitudes toward the implemented technology, and of success outcomes like user satisfaction (e.g., Agarwal & Prasad, 1999; Agarwal, Sambamurthy, & Stair, 2000; Szajna, 1996; Taylor & Todd, 1995; Thompson, Higgins, & Howell, 1994). Mahmood et al.'s (2000) meta-analysis revealed that the combined effect sizes from studies of user skills and prior experience ranked third and eighth, respectively, among other predictors of user satisfaction with IT. Another important type of experience that has been considered is experience with the implemented IT itself (Venkatesh & Davis, 2000; Venkatesh et al., 2003). Several findings suggest that relationships between variables that predict implementation success and measures of success (e.g., intentions to use) change with increases in such experience. Other work suggests, however, that more tenured workers are more resistant to change and less accepting of new IT than are workers with less company tenure (e.g., Adamson & Shine, 2003; Harrison & Rainer, 1992; Majchrzak & Cotton, 1988).

The degree to which individuals feel themselves capable of using IT to achieve results (i.e., self-efficacy) has been posited in an adaptation of Social Cognitive Theory and elsewhere to positively influence IT usage, as well as ease of use, perceived usefulness, and attitude and anxiety toward behaviors involving IT (Adamson & Shine, 2003; Compeau & Higgins, 1995; Lewis et al., 2003; Venkatesh & Davis, 1996; see Agarwal et al., 2000 for a review). Anxiety toward technology use, according to SCT, also predicts success (Compeau & Higgins, 1995). Anxiety and computer aptitude may interact to influence perceptions of ease of use (Venkatesh et al., 2003). The precursors of anxiety itself are not well described, though anxiety may be determined by one's level of education (more education leads to less anxiety Igbaria & Parsuraman, 1989).

ORGANIZATION FACTORS

Consonant with our systems framework, Edmondson (2003) suggests that IT implementation is not driven strictly by technology factors; instead, and quite logically so, implementation is determined largely by the interaction of technology and the organization (see also Lorenzi et al., 1997; see Table 3). For instance, training and support provided by the organization may counteract technology factors such as difficulty and complexity, as well as person factors such as skills and experience.

IT implementation can be a large and important change. Actions taken by organizations to accommodate such change have been proposed to affect implementation success (Gallivan, 2001). Two of the most studied organizational factors affecting success are training and organizational support. Both the amount and type of training have been implicated as direct and/or indirect predictors of success. Reported use of spreadsheet technology by students with one year of work experience, for example, was found to be directly affected by the total amount of various kinds of training (Al-Gahtani & King, 1999). Training has also been reported to predict both perceived usefulness (Agarwal & Prasad, 1999; Igbaria, Guimaraes, & Davis, 1995) and perceived ease of use (Igbaria et al., 1995; Venkatesh & Davis, 1996). Additionally, some research suggests the favorable effect of game-based (Venkatesh, 1999), as well as hands-on and exploratory training (Compeau & Higgins, 1995; Pisano et al., 2001) on implementation success.

Some literature also explores technical and organizational support and organizational commitment. Mahmood et al.'s (2000) review of the literature found that support (including both technical support and management's support of implemented IT) has had the fourth largest effect size in determining user satisfaction with IT. Additionally, facilitating conditions—support structures available that facilitate IT use—are thought to positively affect use (Thompson et al., 1991). Facilitating conditions may be especially important when the implemented IT is complex and difficult to use (Edmondson, 2003; Venkatesh et al., 2003). Top and local management support of/commitment to the implemented IT has also been suggested to indirectly affect usage (e.g., Klein, Conn, & Sorra, 2001; Lewis et al., 2003).

Two predictors of success relate to managerial decisions made about the voluntariness of implemented IT and user participation in its development. In certain cases individuals have no other choice but to use implemented IT, either because the old method has become obsolete or inaccessible, or due to organizational policy mandating its use. The extent to which individuals in an organization are free to choose whether or not to use new IT is referred to as voluntariness (e.g., Hartwick & Barki, 1994; Venkatesh & Davis, 2000). Voluntariness may alter relationships found in models like TAM. In studies examining mandatory IT, ease of use was found to be a much stronger predictor of intention to use than was perceived usefulness; perceived usefulness and ease of use were not found to predict behavioral intention indirectly through an attitude measure; and perceived behavioral control and subjective norms (social influence) were, however, directly predictive of intention to use (Brown et al., 2002); we discuss voluntariness as a moderator of social influence in the next section on social and cultural factors. In addition, user involvement, also described as end-user participation, in development and implementation of IT has been widely studied and has been linked to increased satisfaction with, and positive feelings toward, the IT, less resistance to technological change, and increased user performance, especially in the presence of technological complexity (Baronas & Louis, 1988; Franz & Robey, 1986; Karsh, 1997; Korunka, Weiss, & Karetta, 1993; McKeen, Guimaraes, & Wetherbe, 1994; see Mahmood et al., 2000 for a review). Mahmood et al.'s (2000) meta-analysis of implementation literature found that the effect size of user involvement in new technology development was the largest among all investigated predictors of user satisfaction with IT, including perceived usefulness.

A final organization factor that may be related to implementation success is the type of organization where the IT is implemented. To our knowledge, there has been no explicit comparison of organization type in determining success. Suffice it to say, support for models of implementation success has been achieved in a large variety of organizational settings, including health care, finance, retail, and the computer industry. We discuss studies on implementation specifically in health care later in this chapter.

SOCIAL/CULTURAL FACTORS

Our categorization of implementation success predictors concludes with a discussion of social factors, including such social constructions as group and culture (see Table 4). By "social," we refer to factors defined by implicit and explicit interactions between individuals that are learned and constructed over time.

Recent implementation models—like TAM2, an extended version of TAM—have included the construct of social influence (e.g., Venkatesh & Davis, 2000). Social influence refers to the real or perceived pressure to act in a certain way that is exerted on individuals by others. When this influence is explicit, it can be referred to as a subjective norm: the degree to which people who are important to an individual (e.g., co-workers and superiors) believe that the individual should act in a certain way. Within implementation literature, subjective norm has been found to influence perceived usefulness and intention to use IT (Adamson & Shine, 2003; Brown et al., 2002; Venkatesh & Davis, 2000; Venkatesh et al., 2000; Venkatesh et al., 2003). Similar social influence factors include the IDT constructs of visibility (i.e., the extent to which one can see others using the IT) and image (i.e., the degree to which IT use can "enhance one's image or status in one's social system;" Moore

& Benbasat, 1991, p. 195), and the construct of social factors in Thompson et al.'s (1991) Model of PC Utilization. Social influence has been found to be especially important in predicting intention to use (and thus, perhaps to accept) implemented IT when new IT use is mandatory (e.g., Brown et al., 2002; Hartwick & Barki, 1994; Venkatesh & Davis, 2000). This may suggest that measures of social influence reflect individuals' wishes to secure approval or avoid consequences associated with use or non-use of new IT, respectively (Taylor & Todd, 1995), or similarly, individuals are complying with important others for the purpose of social gain (Venkatesh & Davis, 2000). We note one exception: Chau and Hu (2001; 2002a, 2002b) found that physicians' intention to use new IT was seemingly uninfluenced by subjective norms. The authors discussed this finding in terms of physicians' independence compared to typical end users.

A final social/cultural factor theorized to affect technology implementation is group differences. As opposed to individual differences within groups, group differences imply differences in culture between two or more well-defined entities (e.g., nurses, pharmacists, physicians). Edmondson (2003) and Gallivan (2001) report on qualitative studies revealing cultural differences between different hospitals and between different groups within the same telecommunications company, respectively, that lead to differences in feelings and actions regarding new IT. Other work shows how different—and sometimes inconsistent—views regarding technology may be between different groups of individuals, for example, between IT professionals and IT users (Agarwal & Prasad, 1999). An extreme case of between-groups cultural differences in terms of IT use is discussed by Demeester (1999), who frames technology-related decision making as partially a function of international culture.

IMPLEMENTATION RESEARCH IN HEALTH CARE

Very little overall IT implementation research has been conducted in a health care context. Even less has explicitly addressed how differences between health care and non-health care organizations may impact models of IT implementation. This fact is rather problematic, given the unique characteristics of health care organizations. Consequently, the applicability of implementation theory to health care is not very well established, and further work is needed to understand the factors that contribute to implementation success in health care.

Several studies have, however, addressed implementation specifically in health care settings. We briefly describe them here. Chau and Hu (2001) tested the Technology Acceptance Model and two models of the Theory of Planned Behavior using physicians' self-reported intention to use telemedicine IT as the dependent measure. Their findings were inconsistent with both models. Namely, neither TAM's perceived ease of use nor TPB's subjective norms directly or indirectly predicted intention to use (see also Hu et al., 1999). In contrast, another study of telemedicine found that a composite normative factor (comprising social and personal influence) and a measure of self-identity together explained 81% of the variance in physicians' intention to use telemedicine IT (Gagnon et al., 2003). Ash et al. (2001) conducted a qualitative study comparing three hospitals that successfully implemented computerized provider order entry. Based on observation, focus groups, and formal interviews, the authors found several factors important to successful CPOE implementation including quality, compatibility, and customizability of the system, as well as the training and technical support provided to users. The authors were able to describe their results in the IDT framework. Edmondson's (2003) qualitative work comparing 16 hospitals' implementation of cardiac surgery technology suggests the importance of frameworks—the different ways in which new technology may be viewed by potential end users. Using a broader level of analysis, Hu, Chau, and Sheng (2002) investigated factors leading public health care organizations in Hong Kong to adopt and successfully implement telemedicine IT. Physicians' collective attitudes toward IT predicted adoption (organizations whose physicians had more positive attitudes were more likely to adopt), as did perceptions of risks to patient care as a result of telemedicine use (less perceived risk was related to the tendency to adopt). However, the authors reported an unexpected negative relationship between perceived ease of use and adoption, and no significant relationships between a number of other predictors (e.g., perceived service benefits, perceived technology safety) and adoption (Hu et al., 2002).

Finally, some recent work on IT implementation in health care has adopted a socio-technical systems framework (Berg, 1999; Berg, Langenberg, vander Berg, & Kwakkernaat, 1998; Stricklin & Struk, 2003). This framework regards health care work as

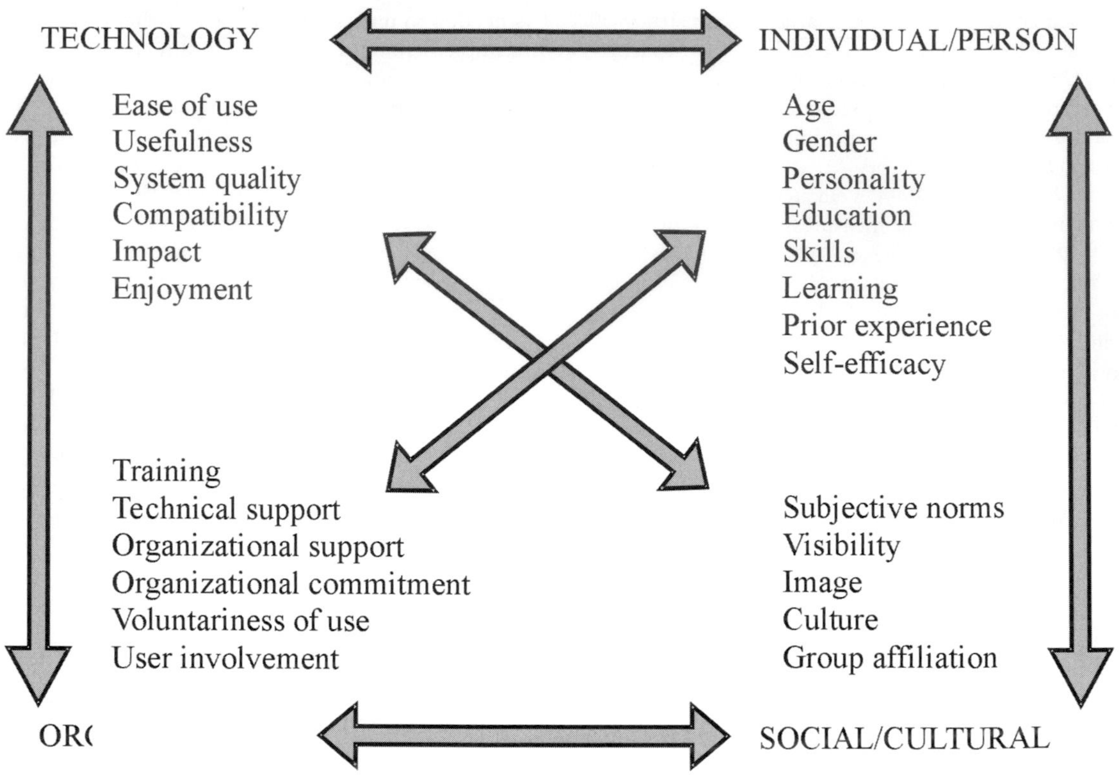

Figure 25–3. Whole-System Organization of Predictors of IT Implementation Success (based on Smith and Sainfort, 1989)

fluid and complex. Thus, in order to successfully implement IT in a health care setting, a bottom-up approach must be taken to understand the complexity of the system pre-implementation. This understanding should be used to implement IT that supports, enhances, fits with, and does not disrupt extant care processes (Berg et al., 1998). To achieve this, end users must be involved in an iterative design and implementation process that continually searches for and addresses concerns about user-IT misfit (Berg, 1999; Berg et al., 1998; Stricklin & Struk, 2003). Stricklin and Struk (2003), in a special section on socio-technical approaches to health care IT, describe several quantitative and qualitative studies of health care (Point of Care) IT implementation. The authors discuss four principles for implementation success: training designed to accommodate the needs and concerns of end users; a system design that is functional and adjustable in response to usability issues; organizational support of change, partially achieved by iteratively implementing, evaluating, and, when necessary, adjusting the IT; and management of end-user satisfaction by involving users in the entire process and being responsive to their needs.

SUMMARY OF IMPLEMENTATION THEORY

Our presentation of the implementation literature condenses a considerably large field of study into four distinct types of factors predicting implementation success (see Figure 25–3). Technology factors such as perceived ease of use and system quality describe characteristics (perceived or real) of the specific IT being implemented. Individual/person factors include demographics and traits like age and personality, as well as individual factors like skills and experience that are instrumental to IT use. Organization factors like training and support reflect the decisions made and the structures created by the organization concerning the development and implementation of IT. Social/cultural factors concern the influences of others (i.e., social influences) and of the groups to which one belongs.

TABLE 25–3. Summary of Organizational Factors

Factor	*Example references*
Organizational support	Mahmood et al., 2000; Thompson et al., 1991
Training	Al-Gahtani & King, 1999; Compeau & Higgins, 1995; Pisano et al., 2001
Voluntariness	Brown et al., 2002
End user participation	Baronas & Louis, 1988; Karsh, 1997; Korunka et al., 1993

TABLE 25–4. Summary of Social/Cultural Factors

Factor	*Example references*
Social influence	Adamson & Shine, 2003; Venkatesh & Davis, 2000
Group differences	Demeester, 1999; Edmondson, 2003; Gallivan, 2001

TABLE 25–5. Design Recommendations

Top management commitment	Simulation before, during and after implementation	Develop clear communication structures
Design to accommodate the range of user needs	Create readiness for technology change	Implement at a pace that facilitates learning
End-user participation	Training	Pilot test
Benchmark	Study the horizon	Utilize feedback systems
Structured program		

Although we have established these four groups, this was not to imply that each described factor falls neatly into any one group. In fact, a factor like attitude toward technology could be categorized as an individual or a technology factor; more likely, it is a combination of both. Additionally, other groups of factors that are not treated here exist. These include, for example, task factors (i.e., those related to the task for which the IT is to be used, e.g., task complexity) and environmental factors (i.e., those external to the organization, e.g., the economy). Finally, we stress the importance of a systems approach to understanding the implementation literature. That is, while few studies address all the factor groups jointly, we have gone out of our way to demonstrate that factors in each group have influence on implementation success. Figure 25–3 illustrates our whole-systems framework, wherein predictors of success are organized in four categories that interact with one another. Accordingly, as we present practical suggestions for implementation success in the following section, we urge a systems approach to new IT implementation.

PRACTICAL SUGGESTIONS FOR IMPLEMENTING NEW TECHNOLOGY IN HEALTH CARE

The review of empirical evidence from research on technology acceptance, technology satisfaction, and new technology implementations identified specific variables that have been shown to impact successful implementation. Those variables can be turned into implementation process design recommendations that, if followed, can help to promote successful IT implementation (see Table 5).

As will be explained, the principles must themselves be designed carefully to promote success and reduce the many possible negative outcomes such as fear, uncertainty, and technology rejection. Collectively, these design recommendations target the predictors of success identified in the literature, such as self-efficacy, perceived ease of use, perceived usefulness, and control while also accommodating for the range of user characteristics that impact implementation success, such as prior experience with technology and age.

Top management commitment, a ubiquitous recommendation for successful change (Smith & Carayon, 1995), must extend beyond statements to actions. True top management commitment makes resources available to the project such as people, money, and time in sufficient quantities so as to ensure a successful implementation. True top management commitment also enables and promotes each of the other strategies of implementation design. Top management commitment is therefore necessary but not sufficient for success. The remaining design recommendations enable success.

To ready the organization for the new IT, a number of steps should be taken. The current system into which the new IT will be implemented should be analyzed to identify (a) the changes in processes that will be caused by the new IT, (b) new support structures that will be necessary to support the change, (c) new processes that might need to be developed because of the change, (d) how jobs will be affected, (e) how the environment might need to be adjusted, and (f) other technologies that might be affected or will need to be added. The above analysis is referred to as the socio-technical systems approach, discussed previously in our review of implementation theory. This analysis is crucial for making sure that the impacts of the new IT can be dealt with proactively through design, as opposed to reactively in response to user and manager performance problems. It goes without saying that implementation success begins with the proper selection or design of the IT. A proper socio-technical systems analysis can provide the information needed to inform design.

Once the analysis is complete, the implementation team should begin to design the change process. That will include designing the support structures necessary to support the change. These structures might include a help desk, accessible via telephone, e-mail, the Web, and walk-in; direct access to vendor support; and direct access to implementation team members. Such structures might also include the designation of lead or expert employees who can serve as resources for the others. The nature of the support will vary on a case-by-case basis, but the minimum requirement is to have technical support easily available at all times when users might need it.

To further anticipate what kinds of changes will take place and be needed, the implementation team should seek out organizations that underwent similar IT implementations. Such benchmarking data will complement that which is learned through internal analysis. Scanning and studying the environmental, technological, organizational, and regulatory horizon is yet a third way to plan for changes down the road that could affect or be affected by new IT implementations. For example, prior to deciding on whether to implement a particular IT, members of the project team could investigate whether or not superior technologies are on the horizon that could offer even better returns on investment. Similarly, the project team should investigate regulatory bodies such as the JCAHO to find out whether any new regulatory policies might be on the horizon that may affect how or whether the organization uses the IT under consideration.

A final way to ready the organization is to design a structured program for the implementation that details the steps necessary for successful implementation and the people responsible for the steps (Da Villa & Panizzolo, 1996; Hunsucker & Loos, 1989; Smith & Carayon, 1995). This provides end users with a road map, which can reduce uncertainty and promote feelings of control and organizational justice. It may reduce many of the fears associated with new IT because end users will know why the change is coming, what to expect, when to expect it, where to expect it to happen and who will be in charge.

Simulation is a tool that should be designed into the implementation process as a means for planning, training, and continuous learning in order to promote self-efficacy and demonstrate ease of use and usefulness of the new IT. After the current work system is analyzed to anticipate how the new IT will impact the system, a new simulated work system with the new processes and structures can be created to study how the new work system functions. Simulation should also be used in deciding which vendor and product to select. Once a product is selected, the integration of the product into the current work system can be simulated to further study potential changes caused by the introduction of the new IT, as well as changes necessary to support the new IT. The most obvious time for simulation is during the user training program, but simulation should not end there.

Even once the new IT is implemented and employees are actively using it, simulators can provide helpful learning experiences. For the first few months following implementation, it is a good idea to have a simulator available to employees for trying out functions without fear of causing patient, employee, or technology harm. Such a simulator could be as simple as an on-site room with computers simulating the new IT. Employees could come into this room at any point during their shift to try out new ideas or new functions and learn how they work. This promotes learning as well as control and may help to reduce the workload of technical support. To be effective, though, the simulator must show the employee the impact of his or her actions. For example, if a physician wants to try out a new function on a new computerized order entry system, the simulator must not only allow the physician to determine if the function is allowable, but also show the physician how the entry is received by pharmacy. That way the physician knows the true impact of the function and can make an informed decision as to how or when to use it.

Training is a well-studied and effective intervention (Agarwal et al., 2000; Cannon-Bowers, Salas,

Tannenbaum, & Mathieu, 1995; Cannon-Bowers, Tannenbaum, Salas, & Converse, 1991; Salas & Cannon-Bowers, 2001) because it can improve self-efficacy and demonstrate ease of use and usefulness of new IT. However, training is typically not designed according to scientific principles and therefore often falls well short of its potential. Training should be designed to promote the transfer of knowledge and skills into work practice. To do that successfully, the content of the training and method of delivery must match the goals. Evaluations must be used to determine if the audience learned the material. Furthermore, barriers to successfully transferring what was learned into work practices must be identified and removed. For example, if certified nursing assistants in nursing homes are trained in a two-person patient transfer technique, but on-the-floors nursing assistants are not available to help one another, successful transfer of the learned technique will not occur.

End-user participation throughout the implementation of a new IT is important (Hurley, 1992; Hyclak & Kolchin, 1986; Korunka et al., 1993), but like the other design recommendations, it must be designed appropriately. If it is not designed appropriately, a myriad of bad outcomes, such as delayed decisions, poor decisions, and angry users, can ensue. To design end-user participation appropriately, an organization must be sure to provide the right type of participation, for the right group of users, at the right time in the process. If that is achieved, end-user participation in decision-making may enhance self-efficacy and control and alleviate fears. For example, during the IT selection phases, it is possible that no end users will have sufficient expertise to actually decide which technologies to select. A more appropriate type of participation at this stage may instead be to involve the end users in defining design requirements based on their expertise as users in the system. The end users chosen to participate in this phase should represent all possible groups of end users that could be affected by the IT. Trying to involve users in decisions that are beyond their skill or well beneath their potential level of contribution will likely result in angry end users. Another key design consideration for end-user participation is the treatment of end-user input. Organizations wanting to involve end users in IT implementation decisions need to design a system for responding to their input. If inputs are ignored, users will stop participating and perceive that management does not really want their opinions. Management must design a system (e.g., through direct communication, emails, newsletters, etc.) that responds to end-user input by explaining which ideas are being used and how, as well as which inputs are not being used and why.

Communication structures and networks, which transmit communication among users, managers, technical support, and others, should be designed to meet a number of different goals. First, they should promote the transmission of information about the implementation to end users to reduce uncertainty about the change and thus reduce fear and stress. Second, they should be designed to allow end users to obtain quick feedback to questions about the new IT prior to, during, and after the implementation. Third, they should be designed such that users know exactly who to go to with different types of questions. By targeting these goals, communication can promote end-user control as well as perceptions of organizational justice.

Pilot testing of new technologies is also important because it facilitates the identification of additional problems that were not uncovered during the proactive system analysis, benchmarking, or horizon scanning. It also promotes self-efficacy and end-user control and may demonstrate ease of use and usefulness of the IT. One shortfall of pilot testing is that it often occurs only in select areas of the organization that do not necessarily generalize to other areas of the organization that will be using the IT. To prevent this common problem, locations for pilot testing should be selected to be representative of all areas that will be affected by the new IT. That typically means pilot testing in a number of locations throughout the organization. Pilot testing should not only be done of the new IT itself, but also of the communication structures and other support systems designed to facilitate successful implementation.

The pace of an IT implementation also deserves careful consideration. The pace must be such that it facilitates learning and does not create stress among the end users. To achieve those goals, it is often a good idea to temporarily suspend throughput goals or even lower them in the early stages of implementation and then only gradually increase the production goals to pre-implementation levels as users become comfortable with the new IT. In this way, users can focus on incorporating the new IT into their jobs without fearing the need to rush or to skip important learning opportunities.

All aspects of the design of the implementation processes for a new IT must be designed to accommodate for the range of potential users and their needs. This is a basic tenet of human factors engineering (Sanders & McCormick, 1993). That means, for example, that training should accommodate the range of competences and self-efficacy; simulators should accommodate for the range of technical, visual, and psychomotor skills among the users; and communication modes should accommodate the range of user preferences. If the implementation processes are not designed to be sufficiently flexible so as to accommodate for the range of user needs, it will be more difficult to achieve a successful implementation.

LIMITATIONS AND FUTURE DIRECTIONS

A number of limitations of the theoretical and empirical work on new IT implementation limit the type and precision of practical suggestions. We therefore suggest directions for the future of IT implementation research. In this chapter, we presented implementation theory using a whole-systems framework and encouraged a systems approach to implementation. Accordingly, future empirical work must address implementation as existing in a complex system. Numerous factors of all sorts should be studied together as should interactions between these factors and the relative importance and efficacy of these factors in predicting implementation success. More research needs to focus on task-related predictors of success, including task difficulty and complexity and task demands and workload. Modelling more objective measures of technology-related predictors, such as usefulness and system quality, can compliment user perceptions and add a deeper understanding of the specific role of technology characteristics in determining implementation outcomes. Future implementation research must explore differences between health care and non-health care organizations, as well as other interorganizational differences (e.g., service vs. manufacturing). Cultural, group, and social factors need to be quantified in order to better understand their contribution to implementation success.

To avoid implementation failure, investigations should not stop at success and should additionally seek out factors that contribute to negative outcomes like rejection of IT. Research should address the importance of success outcomes such as use, acceptance, and satisfaction: are they associated with financial gains, lower turnover, or performance improvement? Could technology use and acceptance have negative consequences as well? Finally, a fruitful approach to understanding IT implementation must include awareness of the individual differences that affect learning, performance, attitude, and perceptions toward new technology. We propose that a deeper understanding of psychological constructs, such as intrinsic and extrinsic motivation (Sansone & Harackiewicz, 2000), affective disposition (Watson, Clark, & Tellegen, 1988), personality (Hogan, Johnson, & Briggs, 1997), intelligence (Sternberg, 1997), and executive (cognitive) control (Kane & Engle, 2002) will provide invaluable direction for design and implementation decisions concerning new IT.

CONCLUSION

Technologies will continue to be implemented into health care organizations at a rapid pace for the foreseeable future. These technologies hold the promise of improving the safety and quality of care delivered to millions of patients. However, positive outcomes are in no way guaranteed and may even elude health care organizations unless the manner in which they are implemented is carefully designed. In this chapter we have presented theoretical and empirical evidence showing that there are known variables that impact the success of technology implementation. We have also translated the predictors of success into practical design recommendations, which if followed, should facilitate successful technology implementation. The pitfalls of poorly designed implementations are great, but through science-based design, they can be avoided.

References

Adamson, I., & Shine, J. (2003). Extending the new technology acceptance model to measure the end user information systems satisfaction in a mandatory environment: a bank's treasury. *Technology Analysis & Strategic Management, 15*(4), 441–455.

Agarwal, R., & Prasad, J. (1997). The role of innovation characteristics and perceived voluntariness in the acceptance of information technologies. *Decision Sciences, 28*(3), 557–582.

Agarwal, R., & Prasad, J. (1998). A conceptual and operational definition of personal innovativeness in the domain of information technology. *Information Systems Research, 9*(2), 204–215.

Agarwal, R., & Prasad, J. (1999). Are individual differences germane to the acceptance of new information technologies? *Decision Sciences, 30*(2), 361–391.

Agarwal, R., Sambamurthy, V., & Stair, R. M. (2000). Research report: the evolving relationship between general and specific computer self-efficacy—an empirical assessment. *Information Systems Research, 11*(4), 418–430.

Al-Gahtani, S. S., & King, M. (1999). Attitudes, satisfaction and usage: factors contributing to each in the acceptance of information technology. *Behaviour & Information Technology, 18*(4), 277–297.

Ash, J. S., Lyman, J., Carpenter, J., & Fournier, L. (2001). A diffusion of innovations model of physician order entry. *Journal of the American Medical Informatics Association*, 22–26.

Baronas, A. M. K., & Louis, M. R. (1988). Restoring a sense of control during implementation—how user involvement leads to system acceptance. *Mis Quarterly, 12*(1), 111–123.

Bates, D. W. (2000). Using information technology to reduce rates of medication errors in hospitals. *British Medical Journal, 320*, 780–791.

Berg, M. (1999). Patient care information systems and health care work: a sociotechnical approach. *International Journal of Medical Informatics, 55*(2), 87–101.

Berg, M., Langenberg, C., vander Berg, I., & Kwakkernaat, J. (1998). Considerations for sociotechnical design: experiences with an electronic patient record in a clinical context. *International Journal of Medical Informatics, 52*(1–3), 243–251.

Bostrom, R. P., Olfman, L., & Sein, M. K. (1990). The importance of learning style in end-user training. *Mis Quarterly, 14*(1), 101–119.

Brown, S. A., Massey, A. P., Montoya-Weiss, M. M., & Burkman, J. R. (2002). Do I really have to? user acceptance of mandated technology. *European Journal of Information Systems, 11*(4), 283–295.

Cannon-Bowers, J. A., Salas, E., Tannenbaum, S. I., & Mathieu, J. E. (1995). Toward theoretically based principles of training effectiveness—a model and initial empirical investigation. *Military Psychology, 7*(3), 141–164.

Cannon-Bowers, J. A., Tannenbaum, S. I., Salas, E., & Converse, S. A. (1991). Toward an integration of training theory and technique. *Human Factors, 33*(3), 281–292.

Carayon, P., & Karsh, B. (2000). Sociotechnical issues in the implementation of imaging technology. *Behaviour and Information Technology, 19*(4), 247–262.

Chau, P. Y. K., & Hu, P. J. H. (2001). Information technology acceptance by individual professionals: a model comparison approach. *Decision Sciences, 32*(4), 699–719.

Chau, P. Y. K., & Hu, P. J. (2002a). Examining a model of information technology acceptance by individual professionals: an exploratory study. *Journal of Management Information Systems, 18*(4), 191–229.

Chau, P. Y. K., & Hu, P. J. H. (2002b). Investigating healthcare professionals' decisions to accept telemedicine technology: an empirical test of competing theories. *Information & Management, 39*(4), 297–311.

Compeau, D. R., & Higgins, C. A. (1995). Computer self-efficacy: development of a measure and initial test. *MIS Quarterly, 19*, 189–211.

Compeau, D. R., Higgins, C. A., & Huff, S. (1999). Social cognitive theory and individual reactions to computing technology: a longitudinal study. *Mis Quarterly, 23*(2), 145–158.

Da Villa, F., & Panizzolo, R. (1996). Empirical study of the adoption and implementation of advanced technologies in the Italian public sector. *International Journal of Technology Management, 12*(2), 181–198.

Davis, F. D. (1989). Perceived usefulness, perceived ease of use, and user acceptance of information technology. *Mis Quarterly, 13*(3), 319–340.

Davis, F. D. (1993). User acceptance of information technology system characteristics, user perceptions and behavioral impacts. *International Journal of Man-Machine Studies, 38*(3), 475–487.

Davis, F. D., Bagozzi, R. P., & Warshaw, P. R. (1989). User acceptance of computer technology a comparison of 2 theoretical models. *Management Science, 35*(8), 982–1003.

Davis, F. D., Bagozzi, R. P., & Warshaw, P. R. (1992). Extrinsic and intrinsic motivation to use computers in the workplace. *Journal of Applied Social Psychology, 22*(14), 1111–1132.

DeLone, W. H., & McLean, E. R. (1992). Information systems success: the quest for the dependent variable. *Information Systems Research, 3*(1), 60–95.

Demeester, M. (1999). Cultural aspects of information technology implementation. *International Journal of Medical Informatics, 56*(1–3), 25–41.

Eason, K. (1997). Understanding the organizational ramifications of implementing information technology systems. In M. Helander, T. K. Landauer & P. Prabhu (Eds.), *Handbook of human–computer interaction* (pp. 1475–1495). New York: Elsevier Science.

Eason, K. (2001). Changing perspectives on the organizational consequences of information technology. *Behaviour and Information Technology, 20*(5), 323–328.

Eason, K. D. (1982). The process of introducing information technology. *Behaviour and Information Technology, 1*(2), 197–213.

Eason, K. D. (1990). New systems implementation. In J. Wilson & E. N. Corlett (Eds.), *Evaluation of human work: a practical ergonomic methodology* (pp. 835–849). London, U.K: Taylor and Francis.

Edmondson, A. C. (2003). Framing for learning: lessons in successful technology implementation. *California Management Review, 45*(2), 34ff.

Food and Drug Administration. (2004). Bar code label for human drug products and blood. *Federal Register, 69*(38), 9119–9171.

Franz, C.-R., & Robey, D. (1986). Organizational context, user involvement, and the usefulness of information systems. *Decision Sciences, 17*, 329–356.

Gagnon, M. P., Godin, G., Gagne, C., Fortin, J. P., Lamothe, L., Reinharz, D., et al. (2003). An adaptation of the theory of interpersonal behaviour to the study of telemedicine adoption by physicians. *International Journal of Medical Informatics, 71*(2–3), 103–115.

Gallivan, M. J. (2001). Meaning to change: how diverse stakeholders interpret organizational communication about change initiatives. *IEEE Transactions on Professional Communication, 44*(4), 243–266.

Goodhue, D. L. (1995). Understanding user evaluations of information systems. *Management Science, 41*(12), 1827–1844.

Goodhue, D. L., & Thompson, R. L. (1995). Task-technology fit and individual-performance. *Mis Quarterly, 19*(2), 213–236.

Harrison, A. W., & Rainer, R. K. (1992). The Influence of individual differences on skill in end-user computing. *Journal of Management Information Systems, 9*(1), 93–112.

Harrison, D. A., Mykytyn, P. P., & Riemenschneider, C. K. (1997). Executive decisions about adoption of information technology in small business: theory and empirical tests. *Information Systems Research, 8*(2), 171–195.

Hartwick, J., & Barki, H. (1994). Explaining the role of user participation in information-system use. *Management Science, 40*(4), 440–465.

Hogan, R., Johnson, J., & Briggs, S. (1997). *Handbook of personality psychology*. San Diego: Academic Press.

Hu, P. J. H., Chau, P. Y. K., & Sheng, O. R. L. (2002). Adoption of telemedicine technology by health care organizations: an exploratory study. *Journal of Organizational Computing and Electronic Commerce, 12*(3), 197–221.

Hu, P. J., Chau, P. Y. K., Sheng, O. R. L., & Tam, K. Y. (1999). Examining the technology acceptance model using physician acceptance of telemedicine technology. *Journal of Management Information Systems, 16*(2), 91–112.

Hunsucker, J. L., & Loos, D. (1989). Transition management: an analysis of strategic considerations for effective implementation. *Engineering Management International, 5,* 167–178.

Hurley, J.-J. (1992). Towards an organisational psychology model for the acceptance and utilisation of new technology in organisations. *Irish Journal of Psychology, 13*(1), 17–31.

Hyclak, T. J., & Kolchin, M. G. (1986). Worker involvement in implementing new technology. *Technovation, 4*, 143–151.

Igbaria, M., & Parsuraman, S. (1989). Path analytic study of individual characteristics, computer anxiety, and attitudes toward microcomputers. *Journal of Management, 15*(3), 373–388.

Igbaria, M., Guimaraes, T., & Davis, G. B. (1995). Testing the determinants of microcomputer usage via a structural equation model. *Journal of Management Information Systems, 11*(4), 87–114.

Igbaria, M., Schiffman, S. J., & Wieckowski, T. J. (1994). The respective roles of perceived usefulness and perceived fun in the acceptance of microcomputer technology. *Behaviour & Information Technology, 13*(6), 349–361.

Institute of Medicine (Ed.). (2001). *Crossing the quality chasm: a new health system for the 21st century*. Washington DC: National Academy Press.

Kane, M. J., & Engle, R. W. (2002). The role of prefrontal cortex in working-memory capacity, executive attention, and general fluid intelligence: an individual-differences perspective. *Psychonomic Bulletin & Review, 9*(4), 637–671.

Karahanna, E., Straub, D. W., & Chervany, N. L. (1999). Information technology adoption across time: a cross-sectional comparison of pre-adoption and post-adoption beliefs. *Mis Quarterly, 23*(2), 183–213.

Karsh, B. (1997). An examination of employee participation during new technology implementation. *Proceedings of the Human Factors and Ergonomics Society 41st Annual Meeting, 2*, 767–771.

Kaushal, R., & Bates, D. W. (2001). Computerized physician order entry (CPOE) with clinical decision support systems (CDSSs). In K. G. Shojania, B. W. Duncan, K. M. McDonald & R. M. Wachter (Eds.), *Making health care safer: a critical analysis of patient safety practices.* (59–69) Rockville, MD: Agency for Healthcare Research and Quality.

Kaushal, R., Barker, K. N., & Bates, D. W. (2001). How can information technology improve patient safety and reduce medication errors in children's health care? *Archives of Pediatrics & Adolescent Medicine, 155*(9), 1002–1007.

Kaushal, R., Shojania, K. G., & Bates, D. W. (2003). Effects of computerized physician order entry and clinical decision support systems on medication safety: a systematic review. *Archives of Internal Medicine, 163*(12), 1409–1416.

King, W. J., Paice, N., Rangrej, J., Forestell, G. J., & Swartz, R. (2003). The effect of computerized physician order entry on medication errors and adverse drug events in pediatric inpatients. *Pediatrics, 112*(3 Pt 1), 506–509.

Klein, K. J., & Ralls, R. S. (1995). The organizational dynamics of computerized technology implementation: a review of the empirical literature. In L. R. G.-M. M. W. Lawless (Ed.), *Implementation management in high technology* (pp. 31–79). Greenwich, CT: JAI Press.

Klein, K. J., & Sorra, J. S. (1996). The challenge of innovation implementation. *Academy of Management Review, 21*(4), 1055–1080.

Klein, K. J., Conn, A. B., & Sorra, J. S. (2001). Implementing computerized technology: an organizational analysis. *Journal of Applied Psychology, 86*(5), 811–824.

Kohn, L. T., Corrigan, J. M., & Donaldson, M. S. (Eds.). (2000). *To err is human: building a safer health system.* Washington DC: National Academy Press.

Korunka, C., Weiss, A., & Karetta, B. (1993). Effects of new technologies with special regard for the implementation process per se. *Journal of Organizational Behavior, 14*(4), 331–348.

Kukafka, R., Johnson, S. B., Linfante, A., & Allegrante, J. P. (2003). Grounding a new information technology implementation framework in behavioral science: a systematic analysis of the literature on IT use. *Journal of Biomedical Informatics, 36*(3), 218–227.

Leape, L. L. (1997). A systems analysis approach to medical error. *Journal of Evaluation in Clinical Practice, 3*(3), 213–222.

Leapfrog Group. (2004). *The Leapfrog Group for patient safety: rewarding higher standards.* Retrieved March 6, 2004, from www.leapfroggroup.org.

Leonard-Barton, D., & Deschamps, I. (1988). Managerial influence in the implementation of new technology. *Management Science, 34*(10), 1252–1265.

Levine, L. (2001). Integrating knowledge and processes in a learning organization. *Information Systems Management, 18*(1), 21–33.

Lewis, W., Agarwal, R., & Sambamurthy, V. (2003). Sources of influence on beliefs about information technology use: an empirical study of knowledge workers. *Mis Quarterly, 27*(4), 657–678.

Liker, J. K., Haddad, C. J., & Karlin, J. (1999). Perspectives on technology and work organization. *Annual Review of Sociology, 25*, 575–596.

Lorenzi, N. M., Riley, R. T., Blyth, A. J. C., Southon, G., & Dixon, B. J. (1997). Antecedents of the people and organizational aspects of medical informatics: review of the literature. *Journal of the American Medical Informatics Association, 4*(2), 79–93.

Lu, H. P., Yu, H. J., & Lu, S. S. K. (2001). The effects of cognitive style and model type on DSS acceptance: an empirical study. *European Journal of Operational Research, 131*(3), 649–663.

Mahmood, M. A., Burn, J. M., Gemoets, L. A., & Jacquez, C. (2000). Variables affecting information technology end-user satisfaction: a meta-analysis of the empirical literature. *International Journal of Human-Computer Studies, 52*(4), 751–771.

Majchrzak, A. (1992). Management of technological and organizational change. In G. Salvendy (Ed.), *Handbook of industrial engineering* (pp. 767–797). New York: John Wiley.

Majchrzak, A., & Cotton, J. (1988). A longitudinal study of adjustment to technological change—from mass to computer automated batch production. *Journal of Occupational Psychology, 61*(1), 43–66.

McKeen, J. D., Guimaraes, T., & Wetherbe, J. C. (1994). The relationship between user participation and user satisfaction—

an investigation of 4 contingency factors. *MIS Quarterly, 18*(4), 427–451.

Mekhjian, H. S., Kumar, R. R., Kuehn, L., Bentley, T. D., Teater, P., Thomas, A., et al. (2002). Immediate benefits realized following implementation of physician order entry at an academic medical center. *Journal of the American Medical Informatics Association, 9*(5), 529–539.

Merli, R., & Stickler, M. (2004). *E-prescriptions promise patient safety, higher drug sales.* Retrieved March 6, 2004, from www.kpm-ginsiders.com/display_analysis.asp?cs_id=96118.

Miller, A. S. (2000a). Prescriber computer order entry: a work in progress. *Hospital Pharmacy, 35*(7), 714–720.

Miller, A. S. (2000b). Prescriber computer order entry: system design. *Hospital Pharmacy, 35*(9), 1008–1010.

Miller, A. S. (2000c). Prescriber computer order entry: team structure and physician impact. *Hospital Pharmacy, 35*(8), 822–824.

Moore, G. C., & Benbasat, I. (1991). Development of an instrument to measure the perceptions of adopting an information technolgy innovation. *Information Systems Research, 2*(3), 192–222.

Moore, G. C., & Benbasat, I. (1996). Integrating diffusion of innovations and theory of reasoned action models to predict the utilization of information technology by end-users. In K. Kautz & J. Pries-Heje (Eds.), *Diffusion and adoption of information technology* (pp. 132–146). New York: Chapman-Hall.

Morris, M. G., & Venkatesh, V. (2000). Age differences in technology adoption decisions: implications for a changing work force. *Personnel Psychology, 53*(2), 375–403.

Nolan, T. W. (1998). Understanding medical systems. *Annals of Internal Medicine, 128*(4), 293–298.

Nolan, T. W. (2000). System changes to improve patient safety. *British Medical Journal, 320*(7237), 771–773.

Ortiz, E., Meyer, G., & Burstin, H. (2001). The role of clinical informatics in the agency for healthcare research and quality's efforts to improve patient safety. *Journal of the American Medical Informatics Association, Suppl S*, 508–512.

Pisano, G. P., Bohmer, R. M. J., & Edmondson, A. C. (2001). Organizational differences in rates of learning: evidence from the adoption of minimally invasive cardiac surgery. *Management Science, 47*(6), 752–768.

Potts, A. L., Barr, F. E., Gregory, D. F., Wright, L., & Patel, N. R. (2004). Computerized physician order entry and medication errors in a pediatric critical care unit. *Pediatrics, 113*(1 Pt 1), 59–63.

Prabhu, M. (2003). *Technology helps docs aid patients.* Retrieved March 6, 2004, from www .chronicle.duke.edu/vnews/display.v/ART/2003/11/17/3fb8be2da6bd9.

Puckett, F. (1995). Medication-management component of a point-of-care information system. *American Journal of Health-System Pharmacy, 52*(12), 1305–1309.

Reason, J. (2000). Human error: models and management. *British Medical Journal, 320*, 768–770.

Rogers, E. M. (1983). *Diffusion of innovations*, 3rd ed. New York: Free Press.

Rogers, E. M. (1995). *Diffusion of innovations*, 4th ed. New York: Free Press.

Rothstein, R. I., & Littenberg, B. (1995). Disposable, sheathed, flexible sigmoidoscopy: a prospective, multicenter, randomized trial. The Disposable Endoscope Study Group. *Gastrointestinal Endoscopy, 41*(6), 566–572.

Sainfort, F., Karsh, B.-T., Booske, B., & Smith, M. J. (2001). Applying quality improvement principles to achieve health work organizations. *Journal on Quality Improvement, 27*(9), 469–483.

Salas, E., & Cannon-Bowers, J. A. (2001). The science of training: a decade of progress. *Annual Review of Psychology, 52*, 471–499.

Sanders, M. S., & McCormick, E. J. (1993). *Human factors in engineering and design,* 7th ed. New York: McGraw-Hill Science.

Sansone, C., & Harackiewicz, J. M. (2000). *Intrinsic and extrinsic motivation: the search for optimal motivation and performance.* New York: Academic Press.

Scholtz, J., & Weidenbeck, S. (1990). Learning second and subsequent programming languages: a problem of transfer. *International Journal of Human-Computer Interaction, 2*(1), 51–72.

Shojania, K. G., Duncan, B. W., McDonald, K. M., & Wachter, R. M. (Eds.). (2001). *Making health care safer: a critical analysis of patient safety practices.* Rockville, MD: Agency for Healthcare Research and Quality.

Smith, M. J., & Carayon, P. (1995). New technology, automation, and work organization: stress problems and improved technology implementation strategies. *International Journal of Human Factors in Manufacturing, 5*(*1*), 99–116.

Smith, M. J., & Sainfort, P. C. (1989). Balance theory of job design for stress reduction. *International Journal of Industrial Ergonomics. 4*(1), 67–79.

Sternberg, R. J. (1997). The concept of intelligence and its role in lifelong learning and success. *American Psychologist, 52*, 1030–1037.

Stricklin, M. L. V., & Struk, C. M. (2003). Point of care technology: a sociotechnical approach to home health implementation. *Methods of Information in Medicine, 42*(4), 463–470.

Szajna, B. (1996). Empirical evaluation of the revised technology acceptance model. *Management Science, 42*(1), 85–92.

Taylor, S., & Todd, P. (1995). Assessing IT usage: the role of prior experience. *Mis Quarterly, 19*(4), 561–570.

Thompson, R. L., Higgins, C. A., & Howell, J. M. (1991). Personal computing—toward a conceptual-model of utilization. *Mis Quarterly, 15*(1), 125–143.

Thompson, R. L., Higgins, C. A., & Howell, J. M. (1994). Influence of experience on personal computer utilization: testing a conceptual model. *Journal of Management Information Systems, 11*(1), 167–187.

Venkatesh, V. (1999). Creation of favorable user perceptions: exploring the role of intrinsic motivation. *Mis Quarterly, 23*(2), 239–260.

Venkatesh, V., & Davis, F. D. (1996). A model of the antecedents of perceived ease of use: development and test. *Decision Sciences, 27*(3), 451–481.

Venkatesh, V., & Davis, F. D. (2000). A theoretical extension of the technology acceptance model: four longitudinal field studies. *Management Science, 46*(2), 186–204.

Venkatesh, V., & Morris, M. G. (2000). Why don't men ever stop to ask for directions? gender, social influence, and their role in technology acceptance and usage behavior. *Mis Quarterly, 24*(1), 115–139.

Venkatesh, V., & Speier, C. (1999). Computer technology training in the workplace: a longitudinal investigation of the effect of mood. *Organizational Behavior and Human Decision Processes, 79*(1), 1–28.

Venkatesh, V., Morris, M. G., & Ackerman, P. L. (2000). A longitudinal field investigation of gender differences in individual technology adoption decision-making processes. *Organizational Behavior and Human Decision Processes, 83*(1), 33–60.

Venkatesh, V., Speier, C., & Morris, M. G. (2002). User acceptance enablers in individual decision making about technology: toward an integrated model. *Decision Sciences, 33*(2), 297–316.

Venkatesh, V., Morris, M. G., Davis, G. B., & Davis, F. D. (2003). User acceptance of information technology: toward a unified view. *Mis Quarterly, 27*(3), 425–478.

Wald, H., & Shojania, K. G. (2001). Prevention of misidentifications. In K. G. Shojania, B. W. Duncan, K. M. McDonald & R. M. Wachter (Eds.), *Making health care safer: a critical analysis of patient safety practices*. Rockville, MD: Agency for Healthcare Research and Quality.

Watson, D., Clark, L. A., & Tellegen, A. (1988). Development and validation of brief measures of positive and negative affect—the panas scales. *Journal of Personality and Social Psychology, 54*(6), 1063.

Zmud, R. W. (1979). Individual differences and MIS success: a review of the empirical literature. *Management Science, 25*(10), 966–979.

26

ROBOTICS IN HEALTH CARE: HF ISSUES IN SURGERY

Caroline GL Cao & Gary Rogers
Tufts University

Technology is ubiquitous in health care today. Without a doubt, automation and use of robotics will play an ever-increasing role as medicine continues to evolve in technological sophistication. Increasingly, automation and use of robotics in the laboratory, such as for blood analysis and cell-sorting, is commonplace; medical research depends on this technology for genomics, proteinomics, and drug discovery processes. State-of-the-art biotechnology enterprises use robots to feed and manage tissue cultures for growing human skin—the source of tissue used for grafting extensive burn victims. Home health care and long-term disease management (e.g., diabetes, kidney dialysis, etc.) are beginning to utilize semi-intelligent and automatic drug delivery and monitoring devices. The field of medicine is in the very early stages of applying robotics to health care. Successful robotic application in health care depends on innovative technology and effective integration of this technology into the complex human–machine interactions in the health care environment. Nowhere is human factors (HF) consideration more critical for the success of technology application than the use of robotics in the clinical setting. In fact, the use of robotics is more prevalent in surgery than in any other specialty of health care. This chapter reviews the key HF issues in applying robotics to medicine, using minimally invasive surgery as a context for discussion. The discussion of HF issues in surgery is divided into two levels: the human-robot level of interaction, and the team-robot level of interaction.

OVERVIEW OF ROBOTICS IN SURGERY

The Robot Institute of America's definition of a robot is "a reprogrammable multi-functional manipulator designed to move material, parts, tools, or specialized devices through variable programmed motions for the performance of a variety of tasks" (Sheridan, 1992, p. 3). Much excitement has been generated about the development of robotic systems in surgery, where technology seems to have the highest potential impact on patient care. For example, robotics promises to enhance surgical performance by replacing tedious and repetitive tasks, increasing dexterity, or automating surgical tasks that are difficult to perform (Garcia-Ruiz, Gagner, Miller, Steiner, & Hahn, 1998; Ben-Porat, Shoham, & Meyer, 2000). Robotics can increase the precision and accuracy of the surgeon's fine motor performance by stabilizing tremor (Taylor et al., 1999), or by scaling down the motions made by a surgeon's hands. There is also the potential to enhance the surgeon's sense of touch through haptic interfaces (Rosen, Hannaford, MacFarlane, & Sinanan, 1999a).

Robotic systems for surgery have seen rapid development but only tentative acceptance by the general public. Initially, the most common use of robots in minimally invasive surgery (MIS) is for positioning the laparoscope using voice control (e. g. AESOP). The benefit of using a robot for this task is that it does not fatigue, does not shake, and

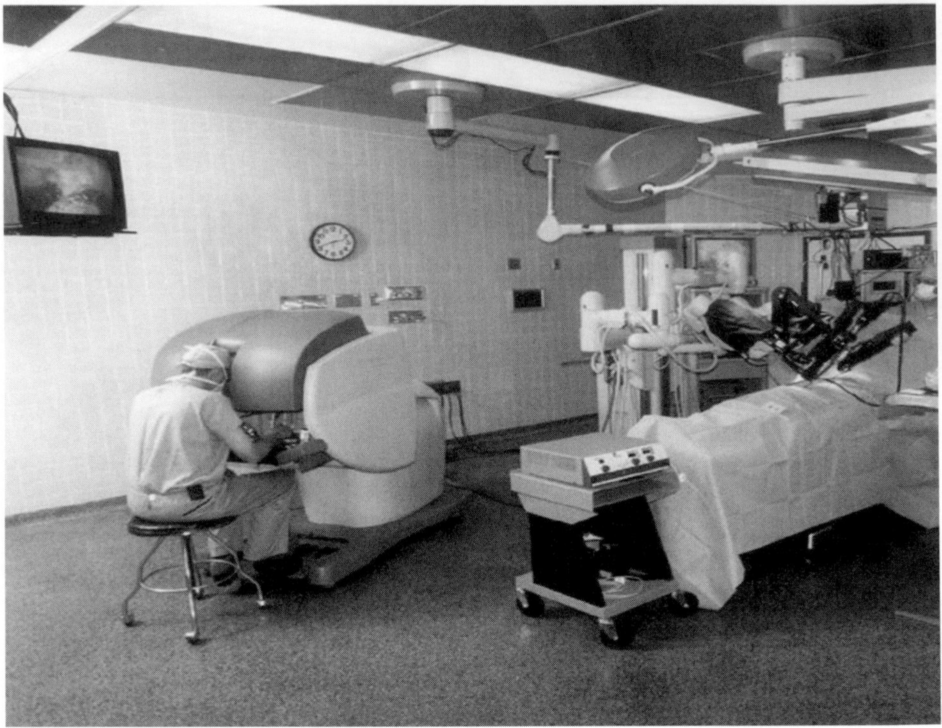

Figure 26–1. The da Vinci Robotic System by Intuitive Surgical

can be controlled verbally by the primary surgeon. This greatly increases the efficiency of the visualization process, as the surgeon controls exactly what is being viewed and brings the procedure closer to a cost-effective "solo-operation," where fewer surgeons are needed in the operating room (Omote et al., 1999). More sophisticated robotic systems have been developed that play a larger role in the surgical process.

Current robotics applications in surgery include positioning devices such as AESOP/Sydne (formerly Computer Motion, Inc., now Intuitive Surgical, Inc.); image-guided radiosurgical systems such as CyberKnife (Accuray, Inc.); and telesurgical manipulators for performing surgery, such as the RoboDoc (Integrated Surgical Systems), Acrobot (Acrobot Company Ltd.), NeuroMate (Integrated Surgical Systems), and Zeus and Da Vinci (Intuitive Surgical, Inc.) (See Figure 26–1). The clinical areas where robots are currently in use are general surgery, gynecological surgery neurosurgery, orthopedics, urology, maxillofacial surgery, ophthalmology, and cardiac surgery.

Because these robotic systems are novel to most researchers working in the field of HF and ergonomics, very little is known about their performance and efficacy from a HF perspective. Few studies have been conducted to compare the performance of robotic surgery to "traditional," MIS (Garcia-Ruiz et al., 1998; Melvin, Needleman, Krause, Schneider, & Ellison, 2002; Nio et al., 2002; Webster, 2004). Case studies and institutional reviews have been published on robot-assisted laparoscopic surgery (Cao & Taylor, 2004; Mack 2001; Marescaux et al., 2001). The successful completion rates were similar to those of human laparoscopic surgery. However, the robot was almost never faster than its human counterpart (Garcia-Ruiz et al., 1998; Marescaux et al., 2001; Nio et al., 2002). Much public attention has also been directed at the failures of robotic surgery (Brink, 2002).

These studies barely scratch the surface of the application of robotics to surgery and highlight the institutional and psychological barriers to the entry of robotic technology into the operating room and to health care in general. For robotics, or any new technology for that matter, to be accepted, the instrumentation must fit into the environment without disrupting familiar routines at the level of individual tasks and functions, at the level of team coordination and functions, and at the level of institutional management. The successful coupling of technology with individuals and teams is determined by the system constraints and requirements, which vary depending on the system environment.

HF AT INDIVIDUAL-ROBOT LEVEL

The use of robotics in MIS is a natural progression following the evolution and development of surgical technology. Indeed, it is a promising solution to the limitations and constraints of MIS technology and requirements of surgical therapy. However, while some of the HF issues associated with MIS technology at the individual level (such as physical and perceptual/cognitive ergonomics and usability) can be addressed by robotics, new ones are also created, often at higher levels of interaction.

Compared to traditional open surgery, MIS procedures impose additional safety concerns and precision requirements, as well as greater physical and visual-motor constraints on the surgeon (Berguer, Forkey, & Smith, 1999; Cao, 1996; Cao, MacKenzie, & Payandeh, 1996; Cuschieri, 1995). These HF issues are due to restricted access to the operative site within the respective body cavities, with respect to visual, tactile, and motor skills. In other words, the very tools that allow surgical operations to be performed with minimal invasiveness represent both major physical and perceptual barriers to the surgeon. Physically, the surgeon's hands and arms are elevated throughout the surgical procedure, leading to increased muscular discomfort and fatigue (Berguer et al., 1999). Visually, information regarding the surgical site is altered. The lack of stereoscopic view and adequate depth cues results in longer performance times (Cao & MacKenzie, 1997; Crostwaithe, Chung, Dunkley, Shimi, & Cuschieri, 1995; Tendick, Jennings, Tharp, & Stark, 1993) and more wasteful movements with the surgical instruments (Kim, Ellis, Tyler, Hannaford, & Stark, 1987). In addition to a restricted view of the operative site and loss of depth perception, other visual-motor constraints include displaced visual space; image magnification, coupled with a separation of hand space from workspace (operative site); and, frequently, a rotation of the display space relative to the operative site. Holden, Flach, and Donchin (1999) have shown that changing the camera position or the surgeon's position with respect to the task space disrupted performance, but when the position of the camera and surgeon changed together while retaining relative orientation, skill performance was maintained. These findings are consistent with the motor control/learning literature (Cunningham & Welch, 1994; Flach, Linter, & Larish, 1990), which suggests that surgical skill depends on a consistent mapping between the virtual hands (displayed tool end-effectors) and the eyes but not the particular visual or motor orientations. This has implications for the design of user interfaces for robotic systems.

In MIS, tactile sensation from the tissues and surgical tools is reduced, while manipulation of the endoscopic tool is usually restricted to 4 degrees of freedom. The control of the surgical tools is further complicated by the fact that the tools rest on a fulcrum at the entry port into the body cavity. Therefore, hand movement direction and tool end-effector directions are reversed. Furthermore, because the camera is controlled by an assistant, it often requires some adjustments guided verbally by the operating surgeon. This requires experience also on the part of the assistant, who ideally should understand the intentions of the operating surgeon, and even anticipate them where possible.

The combination of the physical, precision, safety and visual–motor constraints makes MIS a very difficult task to learn and to master. This is evidenced by the large number of medical errors associated with this technology (Bateman, Kolp, & Hoeger, 1996; Schafer, Lauper, & Krahenbuhl, 2000; Cooper & Fisher, 1993; Shea et al., 1996; Puljiz et al., 2003; Targarona et al., 1997; Way et al., 2003). Some of these "medical errors" have been directly related to poor design of technology and rapidly changing technology (Cook & Woods, 1996; Joice, Hanna, & Cuschieri, 1998; Tendick et al., 2000; Way et al., 2003), as well as insufficient knowledge or training of surgeons (Joice et al., 1998; Katz, 1999; Ostrzenski & Ostrzenska, 1998; Watson, Baigrie, & Jamieson, 1996; Wu, Ou, Chen, Yen, & Rowbotham, 2000;). Some of these injury-causing errors do not exist in open surgery (Cuschieri, 1995), while others are associated with the use, or misuse, of unfamiliar instruments, such as applying too much force with the graspers, or tearing the gallbladder.

In robot-assisted surgery, these same complications persist, although many can be mitigated by

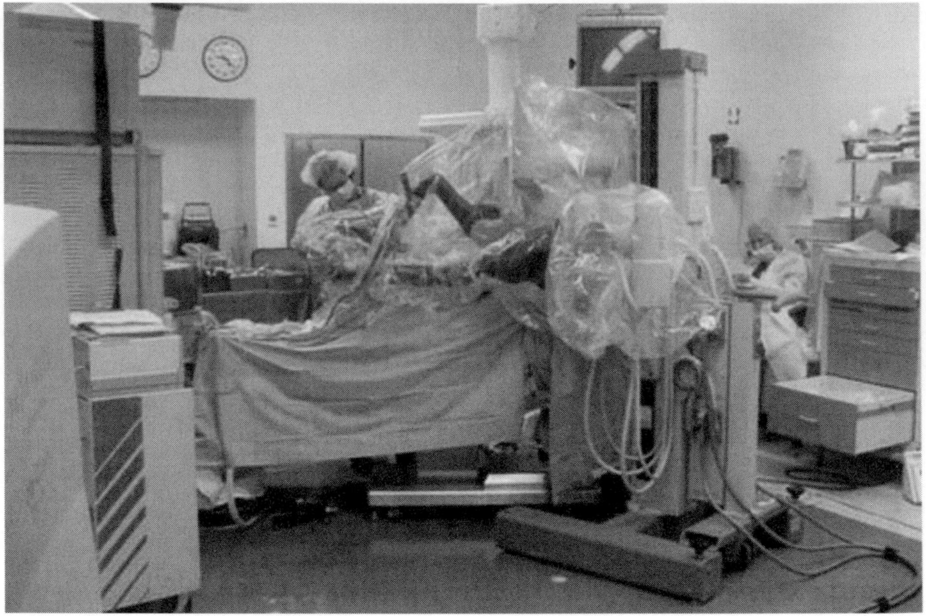

Figure 26–2. The da Vinci robotic system draped at bedside.

implementing 3-D displays (Gulbins et al., 1999), 6 degree-of-freedom manipulators (Guthart & Salisbury, 2000), and a seated posture for the surgeon (Menozzi, von Buol, Krueger, & Miege, 1994). However, the development of haptic interfaces for robotic surgery remains a difficult objective to achieve. The promise of robotic surgery is still hindered by challenges in providing tactile and haptic feedback to the surgeon through the robotic interface (Carrozza, Lencioni, Magnani, D'Altanasio, and Dario, 1997), even though it has been shown that force feedback can increase speed and accuracy in teleoperation tasks (Massimo & Sheridan, 1994; Salcudean, Ku, & Bell, 1997). Thus far, most researchers have dealt primarily with using force feedback to augment the visual feedback (Kennedy, Hu, Desai, Wechsler, & Dresh, 2002; Ortmaier et al., 2001; Rosen, MacFarlane, Richards, Hannaford, & Sinanan, 1999b). These approaches attempt to translate forces, as measured by the force sensors in the instrument, through a servomechanism to the fingers of the surgeon. The force information alone is not representative of the rich information normally perceived by the haptic sense, which includes tactile, kinesthetics, proprioception, as well as texture, temperature, and contour information, but it is better than no force information at all (Cao, Webster, Perreault, Schwaitzberg, & Rogers, 2003).

Unfortunately, these innovations are still not enough for robotics to be accepted by both surgeons and patients. The design of the current robotic systems is modeled after the design of the laparoscopic instruments. In addition to the same HF issues associated with traditional MIS technology (i.e., physical constraints, visual-motor coordination issues, precision and safety requirements), the robots are bulky and difficult to use. Their bulky presence displaces other instrumentation and OR personnel in the room, resulting in re-arrangement of the OR physical space (see Figure 26–2). In addition, the large size of the robotic arms can sometimes result in external collisions of the arms, especially when working on a small patient.

An unexpected outcome of using a robot for laparoscopic surgery is the additional physical and cognitive demands placed on the surgeon. The robot changes the surgeon's tasks and responsibilities (Nio, Bemelman, Busch, Vrouenraets, & Gouma, 2004; Webster, & Cao, in press). Not only is the surgeon responsible for performing the surgery (i.e., cutting and suturing), but she is now also responsible for driving the robot. For the latter task, a high degree of coordination and integration of information from various sources and locations is required. Figure 26–3 shows the difference in the number of steps performed in a surgical task (i.e., changing a tool) between a conventional laparoscopic procedure and a robot-assisted laparoscopic procedure. For example, in laparoscopic surgery, when a surgeon needed to have a tool changed, the surgeon informed the nurse which tool was needed, pulled out the current one, handed it to the nurse in exchange for the new tool, took the new one, and inserted it into the patient (see Figure 26–3, top). Often, the nurse anticipated the need for a new tool and had it ready.

In robot-assisted surgery, this process was much more complicated (see Figure 26–3, bottom). First, the surgeon informed the nurse what tool was needed. Then, using the robotic handles to navigate the menu on the robotic console, the surgeon disabled the robotic arm by selecting an item on the pull-down menu. Because of OR setup (the system console blocked the surgeon's view of the patient and the nurse at the side of the operating table, and vice versa), the nurse could not see the menu screen that indicated that the arm had been disabled. The surgeon had to verbally communicate the state of the robot to the nurse who then removed the old tool. When the new tool was inserted, the surgeon could see the image of the tool end-effector on the monitor at the console, even before the tool can be secured externally. Sitting at the console, the surgeon could not see whether the tool had been secured at the patient site. The surgeon had to rely on the nurse for this information. After being informed that the tool had been secured, the surgeon navigated the menu to enable the arm and then continued with the surgical procedure.

Not only is the number of steps increased, the information required to accomplish the task is distributed and sometimes unavailable. This means that even an expert surgeon must first learn to drive the robot before he or she is able to do their job (i.e., perform surgery) and possibly learn new ways of performing the surgery due to the new tool. This has implications for the safety of the patient and the success of the surgical procedure.

HF AT TEAM-ROBOT LEVEL

The introduction of robotics into the OR has improved the technical performance of surgical procedures, but it has also led to unexpected interactions within the surgical team and new forms of errors (Cook & Woods, 1996; Reason, 1990). Several studies have shown that when new technology is introduced into the operating room, the goals, tasks, and responsibilities of the surgeon and nurses change (Edmonson, Bohmer, & Pisano, 2001; Nio et al., 2004; Webster, 2004). For example, prior to adopting a new, minimally invasive cardiac surgery system, the surgeon and nurses often communicated nonverbally; the nurse could read by the surgeon's body language what was needed and when it was needed. Everyone in the OR also had visual access to the information source (the heart) so could anticipate the next steps (Edmonson et al., 2001). With the new MIS system, information was located in new places, many of the visual cues were removed, and the nurses became responsible for providing critical information to the surgeon (Edmonson et al., 2001). Studies of OR team performance have revealed that preventable medical errors are related not to technical competence, but to interpersonal aspects of the OR team functioning (Helreich & Schaefer, 1994; Zinn, 1995), indicating a need for better communication to improve safety, efficiency, and team morale.

A recent study conducted to examine the changes in performance and communication patterns within the OR team as a result of the introduction of a surgical robot, the LaproTek (see Figure 26–4), revealed large disparities in terms of the amount and type of information required by the surgeon to perform the surgical procedure (Webster & Cao, in press). The introduction of a robotic surgical system into the OR changed the flow of information, as well as the point of access to the information and how that information is shared (see Figure 26–5). Figure 26–5 shows that in laparoscopic surgery, the surgeon's informational needs were fulfilled through interaction with the patient, whereas in robot-assisted surgery, the information was distributed between the nurse and the robotic console. Adjustments in team communication were necessary to accommodate the novel technology, new procedures, and altered roles of the OR personnel. Being removed from the surgical site, the surgeon could not receive the full range of sensory information normally obtained through vision, audition, the vestibular apparatus, haptics, and the olfactory senses. Rather, most of the available information was received through the visual channel. Even that modality, however, was based on video images of the remote surgical site, which provided a restricted field of view and limited depth information from a frequently poor vantage point. Moreover, the physical barrier imposed by the robotic technology also represented a perceptual barrier for the surgeon, increasing the uncertainty

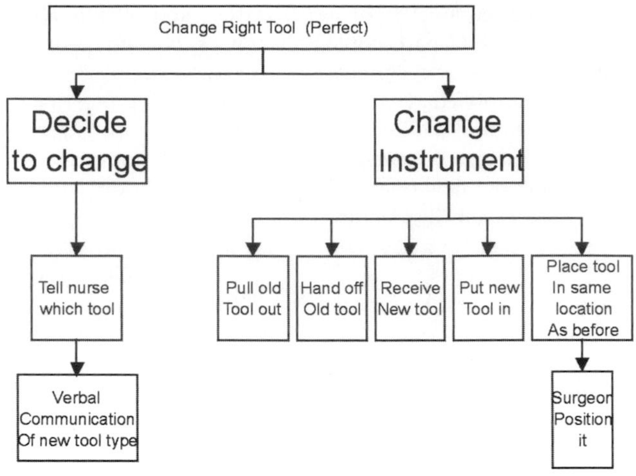

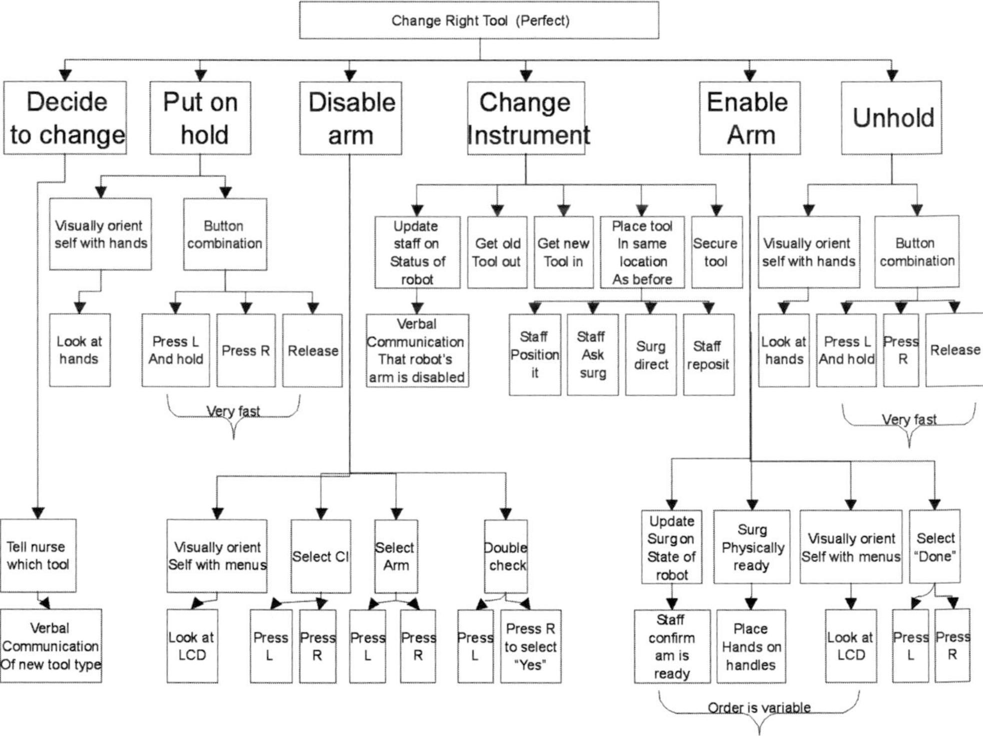

Figure 26–3. Comparison of a Tool-Changing Task in Traditional Laparoscopic Surgery (top) with the Same Task in Robot-Assisted Surgery (bottom)

regarding the status of the remote system, and increasing information processing demands. If the surgeon and nurse could see each other's actions, as is the case with conventional laparoscopic surgery, they would better be able to coordinate their actions, thus reducing down time and potential confusion (Segal, 1995). Confusion about who should provide information and when that information should be provided forms the basis for most of the potential errors. Thus, it is clear that the addition of a robot

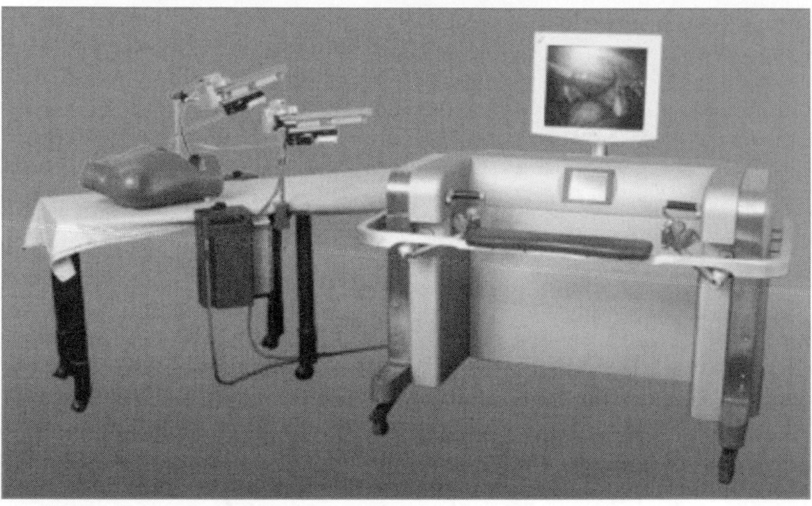

Figure 26–4. Laprotek by endoVia Surgical

places many additional information input and output requirements on the surgeon, increasing the overall communication load on the surgical team.

The dynamic and complex socio-technical environment of an operating room, where each participant has a task to perform and individual expertise relevant to a multitude of tasks, underscores its complexity in team interaction. When new technology that affects work practice in the OR is introduced, adjustments must be made to accommodate the change with greater potential for miscommunication and inefficiency. Adjustments in team communication must be made to accommodate the novel technology, new procedures, or altered roles of the OR personnel in order to establish common ground (Clark, 1996; Clark & Schaefer, 1989; Wilkes-Gibbs & Clark, 1992). As roles change when technology is introduced, people are less familiar with their new roles (as seen with the introduction of a robot). When new technology is introduced, new forms of errors are also possible, as well as interruptions from the use of the technology itself (Coiera & Tombs, 1998; Moss & Xiao, 2004). Given the importance of communication efficiency and safety in surgery, and in particular, the potential of the errors when adapting to working with new technology, it is important to implement training for the team before working with patients.

The operating room environment is a cultural hierarchy with conventions that have evolved over the past century. The surgeon and the assistants must have direct access to the patient and must be able to view each other, the patient, and the anesthesiologist during the procedure. Change in the operating room environment must be gradual, with new technology integrated relatively seamlessly into current routine processes.

The ultimate goal for the designer of robots is to make the robotic technology transparent to the user, just as the complex technology of a desktop computer is not apparent to the person typing a letter. The surgeons should not feel that they are using a robotic system, rather, that they are holding an instrument as an extension of their hands to enhance their performance. The instrumentation must be easy to use. New technological advances that require a steep learning curve will be adopted only gradually, if at all, as younger surgeons trained in their use during residency mature in their practices. On the other hand, robotic instrumentation that is transparent will be adopted much more quickly, even by seasoned surgeons who are expert in using the "old" technology. Therefore, the burden of adaptation must be on the technology and its designers, not on the surgeon.

HF LESSONS FOR IMPLEMENTING ROBOTICS IN HEALTH CARE

Thus far, the robots do not add much to the manual method of surgery, and they are very expensive.

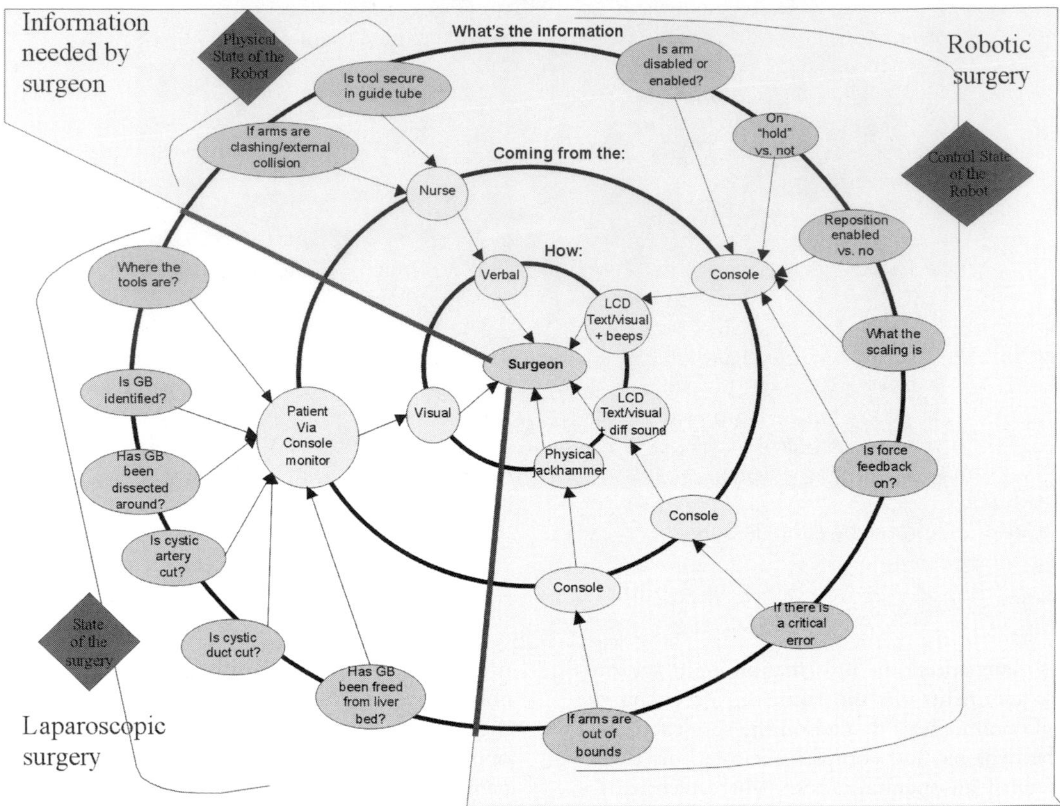

Figure 26–5. Information Needed by the Surgeon to Complete the Surgery, with (shaded area) and without the Robot (unshaded area)

Furthermore, they require refitting of the OR, and a change in the OR culture. For robots to become sufficiently useful in the OR to be generally adopted, and accepted by both surgeons and patients, they must overcome the problems related to safety, usability, and appropriateness and fit within the complex socio-technical environment of the OR. Robots must provide a quantum leap in benefit, not a marginal improvement. Until this occurs, they will not be worth the cost and potential causes of error introduced by their use.

In the aerospace industry, robots have been developed that provide this quantum leap. With increasingly crowded skies in commercial aviation, the autopilot, which can respond to neighboring aircraft position and airport controllers, is in general use, leading to safer air travel. Current technology in aeronautics also allows aircraft to be landed in zero-visibility weather. This increases the utility of the aircraft and extends the pilot's abilities beyond that which he or she would be able to accomplish without technological aid.

The success of technology development and utilization in aviation is due in large part to a great deal of HF research and the design of effective human-machine interfaces. In the case of robotic technology used in medicine, similar efforts are needed. The coupling between the medical team and the robot can greatly affect the usefulness of the enabling technology. Much can be learned from the work already done in the aviation industry and adapted to the complex medical domain.

The ideal interaction between robot and human in the health care environment is one in which there is an optimal "division of labor," that is, the robot is assigned to the task it is good at, while the human is assigned to the task he or she is good at. It is unlikely that the medical robot will ever be autonomous. Therefore, it is in the balance of robotic and human skills that the benefit will be maximized. Prudence must be exercised in the design and adoption of robotics into the health care environment. Again, the analogy to automation in the aviation industry is useful. Automation in the cockpit has increased the safety of air travel and extended its use into circumstances in which the aircraft would be grounded without it. However, pilots of airliners are often resentful of the extent to which automation has taken over their work. While the pilot is still ultimately responsible for all decisions and activities related to flying, he or she no longer performs many of the routine tasks involved in flying. These are performed by an ever-increasingly sophisticated autopilot. This can lead to many problems typical in automated industries, such as loss of control and increased unfamiliarity, resulting in inattention and degradation in performance (Sheridan, 1992). It is important that the analogous situation be avoided in the health care environment. The health care provider must never be in a situation where he or she abdicates control. Rather, robotics should enable the health care provider to enhance his or her skills, and if possible, extend them beyond human capabilities and relieve him or her of physically strenuous and repetitive tasks.

As with any other technology, usable robotics for health care must address the system requirements and constraints at the physical level (i.e., what the robot contains), functional level (i.e., what the robot does), and operational level (i.e., how the robot is used). In addition, its interactions with the individual user and the team must be examined to prevent unintended outcomes. For sophisticated robotic systems, specialized training for the individuals and teams should be implemented. A good coupling between health care system and robotic technology depends on the fit between the requirements of the system and the capabilities of the technology, without imposing additional demands on the system.

References

Bateman, B. G., Kolp, L. A., & Hoeger, K. (1996). Complications of laparoscopy—operative and diagnostic. *Fertility and Sterility, 66*, 30–35.

Ben-Porat, O., Shoham, M., & Meyer, J. (2000). The effect of hand-controller configuration on task performance in endoscopic tele-operation. *Presence, 9*, 256–267.

Berguer, R., Forkey, D. L., & Smith, W. D. (1999). Ergonomic problems associated with laparoscopic surgery. *Surgical Endoscopy, 13*(5), 466–468.

Brink, G. (2002). Patient dies in robot-aided surgery. *St. Petersburg Times*, October 30, P. I. B.

Cao, C. G. L. (1996). *A task analysis of laparoscopic surgery: Requirements for remote manipulation and endoscopic tool design.* Master's thesis, School of Kinesiology, Simon Fraser University, Burnaby, B.C., Canada.

Cao, C. G. L., & MacKenzie, C. L. (1997). Direct, 2-D vs. 3-D endoscopic viewing and surgical task performance. Paper presented at the symposium on evolving technologies: surgeons' performance of surgical tasks. NASPSPA/ACSM meeting, Denver, Colorado, May 28–June 1.

Cao, C. G. L., & Taylor, H. A. (2004). Effects of new technology on the operating room team. *Proceedings of the 7th international conference on work with computing systems*, special session on IT in health care. Kuala Lumpur, Malaysia, June 29–July 2, 2004. 309–312.

Cao, C. G. L., MacKenzie, C. L., & Payandeh, S. (1996). Task and motion analysis in endoscopic surgery. *Proceedings of the ASME dynamic systems and controls division, DSC-vol. 58* (5th annual symposium on haptic interfaces for virtual environments and teleoperation systems), New York, NY: ASME Publishing. 583–590.

Cao, C. G. L., Webster, J., Perreault, J., Schwaitzberg, S. D., & Rogers, G. (2003). Visually perceived force feedback in simulated robotic surgery. *Proceedings of the 47th annual meeting of the human factors and ergonomics society,* Human Factors and Ergonomics Society, Santa Monica, CA, 1466–1470.

Carrozza, M. C., Lencioni, L., Magnani, B., D'Attanasio, S., Dario, P., Pietrabissa, A., & Trivella, M. G. (1997). The development of a microrobot system for colonoscopy. *First joint conference computer vision, virtual reality, and robotics in medicine and medical robotics and computer-assisted surgery (CVRMed-MRCAS-97),* Grenoble, France, March 19–22, 1997.

Clark, H. H. (1996). Communities, commonalities, and communication. In J. J. Gumperz & S. C. Levinson (Eds.), *Rethinking linguistic relativity. Studies in the social and cultural foundations of language,* no. 17. New York, NY: Cambridge University Press, 324–355.

Clark, H. H., & Schaefer, E. F. (1989). Contributing to discourse. *Cognitive Science, 13*(2), 259–294.

Coiera, E., & Tombs, V. (1998). Communication behaviors in a hospital setting: an observational study. *British Medical Journal, 316*, 673–676.

Cook, R., & Woods, D. (1996). Adapting to new technology in the operating room. *Human Factors, 38*(4), 593–611.

Crostwaithe, G., Chung, T., Dunkley, P., Shimi, S., & Cuschieri, A. (1995). Comparison of direct vision and electronic two- and three-dimensional display systems on surgical task efficiency in endoscopic surgery. *British Journal of Surgery, 82*, 849–851.

Cunningham, H. A., & Welch, R. B. (1994). Multiple concurrent visual-motor mappings: Implications for models of adaptation. *Journal of Experimental Psychology: Human Perception and Performance, 20*(5), 987–999.

Cuschieri, A. (1995). Visual displays and visual perception in minimal access surgery. *Seminars in Laparoscopic Surgery, 2*(3), 209–214.

Edmonson, A., Bohmer, R., & Pisano, G. (2001). Disrupted routines: team learning and new technology implementation in hospitals. *Administrative Science Quarterly, 46*, 685–716.

Flach, J. M., Lintern, G., & Larish, J. F. (1990). Perceptual motor skill: a theoretical framework. In R. Warren and A. H. Wertheim (Eds.), *Perception and control of self-motion*. Hillsdale, NJ: Lawrence Erlbaum Associates, 327–355.

Garcia-Ruiz, A., Gagner, M., Miller, J. H., Steiner, C. P., & Hahn, J. F. (1998). Manual vs. robotically assisted laparoscopic surgery in the performance of basic manipulation and suturing tasks. *Archives of Surgery, 133*, 957–961.

Gulbins, H., Boehm, D. H., Reichenspurner, H., Arnold, M., Ellgass, R., & Reichart, B. (1999). 3D-visualization improves the dry-lab coronary anastomoses using the Zeus robotic system. The *Heart Surgury Forum, 2*(4), 318–325.

Guthart, G. S., & Salisbury, J. K. (2000). The intuitive telesurgery system: overview and application. *Proc. IEEE ICRA* '00. IEEE Publishing, Piscataway, NJ., 618–621.

Helreich, R. E., & Schaefer, H. G. (1994). Team performance in the operating room. In M. S. Bogner (Ed.), *Human error in medicine* (pp. 225–253). Hillsdale, NY: Lawrence Erlbaum Associates.

Holden, J. G., Flach, J. M., & Donchin, Y. (1999). Perceptual-motor co-ordination in an endoscopic surgery simulation. *Surgical Endoscopy, 13*, 127–132.

Joice, P., Hanna, G.B., & Cuschieri, A. (1998). Errors enacted during endoscopic surgery—a human reliability analysis. *Applied Ergonomics, 29*(6), 409–414.

Katz, P. (1999). *The scalpel's edge: the culture of surgeons*. Boston: Allyn & Bacon.

Kennedy, C. W., Hu, T., Desai, J. P., Wechsler, A. S., & Dresh, J. Y. (2002). A novel approach to robotic cardiac surgery using haptics and vision. *Cardiovascular Engineering: An International Journal, 2*(1), 15–22.

Kim, W. S., Ellis, S. R., Tyler, M. E., Hannaford, B., & Stark, L. W. (1987). Quantitative evaluation of perspective and stereoscopic displays in three-axis manual tracking tasks. *IEEE Transactions on Systems, Man, and Cybernetics, SMC17*, 61–72.

Mack, M. J. (2001). Minimally invasive and robotic surgery. *Journal of the American Medical Association, 285*, 568–572.

Marescaux, J., Smith, M. K., Folscher, D., Jamali, F., Malassagne, B., & Leroy, J. (2001). Telerobotic laparoscopic cholecystectormy: initial clinical experience with 25 patients. *Annals of Surgery, 234*, 1–7.

Massimino, M. J., & Sheridan, T. B. (1994). Teleoperator performance with varying force and visual feedback. *Human Factors, 36*(1), 145–157.

Melvin, W. S., Needleman, B. J., Krause, D. R., Schneider, C., & Ellison, E. C. (2002). Computer-enhanced vs. standard laparoscopic antireflux surgery. *Journal of Gastrointestinal Surgery, 6*(1), 11–16.

Menozzi, M., von Buol, A., Krueger, H., & Miege, C. (1994). Direction of gaze and comfort: discovering the relation for the ergonomic optimization of visual tasks. *Ophthalmic and Physiological Optics, 14*, 393–399.

Moss, J., & Xiao, Y. (2004). Improving operating room coordination. *Journal of Nursing Administration, 34*(2), 93–100.

Nio, D., Bemelman, W. A., Busch, O. R. C., Vrouenraets, B. C., & Gouma, D. J. (2004). Robot-assisted laparoscopic cholecystectomy versus conventional laparoscopic cholecystectomy: a comparative study. *Surgical Endoscopy, 18*(3), 379–82.

Nio, D., Bemelman, W. A., Boer, K. T., Dunker, M. S., Gouma, D. J., & Gulik, T. M. (2002). Efficiency of manual versus robotic (Zeus) assisted laparoscopic surgery in the performance of standardized tasks. *Surgical Endoscopy, 16*(3), 412–415.

Omote, K., Feussner, H., Ungeheuer, A., Arbter, K., Wei, G. Q., Siewert, J. R., & Hirzinger, G. (1999). Self-guided robotic camera control for laparoscopic surgery compared with human camera control. *American Journal of Surgery, 177*, 321–324.

Ortmaier, T., Groeger, M., Seibold, U., Hagn, U., Boehm, D., Reichenspurner, H., & Hirzinger, G. (2001). Autonomy and haptic feedback in minimally invasive robotic surgery. Paper presented at the fourth annual scientific meeting of the International Society for Minimally Invasive Cardiac Surgery, Munich, Germany, June 27–30.

Ostrzenski, A.,& Ostrzenska, K. M. (1998). Bladder injury during laparoscopic surgery. *Obstetrical and Gynecological Survey, 53*, 175–180.

Puljiz, Z., Kuna, T., Franjic, B.D., Hochstadter, H., Matejcic, A., & Beslin, M. B. (2003). Bile duct injuries during open and laparoscopic cholecystectomy at Sestre Milosrdnice University Hospital from 1995 till 2001. *Acta Clinica Croatica, 42*(3), 217–223.

Reason, J. (1990). *Human error.* Cambridge, UK: Cambridge University Press.

Rosen, J., Hannaford, B., MacFarlane, M. P., & Sinanan, M. N. (1999a). Force controlled and teleoperated endoscopic grasper for MIS- Experimental performance evaluation. *IEEE Transactions on Biomedical Engineering, 46,* 1212–1221.

Rosen, J., MacFarlane, M., Richards, C., Hannaford, B., & Sinanan, M. (1999b). Surgeon-tool force/torque signatures—evaluation of surgical skills in minimally invasive surgery. *Proceedings of Medicine Meets Virtual Reality* (*MMVR*; pp. 290–296). San Francisco: IOS Press.

Salcudean S. E., Ku, S., & Bell, G. (1997). Performance measurement in scaled teleoperation for microsurgery. Paper presented at the first joint conference computer vision, virtual reality, and robotics in medicine and medical robotics and computer-assisted surgery (CVRMed-MRCAS-97), Grenoble, France, March 19–22, 1997.

Schafer, M., Lauper, M., & Krahenbuhl, L. (2000). A nation's experience of bleeding complications during laparoscopy. *The American Journal of Surgery, 180*, 73–77.

See, W. A., Cooper, C. S., & Fisher, R. J. (1993). Predictors of laparoscopic complications after formal training in laparoscopic surgery. *Journal of the American Medical Association, 270*, 2689–2692.

Segal, L. (1995). Designing team workstations: the choreography of teamwork. In P. Hancock, J. Flach, J. Caird, and K. Vicente (Eds.), *Local applications of the ecological approach to human-machine systems* (pp. 392–415). Hillsdale, NJ: Lawrence Erlbaum Associates.

Shea, J. A., Healey, M. J., Berlin, J. A., Clarke, J. R., Malet, P. F., Staroscik, R. N., Schwartz, J. S., & Williams, S. V. (1996). Mortality and complications associated with laparoscopic cholecystectomy: A meta-analysis. *Annals of Surgery, 224*, 609–620.

Sheridan, T. B. (1992). *Telerobotics, automation, and human supervisory control.* Cambridge, MA: MIT Press.

Targarona, E. M., Marco, C., Balague, C., Rodriguez, J., Cugat, E., Hoyuela, C., Veloso, E., & Trias, M. (1997). How, when, and why bile duct injury occurs. *Surgical Endoscopy, 12*, 322–326.

Taylor, R., Jensen, P., Whitcomb, L., Barnes, A., Kumar, R., Stoianovici, D., Gupta, P., Wang, Z. X., deJuan, E., & Kavoussi, L. (1999). A steady-hand robotic system for microsurgical augmentation. *Journal of Robotics Research, 18(*12), 1201–1210.

Tendick, F., Downes, M., Goktekin, T., Cavusoglu, M. C., Feygin, D., Wu, X., Eyal, R., Hegarty, M., & Way, L. W. (2000). A virtual environment test bed for training laparoscopic surgical skills. *Presence, 9*(3), 236–255.

Tendick, F., Jennings, R. W., Tharp, G., & Stark, L. (1993). Sensing and manipulation problems in endoscopic surgery: experiments, analysis, and observation. *Presence, 2*(1), 66–81.

Watson, D. I., Baigrie, R. J., & Jamieson, G. G. (1996). A learning curve for laparoscopic fundoplication: definable, avoidable, or a waste of time? *Annals of Surgery, 224*, 198–203.

Way, L. W., Stewart, L., Gantert, W., Kingsway, L., Lee, C. M., Whang, K., & Hunter, J. G. (2003). Causes and prevention of laparoscopic bile duct injuries—analysis of 252 cases from a human factors and cognitive psychology perspective. *Annals of Surgery, 237*(4), 460–469.

Webster, J., & Cao, C. G. L. (in press). Lowering communication barriers in operating room technology. *Human Factors.*

Wilkes-Gibbs, D., & Clark, H. H. (1992). Coordinating beliefs in conversation. *Journal of Memory & Language, 31(2),* 183–194.

Wu, M. P., Ou, C. S., Chen, S, L., Yen, E. Y., & Rowbotham, R. (2000). Complications and recommended practices for electrosurgery in laparoscopy. *The American Journal of Surgery, 179*, 67–73.

Zinn, C. (1995). 14,000 preventable deaths in Australian hospitals. *British Medical Journal, 310*, 1487.

27

HUMAN–COMPUTER INTERACTION IN HEALTH CARE

Frank A. Drews and Dwayne R. Westenskow
University of Utah, Salt Lake City

A STORY OF INJURY AND DEATH

After Mary Smith,[1] a technician at the East Texas Treatment Center in 1986, had welcomed the cancer patient who was scheduled for a 25 MeV electron beam treatment of 180 rads, she helped him onto the treatment table where he positioned himself comfortably. After situating the patient, the experienced technician left the treatment room and entered the treatment data on the console of the Therac–25 (T–25) dual-mode accelerator. As she entered the data she mistakenly typed an "X" for X-ray treatment instead of an "X" for electron treatment. This slip had happened before, because the majority of treatments were X-ray treatments. Being familiar with the necessary steps, it took her only a few seconds to edit the incorrect entry by moving up the cursor to the entry field and correcting it. The interface of the accelerator was a full-screen interface on a DEC VT100 terminal. Mary proceeded by hitting the enter key several times to verify the previous entries and last hit "B" to begin the treatment. Almost immediately the machine went into treatment pause mode and displayed the error message "Malfunction 54." This happened fairly frequently on a daily basis, so Mary was not worried. The accelerator had two malfunction modes: If the condition was serious, the treatment was suspended and the accelerator required a complete reset; if it was a less serious error, a treatment pause was initiated. Mary knew that "Malfunction 54" initiated a treatment pause, so she pressed "P" to proceed and resume the treatment. Because of the cryptic nature of the error messages, she did not know what the error meant and assumed that it was not a serious problem because it could be fixed easily by pressing "P." Mary hit "B" again to start the treatment, and again the accelerator went into the treatment pause mode. Meanwhile, the patient in the treatment room had experienced two extremely painful shocks when Mary had hit the "B" key. He could not notify her because the video and audio monitoring equipment was malfunctioning. During this treatment, the patient had received a massive overdose (possible doses of up to 25,000 rads in less than 1 second over an area of about 1 cm^2). He died several months after the incident. The error investigation indicated that the accident was caused by the technician who entered, too rapidly, an unexpected sequence of commands, and not by the software developer, who did not expect operators to be this proficient. The operator's rapid entry tripped the computer, which retracted the metal plate used during X-ray mode to absorb radiation but left the power setting on maximum. The terminal display showed the beam mode as active, while it was set to deliver X-ray blasts of 25,000 rads. Over several months, six accidents were reported in which patients received massive overdoses of rads, and several of them died as a consequence.

Good interface design contributes to safer medical equipment. The T–25 incident demonstrates how software failure leads to serious problems in operating equipment (Leveson & Turner, 1993).

[1]Name changed

Multiple factors contributed to the incident: First, the interface allowed the operator to exceed the computer's capabilities by entering the changes too fast, resulting in the computer's slipping into an unexpected mode. Second, the malfunction messages were coded cryptically as "Malfunction 54," making the identification of problems difficult for the user. Third, malfunctions occurred frequently on a daily basis, thus creating a mindset in the operator that lead to a trivialization of these malfunctions. Fourth, from a user's perspective, the problem was "fixable" by pushing the "P" button, which did not reflect the real system status and added to the impression that the problem was nothing serious. And finally, the interface provided no information about the mode the accelerator was operating in, making the system extremely intransparent for the user.

All these contributing factors illustrate the importance of good interface design in health care, where problems can be disastrous. Good interface design has an impact across multiple dimensions: It makes software easier to learn, improves performance speed, increases user satisfaction, and reduces error (Shneiderman, 1992), while "poor user interface design greatly increases the likelihood of error in equipment operation" (Kaye & Crowley, 2000). The health care industry is challenging for interface design because it is a complex, uncertain, highly demanding, and highly stressful environment (Gosbee & Richie, 1997). Problems with interface design are acerbated because the health care professional may not be able to compensate for bad interface design.

Poor interface design in health care is a serious problem. The U.S. Food and Drug Administration (FDA) reports that up to 50% of device recalls from 1985–1989 were due to poor product design, including problems with software and user interfaces. As a result, in 1990 Congress passed the Safe Medical Devices Act (SMDA) requiring device producers to follow good manufacturing practices.

The goal of this chapter is to provide a guide to the topic of human–computer interaction in health care, that is, to outline what "good" practices are in human-computer interaction in health care. Error in medicine is widespread (Institute of Medicine, 2000; see also section 6), and among the contributing factors are the growing complexity of science and technology on the one hand, and constraints of exploiting information technologies on the other hand. Well-developed information technology can help leverage the advances of science and technology, but badly implemented, it will handicap these advances and decrease patient safety (e.g., Cohen, 1999).

This chapter starts with a discussion of the types of interfaces being currently used in health care. Next, we introduce a range of interface design principles using a systems perspective. Then we describe some methods to evaluate user interfaces in health care, and finally, we give an example for good user interface design.

TYPES OF INTERFACES

Computers and software operate in a highly invisible way: Often, the user receives only limited information about the operational or organizational state of the system (Norman, 1993). The user interface provides the operating environment for the user in order to allow interaction with the system and to receive feedback about the system's status. Mayhew (1999, p. 159) defines the user interface as "the way displays and feedback are designed (the application-to-user language) and the way users indicate to the application what they want to do next through interactions with display elements via input devices".

Several types of interfaces can be distinguished. One type of user interface that still can be found in institutional health care environments are character-based interfaces (DOS and UNIX user interfaces). These legacy systems were developed to meet very specific aims of a particular organization. One challenge of these interfaces to novice and intermediate users is the high cognitive load they impose: The user has to remember the command syntax, the spelling, and the procedures of the system. With these interfaces, effortless and fast interaction with the computer relies on knowledge in the head (Norman, 1988) is the only way to interact with the computer.

A more complex variant of character-based interfaces are full-screen interfaces. The user can switch between entry fields all over the screen. Full-screen interfaces can be found, for example, in patient documentation systems, where a form-filling dialogue is used to enter the patient information. The interaction with these displays utilizes menus and function keys. A problem associated with full-screen interfaces is that the menu structure has to be optimized for the user's needs, and the analysis to identify those needs is often not performed. Another problem with these systems is that moving through entry boxes can be time consuming, and as a consequence, interacting with full-screen interfaces is not always optimal. In addition, sometimes functionality such as auto completion of entries increases the likelihood of error (see T–25 example). Because of their lack of user-friendliness, character-based interfaces were abandoned in favor of graphical user interfaces (GUI).

GUIs often consist of windows, icons, menus, and a pointing device (WIMP systems). GUIs allow the user to directly manipulate visual representations of the dialogue objects on the screen. Using a GUI supports the user in multiple ways. The interface supports recognition, whereas character-based interfaces require recall. One of the important features of GUIs is that they can use interface metaphors (e.g., desktop metaphor). A user can apply knowledge about the real objects represented in the metaphor to directly interact with the interface (e.g., the wastepaper basket disposes of documents).

Finally, hypertext-based interfaces allow the user to access information by clicking on hyperlinks. Internet browsers use this type of interface, allowing users to navigate by clicking on hyperlinks on web pages. In health care, a growing number of applications use this approach.

One facet of the development of GUIs or hypertext-based systems is that they are often developed for specific applications and environments, but follow some general conventions (Microsoft, 1995), thus allowing some standardization across applications.

The description of interfaces in the previous section focused mostly on displays, that is, providing information to the user using the visual modality. However, interfaces can use other modalities, for example, the auditory modality (e.g., the pulse oximeter). Another aspect of interface design is related to the choice of the input device that allows the user to interact with the computer. Most systems use keyboards or computer mice; however, alternatives are touch-screen displays, light pens, and voice commands. We do not discuss issues related to input devices here, but recommend Baber (1997) for the interested reader.

DEVICES AND INTERFACES

There is more and more evidence that hazards resulting from medical device use might by far exceed hazards based on device failures. These hazards can be attributed partly to professionals dealing with the interfaces of these devices and illustrate further that effective human–computer interaction design needs to be part of the device development process. According to Leape et al. (1998, p. 1444) one of the main problems is that many systems are not designed for safety, but rely on "error-free performance enforced by punishment." Sustained error-free performance in high-stress and high-stakes environments is impossible (Reason, 1990), but human error can be reduced by good interface design. To have a better understanding of some of the devices used in health care, we describe different functions of devices in health care that pose different requirements for the interface designer. We focus on medical monitoring, computer-controlled devices, and clinical data systems.

Medical Monitoring

Interfaces of devices designed to support patient monitoring provide feedback about the status of the patient. Patient monitoring can be defined as "repeated or continuous observations or measurement of the patient, his or her physiological function, and the function of life support equipment, for the purpose of guiding management decisions" (Gardner & Shabot, 2001, p. 443). Patient monitoring can be found in different contexts (e.g., intensive care, perinatal care, peri-operative care, postsurgical care, and coronary care) utilizing a broad variety of devices. Devices range from patient monitoring systems in the operating room (e.g., anesthesia monitors) to devices measuring blood oxygen saturation in perinatal care. Outside the clinical environment, monitoring devices can be found in home health care contexts (blood pressure and heart rate monitors, glucose meter). Monitoring systems provide the caregiver with information about the patient's current status, thus supporting detection and diagnosis of a problem and planning of treatment. Monitoring devices do not perform treatment; the change of the patient's status is achieved through means ranging from noninvasive to invasive means (e.g., changing the patient position, administration of drugs).

Patient monitoring interfaces frequently follow the single sensor, single indicator approach (Goodstein, 1981), where for each sensor used, a single variable is displayed. Such designs result in sequential, piecemeal data gathering, making it difficult and laborious to develop a coherent understanding of the relationships and their underlying mechanisms (Vicente, Christoffersen, & Pereklita, 1995). An alternative to this approach is using interfaces that support the clinician by providing patient information in an integrated way, thus allowing rapid detection, diagnosis, and treatment. Drews et al. (2001) developed a cardiovascular display that incorporates anesthesiologists' mental model of the cardiovascular system. In this object/configural display, symmetry indicates normal values, and asymmetry shows deviations from normal. The authors incorporated emerging features and implemented patterns that matched particular diagnoses.

Computer-Controlled Devices

Another type of interface is used to control computer-operated equipment (e.g., T–25 interface). An infusion pump interface, for example, provides information about the pump's current status (e.g., mode of operation) and the means to change its status. Computer-controlled device interfaces aim at making interaction with the device simple, fast, and error free. Equipment interfaces should provide information (e.g., a menu) that reflects the structure of the task, the work flow, or the procedures that represent standard clinical practice. Because the clinician interacting with the device often simultaneously performs other tasks or is interrupted by tasks with higher priority (Drews & Westenskow, 2006), information about the current or last operational status should be provided. In addition, equipment interfaces should allow for easy and natural navigation, provide information about performed actions, have an option to reverse and cancel actions, and facilitate correct actions and prevent or discourage potentially hazardous actions.

Both patient monitoring and equipment interfaces practice medicine in a technical sense and are regulated by the FDA. The software of these interfaces is subjected to special verification methods, and the required safety certification ensures safe technical operation of the equipment. For more information on medical devices, see chapter 25.

Clinical Data Systems

Clinical data systems are not considered safety critical, and in the case of medical patient record systems, their main function is to maintain the legally binding patient record. Following are descriptions of two examples of clinical data systems: patient-centered information systems and computerized physician order-entry systems.

Patient-centered information systems can be found in different environments. Here we focus on the primary care environment. In this setting, primary care information systems include different modules such as a data module, a basic module, a medical module, a pharmacy module, and so on. Each module serves a different purpose and uses different data structures. The data module, for example, contains demographic information about the patients, registrations of visits, examinations, laboratory tests, and so forth. The medical module contains diagnosis and reasons for encounter based on the International Classification of Primary Care. The interfaces can be implemented as character-based interfaces or as graphical interfaces.

Electronic medical records are slowly implemented for use in primary care practices. Among the obstacles for implementation are cost, lack of clinically tested systems, and problems with data entry reflecting a lack of user-friendly interface design.

One of the more serious issues in health care is related to the administration of drugs (Barker, Flynn, Pepper, Bates, & Mikeal, 2002). Computerized physician order-entry systems reduce the number of some drug-related errors (Bates et al., 1998), because not only can physicians write the orders online, but the systems provide assistance in prescription decisions by displaying relevant data and by checking prescribed drugs for interactions with other drugs and counterindications (Bates et al., 1999). User satisfaction with implemented systems correlates with the ability to execute tasks in a "staightforward" manner (Murff & Kannry, 2001) and the response time and ease of use of the system (Lee, Teich, Spurr, & Bates, 1996). Both issues are a function of interface design.

Figure 27–1 shows the relationships between clinician, patient, and computer in computerized patient monitoring and patient data management. The clinician receives information about the patient status via human–computer interface on a display that is connected with the computer. The computer receives information (e.g., vital signs such as heart rate) from the patient by transducers and equipment that can receive commands from the computer as well (e.g., alarm threshold settings). The clinician interacts with the equipment via input devices and the computer itself. Finally, the computer receives information from the patient data management system, where patient reports that were entered by the clinician into the computer are stored and can provide important information on the display to support clinical decision making.

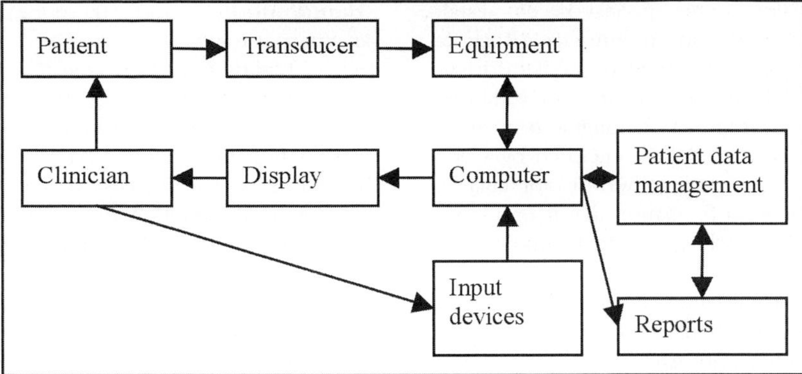

Figure 27–1. Relationships in monitoring patient variables.

PRINCIPLES OF GOOD INTERFACE DESIGN

Often, the design of user interfaces is described as a process that incorporates principles that focus on the user and the user's task. Because of the collaborative nature of health care work, a broader systems perspective on interface design is needed. The following sections describe general principles of good interface design.

THE USER

Several cognitive stages of information processing are involved when a user interacts with a computer interface. Following a standard information processing framework, these are the perceptual stage, cognitive stage, and response stage (see Fig. 2). The following design recommendations loosely follow these three stages, though in some cases the principles affect more than one stage.

Perception

Interface design, especially for patient monitoring and device interfaces, deals with an interesting challenge: On the one hand, the design has to support rapid perception of information, because action has to be taken fast, and on the other hand, action also has to be implemented without error, under stress and/or with concurrent cognitive demand from other tasks. The following principles of interface design can support visual perception of information.

Visibility. Good interface design supports visibility of functionality. Visibility helps to identify the required actions to perform the intended operations (Norman, 1988). GUIs support the principle of visibility by showing visible features: The user can perceive what to do to reach a goal. In contrast to a visible feature, an invisible feature forces the user to remember the existence of a particular feature. Perception provides no guidance.

Affordance. The concept of affordance refers to attributes of objects that allow people to know how to use the object (e.g., a mouse button invites pushing it). Norman (1988) defines affordance as providing a clue about which interaction with the interface is needed to perform a particular operation. Graphical elements such as buttons and icons afford different, but specific activities. Good interface design creates affordance in an unambiguous way.

Both visibility and affordance help to create an interface that is highly transparent.

Redundancy. Redundant coding exploits the redundancy gain, that is, information coded on more than one dimension (e.g., a light that is red and in the highest location on a traffic light) is easier to identify than when coded on one dimension only. In addition, redundantly coded display elements facilitate fast recognition and can help address the issue of color-blind users. Thus redundant coding is an important component of good interface design. However, when using redundancy in color, it is important to follow guidelines for color use (Rice, 1991; Travis, 1991).

Gestalt Principles. The application of gestalt principles when designing an interface (Rock & Palmer, 1990) supports fast perception of information by the user. The gestalt principles indicate that humans tend to see things as belonging together if they are located proximally; are enclosed by a box or lines; move or change together; or look alike in color, size, or shape. Information that is organized following the gestalt principles can be understood better, because the user can extract the organizational structure of the interface easier. Thus interface design should implement the gestalt principles to facilitate rapid perception and comprehension of the display organization.

Saliency and Discriminability. A well-designed interface makes important information salient. This can be done by having information "pop out," which facilitates searches for information. Elements that pop out of a display can be identified faster, because a parallel visual search process can be conducted. Without high saliency, a slower serial search process has to be performed, requiring inspection of each individual element until the target is found (Treisman & Gelade, 1980). The serial search process is negatively affected by similarity of display elements. Discriminability of elements on a display is important because similar elements increase the search time due to additional processes needed to distinguish between these elements. In addition, similar display elements increase error in performance, especially under stress and time pressures. As a consequence, good interface design that supports the user makes important information salient and easy to discriminate.

Visual Clutter. Clutter can be defined as having too many and badly organized elements on a display that have no utility but distract the user. Visual clutter in a display will lead to an increase in the time for search and identification of information. Good human–computer interface design reduces the use of graphical elements to a minimum to avoid visual clutter.

Cognition/Planning

Perceived information has to be processed and integrated into existing knowledge, following either a fast and effortless pattern-matching process or a serial, slow, and cognitively demanding process (Drews & Westenskow, 2006).

Mental Model. A mental model (Craik, 1943) is an internal representation of some aspects of the external world used to make predictions and inferences. The interface designer has to understand a user's mental model to identify goals, actions, and information needs. An interface reflecting a user's mental model can increase performance by reducing time for interaction and learning, by facilitating understanding of the interface structure, and by reducing error. Currently, interfaces often reflect the mental model of the developer and not the user, reducing the likelihood of successful and efficient human computer interaction.

The use of metaphors in graphical interface design attempts to exploit a user's existing domain independent of mental models. For example, common metaphors used in health care could guide interface development. This is in contrast to home health care devices. Here, the user usually has no mental model about the domain, and metaphors implemented to support interaction with the devices have to be very general. A solution is to provide common metaphors that are easily comprehensible. An example of such a general and successfully used metaphor is that of the desktop, which helps a user to understand how a computer works by linking computer operations to the functions of a desktop. Thus good interface design tries to establish a fit between a display and a user's conceptual knowledge, which then facilitates interaction with the display (Carroll, Mack, & Kellogg, 1988; Wozny, 1989).

Proximity Compatibility Principle. Wickens and Carswell (1995) introduced the proximity compatibility principle, which reflects the fact that cognitive integration of information is facilitated when related information is presented in close proximity. For example, to make an assessment of the cardiovascular status of a patient, information from several cardiovascular variables has to be cognitively integrated. If the relevant variables are in proximity on a patient monitor, the information can be integrated faster and with less cognitive demand (see Drews et al. 2001). Thus an interface that follows the proximity compatibility principle has the potential to improve performance by reducing the user's cognitive demand.

Memory: Knowledge in the World versus Knowledge in the Head. Another important design principle is based on Norman's (1988) distinction between knowledge in the world (the interface provides relevant information) and knowledge in the head (the user has to recall how to interact with the interface). That it is easier to deal with a system that provides knowledge in the world is reflected in the popularity of GUIs. Infrequently used systems need to provide information about how to use them because the user likely will have forgotten instructions and procedures of how to use the system.

Response Processes/Execution of Plans

After processing the information provided by an interface and generating a plan, the health care professional has to implement a response. Several principles can support the fast and error-free implementation of action. Because graphical device interfaces and patient information systems frequently allow direct interaction and manipulation, these design principles apply more to these interfaces than to pure medical monitoring systems.

Consistency. Good interface design is consistent design. Consistency refers to the idea that similar operations performed with an interface use similar elements to achieve similar tasks. If a system interface is designed consistently, it allows the user to learn faster to interact with this system, and the system is easier to use and less likely to lead to user error. Consistency of design applies to two levels: within an individual interface and between interfaces of devices, although design that focuses on a single device only tends to ignore the latter. Good interface design creates consistency within a display and attempts to be compatible with other interfaces that are likely to be used in the same environment.

Mapping. Mapping refers to the relationship between controls and their effects in the world. The principle of pictorial realism (Roscoe, 1968) states that the information on a display should have the appearance of the variable represented by it. The principle of moving parts (Roscoe, 1968) describes how to produce a good mapping between controls and effect. According to this principle, dynamically displayed information should reflect the user's mental model about the movement in the real system. Both principles can be used to ensure consistent mapping in a display, which is important for good interface design.

Direct Manipulation. Direct manipulation allows the user to perform actions on the visible objects of the GUI. Direct manipulation has several advantages: It is easy to learn and can be consistent with a user's mental model of a task, thus facilitating performance. Good interface design supports direct manipulation.

Response Alternatives. Providing shortcuts for experienced users allows them to respond faster than does direct manipulation. The issue with shortcuts is that it requires time to learn them, and some users never acquire this knowledge because they make a trade-off between the effort of learning shortcuts and decrease in response times.

A problem related to providing response alternatives is customization of the interface. In general, customization provides the user with more flexibility, which can be beneficial. But customization can be confusing when several people interact with the same system. For example, in the case of a customized patient monitoring system, valuable time could be spent figuring out the controls, diverting attention from the monitoring task. However, Randell (2003) showed how customization in the context of an intensive care unit can have benefits for the users.

Providing a Natural Dialogue. Interfaces should be as simple as possible, providing information about computer concepts to users in a natural way. It is therefore necessary to analyze the user's tasks and to use task terminology for labeling when designing an interface. Related to this is the goal to minimize the effort of navigating through a menu structure by making its logic easy to understand. Thus good interface design uses task-based terminology and creates menu items that reflect the task structure.

Feedback. The concept of feedback is related to visibility. Feedback about what action has been done and what was accomplished is important for interaction. Delays in feedback or nonexistent feedback have negative consequences on behavior. Interfaces frequently use visual or auditory feedback, though verbal, tactile, or combinations of all of these are possible. In the T–25 incident, the feedback provided to the operator was ambiguous: The technician could not know which beam mode was activated when the interface asked, "Beam ready?"

Feedback should be provided within a particular time frame. Miller (1968) and Card, Robertson, and Mackinlay (1991) recommend times for interface feedback: 0.1 seconds is the limit for the user to feel that the system responds immediately, and no special feedback is needed. With an increase in system response time, feedback about the current computer operations has to be provided (after 10 seconds users often want to perform a different task while waiting for the computer to finish).

Another important element of feedback is to provide informative error messages. Cryptic error messages, as the error message "Malfunction 54" in the T–25 example, often require further investigation by the user. Such investigation is unlikely if the error appears to be inconsequential. Good error messages provide information about the source of the error, information about the consequences of the error, and information on how to fix the error using naturalistic language that is comprehensible to the user.

Finally, helping the user to find out how to achieve desired goals is another important issue of feedback. Good interface design allows the user to get specific help information fast and instructs, in simple steps, how to reach the desired goal.

Constraints. Constraints refer to determining ways of restricting the kind of user interaction that can take place at a particular moment. Deactivation of menu options while operating in a particular mode creates such constraints. Restricting the user under certain circumstances reduces the likelihood of error. Norman (1999) distinguishes between three constraint types: physical, logical, and cultural constraints. Physical constraints refer to the way that physical objects constrain the movement of objects. Logical constraints rely on people's understanding of how the world works, that is, people's common sense about actions and consequences. Cultural constraints rely on learned conventions, for example, specific colors that express warnings (red) or desaturated blood (blue). Ecological interface design (Vicente, 2002) emphasizes the importance of making constraints visible.

Error Tolerant Systems and Ease of Error Recovery. To err is human, thus interfaces should minimize the negative consequences of errors and allow fast error recovery. Error-tolerant interfaces prevent or mitigate dangerous and disastrous consequences when error occurs. For the interface designer, it is important to identify potential user errors during the design and evaluation process and to implement solutions for making the system error tolerant. Good interface design approaches provide an undo option and make verification requests in case of irreversible commands.

Systems with multiple operation modes are prone to human error. Two ways of implementing modes can be distinguished: real modes and user-maintained modes (Sellen, Kurtenbach, & Buxton, 1992). An example of a real mode can be found in older text editors, where the user chooses between insert mode and edit mode and where, depending on the current mode, the same user action has very different outcomes. User-maintained modes differ, because here the user constantly pushes a key to keep the mode activated. User-maintained modes create awareness that the particular mode is activated. In the T–25 example, one key (B) activated different treatments depending on the current mode. Ramundo and Larach (1995) report a case where physicians noticed a mismatch between their patient observations and the variables on the patient monitor: The patient's blood pressure was constantly 120/70. Wondering about the constant values, the physicians realized that the monitor was in demonstration mode and not displaying the patient's data.

Figure 27–2 summarizes the stages of information processing and the design recommendations for each stage.

Individual User Characteristics

An important issue regarding the development of efficient user interfaces is to address individual user characteristics. For example, given the fact that the population of anesthesiologists is predominantly male (90%), display designers should be concerned about the high percentage of color-blind users (8% of the male population). Another important user characteristic is age. Aging affects perceptual, cognitive, and response processes, though the changes are gradual and start as early as the 20s and 30s, becoming more serious with increasing age. Other individual user characteristics to be considered are user experience, level of education, work experience, and previous computer experience.

The Task and Its Context

Behavior of experts is goal oriented and structured (Hollnagel, 2000). Thus analyzing behavior in its context helps to identify important aspects and constraints of the user's task. For example, when

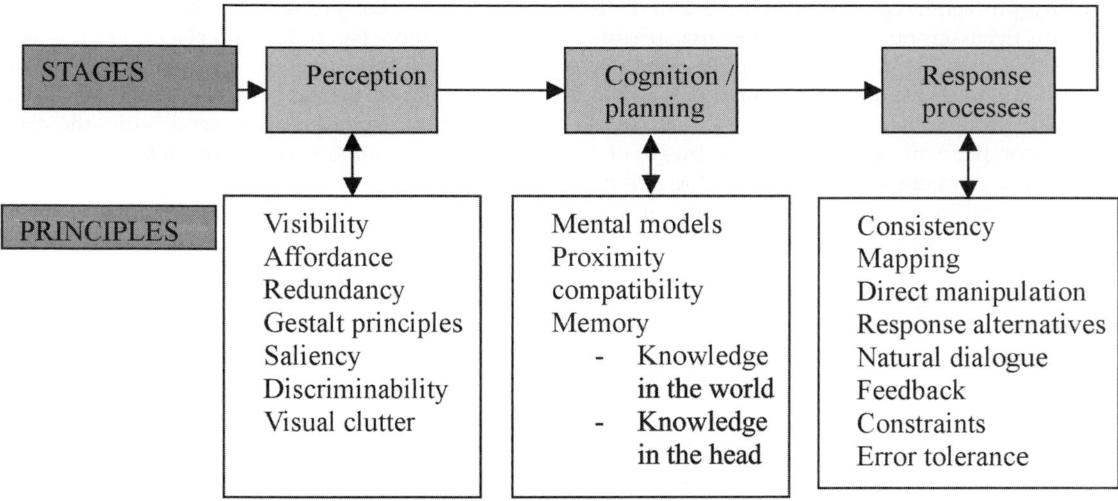

Figure 27–2. Stages of processing and design principles.

developing an infusion pump interface, it is important to analyze how infusions are administered, what steps and procedures are performed before and during administration, and what the current standards of best practice are. In addition, what tasks may interrupt the interaction and how interruptions affect task performance should be identified. Ignoring these influences of the task environment on task performance will overestimate performance in a real context and make error more likely.

Another important facet of the task environment in health care is that health care professionals deal with high levels of stress: Decisions with potentially far-reaching implications have to be made fast, and interventions have to be implemented rapidly. Under stress, human information processing is often suboptimal, that is, not all task-relevant information is processed and included in a decision (Hockey, 1986). Thus interface design in health care has to take into account these task-related circumstances.

Understanding the Task

A task analysis helps to understand the task and its context. Task analysis investigates the situation of the user performing a task with a focus on analyzing the rationale and the purpose of a person's actions. The results help to identify how the task is executed. In the next section we discuss the following techniques used to perform a task analysis: hierarchical task analysis, GOMS and cognitive task analysis (for a more detailed description see chapter 37; Kirwan, & Ainsworth, 1992).

Hierarchical task analysis (Annett & Duncan, 1967) involves the identification of the task and its decomposition into smaller, less complex subtasks. At the lowest level is the description of the user's physical operations. At the end of the analysis, the lowest level subtasks are grouped together as plans specifying how the task might be performed. Thus the hierarchical task analysis focuses on the physical and observable actions of a user.

Another well-known task analysis method is goals, operators, methods, and selection rules (GOMS; Card, Moran, & Newell, 1983). GOMS helps the designer to understand potential problems when a user interacts with interfaces. GOMS assumes hierarchical organization of user action: A user formulates a goal and subgoals and attempts to achieve them by applying methods and operators. A method is a sequence of steps that are perceptual, cognitive, or motor operations. Selection rules guide the user in selecting the method to reach the goal. GOMS can be used to describe software functionality and interface characteristics, which can lead to a systematic analysis of potential usability problems. One problem with GOMS is based in the assumption of error-free user performance (Nielsen, 1993), which is unrealistic and can lead to an underestimation of problems when interacting with an interface.

To complement the previously mentioned approaches, a cognitive task analysis (CTA) should be conducted to allow the identification of the cognitive characteristics of the user's task (see also chapter 37). Cognitive task analysis identifies a user's skills and knowledge when interacting with a system, thus allowing the interface development to focus on specific users and their background. Typical phases of CTA are knowledge elicitation, analysis, and knowledge representation. During the knowledge elicitation phase, information about cognitive events, structures, or mental models is elicitated from domain experts. The analysis phase structures the elicitated knowledge and analyzes its meaning. Finally, the knowledge representation phase shows relationships between important constructs and describes the meaning of the collected data.

Methods to Identify Task Requirements

Taylor (2000) analyzed the success and failure of information technology (IT) projects and found that a major contributor to failure is the lack of clear objectives and requirements. Good interface development identifies the user's needs and requirements as a first step. Requirements describe what the intended interface should do and how it should perform in a particular environment (Robertson & Robertson, 1999). Types of requirements are, for example, functional requirements (what the systems should do), data requirements (what data have to be used), environmental requirements (circumstances under which the product will be expected to operate), user requirements (characteristics of the intended user group), and usability requirements (usability goals).

Multiple methods can be used to gather the information about the requirements. For example, the designer can use questionnaires; structured, semistructured, or unstructured interviews; and focus groups and workshops, and he or she can perform analysis of documents that describe procedures, standards, and regulations or conduct naturalistic observations of the user. Naturalistic observations complement the other techniques by identifying behavior that experts are often not able to verbalize.

Both task analysis and analysis of requirements are of importance for interface development because they help to understand the user's goals as they approach the task, identify the information needs, and help to understand how the user deals with routine events, nonroutine events, or emergencies.

THE SYSTEM

In this section we focus on two important elements of the health care system that have to be considered in the context of human–computer interaction, that is, the heterogeneity of devices and their interfaces and collaboration and distributed cognition.

Heterogeneity of Devices and their Interfaces

Modern health care is provided in a highly interconnected, complex system. Health care professionals are not limited to working in a single setting, but often switch back and forth between settings (e.g., when transitioning a patient). These settings are equipped differently, and the heterogeneity of interfaces has the potential to negatively impact performance of clinicians. For example, the necessity to interact with two different infusion pump interfaces is likely to increase the cognitive demand when programming the devices, the time required to program the pumps, and the frequency of error in programming the pumps.

But even within a particular setting, different devices from different manufacturers are present and professionals are expected to be able to use them proficiently. As described earlier, inconsistent response mapping affects behavior negatively, and interface standardization is likely to increase performance and reduce human error. The need for "a national consensus on comprehensive standards for the definition, collection, coding and exchange of clinical data,"—and this includes condition, procedures, medications, and laboratory data—is described in the Institute of Medicine report (2001). Obviously, this need for standards extends to the design of interfaces.

Interestingly, such standards for the development of graphical interfaces exist already on the national and international levels. On the inter national level, there is ISO–9241 from the International Organization for Standardization

(ISO, 1992) providing guidelines for interface development. Examples for standards on the national level are the American National Standard for Human Factors Engineering of Visual Display Terminal Workstations (ANSI/HFS 100–1988), and the Human Factors Design Standard (HF-STD–001) from the Federal Aviation Administration (Ahlstrom & Longo, 2003).

Collaboration and Distributed Cognition

Another important systems aspect of the health care environment is that it is a highly collaborative environment. This has important implications for interface design. Hutchins (1995) proposed a distributed cognition approach, emphasizing interaction between people as a central part of successful system operation. In this context, it is important to describe how interfaces support the dissemination of information between collaborators.

Multidisciplinary Collaboration: Nurse Information Systems. Teamwork is essential, and information about the status of patients and the performed or planned treatment has to be distributed to all team members. In this section we describe nurse information systems as an illustration of the requirements for human–computer interfaces in the context of multidisciplinary collaboration.

Nurses collaborate with other nurses and other clinical colleagues. Information systems for nurses have to reflect this multidisciplinary collaboration. Nurses use information from different sources. At the bedside they may record particular patient information that is similar or identical to the information physicians use, although the data may have different implications and are used differently by both groups. As a consequence, interface design has to ensure that data are presented consistently with a health worker's mental models, but should also reflect the communalities and disparities between different disciplines in terms of their information requirements.

Facilitating Shared Cognition. A nurse who monitors a patient may call a physician based on his or her assessment of the patient's status. After exchanging information and discussing potential diagnoses they may consult other physicians. Thus collaborative work, which frequently requires distributed cognition, is part of the work conditions in health care. Human–computer interfaces have to take into account this aspect of collaboration and support cognition and problem-solving of teams. Thus one requirement of interface development is that the interaction between collaborators has to be facilitated.

DESIGN METHODOLOGIES FOR USER INTERFACES

Human–computer interface design in health care can utilize a variety of design methodologies (Meister & Enderwick, 2001). In this section we discuss rapid prototyping and iterative design, user-centered design, sociotechnical design approaches, and scenario-focused approaches (see also chapter 37). The methodologies are not mutually exclusive, but complementary.

Rapid Prototyping and Iterative Design

This design methodology includes the fast and early implementation of interface models or mock-ups (Mayhew, 1999). The models are used to perform typical tasks, and the results feed back into the design process. The advantages of this approach are that design models can be developed quickly and comparatively cheaply and that the intended user group evaluates them. A result of this method is that the early presentation to the user influences the overall design, and early design modifications are facilitated. Another advantage of this method is that design changes that affect usability can be identified early in the process and then be revised.

User-Centered Design Approaches

An important aspect of interface development is the user involvement during the design process (Gould & Lewis, 1985). Several user-centered methods can be distinguished ranging in terms of user participation from low (ethnographical methods) to medium (contextual design) to high (participatory design). Participatory design involves the user in the interface development. Central to this approach is the assumption that the user knows best how to optimize an interface so that it fits the needs. Thus the continuous input during the design process ensures that the end product is likely to be user-friendly. The user's function in this context is to serve as a consultant that is involved in all aspects of the design process.

Sociotechnical Design Approaches

The main focus of sociotechnical design approaches is to develop a coherent human–computer system. Direct participation of users is an important part of the design process (Scacci, 2004), but sociotechnical design approaches include, in addition, the network of users, the developers, the current information technology, and the environment in which the system will be used. The basic assumption is that understanding both the social and the technical systems will allow development of interfaces that are highly integrated and support the user in the complex work system. Historically, this approach assumed that a design can be correct, consistent, and complete prior to its implementation, though more recently it has been acknowledged that development is an incremental and iterative process. Unlike most other design approaches, sociotechnical design uses the work system as the unit of analysis, while user-centered approaches focus on the task as a level of analysis (Dillon, 2000).

Scenario-Focused Design Process

This approach involves the identification of particular scenarios in which the interface has to be used (Carroll, 2000). Domain experts participate in the development of the scenarios and the evaluation of performance when testing the interface in the context of these scenarios. Scenario-focused design, for example, has been used to design health care workstations and anesthesia displays.

EVALUATION OF USER INTERFACES

In evaluating an interface, an important trade-off has to be considered: The more frequently evaluations are performed, the more likely it is that problems will be identified and can be addressed. Landauer (1995) argued that each evaluation–redesign cycle performed on an interface has the potential to increase performance by approximately 50%. The maximum cost/benefit ratio for a software system that will be used by 1,000 people has its maximum between three and five usability cycles. In terms of costs, usability assessment can be worth 500 times the expense for large systems and up to 5 times the expense for small systems when choosing the optimal number of cycles.

Interface evaluation focuses on several aspects of the interface interaction and the user's subjective impression of this interaction. Variables for interface evaluation are the effectiveness of an interface, the efficiency with which a task can be performed, and the user's satisfaction when using a particular interface. Focusing only on subjective measures has the danger that subjective assessments do not necessarily correlate with performance. For example, users usually express a preference for color screens and rate their effectiveness higher than monochrome displays. However, when Krupinski and Roehrig (2002) measured actual performance in a pulmonary diagnosis task, participants performed better with monochrome displays.

In the following sections we discuss several methods that can be used to evaluate interfaces (see also chapter 42). These methods are usability testing, field studies, and heuristic evaluation. One less formal way to evaluate an interface is to conduct rapid and less well-controlled evaluations. This approach can be used to get informal feedback from users and has the advantage that it can be done quickly and at almost any stage of the development process.

Usability Testing. Usability testing involves measuring the user's performance on typical tasks. Error rate and time until completion are the measures used to evaluate performance. The performance of a user is recorded on video, and interactions with the system are logged by software. Another element of usability testing is the administration of questionnaires at the end of testing session, where measures such as user satisfaction, cognitive and physical effort, and task difficulty are assessed. Frequently the NASA-TLX is used to assess subjective task load (Hart & Staveland, 1988). Mayhew (1999) points out that usability testing uses a laboratory situation with high levels of control on the side of the evaluator. The level of artificiality of the environment can range widely: On one extreme is the analysis of specific interactions with an interface, without any contextual information; on the other extreme is high-fidelity simulation, with confederates interacting with the expert who uses the new interface.

Field Studies. The advantage of field studies is that they are performed in the naturalistic task environment. The equivalents in the health care environment are clinical studies of interface evaluation. Field studies use a variety of methods to evaluate interfaces. Frequently these methods include audio or video recordings of interactions, verbal

TABLE 27–1. Overview of Interface Evaluation Methods

Method	*Task*	*Participants*	*Costs/Time*	*Control/Fidelity of Evaluation Setting*
Usability testing	Typical task	Typical user	Relatively high/slow	High/range from low to high
Field study	Wide range of tasks	Typical user	Relatively high/very slow	None/highest (naturalistic setting)
Heuristic evaluation	Typical task	Expert – applies knowledge of typical user but heuristics may not apply	Low/fast	High/range from low to high

descriptions of specific events, interviews of experts and novices, or having experts comment on the task while performing it (think-aloud method). The recordings of the observations are analyzed, for example, by focusing on behavior (e.g., a content analysis categorizes observations into categories) or verbalizations (e.g., think-aloud protocols are analyzed using discourse analytic methods). All methods imply that the evaluator is not involved in the interaction in the field, that is, the evaluator takes a clear outsider perspective. An alternative approach is to have the evaluator participate in the interactions as proposed by ethnographic approaches. Ethnographic approaches require the evaluator to be a participant and try to identify what is happening in a particular social setting, for example, work processes (Shapiro, 1995).

Heuristic Evaluation. The goal of heuristic evaluation is to have experts (usually three or more) apply their knowledge of the typical user to predict potential usability problems. The expert walks through the interface and identifies violations of a previously defined catalog of usability heuristics (e.g., the previously mentioned principles of good interface design). The advantage of heuristic evaluation is that regular users are not required during the evaluation process. As a consequence of interviewing only a small number of experts, the evaluation process is fast and comparatively inexpensive (for an example in health care, see Zhang, Johnson, Patel, Paige, & Kubose, 2003). But heuristic evaluation had some serious limitations: The most important of these limitations is that the heuristics that guide the expert during the evaluation might not apply to the specific context. For example, the identification of heuristics for very new and conceptually different products requires more time and resources and may reveal that the heuristics used by experts do not apply, because of the very different nature of the new design approach. Table 27–1 provides a summary of the interface evaluation methods discussed.

GOOD INTERFACE DESIGN IN MEDICINE: AN EXAMPLE

An example of how interfaces in medicine can be improved by good interface design was given in a study by Lin and colleagues (1998) about patient-administered analgesic pumps (PCAs). PCAs allow patients to administer an analgesic to treat postoperative pain. Nurses program the PCAs, and one important step of the setup procedure is to enter the concentration of the analgesic. Several studies report cases where patients died from an overdose of the analgesic (Grover & Heath, 1992; Notcutt, 1992), as a result of incorrect concentration programming (e.g., 1 mg/ml instead of 5 mg/ml). One of the problems with the older PCA's interface was that it used the minimal drug concentration as a default. Thus, if not adjusted, a higher concentrated drug may be administered mistakenly at the higher flow rate of a lower concentrated drug. The consequence is overdosing that can be lethal for the patient. Lin et al. (1998) evaluated whether an improved version of a PCA pump interface would reduce the number of misadministrations and the effectiveness of interacting with the pump. They used CTA to assess pump usage data in a clinical environment. Based on observations and interviews, they designed an alternative pump interface following basic design principles. In their evaluation they compared error rates and time for programming for both the new and the old interfaces. Their results showed that the error rate was reduced by 55% and that the programming time was reduced by 18% when using the new interface. Thus, if fundamental principles of interface design are applied when developing medical device interfaces, patient safety will be improved.

CONCLUSIONS

Successful interface design in health care follows the principles of good interface design and uses the methods described to evaluate user performance in a systematic and realistic way.

But user interface design in health care also has to overcome several challenges to successfully improve quality of patient care. One of these challenges is related to the fact that the health care work environment is a highly stressful, high-risk, and high-stake environment. This affects the health care professional's ability to process information and to make decisions. As a consequence, it should not be assumed that information processing is at the optimal level. Interface design has to take this into account. Another important challenge for the interface designer is related to the fact that health care is a highly cooperative environment where communication and shared cognition are essential for success. Display design that does not take this into account has the potential to exclude team members from access to information and therefore negatively impact efficient teamwork. Another facet of the health care environment is that the professional has to interact with many different devices from multiple manufacturers using very different approaches for user interfaces. One of the potential consequences of such a lack of standards is that performance can be suboptimal, because time has to be spent in identifying the particular interface and the way in which interaction is structured.

All these challenges can be answered, and there is significant potential to improve health care when interface designers are facing these challenges. As identified by the Institute of Medicine (2000), one of the contributing factors to error in medicine is the growing complexity of science and technology on the one hand, and constraints of exploiting information technologies on the other hand. Interface design that realizes the potential of information technologies can help to make the complexity of science and technology manageable and utilize this potential to improve health care.

References

Ahlstrom, V., & Longo, K. (2003). Human Factors Design Standard (HF-STD–001). Atlantic City International Airport: Federal Aviation Administration, William J. Hughes Technical Center.

American National Standard for Human Factors Engineering of Visual Display Terminal Workstations. ANSI/HFS 100–1988. Santa Monica, CA: Human Factors Society.

Annett, J., & Duncan, K. D. (1967). Task analysis and training design. *Occupational Psychology, 41,* 211–221.

Baber, C. (1997). *Beyond the desktop.* San Diego, CA: Academic.

Barker, K. N., Flynn, E. A., Pepper, G. A., Bates, D. W., & Mikeal, R. L. (2002). Medication errors observed in 36 health care facilities. *Archives of Internal Medicine, 162,* 1897–1903.

Bates, D. W., Leape, L. L., Cullen, D. J., Laird, N., Petersen, L. A., & Teich, J. M. (1998). Effects of computerized physician order entry and a team intervention on prevention of serious medication errors. *Journal of the American Medical Association, 280,* 1311–1316.

Bates, D. W., Teich, J. M., Lee, J., Seger, D., Kuperman, G. J., Ma'Luf, N., et al. (1999). The impact of computerized physician order entry on medication error prevention. *Journal of the American Medical Informatics Association, 6,* 313–321.

Card, S. K., Moran, T. P, & Newell, A. (1983). *The psychology of human–computer interaction.* Hillsdale, NJ: Lawrence Erlbaum Associates.

Card, S. K., Robertson, G. G., & Mackinlay, J. D. (1991). The information visualizer: An information workspace. *Proceedings of the Association for Computing Machinery: Computer Human Interface '91 Conference* (New Orleans, LA), 181–188.

Carroll, J. M. (2000). *Making use: Scenario-based design of human computer interactions.* Cambridge, MA: MIT Press.

Carroll, J. M., Mack, R. L., & Kellogg, W. A. (1988). Interface metaphors and user interface design. In M. Helander (Ed.), *Handbook of human–computer interface* (pp. 74–102). Cambridge: Cambridge University Press.

Cohen, M. (1999). Overreliance on pharmacy computer systems may place patients at great risk. *ISMP Medication Safety Alert! 4,* 1–3.

Craik, K. J. W. (1943). *The nature of explanation.* Oxford: University Press, Macmillan.

Dillon, A. (2000). Group dynamics meet cognition: Combining socio-technical concepts and usability engineering in the design of information systems. In E. Coakes (Ed.), *The new socio tech: Graffiti on the long wall* (pp. 119–126). London: Springer.

Drews, F. A., Agutter, J., Syroid, N. S., Albert, R. W., Westenskow, D. R., & Strayer, D. L. (2001). Evaluating a graphical cardiovascular display for anesthesia. In *Proceedings of the 41st Human Factors Meeting.* Human Factors and Ergonomics Society, Santa Monica, CA., p. 1264.

Drews, F. A., & Westenskow, D. R. (2006). Display design in Anesthesia. *Human Factors, 48,* 59–71.

Gardner, R. M., & Shabot, M. (2001). Patient monitoring systems. In E. H. Shortliffe, L. E. Perrault, G. Wiederholt, & L. Fagan (Eds.), *Medical informatics: Computer applications in health care and biomedicine* (2nd ed., pp. 443–484). New York: Springer.

Goodstein, L. P. (1981). Discriminative display support for process operators. In J. Rasmussen & W. B. Rouse (Eds.),

Human detection and diagnosis of system failure (pp. 433–449). New York: Plenum.

Gosbee, J. W., & Richie, E. M. (1997). Human–computer interaction and medical software development. *Interactions, 4*(4), 13–18.

Gould, J. D., & Lewis, C. (1985). Design for usability: Key principles and what designers think. *Communications of the ACM, 28*(3), 300–311.

Grover, E. R., & Heath, M. L. (1992). Patient-controlled anesthesia: A serious incident. *Anesthesia, 47,* 402–404.

Hart, S. G., & Staveland, L. E. (1988). Development of NASA-TLX (Task Load Index): Results of empirical and theoretical research. In P. A. Hancock & N. Meshkati (Eds.), *Human mental workload* (pp. 139–183). Amsterdam: North-Holland.

Hockey, G. R. (1986). Changes in operator efficiency as a function of environmental stress, fatigue, and circadian rhythms. In K. R. Boff, L. Kauffman, & J. P. Thomas (Eds.), *Handbook of perception and human performance* (Vol. 2, pp. 1532–1573). New York, NY: Wiley.

Hollnagel, E. (2000). Modeling the orderliness of human action. In N. B. Sarter & R. Amalberti (Eds.), *Cognitive engineering in the aviation domain* (pp. 65–98). Mahwah, NJ: Lawrence Erlbaum Associates.

Hutchins, E. (1995). *Cognition in the wild.* Cambridge, MA: MIT Press.

Institute of Medicine. (2000). *To err is human: Building a safer health system.* L. T. Kohn, J. M. Corrigan, & M. S. Donaldson (Eds.). Washington, DC: National Academy Press.

Institute of Medicine. (2001). *Crossing the quality chasm: A new health system for the 21st century.* Washington, DC: National Academy Press.

International Organization for Standardization. (1992). Ergonomic requirements for office work with visual display terminals (VDTs): Part 3. Visual display requirements. Geneva: International Standards Organization.

Kirwan, B., & Ainsworth, L. K. (1992). *A guide to task analysis.* London: Taylor & Francis.

Krupinski, E. A., & Roehrig, H. (2002). Pulmonary nodule detection and visual search: P45 and P104 monochrome versus color monitor displays. *Academic Radiology, 9*(6), 638–645.

Landauer, T. K. (1995). *Trouble with computers: Usefulness, usability, and productivity.* Cambridge, MA: MIT Press.

Leape, L. L., Woods, D. D., Hatlie, M. J., Kizer, K. W., Schroeder, S. A., & Lundberg, G. D. (1998). Promoting patient safety by preventing medical error. *Journal of the American Medical Association, 280*(16), 1444–1447.

Lee, F., Teich, J. M., Spurr, C. D., & Bates, D. W. (1996). Implementation of physician order entry: User satisfaction and self reported usage patterns. *Journal of the American Medical Informatics Association, 3,* 42–55.

Leveson, N., & Turner, C. S. (1993). An investigation of the Therac–25 accidents. *IEEE Computer, 26*(7), 18–41.

Lin, L., Isla, R., Doniz, K., Harkness, H., Vicente, K. J., & Doyle, D. J. (1998). Applying human factors to the design of medical equipment: Patient-controlled analgesia. *Journal of Clinical Monitoring and Computing, 14*(4), 253–263.

Mayhew, D. J. (1999). *The usability engineering lifecycle.* San Francisco: Morgan Kaufmann.

Meister, D., & Enderwick, T. P. (2001). *Human factors in system design, development, and testing.* Mahwah, NJ: Lawrence Erlbaum Associates.

Microsoft. (1995). *The windows interface guidelines for software design: An application design guide.* Redmond, WA: Microsoft Press.

Miller, R. B. (1968). Response time in man–computer conversational transactions. *Proceedings of the AFIPS Spring Joint Conference, 33,* 267–277.

Murff, H. J., & Kannry, J. (2001). Physician satisfaction with two order entry systems. *Journal of the American Medical Information Association, 8,* 499–509.

Nielsen, J. (1993). *Usability engineering.* San Diego, CA: Morgan Kauffman.

Norman, D. (1988). *The design of everyday things.* New York: Currency-Doubleday.

Norman, D. (1993). *Things that make us smart: Defending human attributes in the age of the machine.* Reading, MA: Addison-Wesley.

Norman, D. (1999, May/June). Affordances, conventions and design. *ACM Interactions Magazine, 6,* 38–42.

Notcutt, W. (1992). Overdose opioid from patient controlled analgesia pumps. *British Journal of Anesthesia, 68,* 50.

Ramundo, G. B., & Larach, D. R. (1995). A monitor with a mind of its own. *Anesthesiology, 82*(1), 317–318.

Randell, R. (2003). User customisation of medical devices: The reality and the possibilities. *Cognition, Technology & Work, 5,* 163–170.

Reason, J. (1990). *Human error.* Cambridge: Cambridge University Press.

Rice, J. F. (1991). Display color coding: 10 rules of thumb. *IEEE Software, 8*(3), 99–111.

Robertson, S., & Robertson, J. (1999). *Mastering the requirements process.* Boston, MA: Addison-Wesley.

Rock, I., & Palmer, S. (1990).The legacy of gestalt psychology. *Scientific American, 263*(6), 84–90.

Roscoe, S. N. (1968). Airborne displays for flight navigation. *Human Factors, 10,* 321–332.

Sawyer, D. (1996). Do it by design. US Department of Health and Human Services. Food and Drug Administration. P. 1. Retrieved January 15, 2004 from http://www.fda.gov/cdrh/humfac/doitpdf.pdf

Scacci, W. (2004). Socio-technical design. In W. S. Bainbridge (Ed.), *The encyclopedia of human–computer interaction.* Great Berrington, MA: Berkshire.

Shneiderman, B. (1992). *Designing the user interface: Strategies for effective human–computer interaction* (2nd ed.). Reading, MA: Addison-Wesley.

Sellen, A. J., Kurtenbach, G. P., & Buxton, W. A. (1992). The prevention of mode errors through sensory feedback. *Human–Computer Interaction, 7*(2), 141–164.

Shapiro, D. (1995). Noddy's guide to … enthnography and HCI. *HCI Newsletter, 27,* 8–10.

Taylor, A. (2000, January). IT projects: Sink or swim. *ITNOW in 2000, 42*(1), 24–26

Travis, D. (1991). *Effective color displays: Theory and practice.* London: Academic.

Treisman, A., & Gelade, G. (1980). A feature-integration theory of attention. *Cognitive Psychology, 12,* 97–136.

Vicente, K. J. (2002). Ecological interface design: Progress and challenges. *Human Factors, 1,* 62–78.

Vicente, K. J., Christoffersen, K., & Pereklita, A. (1995). Supporting operator problem solving through ecological interface design. *IEEE Transactions on Systems, Man, & Cybernetics, 25*(4), 529–545.

Wickens, C. D., & Carswell, C. M. (1995). The proximity compatibility principle: Its psychological foundation and relevance to display design. *Human Factors, 37*(3), 473–494.

Wozny, L. A. (1989). The application of metaphor, analogy, and conceptual models in computer systems. *Interacting With Computers, 1*(3), 273–283.

Zhang, J., Johnson, T. R., Patel, V. L., Paige, D. L., & Kubose, T. (2003). Using usability heuristics to evaluate patient safety of medical devices. *Journal of Biomedical Informatics, 36,* 23–30.

28

RE-PRESENTING REALITY: THE HUMAN FACTORS OF HEALTH CARE INFORMATION

C. P. Nemeth, M. O'Connor, M. Nunnally, and R. I. Cook
The University of Chicago

Health care is a significant service sector[1] that relies heavily on the use of information to perform the daily work of patient care and related technical work. As part of their 1999 strategies to improve patient safety, the Institute of Medicine (IOM, 1999, pp. 177, 183) recommended improving access to accurate, timely information and making relevant information available at the point of patient care. Information displays are essential to each clinician's performance at both the individual patient level and the unit level in an acute care facility. Because "artifacts shape cognition and collaboration," the way that a problem is presented improves or degrades the cognitive work that a clinical care provider performs to solve it (Woods, 1998, pp. 168–169). Information displays include hard copies of notes, checklists, and status boards. They can also take the form of control/display interfaces on individual pieces of electronic equipment and assignment schedules that are shown throughout a facility on computer monitors. Displays that are suited to cognitive work at both the patient and the unit levels can improve the reliability and efficiency of clinical work and patient safety, and minimize gaps in the continuity of care (Cook, Render, & Woods, 2000).

This chapter describes the evolution of thought related to information displays and offers an approach to develop representations that effectively support the cognitive work that health care clinicians perform. At least one third of new large information (IT) systems in 1994 failed, and the remainder fell short of budget and schedule objectives (Carr, 2005). Heeks, Mundy, and Salazar (1999, p. 2) contend that "many—even most—health care information systems are failures." Efforts such as this and others that are described in Nemeth, Cook, and Woods (2004) may help us to begin to understand causes for IT system failure in health care and to foresee opportunities for improvement.

ECOLOGICAL DISPLAYS

Air and ground transportation, nuclear power, the military, and health care are all high-hazard, complex, uncertain sectors. Operator performance in these settings requires the ability to appreciate previous and current states of the system in which they work and to foresee the implications of previous and current circumstances for what is to follow. In complex systems, these activities rely on the use of one or more displays. By *display* we mean a visual or auditory representation of information, which can include many different kinds of media such as paper copies of photos, diagrams, tables, maps, and electronic versions of these items. An anesthesia coordinator workstation could be considered a display, as it incorporates all of these items to support the daily cognitive work of managing anesthesia assignments. Other terms are also related to displays. *Monitor* refers to an electronic device that

[1]$1,424.5 billion in 2001 national health expenditures (HHS, 2004)

presents data from a computer, while a *representation* is a presentation to the viewer or listener in the form of an idea or image. Representations are embodied in cognitive artifacts (Hutchins, 1995) that are intended to portray an abstract or physical concept for the purpose of explanation. The more clearly one understands a concept, the better the representation. Tufte (1997, pp. 110–111) describes visual aspects of representations that enable them to succeed in this regard. Among many examples of representations that he analyzes, he notes that the medical record is not intended to assist with the medical treatment of the patient. As an alternative, he proposes a graphical view of patient status that combines images, data values in short- and long-term scales, images such as X rays, and comments on patient history and diagnosis. He contends that the high resolution summary would significantly improve information transfer by blending quantitative multiples of data, narrative text, and images.

Representations are used to reveal complex elements of information for clinicians to use. Instead of directing clinician decisions (as rule-based programs tried to do), skillfully crafted *representations* facilitate and empower clinician judgment. They do so by portraying essential elements in a work setting and their relationships.

Patient-level representations synthesize different types of data through the use of diagrams that leverage human skills such as pattern recognition. Patterns and symbols can be used to support crucial tasks such as the comparison and contrast of various data and the assessment of trends. *Unit-level* representations reflect the newly developing understanding of technical work. (Cook, Woods, & Miller, 1998). Those data span groups of patients, the current number of patients on the unit, their locations and condition, pending and in-progress diagnosis and treatment, current care providers, and prospective transfer in or out of the unit. Clinicians currently consider both patient and unit circumstances when deciding which patients should remain in an intensive care unit (ICU) and which should be "bumped" to the patient floor to make way for more acute patients.

In high-tempo settings such as the ICU and emergency department (ED), clinicians develop their own intrinsic mental models of patients and unit activity. They rely on discrete cognitive artifacts as part of the distributed cognition that is required to operate in the uncertain, contingent, tentative, and fast-changing circumstances of the unit. Figure 28–1 suggests how these elements relate to each other. The figure portrays the current circumstance in which clinicians develop their own mental model of each patient and of the unit as a whole. Data sources are separate and vary widely. They include a unit status board showing planned procedures and staff assignments, a monitor showing vital signs telemetry for each patient in the unit, patient chart information, and more. The task of pulling together all of the individual elements of data into a coherent mental model falls to the clinician. Mental models for patients and the unit can differ depending on what artifacts each clinician has seen. Figure 28–2 suggests an approach to support for clinician work that synthesizes data on behalf of clinicians. One value of this approach is its depiction of the entire unit based on the past, current, and anticipated data that are related to each patient on the unit. As data related to the patients change, their representations also change. As patient representations change, the representation of the unit changes. In this setting, clinicians have the opportunity to probe for more particular data related to an individual patient or to view the unit as a whole. This is not an "all-in-one" solution. Rather, it is an information ecology that is created to assist the way that clinicians work. Such an approach can be primarily graphical with alphanumeric elements, or primarily alphanumeric with graphical elements. However it is configured, the fundamental significance of a representation is not in its visual qualities. Instead, its value lies in how well its visual qualities correspond to elements in the work domain that it is intended to represent—what Woods and Hollnagel (1987) refer to as its *domain semantics*. It is the transition between data and task that informs the development of representations.

Representation is the point at which scale, relationships, history, and trends are assembled to evoke meaning from the information. To be effective, displays (which can contain many representations) must reflect the work domain's elements. (Rasmussen & Vicente, 1990). Representations synthesize pertinent elements including scale, relationship, and other aspects of cognitive work that clinicians would otherwise have to combine on their own. They also offer the potential to refine the way data and information are presented. For example, uncertainty and contingency are inherent in the ICU and ED settings, but current displays are blind to these issues. Representations that reflect variations in certainty such as low or high confidence in a data source would better reflect domain

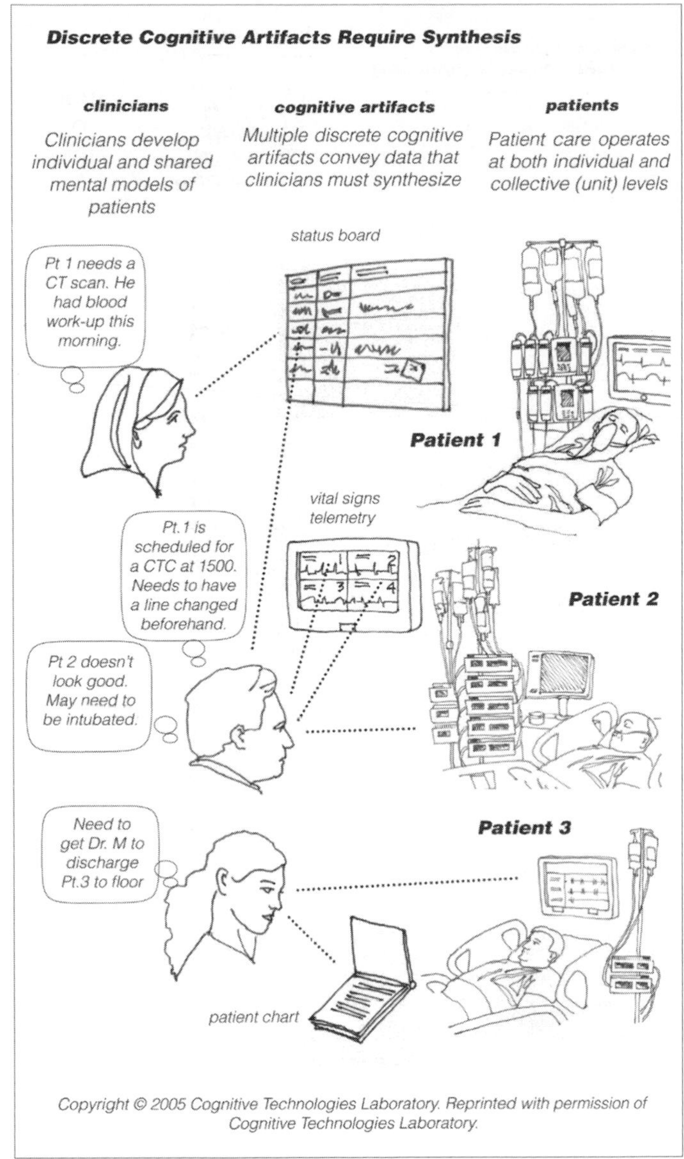

Figure 28–1. Discrete cognitive artifacts require synthesis.

semantics. Effective representations summarize, or abstract (Rasmussen & Pjetersen, 1995), a wealth of discrete elements and thereby spare operators the task of data synthesis. They also enrich operators' ability to contemplate problems and to envision opportunities. Because of this, effective ecological displays and the representations they convey offer substantial potential to benefit clinical care and improve patient safety.

The current understanding of information representation is based on the work of social scientists that has evolved since the 1950s. James J. Gibson first advanced the notion in the 1950s that information shapes an organism's behavior. An individual's ability to act in an environment is influenced by it because each "object offers what it does because it is what it is" (Gibson, 1977, p. 78). Gibson's concept of *affordances* refers to actions a

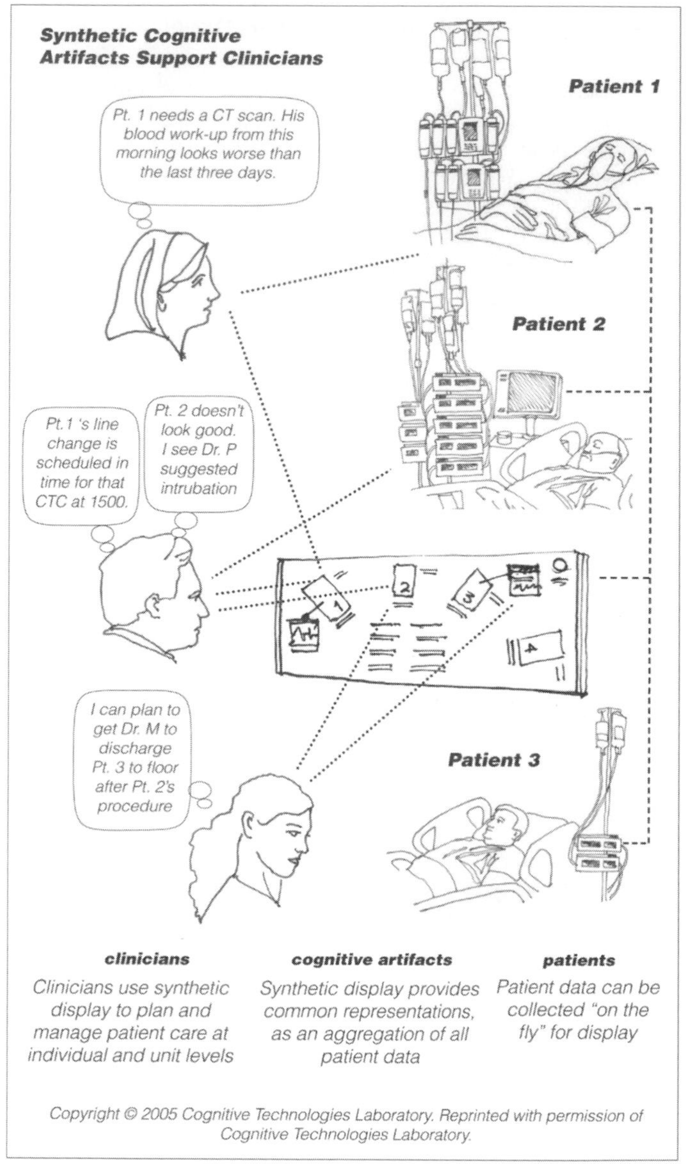

Figure 28–2. Synthetic cognitive artifacts support clinicians.

person can take that are made available by the features that exist in an environment. Those features exist separate and apart from the individual's ability to perceive them (McGrenere & Ho, 2000, p. 179). The concept has since evolved into the "ecological" (Gibson, 1977, p. 78) approach to perception and action. In the ecological approach, individuals are considered to perceive by actively looking, listening, and moving around, rather than passively sensing stimuli. The features of the environment, or ecology, make certain actions possible while concealing others from consideration.

Before the development of digital computing, studies in workstation design were one means to both understand the essential elements that are involved in the management of a complex system and grasp the implications for information display. Shackel (1979), for example, assessed the role of a display as a decision aid for an operator to manage the minute-to-minute activities of an aircraft refueling control center. Shackel's approach followed a path that a current researcher of complex systems would recognize. The study started with a full decision analysis of the controller job, including insights into work domain characteristics such as variable and uncertain demands and limited resources. He then developed a new design for a "stateboard." Results from the study showed how the new display presented information more economically and compatibly than the previous workstation and relieved operator memory load.

Rasmussen's insights into the control of nuclear power generation plants led to the development of improved approaches to manage complex systems. For Rasmussen (1985, p. 236), the issue in system control was to transfer the problem of coordinating a large number of information sources and control actions to a more manageable level that had less resolution. His *abstraction hierarchy* (Rassmussen & Lind, 1981) transformed previous notions of systems as concrete representations that were based on physical properties. The model sorted aspects of systems from a more concrete physical basis (e.g., appearance and anatomy) to more abstract function (e.g., information flow topology) and purpose (e.g., system objectives). The data integration that the system performs spares the operator from spending mental resources to make deductions from a great many variables. It also permits the operator to probe diagnostic pathways from a more abstract function level to deeper levels in order to get a clearer description of the system's state. The hierarchy demonstrated how information, concepts, and structure must necessarily change to represent the state of a function or operation at various levels.

Suchman's (1987) critique of traditional cognitive psychology pointed out the shortcomings of the traditional individual computer display model, showing how it was blind to collaborative work. For Suchman, the sophistication of humans working together far exceeds the software programs that were intended to support that work. Later use of sociological methods including ethnomethodology and conversational analysis derives from dissatisfaction with goal-oriented models of human conduct, as well as the notion that successful design depends on understanding the ordinary work practices, tasks, and situational requirements of users.

Mitchell & Miller (1986) introduced the notion of discrete control models to represent cognitive and decision-making activities that are involved in supervisory control tasks. Dynamic displays of a manufacturing assembly line were used to select and synthesize data that pertained to support decision making, which also minimized the display clutter. The interface for a satellite control workstation (Mitchell & Saisi, 1987) tailored iconic and alphanumeric information to assist what the operator was doing at the time. It used iconic representations, which are more approximate, to monitor system states and detect and diagnose faults. The interface used more detailed alphanumeric data for displays of information related to fault compensation. Jones, Mitchell, and Rubin (1990) probed the potential for an "operator's associate" expert system to infer operator intent and offer intelligent, context-sensitive advice and reminders. Rigorous validation tests found that the operator function model compared favorably with human judgment as a way to formulate real-time hypotheses in dynamic environments. The study concluded that the description and organization of activities for a complex system require that operator activities be modeled within the context of a changing system, easily coded in software, and readily used to describe the semantics that an operator aid would require. The network of finite state automata that results produces an operator function model that links high-level operator functions with low-level operator actions. The network's multiple parallel nodes demonstrate the concurrent nature of supervisory control, and next-state transition functions offer possible actions for the operator to consider. While the study application dealt with a satellite control center, the approach can also inform clinical display design. For example, clinicians may take either diagnostic or therapeutic action at any time depending on the course of activity their judgment dictates. A well-designed display would enable a clinician to monitor and selectively gather necessary data. This makes it possible both to infer (Mitchell & Miller, 1986, p. 352) the patient's current (and possibly prospective) condition and to pursue either diagnosis or therapy.

Among twelve tenets of effective operator interfaces for complex systems, Mitchell (1995) advocates the inclusion of perceptual and conceptual activities beyond physical activities. The decision aids that comply with those tenets are more likely to be understood and accepted by operators, making decision aids preferable to rule-based advisory systems. Traditional task-analytic techniques are normally used to model physical activity. Some other method beyond task analysis would be needed to derive perceptual and conceptual activities.

Norman (1988) expanded on Gibson's initial notion of affordances, contending that they are the result of one's mental interpretation that is based on past knowledge and experience and is applied to our perception of things around us. Norman (1986, pp. 31, 59–61) described *cognitive engineering* as a kind of applied cognitive science that attempts to employ knowledge from psychology and computer science for the design and construction of machines for novice or infrequent users. The notion of *user-centered design* relies on an understanding of user needs to drive development of the system interface, which then drives the system technology.

One of information systems' greatest contributions may be cognitive tools that amplify human powers of conceptualization. This is because "representations of the world that are provided to a problem solver can affect his/her/its problem solving performance" (Woods & Roth, 1988, pp. 25–26). In order to support problem solving, the IT system designer must understand the interactions among representations of the world, the application, and agents involved for both current and prospective changes. Design activity must link semantics that are meaningful for a task to the representation, so that the operator can observe domain semantics directly.

Woods (1994) suggests that representation design maps object or processes data into interrelated visual forms. This mapping is done on behalf of practitioners who are immersed in a goal and task context. Designing for *information extraction* (how well information can be located and understood) relies on mapping the work domain's state and work domain's dynamics to the syntax and dynamics of the visual forms that are intended to represent it. Changes in the domain are reflected in the way that representations of the domain behave and appear. Rather than forcing operators to integrate across successive displays, representations can take advantage of computing systems' ability to behave and change. Following known design principles, such as employing coherent views of processes that are based on patterns or time, allows data to be expressed in a manner that operators can understand or "decode."

Rasmussen and Pjetersen (1995) expanded on Rasmussen (1985) by addressing the broader network of means–ends relations that are part of coping with the requirements and situations in work. Adding whole–part decomposition to means–ends abstraction (Figure 28–3) demonstrated how attention shifts in a trajectory that is primarily along the diagonal of the diagram. Representations vary according to the characteristics of the domain that they are intended to support. Tightly coupled systems such as process control facilities are work driven. Formats for their display are largely influenced by the characteristics of the work system. Loosely coupled systems such as office work domains depend on human activities and responses to constraints. Health care is variably coupled. Certain clinical procedures such as installing an arterial line are tightly coupled. Routine clinical treatment is loosely coupled and allows the clinician to choose and use representations according to the level of patient information (historical to current, the condition of a particular system or organ or tissue) and type of information that is desired or available (consultant opinion, treatment and test activity and results, vital signs).

Regardless of the system's coupling, display format design must rely on an understanding of how individuals and groups perform their work. Vicente (2000) suggests that *work domain analysis* is a means to understand how individuals perform their increasingly complex and varied tasks. Traditional task analysis distills a work role into an efficient but inflexible set of directions. By contrast, the work domain analysis approach produces more of a task map that allows for the worker(s) to meet demands as momentary needs and their preferences dictate. For more information, see chapter 34 on "Cognitive Work Analysis in Health Care" in this handbook.

Hutchins (1995) explained how individuals create a distributed cognition, which is a commonly shared knowledge that benefits a group but cannot be known by any single individual. Distributed cognition makes it possible for a group to perform complex work that is distributed across dimensions such as time and location. In addition to activities, distributed cognition includes artifacts that embody, hold, and share information. In the wake of studies by Suchman (1987) and Hutchins (1995), Heath and Luff (2000) and Luff, Hindmarsh, and Heath (2000) explored how tools and technologies feature in the interaction among people as they work in organizational environments. Their *workplace studies* approach used naturalistic methods, or ethnographies, of the workplace to learn how teams in organizations such as the national health care service and the London Underground accomplish their work. Keen sensitivity to the means that the groups under study use to accomplish goals led to notions of cognitive artifacts such as a small portable electronic device to assist physicians' work

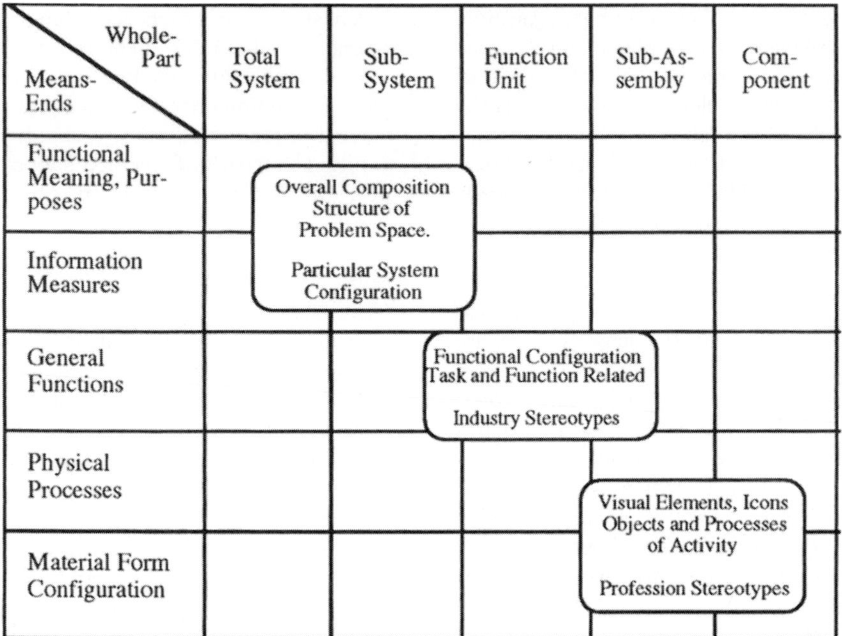

Figure 28–3. System abstraction-decomposition model.

Source: Rasmussen, J., & Pjetersen, A. (1995). Virtual ecology of work. Reprinted with permission of Lawrence Erlbaum Associates.

(Heath & Luff, 2000, p. 238). Hollan, Hutchins, and Kirsh (2000) propose distributed cognition as the replacement for the traditional human–computer interaction (HCI) model that considers the computing system and interface to be separated from cognition, or outside of its boundary. By moving the boundary to be more inclusive, distributed cognition considers all work activity participants together. The symbolic conversion that representations perform can then be viewed as one of many kinds of coordination that occur between users and computer systems.

The creation of ecological displays relies on understanding the work domain in which individuals operate. That understanding flows from ethnographic methods such as work domain analysis. To be effective, displays must reflect a work domain's elements. Representations save operators cognitive work by accounting for a wealth of discrete elements. They also enrich operators' ability to contemplate problems and envision opportunities. Effective ecological displays and the representations they convey offer substantial potential to benefit clinical care.

HEALTH CARE INFORMATION DISPLAYS: AN EXAMPLE AND ITS IMPLICATIONS

Many of the work domains for which cognitive engineering research has been performed are stationary and well bounded. For example, a nuclear power plant is a high-hazard operation, but its condition can be known and its limits can be defined. By contrast, health care provides services to prevent, treat, and manage illness and to preserve mental and physical well-being. The flow of work activities is often much less linear, roles are defined more flexibly and can overlap, and the differences between procedural steps can be less distinct than they are in other sectors (Ash, Berg, & Coheira, 2004, p. 110). Clinicians work on compromised systems (patients) who respond to therapeutic intervention in different ways. Health care is discretionary, involving the ad hoc assembly of information, equipment, and interventions different ways to provide for unique and widely varying patient needs. It is also poorly bounded and subject to the influence of many actors among multiple departments and facilities and considerations of technical work.

Health care activities rely on the acquisition, portrayal, and analysis of therapeutic and diagnostic information as an integral part of individual patient care. Individual elements of information vary enormously in the length of time that they remain reliable, and their weight depends a great deal on their context. The need for accurate, timely information also exists at the unit level, such as the ICU and ED, where the technical work of planning and managing directs who will get care, what type of care will be provided, and when it will be provided. As in other high-hazard settings, expertise (Feltovitch, Ford, & Hoffman, 1997) in health care is the ability to know what is—and what is not—important. The daily work of the clinician requires representations that serve as a map of the ever-changing territory of work that must be successfully navigated (Rassmussen & Pjetersen, 1995, p. 132). What is represented, and how it is represented, depends on the cognitive work that it is intended to support.

Health care relies on the description of process and condition. For each ICU patient, multiple diagnostic and therapeutic processes are underway, about to be started, or being concluded. Each patient's condition can be accounted for by a spectrum of variables that are related, and their interactions exceed the ability of clinicians to perceive them. The internal processes in a patient and the processes that influence how an acute care hospital unit operates cannot be entirely known through direct observation. Even under the best circumstances, there is an irreducible uncertainty that dogs clinicians' ability to fully grasp the phenomena for which they are accountable. For these reasons, representations that accurately depict reality are crucial to successful health care. The presentation of information bears directly on the clinicians' ability to develop an effective mental representation of past, current, and prospective states of patients under their care.

In recent years, electronic displays have been developed to depict information that is available through IT systems. Electronic display use has evolved along with paper and other means of information display, creating what amounts to a hybrid information environment that is intended to support cognitive work in clinical care. The current state of computing system and software support for acute care, or medical informatics, is a result of three developments: information systems, medical records, and decision aids.

Through the 1970s hospital information systems were largely created to handle data related to business matters such as billing, supplies, and operations. Intended benefits such as improvements to reliability have been more recently accompanied by unintended consequences of automating clinical information. The consequences include clinician overtrust of decision support, the imposition of additional work on clinicians (as they develop work-arounds to surmount system shortcomings), and the upset of smooth working relations and communications among care providers (Ash, Berg, & Coheira, 2004, pp. 104–106). Cost considerations also have an effect on IT system development and use. Software systems are often touted as cost-effective, flexible, and scalable. However, Decruyenaere et al. (2005) report that the total cost of ownership for ICU information systems, like other IT systems, is many times more than the initial purchase price. This stems from the unanticipated expenses of installation, use, maintenance, upgrades, changes, and disposal over its lifespan. Once IT systems are implemented, pressures to limit costs force IT staff attention to switch exclusively to the support of individual system functions. Pressures to limit expenditures compel organizations to leverage existing platforms by making only minor variations in order to serve a variety of purposes. For that reason, systems that were developed for one purpose, such as billing, are adapted to serve additional purposes, such as anesthesia assignment scheduling.

Electronic versions of medical records attempted to make the large amount of information that they contain useable. Despite these efforts, clinicians find the records to be a poor match for the kinds of cognitive work that they must perform. This arises from increasing reliance on the medical record to support billing for clinical activity, its configuration to assist billing and not clinical purposes, difficulty in locating critical information among the vast amount of information that it contains, and the inability to use it for important clinical activities such as comparing data. The sign-out sheet is an informal solution to this problem. Created informally by clinicians, the sign-out sheet lists each of the patients on a unit along with critical items of information on their condition and care. There is a need to replace the medical record with representations that are suited to meeting the immediate needs of the patient as well as the collaborative needs of clinicians.

Clinical decision aids sought to help physicians synthesize complex considerations into rule-based guidance on patient care decisions. Berg (1997) described how previous IT approaches to support clinician cognitive work have attempted to create rule-based aids for patient medical care decisions. However, decision support systems need to be constantly monitored to determine whether their suggestions fit a particular case. Also, the number of branching points may become so great to accommodate exceptions that the system is impossible to use and maintain (Ash, Berg, & Coheira, 2004, p. 108). The failure of this approach demonstrates that decision making under clinical conditions is far more complex and less tractable than proponents of these early systems believed.

Recent IT developments have made ever greater amounts of data related to patients available to clinicians. Data availability, however, does not equal data utility. In order to be useful, data must be easy to manage so that it supports clinical decision making. This simple statement belies the depth and complexity it involves. Two levels of acute health care present opportunities to develop representations that are suited for use in health care: the individual patient level and the unit level.

Patient Level

Patients who are admitted to units such as the ED and the ICU are in acute distress due to disease, trauma, or their treatment. Such patients are typically unstable and, as a result, require extraordinary monitoring and attention. For this reason, the ICU and ED require the greatest information flow for individual patients among all acute care settings. Clinical decisions rely on the ability to synthesize many different streams of data. Whether they are an attending, fellow or resident, primary caregiver or consultant, physicians regularly perform the cognitive work of diagnosis, therapeutic intervention, and monitoring. Nurses perform similar activities, although at a different level. Clinicians perform this work by relying on direct observation of the patient's condition, the patient's self-report, past/current vital signs, test results, and expert consultant opinions. In addition to this primary review of data, they also assess data (are they accurate, current, sufficient?) and evaluate prospects (is action indicated, are further data needed, would waiting be prudent?) among other activities.

Even though circumstances are uncertain, the ICU and ED clinician must act on behalf of the clinically ill patient. This compels the clinician to pursue diagnostic and therapeutic interventions that are convenient, rather than sequential. The decision to proceed with a certain treatment relies in part on the trade-offs between what is known about certain courses of treatment and their anticipated harms and benefits. The majority of these activities do not occur in what could be described as familiar territory in which the data are sufficient and the patient's recovery is certain. Instead, patient condition and prognosis are often in the kind of circumstances in which the available evidence on what to do is weak (Sharpe & Faden, 1998, pp. 214–220). Some practitioners contend that much of medical practice takes place where there is little proven knowledge and where anticipated harms and benefits are equivocal (Nemeth, 2005). Trade-offs that are related to anticipated harms and benefits of courses of treatment are directly linked to clinician performance at the bedside. However, the closest information that is available in this regard is the information on medications that is available on personal digital assistant (PDA) devices. While it is a helpful reference, critical cognitive work is still left to the clinician. This includes mapping the drug to the patient's unique physiology, as well as mapping the drug under consideration to medications that are already onboard the patient, that are being introduced at the time, or can be expected to be brought onboard soon.

The need for minute-by-minute monitoring is met by a patient vital signs monitor that depicts data such as electrocardiograph, blood pressure, heart rate, and the percentage of oxygen in the blood. The vital signs monitor may include wave forms from a variety of remote monitors. Which data are displayed, and how they are displayed, largely depends on the information display manufacturer and local hospital convention. Figure 28–4 depicts a patient vital signs monitor and ventilator monitor that are currently in use at a major urban teaching hospital in the United States. Some display options, such as variations in data layout, are available for the clinician to select. In order to do that, the vital signs monitor requires the operator to push a knob, rotate the knob to position an indicator to the preferred position on a pop-up menu, and push again to select the desired option. Making such adjustments can be cumbersome and take time to complete, making them inappropriate to perform during fast-changing circumstances. The vital signs monitor screen is touch-sensitive, but the content of the display and the menus that pop up when the screen is touched are unrelated. As a result, these features are infrequently used. Figure 28–5 depicts the interface of a ventilator that is in use at the same facility. Unlike the monitors in

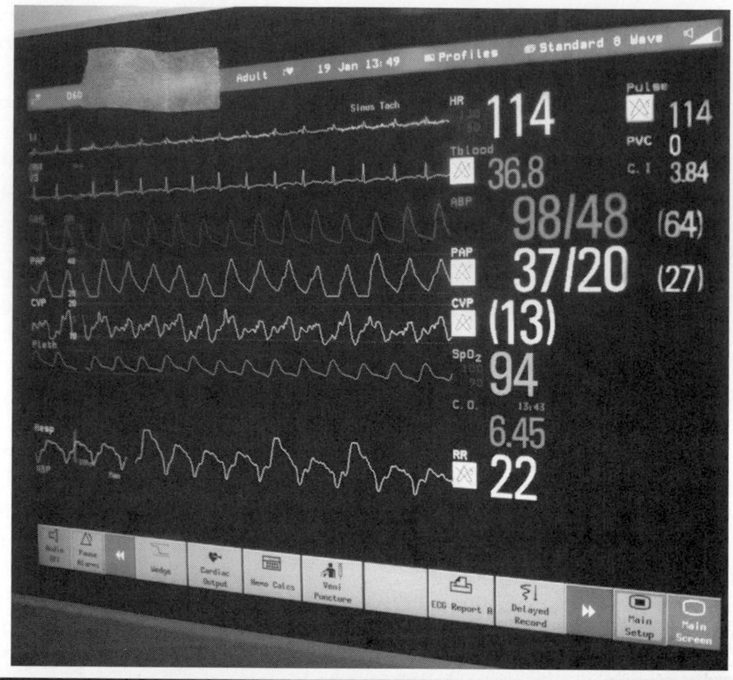

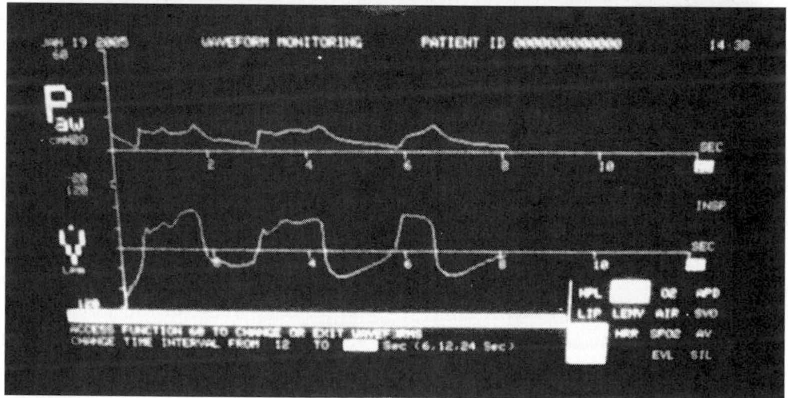

Figure 28–4. Displayed information on a patient vital signs monitor (top) and ventilation monitor (bottom) Copyright 2005 Cognitive Technologies Laboratory. Reproduced with the permission of the Cognitive Technologies Laboratory.

Figure 28–4, some clinicians find it a better match with their daily work activity. A shallow hierarchy makes important functions available to view. Controls map symbols and terms directly to patient data, making them easy to understand (O'Connor, personal communication, November 15, 2004).

The number and kind of equipment that is in use in each room varies depending on each patient's condition. In addition to the vital signs monitor, a patient may also have a ventilator, a separate pulmonary function monitor, a cardiac pacing device, a cardiac balloon assist device, dialysis equipment, a ventricular assist device, warming apparatus, an extra corporeal membrane oxygenation (ECMO) unit, a patient-controlled analgesia (PCA) device, and as many as ten to eleven intravenous infusion devices. Each item has its own unique and separate set of displays, controls, signals, and alarms. Displays typically provide a "keyhole" view into the data, preventing the clinician from reviewing what has occurred earlier and anticipating what will occur if current settings remain. All of the equipment in a room is related to a single patient, yet none of the devices in the same room is able to

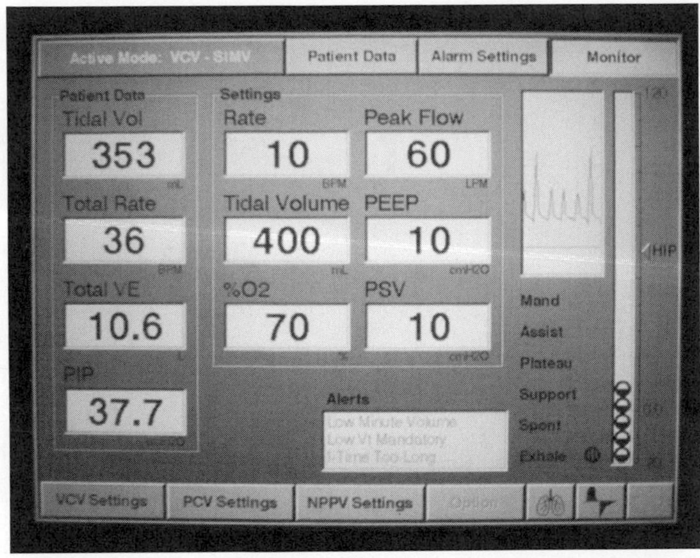

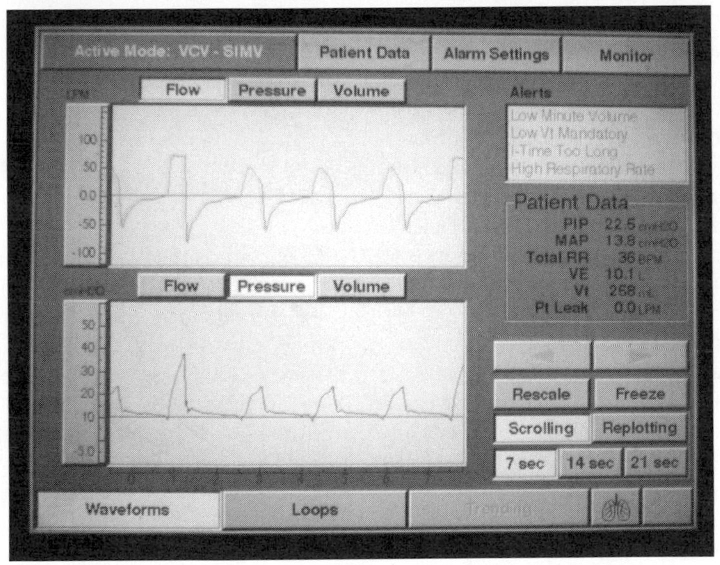

Figure 28–5. Displayed information on a ventilator monitor.
Copyright 2005 Cognitive Technologies Laboratory. Reproduced with the permission of the Cognitive Technologies Laboratory.

share information with another. This forces the clinician to monitor many low-level, discrete displays of patient and equipment data. Display designs are created by widely varying sources, including manufacturers (equipment displays), management (hospital records), and clinicians (locally developed artifacts). This creates what amounts to a culture conflict that the clinicians must resolve.

In addition to equipment, the ICU patient care information ecology also includes test orders, test results, the patient medical record, and sign-out sheet. Albolino and Cook (2005) describe how ICU clinicians use these artifacts for their own work and in concert with others who work on the unit. Their need to use information changes depending on circumstances, which can be routine or emergent. Under routine circumstances, a patient is stable and

Figure 28–6. Telemedicine ICU control station. Reprinted with permission of Sentara Healthcare

clinicians can explore, examine, and evaluate data in different ways. Every day at the start of the a.m. shift, clinicians seek a consensus on the current and prospective state of each patient's course of treatment by conducting "rounds." These informal, structured deliberations assemble many sources of information to inform the group and the attending physician who leads the process. They typically include direct observation of the patient and monitors in the patient room, the flow sheet indicating recent treatment activity, and the informal personal notes written onto the sign-out sheet.

The task load shifts significantly when a patient experiences a critical event such as seizures or cardiac dysfunction. When a patient becomes unstable, clinicians seek the data that are necessary to understand the cause(s) and the likely efforts that will be needed to restabilize the patient. Attention focuses more narrowly on the data that are primarily available through direct observation of the patient and the information that is displayed on certain nearby equipment.

The trend toward computer-supported IT in intensive care can be expected to increase. For example, Breslow et al. (2004) suggest that the use of remote monitoring by intensivists, also referred to as telemedicine, can improve clinical and economic outcomes at hospitals (Piazza et al., 2004). The purpose of such systems is to link one intensivist to multiple remote ICUs by computer-supported data links. The control center–style workstations that intensivists use to interact with remote locations, such as the one shown in Figure 28–6, can be expected to force greater reliance on displays.

Unit Level

Recent studies of displays to support distributed cognition in health care have explored the combined use of video image projection on an operating room (OR) status board (Xiao, Seagull, Hu, Mackenzie, & Gilbert, 2003) and the use of computer-supported media including digital pens and walk-up displays to create hybrid OR workstations that augment paper-based practice (Bång, Larson, & Eriksson, 2003). While these displays use novel media to collect and present lower level information, they still shift the work of synthesis to the clinician.

Planning and management of day-to-day activities, which comprise the technical work on a patient care unit, depend on accurately understanding what occurs there. Few representations of the ICU exist that show its daily fluctuations or the work that is performed. Establishing an accurate, consistent representation of this technical work context is important in itself. This context is essential to compare ICUs and EDs and to assess the meaning of incidents and accidents. In addition, this context serves as a foundation to evaluate claims about the significance of events, to calibrate metrics for the quality of care, and to estimate the impact of factors such as fatigue or work schedule changes.

In the case of ICUs, underlying domain semantics involve matching resources such as nursing assignments and practitioner attention to a range of current and anticipated demands. For example, ICUs and EDs have a status board that is used to account for staff assignments to patients, anticipate procedures and status, and represent resources as they shift through the day. Operating units and hospitals at or near resource saturation (what is termed "100% bed occupancy") places a high premium on the ability of practitioners to anticipate future demands and to predict the clinical course of patients within the unit. Strategies that practitioners use to buffer resource demand and supply include "bumping" and "resource hiding." Such activities may vary substantially from day to day or hour to hour, especially in ICUs with high turnover such as surgical units.

Many recent patient safety initiatives rely heavily on safety management to improve awareness, identify and report adverse events, and suggest interventions to reduce the likelihood they will happen again. However, these issues are far more complex and intractable than safety management alone can address. Patient safety requires methods that get at the actual nature of health care as a work domain. They also require action informed by that understanding in order to develop effective procedures, hardware, and software to assist clinicians' work performance. Practitioner goals and constraints at work are some of the many factors that comprise the technical work context. This complex real-world environment can be studied through ethnographic methods within the naturalistic decision-making (NDM) approach. NDM studies describe how people use their experience in order to arrive at good decisions in such environments without having to compare strengths and weaknesses of alternative courses of action (Klein, 1997, pp. 284–288). Cognitive task analysis (CTA) (Rasmussen, 1986) can be used to identify and describe goals and constraints. CTA has most often been employed in process control industries such as nuclear power control rooms. Using CTA to explain underlying domain semantics provides a pathway for the cognitive engineering of ecological displays (Woods & Hollnagel, 1987). In order to graphically represent work domain semantics. The process tracing (Woods, 1993) identifies durable aspects of the work domain across multiple settings. Critical decision method (Klein, 1997) uses detailed analysis of specific decisions to enumerate the factors that influence decisions and to describe the decision process.

Performing research into this work domain by viewing it from different perspectives can provide powerful insights. The technical work context provides a macro view: a framework to explain the individual critical decisions that are extracted from observation. Critical decision exploration provides a micro view: Details from observation and patient inventories clarify which components of the technical work context are important. The approaches are complementary. Using them together offers the opportunity to describe the nature of ICUs and to understand how their similarities and differences contribute to patient safety.

Research in our lab has shown us that the study of those who make the real world work, care providers, reveals how to aid performance and to increase safety. Ethnographic study examines how experts develop and use cognitive artifacts as a way to deal with real-world problems. This yields insight into what truly aids expertise. For Roth, Patterson, and Mumaw (2002), use-centered systems design pursues five issues: application domain goals and constraints, the range of tasks that practitioners perform, strategies that practitioners currently use to perform those tasks, factors that make tasks complex, and tools that can make it easier for practitioners to accomplish their goals more effectively and easily. This approach can provide significant benefit to the practice of health care. Two projects demonstrate this approach.

A study of hand-offs of care between peer practitioners (Brandwijk, Nemeth, O'Connor, Kahana, & Cook, 2003) analyzed verbal exchanges between practitioners when handing off shifts in an ICU. To support these sign-out observations, a graphical data collection instrument was developed to describe the technical work context. This instrument was used to map the physical layout of an ICU and allowed efficient collection of information regarding patient condition, such as the number of infusion devices being used. Based on these data, several representations of the work context over time were developed. The final version was a dense, highly encoded reflection of the ICU that included high-resolution views of the movement of patients into, out of, and within the ICU. The representation showed the technical work context in the unit that formed the basis for sign-out conversations between shifts. The representation was used to track the use of resources in the ICU and its nearby step-down unit over time. It also used to support the research on between-shift communications that was conducted on that unit.

In another project, Nemeth (2003) described a yearlong observation and analysis of general operating room (GOR) scheduling and activities. The study explains how cognitive artifacts can be used to discover complex, widely varying characteristics of the acute care work domain. It also presents implications and guidance for the development of computer and software support (medical informatics) in the interest of overcoming practitioner resistance to technology and improving the fit between work tools and the tasks that they are intended to support.

DISCUSSION

Computing systems have not been obstacles to the use of IT in health care. The real obstacle has instead been to understand the complex operations of health care systems in which IT systems exist. As early as the 1970s, Shackel (1979, p. 436) noted that operations research techniques such as dynamic programming were available "but fail when the problem involves many criteria with changing priorities." This is precisely the situation that health care providers face. A human's inherent flexibility and adaptability is required to deal with such problems. This suggests that the use of IT in health care should be to represent the relevant domain semantics as the problem space in which practitioners work.

As the theory and study of signs and symbols, wouldn't semiotics offer a means to understand how to create an information ecology? After all, semiotics emphasizes the role of sign systems in the construction of reality: We know things only through the mediation of signs, so we see only what our sign systems allow us to see. While that does sound similar to Gibson's affordances, semiotics has remained a loosely defined critical practice that deals more with the structure and intentions of messages than their results. The approach does not lend itself to quantification and remains more in the realm of philosophy and anecdotal example. As a result, semiotics has not had much of an influence on the more practical issues of real-world information display.

At least two characteristics, drawn from Xiao and Nemeth (2004), are essential to understand collaborative work in high-hazard, complex, uncertain settings and to support that work through effective IT systems. Further study could be expected to reveal even more traits.

Evolution. Clinicians create their work environment by continually tailoring (Cook & Woods, 1996) the workplace to support their collaborative work needs. Cognitive artifacts are developed and modified to efficiently and reliably support cognitive work. Physical artifacts are widely used. Their perceptual and physical properties are adeptly exploited to simplify the effort that is involved in understanding information and to improve workflow. In this way, the collective information that is available to the clinicians evolves in parallel with their work processes.

Power and Utilité. Complexity in acute care spans both social organizations and work processes. This requires some means to represent and convey rich information as well as the subtleties, nuances, and conflicts that are inherent in complex organizations. The flexibility of artifacts such as white boards and paper schedules is not solely due to their being physical objects. It is a result of the artifacts' *power* and *utilité*. By *power,* we mean how closely artifacts represent the actual nature of the relevant domain semantics that they have been created to support. User-created physical artifacts are typically powerful because they express genuine aspects of the work domain. The work domain semantics and the semantics that the artifact represents are nearly the same. There is little else in the artifact other than what matters. We use the French term *utilité* to convey how practical it is to create and use an artifact. Its ease of creation and maintenance, simplicity, and availability all make the artifact useful in a practical sense.

Artifacts that reflect the specific dynamic domain characteristics that are of interest to the clinicians (and no more) are powerful. They demonstrate *utilité* by making it possible for users to easily create meaningful information structures in the workspace as the work evolves. Current collaborative IT systems often fare poorly in both regards. Display configuration has been left primarily to manufacturers and IT staff. Current IT systems often lack power, as little or no effort is invested in understanding the cognitive work that they are intended to support. They lack *utilité* because it is difficult to use them to compose new information structures and to modify structures to suit the tasks at hand. IT systems having low power and *utilité* require clinicians to compensate for system shortcomings by developing "work-around" solutions. Rather than improving clinical work, it causes extra work in addition to clinical care.

Observations of clinician activity indicate initiatives that would make displays of information in complex settings at the unit level more useful. Computer-supported displays can be substantial and flexible. Clinicians have the expertise that IT systems can support. Field research into clinician management of care and of technical work at both patient and unit level can be used to develop representations that have high fidelity to their work domain. Field research can also inform IT system developers about control methods for the data, data presentations, and arrangements that clinicians prefer to use. For example, templates could enable clinicians to quickly choose preferred data and layout to tailor displays to suit how they do their own work. Computing systems are also powerful enough to depict past, current, and anticipated system states. The kind of time-based display that Nemeth and Cook (2004) describe would enable the display of information along various scales by slewing through time from past to present to predicted future. Improved control methods could also support shifts in scale by zooming in to obtain finely grained detail or zooming out to obtain an overview.

For Christoffersen and Woods (2002), the creation of IT systems requires understanding to develop better coordinated teams that are comprised of humans and IT systems. This understanding requires attention to coordination of activity that is *observable* and *directable* among all participants, whether human or machine. Observability involves shared representation of the problem state's nature, difficulty, and priority as well as the nature, status, rationale, and duration of other agents' activities. Directability has to do with who among participating agents really owns how problems are being solved. Power and *utilité* are the kinds of features that strive for the observability and directability that Christoffersen and Woods contend are necessary for IT systems to be team players among acute care clinicians.

CONCLUSION

While all displays are ecological, a good display maps relevant domain semantics (what exists in reality) to the ecological characteristics of the representation (how reality is re-presented). Mere translation of existing artifacts into computer-based versions is insensitive to the sophisticated interactions that operators have developed to engage the complex, uncertain domain that they strive to manage. The opportunity is to derive relevant domain semantics from ethnographic study, to develop concise representations of those semantics, then to use them as the design basis for candidate displays.

Improved information displays will better express the elements of the operational environment that clinicians must manage in their daily work. Ethnographic research to reveal the semantics of the work domain will make it possible to develop display concepts (Nemeth, 2004). Laboratory observation of how candidate displays are used will make it possible to assess their fit with expert preferences. Field observation of clinicians using candidate displays will make it possible to assess their fit with the actual work domain.

The development of both patient- and unit-level representations relies on the collaboration of health care, human factors, and design professionals. (Reid et al., 2005), Applications that would benefit from this approach span all levels of acute care. Carefully developed ecological displays would support clinicians in the ICU and ED, where data change most often and are the basis for crucial decisions. They would be of particular help in the newly evolving practice of remote telemetry ICUs. They would also support the hand-off of patient- and unit-level information between shifts. Clinical improvements in these activities will benefit both health care practitioners and the patients for whom they care.

ACKNOWLEDGMENTS

This work is supported by a grant (HS11816) from the Agency for Healthcare Research and Quality and a grant (LM007947) from the National Library of Medicine, National Institutes of Health.

References

Albolino, S., & Cook, R. I. (2005). Making sense of sensemaking: What are physicians doing on "rounds" in the intensive care unit? Paper presented at the 7th International NDM Conference, Amsterdam, Holland.

Ash, J. S., Berg, M., & Coheira, E. (2004). Some unintended consequences of information technology in health care: The nature of patient care information system-related errors. *Journal of the American Medical Informatics Association, 11,* 104–112.

Bång, M., Larson, A., & Eriksson, H. (2003, November). NOSTOS: A paper-based ubiquitous computing health care environment to support data capture and collaboration. In *27th Annual Symposium* (pp. 46–50). Washington, DC: American Medical Informatics Association.

Berg, M. (1997). *Rationalizing medical work*. Cambridge, MA: MIT Press.

Brandwijk, M., Nemeth, C., O'Connor, M., Kahana, M., & Cook, R. (2003, January). Distributing cognition: ICU handoffs conform to Grice's maxims. *Society of Critical Care Medicine*. San Antonio.

Breslow, M. J., Rosenfeld, B. A., Doerfler, M. D., Burke, G., Yates, G., Stone, D. J., et al. (2004). Effect of a multiple site intensive care unit telemedicine program on clinical and economic outcomes: An alternative paradigm for intensivist staffing. *Critical Care Medicine, 32*(1), 31–38.

Carr, N. G. (2005, January 22). *Does not compute*. Retrieved January 25, 2005, from http://www.nytimes.com/2005/01/22/opinion/22carr.html?incamp'article_popular_4/

Christoffersen, K., & Woods, D. (2002). How to make automated systems team players. *Advances in human performance and cognitive engineering research* (Vol. 2; pp. 1–12). New York: Elsevier Science.

Cook, R., Render, M., & Woods, D. (2000). Gaps in the continuity of care and progress on patient safety. *British Medical Journal, 320,* 792–794.

Cook, R., & Woods, D. (1996). Adapting to new technology in the operating room. *Human Factors, 38*(4), 609.

Cook, R., Woods, D., & Miller, C. (1998). *A tale of two stories: Contrasting views of patient safety*. Chicago: National Healthcare Safety Council of the National Patient Safety Foundation, American Medical Association.

Decruyenaere, J., Danneels, C., Oeyen, S., Colpaert, C., Verwaeren, G., & Myny, D. (2005). Calculation of the total cost of ownership of an intensive care unit system. *Critical Care Medicine, 32*(12), A29.

Feltovich, P. J., Ford, K. M., & Hoffman, R. R. (Eds.). (1997). *Expertise in context: Human and machine*. Cambridge, MA: MIT Press.

Gibson, J. J. (1977). The theory of affordances. In R. Shaw & J. Bransford (Eds.), *Perceiving, acting and knowing: Toward an ecological psychology* (pp. 67–82). New York: Wiley.

Health and Human Services. (2004). Retrieved December 16, 2005, from http://www.cms.hhs.gov/statistics/nhe/projections–2002/t1.asp/

Heath, C., & Luff, P. (2000). *Technology in action*. New York: Cambridge University Press.

Heeks, R., Mundy, D., & Salazar, A. (1999). *Why health care information systems succeed or fail*. Retrieved January 25, 2005, from http://www.man.ac.uk/idpm/idpm_dp.htm#isps_wp/

Hollan, J., Hutchins, E., & Kirsh, D. (2000). Distributed cognition: Toward a new foundation for human–computer interaction research. *ACM Transactions in Computer–Human Interaction, 7*(2), 174–196.

Hutchins, E. (1995). *Cognition in the wild*. Cambridge, MA: MIT Press.

Institute of Medicine (IOM). (1999). *To err is human*. L. Kohn, J. Corrigan, & M. Donaldson (Eds.). Washington, DC: National Academy Press.

Jones, P. M., Mitchell, C. M., & Rubin, K. S. (1990). Validation of intent by a model-based operator's associate. *International Journal of Man-Machine Studies, 33,* 177–202.

Klein, G. (1997). The recognition-primed decision model. In C. E. Zsambok & G. Klein (Eds.), *Naturalistic decision making* (pp. 285–292). Mahwah, NJ: Lawrence Erlbaum Associates.

Luff, P., Hindmarsh, J., & Heath, C. (Eds.). (2000). *Workplace studies: Recovering work practice and informing system design*. New York: Cambridge University Press.

McGrenere, J., & Ho, W. (2000, May). Affordances: Clarifying and evolving a concept. *Proceedings of Graphics Interface 2000* (pp. 179–186). Montreal.

Mitchell, C. M. (1995). The importance of modeling operator functions in complex dynamic systems: A conceptual overview and proposed methodology. *Third International Workshop on Advances in Functional Modeling of Complex Technical Systems*. College Park, MD.

Mitchell, C. M., & Miller, R. A. (1986). A discrete control model of operator function: A methodology for information display design. *IEEE Transactions on Systems, Man, and Cybernetics, 16*(3), 343–357.

Mitchell, C. M., & Saisi, D. L. (1987). Use of model-based qualitative icons and adaptive windows in workstations for supervisory control systems. *IEEE Transactions on Systems, Man, and Cybernetics, 17*(4), 573–593.

Nemeth, C. (2003). *The master schedule: How cognitive artifacts affect distributed cognition in acute care*. Dissertation Abstracts International 64/08, 3990 (UMI No. AAT 3101124).

Nemeth, C. (2004). *Human factors methods for design*. London: Taylor and Francis/CRC Press.

Nemeth, C. (2005). Health care forensics. In I. Noy & W. Karwowski (Eds.), *Handbook of human factors in litigation*. (pp. 31–7). New York, NY: Taylor & Francis.

Nemeth, C., & Cook, R. (2004, August). Discovering and supporting temporal cognition in complex environments. *Proceedings of the National Conference of the Cognitive Science Society*. Chicago.

Nemeth, C., Cook, R., & Woods, D. (2004). Special issue on using field studies to understand health care technical work. *IEEE Transactions on Systems, Man, and Cybernetics-Part A, 34,* 6.

Norman, D. (1986). Cognitive engineering. In D. Norman & S. Draper (Eds.), *User-centered system design*. Hillsdale, NJ: Lawrence Erlbaum Associates.

Norman, D. (1988). *The design of everyday things*. New York: Basic Books.

Piazza, M., Giogino, T., Azzini, I., Stefanelli, M., & Luo, R. (2004, September). Cognitive human factors for telemedicine systems. *MEDINFO2004*. San Francisco: American Medical Informatics Association.

Rasmussen, J. (1985). The role of hierarchical knowledge representation in decision-making and system management. *IEEE Transactions on Systems, Man, and Cybernetics, SMC, 15*(2), 234–243.

Rasmussen, J. (1986). *Information processing and human–machine interaction*. New York: Elsevier Science.

Rasmussen, J., & Lind, M. (1981). *Coping with complexity*. Risø-M–2293. Roskilde, DK: Risø National Laboratory.

Rasmussen, J., & Pjetersen, A. (1995). Virtual ecology of work. In J. Flasch, P. Hancock, J. Caird, & K. Vicente (Eds.), *Global perspectives on the ecology of human–machine systems* (pp. 121–156). Hillsdale, NJ: Lawrence Erlbaum Associates.

Rassmussen, J., & Vicente, K. (1990). Ecological interfaces: A technological imperative in high tech systems? *International Journal of Human–Computer Interaction, 2*(2), 93–111.

Reid, P. R., Compton, W. D., Grossman, J. H., and Fanang, G. (Eds.) *Designing a better delivery system: A new engineering/health-care partnership.* Washington, DC: The National Academies Press.

Roth, E. M., Patterson, E. S., & Mumaw, R. J. (2002). Cognitive engineering: Issues in user-centered system design. In J. J. Marciniak (Ed.), *Encyclopedia of software engineering* (2nd ed.). New York: Wiley-Interscience.

Shackel, B. (1979). Process control: Simple and sophisticated display aids as decision devices. In T. Sheridan & G. Johannsen (Eds.), *Monitoring behavior and supervisory control* (pp. 429–443.). New York: Plenum.

Sharpe, V., & Faden, A. (1998). *Medical harm.* Cambridge: Cambridge University Press.

Suchman. L. (1987). *Plans and situated actions.* New York: Cambridge University Press.

Tufte, E. (1997). *Visual explanations.* Cheshire, CT: Graphics Press.

Vicente, K. (2000). Work domain analysis and task analysis: A difference that matters. In J. M. Schragen, S. F. Chipman, & V. L. Shalin (Eds.), *Cognitive task analysis* (pp. 101–118.). Mahwah, NJ: Lawrence Erlbaum Associates.

Woods, D. (1993). Process tracing methods for the study of cognition outside of the experimental psychology laboratory. In G. Klein, J. Orasanu, R. Calderwood, & C. Zsambok (Eds.), *Decision making in action: Models and methods* (pp. 229–251.). Norwood, NJ: Ablex.

Woods, D. (1994). *Visualizing function: The theory and practice of representation design in the computer medium.* Available from Cognitive Systems Engineering Laboratory. The Ohio State University, Columbus, OH<~>43210.

Woods, D., & Roth, E. (1988). Cognitive systems engineering. In M. Helander (Ed.), *Handbook of human–computer interaction* (pp. 3–43). New York: North Holland.

Woods, D. D. (1998). Designs are hypotheses about how artifacts shape cognition and collaboration. *Ergonomics, 41,* 168–173.

Woods, D. D., & Hollnagel, E. (1987). Mapping cognitive demands in complex problem solving worlds. *International Journal of Man-Machine Studies, 26,* 257–275.

Xiao, Y., & Nemeth, C. (2004). Tailoring workplaces for collaborative work: Artifacts and coordination practices in "Running the ORs." Position paper for Workshop W6: Exploring the role of information, information tools, and information environments in collaboration. *CSCW2004-ACM Conference on Computer Supported Cooperative Work.* Association for Computing Machinery. Chicago.

Xiao, Y., Seagull, F. J., Hu, P., Mackenzie, C. F., & Gilbert, T. B. (2003). Distributed monitoring and a video-based toolset. *Proceedings of IEEE International Conference on Systems, Man, and Cybernetics,* pp. 1778–1782.

Section VI

HUMAN ERROR

·29·

BEHIND HUMAN ERROR: TAMING COMPLEXITY TO IMPROVE PATIENT SAFETY

David D. Woods and Emily S. Patterson
The Ohio State University

Richard I. Cook
University of Chicago

Throughout the patient safety movement, health care leaders have consistently referred to the potential value of human factors research on human performance and system failure (Leape, 2004; Leape, Woods, Hatlie, Kizer, Schroeder, & Lundberg 1998). The patient safety movement has been based on three ideas derived from results of research on human expertise (Feltovich, Ford, & Hoffman, 1997), collaborative work (Rasmussen, Brehmer, & Lepat, 1991), and high-reliability organizations (Rochlin, 1999) built up through investments by other industries:

- Adopt a systems approach to understand how breakdowns can occur and how to support decisions in the increasingly complex worlds of health care,
- Move beyond a culture of blame to create open flow of information and learning about vulnerabilities to failure, and
- Build partnerships across all stakeholders in health care to set aside differences and to make progress on a common overarching goal.

Implementing these ideas requires detailed study, careful assessment, and thoughtful application of approaches that are unfamiliar to health care workers and managers. Success depends on creating durable, informative, and useful partnerships between health care and disciplines with core expertise in areas of human performance. Sustained, substantive partnerships are needed to devise useful approaches to current problems and to anticipate and block the new paths to failure that accompany changes in health care. (Woods & Cook, 2001).

PARALLEL PERSPECTIVES

> Human error in medicine, and the adverse events which may follow, are problems of psychology and engineering not of medicine.—Senders, 1993

Health care organizations and the public demand that we reduce injuries to patients during treatment. From one perspective, medication misadministrations, delayed diagnoses, or wrong site surgeries are events that arise from specific medical issues. From another perspective, these adverse events evolve because of the lawful, predictable effects of factors that affect human performance. Decision making, attentional processes, the use of knowledge, and coping with uncertainty are not medical issues but human performance issues that play out in a health care context.

The research base about human performance was built by studying how practitioners handle routine and challenging situations in aviation, industrial process control, military command and control, and space operations. The studies demonstrate empirical regularities and provide explanatory concepts and models of human performance. These results allow us to see common underlying patterns behind the superficial variability that makes different settings appear unique.

Understanding, predicting, and modulating human performance in any complex setting requires a detailed understanding of both the setting and the factors that influence performance. There are different languages used to describe human performance. These basic themes are the platform—the anatomy and physiology—of human performance in real-world settings. To understand patterns in human *judgment,* one needs to understand concepts such as bounded rationality, knowledge calibration, heuristics, and oversimplification fallacies (Feltovich et al., 1997). To understand patterns in *communication and cooperative work,* one needs to understand concepts such as supervisory control, common ground, communication of intent, and open versus closed work spaces (Clark & Brennan, 1991; Galegher, Kraut, & Egido, 1990; Greenbaum & Kyng, 1991; Rasmussen et al., 1991; Woods & Shattuck, 2000). To understand patterns in *human–computer cooperation,* one needs to understand concepts such as the representation effect, object displays, inattentional blindness, mental models, data overload, and mode error (LaBerge, 1995; Norman, 1993; Rensink, O'Regan, & Clark, 1997; Vicente, 1999; Zhang, 1997).

The complexity and connected nature of health care work ensures that many human performance themes will be found when examining a particular problem and that a single pattern will play out in many different health care settings. Consider the following two medication misadministrations, both of which could be classified as "wrong dose" medication errors. In the first case, an infusion pump was thought to be off when it was actually infusing medication in an unconstrained fashion. In the second case, a patient received two doses of an ordered medication: one from a nurse responding to a verbal order and one from a different nurse on the next shift administering the same medication from the order later entered by the physician into the computerized physician order-entry system. Both of these cases involve some of the same themes, such as poor coordination between multiple providers caring for the same patient, poor observability of the history of recent medication administrations, high workload, and overreliance on memory by frequently interrupted providers. Yet interventions stemming from these cases could be quite different: In the first case, potential interventions could be new software reliability checks, new pattern-based visualizations, automated free-flow protection in infusion devices, and in the second case, the system could highlight potential duplicate orders or enable documentation of administration by a nurse prior to order entry by a physician. This many-to-many mapping of themes in human performance to specific health care topics is one reason that work on patient safety remains difficult: Rather than requiring a particular type of human factors knowledge for progress, these complex problems usually involve the entire range of expertise on human performance. Similarly, it explains some of the challenges in using classifications such as "wrong dose medication error" to prioritize interventions because the classifications are based on failure modes rather than underlying patterns of systemic factors.

To use the human factors knowledge base to support very high levels of human performance in a particular health care setting requires a detailed understanding not only of how these themes play out in the health care setting but also of the technical details that determine medical success and failure. Put another way, to deal effectively with specific problems in health care requires an understanding of both the human performance factors and the medical domain knowledge. Gaining this understanding requires going back and forth between two different perspectives. This is only possible in a genuine collaboration where each party deliberately steps outside of his or her own area of expertise in order to learn from the other's perspective.

Patient safety research requires interdisciplinary synthesis that combines technical knowledge in specific health care areas with technical knowledge of human performance issues that play out in that area as practitioners perform this kind of work in context. Indeed, partnerships of this kind have been and continue to be the engine of progress in work on patient safety to date (see Bogner, 1994), for example, programs such as the Annenberg meetings (see Hendee, 1999) and partnerships at various research labs around the United States and the world.

STARTLING RESULTS FROM THE SCIENCE

One of the great values of science is that, during the process of discovery, conventional beliefs are questioned by putting them in empirical jeopardy. When scientists formulate new ideas and look at the world anew through these conceptual looking glasses, the results often startle us. As a result, we can innovate new approaches to accomplish our goals.

This process has been going on for more than 20 years in the "new look" at the factors behind the label human error (Reason, 1997; Rasmussen, 1990a, 1990b, 2000; Woods, Johannesen, Cook, & Sarter, 1994). Driven by surprising failures in different industries, researchers from different disciplinary backgrounds began to reexamine how systems failed and how people in their various roles contributed to both success and failure. The results often deviated from conventional assumptions in startling ways.

The research found that doing things safely, in the course of meeting other goals, is always part of operational practice. As people in their different roles are aware of potential paths to failure, they develop failure-sensitive strategies to forestall these possibilities. When failures occurred against this background of usual success, researchers found multiple contributors, each necessary but only jointly sufficient, and a process of drift toward failure as planned defenses eroded in the face of production pressures and change. The research revealed systematic, predictable organizational factors at work, not simply erratic individuals. The research also showed that to understand episodes of failure, one had to first understand usual success—how people in their various roles learn and adapt to create safety in a world fraught with hazards, trade-offs, and multiple goals (Cook, Render, & Woods, 2000; Hollnagel, 2004).

Researchers have studied organizations that manage potentially hazardous technical operations remarkably successfully, and the empirical results have been quite surprising also (Rochlin, 1999). Success was not related to how these organizations avoided risks, reduced errors, or prioritized interventions based on probabilities of failure, but rather how these high-reliability organizations created safety by anticipating and planning for unexpected events and future surprises. These organizations did not take past success as an excuse for confidence. Instead they continued to invest in anticipating the changing potential for failure because of the deeply held understanding that their knowledge base was fragile in the face of the hazards inherent in their work and the changes omnipresent in their environment. Safety for these organizations was not a commodity but a value that required continuing reinforcement and investment. The learning activities at the heart of this process depended on open flow of information about the changing face of the potential for failure. High-reliability organizations valued such information flow, used multiple methods to generate this information, and then used this information to guide constructive changes without waiting for accidents to occur.

Perhaps most startling in this research is the finding that the source of failure was not those who are less careful or motivated than those that are more careful or motivated. Instead, the process of investing in safety begins with each person being willing to question his or her beliefs to learn surprising things about how *he or she* can contribute to the potential for failure in a changing and limited resource world.

SYSTEMS ISSUES FOR RESEARCH TO IMPROVE SAFETY

Search for Underlying Patterns to Gain Leverage

From past work, progress has come from going beyond the surface descriptions (the phenotypes of failures) to discover underlying patterns of systemic factors (generic or genotypical patterns). These patterns capture repeated results about how people, teams, and organizations coordinate activities, information, and problem solving to cope with the complexities of problems that arise (Hollnagel, 1993).

The surface characteristics of a near-miss or adverse event are unique to a particular setting and people. Generic patterns reappear in many specific situations. Research in human factors has revealed a wealth of patterns, for example,

- Garden path problems and the potential to fixate on one point of view or hypothesis in problem solving (De Keyser & Woods, 1990; Patterson, Cook, Woods, & Render, 2004; Nguyen, Halloran, & Asch, 2004).
- Missing side effects of an action or change to a plan in highly coupled systems (Rasmussen, 1986).
- Hindsight bias from knowledge of outcome (Fischhoff, 1975).
- Local actors having difficulty tailoring a plan when the situation changes without an understanding of the intent behind an order (Woods & Shattuck, 2000).

- Alarm overload and high false alarm rates leading to missed or ignored warnings (Stanton, 1994).
- Mode errors in computerized devices with multiple modes and poor feedback about device state (Norman, 1988).

A great deal of leverage for improvements is gained by identifying the generic patterns at work in a particular situation of interest (Woods, 2005). For example, we can sample the kinds of difficult situations that can occur in a health care setting and recognize the presence of garden path problems (e.g., in anesthetic management; Gaba, Maxwell & DeAnda, 1987). We may review a corpus of near-misses and note that in several cases a practitioner became fixated on one view of the situation (Cook, McDonald, & Smalhout, 1989). Or we may analyze how people handle simulated problems and see the potential for fixating in certain situations (e.g., as has occurred in crisis training via anesthesia simulators; Howard, Gaba, Fish, Yang, & Sarnquist, 1992; Rudolph, 2003).

Previous work on aiding human and team situation assessment can now seed and guide the development of interventions. To overcome fixation in a garden path problem, one can bring to bear techniques that may break up frozen mindsets such as new kinds of pattern-based displays or new team structures that help broaden the issues under consideration (De Keyser & Woods, 1990; Patterson, Cook, et al., 2004).

Each of the genotypes listed earlier was identified and studied in aerospace, process control, or military settings, but they all also play out in multiple health care settings:

- Fixation as a danger in anesthetic management during surgery (Cook et al., 1989; Rudolph, 2003).
- Missing side effects of planned changes that create new complexities or vulnerabilities (Embi et al., 2004; Patterson, Cook, & Render, 2002).
- Missed warnings or reminders due to high nuisance and false alarm rates in intensive care units (Patterson, Doebbling, et al. 2005; Weinger & Smith, 1993; Xiao, Seagull, Nieves-Khouw, Barzac, & Perkins, 2004).
- Mode errors in computerized infusion devices (Cook, Woods, & Howie, 1992; Cook, Woods, & Miller, 1998; Lin et al., 1998; Nunnally & Cook, 2004).
- Hindsight bias in incident review teams (Caplan, Posner, & Cheney, 1991).

This list is very short and only exemplifies some of the results available to jump-start research and design in health care settings (e.g., for other generic patterns linked to the Columbia accident that also appear in health care organizations, see Woods, 2005b).

Research on patient safety should be using and expanding the set of generic patterns related to breakdowns in human performance that occur in health care settings. Research should focus on developing and testing interventions to reduce these problems. In many cases, previous work has identified the interventions needed (e.g., in the case of mode errors). In other cases, seed ideas exist that need to be further developed given the unique pressures of health care.

Tame Complexity

In the final analysis, the enemy of safety is complexity. In nuclear power and aviation, we have learned at great cost that often it is the underlying complexity of operations that contributes to human performance problems. Simplifying the operation of the system can do wonders to improve its reliability, by making it possible for the humans in the system to operate effectively and more easily detect breakdowns. Often, we have found that proposals to improve systems founder when they increase the complexity of practice (e.g., Xiao et al., 1996). Adding new complexity to already complex systems rarely helps and can often make things worse. This applies to system improvements justified on safety grounds as well.

The search for operational simplicity has a severe catch however. The very nature of improvements and efficiency in health care delivery includes, creates, or exacerbates many forms of complexity. Ultimately, success and progress occur through monitoring, managing, taming, and coping with the changing forms of complexity, and not by mandating simple, one-size-fits-all policies.

This has proven true particularly with respect to efforts to introduce new forms and levels of computerization. Improper computerization can simply exacerbate or create new forms of complexity to plague operations (Woods et al., 1994). The situation is complicated by the fact that new technology often has benefits at the same time that it creates new vulnerabilities.

Again the science startles us. To help people in their various roles create safety, research needs to (Cook et al., 2000; Woods & Cook, 2002):

- Search out the sources of complexity, including the "edges" where simple approaches fail.
- Understand the strategies people, teams, and organizations use to cope with complexity.
- Devise better ways to help people cope with complexity to achieve success.

Adding to a system's or organization's resilience is one of the basic lessons for taming complexity (Carthey, de Leval, & Reason, 2000; Sutcliffe & Vogus, 2003; Woods, 2005b; Hollnagel, Woods & Leveson, 2006).

Adopt Methods for User-Centered Design of Information Technology

When human factors practitioners and researchers examine the typical human interface of computer information systems and computerized devices in health care, they are often shocked. What we take for granted as the least common denominator in user-centered design and testing of computer systems in other high-risk industries (and even in commercial software development houses that produce desktop educational and games software) seems to be far too rare in medical devices and computer systems. The devices are too complex and require too much training to use, given typical workload pressures (e.g., Cook et al., 1992; Nunnally & Cook, 2004; Obradovich & Woods, 1996; Rogers, Mykityshyn, Campbell, & Fisk, 2001).

Computer displays, interfaces, and devices in health care exhibit "classic" human–computer interaction deficiencies. By "classic" we mean that we see these design "errors" in many devices in many settings of use, that these design problems are well understood (i.e., they appear in our textbooks and popular writings; e.g., Norman, 1988; Norman & Draper, 1986), and that the means to avoid these problems are readily available.

We are concerned that the calls for more use of integrated computerized information systems to reduce error could introduce new and predictable forms of error unless there is a significant investment in user-centered design (Winograd & Woods, 1997).

The concepts and methods for user-centered design are available and are being used every day in software houses (Carroll & Rosson, 1992; Flach & Dominguez, 1995; Nielsen, 1993). Focus groups, cognitive walkthroughs, and interviews are conducted to generate a cognitive task analysis that details the nature of work to be supported by a product (Chung, Zhang, Johnson, & Patel, 2003; Garmer, Liljegren, Osvalder, & Dahlman, 2000; Lin, Vicente, & Doyle, 2001; Zhang, Johnson, Patel, Paige, & Kubose, 2003). Iterative usability testing of a system prior to use with a handful of representative users has become a standard, not an exceptional, part of most product development practices. "Out of the box" testing is conducted to elicit feedback on how to improve the initial installation and use of a fielded product. Health care delivery organizations also need to understand how they can use these techniques in their own testing processes and as informed consumers of computer information systems.

Building partnerships, creating demonstration projects, and disseminating the techniques for health care organizations is a significant and rewarding investment to ensure that the health care industry receives the benefits of computer technology while avoiding designs that induce new errors (Kling, 1996).

But there is much more to human–computer interaction than adopting basic techniques such as usability testing (Karsh, 2004). Much of the work in human factors research concerns how to use the potential of computers to enhance expertise and performance. We consider only a few of these issues here.

Study Human Expertise to Develop the Basis for Computerization

The key to skillful rather than clumsy use of technological possibilities lies in understanding both the factors that lead to expert performance and the factors that challenge expert performance (Feltovich, Ford, & Hoffman, 1997). Once one understands the factors that contribute to expertise and to breakdown, one then will understand how to use the powers of the computer to enhance expertise. This is an example of a more general rule—to understand failure and success, first begin by understanding what makes some problems difficult.

The areas of research on human performance in medicine explored in the monograph *A Tale of Two*

Stories (Cook, Woods, & Miller, 1998) illustrate this process. In these cases, progress depended on investigations that identified the factors that made certain situations more difficult to handle and then explored the individual and team strategies used to handle these situations. As the researchers began to understand what made certain kinds of problems difficult, how expert strategies were tailored to these demands, and how other strategies were poor or brittle, new concepts were identified to support and broaden the application of successful strategies. In each of these cases, the introduction of new technology helped create new dilemmas and difficult judgments. In addition, once the basis for human expertise and the threats to that expertise had been studied, new technology was an important means to achieve enhanced performance.

We can achieve substantial gains by understanding the factors that lead to expert performance and the factors that challenge expert performance. This provides the basis to change the system, for example, through new computer support systems and other ways to enhance expertise in practice (Nyssen & De Keyser, 1998).

Make Machine Advisors and Automation Team Players

New levels of automation have had many effects in operational settings. There have been positive effects from both economic and safety points of view. Unfortunately, operational experience, research investigations, incidents, and occasionally accidents have shown that new and surprising problems have arisen as well. Computer agents can be brittle and only able to handle a portion of the situations that could arise. Breakdowns in the interaction between operators and computer-based automated systems can also contribute to near-misses and failures in these complex work environments (Guerlain et al., 1996).

Over the years, human factors investigators have studied many of the "natural experiments" in human–automation cooperation—observing the consequences in cases where an organization or industry shifted levels and kinds of automation. One notable example has been the many studies of the surprising consequences of new levels and types of automation on the flight deck in commercial transport aircraft (Billings, 1996).

The overarching result from the research is that for automation concerned with information processing and decision making to be successful, the key requirement is to design for fluent, coordinated interaction between the human and machine elements of the system. In other words, automated and intelligent systems must be designed to be "team players" (Malin et al., 1991; Roth, Malin, & Schreckenghost, 1997). When automated systems increase autonomy or authority of machines without new tools to support cooperation with people, we find automation surprises contributing to incidents and accidents (Sarter, Woods, & Billings, 1997).

Human factors research has abstracted many patterns and lessons about how to make automated systems team players (Christoffersen & Woods, 2002). Characteristics of successful automation generally include predictability of what the automation will do, inspectability of the basis for action by the automation, high reliability in the most frequent operational situations, observability of the situations in which the automation will fail, direct benefit to the users that are greater than the costs of inputting information for automated processing, and the ability to switch to less automated modes on demand (Klein, Woods, Bradshaw, Hoffman, & Feltovich, 2004; Norman, 1990). One concern with using automated systems is that people have been shown to be more willing to accept even *poor* advice when it comes from a computer and have difficulty revising machine-initiated solutions (e.g., Layton, Smith, & McCoy, 1994). Therefore a useful tactic is to have humans initiate a problem-solving sequence and then use automated systems to remind, suggest, critique, or broaden the factors considered by the human decision maker, even for cases where the computer is unable to generate a good solution on its own (Guerlain et al., 1999).

Many of the developments in computer information systems across health care delivery systems (be they intended to enhance safety or productivity) include embedded forms of automation. Using the lessons from past research to guide the design of automated information processing systems will help avoid new paths to failure and increase the benefits to be obtained from these investments in new technology.

Invest in Collaborative Technologies

> When I order a medication, I think the patient gets the medication directly, but there are many other steps, computer systems, and

hands that intervene in the process.—A physician, 2000

Health care, similar to other settings, is increasingly shifting from paper-based to computer-based systems (Ash, Berg, & Coiera 2004). In the transition, the fundamental distribution of patient care over multiple practitioners, groups, locations, and organizations is easily overlooked or oversimplified by design teams. Collaborative support in prior systems can therefore be unintentionally degraded in new systems, as was experienced with a bar code medication system that initially did not support easy physician access to the complete medication administration record (Patterson et al., 2002).

In addition, opportunities have increased to provide more remote and distributed care through telemedicine (LaMonte et al., 2000; Xiao, Mackenzie, Orasanu, & the LOTAS Group, 1999), web-based consults, and remote surgeries via robotic technology. This change challenges us to coordinate care over these disparate players. We often think that if these different players are connected through new information technology, effective coordination will follow automatically. The exploding field of computer supported cooperative work (CSCW) tells us that achieving high levels of coordination is a special form of expertise requiring significant investment in experience or practice (Clark & Brennan, 1991; Galegher et al., 1990; Greenbaum & Kyng, 1991; Grudin, 1994; Klein et al., 2005). It also tells us that making cooperative work through a computer effective is a difficult challenge. For example, a critical part of effective collaboration is how it helps broaden deliberations and cross-check judgments and actions to detect and recover from incipient failures (Patterson, Cook, et al., 2004). The design of the communication channel and information exchanged can degrade the cross-check function or enhance it, depending on how the channel is designed. A great deal of work is underway to try to identify what factors are important to support good collaborative work to guide investments in technology (e.g., Guerlain et al., 2001; Heath & Luff, 1992; Jones, 1995; Moss & Xiao, 2004; Nyssen & Javaux, 1996; Patterson, Roth, et al., 2004; Xiao et al., 1999).

The field of CSCW is exploding because of the advances in the technology of connectivity—the Internet and telecommunications in general—and because of the potential benefits of being connected. This leads designers to need information about the basis for high levels of skill at coordinated work.

The new technologies of connectivity will transform the face of health care practices, and CSCW should become a core area of expertise, research, and development in health care. Relationships between practitioners will change and new relationships will be introduced. The changes can be designed primarily to support efficiency, or they can be designed primarily to enhance safety. Side effects of these coming changes can also create new paths to failure while they block other paths. Research to understand and direct this wave of change to enhance safety will need to be a critical priority in health care as it is in other high-performance fields.

The development and impact of telemedicine is one example of this process. Achieving continuity of care in the new world of computer connectivity across health care practitioners and providers is another.

Manage the Side Effects of Change

Health care systems exist in a changing world. The environment, organization, economics, capabilities, technology, and regulatory context all change over time. Even the current window of opportunity to improve patient safety is a mechanism for change. And new waves of change are beginning to swell up and move toward shores of health care practices (e.g., reductions to resident work hours). Waves of change are due in part to resource pressures and in part to new capabilities. Uncertainties created by these changes have given rise to public pressures for improving patient safety.

This backdrop of continuous systemic change ensures that hazards and how they are managed are constantly changing. This is important because many of these changes can easily increase complexities of health care delivery. Again, increasing complexity often challenges safety and rarely produces safety benefits without some other investments (Ash, Berg, et al., 2004).

The general lesson is that as capabilities, tools, organizations, and economics change, vulnerabilities to failure change as well—some decay but new forms appear. The state of safety in any system is always dynamic, and stakeholder beliefs about safety and hazard also change. Progress on safety depends on anticipating how these kinds of changes will create new vulnerabilities and paths to failure even as they provide benefits on other scores (Patterson et al., 2002).

For example, new computerization is often seen as a solution to human performance problems. Instead, consider potential new computerization as another source of change. Examine how this change will affect roles, judgments, coordination, and what makes problems difficult. This information will help reveal side effects of the change that could create new systemic vulnerabilities.

Armed with this knowledge, we can address these new vulnerabilities at a time when intervention is less difficult and less expensive (because the system is already in the process of change). In addition, these points of change are opportunities to learn how the system actually functions and sometimes malfunctions.

Another reason to study change is that health care systems are under severe resource and performance pressures from stakeholders. First, change under these circumstances tends to increase coupling, that is, the interconnections between parts and activities, in order to achieve greater efficiency and productivity. However, research has found that increasing coupling also increases operational complexity and increases the difficulty of the problems practitioners can face. Second, when change is undertaken to improve systems under pressure, the benefits of change may be consumed in the form of increased productivity and efficiency and not in the form of a more resilient, robust, and therefore safer system. Thus one linchpin of future success on safety is the ability to anticipate and assess the impact of change to forestall new paths to failure (Rochlin, 1999).

In addition, investments in safety are best timed to coincide with windows of opportunity where change is happening for other reasons as well. A great deal of leverage may result from projects designed to show health care organizations how to take advantage of change points as windows of opportunity where they can rethink processes, work flow, and new modes of collaboration to reduce the potential for breakdowns (Woods & Cook, 2002).

The System of Health Care Delivery Is Changing to Include Patients in New Ways

Patients are becoming involved in their own care (or their family member's care) in new ways. Patients have new access to information, allowing them to take an active role in treatment decisions. Technology and other factors are shifting care from in-patient settings to home settings where patients become a provider of their own treatment (e.g., Klein & Meininger, 2004; Obradovich & Woods, 1996).

Patient self-managed treatment represents a large and growing part of the health care system. Human factors can support patient safety in self-managed treatment. Errors can occur when information and interfaces do not fit the patient's capacities or past experiences or the demands of daily life (Klein, 2003). Human factors can focus attention on underlying mechanisms, behavioral patterns, and contextual contributions and provide the methods and tools to design the support systems that match patient needs (Klein et al., 2004; Porter, Cai, Gribbons, Goldmann, & Kohane, 2004).

Studying Human Performance, Human–Machine Systems, Collaboration, and Organizational Dynamics Requires Methods Unfamiliar to Health Care

Ultimately, the study of human performance is in one sense or another the study of problem solving. Since its origins more than one hundred years ago, understanding problem solving, be it a person, human–machine system, distributed team, or organization, has always been the study of the processes that lead up to outcomes—what is seen as a problem-to-be-solved, how to search for relevant data, how to anticipate future events, how to generate hypotheses, how to evaluate candidate hypotheses, and how to modify plans to handle disruptions.

The base data is the process or the *story* of the particular episode—how multiple factors came together to produce that outcome (Dekker, 2002; Klein, 1998; Klein, Orasanu, Calderwood, & Zsambok, 1995). Patterns abstracted from these processes are aggregated, compared, and contrasted under different conditions—different problem demands (scenarios), different human–human and human–machine teams, different levels of expertise, different external tools.

The fields that study one or another type of problem solving have developed sophisticated methods tailored to meet the uncertainties of studying and modeling these processes. They are deeply foreign to medical research communities, but they are the lifeblood of coming to understand human performance in any complex setting, including health care

(Hoffman & Woods, 2000). Ethnography (Hutchins, 1995), interaction analysis (Jordan & Henderson, 1995), protocol analysis (Ericsson & Simon, 1984), critical incident techniques (Flanagan, 1954; Klein, 1998), and cognitive work analysis (Vicente, 1999) are just a few of the techniques to be mastered by the student of human problem solving at work. (To see some of these techniques in action in health care, see the studies in Nemeth, Cook, & Woods, 2004, and Patterson, Rogers, Chapman, & Render, 2005).

One critical resource for the study of problem solving is mechanisms to build or obtain access to simulation environments at different scopes and degrees of fidelity. Much of the progress in aviation safety has depended on researchers having access to full scope training simulators to study issues such as effective human–human and human–machine cooperation (e.g., Layton et al., 1994). This has occurred through research simulators at NASA Ames and Langley Research Centers, through partnerships with pilot training centers (when research and training goals can be synchronized), and through the use of rapid prototyping tools to create simulation test beds. We have already begun to see how the availability of simulator resources in health care (notably for anesthesia and for the operating room) can be a catalyst to learning about the factors that lead to effective or ineffective human performance (Guerlain et al., 1999; Howard et al., 1992; Nyssen & De Keyser, 1998; Weinger at al., 2004).

Using these resources to understand human performance depends on special skills such as problem or scenario design (the design of the problems that study participants attempt to solve) and in analysis techniques such as interaction and protocol analysis (De Keyser & Samercay, 1998). Health care research organizations will need to create, modify, and use simulation resources to provide critical pieces of evidence in the process of finding effective ways to improve safety for patients.

Technology Evaluation

Evaluating changes intended to improve some aspect of human performance is a difficult problem. Human factors has worked with many industries to assess the impact of technology and other interventions (e.g., training systems) designed to aid human performance. Stakeholders have frequently asked us to give them a simple positive or negative result—does this particular system or technology help significantly or not? We refer to such studies as verification and validation evaluations, or V&V.

Despite the surface appeal of such efforts and the desire to provide definitive answers to guide investments, V&V has proved to be a limited tool in other high-risk domains. The short summary of the lessons is that such studies provide too little information too late in the design process, and at too great a cost.

They provide too little information in a variety of ways. There are multiple degrees of freedom in using new technology to design systems, but V&V studies are not able to tell developers how to use those degrees of freedom to create useful and usable systems. The problem in design today is not the question of whether a certain system can be built, but rather what would be useful to build given the wide array of possibilities new technology provides.

Measurement problems loom large because V&V studies usually try to capture overall outcomes. However, the systems are intended to influence aspects of the processes (human expertise, cooperative work, a culture of safety) that are important to outcomes in particular kinds of situations that could arise (Woods, Cook, & Billings, 1995). As a result, global outcome measures tend to be insensitive to the operative factors in the processes of interest or wash out differences that are significant in restricted kinds of situations.

New systems and technology are not unidimensional, but multifaceted, so that problems of credit assignment become overwhelming. Introducing new technology is not manipulating a single variable, but instead a change that reverberates throughout a system, transforming judgments, roles, relationships, and weightings on different goals (Carroll & Rosson, 1992). This process, called the task-artifact cycle, creates the envisioned world problem for research and design (Dekker & Woods, 1999; Hoffman & Woods, 2000; Woods & Dekker, 2000): How do the results of studies and analyses that characterize cognitive and cooperative activities in the current field of practice inform or apply to the design process, because the introduction of new technology will transform the nature of practice, what it means to be an expert, and the paths to failure? Health care specialists need only consider the many reverberations of the change to laparoscopic surgery or the introduction of new information systems to see these processes play out (e.g., Ash, Gorman, et al., 2004; Cook et al., 1998; Dominguez, Flach, Lake, McKellar, & Dunn, 2004; Patterson et al., 2002; Patterson, Rogers & Render, 2004).

V&V studies occur too late in the design process, especially given their great cost, to provide useful input. By the time the V&V results are available, the design process has committed to certain design concepts and implementation directions. These sunk costs make it extremely difficult to act on what is learned from evaluation studies late in the process. In addition, it is difficult to generalize the results from a single study of one system to other systems at other organizations due to the myriad design and implementation factors that differ between systems, or even from a single study of one system to the same system in a changed environment. (Han et al., 2005; Koppel et al., 2005; Wears & Berg, 2005; Nemeth & Cook, 2005).

The advent of rapid prototyping technology has revolutionized evaluation studies. While late V&V studies still have a role, the emphasis has shifted completely in many different work domains to early, generative techniques such as ethnography, envisioning techniques, and participatory design (Carroll & Rosson, 1992; Greenbaum & Kyng, 1991; Sanders, 2000; Smith et al., 1998). Health care needs to build on this experience and learn the use of these new techniques.

HIGH RELIABILITY ORGANIZATIONS AND REACTIONS TO FAILURE

One of the most productive areas of work on human error in the last 10 years has been about reactions to failure (Dekker, 2002; Rochlin, 1999; Woods et al., 1994). In this work we have come to understand more about what characterizes high reliability organizations (Adamski & Westrum, 2003; Grabowski & Roberts, 1997; Roberts, this volume; Rochlin, La Porte, & Roberts, 1987) and about the common oversimplifications or fallacies about error that block forward progress (Cook et al., 1998; Dekker, 2003; Dekker, 2005).

High-reliability organizations create safety by anticipating and planning for unexpected events and future surprises. These organizations continue to invest in anticipating the changing potential for failure, regardless of past success, because they appreciate that their knowledge is imperfect and that their environment continues to change. The heart of this process is learning activities that depend on open flow of information about the changing threats and about the changing effectiveness of their failure-sensitive strategies. For these organizations, safety is a value, not a commodity.

Escape From Hindsight Bias

There are a variety of factors that block or inhibit the learning processes central to a high-reliability culture. One is the hindsight bias (Fischhoff, 1975; Woods & Cook, 1999; Woods et al., 1994). The hindsight bias is one of the most reproduced research findings relevant to accident analysis and reactions to failure. Knowledge of outcome biases our judgment about the processes that led up to that outcome.

In the typical study, two groups of judges are asked to evaluate the performance of an individual or team. Both groups are shown the same behavior; the only difference is that one group of judges are told that the episode ended in a poor outcome, while other groups of judges are told that the outcome was successful or neutral. Judges in the group told of the negative outcome consistently assess the performance of humans in the story as being flawed in contrast with the group told that the outcome was successful. Surprisingly, this hindsight bias is present even if the judges are instructed beforehand not to allow outcome knowledge to influence their judgment.

Hindsight is not foresight. After an accident, we know all of the critical information and knowledge needed to understand what happened. But that knowledge is not as easily available to the participants before the fact. In looking back, we tend to oversimplify the situation that the actual practitioners faced, and this oversimplification tends to block our ability to see the deeper story behind the label *human error* (Cook et al., 1998; Rasmussen, 1990b; Woods & Cook, 2003).

Researchers use methods designed to remove hindsight bias to see the multiple factors and contributors to incidents, to see how people usually make safety in the face of hazard, and to see systemic vulnerabilities before they contribute to failures (Woods, 2005b). Research has developed a variety of techniques to reduce hindsight bias (Dekker, 2002, 2005). These are available to use and modify as necessary in health care research on safety.

Despite widespread communication about factors that block learning about how safety is created, hindsight and related "biases" continue to plague the literature on adverse events in health care. The studies of injury or death rates as a result of error and virtually all incident review procedures used in health care today fail to control for hindsight bias. This should not be considered acceptable by anyone interested in improving safety. It is time to stop repeating this "error" in the study of error.

Resilience

When research escapes from hindsight, studies reveal the sources of resilience that usually allow people to produce success when failure threatens. Methods to understand the basis for technical work shows how health care workers are struggling to anticipate forms of or paths toward failure, actively adapting to create and sustain failure-sensitive strategies and working to maintain margins in the face of pressures to do more and do it quickly (Woods & Cook, 2002, 2003). In other words, doing things safely, in the course of meeting other goals is and has always been part of operational practice. As people in their different roles are aware of potential paths to failure, they develop failure-sensitive strategies to forestall these possibilities. Failures occurred against this background when multiple contributors—each necessary but only jointly sufficient—combine. Work processes do not chose failure but *drift toward it* as production pressures and change erode the defenses that normally keep failure at a distance. This drift is the result of systematic, predictable organizational factors at work, not simply erratic individuals (Woods, 2005b). To understand how failure sometimes happens, one must first understand how success is obtained—how people learn and adapt to create safety in a world fraught with gaps, hazards, trade-offs, and multiple goals (Cook et al., 2000).

The theme that leaps out from these results is that failure represents *breakdowns in adaptations* directed at coping with complexity. Success relates to organizations, groups, and individuals who produce *resilient systems* that recognize and adapt to variations, changes, and surprises (Rasmussen, 1990a; Rochlin, 1999; Sutcliffe & Vogus, 2003; Weick, Sutcliffe, & Obstfeld, 1999). Resilience is the ability to adapt or absorb disturbance, disruption, and change, especially to disruptions that fall outside of the set of disturbances the system is designed to handle. The brittleness (which is the opposite of resilience) of some organizations and systems becomes evident when cases of near failure or adverse events are examined using the concepts and methods described in the earlier sections. Cook and O'Connor (2005) and Patterson, Cook, et al. (2004) provide detailed analyses of how brittle systems in health care break down in the face of change and a concatenation of factors. Cook and Rasmussen (2005) provide a basic overview of brittleness and resilience for safety management in health care organizations.

These results have led to the emerging area of resilience engineering as an alternative to error tabulations (Hollnagel, Woods, & Leveson, 2006). Many of the essential constituents of resilience engineering are already at hand as a result of research on organizational risk factors (Adamski & Westrum, 2003; Carthey et al., 2001; Hollnagel, 2004). The initial steps in developing a practice of resilience engineering have focused on methods and tools:

1. To analyze, measure, and monitor the resilience of organizations in their operating environment.
2. To improve an organization's resilience vis-à-vis the environment.
3. To model and predict the short- and long-term effects of change and line management decisions on resilience and therefore on risk.

Innovate New Models of Accountability

As the new messages about a systems approach to safety circulated and became more visible in health care, they collided with the common belief that practitioners and managers should be "accountable" to patients and to other stakeholders. But accountability in this common belief is operationalized in terms of pursuit of culprits, threats of disciplinary actions, and threats of stigmatization (Woods, 2005a).

These two trends collide in a basic double bind for the patient safety movement: Blame, even if disguised as accountability, drives out information about systemic vulnerabilities, stops learning, and undermines the potential for improvement (Billings, 1999). The challenge for research is to find a way out of the double bind—How do we create a safe environment for learning about the potential for failure in a publicly accountable system of health care delivery (Sharpe, 2004)?

Accountability is emphasized in the debate on patient safety because how decision makers are held accountable is presumed to influence how they make decisions and the quality of those decisions. Social links such as accountability can be powerful forces influencing human decision making, and these relationships have been studied in organizational dynamics, social cognition, and human–machine interaction (e.g., Hirschhorn, 1993; Lerner & Tetlock, 1999; Ostrom, 1990;

Tetlock, 1999). This information can be integrated with ethical and legal scholarship to stimulate the innovation of new systems for managing accountability relationships (Sharpe, 2000).

Practitioner decision making always occurs in a context of expectation that one may be called to give accounts for those decisions to different parties. How and to whom people expect to be called to account affects their performance in implicit and explicit ways. The expectations for what are considered adequate accounts and the consequences for people when their accounts are judged inadequate are critical parts of the cycle of giving accounts and being called to account (Brown, 2005a).

Interestingly, different factors in this reciprocating cycle can support or undermine practitioner performance and systems learning in predictable ways. Note that, from a behavioral science point of view, accountability is a neutral term that only points to the processes in this cycle of giving and being called to give accounts or reasons for a decision. First, past research shows that there is a complex set of factors, relationships, and effects at work in the reciprocating cycle of calling on and giving of accounts. Second, the empirical regularities and relationships are not consistent with motivational accounts, that is, that accountability creates general improvements by increasing task motivation. Third, and most startling, the research demonstrates that some factors in the reciprocating cycles of accountability may degrade decisions, performance, cooperation, and learning, while other relationships in the cycle may enhance these cognitive processes (Brown, 2005a, 2005b; Ostrom, 2003).

For example, under some conditions, the need to give an account for a decision to others can increase critical thinking and attenuate commitment (presumably ways to enhance the decision), while other conditions can increase self-justification and bolster an initial attitude and commitment (presumably ways that reduce the quality of a decision). Some forms of accountability can increase defensive behavior, create adversarial relationships among parties who need to cooperate, or lead people to prefer options that are easier to justify given knowledge of the standards others impose for giving of suitable accounts (Lerner & Tetlock, 1999).

These results allow us to see the label "culture of blame" in a new way. It is a kind of system of accountability, but it is only one way to design and manage such systems. If one analyzes a "culture of blame" in terms of the dynamics of cycles of accountability, we find many of the factors that have been implicated in degrading performance, cooperation, and learning (Brown, 2005a; Woods, 2005a).

The social cognition research on cycles of accountability clearly captures the complexity of the effects and demonstrates the naïveté of the belief that improving safety only requires holding others accountable. The slogan of "moving beyond a culture of blame" is a call to abandon poor systems of accountability, not an environment where accountability is absent. It is a necessary part of our life as social creatures that we explain our actions to others. The systems approach examines the reciprocating cycle of giving accounts and calling to accounts between the sharp end of practice and the blunt end of organizational context to determine the lawful effects of different systems for accountability (Dekker, 2003).

New work is needed to model and describe systems of accountability and their effects in health care. The results from this work should be used to engage all stakeholders in a process to explore new designs of systems of accountability that will produce the desired effects and advance our common goals.

Create and Share Learning Tools

The research on high-reliability organizations emphasizes continued learning about risks and mitigation strategies. In health care, direct learning and improvement from experience with accidents and incidents has been observed to be very limited and narrow. This appears to be partly because of the fear of blame and litigation. In addition, there are few organizational structures that promote learning about paths to failure.

An important area for new work is creating learning tools that function throughout health care organizations (Adamski & Westrum, 2003). Some believe that expanded requirements for notification of adverse events and near-misses will accomplish this. But lessons from aviation indicate that much more is needed in the analysis of cases and in feedback mechanisms to health care practitioners to complete a cycle of learning based on incident reporting (Billings, 1999). New incident reporting systems will need to encompass much more than new forms to fill out and new notification requirements to be effective. Serious issues remain about whether these programs will be effective, including the independence of the review teams, the need to incorporate human performance expertise in order to recognize deeper patterns, the move to analyze sets of cases rather than one case at a time, how to generate meaningful sustained improvements, and how to provide feedback to practitioners about how what was learned is relevant to them in their area of the world of health care.

The difficulties in learning from accidents and learning before accidents occur are particularly vivid in the contrast between two parallel disasters that played out in 2003—the loss of the Columbia space shuttle and the death of Jesica Santillan, a transplant patient at Duke University Hospital. Both events occurred in February 2003, and two very different processes for understanding and learning from these tragedies played out in parallel over the next few months. The contrast between the two investigations highlights the importance of results on how learning after accidents can break down (see Weick et al., 1999; and chap. 6 of Woods et al., 1994).

An independent and highly distinguished technical panel examined the Columbia accident and NASA as an organization (Columbia Accident Investigation Board, 2003). A wide-ranging and detailed set of information on how to improve the organization resulted from the public and the independent examinations. NASA supported the investigation despite the burden of criticism and is acting on what was learned from the public dissemination of its deficiencies as an organization. Systemic changes are widespread, and the only question is whether these new investments can be sustained over the long term. Plus the lessons are public and can be used by all organizations that manage risky processes under intense production pressures (Starbuck & Farjoun, 2005). During the same time period, the death of a seventeen-year-old girl following mistakes in a transplant procedure at the Duke University Hospital also captured public and news attention.[1] However, little is known publicly about the deeper systemic and organizational contributors to the accident. Press releases by the hospital itself provide most of the available information. The legal, institutional, and professional responses that followed from the tragedy are largely invisible to the public.[2] Little is known publicly about what changes transplant medicine as a whole needs to make and who is monitoring the effectiveness of these changes over time. (but see Wailoo, Livingston & Guarnaccia, 2006).

The comparison of the responses to the two tragedies raises serious questions that health care organizations have not begun to confront adequately. Why are there no *independent* investigations of iatrogenic patient injuries? In no other high-risk system does the organization where the adverse event occurred investigate itself. How are organizations learning about systemic vulnerabilities, developing systemic responses, and monitoring changes over time? Press releases about responses from the affected organization itself are not an effective means to reestablish trust and confidence following serious injuries to patients as a result of care (Cook et al., 1998). Deflecting blame is not a characteristic of high-reliability organizations (Brown, 2005b; Weick et al., 1999).

Interdisciplinary Partnerships

The success of NASA programs in aviation safety was built on interdisciplinary partnerships. The success of safety initiatives in health care is very likely to depend on building the same kinds of partnerships across quite different technical disciplines. Considering patient safety inevitably leads to technical issues about human performance, human–computer cooperation, organizational dynamics, software engineering, and other fields normally considered outside health care. The new research efforts will need to be structured to build up these partnerships between technical areas concerned with different aspects of human performance and different medical and practitioner specialties. One important activity will be to develop a new cadre of experts who are skilled at these interdisciplinary projects.

The success of NASA programs was based on a portfolio of research and development activities that included more basic work on human performance (e.g., mental workload), innovation of new directions for aiding human performance (e.g., cockpit resource management), advanced development, and technology transfer projects. In this process, different kinds of work were carried out and sponsored, including field research, simulator studies, concept development, and evaluation studies. The infrastructure at NASA helped cross-stimulate

[1]"Lessons from Jesica," February, 24, 2003, *INSIDE: The Duke University Medical Center Employee Newsletter*, 12(4). Retrieved January 21, 2005 from http://dukemednews.duke.edu/mediakits/detail.php?id=6498#remembered

[2]"Jesica Santillan Remembered," by R. Snyderman & W. J. Fulkerson, February 4, 2004, Duke University Media Kit.

all of these activities around the goal of developing new design directions that would be useful in improving aviation safety. The structure at NASA and NASA's role in the aviation industry provided another essential ingredient for success—independence. The issues underlying safety are potentially controversial. A technically grounded, independent organization whose only purpose is advancing safety can develop a reservoir of technical results and organizational confidence to ride through these controversies with their substantive efforts for safety intact (Woods, 2006).

Health care would do well to study the formal and informal organizational basis of the successes of the NASA Research and Development (R&D) program and to model their own research efforts on patient safety on NASA's human factors programs.

CONCLUSION

We have a window of opportunity for improving safety for patients, but there are many false trails that could consume the energy and resources available. To take advantage of this window we must be prepared to question conventional wisdom and assumptions by building a partnership between different health care specialties and different human performance specialties that intersect at the label *human error*. These human performance specialties are substantive, deep, and unfamiliar to health care. They are the wellspring for techniques, concepts, and systems that will improve human performance in health care, as has been the case in other high-risk domains.

From the past work in human factors, a simple standard emerges for judging success in research on resilience, error and safety. Research is successful to the degree that it helps recognize, anticipate, and defend against paths to failure that arise as organizations and technology change, **before any patient is injured**.

ACKNOWLEDGMENTS

This chapter is based in part on the written testimony developed by D. Woods as past president of the HFES to the *National Summit on Medical Errors and Patient Safety Research* convened by the Quality Interagency Coordination Task Force and organized by the Agency for Healthcare Research and Quality on September 11, 2000.

The material synthesized here was made possible by the authors' activities as part of the National Patient Safety Foundation and as part of the Veterans Administration's Midwest Center of Inquiry on Patient Safety or GAPS Center at the VAMC, Cincinnati, OH. A VA HSR&D Merit Review Entry Program Award supported Emily Patterson.

We thank Marta Render, M.D., director of the VA's GAPS Center, for her leadership and dedication in building an interdisciplinary partnership between human factors and health care to achieve higher safety for patients.

References

Adamski, A. J., & Westrum, R. (2003). Requisite imagination: The fine art of anticipating what might go wrong. In E. Hollnagel (Ed.), *Handbook of cognitive task design*. Mahwah, NJ: Lawrence Erlbaum Associates.

Ash, J. S., Berg, M., & Coiera, E. (2004). Some unintended consequences of information technology in health care: The nature of patient care information system-related errors. *Journal of the American Medical Informatics Association, 11,* 104–112.

Ash, Joan S., Gorman, Paul N., Seshadri, Veena, & Hersh, William R. (2004, March–April). Computerized physician order entry in U.S. hospitals: Results of a 2002 survey. *Journal of the American Medical Informatics Association, 11*(2), 95–99.

Berg, M., & Goorman, E. (1999). The contextual nature of medical information. *Int. J. Med Inform., 56,* 51–60.

Billings, C. E. (1996). *Aviation automation: The search for a human-centered approach*. Mahwah, NJ: Lawrence Erlbaum Associates.

Billings, C. E. (1999). The NASA aviation safety reporting system: Lessons learned from voluntary incident reporting. *Proceedings of Enhancing patient safety and reducing errors in health care*. National Patient Safety Foundation, Chicago IL (held at Annenberg Center for Health Sciences, Rancho Mirage, CA, November 8–10, 1998).

Bogner, M. S. (Ed.). (1994). *Human error in medicine*. Mahwah, NJ: Lawrence Erlbaum Associates.

Brown, J. P. (2005a). Ethical dilemmas in healthcare. In M. Patankar, J. P. Brown, & M. D. Treadwell (Eds.), *Safety ethics: Cases from aviation, healthcare, and occupational and environmental health*. Burlington, VT: Ashgate.

Brown, J. P. (2005b). Key themes in healthcare safety dilemmas. In M. Patankar, J. P. Brown, & M. D. Treadwell (Eds.), *Safety ethics: Cases from aviation, healthcare, and occupational and environmental health*. Burlington, VT: Ashgate.

Caplan, R. A., Posner, K. L., & Cheney, F. W. (1991). Effect of outcome on physician judgments of appropriateness of care. *Journal of the American Medical Association, 265,* 1957–1960.

Carroll, J. M., & Rosson, M. B. (1992). Getting around the task-artifact cycle: How to make claims and design by scenario. *ACM Transactions on Information Systems, 10,* 181–212.

Carthey, J., de Leval, M. R., & Reason, J. (2000, July). Understanding excellence in complex, dynamic medical systems. In *Proceedings of the 44th Annual Meeting of the Human Factors and Ergonomics Society/IEA2000.*

Carthey, J., de Leval, M. R., & Reason, J. T. (2001). Institutional resilience in healthcare systems. *Quality in Health Care, 10,* 29–32.

Christoffersen, K., & Woods, D. D. (2002). How to make automated systems team players. In E. Salas (Ed.), *Advances in human performance and cognitive engineering research* (Vol. 2; pp. 1–12). St. Louis, MO: Elsevier Science.

Chung, P. H., Zhang, J., Johnson, T. R., & Patel, V. (2003). An extended hierarchical task analysis for error prediction in medical devices. *Proceedings of the American Medical Informatics Association,* 165–169.

Clark, H. H., & Brennan, S. E. (1991). Grounding in communication. In L. Resnick, J. Levine, & S. Teasley (Eds.), *Socially shared cognition.* Washington, DC: American Psychological Association.

Columbia Accident Investigation Board. (2003, August). Columbia Accident Investigation Board (2003, August). Report, 6 vols. Government Printing Office, Washington, DC. Retrieved August 30, 2003, from http://caib.nasa.gov/

Cook, R. I., McDonald, J. S., & Smalhout, R. (1989). *Human error in the operating room: Identifying cognitive lock up.* Cognitive Systems Engineering Laboratory Technical Report 89-TR–07, Columbus, OH, Department of Industrial and Systems Engineering, The Ohio State University.

Cook, R. I., & O'Connor, M. F. (2005). Thinking about accidents and systems. In H. Manasse & K. Thompson (Eds.), *Improving medication safety.* Bethesda MD: American Society for Health-System Pharmacists.

Cook, R. I., & Rasmussen, J. (2005). Going solid: A model of system dynamics and consequences for patient safety. *Quality and Safety in Health Care,* 14, 130–134.

Cook, R. I., Render, M. L., & Woods, D. D. (2000, March 18). Gaps in the continuity of care and progress on patient safety. *British Medical Journal, 320,* 791–794.

Cook, R. I., Woods, D. D, & Howie, M. B. (1992). Unintentional delivery of vasoactive drugs with an electromechanical infusion device. *Journal of Cardiothoracic and Vascular Anesthesia, 6,* 1–7.

Cook, R. I., Woods, D. D., & Miller, C. (1998, April). *A tale of two stories: Contrasting views on patient safety.* National Patient Safety Foundation, Chicago IL. Also available at http://www.npsf.org/exec/report.html

De Keyser, V., & Samercay, R. (1998). Activity theory, situated action and simulators. *Le Travail Humain, 61*(4), 305–312.

De Keyser, V., & Woods, D. D. (1990). Fixation errors: Failures to revise situation assessment in dynamic and risky systems. In A. G. Colombo & A. Saiz de Bustamante (Eds.), *Systems reliability assessment* (pp. 231–252). Dordrecht, The Netherlands: Kluwer Academic Publishers.

Dekker, S., & Woods, D. D. (1999). Extracting data from the future: Assessment and certification of envisioned systems. In S. Dekker & E. Hollnagel (Eds.), *Coping with computers in the cockpit* (pp. 7–27). Aldershot, UK: Ashgate.

Dekker, S. W. A. (2002). *The field guide to human error investigations.* Bedford, United Kingdom: Cranfield University Press.

Dekker, S. W. A. (2003). When human error becomes a crime. *Journal of Human Factors and Aerospace Safety, 3*(1), 83–92.

Dekker, S. W. A. (2005). *Ten questions about human error: A new view of human factors and system safety.* Mahwah, NJ: Lawrence Erlbaum Associates.

Dominguez, C., Flach, J., Lake, P., McKellar, D., & Dunn, M. (2004). The conversion decision in laparoscopic surgery: Knowing your limits and limiting your risks. In J. Shanteau, K. Smith, & P. Johnson (Eds.), *Psychological explorations of competent decision making* (pp. 7–39). New York: Cambridge University Press.

Embi, P. J., Yackel, T. Y., Logan, J., Bowen, J. L., Cooney, T. G., & Gorman, P. N. (2004, July–August). Impacts of computerized physician documentation in a teaching hospital: Perceptions of faculty and resident physicians. *Journal of the American Medical Informatics Association, 11*(4), 300–309.

Ericsson, K. A., & Simon, H. A. (1984). *Protocol analysis: Verbal reports as data.* Cambridge, MA: MIT Press.

Feltovich, P., Ford, K., & Hoffman, R. (Eds.). (1997). *Expertise in context.* Cambridge, MA: MIT Press.

Fischhoff, B. (1975). Hindsight≠foresight: The effect of outcome knowledge on judgment under uncertainty. *Journal of Experimental Psychology: Human Perception and Performance, 1*(3), 288–299.

Flach, J. M., & Dominguez, C. O. (1995, July). Use-centered design: Integrating the user, instrument, and goal. *Ergonomics in Design.*

Flanagan, J. C. (1954). The critical incident technique. *Psychological Bulletin, 51*(4), 327–358.

Gaba, D. M., Howard, S. K., & Jump, B. (1994). Production pressure in the work environment: California anesthesiologists' attitudes and experiences. *Anesthesiology, 81,* 488–500.

Gaba, D. M., Maxwell, M. S., & DeAnda, A. (1987). Anesthetic mishaps: Breaking the chain of accident evolution. *Anesthesiology, 66,* 670–676.

Galegher, J., Kraut, R., & Egido, C. (Eds.). (1990). *Intellectual teamwork: Social and technical bases of cooperative work.* Mahwah, NJ: Lawrence Erlbaum Associates.

Garmer, K., Liljegren, E., Osvalder, A., & Dahlman, S. (2000). Usability evaluation of a new user interface for an infusion pump developed with a human factors approach. *Proceedings of the 46th Annual Meeting of the IEA/HFES 2000 Association* (pp. 1128–1131). Santa Monica, CA, Human Factors and Ergonomics Society.

Grabowski, M., & Roberts, K. H. (1997, Summer). Risk mitigation in large-scale systems: Lessons from high reliability organizations. *California Management Review, 39*(4).

Greenbaum, J., & Kyng, M. (Eds.). (1991). *Design at work: Cooperative design of computer systems.* Hillsdale, NJ: Lawrence Erlbaum Associates.

Grudin, J. (1994). Computer-supported cooperative work: History and focus. *IEEE Computer, 27*(5), 19–27.

Guerlain, S., LeBeau, K., Thompson, M., Donnelly, C., McClelland, H., Syverud, S., & Calland, J. F. (2001). The effect of a standardized data collection form on the examination and diagnosis of patients with abdominal pain. *Proceedings of the Human Factors and Ergonomics Society 45th Annual Meeting,* 1284–1288.

Guerlain, S., Smith, P. J., Obradovich, J. H., Rudmann, S., Strohm, P., Smith, J., et al. (1996). Dealing with brittleness in the design of expert systems for immunohematology. *Immunohematology, 12,* 101–107.

Guerlain, S., Smith, P. J., Obradovich, J. H., Rudmann, S., Strohm, P. Smith, J. W., et al. (1999). Interactive critiquing as a form of decision support: An empirical evaluation. *Human Factors, 41*(1), 72–89.

Han, Y. Y., Carcillo, J. A., Venkataraman, S. T., et al. (2005). Unexpected increased mortality after implementation of a commercially sold computerized physician order entry system. *Pediatrics,* 116, 1506–1512.

Heath, C., & Luff, P. (1992). Collaboration and control: Crisis management and multimedia technology in London underground line control rooms. *Computer-Supported Cognitive Work, 1,* 69–94.

Heath, C., & Luff, P. (2000). *Technology in action.* Cambridge: Cambridge University Press.

Hendee, W. (Ed.). (1999). *Proceedings of Enhancing Patient Safety and Reducing Errors in Health Care.* National Patient Safety Foundation, Chicago IL (held at Annenberg Center for Health Sciences, Rancho Mirage, CA, November 8–10, 1998).

Hirschhorn, L. (1993). Hierarchy vs. bureaucracy: The case of a nuclear reactor. In K. H. Roberts (Ed.), *New challenges to understanding organizations.* New York: Macmillan.

Hoffman, R., & Woods, D. D. (2000). Special section on studying cognitive systems in context. *Human Factors, 42*(1).

Hollnagel. E. (1993). *Human reliability analysis: Context and control.* London: Academic.

Hollnagel, E. (2004). *Barrier analysis and accident prevention.* Aldershot UK: Ashgate.

Hollnagel, E., Woods, D. D., & Leveson, N. (2006). *Resilience engineering: Concepts and precepts* (results from the International Symposium on Resilience Engineering, Soderkoping, Sweden, October 20–25, 2004). Aldershot, U.K.: Ashgate.

Howard, S. K., Gaba, D. M., Fish, K. J., Yang, G. S., & Sarnquist, F. H. (1992). Anesthesia crisis resource management training: Teaching anesthesiologists to handle critical incidents. *Aviation, Space, and Environmental Medicine, 63,* 763–770.

Hutchins, E. (1995). How a cockpit remembers its speeds. *Cognitive Science, 19,* 265–288.

Jones, P. M. (1995). Cooperative work in mission operations: Analysis and implications for computer support. *Computer Supported Cooperative Work, 3,* 103–145.

Jordan, B., & Henderson, A. (1995) Interaction analysis: Foundations and practice. *The Journal for the Learning Sciences, 4*(1), 39–103.

Karsh, B. (2004). Beyond usability for patient safety: Designing effective technology implementation systems. *British Medical Journal: Quality and Safety in Healthcare, 13*(5), 388–394.

Klein, G. (1998). *Sources of power: How people make decisions.* Cambridge, MA: MIT Press.

Klein, G., Feltovich, P. J., Bradshaw, J. M., & Woods, D. D. (2005). Common ground and coordination in joint activity. In W. R. Rouse & K. B. Boff (Eds.), *Organizational simulation.* (pp. 139–178). New York, NY: Wiley.

Klein, G., Orasanu, J., Calderwood, R., & Zsambok, C. (Eds.). (1995). *Decision making in action: Models and methods.* Norwood, NJ: Ablex.

Klein, G., Woods, D. D., Bradshaw, J., Hoffman, R. R., & Feltovich, P. J. (2004, November/December). Ten challenges for making automation a "team player" in joint human-agent activity. *IEEE Intelligent Systems,* November/December, 91–95.

Klein, H. (2003, Spring). Making medication instructions usable. *Ergonomics in Design,* 7–12.

Klein, H., & Meininger, A. R. (2004). Self-management of medication and diabetes: Cognitive Control. *IEEE Transactions on Systems, Man and Cybernetics, Part A, 34*(6), 718–725.

Kling, R. (Ed.). (1996). *Computerization and controversy: Value conflicts and social choices.* San Diego, CA: Academic Press.

Koppel, R., Metlay, J., Cohen, A., Abaluck, B., Localio, A. R., Kimmel, S., et al. (2005). Role of computerized physician order entry systems in facilitating medication errors. *JAMA, 293,* 1197–1203.

LaBerge, D. (1995). *Attentional processing: The brain's art of mindfulness.* Cambridge, MA: Harvard University Press.

LaMonte, M. P., Xiao, Y., Mackenzie, C. F., Hu, P., Gaasch, W., Cullen, J., et al. (2000, May–June). Tele-BAT: Mobile telemedicine for the brain attack team. *Journal of Stroke & Cerebrovascular Diseases, 9*(3), 128–135.

Layton, C., Smith, P. J., & McCoy, C. E. (1994). Design of a cooperative problem-solving system for en-route flight planning: An empirical evaluation. *Human Factors, 36,* 94–119.

Leape, L. L. (2004). Human factors meets health care: The ultimate challenge. *Ergonomics in Design, 12*(3), 6–12.

Leape, L. L., Woods, D. D., Hatlie, M. J., Kizer, K. W., Schroeder, S. A., & Lundberg, G. A. (1998, October 28). Promoting patient safety by reducing medical errors. *Journal of the American Medical Association, 280,* 1444–1447.

Lerner, J. S., & Tetlock, P. E. (1999). Accounting for the effects of accountability. *Psychological Bulletin, 125,* 255–275.

Lin, L., Isla, R., Doniz, K., Harkness, H., Vicente, K., & Doyle, D. (1998). Applying human factors to the design of medical equipment: Patient controlled analgesia. *Journal of Clinical Monitoring, 14,* 253–263.

Lin, L., Vicente, K., & Doyle, D. J. (2001). Patient safety, potential adverse drug events, and medical device design: A human factors engineering approach. *Journal of Biomedical Informatics. 34(4),* 274–284.

Malin, J. T., Schreckenghost, D. L., Woods, D. D., Potter, S. S., Johannesen, L., Holloway, M., et al. (1991). *Making intelligent systems team players* (NASA Technical Memorandum 104738). Houston, TX: NASA Johnson Space Center.

Moss, J., & Xiao, Y. (2004). Improving operating room coordination: Communication pattern assessment. *The Journal of Nursing Administration, 34*(2), 93–100.

Nemeth, C., Cook, R. I. (2005). Hiding in plain sight: What Koppel et al. tell us about healthcare IT. *Journal of Biomedical Infomatics, 38,* 262-263.

Nemeth, C., Cook, R. I., & Woods, D. D. (2004). Messy details: Insights from the study of technical work in healthcare. *IEEE Transactions on Systems, Man and Cybernetics, Part A, 34*(6), 689–692.

Nielsen, J. (1993). *Usability engineering.* Boston: Academic.

Norman. D. A. (1988). *The design of everyday things.* New York: Basic Books.

Norman, D. A. (1990). The "problem" of automation: Inappropriate feedback and interaction, not "over-automation." *Philosophical Transactions of the Royal Society of London, B, 327,* 585–593.

Norman, D. A. (1993). *Things that make us smart.* Reading, MA: Addison-Wesley.

Norman, D. A., & Draper, S. W. (Eds.). (1986). *User centered system design: New perspectives on human–computer interaction.* Hillsdale, NJ: Lawrence Erlbaum Associates.

Nunnally, M., & Cook, R. I. (2004). Lost in menuspace: User interactions with complex medical devices. *IEEE Transactions on Systems, Man and Cybernetics, Part A: Systems and Humans, 34*(6), 736–742.

Nyssen, A.-S. (2004). Integrating cognitive and collective aspects of work in evaluating technology. *IEEE Transactions on Systems, Man and Cybernetics, Part A: Systems and Humans, 34*(6), 743–748.

Nyssen, A. S., & De Keyser, V. (1998). Improving training in problem solving skills: Analysis of anesthetists' performance in simulated problem situations. *Le Travail Humain, 61*(4), 387–402.

Nyssen, A. S., & Javaux, D. (1996). Analysis of synchronization constraints and associated errors in collective work environments. *Ergonomics, 39,* 1249–1264.

Obradovich, J. H., & Woods, D. D. (1996). Users as designers: How people cope with poor HCI design in computer-based medical devices. *Human Factors, 38*(4), 574–592.

Ostrom E. (1990). *Governing the commons: The evolution of institutions for collective action.* New York: Cambridge University Press.

Ostrom, E. (2003). Toward a behavioral theory linking trust, reciprocity, and reputation. In E. Ostrom & J. Walker (Eds.), *Trust and reciprocity: Interdisciplinary lessons from experimental research.* New York: Russell Sage Foundation.

Patterson, E. S., Cook, R. I., & Render, M. L. (2002). Improving patient safety by identifying side effects from introducing bar coding in medication administration. *Journal of the American Medical Informatics Association, 9*(5), 540–553.

Patterson, E. S., Doebbeling, B. N., Fung, C. H., Militello, L., Anders, S., & Asch, S. M. (2005). Identifying barriers to the effective use of clinical reminders: Bootstrapping multiple methods. *Journal of Biomedical Informatics, 38*(3), 189–199. Special Issue on Human-Centered Computing in Health Information Systems.

Patterson, E. S., Nguyen, A. D., Halloran, J. M., & Asch, S. M. (2004). Human factors barriers to the effective use of ten HIV clinical reminders. *Journal of the American Medical Informatics Association, 11*(1), 50–59.

Patterson, E. S., Cook, R. I., Woods, D. D., & Render, M. L. (2004). Examining the Complexity Behind a Medication Error: Generic Patterns in Communication. IEEE Transactions on Systems, Man and Cybernetics, Part A, *34,* 749–756.

Patterson, E. S., Rogers, M. L., Chapman, R. J., & Render, M. L. (2006). Compliance with intended use of bar code medication administration in acute and long-term care: An observational study. *Human Factors. 48,* 15–22.

Patterson, E. S., Rogers, M. L., & Render, M. L. (2004, July). 15 best practice recommendations to improve the effectiveness of bar code medication administration in the Veteran's Health Administration. *Joint Commission Journal on Quality and Safety.*

Patterson, E. S., Roth, E. M., Woods, D. D., Chow, R., & Gomes, J. O. (2004). Handoff strategies in settings with high consequences for failure: Lessons for health care operations. *International Journal for Quality in Health Care, 16*(2), 125–132.

Porter, S. C., Cai, Z., Gribbons, W., Goldmann, D. A., & Kohane, I. S. (2004). The asthma kiosk: A patient-centered technology for collaborative decision support in the emergency department. *Journal of the American Medical Informatics Association, 11,* 458–467.

Rasmussen, J. (1986). *Information processing and human-machine interaction.* New York: North Holland.

Rasmussen, J. (1990a). The role of error in organizing behavior. *Ergonomics, 33,* 1185–1199.

Rasmussen, J. (1990b). Human error and the problem of causality in analysis of accidents. *Philosophical Transactions of the Royal Society London, B, 327,* 449–462.

Rasmussen, J. (2000). The concept of human error: Is it useful for the design of safe systems in health care? In C. Vincent & B. de Mol (Eds.), *Safety in medicine.* New York: Pergamon.

Rasmussen, J., Brehmer, B., & Lepat, J. (Eds.). (1991). *Distributed decision making: Cognitive models for cooperative work.* Chichester, England: Wiley.

Reason, J. (1997). *Managing the risks of organizational accidents.* Brookfield, VT: Ashgate.

Rensink, R. A., O'Regan, J. K., & Clark, J. J. (1997). To see or not to see: The need for attention to perceive changes in scenes. *Psychological Science, 8*(5), 368–373.

Rochlin, G. I. (1999). Safe operation as a social construct. *Ergonomics, 42*(11), 1549–1560.

Rochlin, G. I., La Porte, T. R., & Roberts, K. H. (1987, Autumn). The self-designing high-reliability organization: Aircraft carrier flight operations at sea. *Naval War College Review,* 76–90.

Rogers, W. A., Mykityshyn, A. L., Campbell, R. H., & Fisk, A. D. (2001). Analysis of a "simple" medical device. *Ergonomics in Design, 6.*

Roth, E. M., Malin, J. T., & Schreckenghost, D. L. (1997). Paradigms for intelligent interface design. In M. Helander, T. K. Landauer, & P. Prabhu (Eds.), *Handbook of human-computer interaction* (2nd ed.; pp. 1177–1201). New York, NY: North-Holland.

Rudolph, J. (2003). *Into the big muddy and out again: Error persistence and crisis management in the operating room.* Unpublished doctoral dissertation, Boston College.

Sanders, E. B.-N. (2000). Generative tools for co-designing. In S. Scrivener, L. Ball, & A. Woodcock (Eds.), *Collaborative design.* (pp. 3–16). London, UK: Springer-Verlag.

Sarter, N. B., Woods, D. D., & Billings, C. E. (1997). Automation surprises. In G. Salvendy (Ed.), *Handbook of human factors/ergonomics* (2nd ed.). New York: Wiley.

Senders, J. W. (1993, September). On patient injury and death stemming from the design of medical devices. Report, Columbia Falls, ME.

Sharpe, V. A. (2000). Behind closed doors: Accountability and responsibility in patient care. *Journal of Medicine and Philosophy, 25,* 28–47.

Sharpe, V. A. (2004). Accountability and justice in patient safety reform. In V. A. Sharpe (Ed.), *Accountability: Patient safety and policy reform.* Washington, DC: Georgetown University Press.

Smith, P., Woods, D., McCoy, E., Billings, C., Sarter, N., Denning, R., et al. (1998). Using forecasts of future incidents to evaluate future ATM system designs. *Air Traffic Control Quarterly, 6*(1), 71–85.

Stanton, N. (1994). *Human factors of alarm design.* London: Taylor & Francis.

Starbuck, W., & Farjoun, M. (Eds.). (2005). *Organization at the limit: NASA and the Columbia disaster.* Malden, MA: Blackwell, Forthcoming.

Sutcliffe, K., & Vogus, T. (2003). Organizing for resilience. In K. S. Cameron, I. E. Dutton, & R. E. Quinn (Eds.), *Positive organizational scholarship* (pp. 94–110). San Francisco: Berrett-Koehler.

Tetlock, P. E. (1999). Accountability theory: Mixing properties of human agents with properties of social systems. In L. L. Thompson, J. M. Levine & D. M. Messick (Eds.), *Shared cognition in organizations: The management of knowledge* (pp. 117–137). Lawrence Erlbaum Associates.

Vicente, K. (1999). *Cognitive work analysis: Toward, safe, productive, and healthy computer-based work.* Mahwah, NJ: Lawrence Erlbaum Associates.

Wailoo, K., Livingston, J., & Guarnaccia (Eds.) (2006). *A death retold: Jesica Santillan, the bungled transplant, and paradoxes of medical citizenship.* Chapel Hill, NC: University of North Carolina Press.

Wears, R. L., & Berg, M. (2005). Computer technology and clinical work: Still waiting for Godot. *JAMA, 293,* 1261–1263.

Weick, K. E., Sutcliffe, K. M., & Obstfeld, D. (1999). Organizing for high reliability: Processes of collective mindfulness. *Research in Organizational Behavior, 21,* 13–81.

Weinger, M. B., & Smith, N. T. (1993). Vigilance, alarms, and integrated monitoring systems. In J. Ehrenwerth & J. B. Eisenkraft (Eds.), *Anesthesia equipment: Principles and applications.* Malvern, PA: Mosby Year Book.

Weinger, M. B., Gonzales, D. C., Slagle, J., & Syeed, M. (2004). Video capture of clinical care to enhance patient safety. *Quality and Safety in Health Care, 13,* 136–144.

Winograd, T., & Woods, D. D. (1997). Challenges for human-centered design. In J. Flanagan, T. Huang, P. Jones, & S. Kasif

(Eds.), *Human-centered systems: Information, interactivity, and intelligence.* Washington, DC: National Science Foundation.

Woods, D. D. (2003, October 29). Creating foresight: How resilience engineering can transform NASA's approach to risky decision making. Testimony on *The Future of NASA* to Senate Committee on Commerce, Science and Transportation, John McCain, Chair, Washington, DC. Retrieved December 1, 2003, from http://csel.eng.ohio-state.edu/woods/news/woods_testimony.pdf

Woods, D. D. (2005a). Conflicts between learning and accountability in patient safety. *DePaul Law Review.*

Woods, D. D. (2005b). Creating foresight: Lessons for resilience from Columbia. In M. Farjoun & William Starbuck (Eds.), *Organization at the limit: NASA and the Columbia disaster.* Malden, MA: Blackwell.

Woods, D. D. (2006). How to design a safety organization: Test case for resilience engineering. In E. Hollnagel, D. D. Woods, & N. Leveson (Eds.), *Resilience Engineering: Concepts and Precepts.* Aldershot, U.K.: Ashgate.

Woods, D. D., & Cook, R. I. (1999). Perspectives on human error: Hindsight bias and local rationality. In F. Durso (Ed.), *Handbook of applied cognitive psychology* (pp. 141–171). New York, NY: Wiley.

Woods, D. D., & Cook, R. I. (2001). From counting failures to anticipating risks: Possible futures for patient safety. In L. Zipperer & S. Cushman (Eds.), *Lessons in patient safety: A primer.* Chicago: National Patient Safety Foundation.

Woods, D. D., & Cook, R. I. (2002). Nine steps to move forward from error. *Cognition, Technology, and Work, 4*(2), 137–144.

Woods, D. D., & Cook, R. I. (2003). Mistaking error. In M. J. Hatlie & B. J. Youngberg (Eds.), *Patient Safety Handbook.* (pp. 95–108). Sudbury, MA: Jones and Bartlett.

Woods, D. D., Cook, R. I., & Billings, C. E. (1995). The impact of technology on physician cognition and performance. *Journal of Clinical Monitoring, 11,* 92–95.

Woods, D. D., & Dekker, S. W. A. (2000). Anticipating the effects of technological change: A new era of dynamics for human factors. *Theoretical Issues in Ergonomic Science, 1*(3), 272–282.

Woods, D. D., Johannesen, L., Cook, R. I., & Sarter, N. B. (1994). *Behind human error: Cognitive systems, computers and hindsight.* Dayton, OH: Crew Systems Ergonomic Information and Analysis Center, Wright Patterson Air Force Base.

Woods, D. D., & Shattuck, L. G. (2000). Distant supervision: Local action given the potential for surprise. *Cognition, Technology and Work, 2,* 86–96.

Xiao, Y., Hunter, W. A., Mackenzie, C .F., Jeffries, N. J., Horst, R. L., & The LOTAS Group. (1996). Task complexity in emergency medical care and its implications for team coordination. *Human Factors, 38,* 636–645.

Xiao, Y., Mackenzie, C. F., Orasanu, J., & The LOTAS Group. (1999). Information acquisition from audio-video-data sources: An experimental study on remote diagnosis. *Telemedicine Journal, 5*(2), 139–155.

Xiao, Y., Seagull, F., Nieves-Khouw, F., Barczak, N., & Perkins, S. (2004). Organizational-historical analysis of the "failure to respond to alarm" problems. *IEEE Transactions on Systems, Man and Cybernetics, Part A, 34*(6).

Zhang, J. (1997). The nature of external representations in problem solving. *Cognitive Science, 21*(2), 179–217.

Zhang, J., Johnson, T. R, Patel, V., Paige, D., & Kubose, T. (2003). Using usability heuristics to evaluate patient safety of medical devices. *Journal of Biomedical Informatics, 36,* 23–30.

30

MEDICAL FAILURE TAXONOMIES

Bruce Thomadsen
University of Wisconsin, Madison

> Nearly everyone who has published in this field has devised some form of error classification.—James Reason, 1990

Far from just being academic exercises, classification of the types of failures that contribute to events often reveals information that becomes useful in trying to rectify hazardous situations and conditions. Whether retrospectively investigating the causes of an event or prospectively analyzing potential failures, determining steps to prevent future events seldom is clear. Sometimes analyses lead to many possible actions and selection of the one to take is needed. Other times, finding even a single corrective action remains elusive. In both cases, and those in between, classifying the failures and potential failures by various facets of their characteristics can provide guidance. This chapter considers failure classification systems: some history of their development, how they differ in their focus and procedures, and how they can contribute to reduction of risk to patients.

DEFINITIONS AND RATIONALES FOR CLASSIFYING HUMAN PERFORMANCE FAILURES

Frustration From a Database and Root-Cause Analysis

Many organizations face great frustrations dealing with error prevention. Even with very complete database systems and the performance of root-cause analysis for every event, finding effective remedial actions may remain elusive. In part the problem comes from the type of information involved in the analyses. Most often, databases maintain information on the "hard facts" of the event, such as what happened, what conditions existed at the time, who was involved—all very important information. However, while the information can call attention to a commonality, such as a frequent step in a sequence that is prone to error, the data does not usually lead the operator toward a potential solution nor even suggest what the solution might be.

Root-cause analysis (RCA), another very useful tool, distills the actions of an event to the earliest causes over which the facility could have any control (the progenitor causes). Compared to studying the database, RCA at least deals in causes instead of just data. Yet, with RCA also, the practitioner may be left with either no idea of how to address the causes found, or worse, finds that actions that strike directly at the progenitor causes may not make any difference in the frequency of events of the same nature.

One reason that these tools, and other staples in the armamentarium of risk analysis, sometimes fail is that, while they may identify the "cause" of the problem, they neglect the nature of the cause. A nurse may neglect to give a patient medication, but an effective remedial action to prevent this in the future would be different if the nurse were too overworked to get to the patient than if the nurse were too involved watching a television show.

Taxonomies address the nature of the causes behind events.

Definition of Taxonomy

A *taxonomy* is simply a classification or ordering into groups or categories. The term most often applies to placing living organisms into their classification from phyla through species, but any system with a rational structure used for classification would qualify as a taxonomy.

For applications in patient safety, the taxonomies of interest deal with classifications of failures. An important aspect of the definition should include both classification and ordering.

Natures of Failures

To establish any type of classification system for failures requires an understanding of failures. That is a big order, but fortunately a great deal of progress can be made with knowledge of some of the natures of failures.

Definition of Failure

Practically, failure is simply a lack of success. But this hardly captures the essence of failure. Everyone has experienced failure and recognizes that failure seems to have deeper impact. Even success can seem like failure if it is qualified, and even happy endings may contain many aspects of failure. Sometimes success can be failure if the goal turns out to have been less desirable than originally thought. Thus failure may best be thought of as performance that leaves one unsatisfied.

Distinctions Between Failure, Errors, and Mistakes

Less than acceptable performance characterizes failure. Failure often can be expected. Contests among equal competitors will mean failure for all but one. Such failures often are analyzed to improve performance in the future, although sometimes success lies outside the bounds of likelihood. In other settings, failures fall outside normalcy. In health care, the assumption is that everyone involved in a case performs correctly and effectively, and although the patient may not get well, the cause is not that a member of the health care team failed. Situations such as health care, where stakes are high and all parties are expected to perform correctly, are *high-reliability* situations. Failures in these cases form *events*.

A failure can result from not performing adequately, such as not executing a necessary function at the required time. A failure of this type goes by the name *error*. An execution error generally falls into the class of *omission* or *laps*—the lack of a required action—or the class of *slip*—a botched action. Most often, errors happen without much, or any, thought.

On the other hand, planned actions may not deliver the desired end even if executed well. In addition to misguided plans, the actions may not be planned well, or the result of the planned action may not further a long-term goal. Such failures of intention constitute *mistakes*.

These definitions merely form a convention, because dictionary definitions of *error* and *mistake* share many facets and in conversation are often used interchangeably. Although it may seem silly to make such distinctions (and much finer ones are to come), a differentiation of what sort of failure happened gives insight into why it happened.

RATIONALES FOR STUDYING THE NATURE OF FAILURES

The question of justifying study and classification of failures then turns to why one cares why a failure occurred. As noted in the first section, many different causes could motivate the same act. In the previous example of the nurse not giving a patient medication on schedule, say the administration assumed that the nurse simply forgot about it, so they gave the nurse a wristwatch with an alarm to set for medication times. However, the real reason for the omission was that the nurse was grossly overworked and was still giving the medications to patients from three hours before. Obviously the solution, which might have worked for the assumed cause, would not actually improve the situation. *Knowing the nature of failures helps address the causes that led to the failure.*

Establishing the rationale for studying the nature of failures leads to another question: Is studying the

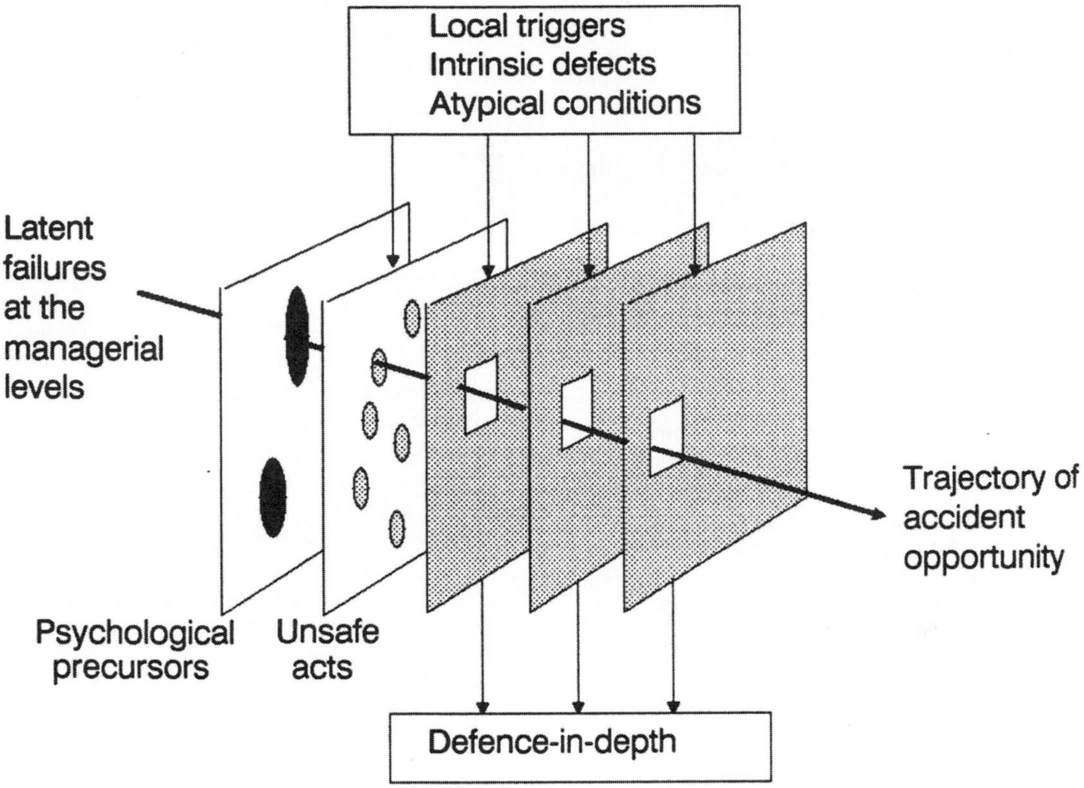

Figure 30–1. Reason's Swiss-cheese model for propagation of errors through holes in the defenses that make up the latent errors.(Reason, 1990, reprinted with permission of the publisher).

nature of failures worthwhile, compared to proactive approaches to risk reduction? Is that like shutting the barn door after the horse runs out? Obviously the proactive approaches serve very important functions and form indispensable tools for setting up protection against failure and promoting risk reduction. However, any event carries the assumption that some of its causes may come back in another event. The very commonly referenced model for events posited by James Reason (1990), called the Swiss-cheese model (see Fig. 30–1), assumes during the proactive phase of risk reduction establishment of many layers of protection against the effects of failures. None of the layers are perfect, however, and each has holes that could allow a failure to propagate. Each time an event occurs, it gives clues as to the procedural locations of the holes. The holes may never again align in the manner that lead to the last event, but new combinations can lead to new and different events.

Understanding of Human Performance Failures

Understanding human performance failures often begins by trying to understand human performance successes, that is, how humans perform functions.

Modern models for human performance not only go back more than a century (models in general go back more than two millennia) but also cover a considerable territory. A comprehensive review will not be presented here, but Reason (1990) presents a good summary for the interested reader.

The types of performances of interest in this chapter—those that influence patient care—mostly are skilled actions that take some training. However, sometimes normal "commonsense" or everyday actions also come into play. Rasmussen (1982) presents a useful model.[1] Most jobs first require some knowledge of the background related to the functions performed. An example for medical personnel would be anatomy classes or infectious disease control principles. There are also particular functions that must be mastered, such as measuring blood pressure or taking blood samples. Learning the functions begins at a level where each step must be carefully considered, possibly following notes or instruction sheets. With practice, the operation becomes routine, and eventually the operator hardly thinks about it during execution. Each operator learns many procedures and rules that dictate the appropriate situation to perform each procedure.

Rasmussen ranks actions using three groups:

1. Skill-based, which when proficient, proceed automatically without intellectual intervention.
2. Rule-based, which consider the situation and determine what skill-based operations to apply.
3. Knowledge-based, which use background information and analytic processes to address situations for which the rules don't apply.

In general, humans tend to try to handle problems in the order given, using a skill if possible, searching for the appropriate rule to tell them what skill to use if that is not obvious, and only as a last resort, trying to figure out the characteristics of the situation and create an original plan to address it. Figure 30–2 presents how inputs, operations and outputs in normal tasks flow through the levels of actions in Rasmussen's model.

Understanding of the Effects of Environments and Other Factors on Performance Failures

How well humans deal with situations requiring actions depends not only on their abilities and training. Surrounding conditions play a big part. While not directly responsible for how well a person performs, these other conditions influence what the person does, and thus they are called *performance shaping factors (PSF).*

The environment, to a great extent, determines what one perceives—both how well sensory inputs are noted and how they are interpreted. For example, a room full of noise, such as from patient monitors and a paging system, may prevent a nurse from hearing the patient tell of an allergy to sulfa. In addition, light levels, angles of computer screens, and smells from a plumbing problem all potentially affect perception. Conditions such as temperature and humidity can add physiological stress that affects performance.

Of a different nature, but still environmental in character, are conditions such as the time of day or day of the week. The quality for human performance during a normal workday follows a pattern of highs and lows as attention waxes and wanes. The end of the workday becomes hazardous as the fatigued worker starts thinking about going home. A similar pattern holds for the workweek, as attention fades on Friday afternoon, the time when most failures occur. The situation worsens for extended workdays (beyond 8 hours) and longer workweeks. The common practice of hospital staff working 12-hour shifts places much of the working time when performance falters. Working on holidays presents a double whammy, first because the workers would rather be elsewhere (where their thoughts are) and second because, if needed, other support staff often are unavailable.

Equipment and instrumentation performance also affects human performance. Equipment malfunctions, while possibly causing a failure or event directly, can also lead to human failure. Equipment may give wrong indications, or signals that are hard to understand or interpret. When equipment fails to work properly, persons that ordinarily depend on them may need to perform functions manually that they are not use to. At the least, equipment failures

[1]This discussion recognizes that models only present useful ways of looking at complex phenomena, rather than actually providing explanations for phenomena. As a result, a model may be useful when considering some aspects of the subject, but obviously completely wrong for others.

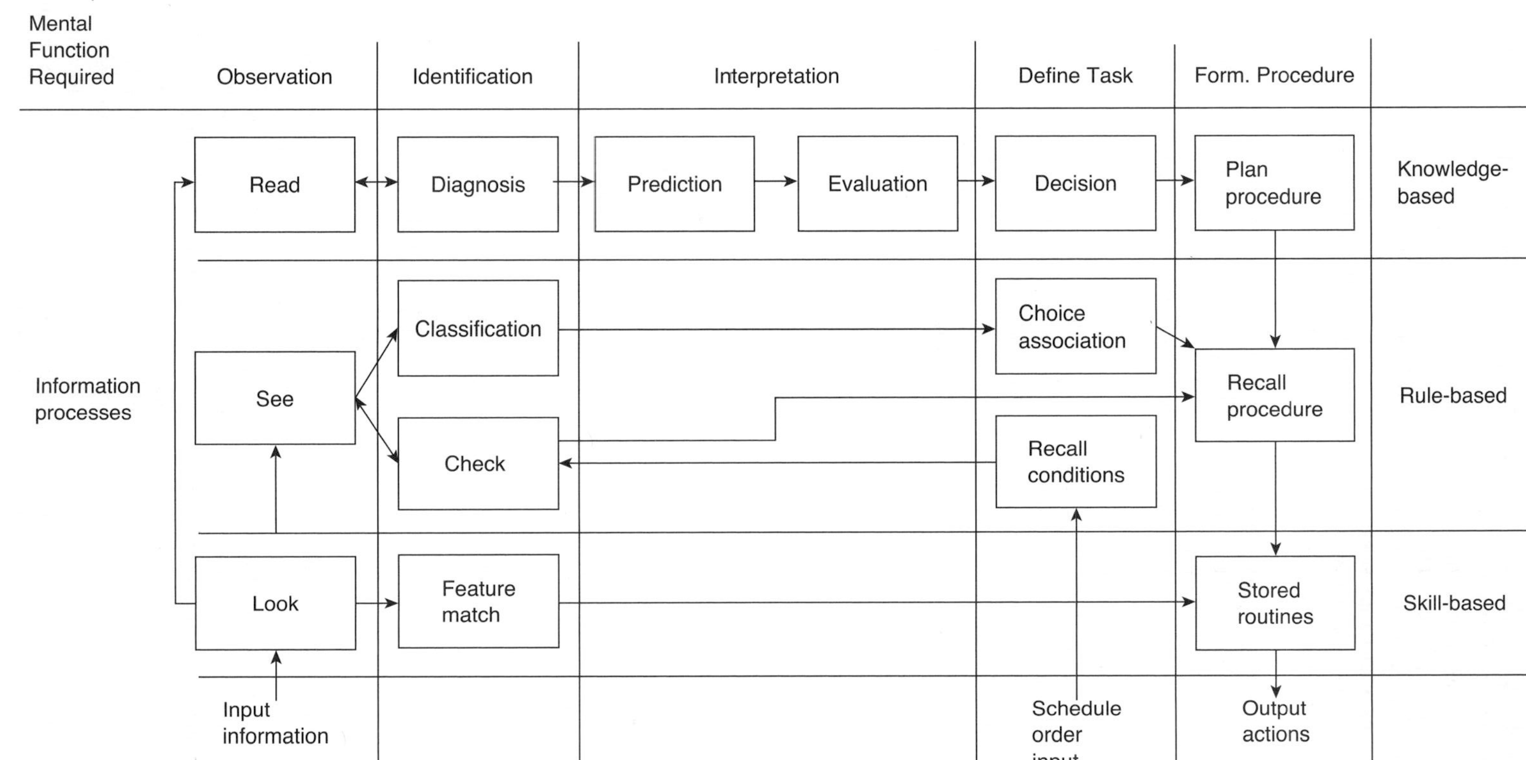

Figure 30–2. The flow of actions in Rasmussen's model. (Rasmussen, 1982). Reprinted with permission of the publisher.

add stress to the work environment, increasing the likelihood of human problems.

Personal factors also affect performance. When personal problems weigh heavily on the mind, other inputs and tasks often get pushed to the back as thoughts about the problems keep creeping to the fore. The same happens with happy occasions and anything exciting. Even the morning news can affect how a person acts during the day.

Many shaping factors stem from an organization's policies. Take, for example, the overworked nurse who failed to give the patient medication on time. The basic problem there was lack of staffing, an organizational decision. Similar decisions involving equipment maintenance could likewise result in failures. The greatest organizational PSF is an administration that simply is not concerned about the quality of care delivered. Such an attitude tends to percolate through the rest of the organization—leading employees who care to leave.

ERROR CODING VERSES TAXONOMIES

As the quote at the beginning of this chapter indicates, error classification systems abound. However, not all classification systems form taxonomies. The key distinction in the previously mentioned definition of taxonomy is that elements are "ordered": There is a structure underlying a taxonomy.

Examples of Error Coding

Many error classification systems, and particularly common in medicine, are error coding systems—basically lists of faults. The lists *do* usually have some organization, such as general categories of errors in medication, errors in surgery, and errors in diagnosis. Under these general categories fall subcategories, for example, the errors in medication may allow designations such as wrong drug, wrong time, wrong amount, wrong patient, and the like. A good example of such an error coding system is the New York Patient Occurrence and Tracking System (NYPORTS). Table 30–1 gives a sample of this classification system.

Error coding systems provide valuable information when assessing what errors take place, identifying procedures or actions that require concentrated resources to improve, or following trends over time. The benefits of NYPORTS can be seen at its Web site (New York State Department of Health, 2001).

Differences Between Error Coding and Classification by Taxonomy

Taxonomies go beyond error coding systems. Error codes may form a first-order base for a taxonomy, but the taxonomy must go deeper into the classification of what happened. The order of a taxonomy places the failures in a logical or functional position in the process of performance as well as in the process performed. That is, not only does a taxonomy consider at which step the failure occurred, but it also considers why or how it occurred. The assumption behind taxonometric theory is that *understanding the nature of failures provides clues for their prevention.*

APPROACHES TO TAXONOMIES—SOME EXAMPLES

Because taxonomies contain an order, they reflect the interests and perspective of their originator. There is no one taxonomy, nor is there likely to be. Taxonomies are just models—as noted before, useful for a purpose but not a true reflection of reality. Taxonomies tend to be the most successful when addressing very narrow conditions or situations and become less robust and useful when made more general and encompassing. The selection of taxonomies considered in the following section only indicates that they provide particular insights into the study (and classification) of taxonomy, and it by no means impugns excluded taxonomies.

A Psychological Viewpoint—Donald Norman

Norman (1981) considered why and in what ways persons say the wrong thing. By considering verbal slips, he developed a list of reasons behind the slips. Table 30–2 shows the list. The order and ranking of errors makes the list a taxonomy rather than an error code. While the particular article addresses verbal slips, it serves as a model of actions in general. In his discussion, and thus in Table 30–2, Norman (1981) does not follow the previously discussed convention in terminology distinguishing errors from mistakes.

Table 30–1. NYPORTS REPORTABLE*

Category/Code	Description
Serious	**Report immediately:**
911	Procedure-related: wrong patient/site—surgical
912	Procedure-related: wrong procedure/treatment—invasive
913	Procedure-related: retained foreign body (sponge pads, guidewires, instruments)
915	All unexpected deaths (including neonate ≥ 28 weeks AND ≥ 1000 grams AND no life threatening anomalies)
916	Unexpected cardiac arrest (BLS/ACLS intervention)
917	Unexpected loss of limb or organ
918	Unexpected impairment/dysfunction of limb
919	Unexpected impairment of body functions
920	Error of omission/delay leading to death/serious injury
921	Crime leading to death/serious injury
922	Suicide or attempt with serious injury
923	Elopement from hospital leading to death/serious injury
938	Malfunction of equipment/defective product leading to death/serious injury
961	Infant abduction
962	Infant discharged to wrong family
963	Rape by another patient or staff
Medication Errors	
108	Medication error with permanent harm
109	Medication error with near death event
110	Medication error with death
Aspiration/Sedation	
201	Aspiration pneumonia (in a nonintubated patient related to conscious sedation)
Intravascular Catheter Related	
301	Catheter-related infection/necrosis
302	Volume overload/pulmonary edema
303	Pneumothorax related to IV catheter
Embolitic & Related Disorders (include readmissions 30 days)	
401	New pulmonary embolism
402	New documented DVT
Laparoscopic	
501	Unplanned conversion: laparoscopic to open procedure
Peri-op Periprocedural related (day of or 1st or 2nd day after procedure including readmissions)	
601	New central nervous system deficit (e.g., TIA, stroke)
602	New peripheral neurologic defect (e.g., palsy, paresis)
603	Cardiac arrest w/successful resuscitation
604	Acute myocardial infarction unrelated to a cardiac procedure
605	Death following: appendectomy, noncardiac angiography, cholecystectomy, endarterectomy, resection of large intestine, hysterectomy, large bowel endoscopy prostatectomy, replacement of joint in lower extremity, spinal fusion
Burns/Falls	
701	Burns (2nd or 3rd degree)
751	Falls with fracture or head injury (subdural or epidural hematoma, SAH)
Procedure related Within 30 days of procedure, including readmissions	
801	Injury requiring repair/organ removal/other procedural intervention
803	Hemorrhage/hematoma requiring drainage, evacuation, or other intervention
804	Anastamotic leak requiring repair
805	Wound dehiscence requiring repair
806	Displacement/breakage of implant
807	Thrombosed distal bypass graft requiring repair
808	Post op wound infection requiring drainage or hospital admission within 30 days
819	Unplanned return to OR, related to primary procedure
851	Postpartum hysterectomy
852	Inverted uterus
853	Ruptured uterus
854	Circumcision requiring repair
Other NYPORTS reportable	
901	Other serious occurrence warranting DOH notification
902	Hospital transfer from diagnostic center
914	Misadministration of radioactive material
931	Strike by hospital staff
932	External disaster affecting hospital operation
933	Termination of vital hospital services
934	Poisoning occurring within the hospital (water, air, food)
935	Hospital fire disrupting patient care/causing harm to patients/staff
936/937	Equipment malfunction w/potential for adverse outcome

* New York State Department of Health, 2001. Used with permission.

Table 30–2. A Classification of Slips Based on Their Presumed Sources *

Slips that result from errors in the formation of the intention
- Errors that are not classified as slips: errors in the determination of goals, in decision making and problem solving, and other related aspects of the determination of an intention
- Mode errors: erroneous classification of the situation
- Description errors: ambiguous or incomplete specification of the intention

Slips that result from faulty activation of schemas
- Unintentional activation: when schemas not part of a current action sequence become activated for extraneous reasons, then become triggered and lead to slips
- Capture errors: when a sequence being performed is similar to another more frequent or better learned sequence, the latter may capture control
  - Data-driven activation: external events cause activation of schemas
  - Associative activation: currently active schemas activate others with which they are associated
- Loss of activation: when schemas that have been activated lose activation, thereby losing effectiveness to control behavior
  - Forgetting an intention (but continuing with the action sequence)
  - Misordering the components of an action sequence
  - Skipping steps in an action sequence
  - Repeating steps in an action sequence

Slips that result from faulty triggering of active schemas
- False triggering: a properly activated schema is triggered at an inappropriate time
  - Spoonerisms: reversal of event components
  - Blends: combinations of components from two competing schemas
  - Thoughts leading to actions: triggering of schemas meant only to be thought, not to govern action

Premature triggering
- Failure to trigger: when an active schema never gets invoked because
  - The action was preempted by competing schemas.
  - There was insufficient activation, either as a result of forgetting or because the initial level was low
  - There was a failure of the trigger condition to match, because either the triggering conditions were badly specified or the match between occurring conditions and the required conditions was never sufficiently close.

* Norman (1981). Reprinted with permission.

Levels of Performance and Types of Failures That Occur at Each Level

The first of three levels of errors, *errors in intention*, falls into our classification of mistakes. Norman considers three types of errors in intention. The first forms high-level errors, that is, errors in cognitive processes: mistakes in thinking about the problem or situation. The second, erroneous classification of the situation, characterizes actions that would be perfectly correct in a usual situation, but, unfortunately, the perpetrator is in another type of situation. The last type of error at this level comes from not giving enough thought to what the intention should be—essentially, going off half-cocked.

The second level is entitled *results from faulty activation of schemas*. A schema, a popular term at the time Norman was writing, refers to a well-practiced mental program. Some schemas run automatically, for example, receptionists answering the phone with the name of a company or many of the actions in driving a car, which almost become autonomic. Others are less reflexive and are just well-known procedures, such as the same receptionist scheduling a patient and asking for the appropriate information. Norman considers two possibilities leading to errors of this type, either when a schema becomes activated when it should not have been or when, after activation, the schema shuts down prematurely. (The possibility that the schema never starts falls into the next level.) Errant schemas form an important and very common classification of errors. The situations in which schemas are likely to start occur frequently. Capture errors happen when the intention is to perform one action, but that action is closely related to a different, more common action, for example, walking into a flim-processing darkroom and, instead of turning on the safelight, turning on the regular white light (and ruining the film). Data-driven activation (a somewhat misleading term) results when some stimulus initiates a schema inappropriately. With the departing salutation, "Have a good day" becoming popular, the common response is "Same to you." This leads to the humorous, and oft-heard exchange with an airline ticket-counter agent telling a traveler, "Have a good flight," followed by the traveler's reply, "Same to you." The response comes well before any thought process formulates. In the last unintentional activation classification, the associative activation, a running schema starts a different schema that often is associated with the first classification, even though in the current context the second is not desired. In medical events, unintentional activation of schemas becomes a common problem. An example could be a unit nurse mentally dealing with a very unresponsive diabetic patient with a blood sugar level of 23 who delivers a food tray to a different psychologically incompetent patient who is not supposed to eat the night before surgery (NPO) simply because the tray came on the cart with all the rest. The mostly automatic delivery of the dinner trays follows closely upon seeing the tray on the cart. With the nurse's thoughts more directed toward the patient with immediate problems, grabbing the tray initiates the delivery schema without thoughts about the NPO status of the patient receiving the tray nor of the possible, life-threatening consequences.

Loss of activation of a schema can also cause errors. Table 30–2 gives several possible ways that schema can stall once started. The last category, *repeating steps in an action sequence,* is not so much the loss of activation of a schema as it is its loss of effectiveness when the operator repeats steps inappropriately (assumedly forgetting that they had been completed once) and the process loops around, such as giving medication to a patient twice during the same round through the ward. The other examples lead to unfinished, or incompletely finished, procedures.

The final of the three major classifications considers schemas that either become triggered at an inappropriate time—essentially by accident, or lack of forethought or present thought—or never get triggered in situations where they ought. The three reasons given for the latter cover a lot of territory. While not tabulated in a study, this category of omissions likely plays a significant role in medical events. Although the first, preemption by another schema, is not so common, the second, the simple lack of activation due to inattention, certainly plays a major role in many events. This should not be thought of as personnel daydreaming and just forgetting a task. Rather, medical caregivers in most settings run just to stay in place. With many other activities going on, the omission of a task more often results from the thought not rising to immediacy. For example, a nurse in a patient ward may forget to change the saline IV bag on a fairly healthy patient while dealing with a number of very sick patients. The fact that the healthy patient has an IV may escape the nurse's consciousness. This differs from the first classification, in that preemption occurs when a schema is about to begin (has been thought of and almost initiated), but a different schema starts instead. The preempting schema likely has a good reason to start in the situation, unlike the capture error, where the schema actually invoked most likely is performed more frequently than intended. The third classification for failure to initiate a schema arises from poor matches between the perceived situation and the situation that would activate the schema. Norman (1981) gives some examples, but many more exist. The reasons may fall along the continuum between problems with the operator, such as ineffective training or perception, to unclear situations. The unclear situation could also lead to mode errors from the first major group.

Complexity of Some Failures

As the discussion in the preceding paragraph highlights, fine lines may separate many of the classifications. Indeed, a given action may carry features of several classifications. Such multiple classifications should not be considered a failure of the taxonomy, but rather an indication of the complexity of human actions. Sometimes the classification assigned to a performance failure reflects the perspective either of the person making the classification or of the resultant event. In most cases with competing classifications, no one is correct and all bring information to the investigation.

An Industrial (Person Faces Machine) Approach—Jens Rasmussen

Rasmussen worked with nuclear power plants. The taxonomy he developed for analysis of events that happened in the plant reflects the work environment (Rasmussen, 1982). Because of the incredible investment in proactive safety in the plants, the plants tend to run very reliably most of the time, with little for the operator to do. Situations generally begin when some system (and its backup systems) fail, or maybe just fall outside of normal ranges. Such a situation is called an initiating event, although it may not qualify yet as an actual "event" as defined earlier, depending on what follows. The operator needs to respond to the abnormal conditions. Inappropriate responses result in failures. Rasmussen's system categorizes the operator's response.

Basis of the Approach and Underpinning Assumptions

Rasmussen's taxonomy focuses on the human performance, divided into categories as in Figure 30–3. The analysis of failure particularly studies three aspects of the operator's response: what failed (or better, where in the required steps the failure occurred), how it failed (what the nature of the fault was), and why the operator failed. Figures 30–4 through 30–6 show the taxonomy's pathways used to find classifications for performance failures. Beginning at the top of the What Failed pathway, the investigator moves into a box with the question "Did the operator recognize the need for activity?" Arrows labeled "yes" or "no" guide the investigator to the next box. Eventually, the path leads to a classification. The procedure is the same for the other two pathways.

The What Failed pathway (Figure 30–4) parallels the steps the operator would follow to execute the appropriate response to the initiating event. The first step is to notice that something happened and that action may be required. Once noticed, the identification of what happened (or is happening) must follow. Before developing a plan of action, the operator needs to have a clear idea of the goal the action should accomplish. For the power plant, this would be stabilizing the reactor state; in health care, it may be to stabilize the patient during a myocardial infarction. The goal leads to a desired target state (for our infarction patient, increasing regular flow to the heart muscle) and that state in turn leads to a task to achieve the target state (e.g., angioplasty). From that task follows a procedure, and finally, the operator executes the planned procedure. Without a failure in at least one of these steps, the response succeeds. This pathway leads through the steps linearly. Upon reaching a box at which a failure occurred, the classification is made.

The How It Failed pathway (Figure 30–5) becomes more convoluted. The main pathway running down the left side progressively takes the investigation farther from normal. Consideration of the classifications in this pathway gives great insights into the types of errors humans make. The first box asks if the situation was normal for which the operator had appropriate skill. That is, things had no particular reason for going astray, except that the operator bungled the execution, which is always a possibility because no one is perfect, or the operator erroneously thought that something was wrong. Two possibilities are given for why the operator might fail in this situation: One is a simple blunder and the other a topographical misorientation, that is, the operator thought the situation was different from the one that actually existed and responded appropriately to the perceived conditions.

Answering "no" in the first box points toward the box below. Whereas in the first box the situation was normal but the operator goes astray, the second box in the column on the left represents a situation that deviates from normal. However, instead of acting appropriately to the abnormal situation, the operator responds as if all were normal. Many medical and nonmedical events contain some aspect of the stereotype fixation. Reactions to warning signals often take this form, such as reaction to the explosion of the SL1 reactor.[2] The first indication that something happened occurred at a sentry post (miles away) where the radiation alarm sounded. The sentinel quickly got on the phone and called the electrician to fix the box, which must be broken, the sentinel reasoned, because the alarm was sounding. More often these types of failures are thoughtless, such as with a worker using a computer program for which a value for a given parameter is so common that just pressing the return key enters that value. A particular patient, however, requires a different value that the operator must calculate. After working hard to calculate the value to enter, the operator, in a moment of mental letdown post-solution, just hits the return key as usually is done accepting the default value, rather than entering the value calculated.

Further along the How It Failed pathway comes an abnormal situation that the operator notes, and

[2]The SL1 was a U.S. Army research reactor in Idaho Falls, Utah. In 1961 an operator removed the central control Rod, apparently intentionally, initiating an explosion.

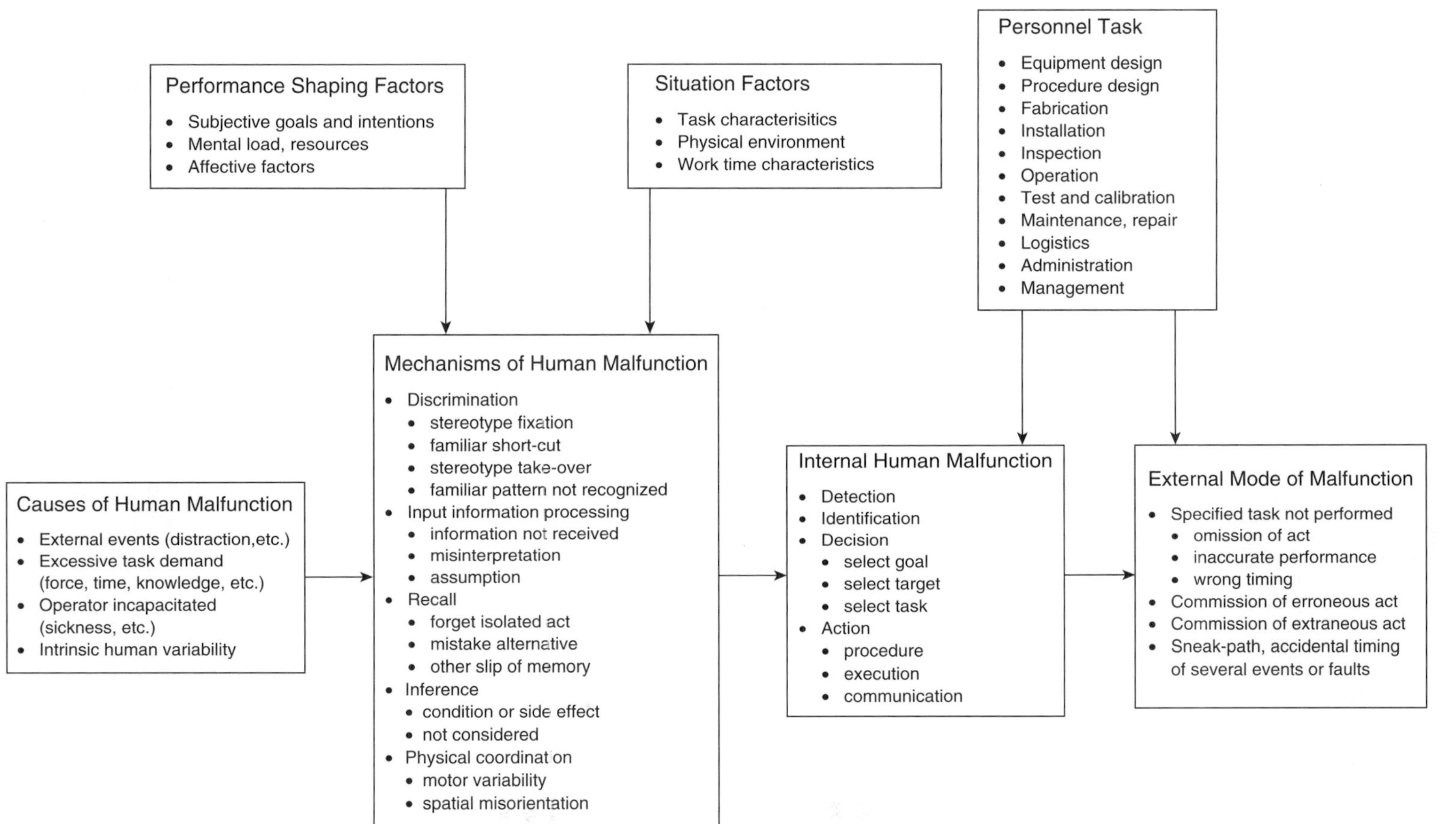

Figure 30–3. Rasmussen's taxonometric organization of failures in human performance. (Rasmussen, 1982, Reprinted with permission of the publisher.)

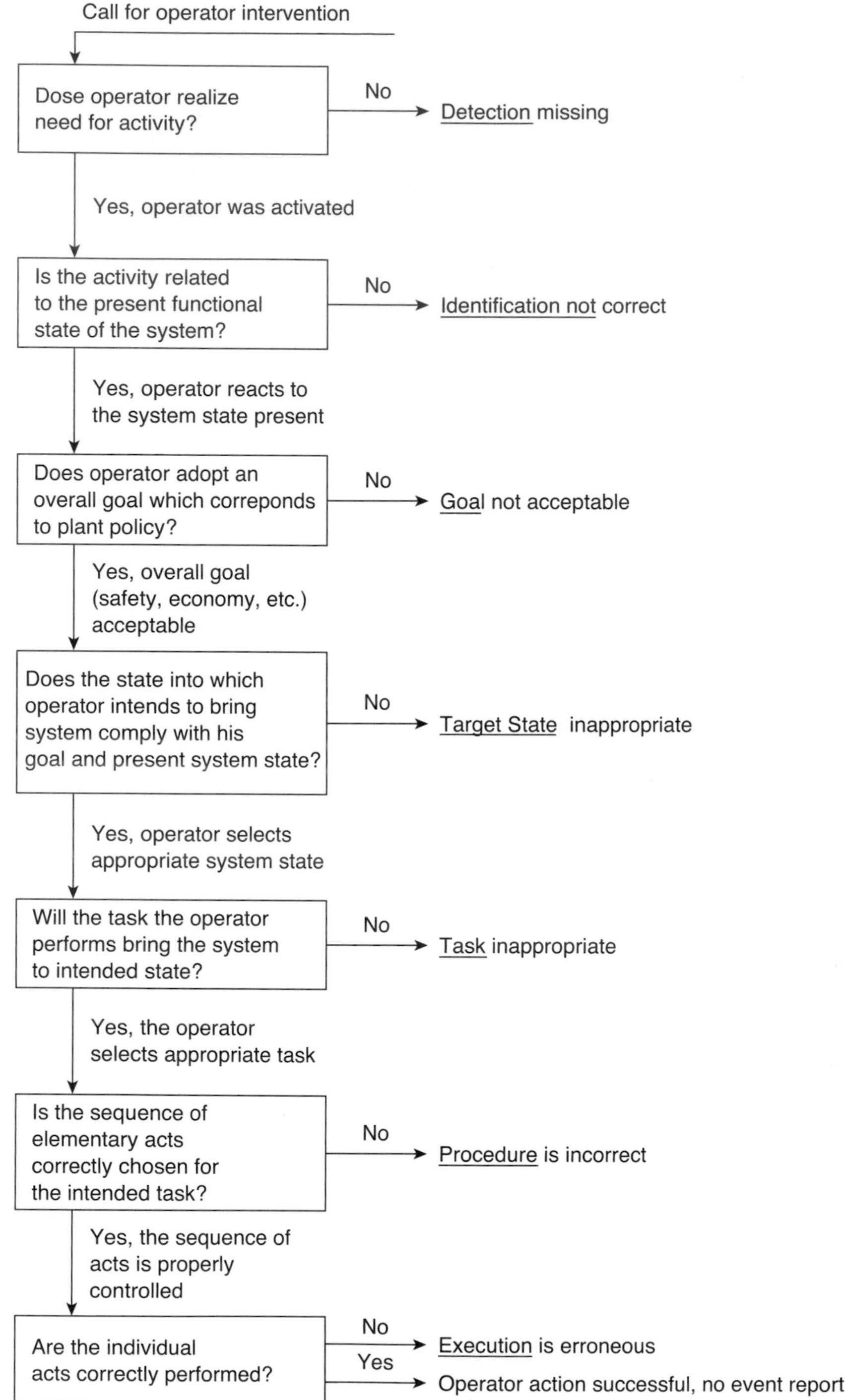

Figure 30–4. Rasmussen's What Failed pathway (Rasmussen, 1982). Reprinted with permission from the publisher.

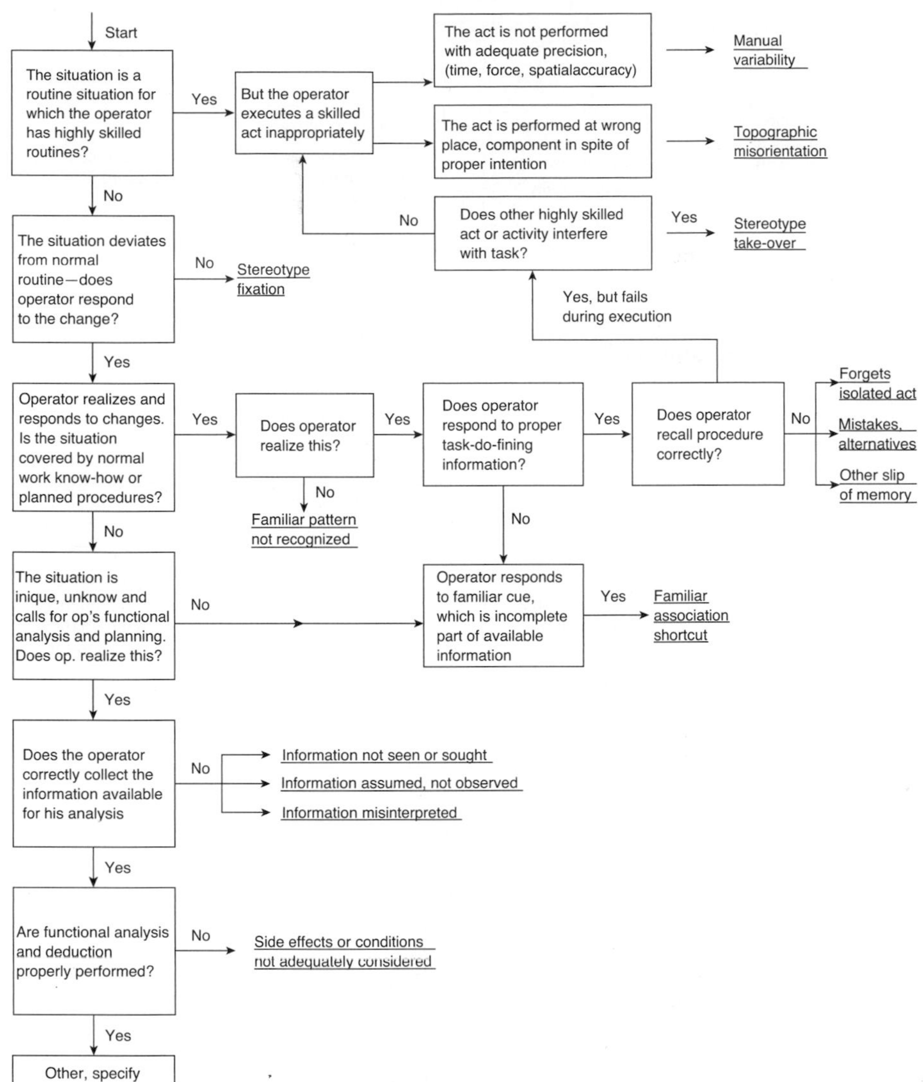

Figure 30–5. Rasmussen's How It Failed pathway (Rasmussen, 1982). Reprinted with permission from the publisher.

one that the operator has procedures to handle. Failure here can result from any of several types of failures. The operator may not recognize the situation as the one covered by the procedures (Familiar pattern not recognized—"to clear the fault on the machine, reset the circuit breaker"). Alternatively, the operator may recognize the situation but forget the seldom-used procedures, select the wrong procedure (also due to infrequent use), or just forget some aspect of what needs to take place. From the box "Does the operator recall procedure correctly?" the arrow "Yes, but fails during execution" leads either to the original execution errors or to a stereotype take-over. Similar to Norman's capture error, the stereotype take-over is where a more commonly used skill preempts a lesser-used one. An example most readers identify with would be planning on picking up milk on the way home from work. One gets into the car with the supermarket in mind and next is aware of pulling into the driveway at home without the milk. The sequence of getting into the car, turning it on, leaving the parking lot at work, and the rest becomes so familiar that we do not usually think about any of the steps—they just flow

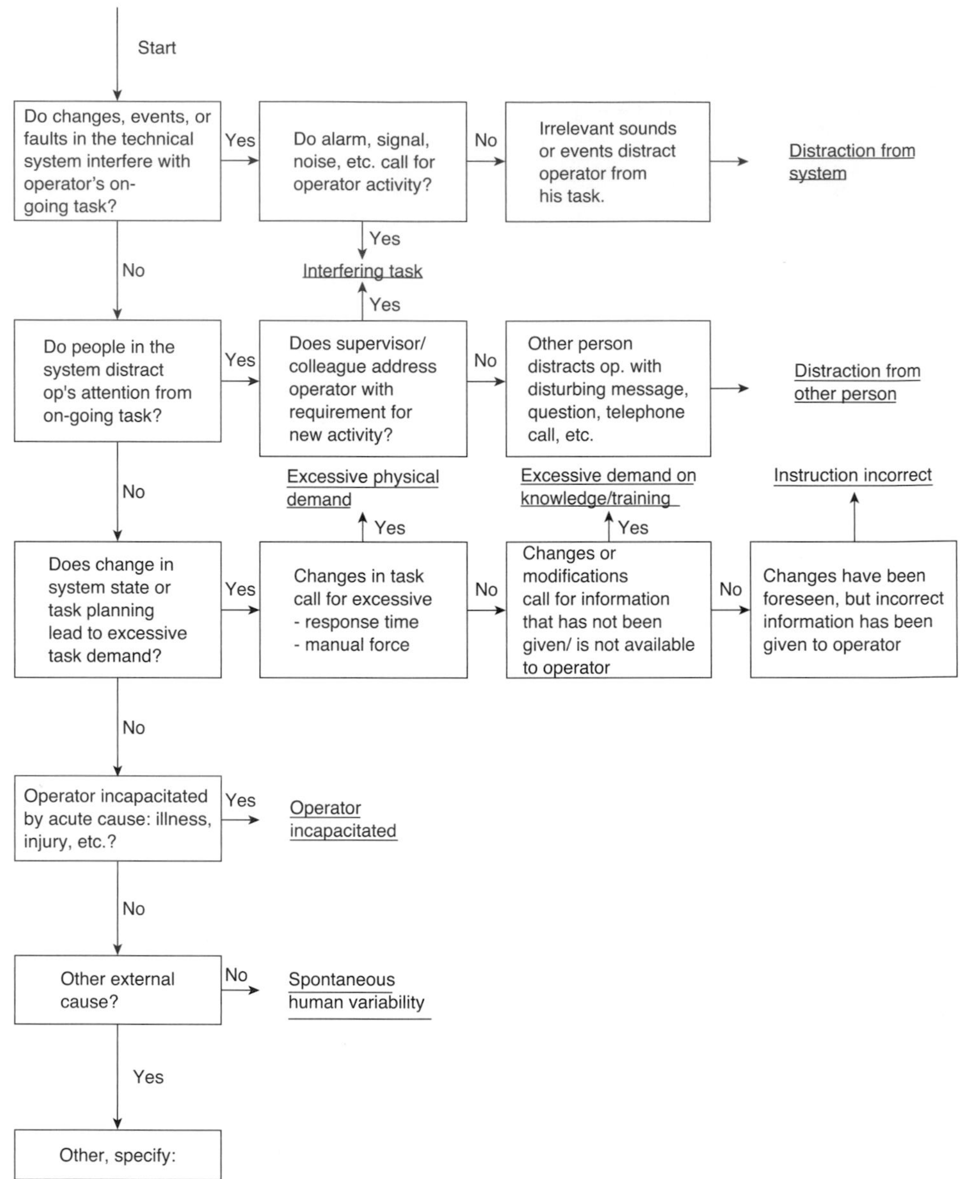

Figure 30–6. Rasmussen's Why It Failed pathway (Rasmussen, 1982). Reprinted with permission from the publisher.

as a well-learned skill: so well learned that they take over if not kept suppressed when changes are required.

Backing up to the "Does the operator respond to proper task-defining information?" box, the "no" alternative indicates that the operator jumps to an action based on premature analysis of the situation. Some aspects of the abnormal condition appear similar to a different, more common situation, and based on the quick match of those aspects, the operator performs the wrong procedures—a familiar association shortcut. This differs from a stereotype fixation, where the operator did not notice that the situation was abnormal. The four classifications, "Stereotype take-over," "Stereotype fixation," "Topographical misorientation," and a "Familiar association shortcut" all share situations in which actions are directed by conditions that do not exist and in which the operator acts without full cognizance.

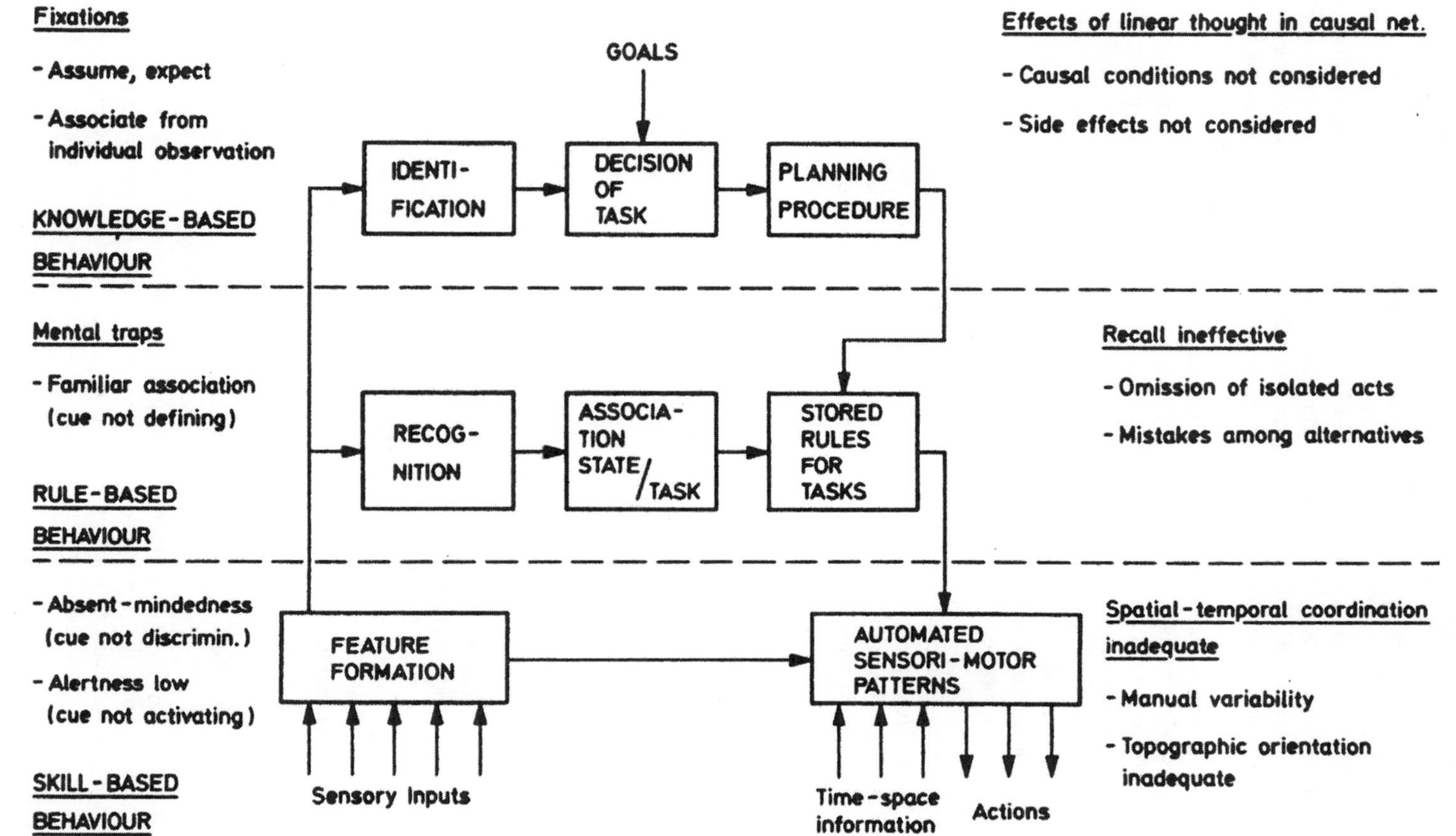

Figure 30–7. Stochastic illustration of categories of human data processes and typical errors. (Rasmussen, 1980). Reprinted with permission.

Information failures come next in line, with various reasons why the operator failed to have necessary information to take the proper actions. Information failures also play significant roles in medical events, particularly when personnel become so rushed that they feel they have no time to gather information. The final classification, "Side effects or conditions not adequately considered," falls in the knowledge-based realm. The situation, following the boxes, is that the operator recognizes the need for action, but no procedures exist to address it, and having gathered the necessary information, the operator must think out the problem. Again, Rasmussen assumes that the operator will work it out correctly unless they fail to consider all important aspects. This is hardly fair, as most unusual situations pose complex problems with interrelated relationships. Indeed, correct conclusions at this point may actually be rare.

The last pathway attempts to identify why the operator failed (Fig. 30–6). Here, the list seems to portray the operator as a mostly effective automaton that would perform the correct act were it not for distractions or incapacitations, or unless the required act exceeds the knowledge, training or physical capacities.[3] The only other category is "Spontaneous human variation."

The layout of the taxonomy's pathways assumes that the first classification encountered is *the* unique classification. That can be argued for the How It Happened pathway. Although an event may result from a failure in identifying the situation and in selecting the correct task, most often the reason the task was inappropriate stems from the misidentification. However, not uncommonly, the Why pathway could easily find multiple causes for the failure. Many events present failures that seem particularly like several of the classifications in the How It Failed pathway. Consideration of the number of classifications describing an event will be picked up in the discussion of the next taxonomy

Analyses Based on the Classifications

In Figure 30–7, Rasmussen (1980) places some of the classifications for failures into the perspective of skill-rule knowledge-based actions. In analyses of nuclear power plant events, Rasmussen (1980) maps the failures into a two dimensional table of the What Failed categories in columns and the How It Failed in rows (Fig. 30–8). "It is necessary to find *what* went wrong rather than *why*," states Rasmussen. This not only explains the relative crudeness of the Why portion of his taxonomy compared to the other two parts but also implies, paradoxically, that the reasons for the events reflect more superficial "causes," whereas the What Failed classifications give insight into the deeper causes. For planning purposes, large flow diagrams following a set of linearly related procedures can be laid out with the potential error modes indicated for each step, marking on the diagram foci where failures have occurred in the past. Figure 30–9 presents a small part of such a chart (Rasmussen, 1982).

Rasmussen (1980) observes that in dealing with abnormal situations, because of memory limitations, understanding a process proves more effective than training for a required skilled procedure. He also notes that most operators have a "point of no return" in making a decision, beyond which observed information becomes unlikely to change their course of action.

Limitations for Use in Medical Settings

Rasmussen's taxonomy as developed over time provides valuable insight into several dimensions of the character of an event. However, the premise of the scenario, that things follow the status quo until an initiating event, while quite appropriate to the environment of a nuclear power plant, applies only to a limited extent in medicine. Of course there are many situations in which the patient will be fine until he or she suddenly takes a turn for the worse, but often the "initiating event" is an action by the health care team. Instead of failing in a reaction, the failure starts the chain of events. For these cases, the questions in the boxes do not always make sense, and the character of erroneous or mistaken actions that originate the event seem unlisted. That being said, patient care forms more of a continuum, with one action always leading to, or meshing with, another. It is rare for a patient to not require some intervention by the staff. In this way, the model applies poorly to medicine. The Why pathway also seems hard pressed to capture the character of medical events, which involve a wider range of reasons for failure.

[3]The title of the work uses the term *human malfunction,* which reviles the Rasmussen's view of the operator.

Table 30–3 Distinctive Features of Failures at Skill-, Rule-, and Knowledge-based Actions

Dimension	*Skill-based Errors*	*Rule-based Errors*	*Knowledge-Based Errors*
Type of Activity	Routine actions	Problem-solving activities	
Focus of Attention	On something other than the task at hand	Directed at problem-related issues	
Control Mode	Mainly by automatic processors (schemata) (stored rules)		Limited, conscious processes
Predictability Of Error Types	Largely predictable "strong-but-wrong" errors (actions) (rules)		Variable
Ratio of Error to Opportunity for Error	Though absolute numbers may be high, these constitute a small proportion of the total number of opportunities for error		Absolute numbers small, but opportunity ratio high
Influence of Situational Factors	Low to moderate; intrinsic factors (frequency of prior use) likely to exert the dominant influence		Extrinsic factors likely to dominate
Ease of Detection	Detection usually fairly rapid and effective	Difficult, and often only achieved through external intervention	
Relationship to Change	Knowledge of change not accessed at proper time	When and how anticipated change will occur unknown	Changes not prepared for or anticipated

From Reason, 1990. Reprinted with permission of the publisher.

Latent Problems—Reason

James Reason, like Jens Rasmussen, is so prolific and such a fundamental force in the field of analysis of human performance that summarizing his works or contributions would require more space than a chapter in this handbook. What will have to suffice is a brief mention of some of the indispensable concepts that Reason cites in his discussion of human performance failures. Anyone doing serious work in the field should become very familiar with Reason's (1990) *Human Error.*

General Approach

Reason (1990) organizes the classification of failures around the mental level of the activity, discussed earlier as skill-based, rule-based, and knowledge-based. Figure 30–10 builds on this model and presents a flow diagram model of the steps leading to an action, called the generic error modeling system (GEMS).

Failures, of course, can take place anywhere along the process, although the characteristics of failures at the three levels differ, as shown in Table 30–3. Table 30–4 lists the potential errors and mistakes by their activity level.

The causes for each type of error are discussed in detail in Reason's book and would take too much space to cover adequately here. The interested reader should take the time to learn the definitions. Reason discusses one important aspect of performance failures not commonly considered. Actions an operator takes occur in a continual flow of inputs and other actions. The history to the time of action strongly affects the operator's performance. For example, because a malfunctioning IV pump began working after the unit was restarted an hour ago, even though the manufacturer's instructions say to retread the line through the unit, a nurse may now first try restarting a failed unit before following the manufacturer's

Table 30–4 Potential Types of Failures That Could Occur According to the Level of the Activity*

Skill-based Performance	
Inattention	*Overattention*
Double-capture slips	Omissions
Omissions following interruptions	Repetitions
Reduced intentionality	Reversals
Perceptual confusions	
Interference errors	
Rule-based Performance	
Misapplication of good rules	*Application of bad rules*
First exceptions	Encoding deficiencies
Countersigns and nonsigns	Action deficiencies
Informational overload	Wrong rules
Rule strength	Inelegant rules
General rules	Inadvisable rules
Redundancy	
Rigidity	
Knowledge-based Performance	
Selectivity	
Workspace limitations	
Out of sight out of mind	
Confirmation bias	
Overconfidence	
Biased reviewing	
Illusory correlation	
Halo effects	
Problems with causality	
Problems with complexity	
Problems with delayed feedback	
Insufficient consideration of processes in time	
Difficulties with exponential developments	
Thinking in causal series, not causal nets	
Thematic vagabonding	
Encysting	

* From Reason, 1990, Reprinted with permission from the the publisher.

instructions. Diagnoses can also be influenced by how recently similar symptoms were seen in a different patient. In addition, working very hard on a problem puts all thought processes into high gear. When a solution is reached, there is a mental letdown, during which vigilance and attentiveness drop temporarily.

Failures, of course, often result from unsafe acts, but unsafe acts do not always lead to failure; if they did, there would be no incentive to perform them. Much of the time, unsafe acts save effort, or at least, cost nothing. However, because of the potential harm, unsafe acts should be discouraged when possible. Figure 30–11 presents a graphical classification of unsafe acts. While the classifications related to the basic error types seem common for failure taxonomies, the bottom classification, Violations, rarely receives notice. In studies of medical events, violations play important parts. Few are sabotage, but many are intentional skipping of a quality assurance step due to a rushed environment (Thomadsen et al., 2003).

Latent Errors

Figure 30–1 showed Reason's Swiss-cheese model of events. Various layers of defenses against failure provide prevention in-depth. However, none are perfect. In fact, some work in reverse and make errors more likely. Decisions made by administrators who never see patients (often referred to as those who operate on the blunt end of the health care

Dec rection of demand	Observation communication	Identification of system state	Goal—strategic decision	Target—tactic system state	Task—determine, select	Procedure—plan, recall	Execution	Various	MENTAL TASK PHASE / ERROR MODE	Distribution across error modes
12		17		5		117	36	2	Distribution across mental 200 task phase	
–	1	1	–	–	–		1	–	Absent-mindedness	3
–	–	5	–	1	–		–	–	Familiar association	6
6	2	–	–	–	–	1	1	–	Alertness low	10
1	–	–	–	–		65	2	–	Omission of functionally isolated act	68
1	2	–	–	1	–	12	1	–	Omision—others	17
–	1	–	–	–	–	1	9	–	Mistakes among alternatives	11
–	4	3	–	–	–	3	–	–	Expect, assume—rather than observe	10
–	–	1	–	1	–	13	–	–	Side effect not adequately considered	15
–	–	5	–	–	–	15	–	–	Latent conditions not adequately considered	20
–	–	–	–	–	–	–	10	–	Manual variability, lack of precision	10
–	–	–	–	–	–	–	10	–	Topographic, spatial orient weak	10
3	2	2	–	2	–	7	2	2	Various; not mentioned	20

Figure 30–8. An analysis of frequency of errors as functions of their classifications by Rasmussen's What Failed and How It Failed pathways. (Rasmussen, 1980). Reprinted with permission.

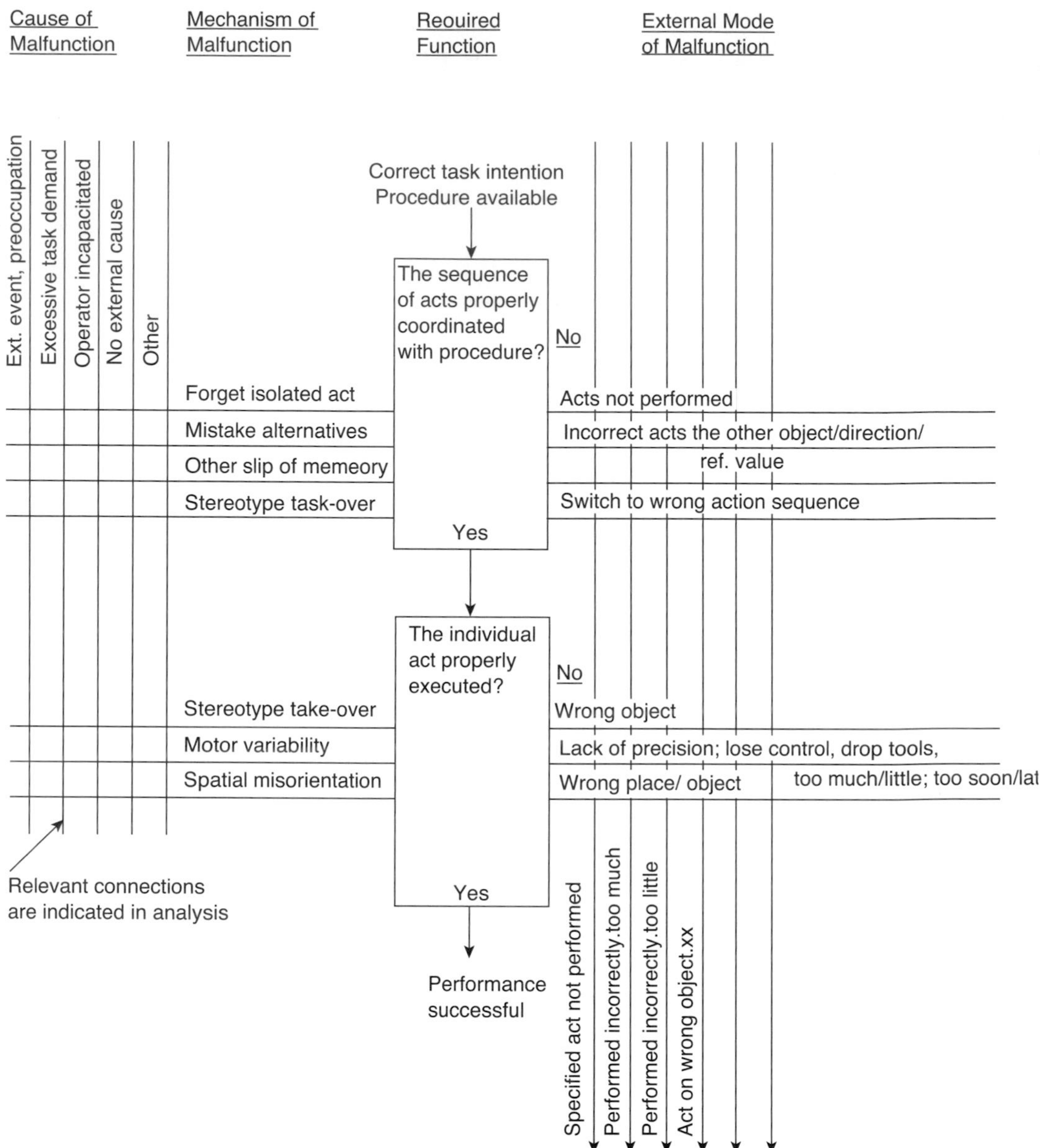

Figure 30–9. Part of a flow chart plotting likely errors based on information in a database organized according to Rasmussen's taxonomy (Rasmussen, 1982). Reprinted with permission from the publisher.

system) can set the stage for failure, for example, by frequently depending on poorly trained temporary nursing staff for regular coverage. The environment affects performance, such as computer screens that have a narrow range of angles from which the contents can be seen. Background holes in the defense comprise the "latent errors," or poor practices that weaken the defense. Most of the time such holes in the defense against failure cause no problem. But an unsafe act by one of the poorly trained nurses combined with the inability to read the computer monitor at the critical time can allow the unsafe act to propagate as an error and cause an event. Despite the importance of eliminating latent errors, the taxonomy does *not* provide classifications for such. Latent errors must be found by back-projecting the human performance failure classifications over several events.

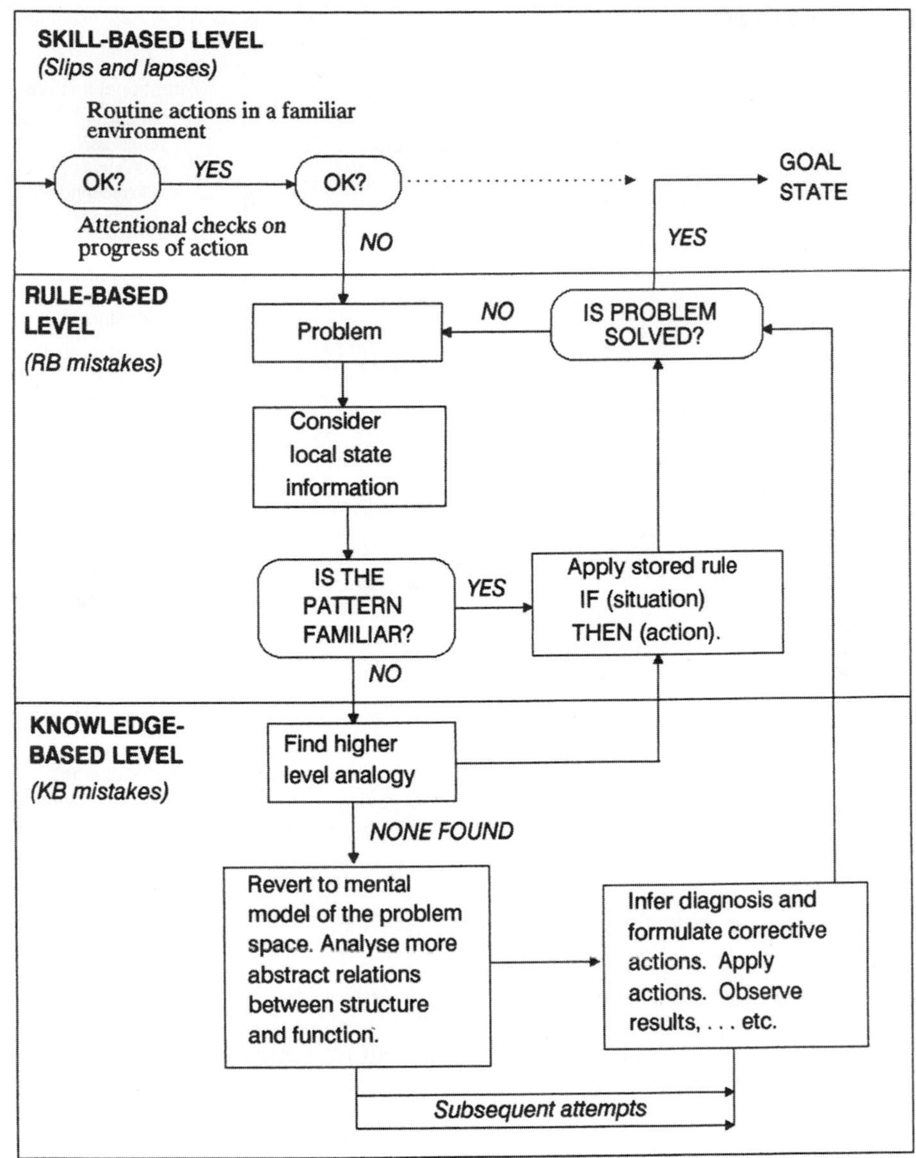

Figure 30–10. Flow diagram depicting the steps taken to determine actions to be taken, according to the generic error modeling system. (Reason, 1990). Used with permission from the publisher.

Limitations

The work of Reason forms a very comprehensive body for classifying human performance failures and assessing the causes, usually the psychological or behavioral causes. The relationships to the latent errors remain more difficult to ferret out, and to that end, the lack of organizational classifications may leave the reader with little guidance. On the whole, however, the book by Reason (1990) presenting this approach to failure analysis provides an excellent discussion of the concepts and should be required reading. Unfortunately, few persons in medical organizations will have the time or inclination to study the book (as delightful reading as it is).

A Systemic Approach—The Eindhoven Taxonomy (Tjerk van der Schaaf)

Van der Schaaf developed his taxonomy in reference to a chemical plant (van der Schaaf, 1992). Similar to the environment of the nuclear power plant, for the

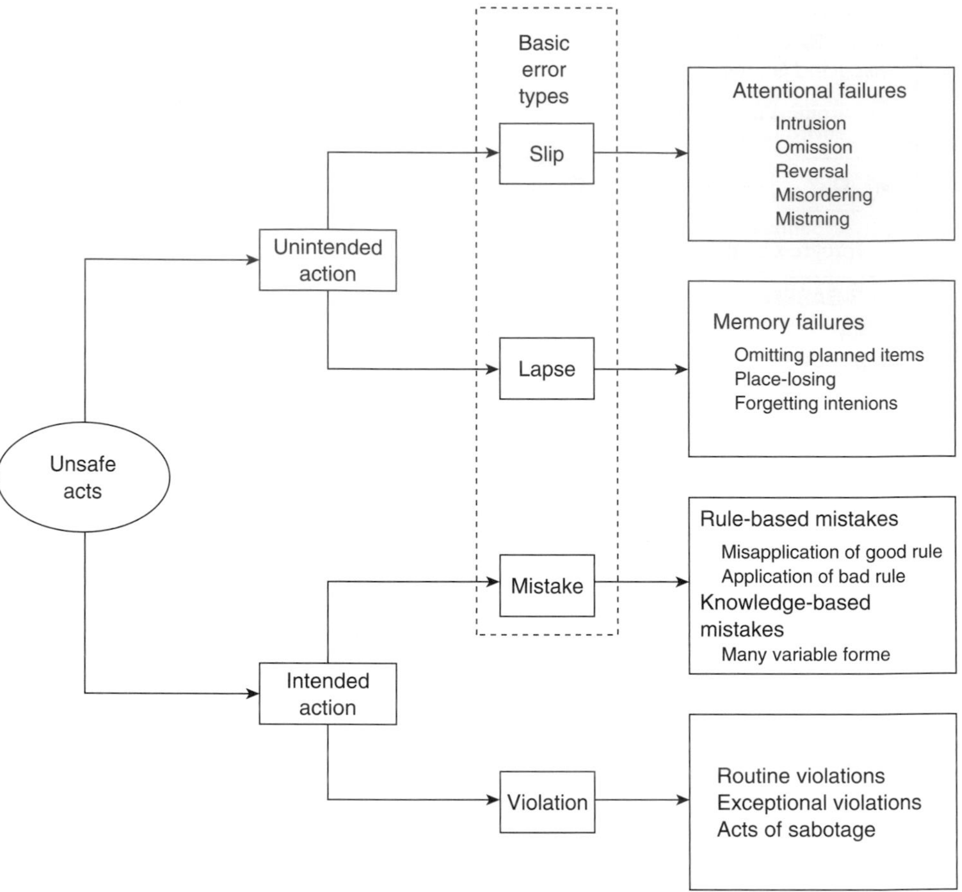

Figure 30–11. Unsafe act classifications. (Reason, 1990) Reprinted with permission from the publisher.

most part the plant does its job fairly autonomously, with human intervention required at specific times or when something goes wrong. However, van der Schaaf takes a more global vantage.

Basis of the Approach and Underpinning Assumptions

This taxonomy, originally called PRISMA, for Prevention and Recovery Information System for Monitoring and Analysis, but reworked for medical applications and renamed SMART (System for Monitoring and Analysis in Radiotherapy—but for wider application in medicine than just that specialty; Koppens, 1997). The overall organization follows Figure 30–12. Van der Schaaf (1992) stresses that for a study of events, near-misses, that is, events that include successful recovery, provide the most useful information. Although studying failing events gives insights into what went wrong, a study of near-misses reveals information on what could prevent a failure following an error.

SMART contains a single pathway with three possible major subpaths (see Fig. 30–13). Beginning at "Start," the first decision is whether technical problems entered into the event, and if so, the subpath gives selections for

- Design
- Construction
- Material problems (i.e., something broke)

The next subpath considers organizational factors:

- Knowledge transfer—inadequate training of the personnel to perform the function that failed.

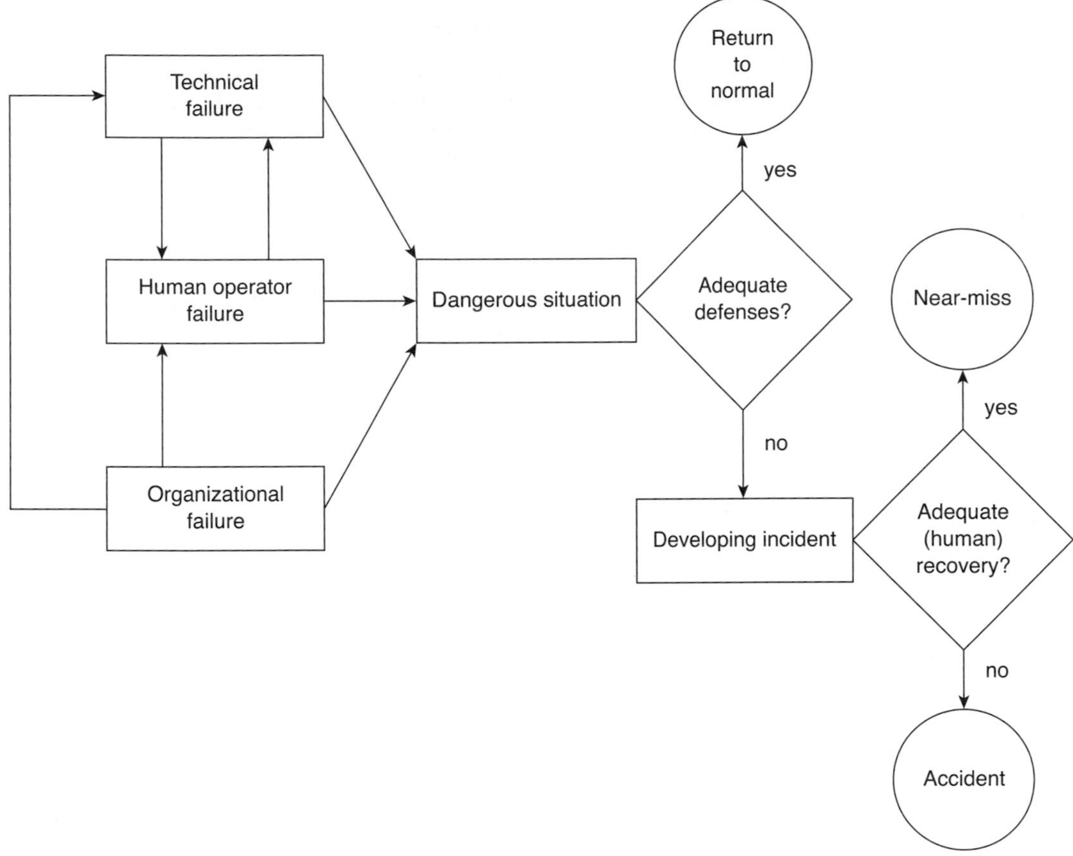

Figure 30–12. An overview of the Eindhoven failure taxonomy. (van der Schaaf, 1992)

- Protocols—inadequate procedures or policies dealing with the situation surrounding the event ("too complicated, inaccurate, unrealistic, absent, poorly presented"; van Vuuren, Shea, & van der Schaaf, 1997).
- Management priorities—management decisions that resulted in situations enabling failures.
- Culture—a pervasive attitude in the organization that places little value on steps to ensure quality performance.

A large subpath for human behavior comes next, subdivided into the three bases for action:

- Knowledge-based, which only contains a self-referential category.
- Rule-based, including
- Qualifications—a mismatch in the operator's knowledge, training, and abilities to perform the task.
- Coordination—a lack of coordinating actions between team members.
- Verification—a failure to completely assess the situation before taking action.
- Intervention—a failure during the action resulting from either poor planning or erroneous execution.
- Monitoring—a failure during the execution or immediately after due to the lack of observing the progress of the task.
- Skill-based, including
- Slips—a fine motor failure.
- Tripping—a large motor failure.

"Patient-related Factors" form the last classification, defined as "failures related to patient characteristics

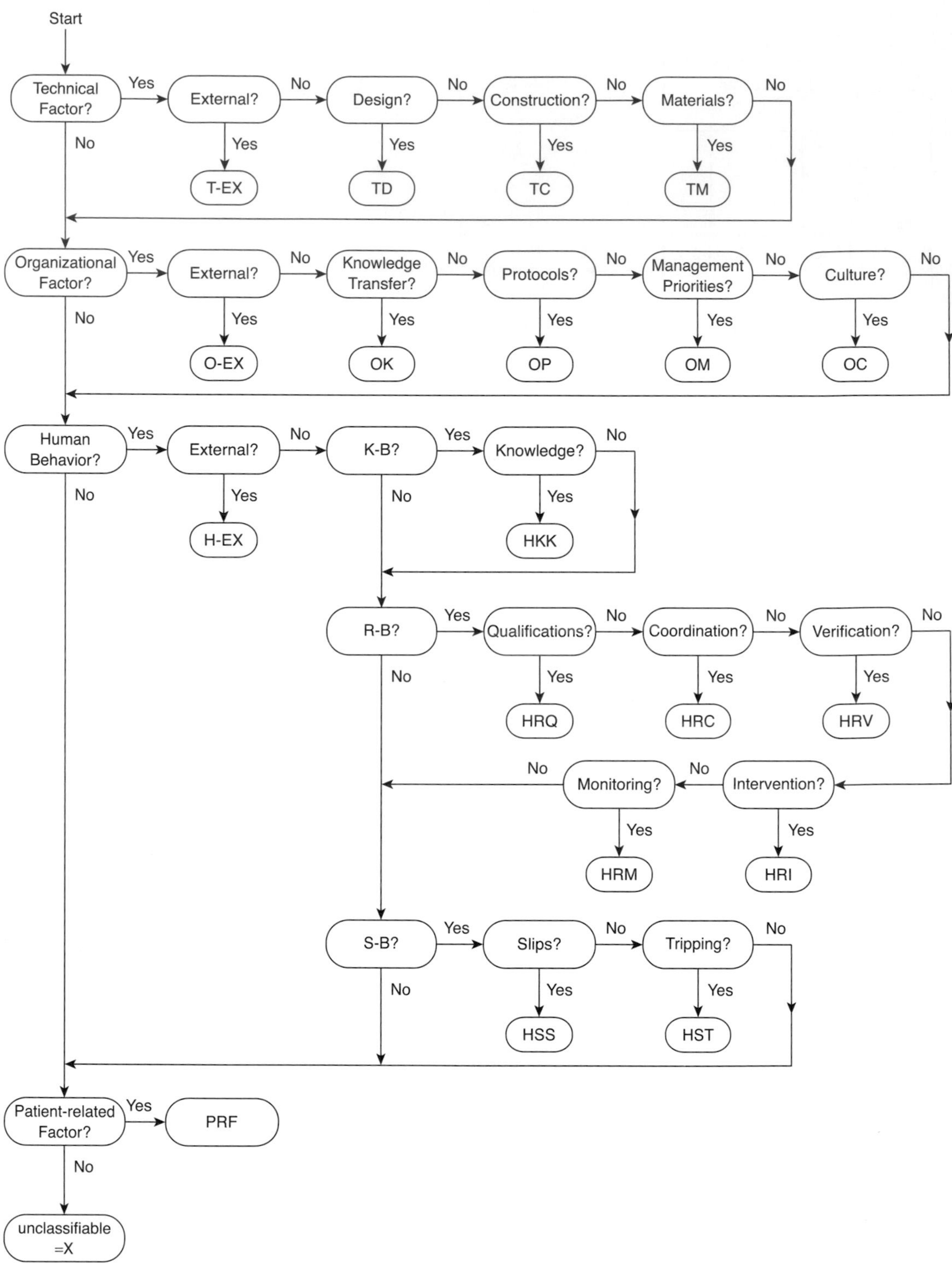

Figure 30–13. The classification pathway for SMART. (Koppens, 1997)

which are beyond the control of staff and influence treatment." (van Vuuren et al., 1997)

Each of the subpaths begins with an "External" classification, each of which applies to failures beyond the control and responsibility of the organization.

Labyrinth or Pinball Approaches

For the designers of the taxonomy, the layout serves more than just a pathway to lead the investigator through the classifications. The order of the classifications also plays an important role. As designed, the first classification encountered becomes *the* classification for the event. For this to make sense, great care goes into the organization of the taxonomy. (This is some of the "ordering" that was so important in the definition.) The rationale offered by the creators explains that any technical problems that result in events should be fixed first. In general, these are the easiest of the categories to address, and if technical problems contributed to an event, removing those problems likely would prevent future events. After technical problems come the sources of latent problems in the system. Latent problems reflect organizational weaknesses, so the organizational categories come second. Only after addressing the technical and organizational problems should the investigator look for human performance causes. This hierarchical approach could be called *labyrinth*, after the game where the player moves a steel ball through a maze with holes in the board. A player's score depends on the distance reached through the maze when the ball falls through a hole.

An alternative to playing labyrinth is *pinball*. In a pinball game, the ball runs the course, falling into holes and racking up points, only to pop out and continue on the course. Only after dropping in many holes (if one is good or lucky) does the ball disappear from play. Pinball players argue that following the whole pathway for every event, noting each classification that contributed to the event, yields more information that could assist in correcting problems and preventing future events. The argument continues that, although addressing the technical problems first and organizational problems next may be the ideal approach, in real organizations, particularly medical organizations, it may be easier to effect changes in the human behavior realm, and thus more effective and efficacious. In practice, both approaches have merit and should be considered in addressing events. During analysis, noting *the* classification for an event along with the other applicable classifications may provide the best set of clues for correcting problems.

In medical settings, the first internal classification—technical design failures—often fall outside of the facility's control. Equipment comes from vendors who control the design, and frequently, only one design exists. Thus, for many events, even the labyrinth players must dig deeper into the pathway for some classification that the facility controls.

Some would argue that the "Patient-related Factors" classification should never find use. Systems should be in place anticipating possible changes in, or actions by, the patient, and if the patient in some way "causes" an event, the real cause lies somewhere in the procedures used with that patient. Consider the case where a patient receives radiotherapy for a bronchial cancer using a radioactive source in a catheter passed through the nose and residing in the bronchus for a day or so. Several events have been recorded where the patient, accidentally or intentionally, removes the catheter before completion of treatment. Although this could be classified as a Patient-related Factor (or action), in truth, the radiation oncologist probably did not fasten the catheter in place well enough to prevent such intervention on the patient's part (an Intervention classification).

Applications

The Eindhoven system has been used in radiotherapy (Koppens, 1997), anesthesiology (Anderson, 1997) and blood bank (Kaplan, Battles, van der Schaaf, Shea, & Mercer, 1998) events. The Kaplan article studied 503 event reports and verified that blood bank events usually resulted from multiple causes, and that latent causes were more important than expected. They also observed that the event classification pattern was similar to that seen in industry. The similarity of events in blood banks to industry may not predict that same finding for other medical setting because blood banks have little interactions with patients and function to a great extent as an industry. However, other studies using the Eindhoven system in patient-intensive settings have also seen this similarity (Thomadsen et al., 2003). Kaplan also noted that the system produced a high degree of agreement among those performing the analyses.

Directive Properties and Limitations

Van der Schaaf (1992) provides a table (similar to Table 30–5) that looks at potential methods of

TABLE 30–5. Suggested Remedial Actions Based on the SMART Failure Classifications

Category	*Equipment*	*Procedures*	*Information and Communications*	*Training*	*Motivation*
TD	X				
TC	X				
(TM)		?			
OK				(X)	
OP		X			
(OM)					
(OC)					
HKK			X		NO!
HRQ				X	
HRC				X	
HRV				X	
HRI				X	
HRM				X	
HSS	X			X	
HST	X				NO!
PRF					NO!

Note. Classification/Action matrix, giving the suggested arenas from remedial actions based on the failure classification in SMART – adapted from van der Schaaf (1992) for PRISMA. Parentheses around an X indicate extrapolation from PRISMA; parentheses about a category indicate that solutions may be elusive.

addressing problems based on the Eindhoven classification system. The table emphasizes that, where technical problems exist, fix them; if procedures are lacking, make them; if a human operative failed in some way, train him or her; and if knowledge was lacking, provide the necessary information, possibly establishing new lines, methods, or modes of communications. Problems with motor skills, fine or gross, require equipment solutions to obviate the human lacking. As the table emphatically states, under no situations do motivational approaches deliver desired results. Posters and rallies seldom make any headway in promoting safety, nor does a policy of picturing the correct result before action or thinking good thoughts.

The usefulness of the suggested correction action, "training," comes under fire. Studies have shown (Rook, 1965; Harris & Chaney, 1969) that retraining human operatives that have already been trained serves no purpose. Quite the contrary, persons sent for retraining usually feel punished, as in the antiquated approach to events of "blame and train." Most medical personnel go through rigorous training in their jobs before becoming clinically responsible for patient care. Failures come from inattention, misunderstanding, or any of the other causes discussed earlier. Retraining seldom has a role in preventing future occurrences. Alternatively, new procedures sometimes do begin before all personnel have been *adequately* trained. In such cases, remedial training, that is, training that should have been completed previously, still is a must.

The problem with Table 30–5 lies in the general nature of the suggestions. Just knowing the general heading of the appropriate type of action does not guide the investigator to the solution—it only limits the universe of solutions.

The reader should note that Table 30–5 offers no suggested solutions for CULTURAL problems. In general, if the organization places little value on error reduction and safety, not only will corrective actions likely be futile, but the organization would not be likely to have an investigator trying to find solutions in the first place.

Scope

Basis of the Approach and Underpinning Assumptions

The System for the Classification of Operator Performance Events (SCOPE) combines the descriptive information on the stage in an action sequence in which an event occurred with identifying the nature of the failure leading to the event (Kapp & Caldwell, 1996). The structure of SCOPE follows the sequence of actions required to react to the need for activity: Detection, Interpretation, Rule selection *or* planning, and Execution. SCOPE also notes failures in linkages between tasks and cross-unit flow of information between organizational teams. The organization and approach build strongly upon the works

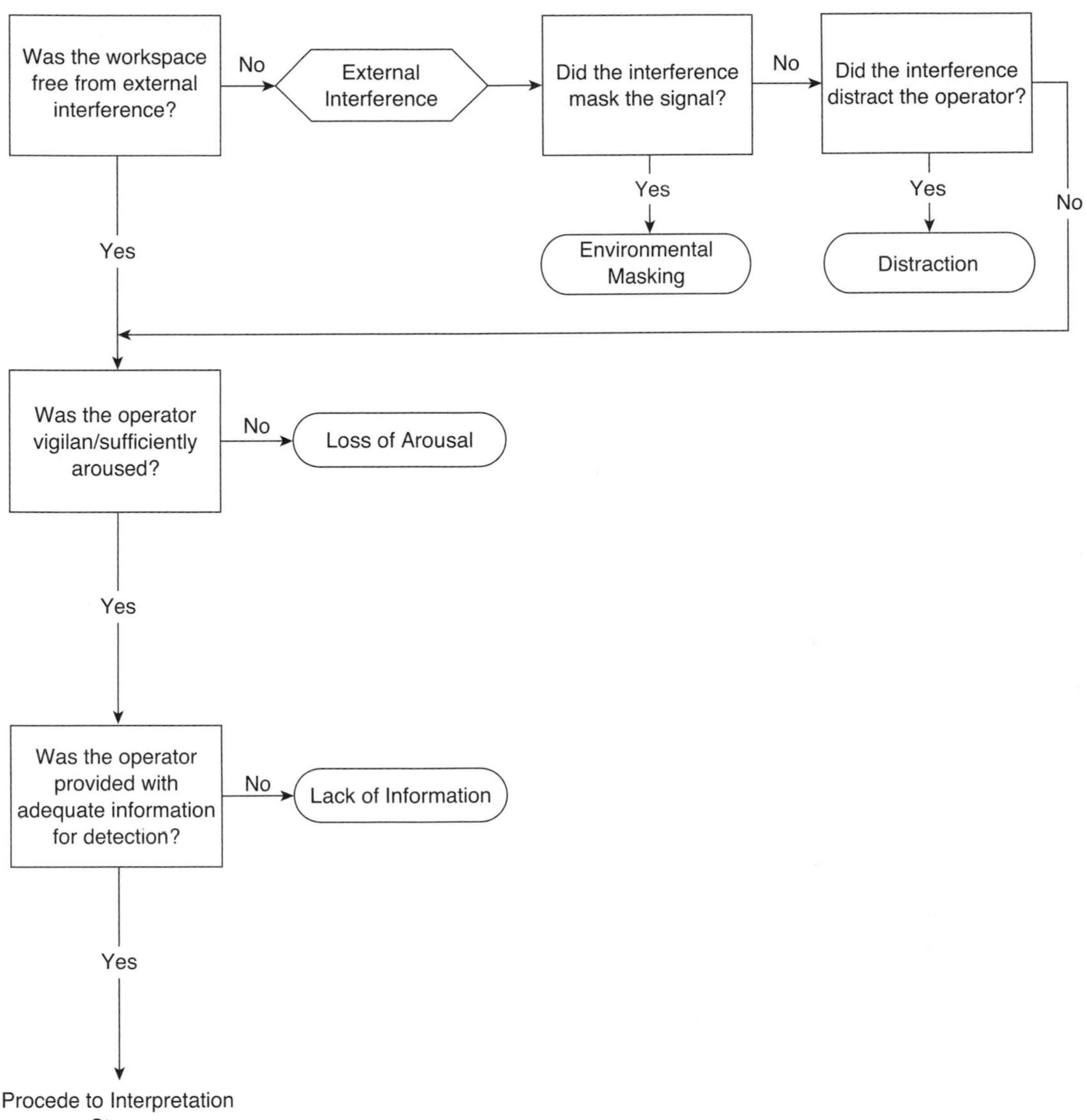

Figure 30–14a. SCOPE failure classification pathway. See also Figures 30–14b, c, d, and e on the following pages.

of Rasmussen (1980, 1982), Reason (1990), and van der Schaaf (1992) cited earlier. As in both of those taxonomies, the classification comes through following a pathway, shown in Figures 30–14a through 30–14e.

The first arena for human performance failures comes with detection of the need for action. Failure of detection may occur in one of four ways, a characteristic of SCOPE for each of its pathways. For each, the classification of the nature of the failure also serves as the designation of the cause. The classification Loss of Arousal implies that the operator was less than alert.

For the detection of the need for action to translate into correct actions, the situation must be interpreted properly. Interpretation failures likewise fall

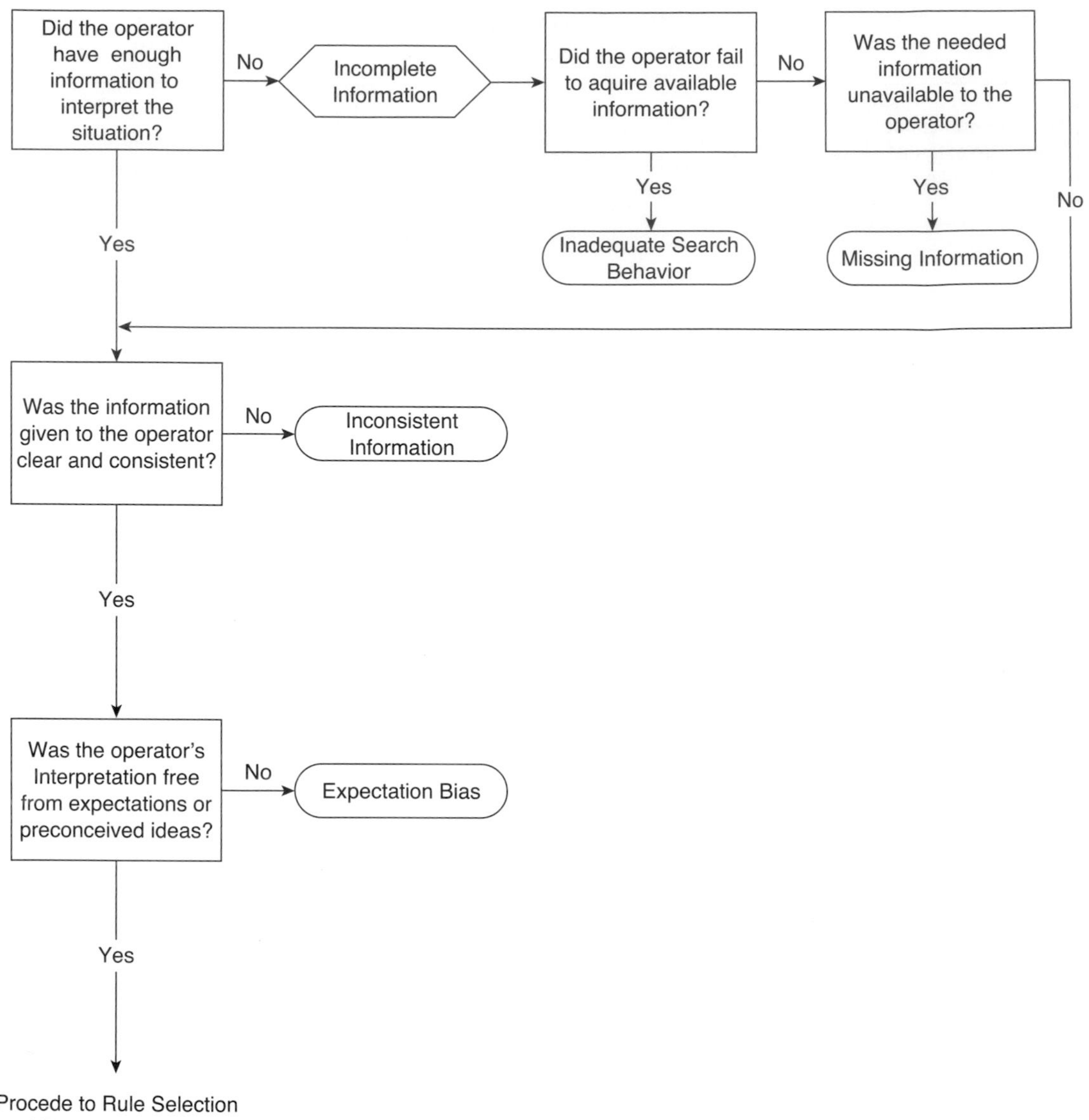

Figure 30–14b.

into four categories. Three of the categories relate to insufficient information to correctly interpret the detected situation: the operator failed to gather sufficient information, all the necessary information was not available, or the information was inconsistent. The final classification, Expectation Bias, plays an important part in a large proportion of medical events (e.g., Thomadsen et al., 2003). An operator exhibits an expectation bias when interpreting a situation as a situation that is expected rather than the actual situation, in spite of receiving information to the contrary. Usually this manifests as the operator assuming that things are normal and making models to explain away the detected indications. Expectation biases relate closely to the Stereotype Fixation, Topographical Misorientation, and Familiar Association Shortcut from the Rasmussen (1982) taxonomy.

Based on the interpretation of the situation, the operator determines a course of action. This decision is based first on selection of a rule that

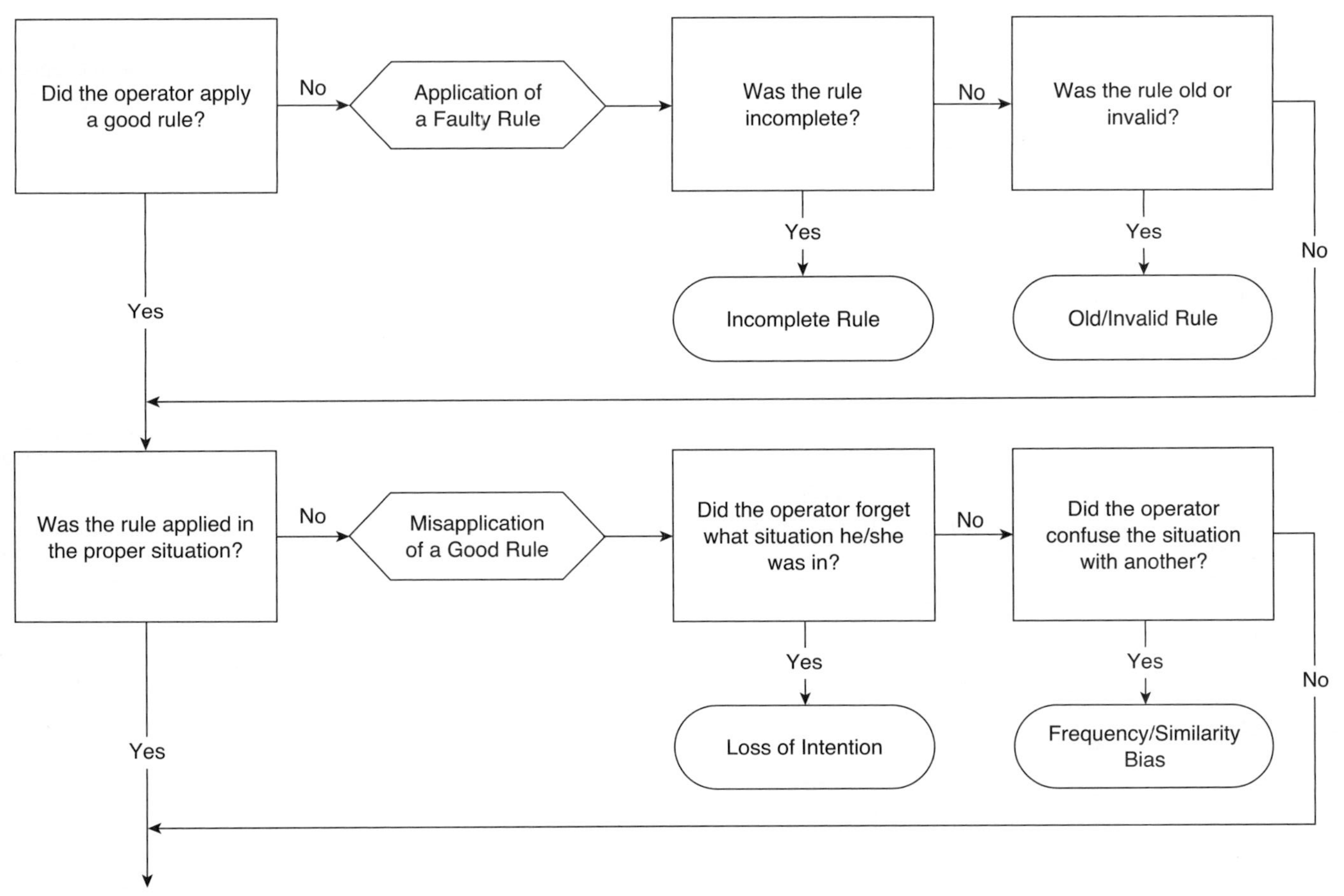

Figure 30–14c.

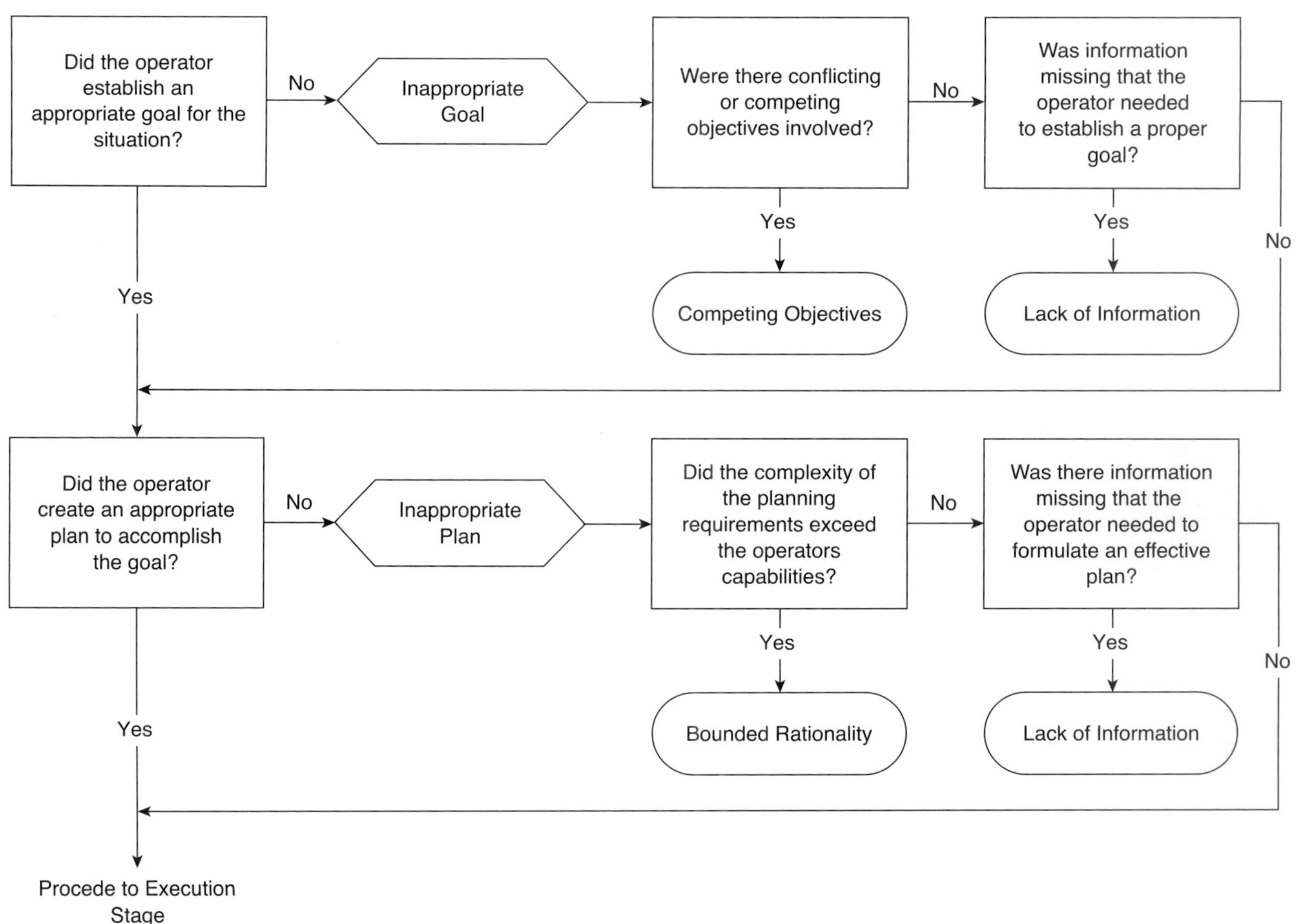

Figure 30–14d.

determines the schemas to activate (treating our operator as an automaton again), if such a rule can be recalled or referenced. However, the operator may select a rule that did not completely cover the situation, an old or outdated rule, or a bad rule—one that never works. Any of those errors lead to failure, as does selecting a good rule at the wrong time, which can happen due to Loss of Intention (not loss of attention), forgetting the actual situation, or confusing the situation with a different, usually more common one (Frequency/Similarity Bias). These last two categories share characteristics with Rasmussen's Familiar Association Shortcut, Topographical Misorientation, and to some extent, Familiar Pattern Not Recognized. Determination of the correct rule leads directly to the execution phase.

Failure to find a rule leads the operator to a knowledge-based action and the Decision and Planning pathway. Failure to develop an appropriate plan comes from either the application of inappropriate goals (due to competing or conflicting objectives or from the lack of information about the goal) or from a failure in the planning proper. An operator may not be able to generate a good plan because the problem may be too difficult for the training and experience level of the operator (Bounded Rationality) or because some necessary information may be missing. Generation of a good plan would lead to the last pathway, Execution.

If all the steps toward actions to this point were failure free, the errors must lie in the execution. Again, the pathway offers four causes for failures. The first is some external interference—the operator would have done fine except something prevented execution. Failure more often results from deviation from a plan, because of entrapment by another rule or plan (Faulty Triggering) or forgetting what to do. Finally, the operator can simply bungle the execution.

The layout of SCOPE intended its use with the labyrinth approach, although its developers later recognized that, because many failures often contribute to events, the taxonomy also lends itself to a pinball approach.

Lack of Information appears as a classification four times in the complete pathway, and Incomplete Information once. The frequency of the classification Incomplete Information indicates the major role receiving correct and complete information plays in error reduction. It is interesting, however, that each of these information deficits actually differ, showing the pervasive nature of information transmission in complex situations.

Limitations

SCOPE, as does the Rasmussen taxonomy, tends to view the operator somewhat as a reacting machine. Although more cognizant of the decision-making function of medical personnel (as seen by the existence of the Decision and Planning pathway) than Rasmussen's, the choice of failure classifications sometimes fails to reflect the richness of errors humans make. The taxonomy also focuses on human performance and, while considering environmental contributing factors, does not code for organizational factors. Such organizational problems may be inherent in the model, for example, a classification of Bounded Rationality automatically raises the question "Why was the operator in the position to be faced with problems beyond their means in the first place?"

Medical Taxonomies

National Coordinating Council for Medication Error Reporting and Prevention

The National Coordinating Council forMedica-tion Error Reporting and Prevention (NCCMERP, 1999), an independent body sponsored by some 25 organizations with interest in medical safety, assembled a reporting system for medical events. This system falls somewhere between an error coding system and a taxonomy (a typical taxonometric problem of trying to fit something into a category). The list of classifications spans 19 pages, and events receive scores in several categories. For example, the initial category gathers information about the event, such as time of day and type of medical facility at which the event initiated as well as where it perpetuated, followed by a free text description of the event.

The next field classifies the level of harm to the patient, followed by a very long section on the products involved. The attention to products makes sense because most errors involve medication administrations, and this taxonomy was mostly designed to address those errors. Personnel involved in the event comes next, and then what type of error occurred.

Up to this point, the system only dealt with objective, noncontroversial facts about the event. Now, however, the attention turns to deeper questions about causes. The first questions do not probe too deeply, for example, offering choices such as "NAME CONFUSION/Proprietary (Trade) Name

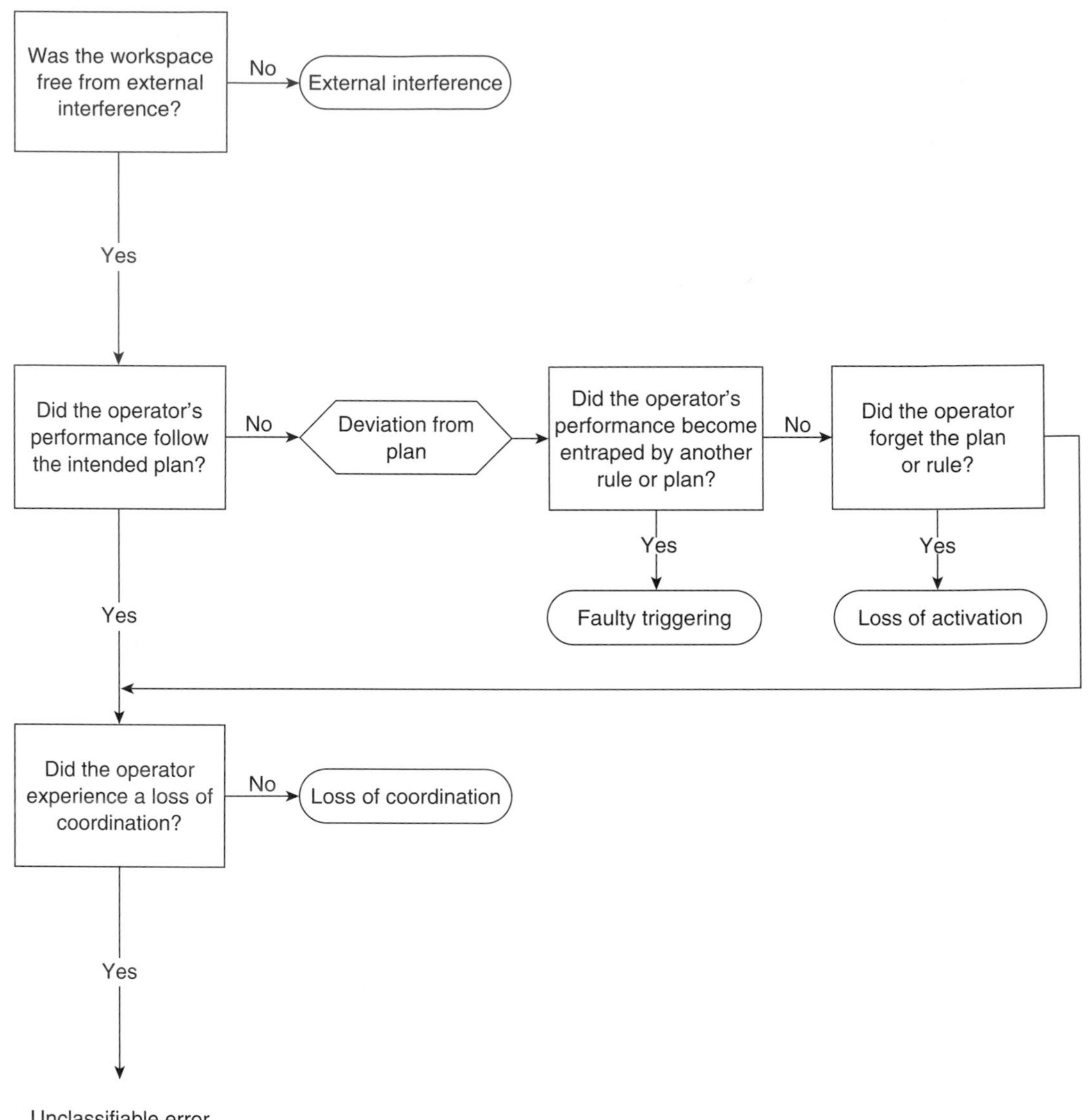

Figure 30–14e.

Confusion; Suffix confusion; Prefix confusion; Sound-alike to another trade name; Sound-alike to an established (generic) name; Look-alike to another trade name; Look-alike to an established name," and so on. The next section looks at the human factors involved in the event. Table 30–6 lists the possible codes in that section. Except for the first (knowledge deficit) and the last three (stress, fatigue, and confrontation), the classifications for human errors also deal only in objective descriptions of actions rather than causes for, or nature of, the actions.

The final category of classification considers contributing factors, or system-related factors (Table 30–7). This last section attempts to provide information on the latent causes of the events. Some of the codes cover broad classifications, such as "Policies and Procedures," while others, for example, "Preprinted medication orders," form narrow, very specific codes.

TABLE 30–6. Human Factors Codes for the Taxonomy of Medication Errors.*

87	Human Factors
87.1	Knowledge deficit
87.2	Performance deficit
87.3	Miscalculation of dosage or infusion rate
87.4	Computer error
87.4.1	Incorrect selection from a list by computer operator
87.4.2	Incorrect programming into the database.
87.4.3	Inadequate screening for allergies, interactions, etc.
87.5	Error in stocking/rsestocking/cart filling
87.6	Drug preparation error
87.6.1	Failure to activate delivery system
87.6.2	Wrong diluent
87.6.3	Wrong amount of diluent
87.6.4	Wrong amount of active ingredient added to the final product
87.6.5	Wrong drug added
87.7	Transcription error
87.7.1	Original to paper/carbon paper
87.7.2	Original to computer
87.7.3	Original to facsimile
87.7.4	Recopying MAR
87.8	Stress (high volume workload, etc.)
87.9	Fatigue/lack of sleep
87.10	Confrontational or intimidating behavior

*Used with permission. (NCCMERP, 2000)

Assumedly, history plays a part in the list, where preprinted medication orders must have been a common problem during the formative period for this taxonomy.

The NCCMERP system orders and organizes the codes, and in that way forms a taxonomy. However, the main thrust seems to be for gathering information for databases, even in the causative sections, making it more of an error coding system. Given the number of major organizations involved with the organization, the system should find widespread use and facilitate generation of large amounts of data, crossing institutions. Such information will be indispensable in improving patient safety.

Edinburgh Taxonomy

The Edinburgh Error Classification System attempts to combine the aspects of both the behavioral (what the operator did) and cognitive (what lead the operator to do it) approaches to classifying types of failures (in the current incarnation, Busse & Wright, 2000). The system also straddles the line separating an error coding system and a taxonomy. The authors state that the classifications focus on proximal cause rather than root cause. The investigation that leads to the classification considers "*what* happened in an incident occurrence (e.g., wrong drug administered), *how* it happened (e.g., drug confusion), and most importantly, *why* it happened (e.g., illegible handwriting on drug container . . .)." In other taxonomies, the examples given could just as well be called the type of event instead of what happened, what happened instead of how it happened, and how it happened instead of why. The distinctions merely emphasize that common meanings of words leaves too much to interpretation to be used without further definition in this specialized field needing fine distinctions between concepts.

Table 30–8 presents the current classification system, expanded from an earlier system used in a study of intensive care unit errors (Wright et al., 1991). As Busse & Wright (2000) point out, the initial classifications mostly involved performance shaping factors, whereas the more recent additions concentrate on "domain-specific, behavioral" categories that, instead of probing "cognitive mechanisms or organizational influences" denote "where in the task sequence a step had been omitted, or had been executed wrongly." Table 30–9 reorganizes Table 30–8 into proximal causes (those immediately leading to the event) and distal causes (those that set the stage for the event.) The developers of the Edinburgh system do not try to capture the latent causes in their classifications, but rather draw that information out from a RCA of the event.

TABLE 30–7. Contributing Factors Codes for the Taxonomy of Medication Errors.*

90	Contributing Factors (Systems Related) [Select as many items as are applicable from this section.]
90.1	Lighting
90.2	Noise level
90.3	Frequent Interruptions and distractions
90.4	Training
90.5	Staffing
90.6	Lack of availability of health care professional
90.6.1	Medical
90.6.2	Other allied health care professional
90.6.3	Pharmacy
90.6.4	Nursing
90.6.5	Other
90.7	Assignment or placement of a health care provider or inexperienced personnel
90.8	System for covering patient care (e.g., floating personnel, agency coverage)
90.8	1 Medical
90.8.2	Other Allied Health Care professional
90.8.3	Pharmacy
90.8.4	Nursing
90.8.5	Other
90.9	Policies and procedures
90.10	Communication systems between health care practitioners
90.11	Patient counseling
90.12	Floor stock
90.13	Pre-printed medication orders
90.14	Other

* Used with permission. (NCCMERP, 2000)

The Edinburgh system has proven useful in analyses of medical events (Wright et al., 1991, and Busse & Johnson, 1999), but because of the specificity of several of the categories, the system may be a little narrow for widespread application through a health care organization. The system also leaves much of the determination of remedial actions to the experience of the investigator.

Nursing Practice Breakdown Research Advisory Panel

The Nursing Practice Breakdown Research Advisory Panel of the National Council of State Boards of Nursing, Inc., investigated 21 events to study the design of a taxonomy for use with nursing-related events (Benner et al., 2002). Although the taxonomy remains under construction, the preliminary structure appears in Table 30–10.

The classification "Lack of agency" refers to failure of the nurse to act as an advocate for the patient, perhaps questioning possibly inappropriate physician orders. "Lack of intervention on the patient's behalf," on the other hand, refers to the failure of a nurse to take nursing actions when such actions are required. This category differs from "Lack of prevention" in that the latter deals with failures involving proactive, instead of reactive, actions. The developers decided that medication errors should be coded using the NCCMERP taxonomy.

Obviously, this taxonomy only descriptively classifies the human performance failures involved in events. Data gathered certainly will be of use in following trends in errors and events and may help point toward functions that need improvement. However, extension of the categories to causes of the failures or even the types of behavior that led to the failures probably would increase its utility.

Madison Taxonomy

The Madison taxonomy (Thomadsen & Lin, 2003) evolved in the 1990s to address the difficulties in applying the existent taxonomies to medical events in radiotherapy (Thomadsen et al., 2003) and to provide a stronger correlation between the taxonometric classification and suggested remedial actions. Drawing heavily from the works of Rasmussen, van der Schaaf, and Kapp and Caldwell, this taxonomy tried to synthesize the most useful features of each. The failure analysis uses the location of the failure on the process and fault trees to determine where in the procedure to address concerns (see next section) and focuses on the nature of the failures.

Figure 30–15 shows the overview of the Madison taxonomy. Unlike many taxonomies that start with

TABLE 30–8. The Categories of the Edinburgh Error Classification System.

Edinburgh Classification of Contributing Factors
Initial categories
1. Inexperience with equipment
2. Shortage of trained staff
3. Night time
4. Fatigue
5. Poor equipment design
6. Unit busy
7. Agency nurse
8. Lack of suitable equipment
9. Failure to check equipment
10. Failure to perform hourly check
11. Poor communication
12. Thoughtlessness
Added categories
1. Presence of students/teaching
2. Too many people present
3. Poor visibility/position of equipment
4. Grossly obese patient
5. Turning the patient
6. Patient inadequately sedated
7. Lines not properly sutured into place
8. Intracranial Pressure Monitor not properly secured
9. Endotracheal tube not properly secured
10. Chest drain tube not properly secured
11. Nasogastric tube not properly secured

Note. The "initial categories" come From the work of Wright et al. (1991) and the "additional categories" from Busse and Wright (2000).

TABLE 30–9. The Edinburgh System Categories Reordered Into Proximal and Distal Causes.

Failure Type Categorization
Proximal causal factors
1. Failure to check equipment
2. Failure to perform hourly check
3. Thoughtlessness
4. Turning the patient
5. Patient inadequately sedated
6. Lines not properly sutured
7. ICP monitor not properly secured
8. En. tube not properly secured
9. Chest drain tube not properly secured
10. Nasogastric tube not properly secured
Distal, causal factors
1. Inexperience with equipment
2. Shortage of trained staff
3. Night time
4. Fatigue
5. Poor equipment design
6. Unit busy (refined by 9 and 10 below)
7. Agency nurse
8. Lack of suitable equipment
9. Presence of students/teaching
10. Too many people present
11. Poor visibility of equipment
12. Poor communication

Note. Table based on data from Busse & Wright (2000).

an initiating event, the Madison approach assumes that, in medical settings, there seldom is a particular condition that upsets an equilibrium. Patients are always in flux and potentially unstable. As long as the patient is under care, there are actions, counteractions, and follow-up actions. And there is always experimentation, testing to find problems, and further testing of possible solutions. Following events, it often becomes difficult to determine precisely when things stopped being normal, in part because no normal may exist—each patient being unique. The beginning of the analysis starts with the last recognizable "normal" condition, or better, the last time that the situation *may* have been under control. As the situation develops, the pathway leads to the box asking if action is required. A "no" response leads to the branching for either action was taken when it should not have been, leading to an event, or no action being taken, leading to a successful outcome or a near-miss. The mechanism that leads to the lack of action separates a successful outcome from a near-miss. Were the decision-making sequence followed according to a well-understood and practiced procedure, the outcome would be a success. If, on the other hand, the decision for action lay in the balance, with questions leading the operator both ways, the correct move, in this case inaction, resulted from a near-miss.

Following the "yes" path from the "Action Required" box leads to "Action Taken," with another yes/no option. Obviously, taking no action at this point leads to failure, while taking action moves the investigation to the question of whether the action was correct. Again, the results of the correct action, success, or near-miss depend on how the action came about.

In Figure 30–15, each of the paths leading to an event moves the analysis to the "A" in a circle, which continues in Figure 30–16. This pathway leads to the general description of where in the process the failure occurred. In this context, the "where in the process" is not a step in the procedure, but a more general view of any process that requires responses to inputs. Any of the inappropriate actions in Figure 30–15 could be described by one of the boxes in Figure 30–16. The first box, "Mistaken Action," applies to an action taken not as a reaction to an immediate situation. Of course, as stated earlier, all actions have antecedents that provide motivation (except, possibly, pathologically initiated actions), but some rely less on response to stimuli than on initiative. For example, a nurse

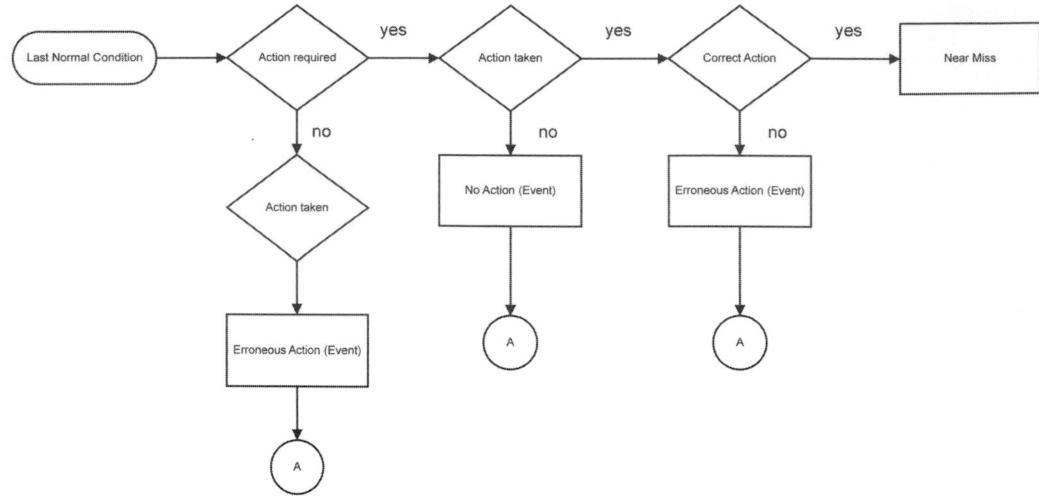

Figure 30–15. Overview of the Madison taxonomy.

TABLE 30–10. The Taxonomy of the Nursing Practice Breakdown Research Advisory Panel

Lack of attentiveness
Lack of agency/fiduciary concern
Inappropriate judgment
Medication errors
Lack of intervention on patient's behalf
Lack of prevention
Missed or mistaken physician or health care provider orders
Documentation errors

decides to change a patient's bed, although the sheets are not dirty and accidentally rolls the patient out of bed. Most events, however, do occur following some need for action. The failures result from either missing that the need arose or not taking the correct action. The former is classified as a detection failure and results from not having procedures to detect the situation, or if procedures exist, they failed in their application. The latter, "Failure of Reaction," results from misinterpreting the information detected, not reacting correctly (for many reasons, considered later), or not taking action when required. Each of the paths leads either to subsequent pathways B or C.

The classification investigation continues in Figure 30–17 at the "B" pathway. The erroneous action could be from a human performance failure (right branch) or a hardware/software failure (left branch). Although purely mechanical failures remain rare in medicine, they do happen. In 1987, the very heavy head of a cobalt–60 radiotherapy unit broke from its gantry and crushed a patient, with no warning signs of weak materials. Equipment failures fall into the classifications taken from the Eindhoven system (van der Schaaf, 1992):

- Hardware failures
  - Design
  - Construction
  - Materials (parts breaking)
  - Maintenance (which may actually be an organizational failure)
- Software failures
  - Design
  - Construction
  - Maintenance (again, may be an organizational failure)

Software failures have no "materials" category because no "parts" break. True, programs may stop working, but the origin of such problems more directly stems from design problems than with hardware.

More commonly, medical events follow the human performance failure pathway. Classification of the nature of the failure takes place at several levels along this pathway (Table 30–11). The first classification simply describes the failure generally, in terms of trips, slips, omissions, blunders, mistakes, or intentional incorrect actions (malfeasance or negligence).

Much of the second section of the human performance classification comes from the works of

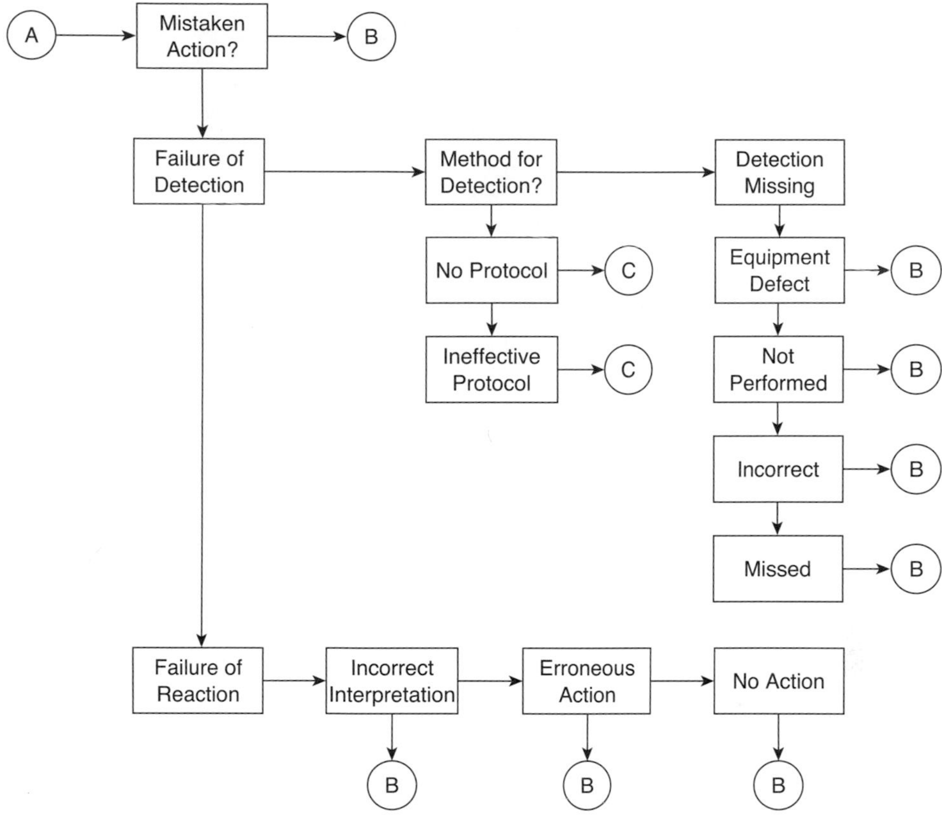

Figure 30–16. General characterization of the failure leading to an event.

Rasmussen and Norman discussed earlier. This subpathway attempts to determine the mechanisms involved in leading the operator to the action actually executed instead of the correct action.

The equipment and human pathways join again leading to "enabling factors" or, more commonly, performance shaping factors. Application of the term "enabling factors" intends to recall the use in psychological settings of actions by one person that facilitate pathological or destructive actions by a second person. "Enabling factors" better project the damaging effects the actions have than the neutral term "performance shaping factors," which could just as well apply to performance improvement as performance degradation.[3] The first of the four general categories of enabling factors reprises the equipment pathway. Although equipment failures seldom directly cause medical events, such failures frequently play important roles in leading the human operators astray. The "External Causes" box leads to no further classifications; being outside of the facility's universe, further classification may not be very useful, and the number of classifications could become massive. Information guiding remedial actions for external causes comes from a fault-tree analysis of the event, rather than from the taxonomy (see discussion that follows).

The two remaining enabling categories usually contain most of the latent causes for the event under investigation. The first, "Environmental Factors" covers a large territory. Table 30–12 gives a partial

[3]In pop psychology, the adjective "empowering" often applies to beneficial actions. It is interesting to note that just as fine distinctions must be made in developing taxonomies between words that in common usage carry equivalent meanings, here also the term "enabling" could have been applied to actions either helpful to recovery or to destruction.

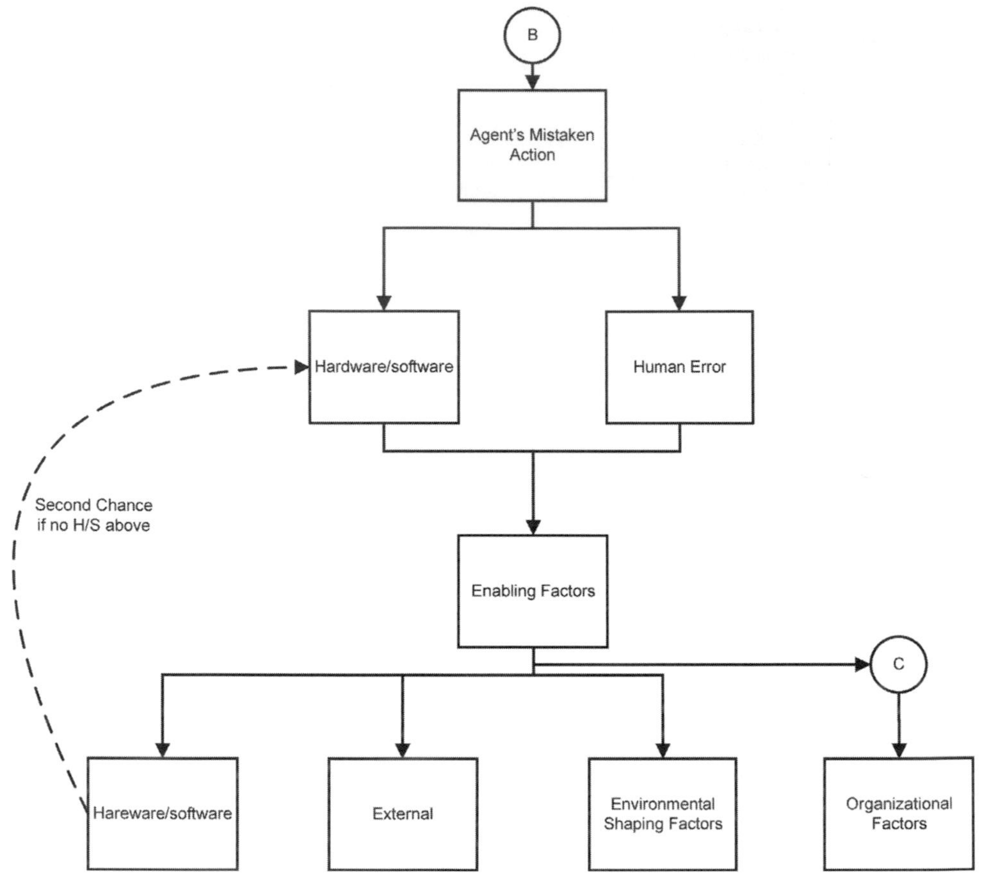

Figure 30–17. Action factors for the Madison taxonomy.

list of possibilities. Some of these classifications apply to the operators, others to the surrounding conditions. As noted previously, immediate attention to environmental factors improves safety immediately with relatively little cost.

The last box, "Organizational Factors," usually contains the preponderance of latent causes. Similar to environmental factors and most often equipment failures, the organizational factors rarely, if ever, directly cause events—they set the stage to make the event more likely, sometimes to the extent of seeming like a trap. Table 30–13 lists some organizational factors.

The nonhierarchical organization of the taxonomy lends itself to the pinball approach, with the basic philosophy that preventing events does not necessarily require addressing the most latent, root causes. In many cases, those types of changes may be the most resistant. All of the classifications give information that could be useful in determining remedial actions.

The text regarding Table 30–15 discusses guidance for the choice of remedial actions based on the failure classifications in the Madison taxonomy in detail.

CLASSIFICATIONS BASED ON ACTIVITY AND ERROR

The taxonomies presented classify failures in terms of types of errors, under the assumption that effective corrective actions depend on such knowledge. However, effective corrective actions also require information on *where* in the process corrective actions can be located. Although the Rasmussen taxonomy partially provides such information (in the What Failed pathway), the information there is intended to

Table 30–11. Classification of Human Performance Failures in the Madison Taxonomy

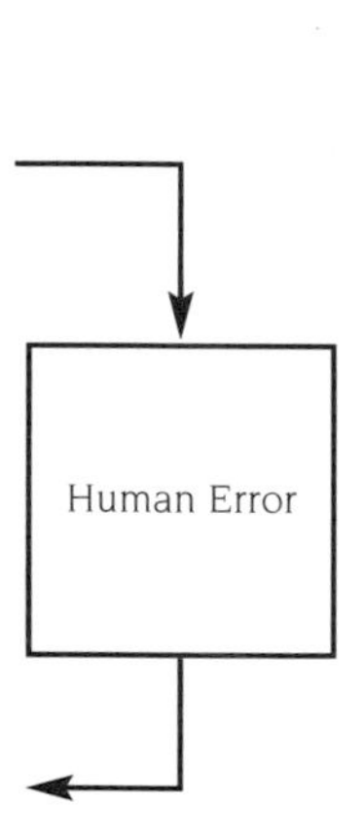

Human Error Subpath
- ❑ Tripping
- ❑ Slips
- ❑ Blunder
- ❑ Error in the intention (mistake)
- ❑ Intentional violation

Human Error How Sub-sub-path
- ❑ Physical coordination
  - Manual variability (SB)
  - Topographic misorientation (SB)
- ❑ Expectation bias
  - Stereotype fixation (SB)
  - Stereotype takeover (SB)
  - familiar association shortcut (RB)
  - familiar pattern not recognized (RB)
- ❑ Mistake choices (recall)
  - Mistake options, mistake consequence (RB)
- ❑ Inference
  - Condition or side effect not considered (KB)
    Information problem (input information processing)
  - Not sought—assumed, negligent omission, not realized it was needed (KB)
- ❑ Vigilance
  - Lack of vigilance (arousal, commitment, complacency)

describe one aspect of the failure, not actually locate the failure in the process. This information comes from other analyses, as discussed next.

Failure Location on a Process Tree

The most direct way to locate where to place corrective actions is using a process tree. A process tree, such as the one shown in Figure 30–18, charts the flow of a procedure. Process trees come in many varieties and sometimes go by the name fishbone diagrams or process flow charts, among others. Locating the position on the chart where the failure happened very directly indicates the step requiring protection. For that, one seldom needs the chart. The chart becomes useful for two functions. The first is when the step requiring protection does not lend itself to any quality management. In such cases, the chart facilitates finding steps downstream in the procedure amenable to some verification or control procedures or, not as likely, positions upstream that could prevent the failures in the noted position.

The second situation noting the position of a failure on the process tree comes when analyzing a large number of errors from a database. Figure 30–18 shows the result of such an analysis. The numbers on the branches of the tree give the number of failures that occurred at that step in the process. Noting where a number of events congregate suggests steps that need particular attention.

Creation of the process tree forms an important part of proactive patient safety but is beyond the scope of this chapter. For additional information on risk management techniques, see the chapter on "Human Factors Risk Management in Medical Products" in this handbook.

Failure Location on a Fault Tree

Similar to the process of gathering information from the process tree, indicating the location of an event on a fault tree provides additional and different information yet. In this case, the tree displays *all possible failures*. Each possible failure *should* be protected by some form of quality management either at the position of the potential fault or downstream, where for an error to propagate, there must be a concomitant failure of the quality management. While not indicating where in the procedure the additional quality management should go, the fault tree helps visualize whether all faults are covered in some way from propagation. Figure 30–19 shows a small portion of a fault tree. As with the process tree, indicating the failures that happened on the

TABLE 30–12. A Partial List of Environmental Enabling Factors

Tangible factors
- (Environmental Controls) Sound control noise (nonhuman)
- (Environmental Controls) Sound control distractions (human)
- (Environmental Controls) Visual control
- (Environmental Controls) Neatening
- (Environmental Controls) Cleaning
- (Environmental Controls) Isolation
- Environmental design

Intangible factors
- Time of day
- Day of week
- Holidays, vacations
- Personal problems
- Health

tree when reviewing an event database shows particular faults that require attention.

REMEDIAL ACTIONS BASED ON CLASSIFICATIONS

Preventative Actions/Remedial Actions

As noted several times, the goal of classifying failures is to direct corrective actions. Such analyses in no way replace the *a priori* approaches for error prevention, such as studying the fault tree and making sure that all error paths are blocked by quality control or quality assurance. Even with good proactive error prevention, some failures are likely and remedial actions necessary. A discussion of guidance for selecting remedial actions based on taxonomic classifications requires a better understanding of the nature of remedial actions themselves.

Categories of Remedial Actions

Van der Schaaf (1992) grouped remedial actions into several large categories (see Table 30–5). The general divisions addressed equipment, individual training, and organizational changes (creation of communication paths or improved procedures). As noted earlier, these categories do not lead the investigator to very specific actions. Any systematic suggestion of remedial actions will be general and descriptive, rather than very concrete, if only because greater specificity would produce unwieldy tables, making it harder to find the appropriate classification and related action.

TABLE 30–13. Organizational Enabling Factors

- Knowledge of leader
- Management priorities
- Communication system
- Knowledge transfer (training inadequate)
- Matching abilities to job
  - Tasks exceed physical ability
  - Tasks exceed mental ability
  - Tasks exceed experience
- Competition for attention
  - Background
    - Lack of staffing or time
    - Other, competitive goals
  - Immediate
    - Other, competitive duties
    - Lack of staffing
    - Other inputs into the system
  - Previous
    - Mental fatigue due to complex work
    - Concern over previous, unaddressed problems

Relation Between Failure Classification and Remedial Actions

The Madison system attempts to provide fairly detailed guidance to the selection of remedial action. Table 30–14 gives general categories of possible remedial action. Of course, the list cannot be exhaustive or static. Table 30–15 correlates some potential actions with the failure classification from the Madison taxonomy. The table lists some of the taxonometric classifications as headings for the rows and potential remedial actions for the columns. An “x” in the box at an intersection of a classification and an action indicates that that action may be applicable for failures with that classification. Blanks at the intersections mark those specified actions that probably would not effectively address the problems

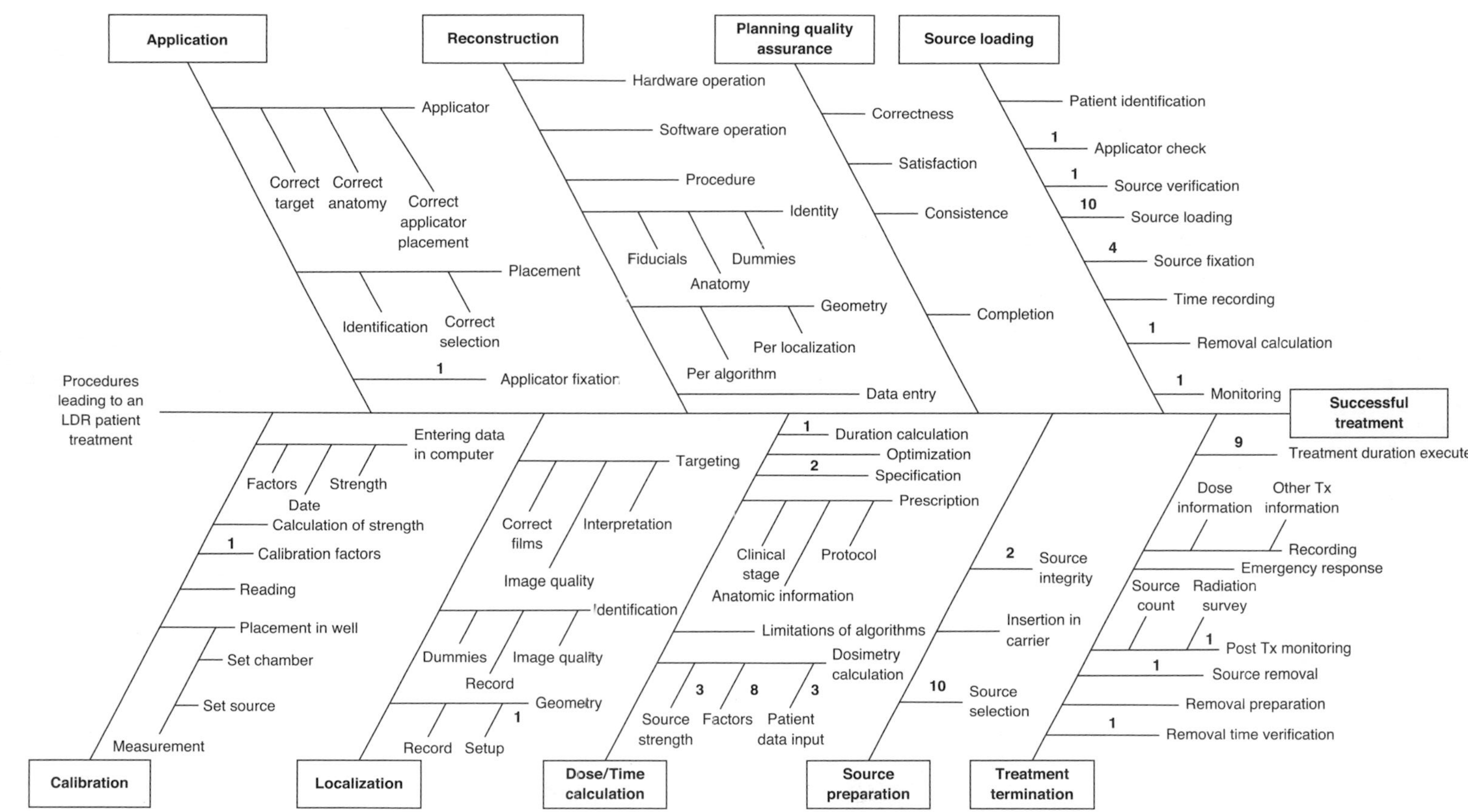

Figure 30–18. Process tree for low dose-rate brachytherapy showing numbers of failures occurring on the branches, based on data in a national database. (Thomadsen et al., 2003)

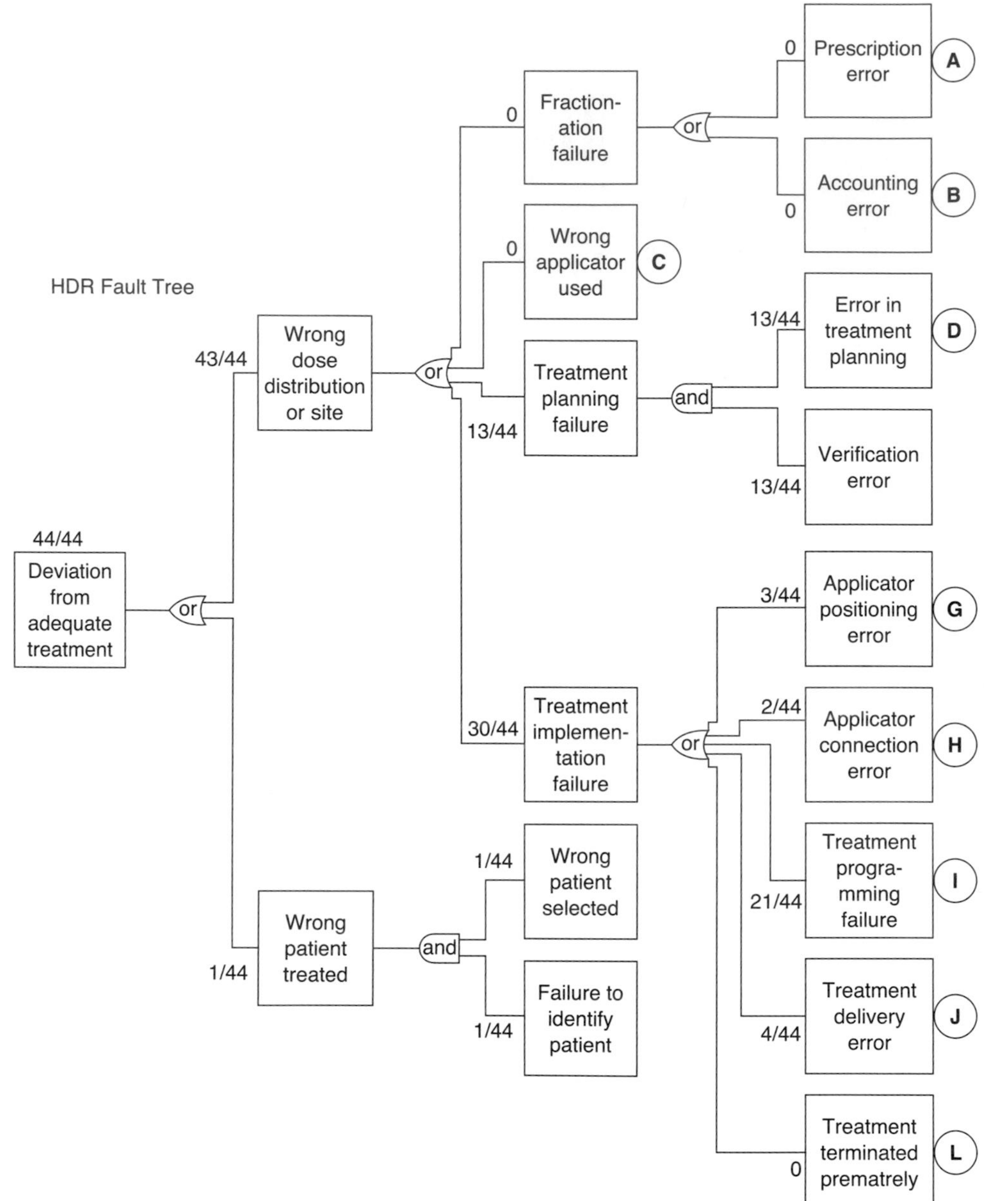

Figure 30–19. A fault tree for high dose-rate brachytherapy, with the number of failures along a branch over the number of total failure reported in a national database. (Thomadsen et al., 2003)

leading to the failure. Question marks indicate actions with undetermined relationships to the failure classification.

Hierarchy of Remedial Actions

Not all remedial actions affect corrections equally. Consider the problem that syringes of the clear solution potassium chloride were sometimes mistaken for syringes of saline, and when injected into a patient's vein, inadvertently produced cardiac arrest. One possible remedial action would be to label the potassium chloride syringes with large letters and a warning, while another approach would be to remove the potassium chloride syringes from the general shelves. The likely effectiveness of each of the two possible solutions differs markedly. Labels, no matter how blatant, sometimes are missed or misinterpreted, while removing the syringes eliminates simply grabbing them by mistake. The Institute for Safe Medical Practices (ISMP) lists a hierarchy for corrective action (Table 30–16).

The Madison taxonomy also ranks the likely effectiveness of possible remedial actions, as given in Table 30–17. Table 30–17 ranks the actions listed in Table 30–14 by power, the likelihood that the selected action will significantly reduce the probability of future failures. It is important to note that the environmental category is not part of the ranking. Most environmental problems (or environmental conditions that facilitated human performance failures) should be corrected without regard for any other actions taken. Usually, such changes in the environment come more easily than changes in practice and may improve the situation markedly with little expenditure of resources. On the other hand, some environments resist changes. Take, for example, the sounds in most intensive care units (ICUs). With many monitors and machines working and producing both white noise (such as from fans cooling computers) and monitor tones (such as heartbeats), alarms from any given machine may go unheeded due to a combination of competition for attention and masking noise. Unyielding environments require other remedial action thrusts. See chapter on "Noise and Alarms in Health Care—An Ergonomics Dilemma" in this handbook for further information.

TABLE 30–14. Selected Remedial Actions Grouped by Category.

Physical tools
Barriers
Interlocks
Alarms
Bar codes
Communication devices
Informational tools
Label
Signs
Reduction in similarity
Measurement tools
Check-off forms
Operational checks
Comparison with standards
Increased monitoring
Automated monitoring
Additional status checks
Redundant measurements
Independent review
Acceptance testing
Computerization
Computerized verification
Computerized order entry (COE)
Computerized order entry with verification and feedback
Knowledge tools
Training
Instruction
Experience
Administrative tools
Mandatory pauses
Increased staffing
Establishing protocol
Clarifying protocol
Establish/clarify communication lines
Reorder administrative priorities
External audit
Internal audit
Environmental actions
Sound control
Sight control (clearing/cleaning)
Sight control (lighting)
Simplification (neatening)
Isolation (removal of competing demands and calls)
Ergonometric improvements
Safety design improvements

INTEGRATED REMEDIAL ACTIONS

Consideration of the Many Dimensions of Information on Failures

Any given remedial action taken may not correct the problems that led to a failure. On this point, the wisdom of the theory that the latent causes require fixing shines. Addressing the human performance problems directly may be the most expedient and the only apparent action to take in a particular setting. However, one may find that the unaddressed latent causes lead to a different human performance problem in a completely different arena. Such a situation illustrates the value of combining a taxonometric approach with a database. The latent causes may not appear during an analysis of single, or several events, but may surface over time. Even multiple events that give the impression of resulting totally from human errors can point to common organizational problems or other enabling factors. Identifying the failures by their location on the process tree or fault tree, while potentially indicating procedures or potential faults that require protection with quality management, cannot direct remedial actions toward latent problems.

Table 30–15 Suggested Remeidal Actions Based on the Failure Classifications in the Madiason Taxonomy

	Physical Tools					*Information Tools*			*Management Tools*									*Knowledge tools*			*Administrative Tools*							*Equip Related*			*Environmental tools*				*Compu-terized*			
	Interlocks	Bar Codes	Alarms	Barriers	Communication Devices	Reduce Similarity	Labels	Signs	Check-off Forms	Operational Checks	Comparison With Standards	Increase Monitoring	Automatic Monitoring	Add Status Check	Redundant Measurement	Independent Review	Acceptance Test	Training	Experience	Instruction	Mandatory Pauses	Staffing	Establish Protocol / Clarify Protocol	Establish / Clarify Communication Lines	Better Scheduling (Reduced Overtime)	(Administrative) Priority	Internal Audits	Repair	PM - Preventive Maintenance Inspection	Establish and Perform QC and QA	(Environmental Controls) Sound Control	(Environmmental Controls) Cleaning	(Environmental Controls) Neatening	(Environmental Controls) Isolation	(Environmental Controls) Visual Control	Computerized Verification	Computerizeed (Order) Data Entry	Computerized Data Entry With Feedback
Mistake action																												O	O	O								
Detection - no protocol																							X															
Detection - ineffective protocol																							X															
Detection - equipment defect										X							X											X	X									
Detection - not performed	X		X						X				X					?	?	?	?															X		
Detection - incorrect					X	X																		X														
Detection - missed	X		X		X				X			X	X	X				?	?	?											X		X		X			
Reaction - incorrect interpretation																X		X	X																		X	X
Reaction - erroneous action																												O	O	O								
Reaction - no action																												O	O	O								
Human error - tripping																																	X		X			
Human error - slips	X	X	X	X	X	X	X	X	X	X	X	X	X	X	X	X																					X	X
Human error - blunder	X	X	X	X	X	X	X	X	X	X	X	X	X	X	X	X																					X	X
Human error - error in the Intension (Mistake)	?		?	?												X																						?
Hardware failure																	X											X	X	X								
Software failure																	X													X								
Enabling factor - hardware									X																													
Enabling factor - software									X								X							X														
Enabling factor - external																																						
Enabling factor - environmental																															X	X	X	X	X			
Enabling factor - organizational																		?	?	?		?	?	?	?	X	X											
Hardware - design																												?										
Hardware - construction																												?										
Hardware - material														X														X	X	X								
Hardware - maintenance																												?										

TABLE 30–15 (Continued)

	Physical Tools					*Information Tools*			*Management Tools*									*Knowledge tools*			*Administrative Tools*							*Equip Related*			*Environmental tools*					*Computerized*		
	Interlocks	Bar Codes	Alarms	Barriers	Communication Devices	Reduce Similarity	Labels	Signs	Check-off Forms	Operational Checks	Comparison With Standards	Increase Monitoring	Automatic Monitoring	Add Status Check	Redundant Measurement	Independent Review	Acceptance Test	Training	Experience	Instruction	Mandatory Pauses	Staffing	Establish Protocol / Clarify Protocol	Establish / Clarify Communication Lines	Better Scheduling (Reduced Overtime)	(Administrative) Priority	Internal Audits	Repair	PM - Preventive Maintenance Inspection	Establish and Perform QC and QA	(Environmental Controls) Sound Control	(Environmental Controls) Cleaning	(Environmental Controls) Neatening	(Environmental Controls) Isolation	(Environmental Controls) Visual Control	Computerized Verification	Computerizeed (Order) Data Entry	Computerized Data Entry With Feedback
Software - design																												?										
Software - construction																												X	X	X								
Software - maintenance																																						
Manual variability (SB)																																						
Topographic misorientation (SB)	X																																					X
Stereotype fixation (SB)	X	X	X	X	X	X	X	X	X	X	X	X	X	X	X	X								X												X		
Stereotype Takeover (SB)	X		X	X		X	X	X	X		X	X	X	X	?	X					X															X		
Familiar Association Shortcut (RB)	X		X	X		X			X				X	X	?					X	X															?		
Familiar Pattern not Recognized (RB)		X			X	X	X	X	XX		X		X		X				X	X	X	?														X		
Mistakes Alternatives (RB)									X							X		X	X	?				X												?		X
Mistake Consequence (RB)									?							X		X	X	?				X												?		X
Forget Isolated Act (RB)	X	X	X	X	X		?	?	X		?		?	?	?	X			?																	X		
Condition or Side Effect not Considered (KB)	X		X	X							X	X		X		X		?	?																	X		
Information not Sought - Assumed, Negligent Omission, Not Reealized It Was Needed	X	X	X	X	X	?	?	?	X		X			X		X							?	?	?	?										X	X	
Lack of Vigilance (Arousal, Commitment, Complacency)	X	X	X	X					?			X	X			X																				X		X
Organizational - Knowledge of Leader																X		X	X																			
Organizational - Management Priority																X																						
Organizational - Communication System						X																				X												
Organizational - Knowledge Transfer																		X		X																		
Organizational - Ability																	X	X	X																			
Competition for Attention (Background) - Lack of Staff or Time	X		X	?					?				X			?							X													X		X

TABLE 30–15 (Continued)

	Physical Tools					*Information Tools*			*Management Tools*									*Knowledge tools*			*Administrative Tools*							*Equip Related*			*Environmental tools*				*Compu-terized*			
	Interlocks	Bar Codes	Alarms	Barriers	Communication Devices	Reduce Similarity	Labels	Signs	Check-off Forms	Operational Checks	Comparison With Standards	Increase Monitoring	Automatic Monitoring	Add Status Check	Redundant Measurement	Independent Review	Acceptance Test	Training	Experience	Instruction	Mandatory Pauses	Staffing	Establish Protocol / Clarify Protocol	Establish / Clarify Communication Lines	Better Scheduling (Reduced Overtime)	(Administrative) Priority	Internal Audits	Repair	PM - Preventive Maintenance Inspection	Establish and Perform QC and QA	(Environmental Controls) Sound Control	(Environmmental Controls) Cleaning	(Environmental Controls) Neatening	(Environmental Controls) Isolation	(Environmental Controls) Visual Control	Computerized Verification	Computerizeed (Order) Data Entry	Computerized Data Entry With Feedback
Competition for Attention (Background) - Other Goals	X		X	?					?				X			?																				X		X
Competition for Attention (Immediate) - Other Duties	?		?	?					?				X			X						X	?			X										X		?
Competition for Attention (Immediate) - Lack of Staff	X		X	?					X				X									X			X											X		
Competition for Attention (Immediate) - Too Many Inputs to System	X		X	X		?			X		?		X			X															X		X		X	X		X
Competition for Attention (Previous) - Fatigue due to Complex Work	X	X	X		X	X	X	X	X	X		X	X	X	X					X				?							X		X		X	X		X
Environment - |Tangible| Noise (Non-Human Environmental Factors)																														X								
Environment - |Tangible| Distraction (Human Related Environmental Factors)																																		X				
Environment - |Intangible| Environmental Problem																															X	X	X		X			
Environment - |Intangible| End of Day, Holiday	?	?	?			X			X				X			X									X											X		X
Environment - |Intangible| Personal Problem	?	?	?			?			X				X			X																				X		X
Others - Exceed Ability - Physical																																						
Others - Exceed Ability - Mental																																						
Others - Lack of Experience																		?	X	X																		

TABLE 30–16. Ranking of the Likely Effectiveness of Types of Actions to Prevent Errors

1. Forcing functions and constraints
2. Automation and computerization
3. Drug protocols and standard order forms
4. Independent double-check systems and other redundancies
5. Rules and policies
6. Education and information

Note. Based on the Institute for Safe Medical Practices's toolbox (ISMP, 1999), listed from the most to the least effective.

TABLE 30–17. Remedial Actions Ranked by Power to Affect Changes.

Forcing functions and constraints	***Rules and policies***
Interlock	External audit
Barriers	Internal audit
Computerized order entry with feedback	Priority
	Establishing/Clarify communication line
Automation and computerization	**Staffing**
Bar codes	Better scheduling
Automate monitoring	Mandatory pauses
Computerized verification	Repair
Computerized order entry	PMI (Preventive Maintenance Inspection)
	Establish and perform QC and QA (hardware and software)
Protocols, standards, and information	
Check-off forms	**Education and information**
Establishing protocol/Clarify protocol	Training
Alarms	Experience
Labels	Instruction
Signs	
Reduce similarity	**Environment (not in ranking)**
	(Environmental controls) Sound control
Independent double-check systems and other redundancies	(Environmental controls) Visual control
	(Environmental controls) Cleaning
Redundant measurement	(Environmental controls) Neatening
Independent review	(Environmental controls) Isolation
Operational checks	Environmental design
Comparison with standards	
Increase monitoring	
Add status check	
Acceptance test	

Note. The environment category is not part of the ranking because any environmental problems should be addressed Independently of other actions taken.

Prevention of Events Rather Than Cure of Root Causes

> The goal is not to prevent error but to prevent events.—Lucian Leape, 2004

While addressing the root causes may be most effective in preventing future events, any effective remedial action is likely to improve the situation. Root causes and latent problems often hide deeply in organizations, making them hard to find and harder yet to root out. Often patients improve even with treatment with placebos. Taking actions, even if not the most effective ones, often brings improvement in patient safety if only because the staff sees the importance of patient safety improvement. Addressing concerns also shows that reporting events brings actions and follow-up. Too often the medical staff see nothing come of reports and become discouraged. Above all, however, remedial actions that directly address human performance problems, without going after the latent problems, frequently still provide sufficient protection against further failures that latent causes are not strong enough to bring about.

References

Anderson, T. (1997). *A case study and conceptual framework for medical event reporting systems.* Unpublished master's thesis, University of Wisconsin–Madison.

Benner, P., Sheets, V., Uris, P., Malloch, K., Schwed, K., & Jamison, D. (2002). Individual, practice, and system causes of errors in nursing: A taxonomy. *Journal of Nursing Administration, 32,* 45–48.

Busse, D., & Johnson, C. W. (1999). Human error in an intensive care unit: A cognitive analysis of critical incidents. In J. Dixon (Ed.), *17th International System Safety Conference* Orlando: Florida Chapter of the System Safety Society.

Busse, D., & Wright, D. (2000). Classification and analysis of incidents in complex medical environments. *Topics in Heath Information Management, 20,* 1–11.

Harris, D., & Chaney, F. (1969). *Human factors in quality assurance.* New York: Wiley.

Institute for Safe Medical Practices. (1999, June 2). Medication error prevention "toolbox." *Medication Safety Alert.* Retrieved June 20, 2006, from http://www.ismp.org/msaarticles/toolbox.html.

Kaplan, H. S., Battles, J. B., van der Schaaf, T. W., Shea, C. E., & Mercer, S. Q. (1998). Identification and classification of the causes of events in transfusion medicine. *Transfusion, 38,* 1071–1081.

Kapp, E. A., & Caldwell, B. (1996). SCOPE Instruction Manual. Unpublished manuscript.

Koppens, H. A. (1997). *"SMART" error management in a radiotherapy quality system.* Unpublished master's thesis, Eindhoven University of Technology, The Netherlands.

Leape, L. (2004). *Where are we going in patient safety?* Presented at the meeting of the Wisconsin Patient Safety Institute, Oconomowoc, WI, November 13, 2003.

National Coordinating Council for Reporting and Preventing Medication Errors. (1999). Retrieved June 20, 2006, from http://www. nccmerp.org/pdf/taxo2001–07–31.pdf

Norman, D. (1981). Categorization of action slips. *Psychological Review, 88,* 1–15.

Rasmussen, J. (1980). What can be learned from human error reports? In K. D. Duncan, M. M. Gruneberg, & D. Wallis (Eds.), *Changes in working life.* (pp. 97–113). New York, NY: Wiley.

Rasmussen, J. (1982). Human errors: A taxonomy for describing human malfunction in industrial installations. *Journal of Occupational Accidents, 4,* 311–335.

Reason, J. (1990). *Human error.* Cambridge, U.K.: Cambridge University Press.

New York State Department of Health. (2001). Retrieved from http://www.health.state.ny.us/nysdoh/commish/2001/nyports/nyports.htm

Rook, L. (1965). *Motivation and human error.* Report SC-TM–65–135. Albuquerque, NM: Sandia Corp.

Thomadsen, B., & Lin, S-W. (2003). *Guidance for remedial actions based on failure classifications.* Wisconsin Patient Safety Institute, November 13, 2003.

Thomadsen, B., Lin, S-W., Laemmrich, P., Waller, T., Cheng, A., Caldwell, B., et al. (2003). Analysis of treatment delivery errors in brachytherapy using formal risk analysis techniques. *International Journal of Radiation Oncology Biology Physics, 57,* 1492–1508.

van der Schaaf, T. W. (1992). *Near miss reporting in the chemical process industry.* Unpublished doctoral thesis, Eindhoven University of Technology, The Netherlands.

van Vuuren, W., Shea, C. E., & van der Schaaf, T. W. (1997). *The development of an incident analysis tool for the medical field.* Eindhoven University of Technology, The Netherlands.

Wright, D., Mackenzie, S. J., Buchan, I., Cairns, C., & Price, L. E. (1991). Critical incidents in the intensive therapy unit. *Lancet, 338,* 676–678.

31

HUMAN FACTORS OF HEALTH CARE REPORTING SYSTEMS

Chris Johnson
University of Glasgow, Scotland

A number of mechanisms can be used to elicit epidemiological information about adverse events in health care. Morbidity and mortality committees provide a primary means of detecting potential problems in the quality of patient care (Wald & Shojania, 2001). Litigation and malpractice statistics focus attention on incidents and accidents. The publication of clinical studies also helps to ensure that the quality of medical practice remains at a high level within particular organizations. However, these epidemiological techniques often provide insights many months and years after the original incidents have occurred. They also are often limited in terms of the insights they provide into mitigation and error reduction strategies. Other techniques such as chart reviews and the use of automated detection systems provide limited information about the causes of adverse events and can provide results that are both partial and biased. This chapter focuses on the role that mandatory and voluntary reporting systems can play in improving patient safety.

INTRODUCTION TO THE HUMAN FACTORS OF ADVERSE EVENTS

Human factors play a dual role in health care. On the one hand, we rely on individual and team decision making to guide most aspects of diagnosis and treatment. We rely on the skill and judgment of clinicians to decide when and when not to intervene. We depend on their vigilance to determine when mistakes have been made or, ideally, to intervene before colleagues make a mistake. Alternatively, as we have seen in other chapters, human factors issues can trigger accidents and incidents. For example, one the one hand, a physician might make a slip if he or she writes down 10 mg of an appropriate medication when the intention was to prescribe 1 mg. On the other hand, he or she might make a mistake by giving a medication that was not intended as part of the patient's treatment. He or she could also lapse by forgetting to deliver an intended drug. Finally, clinicians can commit violations by deliberately ignoring recommended practice. All of these different forms of human "error' have been noted in a range of hospital and primary care settings (Johnson, 2003a).

Estimating the Costs of Adverse Health Care Events

It is difficult to underestimate the significance of human error in health care. We are surrounded by newspaper items and television broadcasts that reinforce concern over a succession of incidents and accidents. The products of research in this area inform much of that media interest. For example, a series of studies have argued that almost 100,000 patients die each year from preventable causes in U.S. hospitals. This annual toll exceeds the combined number of deaths and injuries from motor and air crashes, suicides, falls, poisonings, and drownings (Barach & Small, 2000). It has been estimated that there are 850,000 adverse incidents every year in the U.K. National Health Service (NHS). The U.K. National Patient Safety Agency (NPSA) reinforced this concern when they found more than 24,500 adverse incidents in 28 trusts over a six-month period (BBC, 2002a). Such statistics are, however, very difficult to validate. National figures rely on interpolation from relatively small samples. The biases within these samples further confound interpretation. For instance, some trusts in the NPSA study reported a high number of minor events, such as the misapplication of a bandage, while others reported virtually nothing. The underreporting of adverse events to national monitoring organizations is estimated to range from 50% to 96% annually (IOM, 1999).

The financial costs associated with adverse medical events are slightly easier to determine, although they provide a very indirect measure of the physical and psychological consequences for individual patients. These costs partly stem from the additional treatment that is required in the aftermath of adverse events. They are also associated with litigation. There is a wider perception that rising legal bills are undermining the economic underpinnings of many national and local systems. For instance, the NHS faces a litigation liability in excess of £4.4 billion, a figure that has more than trebled in the last three years (BBC, 2002b). This represents just under one tenth of their annual budget. George W. Bush has responded to these financial costs in a forthright manner: "There are some costs that are unnecessary as far as I'm concerned. And the problem of those unnecessary costs don't start in the waiting room, or the operating room, they're in the courtroom. … And one thing the American people must understand is, even though the lawsuits are junk lawsuits and they have no basis, they're still expensive" (Office of the Press Secretary, 2003).

Tort Reform and an Introduction to Incident Reporting

The rising human and financial costs of adverse health care events have triggered a number of responses. In particular, several governments have proposed a limit on the damages that may be awarded in medical cases. These caps are justified either in terms of limiting the financial exposure of national systems, such as the NHS, or in terms of the "spiraling" insurance costs that individual practitioners must meet in order to protect themselves against such litigation. Most of these limits are inspired by the Californian Medical Injury Compensation Reform Act (MICRA). This was part of a wider initiative to address the financial consequences of health care litigation and resulted in a limit of approximately $250,000 being placed on noneconomic damages in malpractice suits.

Other health care providers have looked beyond a maximum limit for all health care litigation. For instance, Sweden and Norway have moved the burden of insurance from the clinician by developing voluntary insurance schemes for patients. Denmark and Finland rely on mandatory patient insurance. Other proposals have focused on fixed tariffs for specific injuries. Structured payouts instead of large lump sums have also been suggested, as well as noncash compensations, such as home nursing care (Gaine, 2003). Alternative dispute resolution systems have also been proposed. These proposals have been motivated by a number of reports of the inefficiencies that complicate the settlement of claims in the aftermath of adverse incidents. For instance, a report by the U.K. National Audit Office (2001) found that cases can drag on for an average of five and a half years before settlement and that this delay can significantly increase the costs. The same report found that in 44% of cases the final legal bill was substantially higher than the compensation paid to patients and their families. Similarly, legal feels often account for more than one third of compensation paid to injured parties in the United States. For this reason, several states have established "accelerated compensable events." Payments can be made for certain classes of adverse events, mainly in obstetrics where most high-value claims are settled, without requiring that patients and their relatives prove who was to blame for the medical "error."

These U.S. systems illustrate the use of *no-fault liability* as a means of reducing the costs associated with adverse health care events (Vincent, 2003). Both the United Kingdom and the United States rely on the law of tort to resolve most health care claims. Tort law is based on an adversarial process in which the claimant must prove that harm has been caused by a breech of care. This focus on establishing blame may prevent the exchange of information that might prevent future adverse events. Supporters of the current system argue that litigation acts as a deterrent to substandard care. In contrast, the proponents of no-fault liability argue that the claimant must only show a medical error was a causative factor in an injury. They do not need to establish who was to blame for the causative error. In this model, the burden of proof focuses on causative mechanisms rather than establishing the fault of a particular individual or team. The arguments in favor of no-fault liability are counterintuitive. The intention is to reduce the total liability by making it easier to establish a claim. However, the proponents of tort reform argue that lower legal and administrative costs and a lower level of payouts will offset the costs associated with a greater number of claimants.

Just as the advocates of capping point to the success of the MICRA legislation in California, the proponents of no-fault liability also have a number of existing successes to substantiate their arguments. For instance, Virginia and Florida have set up selective forms of no-fault compensation to cover birth-related neurological injuries. New Zealand has established a more sustained system. They replaced a tort-based approach with a form of no-fault litigation following the Woodhouse Commission report in 1972. The initial scheme was criticized by opponents, who felt that it offered undue protection to negligent clinicians. The relevant legislation was then amended to increase the accountability of individual clinicians. This revised act established a model for several other countries. For example, Canada operates a "twin track" approach. Deliberate violations and negligence are separated from the other adverse events that are considered under a no-fault scheme. The parallel approach satisfies the twin demands of economy and of protecting the public through a formal disciplinary process. There are other legal models. For example, French medical negligence claims against the state are handled under administrative rather than civil law and "compensation for hospital mistakes is automatic" (Gaine, 2003).

Although there is a clear dissatisfaction with the current system of tort, it is difficult to find reliable quantitative information that can inform the policy changes being considered in the United Kingdom and in the United States. Davis, Lay-Yee, Briant, and Scott (2003) report that 5.2% of admissions in New Zealand led to a preventable in-hospital event. This rate is similar to that in the United Kingdom. Vincent (2003) argues that this figure also lies in the broad range established by studies in other countries including the United States. He goes on to argue that there is little evidence to support the claim that no-fault systems will encourage the reporting of errors. He provides a useful shift in perspective when he argues that the most important criterion for assessment of any compensation system should be its impact on injured patients and their families, not just in providing appropriate financial recompense where necessary but in ensuring that explanations, apologies, and long term support and care are regarded as the expectation rather than the exception.

USABILITY ISSUES AND MEDICAL DEVICE REPORTING SYSTEMS

Tort reform has been proposed as a means of reducing the costs associated with adverse medical events. These initiatives have been justified by the observation that the value of claims has risen at a time when there is little evidence of an increasing error rate. However, other initiatives have sought to reduce liabilities by reducing the frequency of adverse health care events. In particular, there have been a number of initiatives to establish "lessons learned" and incident reporting systems. These can be used to ensure that information about previous failures and near-misses can be used to inform the subsequent operation of a health care system. Incident reporting systems offer a number of benefits. The most obvious is that they provide a source of information about adverse events. There are further advantages if these schemes capture near-miss information as well as reports of adverse occurrences. These near-misses can be used to find out why accidents *don't* occur. Incident reports also provide a reminder of hazards. They provide means of monitoring potential problems as they recur during the lifetime of an application. They can be used to elicit feedback that keeps staff "in the loop." The data (and lessons) from incident reporting schemes can be shared. Incident reporting systems provide the raw data for comparisons both within and between industries. If common causes of incidents can be observed then, it is argued, common solutions can be found. Incident reporting schemes are cheaper than the costs of an accident. A further argument in favor of incident reporting schemes is that organizations may be required to exploit them by regulatory agencies.

There are many different types of reporting system in health care. One class of applications has been developed for reporting problems with medical devices. For instance, the U.S. Center for Devices and Radiological Health operates a range of schemes that feed into the Manufacturer and User Facility Device Experience Database (MAUDE). For example, the following report describes how the drug calculator of a medication assistant in a patient monitoring application would occasionally round up values to a second decimal place. The users complained that this could easily result in a medication error and that the manufacturer was failing to acknowledge the problem. The manufacturer initially responded that vigilant nursing staff ought to notice any potential problems when calculating the medication. The clinicians countered this by arguing that they had explicitly taught nursing staff to trust the calculation function as a means of *reducing human error.* Subsequent reports from the device manufacturer stressed that clinicians can configure the resolution of medication measurements through a unit manager menu:

> This is best method for clinical staff, it pre-configures drug calculations and allows settings to reflect how drugs are prepared by the pharmacy. Customer was told, drug concentration rounding to nearest hundredths, could be easily addressed in unit manager setup, to reflect higher resolution. Thereby, addressing any concern of a rounding issue. manufacturer has reviewed customer's concern and have determined that "drug calculations" feature is functioning as design. Additionally, manufacturer has reviewed with customer, the user's ability to change units of measure, to achieve desired resolution. The device is performing as designed. (MDR text key: 1601404)

The U.K. Medicines and Healthcare Products Regulatory Agency also provides several mechanisms for reporting adverse health care events including the Manufacturers' On-line Reporting Environment (MORE). These applications help to implement a series of different national and international requirements. In Europe, three directives regulate the marketing and monitoring of medical devices. These are Directive 90/385/EEC-OJL189/20.7.90 on Active Implantable Medical Devices, Directive 98/79/EC-OJ331/7.12.98 on In Vitro Diagnostic Devices, and Directive 93/42/EEC-OJ169/12.7.93, the Medical Devices Directive. Section 3.1 of Annex II of the Medical Devices Directive requires that manufacturers

> institute and keep up to date a systematic procedure to review experience gained from devices in the post-production phase and to implement appropriate means to apply any necessary corrective action. This undertaking must include an obligation for the manufacturer to notify the competent authorities of the following incidents immediately on learning of them: (i) any malfunction or deterioration in the characteristics and/or performance of a device, as well as any inadequacy in the instructions for use which might lead to or might have led to the death of a patient or user or to a serious deterioration in his state of health; (ii) any technical or medical reason connected with the characteristics or performance of a device leading for the reasons referred to in subparagraph (i) to systematic recall of devices of the same type by the manufacturer.

These provisions are important not simply for the reporting of device failures; it can be argued that these reporting obligations extend beyond the reporting of functional system failures to include adverse events that stem from usability or human factors issues during the operation of the device. Each member state within the European Union enacts national legislation to ensure that they conform to the requirements in these directives. For instance, the U.K. regulatory framework is based on the Medical Devices Regulations 2002 (SI 2002, No. 618) and Medical Devices (Amendment) Regulations 2003 (SI 2003, No. 1697). The net effect of all of this is to ensure that incident reports are one of several events that will trigger regulatory intervention and inspection by the Medicines and Healthcare Products Regulatory Agency. In addition, the agency will intervene to inspect a sample of manufacturers who market their devices in the U.K. market whether or not those companies have had any adverse events.

The U.S. Safe Medical Devices Act of 1990 (SMDA) guides the reporting of adverse events involving health care technology. Under the provisions of this act, end users must report device-related deaths to the Food and Drug Administration (FDA) and the manufacturer. Serious injuries must also be reported to the manufacturer, or to the FDA if they do not know how to contact the manufacturer. The FDA established a number of schemes to meet the requirements of the SMDA. These were confirmed under the Medical Devices Amendments of 1992 (Public Law 102–300; the Amendments of 1992) to section 519 of the Food, Drug, and Cosmetic Act relating to the reporting of adverse events. This established a single reporting standard for device user facilities, manufacturers, importers, and distributors. The Medical Devices

Reporting Regulation implements the reporting requirements contained in the SMDA of 1990 and the Medical Device Amendments of 1992. More recently, the 1998 FDA Modernization Act (FDAMA) reduced some of the regulatory burden on manufacturers by removing an obligation to provide annual reports on adverse events. End users could file an annual report instead of semi-annual reports to summarize adverse event reports.

The Canadian reporting system is governed by the Medical Devices Regulations. Australian practice is guided by the Therapeutic Goods Act. Japanese regulations are informed by the Ministry of Health and Welfare. The key point here is to recognize the diversity of different national reporting systems. This can create vulnerabilities if information about adverse events in one country cannot easily be used to inform practice in another. The Global Harmonization Task Force (GHTF) has recently been established to improve systems to support communication about health care incidents and accidents across international boundaries. This is a voluntary group of representatives from medical device regulatory agencies and device manufacturers, distributors, and so on. Figure 31–1 presents the findings of their recent review, which points to considerable differences in the perceived objectives of different reporting systems in different countries.

Figure 31–2 extends this analysis to present the GHTF's assessment of the provision that each of the regions/countries makes for the reporting of human error. As can be seen, there is again considerable diversity. In particular, the U.S. requirement covers a broad range of usability issues, including poor labeling and instruction as well as design flaws. In contrast, European regulations are perceived not to address usability issues except where they stem from manufacturing problems or inadequate labeling. It is also important to acknowledge the wider limitations of these device-related reporting systems. The focus on particular items of equipment implies that many human-related incidents will fall beyond the scope of these national and international schemes. A spate of reports, therefore, argued that these systems be extended to cover, for example, errors that stem from the interaction between different teams of specialists rather than from interaction with particular devices. For example, the 1999 U.S. Institute of Medicine report "To Err Is Human" (IOM, p. 35) identified a broad range of adverse health care events including "transfusion errors and adverse drug events; wrong-site surgery and surgical injuries; preventable suicides; restraint-related injuries or death; hospital-acquired or other treatment-related infections; and falls, burns, pressure ulcers, and mistaken identity." Many of these preventable incidents fall outside the scope of the existing device-related reporting systems. In consequence, it was argued that a new national reporting framework should be established to ensure that as much information as possible is gathered about adverse events in health care. The proposal was to establish a wide-ranging mandatory system for more serious occurrences and a voluntary scheme to elicit information about less serious incidents and near-misses. This multitiered approach was intended to ensure that lessons were both learned from those adverse events that did occur but also more proactively learned from those that were narrowly avoided in the past but that might occur in the future.

Shortly after the Institute of Medicine report, the U.K. NHS (2000, p. IX) expert group on learning from adverse events in health care issued a document titled "Organisation With a Memory." This argued that reporting systems are "vital in providing a core of sound, representative information on which to base analysis and recommendations." It was critical of current reporting practice in the national health care system and made four key recommendations. First, a "unified" mechanism should be developed for reporting and analysis when things go wrong. Second, a more open culture should be established to ensure that errors or service failures can be reported and discussed. Third, techniques should be developed for ensuring that necessary changes are put into practice. Finally, there should be a wider appreciation of the value of the system approach in preventing, analyzing, and learning from errors. A number of authors have challenged the usefulness of this systems view (Johnson, 2003a). For now it is sufficient to observe that these requirements do more than encourage greater reporting of adverse events involving human factors issues. Requirements to improve the reporting "culture" also crucially depend on an appreciation of human factors issues in order to encourage reporting in the first place.

USABILITY ISSUES AND PATIENT SAFETY REPORTING SYSTEMS

The publication of "To Err Is Human" and "Organisation With a Memory" served to increase the prominence of voluntary reporting systems that were already in existence at local or regional levels in several different countries. For instance,

Region	Purpose of Device Reporting
Europe	The purpose of the Vigilance System is to improve the protection of health and safety of patients, users and others by reducing the likelihood of the same type of adverse incident being repeated in different places at different times. This is to be achieved by the evaluation of reported incidents and, where appropriate, dissemination of information, which could be used to prevent such repetitions, or to alleviate the consequences of such incidents. The Vigilance S ystem is intended to allow data to be correlated between Competent Authorities and manufacturers and so facilitate corrective action earlier than would be the case if data were collected and action taken on a State by State basis.
USA	The purpose of the M edical Device Reporting Regulation is to ensure that manufacturers, (including those foreign), and importers promptly inform FDA of all serious injuries, deaths or malfunctions associated with marketed devices. User facilities report deaths and serious inj uries. As the principal US public health agency responsible for ensuring that devices are safe and effective, FDA needs such information to evaluate the risk associated with a device in order to take whatever action is necessary to reduce or eliminate the public's exposure to this risk.
Canada	The purpose of Mandatory Problem Reporting is to reduce the likelihood of recurrence of serious adverse incidents related to medical devices by evaluation of reported incidents and, where appropriate, dissemination of information which could be used to prevent repetitions or to alleviate the consequences of such incidents.
Australia	The purpose of the Incident Reporting and Investigation Scheme is to support the Post market monitoring processes under the Therapeutic Goods Act. Only a small, select group of high -risk, registered devices are evaluated by the TGA prior to being approved for sale on the market, the majority of products being listed on the Australian Register of Therapeutic Goods without evaluation. Postmarket monitoring is considered an important process to evaluate on -going quality, safety and efficacy of therapeutic devices available in the market.
Japan	The purpose is to ensure that safety and effectiveness have been carefully evaluated before approva l time, and expected adverse events and contraindications must be described on the labeling. Before the approval stage, the number of patients is restricted and only narrow ranged group of patients is involved in clinical trial. After approval, the devic e is used for a wide range of patients, and there is the possibility of unexpected adverse events which cannot be foreseen when the device is being approved. Therefore any adverse events must be tracked to ensure safety for marketed device.

Figure 31–1. Global Harmonization Task Force's (2002) review of reporting motivations.

Region	Coverage of User "Error" in Device Reporting
Europe	User errors are generally outside of the adverse reporting system except when; Examination of the device or labeling (inaccuracies in the instruction leaflet or instruction for use include omissions and deficiencies) indicated some factors which could lead to an incident involving death or serious deterioration in health
USA	Use error (errors induced by poor design, poor labeling, poor instruction, etc. which could lead to an incident involving death or serious injury).
Canada	Examination of the device or labeling (inaccuracies in the instruction leaflet or instruction for use include omissions and deficiencies) indicated some factors, which could lead to an incident involving death or serious deterioration in health.
Australia	User error is not specifically defined, but is taken to be: A situation where patient or operator injury, or near injury, is caused by incorrect use, i.e. not following instructions or labeling when these are assessed as a dequate for a "normal" or "reasonable" user. "Off label" use when either the device is not specified for the application or specifically contraindicated within the instructions for use or labeling.
Japan	Recall provisions address inadequate labeling, whic h could lead to an incident involving death or serious injury. There no such definite provisions in adverse incident reporting.

Figure 31–2. Review of user "error" in device reporting. (GHTF, 2002)

the New York State Patient Occurrence and Tracking System (NYPORTS) program was established in 1985. These early state-based schemes tended to focus on more severe accidents that resulted in patient injuries or on facility issues, including structural problems and fire hazards. The early systems also focused on eliciting reports from large secondary health care providers, such as regional hospitals and nursing homes. In Connecticut, 14,000 of the 15,000 reports received in 1996 came from nursing homes. The success of these local systems was very mixed (IOM, 1999). For example, Colorado's program initially received fewer than eight reports per year. However, with a concerted campaign to increase awareness over the benefits of reporting, this increased over a ten-year period to more than 1,000 reports per annum.

There were a number of similar initiatives scattered throughout the United Kingdom. The Edinburgh incident reporting scheme was set up in an adult intensive care unit (ICU) in 1989. It continues to be maintained by Dr. David Wright, an anesthetist ICU consultant (Busse & Wright, 2000). The scale of this system can be illustrated by the observation that the unit has eight beds, with roughly three medical staff, one consultant, and up to eight nurses per shift on the ward. A study of the incidents reported over the first ten years of this scheme found that most fell into four task domains: those relating to ventilation, vascular lines, drug administration, and a miscellaneous group. The scheme encouraged staff to describe adverse events in narrative form, as well as noting contributing factors, detection factors, and grade of staff involved in the event and that of the reporting staff. A number of studies based on this scheme found that approximately one third of the reporters had been involved in the incident that was being reported. Fewer than 10% of the reports were made by medical as opposed to nursing staff.

One of the main problems faced by these early systems was the difficulty of exchanging and aggregating data to determine whether specific incidents formed part of a wider pattern. It was for this reason that the Australian Patient Safety Foundation's system was established in 1989. The work of Runciman (2002) and his colleagues at the Australian Patient Safety Foundation had a profound impact on many health care professionals because it helped to establish a framework for what was arguably the first national voluntary reporting system with a specific remit to elicit information about human factors in adverse health care events. The federal agency for Health Care Research and Quality (AHRQ) and the National Patient Safety Foundation (NPSF) have helped coordinate similar initiatives in the United States. The NPSA fulfils this role in the United Kingdom. These organizations promote a range of initiatives that are intended to reduce "human error" in health care. They are, however, arguably most closely associated with the use of voluntary incident reporting systems as a means of detecting and then addressing common features in adverse events. The NPSF research agenda stresses the importance of "learning about systemic vulnerabilities when incidents and accidents occur; anticipating new areas of concern as change occurs; finding deeper and more generic patterns in failures; developing, prototyping, and evaluating new approaches; and linking the patterns in these to specific health care contexts" (NPSF, 2001, p. 5). The U.K. NPSA is in the process of launching a National Reporting and Learning System (NRLS) across the NHS. The national system is intended to complement local reporting arrangements so that reports entered into a local proprietary system will be automatically forwarded to the NPSA for further processing. The intention is that health care staff will be able to submit anonymous patient safety reports. These will then be "analyzed to identify national patterns, to identify patient safety priorities and to develop practical solutions" (NPSA, 2003).

The NPSA's NRLS was initially intended to help the NHS meet a series of targets. By 2005, the aim was to "reduce by 25% the number of instances of negligent harm in the field of obstetrics and gynecology which result in litigation (currently these account for over 50% of the annual NHS litigation bill); by 2005, reduce by 40% the number of serious errors in the use of prescribed drugs (currently these account for 20% of all clinical negligence litigation); by 2005, reduce to zero the number of suicides by mental health inpatients as a result of hanging from non-collapsible bed or shower curtain rails on wards (currently hanging from these structures is the commonest method of suicide on mental health inpatient wards)" (NHS, 2000, p. 86). Such high objectives must be balanced against a number of prosaic problems that limit the effectiveness of incident reporting systems. For instance, there is a danger that these objectives will act as repositories of information without inspiring direct intervention to correct particular problems. This can lead to incident starvation if potential contributors feel that their reports are being ignored. Further problems stem from the observation that reporting systems often elicit information about known issues, such as maladministered spinal injections or communications problems between particular hospital departments. The collection of such information does little to suggest possible interventions that might be used to address these long-term and deep-seated problems. Many of the issues that jeopardize incident reporting can be directly related to human factors issues. These include the problems of underreporting. The Royal College of Anesthetists' pilot reporting system found that self-reporting retrieves only about 30% of incidents that can be detected by independent audit. Other issues relate more narrowly to the biases that can affect the analysis of incident reports once they have been achieved. Finally the human factors of incident reporting can also complicate the monitoring that must be used to determine whether local and national systems are having any measurable impact on patient safety.

HUMAN FACTORS OF INCIDENT REPORTING

Previous sections have explained recent initiatives to establish incident reporting as a primary means of reducing adverse health care events in several different countries. They have also introduced some of the problems that must be addressed if these initiatives are to be successful. Some of these problems are largely technical, for instance, automated support may be necessary to identify common patterns across the thousands of documents that can be submitted to national schemes (Johnson, 2003a). However, most barriers to the successful application of incident reporting stem from human factors issues. This creates the *"recursive irony of incident reporting"* in which a series of human factors issues must be addressed in order to elicit reports about the underlying causes of, for example, human "error" in medicine. The following pages focus on several aspects of this problem. These include the difficulty of eliciting reports in the first place. This issue can be divided into two subproblems: first, how to persuade potential contributors of the benefits of their involvement and, second, how to ensure that they then provide all necessary information. I also briefly examine the problems of causal analysis; it can be difficult to avoid blaming individuals so that systemic failures can be examined. Equally, there are some incidents in which personal responsibility should not be ignored if external bodies are to believe in the probity of the system. Finally, I consider the human factors issues that arise in the development and implementation of recommendations that are intended to ensure that previous events do not recur.

UNDERREPORTING

A number of attempts have been made to estimate the scale of underreporting in health care systems. For instance, Barach and Small (2000, p. 759) state that the "underreporting of adverse events is estimated to range from 50%–96% annually." The U.K. Royal College of Anaesthetists' concluded that only about 30% of the total number of incidents detected by independent audit will be contributed by voluntary reporting systems. These caveats also affect device-related reporting systems. The U.S. General Accounting Office (GAO; 1997) conducted a study into submission frequencies two years after the requirement was introduced for manufacturers and importers to report all device-related deaths, serious injuries, and certain malfunctions to the FDA. They concluded that less than 1% of device problems occurring in hospitals were reported to the FDA. The more serious the problem, the less likely it was to be reported. A GAO follow-up study concluded that the subsequent implementation of the Medical Device Reporting (MDR) regulation, introduced in previous sections, had not corrected the problems of underreporting (FDA, 2002).

These more general assessments are based on more detailed studies. For example, Mackenzie, Jefferies, Hunter, Bernhard, and Xiao (1996) compared "deficiencies" in the management of patient airways using self-reporting and through exhaustive video analysis. The self-reporting fell into three different categories: anesthesia records constructed during the treatment, retrospective Anesthesia quality assurance reports, and a posttrauma treatment questionnaire that was filled in immediately after each case. Video analysis of 48 patient "encounters" identified 28 deficiencies in 11 cases. These included the omission of necessary tasks and practices that "lessened the margin of patient safety." In comparison, anesthesia quality assurance reports identified none of these incidents. Anesthesia records identified two, and the posttrauma treatment questionnaire suggested contributory factors and corrective measures for five deficiencies. Similarly, Jha and colleagues' (1998) work on adverse drug events compared the efficacy of three different detection techniques: voluntary incident reporting, the computer-based analysis of patient records, and exhaustive manual comparisons of the same data. In one study, they focused on patients admitted to nine medical and surgical units in an 8-month period. Both the automated system and the chart review strategies were independent and blind. The computer monitoring strategy identified 2,620 incidents. Only 275 were determined to be adverse drug events. The manual review found 398 adverse drug events. Voluntary reporting only detected 23.

A number of arguments can be used to explain the wide variation in underreporting within voluntary systems. Differences stem, in part, from the obvious methodological problems that arise when assessing the total number of adverse events that might have occurred but that were not reported. The work of Mackenzie et al. (1996) shows that the retrospective use of patient records will yield different observations of the baseline error rate than the use of more detailed contemporary video analysis. Similarly, the study by Jha et al. (1998) shows that further differences in the baseline rate can be obtained if manual inspections are supported by computer-based search techniques within medical records. It is important not to underestimate the practical consequences of inaccuracies in these baseline rates. For instance, a number of agencies have sought to establish reporting quotas based on estimates of underlying error rates. The ability to meet this quota is then interpreted as a measure of the quality of the reporting system. This then provides an indirect measure of the safety culture in the host organization. In 1998, there was considerable controversy when the U.S. Health Care Financing Administration attempted to place a cap of 2% on the medication error rate in the Medicare Conditions of Participation. This implied that it was "acceptable" if there were errors in up to 2% of medications (Shaw Phillips, 2001). The subsequent controversy also pointed to the difficulty of establishing this 2% figure as a benchmark for adverse medication events. Many different factors could introduce local variations on the underlying error rate. These include differences in the size of health care institutions, their funding profile and equipment provision, the nature and extent of the demands on their services, the profile and characteristics of the population they serve, and so on.

Having raised caveats about the difficult of assessing baseline figures for adverse events, it is still possible to assess changes in the contribution rate over time. However, this is more complex than it might at first appear. For instance, the introduction of a reporting system can encourage a "confessional" phase in which the rate of submissions is temporarily increased by the publicity and availability of a new scheme. It can also be difficult to interpret the cause of any longer term changes in submission rates. For instance, an increase in the number of contributions might reflect a rise in the error rate, and this in turn can be the result of changes in the activities of the reporting organization (Johnson, 2003a). Alternatively, any increase may be due to an increase in the willingness to report adverse events. This ambiguity can have unfortunate implications if risk managers are forced to explain why the reported number of adverse events appears to increase over time. Conversely, any fall in the number of submissions might be due either to specific safety improvements or to a lack of interest in the benefits of contributing to a reporting system. Several other factors can influence submission rates. Most notably, individual participation can depend on individual confidence in the people running the system. For instance, one Scottish hospital-based system received no contributions while the consultant who established the system was on sabbatical. When the consultant returned, the submissions quickly returned to their previous rate of around 120 reports per year (Busse & Wright, 2000).

Having recognized the difficulty of accurately assessing the scale of underreporting, a number of authors have sought to identify the reasons why health care professionals fail to contribute to incident reporting systems. For instance, Lawton and Parker (2000) issued a series of questionnaires to 315 doctors, nurses, and midwives who volunteered to take part in the study from three English NHS trusts. These questionnaires included nine short scenarios describing either a violation of a protocol, compliance with a protocol, or improvisation, where no protocol exists. Different versions of the questionnaire were presented to different volunteers so that each scenario was presented with a good, poor, or bad outcome for the patient. Participants were asked to indicate how likely they were to report the incident described in the scenario to a senior member of staff. The study showed that doctors were particularly reluctant to report adverse events to a superior. The participants were more likely to report incidents with an adverse outcome than those that might be described as "near-misses." They were also more likely to report to a senior member of staff, irrespective of outcome, if the incident involved the violation of a protocol rather than incidents in which a protocol was followed or the clinicians improvised in the absence of such guidelines. The results of this study are, however, difficult to apply across many reporting systems, because the questionnaires and the associated scenarios were drafted to identify the likelihood of report to a "senior colleague" rather than through a confidential or anonymous reporting system.

Van Geest and Cummins (2003) built on this work when they assessed the reasons why physicians fail to report or even detect adverse events. Their work formed part of a NPSF project that was established in 2001 to better understand the physicians' and nurses' experience of health care errors. The intention was then to identify the needs of each group in order to help them "combat" these adverse events. The needs assessment was conducted in two phases. First, the NPSF convened a series of focus groups to determine the origins of, and ways to reduce, health care error. These groups considered the cultural and systemic barriers to identifying, reporting, and analyzing errors in health care. The second phase of this requirement elicitation was conducted through a self-administered mail survey of physicians and nurses. The physician survey utilized a random sample of 1,084 physicians from the American Medical Association's database of all physicians practicing in the United States. The nurse survey used a sample of 1,148 nurses from the American Nursing Association.

The focus group discussions with the physicians helped to reveal a common concern over the growing complexity of many health care systems. This complexity increases the likelihood of adverse events because clinicians may not have a full understanding of the technology that they are expected to operate. Similarly, increasing complexity also stems from the close interaction between varied groups of coworkers. Communication and coordination issues also increase the likelihood of misunderstandings and other forms of adverse events. These problems result in "inefficient therapeutic approaches, lack of follow-up on ordered tests, and failure to monitor medications" (NPSF, p. 8). The physicians' focus groups went on to argue that this complexity can prevent clinicians from identifying and therefore reporting adverse health care events. This observation has recently been confirmed by an independent study of tele-medical incidents (Johnson, 2003b). The focus groups identified further barriers to reporting that help to account for the lack of participation in some systems. In particular, the U.S. physicians argued that the current culture of health care was one of tolerance to error. The authors of the NPSF report argued that denial and complacency were important factors: "Individual egos and marketplace pressures make it unlikely that error will be recognized, let alone addressed" (NPSF, p. 8). A feeling that reporting adverse events will not generate the funds or political support necessary to make sustained improvements within and between health care institutions compounded these underlying cultural issues. In consequence, few were prepared to act as "whistle-blowers" or to question the professional competence of their colleagues. This reluctance to participate was also explained in terms of the previous history of adverse event reporting within local institutions, including the tendency to blame individuals rather than seek more appropriate safeguards. These insights illustrate some of the ambiguities that characterize attitudes toward incident reporting. The physicians perceived that certain forms of error were tolerated while others would elicit a punitive response. In the questionnaire survey, 69% of respondents had identified errors in patient care. However, only 50% reported "working with" (NPSF, p. 10) nonpunitive systems for error reporting and examination. Most of the physicians stated that they knew the proper channels to report safety concerns (61%).

The focus groups involving nurses identified safety more as a "systems issue" than an issue that might be associated with particular individual erroneous actions. This was again explained in terms of the growing integration and complexity of many health care applications. It also reflected the nurses' perception that their individual work was embedded within that of their team of coworkers. However, only 49% of survey respondents agreed that safety was best addressed at the patient level. They argued that safety was better focused at the level of adverse effects on individual patients. The authors of the NPSF study argued that the nurses in the focus group felt that communications failures were one of the most important barriers to the reporting of medical errors. The adverse reporting "culture" was also identified. The nurses explained this tolerance of error in terms of a historical focus on efficiency in health care provision rather than on safety. Nurses also identified a "code of silence" (NPSF, p. 14) that permeates much of the health care system. They felt this to be particularly problematic for nurses who often are the first to identify the consequences of an adverse event but are not "empowered" (NPSF, p. 15) within the medical hierarchy. The focus groups described the sense of isolation that nurses can feel when they either commit or observe an adverse event. Both of these events can alienate them from their coworkers. More than 80% of the survey respondents stated that they had identified error in health care, 35% percent indicated that they had worked with nonpunitive systems for error reporting or examination, 87% of respondent nurses indicated that they knew the proper channels to report safety concerns, and 93% reported discussing patient safety concerns with colleagues and/or supervisors. Only

30% stated that they had read one of the Institute of Medicine's reports on patient safety. However, 72% of nurses were actively engaged in practices to identify medication errors.

Similar evidence for the causes of underreporting can be obtained from Cohen, Robinson, and Mandrack's (2003) survey of 775 nurses across the United States. Although this study focuses on attitudes toward the reporting of medication errors, it reveals a number of more general attitudes and opinions. The survey seemed to provide a consensus in favor of the benefits of incident reporting: 58% of respondents agreed that error reporting is a valuable tool to measure a nurse's medication competency, while 42% disagreed with this statement. In addition, 91% concurred that "A good way to understand why errors occur is through a thorough analysis of information obtained from incident reports" (Cohen, p. 38). Although only 36% agreed with the statement that "during my nursing career, I failed to report one or more medication errors because I thought reporting an error might be personally or professionally damaging," 64% disagreed (Cohen, p. 37). These positive statements in support of incident reporting cannot easily help to explain the problems of underreporting. However, greater insights are provided by the 51% of respondents who observed that "incident reports of my medication errors are placed in my personnel file" (Cohen, p. 38). Individuals may be reluctant to submit reports about their colleagues if it is felt that those reports will adversely affect the career prospects of coworkers. Further insights are provided by the results for the question "I initiate an incident report when I catch" (Cohen, p. 40).

another nurse's mistake
Always (37%), Sometimes (54%), Never (9%)
a pharmacist's mistake
Always (45%), Sometimes (42%), Never (14%)
a physician's mistake
Always (42%), Sometimes (39%), Never (19%)

As can be seen, nurses reveal a slightly greater ambivalence when reporting another nurse's mistake. Cohen, Robinson, and Mandrack then went on to analyze these responses in terms of their respondents' experience and work setting. Nurses working in a hospital were less likely to report another nurse's "error." Those in intensive care (23%) and orthopedic settings (29%) were least likely to report another nurse's mistake compared to other hospital settings. The proportion of nurses stating that they would always report varied from 32% to 53% in these areas. Nurses working in "home health care" were least likely to report a physician's mistake (32%). Furthermore, 67% of student nurses admitted being prepared to initiate a report for a nurse's mistake, compared with 32% of Licensed Practical Nurses and 50% of Registered Nurses. Nurses with less than 1 year (54%) or more than 15 years of experience (50%) are more likely to report a pharmacist's or physician's mistake than nurses with 1 to 15 years' experience. The proportion stating that they would report such incidents in this group varied from 21% to 45%. Finally, this survey also probed some of the ethical issues involved in terms of admitting adverse events to patients and their relatives. Only 18% agreed that they would always tell a patient or their relative if they had made a mistake; 52% would sometimes take this action, and 31% admitted that they would never disclose these details.

These findings and those of previous projects, cited earlier in this chapter, provide insights into the problems of underreporting. They do not, however, suggest immediate solutions. This is regrettable because, unless we address these human factors barriers to reporting, it is unlikely that we will obtain detailed insights into many adverse health care events. One approach would be to make all reporting mandatory rather than voluntary. This could be extended both to near-miss and to minor mishaps as well as the more serious incidents that are covered in existing reporting requirements. For example, the Joint Commission on Accreditation of Health Care Organizations ran a voluntary scheme between 1995 and 2000. Only 798 adverse events were submitted. Two thirds of these came from self-disclosure; however, one third were submitted as a result of media involvement (Minnesota Department of Health, 2000). This level of participation can be contrasted with a mandatory system operating across New York State, where 1,200 mishaps were reported by hospitals in a single year, with approximately 20,000 total submissions. Mandatory systems provide the opportunity to offer a "carrot and stick" approach; the incentive of "no blame" reporting can be combined with legal sanctions for the failure to participate. However, this raises important ethical questions, especially for near-miss or low criticality events. It can be difficult to determine whether or not a clinician had the opportunity to observe an incident or even whether an incident was reportable in the first place. There may, therefore, be a tendency for clinicians to report almost any adverse event that could conceivably be covered by the system, and hence high-reporting frequencies belie the problem that these systems are "cluttered" with relatively minor events. It is for this reason that mandatory systems tend to be used to ensure accountability, but only for more serious mishaps. It can also be argued that the fixation on underreporting misses many important issues in patient safety. Rather than focusing on the information that might not be contributed through a voluntary system, more progress could be made by addressing those concerns that are elicited from health care professionals. In this view, we should focus more on improving safety and less on counting mistakes.

Cohen's (2000) review of mandatory and voluntary reporting systems reiterates many of these points. He identifies the U.S. SMDA of 1990 as an example of a mandatory reporting system that has been "unsuccessful in gaining compliance with reporting requirements for user error" (Cohen, p. 40). As we have seen, the intention behind this federal bill was to state directly that health care facilities and manufacturers must report adverse events related to the failure or misuse of specific medical devices. However, Cohen argues that little action is taken unless the system receives reports about a large number of similar adverse events. He also argues that the state-based mandatory systems are used "almost exclusively to punish individual practitioners or health care organizations." In consequence, mandatory systems often fail to provide insights into the deeper causes of adverse events, which Cohen argues are largely "systemic" rather than "individual." Cohen's view is typical of that put forward by many clinicians working in the area of patient safety. It is well intentioned, but often too narrowly focused on process improvement rather than the patient's concerns, which can also include a desire for greater transparency and equity in the handling of adverse events.

This apparent conflict has been exposed by Robinson et al.'s (2002) study of physician and public opinion on the quality of health care and medical "error." They compared the results of a mail survey of 1,000 Colorado physicians (n = 594) and 1,000 national physicians (n = 304) with a telephone survey of 500 Colorado households. The main aim was to assess their differing attitudes toward some of the main findings in the Institute of Medicine report "To Err Is Human," mentioned earlier. They found that 70% of Colorado physicians believed that the reduction of medical errors should be a national priority. However, only 29% of physicians believed that "quality of care was a problem," compared with 68% of the wider population in this sample. Similarly, only 24% of physicians believed that a national agency is needed to address the problem of medical errors, while 60% of the public agreed with this statement. All of the physicians believed that fear of medical malpractice was a barrier to reporting of errors and that greater legal safeguards are necessary for a mandatory reporting system to be successful. Furthermore, 60% of the physicians agreed that it is difficult to differentiate errors due to negligence from unintended errors.

In April 2000, the U.S. National Academy for State Health Policy conducted an investigation into the state reporting of medical errors and adverse events. They found that 15 states (Colorado, Florida, Kansas, Massachusetts, Nebraska, New Jersey, New York, Ohio, Pennsylvania, Rhode Island, South Carolina, South Dakota, Tennessee, Texas, and Washington) required the mandatory reporting of adverse events from general and acute care hospitals. The levels of participation and the scope of these schemes were very different. The types of events to be reported included unexpected deaths, wrong site surgery, major loss of function, and errors in medication. The National Academy for State Health Policy also conducted a survey of states in February 2000 to examine the way in which various states were addressing the issue of medical error reporting. All 50 states and the District of Columbia responded to the survey. The survey found that most states, including the 10 with mandatory schemes, aggregated the data to identify trends. Nine states administer sanctions and assure corrective action. Eight states issue public reports. In another study of state-based reporting programs by the same agency, several states expressed concerns that their mandatory reporting systems suffer underreporting from hospitals (Raymond & Crane, 2001). The diverse practices identified in these reports motivated a not-for-profit group, known as the U.S. National Quality Forum, to propose a national strategy for health care quality measurement and reporting.

The proponents of mandatory systems argue that some adverse events are so serious that they must be reported in order to reassure the public and ensure that appropriate action is taken. The proponents of voluntary systems, in contrast, point to the problems of underreporting in mandatory systems and to the difficulty in "policing" reporting requirements. They also point to the success that some voluntary systems have had in encouraging participation when health care professionals are offered protection against legal sanction. For example, U.S. Pharmacopoeia and Institute for Safe Medication Practices has established the Medication Errors Reporting Program. This confidential, voluntary medication error-reporting scheme has received around 1,000 error reports each year. Cohen (2000) has argued that the quality of reports made to this voluntary system is just as significant as the number of submissions. He also cites the example of cisplatin. After a series of accidents, the Institute for Safe Medication Practices persuaded manufacturers to include the maximum dose on phial caps and seals.

Few of the proponents on either side of this debate advocate exclusively mandatory or exclusively voluntary schemes as a solution to the problems of underreporting. In contrast, controversy surrounds the extent to which health care professionals should have the discretion to determine what is reportable under each of the various schemes. As mentioned earlier, the Institute of Medicine advocates a national mandatory system for more serious mishaps and a local voluntary system feeding information through regional schemes in the case of less serious adverse events. This architecture is intended to ensure that a national voluntary system is not inundated by a mass of low-risk incidents; local managers help to filter the passage of information up through state schemes to national systems, whereas the more serious events merit a more immediate focus at a higher level. This mixed approach of mandatory and voluntary reporting will only successfully tackle the problems of underreporting if the schemes are supported by legal protection for individual participants. Any breach of confidentiality in general and the (ab)use of voluntary reports in any consequent litigation would undermine confidence in the scheme. For example, the American Medical Association's (2002) recent statement to the Subcommittee on Health, Committee on Energy and Commerce in the U.S. House of Representatives on reducing medical errors argued that Congress must "pass legislation that will encourage reporting of health care errors without the fear of punishment." The primary goal of this legislation would be to facilitate the development of a "confidential, non-punitive, and evidence-based system" for reporting health care errors. They went on to argue that "Congress can help create a culture of safety by allowing medical professionals to convene to discuss patient safety problems and potential solutions without having their discussions, findings, or recommendations become the basis for class action or other lawsuits." Partly as a result of these concerns, a number of initiatives have attempted to reduce underreporting by ensuring that voluntary incident reports are subject to the same legal protection offered by similar schemes in other domains, in particular by NASA and the Federal Aviation Administration's (FAA's) Aviation Safety Reporting System (ASRS).

The ASRS operates an elaborate mechanism whereby reports are initially passed to NASA. They then screen each submission to ensure that information relating to a criminal offense is passed to the Department of Justice and the FAA. Information about accidents rather than incidents is passed to the NTSB (National Transportation Safety Board) and the FAA. All remaining reports fall within the scope of the ASRS and are therefore protected under the following provision. Section 91.25 of the Federal Aviation Regulations prohibits the use of any reports submitted to NASA in any disciplinary action. However, appropriate action can be taken if information about an incident is derived from a source other than the ASRS submission. In addition to the provisions that protect contributors, the action of filing a report is considered to be "indicative of a constructive attitude." Accordingly, the FAA will not seek to impose a civil penalty or suspend a license if the individuals involved submit a report within 10 days of the incident and the violation was inadvertent, if it did not involve a criminal offense, or if it was an accident. These exemptions apply providing that the person has not committed a violation for a 5-year period prior to the date of the incident. These guidelines within the field of aviation are worth citing because they have provided a blueprint for similar protection, which is being offered under health care reporting systems. For example, the Veterans Health Administration's (VA's) National Center for Patient Safety has established two systems. The first is a mandatory reporting scheme for more serious adverse events. This is similar to the FAA and NTSB provision within the ASRS. The second confidential, voluntary Patient Safety Reporting System (PSRS) is closer to the ASRS itself. The PSRS was established in 1999 and allows for local initiatives both to encourage reporting and to initiate remedial actions. This scheme depends on an interagency agreement between the VA and NASA and was modeled on the provisions for the ASRS that were cited earlier. Unlike the ASRS, the VA's PSRS is intended to collect information on adverse events, as well as near-misses. NASA collects the reports and maintains the confidentiality of the system. Under the agreement, the VA may not review any report or data until it has been de-identified. Concern over the inadvertent disclosure of contributor information has led to the decision that these will not be held once the event has undergone an initial analysis. The contributor's identity is also protected under the Privacy Act and recognized exemptions to the Freedom of Information Act. Records created for the VA as part of a medical quality assurance program, such as patient safety reports, have additional protections beyond those of other government agencies. U.S. Code (USC) 5705, with certain exceptions, provides that records and documents created by the VA "as part of a medical quality-assurance program" are "confidential and privileged and may not be disclosed to any person or entity." One recent review of these confidentiality measures raises the caveat that "although federal law appears to provide the VA considerable protection against the discovery and disclosure of data, these unique legal shields are not afforded to non-VA hospitals" (Raymond & Crane, 2001, page 7).

Previous paragraphs have described how a range of human factors issues, including a fear of retribution and concern over the efficacy of any contribution, help to create the problems associated with underreporting. We have also reviewed a wide range of initiatives to address these problems, including the development of mandatory and voluntary schemes as well as the provision of legal protection against disclosure in confidential systems. There are other approaches that help to address the problem of underreporting. In particular, sentinel schemes acknowledge that underreporting will always be a limitation of large scale systems. In contrast, sentinel schemes focus resources more narrowly on a small number of representative institutions or work teams. These groups are given additional training and resources to both encourage and support any reporting. Monitoring systems and exhaustive reviews of patient records may also be used to catch any incidents that are missed. The results from these investigations can then be extrapolated to provide additional insights into the potential scale of any problems at a regional or national level. The FDA was among the first to recognize the potential benefits of this approach (FDA, 1999). They recognized that underreporting was often a feature of what can be described as "passive" national monitoring systems. In contrast, a more active approach would be to take steps that continually remind staff of the benefits derived from incident reporting. Unfortunately, the FDA lacks the resources necessary to train every potential contributor in when and how to submit an incident report. Estimates suggest that there may be 50,000–60,000 "end user" organizations for health care–related devices. They therefore decided to conduct a trial in which a small number of organizations were provided with additional support to explicitly encourage participation in a voluntary reporting system.

Seventeen hospitals and six nursing homes were recruited to participate in a 12-month "DEVICENET" study. Coordinators were identified in each institution; these individuals were typically clinical risk managers. They were offered either a 1-day group training in Washington, D.C., or a slightly shorter course in their own organizations. Videos were also prepared for each of the participant institutions. These were intended for use during in-house staff orientation and in-service training sessions. The video encouraged individuals to follow their facility's internal procedures for reporting of adverse events. After viewing the video, each staff member in the participating institution was given a one-page sheet summarizing the local provision. These sheets also provided information about the confidentiality safeguards offered to participants. Each report had any individual identification information removed as soon as possible after it had been received. After 30 days, the facility ID was removed so that it was no longer possible to link the report to the facility. This period enabled the study team to link the original report with any follow-up reports and provided an opportunity to discuss any questions about the report with the study coordinator. The sentinel trial also enabled participants to contribute anonymous reports. At the end of the year's study, the coordinators had gathered 315 reports, of which 14 were anonymous. They argued that this level of activity was "far above" the average for reporting device-related incidents. By a broad-brush extrapolation, the proponents of this approach suggested that the FDA would receive 100,000 reports per year rather than the 5,000 incidents that were actually filed during the year of the study. However, it is important to note that this study also illustrated some of the limitations of sentinel reporting. A continuing problem for the FDA is that many nursing homes fail to contribute any reports of adverse events even though they operate many of the devices and procedures that give rise to problems in other health care settings. In spite of all of the additional support offered in this trial, none of the 315 reports came from any of the six participating nursing homes.

Sentinel schemes reduce the problems of underreporting by focusing resources on a number of "representative" institutions. A limitation with this is that sentinel schemes may lack the resources to ensure that focused support is provided across all procedures and departments even within one of these favored organizations. In consequence, patient safety organizations have also funded centers to focus on different aspects of patient safety. The VA has established four Centers of Inquiry, with an annual budget of approximately $500,000. For example, Gaba heads a center that studies patient safety in the operating room and is looking at the use of patient simulators in anesthesia (Weeks & Bagian, 2000). Other centers focus on elderly patients. These Patient Safety Centers of Inquiry act as focal points for research and development. They are not primarily intended to support incident reporting. However, it seems clear that their research activities must draw on those adverse events that the VA and other organizations elicit about their main interests. It is also important to recognize that current plans for regional and national reporting systems are often very general. They accept reports from a broad range of health care professionals. There is a risk that they will fail to elicit the support that has been obtained for more specialized systems, such as the pilot scheme promoted by the U.K. Royal College of Anaesthetists (1998). If this is not addressed, then a number of subject-specific sentinel systems are likely to be used by professional organizations to augment the more general national, voluntary schemes being promoted by groups such as the NPSA.

ELICITATION AND FORM DESIGN

The previous section focused on the human factors issues that lead to underreporting. In contrast, this section looks more at the problems of ensuring that adequate information is obtained once a health care worker has decided to submit an incident report. This is not as simple as it might seem. In particular, it may not be possible to interview staff in order to elicit additional details in anonymous schemes. In confidential systems, there is also the danger that any subsequent contacts with managers may inadvertently disclose the identity of the contributor in the process of providing further information. In such circumstances, it is imperative that human factors and human–computer interaction expertise be used to ensure that the design of reporting forms is tailored to support the skills and expectations of potential contributors. The difficulties associated with this task are exacerbated by requirements such as those proposed by the U.S. Quality Interagency Coordination Task Force's report to the president, which strongly urge the provision of facilities for members of the general public to also contribute to health care reporting systems (American Iatrogenic Association, 2000). However, this document focuses on possible improvements to the usability of clinical applications and products. The authors argue that there "is a real need to involve clinicians and other users in the design of systems at an early stage to optimize usability … manufacturers need to ensure that usability testing occurs throughout development, especially in the pre-market design phase of medical device development" (AIA, p. 74). This document neglects the problems of developing usable reporting systems. This is an omission that is common among almost all of the other reports advocating the development of health care incident reporting systems. Those that do consider these issues often make passing comments to the development of web-based interfaces as a panacea for the problems of form design.

Some studies do acknowledge the importance of developing "usable" submission systems for adverse event reports. For instance, a recent roundtable discussion on "Design Considerations for a Patient Safety Improvement Reporting System" organized by the Kaiser Permanente Institute for Health Policy, NASA Aviation Safety Reporting System, and The National Quality Forum recently stressed the role that usability plays in ensuring participation with health care reporting systems (Kaiser Permanente, 2000). This is a laudable aim, but it provides few insights into the mechanisms and techniques that might be used to satisfy such usability requirements. The U.K. NPSA has addressed these concerns through a sustained pilot study for its NRLS. Their aim was to gather feedback on the best ways of reporting to a national system. They evaluated the proposed NPSA standardized method for collecting information. In particular, they studied the usability of the different forms that were "tailored" for each health care sector. Part of this process involved an analysis to ensure that they were asking the right questions to ask NHS organizations and staff to "elicit the maximum amount of meaningful national learning" (NPSA, 2003).

The design of incident reporting forms has remained a focus of debate among the handful of research groups that are active in this area (Johnson, 2000). Meanwhile, hundreds of local, national, and international systems are using ad hoc trial-and-error techniques to arrive at appropriate forms. It is important to stress that there are several different approaches to the presentation and dissemination of incident reporting forms. For example, some organizations provide printed forms that are readily at hand for the individuals that work within particular environments. This approach clearly relies on the active monitoring of staff who must replenish the forms and who must collect completed reports. Other organizations rely on computer-based forms. These can either exist in printable formats such as Adobe's PDF, which must be printed and completed by hand, or in electronic form so that they can be completed online. In either case, there is an assumption that staff will have access to appropriate hardware and software resources. This is not always the case in many health care domains. Many of these machines may also be located in public areas where colleagues and coworkers can observe the submission of an incident report. Each of these different approaches may also be supplemented by, for instance, telephone-based reporting for situations in which forms are unavailable. This plethora of submission techniques is further complicated by the observation that personnel are increasingly expected to file reports through multiple systems. For instance, local voluntary systems such as those proposed by the Institute of Medicine currently operate alongside several mandatory state-based schemes at the same time that federal agencies, including the CDRH (Center for Devices and Radiological Health), also operate national systems. There are also often different parallel schemes for reporting incidents that injure employees rather than patients.

Given this diversity, it can be very difficult to establish under which system to file a report. For instance, many of these schemes define the severity of an incident that should be reported to them. In many cases, however, health care workers may not know what the ultimate outcome of a mishap will be. For instance, medication errors often have uncertain, long-term effects. Should an individual begin by reporting to a local system and then file successive reports to regional and national systems as the results of the incident become more certain? Alternatively, some hospitals have established "one-stop shops" where all reports are filed via a risk manager who ensures that local information is fed into regional and national schemes depending on the nature of the incident. For instance, Spectrum Health facilities (2002, p. 10) have recently introduced a wide-ranging patient safety plan. Employees are required to report any "defect, error, medical accident, near miss, good catch, significant procedural variance, other risk to safety that could result in patient injury, hazardous condition, or risk in the environment of care." Area managers must ensure completion of each report. They must then pass on information about critical events to the vice president, Risk and Compliance, and the vice president, System. The chief operating officer, in consultation with the vice president of Risk and Compliance and the vice president of Quality Improvement, then together determine if the event is reportable to the JCAHO (Joint Commission on Accreditation of Healthcare Organizations) and any additional regulatory bodies. Any approach that relies on such a filtering process must ensure that personnel are confident their reports will be passed on in a timely fashion. They must also be assured of the "just culture" that was mentioned in the previous section if they are to direct all reports via these gatekeepers to the reporting systems.

The design and layout of reporting forms remains a critical issue irrespective of whether individuals report directly to external agencies or via a local safety manager. If potential contributors cannot use the fields of these documents to accurately provide necessary information then there is little likelihood that incident reporting systems will provide an effective tool for "organizational learning." Form design is therefore a critical area for human factors input in the development of most reporting systems. Thus it is surprising that many systems are implemented without even the most cursory forms of user testing (Johnson, 2003a). In consequence, it can be difficult to determine whether underreporting stems from a widespread rejection of the system or from acute frustration with the electronic and paper-based forms that are intended to elicit feedback about past failures. User testing is important because a vast range of different approaches have been used to elicit information about adverse events. For example, Figure 31–3 illustrates a reporting form that was developed for a local system

Critical Incident Study

This is a study that looks at how and why people make mistakes. Information is collected from incident reporting forms (see overleaf) and will be analysed. The results of the analysis and the lessons learnt from the reported incidents will be presented to staff in due course. The reporting forms are anonymous, there is no interest in criticism or blame. We would encourage everyone working in the NICU, at whatever level of experience, to take part. Every incident reported, no matter how trivial, will give information about the way people work and may help to save a life.

When you have completed the form please place it in the Incident Form Box.

Definition of a "Critical Incident"

A critical incident is an occurrence that might have led (or did lead) – if not discovered in time - to an undesirable outcome. Certain requirements need to be fulfilled:

1. It was caused by an error made by a member of staff, or by a failure of equipment
2. It can be described in detail by a person who was involved in or who observed the incident
3. It occurred while the patient was under our care
4. It was clearly preventable

Complications that occur despite normal management are not critical incidents. But if in doubt, fill in a form.

Thank you for your interest.

Critical Incident Reporting Form
(See overleaf for instructions)

The Incident
Description of what happened:

What factors contributed to the incident?

What factors minimised the incident?

The Circumstances
Date: Time: Place:

What procedure was being carried out?

What monitoring was being used?

If equipment failure give details of equipment:

Personnel
Grade of relevant responsible staff: Grade of staff discovering the incident:

Outcome
What happened to the patient?

Prevention
How might such incidents be avoided in the future?

Figure 31–3. Reporting form for a neonatal intensive care unit. (Busse & Wright, 2000)

within a U.K. Neonatal Intensive Care Unit (Busse & Wright, 2000). There are open "free-text" fields for individuals to describe the incident that they have witnessed. Such open-ended questions are appropriate in systems where it is possible for analysts to go back and ask additional questions to clarify any information that is either missing or only partially understood. The benefit of the approach is that it makes only minimal assumptions about the information that the contributor wishes to report. They are not forced to select particular items from a predefined list that may unduly constrain their selections. However, problems arise when analysts must translate the information provided by these "open" forms into the format that is required by regional or national agencies. For example, the NPSA has developed the U.K. NRLS to define the questions and reply options that will be used to collect incident information. One of these questions focuses on "What happened?" Contributors are expected to categorize patient safety incidents according to a large number of predefined choices. These include high-level categories such as incidents involving "Access, admission, transfer, discharge (including missing patient)" or problems involving "Clinical assessment (including diagnosis, scans, tests, assessments)." Each of these high-level categories is further refined into more detailed choices. For instance, more detailed assessment problems: "Assessment—lack of clinical or risk assessment," "Cross-matching error," and "Delay/difficulty in obtaining clinical assistance." Similarly, the high-level category of "Patient accidents" is refined into "Ambulance/patient in road traffic accident," "Collision/contact with an object," "Contact with sharps (includes needle stick)," "Exposure to hazardous substance," "Exposure to cold/heat (inclu des fire)," "Inappropriate patient handling/ positioning," "Slips, trips, falls," or "Other." The need to collect national statistics helps to justify the use of such narrow categories. It is important that data about national trends be gathered in a consistent format. However, this creates problems for local systems such as that illustrated earlier. Clinical risk managers must examine these free-text accounts and then reclassify them according to the guidelines laid down by regional or national organizations such as the NPSA. This can be difficult because it is not always possible to define an ideal match between the details of a particular adverse event and these different categories, particularly if, for example, an incident involved a mixture of events such as a problem in patient access that led to an incorrect clinical assessment.

The NPSA has developed a number of alternatives to the free-text approach for incident reporting. It has collaborated with a number of software developers so that computer-based tools guide potential contributors through the classification process. Participants need never see the hundreds of individual fields in the full taxonomy. Instead, they are only shown those options that are relevant to the incident they are reporting. This relevance is partially determined by the contributor's previous responses to questions about the adverse event. In this limited sense, these computer-based tools are context sensitive. They tailor the elicitation to match the incident that is being described. A limitation, however, is that it can be difficult to ensure that any two contributors will assign the same key words to similar incidents. This is, arguably, more likely to happen when risk managers are trained to classify the free text accounts of their coworkers (Johnson, 2003a). It is difficult to ensure that every potential contributor receives a comparable level of training. Similarly, problems can arise when potential participants cannot find the key words to match the incident that they have witnessed. The NPSA taxonomy addresses this problem in a number of ways. First, the detailed categories usually include the value "other" under each of the high-level terms such as "Patient accidents." This broad classification can act as a catchall. There is, of course, a danger that contributors will too readily use the "other" classification if they do not understand what is meant by the more detailed terms. Second, the NPSA include a question at the end of their classification, which asks contributors to "Please tell us how you think this form could be improved (optional)."

Previous paragraphs have argued that incident reporting systems will be undermined if they embody an incomplete taxonomy; contributors will not be able to provide sufficient information about the incident by ticking appropriate boxes and so forth. This problem can be overcome by extending the taxonomy to ensure that it is broad enough to cover every likely eventuality. However, this creates further problems if users have to navigate hundreds or even thousands of complex terms. Computer-based reporting systems can help to overcome these problems by guiding the users so that the answers to previous questions can help to filter the options that they are presented with. For example, if the user indicates that they he or she has witnessed a patient-related accident, then they are not usually presented with menu options or check boxes that relate to an error in diagnosis. Of course, there may be some unusual incidents that stem from precisely this combination of issues, and thus extensive user testing is required to ensure that users can exploit tool support without becoming so frustrated that they will abandon a submission. User testing can also help to reveal other biases. For instance, there is a tendency for users always to select items from the top of a scrolling list or menu. Few users will scroll to the bottom of long and complex widgets. This can have an unfortunate influence on the findings that may be derived from a reporting system where the position of an item on the display can determine whether or not users recognize it as an attribute of a particular incident. The NPSA has acted to address these problems by conducting a series of field studies into the application of the NRLS between January and May 2003. Thirty-nine organizations from a range of health care settings worked with the NPSA to test reporting methods. Most of this work focused on the completeness and consistency of the taxonomy rather than an evaluation of the computer-based systems being developed by the NPSA's commercial partners. However, it did examine whether an electronic reporting form could "provide a standardised method for collecting information, and the best way to tailor the form to each health care sector" (NPSA, 2003).

The local, paper-based reporting system from an ICU, illustrated earlier, forms a strong contrast with the demands for national reporting, demonstrated by the NPSA's initiatives. There are, however, a number of other reporting systems that do not fit into either of these different stereotypes. For instance, Figure 31–4 illustrates part of the online system that has been developed to support incident reporting within the Swiss Departments of Anaesthesia (Staender, Kaufman, & Scheidegger, 1999). CIRS (Anaesthesia Critical Incident Reporting System) embodies a number of assumptions about the individuals who are likely to use the form. Perhaps the most obvious is that they must be computer literate.

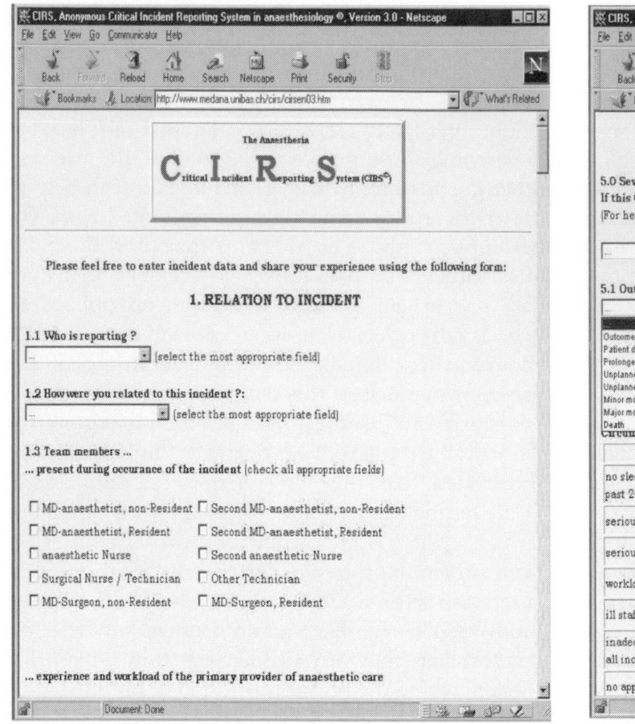

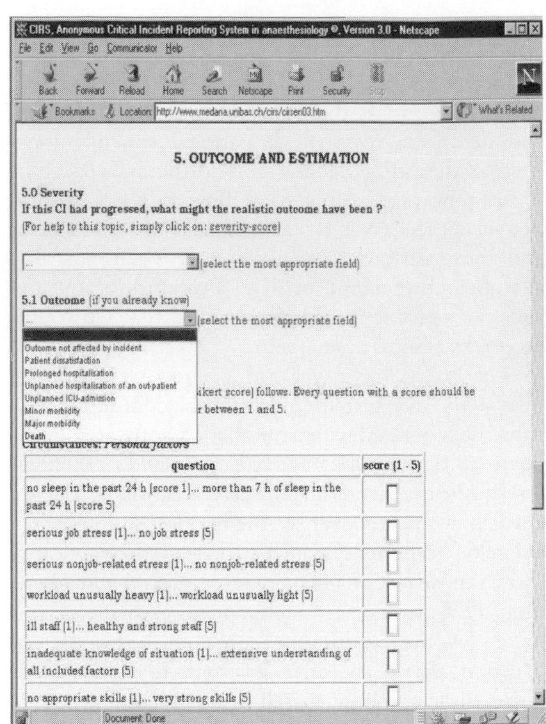

Figure 31–4. The CIRS reporting system. (Staender et al., 1999)

This is significant because CIRS exploits a diverse range of dialogue styles, or interface widgets. These include check boxes and pull-down menus as well as free-text fields. This system is different from the one proposed by the NPSA, because it was established as the result of a self-help initiative from a number of motivated clinicians. It was not set up as part of a government system, although it subsequently attracted this support. Equally, it differs from the local system because it developed beyond a single hospital and hence could not easily be sustained using limited resources and a paper-based approach. As can be seen, CIRS also exploits a number of predetermined categories to characterize each incident. Users must select from 1 of 16 different types of surgical procedure that are recognized by the system. They must also characterize human performance along a number of numeric Likert scales. These are used to assess lack of sleep, amount of work-related stress, amount of non-work-related stress, effects of ill or healthy staff, adequate or inadequate knowledge of the situation, appropriate skills, and appropriate experience. For example, if the individuals involved in the incident had no sleep in the last 24 hours, then the score should be 1. If they had more than 7 hours' sleep, then the score should be 5. Scores between these two extremes should be allocated in proportion to the amount of sleep that had been obtained by the participants. This approach is relatively straightforward when referring to objective amounts of sleep. However, the CIRS workload scale is more difficult to interpret in the same range from 5 (unusually heavy) to 1 (unusually light). The introspective ability to independently assess such factors and provide reliable self-reports again illustrates how many incident reporting forms reflect the designers' assumptions about the knowledge, training, and expertise of the target work force.

One of the most innovative features of the CIRS system is that it is possible for health care professionals to use the Internet as a means of reviewing information about previous incidents. The anonymous cases can be read online and comments can be appended to create a dialogue between individuals either who request additional (anonymous) information or who have experienced similar incidents in other organizations. For example, the following report describes a drug misadministration:

> INCIDENT DESCRIPTION: Female patient 11 y/o was scheduled for tonsillectomy. She was NORMAL as regards the physical examination and lap values. The operation was done as usual without any abnormal events in anesthesia or the recovery. She was discharged awake from PACU to the ward. Shortly after here [sic] arrival, the ward's nurse inject here [sic] by what she was think that is antibiotic. But soon she discovered that this was BROFEN (Ketoprofen) suspension. The poor child developed convulsions and cyanosis at once. She was transmitted quickly again to the OT. The patient was hypotensive (70/40) tachycardic (180) O2 saturation was 75% and the end tidal CO2 was 70.

Another clinician accessing this report left the following observation and request for additional information:

> Sorry about the sad case. Side question: why was an antibiotic given and why afterwards? It seems that some accidents occur as a consequence of an action that wasn't unnecessary in the first place. For example, I once heard of an appendicectomy [sic] case that got an epidural injection where a mix-up also occurred with fatal consequences.

A key point here is that the initial report acts as a focus for further discussion about common factors in previous incidents. From a technical standpoint, this type of facility also requires that the reporting system be extended beyond the forms that elicit information about the initial adverse event to include some mechanism for further dialogue.

FORM CONTENT AND DELIVERY MECHANISMS

Irrespective of whether a reporting system is intended to collect information about local incidents or national statistics about adverse events, it is important that managers consider the range of information that must be elicited in the aftermath of an accident or a near-miss. This includes factual data about what precisely happened. Reporting forms can also prompt potential contributors for more analytical information about what they consider to be potential causes for an adverse event. This section reviews some of the human factors issues that must be considered during the development of reporting forms. It also considers the usability issues that affect the different delivery mechanisms, which enable potential contributors to submit information about near-misses and adverse events.

Delivery Mechanisms

Most of the early reporting systems relied on simple paper-based forms, similar to that illustrated in Figure 31–3. Increasingly, however, as the CIRS and NPSA initiatives show, more schemes are relying on Internet technologies. This approach has numerous benefits. For example, managers do not need to continually check the supply of paper-based forms nor do they have to monitor drop-boxes to check for new submissions. Electronic forms can be revised and then made accessible across many different health care organizations without the overheads associated with conventional distribution networks. The use of appropriate interface design techniques can support users by providing default values for common fields in electronic forms. Inferences can be made to populate these online documents. For example, the date of the report can be set to the day on which the form is accessed unless the end user decides to change it. Similarly, the organization in which an incident occurs might default to the one in which the reporting system is accessed.

A range of problems also complicates the use of computer-based reporting systems. In particular, many organizations have significant concerns about the security of online systems that can be vulnerable to attack from people both inside and outside the reporting organization. These considerations are particularly important given the sensitive nature of confidential and anonymous reporting. There are technical solutions for many of these issues. For instance, digital signatures can be used to authenticate the sender of particular information. Electronic watermarks can be used to ensure that reports are not overwritten or unnecessarily altered after submission. However, many of these technical solutions increase the burdens of system operators. For example, they may be excluded from the system if they do not authenticate their access through the use of an appropriate password. The importance of password protection can be illustrated by a recent mishap reported to the FDA.

> Since the placebo treatment is still active in this version of software (revision 9), it is possible to unintentionally deliver a placebo treatment. This site was not involved in any of the past clinical trials ... and it appears coincidental that the reporter used this particular password ... one possible scenario dicussed [sic] was the x-ray tech operating the unit during this time somehow mistook the default physics password "9999" for "4444," which means the operator would have also confused the treatment password with the physics password. However, this is speculation and could not be confirmed. (MDR text key: 1490034)

Although this incident did not involve password access to an incident reporting system, it does illustrate the generic problems that arise from security mechanisms in health care applications. The use of such a simple numeric password for a placebo is likely to lead to problems in the future. This is illustrated by the potential conflict with the default physics password. The physics department might, in turn, also be criticized for their choice of "9999." This cannot easily be justified as a secure password. There are common examples of such security "failures" across most health care systems (Johnson, 2003a). In consequence, many potential contributors can be dissuaded from participating in a system if they believe that their identity will be compromised by unauthorized access in the future.

Further limitations also affect the use of computer-based forms to elicit incident reports. In particular, it can be difficult to ensure that all potential contributors have easy access to the necessary technical resources. There is a nonuniform distribution of staff with home access to the Internet. Many health care professionals only have work-time access to shared computers in public spaces. They can easily be interrupted or observed as they fill out a confidential or anonymous reporting form online. It can also be possible for other users to access information that their colleagues have entered either by accessing system logs or cached information that has inadvertently been left on disk after a session has ended. Usability issues also determine whether or not potential contributors can easily use computer-based submission systems. The developers of online reporting schemes often assume that end users will have similar technical resources to themselves. For instance, many web-based forms are formatted for large, high-resolution displays. However, U.S. statistics for 2002 indicate the 45% of web users have access to 1024 × 768 pixel displays, 50% use 800 × 600 displays, and 2% are still using 640 × 480 displays. This creates problems because the users of lower resolution equipment will have to scroll through forms in order to access all of the necessary fields. Usability studies have shown that the completion rates for online forms are inversely proportional to the amount of scrolling that users must engage in. For example, many potential contributors quickly become frustrated if they have to move up and down a screen to refer between linked items of information (Johnson, 2003a). The problems of heterogeneous display hardware are exacerbated by software incompatibilities. For instance, cascading style sheets enable designers to ensure that all of the pages in a Web site possess a common look and feel. Changes can easily be propagated through different areas of the site in response to changes in the style sheets rather than forcing manual updates across dozens of pages. However, there is no guarantee that every potential contributor will have access to a machine that implements these style sheets. Early versions of Microsoft's Internet Explorer and Netscape's Communicator do not support this facility. Those that do enable the use of this implementation technique do not all render them in the same manner. It follows that the results of usability tests performed under particular software and hardware configurations will seldom provide a coherent view of the diverse user experience when online systems "go live." The difficulties in ensuring access to computer-based reporting forms have led many national and regional systems to operate hybrid approaches. Online systems are provided alongside paper-based forms or telephone numbers that can be used to leave verbal accounts of adverse events on answering machines.

Preamble and Definitions of an Incident

It is important to provide users with a clear idea of when they should consider in making a submission to the system. The NPSA data set does this through the enumerations that are provided to the user. Each of the items and subsections provides some indication about the types of adverse event that fall within the scope of the system. This guidance is not available within the local system that relies more on free-text fields. In consequence, the local scheme in Figure 31–3 explicitly states that an incident must fulfill the following criteria: "1. It was caused by an error made by a member of staff, or by a failure of equipment. 2. A person who was involved in or who observed the incident can describe it in detail. 3. It occurred while the patient was under our care. 4. It was clearly preventable. Complications that occur despite normal management are not critical incidents. But if in doubt, fill in a form." (Busse & Wright, 2000). It can be surprising that incidents, which occur in spite of normal management, do not fall within the scope of the system. This effectively prevents the system from targeting problems within the existing management system. However, such criticisms neglect the focused nature of this local system, which is specifically intended to *target the doable* rather than capture all possible incidents. CIRS exploits a wider definition of an adverse incident. "Defining critical incidents unfortunately is not straightforward. Nevertheless we want to invite you to report your critical incidents if they match with this definition: an event under anaesthetic care, which had the potential to lead to an undesirable outcome if left to progress. Please also consider any team performance critical incidents, regardless of how minimal they seem" (CIRS, 2006). This could potentially cover a vast range of adverse events. Such a definition would stretch the limited resources of many local or national systems. It also illustrates the way in which the definition of an incident both determines and is determined by the reporting system that is intended to record it. The definition must be broad enough to capture necessary information about adverse events. However, if the definition is drawn too widely, the system may be swamped by a mass of low-risk mishaps and near-misses so that it can be difficult to identify critical events in time to take corrective actions.

It is important also not to forget that the definitions of what should be reported are one of many contributory factors that help to determine whether or not a health care professional will actually submit a report. For example, Lawton and Parker (2002) show that adverse events are more likely to be reported when staff deviate from written protocols. They argue that professionals are unwilling to challenge a fellow professional without strong grounds. They are also reluctant to report behavior that has negative consequences for the patient when the behavior reflects compliance with a protocol or improvisation where no protocol is in place. The key issue here is that such observations about reporting behavior are often orthogonal to the abstract definitions of adverse events that form the basis of many reporting systems (Kaplan, Battles, van der Schaaf, Shea, & Mercer, 1998).

Identification Information

Previous sections have described Spectrum Health's reporting system as an example of a "gatekeeper architecture." Key managers have a responsibility to determine whether reports should be handled locally or whether they should be passed to external agencies. These managers act as the "gatekeepers" to various reporting systems. This approach has numerous benefits in terms of accountability. However, these schemes rely on employees' providing identification information so that risk managers can gather further data if a mishap requires subsequent investigation. These systems, therefore, provide confidentiality rather than anonymity. In contrast, neither the NPSA data set nor the local intensive care system elicits any direct information about the contributor's identity. This anonymity is intended to encourage participation. However, it clearly creates problems during any subsequent causal analysis. It can be difficult to identify the circumstances leading to an incident if analysts cannot interview the person making the report. Without a "gatekeeper" approach, the developers of confidential systems have to provide considerable additional details to ensure that their system does not short-circuit or filter reports that should be submitted under mandatory or regional systems. For instance, the NPSA must distinguish their system from the Serious Untoward Incident (SUI) reporting system that informs Strategic Health Authorities and the Department of Health about incidents that require urgent attention: "They may not necessarily be patient safety incidents, and will often include identifiable information to enable action at a local level. For this reason it is not appropriate to combine the two systems." Similarly, the NPSA is keen to point out that any incidents involving the use of a medical device will be shared with the Medicines and Healthcare products Regulatory Agency (MHRA): "Because the anonymous nature of NPSA information will prevent the MHRA from investigating what happened, you should instead report these incidents directly to the MHRA" (NPSA, 2006a).

The anonymity of a reporting system can be compromised in local reporting systems. Inferences can be made about the identity of a contributor based on shift patterns, on clinical procedures, and on the limited number of personnel who have the opportunity to observe an incident. Clearly there is a strong conflict between the desire to prevent future incidents by breaking anonymity to ask supplementary questions and the desire to safeguard the long-term participation of staff within the system. The move from paper-based schemes to electronic systems raises a host of complex sociotechnical issues surrounding the anonymity of respondents. For instance, each client computer connecting to a Web site will potentially disclose location information through its Internet Protocol (IP) address. This address is not linked to a particular user, but it can be used to trace a report back to a particular machine. If logs are kept about user activity then it will be possible to identify the contributor. Alternatively, many health care organizations routinely log users' keyboard activity; hence there may also be more direct means of identifying the person who contributes an electronic incident report. Balanced against this concern for anonymity and confidentiality, there can also be problems if groups or individuals deliberately seek to distort the findings of a system by generating spurious reports. These could, potentially, implicate third parties. The problems of malicious reporting together with the technical difficulties of providing anonymous reports therefore make it likely that future electronic systems will follow the ASRS approach of confidential submissions.

Time and Place Information

There is a tension between the need to learn as much as possible about the context in which a mishap occurred and the need to preserve the confidentiality or the anonymity of the person contributing a report. A frequent criticism is that in order to protect the identity of those involved in an adverse event, reporting systems also remove information that is vital if other managers and operators are to avoid future mishaps (Johnson, 2003a). As mentioned in the previous paragraph, this is a particular problem in local systems where it may be possible to infer who was on duty if respondents provide information about the time at which an incident occurred. However, if this data is not provided, it may not be possible to determine whether or not night staffing patterns played a role in the incident or whether an adverse event was effected by particular handover procedures between different teams of coworkers. Similarly, if location information is not provided, it can be difficult to determine whether ergonomic issues and the configuration of particular devices contributed to a mishap. If location information is provided, for instance within an ICU, it is then often possible to name a small number of professionals to be identified as responsible for health care provision within that area. The difficulties created by the omission of location information can be seen by the frequent requests for additional unit information in the dialogues that emerge after the contribution of an incident to the CIRS application.

Location information falls into several different types. Geographical information may be important in national and regional systems to detect common patterns within specified areas. These can emerge if local groups of hospitals adopt similar working practices that may contribute to adverse events. Similarly, geographical information can be important to identify "hot spots" that can be created by a batch of medication or other common supply problems. These details need not be explicitly requested from the individuals who contribute a report. They can often be inferred from the delivery mechanism that is used to collect the report. These inferences can be relatively straightforward, for example, if reports from a particular hospital regularly arrive on a specific day of the week. They can also be based on more complex information, such as the IP address of a contributing machine. These assumptions can, however, be unfounded. For instance, problems will arise if contributors work in one location and submit a report from another. This scenario is likely to occur because many health care workers benefit from increasing job flexibility, especially in the delivery of specialist care across a relatively wide geographical area.

As previously mentioned, the location information requested from contributors can take several different forms. Not only do analysts often need to identify geographical patterns within a series of incident reports, but they may also need to locate functional similarities within the different areas of a health care system. For instance, the NPSA's NRLS data set asks "in which service did the patient safety incident occur." As before, contributors must select from a number of predefined fields. The nine options range from "acute/general hospitals" to "general practice" and "learning disabilities service." They also note that individuals may report an incident within a health care service that is different from the one in which they themselves work. The NPSA also asks "in which location did an incident occur?" The FDA takes a similar approach when it offers 30 or more locations in their medical devices reporting system. This taxonomy includes "612 Mobile Health Unit" and "002 Home" as well as "830 Public Venue" and "831 Outdoors." The diversity in the classification reflects the diverse locations in which mass market health care devices might fail.

In confidential systems, location information must be obtained so that analysts can contact reporters in order to follow up any necessary additional details. Even in anonymous systems it can be necessary to provide location information. For example, if a device has failed or if a problem involves a subcontractor, then it may be necessary for the reporter to provide information about the location and identity of the supplier who was involved in the adverse event. This creates considerable opportunity for error in reporting system software. Arguably the most frequent problems center on the misspelling of names. For example, Siemens has been entered into the FDA system under Seimens and Simens. Incidents have also been recorded under Siemens Medical, Siemens Medical Solutions, Siemens Medical Systems, and so forth. Any analysis and retrieval software must cope with such alternative spellings if potentially relevant information is not to be overlooked. An alternative approach is for the system to prevent users from typing in this information. Instead, they are compelled to select a supplier identifier from an enumerated list or menu. This is liable to have several hundred items. Such widgets pose a considerable challenge for human–computer interface design (Johnson, 2003a). They can also generate considerable frustration in users who must scroll through the names of dozens of medical suppliers before they reach the one that they are looking for. This frustration is likely to increase if the supplier identity information is missing from the enumerated list. However, one benefit from this approach is that most software reporting systems can use a supplier list to automatically update address information so that users need not type in their location details. This is a significant benefit, given that many end users will not have this information at hand as they begin to file a report in the aftermath of an adverse event.

Most reporting forms also prompt contributors for the time an incident occurred. As with location information, the elicitation of this information is not as simple as it might appear. As mentioned earlier, temporal information can be used with geographical data to support inferences about the identity of a potential contributor or work group that are implicated in a report. A number of other human factors issues also complicate the elicitation of this information. For example, CIRS prompts the reader with "at what time of the day did the incident happen (1–24)?" In contrast, the NPSA data set asks two related questions. First, contributors must determine the "date on which the incident occurred" by giving the day, month, and year. It is, however, possible for the date to be "unknown." The second related question asks for the "time of the incident." There is slightly more flexibility here than in the previous question. Respondents can supply the precise hour and minute or a time slot from the following list: 08.00–11.29, 12.00–15.59, 16.00–19.59, 20.00–23.59, 00.00–03.59, 04.00–07.59. Alternatively, as before, contributors can also state that the time was "unknown." The NPSA data set enables contributors to specify precise times and also intervals. However, there are additional complexities. For example, some incidents only emerge slowly over a prolonged period. For instance, an infusion device may administer the incorrect medication over several of these intervals. Other mishaps can take place over an even longer time period, even extending to months and years. Similarly, the same adverse event might occur several times before it is detected or might occur at different times to several patients. It is unclear how these different circumstances might be coded within many reporting schemes. For example, contributors might be required to complete a separate form for each instance of the adverse event even though they were strongly connected by sharing the same causes and consequences for the patient. Further complexity arises because the time at which an incident occurs can be different from the time at which an adverse event is detected or reported. These additional details are often elicited so that safety managers can review the monitoring mechanisms that are intended to preserve patient safety. If an adverse event is only detected many months after it has taken place, for example, if a patient returns to report the adverse consequences of a mishap, then many internal quality control and monitoring systems can be argued to have failed.

Detection Factors and Key Events

It is important to determine how adverse events are detected. Many incidents come to light through a combination of luck and vigilance. Unless analysts understand how contributors identified the mishap, it can be difficult to determine whether there have been other similar incidents. CIRS provides an itemized list of detection factors. These include direct clinical observation, laboratory values, airway pressure alarm, and so on. The respondent can identify the first and second options that gave them the best indication of a potential adverse event. The local ICU scheme of Figure 31–3 simply asks for the "Grade of staff discovering the incident." Even though it explicitly asks for factors contributing to and mitigating the incident, it does not explicitly request detection factors.

The reporting of detection factors raises a number of problems. For example, clinical training often emphasizes the importance of "making errors visible." Nolan (2000) identifies the double-checking of physician medication orders (prescriptions) by the pharmacist and the checking of a nurse's dose calculations by another nurse or by a computer as examples of making errors visible. Similarly, if patients are educated about their treatment, they can also play an important role in identifying errors. However, as these checks and balances become more widely integrated into health care practice, it is less and less likely that they will trigger incident reports. There is a paradox that the most effective detection factors are likely to be those that are mentioned least often in incident reports because they are accepted as part of the standard practice. Unfortunately, things are seldom this straightforward. For example, a recent report to the FDA described how the drug calculator of a medication assistant in a patient monitoring application would occasionally round up values to a second decimal place. The users complained that this could easily result in a medication error and that the manufacturer was failing to acknowledge the problem. The manufacturer responded that vigilant nursing staff ought to notice any potential problems when calculating the medication. The clinicians countered this by arguing that they had explicitly taught nursing staff to trust the calculation function as a means of *reducing human error* (Johnson, 2004). This illustrates the recursive nature of incident detection. Manufacturers assume that health care professionals will cross-check any advice to detect potential errors. Conversely, health care professionals increasingly assume that automated systems will provide the correct advice that is necessary to help them detect adverse events.

Many detection factors are focused on the proximal or immediate events that can lead to an adverse outcome. For example, a nurse observing a patient's adverse reaction to a particular medication can trigger a report. It is rare for reports to be filed when health care professionals detect the latent conditions that may eventually contribute to an incident or accident. Many nurses and doctors accept a culture of coping with limited resources and high demands on their attention. Lawton and Parker (2002, p. 17) therefore argue that the U.K. NHS should take a more proactive approach to incident reporting. Individuals and teams must be sensitized so that they are more likely to detect and report the conditions that will lead to error before an error actually occurs:

> Proactive systems work in part by asking people to judge how frequently each of a number of factors such as staffing, supervision, procedures, and communication impact adversely on a specific aspect of their work. So, for example, if nurses in intensive care are experiencing problems with the design of a particular piece of equipment, this will be recorded and action taken to improve the design. This kind of proactive approach allows the identification of latent failures before they give rise to errors that compromise patient safety.

Vincent, Taylor-Adams, and Stanhope (1998) build on this analysis when they identify those latent conditions that should be monitored and detected prior to an adverse event. Their enumeration includes items such as heavy workload, inadequate knowledge or experience, inadequate supervision, a stressful environment, rapid change within an organization, incompatible goals (e.g., conflict between finance and clinical need), inadequate systems of communication, and inadequate maintenance of equipment and buildings. They observe that these latent factors will affect staff performance, can make errors more likely, and will impact on patient outcomes. However, few existing reporting systems or research studies have found concrete means of encouraging respondents to detect and report these latent conditions within health care institutions. Part of the explanation for this may lie in the observation that many of the latent conditions for adverse events are almost characteristic of many modern health care organizations. These include rapid change within an organization and conflict between clinical and financial needs.

Most reporting forms prompt the contributor to explain what happened. In many systems, this field is left open so that individuals and teams can describe the critical events in their own words. Additional prompts are often used to help ensure that contributors provide sufficient detail for subsequent analysis. For instance, the local ICU system breaks down the "what happened' information into a number of different categories. Respondents are first prompted to elicit information about "The Incident." This information is broken down into a "Description of what happened" and "What factors contributed to the incident." Respondents are also asked for mitigation factors; however, a more detailed discussion about this part of the form will follow later. The local system also prompts for other information about what happened. A further section of questions addresses the "Circumstances" of the incident. This includes temporal information, mentioned previously, as well as "What procedure was being carried out?" "What monitoring was being used?" and "If equipment failure give details of equipment." A final section about what happened is intended to elicit general information about the personnel involved. Respondents are asked for the "Grade of relevant responsible staff" and the "Grade of staff discovering the incident." In keeping with the rest of this form, contributors can complete the form using their own terms. They are not expected to select individual items from a predetermined enumeration.

Similar to the local system, the NPSA's NRLS data set also includes a section to elicit information about "what happened." In contrast to the previous example, this scheme exploits a mixed approach using both a predefined taxonomy and open textual responses. For instance, contributors are asked to provide information about "What happened" by categorizing the incident according to a number of choices. These range from problems involving access, admission, transfer, discharge (including missing patient) to problems involving treatment procedures. In addition, respondents can provide a free-text response to describe what happened in their own words. The NPSA also asks a series of more detailed questions about the circumstances in which an incident occurred. These include the location information, described in previous sections. They also include a series of questions about the patients involved in an adverse event. This information is not explicitly prompted for by the local system and arguably reflects the greater diversity of conditions that will be covered by the national reporting system. For instance, respondents are asked, "Does the patient have any of the following known/diagnosed impairments or disabilities?" The options include learning disability(ies), physical disability(ies), sensory impairment(s), or other. Additional questions ask whether the patient was sedated at the time of the incident, whether he or she was being detained under the Mental Health Act, and so forth. The NPSA also elicits contributory factors. Respondents are asked to tick any of the following factors that apply to an incident: Communication factors (includes verbal, written, and nonverbal between individuals, teams, and/or organizations), Education and training factors (e.g., availability of training), Equipment and resources factors (e.g., clear machine displays; poor working order, size, placement, ease of use), Medication factors (where one or more drugs directly contributed to the incident), Organization and strategic factors (e.g., organizational structure, contractor/agency use, culture), Patient factors (e.g., clinical condition, social/physical/psychological factors, relationships), Task factors (includes work guidelines/procedures/policies, availability of decision-making aids), Team and social factors (includes role definitions, leadership, support, and cultural factors), Work and environment factors (e.g., poor/excess administration; physical environment, workload, and hours of work; time pressures), Other, and Unknown. The inclusion of these contributory factors is an important strength in the NPSA's approach because it provides reporters with a means of commenting on the latent factors that may have created the preconditions for an adverse event to occur.

One of the problems with the NPSA's elaborate approach is that respondents may become disillusioned, fatigued, or irritated by the large number of questions that they are being asked. One consequence is that users may spend most of their time completing the free-text description and will therefore pay less attention to ticking the relevant boxes in other areas of a reporting form. Conversely, the tickable boxes may arguably reduce the load on the contributor, who might then feel it unnecessary to provide additional details in the natural language section. It is difficult to determine which of these interpretations will prove correct, as the initial implementation of this approach has been running for a relatively short period of time. However, the NPSA has addressed some of the potential problems by using software suppliers to develop context-sensitive reporting systems. For example, if they state that the incident did not involve a medication error, they would not then need to select options such as the following: Adverse drug reaction (when used as intended), Contra-indication to the use of the medicine in relation, Mismatching between patient and medicine, Omitted medicine/ingredient, Patient allergic to treatment, Wrong/omitted/passed expiration date, and so on. Similarly, the software would only present specific questions about what happened to a particular device once the user had confirmed that the incident did involve a device failure. The dynamic nature of the form content helps to filter out irrelevant questions. This avoids an important limitation of more complex paper-based forms where users are often directed to skip ahead 10 or 20 questions if necessary. Users can become lost as they turn over pages and pages of irrelevant questions. However, the use of context-sensitive software can also create problems if users have to report "hybrid" incidents that bring together complex combinations of, for example, adverse drug reactions and device failures. Again, further experience with applications based on the NPSA data set will be necessary before any sustained analysis can be made of the costs and benefits of this approach within health care organizations.

The CIRS online system also adopts a mixed approach. Initially, the web-based form prompts participants to enter information about what has happened by checking boxes associated with the various team members who were involved in an adverse event. They must then enter the number of hours "on duty without sufficient rest (if known)." This is intended to provide an insight into the workload of the provider of anesthetic care. The subjective nature of this question makes it very difficult to interpret any results. Different individuals can have very different ideas about what represents "sufficient rest" (Johnson, 2003a). This again illustrates the importance of conducting usability studies and of considering the human factors issues when constructing the questions that will be asked as part of a reporting form. Like the NPSA data set, respondents are then asked to provide information about the patients involved. In this case, a radio button widget is used to indicate the sex of the patient. Respondents can type numeric values into an age field. Radio buttons are also used to indicate whether the patient is undergoing an elective or emergency procedure and their ASA status (Classes I–V). The ASA status refers to the American Society of Anaesthesiologists Physical Status Classification, where Class I refers to a normal and healthy patient, and Class V refers to a patient who is unlikely to survive without an operation. This classification provides a crude approximation to the a priori risk involved in anesthesia. The CIRS form assumes that each incident only affects a single patient. This common assumption is often not warranted, for example, if a common "error" is replicated in a number of similar procedures. The key point is that significant testing should be conducted to determine whether such assumptions can be justified within a particular domain. This testing can be performed in several different ways. For example, a sentinel scheme can be used in the manner described in previous paragraphs if representative institutions pilot the system before it is made more widely available. Similarly, potential contributors might be asked to use a prototype system to report information about an incident that they have observed in the past. They can also be given information about stereotypical mishaps and then asked to enter relevant information into the system "as if" they had witnessed such an adverse event.

The CIRS system elicits further information about what happened by asking contributors to select the "Overall anaesthetic technique" during which the incident occurred. A pull-down menu offers nine broad categories of activity that range from general anesthesia to regional anesthesia and care of a multiple trauma patient. These constrained fields are then followed by a number of more open questions. Contributors are asked to describe the incident in their own words. They are warned to be "careful not to present data here, that could identify the patient, the team or the institution." This section elicits information about the events leading to an incident. A second question asks respondents to "describe the management of the situation in your own words" from the moment of occurrence on. In passing, it is worth noting that CIRS warns users that "if you wish to print out this report, please stay in between the margins of the text field." This stems from a problem in the formatting of the online form that does not have a dedicated print function. Such issues are less significant for voluntary reporting systems, where great pains are taken to preserve the anonymity of individual contributors. The need to keep a printed record of particular reports assumes a greater priority within mandatory and confidential schemes.

Consequences and Mitigating Factors

Vincent and Coulter (2002, p. 76) argue that it is vital to consider the patient's perspective when assessing the consequences of an adverse health care event. They refer to the psychological trauma both as a result of an adverse outcome and through the way that an incident is managed. They urge that "if a medical injury occurs it is important to listen to the patient and/or the family, acknowledge the damage, give an honest and open explanation and an apology, ask about emotional trauma and anxieties about future treatment, and provide practical and financial help quickly." They also argue that patients are often best placed to report the consequences of an adverse event. Schemes similar to the Swedish system operated by KILEN, the Consumer Institute for Medicines and Health, should be established to help patients report these incidents. However, most existing reporting systems rely on clinicians to assess the outcomes of adverse events when they submit information about an incident or near-miss.

Most schemes explicitly prompt respondents for information about the consequences of an adverse event. However, this raises a number of complex issues for health care systems. For example, it can often be difficult to determine whether or not a mishap had any appreciable impact on the patient outcome. CIRS asks contributors to respond to this question. They then select an outcome from an enumeration that includes outcome independent from the event, patient dissatisfaction, prolongation of hospitalization, unplanned hospitalization of an outpatient, unplanned admission to an ICU, minor morbidity, major morbidity, and death. This contrasts with the local system that simply provides a free text area for the respondent to provide information about "what happened to the patient?" The NPSA is similar to CIRS, as it also begins with a prompt to state whether or not any patients were actually harmed by an event. If so, then additional information is requested about the degree of harm or severity of the adverse event. The contributor must determine whether the patient suffered a low-severity event that required extra observation or minor treatment, moderate or short-term harm that required further treatment, severe or long-term harm, or death. If more than one patient was affected by an adverse event, contributors should indicate the number of individuals falling into each of these different categories. Further questions probe the nature of any adverse impact on the patient. For example, contributors are asked to state whether the effect was physical, such as an allergic adverse reaction, blood loss, or neurological effect. A free-text field is also provided to record additional details about the nature of psychological or social consequences.

None of these forms are able to resolve the practical problems that arise when individuals have to determine the outcomes of an adverse event. An incident might have no immediate effect. Hence, the distinction between immediate and long-term outcomes is an important issue. Similarly, the administration of an incorrect medication may include side effects that increase the probability of adverse consequences in the future. In such circumstances, it may only be possible to consider the likelihood of an effect rather than commit to a certain outcome. Further problems arise because the individuals who witness an incident may only be able to provide information about the consequences of that event. The lack of clinical audit and of agreed outcome measures in some areas of health care creates additional complexity. Finally, health care professionals can inadvertently compromise the confidentiality of a report by carefully monitoring the progress of particular patients involved in an incident.

Further complexity is created by the need to assess the potential consequences of near-miss incidents. Few health care systems explicitly address this issue. However, it is a common concern in aviation and maritime systems (Johnson, 2003a). Given that no adverse event has actually taken place, it can be argued that the incident resulted in minor or negligible consequences. However, this may ignore the way in which chance occurrences may have intervened to prevent what otherwise might have been a very serious mishap. In consequence, many risk managers adopt the heuristic of "worst plausible consequences" when assessing the severity of a near-miss. The interpretation of "plausible" consequences is subjective and varies from system to system. For example, some air traffic management systems will treat a report of an air proximity violation "as if" a collision had actually occurred if the crew rather than the controllers were forced to initiate an avoiding action.

Voluntary reporting systems are often intended to elicit information about the low-consequence and near-miss incidents that are not covered by mandatory schemes. Van der Schaaf (1992) therefore argues that these reporting systems provide as much information about how to mitigate failure as they do about the causes of adverse events. The local ICU system, introduced in previous paragraphs, simply asks what "minimized" the incident. In contrast, the NPSA illustrates the importance of mitigation information by including a series of questions about the barriers that intervened to protect patient safety. For instance, the subsection titled "Impact on the Patient" includes the question "Did any actions prevent the incident from reaching the patient? (i.e., was this a 'near miss'?)." Contributors are then asked an optional question intended to determine the nature of any preventative actions that were taken. Other questions ask, "Did any actions minimize the impact of the incident on the patient?" and if so, respondents must describe those actions in their own words.

The CIRS system adopts a mixed approach to the elicitation of mitigating factors. Like both the local system and the national NPSA data set, respondents are prompted to describe the management of the situation in their own words. CIRS then provides a number of explicit prompts. The online form asks, "What led you successfully manage the event (recoveries)?" Respondents must select the most important factor using radio buttons that are grouped into a number of categories. Personal factors include knowledge, skill, experience, situation awareness, and use of appropriate algorithms. A further category of mitigation factors focuses on team intervention described in terms of extraordinary briefings, extraordinary team building, extraordinary communication within the anesthetic team, extraordinary communication in the surgical team, and extraordinary communication between the teams. The form also prompts for system factors, including additional monitoring or material, replacement of monitoring or material, additional personnel, and replacement of personnel. Finally, there is an "other" category. As before, this detailed enumeration can help to guide users who may not be used to thinking in terms of "mitigation factors." There is a danger that schemes that ask more open questions may fail to elicit critical information about the ways in which managerial and team factors helped to mitigate the consequences of an incident.

In some incidents, it can be relatively easy to interpret information about the mitigation of adverse events. For instance, one study identified that there were 6.5 adverse drug events for every 100 admissions in a U.S. hospital (Bates et al., 1995). Of those, it was argued that 28% could have been detected and avoided mainly by changing the systems used to order and administer drugs. Similarly, another study showed that computerized monitoring systems were significantly more likely to identify and prevent severe adverse drug events than those identified by chart review (51% vs. 42%, $p = 0.04$; Jha et al., 1998). However, it can be far harder to interpret incident reports where claims are made about human intervention in the mitigation of adverse events. It can be difficult to identify what precisely protected patient safety if another member of staff intervenes to prevent an adverse event. At one level, a safety manager might praise the vigilance of that individual. At another level, they might use this as an example of the success of the monitoring systems within a team of coworkers. Further investigation is required to determine whether such confidence is warranted. Individuals often identify potential incidents in ways that are not directly linked to official monitoring procedures. Conversely, well-developed routines can successfully detect potential incidents even when individuals are tired or operating under an extreme workload (Johnson, 2003a).

Causes and Prevention

The VA's National Center for Patient Safety (2004) uses the VA's vision statement to motivate the development of voluntary reporting systems. They argue that it is necessary to "take advantage of lessons present in *close calls* where things almost go awry, but no harm is done." In order to exploit these lessons it is essential to understand "the real underlying causes (so that) we can better position ourselves to prevent future occurrences." This vision statement also goes on to suggest that people "in the front line" are in the best position to identify the causes of problems and to propose potential solutions. Similar sentiments are expressed by the U.K. NPSA and by the proponents of the CIRS reporting system. Both of these schemes actively seek to elicit information from respondents about what they perceive to be the causes of an adverse event. For example, the NRLS data set includes the question "In your view, what were the underlying causes or events which, if rectified, may prevent another patient safety incident?" This prompts the contributor to provide a free-text explanation of the events leading to an adverse outcome or near-miss. It also raises a host of complex human factors issues. Many issues center on the problems of counterfactual reasoning. Counterfactual arguments lie at the heart of most forms of causal analysis. We can say that some factor A caused an accident if the accident would have been avoided if A had not occurred. This is counterfactual in the aftermath of an adverse event because we know that A did happen and so did the mishap. The NPSA question embodies this counterfactual style of reasoning about causation: "What were the underlying causes or events which, if rectified, may prevent another patient safety incident?" The local system reporting system also asks respondents to suggest "how might such incidents be avoided." This open question is, in part, a consequence of the definition of an incident in this scheme, which included occurrences "that might have led (if not discovered in time) or did lead, to an undesirable outcome." It also provides a further illustration of this counterfactual approach to causal information.

Byrne and Handley (1997) have conducted a number of studies into human reasoning with counterfactuals. They have shown that deductions from counterfactual conditionals differ systematically from factual conditionals. For example, the statement "either the medication was prescribed too late or the disease was more advanced than we had thought' is a factual disjunction. Studies of causal reasoning suggest that readers will think about these possible events and decide which is the most likely. It is often assumed that at least one of them took place. The statement that "had the medication been prescribed sooner or the disease been less advanced then the patient would have recovered" is a counterfactual disjunction. This use of the subjunctive mood communicates information about not only the possible outcome of the incident but also a presumption that neither of these events actually occurred. This theoretical work has pragmatic implications for incident investigation. If factual disjunctions are used, care must be taken to ensure that one of the disjunctions has occurred. If counterfactual disjunctions are used, then readers may assume that neither disjunction has occurred. The key point here is that most reporting systems rely on counterfactual definitions of causation. Human factors studies of counterfactual reasoning have identified systematic biases that make it critical for risk managers to carefully analyze the causal arguments that they receive in response to adverse events.

The close association between causation and counterfactual arguments can also be seen in supplementary questions posed in the NPSA data set. For example, respondents are explicitly asked about potential means of preventing an accident: "In your view, what were the underlying causes or events which, if rectified, may prevent another patient safety incident?" and "please describe any actions planned or taken to date to prevent a reoccurrence." These questions are intended to elicit a natural language response. The relationship to counterfactual reasoning is explicit in the way that responders are asked to think what *might have* prevented the adverse event. The CIRS system adopts a similar approach. However, this reporting form provides an enumeration that is intended to guide the contributor in his or her analysis. This is similar to the way in which CIRS uses a list of potential causes that, arguably, can reduce some of the difficulties associated with informal forms of counterfactual arguments. CIRS asks, "What would you suggest for prevention?" and respondents must select the most important item from a varied list. Potential preventative measures include additional monitoring or material, improved monitoring or material, better maintenance of existing monitoring/equipment, improved management of drugs, or improved arrangement of monitoring/equipment. The CIRS enumeration also provides items for improved training/education, better working conditions, better organization, better supervision, more personnel, better communication, more discipline with existing checklists, better quality assurance, development of algorithms/guidelines, or abandonment of old routine. Finally, there is an opportunity to include other preventative measures, but this time using free-text descriptions.

The NPSA data set also probes for information about factors that did not directly cause an adverse event but that contributed to the course of an incident or accident. Respondents are requested to indicate whether any of the following contributory factors were involved: Communication factors, Education and training factors, Equipment and resources factors, Medication factors, Organization and strategic factors, Patient factors, Task factors, Team and social factors, Work and environment factors, Other, and Unknown. It remains to be seen whether the elements of this taxonomy will have to be revised after prolonged use of the national systems. For instance, most health care mishaps would involve work and environmental factors. It seems likely that additional information would be required to identify specific interventions to address this broad range of contributory factors. A further question asks for any additional "important factors." This question can be interpreted to provide additional details about the context in which an incident occurred. By enumerating potential causes, the list of important factors may avoid some of the potential pitfalls of counterfactual reasoning, mentioned earlier. This list includes the following: Failure to refer for hospital follow-up, Poor transfer/transcription of information between paper and/or electronic forms, Poor communication between care providers, Use of abbreviation(s) of drug name/strength/dose/directions, Handwritten prescription/chart difficult to read, Omitted signature of health care practitioner, Patient/caregiver failure to follow instructions, Failure of compliance aid/monitored dosage system, Failure of adequate medicines security (e.g., missing CD), Substance misuse (including alcohol), medicines with similar looking or sounding names, and poor labeling and packaging from a commercial manufacturer.

The CIRS reporting form mirrors this use of an enumeration rather than a counterfactual approach to causal information. Contributors must select the most "important field" to identify "what led to the incident (cause)." These fields include personal factors such as diminished attention without lack of sleep, diminished attention with lack of sleep, insufficient knowledge, and so on. They also include team factors such as insufficient communication or briefing. System factors include lack of personnel and unfamiliar surroundings. It is important to stress again that the answers to causal questions should be interpreted with care. Although the individuals who directly witness an incident can provide valuable information about how future adverse events might be avoided, they may also express views that are influenced by remorse, guilt, or culpability. Subjective recommendations can also be biased by the individual's interpretation of the performance of their colleagues or their management or of particular technical subsystems. Even if these factors did not obscure their judgment, they may simply have been unaware of critical information about the causes of an incident. These caveats must be balanced against the strengths and weaknesses of alternative analysis techniques. As mentioned earlier in this chapter, previous studies have relied on quantitative approaches for quality improvement using the statistical analysis of nationwide data. These epidemiologic techniques help to analyze the distribution and incidence of adverse events that occur with reasonable frequency and for which it is possible to obtain reliable statistics. However, a number of industries ranging from public transportation to power generation have not begun to complement epidemiological approaches with more qualitative forms of root-cause analysis (Johnson, 2003a). It is important to observe that they provide structured means of minimizing the biases that affect "informal" approaches such as the counterfactual arguments described in the previous paragraphs.

EXPLANATIONS OF FEEDBACK AND ANALYSIS

The human factors issues involved in incident reporting do not end with the submission of a report. Potential contributors must be confident that the information that they submit will be taken seriously. This does not imply that every report must initiate change within the host organization. It is, however, important that contributors know their reports have been successfully received and attended to. In other domains, electronic tracking systems have been introduced so that contributors can monitor who has responsibility for handling their submission from the moment that it is logged (Johnson, 2003a). Such techniques have not yet been introduced within health care applications. In confidential systems it is more usual for contributors to receive an acknowledgment slip in return for their contributions. Such feedback is obviously difficult to provide in anonymous schemes. Most reporting forms provide participants with information about how their contributions will be processed. For example, the local system in Figure 31–3 includes the promise that "information is collected from incident reporting forms (see overleaf) and will be analysed. The results of the analysis and the lessons learnt from the reported incidents will be presented to staff in due course." This informal process is again typical of systems in which the lessons from previous incidents can be fed back through ad hoc notices, reminders, and periodic training sessions. The CIRS web-based system is slightly different. It is not intended to directly support intervention within particular working environments. Instead, the purpose is to record incidents so that anesthetists from different health care organizations can share experiences and lessons learned: "Based on the experiences from the Australian-Incident-Monitoring-Study, we would like to create an international forum where we collect and distribute critical incidents that happened in daily anaesthetic practice. This program not only allows the submission of critical incidents that happened at your place but also serves as a teaching instrument: share your experiences with us and have a look at the experiences of others by browsing through the cases. CIRS© is anonymous" (CIRS, 2006).

The NPSA clearly has a far wider set of responsibilities than either CIRS or the local scheme mentioned earlier. A recent overview of their activities explains to potential contributors that the NPSA will collect reports from across the country and initiate preventative measures, so that the whole country can learn from each case, and patient safety throughout the NHS will be improved every time (NPSA, 2004a). The main mechanisms for achieving this will be by collecting and analyzing information on patient safety incidents from local NHS organizations, patients, and caregivers through the data sets mentioned in previous paragraphs. The NPSA will also use information from other reporting systems. Potential contributors are assured that their data will be used to "learn lessons," "ensuring that they are fed back into health care and (the ways that) treatment is organised and delivered." Potential contributors are also assured that work will be "undertaken on producing solutions to prevent harm, and to specify national goals and establish mechanisms to track progress" where any risks are identified. As can be seen, these health care reporting systems are rather vague on the precise mechanisms that will be used to combat any recurrence of an adverse event. This ambiguity can be explained in a number of ways. First, the NPSA is in the process of establishing their reporting system. They have adopted a step-wise policy of encouraging the establishment of local schemes prior to the development of their overarching national voluntary system. For this reason, they rely on individual health care organizations to develop a complementary mechanism for intervening in the immediate aftermath of an adverse event. Second, it can be difficult to predetermine all of the techniques that might be used to address the vast range of different adverse events covered by this national system. Over time, it is hoped that details will be provided to illustrate the diversity of interventions that will be based on contributions to the reporting system. Third, in the case of the local system, there is little need to provide great detail about the analysis and interventions that will result from a submission, because this information can be directly transmitted through staff meetings, newsletters, and other information sources. The next chapter provides details about the particular interventions recommended as a result of submissions to this local system. Finally, CIRS acts as a medium of exchange rather than an active agent of intervention. Contributors provide information to promote discussion and raise awareness. It is also assumed that they will take local measures to prevent any immediate recurrences through their local system. This illustrates the similarity between aspects of CIRS and the NPSA scheme.

As mentioned previously, the local scheme referred to in this chapter provided feedback to staff through periodic newsletters. CIRS provides feedback in the form of an online dialogue or forum through which professionals can add comments to the various reports that are received. It is likely that the NPSA will also use the Web as a primary means of providing feedback to potential contributors. This approach is justified by the relatively low cost of Web site development. However, it relies on a form of information "pull." Health care professionals have to keep going back to the site to download, or pull, updated information about patient safety initiatives. In contrast, e-mail dissemination provides a form of information "push" to ensure that lessons learned are sent out to health care institutions in a timely fashion. Unfortunately, this approach raises a host of additional human factors issues. For example, the growing problem of spam mail has increased the likelihood that many individuals will overlook or automatically delete messages that have such a mass distribution. Similarly, not everyone has convenient access to e-mail or the Internet. This is less of a problem for systems that elicit information from, and provide feedback to, health care professionals. These groups are likely to have access provided through their workplace. However, electronic dissemination suffers from significant limitations for systems that are intended to provide the general public with reporting facilities. Both in the United States and the United Kingdom, slightly more than 50% of the population currently have regular Internet access either at home or at work (Johnson, 2003a).

The limitations of computer-based dissemination have persuaded other health care reporting systems to explore a range of alternate mechanisms, including telephone- and fax-based applications. A prerecorded message can be used to list all of the most recent recommendations and other documents issued by the reporting system. Callers can then dial another number to request that paper-based copies of the full version are sent to them. This is cheap and simple; the prerecorded messages can be changes frequently and at minimal cost. However, the list of bulletins can become extremely long and tedious to listen to. A further problem is that specialized equipment with multiple inputs must be used so that callers do not find that the call-back lines are frequently engaged. The use of prerecorded messages to provide an index of updates to incident reports still does not address many of the administrative and resources problems that can arise from the paper-based distribution of these documents. At some point, copies of the report have to be printed and shipped to the prospective readers. One solution to these problems is to use fax-servers. These devices automatically ensure that a document is sent to every number on a preprogrammed list. The FDA pioneered the development of a "Facts on Demand" system. The user dials up the service then hears a series of instructions. If, for example, they press "2" on their keypad, they can hear more detailed instructions on how to use the system. If they press

"1," then they can choose to order a document. If they dial "INDX" or "4639" on the keypad, they can order an index of all documents on the system. The only technical requirement for the user of such a system is that they have access to both a fax machine and a touch-tone telephone.

Most incident reporting systems continue to use paper-based dissemination techniques. This situation is gradually changing as a result of financial and administrative pressures. For example, in 1997, the decision was taken to stop printing the FDA's User Facility Reporting Bulletin:

> Ten years ago, our computer capability allowed us to communicate only within FDA. Now, with advanced computer technology we can globally communicate through the Internet and through Fax machines. As you would expect, Congressional budget cuts have affected all parts of government. FDA did not escape these cuts. In the search for ways to reduce our expenses, printing and mailing costs for distribution of publications in traditional paper form have come to be viewed as an extravagant expenditure ... Now, budget restrictions prevent future distribution in paper form. We regret the need to move to this new technology if it means that many of our current readers will no longer have access to the Bulletin. We would like to remind you that you can also obtain copies through our Facts-on- Demand System or the World Wide Web (Wollerton, 1997, page 1).

CONCLUSIONS

This chapter has provided a high-level survey of the human factors issues involved in the reporting of adverse health care events. It began by reviewing recent initiatives from the U.S. Institute of Medicine and a range of national patient safety agencies that have encouraged the development of voluntary and mandatory reporting schemes. Later sections went on to examine the problem of underreporting. Potential contributors are often concerned that they will be blamed for any involvement in an incident or near-miss. Several schemes have arranged for limited legal protection to support participants in voluntary reporting systems.

The middle sections of this chapter introduced a range of different architectures for incident reporting. These included local systems that are designed and operated by individual health care professionals within single units. I also described different regional and national systems. For instance, the FDA has pioneered the use of sentinel reporting to reduce the reporting biases that effect large-scale schemes. This approach focuses training resources and support onto a number of representative institutions so that all staff are sensitized to the importance of incident reporting and hence may be more likely to participate in the scheme. It is typically not possible to provide similar levels of resourcing across all of the thousands of organizations who contribute to less focused national systems. The increasing diversity of mandatory and voluntary reporting systems has made it difficult for many staff to know which scheme they should use after a particular adverse event, particularly when they may not be certain of the ultimate impact on any patients who were involved. As a consequence, an increasing number of hospitals have introduced "gatekeeper" systems where all reports are first submitted to a local safety manager who then assumes responsibility for passing them on to the relevant schemes.

Form design and distribution have a significant impact on the human factors of reporting systems. It can be difficult for individual to access online systems, even once they are persuaded to share information about an adverse event. Conversely, it can be difficult to sustain the levels of funding and management interest necessary to replenish and monitor supplies of paper-based forms. Usability problems can affect online systems if users do not have access to displays with adequate resolution to present increasingly complex forms. Similarly, it can be difficult to ensure that all potential contributors have access to the software that is required for many Internet-based systems. Even if paper-based forms are used, careful consideration must be given to the design of reporting systems. Even apparently simple information requirements, such as the date when an incident occurred, can lead to problems. For instance, many forms provide no means of specifying that the same adverse event recurred on several occasions. This means that several different forms may be submitted if, for instance, an incorrect medication was administered to the same patient over a course of several days. Many forms include questions about why an incident occurred and how it might have been avoided. The closing sections of this chapter have examined recent human factors work that has pointed to the biases that influence causal analysis.

References

American Iatrogenic Association. (2000). *Doing what counts for patient safety: Federal actions to reduce medical errors and their impact report of the quality interagency coordination task force (QuIC) to the president.* Retrieved June 16, 2006, from http://www.iatrogenic.org/library/ presrep.htm

American Medical Association. (2002). Statement for the record of the American Medical Association to the Subcommittee on Health Committee on Energy and Commerce, U.S. House of Representatives Re: Reducing medical errors: A review of innovative strategies to improve patient safety. Retrieved January 10, 2005, from http://www.ama-assn.org/ama /pub/article/6303–6226.html

Critical Incident Reporting System (2006), *The Anaesthesia Critical Incident Reporting System (CIRS©) on the Internet: An experience Based Database.* Retrieved June, 2006, from http://www.medana.unibas.ch/cirs/intreng.htm

Barach, P., & Small, S. D. (2000). Reporting and preventing medical mishaps: Lessons from non-medical near miss reporting systems. *British Medical Journal, 320,* 759–763.

Bates, D. W., Cullen, D. J., Laird, N., Petersen, L. A., Small, S. D., & Servi, D. (1995). Incidence of adverse drug events and potential adverse drug events: Implications for prevention. *Journal of the American Medical Informatics Association, 274,* 29–34.

BBC News. (2002a, June 18). Hospital "near-miss" cover-up denied. Retrieved June 16, 2006, from http://news.bbc.co .uk/1/hi/health/2049453.stm

BBC News. (2002b, April 24). Q&A: The NHS clinical negligence bill. Retrieved June 16, 2006, from http://news.bbc. co.uk/1/hi/health/1948135.stm

Busse, D. K., & Wright, D. J. (2000). Classification and analysis of incidents in complex, medical environments. *Topics in Healthcare Information Management, 20,* 4.

Byrne, R. M. J., & Handley, S. J. (1997). Reasoning strategies for suppositional deductions. *Cognition, 62 (1),* 1–49.

Cohen, H., Robinson, E. S., & Mandrack, M. (2003). Getting to the root of medication errors: Survey results. *Nursing, 33*(9), 36–45.

Cohen, M. R. (2000). Why error reporting systems should be voluntary: They provide better information for reducing errors. *British Medical Journal, 320,* 728–729.

Davis, P., Lay-Yee, R., Briant, R., & Scott, A. (2003). Preventable in-hospital medical injury under the "no fault" system in New Zealand. *Quality and Safety in Health Care, 12,* 251–256.

Food and Drug Administration. (1999). *Final report of a study to evaluate the feasibility and effectiveness of a sentinel reporting system for adverse event reporting of medical device use in user facilities.* Contract #223–96–6052, JUNE 16. http://www.fda.gov/ cdrh/postsurv/medsunappendixa.html

Food and Drug Administration. (2002). *Medical device reporting: General information.* Center for Devices and Radiological Health, Food and Drugs Administration, Washington, DC. Retrieved June 16, 2006 from http://www.fda.gov/ cdrh/mdr/mdr-general.html

Gaine, W. G. (2003). Editorials: No fault compensation systems. *British Medical Journal, 326,* 997–998.

Global Harmonization Task Force. (2002). *Comparison of the device adverse reporting systems in USA, Europe, Canada, Australia & Japan.* GHTF/SG2/N6R3. Retrieved June 16, 2006, from http://www. ghtf.org/sg2/inventorysg2/sg2-n6r3.pdf

Institute of Medicine. (1999). *To err is human: Building a safety health system.* Washington, DC: National Academy Press.

Jha, A. K., Kuperman, G. J., Teich, J. M., Leape, L., Shea, B., Rittenberg, E., et al. (1998). Identifying adverse drug events: Development of a computer-based monitor and comparison with chart review and stimulated voluntary report. *Journal of the American Medical Informatics Association, 5*(3), 305–314.

Johnson, C. W. (2000). Designing forms to support the elicitation of information about accidents involving human error. In P. C. Cacciabue (Ed.), *Proceedings of the 19th European Annual Conference on Human Decision Making and Manual Control* (pp. 127–134). EC Ispra, Research Center.

Johnson, C. W. (2003a). *A handbook of accident and incident reporting.* Glasgow: Glasgow University Press. Retrieved June 16, 2006, from http://www.dcs.gla.ac.uk/~johnson/book

Johnson, C. W. (2003b). The interaction between safety culture and uncertainty over device behaviour: The limitations and hazards of telemedicine. In G. Einarsson & B. Fletcher (Eds.), *Proceedings of the International Systems Safety Conference 2003* (pp. 273–283). International Systems Safety Society, Unionville, VA.

Johnson, C. W. (2004). Communication breakdown between the supplier and the users of clinical devices. *Biomedical Instrumentation and Technology: Journal of the U.S. Association for the Advancement of Medical Instrumentation, 38*(1), 54–78.

Kaiser Permanente. (2000). Roundtable discussion: Design considerations for a patient safety improvement reporting system. Kaiser Permanente Institute for Health Policy, NASA Aviation Safety Reporting System and The National Quality Forum. Held at NASA Ames Research Center, Moffitt Field, California, August 28–29. Retrieved June 16, 2006, from http://www.kpihp.org/publications/docs/ safety_improvment.pdf

Kaplan, H. S., Battles, J. B., van der Schaaf, T. W., Shea, C. E., & Mercer, S. Q. (1998). Identification and classification of the causes of events in transfusion medicine. *Transfusion, 38,* 11–12, 1071–1081.

Lawton, R., & Parker, D. (2000). Barriers to incident reporting in a healthcare system. *Quality and Safety in Healthcare, 11,* 15–18.

Mackenzie, C. F., Jefferies, N. J., Hunter, A., Bernhard, W., & Xiao, Y. (1996). Comparison of self reporting of deficiencies in airway management with video analyses of actual performance. *Human Factors, 38,* 623–635.

Minnesota Department of Health. (2000, December). Medical errors and patient safety: Key issues. Health Economics Program issue paper. Retrieved January 10, 2005, from http://www.mnhealthplans.org//collateral/patient%20 safety.pdf

National Audit Office (2001, May). *Handling Clinical Negligence Claims in England: Report by The Comptroller and Auditor General* (House Of Commons Ref. 403). London: The Stationery Office. Retrieved 15th May 2006 from http:// www.nao.org.uk/publications/nao_reports/00-01/ 0001403 .pdf

National Health Service Expert Group on Learning from Adverse Events in the NHS. (2000). *An organisation with a memory.* Technical report. London, UK. Retrieved June 16, 2006, from http://www.dh.gov.uk/PublicationsAnd Statistics/Publications/PublicationsPolicyAndGuidance/Pub licationsPolicyAndGuidanceArticle/fs/en?CONTENT_ID= 4065083 &chk= PARoiF

National Patient Safety Agency, *National Reporting and Learning System (NRLS) and Dataset,* Retrieved June 16, 2006, from http://www.npsa.nhs.uk/npsa/display?contentId=2389

National Patient Safety Agency. (2004a, March). *About the NPSA.* Retrieved June 16, 2006, from http://www.npsa.nhs.uk/npsa/about

National Patient Safety Agency. (2004b, March). *Healthcare professionals: Frequently asked questions.* Retrieved June 16, 2006, from http://www.npsa.nhs.uk/faq

National Patient Safety Agency (UK), National Reporting and Learning System Update, Retrieved May 15, 2006, from http://www.npsa.nhs.uk/health/reporting/background

National Patient Safety Foundation. (2001). Research agenda. McLean, VA. Retrieved June 16, 2006, from http://www.npsf.org

Nolan, T. W. (2000). System changes to improve patient safety. *British Medical Journal, 320,* 771–773.

Office of the Press Secretary. (2003, January). *President calls for medical liability reform.* Remarks by the president on medical liability reform, University of Scranton, Scranton, PA. Retrieved June 16, 2006, from http://www.whitehouse.gov/news/releases/2003/01/20030116–1.html

Raymond, R., & Crane, R. M. (2001, April). *Design considerations for a patient safety improvement reporting system.* Institute for Health Policy, Kaiser Permanente, Oakland, CA. Retrieved June 15, 2006, from http://www.kpihp.org/publications docs/briefs/safety_ improvement.pdf

Robinson, A. R., Hohmann, K. B., Rifkin, J. I., Topp, D., Gilroy, C. M., Pickard, J. A., et al. (2002). Physician and public opinions on quality of health care and the problem of medical errors. *Archives of Internal Medicine, 162,* 2186–2190.

Royal College of Anaesthetists. (1998, January). Critical incident form. Technical report, Professional Standards Directorate, London, UK. Retrieved June 15, 2006, from http://www.rcoa.ac.uk/docs/CI_record_form.pdf,

Runciman, W. B. (2002). Lessons from the Australian Patient Safety Foundation: Setting up a national patient safety surveillance system—Is this the right model? *Quality and Safety in Health Care, 11*(3), 246–251.

Shaw Phillips, M. A. (2001). National program for medication error reporting and benchmarking: Experience with MedMARx. *Hospital Pharmacy, 36*(5), 509–513.

Spectrum Health. (2002). The Spectrum Health patient safety plan: Setting the course for a broad organizational safety initiative founded on evidence-based practice. *Disease Management and Quality Improvement Report, 2*(6), 1–14.

Staender, S., Kaufman, M., & Scheidegger, D. (1999). Critical incident reporting in anaesthesiology in Switzerland using standard Internet technology. In C. W. Johnson (Ed.), *Proceedings of the 1st workshop on human error and clinical systems,* (pp. 10–12). Glasgow Accident Analysis Group, University of Glasgow.

U.S. General Accounting Office. (1997). *GAO report to Congressional committees: Medical device reporting improvements needed in FDA's system for monitoring problems with approved devices.* GAO-HEHS–97–21, Washington, DC.

van der Schaaf, T. W. (1992). *Near miss reporting in the chemical process industry.* Unpublished doctoral thesis, Technical University of Eindhoven, The Netherlands.

Van Geest, J. B., & Cummins, D. S. (2003). *An educational needs assessment for improving patient safety results of a national study of physicians and nurses, national patient safety foundation.* White paper report, No. 3. Retrieved June 16, 2006, from http://www.npsf.org/ download/EdNeedsAssess.pdf

Veteran's Affair's National Center for Patient Safety. (2004). *Mission statement: Creating a culture of safety.* Retrieved June, 2006, from http://www.patientsafety.gov/vision.html

Vincent, C. (2003). Editorial: No fault compensation: Compensation as a duty of care: The case for "no fault." *Quality and Safety in Healthcare, 12,* 240–241.

Vincent, C., & Coulter, A. (2002). Patient safety: What about the patient? *Quality and Safety in Health Care, 11,* 76–80.

Vincent, C., Taylor-Adams, S., & Stanhope, N. (1998). A framework for analysing risk and safety in clinical medicine. *British Medical Journal, 316,* 1154–1157.

Wald, H., & Shojania, K. G. (2001). Chapter 4: Incident reporting. In Evidence Report/Technology Assessment, No. 43, *Making health care safer: A critical analysis of patient safety practices.* AHRQ Publication No. 01-E058.

Weeks, W. B., & Bagian, J. P. (2000, November/December). *Developing a culture of safety in the Veterans Health Administration: Effective clinical practice.* American College of Physicians. Retrieved June 16, 2006, from http://www.acponline.org/journals/ecp/ novdec00/weeks.htm

Wollerton, M. A. (1997). Last printing of user facility reporting bulletin. FDA User Facility Reporting Bulletin., *(FUSE)* Issue 20, Summer pp. 1–5.

·32·

HUMAN ERROR REDUCTION STRATEGIES IN HEALTH CARE

René Amalberti and Sylvain Hourlier
IMASSA, *Brétigny-sur-Orge, France*

Since the 1980s, there has been an active attempt to increase the health care system's accountability regarding quality, safety, and efficiency. High safety standards are now mandatory for all patients. This situation represents a great challenge since medicine, for many reasons, is laden with potential risks. This chapter tries to offer a comprehensive survey of human error (HE) and related-safety strategies concepts, whether derived from industry or field-dependent. The chapter is divided into three sections. This introduction gives a brief overview of the extent and specifics of risks encountered in the industry, and then in the medical field. The second section defines HE-related issues in medicine. The third section drafts three generic solution spaces for error reduction: a "person approach" (errors of individuals), a "systems approach" (errors of organizations), and a "dynamic approach" (adaptation of error reduction strategies as safety improves). Each subsection suggests a set of related strategies to reduce the errors, and gives directions on how to combine them.

The shock that human error can lead to disasters clearly dates back to the 70s, with the Three Miles Island nuclear incident. Human error was not unknown at the time, but even for psychologists, it was more regarded as a means to characterize behaviors in experiments (good vs. false responses) rather than as a topic of research of its own.

Before Three Miles Island, safety was merely a technical problem. The science of system reliability assessment (SRA) and probability safety assessment (PSA) had been developed in the 60s to accompany the fast growth of technology, providing tools to more effectively and more safely design (see a survey by Cox & Trait, 1991). In other words, achieving better safety was limited to the pursuit of a better and more reliable technology. Humans were hardly considered, and when taken into account, mainly perceived as instruction followers. Unfortunately, engineers suddenly found out after Three Miles Island, that humans were still in the process, and actually, more involved than ever in the control of this process and in charge of decision-making, with their emotional and social skills, and limited rationality. From that date, cognitive ergonomics and cognitive engineering became central concerns to continue in the best possible manner the extreme potential benefit of high technology to the needs and limitations of workers. Nuclear industry and Aviation were pioneering fields, which gradually moved in the 80s and 90s to a set of win–win solutions, combining the extensive introduction of automation and information technology (IT) techniques with the reduction of errors (see for example reviews in Amalberti, 1998; Hollnagel, 1988; Hourlier, Grau, & Valot, 2003).

Human error in medicine only started to be taken into account a decade later, with a few scientific papers and edited books (Brennan, Leape, Laird, Localio, Lawthers, et al., 1991; Bogner, 1994; Leape, 1994; Vincent, 2001), then took another half-decade for this issue to be recognized as major by medical institutions (Kohn, Corrigan, & Donaldson, 1999). Since the late 90s, the number of

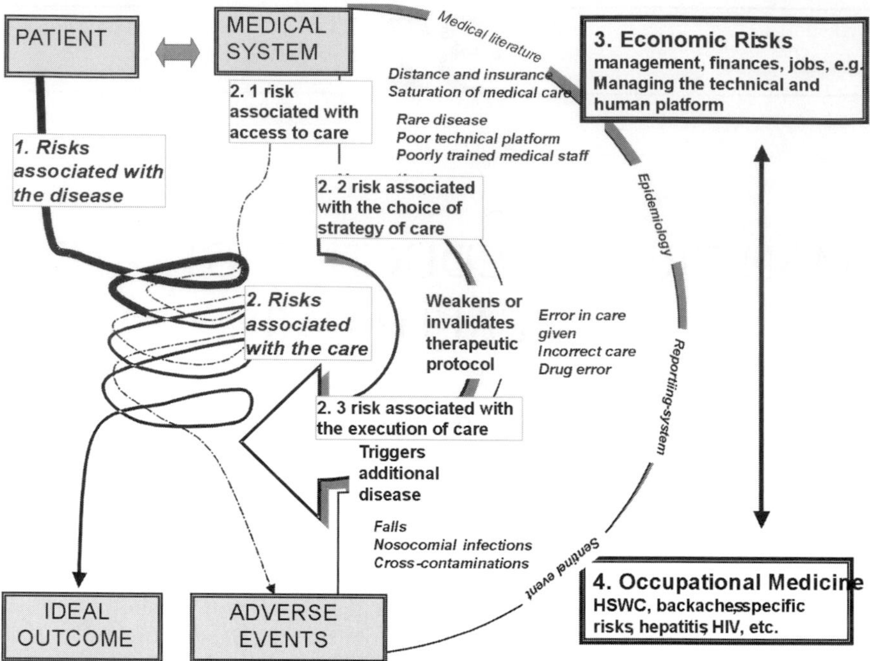

Figure 32–1. Conflicting sources of risks in medicine.

special issues, journals, conferences, and reports dedicated to this topic has continuously grown.

Health care is now following the same pattern regarding the treatment of human error as the one taken by the safest industries a couple decades ago. The same ideas and celebrated safety models are encountered in education on human error management in medicine (with special attention to James Reason's model). These models and ideas tell us to know more about error, to design reporting systems as nonpunitive systems, to add defenses, to reduce contributing factors to HE, to suppress violations, to standardize protocols, to improve quality and traceability, and to create conditions for improved communications and common situation awareness with a better use of IT. Last but not least, the improvement of organizations and culture are considered key factors for long-term success.

However, common sense tells us also that medicine does not strictly behave like industry; patients are not standardized, evolution of disease is not fully predictable. Moreover, for centuries medicine was said to be closer to an art than to a science. Should these differences with industry be true, then the concept of human error in medicine will probably need to be clarified, and additional dedicated strategies will need to be developed to fully meet the challenge of error reduction.

HEALTH CARE AS A LAND OF RISKS AND ERRORS

The Nature and Extent of Risk

Patient safety is a growing concern. However, the risk run by patients remains hard to assess. Risks of different natures are combined: the risk posed by the disease itself, the risk entailed by access to care (waiting time, cost and privilege of individual medical insurance), the risk entailed by the medical decision made, the risk linked to implementing the therapy selected, and last but not least for managers, the managerial and occupational risks which may potentially greatly stress the social system at work (see Figure 32–1).

These risks are generally interconnected and hard to dissociate, and do not generally move in the same direction, therefore their combined effects make it hard to grasp the issue of error. For example, diseases associated with high risks may be tackled

TABLE 32–1. Error Glossary

Generic in work psychology
Error, HE (Senders & Moray, 1991): [Unintentional] human action that fails to meet an implicit or explicit standard. An error occurs when a planned series of actions fails to achieve its desired outcome
Violations (Reason, 1990): [Deliberate] deviations from safe operating practices, procedures, standards, or rules
Active failures, AF (Reason, 1990): Unsafe acts (errors and violations) committed by those at the sharp-end of the system
Latent failures (Reason, 1990): Are the result of decisions taken at the higher echelons of an organization
Near miss (Van der Shaaf, et al. 1991): A situation where an event occurring during the course of action fails to further develop, thus not leading to consequences
Specific to medicine
Adverse Event, AE (Brennan, et al., 1991): Injury that was caused by medical management (rather than by the underlying disease) and that prolonged hospitalization, produced a disability at the time of discharge, or both
Preventable Adverse Event, PAE: All AE that can be avoided by a proper error management strategy. All PAE reflect active failures
Negligence (Brennan, et al., 1991): Care that fell below standards expected of physicians in their community

effectively only by treatments also associated with high risks; high gains for the outcome of the disease may require high risks in the treatment. Patients facing an apparently terminal illness may have a major change in their fatal prognosis, thanks to a daring medical strategy. An example of this is the treatment of hip fractures, previously fatal, which are now successfully replaced in elderly patients even when they are over 90. But, the most daring clinical strategies are also those whose levels of excellence are the least evenly distributed in the profession; they depend on craftsmanship. Inevitably, these complex approaches match those in which errors are the most frequent. Audacity is traded against safety. In short, it is almost impossible to simultaneously successfully reduce the three types of risks: (a) efficient effective disease management; (b) probability of gaining an identical benefit through the use of an advanced treatment strategy, whatever the practitioner and medical centre involved; and (c) achieving the lowest possible rate of error.

The extent of human error and associated risks is also difficult to appreciate in medicine because no system was set up to monitor medical practice as is regularly carried out in the safest industries (see for example systematic after-flight analysis of black boxes, or line-oriented safety audits; Helmreich, 2000); the only observatories that process epidemiological information are not very effective in finding human flaws because they focus on the patient more than on the process of care. Moreover, alerting or sentinel-event systems can only trigger alerts by monitoring large series. Insufficient, sporadic, and scattered cases, which will not pass the alert threshold (even though they eventually often add up to high numbers), are often disregarded (Auroy Amalberti, & Benhamou, 2004). See Table 32–1 for a glossary of error terms used in work psychology and in medicine.

Bearing in mind these contextual reasons for generating human flaws, and the inherent difficulty to characterize and measure human error, medicine made the choice to measure safety with the concepts of adverse event (AE) and negligence. Brennan and colleagues, in the well known pioneering Harvard medical practice study (1991), defines an adverse event as an injury that was caused by medical management (rather than the underlying disease) and that prolonged the hospitalization, produced a disability at the time of discharge, or both. He defines negligence as care that fell below the standard expected of physicians in their community. Preventable adverse events (PAEs) are all AE that can be avoided by a proper error-management strategy. Therefore, PAEs and negligence are the two categories of active failures that medicine considers to be immediately linked to unsafe acts (errors and violations) committed by those at the frontline of the system. However, it is important to note at this point that the two concepts (PAE and HE) should not be totally mistaken for one another. They address the same idea of human failure, but describe the failing process at a different level of abstraction and causality (see Table 32–2).

With these definitions, risks measured in the medical field could be well above anything experienced in safe industries (see Table 32–3).

The percentage of adverse events in the various countries where studies have been conducted ranges from 3 to 12% of hospitalized patients. The proportion of PAEs is at minimum 37%, and may reach 58% in some studies[1]. Operative drug events

TABLE 32–2. Relationships Between Terms Used to Define Human Failures

	Active Failure	*Human Error*	PAE *and Negligence*
Definition	All unsafe acts committed at the sharp end, including human error and violations	Academically, should only refer to unintentional unsafe acts Common usage turns the concept into a synonym of unsafe acts, thus including intentional errors (violations)	Any injury that was caused by medical management and was avoidable by a proper error management strategy
Relations between concepts	All unsafe acts are HE, but not leading to PAEs	All HEs are not leading to PAEs. Most are detected and recovered far before, becoming even a "near miss" or a "close call." Errors are more frequent than near misses, which occur up to 300 times more frequently than AE.	All PAEs are related to at least one unsafe act Neighboring concepts "near miss" and "close-call" are related to potential PAE whose story turned out well
Usual terms used in reports for the description of the type of failure	Reported in terms of generic categories of failure using generic language of ergonomics and reliability (individual failure and organizational malfunction)	Reported in terms of error's mechanisms, error causation, and error contextual shaping factors (in the language of psychology and cognitive ergonomics)	Reported in terms of patients' (potential) consequences and professional's failure (in the language of the medical profession)

(surgery), adverse drug events, and incorrect or delayed diagnosis are the top three areas of PAEs; this result is remarkably stable between studies. Medical errors are estimated to cause more casualties than road accidents or workplace injuries, and to represent a significant percentage of national health expenditures (Gaba, 2000; Kohn, et al., 1999). Moreover, errors are per definition under-reported, since hospital patients only represent a fraction of the total population exposed to medical errors.

Specific Difficulties in Coping With HE in Medicine

Medical professions suffer from various identity crises that make the global system prone to error and the study of errors less easy than in industry.

Risks in Medicine Are far From Being Homogeneous

- On the one hand, a number of sectors display risks of death due to adverse event (whether preventable or not) which exceed 10–2 (1 AE per 100 hospitalized patients, i.e., surgery, internal medicine), and which of course are not all related to medical errors.
- On the other hand, other sectors are extremely safe, and the risk of death due to adverse events is ten to a hundred times lower (i.e., anaesthesiology, radiology, blood transfusion). Furthermore, everything in medical practice encourages heterogeneity. This makes hospitals "sectorial environments," where different safety strategies need to be implemented according to different targets, with heterogeneous levels of regulation governing the various professionals working there, with a reduced capacity for cross-fertilization of lessons learned at local level (strongly limiting the development of a culture of safety).

The Topic of HE *in Medicine* Is *More Difficult to Manage* Than *in Industry*

- The pressure of insurance providers is greater than in any other industrial field.

[1]The apparent differences among studies does not necessarily mean true differences; it may only reflect small variations in the methodology and the inclusion criteria (see Baker, Norton, Flintoft, Blais, Brown, et al, 2004; Michel, Quenon, de Sarasqueta, & Scemama, 2004).

TABLE 32–3. Studies of Adverse Events in Hospital Patients

Study	*Setting (Year)*	*Percentage of Patients With One or More AE*	*Percentage of Preventable AEs*	*Percentage of Fatal Issue When AE*
Brennan et al., 1991 Leape et al., year	Harvard Medical 51 hospitals in NY, N = 30,121	3.7	58	13.6
Thomas et al., 2000	Utah & Colorado 28 hospitals, N = 14,700	2.9	53	6.6
Wilson, Harrison, Gibberd, & Hamilton, 1999	Australia 28 hospitals, N = 14,149	10.6 (after harmonization)	51	4.9
Davis et al: 2001	New Zealand 13 hospitals, N = 6,579	12.9	37	–
Vincent, Neale, & Woloshynowych, 2001	UK 2 London hospitals, N = 1,014	10.8	48	8
Schioler et al., 2001	Denmark, Copenhagen Public Hospitals, N = 1,067	9	40.4	6.1
Baker et al., 2004	Canada 20 hospitals, N = 3,745	7.5	36.9	15.9

The great autonomy of physicians, which is often mentioned as a prerequisite for the profession, also makes frontline actors more individually accountable for errors, with little protection from their organizations when sued and facing justice (Vincent, Young, & Phillips, 1994). Needless to say, this specificity does not encourage disclosure and reporting of malpractice.

- This is one of the areas where actual figures on error are hardest to collect. The figures we have are only extrapolations, and could be discussed, mainly because of the intertwined nature of contradictory dimensions of risks in medicine (see the beginning of this section), and the debatable definition and estimation of PAEs (Hayward & Hofer, 2001). The dilution in medical establishments and wards, the fact that death trickles in, one patient after another without any massive single occurrence in the context of pathologies, all help cover up reality and alleviate any strong reaction of anger from the media. To makes things worse, safety advances uncover safety loopholes which, until then, were discretely hidden away. Because of this, as is the case in all fairly unsafe systems where a culture of denial and hiding is widespread, the first phase of progress is mechanically associated with an increase in known occurrences. This access to such an unbearable truth generates fears and defensive reactions from the system, hindering further safety improvements.
- Nobody can actually really pinpoint the potential gains to be expected. As Marx (2003) pointed out repeatedly in conferences, a good safety management is the result of a comparison between a safety target expected at the phase of design of the system and the reality of occurrences observed in the field. Commercial aviation design expected 1 loss every 1,000,000,000 departures, and the reality is 1 loss every 3,500,000 departures. Space shuttle design expected 1 loss every 212 departures, and the reality is 1 loss every 60 departures. But how can safety be managed in medicine, when there is no expectation, just a reality of losses: 1 loss every 1,000 hospital visits!

In Medicine, Some Conditions of Work Are Behind Usual Industry Standards and Facilitate Errors

- It is one of the few risk-prone areas where direct public pressure is so high that it considerably limits the application of wise (or common-sense) safety-enhancing solutions (i.e., it is difficult to turn down anyone in an emergency ward, to regulate incoming patients to adapt patient flow to the capacity of premises, etc.).

- It is one of the few risk-prone areas where the system is so extensively supported by interns, who are in the process of being qualified for their jobs. First and second year resident interns, as well as nurses in training, often bear huge responsibilities.
- It is one of the few risk-prone areas where there are so many fundamental and obvious sources of human error, and where so little is done to reduce them, because they are considered as being part of the job: excess fatigue on the job because of systematic overtime (Morris & Morris, 2000), overloaded work schedules and chronic staffing shortages (Gander, Millar, & Weller, 2000; Weinger & England 1990), multiplication of quasi-similar products (competing number and forms of medical specialties), reassignment of tasks between the different job categories, and to top it off, staff members acutely sensitive to their own mistakes, with profound psychological consequences, and a closeness to death found in no other profession. (Daugherty, deWitt, & Beverley, 1998; Wu, Folkman, McPhee, & Lo, 1991).

THREE GENERIC SOLUTION SPACES FOR ERROR REDUCTION

Human error in medicine may be studied from various angles: how many (error rate), what type (wrong drug, delay, omission), with what consequence for the patient (from minimal impairment to death), where (surgery, internal medicine, anaesthesiology, etc.), when (week vs. weekend, nighttime vs. daytime), with what mechanism (slips and lapses vs. mistakes, errors vs. violation, active vs. latent human failure), and why (error causation, error-framing factors).

The *how many*, *where*, *when*, and *what* questions (frequency, location, type, and consequences) are key factors to prioritize sectorial efforts (e.g., delay in diagnosis, surgery, and drug events are often considered as first priorities because they represent the three major sources of PAEs). Conversely, the *why* questions are key factors to determine the repertoire of safety strategies to use according to the result of causal analysis.

Because this chapter focuses on error reduction strategies, the following sections give priority to the *why* questions and to the safety strategies to recommend. Reason (1990, 1995, 1997) addresses the problem by clearly opposing on the one hand frontline actors, for example doctors, nurses, or stretcher-bearers, working at the "sharp end" and making "patent errors"; and on the other hand designers, administrators, and system managers working at the "blunt end" and making "latent errors." Design flaws, staff shortage, poor management, or growing incentive to performance are examples of latent errors that can potentially foster patent errors made by frontline actors.

When considering possible solutions, Reason (YEAR) turns the earlier-mentioned opposition into two practical approaches for error reduction: the "person approach" and the "system approach" (Figure 32–2). A third approach turns out to be complementary: the "dynamic approach." This last approach is generally missing or underconsidered in the literature, although it is essential. Strategies adapted to error reduction should change as safety improves.

The Person Approach

Chronologically, the first approach is the person approach. It focuses on the errors and violations of individuals or groups of individuals. Most of the safety strategies relevant to this approach focus on the sharp end. Remedial efforts are directed at people at their work position.

Essential Points

At this level three principles challenge common sense and need to be revisited.

The Total Eradication of HE Should not Be an Objective. However, ergonomic advances in the workplace and facilitating error detection and recovery should be. By definition HEs are nonintentional. The literature distinguishes several types of errors: slips and lapses vs. mistakes (rules based and knowledge based). Slips and lapses are assimilated to routines errors and are most frequent at expert level (Reason, 1990). Their frequency in tightly coupled systems may reach impressive

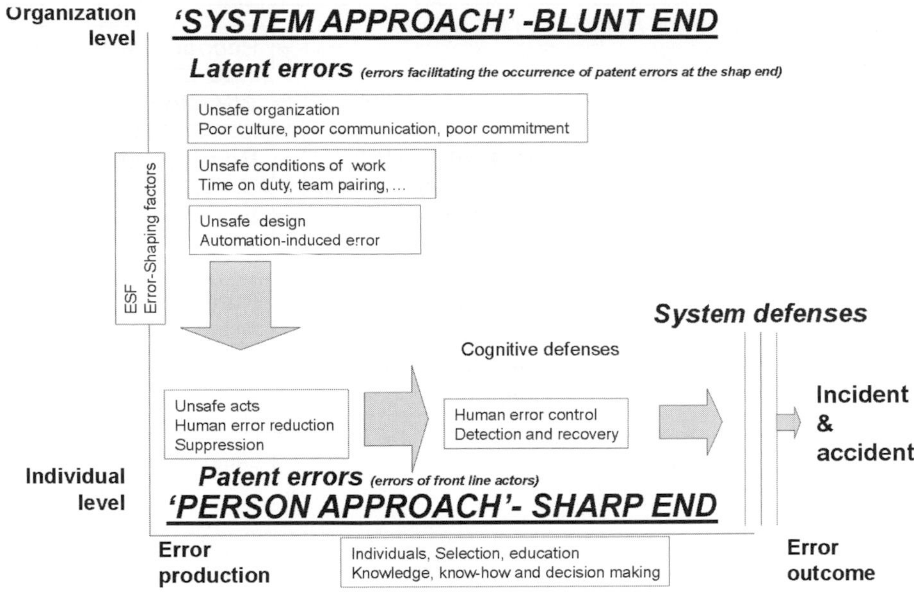

Figure 32–2. Summarizing James Reason's concepts, the person and system approaches.

figures. In aviation, numerous studies show that the minimum rate of error for professional pilots is around one per hour, whatever its outcome and the quality of the workplace design (Amalberti, 1998; Duffey, 1999; Helmreich, 2000). The generic top reasons for why errors occur are complexity of work situations and lack of familiarity with them, time pressure, lack of anticipation, and communications impairment.

It Is More Important to Consider Detection and Recovery than Error Itself. Allwood (1984) even considers that detection performance is the true marker of expertise, while error production is not. The rate of self-detection is then very high, above 70%, and is integrated in the natural cognitive resources that individuals have to manage. Routine errors are better detected than mistakes. Detection and recovery are sensitive to high workload, task interruptions, and system time management. In aviation, a priority principle in design is to tolerate at least one error while giving time to detect it and recover from it before loss of control (fault-tolerant system).

Violations Are Intentional and Should not Be Treated as Errors. Figure 32–3 shows the different types of errors and violations. Violations happen as often as the rest of errors but generally have not been considered as having consequences on safety (Helmreich, 2000). However, additional analysis led to a different conclusion: chronic violators also have increased rates of error, specifically mistakes (Helmreich, 2000). Violations, therefore, are two-faced. On the one hand, they occur frequently but without consequences on safety. On the other hand, violations can also represent the historical source of dangerous behaviors. This duality reinforces the importance of the debate in the scientific community on the safety strategies that allow one to cope with violations. To model and explain why violations occur, Amalberti extended Rasmussen's model of migrations (Amalberti, 2001; Rasmussen, 1997), offering to distinguish a three-phase model (see Figure 32–4).

- The first phase occurs during the initial design of the work situation. At that stage, systems are designed to accommodate a threefold pressure: (a) compliance with sociological rules, (b) use of available technologies, and (c) allowance for economic performance constraints. All the constraints describe a possible space of action, which is itself limited and enclosed by strong barriers (failsafe systems), physically blocking-off various maneuvers or accesses, as well as by

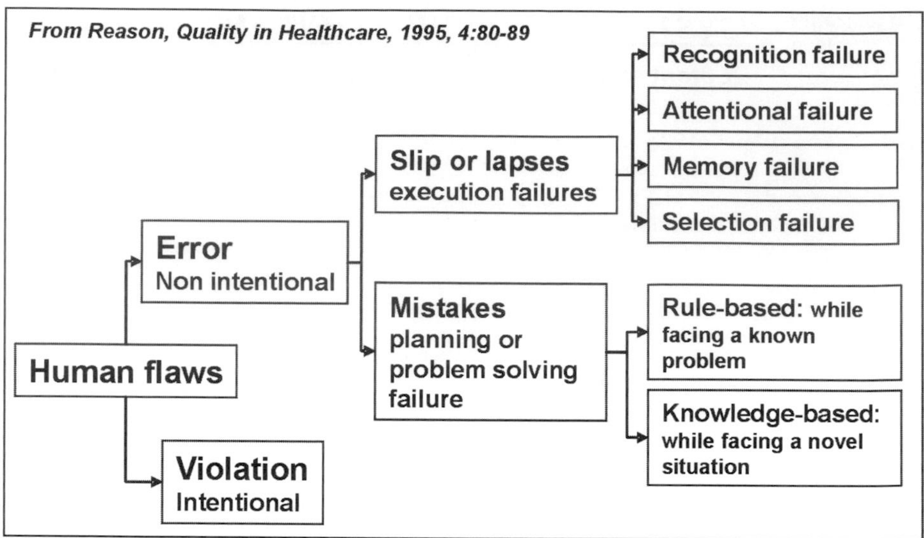

Figure 32–3. Relations between types of errors and violations.

virtual barriers (i.e., rules, protocols, state of the art, diverse operational limitations).

- The second phase occurs when the system is commissioned and must continuously adapt to new social and technical demands. The system migrates towards greater performance (horizontal axis), and additional secondary benefits for the individual (vertical axis). Barriers are quickly bypassed under the pressures of real life. The first transgressions always occur at the senior management level, which, even though it is strapped for resources, is required to provide higher performance than originally expected (e.g., having the same amount of work done with less staff, missing or out-of-order equipment, etc.). Once these transgressions are achieved, the next quasi-immediate move is a second migration. This time, it is for the benefit of individuals who grant themselves secondary rights in payment for efforts made to work "officially illegally" and to routinely "officially transgress" the established rules, in view of delivering the demanded performance. The result is that the system migrates towards a "normal illegal" area of stabilized operation. At that stage, violations are better termed borderline tolerated conditions of use or BCTU (Polet, Vanderhaegen, & Amalberti, 2003), and may be viewed as providing management and individuals with the maximum benefit for the minimum and accepted probability of harm.
- The third and last phase occurs after a certain laps of time. The same violations may be committed as in the second phase, but they are now old, amplified (amplitude of deviation), and made gradually invisible to peers' and managements' inspections (part of usual collective routines). These violations also represent a major safety hazard, as there is a tendency to continue the migration in solo (less dependence with demands) until the incident or accident occurs.

It goes without saying that violation prevention must focus on the appropriate phase.

Recommended HE Reduction Strategies at the Person Approach Level

The best way to achieve the proper error reduction is to apply the "four S" recommendations: standardize, simplify, staff and share, supervise.

Standardize. This requires reducing the span of medical procedures and technologies. Enlarging the span of possible solutions induces a greater risk of error. Evidence-based medicine is the main attempt

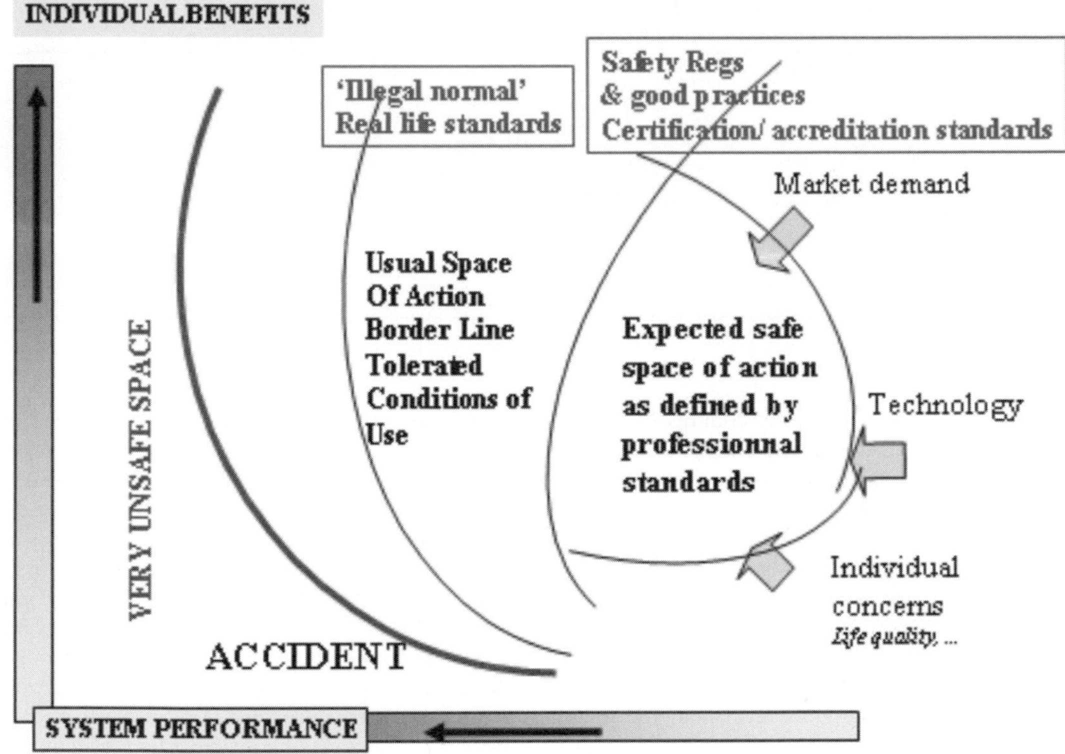

Figure 32–4. A model of systemic migrations to boundaries and violations.

to accomplish this standardization at the level of the profession, although it leads to more protocols being implemented at the ward level. Such standardization implies a significant effort in initial and recurrent training on the selected standards to increase the number of equivalent actors (professionals sharing the same core of knowledge and know-how) and facilitate the round-the-clock coverage of patient care.

Simplify. This is a true challenge for medicine. It comes in addition to standardizing, with the systematic selection of the simplest solution whenever possible: technical devices, procedures, labeling, or patient administration. Some solutions presented as simplifying systems may have drawbacks. Extensive automation, for instance, often has a negative potential, at least when not coherently integrated. In addition to choosing the simplest, the concept of simplifying also encompasses all means to reduce opportunity to make errors and facilitate detection and recovery. The fight against confusion errors is the first priority calling for unambiguous drug labeling (Christie, 2002; Lambert, Ken-Yu, Chang, & Prahlad, 2003). The facilitation of detection relies on security redundancies, hard protections, and reminders (Reason, 2002). Medical instruments commonly have poorly designed user-interfaces that promote human errors with life-threatening consequences. For example, this is particularly frequent with emergency instruments, such as electronic syringes, patient-controlled analgesia pumps (Lin, Isal, Donitz, Harkness, Vicente, et al., 1998). The same errors that occurred a long time ago in aviation (poor error analysis and techno-centered automation), are now being repeated in medicine (Cook & Woods, 1996; Woods, 2002; see Table 32–4)

Staff and Share. This has long been recognized as essential in aviation. Communication through briefings and debriefings is central to facilitating teamwork, and avoiding uncontrolled courses of action. mortality and morbidity conferences (M&Ms) are recommended and often effective. But the concept

TABLE 32–4. Woods and Cook's Nine Steps to Move Forward From Error.

1. Pursue second stories underneath the surface to discover multiple contributors.
2. Escape the hindsight bias.
3. Understand work as performed at the sharp end of the system.
4. Search for systemic vulnerabilities.
5. Study how practices creates safety.
6. Search for underlying patterns.
7. Examine how change will produce new vulnerabilities and paths to failure.
8. Use new to technology to support and enhance human expertise.
9. Stimulate innovation.

of staff and share requires greater continuous interactions to synchronize, explain, and acknowledge decisions, situation awareness, and courses of actions of medical teams. The transfer of aviation-based teamwork skills (crew resource management training) to medicine is gaining ground to educate medical staff in nontechnical skills (Grogan, Stiles, France, Speroff, Morris, Nixon, et al., 2004; Sexton, Thomas, & Helmreich, 2000). IT is taking a rapid central position in securing the process of communication (e.g., electronic patient files). It gives permanent and resident access to patient's data in a nonambiguous and standardized way, yet should never replace interpersonal communication, briefings, and staffs. Paradoxically, in addition to cutting by an average 50% the risk of being sued, sharing more information on error with patients (error disclosure) is also a means to change attitudes in wards and is a first step towards enhancing team communication.

Supervise. This means designing a process to collect and analyze errors, then making ad hoc decisions for improvement of priorities and actions in error reduction, and drawing lessons. Reporting systems are becoming extremely popular. To be efficient, it is commonly accepted that they should remain anonymous or blame-free, and should address near misses are well as misses. Feedback to sharp-end actors guarantees longevity and motivation to report. The technique of analysis of errors is also a challenge. Several methods have been offered (see for example the ALARM method recommended by Vincent, Adams, & Stanhope, 1998, or the root cause analysis method of Battles & Shea, 2001). All of these methods need to go beyond error to evidence the latent causes. Continuous audit and practice supervision are required to control violations. Indeed, solutions to cope with violations are not trivial. On the one hand, a certain amount of flexibility with regulations and standards, resulting in the presence of BTCUs, is probably required in complex social–technical work, to make the system efficient and adaptive. On the other hand, abuses and loss of control may represent true dangers. A compromise attitude between two extreme positions—tolerant versus punitive approaches—is still a matter of debate. Neither mandatory nor voluntary incident reporting systems are very good tools to monitor Phase II violations. These reporting systems only provide necessary information when it is too late for intelligent and smart action and control. Realistic and repetitive short periods of systematic observation of practice in wards (field audit), with a continuous dialogue about practice with peers, are probably the best methods of elegant and intelligent control of deviations and migrations.

The System Approach

The second approach is the system approach. It traces the causal factors back into the system as a whole. Remedial efforts are directed at situations, defenses, and organizations. There are schematically two major sources of error-shaping factors: bad system design implying undue work constraints (general architecture, procurement, and personnel management), and bad organizational design (governance, managerial policies, commitment, safety culture).

Essential Points

Monitoring Changes Is Mandatory. New regulations deeply impact all structures and create instability.

Management Is the Key Actor for Safety Culture. It is essential to promulgate a safety culture by making sure Management adopts it and then, turns it into a priority.

Patient Safety Starts with a Good Hospital Layout. The design of the hospital and of the global care process, recruiting process of staff members, as well as the procurement and maintenance of technologies, are key latent factors that may contribute in time to the emergence of errors. For instance, an undue distance between operating room, emergency, and surgery wards may cause transfer delays, transport injuries, as well as information losses between medical teams.

Time Management Is Crucial. The problem is twofold: first, with poor management of working hours or overtime (e.g., noncoherent closing time for support services and wards), and second, with accumulation of fatigue. A great number of recent papers address the fatigue of residents, showing that eliminating interns' extended work shift in an intensive care unit significantly increases sleep and decreases attentional failures during night work hours (Landrigan, Rothschild, Cronin, Kaushal, Burdick, et al., 2004; Lockley, Cronin, Evans, Cade, Lee, et al., 2004; Sim, Wrigley, & Harris, 2004).

Recommended HE Reduction Strategies at the System Approach Level

Use Macro-Level System Design. A series of ergonomics guidelines and recommendations were published to improve the design of medical systems at a high-abstraction level (Carayon & Smith, 2000; see also a brochure from Reiling, 2003, showing an actual example of a safe hospital design). An example of this is to provide links between the structural level (architecture) and the functional properties and demands of health care processes. Task analysis at ward and hospital levels, addressing tasks, constraints, and products of the administration staff, is a prerequisite for a good macro-ergonomics approach. Participatory ergonomics, asking the workers to participate into the design loop, namely to elaborate relevant scenarios for testing the design, is a second prerequisite (see Kuutti, 1995, and Noro & Imada 1991).

Adapt Governance. Reason (1997) recommends assessing institutional resilience and higher management support for safety by means of three concepts: commitment, cognizance, and competence.

Commitment: In the face of ever-increasing production pressures, do you have the will to make safety management (SM) tools work effectively? Arbitration between economic drivers and safety issues needs to occur.

Cognizance: Do you understand the nature of the safety war, particularly with regards to human and organizational factors?

Competence: Are your safety management techniques, understood, appropriate, and properly utilized?

Share Responsibilities and Professionalism at the Right Level. It was recommended in the person approach paragraph to go beyond errors at the sharp end, and to consider systems as the potential cause of errors of frontline actors. At this point, some people jump forward to conclusions and say that if the system is the root-cause of accidents, the behavior of individual frontline operators is relatively unimportant: Each of them may then feel that what he or she does has little importance because he or she is controlled by the system. If something goes wrong, the blame is inevitably on the system or, possibly, on managers. Others take exactly the opposite view. Here comes the big word: responsibility. So who must be blamed in case of accidents? Those involved in the system, or the system which employs them? These two positions seem impossible to reconcile and may stay that way as long as the idea remains to find a culprit. These two apparently irreconcilable views are actually reconciled in the common term of *professionalism*. This term describes an entire set of qualities specific to an *individual*, including his or her determination, his or her ability to understand, his or her commitment to safety and, eventually, his or her ability to positively express his or her individual freedom. For example, flying an aircraft can never be safe unless all participants are personally totally committed to making it so. But at the same time, professionalism is also the product of a certain education, the personal appropriation of a surrounding culture, and the personal implementation of a collective skill. It is globally the result of social standards, that demand safety. Whatever its mode of action on individuals—fear of reprobation, a means of obtaining recognition from colleagues, corporate culture, or anything else—professionalism bridges, first and foremost, the psychological and social dimensions of individuals (Amalberti, Masson, Merritt, & Pariès, 2000).

Implement a Safety Culture. The concept of safety culture has become a gospel in all complex technical systems. A series of celebrated models were recommended to facilitate the adoption of a safe organization. The best known model is the high reliability organization (HRO; Weick & Sutcliffe, 2001). Weick and Sutcliffe recommend organizations acquire "meaningfulness." By meaningfulness, they mean that these organizations are

- Preoccupied with failure: They treat any lapses as a symptom that something is wrong with the system.
- Reluctant to simplify interpretations: They take deliberate steps to create a more complete and nuanced picture.
- Sensitive to operations: They recognize that unexpected events usually originate in latent failures. Normal operations may reveal these lessons but are visible only if they are attentive to the frontline, where the real work gets done.
- Committed to resilience: They develop capabilities to detect, contain, and bounce back from those inevitable errors that are part of an indeterminate world.
- Deferent to expertise: They encourage decisions to be made at the frontline and migrate authority to the people with the most expertise, regardless of rank

Most of these traits depend on how the organization will react to a danger signal, including weak signals. In complement to the HRO approach, Ron Westrum (1995) suggested a very useful classification of typical ways in which an organization responds to information indicating danger. Westrum described three organization modes:

- The pathological mode: Insufficient account is taken of risks, even in a normal situation. Safety rules are regularly set aside in the interest of economy and danger signals are ignored or concealed. Messengers are shot.
- The bureaucratic mode: The organization applies standard methods conscientiously and abides by the rules. Its does not suppress danger signals, or only very slightly, but does nothing to detect them outside the standardized channels. It minimizes the risks in a normal situation and loses control if important unforeseen events occur.
- The generative mode: The organization encourages inventiveness at all levels. The pursuit of its objectives includes a large proportion of nonstandard activities which constantly reveal new risks. However, because even very low-level staff is allowed to observe and act autonomously, risks are quickly discovered and corrected.

To end this section, it is important to acknowledge the long-term continuous effort needed to succeed in installing a safety culture of transparency. It took years for aviation to install such a culture, and the personnel involved were not as diverse. Medicine will probably require more time. It is vital to continue the effort and investment, even in the absence of immediate payback. Paradoxically, and that has been already said in this chapter, the very first signs that something is changing will probably be the increase in incidents reported (access to transparency). The system and the media must be educated, to have a good and clear understanding of this paradox.

The Dynamic Approach

The third approach, the dynamic approach is generally missing or underconsidered in the literature, although it is essential. The very nature of HE and the effectiveness of safety strategies change with time and systems improvements. Because of this, any safety strategy is sensitive to a specific level of risk, and may lose all effectiveness for another level of risk (Amalberti, 2006).

Essential Points

Effective safety improvement solutions vary with the degree of safety achieved:

- When the system is unsafe, the most effective safety strategies aim at increasing the constraints placed on stakeholders, while providing rapid technological enhancements: increased training, more rules, more protocols, and maybe even a strict sanction policy towards rule-breakers. Technological progress is the other high-priority goal, since it is commonly accepted that it eventually contributes more to improving safety than repressive measures. At this stage, feedback on accidents and serious incidents is sufficient to foster progress. On the one hand, these accidents or incidents are numerous enough to monitor risks, and on the other hand, they quite adequately represent future risks: At that stage, yesterday's accident will be tomorrow's accident, if nothing is done.

- For better systems, safety strategies must enhance the acquisition and generalization of a safety culture. Accidents have become few and far between, and databases now collect close calls or voluntary reports. At this stage, any accident or incident experienced will not repeat itself, it will come back as a recombination of various episodes belonging to different accidents. However, since accident precursors are the only common occurrence, they need to be clearly identified. At this phase, the system is highly regulated. New rules can be developed, adapted to new contexts or new technologies, and older rules which have become obsolete must be pruned. Beside technical innovations, progress will mainly come from generalizing and effectively complying with safety measures, making sure that all stakeholders, in all areas of practice, actually follow the same rules (acquiring and generalizing a culture of safety). In a word, at this stage, safety is not jeopardized by the lack of rules, but by inconsistencies in practice.
- At the highest level, the nature of safety changes once again. Feedback gradually loses its predictive capabilities. Focus must be placed on a more systemic approach, while preserving enough flexibility in the system to avoid being straight-jacketed in a posture that would reduce adaptation to technological progress.

Safety improvements will change professional and technical systems:

Eventually, any continuous safety improvement will add constraints and supervisory systems on actors, standardize performance among clinicians, increase accountability, introduce more automation, and therefore change many standards of the profession.

Recommended Dynamic HE Reduction Strategies

An adaptive strategy must be chosen to address the same problem as the system becomes safer (Figure 32–5). The big picture can be captured with a three-tier model, based on three dedicated actions: standardize, audit, and supervise. These keywords have already been cited in the person approach, yet here they are used in a more systemic point of view. The three tiers are to be considered as a non-end cumulative model. The first tier reaches a threshold allowing for the second tier to start, and at the same time, this first tier continues to improve, moving from one set of techniques to another, as the second tier also starts improving, and so on.

First Tier. When the system is relatively unsafe, the priority is to standardize people (competence), work (procedure), and technology (ergonomics). Recommended procedures in design (ergonomics) and operations (guidelines, protocols) are the main generic tools used during this phase (see person approach). The tools for standardization then move on to more official prescriptions (health care ministry policies) during the second tier, and ultimately will be turned into new federal or national laws in the last tier (patient rights, etc.). Moving from that tier to the second tier does not necessarily depend on reaching a given threshold in safety improvement, but merely on the gradual disclosure of problems to media, which in turn induces a shift in the way the safety level is politically managed. The governance of safety moves from being a local, technical, and domain-dependent professional business to being submitted to new regional or national agencies, reactive to the public's and politicians' wills and fears.

Second Tier. With the new governance, safety strategies benefit from new impulses. A continuous audit is required to address and control residual problems. Now is the time to extensively develop monitoring tools: in-service experience, reporting systems, sentinel events, M&M conferences. It is also time to consider safety at a systemic level,

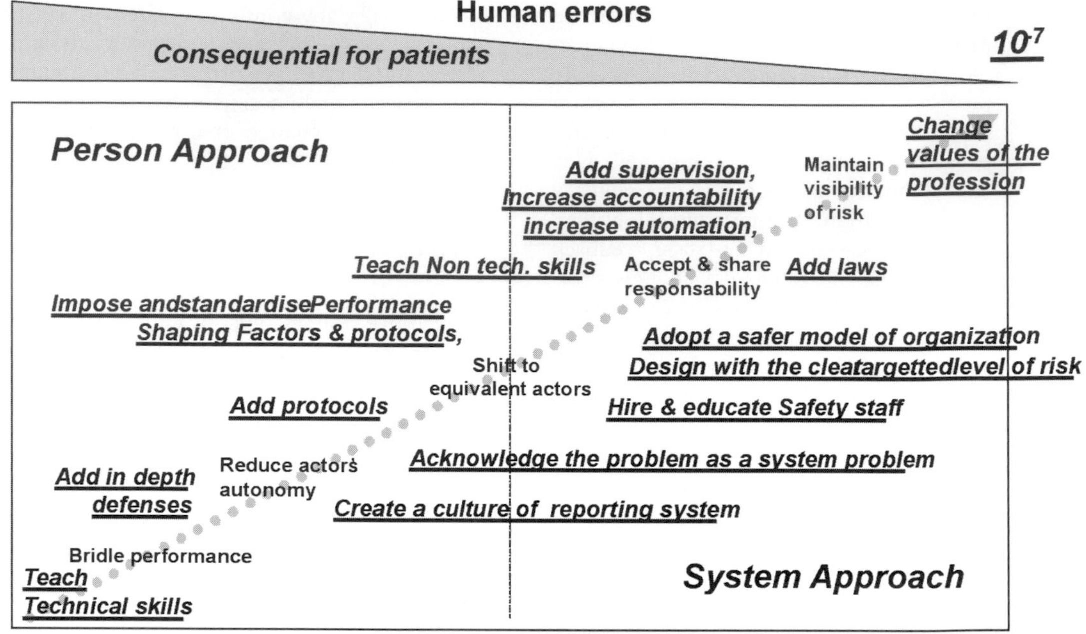

Figure 32–5. The dynamic approach: Variation along time of the safety strategies in the person and system approaches.

enhancing communication and the safety culture at all levels, including the management level. The teaching of nontechnical skills (CRM) becomes a priority to get people to work as a team. Macro-ergonomics tend to replace local ergonomics, and policies tend to replace simple guidelines (see system approach). The audit procedure shows a series of recurrent system and human loopholes which resist safety solutions based on standardization, education, and ergonomics. All solutions follow an inverted U-curve in efficiency. When the cost–benefit of a solution becomes negative (e.g., ratio of time spent on recurrent education to time effective in wards), the governance of the system has to envisage transitioning to the third tier.

Third Tier. Here the watchword is supervision. Supervision means full traceability by means of IT techniques to force standardization and personal accountability for errors, and growing automation to progressively change the role of technical people, freeing them from repetitive, time-consuming, and error-prone techniques, and making them more focused on decisions. For instance, advances in IT and computers provided aviation with an end-to-end supervisory system using systematic analysis of onboard black boxes, and leading to the eventual deployment of a global-sky centralized automated management system called data-link. The impact on people and their jobs, whether pilots, controllers, or mechanics is far-reaching.

Most sectors of health care are today in the second tier of the model. The medical field must be aware that the conditions to enter into the third tier are not yet fully satisfied. The move is just starting with the introduction of ITs. However, it is important not to rush into the extreme solutions specific to the third tier (full monitoring and black boxes) when the top of the second tier has not even been reached. Too much haste could result in professionals severely rejecting innovations (the system being immature), and to an uncontrolled negative feedback effect in terms of payback and effectiveness of the total system. With time, this step eventually occurs, but medicine is at least 10 years behind aviation, and given the inherent specificity and complexity of the medical domain, this move might require many more decades. Possibly, a number of areas in medicine could never transition to this ultimate level, because the potential for creativity and reactivity needed to face the variety of patients may need to be preserved.

CONCLUSION

To paraphrase medical language, at this point, the reader should be convinced that human error is not a disease, it is merely a symptom. To err is human and, as such, is part of natural human behavior, it is just a by-product of performance. Bearing this in mind, proper risk management should not limit itself to errors at the sharp end, for this would jeopardize global success in the long run. Because risks are numerous and diverse, and because the efficiency of risk-management techniques vary according to the safety level at hand, it is best to adopt a dynamic approach to risk management. Last but not least, the first major outcome of successful risk management is a distressing increase of reported events. To avoid a counterproductive interpretation of these numbers, the media (in and outside hospitals) should be educated to the risk-management process underway.

The six high-level dynamic principles that need to accompany health care improvements are as follows:

Principle 1: Audit the Safety Level Preliminary to the Choice of a Strategy. A safety audit includes not only safety figures, but the comprehensive analysis of the source of errors. Go beyond errors.

Principle 2: Adopt a Safety Strategy Consistent with Your Level of Safety. Safety is a nonstop incremental process. Respect the order of progression to make comprehensive changes acceptable and affordable by workers in the horizontal and vertical axes. Do not add laws when even recommended protocols do not exist.

Principle 3: Show Benefits of Change for Private Interest of Workers, and not Benefits for the Institution with Only Inconvenience for Workers. Respect as long as possible accountability, transfer of routines, and ecology of human strategies during transition times. Brusque transition times breaking all routines, destabilize human systems, and generally cost additional losses. Take the win without the losses.

Principle 4: Define and Orient Desirable Future Activity. An envelope of acceptable behaviours should be the aim, instead of setting a rigid but narrow target with yes–no criteria of success. Humans are making errors, but they are also clever managers of contradictory dimensions of risk. Give them instructions for change that have the potential to be equally powerful to cope with contradictory dimensions of the demand (save time, save money, save fatigue, do more, safer). Be careful with new tools that are intended to do—or assist to do—a much better normal job, but fail when the job becomes nonstandard. Real situations are scarcely fully standard. This principle is also preeminent when seeking criteria to comply with new standards, rules, or laws: Do you expect 100% fit with the instruction? Less? How far can your instruction be interpreted accordingly to the context? Think about expectation before introduction.

Principle 5: Adopt a Global Strategy for Change Management. Do not impose change at the sharp end without changing the blunt end.

Principle 6: Monitor Results with Relevant Monitoring Tools and Indicators; Be extremely cautious with indicators (they are a representation of the world, not the world itself): Add new indicators over time to avoid considering progresses only under a few lighting spots, with the risk of disregarding the very reality of field going out of these spots. Monitor the change at the blunt end as well as the change at the sharp end.

References

Allwood, C. (1984). Error detection processes in statistical problem solving. *Cognitive science, 8*, 413–437.

Amalberti, R. (1998). Automation in Aviation: A human factors perspective. In D. Garland, J. Wise, & D. Hopkin (Ed.), *Aviation human factors* (pp. 173–192). Hillsdale, NJ: Lawrence Erlbaum Associates, Inc.

Amalberti, R. (2001). The paradoxes of almost totally safe transportation systems. *Safety Science, 37*, 109–126.

Amalberti, R., Masson, M., Merritt, A., & Pariès, J. (Eds.). (2000). *Briefings* (2nd ed.). Paris: IFSA-Dedale.

Amalberti, R., Auroy, Y., Berwick, D., & Barach, P. (2005). Five systemic barriers keeping health care from becoming ultra safe: A conceptual framework for organizational safety. Manuscript submitted for publication.

Amalberti, R. (2006). Optimum system safety and optimum system resilience: agonist or antagonists concepts? In E. Hollnagel, D. Woods & N. Levison (Eds.), *Resilience engineering: concepts and precepts* (pp. 238-256). Aldershot, U.K.: Ashgate.

Auroy, Y., Amalberti, R., & Benhamou, D. (2004). Risk assessment and control require analysis of both outcomes and process of care. *Anesthesiology, 101*, 815–817.

Baker, R., Norton, P., Flintoft, V., Blais, R., Brown, A., Cox, J., et al. (2004). The Canadian adverse events study: The incidence of adverse events among hospital patients in Canada. *Canadian Medical Association Journal, 170*(11), 1678–1686.

Battles, J., & Shea, C. (2001). A system of analysing medical errors to improve GME curricula and programs. *Academic Medicine, 76*(2), 124–133.

Bogner, M. (Ed.). (1994). *Human error in medicine.* Hillsdale, NJ: Lawrence Erlbaum Associates, Inc.

Brennan, T., Leape, L., Laird, N., Localio, A., Lawthers, A., Newhouse, J., et al. (1991). Incidence of adverse events and negligence in hospitalized patients: Results of the Harvard medical practice survey study I. *New England Journal Medicine, 324*, 370–376.

Carayon, P., & Smith, M. J. (2000). Work organization and ergonomics. *Applied Ergonomics, 31*(6), 649–662.

Cook, R., & Woods, D. (1996). Adapting to new technology in the operating room. *Human Factors, 38*(4), 593–613.

Cox, S., & Trait, N. (1991). *Reliability, safety and risk management: An integrated approach.* Oxford: Butterworth Heinemann.

Christie, W. (2002). Standardized colour coding for syringe drug labels: A national survey. *Anaesthesia, 57*, 793–798.

Daugherty, S., deWitt, B., &Beverley, R. (1998). Learning, satisfaction, and mistreatment during medical internship: A national survey of working conditions. *Journal of the American Medical Association, 279*, 1194–1199.

Davies, P., Lay-Yee, R., Briant, R., Shug, S., Scott, A., Johnson, S, et al. (2001). *Adverse events in New Zealand public hospitals: principal findings from a national survey.* Wellington, New Zealand: Ministery of Health.

Dekker, S. (2004). *Ten questions about human error: A new view of human factors and system safety.* Aldershot, U.K.: Ashgate Avebury.

Duffey, R. S. J. (1999). On a minimum error rate in complex technological systems. In F. S. Foundation (Ed.), *Enhancing safety in the 21th century* (Vol. 289–301). Rio de Janeiro, Brazil: Flight Safety International.

Gaba, D. (2000). Structural and organizational issues in patient safety. *California Management Review, 43*(1), 83–102.

Gaba, D., & Howard, S. (2002). Fatigue among clinicians and the safety of patients. *New England Journal of Medicine, 347*(16), 1249–1255.

Gander, P. M. A., Millar, M., & Weller, J. (2000). Hours of work and fatigue-related error: a survey of New Zealand anaesthetists. *Anaesthesia and Intensive Care, 28*(2), 178–183.

Grogan, E., Stiles, R., France, D., et al. (2004). The impact of aviation-based teamwork training on the attitudes of health-care professionals. *Journal of the American College of Surgery 199*, 843–848.

Hayward, R., & Hofer, T. (2001). Estimating hospital deaths due to medical errors: Preventability is in the eye of reviewer. *Journal of the American Medical Association, 286*(4), 415–420.

Helmreich, R. (2000). On error management: Lessons from aviation. *British Medical Journal, 320*, 721–785.

Hollnagel, E. (1993). *Human reliability analysis context, and control.* London, U.K.: Academic Press.

Hollnagel, E. (1998). *Cognitive Reliability and Error Analysis Method: CREAM.* Amsterdam: Elseiver North Holland.

Hourlier S., Grau, J. Y., & Valot, C. (2003). A human factors approach to adaptive aids. In M. Haas & L. J. Hettinger (Eds.), *Psychological issues in the design and use of virtual and adaptive environments* (pp. 461–482). Hillsdale, NJ: Lawrence Erlbaum Associates, Inc.

Kohn, L., Corrigan, J., & Donaldson, M. (1999). *To err is human: Building a safer health system. committee on quality in America.* Washington, DC: Institute of Medicine, National Academic Press.

Kuutti, K. (1995). Work processes: scenarios as a preliminary vocabulary. In J. Carroll (Ed.), *Scenario-based design* (pp. 19–36). New York: Wiley.

Lambert, B., Ken-Yu, Y., Chang, C., & Prahlad, G. (2003). Effects of frequency and similarity neighborhoods on pharmacist visual perception of drug names. *Social Science & Medicine, 57*, 1939–1955.

Landrigan, C., Rotshchild, J., Chronin, J., Kaushal R., Burdick E., Katz J., et al. (2004). Effect of reducing interns' work hours on serious medical errors in intensive care units. *New England Journal of Medicine, 351*, 1838–1848.

Leape, L. (1994). Error in medicine. *Journal of the American Medical Association, 272*(23), 1851–1857.

Lin, L., Isal, R., Donitz, K., Harkness, H., Vicente, K., & Doyle, J. (1998). Applying human factors to the design of medical equipment: Patient-controlled analgesia. *Journal of Clinical Monitoring and Computing, 14*, 253–263.

Lockley, S., Cronin, J., Evans, E., Cade, B. E., Lee, C. J., Landrigan, C. P., et al. (2004). Effect of reducing interns' weekly hours on sleep and attentional failures. *New England Journal of Medicine, 351*, 1829–1837.

March, J., Sroull, L., & Tamuz, M. (1991) Learning from samples of one or fewer. *Organ Science, 2*, 1–13.

Marx, D. (September, 2003). *Design of safety culture.* Paper presented at the the second United States–United Kingdom Patient Safety Research Methodology Workshop: safety in design, Rockville, MD.

Michel, P., Quenon, J. L., de Sarasqueta, A. M., & Scemama, O. (2004). Comparison of three methods for estimating rates of adverse events and rates of preventable adverse events in acute care hospitals. *British Medical Journal, 328*, 1–5.

Morris, G., & Morris, R. (2000). Anesthesia and fatigue: An analysis of first 10 years of the Australian Incident Monitoring study, 1987–1997. *Anesthesia and Intensive Care, 28*, 300–304.

Noro, K., & Imada, A. (Eds.). (1991). *Participatory ergonomics.* London: Taylor & Francis.

Polet, P., Vanderhaegen, F., & Amalberti, R. (2003). Modeling the borderline tolerated conditions of use. *Safety Science, 41*(1), 111–136.

Rasmussen, J. (1997). Risk management in a dynamic society. *Safety Science, 27*(2–3), 183–214.

Reason, J. (1990). *Human error.* Cambridge, UK: Cambridge University Press.

Reason, J. (1995). Understanding adverse events: Human factors. *Quality in Health Care, 4*, 80–89.

Reason, J. (1997). *Managing the risk of organizational accidents.* Aldershot, U.K.: Ashgate Avebury.

Reason, J. (2002). Combating omissions errors though task analysis and good reminders. *Quality and safety in Health Care, 11*, 40–44.

Reiling, J. (2003). *Designing a safe hospital.* Publications series 1, Center for the Study of Health Care Management, Carson School of Management, University of Minnesota. Retrieved September 20, 2004, from http://www.csom.imn.edu/Assets/4220.pdf

Schioler, T., Lipczac, H., Pedersen, B., & Pedersen, B., (2001). Incidence of adverse events in hospitals: A retrospective study of medical records. *Ugeskr Laeger, 163*, 5370–5378.

Senders, J. & Moray, N. (Eds.). (1991). *Human error: Cause, Prediction and Reduction.* Hillsdale, NJ: Lawrence Erlbaum Associates.

Sexton, B., Thomas, E., & Helmreich, R. (2000). Error, stress, and teamwork in medicine and aviation: cross sectional surveys. *British Medical Journal, 320*, 745–749.

Sim, D., Wrigley, S., & Harris, S. (2004). Effects of the European working time directive on anashetic training in the UK. *Anaesthesia, 59*, 781–784.

Thomas, E., Studdert, L., Burstin, H., Orav, E., Zeena, T., Williams, J., et al. (2000). Incidence and types of adverse events and negligent care in Utah and Colorado. *Medical Care, 38*(3), 261–271.

Valot, C., & Amalberti, R. (2001). Ergonomics in aviation. *Le Travail Humain, 64*(3), 193–196.

Van der Schaaf, T., Hale, A., Lucas, D. (Eds). (1991). *Near miss reporting as a safety tool.* Oxford, U.K.: Butterworth-Heineman.

Vincent, C. (Ed.). (2001). *Clinical risk management* (2nd ed). London: British Medical Journal Publications.

Vincent, C., Adams, S., & Stanhope, N. (1998). A framework for the analysis of risk and safety in medicine. *British Medical Journal, 316*, 1154–1157.

Vincent, C., Neale, G., & Woloshynowych, M. (2001). Adverse events in British hospitals: Preliminary retrospective record review. *British Medical Journal, 322*, 517–519.

Vincent, C., Young, M., & Phillips, A. (1994). Why do people sue doctors? A study of patients and relatives taking legal action. *The Lancet, 343*(June, 25), 1609–1613.

Visciola, M., Armandi, A., & Bagnara, S. (1992). Communication patterns and errors in flight simulation. *Reliability Engineering System Safety, 36*, 253–259.

Weick, K., & Sutcliffe, K. (2001). *Managing the unexpected: Assuring high performance in a range of complexity.* San Francisco: Jossey-Bass.

Weinger, M., & Englund, C. (1990). Ergonomics and human factors affecting anesthetic vigilance and monitoring performance in the operating room environment. *Anesthesiology, 73*, 995–1021.

Westrum, R. (1995). Organisational dynamics and safety. In N. McDonald, N. Johnston, & R. Fuller (Eds.), *Application of psychology to the aviation system* (pp. 75–80). Aldelshot, U.K.: Avebury Aviation.

Wilson, R., Harrison, B., Gibberd, R., & Hamilton, J. (1999). An analysis of the causes of adverse events from the quality in Australian Health Care Study. *Medical Journal of Australia, 170*(May), 411–415.

Woods, D., & Cook, R. (2002). Nine steps to move forward from error. *Cognition, Technology, & Work, 4*, 137–144.

Wu, A., Folkman, S., Mc Phee, S., & Lo (1991). Do house officers learn from their mistakes? *British Medical Journal, 265*, 2089–2094.

33

COMMUNICATING ABOUT UNEXPECTED OUTCOMES AND ERRORS

James W. Pichert and Gerald B. Hickson
Vanderbilt University School of Medicine, Nashville, TN

Charles Vincent
Imperial College, London

A case of a retained sponge: "JS" is a 60-year-old female with adenocarcinoma of the colon. Dr. ABC is her surgeon, relatively new in practice and trying to make a good impression in the community. Almost two weeks ago her surgeon performed on Ms. S a right hemicolectomy. Ms. S's post-op course initially seemed uneventful. Unfortunately, three days after discharge she developed abdominal pain and episodes of vomiting. She called and Dr. ABC instructed her to come in for evaluation. The exam revealed an elevated temperature and diffuse abdominal tenderness. An abdominal film suggests a retained sponge. Ms. S needs surgery to remove the sponge. Dr. ABC has asked you for advice on how best to explain to Ms. S her need for a second surgery. What guidance is available?

This chapter begins with discussion of general issues related to error disclosure. These include recent disclosure-related standards and recommendations, effects of errors on health professionals, and how medical errors impact patients and families. The medical liability environment is briefly assessed, and selected research on medical malpractice is summarized in the second section. The third section presents a series of specific cases designed to highlight a range of challenges associated with discussing adverse outcomes when errors may or may not have contributed. A way to think through a range of disclosure options under various circumstances is presented. The chapter concludes with stories that highlight our strong conviction that appropriate error disclosure is essential in health care, but is only the first step in dealing with harm-causing errors. While the authors claim no special expertise in human factors engineering, analysis or research, we nevertheless try throughout the chapter to suggest where colleagues in those fields have opportunities to help create psychologically safe systems for appropriate and effective error disclosure.

GENERAL ISSUES RELATED TO ERROR DISCLOSURE

The Landscape of Disclosure

The 17th century English Poet Francis Quarles was perhaps only partly exaggerating when he penned

> Physicians, of all [persons], are most happy: whatever good success soever they have, the world proclaimeth... and what faults they commit, the earth covereth" (Francis Quarles, 17th century).

Perhaps Quarles would be surprised, then, that disclosure of medical error is now advocated, even

TABLE 33–1. Recommendations for error disclosure

Joint Commission for Accreditation in Healthcare

The "Elements of Performance" for RI.2.90 require that at ***minimum***, the practitioner or designee informs the patient (and family) about:

1. Outcomes that the patient (or family) must know about to participate in current and future decisions affecting care, treatment, or services
2. Unanticipated outcomes considered reviewable sentinel events
3. Unanticipated outcomes of care, treatment, and services

(Joint Commission on Accreditation of Healthcare Organizations (2006). Comprehensive accreditation manual for hospitals: The official handbook (March, 2006 Update 1). Chicago, IL: Joint Commission Resources, Inc., p. 185.)

Australian Council for Safety and Quality in Health Care

Open disclosure is the open discussion of incidents that result in harm to a patient while receiving health care. The elements of open disclosure are an expression of regret, a factual explanation of what happened, the potential consequences and the steps being taken to manage the event and prevent recurrence. ... While open disclosure is occurring in many elements of the health system the Standard facilitates more consistent and effective communication following adverse events. (Australian Council for Safety and Quality in Health Care. Open disclosure standard: A national standard for open communication in public and private hospitals following an adverse event in health care. Commonwealth of Australia, July 2003. Retrieved on June 4, 2006, from http://www.safetyandquality.org/OpenDisclosure_web.pdf.)

United States, the National Patient Safety Foundation

When a health care injury occurs, the patient and the family or representatives are entitled to a prompt explanation of how the injury occurred and its short- and long-term effects. When an error contributed to the injury, the patient and the family or rep should receive a truthful and compassionate explanation about the error and the remedies available to the patient. (National Patient Safety Foundation Board of Directors (November 14, 2000). Talking to Patients About Health Care Injury: Statement of Principle. Retrieved on June 4, 2006, from http://www.npsf.org/html/statement.html)

The American Medical Association (AMA)

Patients have a right to know their past and present medical status and to be free of any mistaken beliefs concerning their conditions...the physician is ethically required to inform the patient of all the facts necessary to ensure understanding of what has occurred. ...This obligation holds even though the patient's medical treatment or therapeutic options may not be altered by the new information. (Morse, L. J. (2003). Ethical Responsibility to Study and Prevent Error and Harm in the Provision of Health Care. Report of the Council on Ethical and Judicial Affairs #2-A-03. 2003, p. 3. Retrieved on June 4, 2006, from http://www.ama-assn.org/ama1/pub/upload/mm/369/2a03.pdf)

expected worldwide. In July 2001 the Joint Commission on Accreditation of Health Care Organizations (JCAHO) set the stage by stating that: "Patients and, when appropriate, their families are informed about the outcomes of care, including unanticipated outcomes" (JCAHO, 2001). While error disclosure to patients and families is not specifically mentioned, this standard does require hospitals to tell patients when they have been harmed during treatment. In Australia the Council for Safety and Quality in Health Care has established a standard for open disclosure of errors and harm, and the British National Patient Safety Agency has promulgated a similar document. Some examples of the statements and standards from professional and government organizations are shown in Table 33–1.

The standards and recommendations refer to the disclosure of harm and, in some cases, the disclosure of errors as well. Minor errors occur frequently in all walks of life, and health care is no exception. To operate a full disclosure process for any small error, such as a missed dose of a drug which had no consequences, is clearly unnecessary. In this case open disclosure would simply imply letting the patient know with a simple, brief apology that a dose had been missed and that missing the dose should have no consequences, but monitoring would continue. At the other end of the scale, when a patient has been seriously harmed, open disclosure involves not only the initial meeting to let the patient and family know what has happened, but most probably subsequent meetings and support of various kinds in the longer term. Ms. S's retained foreign body clearly falls into the latter category because the flawed counting system resulted in harm: the need for repeat surgery and potential loss of trust. The nature of the disclosure and subsequent support will vary from case to case, as we discuss later. We also consider some complicated cases in which, for example, the cause of an adverse outcome is not yet known, but patients or families want answers.

Open disclosure is sometimes presented as a defined event which is largely confined to a single, carefully and caringly planned meeting when disclosure occurs; for relatively straightforward cases where any harm is short lived and not serious, this may be entirely reasonable. The impact of disclosures in very serious cases needs to be considered more carefully. A clinician discussing the causes and likely consequences of a serious error is making an important intervention with the patient and family. All intend for such an intervention to be beneficial, but if poorly handled without thought for the long term, it can make matters worse. In serious cases a brisk and straightforward disclosure of the facts, which would be fine for minor errors and some families, may not be appropriate in a first meeting with others. For example, if the patient has been seriously traumatized, some patients and families are likely to need a more gentle approach in which they will be helped to gradually face the full implications of the harm and the consequent change(s) in their lives. But clinicians must not overgeneralize: some patients and families will want all the facts immediately. These considerations also apply to staff involved who, in serious cases, may also be quite distressed and disturbed by what has occurred.

Human factors can clearly affect and be affected by error disclosures. Such disclosures have important psychological impacts on clinicians, the usual focus of "user-centered design." But where patients and their families are involved, "family-centered design" is required (France et al., 2005). Before thinking about disclosure, therefore, we need to consider the reactions of those involved and how health care systems can be designed to support the kind and quality of disclosure that serves not only clinicians, but also patients who have suffered harm and their families.

The Impact of Errors and Mistakes on Clinicians

Error disclosure is a crucial first step in helping patients (and their families) who have been harmed or distressed during their treatment. Yet, staff may also be distressed by harm-causing errors and must be supported, both for their own sake and for the patients in their care. Clinicians working while distraught about an error may be so deeply affected by this "performance shaping factor" that they simply cannot provide the kinds and amounts of support required by the affected patient and family (Bogner, 1994, 2004). Such circumstances require that the local health care system be designed in ways that first encourage the error to be acknowledged and disclosed to an appropriate institutional representative. Unfortunately, the medical–legal system in the United States presently can discourage such early disclosure, especially if clinicians are not educated about how to report the event to appropriate risk-management or quality-improvement leaders in ways that protect the report from legal discovery, yet nevertheless appropriately support the patients and the clinicians involved.

Making an error, particularly if a patient is harmed because of it, may have profound consequences for the staff involved, especially if they are seen, rightly or wrongly, as somehow responsible for the outcome. When asked to describe a recent stressful event British junior doctors singled out making mistakes, together with dealing with death and dying, relationships with senior doctors, and overwork (Firth-Cozens, 1987). The typical reaction has been well expressed by Albert Wu in his aptly titled paper, "The second victim":

> Virtually every practitioner knows the sickening realisation of making a bad mistake. You feel singled out and exposed—seized by the instinct to see if anyone has noticed. You agonise about what to do, whether to tell anyone, what to say. Later, the event plays itself over and over in your mind. You question your competence but fear being discovered. You know you should confess, but dread the prospect of potential punishment and of the patient's anger (Wu, 2000, p. 726).

The impact of mistakes was explored in interviews with 11 doctors by Christensen, Levinson, and Dunn (1992). Although this small study did not assess the overall importance of mistakes, a number of very important themes are discussed, in particular: the ubiquity of mistakes in clinical practice; the infrequency of self-disclosure about mistakes to colleagues, friends, and family; the degree of emotional impact on the physician, so that some mistakes were remembered in great detail even after several years; and the influence of beliefs about personal responsibility and medical practice. A variety of mistakes were discussed, all with serious outcomes, including four deaths. All clinicians were affected to some degree, but four described intense agony or anguish as the reality of the mistake had sunk in:

- "I was really shaken. My whole feelings of self-worth and abilities were basically profoundly shaken".
- "I was appalled and devastated that I had done this to somebody".
- "My great fear was that I had missed something, and then there was a sense of panic".
- "It was hard to concentrate on anything else I was doing because I was so worried about what was happening, so I guess that would be anxiety. I felt guilty, sad, had trouble sleeping, wondering what was going on" (p. 246; from Christensen et al., 1992).

After the initial shock the clinicians had a variety of reactions that had lasted from several days to several months. Some of the feelings of fear, guilt, anger, embarrassment, and humiliation were unresolved at the time of the interview, even a year after the mistake. A few reported symptoms of depression, including disturbances in appetite, sleep, and concentration. Fears related to concerns for the patient's welfare, litigation, and colleagues discovery of their incompetence.

The impact on clinicians of serious mistakes was highlighted in a much larger study by Wu and colleagues; they sent questionnaires to 254 interns in the United States asking the respondents to describe the most significant mistake in patient care they had made in the last year, which had serious or potentially serious consequences for the patient (Wu, Folkman, McPhee, & Lo, 1991). Various types of error were reported, most frequently missed diagnoses (30%) and drug errors (29%). Almost all the errors had serious outcomes and almost a third involved a death but, at that time, less than a quarter of the mistakes were discussed with the patient or patient's family. More than a quarter of house officers feared negative repercussions from the mistake. Feelings of remorse, anger, guilt, and inadequacy were common. Therefore, when considering how best to disclose error to patients it is also necessary to consider how the staff involved have been affected and how this in turn may affect the patient and family. Most obviously, if the staff are very affected, the disclosure conversation may need to be led by a senior clinician who has not been directly involved in the patient's care.

Patient and Family Responses to Medical Injury and Error Disclosure

Errors and their disclosures can pose even greater challenges for patients and families. Patients are often vulnerable psychologically, even when diagnosis is clear and treatment goes according to plan. Even routine procedures and normal childbirth may produce posttraumatic symptoms (Clarke, Russell, Polglase, & McKenzie, 1997; Czarnocka & Slade, 2000). When patients experience harm or misadventure, their reaction may be particularly severe. Their families' experiences and reactions must also be considered because families will likely be involved in error communications, may influence the injured party's intentions to pursue a claim, and, in many cases, may provide follow-up care.

Traumatic and life-threatening events produce a variety of symptoms, over and above any physical injury. Sudden, intense, dangerous, or uncontrollable events are particularly likely to lead to psychological problems, especially if accompanied by illness, fatigue, or mood disturbances (Brewin, Dalgleish, & Joseph, 1996). Awareness under anesthesia exemplifies such an event. When people experience such a terrifying, if short-lived, event they may subsequently suffer from anxiety, intrusive and disturbing memories, emotional numbing, and flashbacks. Almost everyone experiences such memories after stressful events, such as a divorce or a bereavement, and while unpleasant they gradually die down. However they can be intense, prolonged, and cause considerable suffering. In severe cases the person may suffer from the full syndrome of posttraumatic stress disorder, discussed later.

The full impact of some incidents only becomes apparent in the longer term. A perforated bowel, for example, may require a series of further operations and time in the hospital. The long-term consequences may include chronic pain, disability, and depression, with a deleterious effect on family relationships and ability to work (Vincent, Pincus, & Scurr, 1993). Depression appears to be a more common long-term response to medical injury than posttraumatic stress disorder (Vincent & Coulter, 2002), although there is little research in this area. Whether people actually become depressed and to what degree will depend on the severity of their injury, their personality, the support they have from family, friends, and health professionals, and a variety of other factors (Kessler, 1997).

When a patient dies the trauma is obviously more severe still, and may be particularly severe after a potentially avoidable death (Lundin, 1984). For instance, many people who have lost a spouse or child in a road accident continue to ruminate about the accident and what could have been done to prevent it for years afterwards. They are often unable to accept, resolve, or find any meaning in the loss (Lehman, Lang, Wortman, & Sorenson, 1989). Relatives of patients whose death was sudden or unexpected may therefore find the loss particularly difficult to bear. If the loss was avoidable in the sense that poor treatment played a part in the death, their relatives may face an unusually traumatic and prolonged bereavement. They may ruminate endlessly on the death and find it hard to deal with the loss.

Injury from Medical Treatment Differs from Other Injures

The impact of a medical injury differs from most other accidents in two important respects. First, patients have been harmed, unintentionally, by people in whom they placed considerable trust, and so their reaction may be especially powerful and hard to cope with. Secondly, and even more important, they are often cared for by the same types of professionals, and perhaps the same people, as those involved in the original injury. As they may have been very frightened by what happened to them and have a range of conflicting feelings about those involved, this too can be very difficult, even when staff are forthcoming, sympathetic, and supportive.

Patients and relatives may therefore suffer in three distinct ways from an injury: initially or long-term from the injury itself, from the way the incident is handled afterwards, and from the loss of trust in their community's health care system and its providers. Many people harmed by their treatment suffer further trauma through the incident being insensitively and inadequately handled. Conversely, when staff come forward, acknowledge the damage, and take the necessary action the overall impact can be greatly reduced. Injured patients need an explanation, an apology, to know that changes have been made to prevent future incidents, and oftentimes practical and financial help. This is not to say that they necessarily need unusual treatment or that staff should be wary of talking to them—in fact, staff avoidance may result in patient fear of abandonment. The problems tend to arise when ordinary feelings are blunted by anxiety, shame, or just not knowing what to say. Human factors engineering must therefore be applied not only to user-centered (i.e., clinician) design, but also to patient- and family-centered design. Later in the chapter we will suggest that identifying and correcting systems contributions to the error and communicating those changes to the patient and family can be very helpful as error disclosure conversations continue over time.

THE MEDICAL MALPRACTICE ENVIRONMENT

Before we turn to the disclosure process itself, the malpractice environment must be addressed. Medical malpractice research reveals that clinicians frequently cite fear of litigation and its associated consequences as barriers to open error disclosure and seeking advice from colleagues when adverse outcomes occur. This is, at first glance, entirely understandable. Medical malpractice research also suggests what (and who) promotes lawsuits. While many medical liability disputes are resolved before trial, when they do go to trial awards can be substantial; national jury awards more than doubled in the United States between 1992 and 2002 (Jury Verdict Research, 2002). Widespread media reporting of high-award cases alongside substantial and ongoing coverage of the Institute of Medicine 1999 report "To Err is Human" have dramatically increased public (and juror) awareness of the potential for error (Kohn, Corrigan, & Donaldson, 1999). Perhaps this awareness has eroded public confidence and trust in the medical profession and medical institutions generally. Physicians whose errors may have caused harm are clearly aware of the social challenges when they consider whether or how to disclose them.

Are Attorneys the Problem?

Attorneys make a convenient target in a legal system whose outcomes appear neither sensitive nor specific; meritorious cases are not pursued, non-meritorious cases are pursued, and the courts seem to reward both about equal proportions of the time. It's little wonder that physicians, patients, and policy makers in the United States find the system of medical–legal dispute resolution so vexing. No one should be surprised that physician concerns about attorneys and courtrooms influence their willingness to disclose or comfort in disclosing. It is easy to understand—if not to pardon—physicians if they sometimes appear to advocate a Shakespearean solution to the malpractice liability climate: "The first thing we do, let's kill all the lawyers." (William Shakespeare, *King Henry VI*, Part II, 1597–1598, act IV, sc. ii).

Tongue-in-cheek sentiments aside, and despite the presence of some ambulance-chasing attorneys, the idea that every error or harm will be followed by litigation is simply not true. First, the incidence of adverse events quoted in major studies (Kohn et al., 1999), far exceeds the extent of litigation experienced. Research based on medical record reviews suggests that adverse events occur in approximately 6% of U.S. hospital stays and that approximately one third of those may be attributed to negligence of some sort (Brennan et al., 1991). Interestingly, however, only about 2% of patients and families whose medical documents revealed a reason to pursue a claim did so (Localio, YEAR). Unfortunately, several times as many patients and families whose medical records did *not* support a claim did so. Lack of social safety nets for persons with significant ongoing medical care requirements and several noneconomic factors appear to drive some families to the courtroom.

Families often find it difficult to identify an attorney who will take their case. Successful plaintiff attorneys report interviewing 50–100 prospective clients for every case they agree to take, and Clayton and colleagues reported that families had to approach more than three lawyers before finding one who would pursue their claim (Clayton, Hickson, Wright, & Sloan, 1993). In a contingency-fee-based system, lawyers' prospects for return on investment mean that some cases with merit do not get filed. On the other hand, questionable cases that have a sympathetic plaintiff and potential for a large award may go forward. Lawyers play a critical role in deciding which cases go forward, but the overall level of claims is not high compared with the overall level of apparent harm.

The Role of Poor Communication

Poor communication itself plays an important role in malpractice claims. Patients and families who file malpractice claims are not necessarily seeking compensation as the primary outcome; they often want explanations, apologies, action to prevent recurrence, and other outcomes (Table 33–2; Beckman, Markakis, Suchman, & Frankel, 1994; Hickson, Clayton, Githens, & Sloan, 1992; Vincent, Young, & Phillips, 1994). Others filed a claim when they came to believe that a medical professional was attempting to mislead them; that is not to say that the patients' physicians intentionally misled or were not forthcoming with information, but that was the families' perception.

TABLE 33–2. Reasons[a] Parents Sued After an Adverse Event Involving their Newborn

Reason	%
Advised to sue by influential other	32%
Needed money	24%
Believed there was a cover-up	24%
Child would have no future	23%
Needed information	20%
Wanted revenge, license	19%

[a]Families could offer more than one reason.

Note. From Hickson et al. (1992)

Research also shows that a small proportion of physicians attract disproportionate shares of malpractice claims (Bovberg & Petronis, 1994; Sloan, Mergenhagen, Burfield, Bovbjerg, & Hassan, 1989). Several studies suggest that physicians who attract more lawsuits than peers do so because they have difficulty establishing and maintaining rapport with patients and families. For instance, in one study interviews were conducted with women who had recently delivered healthy children and who were not involved in malpractice claims (Hickson et al., 1992). Families who saw obstetricians who had, unknown to them, a high rate of previous malpractice claims were more likely to complain about communication failures. For example, one patient reported, "Dr. X offered no information. I felt he was hiding something. He never even tried to talk to my husband." Families seeing high-malpractice-risk physicians were also more likely to complain about problems with access, and feeling that they were not respected as human beings, for example, "Dr. Y was rude. She was nasty that I started labor on the [*national holiday*] … she gave me snappy and smart [*-alec*] answers." A subsequent study linking patient and families' unsolicited complaints about their care experiences to physicians' malpractice histories confirmed the strong relationship between patient dissatisfaction and risk (Hickson et al., 2002). Specifically, the 9% of physician members of a large medical group who accounted for 50% of the group's unsolicited patient complaints—many related to poor communication and lack of apparent concern for the patient—also accounted for a disproportionate share of malpractice claims as well. Consequently, research suggests that the pre-outcome relationship affects how receptive patients and families may be after the consequences are appreciated.

To summarize the malpractice literature, we know that adverse outcomes and errors occur. When patients are injured, good communication and concern are important. Poor communication prompts some patients to sue, compounding the potential impact of any error. Small numbers of physicians attract disproportionate shares of suits, and more patients tend to complain about these physicians' communication. So when adverse outcomes and errors occur, good communication is especially important. Finally, good communication includes appropriate disclosure of adverse outcomes and errors.

So Why Disclose?

Error disclosures are very difficult and uncomfortable for health professionals and patients alike. As health care providers prepare to disclose a clear error, they may reasonably fear the loss of patient trust, injury to reputation in the medical community, and potential for litigation (Kraman & Hamm, 1999; Wu, 1999). Some may even have been taught (directly or via observation of mentors) *not* to disclose errors or to apologize. Others have learned from painful personal or vicarious experience not to discuss errors lest they be publicly humiliated in morbidity and mortality conferences designed more to attribute blame than to consider all potential contributors to an unexpected adverse outcome. It's no wonder then, that several studies report that disclosures are far rarer than errors (Blendon et al., 2001; Wu et al., 1991).

So, if disclosures present difficult challenges and create psychological or legal risks, why disclose? There are a number of reasons, some ethical, some humanitarian, and some practical. First, of course, disclosing an error shows the provider to be honest, forthright, a patient advocate, and someone with nothing to hide. Second, patients generally appreciate and expect appropriate disclosures when errors have occurred. Third, nondisclosure and other communication failures threaten patient–provider relationships, threaten the public's view of health care providers generally, and may promote patient nonadherence to medical regimens. Finally, if patients perceive nondisclosures as willful, they may be driven to plaintiff attorneys, ironically the very thing that influences some physician *not* to disclose errors (Cantor, 2002; Kraman & Hamm, 1999; Liang, 2002).

Once a commitment to disclose errors is achieved by a person or an entire organization, the issue becomes not *whether*, but *how* and *when*. In the sections that follow, we first discuss the general principles of error disclosure applicable when the error is clearly recognized as the cause of harm. We then turn to consider some more complex cases in which the relationship between error and harm is uncertain or disputed, where additional issues arise. In each case the aims are *not* to present scripts, but to share a strategy for critically examining the *how* and *when* of appropriate disclosure-related strategies.

The Process of Error Disclosure

Error disclosure means communicating bad—or at least unsettling—news, about which much has already been written. Generally accepted principles for delivering bad news provide a starting point for error disclosure, and also offer clinicians a foundation and model for the actual discussion and likely reactions. A clinician who, for instance, has had to inform patients that they have a potentially serious condition knows to provide as much information as the patient and family need and desire, to expect a variety of reactions such as initial denial or anger, to pace the discussion carefully and to make sure that the patient is clear about what happens next (Table 33–3). All these basic principles apply to the special case of error disclosure as well.

In addition to employing the general guidelines for delivering bad news listed in Table 33–3, the specific content of an error disclosure must also be carefully considered (Banja, 2001). Before reading on, pause to ask yourself what *you* want discussed if you or a loved one is the patient who has been harmed. Even brief reflection reveals a host of patient and family information needs. A little more reflection suggests a variety of challenges associated with meeting those needs. After all, following the disclosure, patients must deal with the knowledge that their pain or injury might have been avoided, concerns about longer-term consequences of the error, and that the party(ies) responsible for the injury may remain involved in their care (Vincent, 2001).

TABLE 33–3. General Guidelines for Breaking Bad News

Choose a private area; set the stage
Provide a brief review of the course of care: "Ms. S, as you know, we operated on you to remove your cancer about two weeks ago, but since then you've spiked a fever and have been sick to your stomach, so I ordered a film of the area where we operated because we were concerned to learn if those problems were related to your surgery."
Signal what's coming with a "warning shot": "First I want to reassure you that what I have found is not related to your cancer. But I'm sorry to have to tell you that we've found something that will require another surgery very soon."
Be frank but kind in the delivery of the news, pause, pause a little longer, and only then respond: ("Specifically, Ms. S, the film I ordered has clearly revealed that a surgical sponge—a gauze pad we use to soak up blood so we can see where we're cutting—was inadvertently left inside you. That should not have happened, and I'm very sorry that the sponge was not discovered before we closed up. As I said, the main thing now is that we remove the sponge, but of course that will require another surgery, and for that all of us involved in your care also apologize."
Empathize by using statements that signal partnership, empathy, apology, respect, legitimization, support (aka PEARLS) as appropriate
Comfort with silence
Gauge patient or family readiness for information by asking a few questions. Then respond with just enough information to answer, invite a few more questions and respond iteratively until the patient or family appears satisfied
Invite patient or family to ask any questions at any future time as well
Assure family of physician availability (dispel abandonment fears)

Table 33–4 lists error disclosure elements considered important by many patients. Its contents also suggest a few words that might apply to Ms. S's desire for information related to the retained foreign body. Not all 13 elements may apply to every case, or additional considerations may apply in specific instances. And, of course, care providers will choose words and phrasing suited to their best communication style and to the patient and family needs.

Patient and family reactions to bad news and error disclosures cannot be predicted. Many will react with grace, courage, and forgiveness, others with stoic acceptance. Some, who have perhaps suspected that a problem has occurred, will be relieved—at least initially—if their problem is remediable. Others will be distressed and tearful. Still others will appear stunned and numb, unable to respond otherwise at that moment. A few, though, will lash out in anger in their distress (McCord, Floyd, Lang, & Young, 2002). Violent anger, while understandable, poses a real threat of harm to the care provider and others in the clinical environment, and the potential for anger escalating into violence must be considered and anticipated. Generally speaking, most anger can be managed if the providers respond to it in ways that tend to defuse rather than inflame (Table 33–5).

The most straightforward physician–patient conversations about errors occur in cases when the errors are evident and clearly caused injury. Examples include Ms. S's retained sponge, a wrong-sided surgery and a failure to follow up on clearly abnormal test, scan, or procedure in cases where timeliness is important. In such cases, physicians and others involved in patient care should pause to consider general principles for breaking bad news, reflect upon the contents of error discussions desired by most patients and families, and resolve to focus on the patient and family needs (not their own issues during this meeting). But because clinicians and patients and families may discover an error simultaneously and while together, most professionals will need to practice these elements during training. Such training is usually best accomplished when cases are based on real, local incidents, involve personal experiences of clinicians who participate in the training programs, and derive from systematically studied observations of patients and professionals in real circumstances. Human factors researchers therefore have obvious opportunities to contribute to our understanding of how best to disclose errors under various conditions.

Fortunately, while errors occur daily in medical institutions, the need to disclose a harm-causing error will be relatively rare for health care providers. Therefore, most clinicians will benefit from an opportunity, if at all possible, to practice the disclosure in order to think through how best to present it. Such conversations can be very difficult, and made more so by the clinician's emotions and the uncertainty of patient and family reactions and questions. Clinicians should take care to focus on

TABLE 33–4. The Disclosure Process

Element	*How it Might be Conveyed*
1. Apology (precise)	Ms. Smith, I'm sorry to report that the reason you've got a fever and pain is because,
2. Nature of error, harm	while we removed the cancer, we did not remove all the surgical sponges—the gauze pads—we used during your surgery. On behalf of all who participated in your care, I want to apologize that a sponge was left inside of you.
3. When, where error occurred	Surgical sponges are counted three different times in the operating room, including just before and right after we close up. In your case we did the counts and we thought we had all sponges, but we were mistaken. The x-ray we took very clearly shows that a sponge remains inside you.
4. Causes and results of harm	The sponge is most likely causing an infection, so it must be removed very soon. I'm afraid I need to take you back into the OR to get it out.
5. Actions taken to reduce gravity of harm	I've ordered medicines for you to fight the infection, and hoping to remove the sponge quickly, I've scheduled us to do this surgery—with your permission, of course—very soon to minimize any problems.
6. Actions to reduce or prevent re-occurrence	When we go into the operating room this time I will personally oversee the sponge count, and I will ask the persons who assist me to double check the counts.
7. Who will manage ongoing care	If you will allow me to continue your care, your primary care doctor and I will work closely together to manage your condition.
8. Describe error review process, regulatory agencies to be informed	The medical center and I take mistakes like this very seriously. I've already asked that everything we did during your operation be reviewed by experts so that we can learn what caused this error. The results will be reported to me so that if changes in our procedures will make this less likely to happen
9. How systems issues are identified	in the future, we will make them. So if the review shows that there's something we can do to improve the way we do all surgeries, we'll change our systems.
10. Who will do ongoing communications	I will be personally responsible for reporting back to you the outcomes of the surgery to remove the sponge. And, if you wish, I will also be sure you learn the results of the review.
11. Names and contact info for all persons involved in communications	If I'm not available when you have questions or concerns about the sponge or the operation to remove it, please ask one of the nurses here to get hold of me. Or, if you prefer, please call my office, tell them who you are, and ask for my assistant, Ms. J. If she can't help you, she'll know how to get hold of me.
12. Offer counseling, support	I can imagine that this news is very distressing to you [and your family]. The hospital has very good people [in the ___office] who can talk with you, offer emotional support, and get you in touch with any help you might need going through all this. Here's a card with their telephone number on it. Or just ask one of the nurses or me to contact them for you. I encourage you to take advantage of their services. Would you like me to call them or anyone else for you right now?
13. In fee for service systems, address patient's bill for additional care	I want you to know that the review I requested will also result in a recommendation for handling the expenses associated with the second surgery. The important thing to focus on now, though, is removing the sponge and returning you to health.

the patient's and family's reactions, not their own, during error disclosures. This is not the time to share how the adverse outcome has affected them personally (or the team). While such statements may be appropriate later, the attention needs to be on the patient's needs now and in the future, not the clinician's. If or when patients ask who was responsible for the error or harm, it is especially helpful to have rehearsed responses that appropriately convey how individuals and systems may have failed, but that the circumstances will be carefully reviewed (if that is true). Planning ahead, anticipating potential reactions and questions, and having alternatives at hand when the plan goes awry simply increase the odds that the discussion will go as well as possible. Remember that the goal of an error discussion is *not* to make the clinician feel better, but to inform the affected parties. Rarely, if ever, are these conversations entirely satisfactory. That is no reason to avoid them, but it is

TABLE 33–5. If Patient or Family Reacts Angrily

Avoid defensiveness or challenging the patient's anger
Consider the fears or affronts that underlie the anger
Acknowledge the patient's anger directly, don't ignore it
Use empathetic statements
Avoid making excuses, blaming
Use a calm, soft voice; as the angry person dials it up, dial yours down
If an apology is appropriate, make one, but be specific; if one has already been offered, repeat it
If an apology is *not* appropriate, avoid blaming others, the system, and so on
If the anger appears to stem from some point of dispute, reassure the patient or family that it's okay for the physician and patient to disagree (agreeably) on issues of care
May have to call time-out and reconvene for discussion at a later time when emotions have settled.

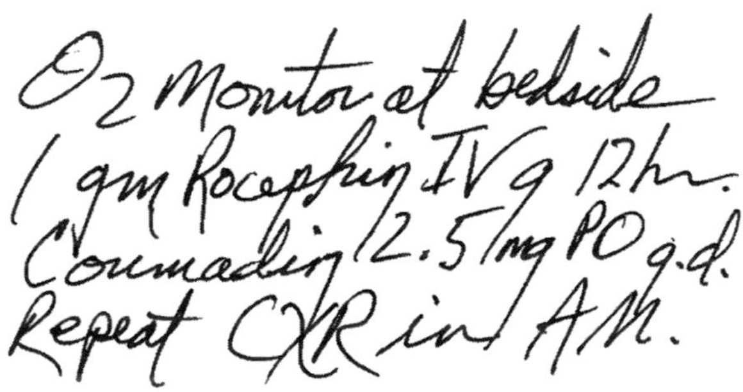

Figure 33–1. Mr. J's bedside drug sheet.

important to have realistic expectations about their outcome.

ERROR DISCLOSURE IN COMPLEX CASES

The general principles outlined in the previous section provide a basis for error disclosure and support in relatively straightforward cases where, even if the outcome is poor, the clinical facts are known and agreed upon. We now consider some more complex cases where the cause is not yet known, there is a dispute about what occurred, and, finally, where an error has occurred, but the family is unaware of the error. These types of cases are experienced daily throughout large medical centers, but how best to handle them are not well researched by the health care and human factors communities. The cases are therefore offered in part as challenges to thoughtful clinical and human factors collaborators to help health care systems cope well when adverse outcomes are associated with real or perceived errors.

Error Disclosure When the Cause of an Unexpected Adverse Outcome is not yet Known You are the general internist for "Mr. J," a 74-year-old man with a history of atrial fibrillation admitted for pneumonia. He was on warfarin (2.5 mg per day) at home to reduce risk of clot formation. His INRs, a measure of anticoagulation, measured 2–3 (checked monthly and just over a week ago). Admission orders included an antibiotic (ceftriaxone) and warfarin (2.5 mg per day). You have just been called at home (early on the 5th hospital day) because Mr. J is no longer able to use the left side of his body. You order stat coagulation studies, a head CT scan, and you proceed to the hospital. When you arrive you discover his INR is 6. Review of hospital lab reports reveals that you had not ordered any coagulation studies since he arrived at the hospital. The omission concerned you because you generally obtain coagulation studies in any patient on both warfarin and a broad spectrum

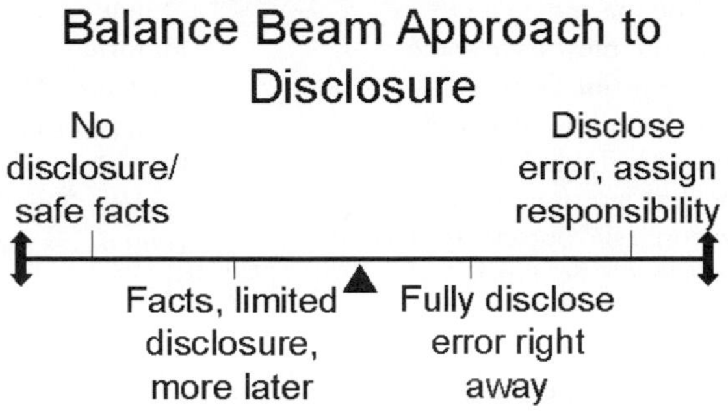

Balance Beam Approach to Disclosure

No disclosure/ safe facts

Disclose error, assign responsibility

Facts, limited disclosure, more later

Fully disclose error right away

Figure 33–2. A balance beam approach to disclosure.

antibiotic. In addition, review of his bedside drug sheet (See Figure 33–1) revealed that he had been getting 12.5 mg of warfarin per day instead of the 2.5 mg that you ordered. You go to Mr. J's room. When you enter his son asks, "Why do you think Dad had a stroke?" How would you recommend the physician respond? Which of the following would you choose?

1. No disclosure, safe facts: "Pneumonia is a serious condition, your father has a bad heart condition. He has several risk factors for a stroke. Under these circumstances strokes just happen sometimes, and I'm sad to see that Mr. J has had one. Now we need to focus on managing …."
2. Facts, more later: "I don't know yet. Mr. J's INR (a measure of blood thinning) was high, which might have contributed. We need to review his condition, the events leading up to the stroke, and the results of his most recent tests to see whether we can find out the cause of the stroke."
3. Disclose error: "INR was high. It looks like Mr. J got too much warfarin (I didn't check his INR since admission), which may or may not have contributed, we need to review his condition, the events leading up to the stroke, and the results of his most recent tests to see whether we can find out the cause of the stroke."
4. Disclose error, assign responsibility: "INR high. Mr. J got too much warfarin. The nurses misread my order (or bad handwriting); I am so sorry that this probably caused Mr. J's stroke."

All four of these responses and their variants have merit. That's what makes the *what*, *when* and *how* of disclosure so challenging under normal conditions of uncertainty, and why disclosure scripts, while sometimes helpful, simply will be neither prudent nor effective in all circumstances associated with apparent or potential harm-causing errors (Pichert & Hickson, 2006; Pichert, Hickson & Trotter, 1998). How clinicians respond in the short term will depend on several factors, including how certain they are that the stroke was caused by the error and their assessment of whether this is the time and place to impart information about a possible serious error. No one response or type of response will suit all situations. Medical professionals are better served weighing the pros and cons of each strategy, factoring in knowledge about themselves and their patients and patients' families, and the particular circumstances associated with the (perceived) adverse outcome and the medical care provided to that point. The strategy we recommend, therefore, may be characterized as a "balance beam" approach to considering what, how and when to disclose (Figure 33–2).

To the left of the fulcrum, no disclosure of any error, real or perceived, is offered. To the right of the fulcrum, the apparent error is disclosed and linked to the adverse outcome. In our experience, medical professionals asked to choose what to disclose in a hypothetical case such as Mr. J's distribute themselves along every point of the continuum and assert strong arguments in favor of their positions. Each selection from "no disclosure" to "disclose the apparent error and assign responsibility" therefore must have some positive features. But a few moments of thought will reveal that each alternative also has some potentially negative features. Before reviewing Table 33–6, see what pros and cons you would assign to each point along the continuum. In addition, ask yourself what follow-up questions might reasonably be anticipated from some families in response to each disclosure strategy.

The questions that appear in the far-right column of Table 33–6 are *not* meant to suggest that one alternative will be better or somehow easier to use than another. The important issue for medical professionals is to consider such questions in advance and, rather than being caught off-guard, be ready to provide answers to those questions as well. So, for example, if the professional postpones disclosure by promising a review, upon the subsequent disclosure some families will ask why the apparent error had not been discussed previously. The health professionals should anticipate the question so they may be prepared to respond. One alternative might be to simply assert that "I was aware of the possibility, but I wasn't sure whether the error caused the stroke. That's why I very much wanted a review of all the circumstances, and why I promised to report the results to you, which is what we're doing now. I just didn't want to speculate." We believe that most (but not all) families will accept this explanation.

When the Physician Deems the Error Unrelated to the Outcome

To highlight another, probably more common problem, consider the case where a physician chooses not to disclose the medication error immediately, but the son asks a nurse who subsequently visits the room, "Why do you think Dad had a stroke?" In this case the physician has judged that the case is to the left of the fulcrum in the balance beam figure, and that a first-encounter disclosure serves no purpose. A useful exercise is to consider the pros and cons if nursing professionals respond to this challenging question by saying:

1. "I don't know. Did you ask your physician? (What did your physician say?)."
2. "Your physician is sad that a coagulation study was not ordered a little sooner. Doing so might have provided us with important information that might have prevented the stroke."
3. "I'm sad to report that the doctor should have but failed to order coagulation studies. This probably would have prevented this stroke."
4. "Unfortunately, your father was given too much of his blood thinner medicine, which probably caused the stroke. Let me show you the doctor's drug order…"

The stroke poses challenges for all of Mr. J's care providers because the medication error may have played a role in causing a bleeding event, but the patient's atrial fibrillation may have caused a thrombotic event. Consequently, even though the drug error is clear, the stroke's *cause* at the time the family asks about it is not. Whether the drug error contributed or not will be unclear at least until CT scan results are read. The delay gives the physician, nurses, and others who care for the patient many opportunities to deliver mixed or even contrary messages.

When the Cause of an Unexpected Adverse Outcome is a Matter of Dispute Between a Physician and Patient or Family

"DA" is a 24-year-old female with ulcerative colitis requiring you (an attending surgeon) to perform a colectomy at your academic medical center. Following good informed consent an intern places a central line under a senior surgical resident's supervision. The surgery goes very well, but DA develops a fever and chills after 2 days. Fearing sepsis, you obtain blood cultures, chest x-ray, start antibiotics just in case, and order regular monitoring. During a follow-up conversation with DA's husband, he asks, "Could a central line infection have caused the fever?" DA's husband goes on to explain that he was in his wife's room when the line was placed and observed something that concerned him. He reports to you that apparently the intern has allergic rhinitis and rubbed his nose on the back side of his gloved hand. After he did, the senior resident declared, "You have just broken sterility," but then the senior resident

TABLE 33-6. Pros and Cons Associated With Disclosures When not all the Facts are Known or may Never be Known.

Alternative Response	*Pros*	*Cons*	*Questions Patients or Families Might ask if Given This Response*
No disclosure or safe facts (at this time, or ever)	Some families won't press you; some just don't want to know or don't need more information; may buy some time to finish an evaluation; may be kind to not share too much at once; others?	It's short of honest; looks BAD when discovered; news gets out various ways; many people have a need to know all the information as soon as possible; may signal a "code of silence"; others?	Do you have any ideas about what may have caused this outcome? Who else can I ask? I just don't understand why this happened, what else can you tell me?
Facts, limited disclosure, more later	It's more honest; buys time for impartial review; may preserve relations; others?	Still may turn out to be short of honest; you know more, may look bad later; news still gets out; premature speculations may create unrealistic expectations or wrong understandings; others?	How long will the review take? Why will it take so long? [Later, if the error was determined to be related to the adverse outcome] Didn't you know about the error when we first spoke? Why did Dr. or Nurse ___ say something different? How can such a thing happen here?
Fully disclose apparent error right away	Shows honesty; many are willing to forgive with apology; *may* cost less later? others?	You might be wrong; it will be challenging to take it back later; may empower legal action; may negatively impact patient confidence in the institution; others?	Who is responsible for this error? Who is going to pay my bills? Why didn't Dr. ___ or Nurse ___ mention the error? How can such a thing happen?
Disclose error, assign responsibility	Shows utter honesty, no "code of silence"; may make you feel better; shows that you're on patient's side; others?	Colleagues don't like jousters; you can't take it back later; jousts drive torts, may cost more money; may be unrelated to the adverse outcome; others?	Can you arrange for the responsible party to apologize to us? We should expect the institution to compensate us, right? Can the responsible parties be disciplined (fired)?

said, "Just go ahead and finish. It'll be okay. We need to get on to the OR." Mr. A concludes, "That's why I'm concerned that the infection was related to the line placement."

Using the same balance-beam approach taken in the previous case, please pause to consider the advantages and disadvantages that might be associated with each of these strategies. How would you as the attending surgeon respond to DA's husband?

1. "No. While it's not impossible, it's highly unlikely. Everyone was masked and gloved during the procedure…line infections just don't happen that way."
2. "It's possible, but very remote. Masks and gloves provide protection. Line infections happen, but I just don't think the intern's slip has caused this infection."
3. "Probably not, but I don't know at present. Let me gather some more information, then I'll come back and share with you whatever I learn."
4. "It's certainly possible … will review … be back to you."

In the meantime, the culture grows MRSA (Multiply Resistant Staph Aureus). When you see the senior resident he explains that they were both gowned and gloved, and that the event occurred with the last stitch to secure the line to the skin. No evidence of infection ever appeared at the stitch site. Later when you enter the room, the husband says: "I haven't seen your resident. Have you talked with the resident today? What have you found out?" You describe your understanding of the circumstances and conclude that the probability that the event caused the infection is less than remote. DA's husband persists in claiming, "But I still think it was the intern." How would you respond at this point?

1. "No, absolutely not."
2. "Possible, but very remote. After all, it happened on the last stitch. Infection should only be at the skin site and there's no evidence for that."
3. "There's only a small uncertainty, but we just can never be 100% sure."
4. "I suppose it's possible. I apologize for the lapse."

No matter how forthcoming or skilled a medical professional may be in disclosing adverse outcomes and their potential causes, not all families will be satisfied with initial explanations. Sources for patient or family nonacceptance are numerous, but must include the nature of medical probability. In the last case, the attending physician's judgment was reasonable: the MRSA has not been transferred from physician to the patient, but he cannot be 100% certain. Families may cling to a belief that an error caused harm, even if the probability is exceedingly small. Reasons for that tenacity might include denial about cause, suspicion of the medical profession in general or the physician specifically (especially with a young physician or intern), direct observation of what the spouse perceived to be an error, overheard jousting by medical professionals, or the patient or family may be correct. Because families ultimately have the final choice to accept or not, it is critical that the medical professional seeks to minimize the "inflammation" that may negatively impact disclosure. In addition, inflammation or anger or dissatisfaction with care has the potential to impact subsequent medical adherence, practice drop-out, complaining to others, and of course, in the face of a prolonged hospital stay, permanent disability or death, and decisions to file suit.

There are numerous reasons why medical professionals have difficulty in walking away from an encounter by simply agreeing to disagree agreeably. Professional training has a way of creating the need in physicians to resolve conflict and always reach an agreed-upon conclusion. Yet consensus is not always going to be possible, and when some families continue to disagree (as in the last case) some physicians will be angry themselves.

When a Subsequently Treating Physician Believes That an Error Occurred, but the Patient and Family are Unaware

"Mr. H" is a 72-year-old smoker with chronic obstructive pulmonary disease and cough with a persistent right lower lobe infiltrate. His family physician orders a chest CT, which reveals a mass in his right lower lobe (RLL), so the family physician refers Mr. H to a specialist for evaluation. A biopsy of the RLL mass is obtained. The pathologist reports out the biopsy as "a poorly differentiated squamous cell carcinoma." Treatment was initiated. Mr. H experienced a very difficult course. Four months into therapy Mr. H developed nausea and abdominal pain, prompting an evaluation including a CT scan of his liver and abdomen. The workup reveals a liver lesion not identified during his previous workup. Mr. H's family decides to transfer care to a comprehensive cancer center in the same city. Following the history, physical exam, and review of medical records, the cancer center oncologist recommends a fine needle biopsy of the liver mass to confirm that it was related to the lung lesion. Pathology report reveals malignant melanoma. When made aware of the unsuspected results, the oncologist requests confirmation of the biopsy report and initiates a request for the previous lung biopsy, concerned whether the new finding represents a second malignancy or whether the initial biopsy's interpretation was incorrect. Within the next 2 days the oncologist receives the first pathology specimen. The specimen is examined and additional stains—standard at the cancer center but not at the first care provider's institution—are performed, revealing that the initial lung lesion was not poorly differentiated squamous cell carcinoma, but malignant melanoma. While both diagnoses are fatal and wouldn't change the outcome, the family continues to ask about the most recent test results saying, "Doc, what did you find? Is it the lung cancer?"

Before considering how the oncologist might respond, who would you have share the new finding with the family (if anyone)?

1. The first oncologist or the first pathologist, or both.
2. The second oncologist or the second pathologist, or both.
3. Representative of the first medical center: that is, an administrator, risk manager, or lawyer.
4. Someone else or some other combination.

Beliefs about who should present the findings affect how the second oncologist would respond to Mr. H's question. Consider the first alternative. What are the pros and cons of engaging the originally treating team to disclose? One pro is that involving them provides the second team an opportunity to provide direct feedback on their decisions, and it may cause them to rethink using special stains. Second, members of the first team likely would want to know the correct data, and some members of the second team might feel professionally motivated to share their newer findings. If the family wants to hear an apology from those who erred, the first team might as well become involved right away. Finally, the first team would not want to first learn about their error from a family member or, worse, a plaintiff attorney.

On the downside, providing the data to the first team and asking them to discuss the results with the family takes time, especially if the first team disputes the findings. In addition, the second team may resent any perceived "highhandedness" and go out of their way to look for payback opportunities. Or some families might not wish to return to that group. The pros and cons for the second alternative are largely opposite those of the first. Finally, involving others can be very helpful if they have outstanding interpersonal skills, experience as bearers of bad news, and knowledge of the institution's policies and procedures regarding disclosure and, if necessary, ability to authorize compensation. One obvious risk, however, includes the potential for making the family suspicious or feeling ganged up upon.

How, then, might Mr. H's second oncologist respond?

1. No disclosure: "I'm sad to report that cancer is in your liver."
2. Principal facts: "It appears to be the original cancer, but the treatment approach needs to change. I recommend…"
3. Set stage for review: "It appears to be the original cancer. With your permission, I'll contact the initial medical team and learn their take on it, then share the findings…"
4. Disclose error: "Both the cancer in your lung and your liver are the result of metastatic melanoma—skin cancer that has spread. I don't know why it was first reported to you as squamous cell carcinoma. I would like to review everything with your first oncologist, then share the findings…"
5. Disclose error, assign responsibility: "The cancer in both your lung and liver is metastatic melanoma. The pathologist who initially diagnosed lung cancer was

> wrong; I am concerned because the treatments you've been getting were also wrong. I would like to review everything with the initial treatment team, then share the findings…"

All five choices have both pros and cons. Consider a few (and add your own). The first two alternatives, non-disclosure and presentation of only the principal facts, may take less time, at least initially, especially if the family chooses not to pursue some action against the first team. In addition, both strategies may be intended kindly; after all, the diagnosis is tough enough for the patient, and the error likely will have no meaningful impact on the eventual outcome, so disclosure at this time might even seem cruel. On the other hand, these strategies may result in the patient or family feeling misled if or when the truth comes out later, and those affected may come to believe that all their health professionals are engaged in a conspiracy of silence. In such cases, the time saved initially may be brief compared with the time and effort required to address such concerns. A con for the second medical team is that those who believe the first team provided genuinely bad care may feel guilty thinking that the family needs to be aware in order to make decisions going forward.

The third option, setting the stage for a review, allows the first team to decide whether and how they wish to be involved in the discussion about the care they provided. In some cases the first team will solicit the second team's assistance, which requires time, but may assure full disclosure. Unfortunately, if the first team wants to handle the disclosure, the strategy may backfire if the family perceives that the second team is simply passing the buck. Perhaps most problematic, however, is that the first team may not disclose the facts or in the fashion the second team believes should occur.

The pros and cons of the fourth and fifth options—those involving disclosure—are essentially the opposites of the first two alternatives. First, they present the truth and show the second oncologist to be an advocate for the truth. Second, by heading off family speculation, disclosure may help the family move through the grief process. In addition, disclosing the error directly may prevent the second team from having to communicate with the first. Unfortunately, collegial relationships might suffer as a result of these choices. Moreover, there's a chance that the second pathologist was wrong, and family anger about the situation may promote litigation.

As we have suggested elsewhere in this chapter, some medical professionals will not like an approach that requires weighing pros and cons. They would prefer disclosure scripts or a structured list of guidelines. Such scripts can be very useful when the circumstances are clear cut: where all the data are known, agreed upon, and everyone involved is a skilled communicator. Unfortunately, experience suggests that the majority of circumstances involving adverse outcomes and apparent errors simply do not meet these tests for implementing scripts. Fortunately, health care professionals have substantial experience weighing pros and cons of treatment alternatives in the face of medical uncertainty. Deciding how best to discuss adverse outcomes associated with apparent or potential errors is just another circumstance in which this skill applies.

Finally, this case highlights the disclosure-related duties of subsequently treating physicians who recognize that patient harm resulted from an error. Once again, the balance beam encourages consideration of alternatives ranging from no disclosure to returning the patient to the initially treating team of physicians, to declaring the error and assigning responsibility for it. The American Medical Association (AMA)'s Council on Ethical and Judicial Affairs states:

> "Even if a physician is not responsible for the harm, that physician still has the ethical obligation of protecting patient welfare in general by disclosing incompetence and promoting operational improvements that enhance patient safety" (AMA, 2003, p. 3).

This admonition assumes that the medical professionals who believe they have identified a harm-causing error indeed possess complete data, have the expertise to evaluate it, and have humbly considered that they might be wrong. Let's be clear: No code of silence is being advocated. Professionals must simply recognize that declarations about the care delivered by a colleague may have serious consequences. These consequences must therefore be considered alongside the equally serious professional duty to honesty and integrity.

DISCLOSURE IS JUST THE BEGINNING

When serious harm has been done, acknowledging, disclosing, and discussing the incident is just the first stage. The longer term needs of patients, families, and staff need to be considered. Injured patients have their own particular problems and needs. Some will require a great deal of professional help. Others will prefer to rely on family and friends. Some will primarily require remedial medical treatment, others primarily psychological assistance. We cannot cover every eventuality, but a few basics bear keeping in mind.

Ask Specific Questions About Emotional Trauma

A common theme in interviews with injured patients is their perception that the professionals involved in their care failed to sufficiently appreciate the depth of their distress. In many cases outright psychiatric disorders were missed. Risk managers, clinicians and others involved with these patients can ask basic questions without fear of making things worse. Other crucial areas of inquiry are feelings of anger, humiliation, betrayal, and loss of trust—all frequently experienced by injured patients.

When something truly awful has happened, health care staff may also be affected. In most clinical situations the need to think clearly and act decisively means that emotions must be kept under wraps. Loss of emotional control is of no help to patients, and may be quite damaging if staff are obviously unable to cope after an adverse event. However this does not mean that staff need to be remote or uninvolved. Many patients have derived comfort from the empathy and sadness of staff involved in tragic incidents describing, for instance, the warmth and support they found in the staff's own sadness at the event (Vincent, 2001).

A proportion of patients who suffer an adverse event will become sufficiently anxious or depressed to warrant formal psychological or psychiatric treatment. While it is important that a senior clinician is involved in giving explanations and monitoring remedial treatment, the staff of say, a surgical unit, cannot be expected to shoulder the burden of formal counseling. They have neither the time nor, in most cases, the necessary training to deal with the more serious reactions. When a referral to a psychologist or psychiatrist is indicated it must be handled carefully. Injured patients are understandably very wary of their problems being seen as psychological or "all in the mind." Therefore, one approach might be to normalize psychological assessment and intervention by suggesting, "When patients are injured as a result of an error, not only do we address the injury and the event, but we also know that the situation can impact you emotionally, so we always recommend that patients see a mental health expert for evaluation."

Continuing Care and Support

Injured patients may receive support, comfort and practical help—usually constructive but sometimes not—from many sources. It may come from their spouse, family, friends, colleagues, doctors, or community organizations. Especially important sources of support will be the doctors and other health professionals who are involved in their treatment. It is vital that the duty of care is paramount.

After an initial mistake most patients find it reassuring to be overseen by a single senior doctor who undertakes to monitor all facets of treatment, even if it involves several specialties (cf. Table 33–3). Where care has been substandard the patient must be offered a referral elsewhere if that is what they wish but if the incident is dealt with openly and honestly then trust may even be strengthened (Clements, 1994).

Inform Patients of Changes

Patients' and relatives' wish to prevent future incidents can be seen both as a genuine desire to safeguard others and as an attempt to find some way of coping with their own pain or loss. The pain may be ameliorated if they feel that, because changes were made, then at least some good came of their experiences. Relatives of patients who have died may express their motives for litigation in terms of an obligation to the dead person to make sure that a similar accident never happens again, so that some good comes of their death. The implication of this is that if changes have been made as a result of the adverse outcome, it is very important to inform the patients concerned. While some may regret that the changes were made too late for them, there will be those who appreciate the fact that their experience was understood and acted upon.

Financial Assistance and Practical Help

Injured patients often need immediately practical help. They may need medical treatment, counseling, and explanations but they may need money as well. They may need to support their family while they are recovering, pay for specialist treatment, facilities to cope with disability, and so on. In less serious cases, relatively modest sums of money to provide private therapy, alterations to the home, or additional nursing may make an enormous difference to patients both practically and in their attitude toward the medical group or medical center. Protracted and adversarial medical–legal negotiations can be very damaging, frustrating, and above all incomprehensible to the patient and their family. One only has to imagine oneself in a similar position to appreciate this. If you were injured in a rail or aviation accident you would expect the responsible organization to help you, not say that "you will be hearing from our lawyers in due course."

Some Good Practices and Positive Outcomes

To close this chapter, consider two examples of the benefits of a positive and proactive approach to error disclosure. Neither had long-term consequences, but both were very frightening to the patients concerned and, if handled badly, might well have affected their recovery and willingness to have future treatment. The examples show that even potentially disastrous events, like awareness under anesthesia, can be handled in a sensitive and innovative way with great benefits to staff and patients alike.

Explanations and Apology After Iatrogenic Cardiac Arrhythmia. "Mrs. A" was admitted for minor day case surgery, expecting to return home later that day. A surgeon requested a weak solution of adrenaline to induce a blood-free field, but was given a stronger solution than requested. As soon as the liquid was applied the patient developed a serious cardiac arrhythmia, the operation was terminated and she was transferred to the Intensive Therapy Unit, where she gradually recovered.

The clinical risk manager was alerted immediately and assessed the likely consequences for the patient and her family. The first task was clearly to apologize and provide a full explanation. However, with both the patient and family in a state of shock, this had to be carried out in stages. The consultant and risk manager had a series of short meetings over a few days, to explain what had happened and keep the family informed about ongoing remedial treatment. Each time the family was given the opportunity to reflect on what they had been told and come back with further questions. A small package of compensation was also arranged, primarily aimed at providing the necessary clinical and psychological support. The whole incident was resolved within 6 months and the patient expressed her thanks to the hospital for the way in which the incident had been handled, particularly the openness about the causes of the incident. Most, but probably not all, families would respond similarly; professionals who attempt to follow this model should be prepared for the instance when the patient and family do not respond well from the very first.

Anesthetic Awareness: Reducing the Fear of Future Operations. A woman was admitted for an elbow replacement. During the operation she awoke, paralyzed and able to hear the discussions amongst the surgical team. She was terrified, in great pain, and absolutely helpless. The lack of anesthetic was fortunately noticed, and she was next aware of waking in recovery screaming.

The risk manager visited the patient at home as soon as practicable, maintained contact, offered psychological treatment for trauma, and advised her on procedures for compensation, including an offer to pay for an independent legal assessment of the eventual offer of compensation. Emotional trauma was the principal long-term concern. In this case a fear of future operations was a major factor, very important in a woman suffering chronic conditions requiring further treatment. So this problem required additional imaginative measures. Specifically, when the patient felt ready, she was given a tour of the operating theatre, and the anesthetic failure was explained in great detail, as were the procedural changes that had been made subsequent to the incident. This was clearly immensely important in reducing her understandable fear of future operations and minimizing the long-term impact of the incident.

CONCLUSION

This chapter began by presenting recent disclosure-related standards and recommendations and discussing effects of errors on health professionals and patients. Selected research on medical malpractice suggested the importance of appropriate error disclosures. A series of specific cases was then used to highlight pros and cons of different approaches to disclosure under various challenging circumstances. Human factors abound, but have not much been analyzed or researched. As a result, experts in human factors engineering, analysis, and research have important opportunities to help assess and create psychologically safe systems for effective error disclosure. We concluded that appropriate error disclosures are essential in health care, but they are merely the beginning of dealing with harm-causing errors and their sequelae.

References

American Medical Association. (2003). *Code of medical ethics.* Council on Ethical and Judicial Affairs. Retrieved on October 3, 2003, from http://www.ama-assn.org/ama1/pub/upload/mm/369/ceja_report_2a03.do

Australian Council for Safety and Quality in Health Care. Open disclosure standard: A national standard for open communication in public and private hospitals following an adverse event in health care. Commonwealth of Australia, July 2003. Retrieved on June 4, 2006, from http://www.safetyandquality.org/OpenDisclosure_web.pdf

Banja, J. (2001). Moral courage in medicine: Disclosing medical error. *Bioethics Forum, 17*, 7–11.

Beckman, H. B., Markakis, K. M., Suchman, A. I., & Frankel, R. M. (1994). The doctor–patient relationship and malpractice: Lessons from plaintiff depositions. *Archives of Internal Medicine, 154*, 1365–1370.

Blendon, R. J., Schoen, C., Donelan, K., Osborn, R., DesRoches, C. M., Scdoles, K., Davis, K., Binns, K., & Zapert, K. (2001). Physicians' views on quality of care: A five-country comparison. *Health Affairs, 20*(3), 233–243 *Association, 272*, 1421–1426.

Bogner, M. S. (1994). *Human error in medicine: Misadventures in health care.* Hillsdale, NJ: Lawrence Erlbaum Associates, Inc.

Bogner, M. S. (2004). *Misadventures in health care: Inside stories.* Mahwah, NJ: Lawrence Erlbaum Associates, Inc.

Brennan, T. A., Hebert, L. E., Laird, N. M., Lawthers, A., Thorpe, K. E., Leape, L. L., et al. (1991). Incidence of adverse events and negligence in hospitalized patients. Results of the Harvard Medical Practice Study I. *New England Journal of Medicine, 324,* 370–376.

Brewin, C. R., Dalgleish, T., & Joseph, S. (1996). A dual representation theory of post-traumatic stress disorder. *Psychological Review, 103*, 670–686.

Cantor, M. D. (2002). Telling patients the truth: A systems approach to disclosing adverse events. *Quality and Safety in Health Care, 11*, 7–8.

Christensen, J. F., Levinson, W., & Dunn, P. M. (1992). The heart of darkness: The impact of perceived mistakes on physicians. *Journal of General Internal Medicine, 7*, 424–431.

Clarke, D. M., Russell, P. A., Polglase, A. L., & McKenzie, D. P. (1997). Psychiatric disturbance and acute stress responses in surgical patients. *Australian & New Zealand Journal of Surgery, 67*, 115–118.

Clayton, E. W., Hickson, G. B., Wright, P. B., & Sloan, F. A. (1993). Doctor–patient relationships. In F. A. Sloan, P. B. Githens, E. W. Clayton, G. B. Hickson, D. A. Gentile, D. A. Partlett (Eds.), *Suing for medical malpractice* (pp. 50–71). Chicago, IL: University of Chicago Press.

Clements, R. (1994). The continuing care of the injured patient. In R. Clements, P. Huntingford, (Eds.), *Safe practice in obstetrics and gynaecology.* London: Churchill Livingstone.

Czarnocka, J., & Slade, P. (2000). Prevalence and predictors of post-traumatic stress symptoms following childbirth. *British Journal of Clinical Psychology, 39*, 35–52.

Firth-Cozens, J. (1987). Emotional distress in junior house officers. *British Medical Journal Clinical Research Ed, 295*(6597), 533–536, 1987.

France, D. J., Throop, P., Walczyk, B., Allen, L., Parekh, A. D., Parsons, A., Rickard, D., & Deshpande, J. K. (2005). *Does patient centered design guarantee patient safety? Using human factors engineering to find a balance between provider and patient needs. Journal of patient safety 1 (3),* 145–153.

Hickson, G. B., Clayton, E. W., Githens, P. B., & Sloan, F. A. (1992). Factors that prompted families to file medical mal practice claims following perinatal injuries. *Journal of the American Medical Association, 267*, 1359–1363.

Hickson, G. B., Federspiel, C. F., Pichert, J. W., Miller, C. S., Gauld-Jaeger, J., & Bost, P. (2002). Patient complaints and malpractice risk. *Journal of the American Medical Association, 287*, 2951–2957.

Joint Commission on Accreditation of Healthcare Organizations (March, 2006). Comprehensive accreditation manual for hospitals: The official handbook (Update 1). Chicago, IL: Joint Commission Resources, Inc.

Jury Verdict Research®. (2002). *Medical malpractice: Verdicts, settlements and statistical analysis.* Palm Beach Gardens, FL: LRP Publications.

Kessler, R. C. (1997). The effects of stressful life events on depression. *Annual Review of Psychology, 48*, 191–214.

Kohn, L. T., Corrigan, J. M., & Donaldson, M. S. (Eds.). (1999). *To err is human: Building a safer health care system.* Washington, DC: Institute of Medicine, National Academy Press.

Kraman, S. S., & Hamm, G. (1999). Risk management: Extreme honesty may be the best policy. *Annals of Internal Medicine, 131*, 963–967.

Lehman, D. R., Lang, E. L., Wortman, C. B., & Sorenson, S. B. (1989). Long-term effects of sudden bereavement: Marital and parent-child relationships and children's reactions. *Journal of Family Psychology, 2*(3), 344–367.

Liang, B. A. (2002). A system of medical error disclosure. *Quality and Safety in Health Care, 11*, 64–68.

Localio, A. R., Lawthers, A. G., Brennan, T. A., Laird, N. M., Hebert, L. E., Peterson, L. M., et al. (1991). Relation between malpractice claims and adverse events due to negligence: results of the Harvard Medical Practice Study III, *New England Journal of Medicine, 325,* 245-251.

Lundin, T. (1984). Morbidity following sudden and unexpected bereavement. *British Journal of Psychiatry, 144*, 84–88.

McCord, R. S., Floyd, M. R., Lang, F., & Young, V. K. (2002). Responding effectively to patient anger directed at the physician. *Family Medicine, 34*, 331–336.

National Patient Safety Foundation. (2003). *Statement of Principle.* Retrieved October 3, 2003, from *http://www.npsf.org/download/*Statement_of_Principle.pdf

Pichert, J. W., & Hickson, G. B. (2006). *Institutional changes following an error disclosure seminar.* Manuscript submitted for publication.

Pichert, J. W., Hickson, G. B., & Trotter, T. S. (1998). Malpractice and communication skills for difficult situations. *Ambulatory Child Health: The Journal of General and Community Pediatrics, 4*(2), 213–221.

Quarles, F. *Hieroglyphics of the life of man.* Retrieved June 4, 2006, from http://www.worldofquotes.com/author/Francis-Quarles/1.

Sloan, F. A., Mergenhagen, P. M., Burfield, W. B., Bovbjerg, R. R., & Hassan, M. (1989). Medical malpractice experience of physicians: Predictable or haphazard? *Journal of the American Medical Association, 262*, 3291–7.

Vincent, C. A. (2001). Caring for patients harmed by treatment. In C.A. Vincent (Ed.), *Clinical risk management: Enhancing patient safety* (pp. 461–479). London: BMJ Publications.

Vincent, C. A., & Coulter, A. (2002). Patient safety: What about the patient? *Quality and Safety in Health Care, 11*, 76–80.

Vincent, C. A., Pincus, T., & Scurr, J. H. (1993). Patients' experience of surgical accidents. *Quality in Health Care, 2*, 77–82.

Vincent, C., Young, M., & Phillips, A. (1994). Why do people sue doctors? A study of patients and relatives taking legal action. *Lancet, 343*, 609–613.

Wu, A. W. (2000). Medical error: The second victim. *British Medical Journal, 320*, 726–727.

Wu, A. W. (1999). Handling hospital errors: *is* disclosure the best defense? *Annals of Internal Medicine, 131*, 970–972.

Wu, A. W., Folkman, S., McPhee, S. J., & Lo, B. (1991). Do house officers learn from their mistakes? *Journal of the American Medical Association, 265*(16), 2089–2094.

Section

VII

HUMAN FACTORS AND ERGONOMICS METHODOLOGIES

·34·

COGNITIVE WORK ANALYSIS IN HEALTH CARE

Michelle Rogers, Emily S. Patterson, David D. Woods, and Marta L. Render
The Ohio State University

Adoption of technology has been proposed as a means to reduce medical error (Kohn, Corrigan, & Donaldson, 1999). This reduction might be derived through automation of tasks, monitoring, and improved flexibility of data handling directed toward better situation assessment (Sarter, Woods, & Billings, 1997). Studies of accidents, however, consistently demonstrate that new computerized systems predictably affect human problem-solving ability in ways that often contribute to accidents. Experience with technology's contribution to new failures is widespread; ranging from personal catastrophes by unintentional but unrecoverable keystrokes that wipe out entire files (Norman, 1983), to software problems that crashed NASA's Mars exploration mission (Young et al., 2000). As health care moves toward increasing dependence on computerized tools—order entry, electronic medical records, and medication administration to name just a few—the design of these computerized tools to support human work (e.g., make it easier, faster, safer, and more accurate) also increases in importance. Recent reports including the failure of computerized order entry at Cedars Sinai (Chin, 2003), unexpected consequences with implementation of patient care information systems reported by Ash, Berg, and Coiera (2004), and bar-coded medication administration (BCMA) by Patterson, Cook, and Render (2002), lends credence to this view. Cognitive work analysis (CWA) is a method that models how environmental, organizational, individual, and technical constraints contribute to work in order to design tools that support work. This methodology, applied to health care systems, could improve safety, efficacy, efficiency, and the acceptance of computerized tools by health care providers.

HUMAN PERFORMANCE IN WORK

High reliability is needed in industries that operate in an unforgiving social and political climate, have risk of serious adverse consequences, have limited opportunity for learning through experimentation, use complex processes and technology, and have potential for unexpected events and surprise (Harper, Hughes, & Shapiro, 1991; Weick & Sutcliffe, 2001). Health care is such a system, with inherent hazard and vulnerability to failure. Workers perceive these hazards or risks for adverse outcomes and adapt individually, in groups, and organizationally to avoid or guard against them in the pursuit of their goals. These continuous adaptive efforts make safety. Feedback on how these adaptations are working or how the environment is changing is a critical factor in preserving safety. Recognition of the boundaries or the vulnerabilities of a system show workers where failures may occur and can guide investment to cope with these paths toward failure. Change within the system may reduce the ability to identify those boundaries and vulnerabilities. The ability of the system to support technical work determines the reliability of a complex system. Such systems, in addition to facilitating completion of the task or goal, must manage access to information, facilitate coordination among team members, and support workload, management of goal conflicts by the workers, deviation from normal, and change (or unexpected events; Woods Johanneson, Cook, and Sarter, 1994). The design of tools for a complex environment must consider operative elements of complexity including those from the work domain and those stemming from humans and organizations.

WHAT IS COGNITIVE WORK ANALYSIS

A CWA describes the goals of a human agent and the means by which they are accomplished considering the entire work system, which includes the individual, organizational, task-related, environmental, and technology factors (Smith & Sainfort, 1989). It is a conceptual framework and methodology developed by Jens Rasmussen and Annelise Mark Pejtersen (1995) and advanced by Kim Vicente (1999). The CWA provides insight into what tasks are cognitively challenging and are therefore candidates for system redesign or tool development.

The method examines how many elements—environment (regulatory, work space), social (organization), individual, and technical constraints—contribute to safe, effective, and efficient work in order to identify the characteristics of these systems that are important for success (see Figure 34–1). The purpose of CWA is to create tools and systems that effectively support human work. The framework and methodology grew out of the failure of the "simple" explanation—that competence in the technical core alone was sufficient for safety and productivity.

OVERVIEW OF CONCEPTUAL FRAMEWORKS TO UNDERSTAND WORK

The theoretical framework and methods of CWA presumes that a sophisticated, deep understanding of human work could facilitate the development and evaluation of tools, training, and systems to improve work and outcomes. In no industry is this more important than health care, where there is high hazard (Kohn et al., 1999), skyrocketing costs, and best practices are inconsistently applied (Corrigan, 2001; Patel, Arocha, & Shortliffe,1998). Three models have been used to examine and improve work. The first, a normative analysis, uses an idealized description of work in isolation, promoting formal stepwise procedures dictating movement through a process to achieve a goal (e.g., policies and procedures). The usefulness of normative approaches in complex industries is limited because of the inevitable discrepancy between how work is actually performed compared to the list of prescribed correct steps. Such a prescriptive approach cannot ever address all the variation in circumstances, goal trade-offs, and management of unexpected events found in work. For instance, a nurse in the in-patient medical unit has written procedures that dictate timely administration of the 9 o'clock medications, review of discharge instructions to a patient, support of a family and patient who have received the worst possible news, resuscitation of an arresting patient, and prevention of a fall by a frail confused patient. Obviously, all of these tasks cannot be achieved in one 30-min time interval at the present rate of staffing. Clinicians use their expertise routinely to choose sequentially how to achieve the most important goals with available resources and time. Rigid adherence to written procedure is not desirable, since all possible combinations of events cannot be articulated or even imagined.

The second approach, descriptive, examines work in a naturalistic setting (observing real people as they perform their work). This approach has the advantage of capturing varied and complex demands imposed on workers, the coordination and communication required between workers to complete a task, how workers use expertise, tools, information, and cues to problem-solve, and the role that contextual and social factors play in promoting or inhibiting problem resolution. Devising systems, training, or tools based solely on this approach, however, unnecessarily limits the design. Some of the strategies and practices discovered from descriptive analysis of work are dependent on the design of existing tools or systems. For example, in the absence of computerized order entry, physicians rely largely on memory for drug–drug interactions, and side-effect profiles. Given the time constraints and effort, it is not possible or even necessary for each physician to look up each ordered drug each time. However, the addition of a computerized textbook or automated checks of drug–drug interaction, if properly designed, might add safety without compromising other goals.

CWA is the third method used to understand technical work. The goal of CWA is to identify the

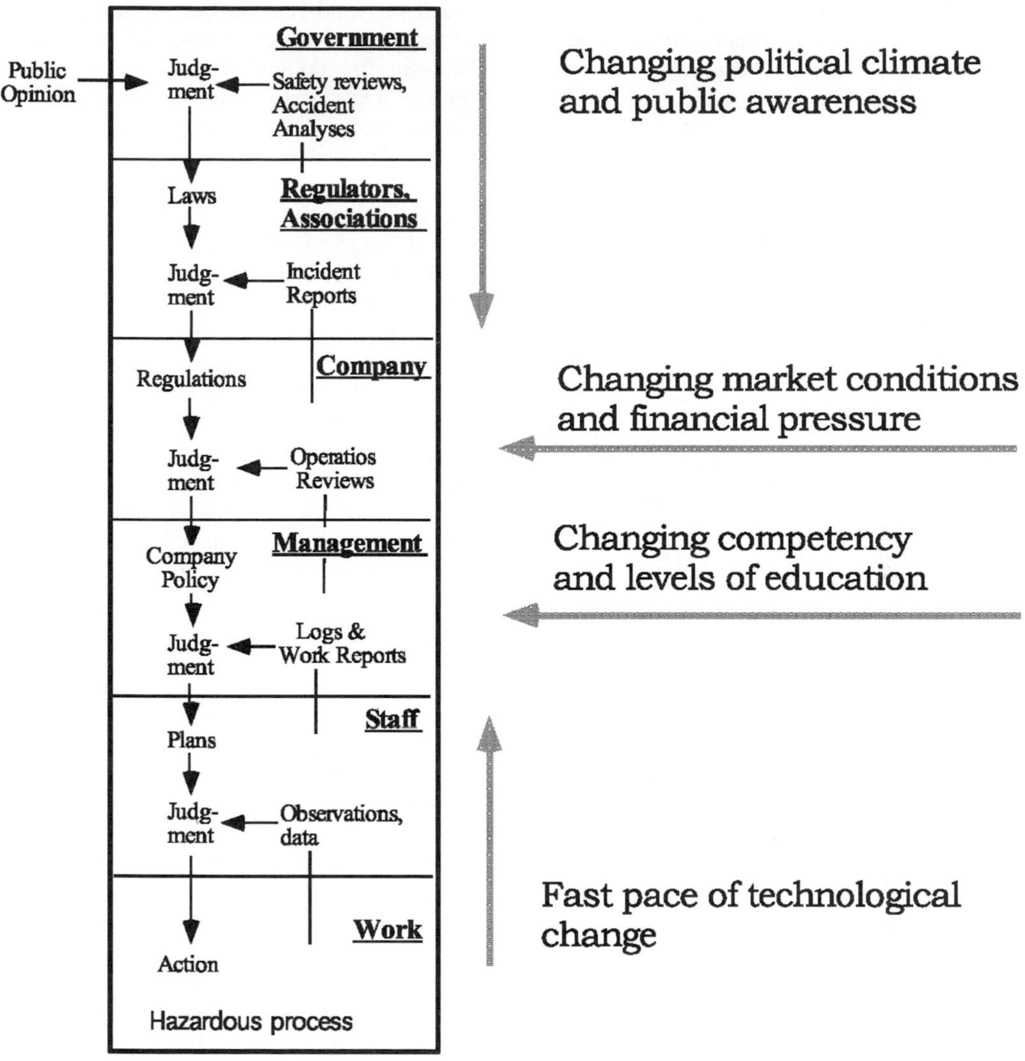

Figure 34–1. Adapted from J. Rasmussen (1997).

technical and organizational requirements necessary in order for a tool to support work effectively. This approach builds on the normative and descriptive models, adding the advantage of adapting to worker practices while allowing innovation that improves support of technical work.

KNOWLEDGE ELICITATION METHODS

CWA provides a way to model the worker adaptation to unique situations (allowing users to capitalize on their expertise) while providing control of the technical system (Vicente, 1999). In order to arrive at these models, data is collected for the domain analyses through the following techniques:

- Ethnographic observations of workers in naturalistic or simulated settings.
- Critical decision method interviews.
- Artifact analysis.
- Usability testing.

The data is then abstracted into two principal domains: technical and social. These categories are subdivided, the technical into three separate conceptual phases—work domain, control tasks, strategies—and the social into two conceptual domains—social organization, and worker competencies. This abstraction makes visible both the intrinsic tasks as well as the constraints of the domains. These constraints are additive, each conferring a separate and unique set of constraints. Through examination of constraints, a tool can be designed with the greatest degree of freedom.

Ethnographic Observations

Ethnographic observation is a methodology derived from anthropology that emphasizes detailed data collection of observable human behavior and interviews. Data collection in ethnographic observational studies involves trained observers sequentially capturing in detail both (a) observable activities and verbalizations, and (b) self-report data about how artifacts (tools) support performance (Hutchins, 1995). One method of analysis of observational data is the process tracing method, developed for the study of problem solving outside of the experimental laboratory, which externalizes internal processes, in order to be able to support inferences about internal cognition or problem solving by applying behavioral interaction protocols (Woods, 1993). Behavioral interaction protocols (Jordan & Henderson, 1995) follow in minute detail the sequence of observed behaviors in order to demonstrate a plan or goal. For example, when a nurse brings the medication cup into the patient, checks his armband, rechecks his armband, goes out to the medicine cart and gets the medication administration record, rechecks the armband against the written record, and then leaves with the medications in her hand, one can assume that she planned to give that patient medications, which on scrutiny proved to be for some other individual.

The process tracing method broadens the data sources beyond verbalizations to include a number of techniques: (a) direct observation of participant behavior, (b) "think aloud" and explanatory verbalizations while performing the task in question, (c) written records of the task performed (print-outs of medication records observed), (d) writing detailed timed notes of actions taken, (e) records of verbal communication among team members or via formal communication media, and (f) interruptions. To further increase reliability and generalizability of observation (Hutchins, 1995), practitioners are observed in their work setting over time so that the effects of being observed in their activities are lessened. In addition, reliance solely on interview data is avoided since it is suspect from memory limitations and bias. Instead, the same types of data are captured from multiple practitioners in the same fashion to enable objective comparison and analyses. Moreover, the focus of the observation is explicitly characterized in order to target the data collection and avoid getting lost in the complexity and missing important details. Finally, data is abstracted in ways that allow comparisons across sites, hospitals, and with other domains such as aviation to increase the generalizability of the findings (Jordan & Henderson, 1995).

Traditional human factors simulation studies are developed to mimic the complex interactions of a cognitive system (a human and a machine in a naturalistic setting; Gaba, 1992). These complex simulations that closely resemble a real-world setting, such as an operating room, reduce concerns about the extreme generalizability of findings to a real-world setting (Gaba & DeAnda, 1988). In designing a simulation study, the focus is normally on stimulus sampling rather than subject sampling because the performance variability is often greater based on the problem to be solved than the subject selection. During data collection, study participants perform challenging, face-valid tasks without interruption. Following the simulation run, there is a debriefing sessions to elicit data about the cognitive processes that were employed. A popular technique is "cued retrospective playback elicitation" (Jordan & Henderson, 1995). This technique aids memory and enforces detailed elicitation around observed actions by playing and pausing video of the simulation performance during an interview immediately following the simulation run.

Critical Decision Method

Interviews to assay knowledge are structured to improve validity and reliability of self-reported information. One approach, the critical decision method (CDM) interview (Klein, Calderwood, & MacGregor, 1989), is a seven-step process that uses a timeline of an event as a forcing function (in the form of an interview) to make it more difficult to stray from personal experience with a specific situation. An interview using the critical decision method follows this format:

- Preparation: Where the interviewer, familiar with the domain, operationally defines goals for knowledge elicitation (i.e., the interviewer might focus on a specific decision involving confusion, change of a plan, interesting use of an artifact).
- Incident selection: Query the interviewee about what makes an incident interesting.
- Incident recall: The interviewee recounts the episode in its entirety.
- Incident retelling: The interviewer tells the story back, asking for details and clarifications.
- Timeline verification and decision point identification: Interviewer reviews incident a second time.
- Deepening: Interviewer, reviewing the incident a third time, employs probe questions that focus attention on aspects of each decision-making event within the incident, usually starting with informational cues and eliciting the meaning of those cues and expectations goals and actions they engender.
- "What if" queries: A fourth sweep through the incident posing hypothetical changes, asking the participant to speculate on what might have happened if an element had been different.

Artifact Analysis

Cognitive artifacts are information tools that are part of the distributed cognition that takes place in technical work (Norman, 1990). They are objects that represent the collective intentions, current state, and past accomplishments across a period of time (Hutchins, 1995). The cognitive artifacts are present to recreate the intention originally made when the document or tool is used. They provide relief for memory and other cognitive functions that can get overwhelmed with high-workload situations. These artifacts are analyzed for their use in the system and the roles they have taken on in the organization. For example, during the medication administration process, depending on the setting, the nurse may use several different paper forms to document the care of patients (e.g., a flow sheet to document inputs and outputs, laboratory communication form, medication administration record, education record). Each of these forms was developed for a specific part of the work process and to meet an information need. As needs change and evolve, seldom are the forms removed but more often than not, additional forms become part of the process.

Usability Testing

Usability testing is iteratively conducted on prototypes of a software product prior to release in order to assess and improve the ability of the target user population to accomplish tasks for which the software was designed easily and efficiently (Rubin, 1994). Test participants are users who meet prespecified criteria. Three to six participants are traditionally used in a single usability test in nonrandomized order. Five test participants are reported to identify 85% of usability problems on average (Nielsen, 1993). During the usability test session, participants interact with a prototype version of the software in a specific testing scenario while thinking out loud. The scenar ios are rooted in the context of the work environment. The verbalizations and actions of the study participants, captured on videotape, are analyzed for confusion, difficulty meeting task goals, time on task, and scenario-based performance metrics. Conclusions are drawn from erroneous assumption or actions, statements indicating surprise or confusion, or tasks taking longer than anticipated. Test participants complete a detailed survey to provide secondary data about the software's usability (Lewis, 1995). Following the scenario execution, there is usually a debriefing session to elicit data about the cognitive processes that were employed.

Triangulation

Using multiple techniques to provide converging evidence attains rigor with these methodologies. In order to consider a phenomenon to have sufficient authenticity or warrant any action, patterns need to be discovered across multiple methodologies, sites, and settings. For example, we use multiple health care sites and settings in our observational studies and compare our findings from the observations with our findings from the critical incident interviews to increase the reliability, validity, and generalizability. Miles and Huberman (1984) define triangulation as a method to check a new finding against another by using other internal indices that should provide convergent evidence. The summaries are compared and contrasted with each other to reach final conclusions in terms of their similarities, differences, and observed patterns. The goal being that the inferences drawn from one source of data would be confirmed by another source of data, that is, one theme is identified and the different data sources are compared to determine what they predict about that theme. Patton (1999) further describes the actual process of triangulation as having four different types: (a) methods, (b) sources, (c) analyst, and (d) theory or perspective. Different kinds of data may yield varying results since the methodologies are sensitive to distinctions of the setting.

PERFORMING A COGNITIVE WORK ANALYSIS

The knowledge elicited is then inspected iteratively and analyzed in five phases: work domain analysis, control tasks analysis, strategies analysis, social organization analysis, and worker competencies. We briefly describe them here, but see Vicente (1999) for a comprehensive discussion of each of the phases in CWA. These domain constructs conceptually move from environmental to cognitive constraints in the work. The analysis moves from the specific descriptive data through modeling tools to characterize intrinsic work constraints. Identification of intrinsic work constraints naturally permits identification of a series of critical elements in the system design and interventions each derived from the five domains. The interventions could result in new social–technical systems and direct future work practices. Because constraints are fixed, they form a basis or boundary for worker adaptation while permitting the maximal flexibility for workers to respond to unexpected demands and adapt to the inevitable continuous changing environment, a critical benefit of CWA (Vicente, 1999). Following review of these phases and the modeling process, we will demonstrate the CWA in a real world example in order to clarify the approach.

Descriptions of the Phases

The *work domain* is the system being examined; independent of workers, tools, events, or goals. It descriptively should "map" the work or display the "lay of the land," often most easily identified as the physical space. For example, the aircraft carrier might represent the work domain of controlling flight. *Control tasks analysis* concentrates on are the products or goals that are to be achieved from the system, that is, what needs to be done independently of how or by whom. The outcome of the analysis can be used to develop policies and procedures for workers to reach those goals. The third phase, *strategies analysis*, includes the processes used in work domain to achieve the product. The *social organizational* phase covers the relationship between agents, either workers or automation, and how responsibility for the processes is distributed across the actors of the system. Finally, the fifth phase, *worker competencies*, recognizes the constraints associated with the worker themselves—the knowledge, physical conditions, and training needed for effective function. Analysis of the constraints from each of the five phases defines the boundaries of action for the workers. These are considered behavior-shaping constraints (Rasmussen & Petersen, 1995) and lead directly to implications for systems design.

Work Domain Analysis. Abstraction of data into the work domain builds a map of the system being acted on or the functional structure. The work domain characterizes two different aspects of work: the parts of the systems and the functional elements that describe the work. To aid in the systematic capture of both elements, Rasmussen (1986) developed a tool called the "abstraction–decomposition space" (Figure 34–2) to break down that map into a hierarchical structure. Decomposition breaks work down into parts of the whole. This category is organized so that the top level is composed of elements from all the levels below. The abstraction hierarchy creates a structural means–end hierarchy. This hierarchy is organized so that each cell in the space represents the same work domain differently. Another way to understand how to build an abstraction hierarchy is that the top level being considered answers the

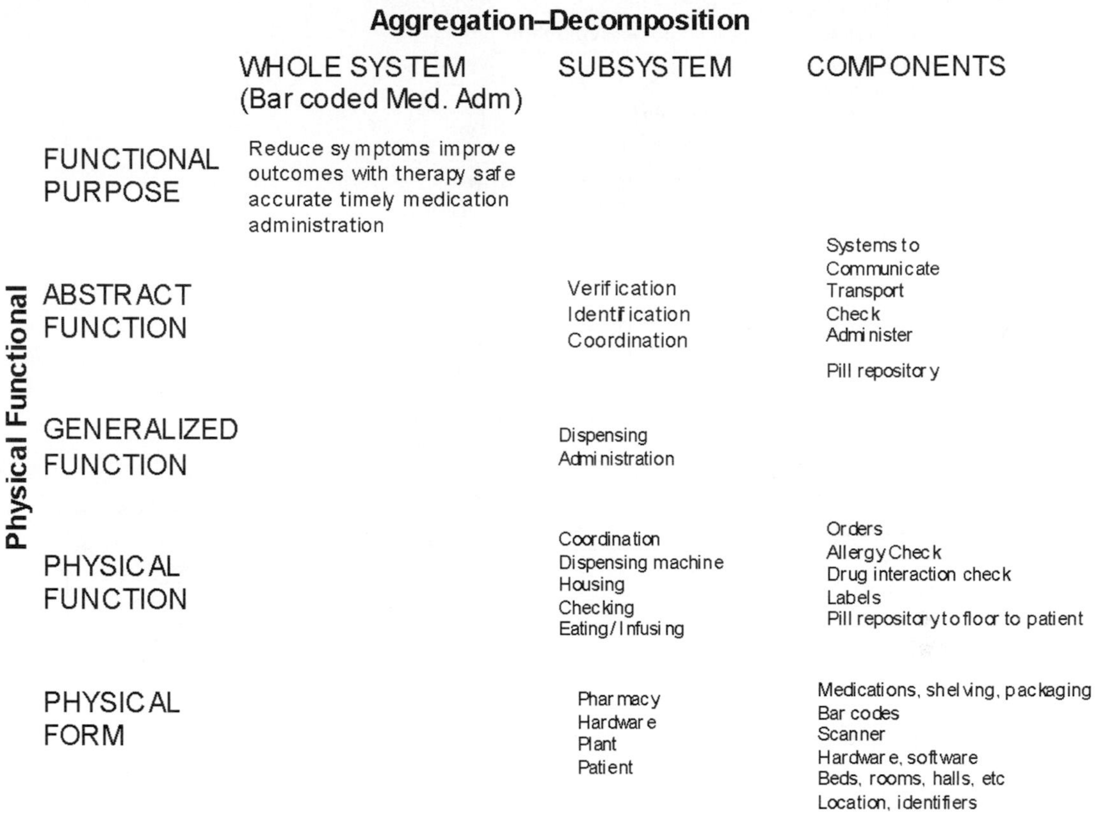

Aggregation–Decomposition

Physical Functional

	WHOLE SYSTEM (Bar coded Med. Adm)	SUBSYSTEM	COMPONENTS
FUNCTIONAL PURPOSE	Reduce symptoms improve outcomes with therapy safe accurate timely medication administration		
ABSTRACT FUNCTION		Verification Identification Coordination	Systems to Communicate Transport Check Administer Pill repository
GENERALIZED FUNCTION		Dispensing Administration	
PHYSICAL FUNCTION		Coordination Dispensing machine Housing Checking Eating/Infusing	Orders Allergy Check Drug interaction check Labels Pill repository to floor to patient
PHYSICAL FORM		Pharmacy Hardware Plant Patient	Medications, shelving, packaging Bar codes Scanner Hardware, software Beds, rooms, halls, etc Location, identifiers

Figure 34–2. Map of bar-coded medication administration process.

question *why* and the level below that answers the question *what*, and the level below that the question *how*. For instance, performance in the ICU (the top *why* level) is measured by risk adjusted mortality and length of stay, staff and patient satisfaction (the middle *what* level). These measures are collected by health information databases, laboratory databases, and surveys (the *how* level). The work domain analysis would then provide information for the control tasks to act on.

To clarify this process, we detail an abstraction decomposition model for medication administration. This map shows only the lay of the land, not the work, the processes, the people or the systems in place. Patterson (2002) conducted a number of ethnographic observations of use of BCMA in three different settings. As a result, the following classification scheme can be structured in the form of an abstraction hierarchy. The representation that resulted (Figure 34–2) identified the information content and structure of the interface. That hierarchy then informs the control task analysis.

In the decomposition hierarchy, the part–whole dimension of the space consists of three levels: components (scanners, barcodes, etc.), subsystems (verification, etc.), and the whole system (the BCMA system). The highest functional analysis recognizes that the goal of the whole system is to improve the symptoms and patient outcomes through drug therapy with accurate and timely administration of medications to people who are ill. The abstract subsystems permit verification, identification, and coordination across systems and people. Notice that the table is filled with nouns, not verbs. Within this abstraction model, links can be made within the categories to identify requirements and constraints.

Control Task Analysis. Control task analysis allows identification of the requirements associated with known recurring types of events. The analysis

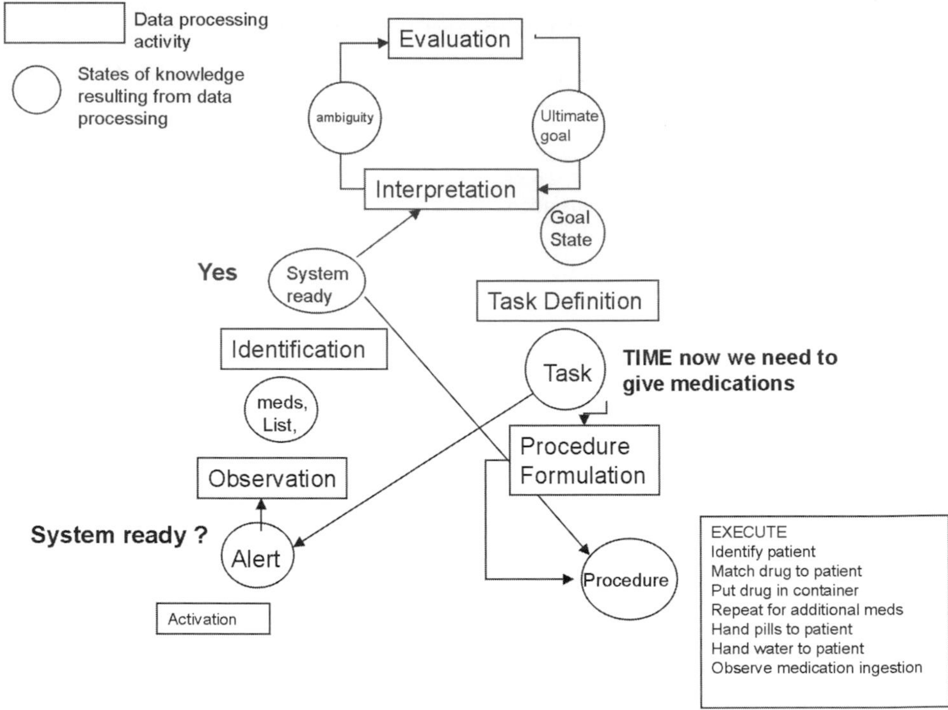

Figure 34–3. Control task analysis decision ladder.

identifies what needs to be done (but not by whom), or the goals of the system, including the variety of ways that each task may be accomplished. Methods adapted from human performance studies can be used to capture how work goals are achieved (Klein et al., 1989; Rasmussen, 1986; Potter, Roth, Woods, 2000).

Structured linear reasoning reflects the processes that novices must navigate to make decisions in the workplace. Experts however, often shortcut the process using their experiences and by matching prior actions to the present situation until one fits. These deviations are hallmarks of expertise and function to economize both on cognition and time. In the control task analysis, a decision ladder (Figure 34–3) represented a folded linear sequence of action. It is used since arrows can represent the shortcuts directly from one stage to another bypassing other stages. Methods of task analysis that have proven more helpful include the decision ladder which is a template identifying requirements associated with information processing activities. These methods de-couple what need to be done from who does it.

Figure 34–3 diagrams only the initiation process of medication administration where the worker begins the task by examining the tools. An expert worker likely follows the bolded script where, beginning with the task, a glance at the tools informs him or her about the readiness of the system (e.g., the list of medications, the medications themselves sorted by patient, and the patient identifiers and location), permitting him or her to begin the medication administration. In contrast, a novice practitioner might need to complete the entire control task sequence, first activating processing, observing the cart specifically for each element, checking them off on a mental checklist, deciding that the cart was ready, evaluating the other tasks waiting for attention against the goals that needed to be achieved, and then beginning the administration of medications. Workers track each element of the list of execution through the control task diagram, bypassing those elements unnecessary to the process at hand based on expert judgment.

Table 34–1 summarizes the control tasks identified from the field studies. Since BCMA is a

TABLE 34–1. Summary of Control Task Analysis of Bar-Coded Medication Administration

Step	*Description*	*Link*	*Ladder Code*	*Abstraction Level*	*Decomposition Level*
1	Determine system readiness (presence of meds, list of meds and patients, lists of patient locations)	Means – ends, part whole	Goal state	Functional purpose	System
2	Select patient	Topological	Define task	Generalized function	Subsystem
3	Identify room	Topological	Formulate procedure	Physical form	Subsystem
4	Match patient and room	Means – ends, part whole	Procedure	Physical function	Subsystem
5	Identify medications (open, pour)		Procedure	Physical form	Component
6	Match medications and patient		Procedure	Physical function	Component
7	Evaluate for discrepancies (blood pressure pulse, glucose, confusion, allergy etc)		Alert	Functional purpose	Subsystem
8	Observe signs and symptoms		Set of observations	Functional purpose	Subsystem
9	Decide medication appropriateness and safety		Interpretation	Abstract function	Subsystem
10	Pour water into cup, or hang IV medication		Execute	Physical form	Component
11	Observe ingestion of medications		Set of observations	Physical form	Subsystem
12	Record medications taken		Task	Generalized function	Subsystem

computer system, the information is communicated by scanning the patient wristband.

Strategies Analysis. Strategies analysis addresses how the task in question can be completed. They are important since the design of a tool is often based on a single strategy (a normative approach), but if multiple strategies are in use, that tool may not work in all environments or conditions of work. The outcome of this analysis will be a description of how the task in question can be done.

Strategies in most industries routinely vary based on situational and workload issues, as it does in medication administration. For instance, nurses administering medication in the nursing home where there can be more than 30 patients assigned to one nurse might be more likely to prepare medications where interruptions are less likely. On the other hand, a nurse administering medication on an acute ward with 4–6 patients might prepare the medications in each individual patient's room. Moreover, medication administration is anything but a straight-line activity. Patients routinely refuse pills because they believe they experienced side effects or because physicians inform them of changes they are officially ordered. Pills may be missing from the medication drawer, discontinued, or initiated continuously. The strategies can be found in careful analysis of the observations by finding consistent patterns of behavior across practitioners. The analysis of strategies also takes into

TABLE 34–2. Examples of Cognitive Artifacts Used by Nurses in Making Decisions

Cognitive Artifacts	*Use as Work Support*	*Specific Usage Examples*	*Theme/Pattern*
Computerized patient record	Off-loading memory	Check dose of medications	Workload complexity/ Distributed resources
Patient record (nurse)	Off-loading memory	Determine last dose given	Communication/ Workload complexity
Flow sheet	Documentation of care	Document patient status	Workload complexity
Test or Lab Communication form	Off-loading memory	Document results	Distributed resources
Medication administration record	Off-loading memory	Track completed and pending meds	Poor observability of team/missing meds
Patient care activities sheet	Documentation of care	Check activities ancillary to medical care	Workload complexity/ Poor observability of patient
Parenteral intake	Documentation of care and proactively monitoring patient status	Monitor fluids administered	Distributed resources
Physician's order form	Documentation of care	Identify changes in therapy	Poor observability of team
Problem progress sheet	Stacking and proactively monitoring patient status	Track interventions across time	Workload complexity/ Poor coordination across service "silos"/Authority
Parent or Patient assessment	Support communication	Timely monitoring of care	Workload complexity
Missing medication sheet	Stacking	Document communication with pharmacy	Poor coordination across service "silos"/Authority
Education record	Off-loading memory	Document communication with family	Workload complexity
Nursing worklist (printed)	Anticipating and Planning	Work progress monitoring	Goal conflicts with safety
Personal ward planning sheet	Stacking	Work progress monitoring	Goal conflicts with safety
Colored ink pen	Off-loading memory	Track changes and progress in work	Workload complexity
Room white-board	Support communication and coordination	Used to track shift change and update vitals	Good observability of system or team

account how workers use information about the process under scrutiny to problem-solve.

Social Organizational Analysis. The social organizational analysis describes how the work can be distributed across human workers and machine automation and how such actors communicate and cooperate. It answers the question "who might perform the tasks?". Any such structure will need to be able to manage change rapidly, with autonomy given to the workers closest to the information source. The redundancy in the system should allow the moving of people and roles as demanded by the situation. Information related to the task must also track with the task, so that workers have the data required available as the task is completed. Rasmussen identified six criteria for dividing work demands: actor competency, access to information or action means, facilitating the communication needed for coordination, workload sharing, safety and reliability, and regulation compliance (Rasmussen, Petersen, & Goldstein, 1994).

The social organizational evaluation of work can use the modeling tools discussed in the previous sections; the abstraction decomposition hierarchy and the decision ladder. The system design interventions would deal with role allocation and organizational structure. The abstraction decomposition space identifies the information set that would be applicable to the entire organization. If the workers are sdistributed across the space, we can identify what subset of the medication administration would be useful for each group. Data collection techniques that can be used include ethnographic observations and the process trace method as they address cooperation and communication issues.

Worker Competencies Analysis. The ultimate objective of this phase is to identify the characteristics an ideal worker would have. To understand the worker competencies, which can be used then to develop training, experts in CWA apply the skills, rules, and knowledge taxonomy (Rasmussen, 1986). Skill-based behavior is automated actions performed without conscious attention, like walking. Rule-based behavior consists of actions based on stored rules derived from prior performance, procedures, or instruction. Rule-based behavior reacts to cues in the environment. In contrast, knowledge-based behavior searches for the correct action to achieve an explicit goal with a clear mental model of the process. Skill-based behavior is cognitively the most economical, rule-based actions require more time and effort, and knowledge-based behavior is the slowest and cognitively the most resource intensive. The behavior used is a function of the skill level of the worker.

Worker competency recognizes that many tasks require varying expertise, making it practical to divide the work along lines of expertise. For instance, in the outpatient clinic, physicians, nurse educators, clerical personnel, and billing clerks all work to provide care to the diabetic patient. Access to information implies that the worker who can see the information that identifies a problem be able to act on that information, or conversely, workers responsible for a task are able to access the information necessary for the task.

To explore what this evaluation means practically, here is an example from BCMA:

- Skill-based knowledge: The worker ideally recognizes names and doses of medications and side effects, the organization of the facility, and room numbers; is able to interact with the tool and the surrounding environment.
- Rule-based knowledge: Worker competencies at this level include icons or signs that support rule-based behavior in medication administration and the awareness of the strategies used to shortcut the process.
- Knowledge-based behavior: This competency measures what will be needed to manage the unexpected changes in medication administration and access to appropriate and rapid information is important in this arena. For instance, the worker administering chemotherapy, a sedative hypnotic, or pain medication must be able to check the dose if unfamiliar and the half-life if the patient remains obtunded. The need for knowledge-based behavior stems from cues that stimulate the worker to question actions from rule-based behavior.

Summary

The Veterans Health Administration (VHA) BCMA system is an example that CWA has practical implications for system design. As a result of the CWA, five unintended side effects> from the introduction of BCMA in VHA were identified, including: (a) missed doses when nurses were unaware of the automated removal of medications after a period of time, (b) reduced access to medication administration data for physicians, (c) nurses employing workaround strategies to increase efficiency during busy periods, (d) nurses prioritizing activities that can be monitored with BCMA data over activities that are not able to be easily monitored, and (e) reduced efficiency for nonroutine activities (Patterson, Cook, & Render, 2002). Others have described nursing "workaround" strategies that reduce the effectiveness of BCMA systems in acute care and long-term care settings (Puckett, 1995; Patterson, Rogers, & Render, 2004).

APPLICATION EXAMPLE

In this section, we provide an example of how CWA was used to understand medication administration. The example identifies the key processes, strategies, and worker competencies in medication administration independent of the technology employed. A study was done of medication administration at a children's hospital considering computerized order entry. Over 90 hours of observations of expert nurses were completed (excluding nurses within 6 months of their degree and recent hires, within 6 months) on three separate pediatric in-patient units.

Medication Administration Without a Technology Interface

Medication administration is a distributed and cooperative activity conducted in a social setting. In human interaction terms, medication administration is a task that requires the nurse to interact with a computerized system with automated features.

Before completing the information system implementation, a CWA was completed to understand the work domain. The first step taken was to document the constraints in the strategies phase.

Strategies Analysis. Adaptations or tailoring strategies develop in all environments that use computer-based devices with variable design deficiencies. Workers use their expertise and experience to find solutions that minimize risk and maximize efficiency to achieve a goal. Attention to strategies used during the data collection phase permits identification of variation in the methods to achieve the tasks identified in the control task analysis.

Several patterns across nurses and nursing units were identified as a result of ethnographic observations and using the critical decision method. Patterns of work that were observed across units include

- Geography of assignments, repetitive travel.
- Disjointed or missing equipment and supply sources.
- Interruptions.
- Waiting on systems or processes to continue or complete care.
- Inconsistencies or breakdown in communication.
- Technology difficulties.
- Trade-offs.
- Managing relationships.
- Managing various organizational rules.

The various strategies have different degrees of resource requirements. For example, the nurses traded accuracy for speed, reduced performance criteria, shed tasks, and deferred tasks to manage interruptions which required a big drain on resources while technology difficulties were managed by maintaining paper backups, a less stressful activity.

Worker Competencies Analysis. The worker competencies analysis was done to support the processes conducted in the different units. We thought of this phase as making visible the otherwise unobserverable goal relevant properties of the domain. Analyzing the artifacts that the workers use did this and supported rule-based behavior.

Supporting Rule-Based Behavior. The nursing staff adeptly utilized a myriad of artifacts (provided by the hospital and self-developed) to manage their activities and the documentation of medication administration. The density of artifacts provides markers for events and system structures that have failed in the past that were redesigned with an additional level of documentation. The conflicting goals of care and time pressure cause the staff to seek ways to manage their environment as much as possible. Table 34–2 describes the artifacts with specific examples of their use in practice.

CONCLUSION

CWA is the framework of choice because it studies the actual work and the system in which it is done. While it is necessary to understand the work of clinicians and practitioners, it is not sufficient because of human performance and limitations in current practice. Similarly, it is not adequate to study only the technical systems because they are not optimized. CWA provides a more complete understanding of the work environment.

CWA brings a powerful tool in health care to design and redesign work, particularly as clinical information systems are implemented. This framework involves identification of the knowledge, mental processes, and decisions that are required to perform technical work. It is commonly used in technical work where the nature of the task, the methods used by practitioners to accomplish the task, and the factors that complicate task performance are less well understood. It is the technical work of physicians, nurses, pharmacists, and technologists that confronts inherent hazards and threats to safety. Practitioners and the system that supports their work in context actually are the means by which patient safety is created and sustained. They are the ones who face the dilemmas and conflicts of practice, they are the ones who receive the demands for production and cope with the complexity of the real world, and they are the ones who bridge the gaps that modern health care products create.

APPENDIX

Description of First-of-a-Kind Implementation of BCMA Software

The Veterans Health Administration (VHA) installed a first-of-a-kind BCMA system in 2000. Currently, BCMA is deployed in all VHA inpatient facilities across the United States. Nurses access BCMA software by using a laptop computer attached to a wheeled medication cart and linked by a wireless network to electronic databases. If the scanned medication bar-code data does not match what is ordered for a patient, who was identified by scanning a bar-coded wristband, the nurse is alerted to the discrepancy by the software.

The canonical medication administration sequence in the Veteran's Administration using the computerized order entry system is

1. A physician orders medications in a computerized provider entry software package, Computerized Patient Record System (CPRS).
2. A pharmacist checks that there are no allergies or drug interactions and dispenses bar-coded medications to a patient's drawer on a medication cart. When necessary, the pharmacist modifies the order, such as by transferring free text orders into BCMA commands.
3. A nurse logs into BCMA with a unique identifier and scans a patient wristband or types in a social security number to access the list of active patient medication orders. The nurse takes the appropriate medications out of the patient's drawer, scans them, opens and pours them into a plastic cup and administers them.

References

Ash, J. S., Berg, M., & Coiera, E. (2004). Some unintended consequences of information technology in health care: The nature of patient care information system-related errors. *Journal of the American Medical Informatics Association, 11*(2), 104–12.

Chin, T. (2003). *Doctors pull the plug on paperless system.* Retrieved on January 2, 2004 from http://www.Amednews.com

Corrigan, J. M. (2001). *Crossing the quality chasm.* Washington, DC: National Academy Press.

Gaba, D. M., (1992). Improving anesthesiologists= performance by simulating reality. *Anesthesiology, 76*(4), 491–4.

Gaba, D. M. & DeAnda, A. (1988). A comprehensive anesthesia simulation environment: Re-creating the operating room for research and training. *Anesthesiology, 69*(3), 387–94.

Harper, R. R., Hughes, J. A., & Shapiro, D. Z., (1991). Harmonious working and CSCW: Computer technology and air traffic control. In J. M. Bowers, & S. D. Benford (Eds.), *Studies in computer supported cooperative work: theory, practice and design* (pp. 225–34). Amsterdam: North-Holland.

Hutchins, E. (1995). *Cognitive in the wild*, Cambridge, MA: MIT Press.

Jordan, B., & Henderson, A. (1995). Interaction analysis: Foundations and practice. *The Journal of the Learning Sciences, 4*, 39–103.

Klein, G., Calderwood, B., & MacGregor, D. (1989). Critical decision method for eliciting knowledge. *IEEE Transactions on Systems, Man and Cybernetics, 19*, 462–472.

Kohn, L. T., Corrigan, J. M. & Donaldson, M. S. (1999). *To err is human: Building a safer health system.* Washington, DC: National Academy Press.

Lewis, J. R. (1995). IBM Computer Usability Satisfaction Questionnaires: Psychometric evaluation and instructions for use. *International Journal of Human-Computer Interaction, 7*(1), 57–78.

Miles, M., & Huberman, A. M. (1984). *Qualitative data analysis: A sourcebook of new methods.* Newbury Park, NJ: Sage.

Nielsen, J., (1993). *A mathematical model of the finding of usability problems.* In the Proceedings ACM/IFIP INTERCHI'93 Conference, Amsterdam, The Netherlands.

Norman, D. (1983). Design rules based on analysis of human error. *Communication of the ACM, 26*, 254–258.

Norman, D. (1990). The "problem" with automation: Inappropriate feedback and interaction, not "over-automation." *Philosophical Transactions of the Royal Society of London, 327*, 585–93.

Patel, V. A. V., Arocha, J. F., & Shortliffe, E. H. (1998). Representing a clinical guideline in GLIF: Individual and collaborative expertise. *Journal of the American Medical Informatics Association, 5*, 467–483.

Patterson, E. S., & Woods, D. D., (2001). Shift changes, updates, and the on-call architecture in space shuttle mission control. *Comput Support Coop Work, 10*(3–4), 317–46.

Patterson, E. S., Cook, R. I., & Render, M. L. (2002). Improving patient safety by identifying side effects from introducing bar coding in medication administration. *Journal of the American Medical Informatics Association, 9*(5), 540–53.

Patterson, E., Rogers, M. L., & Render, M. L., (2004). Simulation-based embedded probe technique for human-computer interaction evaluation. *Cognition, Technology, and Work, 6*(2), 197–205.

Patton, M. Q. (1999). Enhancing the quality and credibility of qualitative analysis. *Health Services Research, 34*(5, Pt. II), 1189–1208.

Potter, S. S., Roth, E. M., Woods, D. D., & Elm, W. (2000). Bootstrapping multiple converging cognitive task analysis techniques for system design. In J. M. C. Schraagen, S. F. Chipman, & V. L.Shalin (Eds.), *Cognitive task analysis* (pp. 317–340). Mahwah, NJ: Lawrence Erlbaum Associates, Inc.

Puckett, F. (1995).Medication-management component of a point-of-care information system. *American Journal of Health-System Pharmacy, 52*, 1305–1309.

Rasmussen, J. (1986). A cognitive engineering approach to the modeling of decision-making and its organization. In *Technical Report No. RIso-M–2589.* Roskilde, Denmark:RIso Laboratory.

Rasmussen, J., Petersen, A., & Goldstein, L. (1994). *Cognitive systems engineering.* New York, NY: Wiley.

Rasmussen, J., & Petersen, A. M., (1995). Virtual ecology of work. In J. Flach, P. Hancock, J. Caird, & K. J. Vicente (Eds.), *Global perspectives on the ecology of the human machine system* (pp. 121–156). Hillsdale, NJ: Lawrence Elrbaum Associates, Inc.

Rasmussen, J. (1997). Risk management in a dynamic society: A modeling problem. *Safety Science, 27*(2–3), 183–213.

Rubin, J. (1994). *Handbook of usability testing: How to plan, design, and conduct effective tests.* New York: Wiley.

Sarter, N., Woods, D., & Billings, C. (1997). Automation surprise. In G. Salvendy (Ed.), *Handbook of human factors and ergonomics* (pp. 1926–1943). New York: Wiley.

Smith, M. J., & Sainfort, P. (1989). A balance theory of job design for stress reduction. *International Journal of Industrial Ergonomics, 4*, 67–79.

Vicente, K. (1999). *Cognitive work analysis: Toward safe productive and healthy computer based work.* Mahwah, NJ: Lawrence Erlbaum Associates, Inc.

Weick, K. E., & Sutcliffe, K. M. (2001). *Managing the unexpected: Assuring high performance in an age of complexity* (1st ed.). University of Michigan Business School Management Series. San Francisco: Jossey-Bass.

Woods, D. D. (1993). Process tracing methods for the study of cognition outside of the experimental psychology laboratory. In G. Klein, J. Orasanu, & R. Calderwood (Eds.), *Decision making in action: Models and methods* (pp. 404–411). Norwood, NJ: Ablex Publishing Corporation

Woods, D. D. (2003) Discovering how distributed cognitive systems work. In E. Hollnagel (Ed.), *Handbook of cognitive task design.* Mahwah, NJ: Lawrence Erlbaum Associates, Inc.

Woods, D. D., et al. (1994). Behind human error: Cognitive systems, computer and hindsight. In *Crew systems ergonomic information and analysis center.* (WPAFB Technical Report). Dayton, OH: Wright Patterson Airforce Base.

Young, T., Arnold, J., Brackey, T., Carr, M., Dwoyer, D., Fogleman, R., et al. (2000). *Concepts and approaches for Mars Exploration.* (Technical Report LPI-Contrib- 1062-Pt-2, 20000701). NASA Center for Aerospace Information (CASI).

35

HUMAN FACTORS RISK MANAGEMENT IN MEDICAL PRODUCTS

Edmond W. Israelski
Abbott Laboratories, Abbott Park, Illinois

William H. Muto
Abbott Laboratories, Irving, Texas

Risk management and its elements of risk analysis, hazard analysis, risk evaluation, and risk control have been used as an engineering tool for many years in industries such as nuclear power, weapons systems, and transportation to identify risks and to control system modes of failure. Recently the health care industry has seen increased attention to risk-management tools since they have the potential to improve patient safety. These tools and related methods have been applied to understanding "use errors" made with medical devices. Use errors are defined as predictable patterns of human errors that can be attributable to inadequate or improper design. Use errors can be predicted through analytical task walk-through techniques and via empirically based usability testing. This chapter explains and discusses the special methodology of use error focused risk analysis and some of its history. Case studies are presented that illustrate the methods of use error risk analysis such as fault tree analysis (FTA) and failure mode effects analysis (FMEA) and some pitfalls to be avoided. The case studies include the Therac–25® computer-controlled radiation therapy system and an automatic external defibrillator (AED).

RISK MANAGEMENT

Risk analysis in the context of "use errors" in medical products and processes has received increasing attention in recent years. Although risk-analysis techniques have been used for decades to assess the effect of human behavior on critical systems such as in aerospace, defense systems, and nuclear power applications, the use of these techniques in medical applications has received comparatively less attention in the published literature. Because human error has been identified as a major contributing cause to patient injury and death, risk-analysis techniques have seen increased attention. Use errors are defined as a pattern of predictable human errors that can be attributable to inadequate or improper design. In the Human Errors section there is a more detailed discussion of use errors including examples. Definitions of relevant risk management terms are included in Appendix A.

HUMAN RELIABILITY AND USE ERROR

Background on Human Reliability

Human reliability and related human errors have been analyzed and studied a great deal in the area of applied psychology, including a great deal of theorizing about error causes and classifications (Bailey, 1983; Miller & Swain, 1987; Reason, 1990, 1997), among others.

Human error may be due to many causes including limitations in

- Attention.
- Perception.
- Memory.
- Cognitive processing.
- Response execution.

Fortunately, a deep understanding of the underlying causes and theories of human error is not critical in applying use-error risk analysis. It is important, however, to be able to systematically identify and analyze potential human errors in the use of a medical product using analytical tools such as FTA and FMEA, which are more fully described in sections on FMWA and on FTA and Other Techniques later in this chapter. The use of these tools leads to a systematic effort to understand and eventually control for these errors and their resulting hazards and harms to patients and other product users. The theory behind human errors plays less of a role compared to the effort to carefully catalog and control for the human errors. It is true that human error theories that would assist in the prediction of error would be beneficial in incorporating effective design mitigations to reduce or eliminate the risks. Again, designers of medical products can use other human factors best-practice tools, such as usability testing that can be used to measure the effectiveness of their design mitigations.

Defining Use Error

Use error is characterized by a repetitive pattern of failure that indicates a failure mode that is likely to occur with use and thus has a reasonable possibility of predictability of occurrence. Use error can be addressed and minimized by the device designer and proactively identified through the use of techniques such as usability testing and hazard analysis. An important point is that in the area of medical products, regulator and standards bodies make a clear distinction between the common terms *human error* and *user error* as compared to *use error*. The term *use error* attempts to remove the blame from the user and open up the analyst to consider other causes including

- Poor user interface design, that is, poor usability.
- Organizational elements, that is, inadequate training or support structure.
- Use environment not properly anticipated in the design.
- Not understanding the users tasks and task flow.
- Not understanding the user profile in terms of training, experience, task performance incentives, and motivation.

There are several taxonomies that attempt to categorize the elements of a use error. For more information, see the chapter on "Medical Failures Taxonomies" (Thomadsen) in this handbook. Figure 35–1 shows one such taxonomy from the International Electrotechnical Commission (IEC) standard for medical devices and in particular the part that focuses on usability (IEC, 2004). This diagram is a formal part of an international standard that focuses exclusively on human factors for medical devices. Human factors professionals are currently debating the details of this model, which is based on Reason's (1997) models of human error. The important point of this model is that it makes the attempt to denote those operator actions that need to be analyzed by medical product manufacturers and those that do not. The dotted horizontal line in Figure 35–1 distinguishes between "abnormal use" and other categories of use, with abnormal use encompassing events such as sabotage and deliberate recklessness. The IEC standard states that abnormal use is outside the scope of the standard. A controversy surrounding Figure 35–1 is that some interpret the standard as implying that manufacturers do not need to consider categories such as off-label use or inadequate training. But, it is common for manufacturers to consider these error categories in risk analysis and offer design mitigations that address them. Figure 35–2 shows one such example, illustrating a plastic lockbox that guards against narcotic substance abuse for a portable ambulatory patient-controlled analgesia (PCA) infusion pump. In the design of this product it became apparent from medical device reports

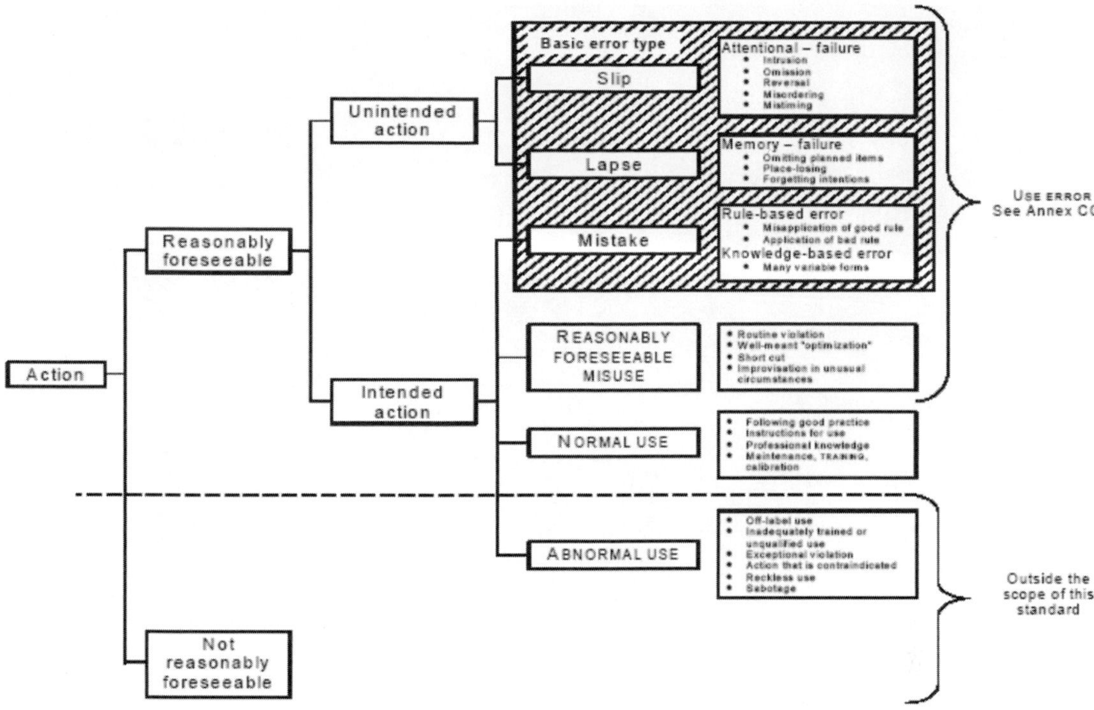

Figure 35–1. A possible taxonomy of use actions and use errors in operation of a medical device (IEC, 2004).

that there could be illegal attempts by narcotic addicts to access vials of pain medication intended to manage chronic patient pain. These actions would fall under the category of abnormal use, but prudent manufacturers of portable PCA pumps have taken proactive steps using a lockbox to reduce the risk of device abuse.

IDENTIFICATION OF HAZARDS AND WHEN RISK MANAGEMENT IS CONDUCTED

Hazard Analysis

An important first step in risk management is to understand and catalog the hazards and possible resulting harms that might be caused by a medical device. Some call this *hazard analysis*. Unfortunately, others use the term in a more general way and use the term *hazard analysis* as a synonym for *risk management*. Hazard analysis is often done as an iterative process with a first draft being updated and expanded as additional risk management methods (e.g., FMEA and FTA) are used. Experts from medical, quality, and product development, among other disciplines, can brainstorm on harms and hazards. Technically, hazards are the potential for harms. Harms are defined as physical injury or damage to the health of people, or damage to property or the environment. Table 35–1 shows examples of harms from hazards for a pen-like automatic needle injector device. Table 35–2 shows similar harms and hazards from an AED.

Where Risk Management Fits in the Development Process

Use-error focused risk analyses, including FMEA and FTA, are particular methods in the user-centered or human factors design process. It is the analytical complement to empirical usability assessment commonly called usability testing. Figure 35–4 shows where use error risk analysis fits into the overall human factors process for medical devices. The figure includes inputs and outputs for the various human factors process steps and also shows the iterative nature of the process as indicated by feedback loops, mainly emanating from the usability testing and post market analysis steps.

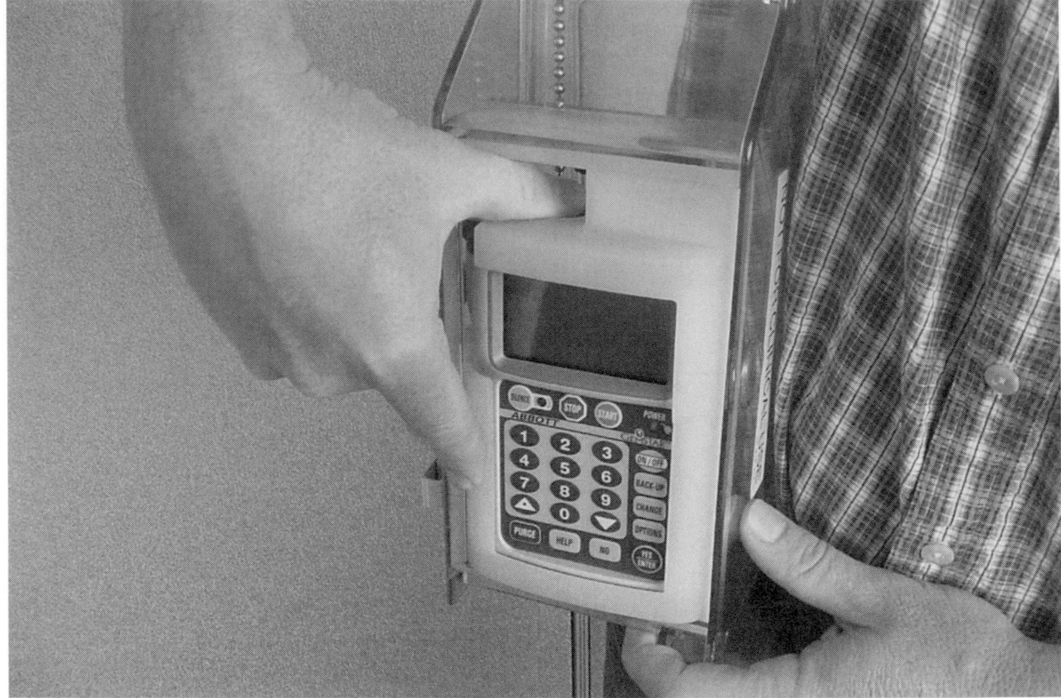

Figure 35–2. A protective plastic lockbox for the prevention of abuse of narcotics used in an ambulatory PCA infusion pump.

Figure 35–5 depicts another view of where risk management fits into the development process as commonly described by design controls. The Code of Federal Regulations (CFR; 2004) regulates the development process in the United States and in these regulations the process is called design controls with distinct development stages such as concept, design input, design output, verification, and validation. In Figure 35–5, risk management begins early in the design process and continues through the end of the process where design validation occurs.

The Examples of Use Error FMEAs and FTAs section and the Criteria for Establishing Acceptable Risk section describe the most-used tools involved in user error risk analysis, (FMEA) and (FTA). Case studies are offered for a radiation therapy system and an AED in later sections, including examples of both FMEAs and FTAs.

FAILURE MODES AND EFFECTS ANALYSIS (FMEA)

The recommended steps for conducting a use-error risk analysis are the same as traditional risk analysis with one significant addition, namely the need to perform a task analysis. Possible use errors are then deduced from the tasks. Each of the use errors are rated in terms of severity of its effects and the probability of occurrence that comprise a risk index used for prioritization. For each of the high-priority items, modes (or methods) of control are assumed for the system or subsystem and reassessed in terms of risk. The process is iterated until all higher level risks are eliminated.

Among the most widely used of the risk analysis tools is FMEA and its close relative, failure modes effects and criticality analysis (FMECA).[1]

[1]FMECA is an extension of FMEA that starts with FMEA elements and further considers ratings of criticality and probability of occurrence. Because of their common basis, FMEA and FMECA are commonly referred to as FMEA. Likewise, in this chapter FMEA and FMECA will be referred to as FMEA.

TABLE 35–1. Possible Harms and Hazards from use of an Automatic Needle Injection Device

Bleeding, bruising, or tearing of skin, leading to a possible infection
Incomplete injection that may to giving another injection, leading to over medication
Under medication
Delay in therapy
Failed therapy due to unsuccessful injection
Pain on injection
Increased bleeding, due to the presence of alcohol
Non delivery, wasted dose
Delivery intramuscularly instead of subcutaneous
Possible infection from microorganisms present on skin

TABLE 35–2. Possible Harms From use of an Automatic External Defibrillator

Non-delivery of defibrillating shock
Delay in delivery of defibrillating shock
Administration of shock when not needed
Bystander shocked when touching patient during delivery
Set victim on fire
Delivery of weak non-effective shock
Ignoring subsequent second episode of cardiac fibrillation
Burns caused by delivery of electrodes touching each other

FMEA is a "bottom-up" design evaluation technique used to define, identify, and eliminate known or potential failures, problems, and errors from the system. The basic approach of an FMEA from an engineering perspective is to answer the question: If a system component fails, what is the effect on system performance or safety? Similarly, from a human factors perspective, FMEA addresses the question, "if a user commits an error, what is the affect on system performance or safety?" A human factors risk analysis has several components that help define and prioritize such faults: (a) the identified fault or use error, (b) occurrence (frequency of failure), (c) severity (seriousness of the hazard and harm resulting from the failure), (d) selection of controls to mitigate the failure before it has an adverse effect, and (e) an assessment of the risk after controls are applied.

A use-error risk analysis is not substantially different from a conventional design FMEA. The main difference is that rather than focusing in on component or system level faults, it focuses on user actions that deviate from expected or ideal user performance.

Basic Steps in Performing Human Factors FMEA

1. *Form a team.* Although risk analyses are occasionally performed by lone individuals, the most effective risk analyses are conducted by a team of critical organizational stakeholders. In a medical device company the critical stakeholders could include: human factors, medical affairs, development (hardware and software engineering where appropriate), and quality assurance. Other stakeholders can include: marketing, regulatory, documentation, and training. The inclusion of a variety of stakeholders maximizes the chance of performing as unbiased an analysis as possible. The objective of the first meeting is to get everyone to understand and agree to the objectives and the process of conducting the FMEA.
2. *Perform a task analysis.* A task analysis is a detailed sequential description (in graphical, tabular, or narrative form) of the tasks performed while operating a device or system. The analysis should cover the major task flows performed by users of the product and would usually include the following (Stamatis, 1995):

   - The person performing the task (in a system that requires multiple operators).
   - The task stimulus; stimulus that initiates the task.
   - The equipment or component used to perform the task.
   - The task feedback.
   - Required human response (including decision making).
   - The characteristics of the task output, including performance requirements.

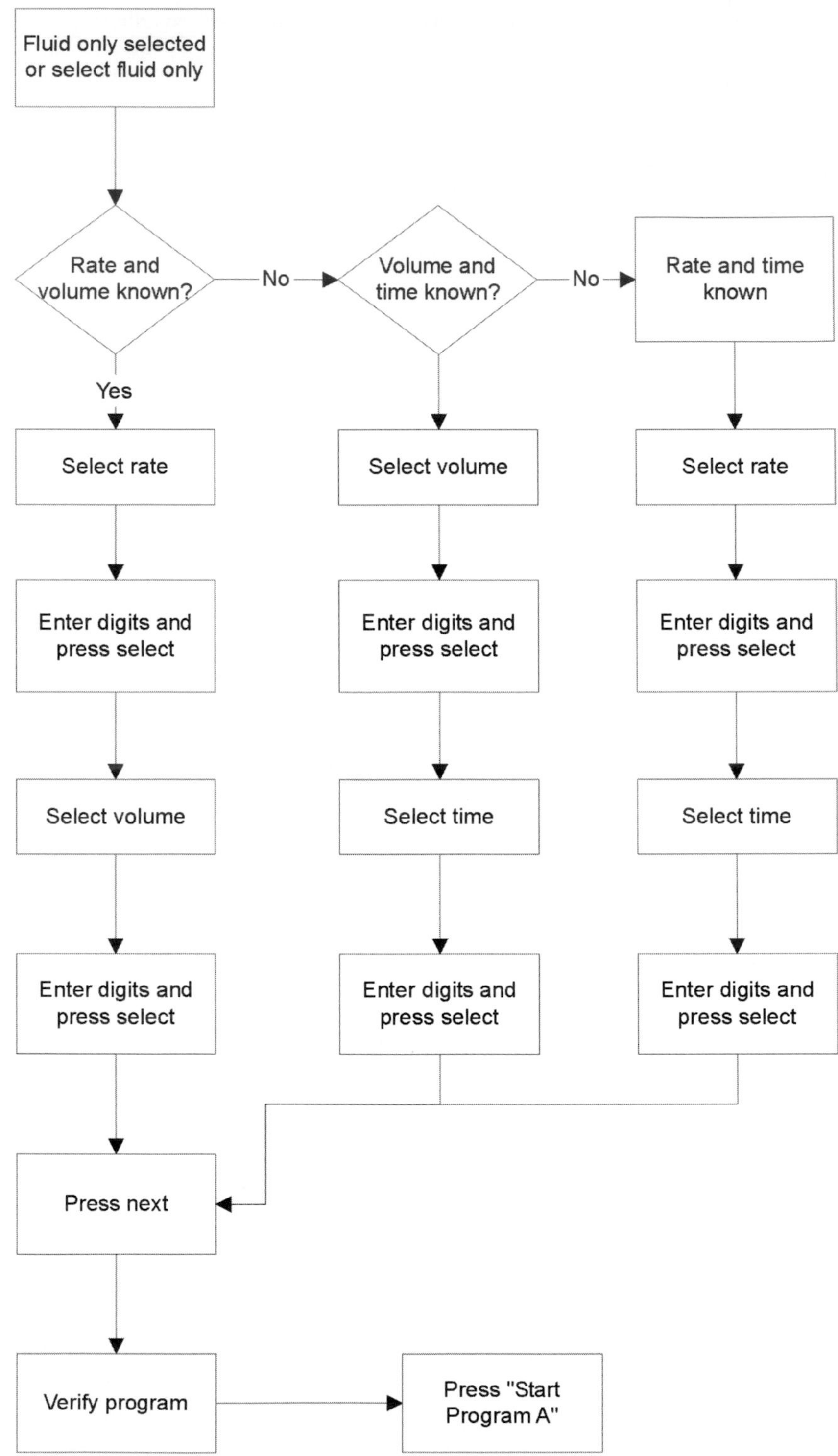

Figure 35–3. Example of a task flow diagram for programming an infusion pump.

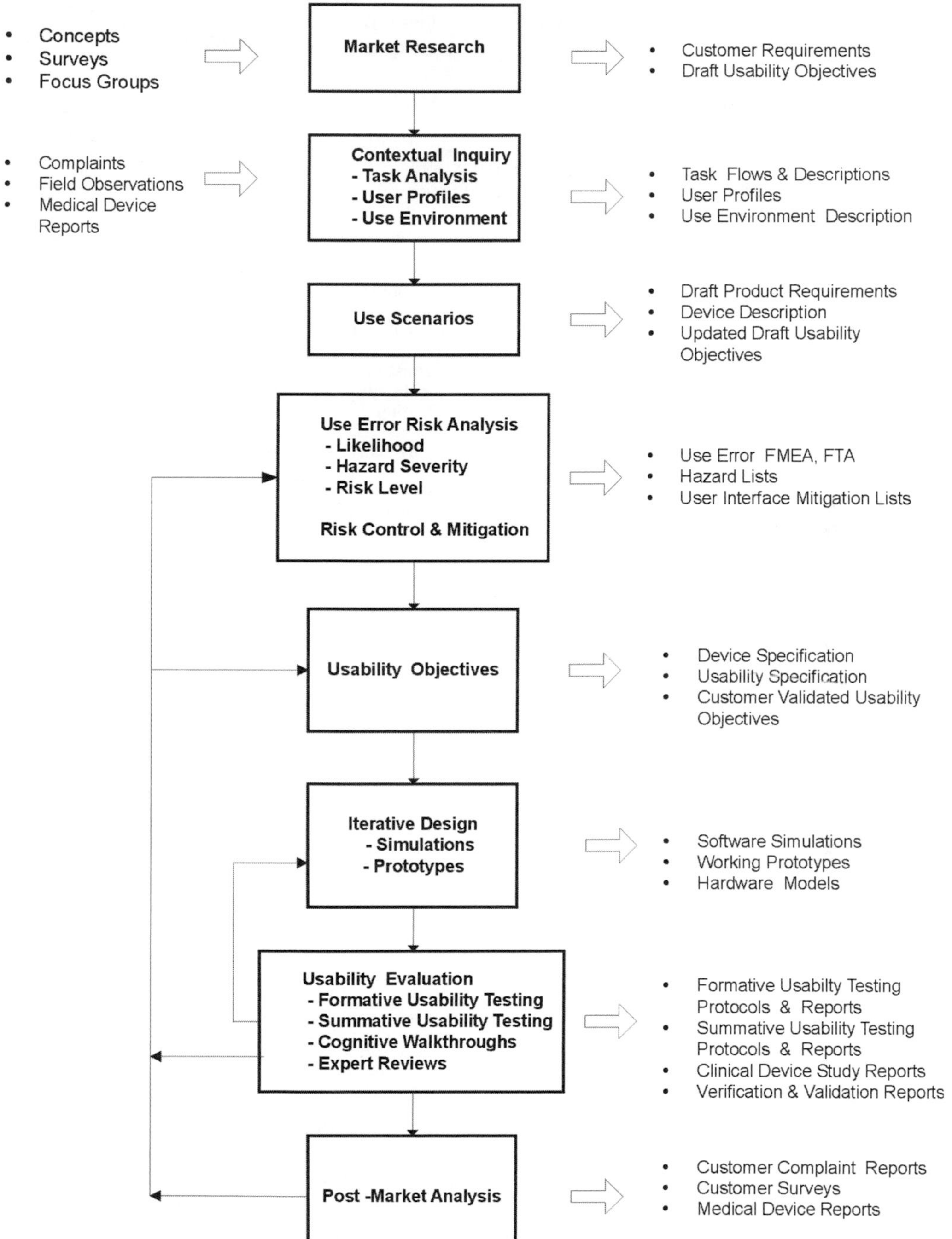

Figure 35–4. Human factors engineering process for medical products.

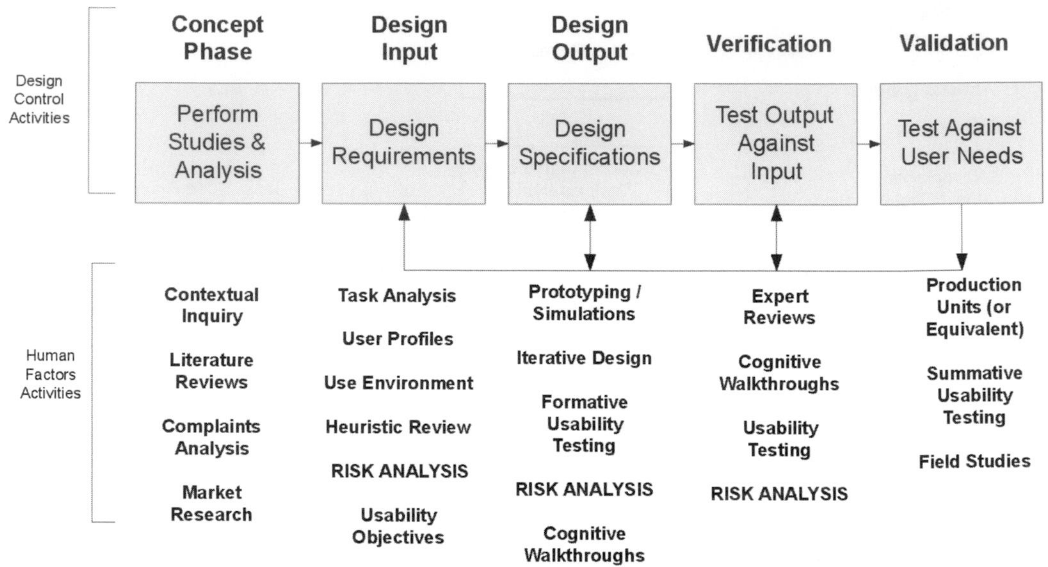

Figure 35–5. Human factors in the design control process.

There are many ways to document a task analysis. Table 35–3 shows a task analysis in the form of a task description table. Figure 35–3 is an example of a task flow diagram. Other forms of task analysis can be found in Kirwan and Ainsworth (1992).

The most commonly recommended methodology for gathering task analysis data is contextual inquiry, an information-gathering process with three components: task analysis, user profiling, and use environment analysis. This method may include the following techniques for data gathering:

- Observations.
- Follow-around studies (shadowing).
- Behavioral checklists.
- Time-slice sampling studies.
- Time and motion studies.
- Usability study repurposed to understand task behaviors.
- Interviews and surveys (one-to-one and focus groups).
- Review of job and task documentation (a draft user instructions manual is one source of user task descriptions).
- Review of training materials, job descripions, job evaluations.
- User-kept diaries and activity logs.

3. *Start a Worksheet.* A use-error focused FMEA is documented using an FMEA worksheet such as the ones shown in Tables 35–13 and 35–14 (Israelski and Muto, 2003). Note that there are many variations of FMEA worksheets. Your company or team may choose variations to better suit you project needs. The worksheet can be developed using overhead projection slides or large paper (e.g., 11" x 17"). Developing FMEAs directly on a computer using dedicated FMEA software tools or generic spreadsheets or word processing (using a "table" function) can be very effective for group interactions, especially when using a computer projector.
4. *Brainstorm potential use errors* (failure modes). For each of the tasks identified in the task analysis, possible operator errors, actions that deviate from expected or optimal behavior, are identified. (These errors, in risk analysis parlance

TABLE 35–3. Example of a Hierarchical Task Description Table

Activity	*Task*	*Subtask*	*Notes*
Device setup	Calibration of unit	Connect sensor to load Measure resistance Adjust sensitivity control	If sensitivity cannot be adjusted go to troubleshooting activity tasks
	Check battery charge state	Read battery meter Verify that reading is greater than 12 volts but less than 15 Volts	If reading is not in range go to battery charging activity
	Check mounting security	Tighten clamp until knob starts to slip Gently pull down on unit to test for no movement	If movement detected than readjust clamp
Start device	Initial start-up	Verify power is on Check that sensor gauge reading is in starting range	
	Monitor status of reading	Check both sensor level and sensitivity readings for in range readings	
Stop device	Press stop button	Check for readings to return to zero	

TABLE 35–4. Example of Numeric Ratings for Severity in FMEA.

Severity Category	*Severity Rating*	*Consequences to Patients, Users or Environment*
Catastrophic	5	May cause death, total loss of operation, or severe impact on the environment that cause death or total loss of operation
Critical	4	May cause permanent injury, permanent occupational illness, major and recoverable damage to operation, or major and recoverable impact on the environment
Significant	3	May cause recoverable injury, recoverable occupational illness, major and recoverable damage to operation, or major and recoverable impact on the environment
Marginal	2	May cause minor injury, minor occupational illness, or minor impact on operation or the environment
Negligible	1	Will not result in significant injury, occupational illness, or a significant impact on the environment. Negligible effect on operation

would be referred to as "faults" or "failure modes".) Team interactions for this phase should follow common brainstorming rules wherein inputs are recorded and the merits of the individual items are not debated. All conceivable errors that might be committed by an operator should be analyzed. Brainstorming for each task and subtask step may generate the following:

- Use errors that may be design-induced or would be anticipated to follow predictable trends or patterns.
- Common human mistakes, lapses of memory and attention that have commonly been called user or human errors (see Figure 35–1 for examples).
- Each distinct use environment should be considered as different environments may lead to error-causing conditions.
- Use-error data from predicate devices. Use errors may be catalogued in customer complaint data or medical device reports (MDRs), which are required in the United States by the Food and Drug Administration.

TABLE 35–5. Example of Qualitative (Descriptive) Ratings for Severity in FMEA.

Severity Rating	*Consequences*
Major	The hazard could directly result in death or serious injury of the patient or operator, or indirectly affect the patient such that delayed or incorrect information could result in death or serious injury to the patient
Moderate	The hazard could directly result in moderate injury to the patient or operator, or indirectly affect the patient such that delayed or incorrect information could result in moderate injury to the patient
Minimal	The hazard is not expected to result in negative medical consequences or any complication

TABLE 35–6. Example of Numeric Ratings for Probability of Occurrence in FMEA.

Frequency Category	*Numeric Rating*	*Frequency of Occurrence*
Extremely likely	5	Very high probability of occurrence
Likely	4	High probability of occurrence
Possible	3	Moderate probability of occurrence
Unlikely	2	Low probability of occurrence
Extremely unlikely	1	Very low or remote probability of occurrence

5. *List potential effects of each failure mode or operator error.* For each of the failure modes, the team identifies the potential effects (harm) of the failure mode if it happens. This step is important for the subsequent determination of risk ratings. Each effect can be thought of as an if–then process (McDermott, Mikulak, & Beauregard, 1996). For most medical devices, the harm will have implications for the patient or the user of the device (e.g., causing death, reversible or non-reversible injury). When there are multiple harms, the convention is to consider the worst-case scenarios.
6. *Assign severity ratings.* A severity rating is a rating of how serious the effects or harm of a given fault would be if it occurs. Depending on the method used, severity can be assigned a numeric value (typically 1–5 or 1–10, with 1 being the lowest severity and 5 or 10 being the highest) or a qualitative descriptive rating. Table 35–4 shows an example list of categories used to assign a numeric rating. Table 35–5 shows an alternative rating using qualitative descriptive ratings. With medical devices the highest severity ratings are typically reserved for faults that have severe consequences for the patient or user, usually involving severe injury or death. The lower ratings would be reserved for minor reversible injuries, such as a scrape or non-injury effect (e.g., customer inconvenience). For many medical devices, ratings of severity should be provided or approved by medical affairs or equivalent.
7. *Assign occurrence ratings.* Occurrence ratings are estimates of the predicted frequency or likelihood of occurrence of a fault. It is best to base occurrence ratings on existing data such as: customer complaint data, telephone support center call data, usability test results, and so on. Where such data are not available, occurrence ratings must come from the collective judgment of a team. As with severity ratings, occurrence ratings can be assigned a numeric value (typically 1–5 or 1–10, with 1 being the lowest frequency of occurrence and 5 or 10 being the highest). Table 35–6 shows a representative example of a numeric occurrence rating scale. An alternative rating scheme, shown in Table 35–7 uses qualitative terms to define the frequency of occurrence where frequency of occurrence is rated as probable, occasional, rare, and improbable.

TABLE 35–7. Example of Qualitative Ratings for Probability of Occurrence in FMEA.

Probability of Occurrence	*Description*
Probable	Likely to occur regularly or many times during the life of the product under specified operating conditions
Occasional	Likely to occur infrequently or several times during the life of the product under specified operating conditions
Rare	Will rarely occur or is very unlikely to occur during the life of the product under specified operating conditions
Improbable	Not expected to occur under specified operating conditions

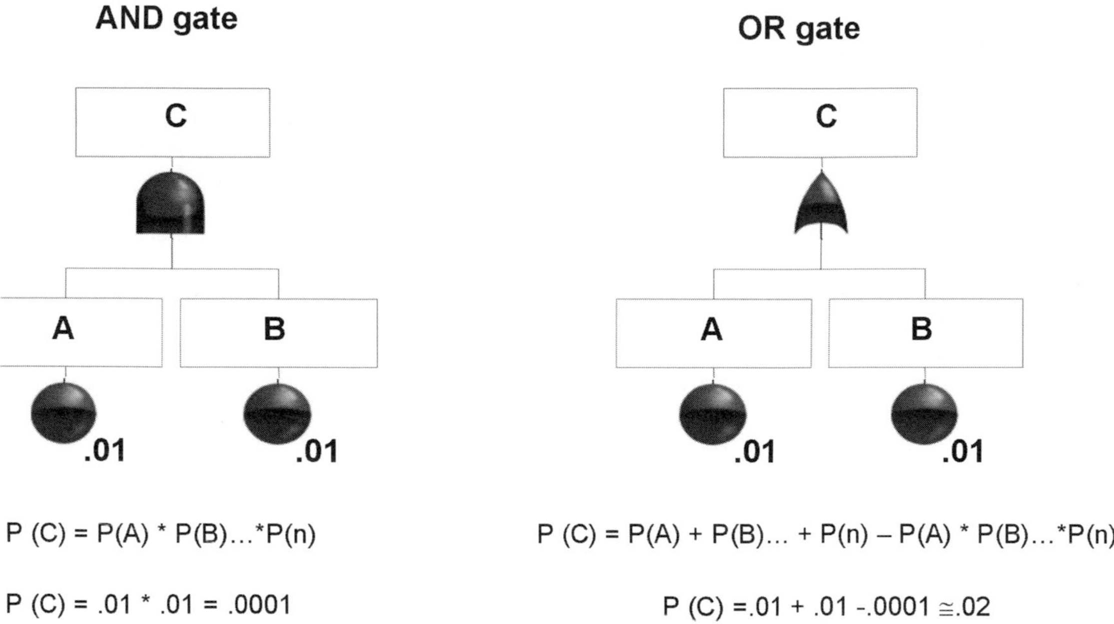

P (C) = P(A) * P(B)…*P(n)

P (C) = P(A) + P(B)… + P(n) – P(A) * P(B)…*P(n)

P (C) = .01 * .01 = .0001

P (C) =.01 + .01 -.0001 ≅.02

Figure. 35–6. Illustration of AND gate and OR gates with 2 events and associated probability calculations.

During this phase the team may decide to eliminate those items that are agreed to be so improbable that their presence in the analysis is judged to have no value.

8. *Derive a risk index.* Using the numeric approach, a "risk index" or "risk level" is calculated by multiplying the severity rating times the occurrence rating

Risk index = Severity × Occurrence

When using qualitative ratings, criteria are developed for how risk levels will be defined based on combinations of severity and occurrence. Table 35–8 is an example of how risk priorities could be defined using qualitative descriptors for severity and occurrence.

9. *Prioritize the risks.* Using the risk index (either numeric or qualitative ratings), the failure modes are sorted according to risk index from the highest to lowest. With the prioritized list, the team will determine

- Failure modes that can be addressed immediately.
- Failure modes that will be addressed at a later date or a later version of the product. Or,
- Failure modes that are determined to require no action.

The criterion used by the team to decide a cutoff line (the threshold below which problems will be designated as needing no action) should be predefined, ideally by standardized procedures. This will ensure that the different project teams will have a uniform means to address risk levels.

TABLE 35–8. Example of a Qualitative Risk Index, Based on Previous Qualitative Occurrence and Severity Ratings.

	Hazard Severity		
Probability of Occurrence	*Major*	*Moderate*	*Minimal*
Probable	High	High	Medium
Occasional	High	Medium	Low
Rare	Medium	Low	Low
Improbable	Low	Low	Low

TABLE 35–9. Example Mitigation Effectiveness Ratings.

Effectiveness of Mitigation	*Rating*
Extremely unlikely	5
Unlikely	4
Possible	3
Unlikely	2
Extremely likely	1

Prioritization may include special rules. For example a team might decide to apply corrective actions to all failure modes with a high severity rating, regardless of the occurrence rating. Such an approach would help eliminate incongruities where a failure mode with a low severity and high occurrence rating is ranked equal to or higher than a failure mode with high severity but with a lower occurrence rating.

10. *Take actions to eliminate or reduce the high priority failure modes.* Using an organized problem solving approach, the team identifies "modes of control" (or "methods of control" or "mitigations") for each of the higher priority failure modes. Modes of control are design or other elements that are intended to reduce the probability of occurrence or the severity of the consequences of the event. Although modes of control can also include warnings or training (often referred to under the broad term *labeling*), such mitigations should be considered less effective and secondary to design modes of control. Where labeling modes of control are applied, care should be taken to validate such measures to ensure that the instructions and training materials are effective and accessible by users.
11. *Assign effectiveness ratings.* For each mode of control, a rating is given for the likelihood that the mode of control will be effective. If the analysis is done early in the design cycle then the effectiveness of potential mode of control will be more difficult to estimate with confidence (formative usability testing and usability inspection methods can aid in these estimates). If done later in the development cycle, then data from summative usability testing will provide better estimates for the effects of the mitigating factors.

Using qualitative descriptors, the effectiveness of the mitigation could be expressed as

- Does not change risk.
- Reduces risk to medium.
- Reduces risk to low.

In a quantitative form, mitigation effectiveness is sometimes given a numeric rating. In the example in Table 35–9, the values are expressed in descending order; wherein the most effective mode of control is given a 1 and least effective, a rating of 5.

12. *Revise the risk priorities.* With the assumed modes of control in place, the numerical or qualitative risk indices are revised or recomputed. Any failure modes with risk indices that remain above the cutoff line should be considered for additional modes of control to further reduce the risk level. The process should be iterated until all use errors have an acceptable level of risk.

FMEA Variations

Numeric vs. Descriptive Ratings. Ratings for severity and occurrence can be given either a numeric or qualitative (descriptive) ratings. As can be seen in the tables shown, whether numeric or descriptive,

TABLE 35–10. Fault Tree Symbols

Symbol	Description	Symbol	Description
 Basic Event	A basic component failure or use error.	 Voting gate	Output occurs if a specified number of eventsoccurs. The input events need not occur simultaneously.
 Undeveloped Event	A component or use error that has not been fully developed due to the lack of information or significance.	 Inhibit gate	A certain condition of the system must exist before one failure produces another. The inhibit condition may be either normal to the system or be the result of failure.
 House Event	A normally occurring event in the system.	 Exclusive OR	Output occurs if any input occurs; specified inputs are mutually exclusive.
 OR gate	Output occurs if any of the input events occur.	 NOR gate	Output occurs when all the input events are absent.
 AND gate	Output occurs if all of the inputs occur.	 NOT gate	The output occurs when the input event does not occur.
 Transfer gate	Symbol used to link elements in a fault tree withot her pages or other segments of the tree.	 NAND gate	Output occurs when at least one of the input events is absent.
		 Priority AND gate	Output occurs when all inputs occur in the sequence specified.

TABLE 35–11. Advantages and Disadvantages of FMEA

Advantages	*Disadvantages*
Risk index/RPN enables prioritization of faults	Difficult to assess combination of events complex interactions (unless explicitly documented)
Explicitly documents modes of control/mitigation	Large documents can be difficult to manage: minimize inconsistencies and redundant items
Format useful for tracking action items	Severity and occurrence ratings are often difficult for individuals or teams to estimate. Much time can be spent in discussions
Easily constructed using hand written spreadsheets or computer-based software tools: Spreadsheets/word processing tables Specialized FMEA tools	Sometimes can be overly conservative. With each fault isolated, failure to consider combinatorial events (as do fault trees) may lead to the false conclusion that every item requires explicit mitigation

TABLE 35–12. Advantages and Disadvantages of Fault Tree Analysis

Advantages	*Disadvantages*
Graphical format enables visualization of combination of events	Drawings can become large and unwieldy in complex systems
Enables estimation of overall probability of failure based on estimates of root causes	Modes of control are not always explicit
Small fault trees can be developed using common flowchart drawing tools	Requires more training than FMEA
	Special software required for rapid development of fault trees

such ratings are almost always categorical rather than precise values for occurrence or severity.

Occurrence ratings are often the most troublesome category for groups to assess. Stamatis (1995) offers benchmarks for how occurrence ratings could be categorized according to probabilistic categories. Although such categories can be useful for mechanistic or component failures, where data are often available, such benchmarks are seldom of use when dealing with human error, mainly because there is a paucity of data related to human reliability, especially when estimating failure in parts per million or even parts per thousand. For this reason, it is incumbent on groups to provide these estimates based on device histories, knowledge of the user, the device, and the use environment.

Range of Numeric Values. When using numeric values, an analysis will incorporate a range of values for estimating severity and occurrence, traditionally 1–10. A common notion is that ratings are more precise with the larger range of numeric values. The implication is that numbers that extend beyond this range will likely not provide benefit. Furthermore, the extended range of numbers will almost certainly require more time and effort to debate and manipulate the resultant values.

Detectability. In addition to ratings for severity and occurrence, it is common practice, especially in manufacturing process settings to include a third rating for "detectability." Detectability is a rating to characterize the ability of the process or user to detect a problem before it occurs. In contrast to severity and occurrence, the values for detecability using numeric ratings are rated in reverse order; where high detectability is given a low numeric value (i.e., 1) and lowest detectability a 5 (or 10). The ratings for severity, occurrence and detectability are multiplied together to form a risk priority number or RPN. Although in common use, the authors offer the following reasons why detectability may not be appropriate for use in human factors FMEAs or perhaps in any design FMEA:

- According to Schmidt (2004), detection of a hazard during use of a device may not assure that harm will be avoided. He delineates three conditions that can determine whether detectability would serve to

TABLE 35–13. Example Qualitative FMEA for Therac-25

Task	*Hazard*	*Failure Mode/ Use Error*	*Use Error Prob*	*Hazard Severity*	*Risk Level*	*Method of Control*	*Effectiveness of Control on Risk Level*	*Risk Acceptability*
Generic Tasks								
Turn on device	Delay in herapy	Fail to hold power button for at least 2 secs	Rare	Min	Low	Training and instructions note that power button requires holding for 2 seconds	Little to no Impact	Accetable
Programming Beam Modes								
	Overdose, single activation	Typed X (high power x-ray mode) instead of E (electron mode) Administer dose.	Occ	Maj	Hi	1. Review screen shows selected dose mode. 2. Enable editing of mode selection.	Reduces to Medium	Acceptable with justification
Enter dose	Insufficient dose, not effective	Typed E instead of X administer dose.	Occ	Mod	Med	Review screen dose mode	Reduces to Medium	Acceptable with justification
	Insufficient dose' not effective	Typed B (beam) command, shutter did not open.	Rare	Mod	Low	Software robust ——— Dosimeter failure.	Little or no impact	Acceptable
Activate beam	Overdose, single activation	Computer selects wrong mode.	Improb	Maj	Low	Software extensively tested	No further control needed	Acceptable
.	Overdose, single activation	Computer selects wrong energy level	Improb	Maj	Low	Software extensively tested	No further controls needed	Acceptable
	Overdose single activation	1. Selected X ray mode. 2. Tungsten filter not properly positioned (software control only- no closed loop feedback 3. Dosimeter failure	Occ	Maj	HI	1. High quality stepper motor. Note to Engineering 2. Can filter position be guaranteed? Filter position sensor could assure that beam will not activate unless in place	Litter or no impact (consider #2) Proper dosimeter operation would indicate Overdose after the fact	Unacceptable
	Over-dose, due to multiple activations	1. Typed B, beam activated 2. User unclear about success. 3. User activates repeatedly.	Occ	Maj	Hi	Software robust ——— Note to engineering: how does the user know beam was actually activated, and at what power level? Could a message (and printout) confirm actual output?	Proper dosimeter operation dosage to warn against multiple exposure (worst case exposure?)	Acceptable with Justification

TABLE 35–14. Example Use-Error FMEA for a Hypothetical Automatic External Defibrillator (AED). Product: AED. Activity: Setup

Task	*Hazard*	*Use Errors*	*Use Error Prob*	*Hazard Severity*	*Risk Level*	*Method of Control**	*Effectiveness of Control*	*Risk level with control*	*Risk Acceptability*
1. Open case	Delay in therapy	Difficulty/unable to open case	3	5	15	Use fabric case with hook-and-loop closures	1	15	Acceptable with review
	Broken/torn fingernail	Use fingernail to open latch	1	2	2	No latch	1	2	Acceptable
2. Tear open electrode package	No therapy delivered	Package missing as result of not being replaced from previous use	3	5	15	Design case with slot positions for accessories Missing item obvious Recommend admin procedures using seals	2	30	Acceptable with review
	No therapy delivered	Tear electrode when attempting to open package	3	5	15	Provide "zipper" closure that allows easy opening of sealed package Construct electrode with non-tear backing	1	15	Acceptable with review
	Delay in therapy	Difficult/unable to open package	3	5	15	Provide "zipper" closure that allows easy opening of sealed package	1	15	Acceptable with review
3. Expose upper chest of patient	Non-delivery of shock	Clothing not adequately removed	2	5	10	Provide scissors Provide pictorial and auditory instructions	2	20	Acceptable with review
	Burn caused by metallic object in clothing	Wire in under garment or metal fastener left in place	2	3	6	Provide pictorial and auditory instructions	3	18	Acceptable with review
4. Peel backing from electrodes	Delay in therapy	Difficulty removing backing	3	5	15	Provide extended tab which allows easy removal of backing	1	15	Acceptable with review
	Non-delivery of shock	Used without moving backing	2	5	10	Detection circuit will alarm because EKG signal will not be detected with nsulated electrodes	1	10	Acceptable with review
5. Apply electrodes to chest	Shock not delivered properly	Improper positioning	3	5	15	Provide pictorial and auditory instructions	3	45	Acceptable with review
	Local burn	Electrodes placed too close together	2	3	6	Provide pictorial and auditory instructions	3	18	Acceptable with review

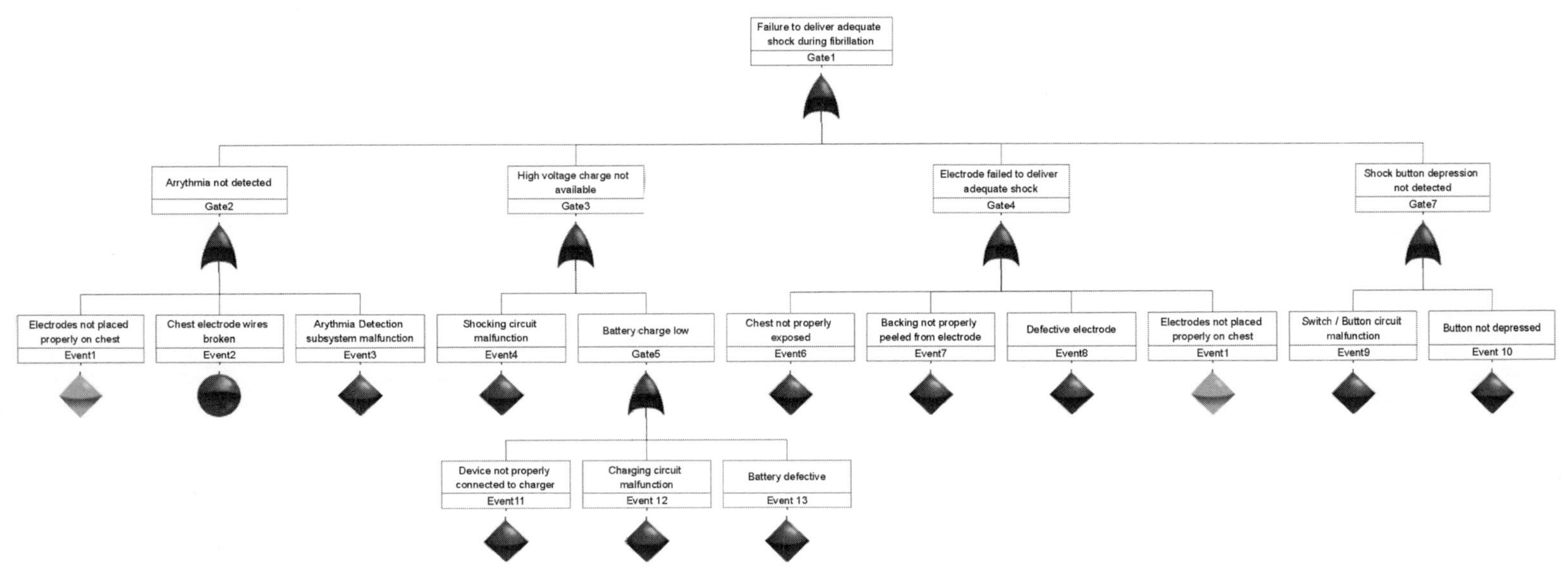

Figure 35–7 Example fault tree analysis for a hypothetical automatic external defibrillator (AED).

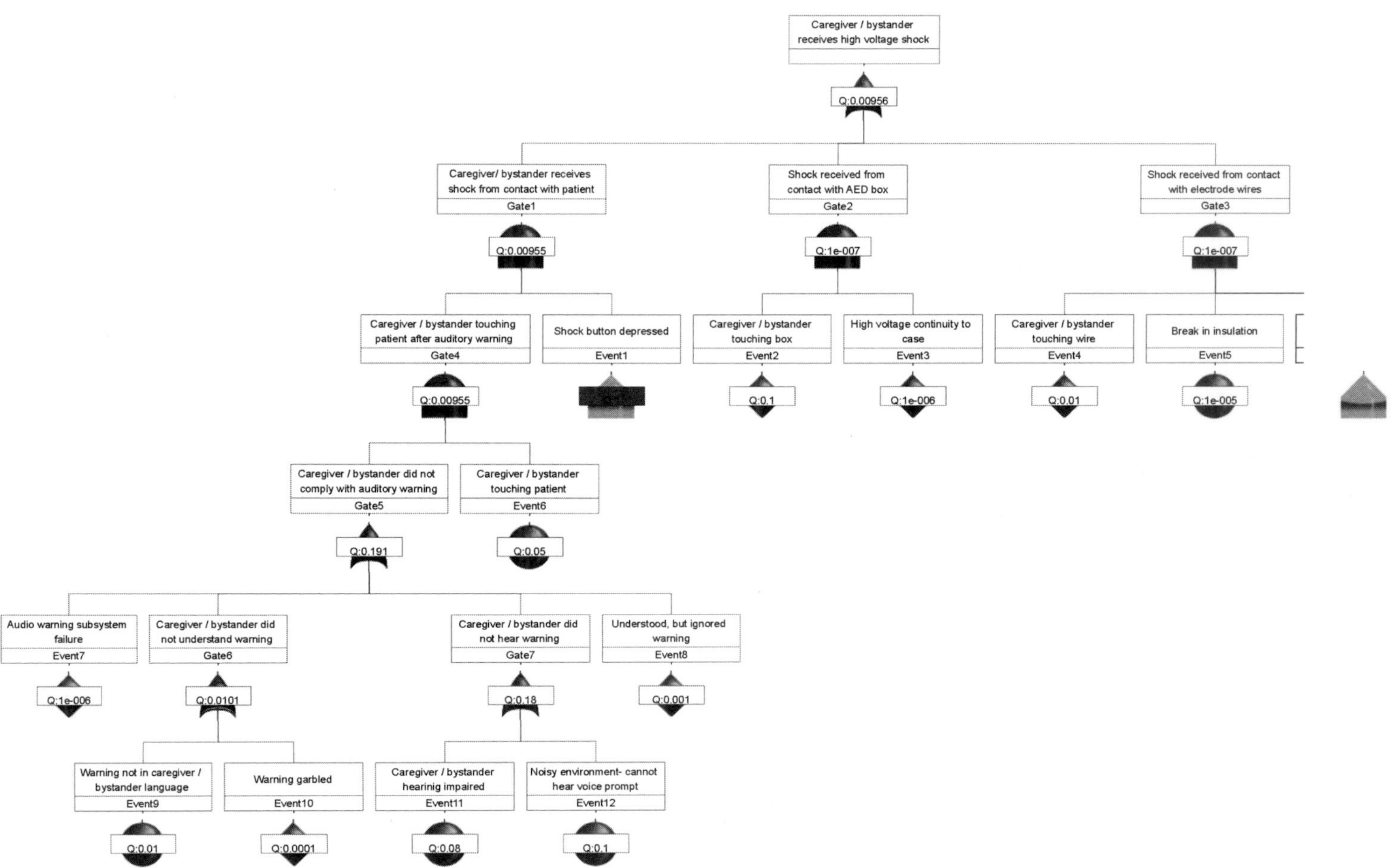

Figure 38–8 Example quantitative fault tree analysis for a fictitious automatic external defibrillator (AED).

TABLE 35–15. Comparison Between Qualitative and Semi-Quantitative (RPN) FMEA Methods

Method	*Risk Severity*	*Fault Probability*	*Risk Level*	*Detect/ Mitigate Risk*	*Outcome (Residual Risk)*
Qualitative	Major	Probable	High	No Change	Unacceptable
	Moderate	Occasional	Medium	Lower	Acceptable w/justification
	Minimal	Rare Improbable	Low	Eliminate	Acceptable
Semi–quantitative risk priority number (RPN)	5	5		1	>45 Unacceptable
	4	4		2	
	3	3		3	
	2	2		4	9 < 45 Acceptable w/review
	1	1		5	1 < 9 Acceptable

prevent harm: (a) Is there enough time to react after detection, (b) is information provided to the user to indicate specific actions to avoid the harm, and (c) will the user have the presence of mind to remember what is done and take action? In many cases, even if the user can detect impending harm, he or she may not be capable of preventing it.

- In situations where a user can detect harm and prevent it, it is the authors' position that this should be factored into the probability of occurrence. If detectability is enhanced or enabled by the system in a way that will reduce the overall probability of occurrence or the severity, this constitutes a mode of control and therefore should be reflected by a reduction in the probability of occurrence. With both detection and occurrence reductions of the RPN value, it is the authors' contention that the risk index is being reduced artificially by overweighting the effectiveness of the detection mode of control.

FAULT TREE ANALYSIS (FTA) AND OTHER TECHNIQUE VARIATIONS

Another commonly used tool for analyzing and predicting failure is FTA. FTA is a "top-down" deductive method used to determine overall system reliability and safety (Stamatis, 1995). A fault tree, depicted graphically, starts with a single undesired event (failure) at the top of an inverted tree, and the "branches" show the faults that can lead to the undesired event—the root causes are shown at the bottom of the tree. For human-factors applications, FTA can be a useful tool for visualizing the effects of human error combined with device faults or normal conditions on the overall system. Further, by assigning probability estimates to the faults, combinatorial probabilistic rules can be used to calculate an estimated probability of the top-level event or hazard. As with FMEAs, fault trees can be developed by teams or by individuals with team review. The following describes basic steps in developing fault trees. For more information, refer to the literature in reliability engineering or systems safety engineering (e.g., Veseley et al., 1981).

Fault trees have often been avoided because the drawings and computations are seen as being labor intensive. In recent years, graphical software programs have been made available for personal computers that enable users to rapidly assemble fault trees by "dragging and dropping" standard logic symbols onto a drawing area and connections are made (and maintained automatically). These tools automatically calculate branch and top-level probabilities based on estimated event probabilities entered. Such tools make FTAs much more accessible and much less labor intensive.

Basic Steps in Conducting a FTA

1. Identify the top-level hazards. The team will brainstorm to identify the top-level hazards (undesired event) to be addressed. A fault tree will be developed for each of these hazards.
2. Identify fault tree events. Identify faults and other events (including normal events) that could result in the top-level undesired

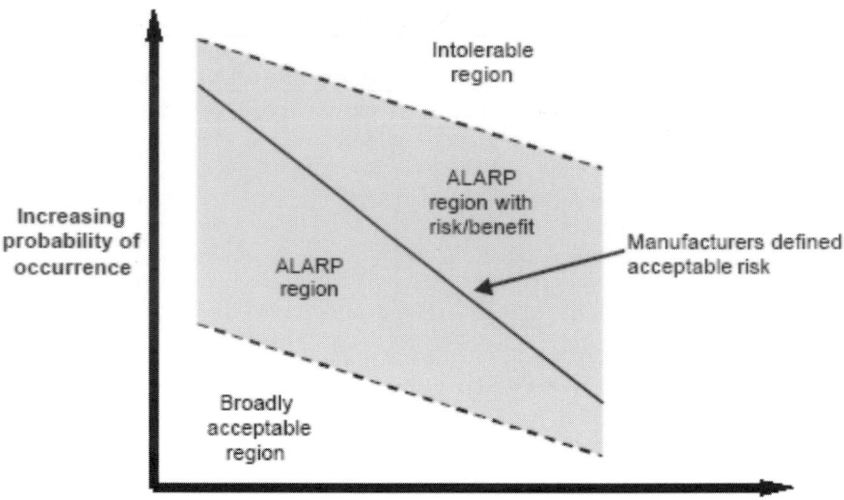

Figure 35–9 Illustration of risk acceptability and the ALARP (as low as reasonably practical) region.

event. These can be documented in a list or on notes posted on a wall.

3. Identify the conditions under which the events can lead to failure. For example

   - Events that may lead directly to harm (single-point failure) or cause another fault without other events occurring.
   - Events that must happen in conjunction with other events to cause failure.
   - Events that must happen in sequence to cause a failure.

4. *Combine the above events into a fault tree.* Table 35–10 shows the symbols (and their usage) that are employed in fault trees. Of these, the more likely to be used are those in the left half of the table that represent individual events ("basic events," "undeveloped events," or "house events") and the most significant logic symbols, the "OR gate" and the "AND gate." Most of the remaining symbols are likely to be used in design-specific analyses or in systems with a high degree of complexity.
5. *Assign probabilities to each event.* At this point, the investigative team may decide that the fault tree sufficiently characterizes the system and human interactions and requires no further development. Other situations, however, may require further quantitative analysis. If so, then the team will assign probabilities to each of the events based on quantitative data or estimates based on expert judgment.
6. *Calculate the probability of each of the branches leading to the top-level hazard.* Fault tree probabilities propagate upward from the individual events. The probability of the individual gates and the overall fault tree probability are computed by using numerical combinatorial rules for various logic gates. Figure 35–6 illustrates two logic gates, OR and AND gates with two events each along with the formulas for their probabilities.

The figure also illustrates the effect of two different logic gates with two identical input events. In computing the probability of failure with the same input probabilities of .01, the OR gate shows a resultant probability of failure of approximately .02 (or 2 in 100), whereas the AND gate shows a probability of .0001 (or 1 in 10,000).

Uses For Fault Tree Probability Estimates

Although fault trees can provide utility without probability estimates, fault tree probabilities can provide benefit in the following ways:

- *Hazard probability*. Probabilities derived from fault trees can be used as a decision tool to assist design teams in determining if further modes of control are needed for further risk reduction. After new modes of control are incorporated into the fault tree, revised probability estimates can be used to determine the effect on overall probabilities. A fault tree may provide a useful supplement to FMEAs that typically do not depict the combinatorial aspect of events.
- *Critical path analysis*. One of the most useful purposes of a fault tree can be realized by tracing high probability pathways through the fault tree. Such a technique, even using qualitative probability ratings may uncover high-risk pathways that require design attention to reduce risks that may otherwise have been overlooked.
- *Sensitivity analysis*. To determine the effect of a given failure on the overall system, one can substitute different probability estimates for individual events or faults and assess the effect on the top-level probability. Such an analysis may be useful for teams performing FMEAs where there is disagreement among team members on the need to mitigate certain failure modes. After constructing a fault tree consisting of the FMEA elements, substituting a range of probabilities (e.g., .1 to 1.0) for selected faults can show to what extent a given failure affects the top-level event probability. A large change in the top-level probability when increasing a single or multiple fault probabilities may indicate where sufficient safeguards are lacking and further modes of control should be considered.

Advantages and Disadvantages of FMEAs and Fault Trees

Because of its tabular format, FMEA is a straightforward method for documenting potential failures in terms of a risk index, enabling explicit documentation of the modes of control or mitigations for each failure mode. For these reasons, FMEA is probably the most commonly used method of documenting risk analyses.

However, because of its bottom-up nature, wherein faults are dealt with one item at a time, it is often difficult to assess how a fault or combination of faults can lead to an undesired event. Further, in complex systems, it is not uncommon for FMEAs to consist of several hundred pages. In such cases, the management of a risk analysis becomes a daunting task; the elimination of inconsistency and redundancy can be challenging. Table 35–11 summarizes advantages and disadvantages in using FMEAs.

The graphical representation of events with associated logic gates of fault trees can provide unique insight into how individual events can lead to or prevent an undesired event.

In large systems, fault trees can become very large and unwieldy. In such cases, it may also be difficult to visualize complex system interactions. The use of fault trees to compute the probability of failure enables teams to compute the probability of selected hazards as well as overall system failure. Lack of data or time, however, will often prevent teams from fully developing fault trees except for the most critical of systems. Table 35–12 summarizes the advantages and disadvantages of fault trees.

Both FMEA and FTA offer distinct advantages and disadvantages. For these reasons we believe that both techniques are complementary.

Other Risk Management Methods (Health Care FMEA, Health Hazard Assessment)

There are many variations on these techniques and they appear under various names, such as

- Health care failure modes and effects analysis (HFMEA™; DeRosier, 2002): A process oriented FMEA focused on health care processes, such as medication ordering in a hospital that was developed by the Veterans Affairs National Center for Patient Safety (VA). HFMEA™ is defined by the VA as
- A prospective assessment that identifies and improves steps in a process thereby reasonably ensuring a safe and clinically desirable outcome.
- A systematic approach to identify and prevent product and process problems before they occur.
- Hazard analysis and critical control points (HACCP): This is a process endorsed by the Food and Drug Administration (FDA) and focuses on hazard analysis of food safety by

examining critical control points in the food-processing flow for fault potential.

- Hazard and operability studies (HAZOPS): Systematic analysis of industrial processing flows. It has been used for identifying potential hazards and operability problems caused by deviations from the design intent of industrial processing plants.
- Health hazard assessment (HHA): A systematic analysis to anticipate, identify, quantify, and recommend controls for health hazards associated with the occupational environment.
- Root cause analysis (RCA): This is a series of analysis techniques that are typically done after an adverse or sentinel event. It is therefore a look back, or retrospective analysis in contrast to FMEA, HFMEA™, FTA, and the other forward-looking prospective techniques. RCA is required by the Joint Commission on Accreditation of Health Care Organizations, guidelines for health care facility accreditation, and includes
- Affinity diagrams: Shows clustering of possible causes of problems arranged in a hierarchy of root-cause groups that resemble an organizational chart.
- Fishbone diagrams: Sometimes they may be referred to as a cause-and-effect diagram. The diagram looks like the skeleton of a fish and hence it is often referred to as the fishbone diagram. See Figure 35–7 for a example.

The steps in the retrospective creation of a fishbone diagram are basically the same as the basic steps of RCA and include (North Carolina Department of Environmental and Natural Resources, 2002) the following:

1. Draw the fishbone diagram.
2. List the problem or issue to be studied in the "head of the fish."
3 Label each "bone" of the fish. The major categories typically utilized are

- The 4 Ms: methods, machines, materials, manpower.
- The 4 Ps: place, procedure, people, policies.
- The 4 Ss: surroundings, suppliers, systems, skills.

Note: You may use one of the four categories suggested, combine them in any fashion or make up your own. The categories are to help you organize your ideas.

4. Use an idea-generating technique (e.g., brainstorming) to identify the factors within each category that may be affecting the problem, issue, or effect being studied. The team should ask "What are the machine issues affecting or causing…?"
5. Repeat this procedure with each factor under the category to produce subfactors. Continue asking "Why is this happening?" and put additional segments under each factor and subsequently under each subfactor.
6. Continue until you no longer get useful information as you ask "Why is that happening?"
7. Analyze the results of the fishbone after team members agree that an adequate amount of detail has been provided under each major category. Do this by looking for those items that appear in more than one category. These become the "most likely causes."
8. For those items identified as the most likely causes, the team should reach consensus on listing those items in priority order with the first item being the most probable cause.

EXAMPLES OF USE ERROR FMEAS AND FTAS

Qualitative Use Error FMEA: Therac–25™ Radiation Therapy System

The Therac–25™ was a radiation therapy machine (a medical linear accelerator) designed for the treatment of tumors. The minicomputer-controlled system was activated through commands entered via a keyboard. Typing an "E" command selected a lower energy electron beam, and an "X" activated a higher energy "X-ray mode." The X-ray beam with an output of 25 Mev (million electron volts) was attenuated to proper dose levels by mechanically inserting a beam-flattening filter into the radiation beam path. This attenuating filter was mounted on a turntable. The filter was designed to be put into the beam path by computer control only when the system was in the high-energy X-ray mode. When using the electron beam mode, the attenuating filter was not used and was rotated away from the beam outlet. The placement of this filter is a key factor in this case (Leveson, 1995).

During this particular event, as the operator mistakenly entered the command "X" instead of "E," she quickly realized her error and pressed the up-arrow key to move the cursor to the beam selection field and then typed the correct command "E." One of the causal problems was that an 8 sec timer was activated in the software by this irregular command sequence and that any edits made before the 8 sec elapsed were ignored. Although the screen displayed the E beam mode, the device beam selection state remained on X. Another, even more serious problem was that the device failed to command the turntable filter to move in place. After the operator commanded "B" for beam on, the system delivered the high-power X-ray energy mode without the beam flattening attenuating filter in place. To further compound the tragedy, the system presented a cryptic error message, "Malfunction 54," after the first dose, which was unrecognizable to the operator. (This is a classic case of poor system error messages that have no meaning to typical users, but do have meaning to the software designers.) Thinking that no dose was delivered, the operator pressed "P" to proceed. The same cryptic error message appeared and the screen displayed no unusual amounts of delivered radiation, so the operator entered the command to activate the beam again, activating a third dose. The repeated doses of the unattenuated beam resulted in the patient receiving over 16,000 rads instead of the intended 180 rads. The patient died about 5 months later from the massive radiation overdose.

This incident, as well as several others, involved a combination of technical failures (involving software and hardware) combined with human behavior resulting in catastrophic radiation overdoses. According to Leveson (1995), AECL, the manufacturer of the Therac–25™ had conducted hazard analyses in the form of FTA. If so, what went wrong? Why were these hazards not anticipated and eliminated through design? According to Leveson, AECL did not consider possible software failures in their consideration of hazards. AECL's reasoning for this was that the software had been extensively tested and that the software was judged not to be subject to risk analysis because software does not wear, fatigue, or fail from reproduction processes. We also know that use-error focused risk analysis was not common in 1985. If such a system were being developed today, current best practices would guide manufacturers to consider all of the system-related considerations in the assessment of risk, including the user and environment as well as system hardware and software. Table 35–13 shows an abbreviated, hypothetical FMEA of the Therac–25™ showing potential faults (use errors). The ratings are qualitative. We propose that if this kind of analysis were performed at the time, the identification of high-risk items would have lead the design team to consider redesign of certain elements to mitigate the high-level risks.

Quantitative Use Error FMEA: AED

An AED is a device used to treat sudden cardiac arrest due to ventricular fibrillation. The typical AED will detect ventricular fibrillation through chest electrodes applied by a caregiver, and automatically administer a defibrillatory shock (after the caregiver pushes a button) through the same electrodes. In a study by Andre, Jorgenson, Froman, Snyder, and Poole (2004), four devices were tested for usability among a group of untrained participants. Although the authors concluded that the AEDs in the test were sufficiently easy to use so that nonmedical people could effectively use them without prior training, the investigators had observed several use-error related problems. Table 35–14 depicts an example of an abbreviated FMEA using quantitative numeric ratings for hazard severity (from Table 35–4), probability of use-error (Table 35–6), and effectiveness of the method of control (Table 35–9).

Nonquantitative Fault Tree for an AED.

One of the features of fault trees is that they can illustrate multiple pathways to an undesired event. Figure 35–7 depicts an example fault tree for an AED in which the top-level undesired event is non-delivery of defibrillatory shock. From this example, it appears that there are a number of potential sources of failure listed in the fault tree, some of which are technical failures and others are use-error related. The fact that all of these items are attached to OR gates implies that any one of these events could lead to the top-level event, the non-delivery of defibrillatory shock. Such a structure generally implies higher risk. This FTA example has no quantitative probability estimates.

Quantitative Fault Tree for an AED

Figure 35–8 depicts an example of a fault tree with quantitative probability estimates for an AED. This example portrays the top-level undesired event as the caregiver or bystander receiving a high-voltage shock. The probability of occurrence of almost .01 is a value that would probably be considered unacceptably high in perhaps most medical devices. Most teams (after checking for errors) would probably begin looking for the cause or causes of the high failure probability. One of the techniques for identifying vulnerable areas is to perform a "critical path analysis." This is done by starting with the top-level event and identifying the gate (or event) with the highest probability. That gate will be traced further to determine which events or gates have the highest probability. This process is repeated until reaching the bottom of the fault tree. The highest probability branches through the tree will constitute a "critical path" in the fault tree, signifying a combination (or sequence) of events that are most likely to lead to the top-level event.

Comparison of Quantitative versus Qualitative Risk Management

Qualitative and semiquantitative methods for FMEAs are very similar. Table 35–15 shows a comparison of the two methods from which it is easier to see how similar they are. When using RPN, an explicit risk index is usually not calculated. Commonly RPN is the result of multiplying three variables; fault frequency × hazard severity × effectiveness of mitigation or detection. Of course, it is feasible to obtain a risk index or risk level quantitative value from the product of the first two variables, that is, fault frequency and hazard severity. It is the authors' recommendation that a risk index be calculated both before and after risk mitigations are considered for both methods. The outcome or residual risk is practically the same comparing the two methods. As noted earlier, many analysts do not endorse the use of a detectability rating for medical device FMEAs.

CRITERIA FOR ESTABLISHING ACCEPTABLE RISK

The end-point for FMEA and other risk-management techniques is whether residual risks remaining after mitigation are acceptable. Three regions of residual risk are discussed

- Acceptable.
- Acceptable with review or justification.
- Unacceptable or intolerable.

The second category of *acceptable with review* is a gray area called ALARP (as low as reasonably practicable) that as shown in Figure 35–9, a manufacturer can further divide into two subregions. A criterion for defining this upper-ALARP region is when a risk versus benefit tradeoff is positive. What is reasonably practicable requires the risk versus benefit tradeoff be explicitly stated and estimated. Lives saved or lives improved versus cost and potential harm must be balanced in favor of the more positive outcomes. An important consideration in making this tradeoff is whether the benefit is proportionate to the cost. An example is heart bypass surgery. The risk of death during surgery and early recovery has been estimated around 3%, which is offset by the fact that death from a heart attack is almost certain without the surgery.

According to the International Organization for Standardization publication 14971 (ISO; 2000), practicability has two areas of consideration:

- Technical practicability, which refers to the ability to reduce the risk regardless of cost, such as having so many warnings and cautions in the labeling that normal operation is hampered.
- Economic practicability, which "refers to the ability to reduce the risk without making the provision of the medical device an unsound economic proposition. Cost and availability implications are considered in deciding what is practicable to the extent that these impact upon the preservation, promotion, or improvement of human health. Major risks should normally be reduced even at considerable cost" (ISO 14971, 2000, p. 200).

ALARP is a difficult concept to incorporate in designing a medical device. Of course, no one wants to put a price on safety or human lives, which is something a true cost–benefit analysis sometimes requires. The ISO 14971 (2000) standard does introduce the difficult concept of costs, that is, a device cannot be made so expensive that it becomes unaffordable for medical product consumers and clearly manufacturers can only exist if they make a profit for their shareholders and employees.

COMMON MISTAKES IN PERFORMING USE ERROR RISK ANALYSIS

For those focusing on use-error risk analysis for the first time, care should be taken to avoid the mistakes often observed among teams:

1. *Blaming the user.* Although there will always be errors that are unpredictable, human-factors literature can provide many examples where the design of products can be shown to induce or increase the probability of error. If a user commits an error that is caused by the design, where a statistical pattern of errors is exhibited among users, the manufacturer will likely be held accountable despite the manufacturer's notion that "it's the user's fault, not ours."
2. *Treating risk management as a necessary evil rather than a development tool.* Although risk analysis has been used to analyze and prevent failure in a myriad of systems, it is not uncommon for development teams to resist performing them. Rather than employing risk analysis as a development tool, it is often treated as a perfunctory exercise at the end of the project, with little or no chance for impact on the design. If significant use-error related defects are found in later development stages, timelines and cost become formidable obstacles to any design changes. Instead, risk analysis (including use-error analyses) should be considered to be an essential tool used early and throughout the development lifecycle.
3. *Underestimating risk probability and severity.* When there are no objective data, practitioners must depend on estimates of fault probability and severities to estimate risk. Individuals and groups are subject to many sources of bias, especially when schedule and cost pressures are evident. Consequently, groups may unintentionally underestimate the probability or severity of various events. The first line of defense for this kind of bias is to assemble teams that include people who have had experience dealing with users of the same or similar devices, and who can represent the users perspective (e.g., human factors, customer support, customer training, and marketing). Secondly, organizational independence from the development organization could be important for minimizing biases in risk estimation. Management support of the group and their output and recommendations is essential. Third, to minimize the effects of group dynamics, techniques for gathering unbiased group estimates (e.g., Delphi method) should be considered.
4. *Overlooking critical sources of failure.* According to Leveson (1995), the developers of the Therac–25™ failed to include software in their considerations of risk. By doing so, they overlooked key fault sources that should have been considered. It is clear from the literature that typical human factors considerations and use errors were not well-considered in the design. These failures allowed critical design flaws in the design of the device that had tragic consequences. One way to minimize the chance of overlooking potentially critical failures is to assemble teams that represent a wide array of relevant stakeholders that would include personnel from human factors, engineering, manufacturing, training, field service, and so on. These teams should have the autonomy to make independent assessments of the overall design. These analytical methods should always be supplemented with empirically based usability testing.
5. *Failure to document low-level risks.* When dealing with user behavior and use errors, it is not uncommon for individuals in a group to declare: "users would never do that" or "users are not that stupid."

Attitudes associated with such statements may lead groups to omit such faults from the analysis documents. Another problem is for groups to assume that certain risks will be (or always have been) well controlled, and therefore not the subject of concern. The problem with both of these scenarios is that if a situation should arise where the development team is queried as to whether a certain risk was considered (as might occur during audits or as the result of litigation) there would be no evidence that the team ever considered the problem. Although common sense should be applied here, it is recommended that teams document all conceivable hazards, and rate those that would have been omitted as having low probability.

6. *Overreliance on training or labeling.* When considering operator errors, a frequently proposed solution is to put a label on it or train users not to do that. Although there are situations where training or labeling may indeed be the solution, the human-factors literature is replete with examples of cases where well-trained operators (with volumes of training documents) commit errors. A main consideration should be that if a design is not consistent with good design practices or conflicts with the users' expectations, training and labeling are not a reliable, sustainable methods for eliminating errors.
7. *Failing to validate modes of control.* Whether the mitigations for an identified use error are elements of design, labeling or training, it is incumbent on the manufacturer to show evidence that the mitigations are effective. The common method for validating a user interface is usability testing, which involves testing representative users of the device performing representative tasks on the actual device or reasonable facsimile. Validation will depend on meeting selected usability objectives. For more information, refer to sources on human factors processes (e.g., American National Standards Institute/Association for the Advancement of Medical Instrumentation [ANSI/AAMI], 2001) and usability testing. For example, see the chapter on "Usability Evaluation in Health Care" (Gosbee & Gosbee) in this handbook.

STANDARDS AND REGULATIONS

Standards for Risk Management

There are ISO standards on the use of risk analysis in the development of medical devices. The current standard is *ISO 14971:2000 Medical Devices: Application of Risk Management to Medical Devices* (International Standards Organization, 2000). This standard describes the FMEA and FTA processes in general. The concept of human factors or use-error focused risk analysis is also endorsed in *ANSI/AAMI HE –74:2001: Human Factors Design Process for Medical Devices* (Association for the Advancement of Medical Instrumentation, 2001), which is a human factors best practices process standard. *IEC 60601–1–6 Collateral Standard on Usability* (International Electrotechnical Commission, 2004) includes "HE 74" as an attachment and also endorses the use of humanfactors risk analysis, evaluation, and control.

Regulations

In the United States, FDA guidance on human factors is very clear on the importance of performing use-error focused risk analysis (FDA, 2000). The guidance urges manufacturers to perform both analytical (e.g., risk analysis) and empirical analyses (e.g., usability testing) during product development to identify and control for use errors.

International regulators commonly endorse international standards and look very favorably on medical device submissions for approval that conform to recognized standards from IEC and ISO. For human-factors risk analysis that would include *IEC 60601–1–6: 2004* (International Electrotechnical Commission, 2004) for usability and *ISO 14971: 2000* (International Standards Organization, 2000) for risk management.

IMPLEMENTATION CONSIDERATIONS

Team Membership

It is recommended that a group perform the use error risk analysis consisting of individuals from the product development team with detailed knowledge of the product's user-interface design. This group should include

- Human factor's specialist (internal or an approved outside consultant).
- Quality assurance.
- Product development, engineering
- Medical, clinical affairs.
- Marketing.

Estimation Techniques

Brainstorming. Brainstorming is the recommended method of examining far-ranging, but feasible, use errors that could be committed for each task and meaningful subtask. Other investigation methods for generating possible use errors are

- Examination of adverse events reported to regulators such as the FDA's Medwatch program.
- Analysis of customer complaints with a focus on use errors attributed by customers to inadequate product design.

Brainstorming needs to follow a few rules to be effective in creatively generating reasonably feasible use errors. To conduct a productive brainstorming session

- Verify that all team participants understand and are satisfied with the main question before you open up for ideas.
- Give everyone a few minutes to personally record a few ideas before getting started. (Give them a pre-session assignment).
- Start with going around the table, giving everyone a chance to voice their ideas or pass. After a few rounds, open the floor to everyone.
- More ideas are better. Encourage radical ideas and piggybacking and some limited tangents.
- Avoid the temptation to evaluate and critique ideas.
- Board and record exactly what is said. Avoid paraphrasing ideas. Clarify only after everyone is out of ideas.
- Do not stop until ideas become sparse. Allow for late-coming ideas.
- Eliminate duplicates and ideas that are not relevant to the topic.
- Combine ideas that fit together.

A typical brainstorming session should start with a simple question, and ends with an unedited list of ideas. It gives you a raw list of ideas. The quality will vary and some ideas will be good, and many will not be. You will spoil the session if you try to judge and evaluate ideas during the session. Just wait. At a later time, you can analyze the results of a brainstorm with other quality improvement tools, such as affinity diagrams or fishbone diagrams, as described earlier in the Other Risk Management Methods section.

Achieving Consensus. The main goal in consensus-building in risk management is to achieve agreement on such things as ratings for fault frequency or occurrence, hazard and harm severity, risk index, effectiveness of mitigations, and level of risk acceptability.

One software tool for facilitating group consensus is PathMaker's Consensus Builder tool (Skymark, 2004). According to PathMaker, this tool uses two main methods to support consensual decision-making:

- One is structured discussion, in which decisions are carefully framed, alternatives systematically discussed, and notes taken.
- The second method involves the effective use of voting—rating systems and multivoting—to reduce lists and quantify opinions. Multivoting allows group members to assign numerical ranks to all alternatives. The ratings are averaged and the values are used to produce a final ranking based on the averages of individual rankings.

The recommended Skymark PathMaker™ consensus building steps are

1. Carefully document by writing out the issues and alternatives.
2. Suggest many alternative candidate answers.
3. Reduce a long list (10+ items) using multivoting.
4. Carefully discuss the remaining candidates. Take notes on each.

5. Decide which criteria will be used to evaluate the alternatives.
6. Perform a rating vote.
7. Look at areas of disagreement and discuss them further.
8. Vote again, if necessary.
9. Discuss the outcome of the vote. Has everyone been heard?
10. Has consensus been achieved? If not, iterate the process.

Delphi Techniques. The Delphi technique is a group forecasting methodology generally used for future events such as technological developments. It uses estimates from experts and feedback summaries of these estimates that allow for additional iterative estimates by these experts until reasonable consensus occurs. The authors propose its use for achieving consensus on risk-management related ratings and decisions. Details of the Delphi technique are described by the Carolla group (Cline, 2004).

According to Cline (2000). The Delphi technique was developed by the RAND Corporation in the late 1960's as a forecasting methodology. Later, the U.S. government enhanced it as a group decision-making tool with the results of Project HINDSIGHT, which established a factual basis for the workability of Delphi. That project produced a tool in which a group of experts could come to some consensus of opinion when the decisive factors were subjective, and not knowledge-based. The general steps in performing a Delphi technique are as follows:

1. *Pick a facilitation leader.* Select a person that can facilitate, is an expert in research data collection, and is not a stakeholder. An outsider is often the common choice.
2. *Select a panel of experts.* The panelists should have an intimate knowledge of the projects, or be familiar with experiential criteria that would allow them to prioritize the projects effectively. In this case, the department managers or project leaders, even though stakeholders, are appropriate. See section 15.1 for recommendations on stakeholders for risk management Delphi panelists.
3. *Identify a strawman criteria list from the panel.* In a brainstorming session, build a list of criteria that all think appropriate to the projects at hand. Input from non-panelists are welcome. At this point, there are no correct criteria. However, technical merit and cost are two primary criteria; secondary criteria may be project-specific.
4. *The panel ranks the criteria.* For each criterion (e.g., risk management related ratings) the panel ranks it as 1 (*very important*), 2 (*somewhat important*), or 3 (*not important*). Each panelist ranks the list individually, and anonymously if the environment is charged politically or emotionally. (This could be done via e-mail.)
5. *Calculate the mean and standard deviation.* For each item in the list, find the mean value and remove all items with a mean greater than or equal to 2.0. Place the criteria in rank order and show the (anonymous) results to the panel. Discuss reasons for items with high standard deviations. The panel may insert removed items back into the list after discussion.
6. *Rerank the criteria.* Repeat the ranking process among the panelists until the results stabilize. The ranking results do not have to have complete agreement, but a consensus such that the all can live with the outcome. Two passes are often enough, but four are frequently performed for maximum benefit. In one variation, general input is allowed after the second ranking in hopes that more information from outsiders will introduce new ideas or new criteria, or improve the list.

USES FOR RISK MANAGEMENT

1. *Tradeoffs regarding design safety.* The goal of every manufacturer is to design and manufacture a quality product that is safe. In so doing, the manufacturer hopes to have high product sales with resultant profits. The question arises, how much quality and safety are enough? A product that has little attention paid to safety will likely fail if word-of-mouth reports the same, or if the product is subject to

recalls due to safety. A product that is the subject of extreme over-design in terms of safety may not may not succeed because it costs significantly more than competitive products, or because it took too long to develop and missed its optimum marketing time window. The risk-management tools in this chapter are useful to help identify areas of risk, and provide the means to assess risk in terms of probability and consequences. Explicit identification of these issues will allow companies to assess these risks in terms of cost-to-benefit analyses.

2. *Decision-point for usability testing.* Because usability test sessions can be time consuming, especially when dealing with complex systems, usability tests are seldom comprehensive assessments of the entire user interface. Among the tasks that should be assessed in a usability test would be those that are determined to have a high risk in terms of possible use errors.
3. *Labeling and training.* As stated earlier, it is common to apply labeling and training as a means to mitigate use errors. Whether or not labeling or training is appropriate should depend on risk levels found in risk analysis. If a risk level is high, labels and training should probably not be used as the sole means to reduce these risks.
4. *Input to corrective and preventative action (CAPA) systems.* When customer complaints are reported for a given product, the manufacturer typically must respond in terms of identified causes of the problem and the solutions. Statistical trend analyses are typically done to determine how common the problem reported. However, when dealing with a small number of occurrences, risk-analysis tools can also be applied to situations involving CAPA, to better understand assess the risks of occurrence (and reoccurrence). Those items involving frequent or even infrequent customer complaints, but which have been identified as having high risk in risk analyses, are issues that should be considered for additional modes of control.

FUTURE DIRECTIONS FOR USE ERROR RISK MANAGEMENT

The focus on patient safety and use errors will likely continue until accident and near miss-rates decline significantly. The following are predicted elements in that focus:

1. *Increased awareness of use errors related to patient safety.* The recent focus on human error in medicine has increased the awareness among manufacturers and health care givers of their responsibilities in this area. A change of attitude toward the following notions will have an impact on reducing error and improving patient safety:
   - Manufacturers may be held responsible for adverse events if designs reflect little or no consideration of possible use errors.
   - Human errors can be predicted and reduced or eliminated by design.
   - Risk-management tools used for traditional engineering and safety analysis are useful or essential for human factors related problems.
2. *Increased focus on the use environment.* Some of the problems and risks with devices reside not in the devices, per se, but in the context of other equipment and the environment. Some examples include
   - Many devices in a hospital room (e.g., patient monitors, infusion pumps, ventilators) have auditory alarms designed to alert caregivers that immediate attention is needed. However, one of the common complaints regarding device alarms is that one device's alarms are not distinguishable from another. The possible confusion of alarms has been an area of concern for several human factors standards bodies (IEC, 2003).
   - Electrical connections for various devices are easily confused and have allowed unintended connections to other devices or sources of power (e.g., inadvertent connection of EKG leads to 120 VAC).

- User interfaces among similar devices are most often not consistent, leading to increased risk of errors, increased time to administer treatment because of confusion, or increased training costs.

  Various standards bodies (e.g., AAMI, IEC, ISO) have been attempting to develop user-interface standards aimed at harmonizing various user-interfaces elements (such as alarms, symbols, etc.) among diverse devices. Attempts to design comprehensive common user interface standards so far have failed for at least two reasons:

- Many manufacturers believe their user interfaces have a competitive advantage over the competitors, and are reluctant to share their ideas as part of a standard.
- Software user-interface technologies have been changing continually over the past several years. Software user interfaces developed during such times may provide standards that are considered obsolete.

3. *Improved medical error data.* One of the biggest challenges in dealing with use errors is the lack of data to make informed decisions regarding risk levels. Consequently, teams are forced to use their own judgments (and imagination) to uncover potential failures and assess designs. In recent years, the FDA, the US Veterans Administration, and others have initiated computerized event-recording systems that enable caregivers and health care facilities to anonymously report accidents and near misses to a database. One of the anticipated benefits of this information is that these sources will eventually become a means for better estimating human error on a wide array of medical devices.
4. *Improved tools.* The emergence of software tools to enable the more rapid construction of FMEAs, fault trees, and RCA documents have greatly improved the ability of small teams to perform risk-analysis procedures in much less time than using manual methods. With similar advances in the next few years, it is likely that future tools will provide even greater ease of use and extended functionality that will assist users in identifying faults (in similar devices), and can automatically monitor for redundancy, automatic tracing of requirements or actions, and so on.

References

Andre, A. D., Jorgenson, D. B., Froman, J. A. Snyder, D. E., & Poole, J. E. (2004). Automated external defibrillator use by untrained bystanders: Can the public-use model work? *Prehospital Emergency Care*, 8(3), 284–291.

American National Standards Institute/Association for the Advancement of Medical Instrumentation (2001) *Human Factors design process for medical devices.* HE 74:2001. (AAMI guidance document)

Bailey, R. W. (1983). *Human error in computer system.* Upper Saddle River, NJ: Prentice Hall.

Code of Federal Regulations (2004, April). *Quality system Regulations, Medical Devices.* 21CFR820, Title 21, volume 8.

Cline, A. (2004) *Prioritization Process Using Delphi Technique.* Retrieved on Feb 1, 2004 from http://wwww.carolla.com/wp-delph.htm.

DeRosier, J. (2002). Using health care failure mode and effect analysis: The VA National Center for Patient Safety's prospective risk analysis system. *The Joint Commission Journal on Quality Improvement, 28*(May), 248–267.

Food and Drug Administration. (2000). *Medical device use safety: Incorporating human factors engineering into risk management*, FDA Guidance document 1497, July 18, 2000.

International Electrotechnical Commission (2004). International Standard 60601–1–6:2004 Medical electrical equipment—Part 1- General requirements for safety: Collateral Standard: Usability

International Electrotechnical Commission (2003). International Standard 60601–1–8:2003 Medical electrical equipment—Part 1–8 General requirements for safety: Alarm Systems—requirements, tests and guidelines—General requirements and guidelines for alarm systems in medical electrical equipment and in medical electrical systems.

International Organization for Standardization (2000). 14971:2000 *Medical devices—Application of risk management to medical devices.*

Israelski, E. W., & Muto, W. H (2003). *Use-error focused risk analysis for medical devices: A case study of the Therac–25 Radiation Therapy System.* Proceedings of the 47th Annual Meeting of the Human Factors and Ergonomics Society. San Diego, CA:

Kirwan, B., & Ainsworth, L. K.(Eds.). (1992). *A guide to task analysis.* New York: Taylor and Francis.

Leveson, N. (1995). *Safeware: System safety and computers.* Boston, MA: Addison-Wesley.

McDermott, R. E., Mikulak, R. J., & Beauregard, M. R.(1996). *The basics of FMEA*. Portland, OR: Productivity, Inc.

Miller, D. P., & Swain, A. D. (1987). Human error and human reliability. In G. Salvendy (Ed.), *Handbook of human factors* (pp. 219–250). New York, NY: Wiley.

North Carolina Department of Environmental and Natural Resources, Office of Organizational Excellence. (2002). Retrieved on Febrauary 1, 2004, from http://quality.enr.state.nc.us/tools/fishbone.htm

Reason, J. (1990). *Human error*. Cambridge, UK: Cambridge University Press.

Reason, J. T. (1997). *Managing the risks of organizational accidents*. Hampshire, UK: Ashgate Publishing Co.

Schmidt, M. W. (2004). The use and misuse of FMEA in risk analysis. *Medical device and diagnostics industry*, *26*(3), 56–58.

PathMaker Consensus Builder Too. Retrieved on Feb.1, 2004 from http://www.skymark.com

Stamatis, D. H.(1995). *Failure mode and effect analysis*. Milwaukee, WI:ASQ Quality Press.

Vesely, W. E., Goldberg, F. F., Roberts, N. H., & Haasl, D. F. (1981) *Fault tree handbook* (NUREG–0492). Washington, DC: US Nuclear Regulatory Commission.

APPENDIX 35–A

DEFINITIONS AND ACRONYMS FOR RISK MANAGEMENT

Abnormal use.	Sabotage, recklessness, unqualified users, off-label use or unforeseeable use.
AER.	Adverse Experience Report: Description of an medical adverse event that resulting in death or serious injury.
ALARP.	As Low As Reasonably Practical: Region of risk acceptability with justification.
BAT.	Best Available Technology: The most current technology for mitigating design risks.
Contextual inquiry.	Process of observing and working with users in their normal environment to better understand the tasks they do and their workflow.
Detection.	The estimation of whether a particular hazard or defect will be discovered and mitigated prior to finished product release, distribution, or use.
FMEA.	Failure Mode and Effects Analysis: A systematic approach that identifies potential failure modes in a system caused by either design or process deficiencies. It also identifies critical or significant design or process characteristics that require special controls to mitigate or detect failure modes. The procedure involves making estimates for the likelihood of each potential failure mode and to classify the severity of each resulting hazard on user or patient safety.
FMECA.	Failure Mode and Effects Criticality Analysis: Same as FMEA with an emphasis on analyzing criticality.
Foreseeable misuse	Recognized potential misuse or misapplication of product.
Formative usability testing.	Usability testing that is performed early with simulations and first working prototypes and explores whether usability objectives are attainable, but does not have strict acceptance criteria.
Frequency.	The estimation of the likelihood that a particular hazard or defect will occur.
FTA.	Fault Tree Analysis: A deductive, top-down method of analyzing system design and performance. It involves specifying a top event to analyze, followed by identifying all of the associated elements in the system that could cause that top event to occur. Fault trees provide a convenient symbolic representation of the combination of events resulting in the occurrence of the top event. FTAs are generally performed graphically using a logical structure of AND and OR gates.
Harm.	Physical injury or damage to the health of people, or damage to property or the environment
Hazard.	A potential source of harm.
HHA.	Health Hazard Assessment: A systematic analysis to anticipate, identify, quantify, and recommend controls for health hazards associated with the occupational environment.
HA. Hazard Analysis:	Analysis of sources of harm to patients and other users of a medical devices.

HACCP.	Hazard Analysis and Critical Control Points: Used to analyze critical process control points in food processing.
HAZOPS.	Hazard and Operability Studies: Systematic analysis of industrial processing flows for deviations from design intent.
Intended use, intended purpose.	The objective intent of the manufacturer as described in labeling, advertising materials, or oral or written representations.
Prior identified criteria.	Documented standards or regulations related to acceptable risk limits.
Prototyping and iterative design.	Iterative design involves the rapid turnaround of user-interface prototypes or simulations that are usability tested and improved in an iterative cycle until usability objectives are attained.
Residual risk.	Risk remaining after protective measures has been taken.
Risk.	(RI) Risk Level or Risk Index: Combination of the probability of occurrence of harm and the severity of that harm.
Risk acceptability.	An analysis of effectiveness of design controls and mitigations on the acceptability of remaining product risk leading to further design actions or the conclusion that a product is safe.
Risk analysis.	Systematic use of available information to identify hazards and to estimate the risk.
Risk control.	Process through which decisions are reached and protective measures are implemented for reducing risks to, or maintaining risks within, specified levels.
Risk evaluation.	Judgment, on the basis of risk analysis, of whether a risk that is acceptable has been achieved in a given context based on the current values of society.
Risk management	The systematic application of management policies, procedures and practices to the tasks of analyzing, evaluating, and controlling risk.
RPN.	Risk Priority Number—The result of multiplying quantitative estimates of three parameters: fault frequency, hazard severity and detection.
Severity.	Measure of the possible consequence of a hazard.
Summative usability testing.	Usability testing that is performed in the late stages of design. These tests include verification and validation and it is a recommended best practice to have formal acceptance criteria (e.g., usability objectives for human performance and satisfaction ratings).
Task analysis.	Task Analysis is a family of systematic methods that produce detailed descriptions of the sequential and simultaneous manual and intellectual activities of personnel who are operating, maintaining or controlling devices or systems.
Usability inspection methods.	Inspection methods involve analytical reviews and systematic walkthroughs of user interactions with simulated or working user interface designs looking to uncover usability problems.
Usability objectives.	Usability Objectives (goals) are a desired quality of a user device interaction that may be expressed in written form, stipulating a particular usability attribute (e.g., task speed) and performance criteria (e.g., number of seconds).
Usability testing	(UT). UT Procedure for determining whether the usability goals have been achieved. Usability tests can be performed in a laboratory setting, in a simulated environment or in the actual environment of intended use.
Use environment analysis.	The actual conditions and settings in which users interact with the device or system.
Use error.	Use error is characterized by a repetitive pattern of failure that indicates a failure mode that is likely to occur with use and thus has a reasonable possibility of predictability of occurrence. Use Error can be addressed and minimized by the device designer, proactively identified through the use of techniques such as usability testing and hazard analysis.
Use error risk analysis.	Analysis focused on the use error component of fault and hazard analysis for medical devices.
User.	The person who interacts with the product.

User error. User error is characterized by an isolated pattern of failure that indicates a failure mode that is due to fundamental errors by humans and has no reasonable possibility of being predicted. User error is non-preventable and non-addressable by the device designer.

User group. Subset of intended users who are differentiated from other intended users by factors such as age, culture, or expertise that are likely to influence usability.

User interface. The hardware and software aspects of a device that can be seen (or heard or otherwise perceived) by the human user, and the commands and mechanisms the user uses to control its operation and input data.

User profiles. Summary of the mental, physical and demographic traits of the end-user population as well as any special characteristics such as occupational skills and job requirements that may have a bearing on design decisions.

36

WORK SYSTEMS AND PROCESS ANALYSIS IN HEALTH CARE

Ingo Marsolek & Wolfgang Friesdorf
Technical University of Berlin, Germany

CHARACTERISTICS OF WORK SYSTEMS IN HEALTH CARE

Momentary Situation

The economic dimension of the health care sector is becoming an enormously influential factor in industrialized countries all over the world. In Germany and the United States the health care sector represents more than 10% of the entire gross domestic product (Organisation for Economic Cooperation and Development, 2003). The following four elements characterize recent developments in health care:

1. *A steadily increasing cost pressure.* For example, in Germany the overall costs for the largest single health care expense, that is, the patient treatment in hospitals, have increased by more than 35% between 1991 and 1999. This large increase in costs has occurred despite a systematic reduction of hospital beds and staff. Reasons for the increasing cost pressure include a continuously growing number of patients combined with growing costs per treated case. Both reasons result from great medical and technological progress and major demographic changes, including increasing number of elderly, multi-morbid, and chronically ill patients (Arnold, Litsch, & Schellschmidt, 2002).

   Nearly all industrialized nations are unable to continue to cope with this large increase in health care costs. Therefore, the introduction of performance-based health care financing systems is witnessed all over the world. These systems aim to motivate health care institutions towards a more efficient usage of available resources. However, it is important that efficiency efforts do not overshadow patient safety, which should have the highest priority (Gorschlueter, 1999).

2. *Rising quality demands.* Quality assurance has a long tradition in medicine. For example, ward rounds and the consultation of medical colleagues have been used as internal quality-assurance procedures in Germany (Eichhorn, 1997). Voluntary external quality comparisons are widely accepted within medical specialties, such as perinatology and neonatology in Germany since 1975 (Jaster, 1997). In many industrialized countries, the increasing cost pressure and the emphasis on performance-based health care financing systems have led to legislative efforts towards more systematic external quality control. Within the health care sector this development has

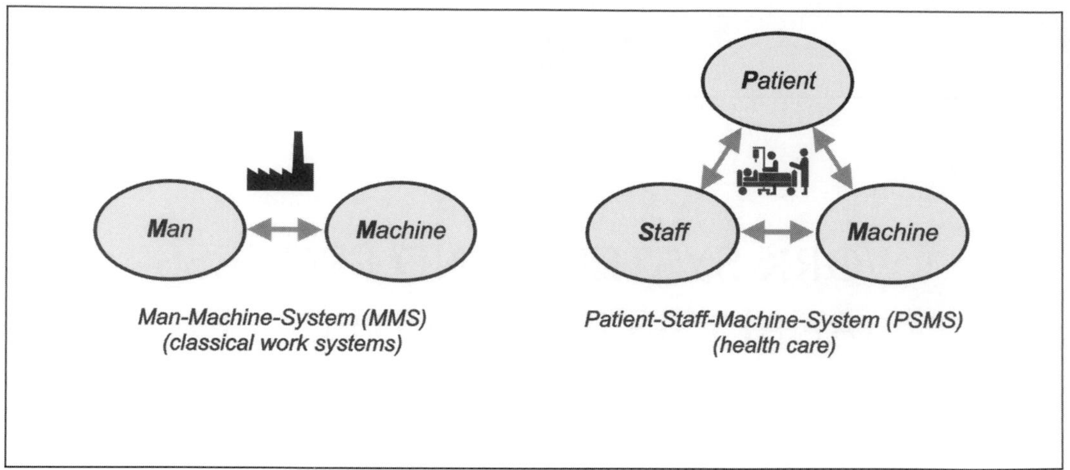

Figure 36–1. The classical man–machine–system (MMS) versus the patient–staff–machine–system (PSMS) for health care.

also resulted in a paradigm change in the understanding of quality; this change of paradigm has already revolutionized work systems in the industrial sector (Vorley & Tickle, 2001). We have also observed a growing emphasis on total quality management concepts, which are based on systematic implementation of "continuous quality improvement" instead of the traditional "quality control" (Eichhorn, 1997).

3. *Continuously growing patient demands.* Over the past few years there has also been a significant increase of patient expectations concerning treatment processes. Reasons for this include a growing number of media reports on medical innovations (Viethen, 1995), a stronger critical reflection about medicine itself (Elfes, 1996), steadily increasing health care insurance costs, and willingness of people to pay for additional health care costs on a private basis.

The health care sector seems to evolve from a sellers' market to a buyers' market, in which the patients and the insurance companies become the customers of the service provider hospital. This leads to growing competition within the health care sector, not only based on the medical patient treatment but also on adequate customer relationships and hotel-like hospital services (Gorschlueter, 1999).

4. *Increasing complexity.* The specificity of patients, ethical issues, and the dynamic characteristics of patient treatment make health care a complex system. Complexity increases through continuous growth of medical and technological innovations and changes as well as the demographic changes described above (Friesdorf, Gross-Alltag, Konichezky, & Schwilk 1993). Three different attempts can be identified for dealing with this complexity: (a) professional specialization of the medical staff (e.g., to surgeons, anesthesiologists, OR nurses etc.), (b) diagnosis-oriented treatment standards for specific diseases (defining *what* has to be done; e.g., critical pathways, clinical pathways, etc.), and (c) task-oriented process standards for specific assignments (specifying *how* it has to be done; e.g., standard operating procedures, etc.; Graybeal, Gheen & McKenna, 1993; Pearson, Goulart-Fisher & Lee, 1995; Ramos & Ratliff, 1997, and others).

Work Object: Patient

The work object in classical work systems (e.g., industry) is a product that has to be produced, maintained, or repaired, and for which the classical man– machine–system can be used. In contrast to that, the work object in health care is the treated patient, who is not comparable to a classical product. For its description the patient–staff–machine–system has to be used as developed by Friesdorf and colleagues (1993) and shown in Figure 36–1.

Therefore, in health care the work object patient does not only result in a much closer costumer–supplier relationship but also ethical aspects directly related to the patient treatment. In addition to that, the treatment of a multiple number of different patients with many different diseases, as well as an individual health conditions, results in a much higher task complexity than in classical work systems. In industry, normally only a very specific number of product alternatives have to be handled.

Organizational Structure

Today's work organization in health care has been mainly influenced by progress in medicine and technology within the past decades. The dramatic increase of medical knowledge has resulted in the need for professional staff specialization, and has led to a corresponding change in the organization, that is, an organization of highly specialized health care institutions and hospital departments (Friesdorf et al., 1993).

The increased medical specialization has occurred beside the traditional separation between physicians, nurses, and administrative personnel. This often results in open conflicts, not only at the management level but also within direct patient treatment processes.

Growing complexity of patient treatment processes results in the need for a systematic integration of many different medical professions for the treatment of a single patient. Therefore, flexible intensive cooperation is necessary between different medical professions during the entire patient treatment (e.g., interdisciplinary team work in the ICU and OR as described by Weissauer & Hirsch, 1994, or Gaba, Fish & Howard, 1994). Unfortunately cooperation is typically restricted to a specific event or to a specific patient status and therefore is not represented in the official organizational structure of medical work systems. Interdisciplinary, autonomously run treatment centers responsible for the entire patient treatment process are rare.

Process Organization

Within medical work systems three different types of work processes can be differentiated (Friesdorf, Marsolek & Goebel, 2002):

- Primary work processes: The direct patient treatment processes (e.g., a surgical intervention, an intensive care treatment).
- Secondary work processes: Processes directly supporting the patient treatment processes (e.g., the OR management, laboratory diagnostics).
- Tertiary work processes: Processes indirectly supporting the patient treatment processes (e.g., the sterilization, cleaning, catering).

Because secondary and tertiary work processes have low complexity, normal analysis and optimization strategies for classical work systems (like standardization, statistical process control, outsourcing, etc.) can be used. But the complexity of the work object patient does not allow an organization of the direct patient treatment processes similar to the well known "flow (or line) production" of modern industry or service companies. Instead it is still organized like a rather traditional "workshop production." Sometimes the work object patient has to be moved between different locations (e.g., ER, OR, ICU, radiology, etc.). But sometimes also specific treatment procedures have to be performed at the work object patient, which is kept within the same location (e.g., an x-ray in the ICU).

The treated patients themselves have a double role. For a successful patient treatment they have to be seen not only as passive work objects but active "co-workers" at the same time.

In all primary work processes it is always the work object patient with its diagnosis (the initial patient status), that defines the "task objective" and therefore what has to be done (see also Figure 36–2). But an optimal task completion (the desired final patient status) can only be achieved by a process organization that is free of process deficits ("how to do things," wherefore specific resources are needed).

Using this model, the "process quality" of the patient treatment is clearly defined as the proportion between task completion and task objective, while the "process efficiency" is the proportion between the produced process quality and the used resources. As soon as it is clearly defined what has to be done with a patient, the loss of process quality or process efficiency can only have one and the same cause: The existence of process deficits, which have to be systematically eliminated for a safe and efficient patient treatment.

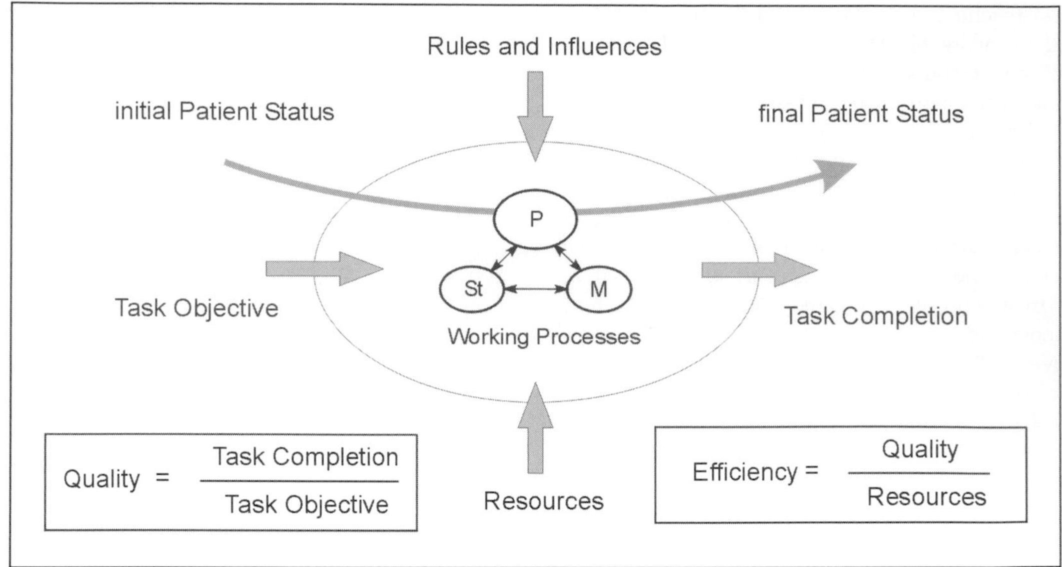

Figure 36–2. Medical work processes and their context in a work system.

Characteristic Problems

Two characteristic problems dominate medical work systems.

First, the high complexity of the direct patient treatment requires a process organization that is individually adapted to the disease and patient requirements and the potential variations and changes in the patient treatment itself. As a result, medical work processes are defined by a much higher individuality and dynamic variation than processes in classical work systems, which can easily be predefined as process chains and controlled on a statistical basis with the help of characteristic benchmarks (Gaitanides, Scholz, Vrohlings, & Raster, 1994; Vorley & Tickle, 2001).

Medical work systems are mainly defined by so called process flows (see Figure 36–3). Even work processes for a planned patient treatment can often not fully be predicted and predefined right from the initial patient status because of the complexity of the work object patient. For instance, a new disease requires the reconsideration of the previously used treatment strategy. Additionally a large number of unpredictable incidents dominates medical work systems. See, for example, the crisis situations in anesthesiology described in detail by Gaba, and colleagues (1994). Both facts result in a high number of process flow distributions with according process flow alternatives, which often have to be adapted again to patient-group-specific or even individual requirements.

This does not mean that medical treatment processes are not preplanned as systematically as possible and controlled by statistical benchmarks. This can be accomplished, for instance, based on diagnosis-related measurement and interpretation of the patient's physiological data. But we are still far from using a single statistically based process control for all patients because of significant differences within the medical process flows of patients with various diseases and individual variations.

The number of process intersections (caused by the characteristic workshop production described earlier) leads to the risk of information losses. Nearly all patients require disease- or patient-specific resources with physicians and nurses of different medical specialties. Therefore, medical process flows have a high number of process intersections (see also Figure 36–3). Either the patient or patient information has to be transferred within those process intersections between different departments or clinical staff members. The process intersections can therefore lead to a loss of information.

In addition, it is extremely difficult to protect all of those process intersections systematically against information losses. Again, the reason for this is the complexity of patient treatment processes, which often require the design of disease- or

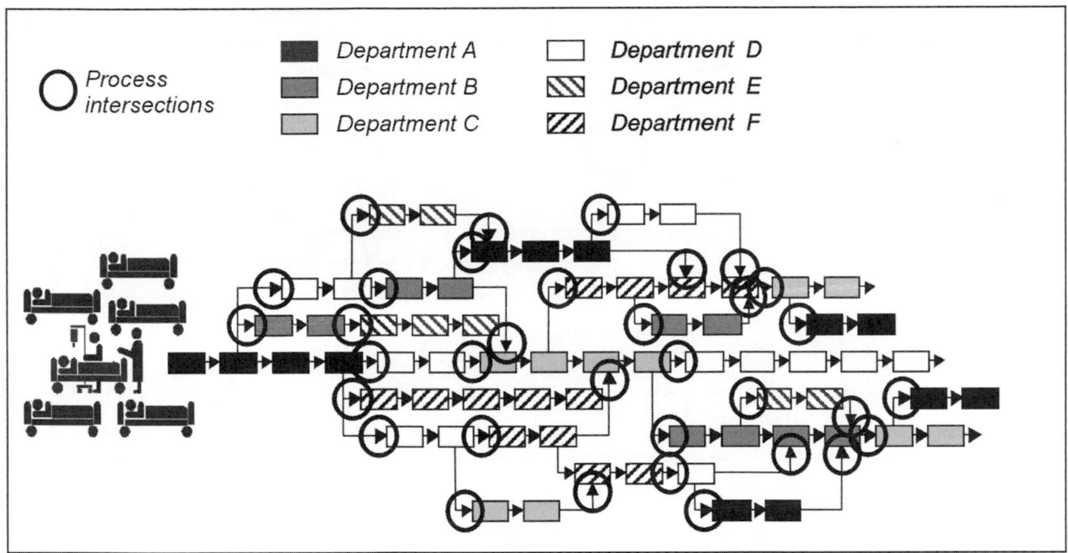

Figure 36–3. Characteristic problems of work systems in health care: Process flows and process intersections.

patient-specific process intersections (Friesdorf et al., 1994).

GUIDELINES FOR THE ANALYSIS AND OPTIMIZATION OF MEDICAL PROCESSES

Rising cost pressure, growing quality and patient demands, as well as an increasing complexity characterize work systems in health care. Therefore, improvements of efficiency are necessary, while quality and safety of the patient's treatment should still remain the highest priority.

But for direct patient treatment processes a top-down oriented tayloristic rationalization approach is not possible. The flexible self-organization of medical expert teams is necessary for mastering the existing complexity in medical work systems. Classical rationalization attempts, such as pure standardization and outsourcing of specific work processes, can only be used for less complex work tasks than the direct patient treatment (e.g., sterilization, cleaning, and catering processes). A different approach is needed for optimizing primary work processes—especially complex process intersections—in medical work systems.

The best task definition (*what* has to be done) cannot lead to the desired result if it is not realized with absolutely correct work processes (*how* to do things). With each incorrectness (in a decision, communication etc.) a loss of quality or efficiency is more likely to occur (see also Figure 36–2). Therefore the existing medical task orientation has to be systematically supported by building up an additional management competence within the medical staff based on the already existing interdisciplinary team work, that is necessary to deal with the high complexity of the patient treatment. Besides the classical work organization (top-down) the establishment of a sustainable process orientation becomes more important. This can only be achieved by using an additional bottom-up oriented approach, similar to the idea of business process reengineering used for industrial work processes (Hammer & Champy, 1993). But this approach has to be carefully adapted to the requirements of medical work systems. A systematic reduction of process complexity has to be the basis for a sustainable optimization of medical work processes. Furthermore an additional management competence has to be built up within the involved staff members besides their professional specialization. Interdisciplinary staff teams have to learn how to improve their processes and process intersections on their own as a basis for a continuous process improvement.

For this aim we have developed and tested the following seven project steps in a variety of applications to implement this process view of medical work systems (Marsolek & Friesdorf, 2002).

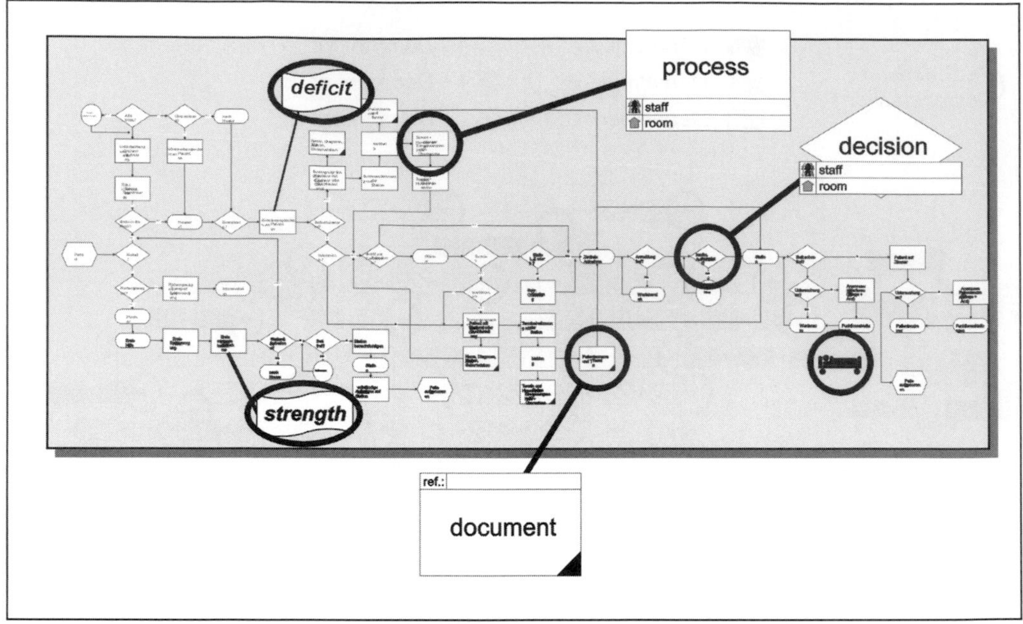

Figure 36–4. Visualization of the process flow as a process flow diagram.

Project Preparation and Staff Participation

- **Define the project objectives** to be achieved depending on the momentary situation of the work system together with the hospital management (e.g., establishing process transparency, increasing process efficiency, improving process quality, improving staff or costumer satisfaction, initiating process standardization, etc.).
- **Define the boundaries of the work process** to be analyzed and distinguish between process flows with different work objects (different patients; see also Figures 36–1 & 36–2) for all primary work processes (i.e., direct patient treatment).
- **Organize kick-off meetings** for all directly involved medical divisions and staff members to integrate the knowledge and improvement ideas of all process experts as early as possible in the process flow optimization. At the same time, this systematic staff integration increases the later acceptance for organizational and technological changes (Fuermann & Dammasch, 1997; Gaitanides, Scholz, Vrohlings & Raster, 1994; Vorley & Tickle, 2001).
- **Collect all existing information** about the analyzed work system.
- **Describe the physical layout** of the medical work system.

Process Flow Visualization and Verification

- **Observe the existing process** flow to get a first overview.
- **Describe and visualize the existing process flow** on large sheets of paper as shown in Figure 36–4. The timely order of sequential process steps is following the horizontal time axis from left to right, while all simultaneous (parallel) or alternative activities are displayed above or below. The aim of the visualization is the establishment of process transparency for all staff members involved. The symbols used have to be understood as the vocabulary of a process language, that should be learned as easy and fast as possible (Fuermann & Dammasch, 1997; Gaitanides et al., 1994).

Therefore the symbols used are kept as simple as possible (similar to the European Norm and International Organization for

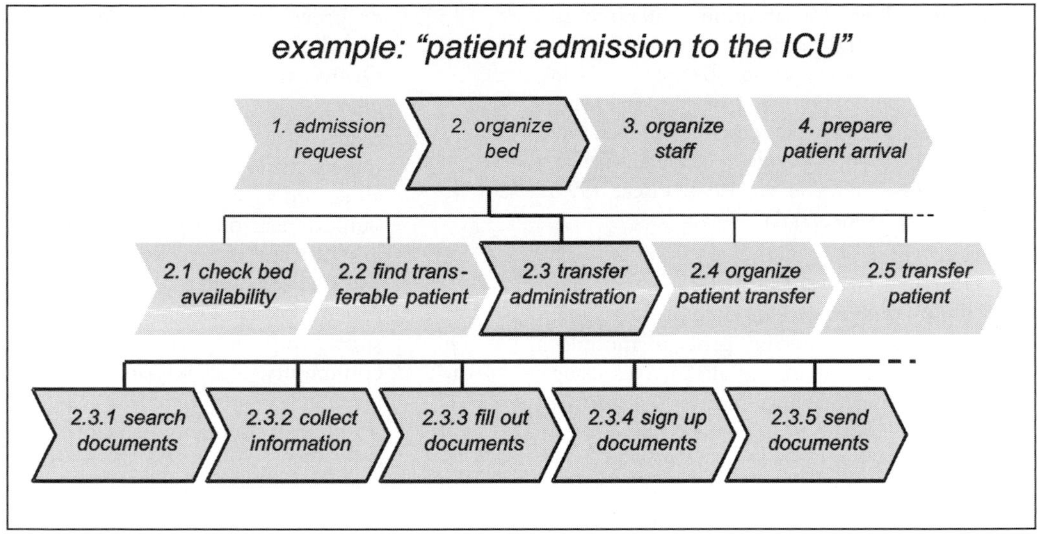

Figure 36–5. Hierarchical structuring of the process flow into process modules.

Standardization proposals), but also especially designed for the requirements of medical work systems (e.g. using additional symbols for the patient's presence).

- **Encourage participation** of as many interdisciplinary staff members as possible. In this context, one-person interviews can be used as well as moderated group discussions. A public display of the process flow diagram allows all process experts to participate anonymously.
- **Verify and revise the process-flow diagram constantly.** The correctness of the process-flow diagram has to be constantly checked together with the involved staff in one-person interviews and moderated group discussions. All necessary corrections identified by the staff have to be integrated into the process flow diagram again. This has to be done until a process-flow diagram is achieved that realistically displays the existing process flows and has as many details as necessary for the understanding of all staff members involved.

Hierarchical Structuring and Process Flow Quantification

- **Define process modules** by grouping all existing activities (processes and decisions) within the verified process-flow diagram to superior process tasks or modules as shown in Figure 36–5 (Gaitanides et al., 1994).
- **Observe and estimate the proportional distributions** after each decision or parallel distribution and documentation of the percentages in the verified process-flow diagram.
- **Elaborate a catalogue with all documents** used for the verified process-flow diagram.
- **Quantify all process modules** (process tasks) systematically by using the seven dimensions ("Ms") of the Ishikawa diagram (Vorley & Tickle, 2001) for each identified process module: men, machines, measurement, methods, milieu, material, and management.

Identification of Process Strengths and Weaknesses

- **Document all characteristic process strengths and weaknesses** that are directly influencing the efficiency and quality of the analyzed work process (see also Figure 36–2), together with the staff involved. Consider patient needs not only from the medical point of view but also from the costumer perspective by interviewing representative patient groups. Use the process-flow visualization, hierarchical structuring, and process flow quantification (described above) as an objective basis in combination with some of the following optimization strategies described in literature (Chang, 1996; Fuermann & Dammasch, 1997; Gaitanides et al., 1994; Plsek, 1999; Vorley & Tickle, 2001):

1. *Process data analysis.* Identify changes within one process module over the time or compare specific process modules from different process alternatives using process performance metrics (e.g., times, costs, quality, costumer satisfaction, etc.) for the identification of unexpected or unwanted differences. Also use specific data analysis tools such as process- or problem-specific check sheets, control charts, scatter diagrams, and so on.
2. *Process benchmarking.* Use the results from a process data analysis for the comparison of process module alternatives from different departments or medical institutions for the identification of a best practice that is then adapted to all other work systems (Camp, 1989, 1995). In this context the existing differences between various departments and medical institutions (e.g., the number and constellation of treated patients, architectural and technological possibilities, etc.) do not always allow a definite identification of one best practice, which makes a systematic adaptation to work system specific requirements even more essential.
3. *Information flow analysis.* Document the needed, checked, and generated information for all activities (process modules) within the analyzed process flow for an elimination of all redundant or unnecessarily generated or checked information.
4. *Value assessment.* Check the added value for all activities (process modules) within the analyzed process flow using such dimensions as time, cost, quality, and costumer satisfaction for the next internal or external costumer. The goal is to eliminate all those activities that are either causing extra work because of avoidable mistakes or are not leading to an added value (not fulfilling a superior work task or process module; also see the Hierarchical Structuring and Process Flow Quantification section).
5. *Output assessment.* Analyze the requirements of all external and internal costumers within the entire process flow for an according adaptation of each activity's output, therefore harmonization of existing process intersections. Examples of external costumers of a hospital are patients, patient families, health care insurance companies, scientific societies, the community and so on. Examples of internal costumers of a hospital are hospital accountants, quality assurance, costumer relations, and so on. Systematically collect their opinions through standardized or nonstandardized interviews and feedback questionnaires and make this information accessible for all involved staff members. Initiate moderated workshops for improving the current situation.
6. *Process check lists.* Use the following proposal from Fuermann and Dammasch (1997) for all process modules, which are mainly based on the approaches described already:

    1. Avoid all unnecessary activities.
    2. Avoid unnecessary external controls and introduce self-control.
    3. Identify unnecessarily divided processes and integrate them into whole tasks.
    4. Arrange processes in parallel when possible.
    5. Standardize processes when helpful, otherwise allow process alternatives.
    6. Identify hazardous work conditions and improve them.
    7. Identify unnecessary stockage and waiting times and reduce them.
    8. Identify unnecessary transportation ways and shorten them.
    9. Identify process times that are too long and shorten them.
    10. Identify shortages in rooms and technology and increase their availability.

- **Visualize the identified process strengths and weaknesses** in the process flow diagram and hierarchical structure.

- **Group all process strengths and weaknesses** considering the desired optimization described in the following section, Development of an Optimized Process Flow:

1. Unnecessary and redundant processes: For a "key process concentration."
2. Process strengths and deficits: For a "process deficit elimination."
3. Stable and unstable processes: For a "process flow stabilization."

Development of an Optimized Process Flow

- **Develop an optimized process flow** together with the staff members involved (including organizational, technical, and architectural changes if necessary) for improving the quality and efficiency of the analyzed work process (see also figure 36–2) together with the involved staff. Also, keep medical patient needs and costumer aspects in mind while using the following strategies:

1. Key process concentration: Avoid all unnecessary and redundant processes, that do not help to fulfill superior process tasks (process modules; also see the Hierarchical Structuring and Process Flow Quantification section).
2. Process deficit elimination: Eliminate all identified process deficits by developing an optimized process flow that still includes the already existing process strengths.
3. Process flow stabilization: Develop an optimized process flow that stabilizes all identified unstable processes but also supports the already existing stable processes.

If the development of an alternative process flow should not be based on a trial-and-error strategy alone, the comparison results from a systematic **process benchmarking** (see also Identification of Process Strengths and Weaknesses section) often seem to be useful at this point. But a classical process benchmarking cannot consider work-system specific restrictions or management goals as already described in the previous chapter. It also risks that the entire process flows of nearly all compared work systems are rejected as "worse practices," while they still might contain partly excellent process segments.

For this reason our preferred optimization strategy for medical process flows is a systematic problem-specific process module comparison used only for those process modules where process weaknesses have been identified. Furthermore, the comparison itself is not reduced to one specific process benchmark and identification of only one best-practice solution. Instead the comparison of the process alternatives from all participating or already analyzed work systems is based on their characteristic process weaknesses and strengths profile. A systematic analysis of all the identified process-flow alternatives, and reflection on their already experienced weaknesses and strengths, is then used for a process-flow optimization, in which the most promising solutions are adapted to the individual requirements of each work system and process module together again with the staff involved (Marsolek & Friesdorf, 2001):

- **List all possible improvement ideas** based on the proposals of as many staff members as possible.
- **Estimate the existing improvement potential** based on the quantified process flow together with the involved staff members.
- **Elaborate realization strategies for defining** the next steps as concrete as possible (e.g., *who* has to do *what* until *when*).

Evaluation of Process Flow Changes and Initiation of Continuous Process Control

Within this project phase, characteristic benchmarks have to be defined that can be collected before and after the actual process redesign is realized; not only for an objective evaluation of the process-flow changes, but also for forcing the actual realization and initiating a continuous process control.

- **Define goal-performance metrics** that are based on the project objectives and only influenced by specific process strengths or weaknesses, but not dependent on other system factors (e.g., the constellation of treated patients). While all benchmarks have to be defined individually for each process optimization, they can normally be chosen from the following four general benchmark categories (similar to the process data analysis described in the Identification of Proccess Strengths and Weaknesses section):

1. Times (e.g., waiting times, process times, usage times, etc.).
2. Costs (e.g., staff costs, room costs, treatment costs, etc.).
3. Quality (e.g., number of mistakes, missing information, complaints, etc.).
4. Costumer Satisfaction (patients, external physicians, internal staff etc.).

- **Define the data collection system,** which is mainly influenced by the possibilities of each work system and the aim either to assess process-flow changes with the help of the staff involved or to measure process-flow changes by hard facts. In general the following four types of collection are possible:

1. External observation and documentation (very objective, but often needs additional resources and is not always accepted by the observed staff).
2. Self-documentation (less objective, but normally does not need additional resources, while it means additional workload for the involved staff).
3. Interviews or questionnaires (allow only a subjective assessment of the achieved process-flow changes, but can be used with little additional workload when standardized; normally well accepted when done anonymously).
4. Automatic documentation (e.g., with an already used data-management system; most objective, while it means only very little additional workload when successfully implemented and accepted by the staff involved).

Initiation of a Continuous Improvement Process

During the entire project, a systematic staff participation is not only necessary for achieving realistic analysis and optimization results, but also for a parallel staff qualification from "process segment experts" to "process partners" and "process managers." Therefore, chosen staff members have to be integrated step-by-step into the analysis as well as optimization work, while their additional qualification in parallel seminars has to guarantee that future optimization projects for other medical work processes will be done with only little or even without external coaching. Along with the project duration, the involved medical work system has to become a self-learning organization.

EXAMPLES OF APPLICATIONS

The seven-step participatory process analysis and optimization approach has been used successfully in a wide variety of medical process flows. Its broad acceptance in very different medical work systems is demonstrated in the first project example. The second project example shows how this participatory approach can be used for improving the architectural planning of medical work systems.

Usability and Acceptance of the Participatory Approach: Results from a Multi-Center Study

For evaluating the usability and acceptance of this participatory process analysis approach the procedure was tested in an international study (Marsolek & Friesdorf, 2003).

Method

In this project the participatory process analysis procedure was used with the help of an external moderator for the optimization of the process intersection concerning patient admission from the OR or ward to the intensive care unit within the following five work systems:

- Charité University Hospital of Berlin (Germany).
- University Hospital of Innsbruck (Austria).

- Kaplan Medical Center Rehovot (Israel).
- Palo Alto Veterans Administration Health Care System Stanford (United States).
- Hokkaido University Hospital Sapporo (Japan).

Together with those clinical project partners, 10 sub-goals were developed for the usability and acceptance evaluation of the participatory process analysis approach:

Five sub-goals describing required project results

1. Staff members to achieve better understanding of analyzed processes.
2. Staff members to be able to bring their own problems into the process analysis.
3. Staff members to achieve better understanding of coworkers' problems.
4. Staff members to achieve better understanding of jointly optimized processes.
5. An optimized process to be developed, that in the opinion of staff members should replace the currently existing one.

Three sub-goals describing procedural requirements

6. Any additional workload acceptable for staff members.
7. Staff members to be able to discuss the linked work processes together.
8. Staff members to be able to participate adequately within the optimization.

Two sub-goals specifying a parallel staff sensitization

9. The results of the project to encourage staff members in the future to consider the analyzed processes critically and if necessary to initiate their improvement.
10. The results of the project to encourage staff members also to analyze and optimize other work processes.

At the end of each project all clinical staff members involved in the five different ICUs had the chance to assess to what extent those 10 project goals were achieved. The project goals were converted into statements and the extent of their accuracy was assessed on a scale between *totally wrong* and *totally correct* using an anonymous questionnaire. All data were transformed into a numerical assessment between 0 and 100 in order to compare the five different projects. The comparison was performed by using the average ssessment provided by the staff of each ICU.

Results

In all five clinical work processes a considerable improvement potential was found. For some process modules this improvement potential was already witnessed during the project itself, while other optimization ideas needed additional time in order to be implemented.

Work System Results. The most important and sustainable process optimization in all five clinical work systems was the elimination of all deficient information transfers between and within different divisions or staff members (e.g., request for patient admissions to the ICU, patient transfer to the ICU, or unscheduled transfer of ICU patients to a general ward). As a result, both efficiency and quality of the process flow were increased simultaneously: The efficiency was improved by eliminating unnecessary and redundant processes (e.g., multiple collection of lost information) and the quality was improved by ensuring a complete stable information transfer right from the beginning.

Impact of Changes on Project Partner. For the evaluation of our procedure, a total of 53 anonymous questionnaires were completed by the clinical staff members involved (16 in the United States, 9 in Israel, 12 in Austria, 8 in Germany, and 8 in Japan). Figure 36–6 shows the assessment results. The average assessment of the clinical staff members involved for the 10 statements is displayed on a scale between *totally wrong* (0) and *totally correct* (100).

For all 10 statements the average results from the 53 clinical staff members involved showed an assessment much closer to *totally correct* than *totally wrong*. Each of the 10 statements became an average assessment within the upper third of our scale (even >> 70). The three statements concerning procedural requirements (Items 6–8) had the lowest approval levels with average results between 72.1 and 76.7. The five statements concerning required project results (Items 1–5) achieved much higher approval with average assessments between 77.8

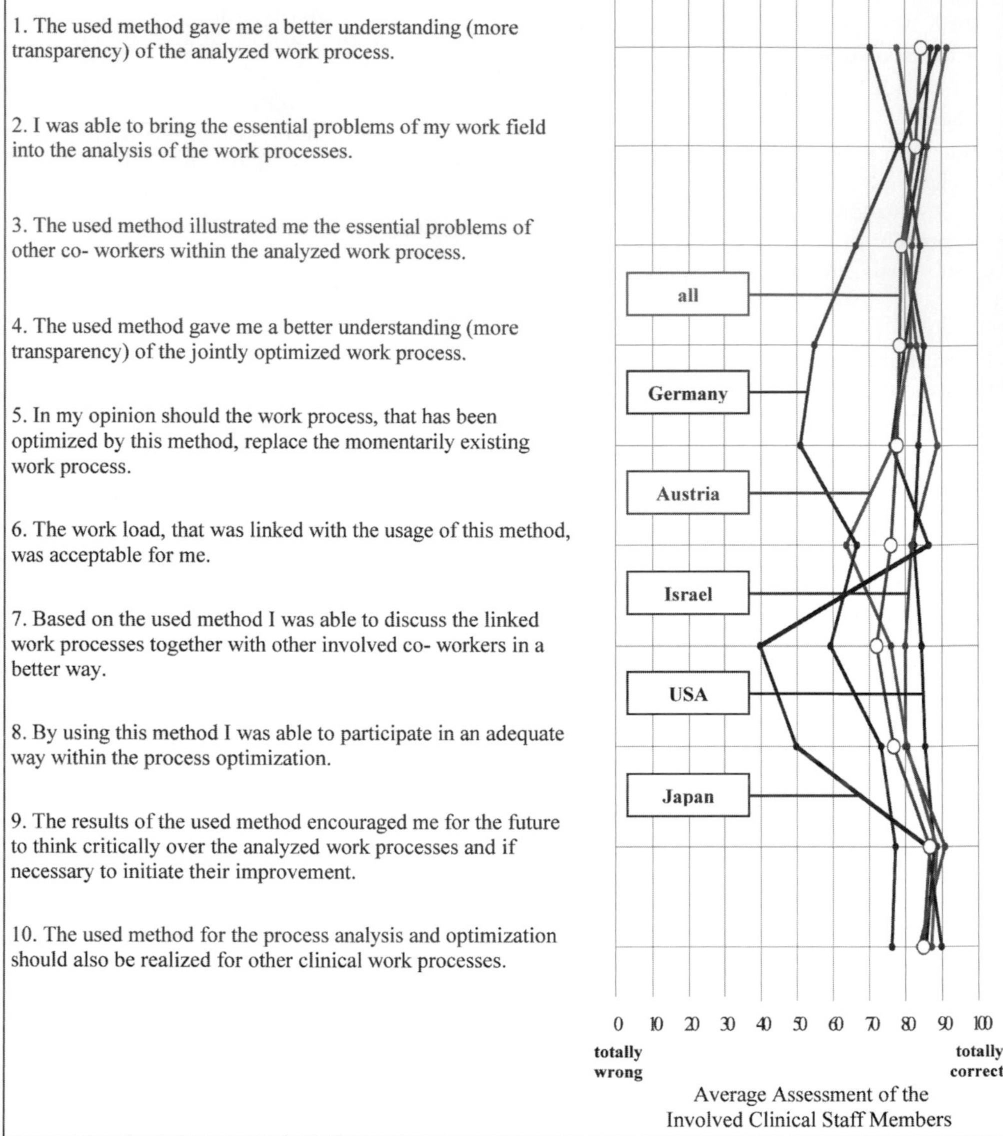

1. The used method gave me a better understanding (more transparency) of the analyzed work process.
2. I was able to bring the essential problems of my work field into the analysis of the work processes.
3. The used method illustrated me the essential problems of other co- workers within the analyzed work process.
4. The used method gave me a better understanding (more transparency) of the jointly optimized work process.
5. In my opinion should the work process, that has been optimized by this method, replace the momentarily existing work process.
6. The work load, that was linked with the usage of this method, was acceptable for me.
7. Based on the used method I was able to discuss the linked work processes together with other involved co- workers in a better way.
8. By using this method I was able to participate in an adequate way within the process optimization.
9. The results of the used method encouraged me for the future to think critically over the analyzed work processes and if necessary to initiate their improvement.
10. The used method for the process analysis and optimization should also be realized for other clinical work processes.
all
Germany
Austria
Israel
USA
Japan
0 10 20 30 40 50 60 70 80 90 100
totally wrong
totally correct
Average Assessment of the Involved Clinical Staff Members

Figure 36–6. Average assessment results for the used procedure.

and 84.2. The highest approval levels were obtained for the two statements concerning staff sensitization (Items 9 and 10), with average assessments comprised between 86.6 and 84.8.

Large differences are observed if one takes a closer look at the assessment results for each project separately. Within some ICUs, some goals regarding required project results (Items 1–5) or procedural requirements (Items 6–8) were not achieved as fully as in other clinical work systems. However, none of the projects showed negative results concerning staff sensitization (Items 9 and 10).

Conclusion

The participatory process-flow analysis allowed for a variety of process flow optimizations in all five clinical work systems. In addition, many different clinical staff members felt they had become more aware of similar problems and projects independent of their professional specialization. Although, in some projects, some goals concerning required project results or procedural requirements were not fully achieved, this did not affect staff sensitization, which is the most important basis for the development of process managers and the establishment of a self-learning organization.

Floor Plan of an OR Area According to Process Flow Requirements

Clinical work flow in the OR is influenced by the architectural design. Therefore the OR floor plan can either support or hinder the accomplishment of optimized work processes in regard to achieving a higher treatment efficiency and quality. The participatory process analysis approach described above was used in a project aimed at supporting the redesign of an architectural OR floor plan to be specifically adapted to existing and future work flow requirements (Marsolek & Friesdorf, 2000).

Method

In this project the participatory process-flow analysis approach was used to help the clinical staff directly involved in the affected work processes (e.g., surgeons, anesthesiologists, nurses) to identify their process requirements and improvement ideas. All members external to the interdisciplinary OR planning team (e.g., architects, engineers, and accountants) used the same procedure in order to develop a better understanding of the work processes within the old OR.

Work flows existing in the old OR (for patients as well as material and staff) and the associated process strengths and deficits were transferred into the first architectural floor plans of the new OR redesign. Potential process flow strengths and deficits of the planned OR redesign were then identified in order to develop an improvement plan together with the clinical staff members involved.

Results

By using this method the following architectural improvements were identified and implemented:

- The creation of a main storage room that can be supplied from the outside in order to reduce transportation times and improve storage inventories.
- Additional entrances for each OR theatre to adjacent storage rooms to reduce transportation times and separate the anesthesia and surgical work flows.
- A redefinition of the OR usage for different medical specialties enabling the introduction of already planned technology (e.g., operation robots for the orthopedic OR needing a larger OR theatre and additional storage rooms).
- A switch between storage rooms (that had windows) and OR staff rooms (that did not have windows), leading to a more comfortable work atmosphere.

Conclusion

By using the participatory process analysis approach process deficits in patient, staff, and material flows were avoided before finalizing the OR redesign plan. Those clinical staff members who were actively involved in the process analysis and optimization were motivated to be involved in future process optimizations upon completion of our project.

References

Arnold, M., Litsch, M., & Schellschmidt, H. (2002). *Krankenhaus-Report 2001*. Stuttgart, Germany: Schattauer Verlagsgesellschaft.

Camp, R. C. (1989). *Benchmarking: The search for industry best practices that lead to superior performance*. Milwaukee, WI: ASQC Press.

Camp, R. C. (1995). *Business process benchmarking: Finding and implementing best practices*. Milwaukee, WI: Quality Press.

Chang, R. Y. (1996). *Process reengineering in action*. London: Kogan Page Ltd.

Eichhorn, S. (1997). *Integratives Qualitaetsmanagement im Krankenhaus*. Stuttgart, Germany: Verlag W. Kohlhammer.

Elfes, K. (1996). Medienbild 1995: Krankenhaeuser am Pranger. *Fuehren und wirtschaften im Krankenhaus, 13*(5), 432–435.

Friesdorf, W., Gross-Alltag, F., Konichezky, S., & Schwilk, B. (1993). Systemergonomie in der Intensivmedizin. In B. Badura, G. Feuerstein, & T. Schott (Eds.), *System Krankenhaus—Arbeit, Technik und Patientenorientierung* (pp. 207–227). Weinheim, Germany: Juventa Verlag.

Friesdorf, W., Konichetzky, S., Gross-Alltag, F., Koller, W., Pollwein, B., Marraro, G., Kari, A., Toro, M. J., Demeester, M., Nahte, M., Samways, S., Polishuk, I., Mueller, I.,

Bonnaire, A., Schraag, S., & Classen, B. (1994). Information transfer in high dependency environments: An ergonomic analysis. *International Journal of Clinical Monitoring and Computing, 11*(2), 105–115.

Friesdorf, W., Marsolek, I., & Goebel, M. (2002). Integrative concepts for the OR. *Journal of Clinical Monitoring and Computing, 17*(7–8), 489–490.

Fuermann, T., & Dammasch, C. (1997). *Prozessmanagement—Anleitung zur staendigen Verbesserung aller Prozesse im Unternehmen.* Muenchen, Germany: Carl Hanser Verlag.

Gaba, D. M., Fish, K. J., & Howard, S. K. (1994). *Crisis management in anesthesiology.* New York: Churchill Livingstone.

Gaitanides, M., Scholz, R., Vrohlings, A,. & Raster, M. (1994). *Prozessmanagement—Konzepte, Umsetzungen und Erfahrungen.* Muenchen, Germany: Carl Hanser Verlag.

Gorschlueter, P. (1999). *Das Krankenhaus der Zukunft.* Stuttgart, Germany: Verlag W. Kohlhammer.

Graybeal, K. B., Gheen, M., & McKenna, B. (1993). Clinical pathway development: The Overlake model. *Nursing Management, 24*(4), 42–45.

Hammer, M., & Champy, C. (1993). *Reengineering the corporation: A manifesto for business revolution.* New York: Harper Business.

Jaster, H. -J. (1997). *Qualitaetssicherung im Gesundheitswesen.* Stuttgart, Germany: Georg Thieme Verlag.

Marsolek, I., & Friesdorf, W. (2000). Ground plan of an OR according to work flow requirements. *Journal of Clinical Monitoring and Computing, 16*(2), 157.

Marsolek, I., & Friesdorf, W. (2001). Comparing clinical processes for a systematic optimization: Patient admissions to the ICU. *Intensive Care in the New Millenium* quoted from the book of abstracts of the 8th World Congress of Intensive and Critical Care Medicine. (p. 167). Sydney, Australia: ICCM Press.

Marsolek, I., & Friesdorf, W. (2002). Remodeling of work processes: The TOPICS method. *Journal of Clinical Monitoring and Computing, 17*(1), 72–73.

Marsolek, I., & Friesdorf, W. (2003). Balanced rationalization in expert systems: Experiences from the analysis and optimization of complex clinical process flows. In K. Landau (Ed.), *Good practice: Ergonomie und Arbeitsgestaltung* (pp. 433–448). Stuttgar, Germanyt: Ergonomia Verlag.

Organisation for Economic Cooperation and Development (2003). *OECD in Figures: Statistics on the Member Countries.* Paris: Author.

Pearson, S. D., Goulart-Fisher, D., & Lee, T. H. (1995). Critical pathways as a strategy for improving care:Problems and potentials. *Annals of Internal Medicine, 123*(12), 941–948.

Plsek, P. E. (1999). Quality improvement methods in clinical medicine. *Pediatrics, 103*(1), 203–214.

Ramos, M. C., & Ratliff, C. (1997). The development and implementation of an integrated multidisciplinary clinical pathway. *Journal of Wound, Ostomy and Continence Nursing, 24*(2), 66–71.

Viethen, G. (1995). *Qualitaet im Krankenhaus.* Stuttgart, Germany: Schattauer Verlag.

Vorley, G., & Tickle, F. (2001): *Quality management: Principles and techniques.* Guildford Surrey, UK: Quality Management & Training Limited.

Weissauer, W., & Hirsch, G. E. (1994). Organisation der Intensiveinheit—Rechtliche Probleme. In H. Benzer, H. Buchardi, R. Larsen, & P. M. Sutter (Eds.), *Intensivmedizin* (pp. 28–35). Berlin, Germany: Springer Verlag.

37

VIDEO ANALYSIS IN HEALTH CARE

Colin F. Mackenzie and Yan Xiao
University of Maryland School of Medicine

Understanding of the strengths and weaknesses of human performance has direct implications for strategies to improve quality of health care and patient safety. However, our knowledge of human performance in real, complex, and dynamic environments, such as those found in clinical care settings, is limited. Studies of health care providers in their natural settings could provide insight into how teams work under time pressure, with constant interruptions, and in suboptimal workplaces.

Primary strengths of ethnographic observations are that they support a discovery process, draw attention to significant phenomena, and suggest new theories whose validity and generality can then be evaluated through additional studies. In comparison, studies carried out in laboratory settings do not recreate many of the real-life variables, such as risk, uncertainty, composition of teams, and the workplace domain. These are significant factors in determining performance during real-world dynamic and stressful events. Studies of decision-making and team performance in the past decade or so have highlighted the importance of understanding human activities in real, complex environments (Klein, Orasanu, Calderwood, & Zsambok, 1993). Many significant variables, such as expertise, risk, uncertainty, and composition of teams, are often difficult to replicate in usual laboratory settings. Studies in real environments and in sophisticated simulation environments with experienced practitioners are required.

A major challenge in carrying out studies in emergency medical settings, either real or simulated, is data collection. Although indirect data such as recalled past incidents can be utilized in one's modeling efforts (e.g., Klein, 1989), direct collection of behavioral data is needed to overcome potential biases in retrospective construction of past events. Tools for collecting behavioral data have become increasingly sophisticated. The most influential among these new tools is probably video recording (Dorwick & Biggs, 1983). Video is perceived as the richest medium to capture the minutest and briefest particulars of human interaction while retaining the context of the event and making it available for analyses by multiple or independent subject matter experts (SMEs). Video recording in the medical environment makes it possible for clinicians to review their own activities and for analysts to extract qualitative and quantitative data.

With video recording, it is potentially possible for analysts to repeatedly examine activities and to extract detailed quantitative data. The person who was recorded can provide comments on his or her covert mental processes cued by video records. Such cognitive approaches to examination of real medical events are a powerful tool to examine performance and identify patient and practitioners safety issues.

Video recording captures rich data that can be reviewed in great detail by different reviewers; the inherent, sub-second timing data in audio and video lends itself to sophisticated time study (Laws, 1989; Mehta, Haluck, Frecker, & Snyder, 2002). Video was used with simulation for medical education (Cooper et al., 2000) and in the analysis of crisis resource management trauma assessment training debriefing after patient simulation (Gaba & DeAnda, 1998; Lee et al., 2003). Video by its nature is a powerful tool for behavioral researchers and its value was recognized soon after its initial consumer availability (e.g., Dowrick & Biggs, 1983; Tardiff, Redfield, & Koran, 1978). The potential utilities of video recording for studying performance in high-risk health care settings are difficult to overstate.

The advances in hardware and software have made video technology a routine tool for research in individual and collaborative performance. An increasing number of research projects include video recording as a key data collection method. How this tool should be exploited methodologically and theoretically is thus a key question for researchers (Xiao & Mackenzie, 2004).

Driven by these and other advantages of video-based data collection techniques, we initiated a research program 11 years ago that included video recording in the fast-paced, highly dynamic health care domain of trauma patient resuscitation. This chapter will review the use of video data as a tool to complement observation in this health care domain.

USE OF VIDEO DATA

The advantage of a video record is that it provides a permanent comprehensive source document. This recording can be reviewed repeatedly by many different analysts. The events in complex environments where time pressure and uncertainty affect performance can now, with the powerful tool of a video record, undergo expanded analyses. Since the audio–video equipment for recording can be miniaturized, events that previously could not be observed due to limited space can now be analyzed through viewing of the video recorded events. Information extracted from the video data can be used for system design, identification of training needs, workload analyses, and performance evaluation and modeling. Due to the recent availability of inexpensive high-quality video cameras and video cassette recorders, there has been a huge increase in use of video recording in many different settings (e.g., Weston, Richmond, McCabe, Evans, & Evans, 1992). Digital video data (DVD) storage and digital cameras have simplified collection analysis and storage (Weinger, Gonzales, Slagle, & Syeed, 2003).

Unlike audio data, which can be readily transcribed (transcription of which is usually adequate), video data contain mostly nonverbal information. The work by deGroot (1965) and Ericsson and Simon (1984) lays the foundation of verbal protocol analysis, but, in the analysis of video data, researchers usually focus on nonverbal data and have relatively fewer previous methodological guidelines. Sanderson and Fisher (1994), based on the ideas presented by Tukey (1977), attempted to build a general framework for analyzing data that are indexed by time. They emphasized the usual exploratory nature in analyzing behavioral data, especially those collected in the real world. In addition, time-indexed data cover not only transcriptions of verbalization, but also events, transactions, non-verbal communications (e.g., gestures and facial expressions), and other observations that one can make in the review of video data. Sanderson and Fisher (1994) describe the steps in analyzing time-indexed, sequential data by 8 Cs: chunking, commenting, coding, connecting, comparing, constraining, converting, and computing. This framework provides a guide to data analysis, yet tools are still needed to increase the efficiency in dealing with video data.

ANALYSIS OF VIDEO DATA

Video analysis is defined as a process by which video data is reviewed and statements (either qualitative or quantitative) on activities and performance are derived. In many aspects video analysis shares similarities with exploratory data analysis, as video recordings contain a rich amount of data on events and activities and there are no simple ways to analyze all relevant aspects. As a consequence, video data analysis tends to be very labor intensive (high sequence-time to analysis-time ratio; Sanderson & Fisher, 1994).

Video records of clinical care are data with multiple applications, including performance feedback, quality improvement, clinician training, as educational tools and for human factors, and ergonomic research.

Template-Based Analysis

A unique characteristic of real-life performance is variations in terms of tasks, duration, workload, and personnel involved. When examined across cases, aggregation methods need to be devised. Another consideration in the process of video analysis is that it can be impossible to efficiently analyze all or even a significant portion of the many facets of recorded cases.

One approach that enables aggregation, and at the same time increases efficiency, in video analysis is through the development of a task template. A task template is a prototypical task sequences with known landmarks. Based on a task template, performance metrics are often possible, such as timing in reaching task landmarks and subjective performance judgment at these landmarks. Development of task templates can be achieved through task analysis by decomposing a task into a sequence of steps. This task template is used to develop review questionnaires, which speed-up video reviewing processes and increased consistency of video reviews across cases.

Through our own experience with video-based performance modeling, it helps greatly if one can define a prototypical task sequence, with a clear starting and ending point. Many tasks can be analyzed in this manner. Medical procedures such as airway management, establishing intravenous lines, and placing chest tubes are examples. Even in a highly changing field of initial trauma patient resuscitation, it is potentially possible to model the team tasks in advanced trauma life support (ATLS) protocols, even though ATLS is formulated to guide a single person's performance.

In our own work, extensive video review questionnaires were developed based on the task templates. SMEs were asked to fill in the questionnaires while reviewing video records. The questionnaires contained timing for major landmarks, subjective ratings of performance in tactic maneuvers and decisions, and judgment of the patient conditions. Results from the completed questionnaires produced the basis for further, often quantitative, analysis. As an example, using this video task-analysis methodology in trauma patient resuscitation identified that in emergency tasks there is compressed and increased workload, increased psychomotor activity, and an increased number of high priority steps in tasks omitted in emergencies compared to elective task performance (Mackenzie, Xiao, & Horst, 2004).

Critical Incident Analysis

Over the course of recording numerous cases, some stand out with unique or interesting characteristics (Xiao & Mackenzie, 1999). The template-based analysis may not capture these characteristics. Critical incident analysis, a technique originally developed by Flanagan (1954), represents a data analysis methodology that focus on individual episodes. These episodes often reflect interesting strategies, team performance patterns, and cognitive errors. With comprehensive data gathering (e.g., including audio–video recordings, system logs, retrospective reviews by participant SME's), one can potentially reconstruct these episodes in great detail. Examples from our critical incident analysis include those on fixation errors (Xiao, Mackenzie, & the LOTAS Group, 1995), communication errors (Mackenzie, Craig, Parr, Horst, & the LOTAS Group, 1994), and team coordination (Xiao, Mackenzie, Patey, & the LOTAS Group, 1998).

Extraction of Qualitative Data on Team Organizational Coordination

As an example of more detailed video data extraction, the qualitative data obtained on team coordination is described. Verbal communications can be viewed as one of many media that the trauma team use to communicate and coordinate. These other types of media include, in addition to utterances and explicit gestures, (a) activities, (b) work space, (c) events, and (d) foci of attention. Use of these media was possible because trauma team members worked in closed physical work spaces. From analysis of video records, several forms of noncommunication team coordination activities were noted in videoanalysis (Xiao & Mackenzie, 1998). These noncommunication activities were

- *Following the protocols.* Established practices (sometimes codified as protocols, such as the Advanced Cardiac Life Support or ACLS protocol), specify task distributions and priorities, immediate goals, and problems to be treated. The tasks to be done by each team member are clear. Without much communication, the surgical, anesthesia, and nursing crews commence their activities after the patient arrives. Clear task distributions were observed among the crews in resuscitation teams at the beginning of each patient admission, despite the uncertainty about the patient's status.
- *Following the leader.* Team or crew members determined what they should do by watching the crew leader. The activities of the crew leader can be viewed in some sense as the medium through which the team leader

passed information (such as instructions) to the rest of the team. If not occupied, team members tended to follow the attention foci of team leaders. Needed materials or help were provided often without explicit solicitation.

- *Anticipation.* The team members were also found to provide unsolicited assistance through the anticipation of the team leader's response to the patient's physiological events. A gagging sound, in one case, led an assistant to offer a suctioning catheter in anticipation that the patient would vomit soon and the anesthesia crew member would have to use suction to clear the patient's airway. Thus the shared physical event space became a medium of communication for the team. The prerequisite, of course, was the ability to understand the significance of patient events. The workspace itself is also a medium through which the teams coordinate. Team members, while not under instruction to perform specific tasks, scanned the workspace and perceived tasks that needed to be carried out.
- *Activity monitoring.* The interdependencies of tasks shared by a team mean that one member's tasks could sometimes only commence after the success of another member's tasks. (For example, surgeons can only begin certain procedures of resuscitation after the patient is anesthetized.) Thus, monitoring the progress of any other member's tasks not only made it possible to compensate for a teammate's performance, but also gave lead information to prepare for the next step. In many cases, the surgical crew did not announce their plans; however, the anesthesia crew inferred what needed to be done from the activities of the other crew. For example, during the review of the video records of a case, one participant in that case revealed that the conversation between two surgical crew members provided cues of what the surgical crew would do next, even though the conversation was not directed at the anesthesia crew.

These strategies improved task coordination, without the use of explicit, verbal or gestural, communications and enabled the resuscitation teams to perform smoothly in most situations. Information flow is an interesting aspect of team coordination, including the explicit, verbal communications regarding situational assessment and future plans, occurring usually when the team was at a decision point. Strategies of team coordination were extracted from video records of trauma patient resuscitation. Breakdowns in team coordination were observed in the following three types of crisis situations:

- *Pressure to seek alternative solutions.* In this type of situation, extreme difficulties or unexpected patient responses were encountered that prevented the implementation of routine procedures. When the patient condition was deteriorating rapidly, the team was under pressure to find an alternative solution and to act immediately.
- *Initiation of unexpected, nonroutine procedures.* When materials or expertise that had not been anticipated in advance by the supporting members of the team were unavailable, coordination breakdowns occurred, such as when no announcement was made about the need to adopt the non-routine approach. As a result, the ability of the supporting members of the team to provide assistance was compromised. Coordination breakdowns in this type of incident were marked by the lack of anticipatory help from the team members, delays in preparing materials, and unnecessary pauses in the team leader's activities to obtain assistance.
- *Diffusion in responsibility.* In critical circumstances during patient resuscitation, a diagnostic procedure or a treatment plan may have to be abandoned if the patient condition is too unstable. Such changes in plans by crew members within the team occur during crises and under great time pressure. The rest of the team may have difficulties in adjusting itself from a diagnostic mode to action mode. The inability of the rest of the team to anticipate this sudden change in plan by crew members prevents them from adjusting their responsibilities accordingly, and results in the omission of critical supporting steps.

The coordination breakdowns that video analysis identified occurred due to: (a) conflicting plans, (b) inadequate support in crisis situations, (c) inadequate verbalization of problems, and (d) lack of task delegation.

Access to Cognitive Function from Video Review

Insights into *team* cognitive function can also be ascertained from video records by review of verbal communications recorded, as in previously published reports (Xiao & Mackenzie, 1997). Many studies of cognitive activities were based on verbal data, but in the real environment, many communications are nonverbal. As illustrated in this chapter, exemplary cognitive analysis can be greatly assisted by video recording of individual cases that are unusual. In the trauma center, multiple video recordings with detailed retrospective accounts and substantiating records have enabled us to assess the usefulness and potential generalizability of this methodology for accessing cognitive aspects of work. Covert cognitive functions can be identified by clinicians when reviewing their own video-recorded clinical care. Visual scanning patterns were used to assess information gathering (Xiao, Mackenzie, & Orasanu, 1998) as a source of access to cognition. The process of identifying task performance factors using a task template is efficient (typically 3:1 ratio of analysis to real video time), and it allows aggregation of multiple occurrences of the tasks. It has been found to be particularly revealing to compare data from the aggregated tasks at two levels of urgency as a means for identifying frequently recurring performance omissions. Exemplary cognitive analysis enables use of individual cases to illustrate how changes in practice can avoid nonoptimal performance factors, and mitigate error evolution.

Video Analysis Tools

Tools are needed to effectively analyze video data. Desirably, these tools should allow basic annotation and event marking or searching. Additionally, tools capable of random access to data, concurrent viewing of multiple time-line documents, touch-coding, assisting iterative and recursive categorization, would certainly help. Document management (in our case, patient records, questionnaires, review annotations, coding results, transcripts, patient physiological data, case logs) made video analysis more productive, as data could be obtained from many different sources.A number of computer systems have been developed in assisting the manipulation and coding of video data:

- MacSHAPA (described in Sanderson, James, & Seidler, 1989). This tool has integrated VCR (video cassette recorder) control, annotation, and coding, with post-coding analysis functionality. Currently it is only available in MacIntosh® platform. It is able to control several major models of VCRs and interface with the time code on videotapes.
- A.C.T. (described in Segal, 1994) provides touch-coding (i.e., one key stroke input) and can be used both in reviewing video tapes and in real-time observation. However it does not provide VCR control. A.C.T. only runs on a MacIntosh® platform.
- OCS Tools (Observational Coding Systems of Tools by Triangle Research Collaborative Inc., NC) is a set of tools that enable VCR control, time code reading, and input of annotation and coding. (It is a commercial product, available at http://depts.washington.edu/chdd/mrd-drc/cores/bsc.html.)
- VANNA (described in Harrison, 1991) is a MacIntosh®-platform utility that allows the analysis of current multiple video sources (e.g., situations in which multiple cameras are used in recording). VANNA can display these video sources along with other time-stamped information on a single computer monitor, thus simplifying video review process.
- VINA. This was a tool developed during the course of our program. The VINA analysis software runs on a UNIX platform and has the following key features (a) manual and scripted VCR control; (b) VCR control by pointing, click-and-drag; (c) automatic highlighting of records in multiple time-line documents by synchronization with the time code on the videotape; (d) touch coding of events and activities; (e) temporal graphic representation of coding; and (f) digital and graphical display of recorded vital signs data synchronization with VCR.
- The Observer®, by Noldus (the Netherlands) is a commercial package available for the

> Microsoft Windows® platform. It can interface with analog media stored on video records, as well as with digital media encoded in MPEG format. The package provides facilities to annotate and logvideo data, to analyze time-line data, and to manage coded data. The package also allows structured coding to capture multiple aspects of video data, such as activities, postures, movements, positions, facial expressions, social interactions, or any other aspect of human or animal behavior.

Several researchers developed their own video analysis tools to meet their unique demands. Guerlain, Turrentine, Calland, and Adams (2004) developed RAVE, a system to view multiple video sources from different cameras in a synchronous manner for the investigation of performance during laparoscopic procedures. Weinger and his colleagues (2003) also developed their own video capturing and analysis tool for studying nonroutine events during anesthesia.

None of these video analysis software tools are ideal for health care use. Weinger's approach is anesthesia-specific, and Guerlain's software is useful for multiple image review. MacSHAPA is used still by many video analysts. Our own software preference now is called JVIDEO and it is simple. Using a video review controller allows half 2×, 4,> and 8× forward speed, and backward speeds together with "jump- back" (e.g., go back to the start of the last 5 sec). This approach allows great flexibility for any type of video and we use it to complete a template needed for a particular video analysis.

STRENGTHS AND WEAKNESSES OF VIDEO ANALYSIS

There are both strengths and weaknesses in analysis of the video record of real patient management for education, quality assurance, human factors, and ergonomic research.

Strengths

The major advantages are that the participant health care providers are not dependent as much on memory, because the video image and audio recording recreate the event, including comments by the team members and alarms that might not have been heard or noted at the time. In our experience, discussions among the participants, care providers, and SMEs, provided additional information that would not necessarily have been revealed without the video image and audio recording. It would have been more difficult to target points where covert events could be revealed by participant SMEs reviewing their own care or where performance could have been changed, without the video image and audio channel. The participants, whose care was video recorded, found video analysis useful and they noted that it allowed them to reflect on their performance in greater detail than is possible without it. They often found it revealing to discuss the video with the nonparticipant SME who could identify that the management was not ideal patient care. It was only on reviewing the video records that it became apparent how their performance could be improved. Qualitative and quantitative data analysis, aggregated over multiple performances of the same task or process, produces powerful data for clinicians on their own workplace safety, performance and ergonomic practices. An extremely important use of video records (in the form of 30 second to 2 min video clips) is in training and education of health care providers. These video abstracts from continuous video records have facilitated buy-in from care providers and have, in our institution, convinced senior staff of the need to change existing practices.

Weaknesses

The weakness of video analysis is that it is tedious and time consuming. It is difficult to accurately estimate how much time was required to analyze each of these video records because much work was spent in development of the analysis system and database necessary to facilitate video analysis of these cases and others (Mackenzie & Xiao, 2003). As an example, about 10 hours were spent in discussion, viewing, and transcription of the 8.5 min of video of prolonged uncorrected esophageal intubation, reported in great detail (Mackenzie, Martin, Xiao, & the LOTAS Group, 1996), whereas, aggregated data obtained from the template analysis took about 30 min for 10 min of video record.

Other weaknesses of video analysis are that even with a two-microphone system, it was difficult to pick up all the utterances and the audio record could be improved. However, it would be more obtrusive to equip care providers with microphones. The video images do not include the entire field of view and cannot identify events occurring off the screen, though the audio channel can be helpful. In analyzing a video image with the physiological data overlaid there is a tendency to think that the participant care providers were aware of these data, when in reality, this is unlikely because of selective attention to other aspects of clinical care. Because analysis occurs after-the-fact, the care providers have time to rationalize the decisions made, as they are aware of the outcome. Thinking aloud, and interviews conducted in the middle of case-management have been used in simulated cases to overcome this problem (Gaba & DeAnda, 1988; Howard, Gaba, Fish, Yang, & Sarnquist, 1992). Video analysis of such real events will be important in development of the database necessary for simulation and to improve clinical practice and equipment in the future (Mackenzie, Hu, Horst, & the LOTAS Group, 1995). We believe it is by systematic study of such real critical incidents that the mechanisms involved in their genesis will be understood and from these analyses preventive measures or particular approaches to training may be devised. This level of effort to collect and analyze video records may only be warranted for important research questions. The successful use of this technique is dependent on careful and extensive study design before the data is collected to insure the research questions are adequately addressed. However, in spite of the difficulties associated with this data-collection strategy, it is a powerful technique that allows for exploration not possible with other techniques when applied in the real, dynamic, and stressful workplace.

CONSENT, SOCIAL–LEGAL, PRIVACY AND CONFIDENTIALITY OF VIDEO RECORDING

A major concern of those included in video records in health care is privacy and the confidentiality of these records. The ubiquitous use of video surveillance in the community has made health care workers and patients well aware of the detailed information that can be extracted from video. In order to enable the potential benefits of video analysis to be realized, while protecting the rights and privacy of individuals creates a dilemma for which there is no single answer.

Consent For Video Recording

The research participants were the clinical care providers in a trauma center. The participants gave consent to be video recorded. Protection of human participants (both research participants and patients) was secured through a formal approval process by the Institutional Review Board (IRB). For studies analyzing human performance carried out in the real workplace, the IRB agreed to allow video recording without patient consent because it was not thought feasible to obtain consent consistently in the emergency circumstances in which the video records were made, and precautions were made to ensure patient privacy by procedures such as masking patient face in video images (Mackenzie & Xiao 2003). The original video records were retained until video analyses were completed and then erased. Video images were acquired from a ceiling-mounted camera and directed so as to minimize patient identifying features (Mackenzie, Hu, Xiao, & Seagull, 2003). The research participants knew when events were being video recorded and they consented to publication of details of video analyses and to use of video clips (30 sec to 3 min duration) for presentations or in training materials.

However, video is a very powerful medium. A concern is that even when researchers obtain a participants's consent, it is not always clear that the participant understands the implications of that consent. We have used the process described in Table 37–1 as an extension of fulfilling the obligatory need to have a signed consent form. The consent process explains the implications to as many as possible of those affected by or likely to be included in video recording. It is time consuming, but rewarding, because participants understand what is happening.

It should be noted that fewer research participants are needed than might be thought to obtain video data. In our first project (Mackenzie et al., 1994) the majority of the 120 video recordings were made by just 6 research participants. No consent forms were signed by other colleagues in the clinical workplace or the patient.

Instead the consent process described in Table 37–1 for non-research participant clinicians was followed. During the first 3 years only 2 research

TABLE 37–1. Research Hypothesis and Plan Implementation

↓	
Research participants	
↓	
Input into development of research protocol	
↓	
Submit to Research Committee and Institutional Review Board	
↓	
APPROVAL →	Educational process for non-research participant clinicians includes:
↓	
Participants sign consent form for video recording research	(a) Availability of research protocol
	(b) Discussion at specialty staff and one-on-one meetings
↓	(c) Institution wide meeting to answer repeatedly asked questions
Participant reviews raw data on video recorded project	(d) Showing of video abstract and identification of area included in video image work
↓	
Participant reviews "masked" video abstracts and signs consent for retention	(e) Efforts to protect confidentiality and privacy of care provider and patients
	(f) Identification of possible role as subject matter expert
	(g) How the research results will be disseminated

participants would not consent to participate or be included in video recordings. One additional participant agreed to participate but only if their image was blurred in any retained video abstracts.

Social–Legal Issues of Video Recording in Real Environments

Despite advantages of video recording, challenges abound, such as gaining support from those being recorded, securing patient confidentiality, overcoming medical–legal obstacles, and effectively using the medium. A major challenge with video recording is social acceptance and potential impact on the recorded behavior. The primary concern for many people being recorded is possible legal implications of video recording. Formal reports (e.g., Hoyt, Shackford, Fridland, Mackersie, Kansbrough, et al., 1988) and informal communications have documented the difficulties in resolving the legal concern of those who are video recorded.

Among informal communications, several e-mail messages circulated in March 1996 regarding other medical centers' experience with video recording. These can be summarized as related to ethics, risk management, medical records, legal counsel, consent and confidentiality, and staff acceptance. Some authors describe video recorded events (e.g., patient fell off the stretcher during transfer onto a resuscitation gurney undergoing full flexion and extension of the neck; these authors do not identify whether any permanent injury resulted). In another event, the relatives of a recently bereaved family saw the video record of their mother's resuscitation played on the waiting room television, including the difficult airway management resulting in nose bleeding, cardiac arrest, and 40 min of cardiopulmonary resuscitation. Most emergency rooms that used video recording erased the records shortly after viewing and completion of any analysis.

In one trauma center in Australia, to gain staff acceptance, the video record was labeled and treated as an aid to memory. There was a considerable sentiment that video recording, if it was erased shortly after the event, was quite acceptable. Some authors made the point that similar confidential information is displayed on a white board prior to patient admission. This white board is erased when the patient is admitted and resuscitation completed. One problem perceived with repeated reviewing of an occurrence is that mistakes would be imprinted in the health care workers minds. Should a medical–legal case arise, every detail of the error would be remembered when time came for a deposition statement.

Due to the nature of consent, the awareness of video recording can change the very behavior being recorded. So far the impact of video on behavior has been deemed insignificant (e.g., Pringle & Stewart-Evang, 1990).

Privacy and Confidentiality

Every effort was made to preserve privacy by using camera angles and tight image border control to avoid recognition of individuals. Patient identifiers were removed from paperwork associated with video records. To preserve confidentiality, only care providers and researchers were given access to the video records which were kept secured under two sets of locks.

Technical approaches used to preserve confidentiality and maintain privacy of the video recorded individuals included video masking with blurring of the face or other distinguishing features of subjects or patients. Voices can be disguised, but these digital manipulations can impair video data analysis if qualities of speech or gaze are being analyzed. The key to the consent process and confidentiality is, in our opinion, the development of trust by those who are video recorded, that the investigators will not abuse the privilege of being allowed to acquire video data for research purposes.

Generally the original video records were destroyed by degaussing within 4–6 weeks of collection. A sign at the entrance to the operating rooms and inside the TRU was posted to indicate image recording was occurring. The wording "Be aware, filming is underway" complies with Joint Commission of Accreditation of Hospital Organizations (JCAHO) regulations for video recording in hospitals. Our experience with 11 years of video recording is that there have been no medical–legal subpoenas and no employment-related or liability issues have resulted.

LOGISTICS OF ACQUISITION AND STORAGE OF VIDEO

Acquisition of video can use home-video camera technology with only minimal expense. However, there are advantages in a more robust system to gather video data in a systematic manner. The system we have had in use for the past 12 years (with interval updates and expansion) is described below.

Video Acquisition System Network (VAASNET®)

Important features of the design and function of equipment used in audio–video acquisition in the clinical domain include unobtrusiveness, "turn key" operation and 24 hours per day, 7-days per week availability, and an interface with the patient vital-signs data. The equipment must also be rugged and reliable enough to withstand use in emergencies, and by multiple users. Because of patient occupancy of locations where the equipment is used, servicing is frequently difficult and industrial grade equipment should be used to withstand the constant wear and tear (Mackenzie et al., 2003).

In our trauma center a Telecontrol Center "command and control" location was built (Figure 37–1) to integrate multiple sources of signal input into one location and to allow use of the system for human factors, ergonomics, and patient safety research, and as a telemedicine testbed. An infrared wireless communication system was installed for mobile audio communication among trauma team members. Each resuscitation bay (among 10) and each operating room (of six) was cabled with four duplex audio, five video, four fast-speed data and four multi-mode fiber connections (Figure 37–2). In addition, similar cabling was installed in the post-anesthesia care unit (PACU), all of which were located on the second floor of the trauma center. The Telecontrol Center was also linked with three ISDN lines and a web-based and cell phone imaging systems that provided multiple inputs from moving emergency vehicles, home-based physician consultants, other hospitals, distance education centers, and pre-hospital care providers (Mackenzie et al., 2003).

Video Acquisition Software

The first version of the video acquisition software was completed in 1991 (Mackenzie et al., 1995). The main data-acquisition system runs under a Microsoft Windows® platform. Data acquisition was in five distinct modules: (a) system control logic, (b) interface software for vital-signs monitor handshake (including Marquette, Mennen, CritiCare, Nellcor, SpaceLabs, Propaq), (c) VCR communication software, (d) definition of the physiological data file sequence from multiple simultaneous sites, and (e) software to control the real-time network communication with the campus-wide fiber-optic system and connections with secondary and auxiliary local area networks.

Figure 37–1 Telecontrol Center shows 3 full screen displays above 28 small screen individual camera images. The "quad" view is shown on the right screen and head mounted video camera (HMVC) is displayed on 1 stand out screen and vital signs waveforms on another. The pan/tilt/zoom (PTZ) camera control panel and image are also indicated. The time code generator and patch board and videocassette recorder are shown to the left of the video technician. The HMVC camera provides detail of procedures such as chest tube insertion or instrument tray use. Ceiling-mounted camera images may be obstructed by care providers. The machine readable time code allows these multiple images of the same event to be synchronized for simultaneous viewing during video analysis.

Camera Systems in VAASNET®

An environmental video camera was positioned on the ceiling of each resuscitation bay in the 10 bays of the trauma resuscitation unit (TRU), to provide an overview and allow remote monitoring of general activities (e.g., patient arrival, team readiness, staff present, etc.) and video recording in each location.

A second camera had pan-tilt-zoom capability (PTZ camera) and allowed remote viewing, video recording, and PTZ control of detailed activities in each bay, for example a specific area on which the trauma team was working such as the chest. The resolution was such that settings used on a mechanical ventilator, O_2 flow meters, intravenous infusion drip rates, and so on, could be determined. The PTZ camera nearest the TRU white board, which recorded information about incoming patients, was also used to distribute the white board information to other locations when not in use for monitoring resuscitation bay activities. The benefits, for example, to the anesthesiology team who covered both resuscitation and operating room activities, was that this white board image saved a 200–300 foot walk to the TRU to obtain details of impending trauma patient admissions.

The third camera system was a wireless head mounted video camera (HMVC). This was worn by an operator during resuscitation and performance of invasive procedures (e.g., chest tube insertion to relieve blood or air trapped inside the rib cage compressing the lung). Four video transmitters placed around the TRU sent wireless video images from the HMVC to the Telecontrol Center. The HMVC images were shown in a "quad" display as a single screen that combined the environmental camera view, the PTZ image, the HMVC image, and the patient vital signs waveforms. This quad display could be video recorded and synchronized with the high resolution PTZ image that was video recorded as a full-screen display. Such a combination gave comprehensive imaging capabilities for recording events during which multiple personnel were performing simultaneous resuscitation tasks, some of which obstructed a single ceiling mounted camera view (Figure 37–3).These upgrades to the

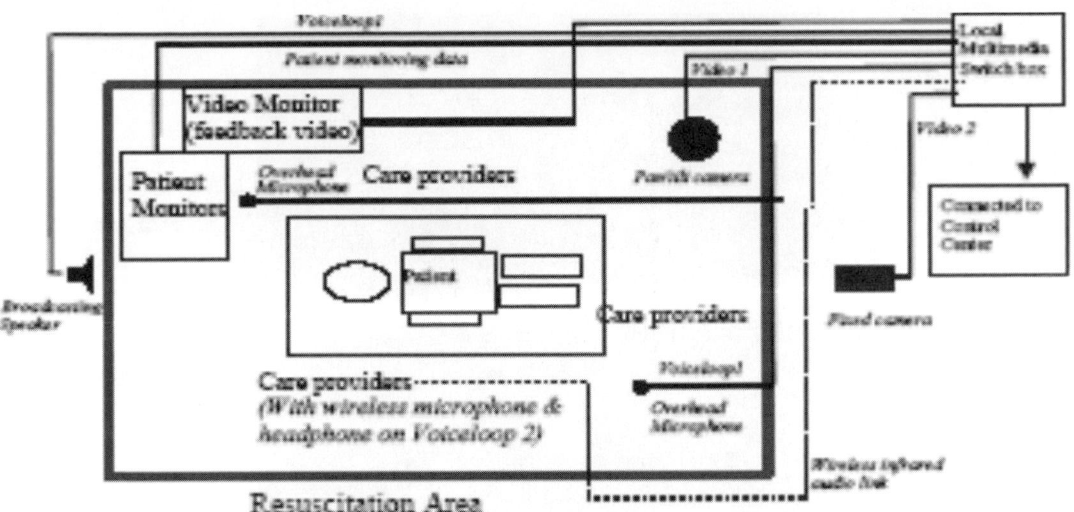

Figure 37–2 The multimedia connections are shown in a TRU bay in relation to care providers and the patient gurney. The wireless audio links enable surgeons, anesthesiology and nursing leaders in the resuscitation team to be in a voice loop with remote care providers. Additional duplex links are available for HMVC images or FAST scans. A surgeon team leader can sit in the Telecontrol Center and view the TRU bay using several images while being in constant duplex audio communication with the team who are using the hands free mobile wireless system shown in Figure 37–4.

original video acquisition system are described in Mackenzie and colleagues (2003).

Core Support Group

A key advantage of studying real-life performance is the access to SMEs. There are four basic elements in this access: (a) data gathering, (b) data analysis, (c) development of measurement tools, and (d) interpretation of data. One should anticipate an enormous amount of effort from SMEs. To achieve these four elements, a core group of SMEs is essential.

In conjunction with the equipment shown in Figures 37–3 and Figure 37–4, a living laboratory for human factors research was created in a real clinical environment.

Events or tasks not associated with the original analysis may be detected during analysis of video records. Whereas with observation, the data-parsing occurs at the time of observation, not later, so only the recorded observations can be analyzed. A great strength of video recording over observation was the ability to document and understand fleeting events, simultaneous interventions, or brief communications, because video allows expanded analysis of time-critical, but brief or uncertain events, whereas such episodes are very difficult to observe accurately in context. Lastly, video was found to be a powerful feedback and training tool that identified problems with existing practices in safety performance and ensured buy-in to participate in change from even the most experienced shock– trauma care provider.

The VAASNET® System became a testbed for assessing remote teleconsultation skills and information gathering practices of the shock–trauma clinicians (Xiao et al., 1995). Using an eye tracker to measure gaze-patterns and dwell times, we documented differences in video display scanning of identical images between anesthesiologists, surgeons, nurses, and untrained observers. We concluded that training for telemedicine will require prompts and decision aids to ensure that all events are noted. High-performing trauma clinicians are not necessarily going to perform well when using audio–video– data links as clinical information without such assistance.

Quality management measures of safety performance deficiencies were compared with video analysis of actual performance to identify the strengths of video recording (Xiao & Mackenzie, 1998). The anesthesia record, the anesthesia quality management report, and the other self-reports did

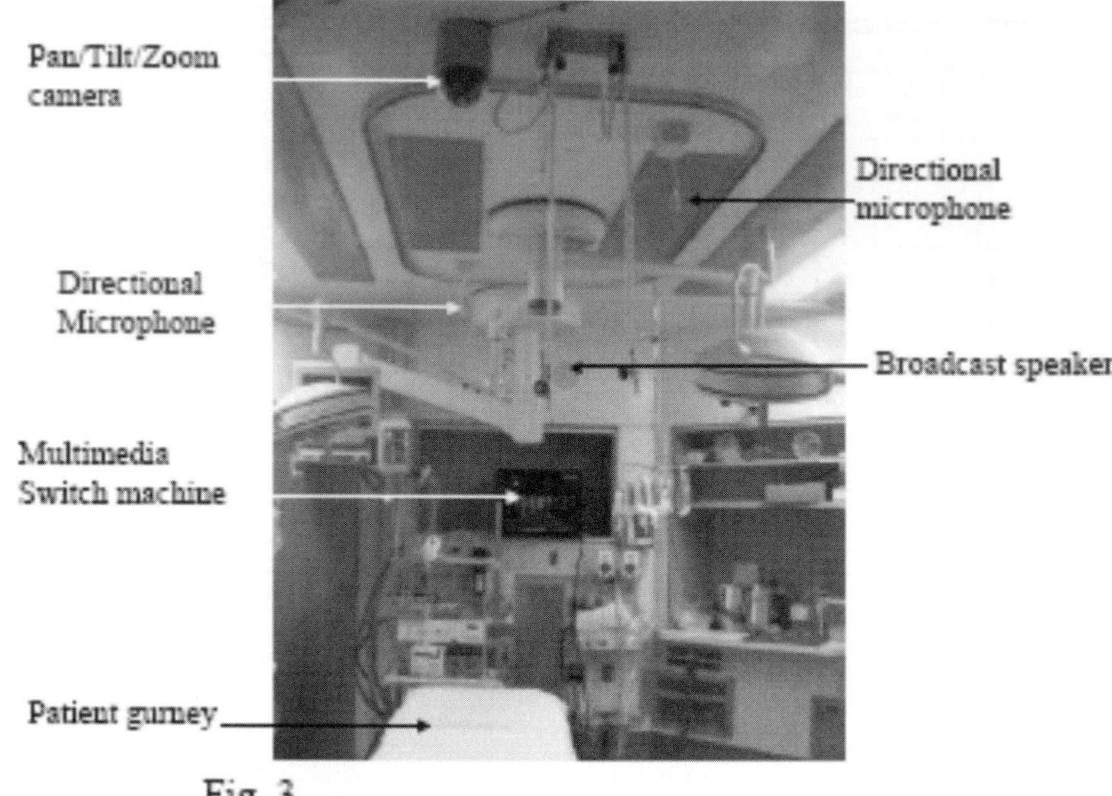

Figure 37–3: Shows the VAASNET® installation in a TRU bay. Pan/tilt/zoom camera provides detailed information, directional microphones at the front and back of the bay gather audio communication from team members. The multimedia switchbox sits on a shelf at the back of the bay beneath the broadcast speaker. There are 4 duplex audio, video and data cable connections to and from the Telecontrol Center to each of the ten resuscitation bays and 6 operating rooms in the trauma center.

not detect 23 of 28 systems and care provider performance deficiencies that were identified from video records. Video recording, therefore, revealed gaps in current information collection about the factors contributing to systems problems as well as their impact on care provider performance.

An ergonomic analysis of the resuscitation workplace simplified the layout (Harper, Mackenzie, & Norman, 1995) by repositioning equipment such as the O_2 flow meter, noninvasive blood pressure monitor and suction system. A more user-friendly resuscitation bay layout was achieved. With the redesign there was space to walk around the patient without the need to step over cables and device connections with the walls. With the revised layout all the patient monitoring, O_2 supply, suction, and equipment necessary for airway management were positioned so that they interfaced with the patient in one place. This interface was beneath the patient vital signs monitor and included vital signs monitor cabling, the ventilator tubing connections, suction, and O_2 supply (now brought from the roof rather than from connections with the walls).

An analysis of chest tube insertion instrument content and tray position showed the problems with existing practices (Seagull, Mackenzie, & Bogner, 2002). Video analysis of chest tube insertion allowed us to identify the difficulties and the time trauma surgeons underwent searching for the few correct instruments among a tray of 50 or more. In addition, instrument-tray position and the number of instrument trays used during insertion of chest tubes were found to be important factors in contamination of the surgical site and of operator injury

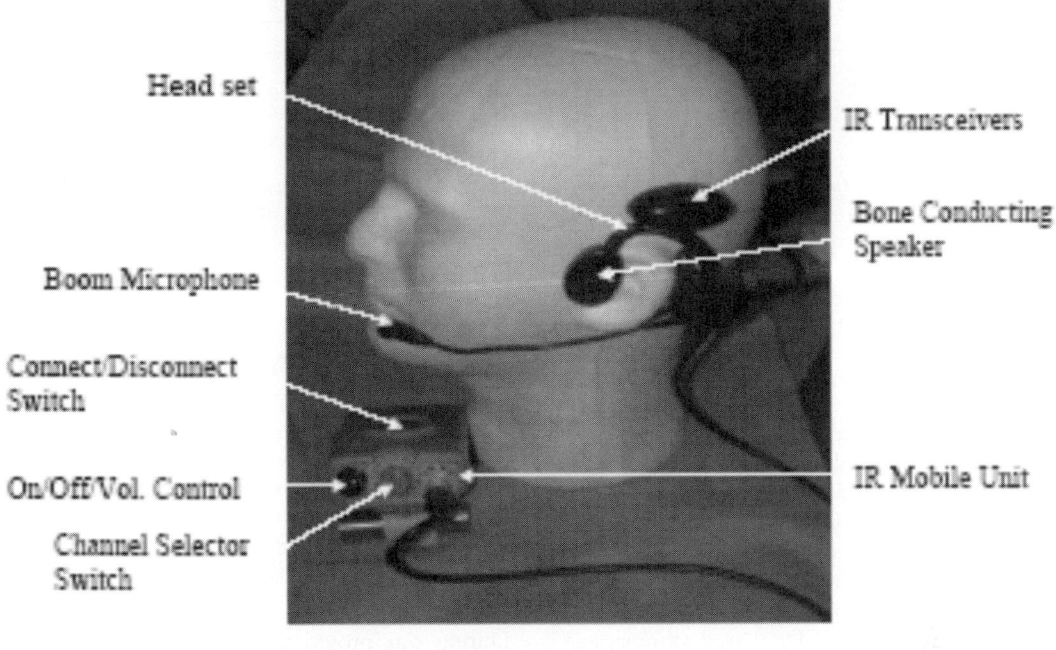

Figure 37–4 IR System Head set and mobile unit are shown. The headset has transceivers, bone conducting speakers (bilateral) and a boom microphone. IR mobile unit has control switches on top and connect/disconnect button on side. The mobile unit works very well in noisy environments (it was designed for military use in tanks). Phone calls be received and made while walking with hands free operation. Line of sight is required with 48 ceiling mounted IR transponders.

due to needle sticks or knife-blade cuts. A new instrument tray was produced for chest tube insertion and its use is currently being evaluated by further video recording. Abstraction of video clips identified best practices for chest tube insertion that could be contrasted with nonoptimal performance. This was powerful for training material and in supporting the quality management process.

Video recording in the clinical domain has been used by others as a tool for education (Townsend, Clark, Ramenofsky, & Diamond, 1993; Hoyt et al., 1988), quality assurance (Jeffries et al., 1994; Townsend et al., 1993), human factors (Burchard & Rowland-Morin, 1990; Goldman, McDonough, & Rosemond, 1972; Mallet, 1990), and ergonomic research (Boquet, Bushman, & Davenport, 1980; Goldman et al., 1972; McDonald, Dzwonczyk, Guptas, & Dahl, 1990; Smith, Macintosh, Sverrisdottir, & Robertson 1992), and to improve time utilization and resuscitation skills during trauma patient management (Townsend et al., 1993; Smith et al., 1992). Video recording has also examined behavioral and educational functions of clinicians, including interpersonal skills of clinicians in simulated patient encounters (Burchard & Rowland-Morin, 1990; Cooper et al., 2002), analysis of communications between nurses and post-anesthesia patients (Mallet, 1990), crisis resource management training of anesthesia care providers (Howard et al., 1992), and assessment of clinical skills (Tardiff, 1981) and teaching in neurology (Kaufman & Kaufman, 1983), family practice (Scheingold & Smith, 1980), emergency medicine (Shesser et al., 1985), and psychiatry (Roeske, 1979; Tardiff et al., 1978).

We found that VAASNET® allowed all of the uses reported by others (Dorwrick, 1991). In addition, because we used the findings from video analysis to assist the clinical care providers rather than as a punitive surveillance tool, we had buy-in for participation.

TABLE 37–2. Key Messages About Video Recording in Healthcare

Key messages about video recording in healthcare
It is important to develop trust with video recorded subjects
Clinician feedback should be obtained on introduction of a new protocol or line of investigation
Aggregated video recorded data should be reported and clinician reviews used for feedback
Task analysis at two levels of task urgency is powerful methodology for brief and risky but beneficial tasks
Multidisciplinary experts in surgery, anesthesiology, and nursing should be involved
Audio records of participants should be used to explain cognitive aspects of events or covert processes
Where events are uncertain or verbal interactions unclear, participant input is needed for clarification
Single critical events may reveal underlying systems failures
Video records detect quality assurance occurrences not identified by self-reports
Video provides powerful feedback and video clips are important training tools

Storage of Video

Initially video were stored on analog VHS tapes. A copy was made to be used for video analysis, annotation, and communication. Other documents accompanying video included questionnaires completed by participants at time of video acquisition, audio records of comments made by participant SMEs and nonparticipant SMEs when viewing the video record. Discharge summaries, operating room and resuscitation area records, laboratory value data, and a description of the role of each person around the patient gurney in the video record were also available.

For the purposes of obtaining interrater reliability data the video records were digitized and clips (short sequences of 30 sec to 5 min duration) were copied onto CDs for distribution to SMEs. These SMEs rated their assessment of clinical performance shown on video clips. As many as 70 video clips could be stored on a single CD which was then reviewed by SMEs at their own convenience at work or on their home computers. A video library was developed after the video records were digitized and any patient facial or body identifiers blurred and identification numbers removed.

Weinger and colleagues (2003) identified the design requirement for a clinical video data-analysis system. They describe a video-capture system that can be moved from one clinical location to another, requiring some set-up time. Weinger and colleagues (2003) archive video on DVDs directly on capture, using stand-alone dedicated DVD recorders that produce an MPEG–2 video stream on a standard DVD video format disc ($1.88 per hour for DVD versus $11.67 for digital videotape). The DVD video standard permits multiple channel surround sound, audio, choice of screen format (wide, letterbox, and pan-and-scan) eight tracks of separate synchronized audio, menus and random access for user interactivity, up to nine camera angles, and digital and audio copy protection. To synchronize all these data streams, a time-code generator with a drop frame longitudinal SMPTE time code was sent to the video recorder.

CONCLUSION

During our program, a considerable amount of time and effort was spent in extracting quantitative data from video data. Some of the efforts were judged successful (e.g., timing data of task landmarks template data extraction), others less so. One example of an effort to obtain quantitative data was collection of subjective stress ratings. In the video analysis of many recorded cases, SMEs were asked to provide ratings on the perceived stress at 1-min intervals. Although valuable data were obtained, it was judged in retrospect that the efforts in collecting such ratings could have been reduced and more exploratory studies could have been accomplished. The template data extraction appears to be the most flexible and generalizable approach to obtaining clinically useful data from video records. Video has significantly more utility for capturing details of events than observation, but it has added cost, and privacy and confidentiality concerns.

This review of video data extraction and analysis techniques used in a clinical environment may help other researchers in formulating their research plans and in data analysis when video recording is considered and used. This belief is based on published reports on video analysis (Goldman et al., 1972; Mackenzie, Jeffries, et al., 1996; Mackenzie & Lippert, 1999; Mackenzie & Xiao, 2003; Laughery, 1984), as well as comments, feedback, and personal communications received from a wide variety of sources. This chapter describing a video data-collection strategy supplements other information including the excellent special issue in October 1989, edited by Mackay and

Tartar (Mackenzie& Xiao, 2002) on video as a research and design tool, and a book chapter by Nardi, Kuchinsky, and Whittaker (1997). The important key messages about video recording in health care are summarized in Table 37–2.

ACKNOWLEDGMENT

The work reported was supported by Office of Naval Research Grant N00014–91-J–1540, NASA NCC–2921, ARL DAAL 01–96-C0091, AHRQ grants and NSF grants. The opinions expressed in this paper are the authors and do not reflect those of the funding agencies.This work would not have been possible without the following group collaborators. The first known as the *LOTAS Group consisted of: Colin Mackenzie (Chair), William Bernhard, Clifford Boehm, John Blenko, Hal Cline, Delores Donnelly, Richard Dutton, Peter Hu, Mahmood Jaberi, Murray Kalish, Kim Mitchell, and Yan Xiao. Our thanks are due to the many care providers in the University of MarylandShock Trauma Center who made the events reported in this chapter possible. A later group, the Invasive Procedure Outcome (*IPO)-Group was formed and included: Colin Mackenzie, MD (Chair), Grant Bochicchio, MD, Bill Chiu, MD, Carnell Cooper, MD, Jim Haan, MD, Sharon Henry, MD, Kerry Kole, DO, Jim O'Connor, MD, Tom Scalea, MD, Amy Sisley, MD, Jennifer Perry, RN, Lynn Gerber-Smith, RN, Ellis Caplan, MD, Manjari Joshi, MD, Steve Johnson, MD, Richard Dutton, MD, Mahmood Jaberi, MD, Yan Xiao, Ph.D, Peter Hu, MS, Jake Seagull, Ph.D.

References

Boquet, G., Bushman, G. A., & Davenport, H. T. (1980). The anaesthetic machine: A study of function and design. *British Journal of Anaesthesia, 52*, 61–67.

Burchard, K. W., & Rowland-Morin, P. A. (1990). A new method of assessing the interpersonal skills of surgeons. *Academic Medicine, 65*, 274–276.

Cooper, J. B., Barron, D., Blum, R., Davison, J. K., Feinstein, D., Halacsz, J., Raemer, D., & Russell, R. (2002). Video teleconferencing with a realistic simulation for medical education. *Journal of Clinical Anesthesia, 12*, 256–61.

deGroot, A. D. (1965). *Thought and choice in chess.* Hague, Netherlands: Morton.

Dorwick, P. W. (1991). *Practical guide to using video in behavioral sciences*, (pp. 17–29). New York: Wiley.

Dorwick, P. W., & Biggs, S. J. (1983). *Using video psychological and social applications.* Chichester, UK: Wiley.

Ericsson, K. A., & Simon, H. A. (1984). *Protocol analysis: Verbal reports as data.* Cambridge MA: MIT Press.

Flanagan, J. C. (1954). The critical incident technique. *Psychological Bulletin, 51*, 327–358.

Gaba, D. M., & DeAnda, A. (1988). A comprehensive anesthesia simulation environment: Recreating the operating room for research and training. *Anesthesiology, 69*, 387–394.

Goldman, L. I., McDonough, M. T., & Rosemond, G. P. (1972). Stress affecting surgical performance and learning I: Correlation of heart rate, electrocardiogram and operation simultaneously recorded on videotapes. *Journal of Surgical Research, 12*, 83–86.

Guerlain, S., Turrentine, B., Calland, J. F., & Adams, R. (2004). Using video data for analysis and training of medical personnel. *Cognition, Technology, and Work, 6*, 131–138.

Harper, B. D., Mackenzie, C. F., & Norman, K. L. (1995). *Quantitative measures in the ergonomic examination of the trauma resuscitation unit anesthesia workplace.* Proceedings of the Human Factors and Ergonomomics Society Annual Meeting.

Harrison, B. (1991). Video annotation and multimedia interfaces: From theory to practice. In: Proceedings of the Human Factors and Ergonomomics Society Annual Meeting.

Howard, S. K., Gaba, D. M., Fish, K. J., Yang, G., & Sarnquist, F. H. (1992). Anesthesia crisis resource management training: Teaching anesthesiologists to handle critical incidents. *Aviation, Space and Environmental Medicine, 63*, 763–770.

Hoyt, D. B., Shackford, S. R., Fridland, P. H., et al (1988). Video recording trauma resuscitation: An effective teaching technique. *Journal of Trauma, 28*, 435–440.

Jeffries, N. J., Hunter, A., Bernhard, W., Mackenzie, C. F., Horst, R., & the LOTAS Group (1994). Incidence of procedural errors and untoward occurrences associated with tracheal intubation assessed from videotapes and self-reports. *Anesthesiology, 81*, A1211.

Kaufman, D. M., & Kaufman, R. G. (1983). Usefulness of video-tape instruction in an academic department of neurology. *Journal of Medical Education, 58*, 474–478.

Klein, G. A. (1989). Recognition-primed decisions. In W. B. Rouse (Ed.), *Advanced in man–machine systems research* (Vol. 5, pp. 47–92). Greenwich, CT: JAI Press.

Klein, G. A., Orasanu, J., Calderwood, R., & Zsambok, C. E. (Eds.). (1993). *Decision making in action: Models and methods*, Norwood, NJ: Ablex.

Laughery, K. R. (1984). Computer modeling of human performance on microcomputers (microSAINT). *Proceedings of Human Factors and Ergonomic Society.* 884–888.

Laws, J. V. (1989). Video analysis in cognitive ergonomics methodological perspective. *Ergonomics, 32*, 1303–1318.

Lee, S. K., Pardo, M., Gaba, D., Sowb, Y., Dicker, R., Straus, E. M., et al (2003). Trauma assessment training with a patient simulator: A prospective randomized study. *Journal of Trauma*, 55: 651–7.

Mackay, W. E., & Tartar, D. E. (1989). Introduction to special issue on video as a research and design tool. *Special Interest Group on Computer and Human Interactions (SIGGH) Bulletin, 21*, 48–129.

Mackenzie, C. F., Craig, G. R., Parr, M. J., Horst, R., & the LOTAS Group (1994). Video analysis of two emergency tracheal intubations identifies flawed decision making. *Anesthesiology, 81*, 911–919.

Mackenzie, C. F., Jefferies, N. J., Hunter, A., Bernhard, W., Xiao, Y., the LOTAS Group, & Horst, R. L. (1996). Comparison of self reporting of deficiencies in airway management with video analyses of actual performance. *Human Factors, 38*, 623–635.

Mackenzie, C. F., Hu, P. F. -M., Horst, R. C., & LOTAS Group (1995). An audio–video acquisition system for automated remote monitoring in the clinical environment. *Journal of Clinical Monitoring, 11*, 335–341.

Mackenzie, C. F., Hu, P. F. -M., Xiao, Y., & Seagull, F. J. (2003). Video acquisition and audio system network (VAASS-NET®) for analysis of workplace safety performance. *Biomedical Instrumentation and Technology*, 221–227.

Mackenzie, C. F., Martin, P., Xiao, Y., & the LOTAS Group (1996). Video analysis of prolonged uncorrected esophageal intubation. *Anesthesiology, 84*, 1494–1503.

Mackenzie, C. F., & Lippert, F. K. (1999). Emergency department management of trauma. *Anesthesiology Clinics of North America, 17*, 45–61.

Mackenzie, C. F., & Xiao, Y. (2003). Video techniques and data compared with observation in emergency trauma care. *Quality and Safety in Health Care, 12*, i51–i56.

Mackenzie, C. F., Xiao, Y., & Horst, R. (2004). Video task analysis in high performance teams. *Cognition Technology and work 6*, 139–147.

Mallet, J. (1990). Communication between nurses and post-anaesthetic patients. *Intensive Care Nursing, 6*, 45–53.

McDonald, J. S., Dzwonczyk, R., Guptas, B., & Dahl, M. (1990). A second time study of the anaesthetists intraoperative period. *British Journal of Anaesthesia, 64*, 582–585.

Mehta, N. Y., Haluck, R. S., Frecker, M. I., & Snyder, A. J. (2002). Sequence and task analysis of instruments used in common laparoscopic procedures. *Surgical Endoscopy, 16*, 280–285.

Nardi, B. A., Kuchinsky, A., & Whittaker, S. (1997). Video-as-data: Technical and social aspects of a collaborative multimedia application. In K. E. Finn, A. J. Sellen, & S. B. Wilbur (Eds.), *Video mediated communication* (pp. 487–517). Mahwah, N.J.: Lawrence Erlbaum Associates, Inc.

Pringle, M., & Stewart-Evang, C. (1990). Does awareness of being video recorded affect doctors' consultation behavior. *British Journal of General Practice, 40*, 455–458.

Roeske, N. C. A. (1979). The medium and the message: Development of videotapes for teaching psychiatry. *American Journal of Psychiatry, 136*, 1391–1397.

Sanderson, P. M., & Fisher, C. (1994). Exploratory sequential data analysis: Foundations. *Human Computer Interaction,* 9, 251–317.

Sanderson, P. M., James, J. M., & Seidler, K. S. (1989) SHAPA. An interactive software environment for protocol analysis. *Ergonomics, 32*, 1271–1302.

Scheingold, L., & Smith, S. (1980). The use of videotape for teaching internal medicine in a family practice residency. *Journal of Family Practice, 11*, 467–468.

Seagull, F. J., Mackenzie, C. F., & Bogner, S. M. (2002). *Combining experts and video clips: Ergonomic analysis for safer medical instrument trays.* Proceedings of the Human Factors and Ergonomics Society Annual Meeting, Denver, CO.

Segal, L. D. (1994). *Action speak louder than words: How pilots use non verbal information for crew communications.* In the Proceedings of the Human Factors and Ergonomic Society 38th Annual Meeting, Boston, MA.

Shesser, R., Smith, M., Kline, P., Rosenthal, R., Turbiack, T., & Chen, H. (1985). A cost effective emergency medicine clerkship. *Journal of Medical Education, 60*, 288–292

Smith, H., Macintosh, P., Sverrisdottir, A., & Robertson, C. (1992). The ergonomic analysis of a trauma resuscitation room. *Health Bulletin, 50*, 252–258.

Tardiff, K. (1981). A videotape technique for measuring clinical skills. Three years of experience. *Journal of Medical Education, 56*, 187–191.

Tardiff, K., Redfield, J., & Koran, L. M. (1978). Evaluation of videotape technique for measuring clinical psychiatric skills of medical students. *Journal of Medical Education, 53*, 438–440.

Townsend, R. N., Clark, R., Ramenofsky, M. L., & Diamond, D. L. (1993). ATLS-based videotape trauma resuscitation review: Education and outcome. *Journal of Trauma, 34*, 133–138.

Tukey, J. W. (1977). *Exploratory data analysis.* Reading, MA: Addison-Wesley.

Weinger, M. B., Gonzales, D. C., Slagle, J., & Syeed, M. (2003). Video capture of clinical care to enhance patient safety. *Quality and Safety in Health Care, 13*, 136–144.

Weston, C. F., Richmond, P., McCabe, M. J., Evans, R., & Evans, R. (1992). Video recording of cardiac arrest management: An aid to training and audit. *Resuscitation, 24*, 13–15.

Xiao, Y., & Mackenzie, C. F. (1997). *Uncertainty in trauma patient resuscitation.* In the Proceedings of the Human Factors and Ergonomics Society 39th Annual Meeting, Albuquerque N. M.

Xiao, Y., & the LOTAS Group (2001). Understanding coordination in a dynamic medical environment: Methods and results. In M. McNeese, E. Salas, & M. Endsley (Eds.) *New trends in collaborative activities: Understanding system dynamics in complex environments* (pp. 242–258). Santa Monica, CA: Human Factors and Ergonomics Society.

Xiao, Y., & Mackenzie, C. F. (1999, October). *Micro-theory methodology in critical incident analysis.* In the Proceedings of IEEE International Conference on Systems, Man and Cybernetics, Tokyo, Japan.

Xiao, Y., & Mackenzie, C. F. (2004). Introduction to the special issue on video-based research in high risk settings: Methodology and experience. *Cogn. Tech. Work, 6*, 127–130.

Xiao, Y., Mackenzie, C. F., & Orasanu, J. (1998). Visual scanning patterns during remote diagnosis. *Proceedings of the Human Factors & Ergonomics Society 42nd* Ann Meeting, Santa Monica, CA. 850–854.

Xiao, Y., Mackenzie, C. F., Patey, R., & the LOTAS Group (1998). *Team coordination and breakdowns in a real-life stressful environment.* In the Proceedings of the Human Factors and Ergonomics Society 42nd Annual Meeting, Santa Monica, CA.

Xiao, Y., Mackenzie, C. F., & the LOTAS Group (1995). *Decision making in dynamic environments: Fixation errors and their causes.* In the Proceedings of the Human Factors and Ergonomics Society 39th Annual Meeting, San Diego, CA.

38

USABILITY EVALUATION IN HEALTH CARE

John Gosbee & Laura Lin Gosbee
National Center for Patient Safety, Veterans Health Administration, Ann Arbor

As in many domains, the lack of human-factors engineering (HFE) process can lead to some very unfortunate consequences in health care. A plug for pediatric EKG leads looked like it should have been plugged into an electrical socket, and was. Labels for a medication vial have "5mg/ml" in large bold type, with "5ml" as total volume in small type at the bottom of the vial. The nurse was asked to draw up and give 5mg, and gave the whole vial. Instructions for defibrillator leads on automated external defibrillator (AED) were located on the peel-away paper covers, and the wind blew them away while policeman were using the AED outdoors.
Consider the following quiz:

1. If it is too hard to assemble life-saving equipment in a timely fashion, whom should you call and let know?
   a. The procurement committee.
   b. The engineer from the company who designed and marketed the equipment.
   c. The local newspaper.
   d. No one. Changing the design of anything is nearly impossible.
   e. a and b.
2. If someone moved the leads on the defibrillator to the opposite side?
   a. A few people would notice, but it would not increase accidents.
   b. It would have no effect.
   c. It would have a measurable effect with an increased accident rate.
   d. A few people who are day-dreaming would mix up the leads, but not those who cared and were paying attention.
   e. Caution labels would solve any confusion that arose.
3. If someone suggests using a computer to solve a systems issue in your hospital:
   a. Embrace the technology and stop being a Luddite.
   b. Go to the training courses, computerization is inevitable.
   c. Ask the developers or vendors to show the usability testing data.
   d. Find out if the people implementing the system have done a simulation to see if there are unintended consequences.
   e. c and d.

(Answers: 1. e, 2. c, 3. e.)

There is much advice about the ways that the medical device industry should embrace HFE (Wiklund, 1995). However, it has been recommended that human factors evaluation should also play a key role in health care delivery organizations (Welch, 1998). It is not entirely clear how a health care system could incorporate such human factors techniques, such as usability evaluation. To address

TABLE 38–1 Summary of Patient Safety Activities That can Incorporate HFE[a]

Candidate Activities for HFE	HF *Concepts/Methods That Should be Applicable*	*Purpose of Applying* HF *Methods*
Development activities		
In-house device or software design and development	User-centered design	Output from applying HF: provide functional requirements, guides design concepts, validatesdesign concepts,
Policy or guideline design and development	Human factors evaluation	promotes user acceptance
Training and education curriculum design and development	Human factors analyses (see Table 2. Overview of Different HFE Methods	Overall goal: Ensure usability (efficient, functional, easy to learn, easy to use, low mental workload
Paper forms (e.g., labels, order forms charts, instruction sheets, etc.)		
Evaluation activities		
Procurement (of devices, software, training progras)	Usability evaluation	Output from applying HF: usability data for comparing products from competing
Adverse event investigation: RCAs, FMEAs and reporting systems	Human factors analyses (see Table 38-2. Overview of Different HFE Methods)	vendors, identification of HF issues or user interaction problems that may system be causing errors
		Overall goal: Identify usability issues with prospective or existing equipment, software or training programs, or existing policies, or procedures that may lead to errors

[a]From Gosbee, 2004

this, the following chapter is filled with practical information about the role of usability evaluation for health care providers.

Table 38–1 provides an overview of patient safety activities that can incorporate HFE. The two rows represent the two broad categories of activities: development (creating and writing policies), and evaluation (e.g., root cause analysis). For each of those categories, Table 38–1 provides one column that lists the concepts and methods that are applicable and another column describing the purpose or rationale for applying HFE.

To incorporate HFE into any organization, one of the first steps is awareness and training. In health care, there has been a huge increase in awareness of errors and the role HFE could play (Murff, Gosbee, & Bates, 2001). There is much less understanding and agreement on the best way to educate and implement HFE principles and tools.

There has been some literature on successful incorporation of HFE and usability testing into health care organizations. Welch (1998) proposed avoiding troublesome devices and software by both performing usability testing of products before purchase and demanding usability testing data from manufacturers. The Mayo Clinic in Rochester, Minnesota built usability (i.e., human factors) labs to evaluate many software products before purchase, as well as to use them for their own development projects (Claus, Gibbons, Kaihoi, & Mathiowetz, 1997). Some in the patient-safety improvement arena have cited the importance of usability testing before purchase or during implementation of a new product (Gosbee & Lin, 2001). It is also recommended that usability testing be integrated into in-house design and development of software and complex work areas.

There are many practical examples of how health care organizations have incorporated usability testing activities into their health care organization, including: (a) providing crucial information to support procurement decisions; (b) understanding where to focus training and other implementation efforts, and (c) creating teachable moments for human factors concepts and framework. These efforts have also helped organizational culture issues in health care that occasionally stand in the way of implementing patient safety overall.

OVERVIEW OF METHODOLOGIES

There are many human factors evaluation techniques and many derivations of each type (see Table 38–2.). Handbooks have been developed for all types of industries, who are struggling to learn and use these helpful techniques (Nielsen, 1993; Wiklund, 1995). The reader is referred to the standards document, *Human Factors Design Process for Medical Devices* (American National Standards Institute/Association for the Advancement of Medical Instrumentation, 2001). This comprehensive document provides an overview of these techniques, key references, and many medical device examples.

Table 38–2 contains a short overview of the various HFE methods one might use in a health care setting. Each row lists the name of the method (activity), a short description, and the type of product one would expect from that activity. The first column describing activities gives the reader a sense of the scope of human factors. While the last column, products, gives specific products that can be inputs or adjuncts to patient safety or health care analyses.

To provide illustrative human factors evaluation samples and a background to subsequent case studies for this chapter, usability testing and heuristic evaluation techniques will be described in more detail. Applicable standards documents will be briefly described, as most health care industry and government leaders will seek HFE guidance from them.

Usability Testing

Usability testing involves putting a system to the test with actual end-users. The key components include the system itself, the setting, the end-user, the set of tasks, and data-collection approaches. Typically, several end-users are brought into a testing situation one at a time and asked to accomplish certain tasks while various data are gathered. The more difficult parts of a usability test often include determining the end-user population(s), defining realistic and representative tasks, and identifying performance measures that will inform designers or decision makers. Performed correctly, researchers have determined that up to 90% of all design problems can be uncovered with as few as four end-users (Nielsen 1993).

The System. The system includes many things beyond the physical device, including accessories, labeling, instruction manuals, and related protocols. The focus of usability testing may span from the micro level through the macro level. At the micro level, tests evaluate very specific characteristics or elements of an interface (e.g., color coding of display elements, readability of text messages, etc.), whereas at the macro level the focus of a test is broader and encompasses the integrated system (e.g., interactions with other equipment, performance under noisy and cluttered conditions).

The Setting. Testing situations vary in realism (i.e., fidelity). For instance, low fidelity can include nonfunctional paper prototypes, and high-fidelity may involve full-scale simulations that involve many people, devices, and rooms. Each of these test situations will provide performance data from direct observation, video recording, or electronic capture of device data.

The End Users. The importance of involving the various people who will actually be using the product or carrying out the process cannot be overstated. One or more end-user populations will have various sets of characteristics: age, gender, education, training, knowledge, experience, and end goals. The users might be new to the product, or be highly experienced with it. They may work individually or rely on teamwork. These characteristics influence how the users interact with a system.

The Set of Tasks. The tasks for usability testing might include entering, updating, or retrieving information on a patient, setting up a system for use, or trouble-shooting a system. The tasks chosen for testing may be individual tasks or a collection of tasks embedded in a longer realistic scenario. For example, infusion pump tasks might include: change pump infusion rate, attach infusion pump to pole, and change alarm volume. There is also an opportunity to select tasks that can be completed by an individual or by a team. One may also choose to add a second task or provide realistic interruptions to observe a more representative scenario where users are dividing their time between multiple goals.

Data Collection. Data collection techniques may involve a range of activities, from an observer taking notes and using a stopwatch to video recording.

TABLE 38–2. Overview of Different HFE Methods [a]

HF *Analysis Activity*	*General Description*	*Analysis Products*
Field observations	Unobtrusively observe actual users in the typical work environment carrying out typical tasks. Taking note of how work is carried out, who carries it out, what they use to carry it out, who they interact with, environmental factors (light levels, noise, crowding from equipment or people, etc.)	Characterizes typical work environment, identifies factors that might impact how clinicians perform (e.g., limitations of equipment, low light levels, high risk or time pressure, frequent distractions, etc.)
Simulation or bench tests	Simulate a process or operation of a device, using different scenarios (e.g., different tasks, time pressure, lighting, errors that one must recover from, etc.). Simulation involves end users, whereas bench tests can be performed by the analyst	Mapping of the system structure (e.g., where do all the menus in a software program lead to? Where does the pharmacist have to go to retrieve XYZ?)
Information requirements or functional needs assessment	Information requirements analysis identifies what information a user needs to carry out specific tasks or activities (from micro to macro. How can a user tell he must push that button next, how does a user know he must perform that task next, how does he know who to contact to relay information?). Similarly, a functional needs-assessment identifies what tools or information a user requires in order to accomplish a task	Information and functional needs of the user Identifies task-related activities that depend on short or long-term memory, identifies where information should be supplied, how it should be supplied, identifies what tools a user needs to accomplish a task
Heuristic evaluation	Evaluates equipment or a process against a set of human factors principles. These principles prompt such questions as: Does the software provide functionality needed by the user? Are buttons grouped in a logical fashion? Is there sufficient feedback to tell the user he has completed a task correctly? Is it obvious what a user must do next?	Identifies areas where human factors principles are violated, which may lead to unwanted consequences such as: frequent user errors, slips, high mental workload, user frustration, inefficient or inaccurate task completion, misunderstanding of policies, and deviation from prescribed guidelines or procedures
Cognitive walkthrough	A user is asked to demonstrate or walk through a device or process, thinking out loud or providing commentary on what he is doing and thinking at each step of the task or process.	Characterizes where human decision-making is involved in a task, factors that influence decision-making, including: expertise a user might rely upon, where information is retrieved from, strategies adopted, work-arounds invented to circumvent a deficiency, and so on.

[a] Adapted From Gosbee, 2004

The quantitative categories of data include time on task, error type and rate, recovery type and speed, and mental workload (e.g., NASA-tlx is an example of a well-accepted measure of workload; see Wickens, 1992, for other methods of assessing mental workload). The qualitative data includes listening to words and observing actions that indicate confusion, frustration, or enjoyment. Surveys and interviews also have a role. The data is gathered towards larger testing goals to understand usability, learnability, and acceptability.

The data collection should match the goal or sub-goals of the usability test. If avoidance of crucial error patterns is key, direct observation might suffice. If serial sets of tasks have to be accomplished in a timely manner (e.g., code situations), then time on task and recovery from error speed would be central to the usability test plan.

Heuristic Evaluation

Heuristic evaluation (i.e., expert evaluation) is the systematic application of HFE principles or guidelines to a system (Kaufmann, Thompson, Patel, Page, & Kubose, 2003; Nielsen & Mack, 1995). As with usability testing, the system can be simple paper forms, single devices or computer screens, or a complex set of work areas. Typically, the HFE expert evaluates all aspects of a system and determines how these features deviate from a list of principles. More than one expert is used, but the research on effectiveness varies from 90% of design issues discovered by two experts to 50% found by four experts (Jacko & Sears, 2003). The inspection can be visual or interactive with the system. There are many variants of heuristic evaluation. Some are listed in Table 38–2 or can be found in Nielsen and Mack (1995).

Next is a partial list of HFE principles from Nielsen (1993), which provides some relevant examples and definitions.

Simple and Natural Dialogue. This refers to the display or provision of visual, auditory, or tactile item information. The dialogue should be as natural as possible. The terms, icons, alarms, and other system features should be familiar to the user and easily and correctly interpreted. The dialogue should also reinforce how the system works and be in line with user expectations.

Minimize User Memory Load. Minimizing user memory load can be accomplished with such design features as: (a) making relevant information apparent to the user, (b) making it obvious how to navigate to a desired location or page in a paper manual, and (c) automatically prefilling information in fields where user must otherwise retype memorized information.

Consistency. Keeping elements of a system consistent refers to obvious features like visual displays, and deeper features like functionality of the system. Elements of a system should behave in a consistent and predictable manner from one component to another.

Feedback. Feedback is the method by which a user is informed about what a system or device is doing, whether the user intended commands were executed correctly and successfully, whether an error has occurred, or how to recover from errors.

Human Factors and Health care Standards Documents

HFE and medical device standards documents are important to know about, since most health care industry and government will seek human factors engineering guidance from them. If your hospital is discussing HFE procurement issues with industry representatives, it will help you to know basic content and scope of ANSI/AAMI HE–8 (1993). This set of guidelines offers specific HFE design specifications for many device and system features (e.g., knob and button specifications). If your health care organization is overhauling your in-house product development process, you can gain key insights from ANSI/AAMI HE–4 (2001). This standard, *Human Factors Design Process for Medical Devices* provides detail about how to bring together risk assessment and control and HFE methodologies during development cycles (U.S. Food and Drug Administration, 1998).

For more information about national and international HFE standards in this book, see Chapter 25, "New Technology Implementation in Health Care" (Karsh & Holden).

PRACTICAL GUIDELINES: APPLICATION TO HEALTH CARE

There is no shortage of candidate products or processes to which human factors can be applied. In this section, we outline various hospital activities that can benefit from HFE principles and methodologies. These can also be adapted for the outpatient setting. Devices range from consumables like syringes to room-sized robotic surgery systems. The work areas range from familiar office-style ward clerk stations to complex and changing anesthesia stations. The architectural aspects range from mundane aspects of bathroom door placement to the colliding needs of privacy and visibility in intensive care units.

In general the activities will have one or more of the following purposes:

1. Determine why and how errors occur (e.g., close call investigation, adverse event, investigation, incident investigation).
2. Determine the usability of any device or process (e.g., compare two devices, information systems, or policies or protocols, prospective risk assessment on existing equipment, prospective risk assessment on policies or protocols, evaluate efficiency of workplace layout, evaluate effectiveness of signs, warnings, labels).
3. Develop a product (e.g., education program, training material, policy or protocol, software or information system, labels warnings or alerts, device or tool, facility or work area).
4. Improve a process or product (e.g., improve a training program, revise a policy or protocol, introduce a new policy or protocol, develop learning aids, develop cognitive aids, such as reminders, cheat sheets, guides, etc., reorganize storage areas, redesign a work area, improve labeling or warning signs).

Activities that have any of the explicit or embedded purposes defined above are candidates for incorporating HFE. We outline seven activities that fall into these categories, which offer leverage for making patient safety improvements in health care organizations: root cause analysis, failure modes and effects analysis, procurement, in-house software development, policy and protocol development, training and education development, and facility design.

Root Cause Analysis

When analyzing the root cause of an accident or adverse event, an understanding of HFE can provide insight into the factors that may have contributed to it. Because health care workers must contend with a work environment that is rife with interruptions, shift work, complex equipment, a multitude of policies and protocols, time pressure, and other stressors, errors are bound to occur in these imperfect conditions. A poorly designed device, for instance, represents a latent failure, or an accident waiting to happen, in the right conditions. The same can be said about a poorly designed workspace or a complex policy that is reinforced with difficult-to-decipher written material or an information system whose functions make it tedious or time-consuming to follow the policy. Incorporating HFE methods and tools into the analysis of actual adverse events or close calls provides a means to identify such factors, directing corrective measures and interventions to the underlying systems issues and not the user, who is usually only the final trigger in a string of events or failures. (Gosbee & Andersen, 2003; Schneider, 2002)

Root cause analysis (RCA) teams are in an excellent position to incorporate human factors methods into their process. Heuristic evaluation, discussed in the previous section, and other analysis methods can be conducted to identify vulnerabilities that contributed to an adverse event or close call. This might involve a human factors engineer trained in these methods working closely with front-line personnel. Also, usability testing can be used to identify the types of errors that might occur as well as the conditions under which their likelihood of occurring increases.

A heuristic evaluation can have a very narrow focus or a broad one. A narrow focus might include an evaluation of specific products or equipment. A broader focus would look at all the related products or processes that surround an activity, for instance, all the equipment that is used in the process of following a protocol, along with the interpersonal communication that occurs, the paperwork that is filled out, the workspace layout, the storage and retrieval of products, and the organizational factors. A broader focus may examine a widevariety of issues that may have an impact on the activity. For example

- What were the working conditions at the time of the event?
- What aspects of the user tasks or working environment may have interfered or influenced the event?

- What aspects of the product or equipment are counterintuitive or are incongruent with the user knowledge, training, education, or experience?
- How could the device or paper forms be redesigned to better accommodate the user task?
- What information was missing for the user to make an informed decision?
- Does the training on one device hinder the learning of another?
- Do the functions of the software support the user task for following a procedure?
- Are there too many functions on the device or software, adding too much complexity to typical tasks?
- What tools should be available to the user to help them follow a policy?

Failure Modes and Effects Analysis

Another formal method in patient safety where human factors evaluation should be used is called failure modes and effects analysis (FMEA). The details of this proactive risk assessment technique are beyond the scope of this chapter. The reader is referred to Stalhandske, DeRosier, Patail, and Gosbee (2003) for a discussion of this process and another chapter in this book, "Human Factors Risk Management in Medical Products" (chapter 35, Israelski & Muto) One activity in the process of FMEA that can benefit from HFE is that of identifying failure modes, or things that can go wrong with a system of interest. The system could include a device or a process such as a collection of activities. Usability evaluation can be used to make the FMEA robust. A heuristic evaluation can be used to identify failure modes of a device that may cause user errors. It can also identify failure modes of a process, that arise from human factors issues, such as lack of information for decision-making, poor labeling, insufficient lighting, interruptions, and so on.

Usability testing can also be used as a means for identifying various failure modes, by creating test conditions (task, environment, etc.) that reflect the conditions and situations under which a process is carried out or device is used. For instance, a FMEA team might not brainstorm or identify the failure mode of "broken infusion pump after being dropped." Such a failure mode could either occur or nearly occur when 5 nurses grasp the new infusion pump with unwieldy handles and try to attach the pump to a pole.

Procurement

Procurement decisions often consider a multitude of factors, one of which is usability. Usability itself can be defined as having multiple criteria. These can be generalized into the following categories:

1. Functionality: Does it meet functional needs of end-users, does it create new needs or tasks based on limitations or excessiveness in it functionality, does it present functionality rarely used, is it used for different functions depending on the user? Can it be used in the work environment that it is intended for (e.g., portability)?
2. Interface Design: Does it present high risk for user errors? Is the interface or method of operating intuitive or confusing? Are labels and warnings easily perceived and understood? Does it require daily use to become a competent user? Is it easy to tell what the device is doing? Are there annoying features that in some situations become a risk for misuse?
3. Training and Learning: Does it pose training challenges for novice users, users with extensive experience on another system (transfer of training issues), users who only occasionally use the device or software, different user groups who may not have the same goals and thus use it differently?

The HFE methods covered in the Overview of Methodologies section (also in Table 38–1) are useful for ascertaining levels of usability of devices or software prior to its purchase. (Gosbee & Lin, 2001; Schneider, 2002). There are a growing number of health care organizations adopting this practice. Allina Health System in Minneapolis, Minnesota for instance incorporated human factors in their patient safety program by using HFE concepts in their workplace analysis, accident investigation, and incident reporting program. Staff training on how to conduct usability testing has led to the integration of user testing in their product selection process. Desert Samaritan Medical Center in Mesa, Arizona performed user testing in order to anticipate problems with training and implementation of new IV pumps (Gosbee, Arnecke, Klancher, Wurster, & Scanlon). User testing can uncover problems ranging from inappropriate machine default settings, obscurity of critical functions, ineffective or nuisance alarms (e.g., see Stanton, 1994, for human factors issues related to alarms), poor labeling, or ambiguous machine dialogue.

Beyond a usability evaluation in the intended work environment, HFE can also be used to examine maintenance issues that would be encountered during the life cycle of the device. Maintenance issues might include the usability of the device from a biomedical engineer perspective. This points to the importance of identifying all the different groups of end users. For instance, a biomedical engineer who is responsible for setting defaults for various pumps, may encounter several usability issues. The sequence of steps for doing this may be complex and prone to error. The default values may vary depending on whether the pump is used in the recovery room versus on the floors. The preferred default values may also be unknown to the engineer maintaining the pump. When there are software updates, this may impact how the default settings are set up or changed. These issues should be evaluated when assessing a device usability.

In-House Software Design and Development

When making product selections, HFE can help identify potential risks for error. In developing in-house products, developers should be no less vigilant about usability issues. (Gosbee & Gardner-Bonneau, 1998; Zhang, et al., 2003). HFE can provide a framework for the development cycle (user-centered design and iterative design and testing). Such a framework provides opportunity for periodic evaluation so that usability issues that have the potential to become latent errors are resolved during development and not after the product is rolled out.

In a user-centered design framework, the development cycle becomes iterative in nature. That is, design and testing are conducted iteratively (repeated at various intervals throughout the development cycle of the product) so that user testing and design refinements start early in the design process and continue throughout the design cycle. Design concepts and early prototypes should undergo user testing for validation and to identify any aspects of design that don adhere to HFE principles. What is learned from this early user testing guides any improvements to the design. User testing and design refinement are then repeated throughout the design and development.

In addition to a framework for the development cycle, HFE evaluation can be used to determine functional requirements and information requirements. These requirements influence a variety of aspects of the products design including: the organization and layout (of information, controls, etc.), structure and sequence of tasks, level of customization, level of detail of information, organization and layout of the menu structure (for software), appropriate graphics, on-line training features or help functions, and so on.

There are numerous examples of in-house software products. They include computerized provider order entry (CPOE) systems, bedside barcode medication administration (BCMA) systems, pharmacy information systems, clinical decision support systems, clinical data repositories, patient registration, and medical record systems. Although different end-users of these information systems may have similar information needs for their decision-making, users may differ vastly in the ways in which they search for information, in the context of their decisions, in their need for specific representations or organization of that information, in their need for detail, and in the end goals that drive the way in which they interact with the system. These sometimes subtle but crucial differences determine whether a software product is functionally appropriate and easy to use for any user group. An HFE analysis can help to identify the functional and information needs of the various user groups and usability testing can help to verify that the software being developed meets these needs in an efficient and usable manner.

Policies and Protocols

Up to this point, the discussion has centered around conducting usability evaluations on commercial or in-house products. It is important not to overlook the fact that policies and protocols are products that can be evaluated as well. HFE can be used to help assess or develop policies, protocols or guidelines. A HFE evaluation can help to identify any special training needs with respect to specific policies or protocols that might be challenging due to poor design (of equipment or software used to help carry it out), complexity of the protocol, or requisite knowledge or expertise in an area needed to carry it out.

The task of developing policy or protocol material is a two-tiered challenge. The first is to address the problem or issue for which the policy or protocol will alleviate, and the second is to avoid creating additional problems or issues that might arise from the introduction of a new policy or protocol. To tackle these two challenges, it is helpful to first gain an understanding of the nature of the activities involved in carrying out the policy or protocol, and the information, material, and machines that the user must come in contact with the carry it out.

To gain this understanding, the designer or developer of policies or protocols should carry out human factors evaluations. Heuristic evaluation and testing of the policy or protocol can be conducted to evaluate its readability, ease of learning, prominence of critical steps or information, complexity, mental workload, and so on. Similar to in-house software development, this can be done iteratively to guide development of policy content and the method of conveying it the end-users.

When a policy or protocol is perceived to be complex or requires specialized knowledge to carry it out, HFE can also be used to help identify portions of the process where supplemental material is needed. Supplemental material includes any aids such as cognitive aids (e.g., "cheat sheets"), that either help learning or to help guide users (in decision-making, device operation, following steps, etc) of a policy or protocol. Aids may also come in the form of risk communication (e.g., warning signs), whose design and evaluation can be aided by the human factors research in the area of warnings and risk communication (Isaacson, Klein, & Muldoon 2001; Lehto, 2001; Wolgater, DeJoy, & Laughery, 1999).

The issues that are dealt with in a heuristic evaluation or usability testing encompass how a policy or procedure is conveyed (e.g., readability, availability, complexity), how a policy or procedure may interact or interfere with other work processes, what portions of the process should and can be standardized or automated, impact on mental or physical workload or timeliness of other time-critical activities (added paperwork, tedious or repetitious procedures).

Training and Education Development

Training and education programs are also products that should be developed with the aid of HFE evaluation methods. Doing so can help developers identify specific training needs with respect to specific devices, software, or policies that might be challenging due to poor design or complexity. This in turn would provide focus areas for training material being developed and a means to evaluate it for effectiveness. Other aspects of training and HFE can be found in chapter 44, "Teamwork Training for Patient Safety: Best Practices and Guiding Principles" (Salas, Wilson-Donnelly, Sims, Burke, & Priest).

The approach is the same as that used for any other in-house product discussed in the In-House Software Designs and Development and the Policies and Protocols sections. The goal of the end-product is to teach users, providing knowledge or guidance, that in turn will help them carry out a physical or cognitive task. Developers of training and education curriculum can adopt the same methods and tools (user testing and heuristic evaluation, or other methods in Table 38–2) in order to determine the functional requirements and information requirements for the training material. The product may be paper-based (e.g., user manual or guide) or electronic based (educational or training software). A heuristic evaluation can: Help identify issues or requirements for training such as the need for supplemental material, such as cognitive aids (e.g., cheat sheets), that either helps learning or helps guide users (in decision-making, device operation, following protocols, etc.); determine the frequency of refresher training to maintain competency with a product; and determine transfer of training issues (e.g., clinicians have extensive training with a specific software; how does this negatively or positively influence whether and how they gain competency in a new prospective system or system upgrade). These issues apply to trainees who are either health care workers or patients. Patient education programs can be designed based on what is learned in an HFE evaluation. For the outpatient setting, this is especially important if a home medical device poses usability issues (e.g., Gosbee, L., 2004).

Facility Design

When applying HFE to the remodeling or construction of a new facility or work area, or to the purchase of office furniture and storage equipment, attention must be paid to the degree to which it conforms to or supports the work that is carried out there (e.g., Harper, Mackenzie, & Norman, 1995; Scilafa, Laberge, & Ho, 2004). HFE can help determine the requirements for layout of a new facility based on an analysis of the work flow in different areas, functional needs of the task and the users, and ergonomic specifications. HFE can also help to inform requirements for furniture or storage equipment based on the nature of the task and user characteristics or functional needs. Finally, user testing can help evaluate the efficiency that the workspace affords and the adherence to the functional needs uncovered in a human factors evaluation (e.g., distance and accessibility between related work areas, width of pathways to accommodate equipment and personnel, level of lighting required for specific tasks, height of shelves so that 10th percentile female can read the labels and reach stored products, etc.) This topic is discussed in more detail in chapter 19 "The Physical Environment in Health Care" (Alvarado).

CASE STUDIES

Code Cart

In a Salt Lake City hospital they had many issues with code teams involving confusion and delay with medication drawer retrieval. When a patient had a cardiac or respiratory arrest, the code or emergency team is assembled in a room with a code cart containing medications and devices to perform advanced life support. All systems and personnel have to fit together and work together well, because seconds count.

McLaughlin (2003) and his colleagues focused on the HFE aspects of the code-cart drawer containing life-saving medications. With upper-management support, they developed a test plan with an aim to redesign the drawer for efficient and accurate retrieval by nurses given verbal orders. The scope of redesign was limited to the drawer that already fit within the several code carts deployed around the hospital.

The test plan included: (a) 9–11 nurses who respond to codes were the end-users for each of five design iterations (versions), (b) the actual drawer with real medications and packaging was used, (c) the tests were performed on tables or other convenient surfaces near the nurses work area, (d) each of 10 medications was called out one at a time for retrieval (not embedded in a full resuscitation scenario), and (e) the key data collection was time on task with accurate retrieval of all 10 medications.

The initial baseline drawer resembled a laundry hamper with several loose items placed in various places. Boxed injection systems were aligned together, as were the vials and other tubes of medications. There were also irregularly shaped items stacked to the side or on top of other items. Results of this drawer showed a total time ranging from 2:43 min to 3:58 min, and an average of 3:07 min. The general usability or design goal was to retrieve them all within about 1 min.

Future versions of drawers included various inserts, cutouts, and labeling (or not). Some versions had adequate average time on task (1:09), but the range went as long as 1:52 min. One version had labels for each cutout area within a blue foam, but the retrieval time actually slowed, since nurses did not trust that slots were filled correctly and looked at the labels on the vial or box. The final version, which was subsequently introduced into all code carts in the hospital, provided a retrieval time range of :55 min to 1:25 min, and an average of 1:08.

One bonus to this process of redesign was the interest level it generated for change. Change in health care settings is often hard to encourage or maintain in dynamic settings with a lot of traditional thinking. McLaughlin (2003) found that the usability testing with so many thought leaders helped decrease the time and effort for implementation planning and change management to almost zero.

IV Pumps

The Department of Veterans Affairs National Center for Patient Safety worked with American Institutes for Research to complete a comparative usability test of intravenous infusion pumps (U.S. Department of Veterans Affairs, 2002). The purposes were: (a) to assess the value of usability testing as part of an operational patient safety program, (b) to generally determine whether data regarding the comparative usability of infusion pumps (i.e., IV pumps) had the potential to be useful for procurement, (c) to assess the value of the findings with respect to findings and issues resulting from RCA involving IV pumps, and (d) to make a rudimentary comparison of the value of usability testing in a ICU simulator versus a usability testing laboratory.

Seventeen registered nurses participated in the usability test. Four nurses participated in a standard usability testing lab. Thirteen participated in an ICU simulator, which was equipped with two patient manikin simulators, patient monitors, a ventilator, and many other examples of ICU equipment and furnishings.

Three single-channel IV infusion pumps that offered comparable functionality were used. Test participants used the devices for the first time following a brief, in-service style training session. The test participants lack of prior experience using the pumps enabled us to focus the test on initial ease of use, that is, learnability (Nielsen, 1993), a strong indicator of long-term ease of use.

Each participant underwent two scenarios with two of the pumps for 30–40 min each. Performance measures included an assessment of task correctness, task times, ease-of-use ratings, confidence ratings, and an assessment of the need for design improvement ratings, overall preference, and scoring according to a 10-question usability rating form. Participants performed eight major tasks, some of which included several subtasks that fit into an overall patient-care scenario.

An analysis of both the objective and subjective data suggested a relatively distinct ordinal ranking among the three pumps with regard to their initial usability. Averaging all of the simulator-based test results, the highest ranked pump demonstrated superior performance in all major performance dimensions. Not surprisingly, the subjective impression of the participants was often in discordance with objective performance measures. That is, occasionally the task was not completed or completed correctly, but the nurse was confident the task was completed without issue.

Whether the purposes of the pilot test were realized is mixed and still in progress. The pragmatic difficulties of the usability tests have already informed design and scope of future testing for patient safety at VA. The utility for procurement is still under review and continues to evolve. Some patient safety findings from the testing correlate with those of RCA, but detail and specificity are lacking in both data sets to arrive at a definite conclusion. Finally, there were some inconsistencies in the findings from the simulator and usability laboratory test sessions, but without statistically significant differences (full results yet to be published).

Patient-Controlled Analgesia

A University of Toronto study looked at improving medical device design through the application of HFE (Lin, 1998). Human factors engineers focused on a commonly used device as the testbed, the patient-controlled analgesia (PCA) infusion pump. According to FDA incident reports reviewed in the study, human error was found to be responsible for a majority (68%) of fatalities and serious injuries associated with the PCA in a randomly chosen year. This is similar to what was found in other studies (Callan, 1990). Of the reports where human error was involved, nurse programming errors were found to be the most common type of human error in PCA use. Furthermore, the majority of programming errors involved setting an incorrect drug concentration, all of which led to an overdelivery of medication. Taken together, there was strong evidence that the design of the PCA infuser had many latent errors and could be improved with HFE.

As the first step towards demonstrating the utility of HFE, the engineers develop a redesigned PCA pump based on human factors analytical techniques and design principles, including cognitive task analysis (field studies, bench tests, and design checklist). In the field studies, for instance, nurses were observed in the recovery room doing many tasks at once and constantly being interrupted while programming the PCA. During bench tests, many difficulties were encountered even while operating the equipment under ideal conditions with the help of a users manual. Finally, the design checklist helped the engineers find where design principles were violated. The redesigned PCA pump included such features as logical grouping and labeling of controls, simplified and more natural language in the displayed messages, and improved status display and feedback.

The second step was to empirically evaluate the design with users. Two user groups participated: novice users (nursing students) and experienced PCA users (recovery room nurses). Participants were given a PCA order form (prescription) and their task was to program the pumps accordingly using each interface. The evaluations included performance metrics such as number of errors, time to complete the task of programming, and subjective workload measures (i.e., mental demand, frustration, etc.).

In both groups, there were marked improvements with the redesigned PCA pump. First, there was a reduction in number of errors recorded, 50% and 55% reduction for nursing students and nurses, respectively. Furthermore, there were no errors in setting the drug concentration with the redesigned system. This demonstrated that the new design had a degree of resistance to the most culpable error found in the medical device reports.

Accompanying the reduction in programming errors with the redesigned system was a statistically significant improvement in task completion time. Nursing students were able to complete programming tasks with the redesigned system 15% faster. The nurses showed 18% faster completion times despite having no prior experience with the new system, compared to several years of experience with the existing Abbott pump. This improvement can be attributed to, among other things, the fact that significantly fewer programming errors being made, and thus less time was wasted recovering from errors.

The subjective workload associated with the redesigned interface was found to be lower for both user groups using the redesigned pump compared to the existing pump: 53% and 14% lower for nursing students and nurses, respectively. Finally, post-experiment interviews with the nurses and nursing students showed that an overwhelming majority (100% of nursing students and 90% of nurses) preferred the redesigned system over the current system.

HFE AS TEACHING AND SELLING TOOL

HFE might provide a less-than-obvious benefit beyond those listed already. During the training for participation in safety activities, usability testing might be an effective tool to train a HFE mindset for clinicians and management (Gosbee, 1999).

The HFE mind-set (turning people brains around 180 degrees) is rarely accomplished with didactic approaches (Shapiro & Fox, 2001). This is mainly true because it is hard to teach the concept of–learned intuition.–Briefly, learned intuition is the phenomena whereby you cannot recall ever not knowing how to use something, and you cannot imagine someone else would not know how it should be used. Educators have created "teachable moments" about learned intuition by having students grapple with human factors design problems and having them develop and test remedies (Sojourner, Aretz, & Vance, 1993). Also, incorporating usability testing into resident, medical, pharmacy, and nursing student training has been used with success (Gosbee & Gardner-Bonneau, 1998).

One practical exercise that the authors have used dozens of times, include many aspects of interactive training tools from other HFE coursework. Groups of 3–4 people sit around a common table or work surface. One person is assigned to be the "Director," who reminds the end-user to think aloud, and leads the team in discussion about issues and redesign *after* usability test is complete. Another trainee is assigned to be the "End-User," who uses the system as they would (without interference or help). The remaining learners are "Observers" who document end-users actions and words, facial expressions, swearing, and so on. Once the person has completed the task of using the system (without guidance), the team summarizes issues related to observed behavior, comments, and other observations, and the team recommends redesign related to these issues. During the larger group discussion, the facilitator leads the discussion and team reports towards common themes for design deficiencies (HFE flaws).

In practice, many kinds of items and levels of complexity can be used for this exercise. Simple everyday items with design features that lend themselves to teaching points can be very useful. A generic list includes various types of juice pouches or travel package size items. Some home medical devices found in pharmacies or portable medical devices in hospitals can also be used with significantly more preparation and time allotment.

SUMMARY

This chapter focused on HFE methodologies that can be adopted in the health care setting to improve patient safety. Two methods were discussed as examples: heuristic evaluation and usability testing. These evaluation methodologies can be applied to a number of activities in a hospital setting or outpatient setting including: RCA, FMEA, procurement, in-house software design and development, policies and protocols, training and education, and facility design. In all these activities, HFE can be used to gain insight into human–system interaction issues that we also call usability issues, or latent failures that are "accidents waiting to happen." The intent of this chapter was to provide guidance to the practitioner wishing to employ human factors in their patient safety program or to teaching and demonstrating the concept or human factors mindset to clinicians and management.

References

American National Standards Institute, Association for the Advancement of Medical Instrumentation (1993). *Human factors engineering guidelines and preferred practices for the design of medical devices* (ANSI/AAMI HE–8). Arlington, VA: Association for the Advancement of Medical Instrumentation.

American National Standards Institute, Association for the Advancement of Medical Instrumentation (2001) *Human factors design process for medical devices* (ANSI/AAMI HE–74. Arlington, VA: Association for the Advancement of Medical Instrumentation.

Callan, C. M. (1990) Analysis of complaints and complications with patient-controlled analgesia. In F. M. Ferrante, G. W. Ostheimer, B. G. Covino (Eds.), *Patient-controlled analgesia* (pp. 139–150). Boston: Blackwell.

Claus, P. L., Gibbons P. S., Kaihoi B. H., & Mathiowetz, M. (1997). Usability lab: A new tool for process analysis at the Mayo Clinic. *Health care Information Management Systems Society Proceedings, 2*, 149–159.

Gosbee, L. L. (2004). Nuts! I can't figure out how to use my life-saving epinephrine auto-injector. *The Joint Commission Journal on Quality and Safety, 30*(4), 220–223.

Gosbee, J. W. (1999). *Human factors engineering is the basis for a practical error-in-medicine curriculum.* In the proceedings of the First Workshop on Human Error in Clinical Systems. Glasgow, Scotland.

Gosbee, J. W. (2004). Human factors engineering and patient safety: Starting a new series. *The Joint Commission Journal on Quality and Safety, 30*(4), 215–219.

Gosbee, J. W., Anderson, T. (2003). Human factors engineering design demonstrations can enlighten your RCA team. *Quality & Safety in Health Care, 12,* 119–121.

Gosbee, J. W., Arnecke, B., Klancher, J., Wurster, H., & Scanlon, M. (2001). *The role of usability testing in health care organizations.* In the Proceedings of the Human Factors Society 40th Annual Meeting, Santa Monica, CA.

Gosbee, J. W., & Gardner-Bonneau, D. (1998). The human factor: Systems work better when designed for the people who use them. *Health Care Informatics, 15*(2), 141–144.

Gosbee, J. W., & Lin, L. (2001). The role of human factors engineering in medical device and medical system errors. In C. Vincent (Ed.), *Clinical risk management: Enhancing patient safety* (2nd ed.). London: BMJ Press.

Harper, B. D., Mackenzie C. F., & Norman K. L. (1995). *Quantitative measures in the ergonomic examination of the trauma resuscitation units in anesthesia workspace.* In the Proceedings of the Human Factors and Ergonomics Society 39th Annual Meeting, Santa Monica, CA.

Jacko, J. A., & Sears, A. (2003). *The Human-computer interaction handbook: Fundamentals, evolving technologies, and emerging applications.* Mahwah, NJ: Lawrence Erlbaum Associates, Inc.

Isaacson, J. J., Klein, H. A., & Muldoon, R. V. (2001). Prescription medication information: Improving usability through human factors design. In M. S. Wogalter, S. L. Young, & K. R. Laughery (Eds.), *Human factors perspectives on warnings* (Vol. 2, pp. 104–108). Santa Monica, CA: Human Factors and Ergonomics Society.

Kaufman, D. R., Thompson, T. R., Patel, V. L,, Page, D. L., & Kubose T. (2003). Using usability heuristics to evaluate patient safety of medical devices. *Journal of Biomedical Informatics, 36*, 23–30.

Lehto, M. R. (2001). Determining warning label content and format using EMEA. In M. S. Wogalter, S. L. Young, & K. R. Laughery (Eds.), *Human factors perspectives on warnings* (Vol. 2, pp.143–146). Santa Monica, CA: Human Factors and Ergonomics Society.

Lin, L., Isla, R., Harkness, H., Doniz, D., Vicente, K. J., & Doyle, D. J. (1998). Applying human factors to the design of medical equipment: Patient-controlled analgesia. *Journal of Clinical Monitoring and Computing, 14*, 253–263.

McLaughlin, R. C. (2003). Redesigning the crash cart: usability testing improves one facility's medication drawers. *American Journal of Nursing, 103*(4), 64A–64F.

Murff, H. J., Gosbee, J. W., & Bates, D. W. (2001). Human factors engineering and medical devices. In K. G. Shojania, B. W. Duncan, K. M. McDonald, & R. M. Wachter (Eds.). *Making health care safer: A critical analysis of patient safety practices.* (pp. 459–470), Retrieved June, 2006, from http://www.ahrq.gov/CLINIC/PTSAFETY/index.html

Shojania K. G., Duncan B. W., McDonald K. M., & Wachter R.M. (Eds.). (2001, July). Evidence report/technology assessment No. 43 (AHRQ Publication No. 01-E058). Retrieved June, 2006, from http://www.ahrq.gov /CLINIC/ PTSAFETY/ index.html

Nielsen, J. (1993). *Usability engineering.* Cambridge, MA: AP Professional.

Nielsen, J., & Mack, R. L. (1995). *Usability inspection methods.* New York: Wiley.

Ritchie, E. M., & Gosbee, J. W. (1997). Design of a web site for rural health practitioners. *Health care Information Management, 11*(3), 97–112.

Schneider, P. J. (2002). Applying human factors in improving medication-use safety. *American Journal of Health-System Pharmacy, 59*(12), 1155–1159.

Scilafa, C., Laberge, J., & Ho, G. (2004). A wayfinding system for long term care facilities. *Ergonomics in Design, 12*(2), 5–11.

Shapiro, R. G., & Fox, J. E. (2001). *Games to explain human factors.* In the Proceedings of the Human Factors and Ergonomics Society 45th Annual Meeting, Santa Monica, CA.

Sojourner, R. J., Aretz, A. J., & Vance, K. M. (1993). *Teaching an introductory course in human factors engineering: a successful learning experience.* In the Proceedings of the Human Factors and Ergonomics Society 33rd Annual Meeting, Santa Monica, CA.

Stalhandske, E., DeRosier, J., Patail, B., & Gosbee, J. W. (2003). How to make the most of failure mode and effect analysis. *Biomedical Instrumentation and Technology, 37*(2), 96–102.

Stanton, N. (1994). *Human factors in alarm design.* London: Taylor and Francis.

United States Department of Veterans Affairs. (2002). *Pilot comparison usability test of intravenous infusion pumps* (technical report). Ann Arbor, MI: Veterans Affairs National Center for Patient Safety.

United States Food and Drug Administration. (1998). Human factors implications of the new GMP rule: New quality system regulation that apply to human factors. *Selections of center for devices and radiologic health guidance documents*, Retrieved from http://www.fda.gov/cdrh/humfac/hufacimp.html

Welch, D. L. (1998). Human factors in the health care facility. *Biomedical Instrumentation and Technology*, March/April, *32*(3), 311–316.

Wickens, C. D. (1992) *Engineering psychology and human performance (2nd ed.)*. New York: Harper Collins.

Wiklund, M. (1995). *Medical device and equipment design*. Buffalo Grove, IL: Interpharm Press.

Wogalter, M. S., DeJoy, D. M., & Laughery, K. R. (1999) *Warnings and risk communication*. London: Taylor and Francis.

Zhang, R., Patel, V. L., Hilliman, C., Morin, P. C., Pevzner, J., Weinstock, R. S., Goland, R., Shea, S., & Starren, J. (2003). Usability in the real world: Assessing medical information technologies in patients' homes. *Journal of Biomedical Informatics*, *36*, 45–60.

39

ASSESSING SAFETY CULTURE AND CLIMATE IN HEALTH CARE

Marlene Dyrløv Madsen and Henning Boje Andersen
Risø National Laboratory, Roskilde, Denmark

Kenji Itoh
Tokyo Institute of Technology, Tokyo, Japan

THE ROLE OF SAFETY CULTURE AND CLIMATE ASSESSMENT IN PATIENT SAFETY IMPROVEMENT

In recent years, the promotion of a culture of safety has, in many countries, become one of the key issues in patient care. It has been argued that a positive safety culture (or climate) is essential for minimizing the number of preventable patient injuries and their overall cost to society (e.g., Kohn, Corrigan, Donaldson, 1999; Nieva & Sorra, 2003; Zhan & Miller, 2003). At the same time, there is also an increasing recognition that it is necessary to determine the relationship between the effects of safety culture on health care outcome. (Gershon, Stone, Bakken, & Larson, 2004; Scott, Mannion, Davies, & Marshall, 2003a; Scott, Mannion, Marshall, & Davies, 2003b;). Efforts in this direction are, however, hampered in two respects: First, patient safety outcomes are hard to establish and validate across different patient populations and health care services; and second, there is no generally accepted model of safety culture and climate, identifying its components and their interrelationships (Collins & Gadd, 2002; Flin, Mearns, O'Connor & Bryden, 2000; Guldenmund, 2000; see chapter on "Safety Culture in Healthcare" in this handbook).

If safety culture does indeed play a significant role in patient safety—over and above the contribution of material, professional and organizational resources—then it is important to identify which elements of safety culture correlate with safety outcomes and to develop reliable methods and techniques for determining the type and nature of the safety culture and climate of individual hospitals, departments, and wards.

Adapting suggestions by Nieva and Sorra (2003), we propose that safety culture assessment for a given organization, ward or department, may serve a number of objectives:

1. Profiling (diagnosis): An assessment may aid in determining the specific safety culture or climate profile of the unit, including the identification of "strong" and "weak" points.
2. Awareness enhancement: It may serve to raise staff awareness, typically when conducted in parallel with other staff-oriented patient safety initiatives.
3. Measuring change: Assessment may be applied and repeated over time to detect changes in perceptions and attitudes, possibly as part of a "before-and-after-intervention" design.

4. Benchmarking: It may be used to evaluate the standing of the unit in relation to a reference sample (comparable organizations and groups).
5. Accreditation: It may be part of a possibly mandated safety management review or accreditation program.

Based on the findings of pervasive medical error in both the United States (Brennan et al., 1991; Thomas et al., 2000;) and Australia (Wilson et al., 1995), several research programs were established in the late 1990s to investigate the effects of climate and culture on patient safety. The U.S. Institute of Medicine formed the Quality of Healthcare Committee who, in their influential report *To Err is Human* (Kohn et al., 1999), suggested a strategy for improvement of the overall quality of patient care in the United States. In addition, the Agency for Healthcare Research and Quality (AHRQ) funded 21 studies examining specific work conditions, and 14 (66%) of these studies involved measures of organizational culture and climate (et al., 2004). A majority of the studies, the results from which have now begun to be published, used quantitative methods, and we include in our review of instruments one of the tools developed by the program mentioned earlier (Nieva & Sorra, 2003). In the UK, a similar historical development took place, and patient safety and safety culture were put on the agenda, reflected in the reforms of the National Health Service (NHS; Scott et al., 2003a). A highly significant publication was the report put out by the Department of Health (2000), *An Organization With a Memory.*

Both of these reports, *To Err is Human* and *An Organization With a Memory,* signaled a discussion—not only in the English-speaking world but internationally—about the role of organizational culture in the occurrence of preventable adverse events in health care. Prompted in large measure by the experience from other domains, especially aviation and the nuclear industry, new conceptions of human error were suggested to health care that stressed a systems-based and organizational perspective (Helmreich, 2000; Reason, 2000; Sexton, Thomas, & Helmreich, 2000). As an alternative to reactive strategies to error management, a systems-approach based on proactive strategies—and thus involving systematic reporting of errors and adverse events—was recommended in order to identify and ultimately control so-called "latent conditions" (Reason, 1997). For additional information, see the chapters in the section on "Macroergonomics and Systems" in this handbook.

In this chapter, we briefly review (Section 2) the distinction between safety culture and safety climate; then (Section 3) we present an integrative model of safety culture and safety management structure, stressing the link between safety culture and the traditional *human factors* links to organizational factors that influence performance and safety outcomes. In Section 4, we discuss some different methods and techniques for assessing safety culture and climate, and in Section 5, we review a number of criteria that may be used to select a survey tool suitable to the user's specific needs and wants. In Section 6, we describe a few illustrative examples of assessment, and in Section 7, we summarize some guidelines for using the results of assessments. Finally, in the concluding section, we briefly discuss problems and prospects for use of safety culture assessment and future research in this field.

SAFETY CULTURE VERSUS CLIMATE

The term "safety culture" is of surprisingly recent origin, having been coined in the late 1980s in the aftermath of the Chernobyl nuclear accident to characterize a poor and risky mindset among management and staff at the nuclear power plant (see chapter on Safety Culture in Health Care in this handbook). The concept and the term quickly caught on, not least because it appeared to be a natural focusing of the familiar notion of *organizational culture.* This well-established and broader notion had been in use for some years to characterize the shared view of "how we do things here" in organizations—but with no emphasis on safety critical domains. *Safety culture* was thus introduced to connote a set of special aspects of organizational culture—namely, shared values and attitudes relating to safety. In turn, organizational theorists had borrowed the notion of culture from anthropology. So it is only natural—although possibly not always so productive—that a number of authors feel obliged to seek the roots of the concept of safety culture in the anthropological and organizational theories of culture (see e.g., Guldenmund, 2000; Wiegmann, Zhang, von Thaden, Sharma, & Gibbons, 2004 for concise overviews).

However, the closely related notion of *safety climate,* introduced by Zohar (1980), also refers to attitudes and perceptions relating to safety. Safety

TABLE 39–1. Edgar Schein's Three Levels of "Culture" Adapted to Medicine

Levels of culture	*Characteristics*	*Examples in medicine*	*Levels of interpretion and methods*
Artifacts (cultural symptoms)	Visible artifacts,objects and behavior; Changeable, context dependent, but often difficult to decipher	Equipment, procedures, communication routines, standard, services alarms, dress code hierarchical structures	Climate/ Quantitative methods
Espoused values (and attitudes)	Official and unofficial policies, norms and values (in/not in accordance with underlying assumptions)	Mission statement, team norms, "Learn from mistakes"	
Basic assumptions	Unconscious beliefs, values and expectations shared by individuals (taken as given), implicit (tacit), relatively persistent, cognitive and normative structures	The "Primum non nocere" creed, the natural science paradigm	Culture Qualitative methods

climate has been characterized as reflecting the surface manifestation of culture; "the workforce's attitudes and perceptions at a given place and time. It is a snapshot of the state of safety providing an indicator of the underlying safety culture of an organisation" (Mearns, Flin, Fleming, & Gordon, 1997).

The distinction between culture and climate is not sharp, but is to a large extent related to somewhat different research traditions: Culture theorists have tended to apply qualitative methods and climate theorists have typically applied quantitative methods including in particular questionnaire survey techniques. The debate about the distinction between climate and culture has not been made any clearer by the fact that neither notion has anything that looks like a standard definition. (Flin et al., 2000; Guldenmund, 2000; HSL, 2002; Scott et al., 2003a; Wiegmann et al., 2004).

Culture and climate can, however, theoretically be distinguished in terms of how *stable, tacit,* and *interpretable* these shared values, attitudes, and so forth are. Thus, culture concerns shared symbolic and normative structures that (a) are largely tacit (implicit, unconscious), (b) are largely stable over time, and (c) can be assigned a meaning only by reference to surrounding symbolic practices of the cultural community. In contrast, climate reflects the largely overt and explicit, context dependent, and most directly interpretable manifestations of culture and the meaning of its expressions can be compared across groups.

One often quoted source for characterization of organizational culture is found in the organizational theorist Edgar Schein' works. Schein distinguished three levels of culture: "basic assumptions," "espoused values," and "artifacts" (Schein, 1985). In Table 39–1, we illustrate (an adaptation of) Schein's three levels of culture, their individual characteristics, and examples from health care. Furthermore, we seek to demonstrate at which levels and by which methods culture and climate may be recognized (confer Section 4).

In the rest of this chapter, however, unless precision is required, we refer to "safety culture and climate."

The contents and variety of issues addressed in both safety culture and climate research may be illustrated by the following list of examples of factors and dimensions probed in safety culture and climate questionnaires (Guldenmund, 2000; HSE, 1999; Nieva & Sorra, 2003; Scott, 2003a; Wiegmann et al., 2004).

- Learning and reporting of incidents.
- Motivation, involvement and trust.
- Accountability and responsibility.
- Communication and cooperation.
- Safety and production priorities.
- Risk perception.
- Perceptions of performance-shaping factors (e.g. fatigue, training, human–machine interface design).

- Management and employee commitment to safety.
- Procedures, compliance and violations.
- Teamwork within and between hospital units
- Perceived causes of incidents.
- Reasons for not reporting.

A widely accepted and often quoted definition of safety culture, partially covering the previous factors, is one proposed by the Advisory Committee on the Safety of Nuclear Installations, UK, [now NuSAC: Nuclear Safety Advisory Committee:

> *The safety culture of an organization is the product of individual and group values, attitudes, perceptions, competencies, and patterns of behavior that determine the commitment to, and the style and proficiency of, an organization's health and safety management. Organizations with a positive safety culture are characterized by communications founded on mutual trust, by shared perceptions of the importance of safety, and by confidence in the efficacy of preventive measures*

(ASCNI, 1993; See chapter on "Safety Culture in Healthcare" in this handbook for further discussion about the notion of safety culture).

All hospitals and wards have *a* safety culture, but some safety cultures are stronger and more mature than others. Thus, current theories of safety culture hold that the safety cultures of individual organizations or units may have different degrees of strength and different profiles, and that assessment of the maturity and profile of the safety culture of a given organization provides a useful, perhaps even essential, basis for working proactively with safety culture in that organization. Accordingly, assessment tools are designed to provide this type of basis.

MODELS OF SAFETY CULTURE / CLIMATE

Having reviewed some concepts of safety culture/climate, we need to address the relation between safety culture and other sociotechnical factors that determine the risk of patient injury. Human factors has traditionally studied how human–machine systems may be designed so that people can accomplish their tasks with efficiency and safety. The focus has not just been on the characteristics of the staff on the front line (skills, competencies, and physiological characteristics) and teams (coordination, shared situation awareness), but also on how factors such as the design of the human–machine interface, the layout of the physical work environment, the quality of training and procedures, work schedules, and fatigue effect human performance (e.g., Rasmussen, 1986; Sanders & McCormick, 1993). The factors that are known to impact on human performance are usually referred to by the umbrella term *performance shaping factors* (PSFs). Gradually, however, PSFs were expanded to include more traditional organizational factors, including management, learning, and organizational culture (e.g., Reason, 1997). See chapter on "An Historical Perspective and Overview of Macroergonomics" by Hendrick in this handbook for a discussion of the system-level human factors variables. Finally, it has been argued that we cannot establish adequate models of accident causation unless we also take explicit account of factors that are outside the control of the individual organization and its management—for example, law making, market forces, public awareness (Rasmussen, 1997; Rasmussen & Svedung, 2000; see also Maurino, Reason, Johnston, & Lee, 1995).

In Figure 39-1, we depict different *types* of factors that may determine the risk of given tasks. This map of the socio-technical factors (indebted to, *inter alia*, the frameworks of Rasmussen (1997) and Reason (1997)), divides the potential causal factors that determine preventable adverse outcomes into four groups. The factors referred to as being "largely outside organizational control" comprise decisions and forces that may have a massive influence on the options available to an organization. The most obvious example, of course, will be the economic forces that may prompt the organization to realign its balancing between productivity and safety. Similarly, the recent introduction of mandatory, national reporting systems in 2004 (in Denmark and in England and Wales) is imposing major changes to the structures and mechanisms for learning from preventable patient injury.

In the cluster of factors that are "largely within organizational control" (Hale, 2000) we distinguish between safety culture and safety management structure. While tightly related, culture and structure are nevertheless quite distinct. Structure is the object of investigation when a safety audit or an accreditation review is made. An audit "determines

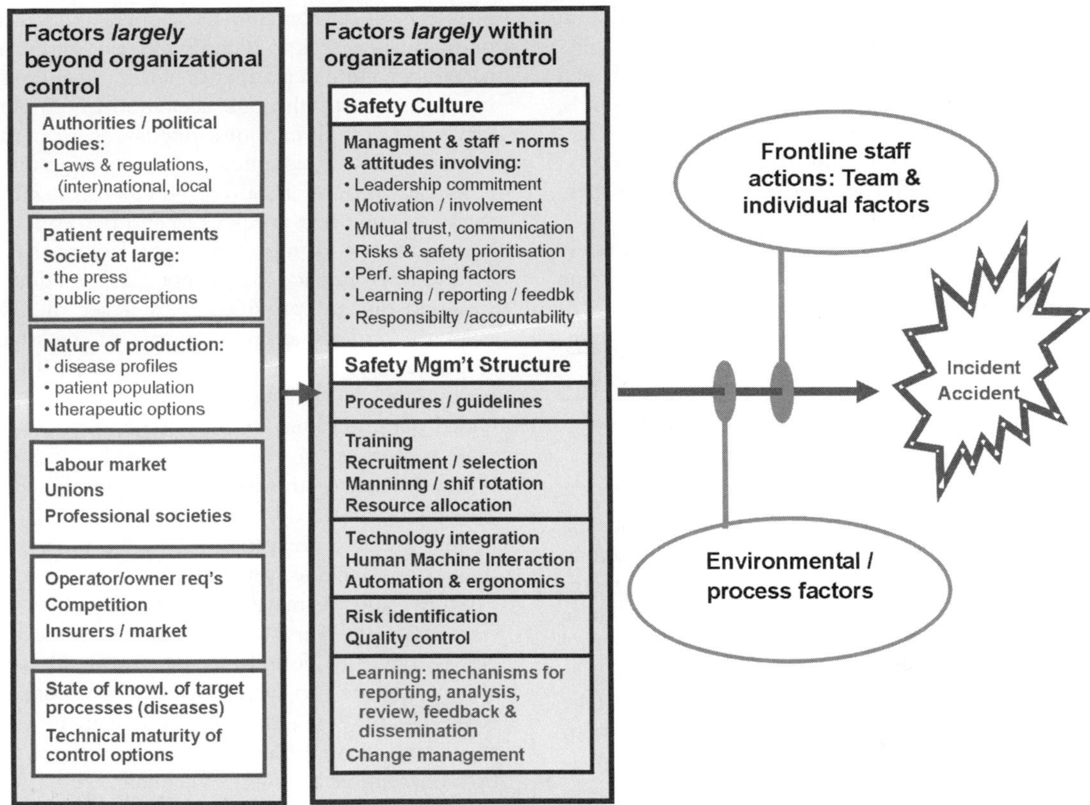

Figure 39-1. Model of performance-shaping factors—cultural and sociotechnical—that may have an impact on patient injury risk

whether there are policies, plans and procedures, whether responsibilities are allocated and communication channels exist and operate, whether risk assessment takes place, design solutions are implemented and monitoring, feedback and learning systems are in place"

(Hale, p. 6). Tools to investigate structure shall tell us whether mechanisms are in place and working (Phillips, 1999). Tools to investigate culture and climate shall tell us whether management and staff attitudes and perceptions are such that they are disposed to act safely. We may expect that the quality of the safety management structure is correlated with the maturity of the safety culture and climate—i.e., we should expect that if the mechanisms are in place *and* working well, then the safety climate is good. Yet, there are examples from industry that show comparable organizations, which apparently have the same safety management structures and yet exhibit significant differences in safety climate *and* safety outcomes (Andersen et al., 2004; Itoh, Andersen, Madsen, & Abe, 2004).

The mechanisms involved in safety management structure may be separated into those that operate as first-order delivery systems and those that operate as higher order or "reflective" ones. The first-order structural mechanisms control the processes of health care delivery (training, procedures, etc.) and therefore embody the *performance-shaping factors* that are under organizational control. In contrast, the higher order mechanisms aim to control the adequate functioning of the former, and they, therefore, include learning (incident reporting) and change management, including continuous quality control optimization.

In Figure 39-1, we have mapped two additional groups of factors that determine risk and possibly the difference between an accident and no accident. "Environmental/process factors" refer to physical conditions and patient conditions (underlying disease, patient characteristics). "individual and team factors" comprise personal and team characteristics that are, to some extent, of course, shaped by organizational decisions (such as skill levels, motivation, involvement), but they also include factors that, in principle and in practice, are beyond organizational influence—personal problems, personalities, and indisposition.

Having reviewed the relationship between safety culture factors and other factors that determine patient safety, we will now describe methods and techniques that are dedicated to measuring safety culture.

METHODS TO ASSESS SAFETY CULTURE AND CLIMATE

In social research, there is a—not too sharp, but still basic—distinction between qualitative and quantitative methods; and similarly, we should classify methods of assessing safety culture and climate into these two groups. In general, qualitative methods are used when the *sense* of the acts and utterances of the subjects studied must be *interpreted* (Taylor, 1986), and the qualitative assessment methods that are most often used to study work practices include in-depth interviews, semistructured interviews (staff and management), focus group interviews, field studies or observations. Quantitative approaches seek, first, to operationalize and, second, to numerically measure safety culture and climate aspects; therefore, quantitative assessment methods typically expose subjects (respondents) to the same set of cues (question items or vignettes) and collect subject responses in terms of fixed response options. Each type of approach has advantages and disadvantages, and many researchers will frequently combine both approaches—for instance, conduct interviews in order to prepare a questionnaire or, after a survey, elucidate the reasons for subjects' responses.

The strengths and weaknesses of the two approaches in successfully uncovering organizational culture are continually being debated (Ashkanasy, Wilderom, & Peterson, 2000; Schein, 2000). Here we briefly describe some of the most often used methods of the two approaches.

Safety Culture / Climate Questionnaire Survey

Safety climate/culture questionnaires are tools for measuring safety culture in organizations operating in safety critical domains. Developers and users of questionnaires seek to identify the level and profile of safety culture in a target organization and possibly in its groups through elicitation of employee views and attitudes about safety issues. In particular, emphasis is assigned to the perception of employee groups regarding their organization's safety system, their attitudes to and perceptions of management and, more generally, factors that are believed to impact on safety (see Section 3). Respondents are provided with a set of fixed-response options, often in terms of rank-based responses on a Likert-type response scale. It is not uncommon to include in questionnaires open-ended questions that prompt respondents to provide responses in their own words, and thus, qualitative data of this type must be interpreted and possibly categorized by the researchers.

Just as there are variations in the themes and dimensions included by different theorists under "culture" and "climate", there is, in similar vein, only moderate consistency—although some overlap is typically found—across different safety climate/culture survey tools in the dimensions they cover (Collins & Gadd, 2002). Several attempts and suggestions have been made to identify emerging themes (Flin, et al., 2000; Wiegmann, et al., 2004) for a common classification system to reduce the general number of dimensions (Guldenmund, 2000). Dimensions are either determined a priori (e.g., reproduced from previous questionnaires) or through survey iteration and refinement of items and dimensions. Dimensions are validated statistically using a measure of reliability (Cronbach's alpha) and some form of multivariate statistical method (e.g., factor analysis or principal component analysis). Still, not all questionnaires are validated statistically. It is important, however, to note that even though it is possible to make a statistical evaluation of the identification of dimensions, the *labeling* of these will always remain subjective.

Focus Group Interviews

This type of interview normally involves 5–8 interviewees and a pair of interviewers. The interviewers ask the interviewees to react to a few open-ended, related issues. The aim of focus group interviews is to establish an informal forum where the interviewees are invited to articulate their own attitudes, perceptions, feelings, and ideas about the issues under scrutiny (Kitzinger, 1995; Marshall & Rossman,

1999). It is a strength of the focus group interview that themes, viewpoints, and perspectives are brought up, which might not otherwise have been thought of. It is important to promote a free exchange of viewpoints among the interviewees, and this is most often done using a prompting technique that resembles semistructured interviews. A semistructured interview is characterized by the use of an interview guide structured by clusters of themes that may be injected with probes. The role of the interviewer is to usher the interviewees through the selected themes and, importantly, to ensure that all participants get a chance to voice their opinion and that no single person gets to dominate the others. Focus group interviews can be especially helpful when developing questionnaires because they provide the researcher with perspectives and views—and even terminology and phrases—related to themes that might otherwise be missed. In addition, these interviews also work very well as follow-up on results from a survey, providing the researcher with "reasons" and background for the data collected. The focus group interview is ideal for employee groups, whereas the management group might be too small or too hierarchical to lend itself to this type of interview. Therefore, upper management representatives will, therefore, often be asked to participate in semistructured interviews (one or two interviewees).

Critical Incident Interview Technique (CIT)

Another interview technique often used when studying human factors issues in safety critical domains is the critical incident technique (CIT) (Flanagan, 1954; Carlisle, 1986). Whereas the semistructured interview is meant to draw a wide and comprehensive picture of the operators' attitudes and values, the CIT focuses on the operators' narratives about specific incidents or accidents in which they themselves have been involved. They are asked to recall a critical incident and talk about what happened; how they reacted, what the consequences were to themselves, to others or to their work environment, what went well and not so well, and what they or others might have learned from the incident. In particular, interviewees are asked to recall and recount the precursors—the contributing and possibly exacerbating factors behind the event—and factors that, if in place, could have contributed to resolving the incident. This technique is highly useful for identifying human factors issues, to understand and provide a basis for possibly planning a change of the "performance-shaping factors"—for example, procedures, training, team interaction guidelines, human-machine interfaces, and workplace redesign. Thus, when a number of interview persons has offered their recalls of specific incidents, the data may reflect strengths and weaknesses in the current safety culture and climate.

For each of the methods described, results may be used in accordance with each of the five goals behind assessment efforts described earlier (Section 1). For instance, data from interviews and/or a questionnaire survey may be used for different purposes. They will, however, nearly always be used diagnostically to profile the strong and weak points of the target groups, and thereby to provide a basis for considering options for addressing weaknesses (see Section 8).

Some authors believe that questionnaires are unsuitable to fully uncover culture. For instance, Schein argued that it is only through iterative in-depth interviews that values and assumptions of organizational members may be revealed, and that it is doubtful whether questionnaires may be capable of exposing values and hidden cultural assumptions. According to Schein (2000), "Culture questionnaire scores do correlate with various indexes of organizational performance, but these measures are more appropriately measures of climate than of culture". Although most will agree that questionnaires are best suited to elicit (explicit) attitudes and perceptions—and therefore climate—not everyone will agree that climate measures cannot illuminate culture. Indeed, to the extent that a successful factor analysis may identify underlying factors behind overt responses to items that might not, on their surface, appear similar, this type of approach can be said to uncover normative or attitudinal structures.

When one has to choose among methods of assessment, one may wish to consider that culture can have several levels of expression within an organization, and each method has strengths and weaknesses. If the aim is to obtain a comprehensive picture of the organizational culture, one should preferably apply both quantitative and qualitative methods. If resources are scarce and the aim may be solely to provide an empirical basis for planning and selecting a limited set of interventions or revisions of safety management mechanisms, it may not be practical to aim for a comprehensive picture of the safety culture. Considering the resources and possibly the time available, the potential user should try to choose the method according to the overall safety aims behind the study under consideration as well as any available prior knowledge. For instance, qualitative methods are not as easily adopted without prior knowledge or experience, as is the case with quantitative methods, where guidelines for use often follow the tools available. Additionally, quantitative results can be benchmarked with other departments and hospitals, and may be repeated to detect the effects of interventional programs.

Finally, although the topic of this chapter is on safety climate and culture with a focus on quantitative methods, it is important to emphasize that culture and climate cannot fruitfully be investigated unless the structure of the safety management system is taken into account. If a given group turns out to nourish negative perceptions of a given structural mechanism and the practices surrounding it—for instance, the reporting of incidents—it can be argued that the threat posed to patient safety lies not with the perceptions per se but, rather, with the fact that the procedures, mechanisms, and practices for reporting are possibly lacking (see Section 3).

REQUIREMENTS AND QUALITIES OF SAFETY CULTURE/CLIMATE TOOLS

There are a number of quality requirements for questionnaire tools that should be considered when selecting a safety climate tool for a given application. In the following, we refer to these requirements as "selection criteria." Generally speaking, the requirements contained in these criteria raise issues about whether a given tool: can be used to measure what it intends to measure (content validity), correlates with safety performance (discriminant validity, external validity), yields consistent results (reliability), covers the safety culture dimensions that the user wants to have covered (relevance and comprehensiveness), is practical to administer (usability), is culturally referenced and tested in environments demographically and culturally much different from the user's (universality) and is targeted at the user's respondent groups (group targeting).

These selection criteria are described in more detail below. It is worthwhile pointing out some of these criteria have to be balanced against each other: for instance, no tool can be rated highly on both comprehensiveness, validity and practicality, since the more comprehensive it is with respect to themes it covers, the more items it must include and, therefore, the lower it will score on usability.

1. Validity. The issue of validity concerns whether the tool may successfully be used to measure what it is supposed to measure. There are four levels of validity that are relevant:

    - Pilot testing. A pilot test serves to identify items that are ambiguous, hard to understand or are understood in ways that differ from what the developers had in mind. No questionnaire should be used as a survey instrument unless it has been thoroughly pilot tested.
    - Consistency (interitem reliability). A questionnaire containing items purporting to probe a number of different underlying factors or dimensions should be tested for internal consistency. Internal consistency means that the items that address the same underlying factor correlate with each other. A widely used measure of this is Cronbach's alpha (Pett, Lackey, & Sullivan, 2003). A high value of internal consistency does not ensure that the purported factor may not consist of subfactors (to be determined by various factor analysis or, in general, multivariate methods).
    - Criterion validity (a)—self-reported safety performance. By measuring external validity by reference to self-reported incidents, the researchers will obtain a measure of the extent to which the attitudes and perceptions that are elicited correspond to a subjective measure of safety performance. This is a common means of validating safety climate tools for industrial applications (Cooper, 2000; Gershon, et al., 2004).
    - Criterion validity (b)—safety performance. To test this type of validity, it is necessary to obtain data about independently assessed safety performance of the organization or units surveyed. At the same time, this is the ultimate test of the usefulness of a safety climate tool as an instrument to differentiate between cultures and

climates that correlate with patient safety. It seems no health care safety climate survey instrument (so far) has been validated in this sense of validation. However, validations have been made in other domains with safety climate survey instruments (Andersen, et al., 2004; Itoh, et al., 2004).

2. Reliability: This means that the survey tool will yield the same result if the same population is surveyed repeatedly (with the same techniques and in the same circumstances). In practice, this is often impossible to establish, because attitudes and perceptions are liable to change over time. A useful version of the requirement of reliability concerns internal consistency (which we previously categorized as validity of Level 2). When a tool has passed a thorough pilot-testing phase and has been validated in terms of internal consistency, it is likely that it is reliable.
3. Relevance and comprehensiveness. We deal with these, in principle, distinct qualities at the same time. Relevance refers to whether a tool seeks to measure the important—or the relevant—dimensions of safety climate. Comprehensiveness refers to the extent to which the tool covers all the (relevant) dimensions.
4. Practicality refers to ease of use with which a given tool may be administered and it includes considerations of length and the time that respondents require to complete it. Also relevant in this respect are considerations about statistical analysis of results.
5. Nonlocality. This refers to the universality or possible cultural bias of a tool: Is it tied to a specific regional or national culture? Users who consider using a questionnaire developed and tested within their own national or ethnic culture need not worry about this requirement.
6. Job orientation and setting. This refers to the types of staff for which the tool has been designed and tested (nurses, hospital physicians, pharmacists, etc). It also refers to the setting for which the tool is developed: Is it for health care or is it developed for another domain? Questionnaires will often require considerable adaptation when they are transferred across work domains.
7. Documentation. This refers to the documentation that is available about a given tool; whether there are sources available that describe the tool in terms of the previous selection criteria as well as the history of its development and use.

EXAMPLES OF ASSESSMENT TOOLS FOR MEASURING SAFETY CULTURE AND CLIMATE IN HEALTH CARE

In this section, we review some assessment tools in order to illustrate their variety and scope in measuring safety culture and climate. (More extensive overviews of assessment tools and recommendations for their use can be found in the review articles listed in the Appendix). The tools have been chosen to illustrate the requirements and criteria described in Section 5. In the selection process, we applied two overall requirements; (a) proven validation of at least Level 2; pilot tested and tested for coherence, and (b) development for and application in health care.

In Table 39–2, we outline the general objective, content, and construct of the tools, the types (professional groups) of health care personnel assessed, the dimensions covered, validation status and availability. In Table 39–2, we assess and evaluate the strengths and weaknesses of the tools in terms of the overall requirements and discuss their applicability.

Overview of the Tools

The Stanford/PSCI Culture Survey (Singer et al., 2003) is a good general tool to assess safety culture/climate across different hospital settings and personnel.

In general, the Safety Climate Survey [Institute of Healthcare Improvement (IHI), 2004] is easy to use, making modest requirements on staff time to fill in their responses. The instrument has been tested in many countries, the authors note. It is not comprehensive, however, leaving out a number of potentially revealing dimensions. It would be a sensible choice if the user wanted a tool that imposes few demands on staff's time and to track changes over time.

The 20-Item Safety Climate Scale (Gershon, et al., 2000) seeks to derive safety climate measures to a specific hospital setting; care workers at risk of blood borne pathogens exposure incidents, which distinguishes it from most of the other climate survey tools. As a consequence, it would require considerable efforts to adapt it to health care settings different from their target environment. The tool illustrates the possibility of creating and validating a specialized tool for measures of safety climate.

The Hospital Survey on Patient Safety Culture (Sorra & Nieva, 2004) is a good choice for a comprehensive measure of climate and culture. It has strong content validity, is well structured, and because it also considers outcome measures, it facilitates an external (criterion) validation. One of the objectives of this tool is to provide feedback to staff to strengthen awareness of patient safety and the importance of reinforcing a positive culture. We find this to be the tool of choice if the user wants to establish a basis for planning an intervention program.

The tools included in this overview all aim at measuring safety climate or culture, but they differ in their focus and comprehensiveness as shown in Tables 39–2 and 39–3. Because the Stanford/PSCI Culture Survey and the Hospital Survey on Patient Safety Culture are directed at a broad range of specialties and work settings, the questions (items) will necessarily be more general and less task focused than the The 20-Item Safety Climate Scale, which is guided by questions related to the specific safety measures of the target staff. A comprehensive tool, however, like the Hospital Survey on Patient Safety Culture, has a greater potential for revealing possible focus points for an interventional program as compared to a short questionnaire like the Safety Climate Survey, which is limited in its scope and liable to be less precise capturing the actual problem areas. Still, shorter questionnaires are obviously more suited for making quick and repeated measures of safety climate, and they may also increase response rate (Edwards, et al. 2002), reducing a potential bias in the data acquired.

Additional Considerations When Conducting Surveys

There are several things to consider when choosing a tool. The user should be clear about the focus and objectives of the assessment (Section 1) and the resources required. This includes the time individual respondents need to fill in their questionnaire, resources for data collection, entry, analysis, interpretation, reporting, and—importantly—feedback to management, safety managers, and staff. The work needed to make a useful and successful survey should not be underestimated, but experience quickly accumulates and the second, third and so forth. survey will be much quicker to run. It is therefore useful to seek collaboration with researchers or consultants. Researchers who have developed a survey tool may frequently have a scientific and personal interest in providing their tool free of charge in exchange for obtaining more data and possibly get a chance to perform comparisons between organizations and/or countries. For instance, Andersen, Madsen, Hermann, Schiøler, and Østergaard (2002) developed a questionnaire for the Danish Ministry of Health in 2001, which has subsequently been translated and applied in several countries. (Itoh, Andersen, Madsen, & Abe, 2004).

For health care staff, it will probably be most relevant to run a self-administered survey (and not a phone interview or personal interview survey). Web-based tools may be a possibility, although availability of web-linked PCs and staff familiarity with IT should be considered. Carayon and Hoonakker (2004) have reviewed studies that used mail surveys and different forms of electronic surveys, summarizing advantages and disadvantages and what to consider when choosing a survey method. Key findings show that mail surveys tend to lead to a higher response rate than do electronic surveys (but Web-based ones are getting better results), whereas the latter tend to yield more completed questionnaires, a greater likelihood of answers to open-ended questions, and in general, a higher response quality. In addition, electronic surveys are of course much easier and cheaper to administer.

Involvement of key stakeholders is the key to obtaining a high response rate from all relevant groups, which, in turn, is necessary to reduce the risk of bias. A low response rate (50% or less) will necessarily invite speculation that respondents may not be representative of the target group. In our experience, most people want to know what the survey will be used for and they want to receive firm assurances about anonymity. This also means that respondents may be reluctant to supply potentially revealing demographic information (e.g., age, position, department, length of employment in current department). It might, though, be helpful if the survey is administered by an independent, reputable research or survey organization that issues a guarantee that data will only be reported to the host

TABLE 39–2. Examples of Assessment Tools: Objective, Content, and Constructs of Tools

Tools	*Stanford*/PSCI *Culture Survey (close ended)*	*Safety Climate Survey Quality and Health Care, Institute for Healthcare Improvement*	*The 20-Item Hospital Safety Climate Scale*	*Che Hospital Survey on Patient Safety Culture Agency for Health Care Research and Quality* (AHRQ)
Objective and the description of tool	"To measure and understand fundamental attitudes toward patient safety culture and organizational culture and ways in which attitudes vary by hospital and between different types of health care personnel." A general tool for assessing safety culture/climate across different hospitals settings and personnel.	"To gain information about the perceptions of safety of front line clinical staff and management commitment to safety." Can show variations across different departments and disciplines; can be repeated to assess impact on interventions. A possible measure to develop improve or monitor changes in the culture of safety.	To measure hospitals commitment to bloodborne pathogen risk management programs and the relation with safety climate. "A short and effective tool to measure hospital safety climate and (1) employee compliance with safe work practice and (2)incidents of workplace exposure to blood and other body fluids." A highly specialized safety climate tool, aimed at external validation to investigate the relationship between safetyclimate, safe work practices and work-place exposure incidents.	"To help hospitals assess extent to which its cultural components emphasize the importance of patient safety, facilitate openness in discussing error, and create an atmosphere of continuous learning and improvement rather than a culture of covering up mistakes and punishment".
Construct & development	Adapted from five former U.S. questionnaires (aviation and health care).	Not clear.	Originally a large 99-item survey that measured 4 major constructs 1) safety climate, 2) demographics, 3) self-reported compliance rates, and 4) exposure history. 46 were safety climate items developed on the basis of intensive qualitative data generating techniques and work site surveys, as well as restructuring and testing of existing safety climate scales.	Developed on the basis of literature pertaining to safety, error and and accidents, and error reporting; review of existing safety culture surveys; interviews with employees and managers in different hospitals; a number of transfusion service survey items and scales with demonstrated internal consistency reliability have been modified and included in the hospital-wide suvey (informed by earlier work by Nieva and Sorra (2004):
Dimensions and items (excluding demographics)	16 dimensions and 30 items.	Dimensions not identified; 19 items.	Six dimensions; 20 items.	12 dimension and 42 items grouped into four levels: unit level, hospital level, outcome variables and other measures.

TABLE 39–2 (Continued)

Tools	*Stanford/PSCI Culture Survey (close ended)*	*Safety Climate Survey Quality and Health Care, Institute for Healthcare Improvement*	*The 20-Item Hospital Safety Climate Scale*	*The Hospital Survey on Patient Safety Culture Agency for Health Care Research and Quality* (AHRQ)
Tested & applied	15 hospitals in California, sample of 6,312, response rate 47.4% overall.	Many hospitals in both U.S. and Europe.	A sample of 789 hospital-based healthcare workers at risk for blood-borne pathogen exposure incidents.	Piloted in 12 hospitals in 2003 in the United States.
Test history and results	After first revision 82 items, final revision 30 items plus demographics, leaving at least 1-2 items per dimension.	Results of resurveying shows that improvements in staff perceptions of the safety climate is linked to decreases in actual errors, patient length of stay, and employment turnover.	Test resulted in a final 20-item hospital safety climate scale, which factored into 6 different organizational dimensions; several significantly related to compliance and workplace exposure incidents. (Key finding; "the importance of the perception that senior management was supportive of the blood-borne pathogen safety program in terms of enhancing compliance and reducing exposure incidents").	Psychometric analysis provides solid evidence supporting 12 dimensions and 42 items, plus additional background questions (originally 20 dimensions) Psychometric analysis consisting of item analysis, content analysis, exploratory and confirmatory factor analysis, reliability analysis, composite score construction, correlation analysis and analysis of variance.

TABLE 39–3. Examples of Assessment Tools: Evaluation of Tools in Terms of the Overall Requirements and Qualities

Tools Assessed in Accordance With Requirements	*Stanford/PSCI Culture Survey (closed ended)*	*Safety Climate Survey Quality and Health Care Institute for Healthcare Improvement*	*The 20-item Hospital Safety Climate Scale*	*The Hospital Survey on Patient Safety Culture Agency for Health Care Research and Quality* (AHRQ)
Validity	Pilot tested, tested for internal consistency and dimensionality. The five factors identified: organization, department, production, reporting/seeking help, shame/self-awareness, are different from the overall dimensions. Not validated against external validation.	Pilot tested, tested for consistency. No further information about validation.	Validated at all levels.	Validated at all levels.
Relevance and comprehensiveness	Covers most relevant dimensions, but each only sparsely (only 1–2 items per dimension).	Covers relevant issues concerning climate, but thinly. Low comprehensiveness.	Covers relevant issues in a narrow field: compliance and workplace exposure.	Large, comprehensive tool covering most relevant safety culture dimensions listed in the literature.
Practicality	Relatively short, 30 items. Analyzing data in "problematic responses" provides a simple overview of data.	Supplied with guidelines; easy to administer and fill out; gives a simple overview of data. Possible to calculate the Overall Mean, Safety Climate Mean, Safety Climate Score, and Percent of Respondents Reporting a Positive Safety Climate	Practical and well instructed; shouldn't take much time to answer; option to add the 14-item Universal Precautions compliance scale and four types of exposure incidents.	Long questionnaire. Analysis of results not available at this time.
Benchmarking	Yes, across hospital settings.	Yes, across hospital settings.	Yes, but only in settings at risk for bloodborne pathogens exposure incidents.	Yes, across hospital settings.
Tracking changes	Possible.	Possible.	Possible.	Possible.
Non-locality	Produced in the United States.	Produced in the United States and tested in both the United States and Europe.	Produced and tested in the US.	Produced and tested in the US.

TABLE 39–3 (Continued)

Tools Assessed in Accordance With Requirements	*Stanford/PSCI Culture Survey (closed ended)*	*Safety Climate Survey Quality and Health Care Institute for Healthcare Improvement*	*The 20-item Hospital Safety Climate Scale*	*The Hospital Survey on Patient Safety Culture Agency for Health Care Research and Quality* (AHRQ)
Job orientation & setting	Seeks broadly to cover all types of hospital personnel and specialties.	Directed at front-line staff.	Only relevant for care workers at risk for bloodborne pathogens exposure incidents.	Seeks broadly to cover most types of hospital personnel and specialties. Targeted at nurses, pharmacist, laboratory staff, therapist etc., and full-time physicians.
Documentation and availability	Published documentation (Singer et al., 2003). Extensive results are availabledirectly from the authors.	Developed as collaboration between University of Texas Center of Patient Safety Research and Practice, University of Texas, Austin, Texas, US. (Sexton, J. B., Helmreich, R., and Thomas, E.) and John Hopkins Hospital (Peter Pronovost). Available at: http://www.ihi.org Inst. Healthcare Improvement, 2004.	Published documen (Gershon (et al., 2000).	Published documentation (Gershon et al., 2000). Developed by Ph.D. Joann Sorra and Ph.D. Veronica Nieva, Westate.The Hospital Survey form and the complete set of Survey Feedback Report templates are available free of charge at: www.ahrq.gov/qual/hospculture/.

hospital and the departments at an aggregate level. Even when staff is encouraged to fill out questionnaires during working time, they may often not feel they can take the time to fill out the survey. Finally, management and department leaders may feel that some items get "too close" and that the survey invites respondents to criticize their superiors. All of these considerations make it necessary for a local survey leader to obtain explicit support from management, local leaders, and employee representatives. Low response rates are not uncommon. For instance Singer et al. (2003) reported a response rate of just 47.4%. Similarly, Andersen et al. (2002) obtained a response rate of just 46% for doctors and 53% for nurses (total of 51%). On the other hand, other surveys in health care have obtained quite high response rates. For instance, Itoh et al.'s (2004) survey of Japanese health care staff had a rate of 91%.

USING RESULTS FROM SAFETY CULTURE SURVEYS

In the introduction, we mention five purposes for using safety culture assessment; profiling or diagnosing safety culture; raising awareness; measuring change, possibly in relation to an intervention program; benchmarking against comparable units and organizations; and finally, as part of an audit or accreditation program. In this section, we focus on how results from safety culture assessment can be used to diagnose, raise awareness, and prepare interventional change to improve safety culture and practice.

Within organizational theory, there is an ongoing discussion of whether culture is something an organization *has* or *is* (Scott, 1998). Even though the discussion may be theoretical, the implications are that if an organization *is* culture, it cannot be changed, whereas if culture is something the organization *has*, it can be shaped and managed. The use of safety culture assessment tools is based on the idea that organizations have a safety culture, and that the culture (to some extent) can be shaped, managed, and is malleable when exposed to intervention. To change culture can be difficult and can take time, whereas safety climate is more easily manipulated (confer Section 2). An illustrative example of successfully managed change of culture was reported when the Danish Air Traffic Control of Copenhagen was able to change the safety culture of the organization within less than one year (Nørbjerg, 2003). The change was prompted first by the introduction of a new law that guaranteed strict confidentiality for controllers who reported incidents and, moreover, made it illegal to use any information thus collected for disciplinary and punitive reasons (so, the police, courts, the press, and the public have no access to individual reports). Second, when the new law on reporting was introduced, management and controller representatives implemented an intensive campaign to encourage reporting and learning. The transition has apparently succeeded (Nørbjerg, 2003), changing the culture from a punitive, nonreporting to a nonpunitive and learning culture. Thus, the "basic assumption" (in Schein's sense—see Section 4) that "you may talk with colleagues about an incident, but you do not write a report about it unless you are required by the pilot to do so" was changed to "if you have an incident you should write about it for the sake of your colleagues."

Schein (2000) has argued that it is not possible to create a climate of openness if history has shown that messengers are "shot" for bad news or making mistakes, and that such changes can only be brought about by modifying "basic assumptions," which at best will be a long-term endeavor. In many cases, this would probably hold true, because change programs are often implemented without regard to the underlying assumptions or nonfunctioning artifacts existing in the organization. However, the example just given shows that changes in artifacts and shared values can in fact change basic assumptions, which means that it is possible to change culture, even by structural means. (See Madsen, 2002 for a short description of the problematic safety culture and barriers towards reporting of an ATC center prior to a major structural change involving new legal and administrative procedures as described in the previous paragraph.)

Most survey instruments make no attempt to uncover "basic assumptions"; and even when they do, safety culture assessment tools will not be able to capture the nature of a professional culture in any detail. For a discussion of professional medical culture, see the chapter by Smith and Bartell on "The Relationship Between Physician Professionalism and Health Care Systems Change" in this handbook. Still, such tools can be used to show how different groups react to and think about safety issues and about factors that impact on safety. Moreover, some survey instruments have succeeded in identifying what may be regarded as basic assumptions, using multivariate statistical methods techniques such as principal component analysis. For instance, Itoh et al., 2004 identified two underlying factors describing reasons for not reporting adverse events and mistakes among hospital staff.

Assessing safety culture is a process that can contribute to a positive change in culture, and from which the first results—if used well—can be the beginning of a path of continuous patient safety improvement.

Different stakeholders may have different—and sometimes conflicting—interests in the knowledge that is acquired from safety culture assessment. The question, therefore, is how and to whom survey results should be conveyed and for which purpose? It is possible and often relevant to use results at the hospital, the department, and ward level. For instance, results may be used locally to address poor levels of safety culture for certain areas; to help staff better understand the mechanisms of safety practice; to facilitate the development of concrete and specific safety practices; or to benchmark one's own department and track changes over time. However, when preparing presentations of results to individual departments (or hospital management teams), it is useful to prioritize among these options before proceeding.

Whatever the purpose of the survey, one should expect some resistance. If assessments are capable of distinguishing individual units and of making comparisons across them, it is not uncommon that local leaders will be hesitant in participating or taking the results seriously. They may find their authority threatened and they may be well aware that safety culture is still a somewhat vague and intangible phenomenon. Equally, some units may find it stigmatizing and threatening to be defined as having an "immature" or a "relatively negative" safety culture, whereas others will welcome the results, even when they are negative. Others will, however, find them much too "qualitative", and will refuse to take seriously anything that bears little resemblance to scientific "evidence" of health care delivery quality.

People who are responsible for conducting and presenting results from safety culture assessments should be prepared for such very different reactions. It is especially important to demonstrate understanding toward those who do not embrace the results or the changes that follow from assessment. Much resistance, however, can be avoided by clearly communicating the purpose(s) of the assessment.

There is no single, optimal way of using survey results for preparing and implementing changes; but there are several heuristics and theoretical frameworks that are useful for guiding change. Inspired by the framework of Kotter (1995), we illustrate an eight-step process for using and managing change based on results from safety culture assessment.

1. Establish a sense of urgency. Employees will be motivated only if they find the change necessary. It is important to convey survey results to staff to initiate an understanding for improvement and change. Sometimes survey results are shown only to top management—this is a grave mistake. Use the survey results as a point of departure and as a basis for dialog, reflection, and constructive discussion; does staff agree or disagree with the results, why and why not? What should be the consequences of the results, and which changes should be initiated?
2. Form a powerful guiding coalition. Get key persons or opinion leaders involved and engaged in taking part in the planning in order for actions to improve patient safety. Management commitment and trust are of course central throughout the process.
3. Create a vision. Focus on what the aim of change is, making the aim guide—rather than having the change direct the focus. The strategy should ensure that the aim is focused. Do not initiate too many changes at the same time. Even though results show many safety cultural dimensions that deserve to be improved, it is important not to be too ambitious, but to focus, instead, on key elements for improvement. For instance, practices will be more easily changed than will norms and values, and "articulating new visions and new values is a waste of time if these are not calibrated against existing assumptions and norms" (Schein, 2000). Make sure that the organization is not undergoing other major changes at the same time—because employees will easily be overwhelmed by too much simultaneous change.
4. Communicate the vision of change. Use all channels to communicate the new

vision and strategy; in groups and units, newsletters, posters, and so forth. It is essential for the process that staff understands the purpose of change.

5. Empower others to act on the vision. Get rid of barriers against change. Change necessary systems or structures. As James Reason has noted (1997), "you can not change the human being but you can change the conditions under which they work." Welcome untraditional ideas and actions and use your positive survey results to engage and rebuild. Ask how we can use our strengths to overcome our weaknesses.
6. Plan and create short-term successes. Generating short-term success is important for continuous motivation, as well as for recognizing and rewarding those who made the changes possible.
7. Consolidate improvements and produce continued change. The biggest mistake is too early to believe the vision has succeeded. The results and effects of the cultural change need to be consolidated to avoid old traditions to reappear. Visible successes will make it more credible to implement further changes, such as employing new personal, or promoting and training employees to implement the sought changes.
8. Institutionalize new approaches. Make efforts to create new and safer work methods and practices within the culture. It is not possible to successfully speak of achieving the vision before all changes are accounted for and consolidated into social norms and values. It takes time, and there is a constant danger of falling back into old traditions and customs, which can be very strong and, arguably, especially so in health care. If one seriously seeks changes, then resources and continuous improvement are needed as well as continuity in management. It is important that management communicates the relation between the improvements, actions, and attitudes, and the effects of these on patient safety culture.

Nieva and Sorra (2003) suggested that action-planning sessions are most successful when using trained line managers rather than top management or external experts. Ultimately, busy professionals such as doctors and nurses may have a low degree of tolerance for naïve and inexperienced facilitators. It is important, therefore, that change processes are carefully planned and facilitated, either with professional guidance or trained in-house personnel.

As a final point, let us repeat that performing safety culture assessments periodically is an effective means of tracking changes, possible improvements, or degradations. Results can be used to measure improvements, review practices, and discuss the direction of further improvements. Alhough self-evident, it should be noted that if a culture is very positive from the outset, it might be difficult to track significant changes over time, whereas a lesser positive culture has a greater potential for change.

PROBLEMS AND PROSPECTS—FUTURE DIRECTIONS

In this chapter, we focus on the background for, requirements to, and examples of safety culture assessment tools, but have touched only briefly on some of the problems of carrying out successful assessment. The state of the art is far from perfect, and there are a number of issues that need to be addressed in future developments in this area. Here we single out five areas that, we believe, will be the focus of much further research and development.

First, so far, there is some amount of vague agreement about the factors (dimensions, scales) that underlie safety culture and climate. Most authors do agree that leadership commitment and involvement, and learning and safety prioritization vis-à-vis production pressure are essential elements (Flin, et al., 2000; Guldenmund, 2000; Wiegmann, et al., 2004). But beyond this, there is at present little sign of a convergence of opinion. Moreover, there are no generally accepted models of how individual candidate factors may influence each other—except agreement that leadership is a primary driver.

This leads us immediately to the second area in which we suggest that research and development are urgently needed, namely validation of candidate factors against actual safety performance (i.e., criterion validity—see Section 5—held up against either self-reported outcomes or independently estimated preventable adverse outcomes). Thus, the requirements for an evidence-based test of whether a presumed safety cultural factor is in fact related to patient safety is easy to describe but hard to carry out. Any such test must demonstrate that otherwise comparable units (comparable in terms of the potential confounders, e.g., patient profiles, stage of disease when admitted, therapeutic regimes, staff skills and experience, and, most importantly, resources) turn out to correlate in terms of safety culture factor measures *and* preventable patient safety outcomes. For instance, the recent Canadian study of adverse events (Baker, et al., 2004) showed that teaching hospitals have a significantly higher rate of adverse events—but not preventable adverse events. So far, it seems that no study has combined results from the considerable efforts devoted to developing quality indicators (Mainz, 2003) with the development of criterion-based safety culture factor identification.

A third, and closely related area of development is the combination of safety culture assessment with assessment of the safety management structure (Section 3). Safety management mechanisms define the policies, plans, and procedures of an organization and the routines and responsibilities for their realization. They are tightly related to the norms, attitudes, and perceptions of staff members, but there is no one-way causation: Structure will impact on culture, but safety cultural forces (internal to the organization or unit) will of course often produce a change in structure. There is something artificial and incomplete about assessing safety culture independent from structural mechanisms: If perceptions are negative, they might be "valid" and reflect a poor delivery system or a poor implementation of this. We believe, therefore, that one of the significant ways forward for developing and applying successful safety culture assessment is to link it to safety management reviews (audits and accreditation efforts).

A fourth area for development is to include the patient perspective in culture assessments. Hospitals in the 21st century are forced to have a more open perspective in terms of patients and other stakeholders, because they depend more and more on their demands. Patients are unavoidably the most important stakeholders in improvements of safety; and there is a growing call for openness, honesty, and disclosure of errors (Cantor, 2002; Hébert, Levin. & Robertson, 2001). If a hospital has a bad reputation, patients will not hesitate to find another. Future assessment tools should seek to include patients' experiences, perspectives, and attitudes toward health care. Klingle, Burgoon, Afifi, and Callister (1995) pointed out that most patients should be viewed as part of the organization, and therefore should also participate in assessing the safety climate. They also noted that patients are surprisingly good at assessing the climate. We therefore recommend developing tools that are able to assess climate from all relevant views—management, staff, and patients. Health care professional's views of what patients want and expect may differ from what patients themselves indicate (Gallagher, Waterman, Ebers, Fraser, & Levinson, 2003; Hingorani, Wong, &Vafidis, 1999). Among the measures of the quality of clinical performance, one can find process measures of patient satisfaction and trust in quality of care. Patient satisfaction surveys focus on issues such as patients' perceptions of information received, being listened to by staff, explanation of care, diagnosis and therapy, involvement in decisions about care, respect for dignity and privacy, and wait times, discharge (e.g., Cohen, Forbes & Garraway, 1996; Idvall, Hamrin, Sjöström, & Unosson, 2002; Jung, Wensing, Olesen, & Grol, 2002; Jenkinson, Coulter, & Bruster, 2002). So far, few patient satisfaction survey instruments have included items about patients' possible experiences of mistakes, their trust in having adverse events disclosed (see Oesterbye, Gut, Petersen, & Freil, 2005). We believe that it will be a useful extension of traditional safety culture and climate survey methods to capture patients' views on trust and openness (see Klingle, et. al., 1995, for one such attempt).

The final and fifth area that we wish to identify as requiring additional research and development efforts concerns the application of culture assessment results for improvements. This has two aspects. First, it would be somewhat useful to have (at least experience-based) guidelines for translating assessment results into recommendations for selecting and prioritizing among possible interventions. But perhaps more importantly, there is currently little systematic, general evidence that will tell users of assessment results which types of interventions will have the greatest intended effects. There is, therefore, an urgent need for collecting, classifying, and comparing intervention program results to provide safety managers at both hospital and department levels with a basis for selecting among the vast set of options.

References

Andersen, H. B., Madsen, M. D., Hermann, N., Schiøler, T., & Østergaard, D. (2002, July). Reporting adverse events in hospitals: A survey of the views of doctors and nurses on reporting practices and models of reporting. In C. Johnson (Ed.), *Proceedings of the Workshop on the Investigation and Reporting of Incidents and Accidents* (IRIA 2002; pp. 127–136). Glasgow, Scotland: University of Glasgow.

Andersen, H. B., Nielsen, K. J., Carstensen, O., Dyreborg, J., Guldenmund, F., Hansen, O. N., et al. (2004). Identifying safety culture factors in process industry. *Loss prevention 2004*. (CD-ROM). *Eleventh International Symposium on Loss Prevention and Safety Promotion in Process Industries* (pp. PG5225-PG5233). Prague: Czech Society of Chemical Engineering.

Ashkanasy, N. M., Wilderom, C. P. M., & Peterson, M. F. (Eds.). (2000). *Handbook of organizational culture & climate* Thousand Oaks, CA: Sage.

ASCNI. (1993). Advisory commitee of the safety of nuclear installations: Study group on human factors *Third Report: Organizing for Safety*. Sheffield, England: HSE Books.

Baker, G. R., Norton, P. G., Flintoft, V., Blais, R., Brown, A., Cox, J., et al. (2004) The Canadian Adverse Events Study: The incidence of adverse events among hospital patients in Canada. *Canadian Medical Association Journal, 170*(11), 1678–1686.

Brennan, T. A., Leape, L. L., Laird, N. M., Herbert, L., Localio, A. R., Lawthers A. G. et al., (1991). Incidence of adverse events and neglience in hospitalized patients: Results of the Havard Medical Practice Study. I. *New England Journal of Medicine, 324*, 370–376.

Cantor, M. D. (2002). Telling patients the truth: A systems approach to disclosing adverse events. *Quality. Safety. Health Care, 11*(1), 7—8.

Carayon, P., & Hoonakker, P. L. T. (2004). Macroergonomics organizational questionnaire survey (MOQS). In N. Stanton, A. Hedge, K. Brookhuis, E. Salas, & H. Hendrick (Eds.), *Handbook of human factors and ergonomics methods* (pp. 76–1/76–10). Boca Raton, FL: CRC Press.

Carlisle, K.E. (1986). *Analysing jobs and tasks*. Englewood Cliffs, NJ: Educational Technology Publications.

Carruthers, A. E., & Jeacocke, D. A. (2000), Adjusting the balance in health-care quality. *Journal of Quality in Clinical Practice, 20*, 158–160.

Cohen, G., Forbes, J., & Garraway, M. (1996) Can different patient satisfaction survey methods yield consistent results? Comparison of three surveys. *British Medical Journal; 313*, 841–4.

Collins, A. M., & Gadd, S. (2002). Safety culture: a review of the literature. Sheffield: Health and Safety Laboratory: An Agency of the Health and Safety Executive.

Cooper, M. D. (2000). Towards a model of safety culture. *Safety Science, 36*, 111–136.

Davies, H. T. O., Nutley, S. M., & Mannion, R. (2000). Organisational culture and quality of health care. *Quality in Health Care, 9*, 111–119.

Department of Health. (2000). *An organization with a memory*. London: The Stationary Office. Retrieved February 1, 2004, www.doh.gov.uk/org.memreport/index.htm

Edwards, P., Roberts, I., Clarke, M., DiGuiseppi, C., Pratap, S., Wentz, R., & Kwan, I. (2002, May). Increasing response rates to postal questionnaires: Systematic review. *British Medical Journal, 324*, 1183.

Flanagan, J. C. (1954). The critical incident teqnique. *Psychological Bulletin, 51(*4), 327–359.

Flin, R., Mearns, K., O'Connor, P., & Bryden, R. (2000). Measuring safety climate: Identifying the common features. *Safety Science, 34:1–3*, 177–193.

Gallagher, T. H., Waterman, A. D., Ebers, A. G., Fraser, V. J., & Levinson, W. (2003). Patients' and physicians' attitudes regarding the disclosure of medical errors. *Journal of the American Medical Association, 289(8),* 1001–1007.

Gershon, R. R. M., Karkashian, C. D., Grosch, J. W., Murphy, L. R., Escamilla-Cejudo, A., Flanagan, P. A. et al. (2000). Hospital safety climate and its relationship with safe work practices and workplace exposure incidents. *American Journal of Infection Control, 28*(3), 211–221.

Gershon, R. R. M., Stone, P. W., Bakken, S., & Larson, E. (2004). Measurement of organizational culture and climate in healthcare. *Journal of Nursing Administration, 34*(1), 33–40.

Guldenmund, F. W. (2000). The nature of safety culture: A review of theory and research. *Safety Science, 34*, 215–257.

Hale, A. (2000). Culture's confusions. *Safety Science, 34*, 1–14.

Health & Safety Executive [HSE]. (1999). *Summary guide to safety climate tools* [Offshore Technology Report 1999/063]. Sudbury, U.K.: HSE Books.

Health & Safety Laboratory (HSL)—Human Factors Group. (2002). *Safety culture: A review of the literature*. HSL/2002/25. Sheffield: HSL.

Hebert P. C., Levin A. V., & Robertson, G. (2001). Bioethics for clinicians:23. Disclosure of medical error. *Canadiun Medical Association Journal, 164*(4), 509–513.

Helmreich, R. L. (2000). On error management: Lessons from aviation. *British Medical Journal, 320*, 781–785.

Hingorani, M., Wong, T., & Vafidis, G. (1999). Patients' and doctors' attitudes to amount of information given after unintended injury during treatment: Cross sectional, questionnaire survey. *Britiah Medical Journal, 318,* 640–1.

Idvall, E., Hamrin, E., Sjöström, B., & Unosson, M. (2002). Patient and nurse assessment of quality of care in postoperative pain management. *Quality and Safety in Safe Health Care, 11*, 327–334.

Institute for Healthcare Improvement [IHI]. (2004). *Safety Climate Survey*. Cambridge, MA: IHT. Retrieved April 1, 2004, from http://www.ihi.org/NR/rdonlyres/145C099B–5FB4–46A–8CFD-D08D3CE9082C/1704/SafetyClimateSurvey1.pdf

Itoh, K., Andersen, H. B., Madsen, M. D., & Abe, T. (2004). *Common factors behind reasons for not reporting in healthcare.* Manuscript submitted for publication.

Jenkinson, C., Coulter, A., & Bruster, S. (2002). The Picker patient experience questionnaire: Development and validation using data from in-patient surveys in five countries. *International Society for Quality in Health Care, 14*(5), 353—358.

Jung, H. P., Wensing, M., Olesen, F., & Grol, R. (2002). Comparison of patients' and general practitioners' evaluations of general practice care. *Quality and Safety in Health Care, 11*, 315–319.

Kitzinger J. (1995). Introducing focus groups, *British Medical Journal, 311*, 299–302.

Klingle, R. S., Burgoon, M., Afifi, W,. & Callister, M. (1995). Rethinking how to measure organizational culture in the hospital setting: The hospital culture scale. *Evaluation & The Health Professions, 18*(2), 166–168.

Kohn, L., Corrigan, J. M., & Donaldson, M. S. (1999). *To err is human—Building a safer health system.*Washington, DC: National Academy Press.

Kotter, J. P. (1995, March/April). Why transformation efforts fail. *Harvard Business Review*, 59–67.

Madsen, M. D. (2002, July). A study of incident reporting in air traffic control—Moral dilemmas and the prospects of a reporting culture based on professional ethics. In C. Johnson (Ed.), *Proceedings of the Workshop on the Investigation and Reporting of Incidents and Accidents* (pp. 161–170). Glasgow: University of Glasgow.

Mainz, J. (2003). Defining and classifying clinical indicators for quality improvement *International Journal of Quality Health Care, 15,* 523–530.

Marshall, C., & Rossman G. B. (Eds.). (1999). *Designing qualitative research.* Thousand Oaks, CA: Sage.

Maurino, D.E., Reason, J., Johnston, N., & Lee, R. B. (1995). *Beyond aviation human factors: Safety in high technology systems.* Aldershot, U.K.: Ashgate Publishing Limited.

Mearns, K., Flin, R., Fleming, M., & Gordon, R. (1997). *Human and organisational oactors in offshore safety* (OTH 543). Suffolk: Offshore Safety Division, Sudbury, U.K.: HSE Books.

Nieva, V. F. & Sorra, J. (2003). Safety culture assessment: A tool for improving patient safety in healthcare organizations. *Quality and Safety in Health Care, 12(*suppl II*)*, ii17–ii23.

Nørbjerg, P. M. (2003). The ceation of an aviation safety reporting culture in Danish air traffic control. *Second Workshop on the Investigation and Reporting of Incidents and Accidents* (pp. 153–164). Retrieved February 1, 2004, from http : // shemesh.larc.nasa.gov/iria03/ iria2003proceedings.pdf

Oesterbye, T., Gut, R., Peterson, M., & Freil, M. (2005). National Danish Survey of Patient Experiences. Unit of Patient Evaluation, Copenhagen County. Glostrup, Denmark. Retrieved April 1, 2005, from http://www.efb.kbhamt.dk/emnekatalog/pdf/English/Survey_20041.pdf

Pett, M. A., Lackey, N. R., & Sullivan, J. J. (2003) *Making sense of factor analysis : The use of factor analysis for instrument development in health care research.* Thousand Oaks, CA: Sage.

Phillips, D. F. (1999) "New look" reflects changing style of patient safety enhancement. *Journal of the American Medical Association, 281*, 217–219.

Rasmussen, J. (1986). *Information processing and human–machine interaction: An approach to cognitive engineering.* New York: North-Holland.

Rasmussen, J. (1997). Risk management in a dynamic society: A modelling problem. *Safety Science, 27*(2–3), 183–213.

Rasmussen, J., & Svedung, I. (2000). *Proactive risk management in a dynamic society*. Karlstad, Sweden: Swedish Rescue Services Agency.

Reason, J. (1997). *Managing the risks of organizational accidents.* England: Ashgate.

Reason, J. (2000). Human error: Models and management. *British Medical Journal, 320*, 768–770.

Sanders, M. S., & McCormick, E. J. (1993). *Human factors in engineering and design* (7th ed.). New York: McGraw Hill.

Schein, E. H. (1985). *Organizational culture and leadership: A dynamic view.* San Francisco: Jossey-Bass.

Schein, E. H. (2000). Sense and nonsense about culture and climate. In N. M. Ashkanasy, C. P. M. Wilderom, & M. F. Peterson (Eds.), *Handbook of organizational culture & climate* (pp. xxiii-xxx). Thousand Oaks: Sage.

Scott, T., Mannion, R., Davies, H., & Marshall, M. (2003a). The quantitative measurement of organizational culture in health care: A review of the available instruments. *HSR: Health Services Research, 38*(3), 923–945.

Scott, T., Mannion, R., Marshall, M., & Davies, H. (2003b). Does organizational culture influence health care performance? A review of the evidence. *Journal of Health Service Research Policy, 8*(*2*), 105–117.

Scott, W. R. (1998). *Organizations—Rational, natural, and open systems* (4th ed.). Englewood Cliffs, NJ: Prentice Hall.

Sexton, J. B., Thomas, E. J., & Helmreich, R. L (2000). Error, stress, and teamwork in medicine and aviation: cross sectional surveys. *British Medical Journal, 320*, 745–749.

Singer, S. J, Gaba, D. M., Geppert, J. J., Sinaiko, A. D., Howard, S. K., & Park, K. C. (2003). The culture of safety: results of an organization-wide survey in 15 California hospitals. *Quality of Safe Health Care, 12*, 112–118.

Sorra, J. ,& Nieva, V. (2004). *Hospital survey on patient safety culture* (AHRQ Publication No. 04–0041). Retrieved April, 1, 2004, from http://www. ahrq.gov.

Taylor, C. (1986) *Philosophy and the human sciences.* Cambridge, New York: Cambridge University Press.

Thomas, E. J., Studdert, D. M., Burstin, H. R., Orav, E. J., Zeena, T., Williams, E. J. et al., (2000). Incidence and risk factors for adverse events and negligent care in Utah and Colorado in 1992. *Medical Care*; *38* (3), 261–271.

Vicente, K. J. (2002). From patients to politicians: A cognitive engineering view of patient safety. *Quality and Safety in Health Care, 11*,. 302–304.

Wiegmann, D. A., Zhang, H., von Thaden, T. L., Sharma, G., & Gibbons, A. M. (2004). Safety culture: An integrative review. *The International Journal of Aviation Psychology*, *14*(2), 117–134.

Wilson, R. M., Runciman, W. B., Gibberd, R. W., Harrison, B. T., Newby, L., & Hamilton, J. D. (1995). The quality in Australian health care study. *Medical Journal of Australia*, *163*, 458–471.

Zhan, C., & Miller, M. R. (2003). Excess length of stay, charges and mortality attributable to medical injuries during hospitalization. *Journal of the American Medical Association*, *290*(34), 1868–1874.

Zohar, D. (1980). Safety climate in industrial organizations: Theoretical and applied implications. *Journal of Applied Psychology, 65*(1), 96–102.

APPENDIX

An Overview of Three Review Articles of Safety Culture Assessment Tools and Their Recommendations

Gershon et al. (2004): *Measurement of Organizational Culture and Climate in Health Care.*

Aim. Instruments are described and characterized in order to reveal the implication for nurse administrators (but can easily be used by administrators as such).

Types and Amount of Tools Assessed. Review of 12 instruments that may have applicability in measuring organizational constructs in health care settings. The focus is on global measures of culture and climate.

Conclusions and Recommendations. Provides guidelines for measuring organizational constructs in health care. Limitations may include that the search strategy may have missed some information—the authors make reference to Scott et al. 2003a that identified other instruments. Recommendations are (a) to adopt and consistently use uniform terminology; (b) to guide all health services organizational studies with a theoretical framework that can be tested; (3) to apply standard and psychometrically sound instruments, possessing content, face, criterion, and construct validity; (c) to ensure that all measures be as specific and targeted as possible; (d) to apply high-level statistical analysis where feasible; including path analysis and multiple regression to verify the relationship between culture, climate and various outcomes.

Scott et al. 2003*a*: *The Quantitative Measurement of Organizational Culture in Health Care*: *A Review of the Available instruments*

Aim. To review the quantitative instruments available to health service researchers who want to measure culture and cultural change.

Types and Amount of Tools Assessed Thirteen tools; 9 with track record in health care organizations; four with potential for use in health care setting.

Conclusions and Recommendations. A range of instruments with differing characteristics are available to researchers interested in organizational culture, all of which have limitations in terms of scope, ease of use, or scientific properties. It is recommended that the choice of instrument should be determined by how the research team conceptualizes organizational culture, the purpose of investigation, intended use of the results, and availability of resources.

Guldenmund, 2000: *The Nature of Safety Culture*: *A Review of Theory and Research*

Aim. Main emphasis is on applied research in the social psychological and organizational psychological traditions, the assumption of which is that a large group of organizational cultures can be described with a limited number of dimensions.

Types and Amount of Tools assessed. Sixteen questionnaires, none of them targeted specifically for healthcare.

Conclusions and Recommendations. Safety climate might be considered an alternative safety performance indicator and research should focus on its scientific validity.

40

INCIDENT ANALYSIS IN HEALTH CARE

Chris Johnson
University of Glasgow, Scotland

CAUSATION

The investigation of adverse events can be decomposed into a number of different activities. For example, data must be collected about the events that led to a mishap. Interviews and the analysis of data logs and charts provide the information that is necessary to understand what happened. Elicitation techniques may also extend more widely into the organizational and managerial context in which an incident occurred. Together these different sources of information can contribute to our understanding of why there was an accident or near miss. Any causal analysis, in turn, helps to guide the identification of recommendations that are ultimately intended to minimize the likelihood of any future recurrence. It is important to stress that these different activities often overlap so that, for instance, it is often necessary to gather additional evidence to support particular causal hypotheses. Similarly, the identification of potential recommendations often forces analysts to reconsider their interpretation of why an adverse event occurred. The U.S. Joint Commission on Accreditation of Healthcare Organizations (http://www.jcipatientsafety.org/show.asp?durki = 9348, 2004) identified similar stages when it argued that a "meaningful improvement in patient safety" is dependent on:

- "Identification of the errors that occur.
- Analysis of each error to determine the underlying factors—the "root causes"—that, if eliminated, could reduce the risk of similar errors in the future.
- Compilation of data about error frequency and type and the root causes of these errors.
- Dissemination of information about these errors and their root causes to permit health care organizations, where appropriate, to redesign their systems and processes to reduce the risk of future errors.
- Periodic assessment of the effectiveness of the efforts taken to reduce the risk of errors."

Other chapters in this handbook describe problems that complicate the various activities involved in the analysis of health care incidents. In particular, the chapter on "Human Factors of Reporting Systems" focuses on the problems of unde-reporting and on the difficulty of eliciting adequate information in the aftermath of an accident or near-miss incident. In contrast, this chapter focuses more narrowly on the problems of determining *why* a mishap occurs. In particular, the following pages consider a number of different perspectives on the role of human error as a causal factor. Several authors have identified the "perfective" approach to incident analysis in health care systems (Helmreich & Merritt, 1998; Johnson, 2003). In the past, many medical adverse events have been "blamed" on the clinicians who were most closely involved in the immediate events leading to an adverse event. This led to recommendations that focused on improvements in operator performance, most often this involved exhortations to "be more careful" or to attend additional training sessions. It is important to stress that health care was not alone in adopting this perfective approach. Wagenaar's (1992) survey of industrial practice in energy production and the transportation industries observed that 80–100% of all incidents were attributed to human failure. More recently, however, attention has shifted away from the individuals at the "sharp end" of an adverse event. Reason (1997) argued that greater attention should be paid to the context in which an incident occurs. Analysts must understand the error-producing conditions that make mishaps more likely. This work has contributed to the popularity of "systemic" theories as explanations of accident causation (Leveson, 2003). In this view, individual errors rarely create the causes of an adverse event. Instead, we must look at the complex conjunction of managerial, regulatory, and even legislative constraints that jeopardized the safety of a more general "system". There are problems with this approach. It arguably undervalues the importance of individual responsibility. It also creates a recursive problem when we must try to understand the circumstances that led, for example, to management error that, in turn, contributed to clinical error (Johnson, 2003).

The Case Study Incident

The previous paragraph provides a deliberately broad overview of causal arguments about health care incidents. Chassin and Becher (2002) provided a more focused example from the health care domain. They studied the causes of an adverse event in which a patient mistakenly underwent an invasive cardiac electrophysiology study. She had struck her head and was found to have two cerebral aneurysms. She was, therefore, admitted for cerebral angiography. The day after admission, this procedure successfully embolized one of the aneurysms. A subsequent admission was planned for surgery to treat the second aneurysm. The patient was, therefore, transferred to the oncology floor prior to discharge rather than to her original bed on the telemetry unit. The next morning, however, the patient was taken for an invasive cardiac electrophysiology study. After approximately 60 mins, it became apparent that this procedure was being performed on the wrong patient. The intended patient had a similar name and had recently been transferred from an outside hospital for a cardiac electrophysiology procedure. She had also been admitted to the telemetry unit. This second patient's procedure had been delayed for 2 days but was now scheduled as the first electrophysiology on the day of the first patient's planned discharge. The electrophysiology nurse used the electrophysiology laboratory computer to check the schedule and saw the second patient correctly listed. She telephoned the telemetry floor, identified herself by name, and asked for the patient by their surname only. The person answering the telephone incorrectly stated that the patient had been moved to oncology when the second intended patient was still on the telemetry floor. The electrophysiology nurse was told that her patient would be transferred from oncology to the electrophysiology laboratory.

The original patient's nurse was nearing the end of her shift but agreed to transport her to electrophysiology even though she had not been told about any change of plan over her expected discharge. When asked about the procedure, the patient told the nurse that she had not been informed of any electrophysiology and that she did not want to undergo the procedure. Her nurse told her that she could refuse the procedure after she arrived in the electrophysiology laboratory. The patient repeated her reservations in the lab and so the attending physician was called. He was surprised to hear of this apparent change in opinion because he mistakenly believed that he was now talking with the same patient that he had beefed about the procedure on the previous evening. He reassured the first patient and prescribed medication to reduce the nausea that partly explained her reluctance to undergo the procedure. The electrophysiology nurse reviewed the patient's chart and noted that no consent had been obtained even though other records indicated that consent had been obtained. She paged the electrophysiology fellow scheduled to do the procedure. He reviewed the chart and was surprised at its relative lack of pertinent information. However, he then discussed the procedure with the patient and had her sign the consent for "EP study with possible ICD and possible PM placement" (EP refers to electrophysiology; ICD refers to an implantable cardiac defibrillator; PM refers to a pacemaker). The charge nurse arrived and was told by the electrophysiology nurse that their first patient had arrived, however, the patient's name was not referred to. A temporary nurse then placed the patient on the table, attached monitors, and spoke to them about the procedure. The patient stated that the original injury to her head had occurred when she had "fainted." The nurse thought that this was a reasonable indication for an electrophysiology procedure.

Meanwhile, a resident from the neurosurgery team was surprised to find that the original patient was absent from her room. He discovered that she had been moved to the electrophysiology laboratory and then demanded to know why the patient was there. Again, the patient's name was not used in the conversation. He was told that the patient had already missed this procedure on two previous occasions and was now being taken as the first case of the day. The resident left assuming that his attending had ordered the study without telling him. An additional electrophysiology nurse and the electrophysiology attending arrived. They stood outside the procedure room at the computer console but could not see the patient's face because her head was draped. The procedure was then started.

A nurse from the telemetry floor telephoned the electrophysiology laboratory to find out why no one had called for the second patient who was correctly scheduled for the electrophysiology. The electrophysiology nurses advised the telemetry nurse to send the second patient down when they estimated that the procedure on the first patient would have been completed. The electrophysiology charge nurse, making patient stickers for the morning cases, noticed that the first patient's name did not match any of the five names listed in the morning log. She raised the problem with the fellow who reassured her that this was "our" patient. The nurse did not want to inquire any further because she was concerned about interrupting such a demanding procedure. An interventional radiology attending went to the first patient's room and was also surprised to find it empty. He called the electrophysiology laboratory to ask why she was undergoing the procedure. At this point, the radiology attending and the electrophysiology charge nurse identified that the first patient's name was similar to the intended patient who was still waiting to be transferred from the telemetry floor.

The importance of Chassin and Becher's (2002, p. 829) study partly lies in their detailed exposition of the type of adverse event that is depressingly familiar within many health care systems. The significance of their work also lies in their subsequent analysis of the active, latent, and environmental conditions that they argued were causes of this incident. The perfective approach, mentioned in previous paragraphs, might focus blame on the individuals involved in the events that led to the electrophysiology procedure. For example, a causal analysis might argue that the nurse who mistakenly brought the first patient to the electrophysiology lab should have checked more carefully that their identity matched the names scheduled for the morning's procedures. Similarly, it can be argued that the attending physician should have introduced themselves more directly to the patient prior to the procedure taking place. Chassin and Becher's analysis identified 17 such instances of individual "failure." However, their interpretation of Reason's work (1997) led them to emphasise the systemic causes. In other words, "the errors of many individuals (active errors) converge and interact with system weaknesses (latent conditions), increasing the likelihood that individual errors will do harm." In particular, they distinguish between environmental factors that are not readily changeable in the short run and latent conditions. These are system faults that can be remedied but if they are ignored, will increase the probability that individual errors will have an adverse effect.

Subsequent analysis revealed a number of environmental factors that form the background to this incident. These included the increasing specialization of medical disciplines, pressures to reduce the number of hospital staff, and the increasing range of procedures being conducted on a "short stay" basis. These environmental pressures act together to make it less likely that the patient will be familiar with the individuals and teams who are responsible for their care. Latent conditions in this incident were identified as including failures of communication, teamwork, and procedures for the verification of the patient's identity. Nurses failed to communicate with their colleagues, physicians failed to communicate with nurses, attendings failed to communicate with residents and fellows and so on.

Previous Studies of Causation in Health Care Incidents

Chassin and Becher's (2002) study of an individual incident has been supported by more sustained studies into the causes of adverse health care events. For example, Van Vuuren's (2000) study of intensive care, accident, emergency, and anaesthesia-related incidents in English hospitals found that poor communication was a major factor among the many organization issues that contributed to adverse events. A recent investigation into the causes of near-miss incidents in an Edinburgh Intensive Care Unit also focused on organizational factors, including poor communication between health care professionals (Busse & Johnson, 1999). This study was based on over 10 years of incident data that was collected by a consultant, Dr. David Wright. Over the lifetime of the reporting system, a study of the human factors literature together with operational experience were used to inform a classification scheme for adverse events. Each report was analyzed to identify "causes," "contributory factors," and "detection factors." Causes included human error and equipment failure. Incidents involving human error were further associated with particular tasks, such as "vascular lines related," "drugs-administration related," or "ventilator related." Contributory factors included: inexperience with equipment; shortage of trained staff; nighttime; fatigue; poor equipment design; busy unit; agency nurse; lack of suitable equipment; failure to check equipment; failure to perform hourly check; poor communication; thoughtlessness; presence of students/teaching; too many people present; poor visibility/position of equipment; grossly obese patient; turning the patient; patient inadequately sedated; lines not properly sutured into place; intracranial pressure monitor not properly secured; endotracheal tube not properly secured; chest drain tube not properly secured; and nasogastric tube not properly secured. Finally, incidents were analyzed to identify detection factors. This illustrates an important issue. It can often be difficult to entirely remove the potential for adverse events; especially when health care professionals rely on information and services from other groups and individuals. In such circumstances, it is particularly important to strengthen those detection factors that have successfully identified potential adverse events in the past. The subsequent study of the Edinburgh reports focused on a comparison of two samples. One included reports that were received between January and February, 1989. The second included reports received between May and November, 1998. The study focused on a random selection of 25 reports from each period.

Table 40–1 summarizes the results from analyzing the 25 incidents in each sample. The totals in each column can exceed the sample size because each causal, contributing, and detection factor can appear more than once in each incident. As can be seen, poor communication remains a cause of adverse events across both samples (14 of 25 incidents in 1989 and 8 of 25 in 1998). However, thoughtlessness has increased considerably as a cause identified in these samples (2 in 1989 and 11 in 1998). It is, however, very difficult to interpret any changes in the distribution of causal, contributory, and detection factors over time. On the one hand, these may be due to differences in the underlying incidents that occurred in the unit being studied. For example, there was a determined training initiative between August, 1995 and August, 1998 to address the recurring problem of dislodged endotracheal tubes. Such initiatives can have the dual effect of reducing the frequency of such incidents and sensitizing staff so that they are now more likely to report these incidents in the first place. Changes in the distribution of causes not only stem from differences in the underlying incidents but they may also reflect changes in the way in which health care professionals performed the classification. In particular, it seems likely that the attitudes toward human error within the unit will change as they become increasingly familiar with some of the changes in human factors research, as summarized at the start of this chapter.

Bias and Causation

The causal analysis of adverse events helps to identify recommendations that are intended to avoid or mitigate any subsequent recurrence. The previous examples have shown, however, that it can be difficult to interpret whether changes in the distribution of causal factors reflect changes in the underlying pattern of incidents or in the interpretation of those adverse events. Further problems arise from the many different forms of bias that can affect the analysis of incidents and near misses (Johnson, 2003). For instance, author bias arises when individuals are reluctant to accept the findings of any causal analysis in which they themselves have not been involved. Confidence bias occurs when individuals unwittingly place the greatest store in causal analyses that are performed by individuals who express the greatest confidence in the results of their techniques. Previous work into eye-witness testimonies and expert judgments has shown that it may be better to place greatest trust in those who do not exhibit this form of overconfidence (Steblay, 1992). Hindsight bias arises when investigators criticize individuals and groups on the basis of information that may not have been available to those participants at the time of an incident (Johnson, 2003). More generally it can be seen as the tendency to search for human error

TABLE 40–1. Categories Used in Analysis of 1989 and 1998 Samples (50 Reports)

Frequency of Cause (*Reported incidents*)			*Frequency of Contributory Factor* (*Reported incidents*)			*Frequency of Detection Factor* (*Reported incidents*)		
Description	1989	1998	Description	1989	1998	Description	1989	1998
Ventilator	10	8	Poor Communication	14	8	Regular 11 checking	9	
Vascular line	6	4	Poor Equipment Design	11	1	Alarm 11	2	
Miscellaneous	5	4	Inexperience with Equipment	5	4	Experienced staff	8	8
Disposable Equipment	4	0	Lack of Suitable Equipment	4		Patient 1 noticed	1	
Drug–administration	3	10	Nighttime	3	3	Unfamiliar noise	0	1
Nondisposable Equipment	2	1	Fatigue	3		Handover check	0	1
			Unit Busy	2				
			Failure to Perform Checks	2	5			
			Thoughtlessness	2	11			
			Endotrach not properly sutured	0	2			
			Patient inadequately sedated	0	1			
			Turning the patient	0	1			

rather than for deeper, organizational causes in the aftermath of a failure. Judgment bias occurs when investigators perceive the need to reach a decision within a constrained time period. The quality of the causal analysis is less important that the need to make a decision and act upon it. Political bias arises when a judgment or hypothesis from a high status member commands influence because others respect that status rather than the value of the judgment itself. This can be paraphrased as "pressure from above." Sponsor bias arises when a causal analysis indirectly affects the prosperity or reputation of the organization that an investigator manages or is responsible for. This can be paraphrased as "pressure from below." Professional bias occurs when an investigator's colleagues favor particular outcomes from a causal analysis. The investigator may find them excluded from professional society if the causal analysis does not sustain particular professional practices. This can be paraphrased as "pressure from beside." Recognition bias arises when investigators have a limited vocabulary of causal factors. They actively attempt to make any incident "fit" with one of those factors irrespective of the complexity of the circumstances that characterize the incident. Confirmation bias arises when investigators attempt to interpret any causal analysis as supporting particular hypotheses that exist before the analysis is completed. In other words, the analysis is simply conducted to confirm their initial ideas. Frequency bias occurs when investigators become familiar with certain causal factors because they are observed most often. Any subsequent incident is, therefore, likely to be classified according to one of these common categories irrespective of whether an incident is actually caused by those factors. Recency bias occurs when the causal analysis of an incident is heavily influenced by the analysis of previous incidents. Weapon bias occurs when causal analyses focus on issues that have a particular "sensational" appeal. For example, investigators may be biased to exclude factors that are considered to be everyday occurrences.

It is difficult, if not impossible, to avoid the many different forms of bias that can affect the causal analysis of health care incidents. One reason for this is that many investigatory techniques provide guidance that is likely to influence the outcome of any causal analysis. On the one hand, this can be seen as beneficial if analysts are directed toward a broader consideration of potential causes. The influence of such direction can also be potentially dangerous if it places undue constraints on an investigation. The latter sections of this chapter presents a range of techniques that can be used to reduce some of the biases that affect the causal analysis of adverse health care events. In contrast, the following paragraphs present some of the underlying theoretical and practical problems that complicate incident investigation. The problems created by these various forms of bias are compounded by the complex nature of causation. For example, the FDA described a case study in which a violent patient in a wheelchair was suffocated through the use of a vest restraint that was too small. The risk manager, JC, used an FDA coding sheet to categorize the causes of the adverse event: She finds the list of event terms,

> which was detached from the rest of the coding manual ... She muses: 'Mr. Dunbar had OBS which isn't listed in these codes; he had an amputation which is listed; he had diabetes which isn't listed; and he had hypertension which is listed.' JC promptly enters 1702 (amputation) and 1908 (hypertension) in the patient codes. She then finds the list for Device-Related Terms ... She reviews the terms, decides there was nothing wrong with the wheelchair or the vest restraint, and leaves the device code area blank.(Weick-Brady, 1996, p. 3)

The resulting classification of 1702 (amputation) and 1908 (hypertension) provided few insights into the nature of the incident.

THEORETICAL APPROACHES TO CAUSATION

The previous example illustrates the practical difficulties that complicate the causal analysis of health care events. There are also a number of theoretical problems about defining what exactly is a cause of any observed effect. For instance, it can be difficult to determine whether or not an incident was *caused* by a particular staffing level on a ward rather than by the actions of individual clinicians. The distinctions between latent and active causes and between environmental and contributory causes can be used to guide such an analysis.

Epidemiological Approaches

Epidemiology has a number of well-considered requirements that must be satisfied in order to conclude that there is a causal connection between two events (Johnson, 2003). These can be summarized in the following manner. First, there is a temporality requirement. Most obviously the cause must precede the effect. However, it may well be that the hypothesized cause and the effect are both the effect of some earlier root cause. Hence, any investigation must consider a range of temporal relations. It is also important to stress that there need not be any causal relationship simply because a hypothesised cause precedes an effect. We can also impose a requirement for reversibility. In epidemiological terms, this implies that the removal of a presumed cause will result in a reduced risk of some adverse consequence, such as a fall in the prevalence of a particular disease. Equally, however, there may be some confounding factor that reduces the observed effect in a manner that is independent of the hypothesized cause. We might also expect that investigators demonstrate the strength of association between a cause and an effect. In particular, we would like to show that exposure to a potential cause has a significant impact on the relative risk of an adverse effect. This requirement is important because there is seldom a deterministic relationship between cause and effect in health care incidents. The fact that a unit is short-staffed, for example, does not guarantee that an incident will occur. However, there may be a strong association between these situations and particular types of adverse event. Epidemiologists also expect some relationship between exposure and response. This requirement is slightly different from the strength of association because it refers to the exposure and dose associated with a cause or risk. Exposure is usually quantified as a product of duration and intensity. For example, the period of time at which a unit is understaffed at a particular level. Dose is a measure of the "infecting agent" that is taken up by the human body. In terms of this chapter, it might be thought of as the number of incidents and near misses that occur in a particular time period. The establishment of a relationship between dose and exposure can provide powerful evidence of a causal relationship, providing it is not due to a confounding factor as previously described. This requirement can also be used constructively to establish risk thresholds; for example, establishing staffing minima that are intended to reduce the negative outcomes to an "acceptable" level.

A number of problems arise when epidemiological approaches are applied to identify the causes of adverse events in health care. In particular, a requirement for reversibility is often difficult to satisfy in more complex incidents. This requires that the removal of a cause would also lead to a reduction in the risk of an adverse event without any confounding factors. This recognizes that the identification of previous causal factors involves a study of future incidents. In the immediate aftermath of an adverse event, this creates a number of practical problems. In particular, it often forces investigators to construct counterfactual arguments of the form "an incident X would not have occurred if causal factor Y had not also occurred." This is counterfactual because we know that Y did happen and then the accident X also occurred. Practical problems arise because these arguments are nontruth functional. In other words, it can be difficult to provide evidence to support assertions that the removal of any single cause would have avoided an adverse event. A number of cognitive psychologists, such as Byrne and Tasso (1999), have studied the problems and paradoxes that stem from counterfactual reasoning. For example, in the incident described by Chassin and Becher (2002), it might be argued that the wrong patient would not have been selected by the electrophysiology nurse if she had referred to the intended patient's name during the initial phone call. However, previous incidents have shown that such verbal procedures can be error prone and that confusion can still occur where patients have similar names (Johnson, 2003). If we were to adopt the reversibility requirement previously described, then it would be sufficient to show that the risk of treating the wrong patient is diminished by ensuring the use of patient names while arranging for transfers between units. This would ignore many of the systemic factors identified as being critical to an understanding of adverse events across the human factors and systems engineering literature.

Primary (Catalytic) Failures

Further problems complicate the application of epidemiological approaches to understanding the causes of accidents and incidents in health care. For instance, it can be difficult to talk about the exposure to a hazard when the risk depends not on an "infecting" agent but on a complex conjunction of technical, social, and managerial precursors. The problems that lead to accidents often form part of a more complex landscape of managerial and regulatory failure, of poor design and equipment malfunctions, of environmental conditions, and of operational inadequacies. Mackie (1993) uses the term *causal complex* to describe this landscape of failure. Although he was looking purely at the philosophy of causation, it is possible to apply his ideas to clarify some of the issues that complicate the investigation of health care accidents and incidents. Each individual factor in a causal complex may be necessary for a mishap to occur but an adverse event will only occur if they happen in combination. Several different causal complexes can lead to the same accident even though only one may actually have caused a particular failure. For instance, initial confusion over the location of the patient from the electrophysiology treatment can be compounded by inadequate confirmation of the

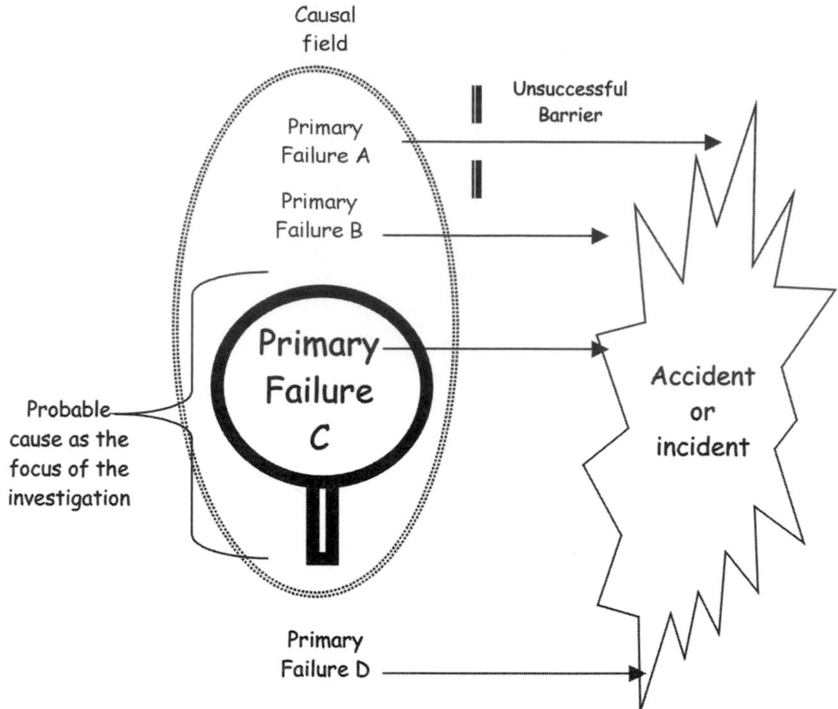

Figure 40–1. Causal fields and primary interaction failures.

patient's identity in the electrophysiology lab or by confusion over whether the patient had already provided consent to result in treatment of the wrong patient. It is for this reason that most accident investigations consider alternate scenarios in order to learn as much as possible about the potential for future failures.

Mackie goes on to argue that we often make subjective decisions about those factors that we focus on within a causal complex. The term *causal field* refers to those factors that an investigator considers relevant to a particular investigation. If a cause does not appear within this subjective frame of reference, then it is unlikely that it will be identified. This philosophical work has empirical support from the findings of Lekberg (1997) who was able to show a strong correlation between the findings of accident investigators in the Swedish nuclear power industry and the subject of their first degree. Human factors graduates were more likely to identify usability issues, process engineers were more likely to find problems in plant design, and so on. Figure 40–1 provides an overview of Mackie's ideas and how they might relate to Reason's (1997) view of accident investigation, mentioned in the opening paragraphs of this chapter. The causal field in this case concentrates on primary causes A, B, and C. Within that, we can focus on particular issues that we raise to the status of "probable causes". This is illustrated by the magnifying glass. For example, an investigator might be predisposed, or biased, to look at the behavioral issues in a particular working group. This would be illustrated by the focus on primary failure C in Figure 40–1. However, the causal field may not encompass a sufficient set of conditions and in this case primary failure D is not within the range of issues being considered by the investigator. For instance, if the investigation focuses on team-based issues, then correspondingly fewer resources will be available to consider other potential problems, including managerial or equipment concerns. In terms of our case study, the focus might be placed on the electrophysiology nurse's use of the patient's second name only during the initial contact with the telemetry department, primary failure A, or on the original patient's nurse's failure to check up on the movement to electrophysiology even though

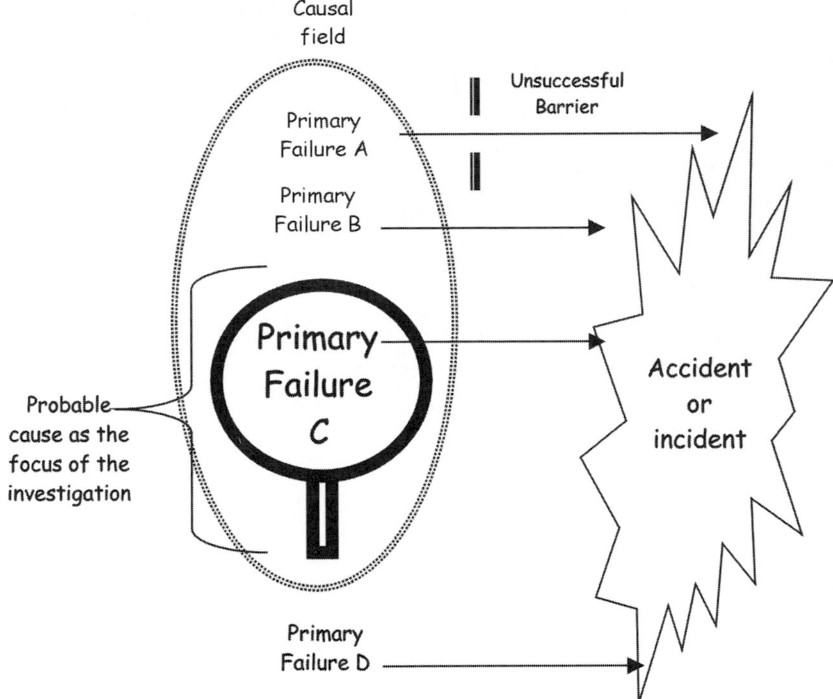

Figure 40–2. Causal fields and secondary failures.

they had not been told about the planned procedure, primary failure B. By focusing on these issues, investigators might neglect the role played by the lack of integration between patient records on the different computer systems operating in the various departments, primary failure D.

Secondary (Latent) Failures

The events leading to an adverse event can be traced back to form what are termed *causal chains*. For instance, secondary or latent problems create the preconditions for primary failures. These events might include particular decisions to reduce staffing levels within the various departments involved in the Chassin and Becher study. These events make it more likely that, for example, the original patient's nurse from the telemetry unit would not double check at the end of their shift before agreeing to move the patient down to the electrophysiology laboratory. Henc,e secondary problems do not directly cause an adverse event but can help to create the conditions in which a mishap is more likely to occur. Figure 40–2 provides an overview of secondary failures. As can be seen, these problems contribute to primary failures.

For instance, the fact that the electrophysiology laboratory's computer system was isolated from the main hospital system, represented by secondary failure 2, can create a situation in which personnel are more likely to make an error over the name and identity of an electrophysiology patient. This is illustrated by primary failure B in Figure 40–2. Alternatively, effective cross-checking procedures, for example, by the electrophysiology fellow prior to commencing the procedure might have discovered the potential adverse event. The successful barrier to secondary failure 1 in Figure 40–2 would illustrate this. An important aim of this chapter is to extend the causal field of accident investigations to consider these secondary or latent causes of adverse events. This is illustrated in Figure 40–2 by moving the magnifying glass to the left. The dotted ellipse used to denote the causal field in Figure 40–1 could also be redrawn to show the extended scope of an investigation in this figure. Our emphasis on secondary problems is intended to guide the

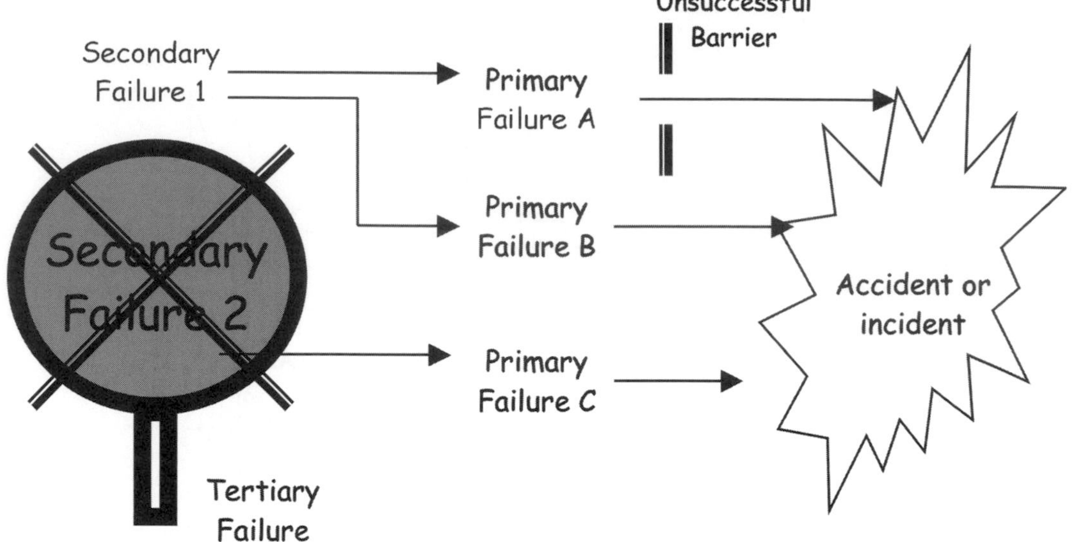

Figure 40–3. Causal fields and tertiary interaction failures.

composition of a causal field, which Mackie argued can be a subjective and arbitrary process. These latent failures are an increasingly common factor in the assorted lists of "contributory factors" that appear in accident reports. We would, therefore, argue that these secondary failures deserve greater and more sustained attention.

Tertiary (Analytic) Failures

First order failures lead directly to an incident or accident. They are cited as the probable cause of an adverse event when, for instance, a clinician performs a particular procedure on the wrong patient. Such primary failures are rare. In contrast, there is a host of secondary failures in most organizations that create the preconditions for an adverse event but for which there are, as yet, insufficient primary failures to trigger an adverse event. Figure 40–3 illustrates a final form of failure that complicates the analysis of adverse events in health care. Tertiary problems stem from the difficulty that investigators face when they use particular analytical tools to identify the primary and secondary causes of adverse events. Some techniques are poorly documented, especially if they were developed for domains other than health care. These techniques are, therefore, often used incorrectly when they are introduced to analyse adverse events in surgeries and hospitals. Other techniques require considerable training and expertise in order to understand and apply the underlying concepts. In particular, the application of the same techniques by different analysts can yield radically different insights into the same adverse event. This can create problems when health care organizations have to establish the priorities that will guide any subsequent recommendations. Figure 40–3 uses a darkened magnifying glass to illustrate the tertiary problems that complicate the use of causal analysis techniques. For example, Chassin and Belcher's (2002) case study was intended to warn against the dangers of the "perfective approach" where investigators focus narrowly on blame and retribution. In such circumstances, risk managers and safety professionals are unlikely to use the analytical techniques described in the rest of the chapter. They are also, therefore, less likely to identify the underlying causes of individual human error such as those described in the previous case study (Johnson, 2003). The remainder of this chapter describes a range of tools and techniques that bare intended to support root cause analysis.

CAUSAL ANALYSIS TECHNIQUES

In the aftermath of adverse events, it is important to identify those hazards that threatened the safety of an application process. Each of these may stem

TABLE 40–2. Results of an Incident Investigation

Hazard	*Root Cause of the Hazard*	*Proposed Remedial Action*	*Responsible Authority*
Hazard 1	Root Causes	Remedial Actions	Person or team to sign off
Hazard 2	Root Causes	Remedial Actions	Person or team to sign off

from numerous causes. These can be catalytic or primary events that triggered the mishap. They can also stem from the secondary, background or latent conditions that emerge slowly over a longer period of time. The identification of these causal factors can, in turn, be jeopardized by a higher order or tertiary form of failure that occurs when biases or technical limitations affect the analysis of an adverse event. This section presents a number of techniques that are intended to reduce the likelihood of tertiary failures in the investigation of accidents and incidents. They are intended to help investigators identify the "root causes" of adverse incidents. Unfortunately, few of these techniques have been specifically tailored to support the analysis of health care accidents. Most stem from work in the transportation and power production industries. The following exposition, therefore, uses the incident described by Chassin and Becher to illustrate how these different approaches might be applied to this domain.

Before introducing the various techniques, it is important to reiterate a point made at the start of this chapter; the output from any causal analysis is not simply the root causes and contributory factors. In particular, the tools and techniques must help investigators to identify recommendations and remedial actions. Table 40–2 illustrates a common format that is used to summarize these products of an incident investigation.

As we have seen from the previous section, Mackie's work suggests that each incident may help to identify a number of different hazards. These can be thought of as different causal chains that are individually sufficient to result in an adverse consequence. For example, the Chassin and Becher incident illustrates the hazards that arise when staff fails to confirm the name of the patient being transferred between units. Similarly, it also illustrates the problems that arise when the computer systems that hold information about patients in different units are not integrated so that cross-checks cannot easily be made on an individual patient as they move between those units. Each of these hazards may be sufficient to cause a mistake over the identity of a patient. Any particular incident can, therefore, involve several different hazards. Each hazard can be the result of several different combinations of secondary causes. For instance, there may be complex managerial and technical reasons for the lack of integration between the computerised records systems. These causes are likely to be very different from those that led to the lack of verbal confirmation for the patient's identity during the transfer. Each of these causes may, in turn, require a range of remedial actions. The following pages introduce techniques that investigators might use to identify the root causes of hazards involving health care systems.

As mentioned, causal analysis forms part of a wider process of mishap investigation. Ideally, investigators and safety managers must ensure the immediate safety of a system and gather all necessary evidence before any attempts are made to identify causal factors. In practice, however, investigators may have preconceived notions about what led to a failure. This can bias the way in which they gather evidence so that they only look for information that supports preconceived theories. From this it follows that the use of a causal analysis technique does not guarantee that appropriate lessons will be learned from adverse events.

Elicitation and Analysis Techniques

A number of causal analysis techniques are tightly integrated into the elicitation of evidence and mishap reconstruction. Investigators who are still considering "what" happened are encouraged to consider a number of possible causal factors so that they gather an appropriate range of evidence about the incident. This is important because, as mentioned previously, investigators' initial causal hypotheses may mean that evidence is only gathered if it supports their preconceptions. Barrier analysis provides an example of this form of causal analysis technique.

Barrier analysis

Previous sections have described how barriers can be created to protect a safety critical system from particular hazards. These barriers include technical

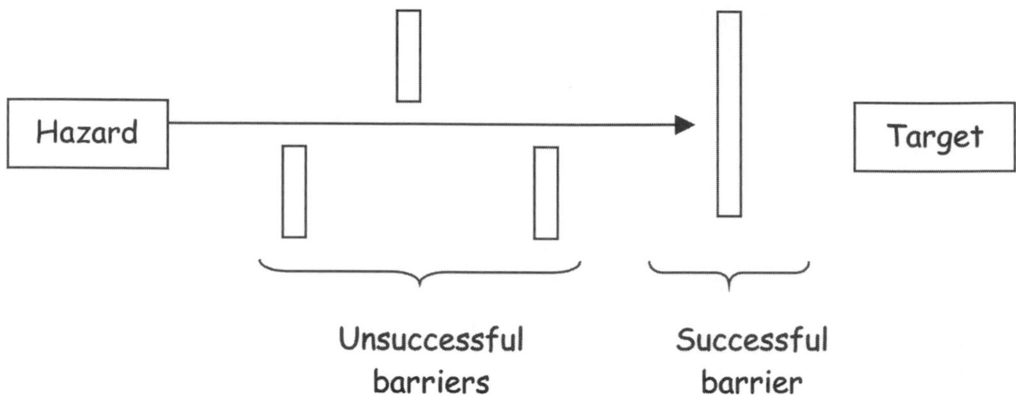

Figure 40–4. Targets, hazards and barriers.

features, such as the safety interlocks that physically prevent laboratory staff from placing their hands inside a moving centrifuge. They also include organization and procedural requirements, such as the standard operating procedures that may require staff to confirm the name of a drug with a colleague before it is administered to a patient. Barrier analysis focuses on the ways in which these measures are undermined during an incident or accident (Johnson, 2003). It traces the way in which an adverse effect stems from a potential hazard and ultimately affects the target. In this context, the target of the hazard is assumed to be the patient. In other incidents, the target might include other members of staff or even systems within a hospital. Figure 40–4 shows how the adverse effects of a hazard must pass through a series of potential barriers before they can reach the ultimate target. In this case, the final barrier prevents the incident from affecting the target. This typifies the way in which a final layer of defences can make the difference between a near miss and an accident. In such circumstances, incident reports provide important insights about both those barriers that failed and those that acted to protect the target from a hazard.

Barrier analysis, therefore, begins by drawing up tables that identify the hazard and the targets involved in an incident or accident. Table 40–3 illustrates these entities for the Chassin and Becher case study. As can be seen, we have extended the targets to also include the second patient who was correctly intended to have the electrophysiology procedure. In this instance, missing the intervention had no apparent effects on their prognosis even though this incident delayed their study for a

TABLE 40–3. Hazard and Target Identification

What?	Rationale
Hazard	Cardiac electrophysiology study per formed on the wrong patient.
Targets	The patient who incorrectly underwent the electrophysiology and the patient who missed their scheduled electrophysiology procedure.

third time. They are, however, included in Table 40–3 to illustrate how an initial barrier analysis should deliberately consider as wide a range of targets as possible.

Analysis progresses by examining the barriers that might prevent a hazard from affecting the targets. Analysts must account for the reasons why each barrier might have failed to protect the target. Table 40–4 illustrates the output from this stage.

Table 40–4 illustrates the way in which barrier analysis can be used to identify potential reasons for the failure of particular protection mechanisms. As can be seen, however, this initial analysis often focuses on the individual actions or primary causes of an adverse event. Further analysis may be required to identify the underlying secondary causes of an incident or accident. For instance, the previous analysis does not explain *why* the attending physician in electrophysiology did not pursue patient's apparent confusion over the procedure that had already been explained. For instance, "obtaining consent is frequently delegated to an overburdened or exhausted physician who has not met the patient previously and does not know the details of the medical history … cultural or social barriers to effective communication

TABLE 40–4. More Detailed Barrier Analysis

Barrier	*Reason for failure?*
Electrophysiology computer system provides patient details on the lab schedule.	No automatic way for telemetry nurse to cross-check her patient's data with the electrophysiology schedule because the systems were incompatible.
	Manual cross-check from the schedule fails because the electrophysiology nurse only uses the patient's surname.
Consent procedure requires patient's explicit permission for the procedure.	Telemetry nurse told patient she could refuse consent in the electrophysiology lab without cross-checking reason for patient's confusion over the procedure.
	Attending physician in electrophysiology did not pursue patient's apparent confusion over the procedure that had already been explained (to the other patient) on the previous evening.
	Nurses notice no consent on the patient's chart even though other records show it had been obtained, pass problem to electrophysiology fellow.
	Electrophysiology fellow briefs patient to obtain consent without clarifying source of confusion.
Electrophysiology clinicians required to perform chart review before conducting procedure.	Staff surprised by lack of relevant information but assume procedure ordered as part of treatment from other department. Original "fainting" mentioned by the patient considered reasonable indication for the electrophysiology procedure.

may be neither appreciated nor overcome" (Chassin & Becher, 2002, p. 828).

The metalevel point here is that causal analysis techniques often identify additional questions about practices and procedures, which are intended to protect patients and staff. Asking questions about why barriers fail can help analysts to look beyond the immediate triggering events that led to the mishap. It can be difficult to predict all of the possible events that might individually contribute to an adverse incident. In contrast, analysts must focus on the protection mechanisms that were in place to prevent those individual events from threatening the safety of the system.

Change Analysis

Change analysis is similar to barrier analysis in that it provides a framework both for causal analysis and also for the elicitation of additional information about an adverse event. Change analysis looks at the differences between the events leading to an incident and "normal" or "ideal" operating practices (Johnson, 2003). As with barrier analysis, this technique was not originally developed to support the investigation of health care incidents. It is, however, possible to use change analysis to analyze aspects of our case study. For example, the manner in which the patient's consent was obtained in this incident can be compared with the hospital's procedures or in the practices recommended by professional and regulatory organizations. The American Medical Association's Ethics Policy states that:

> The patient's right of self-decision can be effectively exercised only if the patient possesses enough information to enable an intelligent choice. The patient should make his or her own determination on treatment. The physician's obligation is to present the medical facts accurately to the patient or to the individual responsible for the patient's care and to make recommendations for management in accordance with good medical practice... (AMA, 2004, p. 73, paragraph E - 8.08)

Similarly, the events leading to this mishap can be compared with the procedures and policies established for the identification of patients prior to treatment. For instance, Louisiana State University requires that:

> Prior to the administration of tests, treatments, medications, or procedures the health care professional providing the care is responsible for verifying the patient's identity by utilizing two identifiers: patient name and patient medical record number. Whenever possible, staff shall also verbally assess the patient to assure proper identification, asking the patient's name and date of birth and matching the verbal confirmation to the written information on the identification ... If the identification band is illegible, missing, or contains information that is incorrect the test, treatment, medication, or procedures will not be done until the patient is properly identified. (Louisiana, 2003, p. 2)

Table 40–5 shows how the first column of a change analysis describes the ideal condition or the condition prior to the incident. This is an important distinction because the causes of an adverse event may have stemmed from inappropriate practices that continued for many months. In such circumstances, the change analysis would focus less on the conditions immediately before the incident and more on the reasons why practice changed from the ideal some time before the mishap. In this case, the analysis focuses more on the prescribed procedures and practices identified by the American Medical Association and by local hospital guidelines. The middle column summarizes the way in which these "ideals" may have been compromised during the incident under investigation. The final column discusses the effects of those changes from established guidelines and procedures. In this example, the result was that consent was obtained from the patient but this approval could not be described as "informed." Similarly, Table 40–5 illustrates the way in which failures in the identification procedure led to a patient with a similar surname receiving the procedure that was intended for the electrophysiology patient. Conversely, the electrophysiology patient did not receive their intended treatment.

An important strength of change analysis is that it can help to identify potential recommendations in the aftermath of an adverse event. In the simple case, it may be sufficient to ensure that the prior or ideal situation should be restored. In many situations, however, it will be important to question the reasons why previous procedures were violated or why norms emerged that might otherwise threaten the safety of an application. For instance, stating that clinicians should follow the AMA guidelines does little to address the systems-level issues identified by Chassin and Besson (2002) or by Reason (1997) and his colleagues. Further problems complicate the application of this technique. The previous example illustrates the manner in which change analysis often focuses on individual violations. The use of published policies and procedures in establishing ideal conditions can influence analysts to look for the individuals who were responsible for breaking those requirements. Finally, it can be difficult to connect this form of analysis to the mass of more immediate events that are, typically, documented in the evidence that is gathered following near-miss events. Event-based causal analysis techniques arguably provide a more convenient bridge to these reconstructions.

Event-Based Reconstruction Techniques

Barrier analysis can help risk managers to distinguish successful protective devices and procedures from those that failed to safeguard the patient or any other associated targets. Change analysis can also be used to elicit information in the aftermath of an adverse event by prompting investigators to identify the particular circumstances that led to a mishap. The analysis focuses on those factors that distinguish a mishap from previously successful procedures or from the ideal circumstances that are often embodied in relevant guidelines. Both of these approaches tend to analyze an incident at a relatively high level of abstraction. They do not usually provide a detailed reconstruction of the events that occurred during an incident or accident. In contrast, event-based techniques are intended to help analysts clarify *what* happened. They are often used in conjunction with a secondary form of analysis that uses these event reconstructions to determine *why* these events took place.

Time lines

Time lines provide arguably the simplest form of event-based analysis technique. They provide a straightforward representation of the ways in which events unfold over time. For instance, many devices provide automatic means of generating a log of system-level events. Table 40–6 recreates part of the alarm log that might have been derived from a monitoring system. These time lines raise a

TABLE 40–5. Change Analysis

Prior/Ideal Condition	*Present Condition*	*Effect of Change*
Patient identity should be confirmed prior to moving them from their current location within the hospital or performing procedure (Hospital guidelines).	Electrophysiology nurse only uses patient's surname when requesting they be moved from telemetry. Electrophysiology attending fails to confirm this is the patient he discussed procedure with on previous evening. Charge nurse and resident do not refer to patient name when discussing the first case of the day. Charge nurse fails to pursue discrepancy noticed when making name stickers ...	Patient with a similar surname receives procedure intended for electrophysiology patient. Electrophysiology patient does not receive their intended treatment.
Patient should provide informed consent prior to any procedure having been provided with adequate information about the risks and benefits. (American Medical Association, Ethical Guidance E8.08)	Oncology patient told nurse she didn't know about the electrophysiology. Patient also told attending electrophysiology physician of her objections. He reassured her. Electrophysiology nurse notices no written consent even though records state it should be there.	Consent is obtained but it cannot be described as informed given that they agreed to an "unnecessary" procedure. Staff had several prompts to inquire further into apparent anomalies over the missing consent forms.

TABLE 40–6. Example Summary From Automated Alarm Log

Point	Time	State of the Alarm	Description	State–Start of Scan	Current Status	State once Scan Complete
BLS_605	11:27:20	Normal	Gas detector	Acknowledged	Reset	Deleted
BLS_605	11:27:37	Beam Blocked	Gas detector	Nominal	Generated	Generated
BLS_605	11:27:40	Normal	Gas detector	Generated	Reset	Reset
BLS_605	11:28:30	Normal	Gas detector	Reset	Acknowledged	Deleted
PLT-23	11:28:34	Loop Fault	F/Disch	Nominal	Generated	Generated

host of practical problems. For instance, most devices have a very limited capacity to record alarm and status information. It, therefore, follows if the logs are not printed or saved in an aftermath of an adverse event, then key records will be overwritten. In consequence, many agencies in other domains such as Air Traffic Management have explicit requirements for supervisors to ensure that any relevant automated logs are protected (Johnson, 2003). There are further problems. Some device manufacturers view these logs as diagnostic tools that should be used during installation and calibration. They are not easily accessible to end-users and may even be disabled after deployment. It can also be difficult to interpret the meaning of these automatic logs without detailed technical support from device manufacturers.

Table 40–6 illustrates the problem that exists in moving from device logs to the more "systemic" forms of causal analysis, described earlier in this chapter. In particular, it can be difficult to incorporate operator interventions and management decision making processes that will only be indirectly represented in the output of such systems. In consequence, most incident investigations construct higher level, graphical time lines to record the events that contributed to an accident or near miss. Figure 40–5 illustrates this approach.

Figure 40–5 illustrates a form of time line that was developed by the U.S. National Transportation Safety Board (Johnson, 2003). The "actors" involved in an incident are named vertically on the left side. For instance, the events involving the oncology nurse can be distinguished from the telemetry nurse and the attending physician. The events that they are then involved in are enumerated on a horizontal time line to their right. Once these events have been mapped out, it is then possible to draw arrows between those events to indicate "informal" causal relationships. These are informal because there are few rules to guide investigators at this stage. For instance, there is no way of determining whether the precursors are sufficient to cause an adverse event or whether there are other ways in which the mishap might have occurred. As we shall see, other techniques provide heuristics or rules of thumb that can be used to support these different forms of causal analysis.

Practical problems arise because in complex incidents, it is relatively easy to run out of space on any single sheet of paper. Using continuation sheets can reduce this problem. However, investigators often adapt this approach by using sticky notes on a large section of wall. These can be placed and replaced as more information becomes available. Further problems complicate the use of time lines in the causal analysis of adverse health care events. First, it can be difficult to obtain exact timings for many different events distributed across several departments. Hence there will often be inconsistencies and contradictory evidence for exact timings. This caveat has persuaded many analysts to represent and reason about adverse events at a more abstract level. The intention is not to model every detailed event that occurred but to sketch critical causal relationships between lesser numbers of more important events.

Accident Fault Trees

Fault trees extend concepts from systems engineering to support the analysis of adverse events (Johnson, 2003). The key idea is that the causes of a complex event can be analyzed in terms of a conjunction of simpler precursors. In other words, an accident is the result of A happening and B happening and C happening, and so on. Disjunctions

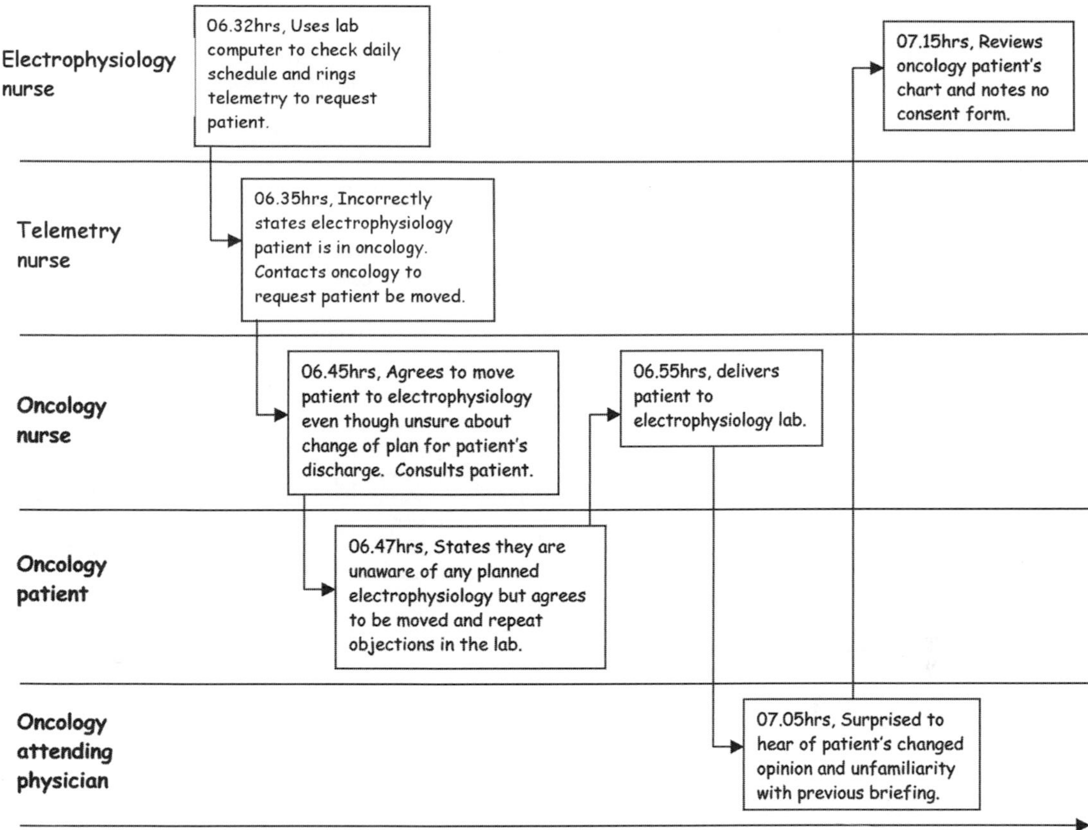

Figure 40–5. High-level time line of the case study incident.

can also be used to represent alternative causes. For instance, an accident is the result of A or B or C and so forth. These distinctions help to clarify the causal relationships that were less apparent in time lines. Sufficient causes are represented by a conjunction of events that lead to an adverse event. There may be several of these conjunctions if disjunctions appear in the tree. Necessary causes are events that appear in every sufficient conjunction. In other words, if a necessary cause is omitted, then the accident will be avoided. Figure 40–6 provides an overview of one form of accident fault tree for the Chassin and Becher's (2002) case study. For additional information on fault tree analysis and other risk management techniques, see the chapter on "Human Factors Risk Management in Medical Products" in this handbook.

The events that contribute to a mishap are represented as rectangles. In this case, the tree only includes conjunctions; these are denoted by the semicircles labeled *AND*. For example, the bottom left subtree illustrates the observation that two patients shared similar names AND the surname was only used when the electrophysiology nurse requested the transfer AND the electrophysiology computer was not linked to the main hospital system AND the telemetry department stated incorrectly that the patient was in oncology. These four observations together are used to conclude that the there was a "failure to identify the correct patient prior to transfer." If any one of the component events were omitted, for example if the patient's forename was used as well, then the identification failure need not have occurred. Similarly, this diagram also includes some of the events that led to the failure to obtain informed consent. This event and the righthand "failure to confirm patient's identity before procedure" illustrate how accident fault trees can be used to incrementally build up the level of detail in an analysis. Subsequent investigation

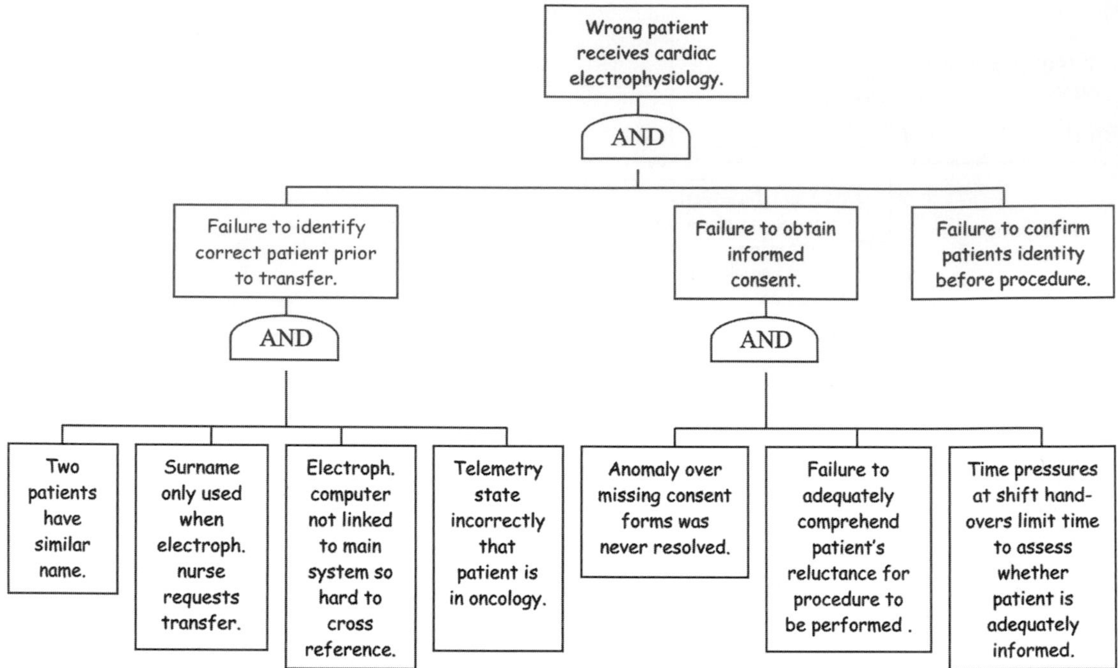

Figure 40–6. Overview of an accident fault tree.

might focus on adding additional information so that these aspects of the diagram mirror the detail devoted to explain the "failure to identify the correct patient prior to transfer."

Although accident fault trees avoid some of the problems that affect time lines, there are a number of additional problems (Johnson, 2003). In particular, it is unclear how to represent the way in which any response to an incident helps to determine the eventual outcome. In conventional fault trees, the analysis stops with a potential hazard. In our example, this is the top-level event "wrong patient receives cardiac electrophysiology." Extending the diagrams to representing different outcomes, from the subsequent mitigation or exacerbation of a mishap, would create considerable differences with the analytical use of fault trees in design. It might also result in complex diagrams that hinder rather than support the analysis of what are often complex failures. Further problems stem from the lack of any explicit temporal information in accident fault trees. There is an implicit assumption that lower level events in the diagram occur before the top-level events and that events in the same level occur from left to right. However, these are informal conventions and there is no explicit relationship to real time that can be critical, for instance, in drug misadministration mishaps. Figure 40–6 also blurs the distinction between events and conditions. For example, the use of the surname only during patient transfer is an immediate event that cannot easily be decomposed any further. In contrast, the lack of any link between the two computer systems and the time pressures between shift handovers are both conditions that could be decomposed to identify the detailed events that led to these problems. This distinction between events and conditions can create confusion about whether any analysis has explored the causes of an incident in sufficient depth.

Failure event tree, ECF *Charts,* MES *and* STEP

As mentioned, fault trees were originally developed to support the design and engineering of complex applications. The previous paragraphs describe how this approach can be extended to support the analysis of adverse events in health care. In contrast, a range of event-based techniques has been specifically developed to support the reconstruction and interpretation of accidents and incidents. There are strong similarities between many of these approaches, including events and causal factors charting (ECF), multilinear events sequencing (MES) and sequential timed event plotting (STEP). For more information on each of these techniques, see Johnson (2003).

Figure 40–7 provides an illustration of these event-based approaches by applying a failure event tree to our case study. In contrast to the previous time line, this diagram focuses on what happened after the patient had reached the electrophysiology department. All timings are indicative and should not be read as an accurate representation of any particular accident. The sequence of events leading to a mishap is denoted by the rectangles at the top of the image. For example, the diagram includes the events "07.35hrs, procedure started" and "07.19hrs, Electro(physiology) fellow reviews chart and had patient sign consent form." Outcomes are denoted by bold rectangles with dotted borders. In this example there are two. One is associated with the performance of the procedure on the oncology patient. The second is associated with the failure to perform the procedure on the intended patient. This represents a slightly unusual application of the failure event tree technique (Johnson, 2003). It could be argued that the decision to halt the procedure at 08.30hrs represents the outcome of this incident. However, the current version of the diagram focuses attention on the adverse outcome rather than on this mitigating event. This decision could be revised during any subsequent investigation of the mishap.

Figure 40–7 also uses rectangles with a double line border to denote direct factors that influence an adverse event. These factors can be thought of as conditions that influence the course of a mishap but that cannot easily be captured by particular events. For example, it would be possible to extend the diagram back many months or years to consider the precise moments when the decisions were made that prevented any cross-referencing between the telemetry and electrophysiology computer systems. However, in the aftermath of many adverse events it can be extremely difficult to identify the particular moments that led to such decisions. It is often easier simply to represent these more detailed precursor events as direct influences on the course of events. Similarly, it is often easier to represent cognitive and human factors influences on decision making as direct factors rather than attempt to trace the individual events that might have affected their actions. As before, we could extend the diagram to represent these factors as events. For example, the observation that the "Electrophysiology attending believes he spoke to patient on the previous evening" could be associated with a particular location in the time line at the top of the diagram. However, it can be difficult to explicitly identify the moment at which such cognitive events occurred. It can also be hard to gather necessary evidence to support such inferences about individual cognition.

Failure event trees can also capture the less direct, or distal, factors that contribute to an incident. Dotted double borders around a rectangle denote these. For example, Chassin and Becher (2002) observed that the underlying causes of our case study include reductions in staffing levels; the increasing number of short-stay patients, and the increased specialization of medical disciplines. Unfortunately, they do not explain precisely how these distal factors affected the direct factors and events in this incident. As can be seen from Figure 40–7, failure event trees provide a means of explicitly representing these relationships between proximal and distal factors. The underlying indirect factors help create the conditions for more direct factors, which in turn contribute to the events leading to this particular mishap. For instance, reductions in staffing levels and increasing numbers of short-stay patients may have combined to increase the pressures on staff that were then less likely to carefully confirm the name of particular patients. Hence, their pressures may indirectly have contributed to electrophysiology nurse's failure to state the forename of the first patient on their schedule. There are, however, no agreed-upon rules for distinguishing events from direct or indirect factors. These distinctions often, therefore, result from a process of negotiation between the participants in an investigation.

Figure 40–7 shows how event-based techniques often "map" out the course of an incident or accident in a manner that is very similar to a time line representation. The key difference is that techniques such as failure event trees and events and causal factors charts also explicitly represent the distal factors that indirectly contribute to an adverse event. Most of these approaches rely on a secondary stage of analysis to identify the root causes of a mishap. The central problem is to

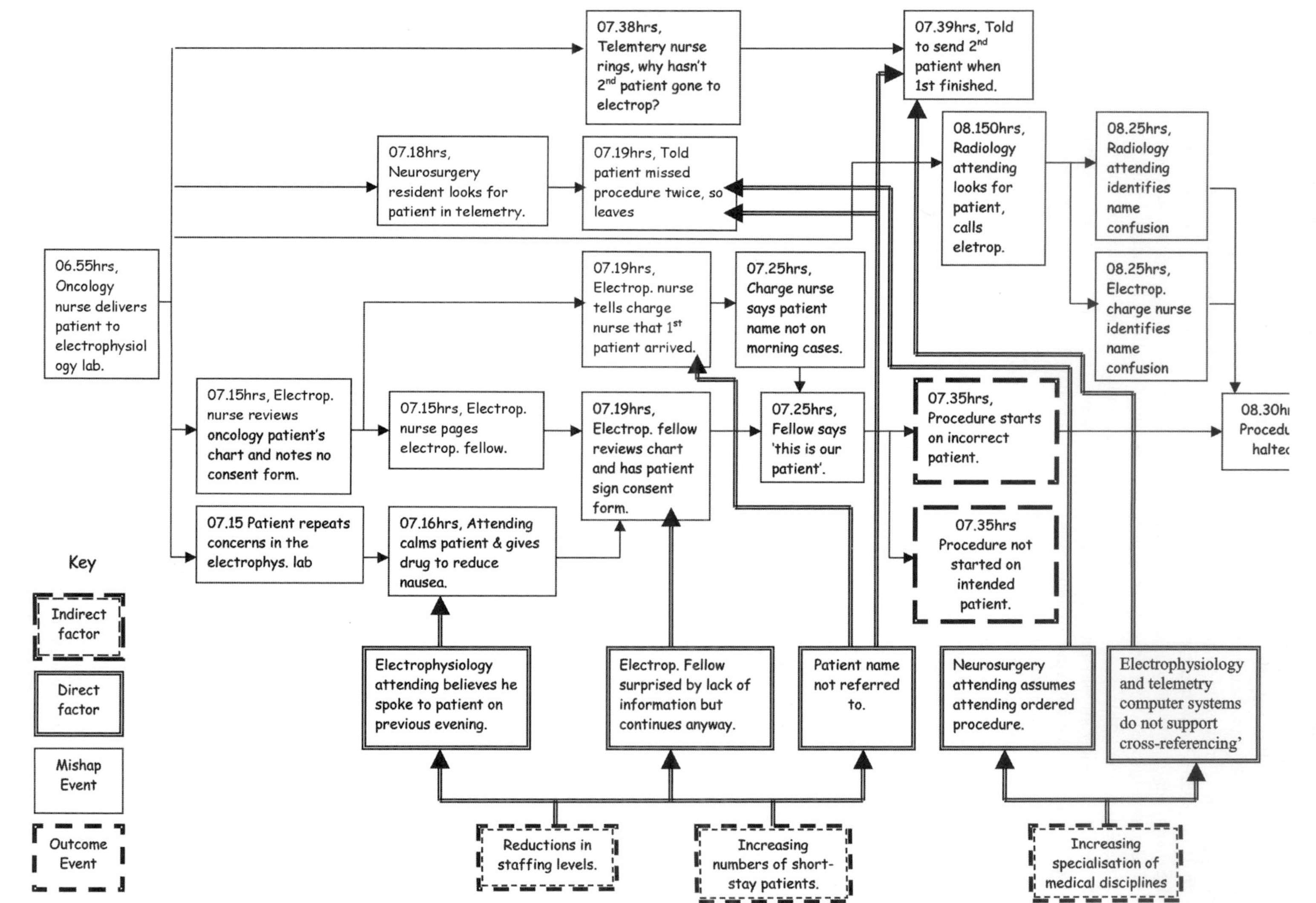

Figure 40–7. A failure event tree

distinguish these critical aspects of an accident or near miss from the other less significant events in the diagram. Counterfactual reasoning is again used to identify these root causes. Analysts begin by looking at the event closest to the incident. In Figure 40–7, we ask would the procedure still have started on the wrong patient if the electrophysiology fellow had not confirmed that "this is our patient." If the answer is yes and the mishap would still have happened, then this event cannot be a candidate root cause of the incident. If the answer is no and the mishap would not have occurred without this event, then we can argue that it was necessary for the incident to occur so it can be considered as a root cause. The process continues for each of the mishap events shown in the diagram. Once potential root causes have been identified, remedial measures can be introduced to address the direct and indirect factors that led to each of the particular mishap events that were identified as the root causes of this mishap. For instance, in our case study, it is conceivable that the procedure would still have started even if the fellow had not made his pronouncement. However, it seems less likely that the problem would have arisen if they had not persuaded the patient to sign the consent form. Hence, the indirect factors associated with this event, including the rising number of short-stay patients and reduced staffing levels, can be considered root causes of the incident as a whole.

A number of caveats complicate the use of event-based analysis techniques. For example, we have already argued that the distinctions between "first class" events, direct and indirect factors can be arbitrary. Further problems arise because there are few heuristics for determining the scope of the analysis. For example, we could have pushed the initial event in Figure 40–7 back to consider the moment when the two patients were admitted or to precursor events such as the acquisition of the hospital computer system. In consequence, it is entirely likely that two investigators might produce different models from the same accident. The subjective nature of these approaches need not be a weakness. These techniques are relatively easy to learn and in consequence, the members of a multidisciplinary team might conduct a semi-independent analysis of the incident. The different perspectives offered by these studies could then be combined into a unified model of an adverse event. However, the arrows between events introduce further confusion. They represent causal relationships. For example, the charge nurses' statement that the patient's name is not on that morning's list causes the eectrophysiology fellow to remark that "this is our patient." Elsewhere they represent the flow of events without any causal information. For instance, there is no immediate causal relationship between the electrophysiology nurse telling the charge nurse that their first patient has arrived and the charge nurse's observations about the names on that morning's list. Such criticisms have resulted in alternative forms of causal analysis techniques such as Leveson's (2003) STAMP and Ladkin and Loer's (1998) WBA, which avoid some of these confusions between temporal sequences and causal relationships. These techniques are described in the later sections of this chapter.

Flow Charts and Taxonomies

One of the main criticisms leveled at elicitation techniques, such as barrier analysis, and event-based approaches, including failure event trees, is that they provide little explicit encouragement for consistency between investigators. Different investigators will produce very different models of the same adverse events (Johnson, 2003). In contrast, flow charts, typically, guide the analysis toward a number of predefined causal factors. This supports the extraction and validation of statistical information from large-scale incident reporting systems. The flow charts help to ensure that analysts consider the same range of causal factors by constraining the scope of their analysis.

MORT

Management oversight and risk trees (MORT) provide a flow-chart approach to the analysis of organization failures (Johnson, 1973). Figure 40–8 provides an abbreviated version of a MORT diagram that is at the heart of this technique. The diagram is built from components that are similar to those in the accident fault tree, introduced in previous sections. For instance, an adverse event is caused because the oversight of an application was less than adequate AND because there were risks involved in the operation of the system. These two components of an incident are related by the conjunction at the top level of Figure 40–8. The causal analysis of an adverse event begins at these top levels of the tree. Investigators must ask themselves whether the mishap was the result of an omission of some management function and whether the

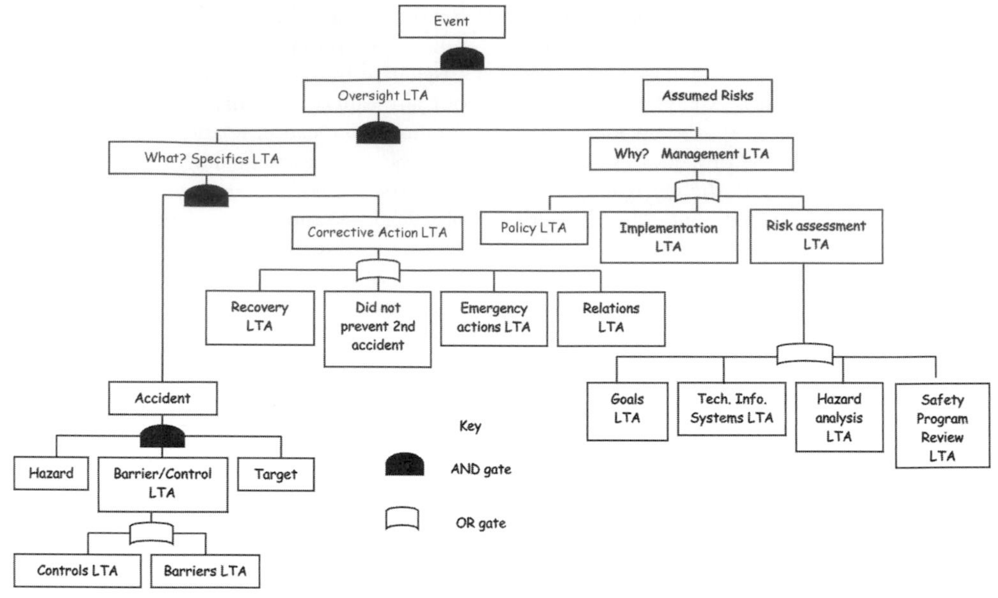

Figure 40–8. Abbreviated form of a MORT diagram.

incident occurred from a risk that had already been recognized. In the tree, the term LTA refers to a "less than adequate" performance of some necessary activity. If there was an oversight problem, then analysis progresses to the next level of the tree. Investigators are encouraged to consider why a mishap occurred. The reasons why an oversight might occur include less than adequate management policy, implementation, or risk assessment. The analysis progresses in this manner until investigators reach a number of terminal nodes, not shown here, that describe the more detailed causes of the incident. In passing, it is important to note that the bottom lefthand branches of Figure 40–8 show how MORT encompasses aspects of barrier analysis, introduced in previous sections of this chapter. This illustrates how the elements of the MORT tree capture high-level categories that can be applied to describe management problems in many different domains. This enables comparisons to be made between the management of adverse events across many different health care sectors. The tree structure also encourages consistency because investigators must use the tree to ask the same analytical questions determined by a left to right traversal of Figure 40–8.

As we have seen, our case study involved risks to the patient. It can also be argued that oversight was less that adequate. Analysis can, therefore, progress to the lower levels of the tree. We must determine whether the oversight during either development or operation was adequate. If it was not, then we can begin to analyze what happened during the incident by going down the far left branch of the figure. This involves the identification of hazards, barriers, and targets in an identical fashion to barrier analysis introduced previously. After having identified these components of what occurred, analysis might go on to consider the right branches including the reasons why management might have been less than adequate. Figure 40–8 encourages analysts to consider whether the policy, the implementation, or the risk assessment in the design and operation of the system might have contributed to the mishap. This might lead an investigation to question whether the policies identified in our previous change analysis had been successfully implemented, for example, in order to obtain informed consent.

Investigators must document the results of such causal analysis. Table 40–7 illustrates one technique that can be used in conjunction with MORT diagrams. A brief argument states the reasons why a mishap was caused by one of the factors that are represented in the nodes of the tree. In this case, the risk assessment was less than adequate because the danger of a loss of control functions after a system trip for the crew and the vessel was not considered in sufficient detail. Such documentation is

TABLE 40–7. Documenting the Products of a MORT Analysis

Branch in Mort Tree	*Node of MORT Tree*	*Incident description*
Risk assessment less than	Hazard	Electrophysiology procedure performed on wrong patient.
	Target	The patient who received the treatment and the patient who should have received it but did not.
Hazard analysis less than	Control operability problems	Failure to identify potential confusion caused by lack of cross referencing between two patient record systems.

important if others within an organization are to understand the reasons why particular causes have been identified in the aftermath of an adverse event or near-miss incident. They can act as a focus of subsequent discussion and can help to direct resources to improve areas where previous management activities have proven to be less than adequate.

As mentioned, MORT is a generic technique intended to help identify management problems across many different industries. It lacks the technical details necessary for example to distinguish some of the more detailed problems that can arise in health care. In particular, it seems ill suited to consider the technical causes of adverse events and their interaction with, for example, failures in teamwork across departmental boundaries. There are several other flow-chart techniques that arguably avoid some of these problems because they have been tailored to particular domains.

PRISMA

The PRISMA technique is similar to some of the event-based approaches in that it consists of a reconstruction phase and an analysis phase. The initial reconstructions develop accident fault trees similar to those seen previously in this chapter (van der Schaaf, 1992). The leaf or terminal nodes on the tree are then classified to identify more generic root causes. This is important because a number of different incidents could yield very different trees. For instance, one might address the communications issues that led to confusion over a particular patient as in our case study. Another subsequent incident within the same hospital might, in contrast, yield an accident tree with nodes describing a drug misadministration. By classifying each of these different nodes, it may be possible to identify common causes, for instance involving staff reduction, that led to both of these apparently different adverse events. A flow chart is, therefore, used to provide a higher level classification of these more detailed causes. The use of this flow chart not only enables investigators to identify common causes between different incidents. It can also encourage a consistent analysis of individual mishaps. Investigators may disagree about the detailed causes of an adverse event but may exhibit greater agreement about the higher-level classification.

Unlike some of the other methods reviewed in this chapter, PRISMA has been trialed and tailored for use in the health care industries. For instance, Figure 40–9 presents a flow chart that was developed for use within the United Kingdom NHS and the Netherlands' health care systems. As can be seen, each terminal node is associated with a particular abbreviation such as OK for an organizational factor related to the transfer of knowledge, this might be applied in our case study to explain the problems in cross-referencing patient names between telemetry, oncology, and electrophysiology. It is also extremely important to stress that the ordering of terminal nodes can bias the insights obtained from any causal analysis. In Figure 40–9, organizational issues appear before technical factors and human behavior. It is therefore more likely that analysis will identify organizational issues before considering these other potential classes of causal factors. When PRISMA was adapted for the process industries, the ordering was altered so that technical issues were considered first to reflect the relative priorities of safety managers in these different industries. A further difference between Figure 40–9 and other variants of the PRISMA flow chart is that it includes "patient-related factors" as a potential cause in health care incidents.

It is important that recommendations can be derived from the findings of an investigation. This

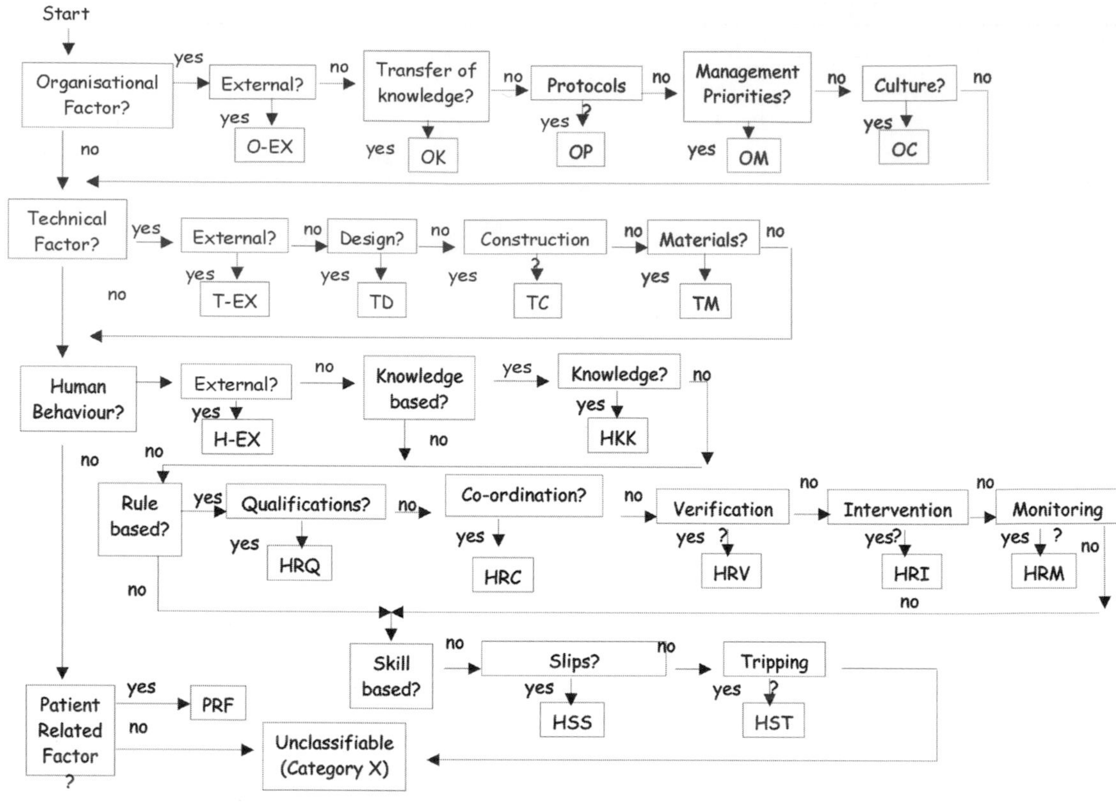

Figure 40–9 PRISMA flow chart (van der Schaaf, 1992; van Vuuren, 1998)

can create problems. Even if a PRISMA flow chart can help to ensure that different investigators agree on the high-level causes of an adverse event, there is no guarantee that they will agree on potential interventions to avoid future incidents. PRISMA, therefore, uses the output from the flow chart to direct the process of identifying recommendations. Table 40–8 illustrates a classification action matrix. This shows that if, for example, an incident were due to problems with management priorities, then subsequent recommendations might focus more on "bottom-up communication." This is intended to ensure that investigators offer a common response to incidents with similar causal factors. If incidents continue to recur with the same set of causal factors, then safety managers might decide to revise the interventions advocated in Table 40–8. In practice, however, it is likely that the elements in Table 40–8 would have to be revised and carefully monitored. For instance, additional details must be provided in order to improve bottom-up communication. In terms of our case study, this might involve changes in the procedures by which staff confirm the patient's identity during shift handovers or between clinical and nursing staff. Measures that are intended to improve bottom-up communication must in turn be assessed to determine whether they are having the outcome intended by the rows in Table 40–8.

Accident Models

Many of the previous techniques assume that investigators can produce complex representations, such as accident fault trees or time line event models, of the adverse events that are reported to them. This assumption may not be justified. For instance, it can be difficult to determine when an accident begins or ends. Our case study could begin when the patient was delivered to the electrophysiology lab, it could equally well begin when the transfer request was made to the telemetry unit, when the two patients were first admitted or even when

TABLE 40–8. Example of PRISMA Classification/Action Matrix[a]

	External Factors (O-EX)	*Knowledge Transfer* (OK)	*Operating procedures* (OP)	*Manag priorities* (OM)	*Culture* (OC)
Inter-departmental communication	X				
Training and coaching		X			
Procedures and protocols			X		
Bottom-up communication				X	
Maximise reflexivity					X

[a]Van Vuuren, 1998

the consent procedures were introduced into the hospital. Similarly, it could end when the procedure was started or if we are to consider mitigating actions, it might also consider the chain of events that led to the error being detected. Similarly, it can be difficult to determine the level of detail to be included in any analysis. In some incident investigations, it is necessary to examine the exact verbal protocols used during shift transitions and the transfer of patients. In our case study, investigators might focus on the ambiguity in the request that was first made by the nurses on the electrophysiology ward to the nurses in the telemetry unit. In other incidents, it may only be necessary to consider higher level communications between individual working groups rather than the actual words and phrases that were used. *Accident models* can help to address some of these problems. In contrast to event-based approaches and flow chart techniques, accident models provide strong guidance about what causes an adverse event. They enforce a particular viewpoint on the analytical process.

TRIPOD

The tripod technique builds on a number of "general failure types" that are assumed to cause the majority of adverse events (Johnson, 2003). These generic causes include failures in hardware; maintenance management; design; operating procedures; error-enforcing conditions; housekeeping; incompatible goals; communication; organization; training; defense planning. There are strong similarities between these general failure types and some of the concepts that are embedded in the techniques of previous sections. For instance, the failures in defense planning that are identified by tripod are similar to the inadequate defenses that are identified in barrier analysis. Similarly, operating procedures are considered within the PRISMA flow chart and are included in tripod. However, some general failure types are not explicitly considered by the other techniques that we have introduced. These include maintenance management. Conversely, it can be argued that additional general failure types should be introduced into tripod. In particular, it seems odd that hardware should be considered as a cause but not software.

Figure 40–10 illustrates the graphical model that tripod provides to represent the way in which general failure types combine to create the conditions for an adverse event. As can be seen, tripod uses many of the concepts that were introduced as part of barrier analysis. Defenses fail to protect a target from a hazard. In our case study, there is a danger that cardiac electrophysiology procedures will be performed on the wrong patient. This hazard threatens both the patient on which the procedure is performed and the intended recipient who inadvertently missed their operation. In this case, the defenses included handover procedures that are intended to establish patient identity prior to a transfer and the requirement for informed consent to be obtained prior to any procedure being performed. It is possible to identify a number of other barriers that failed in this incident, including the need to establish patient identify before the procedure is performed, however these are omitted for the sake of brevity and could be included in any subsequent analysis. Active failures are associated with each of the defenses that did not protect the target. They are made more likely by a number of preconditions. For instance, the reliance on patient surnames was made more likely by a divergence from recommended practices and procedures. This

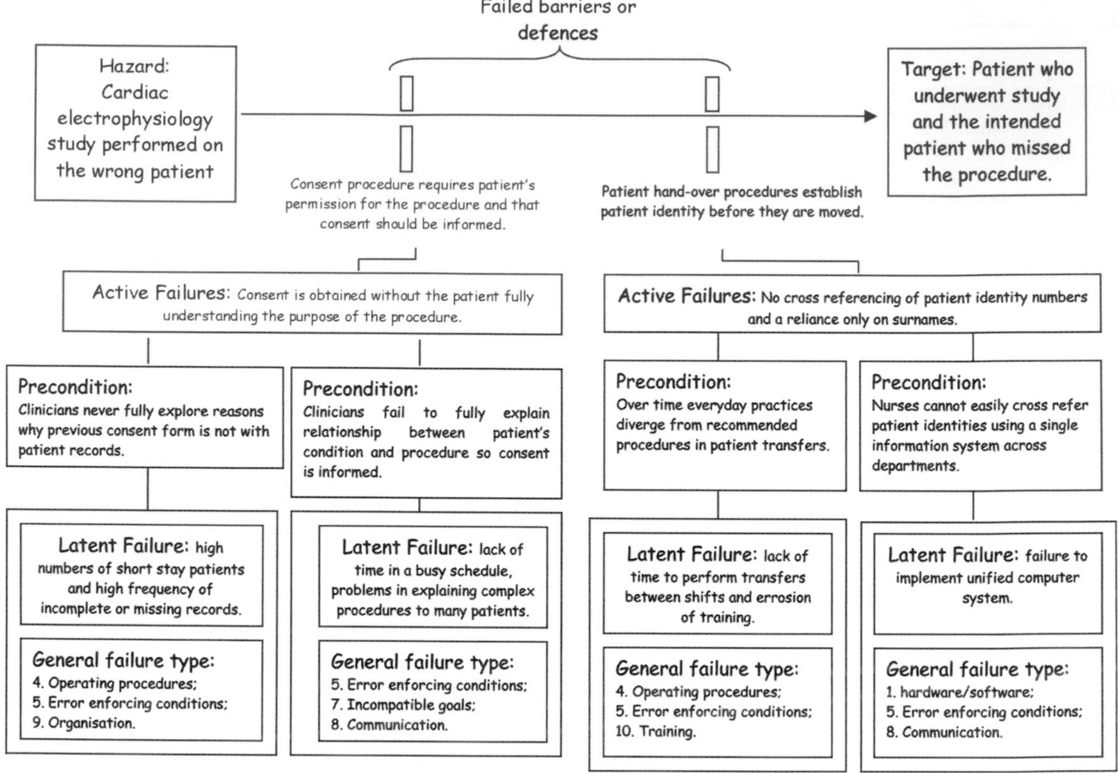

Figure 40–10 Example application of TRIPOD general failure types.

precondition was, in turn, satisfied by a latent failure in terms of the time pressures that affected transfers between shifts. As with flow chart techniques, the intention is to move away from the specific events that led to a mishap so that greater attention is paid to the generic or systemic failures that are likely to threaten future operation. It is unlikely that an identical failure will recur in the immediate future. Most safety-critical organizations can ensure that recommendations are put in place to prevent a pattern of identical failures. However, other similar problems, for instance to do with limited time at shift handovers, may manifest themselves in future incidents unless these latent causes are addressed.

TRIPOD builds on concepts from barrier analysis to develop a general model of how accidents occur. This model links general failure types, latent causes, preconditions and active failures to enable analysts to trace back from immediate events to the underlying causes of a mishap. A range of computer-based tools can also be recruited to support the application of this approach (Hudson, et. al., 1994). Such support is critical because it can help to reduce the administrative costs associated with each analysis. It can be a nontrivial task to construct TRIPOD diagrams for relatively complex incidents. Tool support can be used to automatically perform certain consistency checks. It is also often possible to search for incidents that stem from similar patterns of active and latent failure (Johnson, 2003).

STAMP

The systems theory accident modeling and process (STAMP) provides less supporting infrastructure than TRIPOD. It avoids any strong assumptions about the relationship between latent and active failures or between barriers and hazards. Instead, STAMP exploits elements of control theory to help identify causal factors (Leveson, 2002). The motivations for using control theory resemble some of the arguments behind barrier analysis. Control theory is used because mishaps occur when external disturbances are inadequately controlled. Adverse events can also arise when failures go undetected or when the individuals that might respond to such a failure are unsuccessful in their attempts to control any adverse consequences. STAMP can be distinguished from more general techniques, such as accident fault trees, because it develops a particular view of adverse events in terms of "dysfunctional interactions" between system components. These can arise if, for example, one subsystem embodies inappropriate assumptions about another process component. In our case study, the electrophysiology nurse might have assumed that the telemetry unit would double check the identity of the patient being recalled from the oncology unit. Similarly, the neurosurgery resident assumed that another member of her care team had ordered the electrophysiology procedure for the cerebral angiography patient and that he had not been informed. This view of incidents as the product of suboptimal interactions is based on the argument that previous analysis techniques have been too narrowly focused on event-based models. In contrast, STAMP focuses on relationships and the constraints that hold between system components. Safety is, therefore a dynamic property because the degree to which a system satisfies any constraints will evolve over time. For example, the interventional radiologist assumed that he would have been informed of any additional treatment for his patient. Unlike the neurosurgeon, he was anxious to determine why this usual constraint had been violated with the decision to conduct the electrophysiology procedure.

STAMP analysis begins by developing a control model of the relationships between entities in the system. Figure 40–11 shows how arrows can be used to represent communication and control flows. The rectangles are entities, including people, systems, and organizations; they do not represent the events shown in the failure event tree of Figure 40–7. As can be seen from Figure 40–11, the STAMP control analysis extends from individual staff members, such as the electrophysiology staff nurse, and the systems under their immediate control, including the electrophysiology computer systems, to also consider the relationships between units. These different relationships must be captured in any analysis because they have a profound influence on the operation of safety-critical systems. However, this is a partial diagram. Most previous examples of the STAMP technique would extend the diagram to consider higher levels of management. The scope of this analysis, typically, would also include the role played by governmental and regulatory agencies. This might cover, for instance, the manner in which procedures for informed consent were specified and monitored to ensure both conformance and quality control. These relationships are missing from the STAMP analysis because Chassin and Becher only consider these issues in passing. They provide few details that would be sufficient for any extended investigation of the higher-levels of control within hospital management. This information could, of course, be added as part of a more sustained analysis of the circumstances surrounding this and similar mishaps.

Figure 40–11 traces the control and communication flows that led to the adverse event in our case study. As can be seen, information was widely distributed across several departments and between a number of key individuals. In particular, the diagram makes clear that the cerebral angiography patient was faced with a succession of clinical staff as they questioned the decision to undergo the electrophysiology procedure. This diagram also illustrates how the interventional radiologist's decision to pursue the missing patient from one department to the next finally led to the discovery of the mishap. After having conducted the extended form of control analysis illustrated in this diagram, the STAMP technique considers each of the control loops that are identified in the sociotechnical system. Potential mishaps stem from missing or inadequate constraints or from the inadequate enforcement of a constraint that contributed to its violation. Table 40–9 illustrates the general classification scheme that guides this form of analysis. It helps to identify potential causal factors in the control loops that exist at different levels of the management and operation hierarchy characterized using diagrams similar to that shown in Figure 40–11.

STAMP is similar to other techniques, such as failure event tree analysis and PRISMA, because it also uses a second stage of analysis to identify the potential causes of adverse events. In this case, each of the arrows in the control model is examined to determine whether any of the flaws in Table 40–9 can be identified in the relationships that they represent. It might be argued that there were unidentified hazards in the control loop between the electrophysiology computer and the electrophysiology nurse. Similarly, subsequent

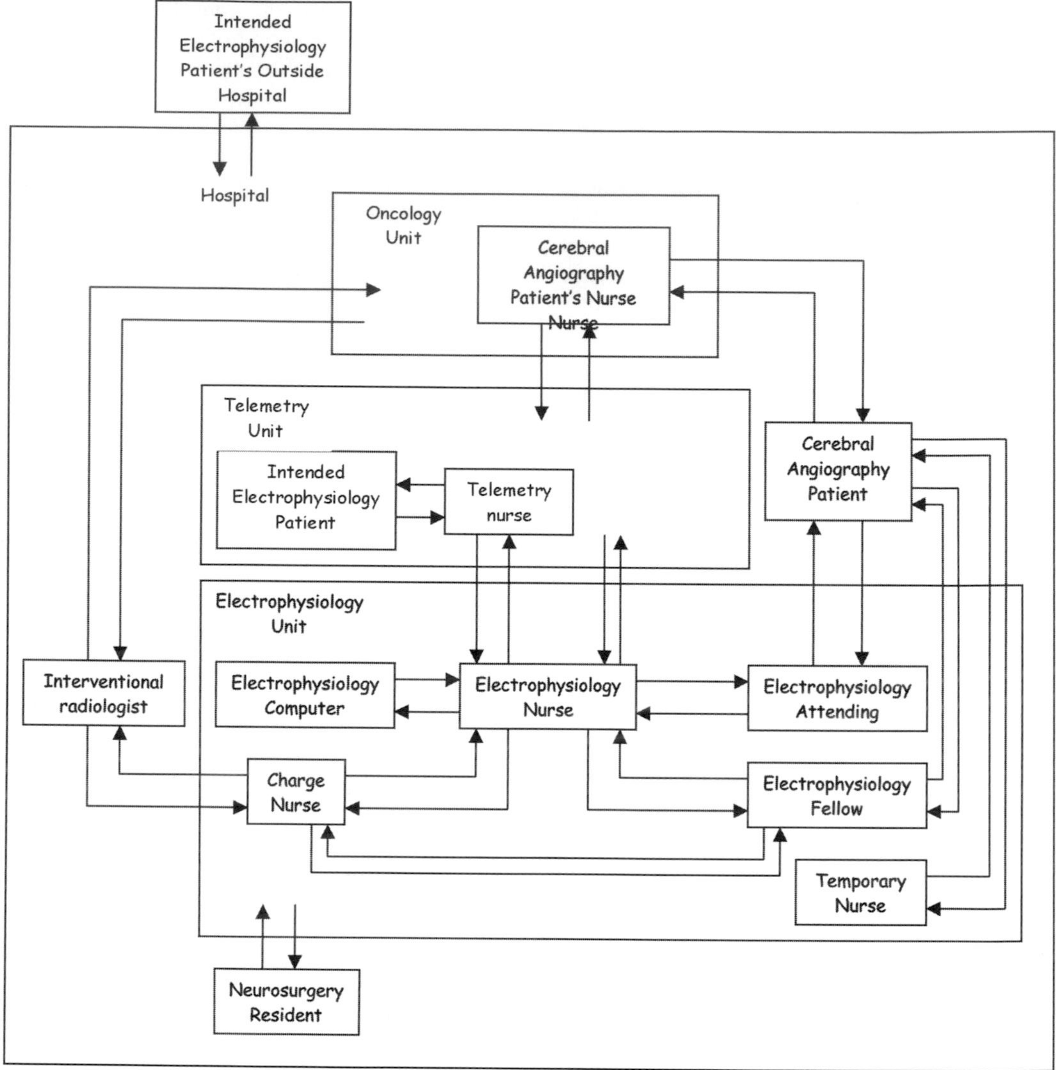

Figure 40–11 Control model of the example case study from a STAMP analysis.

investigation might identify flaws in the creation process that led to the computer systems presentation of patient information. Subsequent analysis might also identify communication flaws in the passage of information between the electrophysiology nurse, unnamed individuals in the telemetry unit and the original nurse for the angiography patient. As mentioned, this second stage of analysis is similar to several other techniques. For example, both PRISMA and MORT rely on taxonomies of general causal factors that have some similarities to the control flaws used to guide a STAMP analysis. The list of potential problems helps investigators to focus their analysis.

Argumentation Techniques

The previous sections in this chapter have reviewed a range of different causal analysis techniques. Barrier and change analysis can be used to determine why an incident occurred by looking at the differences between what happened actually did happen and what should have happened according

TABLE 40–9. Control flaws leading to hazards [a]

1.	Inadequate Enforcements of Constraints (Control Actions)
1.1	Unidentified hazards
1.2	Inappropriate, ineffective or missing control actions for identified hazards
1.2.1	Design of control algorithm (process) does not enforce constraints – Flaws in creation process – Process changes without appropriate change in control algorithm (asynchronous evolution) – Incorrect modification or adaptation.
1.2.2	Process models inconsistent, incomplete or incorrect (lack of linkup) – Flaws in creation process – Flaws in updating process (asynchronous evolution) – Time lags and measurement inaccuracies not accounted for
1.2.3	Inadequate coordination among controllers and decision makers
2	Inadequate Execution of Control Action
2.1	Communication flaw
2.2	Inadequate actuator operation
2.3	Time lag
3.	Inadequate or Missing Feedback
3.1	Not provided in system design
3.2	Communication flow
3.3	Time lag
3.4	Inadequate sensor operation (incorrect or no information provided)

[a] Leveson, 2002

to previous practices and procedures. Event based techniques provide general modeling tools that can be used with secondary forms of analysis, such as counterfactual arguments, to distinguish root causes from contributory factors. Flow charts and accident models provide additional guidance to analysts by restricting the way in which an incident is represented and the causes are identified. They often make strong assumptions about the ways in which safety constraints are violated during adverse events. All of these different approaches can be used to support arguments about the causes of an incident. For instance, Barrier analysis can be used to show how safety measures failed to prevent a hazard from threatening a target. STAMP control models can be used to argue that there were problems in the relationships between systems, operators and management organizations. Few of these techniques provide explicit support for the development of causal arguments. This is an important issue because rhetorical techniques can be used to bias the findings that are derived from any causal analysis (Johnson, 2003). It is perfectly possible to develop a partial failure event tree or PRISMA model to show that individual human error is the root cause of an adverse event even if managerial and organizational issues created the necessary preconditions for that failure. In contrast, a small number of alternate analysis techniques explicitly consider the structure of arguments that are made about the causes of incidents and accidents.

WBA

Why–because analysis begins like many other causal analysis techniques by developing a time line (Ladkin & Loer, 1998). Angled arrows are used to denote the sequence of events leading to an incident. For example, Figure 40–12 uses one of these arrows to denote that the cerebral angiography patient arrives in the electrophysiology unit before the nurse identified that there was no consent on the patient's records. It is important to stress, however, that this "occurs before" relationship does not imply causation. We cannot in this early stage of the analysis say that the arrival of the patient actually triggered the nurse's observations about the consent form. In order to make this argument, we must demonstrate that we have identified sufficient causes for an "effect" to occur. For example, there may have been other factors including preexisting procedures that motivated the nurse to check for the consent documents in the patient record. A more sustained causal analysis must consider these additional issues in order to fully explain the reasons why each event occurred.

Figure 40–12 shows the results of applying a further stage of analysis to an initial WBA time line.

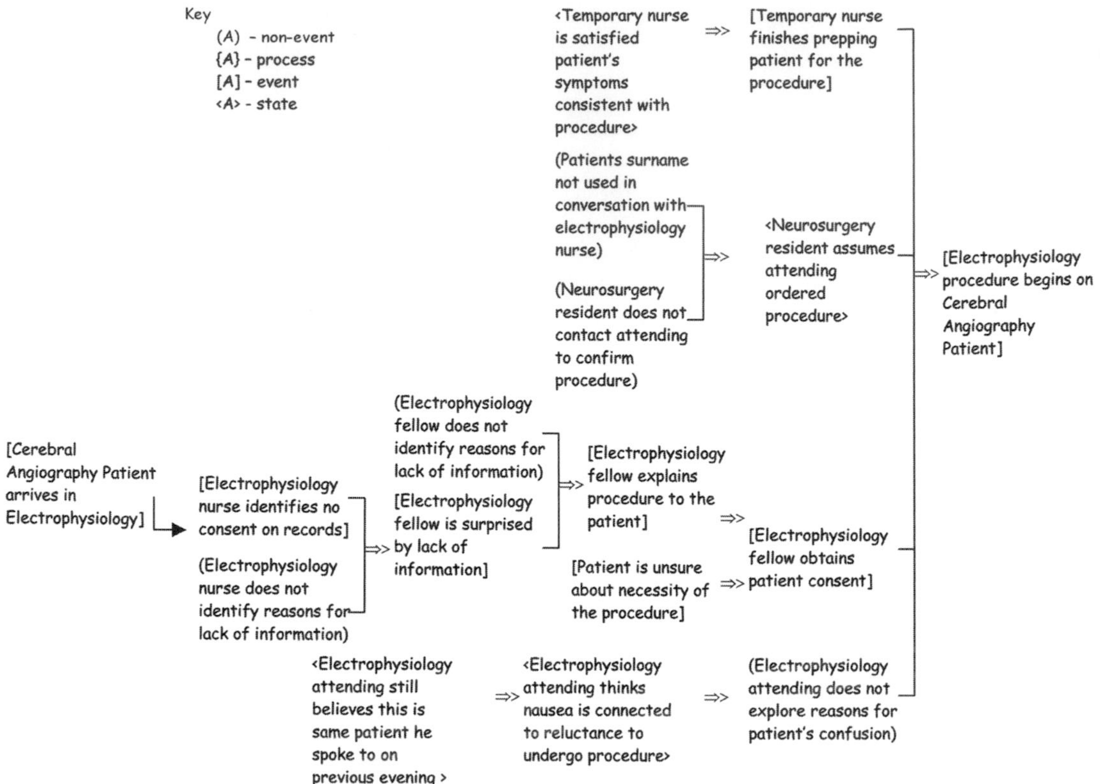

Figure 40–12 Example WBA diagram.

Each node in an initial diagram is considered in turn. Analysts must ask "why did this happen." Each reason or cause is then added to the diagram and denoted using double-headed arrows. This transition from temporal sequences to more rigid causal relationships produces insights that are not apparent in purely event-based approaches, such as time lines. For example, Figure 40–12 arguably provides a more explicit representation of the non-events or omissions that led to the case study. These include the failure to use the patient's surname, the resident's failure to confirm the attending physician's decision to order the electrophysiology procedure and so on. Figure 40–12 also illustrates that the electrophysiology procedure begins because the neurosurgery resident assumed that the attending ordered the procedure and the temporary nurse prepared the patient for the procedure and the fellow obtained consent and the attending did not explore the reasons for the patient's confusion over the procedure. If we explore one of these causal arguments back, we can find that the attending did not explore the patient's confusion because they believe that this is part of a reluctance to undergo the procedure.

WBA provides a set of mathematically based procedures that analysts must follow in order to replace the angled arrows of a temporal sequence with the double-headed arrows of the causal relationship. They are based on counterfactual arguments of the form "A causes B" if we know that A and B occurred and that if A had not occurred, then B would not have occurred. However, analysts must explicitly consider the level of formality that is appropriate to their needs. It would not be cost effective to conduct this form of mathematical modeling for many of the "slips, trips, and falls" that characterize the majority of adverse events in health care. Investigators must typically choose between restricting their analysis to the more accessible aspects of the informal graphical reasoning, based on Figure 40–12, and more complete forms of WBA involving the use of discrete mathematics. It remains an open question as to whether the

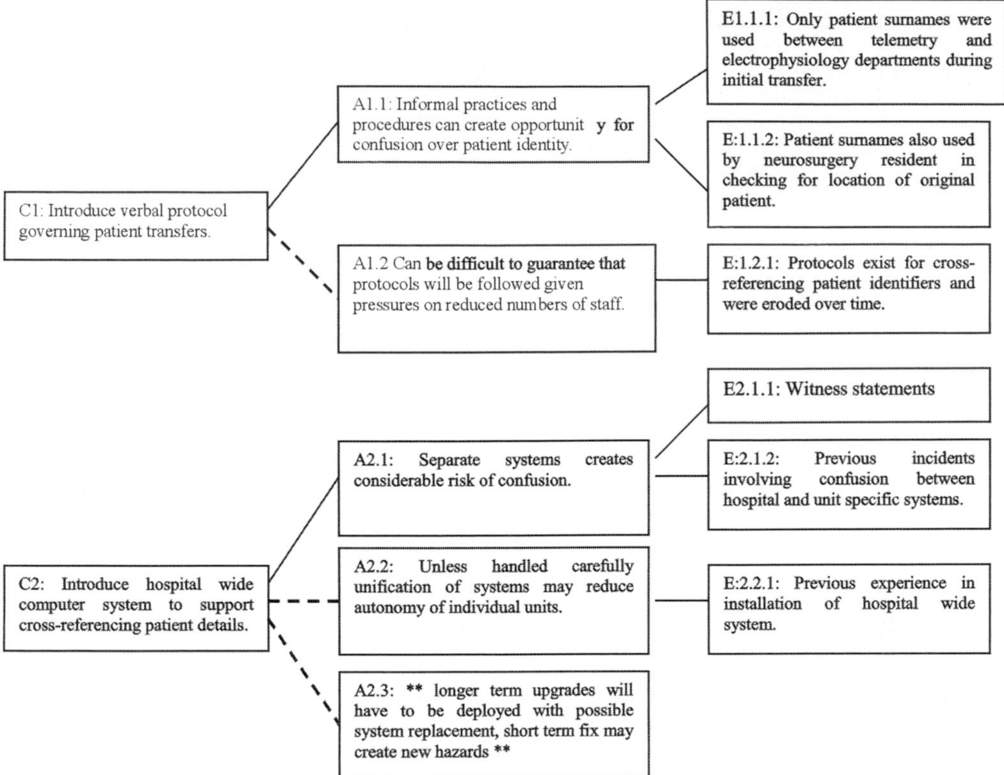

Figure 40–13 Example of a CAE diagram.

additional assurance provided by the formal reasoning would ever justify the additional costs in analyzing adverse health care events using this approach.

CAE *Diagrams*

It is important that the output from any causal analysis should inform the identification of appropriate recommendations. Analysts must help risk managers to reduce the likelihood of future failures or to mitigate their consequences. Table 40–10 presents a simple tabular form that can be used to associate root causes with potential recommendations. For instance, a recommendation to introduce new verbal protocols governing patient transfers is supported by the argument that informal practices and procedures create the opportunity for confusion over patient identity. There is evidence to support this argument from the manner in which only surnames were used between the telemetry and electrophysiology departments during the initial transfer and by the neurosurgery resident when they checked on the location of the original patient.

It can, however, be difficult to construct such tables for complex incidents. There may be many hundreds of items of evidence in complex failures. Similarly, there can be competing arguments that undermine particular recommendations. For instance, any decision to introduce a hospitalwide computer system for cross-referencing patient identities might force major changes in the existing infrastructure. These changes could introduce further problems and may undermine the autonomy of individual units who tailor their computational support for their particular requirements. Conclusion, analysis and evidence (CAE) diagrams can help designers to map out the competing arguments for and against particular conclusions or recommendations in the aftermath of a complex incident. These diagrams are simpler than many of the techniques we have described (Johnson, 2003). CAE provides a graphical representation with less

TABLE 40–10. General Format for a Recommendation Table

Conclusion/Recommendation	*Root Cause (Analysis)*	*Supporting Evidence*
C1. Introduce verbal protocols governing patient transfers.	A1.1 informal practices and procedures can create the opportunity for confusion over patient identity	E.1.1.1 Only patient surnames were used between telemetry and electrophysiology departments during initial transfer.
		E.1.1.2 Patient surnames were also used by neurosurgery resident in checking location of original patient.
C2. Introduce hospital wide computer system to support cross-referencing of patient transfers.	A2.1 Separate systems create considerable risk ofconfusion.	E2.1 Witness statements.
		E2.2 Previous incidents involving confusion between hospital and unit specific systems.

strict rules on how to conduct the analysis. This arguably reduces the costs and increases the flexibility of the approach.

Figure 40–13 illustrates a conclusion, analysis, and evidence (CAE) network. As the name suggests, the rectangles labeled with a C are used to denote conclusions or recommendations; those labeled with an A are lines of analysis; and the E rectangles denote evidence. Lines are drawn to show those lines of analysis that support particular conclusions.

For example, the recommendation to introduce verbal protocols governing patient transfers (C.1) is supported by the argument that informal practices and procedures create the opportunity for confusion over patient identity (A1.1). The evidence for this assertion is provided by the manner in which surnames only were used in the initial patient transfer (E.1.1.1) and by the neurosurgery resident's initial attempts to locate the original patient (E.1.1.2). Figure 40–13 also captures contradictory arguments. For instance, the dotted line in the first of the networks denotes that the introduction of protocols recommended in C.1 cannot guarantee staff will follow them under their everyday pressure of work (A.1.2). Evidence for this is provided by the manner in which the effectiveness of existing protocols had been eroded over time (E1.2.1).

The lower of the two networks in Figure 40–13 illustrates the argument that a hospitalwide computer system should be introduced to help cross-reference patient details (C2). This recommendation is based on the observation that separate systems create considerable risk of confusion between different units (A2.1). This argument is based on evidence of witness statements (E.2.1.1) and on evidence from previous incidents involving the transfer of information between hospital and local systems (E.2.1.2). The recommendation to introduce a unified computer system is weakened by an argument that it may compromise the independence of individual units (A.2.2) given previous experience from centralized hospital systems (E.2.2.1). It is also weakened by the introduction of further hazards from the short-term expedients that must support the longer term development of a more unified system (A.2.3). As can be seen from Figure 40–13, CAE diagrams capture general arguments about incidents and accidents. For example, a conclusion might refer to a recommended action rather than a causal relationship. It is also important to mention that this technique was specifically developed to enable investigators to sketch out the arguments that might appear in an incident report. This helps to ensure that any document avoids contradictory arguments prior to the publication of a report.

RECOMMENDATIONS AND MONITORING

The previous section argues that the output from causal analysis techniques must help to identify the recommendations that will reduce the likelihood or consequences of any recurrence. Unfortunately, there has been relatively little research or even applied methods that can be used to generate recommendations from the products of a causal analysis (Johnson, 2003). Most of the available techniques, such as the tabular forms and diagrams shown in the previous section, can be criticized for being very simplistic. However, it is critical that there be some documented justification that links proposed interventions to the identified causal factors. If this link cannot be established, then there is a danger that managers will fund changes that do not address the underlying problems or that even exacerbate any previous safety threats. The need to justify and document proposed recommendations is particularly important given that there is a traditional separation between the teams who derive recommendations from a causal analysis and those who must choose to implement them. For instance, the NTSB can propose regulatory changes that must then be implemented by the FAA. Similarly, the U.K. Health and Safety Executive might propose changes to the particular governmental departments dealing with energy production. There are good reasons why different people make and accept recommendations. For example, if the same individual is responsible for making recommendations and funding them, then they may choose not to propose interventions that cannot be funded in the short term. It is often better for safety managers to make recommendations independent of whether or not they can be funded from immediate resources. From this it follows that particular attention must be paid to documenting the reasons why a causal analysis has identified particular recommendations. Without such a motivation, then, it will be difficult for managers outside of an investigation to understand why funds should be spent on necessary improvements,

A number of more prosaic problems can affect the implementation of recommendations in the aftermath of health care incidents. It can be difficult for managers to obtain independent or expert advice about particular recommendations, for example those involving human factors issues. Similarly, it can be difficult to obtain the funds that are necessary to implement longer term changes. This creates a cycle in which the limited scope of potential recommendations can influence the course of any causal analysis. There can be a tendency to blame incidents on inadequate attention or on poor staff performance because relatively cheap remedies can be found. For example, a local reporting system identified a class of incidents in which the staff had failed to realign the taps that are used to control the flow of medication to a patient when bags are being changed. It was, therefore, recommended that acronyms be used to remind staff to perform particular actions. TAP stood for tap aligned properly (Johnson, 2003). Such advice provides short-term protection against certain classes of adverse events. However, their effectiveness declines rapidly over time. It can also be difficult to ensure that new staff are taught to use these reminders. A subsequent study of many of the incidents that helped to generate these acronyms revealed that they were often "work arounds" that were intended to support the use of poorly designed or faulty equipment.

Such examples illustrate the importance of linking incident reporting into the wider risk assessment practices that are being introduced into health care. The decision of whether or not to fund a recommendation will be made partly on the perception of risk associated with any recurrence of an adverse event. It is also important to monitor the success or failure of a recommendation. If it is ineffective and an adverse event continues to recur, then additional measures will be necessary. The human factors issues here relate to the reliance on short-term fixes and coping strategies in health care. As the previous example shows, many organizations respond to incidents by reminding staff to do better. This perfective approach often fails to secure the longer term safety of a system because training cannot easily sustain long-term changes in behavior unless it is repeated and invested in over similar periods of time.

The Agency for Healthcare Research and Quality's Review of patient safety practices includes a chapter on root cause analysis (Wald & Shojania, 2001). Unlike the analysis in this document, the AHRQ report focused less on explaining root cause analysis techniques and more on a critical assessment of the costs and benefits of incident analysis in general. It is argued that "there is little published literature that systematically evaluates the impact of formal RCA on error rates." They do, however, present the results of a study (Rex, Turnbull, Allen, Vande Voorde, & Luther, 2000) in a tertiary referral hospital in Texas that applied root cause analysis to all adverse drug events over a 19-month trial. This study reported a 45% decrease in the rate of voluntarily reported serious incidents between the 12-month study interval and a 19-month follow-up period (7.2 per 100,000 to 4.0 per 100,000 patient days, $p < 0.001$). The authors of the study argued that a "blame-free" environment helped senior management to focus on the underlying causes of these incidents. Procedures were revised and changes were made in organizational policies including staffing levels. As we have seen in the chapter on "Human Factors of Reporting Systems" in this handbook, the decline in reporting does not, however, guarantee any related decline in adverse events. It may reflect a decline in reporting behavior as staff become disillusioned with the system. There is also a common phenomena known as the "confessional period" in which reporting systems will initially elicit very high levels of participation as staff use these schemes to contribute information about long-standing concerns (Johnson, 2003). Hence, it is to be expected that contribution rates would decline after this initial uptake. The AHRQ analysis argued that the results of the Texas study are "unclear." In particular, it was suggested that "as the study followed a highly publicized, fatal adverse drug event at the hospital, other cultural or systems changes may have contributed to the measured effect.". Further caveats related to the absence of any control group and to the lack of any longer term data on adverse drug events from the hospital. Finally, there was no attempt to determine the precise benefits that a formal root cause analysis might provide over less formal analysis.

A similar study reported in the British Medical Journal looked at the reasons why clinicians change their practice (Allery, Owen, & Robling, 1997). A random sample of 50 consultants and 50 general practitioners were interviewed about key events that had caused them to revise their practice. The sample yielded 361 changes in clinical practice. These included major organizational changes, such as setting up an asthma clinic. They also included more specific changes in clinical practice, such as the decision to change from cefotaxime to cefuroxime for chest infections. The interviews probed for the reasons behind these changes. This yielded an average of three justifications per change. Of these, the most frequently mentioned reasons were organizational factors, education, and contact with professionals. These justifications accounted for 47.9% of the total number of reasons for change. Education accounted for 16.9% and was involved in 37.1% of the changes. This was defined to include "reading medical journals, attending an organised educational event, participation in research and audit, and the provision of clinical guidelines." Neither root cause analysis nor incident reporting was included amongst the list of 12 factors that were used to classify the results. The findings from such studies only had an indirect influence on clinical change through, for example, the education activities previously mentioned. This provides another reminder that root cause analysis is a limited tool. It guides the recommendations that help to prevent any future recurrence. However, the lessons learned within particular clinical organizations will be of little benefit unless they are disseminated and acted upon. Studies such as that conducted by Allery et al. (!997) help to identify the most effective mechanisms for change in health care.

CONCLUSIONS

This chapter introduces incident analysis within the context of health care related incidents. The opening sections present information about the distribution of causal factors that have been identified in a number of previous studies. In particular, we distinguish between primary or catalytic causes and secondary or latent failures. Primary causes describe immediate failures that trigger an adverse event, such as individual human error. Secondary failures include the conditions that make these primary incidents more likely. These often stem from managerial and organizational factors such as changes in staffing levels.

Many previous reporting systems used informal and qualitative causal analysis techniques. This can create problems of individual and organizational bias. We term such problems tertiary failures. Rather than affecting the immediate operation of a health care system or the factors that contribute to an incident, these problems can impair effective learning from an adverse event. It can also be important to document the process that was used to identify particular causes and contributory factors. Subsequent sections, therefore, introduce a range of techniques for root cause analysis. These include elicitation approaches such as barrier and change analysis. Such methods can be used to compare what happened in an adverse event with what should have happened according to previous practice and procedures. These techniques can also assist in the elicitation of information after an incident because they encourage investigators to determine both what did happen and what should have happened.

A second class of analysis techniques support incident reconstruction. For instance, time lines can be used to map out the sequence of events leading to an incident. This approach has important limitations. It can be difficult to represent the many subtle constraints that make particular events more likely. Organisational factors can affect many different individual actions and it is unclear how to represent this influence on a time line. A number of other techniques, such as accident fault trees, suffer from similar limitations. Fortunately, there is a range of causal analysis techniques that provide means of representing these factors within a reconstruction. In particular, we show how failure event trees can be applied to represent and reason about an example health care incident. Many of these techniques use a second stage of analysis based on counterfactual reasoning to distinguish root causes from contributory factors. Analysts must ask for each event whether an incident would still have occurred if the event had not occurred. If the incident would have happened anyway, then that cause was not necessary for the accident to occur and so it cannot be a root cause.

Most reconstruction techniques can be applied to analyze a wide range of different adverse events. This flexibility can create problems. In particular, there are few guarantees that different analysts will reach the same conclusions about the same incident. Flow chart techniques help to encourage consistency by guiding analysts to reach a limited number of conclusions about the causes of an adverse event. In particular, we present the MORT and PRISMA techniques. MORT provides a general scheme for the analysis of management-related incidents, focusing on risk assessment. PRISMA has been specifically tailored to support the analysis of health care related incidents. These techniques provide pragmatic tools for causal analysis but can still suffer from particular forms of bias. For example, analysts often tend to focus around the initial stages in any flow chart. Additional effort and concentration is required to pursue the causes of an incident through the details of embedded branches in the tree of options that these techniques present.

Flow charts and reconstruction techniques make minimal assumptions about the nature of adverse events. In contrast, accident-modeling techniques embody particular ideas about the manner in which accidents are caused. For instance, the TRIPOD method insists that analysts trace the way in which barriers fail to protect a target from a hazard. In addition, the preconditions for the failure of any barrier must be tied back to a limited number of general failure types. One problem with all of these techniques is that it can be difficult to ensure that the output of a causal analysis informs the recommendations that are made in the aftermath of an adverse event. The final sections of this chapter, therefore, introduce argumentation techniques for causal analysis. Why–because analysis and conclusion, analysis, and evidence diagrams look at the reasons why particular causal findings are made. These approaches also, therefore, help to document the reasons that support particular recommendations.

This chapter closes by arguing that causal analysis is not an end in itself. It is important to monitor whether the findings from any investigation are having any impact on the patient care. This is a notoriously difficult problem; participation levels in a reporting system provide extremely ambiguous signals about the underlying incident rate. Root cause analysis will never yield "zero incidents" and so questions must be asked about the cost effectiveness of both the recommendations that are produced and of the analysis that is performed to identify those recommendations. More generally, the identification of a root cause and the development of a subsequent recommendation do not guarantee that appropriate changes will be made in clinical practice.

References

Allery, L. A., Owen, P.A., & Robling, M. R. (1997). Why general practitioners and consultants change their clinical practice: A critical incident study. *British Medical Journal, 314*, 870.

American Medical Association. (2004). *Ethical Guidance Statement E8.08*. Retrieved July 1, 2006 from http://www.ama- assn . org/ama/ noindex/category/11760.html

Busse, D. K., & Johnson, C. W. (1999, August). Human error in an intensive care unit: A cognitive analysis of critical incidents. In J. Dixon (Ed.), *Proceedings of the 17th International Systems Safety Conference* (pp. 138 –147). Unionville, VA: The Systems Safety Society.

Byrne, R. M. J., & Tasso, A. (1999). Deductive reasoning with factual, possible and counterfactual conditionals. *Memory and Cognition, 27,* 726–740.

Chassin, M. R., & Becher, E. C. (2002). The wrong patient. *Annals of Internal Medicine, 136*, 826–833. Retrieved July 1, 2006, from http://www.annals.org/issues/v136n11/full/200206040–00012.html

Helmreich, R. L., & Merritt, A. C. (1998). *Culture at work in aviation and medicine*. Aldershot, UK: Ashgate Press.

Hudson, P., Reason, J., Wagenaar, W., Bentley, P. , Primrose, M., & Visser, J. (1994). Tripod-delta: Pro-active approach to enhanced safety. *Journal of Petroleum Technology*, vol 46, 58–62.

Johnson, C. W. (2003). *A handbook of incident and accident reporting.* Glasgow UK: Glasgow University Press.

Johnson, W. G. (1973). *MORT: The management oversight and risk tree analysis* (Tech. Rep. SAN 8212). Washington, DC: Atomic Energy Commission.

Ladkin, P., & Loer, K. (1998). *Why–because analysis: Formal reasoning about incidents* (Document RVS-Bk–98–01). Bielefeld, Germany: Technischen Fakultat der Universitat Bielefeld.

Lekberg, A. K. (1997). Different approaches to incident investigation: How the analyst makes a difference. In S. Smith & B. Lewis (Eds.), *Proceedings of the 15th International Systems Safety Conference* (pp. 178–193). Unionville, VA: The Systems Safety Society.

Leveson, N. (2002). A systems model of accidents. In J. H. Wiggins & S. Thomason (Eds.), *Proceedings of the 20th International System Safety Conference* (pp. 476–486). Unionville, VA: International Systems Safety Society.

Leveson, N. (2003). A new approach to hazard analysis. In G. Einarsson & R. Fletcher (Eds.), *Proceedings of the 21st International Systems Safety Conference* (pp. 498–507). Unionville, VA: International Systems Safety Society.

Louisiana State University Health Sciences. (2003). *Patient Identification, Policy number: 5.10.1.* Retrieved from http://www.sh.lsuhsc.edu/policies/policy_manuals_via_ms_word/hospital_policy/Patid.pdf

Mackie, M. L. (1993). Causation and conditions. In E. Sosa & M. Tooley (Eds.), *Causation and conditions* (pp. 33–56). Oxford, England: Oxford University Press.

Reason, J. (1997), *Managing the risks of organisational accidents.* Brookfield: Ashgate Press.

Rex, J. H., Turnbull, J. E., Allen, S. J., Van de Voorde, K., & Luther, K. (2000), Systematic root cause analysis of adverse drug events in a tertiary referral hospital. *Journal of Quality Improvement, 26*, 563–575.

Steblay, N. M. (1992). A meta-analytic review of the weapon focus effect. *Law and Human Behaviour*, 413–424.

U.S. Joint Commission on Accreditation of Healthcare Organizations. (2004). *Reporting of Medical/Health Care Errors: A Position Statement of the Joint Commission on Accreditation of Healthcare Organizations.* Retrieved http://www.jcaho.org/accredited+organizations/patient+safety/medical+errors+disclosure

van der Schaaf, T. W. (1992). Near *miss reporting in the chemical process industry.* Unpublished doctoral dissertation, Technical University of Eindhoven, Eindhoven, The Netherlands.

van Vuuren, W. (2000). Organisational failure: An exploratory study in the steel industry and the Medical domain. Unpublished doctoral dissertation, Institute for Business Engineering and Technology Application, Technical University of Eindhoven, Eindhoven, The Netherlands.

Weick-Brady, M. (1996). Those codes! Food and Drug Administration. *User Facility Reporting Bulletins, 18.* Retrireved. http://www.fda.gov/cdrh/issue18.pdf.

Wagenaar, W. A. (1992), Risk taking and accident causation. In J. F. Yates (Ed.), *Risk taking behavior.* Chicester, England: Wiley.

Wald, H., & Shojania, K. G. (2001). Root cause analysis. In K. G. Shojania, B. Duncan, K. McDonald, & R. Wachter (Eds.), *Making health care safer: A critical analysis of patient safety practices* (chap. 5,. Evidence Report/Technology Assessment No. 43). Agency for Healthcare Research and Quality.

Section

VIII

HUMAN FACTORS AND ERGONOMICS INTERVENTIONS

41

ERGONOMICS PROGRAMS AND EFFECTIVE INTERVENTIONS

Michael Smith
University of Wisconsin, Madison

WHAT IS ERGONOMICS?

Ergonomics is the science that matches the capabilities of people and the work they perform. The primary interest is an understanding of how the design of work affects people's safety, health, performance, and productivity. The purpose is to match the capabilities of employees with the requirements of work by designing work processes and tasks to provide the best fit for each employee. Ergonomics takes knowledge from many fields including engineering, physiology, medicine, psychology, anthropology, sociology, and business management to develop an understanding of how to make effective and healthy working conditions. Of specific interest is the prevention or reduction of musculoskeletal injuries produced by working.

WHY IS ERGONOMICS IMPORTANT?

Ergonomics is important because research and experience has shown that employees can be injured because their work exceeds their capacities. Ergonomics provides a way to evaluate an employee's capacities as well as the demands that working puts on her or him. Ergonomics helps employers determine if working conditions are well designed to protect employees' health and to enhance their performance. It provides tools to assess whether there is a misfit between employee capabilities and the work demands. Ergonomics also provides guidance for improving the fit between the capabilities of employees and the demands of work. This leads to fewer worker injuries, better worker performance and productivity, and lower worker's compensation and hospitalization costs. It is also believed that this will bring about better patient care and safety because the staff will be performing at a higher level with less stress and strain to detract from their performance.

PURPOSE AND CONTENT OF THIS CHAPTER

The purpose of this chapter is to provide general guidance in how to establish an ergonomics program to reduce the risk of musculoskeletal injuries. In terms of health care operations and patient safety, an ergonomics program is important because health care workers who are injured, who are experiencing discomfort, or who are fatigued have a higher probability of making errors that can lead to patient injuries. Other chapters in this book provide specific information about the nature of the ergonomic risks and safety risks that occur in health care settings, so this chapter discusses such risks as examples in need of ergonomic interventions. Thus, this chapter does not cover the entire range of ergonomic risks or the entirety of specific ergonomic interventions for particular risks. Rather, it provides an approach for establishing an ergonomics program, for defining ergonomic risks (hazards), for evaluating the seriousness of the risk, and general guidance for reducing adverse ergonomic exposures (hazards). Particular health care situations and settings will be used as examples to show how to use the general approach for controlling ergonomic risks.

A BRIEF OVERVIEW OF WORKPLACE MUSCULOSKELETAL INJURIES

The scientific literature has established that there is a relationship between working in certain occupations or particular types of job exposures and the occurrence of work-related musculoskeletal disorders (WMSDs; see Hagberg et al., 1995; NIOSH, 1997; NRC, 2001). For example, warehouse workers have a higher prevalence of low back injury, meat cutters have a higher prevalence of carpal tunnel syndrome (CTS), and painters have a higher prevalence of shoulder pain. However, the research evidence necessary to define the specific levels of workplace risk factors that cause each of these WMSDs is very limited, and makes it hard to establish a risk threshold. In addition, the evidence examining the effectiveness of specific interventions to control the ergonomic risk factors and to reduce injuries has mixed findings. Thus, the control of WMSDs using ergonomic interventions is best done using a process similar to quality improvement where incremental changes are made and results evaluated that then lead to a stable situation or additional interventions as the results dictate.

WMSDs are unique among occupational injuries in the complexity of their causation and in the time course of their development. Many only occur after a long period of exposure to ergonomic risk factors, whereas others occur due to acute events. In addition, many of the hypothesized "pathomechanisms" for the cumulative causation of WMSDs are not completely understood. There are several plausible and reasonable theories that provide the basis for defining possible ergonomic hazards that are likely to increase the risk for WMSDs such as microtrauma to soft tissue that accumulates over time. What appears clear from the scientific literature is that WMSDs are more likely to occur when there are multiple ergonomic risk factors present.

There is insufficient evidence in the scientific literature to establish dose–response relationships among particular ergonomic risk exposures (for instance, the number of repetitions of motions, or the weight of the load being handled) and the specific medical conditions that may result such as low back pain or CTS. In addition, it is not possible to define the exact duration of such exposures that will produce specific adverse medical outcomes (hours per day, weeks, months, years) in a particular person. Therefore, there is no definitive basis to establish the specific level(s) of a particular ergonomic risk exposure that is hazardous, or conversely the level of exposure that is safe. Due to this lack of definitive evidence, ergonomic intervention programs for preventing WMSDs often use very conservative ergonomic risk criteria to establish the limits of exposure. For this reason, it is not unreasonable to customize these criteria for specific companies, operations, or a particular workforce, or for a particular employee to account for specific circumstances of exposure and employee susceptibility, and to provide reduced ergonomic risk.

Jobs are an integration of many work demands that, taken together, put loads on the person's body (and the mind). It is the combined work demands that produce the totality of ergonomic exposures that can affect the risk of developing WMSDs. No single job exposure (frequency of back or hand movements, weight of objects being lifted, posture of the back or wrists) can define the entirety of the risk for developing WMSDs. Most often, employees are exposed to multiple ergonomic risk factors simultaneously. In addition, there are personal susceptibilities, work habits, and individual work methods that also contribute to the development of WMSDs. A "holistic" approach to an ergonomics program accounts for the entirety of exposures and risks through an understanding of the combination of the ergonomic risk exposures, employee susceptibilities, and employee work habits (Carayon, Smith, & Haims, 1999; Smith & Carayon, 1996). The ergonomics program must look for ways to improve all of these considerations if the best results are to be attained. This is discussed in more detail later.

Recognized Work-Related Ergonomic Risk Factors

The traditional workplace ergonomic risk factors that have been correlated with the development of WMSDs of the upper extremities in industrial, food processing, and service settings are; (a) a high frequency of sustained repetitive motion, combined with (b) heavy exertion due to loads from the weight of materials being handled or tools held, and/or combined with (c) extreme postures of the joints, limbs, shoulders, and neck from neutral positions. In addition, mechanical vibration may also increase the risk, especially in combination with any of the aforementioned factors (primarily high repetition; Armstrong & Chaffin, 1979; Armstrong et. al., 1993; deKrom, Kester, Knipschild, & Spaans, 1990; Hagberg, Morgenstern, & Kelsh, 1992; Hagberg, et al., 1995; NRC, 2001; Nathan, Kerniston, Myers, & Meadows, 1992; NIOSH, 1997; Nordstrom, Vierkant, DeStefano, & Layde, 1997; Silverstein, Fine, & Armstrong., 1987; Stock, 1991).

The traditional workplace ergonomic risk factors for low back injury that have been identified in a wide variety of work settings include (a) handling a very heavy weight that puts too much load on the spine or back muscles; (b) moving heavy loads a long distance in poor postures; (c) the proximity of the load away from the body's center of gravity, which overloads the spine; (d) overexertion in handling the load caused by the load being too heavy or by the inability to grip the load securely; (e) a very high frequency of lifts per work shift for prolonged periods of time; (f) the improper posture of the back, legs, and shoulders during the lift (materials handling), which loads the spine, shoulders, and legs; (g) certain situations of prolonged whole body and/or segmental vibration exposure; and (h) certain situations of prolonged static seated posture (Andersson, 1981; Bigos et. al., 1986a , 1986b; Bigos et al., 1991; Boshuizen, Bongers, & Hulshof, 1990; Burdorf, Naaktgeboren, & deGroot, 1993; Chaffin & Park, 1973; Kelsey, 1975; NRC, 2001; NIOSH, 1997; Svensson & Andersson, 1983).

The etiology of WMSDs is very complex, and factors independent of workplace exposures are also substantive risk factors for WMSDs. For instance, for lower back pain or for carpal tunnel syndrome (CTS) some of these nonoccupational factors include personal constitution, health status, heredity, age, gender, weight, and psychological status, as well as activities that are not occupational (de Krom et al., 1990; Hanrahan, Higgins, Anderson, & Smith, 1993; Nordstrom, 1996; Nordstrom et. al., 1997; NIOSH, 1997; NRC, 2001). The nonoccupationally related cases of low back pain and CTS are as prevalent in the general population as occupationally related cases. In other words, working is not the only cause of low back pain, CTS, shoulder tendonitis, elbow epicondylitis, thoracic outlet syndrome, shoulder pain, or neck pain.

Some General Resources for Understanding and Dealing With WMSDs

Table 41–1 defines some general resource documents that can be helpful in recognizing and dealing with work-related musculoskeletal disorders, and for developing an ergonomics program. The table provides the authors of the document, the source of the document, and a short statement about the information in the document.

THE NEED FOR A "HOLISTIC" APPROACH TO ERGONOMICS

Ergonomics and human factors have been around for approximately 60 years. During WWII, there was a need to deal with employee fatigue, stress, injuries, and poor performance due to mismatches between people and technology, unusual work schedules, and demanding and threatening working conditions. Over the ensuing 60 years, ergonomic researchers and practitioners of ergonomics have learned that the best way to address ergonomic concerns is to take a broad view of the problems encountered combined with focused solutions. The broad view of a work process and improvements addresses the need to make sure that improvements in one area do not lead to problems in another area. It accounts for the issues that often occur elsewhere in the work system when just a focused approach is used in one area. This has led to the understanding that a bigger view allows the ergonomics practitioner who is trying to solve a problem to address both the specific problem (the "microergonomics"), and to also deal with the ramifications in the entire work process (the "macroergonomics").

Microergonomic factors deal with design characteristics of tasks, or tools/technology, or the environmental conditions, or the capacity/knowledge of each individual employee. It addresses the specific ergonomic risk factors in a particular task, job, or operation. Macroergonomic factors deal with larger issues such as the organization of the work process, the coordination of tasks and activities among employees, the supervision of processes and

TABLE 41–1. Ergonomic Program Resource References

Author/Year	*Title/Source of Publication*	*General Material*
Chaffin, Don 1987	Manual materials handling and the biomechanical basis for prevention of low back pain in industry—an overview. *American Industrial Hygiene Association Journal*, 48, 989–996	Advice on how to control back injuries
Kilbom, A. 1994a	Repetitive work of the upper extremity: Part I-Guidelines for the practitioner. *International Journal of Industrial. Ergonomics*, 14, 51–57 (also see pages 59–86)	Advice on how to control upper extremity injury risk
Konz, Stephen & Johnson, Steven, 2000	*Work Design: Industrial Ergonomics.* Scottsdale, AZ: Holcomb Hathway Publishers.	General ergonomic advance on workplace evaluation and improvement
Kroemer, Karl & Grandjean, E., 1997	*Fitting the task to the human.* Bristol, PA: Taylor & Francis.	General ergonomic advance on workplace evaluation and improvement
Moore, Steven & Garg, Arun 1995	The strain index: A proposed method to analyze jobs for the risk of distal upper extremity disorders *American Industrial Hygiene Association Journal*, 56, 443–458	A way to measure the risk for upper extremity disorders
NRC/IOM 2001	*Musculoskeletal disorders and the workplace.* Washington, DC: National Academy Press.	Review of scientific literature on the causes and cures of work related musculoskeletal disorders
NIOSH (1988 Putz-Anderson, V.)	Cumulative trauma disorders. Bristol, PA: Taylor & Francis.	Overview of the causes and ways to control workplace musculoskeletal disorders
NIOSH 1981	*Work practices guide for manual lifting* (NIOSH Publication 81–122). Cincinnati, OH: NIOSH	Guidance for lifting
NIOSH (Bernard, B.) 1997	*Musculoskeletal disorders and workplace factors.* (DHSS (NIOSH) Publication No. 97–141) Cincinnati, OH: NIOSH	Review of the scientific literature on causes of workplace musculoskeletal disorders
OSHA 1990, 1999, 2000, 2002, 2003 2004/2005	www.osha.gov/SLTC/ergonomics/index.html Washington, DC: Occupational Safety and Health Administration	Provides guidance on ergonomics programs
Salvendy, Gavriel 1997, 2006	*Handbook of human factors and ergonomics.* New York, NY: Wiley Interscience Publication	Provides expert advice on ergonomics stopics
Snook, Stover 1988	Approaches to the control of back pain in industry: job design, job placement and education/training. *Professional Safety*, 33, 23–31	Suggestions for reducing back injuries at work
Snook, Stover 1991	The design of manual handling tasks: revised tables of maximum acceptable weights and forces. *Ergonomics*, 34, 1197–1213.	Advance on design of work tasks and lifting loads
Waters, T., Putz-Anderson, V. & Garg, A. 1993	Revised NIOSH equation for the design and evaluation of manual lifting tasks. *Ergonomics*, 36,. 749–776	Revised NIOSH guidance on lifting tasks and weight limits

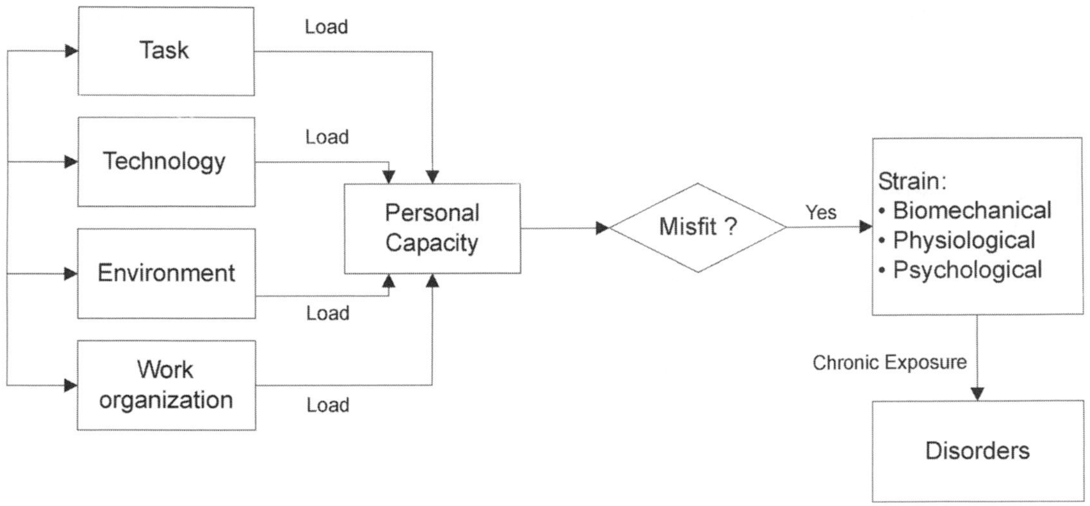

Figure 41–1. Holistic model of workplace.

employees, and how people, technology, tasks, and environmental features are integrated. Ergonomic improvements and interventions must account for both levels and their effects on the work system to be successful.

The Balance Model as a "Holistic" Approach

As previously indicated, it is important for an employer to recognize that there are many personal and workplace factors that interact together to produce situations that lead to musculoskeletal injuries. Therefore, any strategy to control these situations should consider a broad range of factors and their influences on each other. One holistic model of human workplace interaction is the "balance theory that is presented in Figure 41–1 (Smith & Carayon, 1995; Smith, Karsh, & Moro, 1999; Smith & Sainfort, 1989). Each element of this model can present demands or loads on the worker that can lead to musculoskeletal injury. In combination, their effects can produce even greater injury. Each one of these elements of the model has specific features that can lead to stress and strain in the person, and this stress and strain may produce injury. In addition, each element of the model can be improved to reduce the risk of WMSDs, or a combination of elements may need improvement to obtain the desired benefits.

The Person There is a wide range of individual attributes that can affect injury potential. These include perceptual-motor abilities, physical capabilities and capacity such as strength and endurance, current health status, susceptibilities to disease, personality, intelligence, and behavior (work practices, task technique). Perceptual/motor skills level can affect the ease and quality with which a task is carried out, physiological energy expenditure, posture, balance, and fatigue. Personal physiological considerations such as strength, endurance, stress tolerance, and disease status affect susceptibility to injury. Intelligence affects the ability for hazard recognition and training in hazard recognition and elimination. It is critical that a proper "fit" be achieved between employees and other characteristics of the model to achieve proper balance. Each one of the personal attributes and characteristics can be improved to reduce the risk of a WMSD. For example, a person can become stronger or better conditioned physically. A person can be trained to improve the work practices used to handle loads. A person who has a strain can be given light duty tasks or excused from work until the strain heals. A person who is injured can be given proper treatment to provide the maximum possible healing.

Machinery, Technology, and Materials. There are characteristics of machinery, tools, technology, and materials used by the employee that can influence the potential for injury. One consideration is the extent to which machinery and tools influence the use of the most appropriate and effective perceptual/motor skills and energy resources. The relationship between the controls of a machine and the action of that machine dictates the level of perceptual/motor skill necessary to perform a task, the repetition level of motions, the postures of the joints and back, and force necessary for activation. The proper design considerations for the technology to accommodate the human factors issues is well established and identified in ergonomic books such as the book by Kroemer and Grandjean (1997), and by the U.S. Military Standard as defined in the document Human Engineering Design Criteria for Military Systems, Equipment and Facilities (MIL-STDB1472, 1981)). In addition, technology can be used to reduce loads on the employees by providing leverage, physical power, and mental power (decision-making tools). For example, lifting aids can be used to reduce back strain, and conveyors can be used to reduce materials handling and excessive walking. Software programs can be used to determine lifting frequency and weight limits for lifting.

Task Factors. The demands of a work activity and the way in which tasks are conducted can influence employee strain and subsequent injury. Work task demands can be broken into the physical requirements, the mental requirements, and psychological considerations. The physical requirements influence the amount of energy expenditure necessary to carry out a task, and the forces imposed on the body. Excessive physical requirements can lead to fatigue, both physiological and mental, which can then lead to employee strain. Employees can only tolerate very high physical demands for short periods. Longer exposure to heavy workloads or multiple exposures to shorter duration heavy workloads can reduce an employee's capacity to respond normally. This can lead to strain and injury. Other physical task demands include the pace or rate of work, the amount of repetition in task activities, and work pressure due to production demands. Task activities that are highly repetitive and paced by machinery, rather than employee paced, tend to be stressful.

Psychological task demands are tied to the amount of control over the work process, participation in decision making, the ability to use knowledge and skills, the amount of esteem associated with the job, and product identity. These can influence employee job satisfaction, motivation, and attention. They also can cause job stress that can affect the employee's sensitivity to discomfort and pain, and increased muscle tension.

Ergonomic improvements can be achieved by setting the appropriate workload using industrial engineering work methods analysis procedures. In addition, establishing the proper work methods provides the basis for proper work practices to be used by the employees. With adequate employee training, the improved work practices will replace the bad habits that employees develop over months and years of working.

The Work Environment. The work environment exposes employees to climates, materials, chemicals, and physical agents that can cause harm or injury if the exposure exceeds safe limits. Such exposures vary widely from job to job, and from task to task. Hazard exposure in the work environment influences the probability for an injury or illness, and the extent of exposure often determines the seriousness of injury. Climatic conditions can also hamper the ability of an employee to use her or his senses properly (poor lighting, excessive noise, too hot). Cold temperatures make muscles more susceptible to strain, and hot temperatures affect the extent of energy expenditure and body fluid balance. The environment should be compatible with worker perceptual/motor skill requirements, energy expenditure, and motivational needs to encourage proper work methods.

Organizational Structure. Many aspects of organizational management can affect health and safety performance. These include management policies and procedures, the organization of tasks, the style of employee supervision, the motivational culture in the plant, the amount of socialization and interaction between employees, the amount of social support employees receive, and management attitude toward safety and ergonomics. The latter point, management attitude, has often been cited as the most critical element in a successful safety program (Cleveland Cohen, Smith & Cohen, 1979; Cohen, 1977; Smith, Cohen, Cohen, & Cleveland, 1978). If the individuals that manage an organization are not interested in good ergonomics, then employees tend to be less motivated to work safely. Conversely, if the management attitude is one in which safety and ergonomic considerations are important, then managers, supervisors, and employees will put substantial interest and energy into positive ergonomic efforts.

A consistent factor in injury causation is management pressure for higher production, or faster output, or a quick response to production problems. Technology malfunctions, insufficient staffing, and improper workload standards can exacerbate work pressure caused by management's insistence on greater production. Management emphasis on reducing costs, and enhancing profits may stretch the limits of the capabilities of the workforce and the technology. When breakdowns occur or operations are not running normally, employees tend to work harder to keep production up to speed. It is during these heavy load situations that many ergonomic injuries occur.

Management must provide adequate resources to meet production goals and to accommodate heavy demand operations. This means adequate staff, effective technology, proper process design, and fair work standards. Management must also establish policies to ensure that employees and supervisors accept the importance of proper ergonomics.

A GENERAL APPROACH TO AN ERGONOMICS PROGRAM

An ergonomic program standard proposed by Federal Register 1999 provided guidance that companies can follow to develop an effective ergonomic program (See *www.osha.gov*). Unfortunately this standard was rescinded by the U.S. Congress in 2001. The first step in establishing an ergonomics program is to have a corporate policy statement that defines the business' commitment and approach to ergonomic improvements. This is then followed by a structured process for defining ergonomic risks, developing solutions, implementing solutions, and assessing the effectiveness of the interventions. As stated earlier, it is essential that management be committed to and active in ergonomic issues if an ergonomic program is to succeed. In addition, ergonomic programs are more effective when employees have an active role in defining ergonomic risks and potential solutions, and when they can participate in the improvement process.

The OSHA ergonomics program standard defined critical areas that should be addressed in an ergonomics program. These were (a) management leadership and employee involvement, (b) hazard information and reporting, (c) job hazard analysis and control, (d) training, (e) WMSDs management, and (f) program evaluation.

Management leadership calls for companies to demonstrate leadership for the ergonomics program. This recommendation recognizes the research on safety programs by Cohen (1977) and Smith et al. (1978) that was discussed earlier showing the importance of management commitment of time and resources, and active management involvement in the program. In addition, the OSHA ergonomics program directive specified that companies must provide mechanisms for employees and unions to report ergonomics hazards and WMSDs symptoms and signs. The company policies must not discourage employees from reporting injuries and hazards, or from participating in the ergonomics program. To achieve these goals, the company must assign the responsibility for setting up and managing the ergonomics program to managers and supervisors. There must be resources for operating the ergonomics program. There must be ways for employees to be involved in the development, implementation, and evaluation of all aspects of the ergonomics program. The company should periodically communicate with employees about the ergonomics program and WMSDs.

In terms of hazard information and reporting, the company must provide ways for employees to (a) be aware of WMSDs hazards and the etiology of WMSDs, (b) report WMSDs and symptoms, and (c) get a prompt response about their reporting. This requires a central reporting location that will aggregate symptoms and injuries, and to report the findings back to employees periodically.

The hazard analysis requires the company to use ergonomic methods to define hazards for WMSDs. This calls for a job hazard analysis for any person or occupational group in which employees are reporting WMSDs and/or symptoms. Some areas of evaluation in the job hazard analysis that OSHA recommended be examined are: (a) high exertion activities, (b) highly repetitive motions, (c) the absence of breaks in work when there is repetition, (d) tasks with long reaches, (e) improper working surface height, (f) static postures, (g) sitting too long, (h) use of hand tools and powered tools, (i) contact with vibrating machines or surfaces, (j) compression of tissue on surfaces, (k) contact stresses on the hand, (l) gripping/holding objects with the hands or body while working, (m) using gloves that fit poorly, (n) moving heavy objects or people, (o) long horizontal reaches away from the body, (p) vertical reaches below the knees or above the shoulders, (q) moving objects a long distance, (r) bending or twisting the back, (s) slippery objects or objects without grips/handles, and (t) uneven, slippery, or sloped floor surfaces.

Hazard control efforts require the company to ask affected employees about their recommendations for improvements, to identify controls that will eliminate or reduce the hazards, to track the progress of the controls in eliminating or reducing the hazards, and to evaluate the potential WMSDs hazards whenever any workplace design change is made or new equipment purchased.

OSHA proposed the use of any combination of engineering solutions, administrative changes, or work practice improvements that eliminate or reduce the ergonomic hazards. Engineering controls require physical changes to the work environment or equipment to reduce the hazard level; for example, limiting the amount of weight handled by one person to 45 lbs, or providing a lifting device for loads over 45 lbs. Administrative changes deal with modifying the work process to reduce the hazard level; for example, forbidding individual employees from handling a patient, and requiring that two or more employees handle a patient. Or limiting the number of continuous hours worked to a maximum of 12 hours per 24-hour period, and the number of consecutive days worked to no more than 6 per week. Work practice improvements reduce hazards by eliminating unsafe behaviors used by employees.

Regarding training, OSHA recommended employees know about WMSDs hazards, the company ergonomics program, and how to reduce WMSDs risks. Employees in problem jobs, their supervisors, and those employees who are responsible for the ergonomics program must be trained. Retraining should occur at least every 3 years.

The company is responsible for promptly making available WMSD medical management for any injured employee at no cost. An injured employee should have work restrictions imposed to protect her or him from exposure to ergonomic hazards until they have recovered from the injury. The basic concept is early intervention in the injury process to limit damage and speed up recovery.

Lastly, OSHA recommended periodic evaluation of the effectiveness of the ergonomics program at least every 3 years. This can involve a compilation of the ergonomic hazards identified and controlled and a determination of a reduction in WMSDs frequency and severity.

General Guidelines for an Occupational Ergonomics/Safety Program

There are a number of other elements to consider in developing an ergonomics program or in upgrading your current program. These include organizational policies, managing various elements of the program, motivational practices, hazard control procedures, dealing with employees' concerns, injury investigations, and injury recording and surveillance. There has been considerable research into the necessary elements for a successful safety program (Cohen, 1977, Smith et al., 1978) and how these elements should be applied. These can be used as a base for organizing an ergonomics program. In addition, OSHA first published guidelines for an ergonomics program in 1999 (Federal Register, 1999), and promulgated the standard in 2000 (OSHA, 2000).

One primary factor that emerges from every study of successful company safety programs and a major aspect of the recommended OSHA ergonomics program is that the program will not be successful unless there is a commitment to the program by top management. This indicates that there needs to be a written organizational policy statement on the importance of safety and ergonomics, and the policy should specify the general approach and procedures the organization intends to use. Having such a policy is the first step toward effective management commitment and a successful ergonomics program.

Even so, Smith et al. (1978) have shown that it takes more than just a written policy to ensure a successful safety program and good safety performance. It also takes involvement in the program by all levels of management. For the top managers, it means they must periodically get out to where employees are working to show their concern, to talk to employees about working conditions, and to view safety and ergonomic problems. This has been shown to be more effective when done on an informal basis. For middle managers and supervisors, there is a need to participate in safety and ergonomic program activities with employees. These safety and ergonomic activities put middle managers in touch with employee concerns about potential hazards and educate them to operational problems. Management involvement demonstrates to the employees that management cares about their safety and health and that ergonomics is important. This reinforces employee commitment to safe work practices.

Another aspect of management commitment is the level of resources that are made available for safety and ergonomics programming. Cohen (1977) found that organizational investment in full time safety staff was a key feature to good plant safety performance. The effectiveness of safety and health staff was greater the higher they were positioned in the management structure. However, Smith, et. al. (1978) found that safety staff was not as critical to program effectiveness as management and employee active involvement in the safety program efforts. A full-time ergonomics staff may not be as necessary when the supervisors and employees are actively involved in ergonomic efforts.

An organization's motivational practices will influence employee safety behavior and compliance with safe work practices. Research has demonstrated that organizations that exercise participative management approaches have better safety performance. These approaches encourage employee involvement in defining the risks and in developing solutions such as improved work practices. Employee involvement leads to a greater awareness of ergonomics, and leads to higher motivation levels conducive to proper employee behavior and work practices. Research has shown that organizations that use punitive motivational techniques for influencing safety behavior have poorer safety records than those using positive approaches (Cohen, 1977; Smith et al., 1978).

Another important motivational factor is encouraging communication between various levels of the organization (employees, supervisors, managers). Such communication increases participation in ergonomic issues and builds employee and management commitment to safety and ergonomic goals and objectives. Often informal communication between supervisors and employees is a potent motivator and provides more meaningful information for ergonomic hazard control.

All safety and ergonomic programs should have a formalized approach to hazard detection and control. This often includes an inspection system to define ergonomic workplace hazards, injury investigations to identify critical causes, recordkeeping to monitor progress, the review of new purchases to ensure compliance with safety and ergonomic guidelines, and good housekeeping for a comfortable environment. All of these elements contribute to a positive "safety climate" that encourages employee compliance with proper safety and ergonomic procedures. The importance of formalized hazard control programs is that they establish the groundwork for other programs such as employee training for work practices improvements and equipment purchasing requirements.

Defining How Well a Company is Doing Using Injury Data

The OSHA-proposed ergonomic standard provides some general guidance for hazard analysis, and specific areas to examine. The following details an approach for assessing WMSDs experience and for conducting hazard analysis. The first step in assessing ergonomic risks is to define where WMSDs are occurring in your organization. This requires an injury surveillance system. The first place to start such a surveillance system is by using the records required by OSHA for injury reporting. Occupational injuries are recorded on the OSHA 300 Injury Log. More detailed information about an injury is recorded on the supplemental injury form (OSHA 301 reports or worker's compensation reports). In addition, there may be medical department or infirmary reports that can be used. These injury reports can serve as the basis for looking at the number, type, and seriousness of injuries in specific departments or jobs.

There are four main uses of injury statistics; (a) to identify high-risk jobs or work areas, (b) to evaluate company health and safety performance, (c) to evaluate the effectiveness of hazard abatement approaches, and (d) to identify factors related to illness and injury causation. An illness and injury reporting and analysis system requires that detailed information be collected about the characteristics of illness and injuries, and their frequency and severity.

The 1970 OSHA Act requirements specify that any illness or injury to an employee that causes time lost from the job, treatment beyond first-aid, transfer to another job, loss of consciousness or an occupational illness must be recorded on a daily log of injuries (OSHA 300 form). This log identifies the injured person, the date and time of the injury, the department or plant location where the injury occurred, and a brief description about the occurrence of the injury highlighting salient facts such as the chemical, physical agent, or machinery involved and the nature of the injury. The number of days that the person is absent from the job is also recorded upon the employee's return to work. In addition to the daily log, a more detailed form (OSHA 301) is filled out for each injury that occurs. This form provides a more detailed description of the nature of the injury, the extent of damage to the employee, the factors that could be related to the cause of the injury, such as the source or agent that produced the injury, and events surrounding the injury occurrence. A worker's compensation form can be substituted for the OSHA 301 form, as equivalent information is gathered on these forms.

The OSHA Act injury and illness system specifies a procedure for calculating the frequency of occurrence of occupational injuries and illnesses, and an index of their severity. These can be used by companies to monitor their health and safety performance over time, and to compare their experience to those companies in the same industry category. National data by major industrial categories is compiled by the U.S. Bureau of Labor Statistics annually and can serve as a basis of comparison of individual company performance within an industry sector in the United States. Thus, a company can benchmark its injury rate to see how it compares with other companies in its industry. This industrywide injury information is available on the OSHA Web site (*http://www.osha.gov*).

The OSHA system uses the following formula in determining a company's annual injury and illness incidence. The total number of recordable injuries are multiplied by 200,000 and then divided by the number of hours worked by the company employees. This gives an injury frequency per 100 person-years of work (injury incidence). These measures can be compared to an industry average.

> Incidence Rate = Number of WMSDs multiplied by 200,000 ÷ (divided by)
> The number of hours worked by company employees
> The equation for calculating the incidence rate is:IR = # WMSDs × 200,000 ÷
> # person-hours where:
>
> 1. The number of recordable injuries and illnesses is taken from the OSHA 300 daily log of injuries.
> 2. The number of hours worked by employees is taken from payroll records and reports prepared for the government.

It is also possible to determine the severity of a company's injuries. Two methods are typically used. In the first, the total number of days lost due to injuries are compiled from the OSHA 300 daily log and are divided by the total number of injuries recorded on the OSHA 300 daily log. This gives an average number of days lost per injury. In the second approach, the total number of days lost is multiplied by 200,000 and then divided by the number of person-hours worked by the company employees. This gives a severity rate per 100 person-years of work. These measures can also be compared to an industry average or across departments within a facility, or among jobs in a facility to identify high risk areas.

Evaluating Ergonomic Risk Factors (Hazards) as a Basis for Improvement

A good understanding of the demands of specific jobs and tasks provides the base for developing ways to better match and fit tasks and activities to employees. Knowledge of the loads and forces imposed on employees, the energy requirements of work, the mental demands, and the psychological stress leads to an evaluation of the potential risks for employee injury or reduced performance. In order to successfully control occupational hazards that lead to illness and injuries, it is necessary to define their nature, and predict when and where they will occur. This requires a hazard detection process that can define the frequency of the hazards, their seriousness, and their amenability to control.

Traditionally, two parallel systems of information have been used to define occupational hazards. One system is hazard identification, such as plant inspections, fault-free analysis, and employee hazard reporting programs, which are used to define the nature and frequency of the hazards. With these approaches, preventive action can be taken before an injury or illness occurs. The second system is after the fact because it uses employee injury and company loss control information to define company problem spots based on the number of injuries and the costs to the organization. When pre and post injury systems are integrated, they can be used to define high-risk jobs, tasks, plant areas, or working conditions where remedial programs can be established for hazard control.

Hazard identification prior to the occurrence of an occupational injury is a major goal of a hazard inspection program. In the United States, such programs have been formalized in terms of federal and state regulations that require employers to monitor and abate recognized occupational health and safety hazards defined by the standards. These standards of unsafe exposures define the benchmarks that must be achieved by companies. Unfortunately there are no U.S.A. federal or state standards for ergonomic hazards that define the upper limits of specific exposures that are dangerous.

Research has shown that formal inspections are most effective in identifying permanent, fixed physical and environmental hazards that do not vary over time. Inspections are not very effective in identifying transient physical and environmental hazards, or improper employee behaviors, as these hazards may not be present when an inspection is taking place (Smith et. al, 1971). A major benefit of conducting an inspection, beyond the definition of serious hazards, is the positive motivational influence on supervisors and employees. Inspections demonstrate that management is interested in the health and safety of employees, and to a safe working environment. To capitalize on this positive motivational influence, an inspection should not be a punitive process. Indicating the good aspects of a work area and not just the hazards is important in this respect. It is also important to have employees participate in hazard inspections as this increases hazard recognition skills and increases motivation for safe and proper behavior.

The first step in an inspection program is to develop a checklist that identifies all potential hazards. A good starting point for an ergonomic checklist is to examine the references cited earlier (Kroemer & Grandjean, 1997; NRC, 2001; NIOSH, 1997). In addition, the topical areas defined in the proposed 1999 OSHA ergonomics program standard (see page 7) delineate important areas to evaluate. Many insurance companies have developed general checklists that can be tailored to a particular company. Once the checklist is completed, a process for conducting inspections can be established.

A systematic inspection procedure is preferred. This requires that the inspectors know what to look for, where to look for it, and have the proper tools to conduct an effective assessment. Aspects of the inspection that need to be determined are (a) the factors to be examined, (b) the frequency of inspection necessary to detect and control hazards, (c) the individuals who should conduct and/or participate in the inspection, and (d) the instrumentation needed to make measurements of the hazard(s). The factors to be inspected include the technology, machinery, tools and materials; the environmental conditions; task risk factors; and employee work practices.

The frequency of inspections should be based on the nature of the hazards being evaluated. Intermittent hazards may need to be evaluated more often than fixed physical hazards. Random spot-checking is a useful method that "catches" transient hazards. These intermittent hazards require more frequent inspection. In many cases, monthly inspections are warranted, and in some cases daily inspections are reasonable. Many times the transient hazards are poor work practices that occur due to specific circumstances such as handling a very heavy bedridden patient alone rather than getting help. When such work practices are observed, they need to be called to the attention of the employee and instructions given on proper procedures.

In contrast to the periodic, formal inspection process, Smith, Karsh, Carayon, & Conway (2002) presented a hazard survey approach that is based on continuous reporting of potential hazards by employees as they encounter a hazard. This approach has formal methods for reporting and evaluating the hazards to bring about a quick response to the problem. One weakness of the approach is that it is unlikely that employees will report poor work practices and behaviors in themselves or peers. One approach for effective identification of these behavioral hazards could be periodic observations by supervisors. This assumes that supervisors have knowledge of which behaviors are hazardous, that they will report/record the behaviors, and that they will take action to improve employee behaviors.

SOME EXAMPLES OF ERGONOMIC INTERVENTIONS FOR HEALTH CARE ACTIVITIES

Lifting and Moving Patients and Objects

Materials handling causes the greatest number of lower back injuries in manufacturing, warehousing, long shoring, mining, and construction, and this is also true in health care. Lifting, adjusting, pulling, and moving people who are sick and fragile is very difficult and risky for a worker's back (OSHA, 2000, 2002). There are several factors that make people handling so difficult and risky such as a patient's size (weight, height, girth), the "dead" weight of the patient, the lack of good places to grip the patient, poor postures in accessing the patient or when moving the patient, the need to be gentle when handling the patient, potential exposure to pathogens that affect how/where the patient is handled, and some patients are connected to technology/medication systems that cannot be removed. These and other factors make it very difficult to handle the patient without having some exposure to risk of back injury. What can be done to reduce this risk?

Factors that cannot be changed are the patient's weight and immobility. What can be changed is how the weight is handled. First, when handling a patient, the health care worker should assess whether the patient has the capability to move herself, or to provide assistance in the move. Where the patient has the capability to move herself or provide assistance in moving herself, the health care worker should take advantage of this help. Second, it is best to limit the amount of the patient's weight that is handled. Never lift the entire weight. Always use patient handling techniques where the patient can be slid, pulled, pushed, and/or swung onto/into a patient transfer and/or transport device. This greatly reduces the amount of weight that loads the health care worker's back and the amount of force generated by the health care worker to make the move. When the patient is a dead weight, this is not possible, and then it is best to use technology to handle the patient to make the move. A height adjustable transfer table (gurney) can be aligned with the bed height to slide or push the patient onto the transfer table, or lifts can be used to move the patient from the bed to the transfer table. When technology is not available, then teams of health care workers can work together to handle the patient when making a move. The primary principle is to reduce the load on the back, shoulders, and hands of the health care worker by using good material handling methods, technology, and/or multiple workers.

An example provides some understanding problem and the some ideas for dealing with it. A very heavy patient (~700 lbs.) was admitted to the hospital in a coma. The patient was placed on an air mattress bed to reduce the risk of bedsores. This air mattress bed did not provide the surface resistance that is useful for the health care worker to gain leverage when moving the patient in the bed and on/off of the bed. It was also very hard to get a solid grip on the patient. A standard lifting device did not have the capacity to lift the patient, and thus a team of health care workers was required when the patient was moved. Even so, several health care workers strained their backs when handling this patient even when working as a team. A solution was found when the head nurse contacted a local livestock veterinary clinic and borrowed a "horse lift" that had the capacity to lift this patient.

A less extreme example also deals with patient handling. A quadriplegic patient, who weighed 160 lbs., needed to be transferred daily from the nursing home bed to a wheelchair to be taken to physical therapy. The patient could be pulled upright, positioned onto the edge of the bed, and swung into the wheelchair. This method of patient handling did not require the health care worker to support the full weight of the patient during any of the steps of positioning and moving (swinging). The patient was delivered to the physical therapy department where he was swung from the wheelchair onto a "stool" that had a powered height adjustment that could be lowered to the floor or raised up to table height. The stool also had a powered drive to transport the patient to a mat or to a therapy table where the patient could be slid to the working surface.

A third example deals with equipment design and portable equipment. A hospital had a portable x-ray machine that was wheeled to patient rooms for patients too ill to be moved to the regular x-ray laboratory. The manufacturer had installed small casters on the machine that supported the heavy weight well. However, the machine was hard to push and pull across the floors particularly when maneuvering in the small hospital rooms. This caused strain on the back, shoulders, and arms of the x-ray technician(s) when transporting and positioning the machine. They complained to the maintenance supervisor who contacted the engineering department. The engineers found suitable 4 in. diameter wheels with ball bearings that could handle the weight of the machine and that provided much easier rolling of the machine. This reduced the force necessary to push and pull the machine and eliminated the strain on the x-ray technicians.

These examples show that by first defining the cause of the strain and then by using simple solutions, many ergonomic risks can be controlled. However, not all ergonomic risks are simple to control as is seen later. This is because the risk factors are an inherent part of the work tasks. When the risk factor cannot be removed as in the case of inherent ergonomic risks, then the solution deals with limiting the amount of exposure to these risks.

Prolonged Standing and Walking

Like many active jobs, most health care jobs require workers to stand up a great deal of the time during the day. This places strain on the muscles of the back, shoulders, neck, and legs. It also puts loads on the spine, knees, and feet. Studies of factory workers doing assembly work at a fixed work station have shown that static standing for many minutes and even hours leads to fatigue and discomfort in several parts of the body including the neck, shoulders, back, and legs. For most health care workers, the standing is dynamic and includes walking from patient to patient or work station to work station. This is different from the factory assembly workers as it reduces the static standing, but it adds the increased energy expenditure requirement of the walking. The long periods of standing combined with walking not only put a strain on the torso and legs, but also on the feet. Because the nature of the job tasks requires standing and walking, there is no way to eliminate the ergonomic risk factors. The solution must be to reduce the time of exposure to these ergonomic risk factors. That means that time limits should be placed on these exposures with breaks of activities between exposures that do not strain the torso or feet. In other words, there needs to be periods of sitting interspersed with the standing and walking. Based on personal experience, it is a reasonable rule of thumb that health care workers when standing and walking a majority of their day should have a sit-down break at least every 2 hours. This could be a rest break or alternative work done while sitting.

Prolonged Sitting

Static sitting for too long is just as problematic from an ergonomic perspective as standing and walking for too long. Prolonged static seated postures produce fatigue, discomfort, and pain in the neck, shoulders, back, and buttocks. This affects office staff, receptionists, lab technicians, administrators, information technology staff, and others who sit the majority of their working day.

In any work station, postural support is essential for controlling loads on the spine and appendages. Studies have revealed that the sitting position, as compared to the standing position, reduces static muscular efforts in legs and hips, but increases the physical load on the intervertebral discs in the lumbar region of the spine. Research by Nachemson and Elfstrom (1970) and Andersson and Ortengreen (1974) offers some guidance about proper seated posture. When the chair's backrest angle was increased from 90 (straight up) to 120 degrees (leaning back), subjects exhibited an important decrease of the intervertebral disc pressure and of the electromyographic activity of the back muscles. These researchers concluded that a sitting posture with reduced disc pressure is more healthy and desirable. The results of these studies indicated that leaning the back against a backward leaning backrest transfers some of the weight of the upper part of the body to the backrest. This reduces considerably the physical load on the intervertebral discs and the static strain of the back and shoulder muscles.

Poorly designed chairs can contribute to discomfort. Chair adjustability in terms of height, seat angle, backward tilt, and lumbar support helps to provide trunk, shoulder, neck, and leg postures that reduce strain on the muscles, tendons, ligaments, and discs. The "motion" of the chair helps encourage good movement patterns. A chair that provides swivel action encourages movement, whereas backward tilting increases the number of postures that can be assumed. The chair height should be adjustable so that the computer operator's feet can rest firmly on the floor with minimal pressure beneath the thighs. To enable short users to sit with their feet on the floor without compressing their thighs, it may be necessary to add a footrest.

The seat pan should be wide enough to permit operators to make slight shifts in posture from side to side. This not only helps to avoid static postures, but also accommodates a large range of individual buttock sizes. The front edge of the seat pan should be well rounded downward to reduce pressure on the underside of the thighs that can affect blood flow to the legs and feet. The seat needs to be padded to the proper firmness that ensures an even distribution of pressure on the thighs and buttocks.

The tension and tilt angle of the chair's backrest should be adjustable. Inclination of chair backrest is important for operators to be able to lean forward or backward in a comfortable manner while maintaining a correct relationship between the seat pan angle and the backrest inclination. A backrest inclination of that allows adjustment of up to 125 degrees is considered appropriate. The backrest tilt adjustments should be accessible and easy to use. Chairs with high backrests are preferred because they provide support to both the lower back and the upper back (shoulders).

Another important chair feature is armrests. Armrests can provide support for resting the arms to prevent or reduce arm, shoulder, and neck fatigue. Removable armrests are an advantage because they provide greater flexibility for individual operator preference. For specific tasks such as using a microscope, removable armrests can be beneficial. Many chairs have height-adjustable armrests that provide better positioning of the arms when resting on the armrest.

Work Station and Technology Design

The work station where the health care worker carries out tasks needs to be designed to minimize the ergonomic risk factors. As general guidance, Grandjean (1984) proposed the consideration of the following features of work station design:

1. The furniture should be as flexible as possible with adjustment ranges to accommodate the anthropometric diversity of the users in sitting and standing.
2. Controls for work station adjustment should be easy to use.
3. There should be sufficient knee space for seated operators, and leg space for standing users.
4. The chair should have an elongated backrest with an adjustable inclination and a lumbar support.
5. The desk surface should provide adequate space to properly situate equipment, materials, and papers being used by the operator.

The recommended size of the work surface is dependent on the task(s) and the characteristics of the technology being used (dimensions of technology devices, tasks being performed, reference materials, products, storage requirements). Work stations are composed of primary work surfaces, secondary surfaces, storage, and postural supports. The primary working surface should allow tasks to be carried out with good motion patterns that do not constrain the postures of the limbs, neck, shoulders, and back. There should be the possibility to adjust the height and orientation of the working surfaces to achieve proper postures. There should be adequate knee and leg-room for the user to move around while working. It is important to provide unobstructed room under the working surface for the feet and legs so that employees can easily shift their seated or standing postures. When adjustable working surfaces are used, the ease of adjustment is essential.

Now let us consider an unstructured work station situation. Doctors and nurses often have to record information on a patient chart without a working surface to rest the chart on. Often a handheld PDA, laptop computer, or other electronic recording device is used as the doctor or nurse is standing in the patient's room. This lack of support surfaces and postural supports leads to strain on the back, shoulders, and neck (and maybe the arms and wrists as well). Although the convenience and effectiveness of lightweight portable computing is very high for recording or retrieving patient information, the comfort and health factors are often very low because the person uses the laptop or PDA in all manner of environments, work stations, and tasks that diminish the consistent application of good ergonomic principles.

In situations where there is not a fixed work station, the device is typically positioned wherever is convenient. Very often such positioning creates bad postures for the legs, back, shoulders, arms, wrists/hands, or neck. In addition, the smaller dimensions of the manual input devices (touch pad, buttons, keyboard, joy stick, roller ball) make motions much more difficult, and these often produce constrained postures. If the devices are used continuously for a prolonged period (such as one hour or more), muscle tension builds up and discomfort in joints, muscles, ligaments, tendons, and nerves can occur. To reduce the undesirable effects of the poor work station characteristics that lead to the discomfort, the following recommendations are given:

1 If you are using a laptop on your lap, find a work area where you can put the laptop on a table (rather than on your lap). Then arrange the work area as closely as possible with the recommendations presented for a standard office.

2. If you are using a handheld PDA, you should position yourself so that your back is supported. It is preferable to use the device sitting down. Of course, if you are using the PDA as you are walking, then this is not possible. If the PDA

has a voice interface, then use an earpiece and a microphone so that you do not have to be constantly gripping the PDA in your hand.

3. Never work in poor postural conditions for more than a few minutes continuously.
4. Buy equipment that provides the best possible input interfaces and displays (screens, headphones, typing pads). Because these devices are small, the perceptual motor requirements for their use are much more difficult (sensory requirements, motion patterns, skill requirements, postural demands). Therefore, screens should provide easily readable characters (large, understandable), and input buttons should be easy to operate (large, properly spaced, easily accessible).
5. Only use these devices when you do not have access to fixed work stations that have better ergonomic characteristics. Do not use these devices continuously for more than 30 minutes.

Health care has unique work station considerations that need to be addressed. We can start by looking at health care workers and patient interactions at the hospital bed. Many long-term care facilities have beds similar to hospital beds, but may also have more traditional beds found in residences. The hospital bed has some positive features that can help when handling a patient. For instance, the ability to raise or lower the head and foot of the bed to move the patient into a better position for grasping or sliding or lifting. In addition, the bed is on wheels that make it possible to position the bed for easier access to the patient. Keep in mind the previous example about the portable x-ray machine and be sure the beds have wheels that provide ease of wheeling. The beds also have height adjustability that can reduce the extent of back bending to reach the patient and can facilitate sliding the patient from the bed to a hospital gurney or wheelchair. But the hospital beds are large and very heavy, and they sometimes have technology attached to them that can impede access to the patient. Technology can be hung on the bed to facilitate moving with the patient, but this adds to the weight of the bed and difficulty of patient access. Proper work practices need to be employed to deal with these impediments. For example, when the access to the patient is obstructed, then more than one employee may be necessary to handle the patient to reduce the strain.

Given the characteristics of the hospital bed, it is important for good ergonomic practice that health care workers take advantage of the adjustability of the bed to reduce musculoskeletal strain. This means using good practices when interacting with patients in bed that capitalize on the beds capabilities. The enemy of using good work practices is too much workload and a lack of time to do tasks properly. Health care workers have very heavy work schedules and very often do not take the time to use the features of the bed. For example, to raise or lower the bed height to reduce a poor posture that can cause back strain when handling a patient, or to position a patient using the bed's head/foot adjustments to reduce the force necessary to handle the patient. Technology like the adjustability characteristics of the hospital bed can be helpful in reducing strain, but only when good work practices are employed by workers using the technology.

Residential beds do not have the adjustable features or the mobility of hospital beds. For long-term care residents that are ambulatory, the residential beds are preferable because hospital beds are hard to get into and out of easily by people with limited mobility. But residential beds can be problematic for health care workers who have to handle residents that are not mobile. A resting long-term care resident who needs help getting up is positioned lower to the ground in a residential bed, and therefore the effort exerted by the health care worker to transfer the resident is higher than from a hospital bed. In addition, the residential bed cannot be positioned for patient transfer or access as easily as a hospital bed.

Examining rooms often are small and filled with the examining table, chairs, cabinets, shelves, and technology. This makes them crowded and cramped for space, and can make it difficult to handle frail patients. In particular, the examining tables are very often at a fixed height that makes it hard for frail patients to get onto the table. More and more, adjustable-height tables are being used that go down low enough for the patient to sit on the table and then be elevated to a comfortable height for the health care worker to examine the patient. This eliminates the back and shoulder strain from lifting or assisting patients up onto the examining table, and provides a comfortable height with less back flexion when examining the patient. It also allows the health care worker to sit down to examine the patient by lowering the examining table to a suitable height.

Examining rooms that use equipment such as sonograms many times are so small that the technician must use extreme postures of the shoulders, arms, hands, and wrists to position the instruments. To reduce this problem, rooms need to be of adequate size for technicians to get in comfortable postures during the medical evaluation process. Technicians should be able to either sit or stand at their discretion when conducting the medical evaluation.

CONCLUSIONS

An overall ergonomics program is necessary to be able to develop good specific ergonomic interventions to reduce the risk of WMSDs in your employees. This is based on the belief that a holistic approach is needed in order to be successful. This program should include appropriate management commitment and involvement, and employee involvement. There should be mechanisms established for defining ergonomic hazards that can cause WMSDs, to prioritize the importance of the ergonomic hazards, and to define interventions to remove or control the risks due to the serious hazards. Interventions might require engineering redesigns or new technology, but might also use administrative procedures and/or work practice improvements to resolve the risk due to the ergonomic hazards. All of these aspects require a structured approach for managing the ergonomic program, hazard identification and control, and assessment of success.

There are many resource books that can provide guidance in developing specific ergonomic interventions (microergonomics) that are identified in this chapter. Of particular significance is the *Handbook of Human Factors and Ergonomics* (Salvendy, 1997), *Fitting the Task to the Human* (Kroemer & Grandjean, 1997), and several ergonomic guidelines for specific industries published by OSHA (see *www.osha.gov*).

Participative approaches for ergonomic hazard definition and ergonomic hazard control are the most effective. These provide critical information about the sources of hazards, motivation to employees to be actively involved in ergonomics improvements, and motivation for developing improved work practices. In addition, ergonomic approaches that employ multiple interventions for ergonomic hazard reduction are the most successful (Karsh, Moro, & Smith, 2001; Smith et al., 1999). A holistic approach is the best approach.

References

Andersson, G. B. J. (1981). Epidemiologic aspects of low back pain in industry. *Spine, 6,* 53–60.

Andersson, G. B. J.,. & Ortengren, R. (1974). Lumbar disc pressure and myoelectric back muscle activity during sitting. *Scandinavian Journal of Rehabilitation Medicine, 3,* 115–121.

Armstrong, T. J., Buckle, P., Fine, L. J., Hagberg, M., Jonsson, B., Kilbom, A., Kuorinka, I. A. A., Silverstein, B. A., Sjogaard, G., & Viikari-Juntura, E. R. A. (1993). A conceptual model for work-related neck and upper-limb musculoskeletal disorders. *Scandinavian Journal of Work Environment Health, 19,* 73–84.

Armstrong, T. J., & Chaffin, C. B. (1979). Some biomechanical aspects of the carpal tunnel. *Journal of Biomechanics, 12,* 567–570.

Bigos, S. J., Battie, M. C., Spengler, D. M., Fisher, L. D., Fordyce, W. E., & Hansson, T.H. (1991). A prospective study of work perceptions and psychosocial factors affecting the report of low back injury. *Spine, 16,* 1–6.

Bigos, S. J., Spengler, D. M., Martin, N. A., Zeh, J., Fisher, L., & Nachemson, A. (1986a), Back injuries in industry: A retrospective study. II: Iinjury factors. *Spine, 11,* 246–251.

Bigos, S. J., Spengler, D. M., Martin, N. A., Zeh, J., Fisher, L., & Nachemson, A. (1986b), Back injuries in industry: A retrospective study. II: Iinjury factors. *Spine, 11,* 252–256.

Boshuizen, H. C., Bongers, P. M., & Hulshof, C. T. J. (1990). Long-term sick leave and pensioning due to back disorders of tractor drivers exposed to whole-body vibration. *International Archives of Occupational Environmental Health, 62,* 117–122.

Burdorf, A., Naaktgeboren, B., & deGroot, H. C. (1993). Occupational risk factors for low back pain among sedentary workers. *Journal of Occupational Medicine, 35,* 1213–1220.

Carayon, P., Smith, M. J., & Haims, M. (1999). Psychosocial aspects of work-related musculoskeletal disorders. *Human Factors, 41*(6) 644–663.

Chaffin, D.B. (1987). Manual materials handling and the biomechanical basis for prevention of low-back pain in industry Can overview. *American Industrial Hygiene Association Journal, 48,* 989–996.

Chaffin, D. B., & Park, K. S. (1973). A longitudinal study of low-back pain as associated with occupational weight lifting factors. *American Industrial Hygiene Association Journal, 34*, 513–525.

Cleveland, R. J., Cohen, H. H., Smith, M. J., & Cohen, A. (1979). *Safety program practices in record holding companies*, [DHEW (NIOSH) Publication No. 79–136]., Washington, DC: United States Government Printing Office.

Cohen, A. (1977). Factors in successful occupational safety programs. *Journal of Safety Research, 9,* 168–178.

deKrom, M. C. T. F. M., Kester, A. D. M., Knipschild, P. G., & Spaans, F. (1990). Risk factors for carpal tunnel syndrome. *American Journal of Epidemiology, 132,* 1102–1110.

Federal Register. (1999, November 23). *Part 1910, Subpart Y–Ergonomic Program Standard, 64*(225), 66067–66078.

Grandjean, E. (1984). Postural problems at office machine work stations. In E. Grandjean (Ed.), *Ergonomics and health in modern offices* (pp. 445–455). London: Taylor & Francis. Ltd..

Hagberg, M., Morgenstern, H., & Kelsh, M. (1992). Impact of occupation and job tasks on the prevalence of carpal tunnel syndrome: A review. *Scandinavian Journal of Work Environment Health, 18,* 337–345.

Hagberg, M., Silverstein, B., Wells, R., Smith, M. J., Hendrick, H. W., Carayon, P., Perusse, M., Kuorinka, I., & Forcier, L. (1995). *Work related musculoskeletal disorders (WMSDs): A reference book for prevention.* London: Taylor & Francis.

Hanrahan, L. P., Higgins, D., Anderson, H., & Smith, M. (1993, December). Wisconsin occupational carpal tunnel syndrome surveillance: The incidence of surgically treated cases. *Wisconsin Medical Journal,* 685–689.

Karsh, B.-T., Moro, F. B. P., & Smith, M. J. (2001). The efficacy of workplace ergonomic interventions to control musculoskeletal disorders: A critical analysis of the peer-reviewed literature. *Theoretical Issues in Ergonomic Science, 2,* 23–96.

Kelsey, J. L. (1975). An epidemiological study of the relationship between occupations and acute herniated lumbar intervertebral discs. *International Journal of Epidemiology, 4,* 197–205.

Kilbom, A. (1994a). Repetitive work of the upper extremity: Part I. Guidelines for the practitioner. *International Journal of Industrial Ergonomics, 14,* 51–57.

Kilbom, A. (1994b). Repetitive work of the upper extremity: Part II. The scientific basis (knowledge base) for the guide. *International Journal of Industrial Ergonomics, 14,* 59–86.

Konz, S., & Johnson, S. (2000). *Work design: Industrial ergonomics* (5th ed.). Scottsdale, AZ: Holcomb Hathaway.

Kroemer, K. H. E., & Grandjean, E. (1997). *Fitting the task to the human* (5th ed.). Bristol, PA: Taylor & Francis.

Moore, J. S., & Garg, A. (1995). The strain index: A proposed method to analyze jobs for the risk of distal upper extremity disorders. *American Industrial Hygiene Association Journal, 56,* 443–458.

Nachemson, A., & Elfstrom, G. (1970). Intravital dynamic pressuremeasurements in lumbar discs. *Scandinavian Journal of Rehabilitation Medicine 1(suppl),* 1–40.

Nathan, P. A., Kerniston, R. C., Myers, L. D., & Meadows, K. D. (1992). Longitudinal study of median nerve sensory conduction in industry: Relationship to age, gender, hand dominance, occupational hand use, and clinical diagnosis. *The Journal of Hand Surgery, 17A,* 850–857.

NIOSH. (1981). *Work practices guide for manual lifting* Cincinnati, OH: NIOSH.

NIOSH. (1997). *Musculoskeletal disorders and workplace factors.* (DHSS/NIOSH Publication No. 97–141). Cincinnati, OH: Author.

Nordstrom, D. L. (1996). A population based, case control study of carpal tunnel syndrome (Doctoral dissertation, University of Michigan). *Dissertation Abstracts International*

Nordstrom, D. L., Vierkant, R. A., DeStefano, F., & Layde, P. (1997). Risk factors for carpal tunnel syndrome in a general population. *Occupational and Environmental Medicine, 54*: 734–740.

NRC/IOM. (2001). *Musculoskeletal disorders and the workplace.* Washington, DC: National Academy Press.

OSHA. (1990). *Ergonomics program management guidelines for meatpacking plants.* (OSHA Publication No. 3121). Washington, DC: OSHA.

OSHA. (2000). *Final ergonomics program standard—1910.900.* Retrieved November, 2002 from www.osha.gov.

OSHA. (2002). *Ergonomic guideline for nursing homes.* Retrieved from http://www.osha.gov.

OSHA. (2003). *Draft ergonomic guideline for retail grocery stores.* See *www.osha.gov.*

OSHA. (2004/2005). Retrieved from www.osha.gov/SLTC/ergonomics/index.html.

Putz-Anderson, V. (Ed.). (1988). *Cumulative tauma disorders— Manual for musculoskeletal diseases of the upper limbs.* London: Taylor & Francis.

Salvendy, G. (Ed.). (1997). *Handbook of human factors and ergonomics.* New York: Wiley.

Salvendy, G. (Ed.). (2006). *Handbook of human factors and ergonomics.* New York, NY: Wiley.

Silverstein, B. A., Fine, L. J., & Armstrong, T. J. (1987). Occupational factors and carpal tunnel syndrome. *American Journal of Industrial Medicine, 11,* 343–358.

Smith, M. J., et al. (1971). *Wisconsin inspection effectiveness projec,* (Contract Rep. L71–171). Washington, DC: U.S. Department of Labor.

Smith, M. J., & Carayon, P. (1995). New technology, automation and work organization: Stress problems and improved technology implementation strategies. *International Journal of Human Factors in Manufacturing, 5,* 99–116.

Smith, M. J., & Carayon, P. (1996). Work organization, stress and cumulative trauma disorders. In S. Moon & S. Sauter (Eds.), *Beyond biomechanics: Psychosocial aspects of cumulative trauma disorders* (pp. London: Taylor & Francis, 23–42.

Smith, M. J., Cohen, H., Cohen, A., & Cleveland, R. (1978). Characteristics of successful safety programs. *Journal of Safety Research, 10*(2), 5–15.

Smith, M. J., Karsh, B-T., Carayon, P., & Conway, F. T. (2002). Controlling occupational safety and health hazards. In J. C. Quick & L. E. Tetrick (Eds.), *Handbook of occupational health psychology* (pp. 35–68). Washington, DC: American Psychological Association.

Smith, M. J., Karsh, B.-T., & Moro, F. B. (1999). A review of research on interventions to control musculoskeletal disorders *Work-related musculoskeletal disorders* (pp. 200–229). Washington, DC: National Research Council.

Smith, M. J., & Sainfort (1989). A balance theory of job design for stress reduction. *International Journal of Industrial Ergonomics*, *4*, 67–79.

Snook, S. H. (1988). Approaches to the control of back pain in industry: job design, job placement and education/training. *Professional Safety, 33,* 23–31.

Snook, S. H. (1991). The design of manual handling tasks: Revised tables of maximum acceptable weights and forces. *Ergonomics, 34*(9), 1197–1213.

Stock, S. (1991). Workplace ergonomic factors and the development of musculoskeletal disorders of the neck and upper limbs: A meta analysis. *American Journal of Industrial Medicine*, *19*, 87–107.

Svensson, H., & Andersson, G. B. J. (1983). Low-back pain in 40–47-year-old men: Work history and work environment factors. *Spine*, *8*, 272–276.

U.S.MIL. (1981). *Military standard 1472: Human engineering design criteria for military systems, equipment and facilities.* Washington, DC: U.S. Department of Defense.

Waters, T. R., Putz-Anderson, V., & Garg, A. (1993). Revised NIOSH equation for the design and evaluation of manual lifting tasks. *Ergonomics, 36,* 749–776.

42

QUALITY IMPROVEMENT IN HEALTH CARE

Todd D. Molfenter and David H. Gustafson
University of Wisconsin, Madison

OPPORTUNITY AND PARADOXES

Quality Improvement (QI) has become part of the health care lexicon. It is part of licensing requirements, a focus of organizational committees, and is seen as a component of health care delivery by payers, regulators, and administrators. Yet, the quality improvement discipline in the health care industry is underdeveloped and inadequate. It is rather ironic that a field focused on making people "better" is ill equipped to make itself "better" (Blumenthal & Kilo, 2000; Solberg et. al., 2000). Health care delivery costs continue to increase; errors are common and result in 44,000 to 98,000 deaths per year; and the United States compares poorly to most Western countries in benchmark public health measures like infant mortality rates and death rates for coronary heart disease (Institute of Medicine, 2000; World Health Organization, 2003.)

For each of these shortcomings, examples exist within and outside of health care that the health care system can do better. Policymakers and administrators are aware of the gap between current and potential performance. The inability to make quality improvements suppresses attainment of superior efficiency, effectiveness, and clinical outcomes, and is viewed as one of health care's greatest challenges (IOM, 2000).

Quality improvement in organizations is performed by people. This makes the quality improvement discipline seem to be "an art form," hard to document, quantify, and replicate. To the contrary, there is evidence describing what factors make quality improvement and other types of organizational innovation more likely to succeed (Frambach & Schillewaert, 2002; Gustafson & Hundt, 1995). Moving beyond science to practice, in many industries an individual's ability to effectively lead quality improvement projects is a prerequisite for executive level promotions. Organizations within these industries have used quality improvement principles to set new levels of performance, and in doing so, have become the standard for others to follow. For example, quality improvement principles made Alcoa a dramatically safer place to work, Toyota the most reliable car, and so forth. (Consumer Report Buying Guide, 2003; Varley, 1994).

The relationship health care has with quality improvement is paradoxical. On one hand, quality improvement and related research practices are accepted as fundamentally important to the development of life-saving medical devices and pharmaceuticals. On the other hand, quality improvement has not been effectively used to improve the very processes that deliver these therapies. A critical element in bringing about any change, including a commitment to quality improvement, is the will to do so. Here we have a significant paradox. The traditional values of caring and curing would lead us to expect that health care organizations would do anything they could to improve the care they deliver. But organizational cultures, financial incentives, evolving regulatory requirements push the field in exactly the opposite direction. For example, no other field gets paid to fix their mistakes. However, a hospital can often make more money if a patient develops an infection caused by staff negligence while they are in the hospital, because the hospital gets paid for that patient's extra inpatient days. Hence, the health care system faces very powerful incentives not to improve the care they deliver and not to reduce the mistakes they make. These paradoxes impact quality improvement as well, for some of the historic efforts to improve the quality of care and services have impeded quality whereas others have not.

Understanding the residual impact key historic programs and initiates have had on quality improvement deepen our understanding of the cultural and structural barriers that will impede or facilitate future quality improvement efforts. Past experience as well as recent events offer four opportunities to facilitate the "improvement" of quality improvement. They are: (a) an understanding of how historical trends have affected health care culture and how these trends offer strengths, threats, misconceptions, and opportunities for future quality improvement practice; (b) the use of science to guide quality improvement practice and policy; (c) a quality improvement framework, built on evidence, that provides a common definition for quality improvement and can guide quality improvement practice, education, and research; and (d) taking greater advantage of human factors principles and procedures to guide how quality improvement performed and what it should achieve.

This chapter is organized to help the reader better understand how quality improvement is defined, followed by a discussion of each of the four opportunities. It concludes with projections of how quality improvement can be advanced into the future so patient needs are met, errors are reduced, and better health outcomes achieved.

QUALITY IMPROVEMENT: A DEFINITION

Quality improvement is a set of organizational processes and tools used to design and deliver products and services so customer and business priorities are simultaneously met and exceeded. For instance, this means that understanding, meeting, and exceeding customer needs will build consumer loyalty and increase organizational revenues; that reducing waste and rework and streamlining processes will decrease consumer as well as employee frustration and lead to increased productivity; and that error reduction will improve employee safety, customer safety, and avoid lawsuits. It also means that the definition of "customer" in the health field needs to be carefully understood; indeed, it can be argued that there are many customers. Certainly the patient comes to mind as the primary customer. But there are others, not the least of whom are (a) the physician, who frequently decides what care will be delivered and where a patient will go for care, and (b) the health care payer (e.g,. Medicare, HMO, etc.) because they should be concerned about the costs of care and in many cases limit access to care. In many cases, the interests of the physician, the HMO and the patient conflict, creating a dilemma for people interested in improving processes. However, one key customer is almost always ignored; the family. The family often determines whether a patient will recover and how fast, and they will determine how well a patient adheres to behavior changes regarding diet, smoking, and so forth. The family also will be strongly affected by the disease and its treatment. Yet it is rare to find a quality improvement initiative focused on helping the family meet their needs.

HEALTH CARE QI STRENGTHS, WEAKNESSES, AND MISNOMERS: AN HISTORICAL PERSPECTIVE

This section describes how historic quality improvement trends impact current practice (Table 42–1). In Donabedian's (1966) classic structure/process/outcome model, he described quality in health care as being composed of these three elements. This work effectively summarized pre–1966 health care quality improvement in thinking, policy, and practice and this model continues to have some relevance today.

Donabedian defined structure as the professional and organizational resources associated with clinical care, such as professional credentials and facility operating policies. Structure became part of health care culture early in the 20th century, beginning with the Flexner report of "Medical Education in the United States and Canada" (Flexner, 1910). The Flexner report was a response to substantial inconsistencies observed in physician education. This report forwarded a set of principles and practices that led to standardized medical school education, and in turn, that improved the quality of physician care. The strength of the Flexner report is in creating a legacy that certain professional requirements, or licensure, should be

TABLE 42–1. Quality Improvement in Health Care Chronology

Time Period	*Event*
Early 1900s	Flexner report (Flexner, 1910) Hospital Standardization Study(Codman, 1914)
1960s	Donabedian's Structure-Process-Outcomes Theory (Donabedian, 1966)
1970s	Quality by Inspection Movement (Berwick et al.,1990) Industrial Quality Movement (Deming, 1986)
1980s	National Demonstration Project (Berwicket al., 1990)
1990s	Total Quality Management (Shortell et al., 2000) Clinical Guidelines (Wennberg & Cooper, 1996)
2000s	Patient Safety Movement (IOM, 2000) Broad use of Human Factors Methods (Gosbee, 1999)

met before a physician is allowed to treat patients. Nursing, physical therapy, pharmacy, and most other health care professions have adopted the licensure concept since then. Conversely, a weakness arising from licensure requirements is that health professionals can be partial to their licensure area, and may not be receptive to issues beyond their specialty. This inhibits the teamwork needed to make improvements to patient care processes that span across job functions. During the same time period as the Flexner report, the Hospital Standardization Program was published, stressing the need for sanitary conditions and patient outcome measurement in hospitals.

Similar in impact to the Flexner report, the Hospital Standardization Program led to licensure requirements. These requirements held hospitals accountable to a set of facility requirements that reduced the spread of infection and enhanced patient safety. This program provided the foundation for the accrediting practices of the Joint Commission of Accreditation of Health Care Organizations (JCAHO). Since then, licensure requirements have begun to regulate other health care entities like outpatient clinics, home care agencies, and nursing homes. However, this licensure approach has also led to a misperception that to assure quality services are being provided, facilities need to be solely concerned about addressing issues such as whether or not all nurses are certified, all fire extinguishers are changed, medical records follow a prescribed documentation format, and so forth. This mindset ignores the role process plays in outcomes.

Donabedian defined process as the clinical actions done to and for patients by practitioners in the course of treatment. Prior to 1966, the typical method for quality improvement was the scientific method. This approach guides all scientific inquiry and is based on four steps; (a) observation and description of a phenomenon or group of phenomena; (b) formulation of an hypothesis to explain the phenomena; (c) use of the hypothesis to predict the existence of the phenomena; and (d) performance of experimental tests of the prediction by several independent experimenters using properly performed experiments (Bright, 1952).

The next advance in quality improvement was based on expanding the definition of process to include organizational as well as clinical processes. In 1914, Ernest Codman noted that organizational processes impacted clinical outcomes. This was a novel concept, for it was culturally accepted that clinicians delivered medical care, that the hospital or clinic was simply their workshop, and how that "workshop" operated had little significance on outcomes. It took nearly a century for Codman's notion to become more widely recognized and accepted. Two more recent key events that facilitated the appreciation of organizational processes were (a) the movement away from the quality by inspection philosophy and (b) the National Demonstration Project (Berwick, Godfrey, & Roessner, 1990).

In the 1970s and into the 1980s, groups of regulators responsible for assuring quality focused on finding the individuals who were performing poorly; the "bad apples." These entities were called peer review organizations. As health care was learning lessons from industrial quality improvement efforts, regulatory review processes began to emerge that focused the blame for errors on process inadequacies rather than on the individual. The National Demonstration Project, led by Don

TABLE 42–2. TQM Concepts[a]

Constancy of purpose implies that organization's leadership has developed a set of key long-term strategic priorities and their actions continually emphasize and do not contradict these priorities. Thus, the people in the organization know why the organization exists and in turn, operate within the bounds of the fundamental mission (Deming, 1986).
Customer mindedness calls for a continued commitment to understanding the needs and expectations of purchasers/users of products and/or services produced by an organization.
Quality mindedness defines quality as the design and production of products and services that meet customer needs.
Process mindedness means improving products and services and occurs by improving processes and not identifying and blaming people.
Statistical mindedness means that data should guide decisionmaking and evaluate the effectiveness of a quality improvement.
Employee mindedness argues that employees who are directly involved with the process are highly qualified to make suggestions on how to improve it and their buy-in to an improvement will impact the sustainability of the change.
Management and leadership indicate successful improvements will only occur when leadership provides clear expectations, stays committed to these expectations, and supports these expectations by providing resources and removing barriers.

[a] Adapted from Gustafson & Hundt, 1995 & Batalden & Nolan, 1994)

Berwick, M.D., paired a health care provider with a quality improvement expert from industry to use industrial quality improvement methods and tools to improve processes in several hospitals. This project's success led to a gradual increase in the use of Total Quality Management (TQM) in health care during the early 1990s. As time went on, process improvement began to gain acceptance in regulatory and educational arms of health care. A key accreditation body, the Joint Commission on Accreditation of Healthcare Organizations (JCAHO) incorporated many of these concepts into their regulatory requirements. Although TQM as a strict methodology faded over time, certain aspects continue as important foundations for process improvement in health care today (Table 42–2). The concepts can still be found in accreditation criteria and some organizations continued to apply TQM principles and techniques. The primary reason for the fading of the TQM movement was that it did not have a greater impact on clinical and business results.

After TQM was introduced, three subsequent major developments have influenced how health care quality improvement is approached and applied; the use of clinical guidelines; the patient safety movement; and broader use of human factors.

Movement #1 (Clinical Guidelines). In 1996, John Wennberg, MD released research that demonstrated wide variance in physician clinical practice patterns based solely on where patients were located geographically. For instance, he demonstrated how an individual in one zip code was five times more likely to have a knee replacement than an individual with similar demographic characteristics in a zip code located just 20 miles away. This evidence also suggested that provider practice patterns were significantly influenced by where they were trained and the practices of their peers. These practice patterns at times ignored reigning clinical evidence. In response to these findings, a clinical standardization movement began that emphasized use of clinical guidelines (or clinical care maps). These guidelines outlined the steps that should be followed in treating a given disease. Leading professional bodies such as the Agency for Health Care Quality and the American Medical Association published practice guidelines. Within individual organizations, guidelines were developed as well. These guidelines often had little impact on how care was actually delivered—highlighting the difficulty in "legislating" standardized processes into clinical decision making. Yet, standardization is often a central element in industrial improvement practices, because it standardizes practices and limits variation. Although application of guidelines is slowly gaining influence, the lack of standardization of key processes represents a challenge and opportunity into the future.

In the early 1990s, evidence began to emerge that patient safety was a problem. In a study of two large Boston teaching hospitals, patient injuries resulting from errors in dispensing medication were experienced by 2% of hospital patients (Brennan et. al., 1991). The Institute of Medicine's 1999 report, "To Err is Human," shocked the nation with estimates that mistakes and system failures in medicine result in 44,000 to 98,000 deaths in hospitals each year. This ranked medical mistakes among the leading causes of death in the United States (IOM, 2000). Overnight, the IOM report made patient safety a major health policy issue. The leading cause of these errors was system (or process) failures.

Preventing harmful errors forced health care to address two issues in a meaningful way for the first time. The first was the prevention of low frequency, but catastrophic events. The second was that medical care was becoming progressively more reliant on technology, which was also contributing to medication errors due to failed interactions between people and machines. Both issues called upon the discipline of human factors for insights and solutions.

THE FOCUS ON HUMAN FACTORS

A historic weakness of health care is that performance successes and failures are assumed to be due primarily to provider skill. This is a dangerous precedent in today's technologically dependent health care environment. Slowly, health care is recognizing that interface between human and machine may be the root cause of many quality errors, and has looked to the human factors field for answers. James Reason, a human factor's pioneer, who focused on human errors and safety, advanced the study of the interaction between people and mechanical equipment by examining the factors that affect the outcome of that contact. Although new to health care, the study of low frequency, but severe errors that occur between human and machine had been successfully addressed elsewhere. For example, analysis of World War II Air Corps (today's Air Force) accidents discovered that poorly designed cockpits caused pilots to make mistakes. They were highly trained personnel, but the design and layout of controls did not take into account the effects of intense stress on their perception and cognition. The analysis led to cockpit design improvements, created a new era of dramatically safer airplanes, and served as an important initial step in human factors science. Since then, the human factors field has advanced considerably and guides safety efforts in manufacturing, food processing, health care, and many other industries (Gosbee, 1999).

The human factors field has a newly developed set of tools intended to prevent failures by identifying hazards and then developing countermeasures to reduce their risk. The hazards can be purely mechanical in nature (e.g., a malfunctioning computer that causes lethal amounts of an intravenous drip), but most failures in medicine are due to the actions of humans, because few health care systems are automated. This is why human factors tools have begun to experience widespread implementation in health care. As with other improvement approaches, the goal of hazard identification is to redesign the system, not to find "bad apples." Some tools of hazard identification are described later. For additional information on some of these tools, see the chapters on "Human Factors Risk Management in Medical Products" and on "Incident Analysis in Health Care" in this handbook.

Failure mode effect analysis (FMEA) was developed in the U.S. military in 1949 to determine effects of system and equipment failures (Military Procedure MIL-P-1629). The output of an FMEA analysis is to identify failures, their causes, their effects, and then identify actions to mitigate the failures. A crucial step in FMEA is anticipating what can go wrong in a system. Minierrors or latent errors are those errors that make the larger, more detrimental error more likely. In a FMEA analysis, the latent errors are identified and corrected so the "chain of errors" is prevented.

Hazard analysis and critical control points (HACCP) is a systematic methodology for the identification and control of hazards. The approach, developed by the National Advisory Committee on Microbiological Criteria for Foods (1997), has been almost exclusively used by the food industry, but is applicable to manufacturing, distribution, and use of any product or service. A "critical control point" is a point, step, or procedure at which control can be exercised to prevent, minimize, or eliminate a hazard. Seven principles form the core of the HACCP method (National Advisory Committee on Microbiological Criteria for Foods; 1997):

1. Conduct a hazard analysis by preparing a list of steps in a process where hazards that are likely to cause injury, illness, or some other hazard if they are not effectively controlled.
2. Determine critical control points (CCP) at which control can be applied and that

is essential to prevent or eliminate a hazard or reduce it to an acceptable level.

3. Establish critical limits that are maximum and/or minimum value at which a parameter must be controlled to prevent, eliminate, or reduce to an acceptable level the occurrence of a hazard.
4. Establish monitoring procedurse to assess whether a CCP is under control and to produce an accurate record for future use in verification.
5. Establish corrective actions to be taken when a critical limit deviation occurs.
6. Establish verification procedures to determine the validity of the HACCP plan and to verify that the system is operating according to plan.
7. Establish record-keeping procedures to document the HACCP system.

Root cause analysis (RCA) is the hazard analysis approach mandated by the Joint Commission on Accreditation of Healthcare Organizations. In 1997, the JCAHO mandated the use of RCA in the investigation of sentinel events in its accredited hospitals. RCA is a qualitative retrospective approach to error analysis widely applied to major industrial accidents. RCA searches out hazards or latent failures that may have caused the accidents.

Probability risk assessment (PRA) estimates the risk of failure of error occurring. Proponents to this approach prefer its ability to model multiple causes of error simultaneously. A weakness in the FMEA and HACCP approaches is that they tend to focus on one hazard at a time, and fail to recognize combinations of failures that could lead to a catastrophic event. Subjective risk assessments have been used to develop predictive models of the uncertainty of medication adherence, organizational success, and health care costs (Gustafson, Cats-Beril, & Alemi, 1992; Molfenter, Gustafson, Kilo, Bhattachrays, & Olsen, 2005). These models allow for multiple attributes to be simultaneously modeled in order to predict the probability of success and failure. Because these models use subjective data, obtained by experts, they provide a practical approach to create valid risk assessment models in the work setting (Gustafson et. al., 1992). The introduction of hierarchical modeling, using fault and event trees, gives risk assessment models the added feature of modeling a series of events. This is important because catastrophic events are not a result of a single hazard but a succession of hazards. The fault and event tree graphically illustrates risk and reliability; the assignment of probabilities allows process owners to establish thresholds of risk and to specify safety and reliability standards (National Research Council, 1994).

The hazard analysis approach relies on steps typically found in quality improvement models to reduce errors. Their uniqueness resides in their ability to identify hazards and root causes of errors as part of the quality improvement process. These hazard identification approaches represent a recent historic trend in health care, and optimistically will aid in the reduction of medical errors.

ORGANIZATIONAL PROCESS INNOVATION MODEL

The history of quality improvement offers an assortment of perspectives on how it can be approached. They range from the structural components of the Flexner report in the early 1900s, to the more in-depth process analysis required in hazard analysis. This section introduces an "organizational process innovation model" that integrates past and current quality improvement theory and practice into an integrated framework. The framework has three general areas; identification of needs and opportunities; process innovation; and sustainability.

Identification of needs and opportunities was chosen because it is present in most of the historic movements. For instance, TQM promoted the use of customer needs to identify opportunities, and the patient safety movement used hazard identification to uncover opportunities. Donabedian (1966) noted that process innovations were at the core of quality improvements. Many historic movements agree, but steps emphasized in pursuing process improvement vary by movement. An evidence-based approach to process innovation will be suggested, based on the research and operational findings from quality improvement and organizational innovation. There is rising concern that organizational researchers are noting once a process improvement is made, there is no guarantee that the improvements will be sustained—in fact, 20+% of process improvements fail to be sustained (Molfenter et. al.,; Øvretveit et. al., 2002). Sustainability is the last part of the organizational process innovation model and for all steps, tips, insights, and when available evidence will be offered.

Identification of Needs and Opportunities

Quality improvement is built upon projects. The project can be delivered by an individual or by a team, but there is a series of tasks that results in a process innovation. Identified projects should meet two critical criteria: (a) "Will the project have strategic alignment with or address a key an organizational priority?" and (b) "Will the project meet a key customer need or requirement?" Strategic alignment between an innovation and the eventual success of that innovation has been documented across research studies (Gustafson & Hundt, 1995). This is an absolutely vital step where a key executive notes the strategic importance of a quality improvement project, and then champions the project by communicating its merits to the board of directors, key customers, workforce, and so forth. The TQM movement and general quality improvement theory advance customer needs and requirements as the single most important factor in guiding an improvement process. Customer needs and requirements can be discovered through a variety of techniques:

- **Spontaneous reports** of customer needs learned by front-line staff during customary transactions. When these reports are valued and used, they become a rich source of customer needs identification and foster a culture focused on the customer. Developing structural mechanisms to capture these reports and reward staff for providing these reports reinforces a customer-focused culture.
- **Focus groups** of customers will uncover customer needs and provide the valuable opportunity for a trained facilitator to conduct an in-depth analysis of customer motivators and their reactions to specific product or service attributes.
- **Interviews** with customers can provide a detailed review of a product or services as well, and tend to be more easily organized than focus groups due to the flexibility of scheduling individual interviews versus coordinating a group of individuals.
- **Critical incident technique** is an interview technique used to identify behaviors or events that contribute to the success or failure of a product or service (Flanagan, 1954). In health care, "this technique involves asking patients to think back to specific stages in their disease experience and describe events that stand out in their mind. For instance, a patient who had a heart attack is asked to think back to specific events in the attack and remember specific experiences during the diagnostic treatments, recovery, and behavior change experiences. The patient is asked to describe things, good or bad, that stand out about each stage." (Gustafson, Arora, Nelson, & Boberg, 2001, p. 83).
- **In customer shopping or service walkthroughs**, an organization's staff member "walks in the customer's shoes." Acting as customer, they can directly see and feel the customer's experience. Directly experiencing a service error can be enlightening and leave a lasting impression. This exercise can be even more informative if the staff member(s) conducting the walkthrough asks the staff members for their improvement recommendations. An example of a walkthrough exercise is documented in Figure 42–1 and describes a client trying to gain access for alcohol addiction treatment.

An important first step many organizations fail to take is to build a culture that values understanding customer needs and requirements. In such cultures, these tools are not solely reserved for special projects; instead, they become part of daily work.

Process Innovation

Determining how to best approach process innovation can become quite a complicated task, because a wide variety of attributes have been found to influence innovation, including:

- External traits such as competitive pressures and industry trends.
- Organization traits such as size, structure, strategy, resources and culture.
- Work group traits such as team structure, team climate, and team member characteristics.

Notes From a Walkthrough

(based at Brandywine Services, Wilmington, Delaware)

The walkthrough exercise was conducted by having one staff member posed as a client attempting to enter methadone treatment for heroin addiction. This employee was selected because she is a former client with 3 years in recovery. A second staff member posed as a family member and observed the process. Walkthrough observations fell into two categories: utilization, and access/retention. Regarding utilization, Brandywine Counseling (BC) has an estimated 40–50 available treatment slots, and yet turned away 12 walk-ins who sought treatment on Day 1 of the simulation. We believe we have created "artificial barriers," such as offering only 12 intake slots per week. Our second major observation is that there are significant barriers to access to, and retention in, our program. The first contact our client had with the agency when calling for an appointment was negative and lacking dignity. Furthermore, our existing triage process for walk-ins does not adequately address the needs they present at that time, nor properly identify which should have priority for an intake slot. Finally, the early morning timing and nearly 6-hour length of our current intake process represent a major barrier to engaging the client.

Figure 42–1. Walkthrough exercise

- Individual traits such as personality, motivation, cognitive ability, and job characteristics, which have been found to influence organizational innovation (Anderson, De Dreu, & Nijstad, 2004).

A series of classic innovation studies investigated the factors that facilitate successful innovation adoption. A meta-analysis of these studies found five factors that consistently influenced successful innovation adoption (Gustafson & Hundt, 1995). These studies investigated 80 factors, 640 companies, and 13 industries.

The five factors that can provide focus and direction to quality improvement efforts are constancy of purpose; customer focus; external influence; leadership; and quality of the innovation. These factors have been supported in more recent reviews of the innovation literature (Frambauch & Schillewart, 2004; Montoya-Weiss & Calantone, 1994; Rogers, 1995). The meta-analysis of these classic studies described each project as successful or unsuccessful, so ratios of success to failure could be calculated.

For projects that possessed **constancy of purpose**, the ratio of success to failure ranged from 3:1 to 1:1. The studies identified successful projects that align with a pressing strategic issue are more likely to obtain needed organizational resources and to be successful.

The principle "customer-mindedness" was overwhelmingly supported by all of the studies that examined it. Understanding customer needs had a favorable success to failure ratio ranging from 33:1 to 1.4:1. These results emphasized the need of knowing customer requirements. Similarly, the central finding of the Montoya-Weiss and Calantrone (1994) meta-analysis of innovation found many new innovations failed either because customer needs were not fulfilled, or the innovation did not provide a superior alternative for the customer. In health care, the customer extends beyond the patient; there is a need for an emphasis on customer–supplier relationships between providers, so the patient experiences a smooth transition between treatment settings.

The research also found organizations need external ideas and pressures to catalyze innovation. Innovations influenced by **external ideas** have ratios of success to failure of 6:1 to 5:1. Specifically, getting ideas for specific projects via peer networking was advantageous to innovation success.

The **quality of the innovation** or the ability of the innovation to be "bug-free," results in few user complaints on initial use and this facilitates user acceptance and improved outcomes. The ratio of success to failure for "bug-free" versus "complaint-prone" was 3:1 to 1.45:1. One key vehicle for reducing bugs is extensive pilot testing of an innovation before it is implemented. A recent formulation of pilot testing is referred to as "rapid cycle improvement," where changes are tested on a very small scale, problems identified and quickly corrected, and tested again and again until the bugs have been eliminated (Langley, Nolan, Nolan, Norman, & Provost, 2005).

Senior leadership was attributed to innovation success, but not directly. It seemed that senior leaders could not dedicate the direct time needed for project success, but needed to be clearly supportive of the innovation project. Their ability to demonstrate and delegate this support was what directly influenced the project. For instance, the senior leader who assigns a change leader who has power, prestige, direct access to senior leadership, and who will act in accordance to the senior leader's wishes is increasing the likelihood of innovation success. Or, the senior leader who (a) emphasizes constancy of purpose, (b) holds organizational staff accountable for assuring that innovations meet customer needs, (c) incorporates external ideas in the innovation process, and (d) ensures the innovation is "bug-free" is more likely to have successful innovations.

Van de Ven (1980) found innovation success was increased when a structured step-by-step model was used. Two types of models have been popular in quality improvement applications; planning models and experimentation models. Which model type to use can be controversial and is a decision for organizations embarking on a quality improvement project.

The TQM movement emphasized use of planning models. These models were characterized by intensive analysis to diagnose the causes of the current problem, a review of internal and external customer needs, a detailed implementation plan, and implementation of a detailed one-time solution. The planning portion of these models took anywhere from one to 12 months to develop, then the solution was implemented. In cases of new system design, this elaborate process seemed to be warranted (Van de Ven, 1980). In cases of process redesign, planning models were criticized for "taking too long" and being too labor intensive. Additionally, improvement to outcomes from these models often was negligible (Blumenthal & Kilo, 1998; Shortell et. al., 2000; Solberg et. al, 2000). A core element in the implementation portion of many of the planning models was the one-time use of Shewhart's plan-do-study-zct (PDSA) model (Van Nostrand, 1931).

In the mid–1990s, a revision was made to how PDSA cycles were applied that led to onset of "experimentation models." The experimentation models made the PDSA cycle the central priority as opposed to diagnosing the problem. With these models, small-scale PDSA experiments are rapidly conducted and the experiments that prove beneficial in the "study phase" are expanded upon in the "act phase (Table 42–3)." As mentioned earlier, these models are often called "rapid cycle" PDSA models (Langley et. al., 2005).

The experimentation models have yielded better improvement results due to the flexibility that the rapid cycle PDSA approach affords, and the results orientation the "aim statement" causes (Clemmer, Spuhler, Oniki, & Horn, 1999; Cretin, Shortell, & Keeler, 2004; Nolan, Schall, Berwick, & Roessner, 1996). Although the rapid cycle PDSA models have demonstrated effectiveness in process redesign, their results have not been as prevalent in new process design. These models are also ill-prepared to address patient safety issues because they are ineffective at hazard identification.

There is not a "best" model, but there seem to be circumstances in which one model type may be preferred over another. A constant need across all models is the use of a team that includes representation from all disciplines that impact the system being designed or improved. A quality improvement decision tree has been formulated to aid model selection decision making (Figure 42–2). If a new process needs to be designed or the existing model is so flawed that an improvement in it would not work, the planning model is recommended because new processes possess more uncertainty and require infrastructure development.

Planning models reduce uncertainty by allowing more time than experimental models for coalition

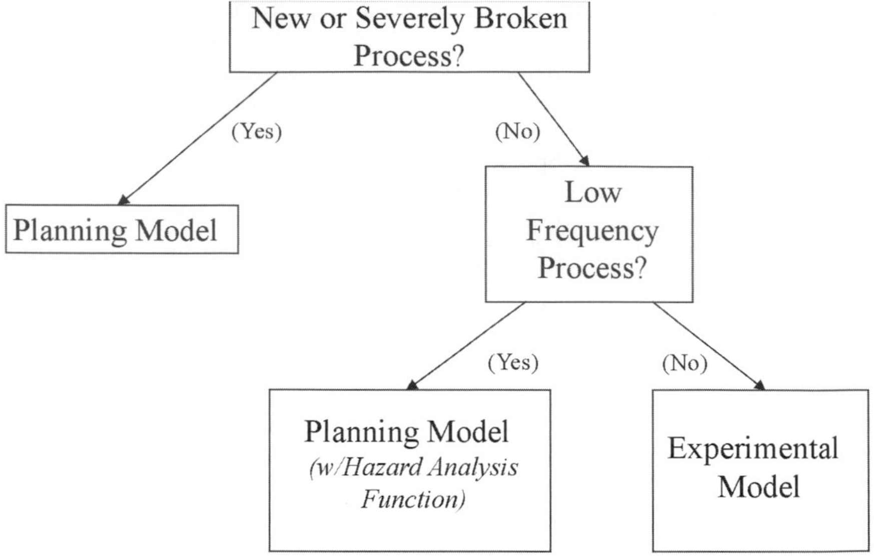

Figure 42–2. Quality improvement decision tree.

building among stakeholders and for understanding customer needs. Experimental models are not recommended for new processes because the investment of time and resources to build the new infrastructure is considerable and the "cost" of a failed experiment is greater. Thus, the resources spent on planning to increase the likelihood of success is justified. If the quality improvement project is addressing a low frequency problem in an existing process, a planning model that includes a hazard analysis should be used (see the earlier "Focus on Human Factors" section for examples of hazard analysis techniques).

The planning model is recommended because low frequency issues are hard to uncover and the attention that planning models dedicate to problem identification is warranted. Additionally, experimental models rely on series of trials and errors to eventually arrive at solutions and low frequency problems do not allow for this repetition of experiments to occur. This is why hazard analysis procedures try to uncover latent errors that lead to catastrophic errors. Latent errors have higher frequency and are amenable to experimental models. Experimental models, and their trial and learning approach on a small scale, are amenable to addressing high frequency problems. The majority of process problems involve high frequency events; this led to the popular use and success of these models over the past decade (Leape, Kabcenell, Berwick, & Roessner, 1998; Nolan et. al., 1996; Weiss, Mendoza, Schall, Berwick, & Roessner, 1997).

Sustainability

Those studying organizational change are realizing what researchers and practitioners of individual change have noted for some time: Adoption of a change does not always lead to sustained use of that change. Two well-studied examples of this phenomenon in individual change behavior are poor sustainability rates with smoking cessation and alcohol and other drug addiction recovery (McLellan, O'Brien, Lewis, & Kleber, 2000; Ockene et. al., 2000;). Rogers (1995) noted that the sustained use of innovations in both individual and organizational settings is a distinct and important step in the diffusion of innovations. Yet, the emphasis on sustainability in organizational settings has been limited and is beginning to emerge as a point of interest in settings like the British National Health Service (http://www.modern.nhs.uk/ improvementguides/sustainability/web page link).

Whether or not an organizational change will be sustained depends on many factors. A panel of theoreticians and practitioners, with representation from business and health care management, family therapy, economics and education, was used to develop a predictive model of sustainability (Molfenter, Bhattacharya, Ford, & Wen, 2005). The model was then tested with 81 individuals experienced with organizational change using a set of hypothetical profiles that included a series of the model's factors. The expert panel rated the probability of success of the profile. The model explained 58% of the variance in the expert panel's profile ratings. The model's nine factors are:

1. Adaptability (so the change can easily accept adjustments from all levels of the organization).
2. Change reversibility (so the nature of the change makes it impossible to revert back to the old system).
3. Champion turnover and new staff development.
4. Ongoing leadership.
5. Political environment (so there is support between departments for maintaining the change).
6. Staff motivation.
7. Resources devoted to change.
8. Evidence of effectiveness.
9. External pressure (regulatory agencies, community leaders, or competitors).

Assuring these sustainability traits will help prevent innovation success from declining. Specific interventions that have been suggested for improving sustainability are:

- **"Signaling" systems**. This has been the fundamental premise behind use of quality control. If a system is continually monitored, sustainability of low levels of variance can be achieved by noting and fixing systems that begin to show variance. The use of quality control is popular on the production line but can be used in all processes. For example, a clinic that wants to assure their addiction treatment clients will see a provider within 48 hours of request for service can track the percentage of patients who do meet the performance expectation. When the percentage passes a certain threshold, process innovations are made to once again meet the performance expectation. A desired goal of these systems is process standardization.
- **Sustainability leaders**. These leaders monitor the "signaling" systems and when sustainability is jeopardized, they are empowered and expected to immediately take the actions needed to assure sustainability.
- **Ease of use**. Venkatesh (2000) noted that ease of use was one of the key indicators of technology acceptance. Technology acceptance is beginning to be hypothesized, at the individual level, to be associated with sustainability (Brassington, Atienza, Perczek, DiLorenzo, & King, 2002). Aspiring to ease of use proposes that systems are designed to be as simple as possible for the user by reducing steps to complete, the need for additional training, and reliance on memory. For additional information, see the chapter on "Usability Evaluation in Health Care" in this handbook.
- **Clear benefits**. The perceived advantage of a new process innovation over the competing alternatives has been central to behavioral change models (Salovey Rothman, & Rodin, 1998; Weinstein, 1993) and to diffusion of innovation theory (Rogers, 1995). Hence, the ability of management and staff to view a process innovation as useful, both initially and over time, will determine the sustainability of a change. And part of a signaling system should be collecting data on how process innovations are being received by staff.

In conclusion, quality improvement success does not stop with sustainability. Overall success resides in renewal, which is not a step in the model, but an occurrence when the identification of needs and opportunities begins anew for a given process.

THE FUTURE (OPPORTUNITY AND NEEDS)

Evolving over the last century, quality improvement has improved, but some challenges remain. With challenge there is opportunity. In 1995, Gustafson and Hundt reported that 18% of attempted quality improvement projects were successful. Similarly,

Shortell, et. al. (2000) published research that TQM had had minimal impact. More recently, quality improvement projects are documenting success rates of 50% or more (Cretin et. al., 2004). What has been learned about process improvement has advanced the field, but a leading health systems researcher has noted: "quality improvement can be viewed as a glass of water that is half-empty or half-full" (S. Shortell, personal communication, August 12, 2002). By following quality improvement's historical progression, the glass is half full. Advances have improved the quality of practitioners and treatment facilities; process has been identified as a source of errors as well as of consumer dissatisfaction; and outcomes are improving as evidenced by increased life expectancy and reduced mortality.

However, the glass remains half empty, for many quality deficiencies still exist due to the high rate of medication errors, poor coordination between care providers, increasing health care costs, and unmet customer needs. Pulling against these needs is a health care system that is becoming less personal and more disjointed due to specialization and billing practices. How can the glass become full? How can the health care system be better prepared to cure itself? A series of suggestions are being offered to address current concerns and anticipated future needs.

Suggestion #1: Place the Patient at the Center of Health Care Delivery System Design, Product Development, and Quality Improvement

The fundamental precept of quality improvement is a customer focus and, for health care, like many other fields, the customer will provide the insights needed to guide dominant innovations. The nature of health care is changing from an emphasis on treatment of acute health care episodes to an emphasis on prevention of chronic disease exacerbations. For example, if heart disease is effectively treated, heart attacks can be prevented. Major disease categories like heart disease, diabetes, and asthma are chronic diseases, are the leading causes of death and disease prevalence, and can most effectively be treated through prevention (CDC, 2004).

Our current health care system is not designed to prevent and treat chronic disease. This shortcoming will become even more prominent with the onset of health risk detection technology. Genetic tests and diagnostic screening tools will detect at young ages, sometimes before birth, health risks that should be addressed to avoid health problems. For instance, a teenager could be informed they are susceptible to heart disease and need to adhere to a low-fat diet in order to reduce their risk of heart disease and premature death. The apparent challenge is getting the teenage to steer clear of fast foods and high fat dietary options. Putting the 15-year-old at the center of the health care system may not guarantee low-fat intake for all high-risk teenagers, but the chances that the health care system can address their needs will be greatly increased with the 15-year-old's insight, guidance, and a simple understanding of how they view the world.

Lack of time to include patient perspectives, possibly provider arrogance, or simply the strong conviction that the provider knows what is right for the patient have all led to a provider-focused health care system. This focus was more appropriate when the providers were administering most of the treatment—as the acute care model required. With the chronic model, the patient and family play a more central role. How they live and how they navigate the health care system will have a pronounced impact on their health. This is not suggesting that providers will no longer have an important role in health care delivery; but instead, they will have a different role. They will continue to be vital to disease diagnosis and treatment, but their role will need to change.

Gustafson, et. al. (2001) conducted a study of the needs of heart attack patients treated in a hospital and found two results. First, the typical needs evaluated in provider-designed patient satisfaction surveys generated a general sense of the satisfaction of the patient's hospital experience, but did not assess the needs the patients identified as most important. Second, the needs patients identified as most important were informational and emotional support. Coincidentally, clinical quality, the focus of the provider surveys, often is simply assumed to be present by the patient and is not the focus of their assessment of the health care experience.

In the future, the ability of the patient to gain provider diagnosis and treatment advice, while having their information and support needs addressed, will be the path to better health in a chronic disease environment. This model must understand patient needs, how patients live, who provides social support—and then integrate these needs and lifestyles into a health care delivery system that sees itself as providing mentoring as well as expertise. It will require experts who can recommend and interpret evidence-based tests, as well as mentors who can integrate patient feedback and test results into a treatment approach that is patient-driven and does not rely on office visits for continued support and success. Future health care design must consider the patient as part of, not the recipient of, the health care delivery process.

Suggestion #2: Use Technology to Meet Patient Needs

An avenue for health care to deliver greater connectivity between provider and patient, while increasing informational and emotional support, is through interactive computer technology often called eHealth systems. These systems are not dependent on provider schedules, and can be accessed at the patient's convenience 7 days a week, 24 hours per day. Application of these systems in coping with severe illness suggests that they can have substantial effects on quality of life and use of health systems (Gustafson et. al., 1998).

One key feature of an eHealth system is electronic discussion groups. These forms of "e-community" offer emotional support that can be critical to effectively coping with disease (McTavish, Pingree, Hawkins, & Gustafson, 2003). eHealth can also provide decision support. Many disease diagnoses offer multiple treatment modalities from which the patient is asked to choose. It is often difficult to make these decisions in a 15-min office appointment without the input of family and friends. eHealth systems both provides the information needed to make an informed decision, and allows the individual the time needed to make the decision. However, human factors should play a much greater role in the design of eHealth systems. Many of the potential users are older with limited vision and eye/hand coordination. They need interfaces and technology that can help them make the most effective use of the content that could be available to them.

Another type of technology that will have a great impact on future health care delivery is "enabling technology" that will assist an individual in meeting their health care goals. For example, the 15-year-old with the potential for heart disease may have a palm pilot where she can list the foods she is eating as well as the serving size. When she is exceeding the "fat allotment" for a meal, the palm pilot will sound an alarm. Although such a solution is theoretically possible, it is not practical because of the effort involved. However, a much less labor-intensive technology would be a "fat sensor" implanted in the stomach lining that would sound an alarm on the palm pilot when the fat allotment had been achieved. Interactive and enabling technology will serve as tools to engage patients in the provision of their care.

Suggestion #3: Use Human Factors Techniques to Develop, Test, and Evaluate New and Existing Technology

Opportunities for human factors engineering challenges will become greater with the increased use of technology to meet customer needs. As mentioned earlier, new technology needs to be easy to use and not create new hazards for the patients as well as for the providers. These are challenges human factors are capable of addressing, but human factors in health care has been underutilized. The human factors field has a history of determining user needs and then designing technology to meet those needs.

One attribute associated with successful innovation is using pilot tests to assure that the quality of the innovation is high and that errors and breakdowns are minimized (Utterback, 1975). Pilot tests are a way to conduct usability analyses, ergonomic studies, begin to examine hazard analysis, and determine the user's acceptance of the technology. Acceptance, as measured by the technology acceptance model (TAM), has been shown to be a link between an initial trial and continued use of a technology (Venkatesh, 2000). This model measures relative benefits and perceived ease of use. This model, as well other types of human factors tools, can increase patient use rates as well as providers' adoption of interactive technology. The latter is particularly salient, because lack of provider acceptance has been a significant barrier to the adoption of interactive technology (Molfenter, Johnson, Gustafson, DeVries, & Veeramani, 2002). For additional discussion of human factors of technology implementation in health care, see the chapter on "New Technology Implementation in Health Care" in this handbook.

Pilot tests can also be used to assess the physical stress the technology will have on the individual through usability and ergonomics testing. These tests can inform the redesign of products and systems to facilitate ease of use and to reduce worker injury rates. From a broader perspective, the usefulness of human factors techniques to reduce worker injury extends beyond technology assessment, for worker injury rates are very high in health care. Applying these techniques to reduce injuries common to patient lifting could serve to reduce injury rates, decrease the worker compensation cost burden on the medical system, increase productivity, and reduce the effects of labor shortages of nurses and other health care professionals whose supply may be less than the demand. For additional information, see the chapters on "Evidence-Based Strategies for Safe Patient Handling and Movement" and "Physical Ergonomics in Health care" in this handbook.

The use of human factor tools in health care is needed for technology acceptance as well as safety enhancement. Thus, individuals should be trained on how to apply human factors techniques, but more importantly, the demonstrated effectiveness of these techniques should be documented for others to follow and also to enhance their credibility and acceptance.

Suggestion #4: Mentor and Value Leaders of Change

The pace in the current health care environment is hectic; daily crises distract from focused work on improvement. Under these circumstances, improvement must rely on leadership. Two leadership roles associated with organizational improvement are that of an executive sponsor and a change leader. The executive sponsor champions the project and the change leader delivers the results.

There are competencies that can make change leaders and executive sponsors more effective at improvement. Fifty-four health care leaders were organized to select executive competencies they needed to possess and execute in order to increase the likelihood of a successful quality improvement project. The attributes they selected were the ability to:

- Strategically align the improvement project with a pressing organizational goal so that the project becomes an integral part of meeting this goal and is not perceived as a barrier.
- Engage an enthusiastic and influential clinical champion—preferably a physician.
- Create the expectation that change will occur.
- Assure the systems are in place to keep the senior leader informed.
- Allocate necessary resources.
- Assure an improvement method or model is used.
- Reward results.

Most of the selected executive sponsor attributes relate to organizational management skills. Change leaders need both organizational management skills and technical skills. The literature suggests that organizational management attributes associated with success are the ability to motivate others, to manage relationships, to set priorities, to manage conflict, to listen, to inform, and to achieve results (Lombardo & Eichinger, 2002).

Technical quality improvement and human factors skills are vital as well. Change leaders need an opportunity to learn and refine their organizational and technical skills simultaneously. Six-sigma training and other types of leadership development programs are being used in many organizations to accomplish this task. These organizations recognize that the ability to effectively manage change is critical to organizational success and that leaders must have organizational as well as technical skills.

These approaches are being used to develop and select management through a training approach that integrates education with application. The training approach, often in 6 sigma programs, has three phases; learn it, do it, and then teach it. Hence, in a training session, the individual learns techniques and skills. In the application section, they are then asked to apply what they have learned by making a successful change in their work area over the next 2–4 months. Once individuals become skilled at making successful changes routinely, they begin to teach others these techniques. This methodology can be used to create the quality improvement and human factors capital in health care as it has in industry.

Suggestion #5: Building the Business Case

The downfall of the popular TQM movement was because TQM-related projects failed to significantly impact the organization's strategic and financial priorities. In industry, although there was a frustration with TQM impact, the need to improve quality continued to be vital due to competitive pressures to deliver high quality products. The answer to this dilemma emerged through the 6 sigma movement. Six-sigma initially became popular through applications at General Electric and Motorola and has since spread throughout the manufacturing industry (Pande & Holpp, 2002). Six-sigma applications continued to use tools and techniques common to TQM, but possessed four critical differences: (a) Quality projects are related to strategic priorities; (b) software has emerged that simplifies the use the powerful statistical tool capable of uncovering "root causes" of quality errors; (c) a focus of quality projects is on achieving 6-sigma variation in product delivery (this means achieving 3–4 errors in every product made); and (d) linking education in quality improvement techniques with application of those techniques, and the expectation that the techniques will be used to deliver business results (described in suggestion #4; Barney & McCarty; 2003).

For quality improvement to become a central business strategy in health care, there needs to be a clear business case as evidenced by linkage to key strategic priorities. Unfortunately, quality improvement is often viewed as something that is nice to do, as something we "should" do, but only after we complete the tasks necessary to "keep the doors open," meet quarterly objectives, and so forth. So the organization should address the important question: "Is quality improvement worth the investment of resources, time, and opportunity costs?"

Improvement is not cost-free; resources need to be dedicated. There are minimum training costs, and time is usually the greatest cost in improvement. The time required for individuals to focus on how processes are performing, to test and measure the impact of the changes, and then to standardize improvement is time intensive and results in an opportunity cost; a cost of what individuals could be doing instead of the quality improvement work. It is the doctor who could be seeing patients, the administrator who could be developing a new business line, or the coverage for a clerk to participate in a change project. A skill or trait that needs to be part of all quality improvement projects is to conduct a cost-benefit analysis that reflects the costs and benefits of a quality improvement project. Perhaps more important, for a business case to be likely, all quality improvement projects should be linked to a key organizational goal.

Toyota has strategically aligned quality improvement projects with a key business priority; reliability. They want to sell reliable products because this will create brand loyalty and allow them to charge higher prices that lead to greater profit margins. Therefore, one in eight Toyota employees is solely dedicated to making improvements. They walk the production line, look for improvement opportunities and side-by-side with other employees, make changes to increase reliability and improve worker safety.

There are no Toyotas in health care. Many of the weaknesses of historic quality improvement practices, such as heavy focus on structure versus process, lack of coordination between levels of care, poor customer need identification systems, and insufficient use of human factors engineering techniques still exist. However, this could change by following some of the suggestions made in this chapter.

Returning to the Flexner report referenced earlier: What made it successful? This report was successful because there was a business case. There was a cost-benefit for medical schools to standardize their medical school curriculum, for those that did not make this investment were avoided by students and eventually had to close. Will history repeat itself? Will those organizations and individuals who cannot make the improvements desired by patients and the public eventually fail, like the medical schools that did not follow the recommendations of the Flexner report? Only the history that is currently unfolding will yield the answer. But, it is easy to envision that those organizations successfully making rapid innovations, meeting customer needs, and offering a safe environment may be better positioned to succeed. This will be both the business case and the desire of all patients whose health depends on this system.

ACKNOWLEDGMENTS

The authors would like to acknowledge Becky Rice for preparing this chapter.

References

Anderson, N., De Dreu, C. K., & Nijstad, B.A. (2004). The routinization of innovation research: A constructively critical review of the state-of-the-science. *Journal of Organizational Behavior, 25*, 147–173.

Barney, M., & McCarty, T. (2003). *The new six sigma: A leader's guide to achieving rapid business improvement and sustainable results.* Englewood Cliffs, NJ: Prentice Hall.

Batalden, P., Nolan, T. W., (1993). Knowledge for the Leadership of Continual Improvement of Health Care. *The AUPHA Manual of Health Services Management,* 22–23.

Berwick, D. M., Godfrey, A. B., & Roessner, J. (1990). *Curing health care: New strategies for quality improvement.* San Francisco: Jossey-Bass.

Blumenthal D., & Kilo, C. (1998). A report card on continuous quality improvement. *Milbank Quarterly, 76*(4), 625–648.

Brassington, G., Atienza, A., Perczek, R., DiLorenzo, T., & King, A. (2002). Intervention-related cognitive versus social mediators of exercise adherence in the elderly. *American Journal of Preventive Medicine, 23*(2S), 80–86.

Brennan. T., & Leape, L. (1991). Adverse events, negligence in hospitalized patients: Results from the Harvard Medical Practice Study. *Perspectives in Healthcare Risk Management, 11*(2), 2–8.

Bright, W. (1952). *An introduction to scientific research.* New York: McGraw-Hill.

British National Health Service. (2004). Retrieved May 4, 2004 from http://www.modern.nhs.uk/improvementguides/sustainability/

CDC. (2004). Retrieved April 15, 2004 from http://www.cdc.gov/nccdphp/overview.htm.

Clemmer, T., Spuhler, V., Oniki, T., & Horn, S. (1999). Results of a collaborative quality improvement program on outcomes and costs in a tertiary critical care unit. *Critical Care Medicine, 27*(9) 1768–1774.

Codman, E. A. (1914). *A study in hospital efficiency* (Reprinted). Chicago: Joint Commission Resources.

Consumer Report. (2003). *Consumer report buying guide*: 2003. New York, N.Y.: Consumer Report.

Cretin, S., Shortell, S. M., & Keeler, E. (2004). An evaluation of collaborative interventions to improve chronic illness care. Framework and study design. *Evaluation Review, 28*(1), 28–51.

Deming, E., (1986). *Out of the Crisis.* Cambridge, Mass.: MIT Center for Advanced Engineering Studies.

Donabedian, A. (1966). Evaluating the Quality of Medical Care. *Milbank Quarterly Fund Quarterly.* 44(1): 166–203.

Flanagan, J. (1954). The critical incident technique. *Psychological Bulletin, 51,* 327–358.

Flexner, A (1910). Medical Education in the United States and Canada: A Report to the Carnegie Foundation for the Advancement of Teaching, *bulletin 4.* New York: The Carnegie Foundation.

Frambach, R., & Schillewaert, N. (2002). Organizational innovation adoption: A multilevel framework of determinants and opportunities for future research. *Journal of Business Research, 55,* 163–176.

Gosbee, J. W. (1999). Human factors engineering is the basis for a practical error-in-medicine curriculum. In C. Johnson (Ed.), *Human error and clinical systems (HECS '99).* Glasgow, UK: University of Glasgow, 135-145 # Technical Report G99-1.

Gustafson, D., Arora, N., Nelson, E., & Boberg, E. (2001). Increasing understanding of patient needs during and after hospitalization. *Journal of Quality Improvement, 27*(2), 81–92.

Gustafson, D., Cats-Beril, W., & Alemi F. (1992). *Systems to support health policy analysis.* Ann Arbor, MI: Health Administration Press.

Gustafson, D., Hawkins, R., Boberg, E., Pingree, S., Serlin, R., Graziano, F., & Chan, C. (1998). Impact of a patient-centered, computer-based health information/support system. *American Journal of Preventive Medicine, 16*(1), 1–9.

Gustafson, D., & Hundt, A. (1995). Findings of innovation research applied to quality management principles for health care. *Health Care Management Review, 20*(2), 16–33.

Institute of Medicine (IOM). (2000). *To err is human: Building a safer healthcare system.* Washington, DC: National Academy Press.

Langley, G.J., Nolan, K., Nolan, T., Norman, C., & Provost, L. (2005). *The improvement guide: A practical approach to enhancing organizational performance.* San Francisco: Jossey-Bass.

Leape, L., Kabcenell, A., Berwick ,D., & Roessner, J. (1998). *Reducing adverse drug events.* Boston, MA: Institute for Healthcare Improvement.

Lombardo, M., & Eichinger, R. (2002). *For your improvement: A development and coaching guide for learners, supervisors, managers, mentors, and feedback givers.* Minneapolis, MN: Lominger.

McLellan, A., O'Brien, C., Lewis, D., & Kleber, H. (2000). Drug addiction as a chronic medical illness: Implications for treatment, insurance, and evaluation. *Journal of the American Medical Association, 284,* 1689–1695.

McTavish, F., Pingree, S., Hawkins, R., & Gustafson, D. (2003). Cultural differences in use of an electronic discussion group. *Journal of Health Psychology, 8*(1), 105–117.

Military Procedure MIL-P–1629. (1949). *Procedures for performing a failure mode, effects and criticality analysis.*

Molfenter, T., Bhattacharya, A., Ford, J., & Wen, K. Y. (2005). *The development and use of hybrid conjoint model to predict sustainability of organizational change.* Unpublished manuscript, May, 2005.

Molfenter, T., Gustafson, D., Kilo, C., Bhattacharya, A., & Ollsen, J. (2005). Prospective evaluation of a Bayesian model to predict organizational change. *Health Care Management Review. 30(3)*, 270–279.

Molfenter, T., Johnson, P., Gustafson, D., DeVries, K., & Veeramani D. (2002). Patient internet services: Creating the value-added paradigm. *Journal of Healthcare Information Management,. 16*(4), 73–79.

Montoya-Weiss, M. M., & Calantone, R. (1994). Determinants of new product performance: A review and meta-analysis. *Journal of Product Innovation Management, 5,* 397–417.

National Advisory Committee on Microbiological Critieria for Foods. (1997). *Hazard analysis and critical control point principles and application guidelines.* Washington, DC: U.S. Food and Drug Administration & U.S. Department of Agriculture.

National Research Council. (1994). *Science and judgment in risk assessment.* Washington, DC: National Academy Press.

Nolan, T., Schall, M., Berwick, D., & Roessner, J. (1996). *Reducing delays and waiting times: Throughout the healthcare system.* Boston, MA: Institute for Healthcare Improvement.

Ockene, J., Emmons, K., Mermelstein, R., Perkins, K., Bonnollo, D., Hollis, J., & Voorhees, C. (2000). Relapse and maintenance issues for tobacco cessation. *Health Psychology, 19*(Suppl.), 17–31.

Øvretveit, J., Bate, P., Cleary, P., Cretin, S., Gustafson, D., McInnes, K., McLeod, H., Molfenter, T., Plsek, P., Robert, G., Shortell, S., & Wilson, T. (2002). Quality collaboratives: Lessons from research. *Quality & Safety in Health Care, 11*, 345–351.

Pande, P., & Holpp, L. (2002). *What is six sigma?* New York: McGraw-Hill.

Reason, J. (1990). *Human error.* New York: Cambridge University Press.

Rogers, E. (1995). *Diffusion of innovations* (4th ed.). New York: Free Press.

Salovey, P., Rothman, A., & Rodin, J. (1998). Health behavior. In D. T. Gilbert, S. T. Fiske, & G. Lindzey (Eds.), *The handbook of social psychology* (4th ed., vol. 2, pp. 633–683). Boston: McGraw-Hill.

Shortell, S., Jones, R., Rademaker, A., Gillies, R., Donovan D., & Hughes, E. F. (2000). Assessing the impact of total quality management and organizational culture on multiple outcomes of care for coronary artery bypass graft surgery patients. *Medical Care, 38*(2), 207–217.

Solberg, L., Kottke, T., Brekke, M., Magnan, S., Davidson, G., Calomeni, C., Conn, S., Amundson, G., & Nelson, A. (2000). Failure of a continuous quality improvement intervention to increase the delivery of preventive services: A randomized trial. *Effective Clinical Practice, 3*(3), 105–115.

Utterback, J. (1975). Successful industrial innovations: A multivariate analysis. *Decision Sciences, 16,* 65–77.

Van de Ven, A. (1980). Problem solving, planning, and innovation: Part I. Test of the program planning model. *Human Relations, 32*(10), 771–740.

Shewhart, W. A. (1931). *Economic Control of Quality Manufactured Product,* New York: D. Van Nostrand Company, Inc. (Republished in 1980 by the American Society of Quality Control.)

Varley, P. (1994). Vision and Strategy: Paul H. O'Neill at OMB and Alcoa. Case C16-94-1134.1, Kennedy School of Government.

Venkatesh, V. (2000). Determinants of perceived ease of use: Integrating control, intrinsic motivation, and emotion into the technology acceptance model. *Information Systems Research, 11*(4), 342–365.

Weinstein, N. (1993). Testing four competing theories of health-protective behavior. *Health Psychology, 12,* 324–333.

Weiss, K., Mendoza, G., Schall ,M., Berwick, D., & Roessner, J. (1997). *Improving asthma care in children and adults.* Boston, MA: Institute for Healthcare Improvement.

Wennberg, J., & Cooper, M. (Eds.). (1996). *The Dartmouth atlas of health care in the United States.* Chicago: American Hospital Publishing, Inc.

World Health Organisation (WHO). (2003). *World health report.* Geneva, Switzerland: Author.

43

WORK ORGANIZATION INTERVENTIONS IN HEALTH CARE

Kari Lindström
Finnish Institute of Occupational Health, Finland

Gustaf Molander
University of Helsinki, Finland

Good interpersonal skills and competence to deliver high-quality care are important characteristics of health care workers and of an entire health care organization. In many countries, health care systems have recently experienced a crisis caused by the rising cost of care, the long patient waits, and the rigidity and slowness of health care organizations to respond to changing external demands and challenges. In such situations, organizational interventions targeting both organizational structures and functioning are needed.

Organizational interventions must be contextualized with respect to the nature of the work and the type of health care organization in which they are being carried out. They can focus on the jobs, the apparent organizational stressors, and the effects of organizational stressors on the well-being and health of personnel. Therefore, a short description of the psychological, social, and organizational factors that should be improved or changed and the well-being and health problems of health care personnel that should be affected is given at the beginning of this chapter.

The main focus of the chapter is the various kinds of organizational intervention processes and methods applied, as well as their results. Emphasis is not only placed on participatory intervention processes such as the survey-feedback method and participatory action research, but also on interventions involving health promotion, job stress, emotions at work, total quality management, and various nursing modes. The aim is to describe organizational interventions that have been successful in health care and to illustrate the main factors contributing to their success.

The issues dealing with the organizational culture of health care organizations are, however, dealt with in the chapter entitled "Changing Organizational Culture in Health Care: Redirecting Traditional Professional Values to Support Safety", by John Carroll and his colleagues.

WHAT ARE THE STAGES OF PREVENTION AND THE GOALS AND TARGET GROUPS OF ORGANIZATIONAL INTERVENTIONS?

Stages of Prevention in Organizational Interventions

The prevention of occupational stress at work can be considered primary prevention. It includes changing work conditions, such as providing well-defined job descriptions, redesigning work content and its organization, applying good ergonomic practices, using joint employee–employer action to improve the work environment and work–life balance, and offering training to workers to help develop their competencies. Often, however, interventions involve secondary forms of prevention (e.g., to help employees control their reactions to stressful work factors). These interventions focus on individuals and their symptoms. Tertiary interventions, in turn, concentrate on disease and disability management. Organizational interventions are seldom tertiary forms of prevention, however (Murphy & Sauter, 2004).

Goals of Organizational Interventions

Organizational interventions and the management of change processes involve the following four general aspects of job perception that should be considered targets when the content and process of an intervention are planned and evaluated; changing job characteristics, role characteristics, leadership behavior, and work group characteristics. Of course, the direct effects on well-being, health, and productivity and the effects on these factors via changes in the aforementioned aspects of job characteristics are also intervention goals.

The targets of organizational interventions can be conceptualized on the basis of either the work climate (e.g., James & James, 1989) or open system theories of organizational behavior. For example, the causal model of organizational performance and change by Burke and Litwin (1992) describes organizational behavior at the transformational (organizational culture) and transactional (group and individual) levels. The process variables in organizational interventions are important, such as how actions are prepared and carried out (Nytrø, Saksvik, Mikkelsen, Bohle, & Quinlan, 2000).

The nature and development of work organizations have been explored using, for example, the concept of organizational healthiness. According to this concept, the psychosocial subsystem of the organization can be described as representing perceived internal functioning with regard to task completion, problem solving, and staff development (Cox & Leiter, 1992). These elements are also relevant with respect to health care organizations.

Target Groups in Organizational Interventions

Most studies and organizational interventions in health care have focused on the work of nurses and physicians. Segregation between various professional groups and their tasks is common in health care, especially in hospital settings. Such an approach easily leads to intervention practices that cannot target the whole work unit. In primary health care, the individual-based working model has been the object of change, and team building among personnel has been the goal. In hospitals, nurses have often tried to enhance their mutual collaboration and the organization of daily patient care routines. One key issue in hospitals has been the tension between the managerial and medical subcultures. Professional autonomy and relationships between colleagues have been emphasized (Loan-Clarke & Preston, 1999).

JOB STRESSORS AND LOWERED WELL-BEING AS TARGETS OF HEALTH CARE INTERVENTIONS

Much research data are available on job stressors and their possible effects on the well-being of health care personnel. Job and organizational stressors are the starting points for organizational interventions. In addition to the traditional job and organizational stressors, functional and structural change processes are also sources of stress in health care organizations (Roald & Edgren, 2001).

According to a comprehensive British literature review (Michie & Williams, 2003), long work hours, work overload and pressure, and their effects on personal life are associated with both ill health and absenteeism among health care professionals. In addition, lack of control over work, lack of participation in decision making, poor social support, and ambiguous management and work roles are related to ill health and absenteeism. Successful organizational interventions have been able to affect and change these factors.

Among nurses, more than in other comparable professional groups, the level of psychological disturbances, from mild emotional exhaustion to suicide, is higher according to a British report (Williams, Mitchie, & Pattani, 1998). The job-related sources of these psychological disturbances among nurses are high workload and its effect on personal life, as well as staff shortages, unpredictable staffing and scheduling, and lack of time for providing emotional support to patients. A common job stressor is also poor leadership (Williams et. al., 1998).

Physicians form another professional group that has been studied in health care to a great extent.

Their main job stressors have also been high workload and pressure at work (Williams et al., 1998). Poor teamwork seems to contribute to the sick leave absences of hospital physicians even more than do traditional psychosocial risks, such as workload and work pressure (Kivimäki et. al., 2001). Long work hours are also typical for physicians, and they lead to problems in the balance between work and private life. In addition, other job factors specific to the field of specialization and the type of organization can be stressors or sources of job satisfaction for physicians. For example, among both British and Swedish psychiatrists, high job satisfaction was explained by a lower workload, a positive view of leadership, low work-related exhaustion, and a sense of participation in the organization. Also noteworthy were the few differences between British and Swedish psychiatrists (Thomsen, Dallender, Soares, Nolan, & Arnetz, 1998). Low organizational justice has been found to increase the risk of psychological distress among male physicians but not among female physicians (Sutinen, Kivimäki, Elovainio, & Virtanen,. 2002). In addition, management and decision-making practices, especially in large hospitals, are often problematic. Finnish studies have shown that fair leadership and fair decision making are important factors for the well-being of health care workers in general (Elovainio, Kivimäki, & Vahtera, 2002).

For other occupational groups in health care, the main stressors are generally at the same level as those for nurses and physicians. High psychosocial workload is expected because health care work requires much social interaction, both with other personnel and with patients and their relatives. For more information, see the chapter "Job Stress in Health Care Workers" by Lawrence Murphy.

Health problems that have been found to be related to the aforementioned job stressors include reduced well-being (usually measured with the General Health Questionnaire), psychological disturbances such as depression and anxiety, and cardiovascular illnesses. In addition, back and joint pain have been strongly associated with sickness absenteeism (Williams et al., 1998).

Even though the stressors in health care work are well known, in order for interventions to be successful, the organizational structure, organizational culture, type of patients, and, finally, the financing model used for activities must be taken into account.

PLANNED ORGANIZATIONAL INTERVENTIONS AND JOB REDESIGN AS DEVELOPMENTAL PROCESSES

Most of the methodologically acceptable interventions among health care workers have aimed at improving general physical and psychological problems. The general health-improving interventions are systemic organizational programs, which also include various kinds of personnel and management training.

Phases of the Participatory Intervention Process

We propose that organizational intervention processes in health care should be participatory, meaning that all stakeholders should be involved in the whole intervention process, from the beginning to the end. The key factors of the intervention process are good planning and clear responsibilities for process management. The impact of the intervention may depend even more on the process factors than on the actual goal of the intervention (Nytrø et al. 2000).

A planned organizational intervention needs a project group that is responsible for the intervention. It should be comprised of representatives of the employees and the employer, as this joint representation guarantees participation and strengthens the commitment of both groups. The process usually needs an internal or external consultant who insures that the process proceeds smoothly.

Organizational interventions usually follow a specific series of steps (Figure 43–1). The intervention process starts with a preliminary problem definition and the expectation of both the employees and the employer (1). The second phase is the commitment of all the stakeholders to the intervention, so that a shared vision is formed of the goals, the possible content, and the process of the intervention (2). Once a shared vision is attained, the first step is to organize a project group responsible for carrying out the program, along with possible external consultative support (3). It is then possible to start the intervention process with an organizational diagnosis, for example, a survey of psychosocial stressors at work (4).

Using a participatory planning process makes it possible to choose the method to be applied, the targets, and the time schedule (5). The implementation

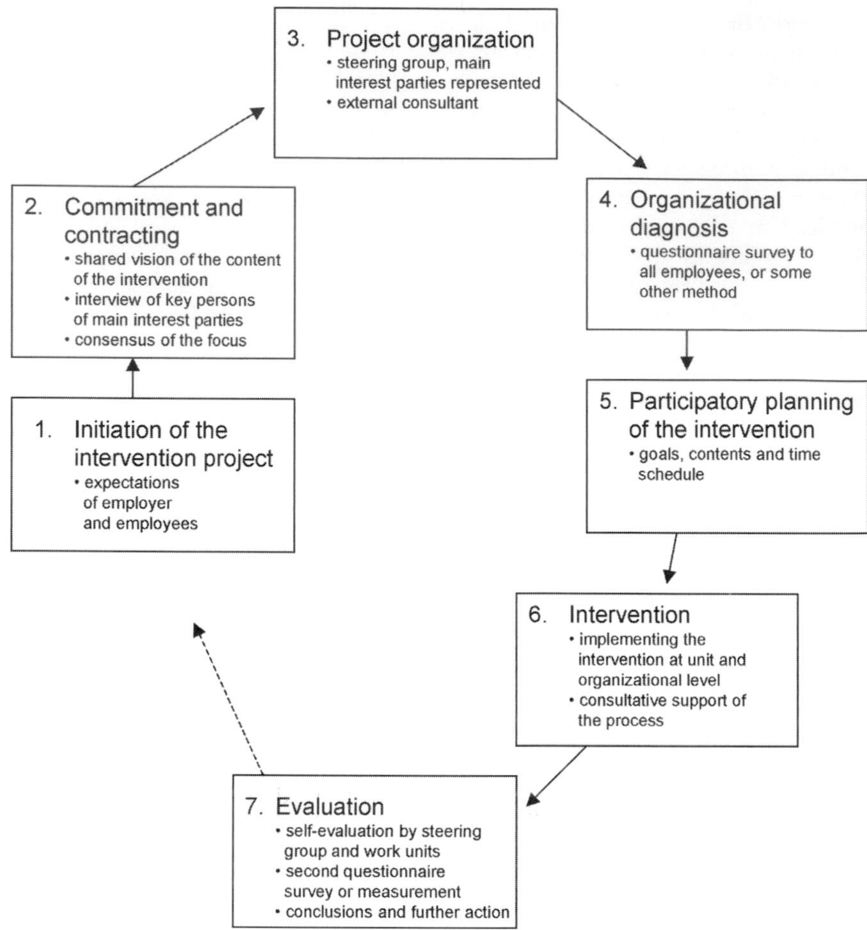

Figure 43–1. Phases of a participatory organizational intervention.

of the planned actions requires the commitment of both the employers and employees and usually consultative support from an external consultant (6). The last phase is to evaluate the process (7), which, at the very least, should include a second measurement with the same survey instrument used in the beginning of the intervention. However, data should be also collected as the intervention process proceeds. These data provide information about the successes and failures of the intervention.

This general process model can include several kinds of group work activities based on methods that promote the active participation of employees. One of them is the conference method, which is based on a democratic dialogue between the participants (Gustavsen & Engelstad, 1986).

The participating interventions described in this section are the survey-feedback method, the conference method, and multilevel organizational interventions. In addition, job stress interventions, management of structural changes, organizational health promotion, and total quality management including employee–employer confidential discussions and various nursing model implementations are described. The target groups of these forms of intervention have varied from small work units to several hospitals or organizations. The length of the intervention varies from several days to 2 or even 3 years.

Survey-Feedback Method Used in Health Care Organization

The survey-feedback method follows the general steps of an organizational intervention process (Figure 43–1). It belongs to the sociotechnological tradition of organizational psychology (Harrison,

1994). Many workplaces, especially in the health care sector, annually carry out work climate surveys that involve self-administered questionnaires. These questionnaires generally cover the main psychological, social, and ergonomic aspects of work and also include measures of the well-being and subjective health of personnel. Questionnaire tools that are useful for these purposes are, for example, the NIOSH Job Stress Questionnaire (Hurrell & McLaney, 1988) or the Nordic Questionnaire for Psychological and Social Factors at Work (QPSNordic; Lindström et al. 2000a).

The results of the questionnaire survey can be used by management and employees of work units for the so-called survey-feedback process that is often accompanied by practical organizational interventions and their evaluation. At the least, the evaluation should be based on a comparison of survey results before and after the intervention.

An example from a municipal health care organization illustrates how the survey-feedback method, which formed the main structure of the intervention, can aim at managing job and organizational stress problems (Lindström & Kivimäki, 1999). The target organization employed about 900 people and consisted of the following five departments; primary health care, general hospital care, dental care, hygiene control, and the supplies and maintenance department.

In the initiation phase, the organization shared the need to manage psychosocial stress and workload and to improve the organizational climate, but the intervention plans of the employer and employees were contradictory. The situation with respect to psychosocial stress and organizational climate had worsened significantly as the organization had to cut personnel costs at a time when the staff already considered their workload to be too high. The employer was interested in saving personnel costs and reorganizing the work. The city mayor, together with representatives of the local trade union of nurses, contacted a consultant in order to implement an organizational intervention program in the entire organization.

A shared vision of the content of the intervention program was created after interviews and joint discussions of the main interested parties. The project was organized and implemented by a steering group that included representatives of the main stakeholders. The problems of the organization were diagnosed on the basis of questionnaire results. The participatory planning of intervention actions was defined in survey-feedback meetings. The intervention plan was implemented with the support of an outside consultant. Actions were taken to improve work organization and to reduce stress and strain. These actions included reducing the work overload and underload peaks by revising the timing of shifts in the hospital wards, establishing work counseling groups to promote professional competence among the health care personnel, and enhancing cooperation between the physicians and nurses in primary health care.

In addition, ergonomic interventions were initiated. For instance, interdepartmental exercise groups were set up to help those suffering from neck and shoulder problems, lifting aids were acquired to facilitate the handling of patients, and training in patient handling and the use of lifting aids was provided. The evaluation included a questionnaire survey similar to the questionnaire used in the diagnosis phase, as well as documentation of the intervention process. All units produced a report on how they had succeeded in their own projects, and the results were discussed in the steering group. The overall effect was that collaboration within the workers' own work group and professional competence were improved, but the climate at the organizational level was still poor.

Conference Method for Maximizing Employee Participation

The conference method, or search seminars, used in organizational interventions is an example of the participatory approach in that the setting of goals and the intervention process are based on a democratic dialogue between the participants in the organization. This method is used especially in Scandinavian countries, in both health care and in other types of organizations (Gustavsen & Engelstad, 1986). The method can be used as such or in combination with other approaches. The conference method facilitates joint discussion in organizational interventions and aids the participatory intervention process.

When this method was used in an organizational merger in the municipal health care organization, the aim was to create and transfer the common goals of the merger to everyday practice in the new organization and to guarantee the availability of good services for all residents of the municipality. Several small projects were also launched to improve work practices (Lindström & Kivimäki, 1999).

A process based on the conference method usually starts with a one-day search seminar with persons from different units, tasks, and hierarchical levels. At the beginning of the seminar, the chair or leader presents the future plans and vision of the organization. Thereafter, discussions are held in small parallel groups. The same topics (i.e., the vision of the organization, potential problems in the realization of the vision, and actions needed to realize these visions and reach the goals) are discussed by the groups. In addition to furthering people's commitment to the project, the dialogue in the working conferences increases mutual understanding and helps generate new ideas. The project group then utilizes these ideas to produce main ideas and goals for action. During the 1-or 2-year project, the management and personnel meet at least once more to exchange experiences in order to check on the progress of the program and to generate new ideas on how to continue. This exchange of experiences and common learning improves the process and helps people commit to it. At the end of the project, a third seminar is organized to evaluate what has been learned during the process. These seminars are forums for exchanging experiences and increasing commitment.

Organizational Participatory Intervention to Reduce Stress at Work

One Norwegian example may be typical of how participatory organizational stress intervention is usually carried out. A short-term participatory intervention to reduce stress at work was implemented in a health care institution. The participants were randomly allocated to an intervention group and a control group (Mikkelsen et al. 2000).

The intervention proceeded as follows. Two weeks before starting the intervention, the researchers informed all the employees and supervisors about the project. An external OD (organizational development) consultant was responsible for carrying out the intervention together with a group comprised of the supervisor, trade union representative, and employees' safety representative. The aim was to start a learning process to identify and solve work problems in order to continuously improve worker health and organizational performance on a long-term basis. The process started with a 6-hour seminar in the form of a participative search conference (Emery & Purser, 1996). The seminar was held to collectively create a plan that the members themselves would implement. The key questions asked were: (1) What are the key factors in the work unit that insure a good work environment and (2) what actions do you see as necessary for reducing the gap between the desired outcome and existing reality? The planned actions were based on the main problems reported by the participants, which were lack of information, poor communication, and insufficient respect between the different professions. There was also a need for professional and personal development. After the initial orientation seminar, the work unit held nine individual group meetings, each lasting 2 hours.

The evaluation of the process was based on the logbooks and written reports of the supervisor and the work groups. The measurement instruments included a questionnaire scale of work-related stress, a health inventory of subjective health, job demands and control, social support, learning climate, and leadership style.

Because the intervention was carried out during work hours, it was difficult to find meeting times that were suitable for everyone (e.g., because of shift work). Initially, the intervention did not have a high priority among the employees. The supervisors of the two intervention units held opposite views, one was skeptical and the other one was enthusiastic. As a result, the intervention had a positive, but limited effect on work-related stress, and psychological work demands improved in the intervention group but not in the control group.

The intervention seemed to have begun a positive change process. In spite of the short-term changes, it did not give a consistent and convincing picture of the role of organizational intervention in reducing work stress (Mikkelsen & Saksvik, 1998). The intervention period was too short, and the work methods were not sufficiently established to encourage the participants to continue the process after the project period. The employees reported that the introduced methods were very useful for problem solving, but that the methods were not used in the normal staff meetings until the end of the project. A more focused strategy concentrating on some of the problems might have worked better. The participation process itself had a positive impact on well-being, and it reduced stress at work.

Managing Structural Changes and Mergers in Health Care

Hospital mergers are typical structural changes that have both positive and negative effects on employee work conditions and well-being. For example, in the 1980s, the goal of most hospital mergers in North America was to strengthen their financial situation (Bazzoli, LoSasso, Arnould, & Shalowitz,. 2002). Overlapping services were discontinued, wards were combined, the number of full-time and especially auxiliary personnel diminished, and the number of part-time employees increased. Later, in the 1990s, the main trend was to downsize the number of personnel in general in order to cut costs. Many of these mergers and downsizing procedures were carried out without the participation of the personnel (Bazzoli et. al., 2002). The negative consequences of the mergers and downsizing have been found to lead to an elevated workload for the remaining employees and, therefore, reduced well-being and job dissatisfaction and increased stress-related sickness absenteeism—and, in the long run, even increased cardiovascular diseases and elevated mortality (Vahtera et al. 2004).

An important focus of organizational intervention in hospitals is how to facilitate structural and cultural changes when wards and hospitals with different cultures are merged (Cavanagh, 1996). Employees' reactions to mergers can vary from neutral to strong resistance.

The management process in a merger in a Finnish hospital with 20,000 employees included the following phases:

- Informing personnel about the coming structural change.
- Meetings for supervisors to help them prepare for the change process.
- Questionnaire survey of all personnel immediately after the merger.
- Feedback discussion about the questionnaire survey results and the implementation of supportive actions.
- Intensified information communication within the organization.
- Special consultation support for units with major changes.

In a prospective study of this merger, negatively perceived changes were explained by the extent of the change, poor management of the change, and poor leadership in general. Low employee participation in the organizational change process was also related to negative attitudes toward the change (Lindström, Turpeinen, & Kinnunen,. 2003). It is therefore crucial to prepare the merger carefully and to inform the personnel (Roald & Edgren, 2001; Schweiger & DeNisi, 1991). The corporation of Swedish hospitals showed that the restructuring proceeded in a neutral manner due to its careful implementation by management and resulted in a realistic preview of the change (Sverke, Hellgren, & Öhrming, 1999).

Multilevel Organizational Health Promotion

Intervention in the form of health promotion has usually been individual oriented, focused on the lifestyle, well-being, and health of the individual (e.g., Hurrell & Murphy, 1996). The Finnish extension of this health promotion strategy is called the maintenance of work ability (MWA). It refers to participatory workplace health promotion interventions targeted toward work and organizational factors, the physical work environment, and the health and professional competence of the employees. The interventions are carried out by the health promotion team with the help of occupational health personnel or an outside organizational consultant (Elo & Leppänen, 1999). Organizational interventions represent one part of this health promotion strategy. Some of the interventions have been evaluated and found to be successful when planned and carried out systematically (Lindström et. al., 2000b). A large-scale organizational health promotion program improved both individual well-being and productivity at the company level. Continuous improvement practices and leadership practices brought about the most positive changes. The most effective actions were those focused on social interaction and leadership.

In a multilevel and multicomponent intervention process of workplace health promotion in a nursing home, the conference method was used to structure the intervention process. The intervention focused on the individual/organization interface at both the group and organization level. All 600 employees of the nursing home participated in the study (Lindström, 2001). The 2 1/2-year process was led by a project group that included the head of the organization, the occupational health physician, a personnel development expert, a union representative, and an outside consultant. The project started by diagnosing the situation on the basis of the results of a questionnaire survey and a search seminar. The intervention was comprised of lectures on patient care and work organization (e.g., learning organization principles), followed by participatory group work in intensive case projects at specific wards. The group work dealt with topics like team building, customer-oriented work, utilizing the competence of employees of various ages, and developing the supervisor's role. Every 6 months a feedback and evaluation meeting was organized at the organizational level, and, at the end of the project, a final evaluation meeting took place. In addition, a questionnaire survey was carried out at the end of the project. The intervention was evaluated from the following four perspectives:

- How it influenced the actual problems in the nursing home.
- How effective it was in improving organizational functioning.
- How it contributed to organizational and personnel learning.
- How it influenced the future perspectives of the nursing home.

The improvements with respect to psychosocial stressors and professional competence were related to the person's frequency of participation in the various intervention events (Lindström, 2001).

The success factors and experiences from this case of organizational health promotion were the extensive participation of both the employers and the employees, the external consultant's facilitation of the process structuring, the motivation of the project group, the established joint-learning process, and the documentation of the whole process. In some units of the nursing home, the simultaneous structural and functional changes in the organization had a negative effect on the results of the intervention.

Stress Management and the Handling of Negative Emotions at Work

Several stress management and burnout interventions have been implemented for health care personnel (Schaufeli & Enzmann, 1998). Stress management interventions can be directed toward the individual, the individual/organization interface, or the levels of organization. Stress and burnout interventions are, however, dealt with in this chapter only with respect to innovative interventions and the handling of negative emotions at work. Burnout and its management is described in more detail in the chapter "Burnout in Health Care" by Wilmar Schaufeli.

Stress management programs have traditionally focused on individuals and their possibilities to cope with stressors at work. The systematic actions taken have included recognition of the job stressors and means of coping with the strain.

Stress management intervention at the individual level can focus on didactic stress management, the promotion of a healthy lifestyle, relaxation and cognitive-therapy techniques. Interventions focusing on the individual/organization interface deal with time management, balancing work and private life, improving peer support, or some type of counseling. At the organizational level, the methods used include stress audits, changes to improve job content and the work environment, conflict management, and various types of organizational interventions.

An innovative climate at work has been found to contribute to satisfaction with leadership among hospital ward personnel (Kivimäki, Kalimo, & Lindströ,. 1994). A British study compared innovative interventions and stress management interventions in health care (Bunce & West, 1996). A traditional stress management intervention and an intervention promoting innovative responses to stressors were carried out among mixed community-based and hospital-based staff. The traditional cognitive-behavioral stress management intervention was comprised of interactive sessions/workshops. The innovation program was based on the idea of innovative coping, and the participants were encouraged to develop innovative responses to stressors through group discussions and individual action planning. At first, the general stress level decreased in the stress management group, but this effect did not last very long. The innovative stress management group displayed more long-lasting effects in the level of innovation, but the result was not evident immediately after the intervention sessions. The general conclusion was that, in a stress management program, the process itself influences the results.

Positive effects were found among health care personnel when short- and long-term stress management interventions were carried out. The long-term intervention included a stress management session that was repeated three times. Exhaustion symptoms decreased and stayed at a lower level among those who attended the repeated stress management sessions. In addition, cynicism was lower in the group with repeated sessions, and professional competence was higher (Rowe, 2000).

Situations involving negative emotions like jealousy and envy have been found to cause strain in nursing. Both individual and work-related issues have been found to provoke these emotions (Vecchio, 1999).

Among the nurses, the emotional workload was related to situations like nurse–patient or nurse–nurse interactions with a lack of social support and fair treatment, high responsibility for the patients, difficult ethical issues, dissatisfaction on the part of patients, and balancing one's work and private life (Molander, 2003). These situations produced feelings of shame or guilt in the nurses. In order to solve and handle emotions in joint discussion groups, a systematic problem-solving procedure is needed.

Problem-solving procedures were developed for these emotionally burdensome situations and their usefulness was tested in an intervention study in a nursing home (Molander, 2003). Two negatively loaded situations that had actually occurred were discussed in a counseling group. They included situations of anger in the team when one of the coworkers became the supervisor and feelings of guilt when an employee had forgotten something that was important for the patient. These feelings of anger and guilt were discussed in joint meetings in which an outside consultant was present as a facilitator. The group situation proceeded through the following steps:

- Presenting the anger- or guilt-provoking event for joint discussion in the group.
- Negotiating and trying to find a joint solution.
- Agreeing about the ways in which to solve the problem.

The group helped the person feeling guilt or anger to define the problem, to put it into a broader context, and to find ways of handling the negative emotions provoked by the incident. The developed process models were then tested by another group. These step-by-step models helped people to cope with emotionally difficult situations, which are impossible to avoid in patient work (Molander, 2003).

Total Quality Management and Employer–Employee Discussion

Total quality management (TQM) can potentially improve the quality of products, the services produced, and different aspects of work design and quality of work life. Its principles are customer focus, continuous improvement, and teamwork. Some aspects are also dealt with at the individual level using annual confidential employer–employee discussions.

TQM has been found to be positive or negative or to produce no effects on work design. In public sector organizations in the United States, the main positive impact of TQM concerned job content, job control, participation, and social relations. At the same time, negative changes were reported for workload and the uncertainty and clarity of duties. The results, however, varied between and within departments (Carayon, Sainfort, & Smith, 1999).

Implementing or improving TQM systems in health care has been one goal of organizational interventions. TQM applied in health care organizations produces an organizational development process that aims to improve the quality of care processes, the work flow and collaboration between various professionals involved in the same care process. There are numerous studies on the relationship between TQM and the realignment of critical process flow, cost effectiveness, client satisfaction, and level of quality (Sommer & Merrit, 1994). Results show that some professional groups have a better collective work orientation than do others, and they also know the impact of their own activity better.

TQM has mainly been applied and studied by consultants, and little research data are available. In addition, there are contradictory opinions about the effects of TQM. One element of TQM is the use of quality circles. People have thought that quality circles increase the innovativeness and participation of personnel, but they have had only a limited effect in changing employees' attitudes and organizational culture (e.g., Adam, 1991). Thus far, results on the relationship between personnel well-being and TQM are inconclusive and contradictory.

A longitudinal case-control study (Kivimäki et. al., 1997) was conducted in a surgical clinic that won an international prize for successful TQM implementation, and two other surgical clinics that served as controls. The support for the potential economic impact of TQM was strong, but the evidence concerning the effects of TQM in promoting the well-being and work-related attitudes of the personnel was weak. Although the proponents of TQM have argued that it creates a cultural change that enhances the well-being of personnel, this assumption was not supported by the study.

Commitment to the new system was high among the employees, except for the physicians. Among the physicians, the implementation of TQM led to a loss of autonomy due to increased teamwork, and there was no perception of their future being connected to the success of the whole organization. TQM systems therefore need continuous monitoring regarding their benefits and drawbacks in the organization.

Annual confidential conversations between a supervisor and an employee are often recommended in order to increase an awareness of goals, to provide feedback, and to enhance innovativeness (Katzell & Thompson, 1990). Typically these conversations last 1or 2 hours and cover topics such as the participatory setting and a clarification of work goals and consideration of what might help employees to attain them, feedback on past performance, and a discussion of innovative ideas. The aim of these discussions is to enhance good leadership and increase employee job satisfaction. In practice, they usually include planning of the personal objectives for the employee for the coming year and getting feedback concerning the past year's accomplishments.

Although the concept for these confidential discussions comes from management by objectives (e.g., Drucker, 1976), the feedback can help develop the employee's professional skills. Research has been conducted to examine whether these conversations improve the perception of work goal clarity, the sufficiency of feedback and innovativeness, and satisfaction with leadership. The perceived work characteristics and supervisory practices of a group that had such discussions was compared with those of a control group with measurements made before and after the discussions. The sufficiency of feedback was significantly better in the group with confidential discussions, but no change was found in goal clarity, innovation, or satisfaction with leadership (Kivimäki, 1996). Therefore, the benefits were rather narrow given the theoretical assumptions of the discussions.

Reorganizing Nursing Work

There are at least three different theoretical models for organizing nursing work (i.e,. functional nursing, team nursing, and primary nursing; Thomas & Bond, 1990). Functional nursing is a task-oriented mechanistic model in which the work is divided into separate tasks and allocated to nurses according to their skills. This model has been further developed, and transitions to team nursing and primary nursing have taken place. In team nursing, every nurse carries out a more comprehensive set of tasks than in individual nursing. Primary nursing, on the other hand, is characterized by horizontal colleague support. In practice, many wards have a work organization that is a mixture of these three models. The amount of stress among nurses has been studied by comparing these models. Patient-focused work allocation, opportunity to write nursing notes, and accountability for patient care contributed to the nurses' satisfaction with supervision and opportunities for personal growth (Mäkinen, Kivimäki, Elovainio, Virtanen, & Bond, 2003).

Interventions in which new models have been implemented have shown some differences in job stressors, such as work overload and interpersonal conflicts. The transition from team and task-oriented nursing to primary nursing has been found to lead to increased autonomy and decreased difficulty with work and work pressure (e.g., Manley, Hamill, & Hanlon, 1997).

However, the results of implementing different nursing models have shown that the changes in perceived job stressors and well-being are not merely dependent on the nursing model. In addition, ward characteristics are important, such as the number of beds, the nurse/patient ratio, and the level of education and skills of the nursing staff. The adoption of a new model may take several years and may therefore lead to additional stress due to change. A recent study has shown that the ways used to organize nursing and nursing staff increased the work motivation and job satisfaction of nurses, but their role in job stress was limited (Kivimäki et al. 1994; Mäkinen et. al., 2003).

The process of redesign and organizational development from a functional to a holistic nursing system was evaluated with regard to its influence on the work stressors and its implications for burnout and interactional stress (Büssing & Glaser, 1999). The longitudinal study design included a comparison between one model hospital and two control hospitals. Both quantitative indicators of various work stressors that were predictors of burnout and qualitative data from group discussions were collected. Work stressors were substantially reduced during the process, whereas emotional exhaustion and depersonalization increased. According to the qualitative data, the results were interpreted to indicate that a holistic nursing system meant an intensification of interactional stress and emotional work.

EVALUATION OF INTERVENTIONS AND CRITICAL FACTORS

Evaluation of Interventions

Work organization interventions are increasingly used to improve the occupational safety and health of health care organizations and their personnel. Usually a research-based evaluation of the impact of the interventions is difficult because traditional experimental designs, by themselves, are not sufficient and relevant. Therefore, it has been difficult to identify effective intervention methods and encourage people to use them (e.g., Goldenhar & Schulte, 1994). The evaluation of an organizational intervention should be both a process evaluation and an evaluation comparing the pre- and post-measurements of expected outcomes. The weakest evidence of an intervention is based on the description and reporting of the process and possible changes. The strongest evidence is based on randomized control studies (Murphy, 1996). Planned study designs in organizational interventions are usually the bases for evaluation, but natural changes inside and outside organizations make the evaluation challenging.

Organizational interventions have different targets. Some aim to change job design and content, some are directed more at social interaction or leadership, whereas others are more focused on the symptoms of psychological distress or health complaints. The evaluation method varies depending on these goals of intervention.

Because organizational interventions are often long processes, their evaluation requires:

- Documentation of the whole intervention process.
- Analysis of the changes at micro- and macrolevels (i.e., the dynamics of the change).
- Contextualization of the intervention process with regard to the target organization and simultaneous changes occurring in society.
- Analysis of the roles of various actors during the process.

In the evaluation, the nature of behavioral change and organizational dynamics should be taken into account. Organizational factors and process issues and people's conflicting preferences in changing or intervening should be considered (Griffiths, 1999; Mikkelsen, Saksvik, & Landsbergis, 2000). Intervention consists not only of the "treatment" or planned intervention, but also of everything related to the target organization and the process of administering the "treatment."

The following two main reasons for clearly describing the intervention process have been recommended for health care organizations:

- To strengthen the evidence that an observed effect was really due to the actions implemented by identifying links in the chain of causes and effects that could have led to the observed outcome
- To recommend which factors should be given special attention in future attempts to arrive at a similar or better result of the evaluation (Wickström, Joki, & Lindström, 2000).

Although qualitative data from the process are important in organizational interventions, quantitative data should not be overlooked.

A Norwegian participatory organization intervention on job stress in community health care institutions is a good example of the importance of documenting the process for evaluation (Mikkelsen et. al., 2000). This short-term intervention was targeted toward the problems perceived by employees. Positive but small effects were found in relation to work-related stress, job characteristics, learning climate, and management style. A beneficial change process was also initiated. The intervention process took time to be established as an integrated process and as a learning process. The intervention period was, however, too short for any conclusions to be drawn about long-term effects. The intervention was not integrated into the normal daily routines of the work units. The participative process itself, however, had a positive effect (Mikkelsen et al. 2000). If important qualitative process elements are not documented, the conclusions about the pre- and postresults may be very superficial, even misleading.

Critical Factors in Planning and Implementing Interventions

There are general success factors for organizational interventions that also fit health care organizations. Because of their specific nature as service organizations dealing with people's lives and their deaths, they have many specific requirements that either promote or prevent the planned improvements or positive changes.

According to the results of European case studies with organizational intervention, several success and failure factors can be identified (Kompier & Cooper, 1999). The national legal framework for preventing occupational stress facilitated organizational development activities at the workplace level. The main prerequisite of a successful organizational intervention was to apply a step-by-step model starting from careful planning and ending in proper evaluation. Also very important was the commitment of supervisors and employees from the very beginning (Murphy, 1999). Top management especially should provide the support needed, both initially and later, in order for the intervention process to survive.

In organizational interventions and structural changes, as in mergers, the patient care itself and the competence of the organization, teams, or individuals needs to be taken into account, along with job stressors and well-being. This requirement was clearly seen in a postmerger situation of a surgical clinic at which the patient waiting lists grew and old work routines and practices did not work. In this case, innovators from inside the organization took the initiative and started to reorganize patient care with the participation of personnel. The problems were solved because this group had management support and personnel participation (Lindström & Turpeinen, 2004).

The combination of a bottom-up (participation) approach and a top-down (top management support) approach promotes the goals of intervention. This combination is especially critical in health care, where organizations are big, and hierarchically multilevel.

It is important to realize that, in health care, there are separate professional subcultures that need to be taken into account when intervention processes are planned. Therefore, the intervention should have a sufficient number of cross-professional goals and subprojects. Involving as many employees as possible in subprojects of the intervention increases motivation.

The main success factor has usually been the application of a participatory approach in which employees have been able to commit and contribute actively to the intervention. Especially in the Swedish hospital corporation project, the positive results were interpreted as being related to the careful preparation and information that resulted in a realistic change process among the employees (Sverke et. al., 1999). In this case, however, the change did not have a major influence on employees.

Well-defined responsibilities are needed for people involved in intervention processes. It is important that the project has an employer who is committed to the plans and actions. The process also needs external facilitators or a process consultant because there are obstacles along the way that must be dealt with.

Factors obstructing stress prevention have been listed in earlier studies (Kompier & Cooper, 1999). Of these factors, most are relevant also for organizational intervention in health care. One of the main problems has usually been too tight a time allocation and a lack of time to participate on the part of personnel. In health care work, time must be reserved for personnel to participate in the intervention; otherwise participation stays low. Head ward nurses or middle managers used to be key persons both in motivating personnel and in organizing the work schedules and allocating work so that personnel has time to participate in joint training and intervention sessions.

Making a decision about the focus of an intervention is critical because questionnaire surveys used as diagnostic tools reveal the frequency of problems, but not necessarily their real importance. The practical and scientific aims of intervention compete because evidence and documentation for practical purposes is not so crucial as that for scientific purposes.

The success factors for organizational interventions in health care are, for the most part, similar to those in other sectors or organizations, but the type of work, especially patient work, and the care culture have specific effects. Often primary prevention should be combined with an individual approach to strengthen individual coping resources.

CONCLUSIONS

Organizational interventions in health care organizations have traditionally dealt mainly with individual job stressors and strain. In addition, various professional groups have their own work culture, and job redesign intervention usually focuses either on implementing various nursing models or, at a higher level, a new management system that does not include all occupational groups at the same time. Implementing a new management system does not necessarily influence the entire care process and take into account parallel goals of well-being. Competing with organizational intervention aimed at improved well-being and a higher quality of work life in a health care organization are the restructuring and mergers taking place to cut costs. These structural changes will continue in the future because of rising health care costs. Therefore multilevel and multitargeted organizational interventions are needed in which well-being and productivity goals can be balanced simultaneously. This balance requires much more management commitment to planned interventions. The participatory approach, however, and employee involvement can be seen as the most important factors when patient work, which is demanding emotionally, socially, and ethically, is being dealt with. Organizational culture and values should receive much more attention.

References

Adam, E. E. (1991). Quality circle performance. *Journal of Management, 17*, 25–39.

Bazzoli, G., LoSasso, A., Arnould, R., & Shalowitz, M. (2002). Hospital reorganization and restructuring achieved through merger. *Health Care Management Review, 27*, 7–20.

Bunce, D., & West, M. A. (1996). Stress management and innovation interventions at work. *Human Relations, 49*, 209–232.

Burke, W. W., & Litwin, G. H. (1992). A causal model of organizational performance and change. *Journal of Management, 18*, 523–545.

Büssing, A., & Glaser, J. (1999). Work stressors in nursing in the course of redesign: Implications for burnout and intreractional stress. *European Journal of Work and Organizational Psychology, 8*, 401–426.

Carayon, P., Sainfort, F., & Smith, M. J. (1999). Macroergonomics and total quality management: How to improve quality of working life? *International Journal of Occupational Safety and Ergonomics, 5*, 303–334.

Cavanagh, S. J. (1996). Mergers and acquisition: Some implications of cultural change. *Journal of Nursing Management, 4*, 45–51.

Cox, T., & Leiter, M. (1992). The health of health care organizations. *Work & Stress, 6*, 219–227.

Drucker, P. (1976). What results should you expect? A users' guide for MBO. *Public Administration Review, 36*, 12–19.

Elo, A-L., & Leppänen, A. (1999). Efforts of health promotion teams to improve the pscyhosocial work environment. *Journal of Occupational Health Psychology, 4*, 87–94.

Elovainio, M., Kivimäki, M., & Vahtera, J. (2002). Organizational justice: Evidence of a new psychosocial predictor of health. *American Journal of Public Health, 92*, 105–108.

Emery, M., & Purser, R. E. (1996). *The search conference. A powerful method for planning organizational change and community action.* San Francisco, CA: Jossey-Bass.

Goldenhar, L. M., & Schulte, P. A. (1994). Intervention research in occupational health and safety. *Journal of Occupational Medicine, 36*, 763–773.

Griffiths, A. (1999). Organizational interventions: Facing the limits of the natural science paradigm. *Scandinavian Journal of Work Environment and Health, 25*, 589–596.

Gustavsen, B., & Engelstad, P. H. (1986). The design of conferences and the evolving role of democratic dialogue in changing work life. *Human Relations, 39*, 101–116.

Harrison, M. I. (1994). *Diagnosing organizations. Methods, models, and processes.* Thousand Oaks, CA: Sage.

Hurrell, J. J., Jr., & McLaney, A. M. (1988). Exposure to job stress: A new psychometric instrument. *Scandinavian Journal of Work Environment and Health, 14*, 27–28.

Hurrell, Jr., J. J., & Murphy, L. R. (1996). Occupational stress intervention. *American Journal of Industrial Medicine, 29*, 338–341.

James, L. A., & James, L. R. (1989). Integrating work environment perceptions: Explorations into the measurement of meaning. *Journal of Applied Psychology, 74*, 739–751.

Katzell, R. A., & Thompson, D. E. (1990). Work motivation: Theory and practice. *American Psychologist, 45*, 144–153.

Kivimäki, M. (1996). Confidential conversations between supervisor and employee as a means for improving leadership: A quasi-experimental study in hospital wards. *Journal of Nursing Management, 4*, 325–335.

Kivimäki, M., Kalimo, R., & Lindström, K. (1994). Contributors to satisfaction with management in hospital bed wards. *Journal of Nursing Management, 2*, 229–234.

Kivimäki, M., Mäki, E., Lindström, K., Alanko, A., Seitsonen, S., & Järvinen, K. (1997). Does the implementation of total quality management (TQM) change the well-being and work-related attitudes of health care personnel? *Journal of Organizational Change Management, 10*, 456–470.

Kivimäki, M., Sutinen, R., Elovainio, M., Vahtera, J., Räsänen, K., Töyry, S., Ferrie, J. E., & Firth-Cozens, J. (2001). Sickness absence in hospital physicians 2 year follow up study on determinants. *Occupational & Environmental Medicine, 58*, 361–366.

Kompier, M., & Cooper, C. (1999). Stress prevention: European countries and European cases compared. In M. Kompier & C. Cooper (Eds.), *Preventing stress, improving productivity: European case studies in the workplace* (pp. 312–336). London: Routledge.

Lindström, K. (2001). Promoting organizational health as a part of the maintenance of work ability (MWA) activity in Finland. In C. Weikert, E. Torkelson, & J. Pryce (Eds.), *Occupational Health Psychology in Europe 2001: European Academy of Occupational Health Psychology Conference Proceedings Series* (pp. 129–133). Nottingham, UK: I-WHO Publications.

Lindström, K., Elo, A-L., Skogstad, A., Dallner, M., Gamberale, F., Hottinen, V., Knardahl, S., & Ørhede, E. (2000a). *User's guide for the QPSNordic: General Nordic questionnaire for psychological and social factors at work.* TemaNord 2000:(603). Copenhagen: Nordic Council of Ministers.

Lindström, K., & Kivimäki, M. (1999). Organizational interventions and employee well-being in health care settings. In P. M. Le Blanc, M. C. W. Peters, A. Büssing, & W. B. Schaufeli (Eds.), *Organizational psychology and health care* (pp. 135–151). München und Mering: Rainer Hampp Verlag.

Lindström, K., Schrey, K., Ahonen, G., & Kaleva, S. (2000b). The effects of promoting organizational health on worker well-being and organizational effectiveness in small and medium-sized enterprises. In L. R. Murphy, & C. L. Cooper (Eds.), *Healthy and productive work* (pp. 83–104). New York, NY: Taylor & Francis.

Lindström, K., & Turpeinen, M. (2004). How to manage patient overflow after a hospital merger. In C. R. Johansson, A. Frevel, B. Geibler-Gruber, & G. Strina (Eds.), *Applied participation and empowerment at work* (pp. 151–161). Lund: Studentlitteratur.

Lindström, K., Turpeinen, M., & Kinnunen, J. (2003). Effects of data system changes in job characteristics and well-being of hospital personnel: A longitudinal study. In D. Harris, V. Duffy, M. Smith, & C. Stephanidis (Eds.), *Human-Centered Computing, Cognitive, Social and Ergonomic Aspects: Proceedings of HCI International 2003* (pp. 88–92). Mahwah, NJ: Lawrence Erlbaum associates.

Loan-Clarke, J., & Preston, D. (1999). Organizational climate and culture issues amongst nurses within a Community Healthcare Trust in England. In P. M. Le Blanc, M. C. W. Peters, A. Büssing, & W. B. Schaufeli (Eds.), *Organizational psychology and health care* (pp. 205–219). München und Mering: Rainer Hampp Verlag.

Manley, K., Hamill, J. M., & Hanlon, M. (1997). Nursing staff's perceptions and experiences of primary nursing practice in intensive care 4 years on. *Journal of Clinical Nursing, 6,* 277–287.

Michie, S., & Williams, S. (2003). Reducing work related psychological ill health and sickness absence: A systematic literature review. *Occupational Environment Medicine, 60,* 3–9.

Mikkelsen, A., & Saksvik, P. Ø. (1998). An evaluation of the implementation process of an organizational intervention to improve work environment and productivity. *Review of Public Personnel Administration, 2,* 5–22.

Mikkelsen, A., Saksvik, P. Ø., & Landsbergis, P. (2000). The impact of a participatory organizational intervention on job stress in community health care institutions. *Work & Stress, 14,* 156–170.

Molander, G. (2003). The emotional burden of work in the care of elderly people, and coping with it. In *Proceedings of XIth European Congress on Work and Organizational Psychology* (p. 51). Lisboa, Portugal: European Association of Work and Organizational Psychology.

Murphy, L. R. (1996). Stress management in work settings: A critical review of the health effects. *American Journal of Health Promotion, 11,* 112–135.

Murphy, L. R. (1999.) Organisational interventions to reduce stress in health care professionals. In J. Firth-Cozens, & R. L. Payne (Eds.), *Stress in health professionals* (pp. 149–162). Chichester, England: Wiley.

Murphy, L. R., & Sauter S. L. (2004). Work organization interventions: State of knowledge and future directions. *International Journal of Public Health, 49,* 79–86.

Mäkinen, A., Kivimäki, M., Elovainio, M., Virtanen, M., & Bond, S. (2003). Organization of nursing care as a determinant of job satisfaction among hospital nurses. *Journal of Nursing Management, 11,* 299–306.

Nytrø, K., Saksvik, P. Ø., Mikkelsen, A., Bohle, P., & Quinlan, M. (2000). An appraisal of key factors in the implementation of occupational stress interventions. *Work & Stress, 14,* 213–225.

Roald, J., & Edgren, L. (2001). Employee experience of structural change in two Norwegian hospitals. *International Journal of Health Planning Management, 16,* 311–324.

Rowe, M. M. (2000). Skills training in the long-term management of stress and occupational burnout. *Current Psychology, 19,* 215–228.

Schaufeli, W., & Enzmann, D. (1998). The burnout companion to study and practice: A critical analysis. *Issues in occupational health,* . London: Taylor & Francis. p. 220.

Schweiger, D. L., & DeNisi, A. S. (1991). Communication with employees following the merger: A longitudinal field experiment. *Academy of Management Journal, 34,* 110–135.

Sommer, S. M., & Merrit, D. E. (1994). The impact of a TQM intervention on workplace attitudes in a health care organization. *Journal of Organizational Change Management, 7,* 53–62.

Sutinen, R., Kivimäki, M., Elovainio, M., & Virtanen, M. (2002). Organizational fairness and psychological distress in hospital physicians. *Scandinavian Journal of Public Health, 30,* 209–215.

Sverke, M., Hellgren, J., & Öhrming, J. (1999). Organizational restructuring and health care work: A quasi-experimental study. In P. M. Le Blanc, M. C. W. Peters, A. Büssing, & W. B. Schaufeli (Eds.), *Organizational psychology and health care (pp. 15–32).* München und Mering: Rainer Hampp Verlag.

Thomas, L. H., & Bond, S. (1990). Towards defining the organization of nursing care in hospital wards: An empirical study. *Journal of Advanced Nursing, 15,* 1106–1112.

Thomsen, S., Dallender, J., Soares, J., Nolan, P., & Arnetz, B. (1998). Predictors of a health workplace for Swedish and English psychiatrists. *British Journal of Psychiatry, 173,* 80–84.

Vahtera, J., Kivimäki, M., Pentti, J., Linna, A., Virtanen, M., & Ferrie, J. E. (2004). Organisational downsizing, sickness absence and mortality: The 10-Town prospective cohort study. *British Medical Journal, 328,* 555–557.

Vecchio, R. P. (1999). Jealousy and envy among health care professionals. In P. M. Le Blanc, M. C. W. Peters, A. Büssing, & W. B. Schaufeli (Eds.), *Organizational psychology and health care (pp. 121–132).* München und Mering: Rainer Hampp Verlag.

Wickström, G., Joki, M., & Lindström, K. (2000). Description of the intervention process. In G. Wickström (Ed.), *Intervention studies in the health care work environment* (pp. 103–111). Arbete och Hälsa nr 2000:10. Stockholm: National Institute for Working Life.

Williams, S., Michie, S., & Pattani, S. (1998). *Report of the partnership on the health of the NHS workforce.* London: Nuffield Trust.

44

TEAMWORK TRAINING FOR PATIENT SAFETY: BEST PRACTICES AND GUIDING PRINCIPLES

Eduardo Salas, Katherine A. Wilson-Donnelly, Dana E. Sims, C. Shawn Burke, and Heather A. Priest
University of Central Florida, Orlando

TEAMWORK TRAINING IN HEALTH CARE: BEST PRACTICES AND GUIDING PRINCIPLES

A resident's shift started at 7 a.m. in the emergency medicine department at a large urban hospital. Around midnight the resident tried to get some rest, but a gunshot victim was rushed into the emergency room not 5 minutes later. No other time to rest presented itself during the shift. It is 5 a.m. the next morning when several victims of a car crash arrive and are severely injured. As one of the victims cries out in pain, the tired resident asks a nurse to administer the pain relieving drug Demerol without checking the patient's chart for allergies. Before administering the drug, the nurse decides to check the chart, which indicates that the patient was given Demerol when admitted to the hospital on a previous occasion that resulted in severely low blood pressure and difficulty breathing. The nurse is concerned that the same allergic reaction may occur again. However, the nurse has been previously reprimanded for questioning this doctor's order and administers the Demerol without discussing this patient's history with the doctor. As a result, the patient dies.

Who is to blame in this example? The nurse for not speaking up to the doctor? The doctor for discouraging nurses from voicing concerns? The organization for fostering a culture that has a closed authoritarian hierarchy in which what the doctor says goes? This hypothetical situation is more realistic than we would like to think, and it is not confined to just the health care community.

Although the previous example shows how health care teams may derail, there are many times when patients' lives are saved as a result of effective teamwork. For example, a father rushes his 1-year-old son into the emergency room and tells the nurse that the child is choking on something but he can't see the object. The attending resident rushes over and attempts to ventilate. Simultaneously, a second doctor is listening for any sign of breathing. The doctor indicates that the child is breathing slightly and orders several instruments (e.g., forceps) to try and remove the object. Nurses rush to get the instruments quickly to the doctor's side. The object is severely lodged and efforts to remove it have failed, yet the team perseveres. A number of suggestions are provided by doctors and nurses of varying specialties as to how to remove the lodged object. Each of these suggestions is seen as viable, and the team tries to implement them without success. Brain damage is of concern as very little oxygen is reaching the child's system. Numerous intubation attempts are unsuccessful and the drained doctors ask for someone else to step in—this time the attempt is successful and the child begins to breathe. After some additional hard work and coordination, the object (the family cat's play toy) is removed from the child's throat. Only time will tell if the teamwork demonstrated will allow this child to lead a normal, healthy life.

WHY SHOULD HEALTH CARE BE CONCERNED ABOUT TEAMS AND TEAMWORK TRAINING?

The hypothetical examples we provided indicate the need for teams and teamwork in health care settings. The latter example demonstrates the importance of teamwork and the reliance of team members on one another to save a patient's life. However, the former example is all too common as well, and errors can lead to catastrophic results. The health care community has for many years overlooked the role that individual and team errors have played in the field (Pietro, Shyavitz, Smith, & Auerbach, 2000). It has been estimated that at least 50% of errors are not reported, and 70% of those reported were deemed preventable (Leape, 1994). The 1999 Institute of Medicine report on medical errors (Kohn, Corrigan, & Donaldson, 1999) shed some light on this delicate subject, and greater attention is now being paid. Reasons as to why errors are ignored can be attributed to several factors. For example, the medical professional is not in any physical danger if an error is committed (e.g., there is no risk of death to the doctor if an error is committed on a patient; this is opposed to the risk of death to a pilot who makes a catastrophic error in the aircraft). Additionally, there is a risk of legal liability that could be brought against both individual health care professionals and health care organizations if errors are documented. Finally, medical errors are often justified and rationalized due to the complex and subjective nature of health care (Pietro et al., 2000).

As medical errors will continue to occur, the need for highly trained teams in health care to recognize errors and to know how to correct errors to improve patient safety is imperative. Although the health care community has long used teams that are highly trained in the technical skills needed for the job, these teams often lack knowledge of teamwork skills. It has been argued that merely requiring a group of experts to work together does not make them an expert team (e.g., Salas, Sims, Klein, & Burke, 2003). Rather, training—team training—is needed to give the team the necessary knowledge, skills, and attitudes (KSAs) to effectively communicate, collaborate, and coordinate as a team.

Therefore, the purpose of this chapter is twofold. First, we offer a brief discussion about what is known about teams, teamwork, team effectiveness, and the design and delivery of team training. A more thorough discussion of the importance of teamwork to patient safety can be found in Baker et al.'s chapter in this volume. Second, we discuss some best practices and principles that training designers, managers, and patient safety officers in health care can use to systematically guide the design, delivery, and evaluation of team training programs.

WHAT DO WE KNOW ABOUT TEAMS AND TEAMWORK?

Teams

Teams are widely used throughout the health care community (e.g., surgery, emergency medicine, fire rescue), the military, industry, and aviation. Health care teams typically consist of team members with different specialties (e.g., anesthesiology, cardiology) and different status (e.g., doctor, nurse). Indeed, all teams are not created equal. And groups are not the same as teams. Although some may think that this is an academic distinction or debate, the fact remains: there is a difference between a collection of individuals in a group and a set of interdependent team members with distributed expertise. For our purpose, we define a team as two or more individuals with specialized roles and responsibilities who must interact dynamically and interdependently and are organized hierarchically to achieve common or valued goals and objectives (Salas, Dickinson, Converse, & Tannenbaum, 1992). Groups are just a collection of individuals joined temporally, for example, to brainstorm, troubleshoot or problem-solve, or discuss an issue. Groups by in large do not have meaningful task interdependency. They may not have a past or a future. Teams do. To function effectively, (a) resources and information must be exchanged between team members; (b) task activities, actions, and events must be synchronized and coordinated; (c) team members must continually adapt to task environmental and cue demands; and (d) members must be engaged in closed-loop communication, back-up behavior, and mutual performance monitoring (Hackman, 1990; Salas et al., 2003). All these are supported by a functional team leader.

Core Components of Teamwork

Teamwork is more than getting the job done or how people interact with each other—it is the ability to create value-added outcomes through a shared understanding of the team's resources (i.e., members' KSAs), the team goals and objectives, and the constraints under which the team works. Teamwork can be conceptualized as the interactions among team members to reach shared goals and adaptation of strategies with changing environmental demands to reach those goals (Salas & Cannon-Bowers, 2000). Furthermore, effective teamwork requires a set of team competencies—cognitions (what team members think), behaviors (what team members do), and attitudes (what team members feel). Although a number of team effectiveness models have been proposed in the literature that discuss what factors promote or deter teamwork (e.g., Campion, Medsker, & Higgs, 1993; Fleishman & Zaccaro, 1992; Marks, Mathieu, & Zaccaro, 2001; Salas et al., 1992; Stevens & Campion, 1994), there is little consensus among researchers regarding teamwork and the factors that comprise it. As such, Salas, Burke, and Stagl (2004) reviewed the team literature and proposed that, at its core, teamwork is comprised of five core factors—the "Big 5 of Teamwork." The five core components proposed are team leadership, mutual performance monitoring, back-up behavior, adaptability, and team orientation. Furthermore, three coordinating mechanisms (i.e., a shared mental model, engaging in closed-loop communication, and holding mutual trust) work around the Big 5 to ensure effective team performance. We now present a brief discussion of each core component and the corresponding coordinating mechanisms and provide examples to illustrate their relevance to the medical community. Additionally, a number of principles associated with the components of teamwork have been offered in the literature and are presented in Table 44–1.

Team leadership. Leaders make or break a team. Team leadership serves to guide and structure the team (Stewart & Manz, 1995). A clear difference should be drawn between the leadership of individuals versus team leadership. Generally one who is leading independent individuals will diagnose a problem, generate possible solutions, and implement the most appropriate solution (e.g., Fleishman et al., 1991). An example of this type of leadership might be seen in a situation in which the leader makes all of the decisions without input from other individuals. Conversely, team leadership does not involve handing down solutions to team members but rather defining team goals, setting expectations, coordinating activities, organizing team resources, and guiding the team toward their goals (Salas et al., 2004). In this case, the goal might be the repair of heart damage and the survival of the patient. The team leader will set the expectations of how the team will meet these goals but is willing to adapt to the situation. Furthermore, the team leader creates, fosters, and maintains shared knowledge among team members, assigns tasks, assesses performance, motivates team members, promotes adaptability, and creates a positive atmosphere (Cannon-Bowers, Tannenbaum, Salas, & Volpe, 1995).

Mutual performance monitoring. The second core teamwork component is mutual performance monitoring. For a team to be effective, it is important that team members maintain an "awareness" of how others (and themselves) are functioning by monitoring each other's work while simultaneously performing their own tasks (McIntyre & Salas, 1995). The purpose of this is to (hopefully) prevent or mitigate the consequences of errors. Certainly, tasks that medical professionals face occur in complex and fast-paced environments with many opportunities for errors. Mutual performance monitoring can, thus, benefit teams by allowing errors or lapses to be caught earlier in their sequence and before they transpire into larger, more catastrophic errors. Thinking back to the first example provided in this chapter: The nurse caught the doctor's error by checking the patient's chart for him. However, due to a lack of teamwork skills (e.g., assertiveness by the nurse, leadership by the doctor), the error was not stopped and deadly consequences occurred.

Probably one of the most high profile cases of medical error involved Willie King in 1995 in Tampa, Florida. Mr. King's left leg was amputated instead of his right. Although much of the media attention focused on the error of the attending surgeon and the subsequent suspension of the doctor's license, medical experts testified that failures within the hospital information management records and errors by other hospital employees played a large part in the ultimate occurrence of the error (Crane, 1997). Specifically, some team

TABLE 44–1. Teamwork Principles

General Principles

Teamwork and taskwork are different components of team performance.
Teamwork is affected by a number of external and internal factors.
Effective teamwork requires that team members amass competencies for their specific team task before receiving team training.
There are a number of teamwork skills that are nonexclusive (generic).
Teams that are motivated, and think about their efficacy, will "stretch" themselves to attain what an individual would not contemplate as possible.
Effective teams optimize resources.
Teams develop and transform over time.
"Mature" teams are composed of members who can foresee one another's needs.
"Mature" teams depend less on overt communication to perform effectively.
Teamwork is depicted as a set of behavioral skills that are suited to situational demands.

Team Leadership-Related Principles

Team performance is influenced by team leadership.
Team leaders can be used as models of teamwork.
Teamwork involves active participation by the team leader and its teammates.
Effective team leaders are respected by team members.
Team leaders have different experience levels in team operations and various readiness levels to lead a team.
Effective team leaders are knowledgeable in their specialty areas and accept suggestions of other team members who are experts in a special area.
Team leaders clearly define social structure, encourage open communications, and exhibit self-disclosure to develop team cohesion.
Leaders should clearly define and encourage team goals and performance expectations to promote commitment and consensus on team climate.
Team coherence is moderated by team leader actions within the task-contingent role.
Effective team leaders plan, structure, and coordinate the team.
Effective team leaders exhibit more initiating structure behavior as the number of team members increases.
Team leaders maintain the team members' focus on their task.
Team leaders use effective communications. They inform the team about matters that affect team performance.
One of the functions of the team leader is to provide feedback.
A team leader's style affects the degree of successful feedback provided to the team.
Effective leaders develop coherent teams.
Leaders adjust their role to match the team's progress.
The leader must define, clarify, and instill team goals and objectives early during team formation.

Performance Monitoring and Back Up-Related Principles

Teamwork involves members overseeing one another's performance.
Teamwork involves members exchanging and accepting feedback from one another.
Teamwork involves being willing, prepared, and inclined to back up teammates during performance.
Teamwork involves "intermember reinforcement" behaviors, for example teammates thanking one another for correcting mistakes and complimenting one another for performing a task well. These behaviors seem to encourage a feeling of task competency among teammates and to some extent decrease the perception of a formal hierarchy that exists between team leaders and its members.

Adaptability-Related Principles

Team members should be adept at adjusting crucial information to the task being performed.
Team members that are versatile can vary the way they perform a task when needed or when requested to do so.
Team members that are versatile can provide information on how to identify and correct mistakes.

Team Orientation-Related Principles

Effective teams exhibit a strong feeling of "teamness".
Team members value and accept the input of others.
Team members enhance individual performance through the coordination, evaluation, and utilization of task inputs from other group members while performing group tasks.

Shared Mental Models-Related Principles

Effective teams share a similar understanding and representation of team goals, individual team member tasks, and how the team will coordinate to achieve their common goals.
Team members share team-related and task-related mental models.
Effective teams do not require the sharing of an exact mental model, but rather a similar, slightly overlapping mental model.

TABLE 44–1. (Continued)

Communication-Related Principles

Team members need to convey information to other members using the appropriate terminology.
Team members need to convey information to other members in the appropriate order.
Effective teams have team members who complement one another's accomplishments.
Effective teamwork entails members exchanging information that usually include closed-loop communication.
The effectiveness of new communication tools (e.g., e-mail) depends on the structure of the team.
On some occasions more communication among team members can lead to effective performance and at other times it will hinder performance, depending on the team structure operating at that time.

Mutual Trust-Related Principles

Team members share the perception that individual team members will perform their tasks.
Effective teams are more willing to accept a certain amount of risk by relying on team members to reach their goals.
Team members accept team leadership behaviors.

Interpersonal Relations-Related Principles

Teammates encourage one another when they make errors.
Teammates who perceive themselves as integral to the team's accomplishments feel more satisfied than those who do not perceive themselves this way.
Successful team motivation and performance are related to team members making positive comments and complimenting one another.
Teamwork means encouraging an attitude of interdependence, so that members recognize that the team's success depends on their interaction.

Coordination/Cooperation-Related Principles

Effective teams exhibit "intermember assistance" behaviors, such as prompting and behaviors that indicate task coordination. This component is characterized by conduct such as prompting and guiding others on the next action to be taken.
Team cooperation can be characterized as team members checking with one another when unclear about what to do next.
Teams that are well coordinated tend to be successful. Team members of effectively coordinated teams can shift easily from one task to another when necessary for task completion.
Team cooperation can be characterized as team members assisting one another to perform a task when such tasks are not part of the teammate's actual responsibility.
Well-coordinated team members that are unoccupied seek to learn about other teammates' responsibilities and assist them when necessary.
Successful team members coordinate with one another to collect information in a systematic manner.
Successful team members tend to assist other teammates who are experiencing difficulties with a task.
Successful team performance is related to team members asking for assistance when needed.

members were not included in the verification process, and others did not feel they were allowed to call attention to the error (Joint Commission Resources, 1998). Mutual performance monitoring, which promotes team members in providing, seeking, and receiving feedback related to the task (Salas & Cannon-Bowers, 2000), might have ameliorated the error. Critical to the success of mutual performance monitoring is the support of the team leader (who will foster the shared understanding of the tasks) and also a sense that the feedback that may result from mutual performance monitoring is valuable.

Back-up behavior. If team members engage in mutual performance monitoring, then they can provide back-up behavior(s). They can step in to help. If when monitoring another team member's performance a decrement is detected or expected (e.g., due to high workload), it is the responsibility of that team member to provide the other member with support or assistance (i.e., backup). Thus, back-up behavior can be defined as a team member's ability to provide resources and task-related assistance to others in times when performance may decline (Porter et al., 2003). Back-up behavior can serve one of three purposes: (a) provide feedback to improve performance, (b) assist in performing a task, and (c) complete a task for another (Marks et al., 2001). Returning to the example presented earlier in which a tired and overloaded resident committed a medication error, mutual performance monitoring and back-up behavior may have prevented this error. McIntyre and Salas (1995) argued that the ability to provide back-up behavior to team members separates a team from a group of individuals who happen to be working colocated.

Adaptability. If team members can provide back-up behavior, then they can adapt. The ability of a team member to monitor another's performance, assist in performing a task, and accept feedback and alter their behaviors appropriately is important to team effectiveness as it demonstrates adaptability (Campion, Medsker, & Higgs, 1993). Although this type of adaptability is at the individual level, the team as a whole must also adapt at times due to changes in the environment or task situation. Adaptability can, therefore, be defined as the ability of the team or individual team members to adjust their strategies based on information gathered from the task environment through flexibility, compensatory behaviors, and reallocation of resources (Salas & Cannon-Bowers, 2000). Because many teams, such as those in health care, operate in a dynamic and stressful environment, the ability of teams to adapt to changing situations is imperative.

The team's operational need for team adaptation is driven both by the complexity of the environments that many teams operate within (i.e., things don't always go as planned) as well as by the interdependent nature of such teams. A vivid example might be easily envisioned in the emergency room. Imagine a team of medical professionals working on a patient with symptoms of an anxiety attack (e.g., dizziness, difficulty breathing, and chest pain). As the treatment progresses, the patient's condition worsens and the team realizes that in fact the patient is experiencing a heart attack. It is, thus, the responsibility of the team to notice the changing medical condition (e.g., nonresponse to traditional anxiety treatment), assign meaning to that change (i.e., heart attack), and finally to develop and successfully carry out a new plan of treatment (i.e., adjust their strategies) while still coordinating. If any step within this process is skipped or breaks down, the patient's chance of survival deteriorates.

Team orientation. The final component essential to teamwork is team orientation. Although the previous dimensions have been behavioral in nature, team orientation is attitudinal. It is a disposition. Team orientation involves not only a preference for working with others but also the tendency to enhance individual performance through the coordination, evaluation, and utilization of task inputs from other members while performing group tasks (Driskell & Salas, 1992). Team orientation is important not only because it improves individual effort and performance within a team (Shamir, 1990), but it has also been found to facilitate overall team performance (e.g., better decision making; Driskell & Salas, 1992). Specifically, team orientation results in increased cooperation and coordination among team members (Eby & Dobbins, 1997), which facilitates increased task involvement, information sharing, strategizing, and goal setting. For instance, Driskell and Salas found that individuals with a team orientation more frequently considered teammate input, even if the input was not always accepted as correct. Team orientation also increased the likelihood of uncovering errors, and the resultant reevaluation of the team product led to higher quality decisions.

In the medical setting, team orientation requires team members to accept and consider the input from others who have knowledge of a particular patient, regardless of status or rank. For example, it is important that doctors and nurses consider each other as members of the same team and therefore recognize the need to share relevant information regarding the patient. We would argue that those with a team orientation would be more likely to share and accept this information.

Coordinating Mechanisms of Teamwork

As noted, having the core components of teamwork is not enough, as teams also require that several coordinating mechanisms operate to produce effective teamwork. Although very little of the literature accurately addresses how these mechanisms are created, we believe that these coordinating concepts are best described as "mechanisms," because they are a process for achieving the goal of teamwork. Although there is still a great deal we do not understand about implementing and validating such concepts (e.g., shared mental models), effective teams are able to develop these mechanisms and coordinate effectively. We next discuss these mechanisms, specifically shared mental models, closed-loop communication, and mutual trust.

Shared mental models. For teamwork to be successful, especially under stress, it is important that team members anticipate and predict the needs of others. This is accomplished through a common understanding of the team's goals and expectations, individual team member tasks and task environment, and the method(s) by which the team will coordinate to achieve their goals (i.e., shared mental models; Cannon-Bowers, Salas, Tannenbaum, & Mathieu, 1995). There are two types of mental models discussed in the literature: team related and task related (e.g., Mathieu, Heffner, Goodwin, Salas, & Cannon-Bowers, 2000). Team-related mental models deal with information pertaining to team functioning and expected behaviors. On the other hand, task-related mental models relate to information dealing with the materials (e.g., equipment) required to complete the task and how to use the materials. Relating to the core components of teamwork, shared mental models should allow for more and better mutual performance monitoring and back-up behavior because team members will have a shared understanding about each other's tasks and what is expected of them. Research supports this. For example, some research suggests that shared mental models lead to more effective communication, improved performance (e.g., Griepentrog & Fleming, 2003; Mohammed, Klimoski, & Rentsch, 2000; Stout, Salas, & Fowlkes, 1997), and willingness of team members to work with others in the future (Rentsch & Klimoski, 2001).

Closed-loop communication. Communication is important in the workplace as well as socially, as this allows us to exchange information that is relevant to a given situation. Health care teams communicate every day—whether it be nurses exchanging information during a shift change, emergency medical technicians providing an incoming patient's stats to emergency department personnel, or a nurse updating a doctor on a patient's status. However, research suggests that team members often interpret the same information in different ways due to differing perspectives and biases (Bandow, 2001). Therefore, effective teamwork requires more than just sending and receiving information. It requires closed-loop communication. Closed-loop communication is the process of sending information, acknowledging receipt of that information, and ensuring the proper understanding of that information (Cannon-Bowers et al., 1995). Without closed-loop communication, the chance of team members misunderstanding the information being shared increases, leading to errors. This is especially important in health care due to the dynamic and complex nature of the environment in which the consequences for errors are high.

Mutual trust. The final coordinating mechanism essential for effective teamwork is mutual trust. Mutual trust is defined as the shared perceptions that team members will perform the necessary actions to complete their task and respect the rights and interests of team members (Webber, 2002). A lack of trust among team members can result in wasted time and energy protecting, checking on, and inspecting other team members (Cooper & Sawaf, 1996). Research suggests that mutual trust in teams leads to improved participation and contribution, better quality of products, decreased cycle times, and retention of team members (Bandow, 2001). In the health care community, doctors (for example) rely heavily on nurses to assist them with a patient. It is important that the doctor trust that the nurse will do the job effectively, so the doctor does not feel the need to direct the nurse to perform certain tasks. Without trust, teamwork will be hindered.

The Big 5 of Teamwork that we proposed are indeed teamwork competencies. That is, training that is well designed, is focused on teamwork, and provides opportunities to practice and get feedback will provide team members (and the teams) with a required behavioral repertoire to perform effectively. We discuss the design and delivery of team training next.

WHAT DO WE KNOW ABOUT TEAM TRAINING?

The teamwork core competencies and coordinating mechanisms discussed in the previous section do not generally come naturally to team members. Rather, they must be trained. Hopefully, the design of the training will be guided from what is known about the science of training and learning (Salas & Cannon-Bowers, 2001, 2000). Team training can be defined as the systematic acquisition of KSAs that lead to improved performance and safety in a specific environment (Salas et al., 1992). There are four necessary components (i.e., tools, methods, competencies, and learning objectives) that come together to shape team training strategies (Cannon-Bowers & Salas, 1998; Salas & Cannon-Bowers, 1997). Team training tools include, for example, team task analysis (see Burke, 2005), task simulation and exercises (see Salas, Wilson-Donnelly, Burke, & Priest, in press), and feedback strategies. These tools aid in collecting information needed to focus the training and structure the design, delivery, evaluation, and transfer of team training. The methods used to deliver the training are driven, of course, by the training objectives and content. There are typically three methods used to deliver team training: information based (e.g., presentation of knowledge via lecture), demonstration based (e.g., presentation of critical events via video), and practice based (e.g., simulators; Salas & Cannon-Bowers, 2000). Taking these components together, instructional strategies focused on teamwork can be determined. Examples of team training strategies relevant to the health care community include team coordination training (also called crisis resource management [CRM] training), cross-training, assertiveness training, simulation-based training, team self-correction, and team leadership training (see Table 44–2 for definitions). These team-based instructional strategies (discussed later) facilitate the competencies necessary for effective teamwork. We should note that CRM training is not the only strategy available for developing effective teams. There are more than we note here. The health care community needs to learn more about these and use them as needed. Given what we know today about the intricacy of teams and team training, a number of guiding principles can be developed to help training designers and developers ensure the effectiveness of their team training program. We present these principles next. For a full discussion of these, see Salas and Cannon-Bowers (2001).

WHAT ARE THE BEST PRACTICES FOR DESIGNING, IMPLEMENTING, AND EVALUATING TEAMWORK TRAINING?

Up to this point we have presented a brief review of what is known about teams and team training. We now focus on how to systematically design, implement, and evaluate team training. Team training is influenced by multiple factors before, during, and following training. Each of these factors must be considered when designing and delivering training to ensure its effectiveness. This is especially critical in communities, such as health care, in which the consequences for errors due to a lack of team performance can be catastrophic. As such, there are a number of guiding principles that training designers and developers should follow when implementing a team training program. We present these principles next.

Best Practices for Designing Team Training (and Teams)

Best Practice 1. Health care organizations should conduct a team task analysis to determine the coordination demands, as well as the requirements for the design of team training.

Glickman and colleagues (1987) have argued that both taskwork (i.e., task-oriented skills) and teamwork skills (i.e., behavioral, attitudinal, and cognitive responses needed to coordinate with fellow team members) are needed for teams to successfully complete their tasks. As such, some researchers have developed and refined the procedure known as team task analysis as a means to identify taskwork and teamwork skills (Bowers, Baker, & Salas, 1994; Bowers, Morgan, Salas, & Prince, 1993; McNeese & Rentsch, 2001).

There are seven key steps to conducting a team task analysis (see Burke, 2004, for a complete discussion of conducting team task analysis; see also Table 44–3). The first step involves conducting a requirements analysis in which the target job is identified, the associated duties and conditions under which the job is to be performed are clarified, information-gathering methods are identified (e.g., observation, questionnaires; Goldstein, 1993), the team task analysis protocol is developed, and subject matter experts are contacted for participation. The next step is to identify the tasks that comprise the target job through subject matter expert interviews (Goldstein, 1993) and to write statements describing the work, what the worker does and how, to whom, and why it is done (Goldstein & Ford, 2002). The purpose of Step 3 is to identify a teamwork taxonomy (e.g., Fleishman & Zaccaro, 1992; Stevens & Campion, 1999) so that tasks relating to taskwork can be distinguished from those relating to teamwork. Step 4 involves conducting a coordination analysis to determine which tasks require team members to coordinate activities to complete their tasks. This information is typically determined using surveys and cluster analyses (e.g., Bowers et al., 1993). Fifth, once the teamwork and taskwork tasks have been identified, it is necessary to determine which of these tasks are most relevant by asking subject matter experts to rate the tasks. The sixth step requires the translation of tasks into KSAs and abilities (see Goldstein & Ford, 2002) by subject matter experts. Finally, the competencies identified in Step 6 must be linked to each of the team tasks by stating whether they are essential, helpful, or not relevant.

Best Practice 2. Health care organizations should consider early on the factors external to the training program that may influence its success.

When designing a team training program (or any training program for that matter), it is necessary to consider factors external to the training program that will influence its success over and above the

TABLE 44–2. Team-based Instructional Strategies Relevant in Health Care

Strategy	*Definition*	*Tools*	*Methods*	*Sources*
Team coordination training	Improves team coordination communication (both explicit and implicit) encourages backup behavior, and provides practice, opportunities for other KSAs that lead to effective coordination	Performance measures Feedback Simulations Learning principles	Information-based: Lecture: Demonstration based Video Practice-based: Guided practice	Bowers Blickensderfer & Morgan, 1998; Entin & Serfaty 1999
Cross training	Team members receive practice in performing other team members' roles and tasks. Leads to a better understanding of other team members' responsibilities and taskwork Leads to enhanced shared mental models and interpositional knowledge.	Performance measures Team task analysis Learning principles Simulations Feedback	Information-based: Lecture and multimedia Demonstration-based: Role modeling Practice-based Guided practice	Salas et al., 1997; Volpe, Cannon-Bowers, Salas, & Spector, 2001
Assertiveness training	Practice and feedback help create and reinforce assertiveness in trainees. Provides opportunities for practice and supplies feedback.		Practice-based: Role play	Smith-Jentsch et al., 1996
Simulation-based training and games	Provides opportunities for trainees to operate in a realistic setting with life-like terrain, interaction, and dynamic situations. Range in fidelity, immersion, and cost. Widely used in business, the military, and research.			Marks, 2000; Tannenbaum & Yukl, 1992
Team self-correction training	Helps individuals correct and evaluate their own behavior to assess th effectiveness of the behavior. Team members learn to assess other team members. Allows constructive feedback and correction of discrepancies	Team task analysis Performance measures Learning principles Feedback	Practice-based: Role play	Blickensderfer et al., 1997a; Smith-Jentsch et al., 1998
Team leadership training		Team task analysis Performance measures Feedback Simulations	Information-based: Lecture and seminar/ workshop Demonstration-based: Video Practice-based: Guided practices and behavior modeling	

Note. KSA = Knowledge, skills, and attitudes.

TABLE 44–3. Steps for Conducting a Team Task Analysis

1. Conduct a requirements analysis.
   - What is the target job?
   - What are the duties and conditions under which the team has to perform?
   - What knowledge gathering methodologies will be used?
   - What is the protocol for conducting the team task analysis?
   - What subject matter experts will be used throughout the team task analysis?

2. Identify the tasks that comprise the target job.
   - What tasks do the teams perform on the job?

3. Task description.
   - How is the task described?
   - What are the task characteristics?
   - What are the task requirements?
     - What does the team do?
     - How does the team do it?
     - To whom/what and why does the team do it?
   - What are the task competencies?
     - What knowledge (e.g., declarative, procedural) is necessary?
     - What skills are necessary?
     - What cognitive abilities are necessary?
     - What attitudes are necessary?
   - How are tasks prioritized?

4. Determine relevant taskwork and teamwork tasks.
   - What are the most relevant tasks?
   - Which tasks can be clustered together?
     - Have subject-matter experts been consulted to help with clustering?

5. Identify teamwork taxonomy.
   - What tasks can be categorized as taskwork related?
   - What tasks can be categorized as teamwork related?
   - What teamwork taxonomy will be used?

6. Conduct a coordination analysis.
   - Which tasks place a requirement on the team to coordinate activities?

7. Translate tasks into KSAs.
   - What are the requisite knowledge, skills, abilities, and attitudes related to the relevant tasks?

8. Link KSAs to team tasks.
   - Which KSAs are essential, helpful, or not relevant to the task?

Note. KSAs = knowledge, skills, and attitude.

content and strategies used. We discuss several factors that we believe will influence team training programs in health care, specifically, the pretraining environment, organizational and individual characteristics, and trainee motivation.

Best Practice 2a. A pretraining environment should be created that prepares trainees for team training.

A key factor influencing a training program's success is the pretraining environment. There are two main characteristics of the pretraining environment that need to be considered: (a) prepractice conditions and (b) pretraining climate (Salas & Cannon-Bowers, 2001). First, prepractice conditions are elements in the pretraining environment whose purpose is to help prepare trainees for practice exercises during training. Because practice is more than the mere repetition of a task and is rather a complex process that leads to skill acquisition (Ehrenstein, Walker, Czerwinski, & Feldman, 1997; Shute & Gawlick, 1995), providing trainees with preparatory information and advanced organizers have been suggested as interventions that can help trainees to better prepare for practice sessions (Cannon-Bowers, Burns, Salas, & Pruitt, 1998).

In addition to prepractice conditions, the pretraining climate will also impact the outcomes of team training. Specifically, how the training is framed (i.e., remedial vs. advanced) will influence trainees' motivation and learning (Quinones, 1995, 1997). Additionally, the training's attendance policy (i.e., voluntary vs. mandatory) has been argued to influence the success of training (Baldwin & Magjuka, 1997). Finally, trainees' previous training experience (i.e., positive vs. negative) is thought to influence trainees' learning and retention (Smith-Jentsch, Salas, & Baker, 1996).

Best Practice 2b. The organization should support and provide the necessary resources for training.

Beyond the pretraining environment, characteristics of the organization (i.e., those present within the organization to which the newly acquired competencies must be performed) have been argued to influence the outcomes of training. Specifically, situational constraints, such as improper equipment, can lead to less than ideal training outcomes. Additionally, the organizational climate (e.g., perceived organizational support, safety culture and policies) has been suggested as having a direct impact on the outcomes of training (e.g., Rouiller & Goldstein, 1993; Tracey, Tannenbaum, & Kavanagh, 1995). In other words, for training to be more successful, it is important that the organization's goals support the training, the climate (i.e., management support) encourages demonstration of the trained competencies, and the organization provides the necessary resources to do this (Goldstein, 1993).

Best Practice 2c. Characteristics of the trainees should be considered when designing training.

In addition to the pretraining environment and organizational characteristics, training outcomes are influenced by characteristics that trainees bring to the training program. Individual characteristics suggested to influence the success of training are trainees' cognitive abilities (i.e., general intelligence, or "g"), self-efficacy (i.e., belief in own ability), expectations, and goal orientation (i.e., mastery vs. performance). First, research has shown that cognitive ability influences trainees' attainment of knowledge about the job (see Colquitt, LePine, & Noe, 2000; Ree, Carretta, & Teachout, 1995) and is a strong determinant of success in training (Ree & Earles, 1991). Next, trainees' self-efficacy has been shown to lead to better performance (see Ford, Kozlowski, Kraiger, Salas, & Teachout, 1997; Martocchio & Webster, 1992; Quinones, 1995) and is influenced by cognitive ability (see Hunter, 1986). Trainees' expectations regarding training has also been a factor influencing the success of training, and some research has indicated that when trainees' expectations are met, they demonstrate more commitment to transferring the learned competencies and improved self-efficacy (Tannenbaum, Mathieu, Salas, & Cannon-Bowers, 1991). Finally, research suggests that trainees' goal orientation will influence training outcomes (Dweck, 1986; Dweck & Leggett, 1988). The orientation that is more successful will depend on the goals of the training. For example, individuals high in mastery orientation aim to acquire new skills and master novel situations (e.g., Ford et al., 1997; Phillips & Gully, 1997), whereas trainees high in performance orientation aim to achieve high performance ratings and to avoid negative ones (see Salas, Burke, Bowers, & Wilson, 2001; Ford, Smith, Weissbein, Gully, & Salas, 1998).

Best Practice 2d. Trainees should be motivated to attend, participate, and be engaged in the team training program.

Finally, trainee motivation, which is influenced by both individual (e.g., self-efficacy) and organizational (e.g., notification) characteristics, influence trainees' willingness to participate in and learn from the training and will, therefore, influence the outcomes of training (e.g., Baldwin, Magjukia, & Loher, 1991; Mathieu, Tannenbaum, & Salas, 1992). Training motivation will influence training in terms of the amount of time and effort invested by trainees, as well as the behaviors exhibited after training on the job (Naylor, Pritchard, & Ilgen, 1980, as cited in Goldstein, 1993). Finally, when trainees believe that the training and its outcomes are relevant to their job performance (Noe, 1986), they will be more motivated to participate and learn.

Best Practice 3. Health care organizations should avoid falling prey to the myths of training.

Despite what is known about training, there are a number of myths or misconceptions that persist regarding the design, delivery, and evaluation of such programs in organizations (Salas, Rhodenizer, & Bowers, 2000). These myths exist because training is often designed based on many unsupported assumptions about how to optimize the acquisition of skills, thus hindering training's effectiveness. We believe that the health care community may fall prey to some training myths (as many others have), diminishing the effects of its team training programs. See Table 44–4 for the relevant myths and their descriptions.

Best Practices for Implementing Team Training

Best Practice 4. Training objectives (i.e., learning outcomes) should be developed based on information obtained in the team task analysis.

TABLE 44–4. Team Training Myths and Associated Realities

Myth	*Reality*
Myth 1. Everyone who has ever learned anything or has gone to training is a training expert and therefore can design it	Training is a complex event involving behavioral and cognitive aspects.Designers must consider scenario design, guided practice, feedback, measurement, and supporting technology. To develop and deliver a successful training program, one must first be trained in the necessary knowledge, skills, and attitudes to do so.
Myth 2. Subject-matter experts should drive the design of training.	Subject-matter experts are a great source of task domain knowledge and can and should articulate the needs and requirements to execute a task. They are neces sary but not sufficient to ensure a sound learning environment. As such, the design of team training requires a partnership between subject-matter experts and learning experts.
Myth 3. The higher the fidelity of the simulation, the better one learns.	The literature available regarding team training in health care has indicated that the community relies heavily on high fidelity simulations to train its teams. Of concern is that much of this training appears to focus more on how to use the simulation rather than the teamwork competencies being trained. It is suggested that simulations designed with low physical and functional fidelity will be as effective as those with high fidelity as long as the psychological fidelity of the system is adequate.
Myth 4. The more one practices, the better one gets.	Practice alone does not improve performance. Practice needs to be guided and requires measurement and feedback. In addition, practice should accompany training to help trainees develop appropriate mental representations of the task.
Myth 5. If you know how well you did during training, learning has taken place.	Feedback must provide more than just a rating of "good" or "bad." It should be diagnostic and constructive, based on the processes performed during training, specific to the trainees' skill performance, and provide an indication of how the trainee can improve.
Myth 6. Positive reactions to what happened during training lead to learning.	Research suggests that just because trainees liked training does not mean that the trained knowledge, skills, and attitudes were learned. It is important that organizations look at training beyond reactions to accurately assess, diagnose, and evaluate the program.
Myth 7. If one learns during training, this will lead to behavior change on the job.	Research suggests that just because trainees learn the trained knowledge, skills, and attitudes, these may not transfer to the job. Transfer requires that the mate rial learned in training be generalized to the job setting and that skills be retained over time.
Myth 8. Team training is just a program.	Team training involves the design and delivery of instructional strategies and requires the integration of tools, methods, and content, as well as organizational support.
Myth 9: Team training equals crisis resource management training.	Crisis resource management training is just one team training strategy. There are a number of other proven team training strategies that can improve patient safety (e.g., cross-training, team leadership training, team self-correction).
Myth 10: Team training is a one-time deal.	Results of longitudinal studies in the aviation community have shown that team work behaviors and attitudes toward teamwork decline over time without refresher training (e.g., Irwin, 1991). These results are likely to be found in the health care community as well. As such, recurrent team training should be offered to patient safety professionals.

The information gained from the team task analysis phase will drive the objectives of the team training program. When developing training objectives, it is important that they be specific, measurable, and task relevant so that they can be evaluated after the completion of training. The training objectives, which guide the training, are important for three key reasons. Training objectives (a) state how trainees should be able to perform after training (i.e., at what level) to be judged acceptable;

(b) describe the conditions during which the performance, stated previously, should occur; and (c) provide a description of acceptable performance criterion (Goldstein, 1993). In sum, training objectives state what competencies trainees are expected to acquire and demonstrate after the completion of training. After clearly defining the training objectives, they are used to guide what instructional strategies should be implemented based on their effectiveness at promoting the task-relevant team competencies as stated in the objectives.

Best Practice 5. Health care organizations should consider a variety of proven team training strategies beyond CRM training.

Once the training objectives have been established, the next step is to determine what instructional strategies will be used during training (i.e., how to train the requisite safe behaviors). There are numerous instructional strategies that have been developed over the past several decades that can prepare both individuals and teams to increase their KSAs, reduce errors, and increase their expertise in performing their tasks—ultimately leading to safe behaviors. When training individuals to perform safe behaviors, we argue that there are three important issues. First, trainees should learn to be adaptable to changing situations and to recognize when things go wrong. By training flexible knowledge structures (i.e., cognitive representations), trainees can adjust their behavior to compensate for any changes. Rigid knowledge structures in a changing environment could lead to errors. Second, we argue that all training strategies must provide trainees with constructive feedback that focuses on the task. Providing feedback to trainees allows them to compensate for incorrect behaviors and readjust or correct their strategy to be more appropriate for a given situation. Finally, training needs to be dynamic (i.e., interactive). A recent report suggests that almost 84% of all companies use classroom-based and instructor-led training (Bassi & Van Buren, 1998). In addition, it was discovered that the most commonly used delivery methods (approximately 90% of the time) were videotapes and workbooks, as compared to only 10% that used interactive, digital technologies. Computer-based or other technology-based training was used less than 35% of the time. Therefore, we argue that training for teamwork must involve providing the trainees with opportunities to practice the teamwork behaviors (e.g., role-play, simulations). Table 44–2 presents some frequently used instructional strategies that may be used to improve safe practices in the workplace. We focus on six: CRM training, simulation-based training, cross-training, assertiveness training, team self-correction, and team leadership training.

Best Practice 5a. CRM training should be used to improve how team members communicate and coordinate effectively.

CRM training is a commonly used instructional strategy to train teams in health care. As teams are sometimes required to switch from explicit to implicit coordination (Entin & Serfaty, 1999; Kleinman & Serfaty, 1989), CRM training has been shown to make this possible (Serfaty, Entin, & Johnston, 1998). CRM training teaches teams to use all available resources (i.e., people, equipment, and information). A recent review by Salas and colleagues (in press) examined the impact of CRM training on patient safety in health care. Due to the infancy of this training strategy in health care, its impact on patient safety, although promising, is not fully known.

Best Practice 5b. Simulation-based training should be offered to provide trainees the opportunity to practice the trained skills.

Simulation-based training is an instructional strategy helpful in training teams to exhibit teamwork behaviors (Fowlkes, Dwyer, Oser, & Salas, 1998). Simulation-based training offers training designers and developers the ability to embed learning events in scenarios (e.g., as determined from critical incidents data), giving trainees a meaningful framework by which to learn (Fowlkes et al., 1998; Salas & Cannon-Bowers, 2000). Furthermore, this instructional strategy provides trainers with valuable tools, including guidelines and six critical steps to achieve training objectives: trigger events, measures of performance, scenario generation, exercise conduct and control, data collection, and feedback (see Figure 44–1). The key to the success of simulation-based training is that this instructional strategy is practice based paired with feedback, giving trainees the opportunity to practice in scenarios (defined a priori) while performance is being evaluated. Additionally, the embedded events serve to trigger the trained behaviors at a predetermined time so that they can best be evaluated and feedback provided.

Best Practice 5c. Cross-training should be used to foster a shared understanding of each team member's roles and responsibilities.

Another instructional strategy used to train teamwork-related competencies is cross-training. Cross-training involves exposing trainees to the

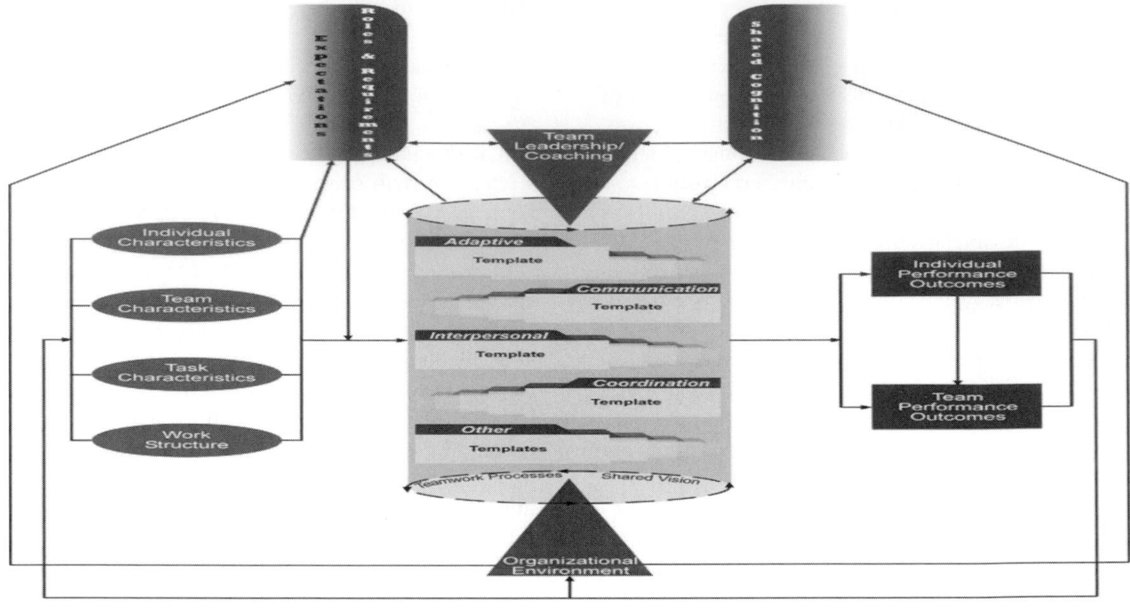

Figure 44–1. Conditions and processes of team performance.

goals, roles, tasks, and responsibilities of other team members within the team (Salas & Cannon-Bowers, 2000). There are two advantages to cross-training: (a) trainees learn and practice the tasks required of team members, allowing them to gain some degree of proficiency to assist with each others' tasks; and (b) a common understanding of the roles and responsibilities of others is created and reinforced. This can lead to a more common shared knowledge structure that can be measured in an effort to approximate the shared mental model of the team (see Cooke, Salas, & Cannon-Bowers, 2000). This allows team members to effectively monitor each others' performance and provide assistance as needed.

Best Practice 5d. Assertiveness training should be offered to ensure that each team member's ideas, opinions, and concerns are heard in an appropriate manner.

Assertiveness training involves teaching team members to clearly and directly communicate their concerns, ideas, feelings, and needs to others (Jentsch & Smith-Jentsch, 2001). Assertiveness training is important not only because it trains less senior team members to feel comfortable providing input (e.g., nurses sharing concerns with doctors), but also because it teaches them to communicate this information in a manner that does not demean others or infringe on their rights. Finally, assertiveness training also teaches more senior team members to accept information from a team member of lower status without feeling threatened. Assertiveness training does not attempt to remove the authority of the team leader (e.g., doctor), but rather its purpose is to ensure that critical information (i.e., through concerns, ideas, and so on) does not go unspoken.

Best Practice 5e. Team self-correction training should offer trainees a means to providing feedback in a timely, constructive manner.

Team self-correction is based on the theory of shared mental models (Blickensderfer, Cannon-Bowers, & Salas, 1997a, 1997b) and requires not only that individual team members have an accurate mental model of the teamwork processes that will influence their performance, but also that the whole team shares the same (or at least an overlapping) mental model (Smith-Jentsch, Zeisig, Acton, & McPherson, 1998). Specific examples of teamwork can be used to facilitate an understanding of what general team skills should look and sound like during exercises, fostering the team's shared mental model (Smith-Jentsch et al., 1998). The team must also be provided with formal training on how to self-correct, monitor one another's performance, and provide constructive, nonaccusatory feedback to team members in a debrief setting, thus creating a natural tendency (Salas & Cannon-Bowers, 2000). There are four critical elements or stages to team self-correction as a part of the debrief: event review or recap (both positive and negative behaviors), error identification and feedback exchange, stating of expectations, and planning for the future. Until self-correction becomes a natural tendency for the team through practice, it is important that a trained instructor or team leader facilitate these self-correction methods.

Best Practice 5f. Team leadership training should be used to help team members reach their goals in an effective manner.

Team leadership has been shown to have a considerable impact on the promotion of the dynamic, teamwork processes (i.e., the KSAs) required for a team to perform its tasks (Komaki, Desselles, & Bowman, 1989; Zaccaro, Rittman, & Marks, 2001) and becomes increasingly important as the complexity of the team and the work environment increases (Jacobs & Jacques, 1987). Without team leadership, the team may fail. Unfortunately, no formal training technique exists for training team leaders. However, there are a number of guidelines available in the literature to help organizations identify the team leader's role. One way in which the team leader can lead the team is through coaching. As a coach, the team leader can help team members develop strategies to overcome process losses that may be occurring (e.g., communication; Kozlowski, Gully, Salas, & Cannon-Bowers, 1996). Martin and Lumsden (1987) offered several strategies for effective coaching; namely, the team leader should (a) praise those processes and efforts that are desirable (e.g., open communication), (b) reward team members when desired behaviors are exhibited, and (c) encourage positive interactions among team members (e.g., avoid stereotyping). When the team leader takes on the role of a coach, he or she will better motivate team members to work together to overcome their differences and to be successful as a team.

Best Practice 6. Team training should be developed such that it facilitates the presentation, demonstration, and practice of teamwork competencies.

Information pertaining to the training of teamwork competencies should be provided to trainees using a number of methods. These methods include presenting information in the form of lectures, demonstration of teamwork behaviors via videos, and practice of behaviors using role-playing or guided practice. Depending on the stage of training, one method may work better than another. For example, lecture-based training is a useful way to first develop an understanding of the teamwork competencies (e.g., what they are, why they are important; Salas & Cannon-Bowers, 2000). Demonstration-based training is useful for developing competencies that are task contingent, whereas as practice-based training is useful for developing team-contingent competencies (Salas & Cannon-Bowers, 2000).

Best Practice 7. Team training should include the means for providing constructive, timely feedback to trainees.

Providing feedback to trainees in a constructive and timely manner is important to the success of a team training program (Cannon-Bowers & Salas, 1997). This requires that several criteria are met. First, feedback should be based on the teamwork processes performed during practice and on the training outcomes. Next, feedback provided to trainees should be specific to the skill performance of trainees but not critical of the individual. Third, feedback should provide trainees with the necessary knowledge that allows them to adjust their learning strategies to meet the expected performance levels. Finally, feedback must be meaningful to trainees and focus on both individual and team performance. Without feedback, breakdowns in performance may go unnoticed by trainees, corrective strategies will not be developed, and errors will likely occur on the job.

Best Practices for Evaluating and Transferring Team Training

The posttraining environment is important in determining whether the competencies learned during training will transfer to the actual job. Regardless of how well the training program was developed, without an environment that encourages the transfer of the learned competencies, it will not be effective. Research suggests that several characteristics of the work environment are essential for the transfer of training: (a) supervisor support, (b) organizational transfer climate, and (c) continuous-learning culture, although supporting empirical evidence is limited (see Baldwin & Ford, 1988; Ford & Weissbein, 1997; Rouiller & Goldstein, 1993; Tracey et al., 1995). Additionally, it is argued that some elements of the transfer climate may facilitate (e.g., rewards, positive transfer climate) or hinder (e.g., lack of peer or supervisor support, lack of resources) the transfer of training (Rouiller & Goldstein, 1993; Tannenbaum & Yukl, 1992). We briefly discuss how supervisor support and organizational transfer climate can promote the transfer of training.

Best Practice 8. Training evaluations should be developed to focus on multiple levels of the training program.

Once the instructional strategy has been chosen and the training program has been implemented, it is imperative that the training (or team training) be evaluated. Few organizations conduct systematic evaluations of their training programs. Although we acknowledge that evaluation can be resource intensive, it is the only way to truly assess training's effectiveness—whether training yielded the expected outcomes. There are numerous methods of assessment that can be applied to training. We argue for the use of a multilevel approach, such as that suggested by Kirkpatrick (1976). Kirkpatrick proposed a method of training evaluation that constituted a multilevel approach to evaluating the outcomes of training programs. He argued that training evaluation should include assessment at four levels: (a) reactions (i.e., what trainees think of the training), (b) learning (i.e., what trainees learned), (c) behavior (i.e., how trainees' behavior changes), and (d) results (i.e., impact on organization). Building on Kirkpatrick's framework, Kraiger, Ford, and Salas (1993) outlined three similar outcomes of training programs: (a) affective (i.e., reactions), (b) cognitive (i.e., learning), and (c) skill-based (i.e., behavior) outcomes. Several training reviews have been conducted recently arguing for the use of a multilevel approach to training evaluation (Alliger & Janak, 1989; Salas et al., 2001).

Best Practice 9. Supervisor support following training should be encouraged to improve transfer of (team) training.

Supervisor support has been argued to influence the transfer of learned teamwork competencies to the actual task environment. Thus, if trainees perceive that the teamwork competencies are valued by the organization and management, they will be more likely to integrate what they learned on the job. Research indicates that discussions held by supervisors prior to and following training (e.g., Huczynski & Lewis, 1980), supervisor sponsorship (e.g., Brinkerhoff & Montesino, 1995), and opportunities to perform the learned skills (Ford, Quinones, Sego, & Speer, 1991) result in transfer of learned skills to the job. However, Baldwin and Ford (1988) argued that there is a lack of understanding regarding the behaviors that lead trainees to perceive support. Tannenbaum and Yukl (1992) stated that supervisor support could include goal-setting activities (e.g., minimize number of accidents), reinforcement (e.g., error reporting), and modeling of trained behaviors (e.g., teamwork behaviors). Finally, it is important that the performance appraisals and reward systems used by supervisors focus on outcomes at the team level, not just the individual level (Cannon-Bowers, Salas, & Milham, 2000; Smith-Jentsch, Salas, & Brannick, 2001). Although these results are encouraging, additional research is needed to determine the true impact of supervisor support on transfer of training.

Best Practice 10. An organizational climate that encourages continuous learning should be created to improve transfer of team training.

Research supporting the role of an organization's climate on the transfer of training is slightly larger than that of supervisor support. Organizational climate can be defined as the interaction among elements within the organizational setting that are observable as well as those that are perceived by trainees (Hellreigel & Slocum, 1974; James & Jones, 1974). Research suggests that trainees who perceive a positive organizational climate (e.g., organizational support, rewards) apply learned competencies on the job (Baumgartel, Reynolds, & Pathan, 1984; Rouiller & Goldstein, 1993; Tracey et al., 1995). In addition, we argue that a continuous-learning culture (a part of the organizational climate) that supports and provides opportunities to acquire and demonstrate teamwork knowledge and skills, reinforces achievement, and encourages innovation and competition (Dubin, 1990; Rosow & Zager, 1988; Tracey et al., 1995) is important. Teams that are subjected to this positive environment understand that learning is a part of their daily work environment, and thus it is accepted. Research conducted by Tracey et al. suggests that trainees who perceived a continuous-learning environment demonstrated more posttraining behaviors. In high consequences environments where errors can be fatal, such as in health care, a continuous-learning environment is necessary to encourage team members to continuously demonstrate teamwork behaviors learned during training and to learn from any errors that do occur.

CONCLUDING REMARKS

This chapter has provided a brief summary of what we know about teamwork and team training, which can be useful and applicable to guide the design, delivery, evaluation, and transfer of team training in health care. Much of what we know, we submit, is applicable. The science of team training has produced tools, techniques, strategies, and lessons learned that can be very useful (once adapted) to enhance patient safety. So there is much that can be directed toward improving teamwork in the operating rooms, emergency rooms, intensive care units, Code Blue calls, and other health care situations.

We hope this chapter helps implement and institutionalize effective team training strategies in health care. We also hope that partnerships are created between those who know team training with those who know health care and together create training systems that improve patient safety. We hope this chapter will motivate additional research in, for example, multicultural medical teams and simulation-based training and that it illustrates that there are more team training strategies than just CRM. Finally, we hope this chapter helps health care organizations implement sound training to reduce errors and improve patient safety.

References

Alliger, G., & Janak, E. (1989). Kirkpatrick's levels of training criteria: Thirty years later. *Personnel Psychology, 42,* 331–342.

Baldwin, T. T., & Ford, J. K. (1988). Transfer of training: A review and directions for future research. *Personnel Psychology, 41,* 63–105.

Baldwin, T. T., & Magjuka, R. J. (1997). Organizational context and training effectiveness. In J. K. Ford et al. (Eds.), *Improving training effectiveness in work organizations* (pp. 99–127). Mahwah, NJ: Lawrence Erlbaum Associates, Inc.

Baldwin, T. T., Magjuka, R. J., & Loher, B. T. (1991). The perils of participation: Effects of choice of training on trainee motivation and learning. *Personnel Psychology, 44,* 51–65.

Bandow, D. (2001). Time to create sound teamwork. *Journal for Quality and Participation, 24,* 41–47.

Bassi, L., & Van Buren, M. (1998). The 1998 state of the industry report. *Training and Development, 52,* 22–43.

Baumgartel, H., Reynolds, M., & Pathan, R. (1984). How personality and organizational-climate variables moderate the effectiveness of management development programmes: A review and some recent research findings. *Management and Labour Studies, 9,* 1–16.

Blickensderfer, E. L., Cannon-Bowers, J. A., & Salas, E. (1997a). Theoretical bases for team self-correction: Fostering shared mental models. In M. Beyerlein, D. Johnson, & S. Beyerlein (Eds.), *Advances in interdisciplinary studies in work teams series* (Vol. 4, pp. 249–279). Greenwich, CT: JAI.

Blickensderfer, E. L., Cannon-Bowers, J. A., & Salas, E. (1997b, April). *Training teams to self-correct: An empirical investigation.* Paper presented at the 12th annual meeting of the Society for Industrial and Organizational Psychology, St. Louis, MO.

Bowers, C. A., Baker, D. P., & Salas, E. (1994). Measuring the importance of teamwork: The reliability and validity of job/task analysis indices for team-training design. *Military Psychology, 6,* 205–214.

Bowers, C. A., Blickensderfer, E. L., & Morgan, B. B. (1998). Air traffic control specialist team coordination. In M. W. Smolensky & E. S. Stein (Eds.), *Human factors in air traffic control* (pp. 215–236). San Diego, CA: Academic.

Bowers, C. A., Morgan, B. B., Jr., Salas, E., & Prince, C. (1993). Assessment of coordination demand for aircrew coordination training. *Military Psychology, 5,* 95–112.

Brinkerhoff, R. O., & Montesino, M. U. (1995). Partnership for training transfer: Lessons from a corporate study. *Human Resource Development Quarterly, 6,* 263–274.

Burke, C. S. (2004). Team task analysis. In N. Stanton, A. Hedge, K. Brookhuis, E. Salas, & H. Hendrick (Eds.), *Handbook of human factors and ergonomics methods* (pp. 56.1–56.8). London: Taylor & Francis.

Burke, C. S. (2005). Team task analysis. In N. Stanton, H. Hendrick, S. Konz, K. Parsons, & E. Salas (Eds.), *Handbook of human factors and ergonomics methods* (pp. 56–1 through 56–8). London, UK: Taylor & Francis.

Campion, M. A., Medsker, G. J., & Higgs, A. C. (1993). Relations between work group characteristics and effectiveness: Implications for designing effective work groups. *Personnel Psychology, 46,* 823–850.

Cannon-Bowers, J. A., Burns, J. J., Salas, E., & Pruitt, J. S. (1998). Advanced technology in scenario-based training. In J. A. Cannon-Bowers & E. Salas (Eds.), *Making decisions under stress: Implications for individual and team training* (pp. 365–374). Washington, DC: American Psychological Association.

Cannon-Bowers, J. A., & Salas, E. (1997). A framework for measuring team performance measures in training. In M. T. Brannick, E. Salas, & C. Prince (Eds.), *Team performance assessment and measurement: Theory, methods, and applications* (pp. 45–62). Hillsdale, NJ: Lawrence Erlbaum Associates, Inc.

Cannon-Bowers, J. A., & Salas, E. (1998). Team performance and training in complex environments: Recent findings from applied research. *Current Directions in Psychological Science, 7,* 83–87.

Cannon-Bowers, J. A., Salas, E., & Milham, L. M. (2000). The transfer of team training: Propositions and preliminary guidance. *Advances in Developing Human Resources, 8,* 63–74.

Cannon-Bowers, J. A., Salas, E., Tannenbaum, S. I., & Mathieu, J. E. (1995). Toward theoretically based principles of training effectiveness: A model and initial empirical investigation. *Military Psychology, 7,* 141–164.

Cannon-Bowers, J. A., Tannenbaum, S. I., Salas, E., & Volpe, C. E. (1995). Defining team competencies and establishing team training requirements. In R. Guzzo, E. Salas, & Associates (Eds.), *Team effectiveness and decision making in organizations* (pp. 333–380). San Francisco, CA: Jossey-Bass.

Colquitt, J. A., LePine, J. A., & Noe, R. A. (2000). Toward an integrative theory of training motivation: A meta-analytic path analysis of 20 years of research. *Journal of Applied Psychology, 85,* 678–707.

Cooke, N. J., Salas, E., & Cannon-Bowers, J. A. (2000). Measuring team knowledge. *Human Factors, 42,* 151–173.

Cooper, R. K., & Sawaf, A. (1996). *Executive EQ: Emotional intelligence in leadership and organization.* New York: Perigee.

Cox, T., Jr., Lobel, S. A., & McLeod, P. L. (1991). Effects of ethnic group cultural differences on cooperative and competitive behavior on a group task. *Academy of Management Journal, 4,* 827–847.

Crane, M. (1997, July). Malpractice is not criminal; mistake is? *The National Law Journal,* A17.

Driskell, J. E., & Salas, E. (1992). Collective behavior and team performance. *Human Factors, 34,* 277–288.

Dubin, S. (1990). Maintaining competence through updating. In S. L. Willis & S. S. Dubin (Eds.), *Maintaining professional competence* (pp. 9–43). San Francisco: Jossey-Bass.

Dweck, C. S. (1986). Motivational processes affecting learning. *American Psychology, 41,* 1040–1048.

Dweck, C. S., & Leggett, E. L. (1988). A social–cognitive approach to motivation and personality. *Psychological Review, 95,* 256–273.

Eby, L. T., & Dobbins, G. H. (1997). Collectivistic orientation in teams: An individual and group level analysis. *Journal of Organizational Behavior, 18,* 275–295.

Ehrenstein, A., Walker, B., Czerwinski, M., & Feldman, E. (1997). Some fundamentals of training and transfer: Practice benefits are not automatic. In M. Quiñones & A. Ehrenstein (Eds.), *Training for a rapidly changing workplace: Applications of psychological research* (pp. 119–147). Washington, DC: American Psychological Association.

Entin, E. E., & Serfaty, D. (1999). Adaptive team coordination. *Human Factors, 41,* 312–325.

Fleishman, E. A., Mumford, M. D., Zaccaro, S. J., Levin, K. Y., Korotkin, A. L., & Hein, M. B. (1991). Taxonomic efforts in the description of leader behavior: A synthesis and functional interpretation. *Leadership Quarterly, 2,* 245–287.

Fleishman, E. A., & Zaccaro, S. J. (1992). Toward a taxonomy of team performance functions. In R. W. Swezey & E. Salas (Eds.), *Teams: Their training and performance* (pp. 31–56). Norwood, NJ: Ablex.

Ford, J. K., Kozlowski, S., Kraiger, K., Salas, E., & Teachout, M. (Eds.). (1997). *Improving training effectiveness in work organizations.* Mahwah, NJ: Lawrence Erlbaum Associates, Inc.

Ford, J. K., Quinones, M., Sego, D., & Speer, J. (1991, April). *Factors affecting the opportunity to use trained skills on the job.* Paper presented at the 6th Annual Conference for the Society of Industrial and Organizational Psychology, St. Louis, MO.

Ford, J. K., Smith, E. M., Weissbeinn, D. A.., Gully, S. M.., & Salas, E. (1998). Relationships of goal-orientation, metacognitive activity, and practice strategies with learning outcomes and transfer. *Journal of Applied Psychology, 83,* 218–233.

Ford, J. K., & Weissbein, D. A. (1997). Transfer of training: An updated review and analysis. *Performance Improvement Quarterly, 10,* 22–41.

Fowlkes, J., Dwyer, D. J., Oser, R. L., & Salas, E. (1998). Event-based approach to training (EBAT). *International Journal of Aviation Psychology, 8,* 209–221.

Glickman, A. S., Zimmer, S., Montero, R. C., Guerette, P. J., Morgan, B. B., Jr., & Salas, E. (1987). *The evolution of teamwork skills: An empirical assessment with implications for training.* Technical report for the Naval Training Systems Center. Orlando: University of Central Florida.

Goldstein, I. L. (1993). *Training in organizations* (3rd ed.). Pacific Grove, CA: Brooks/Cole.

Goldstein, I. L., & Ford, J. K. (2002). *Training in organizations: Needs assessment, development, and evaluation* (4th ed.). Belmont, CA: Wadsworth.

Griepentrog, B. K., & Fleming, P. J. (2003, April). *Shared mental models and team performance: Are you thinking what we're thinking?* Paper presented at the 18th annual meeting of the Society for Industrial/Organizational Psychologists, Orlando, FL.

Hackman, R. (Ed.). (1990). *Groups that work (and those that don't): Creating conditions for effective teamwork.* San Francisco, CA: Jossey-Bass.

Hellreigel, D., & Slocum, J. W. (1974). Organizational climate: Measures, research, and contingencies. *Academy of Management Journal, 17,* 255–280.

Huczynski, A. A., & Lewis, J. W. (1980). An empirical study into the learning transfer process management training. *Journal of Management Studies, 17,* 227–240.

Hunter, J. E. (1986). Cognitive ability, cognitive aptitudes, job knowledge, and job performance. *Journal of Vocational Behavior, 29,* 340–362.

Irwin, C. M. (1991). The impact of initial and recurrent cockpit resource management training on attitudes. In R. S. Jensen (Ed.), *Proceedings of the 6th International Symposium on Aviation Psychology* (pp. 344–349). Columbus, OH: The Ohio State University.

Jacobs, T. O., & Jaques, E. (1987). Leadership in complex systems. In J. Zeidner (Ed.), *Human productivity enhancement Vol. 2. Organizations, Personnel, and decision making,* (pp. 7–65). New York, NY: Praeger.

James, L. R., & Jones, A. P. (1974). Organizational climate: A review of theory and research. *Psychological Bulletin, 81,* 1096–1112.

Jentsch, F., & Smith-Jentsch, K. (2001). Assertiveness and team performance: More than "just say no." In E. Salas, C. Bowers, & E. Edens (Eds.), *Improving teamwork in organizations* (pp. 73–94). Mahwah, NJ: Lawrence Erlbaum Associates, Inc.

Joint Commission Resources (Ed.). (1998). Sentinel events: Evaluating cause and planning improvement, 2nd ed. Oak Brook, IL: Joint Commission Resources.

Kirkpatrick, D. L. (1976). Evaluation of training. In R. L. Craig (Ed.), *Training and development handbook: A guide to human resource development* (2nd ed., pp. 1–26). New York: McGraw-Hill.

Kleinman, D. L., & Serfaty, D. (1989, April). *Team performance assessment in distributed decision making.* Paper presented at the Simulation and Training Research Symposium on Interactive Networked Simulation for Training, Orlando, FL.

Kohn, L. T., Corrigan, J. M., & Donaldson, M. S. (Eds.). (1999). *To err is human: Building a safer health system.* Washington, DC: National Academy Press.

Komaki, J., Desselles, M., & Bowman, E. (1989). Definitely not a breeze: Extending an operant model of effective supervision teams. *Journal of Applied Psychology, 74,* 522–529.

Kozlowski, S. W. J., Gully, S. M., Salas, E., & Cannon-Bowers, J. A. (1996). Team leadership and development: Theory, principles, and guidelines for training leaders and teams. In M. Beyerlein, S. Beyerlein, & D. Johnson (Eds.), *Advances in interdisciplinary studies of work teams: Team leadership* (Vol. 3, pp. 253–292). Greenwich, CT: JAI.

Kraiger, K., Ford, J. K., & Salas, E. (1993). Application of cognitive, skill-based, and affective theories of learning outcomes to new methods of training evaluation. *Journal of Applied Psychology, 78,* 311–328.

Leape, L. L. (1994). The preventability of medical injury. In M. S. Bogner (Ed.), *Human error in medicine* (pp. 13–25). Hillsdale, NJ: Lawrence Erlbaum Associates, Inc.

Marks, M. A. (2000). A critical analysis of computer simulations for conducting team research. *Small Group Research, 31,* 653–675.

Marks, M. A., Mathieu, J. E., & Zaccaro, S. J. (2001). A conceptual framework and taxonomy of team processes. *Academy of Management Review, 26,* 356–376.

Martin, G. L., & Lumsden, J. A. (1987). *Coaching: An effective behavioral approach.* St. Louis, MO: Mosby.

Martocchio, J. J., & Webster, J. (1992). Effects of feedback and cognitive playfulness on performance in microcomputer software training. *Personality Psychology, 45,* 553–578.

Mathieu, J. E., Heffner, T. S., Goodwin, G. F., Salas, E., & Cannon-Bowers, J. A. (2000). The influence of shared mental models on team process and performance. *Journal of Applied Psychology, 85,* 273–283.

Mathieu, J. E., Tannenbaum, S. I., & Salas, E. (1992). An influence of individual and situational characteristics on training effectiveness measures. *Academy of Management Journal, 35,* 827–847.

McIntyre, R. M., & Salas, E. (1995). Measuring and managing for team performance: Emerging principles from complex environments. In R. A. Guzzo & E. Salas (Eds.), *Team effectiveness and decision making in organizations* (pp. 9–45). San Francisco, CA: Jossey-Bass.

McNeese, M., & Rentsch, J. (2001). Identifying the cognitive and social requirements of teamwork using collaborative task analysis. In M. D. McNeese, E. Salas, & M. R. Endsley (Eds.), *New trends in cooperative activities: System dynamics in complex environments* (pp. 96–113). Santa Monica, CA: Human Factors and Ergonomics Society.

Mohammed, S., Klimoski, R., & Rentsch, J. R. (2000). The measurement of team mental models: We have no shared schema. *Organizational Research Methods, 3,* 123–165.

Noe, R. A. (1986). Trainee attributes and attitudes: Neglected influences on training effectiveness. *Academy of Management Review, 4,* 736–749.

Phillips, J. M., & Gully, S. M. (1997). Role of goal orientation, ability, need for achievement, and locus of control in the self-efficacy and goal setting process. *Journal of Applied Psychology, 82,* 792–802.

Pietro, D. A., Shyavitz, L. J., Smith, R. A., & Auerbach, B. S. (2000). Detecting and reporting medical errors: Why the dilemma? *British Medical Journal, 320,* 794–796.

Porter, C. O. L. H., Hollenbeck, J. R., Ilgen, D. R., Ellis, A. P. J., West, B. J., & Moon, H. (2003). Backing up behaviors in teams: The role of personality and legitimacy of need. *Journal of Applied Psychology, 88,* 391–403.

Quinones, M. A. (1995). Pretraining context effects: Training assignment as feedback. *Journal of Applied Psychology, 80,* 226–238.

Quinones, M. A. (1997). Contextual influencing on training effectiveness. In M. A. Quinones & A. Ehrenstein (Eds.), *Training for a rapidly changing workplace: Applications of psychological research* (pp. 177–200). Washington, DC: American Psychological Association.

Ree, M. J., Carretta, T. R., & Teachout, M. S. (1995). Role of ability and prior job knowledge in complex training performance. *Journal of Applied Psychology, 80,* 721–730.

Ree, M. J., & Earles, J. A. (1991). Predicting training success: Not much more than G. *Personality Psychology, 44,* 321–332.

Rentsch, J. R., & Klimoski, R. J. (2001). Why do "great minds" think alike? Antecedents of team member schema agreement. *Journal of Organizational Behavior, 22,* 107–120.

Rosow, J. M., & Zager, R. (1988). *Training the competitive edge.* San Francisco, CA: Jossey-Bass.

Rouiller, J. Z., & Goldstein, I. L. (1993). The relationship between organizational transfer climate and positive transfer of training. *Human Resource Development Quarterly, 4,* 377–390.

Salas, E., Burke, C. S. Bowers, C. A., & Wilson, K. A. (2001). Team training in the skies: Does crew resource management (CRM) training work? *Human Factors, 43,* 641–674.

Salas, E., Burke, C. S., & Stagl, K. C. (2004). Developing teams and team leaders: Strategies and principles. In D. Day, S. J. Zaccaro, & S. M. Halpin (Eds.), *Leader development for transforming organizations: Growing leaders for tomorrow* (pp. 325–355). Mahwah, NJ: Lawrence Erlbaum Associates, Inc.

Salas, E., & Cannon-Bowers, J. A. (1997). Methods, tools, and strategies for team training. In M. A. Quinones & A. Ehrenstein (Eds.), *Training for a rapidly changing workplace: Applications of psychological research* (pp. 249–279). Washington, DC: American Psychological Association.

Salas, E., & Cannon-Bowers, J. A. (2000). The anatomy of team training. In S. Tobias & J. D. Fletcher (Eds.), *Training and retraining: A handbook for business, industry, government, and the military* (pp. 312–335). New York: Macmillan Reference.

Salas, E., & Cannon-Bowers, J. A. (2001). The science of training: A decade of progress. *Annual Review of Psychology, 52,* 471–499.

Salas, E., Cannon-Bowers, J. A., & Johnston, J. H. (1997). How can you turn a team of experts into an expert team?: Emerging training strategies. In C. E. Zsambok & G. Klein (Eds.), *Naturalistic decision making* (pp. 359–370). Mahwah, NJ: Lawrence Erlbaum Associates.

Salas, E., Dickenson, T. L., Converse, S. A., & Tannenbaum, S. I. (1992). Toward an understanding of team performance and training. In R. J. Swezey & E. Salas (Eds.), *Teams: Their training and performance* (pp. 3–29). Norwood, NJ: Ablex.

Salas, E., Rhodenizer, L., & Bowers, C. A. (2000). The design and delivery of CRM training: Exploiting available resources. *Human Factors, 42,* 490–511.

Salas, E., Sims, D. E., Klein, C., & Burke, C. S. (2003). Can teamwork enhance patient safety? *Forum, 23,* 5–9.

Salas, E., Wilson-Donnelly, K. A., Burke, C. S., & Priest, H. A. (2005). Using simulation-based training to improve patient safety: What does it take? *Joint Commission Journal on Quality and Patient Safety, 31*(7), 363–371.

Serfaty, D., Entin, E. E., & Johnston, J. H. (1998). Team coordination training. In J. A. Cannon-Bowers & E. Salas (Eds.), *Making decisions under stress: Implications for individual and team training* (pp. 221–246). Washington, DC: American Psychological Association.

Shamir, B. (1990). Calculations, values and identities: The sources of collectivistic work motivation. *Human Relations, 43,* 313–332.

Shute, V. J., & Gawlick, L. A. (1995). Practice effects on skill acquisition, learning outcome, retention, and sensitivity to relearning. *Human Factors, 37,* 781–803.

Smith-Jentsch, K., Salas, E., & Baker, D. P. (1996). Training team performance-related assertiveness. *Personnel Psychology, 49,* 909–936.

Smith-Jentsch, K. A., Salas, E., & Brannick, M. T. (2001). To transfer or not to transfer? Investigating the combined effects of trainee characteristics, team leader support, and team climate. *Journal of Applied Psychology, 86,* 279–292.

Smith-Jentsch, K. A., Zeisig, R. L., Acton, B., & McPherson, J. A. (1998). Team dimensional training: A strategy for guided team self-correction. In J. A. Cannon Bowers & E. Salas (Eds.), *Making decisions under stress: Implications for individual and team training* (pp. 271–297). Washington, DC: American Psychological Association.

Stevens, M. J., & Campion, M. A. (1994). The knowledge, skill and ability requirements for teamwork: Implications for human resource management. *Journal of Management, 20,* 503–530.

Stevens, M. J., & Campion, M. A. (1999). Staffing work teams: Development and validation of a selection test for teamwork settings. *Journal of Management, 25,* 207–228.

Stewart, G. L., & Manz, C. C. (1995). Leadership for self-managing work teams: A typology and integrative model. *Human Relations, 48,* 747–770.

Stout, R. J., Salas, E., & Fowlkes, J. E. (1997). Enhancing teamwork in complex environments through team training. *Group Dynamics, 1,* 169–182.

Tannenbaum, S. I., Mathieu, J. E., Salas, E., & Cannon-Bowers, J. A. (1991). Meeting trainees' expectations: The influences of training fulfillment on the development of commitment, self-efficacy, and motivation. *Journal of Applied Psychology, 76,* 759–769.

Tannenbaum, S. I., & Yukl, G. (1992). Training and development in work organizations. *Annual Review of Psychology, 43,* 399–441.

Tracey, J. B., Tannenbaum, S. I., & Kavanagh, M. J. (1995). Applying trained skills on the job: The importance of work environment. *Journal of Applied Psychology, 80,* 239–252.

Volpe, C. E., Cannon-Bowers, J. A., Salas, E., & Spector, P. E. (2001). The impact of cross-training on team functioning: An empirical investigation. In R. W. Swezey & D. H. Andrews (Eds.), *Readings in training and simulation: A 30-year perspective* (pp. 115–128). Santa Monica, CA: Human Factors and Ergonomics Society.

Webber, S. S. (2002). Leadership and trust facilitating cross-functional team success. *Journal of Management Development, 21,* 201–274.

Zaccaro, S. J., Rittman, A. L., & Marks, M. A. (2001). Team leadership. *The Leadership Quarterly, 12,* 451–483

45

TILTING THE CULTURE IN HEALTH CARE: USING CULTURAL STRENGTHS TO TRANSFORM ORGANIZATIONS

John S. Carroll and Maria Alejandra Quijada
MIT Sloan School of Management

"Searching for one word to describe the state of mind of the physician in the United States today, we might choose beleaguered" (Berwick & Nolan, 1998, p. 289). Health care professionals have barely enough time and energy to cope with daily problems, leaving few resources for innovation and fundamental change (Edmondson, 2004). Economic pressures and new technologies have shifted the balance of power among physicians, technicians, nurses, hospitals, and insurers (Barley, 1986; Edmondson, Bohmer, & Pisano, 2001; Millman, 1977). Where health care was once a high calling, it is now a regulated business whose customers want permanent good health whereas payers insist on limiting costs and curtailing professional discretion.

The recent movements in health care toward enhanced patient safety and quality of care are directed at systemic changes: "quality problems occur typically not because of a failure of goodwill, knowledge, effort, or resources devoted to health care, but because of fundamental shortcomings in the ways care is organized" (Corrigan, Donaldson, Kohn, Maguire, & Pike, 2001, p. 25). Health care is delivered by individuals with personal and professional identities who work within and across complex organizations structured by our economic–political–legal system. Fundamental changes, such as those advocated by the Institute of Medicine (Corrigan et al., 2001; Kohn, Corrigan, & Donaldson, 2000) and the Institute for Healthcare Improvement[1], require that health care practices, systems, and the cultural assumptions that underlie and reinforce existing behaviors realign in a more effective and sustainable way.

Our purpose in this chapter is to present a somewhat different approach to change, with a particular focus on health care, that emphasizes working with existing cultural strengths to address practical problems in new ways that will gradually shift cultural assumptions and values. In the first section of the chapter, we present definitions and concepts about organizational culture. We then review potential strategies for change, including culture change. Finally, we examine the health care culture and suggest a strategy for redirecting existing cultural strengths to support change.

ORGANIZATIONAL CULTURE

Organizational culture seems to be blamed for both successes and failures, often without a clear understanding of its meaning (Weeks, 2004). For our purposes, a good everyday definition of culture is: **the way we do things around here and why we do them.** This definition emphasizes both visible behaviors and the more subtle values and assumptions underlying them. Culture emerges from successful collective solutions to organizational problems. Culture is pervasive;

[1]http://www.ihi.org

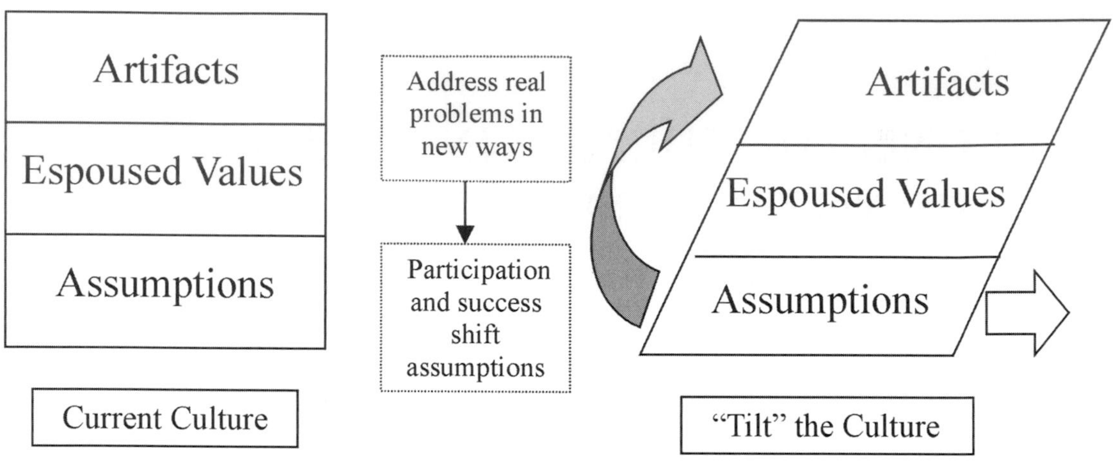

Figure 45–1. Culture change by "tilting" the culture.

understanding it can help us comprehend why things are they way they are, what needs to be changed, and what needs to be preserved for the organization to achieve a desired set of behaviors that will be the foundation of a new culture.

Edgar Schein (1992, 1999), a seminal organizational researcher, articulated the three-level conceptualization of culture shown on the left side of Figure 45–1. At the top or surface level are the observable behaviors and artifacts of the organization. This is what a television camera or documentary crew would capture in a hospital, a doctor's office, a community clinic, a medical research lab, an insurance company, or any other health care setting. An observer could extract patterns of behavior or norms, but it would be difficult to understand what is happening without talking to people about the meaning of these activities. This level "is easy to observe but very difficult to decipher" (Schein, 1992, p. 17).

The second level of culture comprises the beliefs and values that participants espouse, what they are willing and able to verbalize. Whether articulated dutifully by employees or displayed on the walls, this gives us an idea of what people say they do or what they think should be done but not necessarily what is actually done or why. Sometimes espoused values and visible behaviors and artifacts are consistent as people live a coherent and "authentic" culture, but often articulated beliefs and values represent an ideal aspiration or an image carefully crafted to achieve legitimacy with employees, customers, or investors. Although the mission statement may emphasize quality and teamwork, for example, the working reality may be cost reduction and individual reward and blame.

Contradictions between behaviors and espoused values help reveal Schein's (1992) third and deepest level of culture: underlying assumptions, often taken for granted and unarticulated, that have developed over time through successful collective problem solving and then became part of "the way we do things here" to the point where they are no longer questioned. Symbols, myths, and stories embody cultural assumptions and give meaning to everyday events. People continue to enact the culture and teach these ways to newcomers, even when they have outlived their usefulness, and find it strange when the assumptions are challenged.

For example, as a patient in the health care system it is evident that doctors mostly work individually. They rarely ask for help and do not take kindly to someone telling them how to practice medicine (Adler, Riley, Kwon, Lee, & Satrasala, 2003). In many hospitals, doctors are not even employees of the hospital but rather individual or small group practitioners who have privileges to work at one or more hospitals. These cultural artifacts of work organization are accompanied by values around individual autonomy and independence articulated in medical school training, certification processes, awards for excellence, and so forth. Underlying these values are deeply held assumptions that have developed over centuries when individual doctors had to be responsible for their patients under extreme and uncertain conditions. Anything less is a violation of their identity, their oath of professional responsibility, and their understanding of how health care works, even to the point of assuming that asking for help is a sign of incompetence.

An organizational culture is rarely uniform throughout the organization, and there is no single ideal amount of culture strength or fragmentation (Martin, 1992; Schein, 1992, 1999). Hospitals, for example, have particularly strong subcultures defined by professions or occupations such as medicine, nursing, technicians, and administration. These are further divided into surgery, anesthesiology, pharmacy, accounting, marketing, and so forth. Cutting across these occupational groups are local, national, and ethnic cultures. Individual leadership styles can create unique cultures, as Edmondson (2004) found among work groups in the same nursing unit or operating room. An organization leader may try to meld those bits and pieces into a single identity and culture, or a crisis may bring everyone together with common purpose and force some cultural blending in answer to the crisis, but the individuals and groups are likely to retain diverse cultural elements within a more or less uniform organizational culture. We believe effective cultural change will involve participation from diverse groups in addressing common problems and thereby building cultural bridges.

CULTURE AND CHANGE

Change that is not rooted in culture can be ephemeral. People resist changes that undermine their hard-won expertise, status, identity, habits, and understandings (Beckhard & Pritchard, 1992; Kanter, Stein, & Jick, 1992). Early improvement efforts often lose momentum as resistance emerges and attention shifts to other issues. Apparent success in pilot projects may not transfer to the rest of the organization (Keating, Oliva, Repenning, Rockart, & Sterman, 1999). People easily revert to familiar routines and thought patterns with even greater cynicism about the change program of the month.

Real long-term change that becomes integrated with the culture takes time, because the new ways need to be tested and shown successful as new meanings emerge in the form of symbols, heroes, stories, and sayings that become part of the culture. Without a sense of urgency about the need to change (Kotter, 1996), people will not take risks in trying new things, and change efforts will likely fail. Sometimes scandals and emergencies provide that incentive for change. The Institute of Medicine (Kohn et al., 2000) identifying at least 44,000 unnecessary deaths each year in the United States from medical errors created this sense of urgency and desire for change. Yet at the same time, people need to feel comfortable enough to try out new approaches without a paralyzing fear of punishment (Schein, 2002).

Schein (1992, 1999) suggested a variety of strategies for culture change that are related to general approaches to change (Beer & Nohria, 2000a). It is common for senior managers to mandate change by making a new mission statement and changing performance measures and incentives. Sometimes leaders are imported from another organization in the hope that they will shape the culture in the image of their previous culture. For example, nuclear power plants that get in trouble with regulators have a pattern of hiring Navy admirals into senior management positions on the assumption that the Navy has an effective and disciplined way of doing things and the admiral will be able to make the plant more Navy-like. Sometimes the culture to emulate is found within one of the organizational subcultures, and individuals from that ideal subculture are promoted to other positions in the hope that they will spread their assumptions and values to the rest of the organization (Schein, 1992).

However, the research literature suggests that such a "culture war" is successful only if the new leaders gain credibility with the employees and mandate changes that make some sense in the present culture (Dyer, 1986). For example, should hospitals hire from the Navy or the airlines or other hospitals?[2] What are the implications of promoting from another industry or choosing medicine over

[2]For a comparison of aviation and medicine culture see Helmreich and Merritt (1998).

nursing or finance over medicine as sources of leadership? This process is much more likely to succeed if current organization members are involved in helping to design new ways of doing things that solve real problems and thereby engage support from them and others and eventual acceptance of the values and assumptions.

An alternative strategy that could be advantageous for health care focuses on more cooperative and bottom-up change. Schein (1992, 1999) suggested that every culture has strengths that can be drawn on and reframed to be supportive of the desired changes. The immediate objective is not to change the culture but rather to work together on common problems in ways that draw on desirable elements of the existing culture. In this way new behaviors and values can be tied to existing assumptions and values, given legitimacy, and made more acceptable and less threatening. Over time, as these new behaviors are tried out, and if people recognize that they are working well to bring desired results, new assumptions arise to articulate the cultural lessons learned. In the right-hand panel of Figure 45–1, we represent this as a curved arrow using cultural elements to design new behaviors (artifacts) that shift or tilt the top level. As the surface level of culture shifts, whereas the connections to deeper levels are retained, the deeper levels will gradually shift (the second, smaller arrow in Figure 45–1) as people reshape their own culture to give meaning to the links between behaviors and values.

Cooperrider et al. (2000) offered a related idea in arguing for appreciative inquiry, that is, assessing strengths and sources of resiliency rather than focusing on deficiencies and problems. The idea is to use the strengths of the culture for leverage to get people to try out the new solutions, and when these prove more useful than the previous ones the old assumptions will be abandoned and new ones created. Beer and Nohria (2000b) pointed out that top-down and bottom-up approaches to change are not inherently incompatible, and successful change may involve both.

AN EXAMPLE OF "TILTING" THE CULTURE FROM ANOTHER INDUSTRY

In the mid-1990s the Millstone Nuclear Power Station in Connecticut was forced to cease operations by the U.S. Nuclear Regulatory Commission (NRC) for a variety of problems, although management insisted that they were running a safe and effective operation. Employees within Millstone who had raised questions about quality and safety had reported being pressured by management to remain silent. The founding culture of Millstone, which had focused on technical excellence, had been thrust underground when financial crises in the 1980s made heroes out of those who got things done at low cost, even when it meant starting the newest unit with a long list of work to be done. An interviewee at Millstone labeled the management culture as "male … militaristic—control and command" (Carroll, Rudolph, & Hatekenaka, 2001, p. 113). The NRC had been promised by Millstone that there would be changes, but Millstone had not delivered. A few engineers who never bought into the current culture became whistleblowers to the NRC and the media. The NRC then acted to force change, requiring not only technical changes but also the creation of a safety-conscious work environment that would ensure a proper response to employee concerns.

The Board of Trustees' first step was to hire a new senior manager (CEO-Nuclear) to be in charge of Millstone (and the other nuclear plants of this utility). He moved swiftly to communicate his values—high standards, openness and honesty, commitment to do what was right, and two-way communication—through numerous meetings and pronouncements. He quickly solidified his management team with changes, including the hiring of a Navy admiral for oversight (quality) and the admiral's former chief of staff to manage the Employee Concerns Program, intended to provide a safe way for employees to communicate about issues. However, the CEO-Nuclear admitted he had "never encountered a culture as broken" (Carroll & Hatekenaka, 2001, p. 70) in which management actions were regarded automatically with suspicion and distrust. As a result, changes moved very slowly at first, because employees were reluctant to take the risks that change requires.

Trust developed over time because senior management created opportunities for broad participation and personal transformations in an environment that accepted honest mistakes. For example, the CEO put the operations vice president in charge of the safety-conscious work experience initiative. This vice president was a bright and articulate engineer but also a typical old-style manager; he described himself as weary of "whiners" and that "a lot of people were scared of me" (Carroll & Hatekenaka, 2001, p. 73). He admitted that in implementing the initiative he was just "going through the motions … I didn't believe anyone would harass someone who brought forth safety concerns. … After all, I live near here" (Carroll & Hatekenaka, 2001, p. 74) Yet, several months later, when two contractors were terminated for alleged poor performance, the director of the Employee Concerns Program immediately protested (and the CEO reversed the termination pending a review), and a prompt investigation provided evidence that the terminations had been improper; the operations vice president then realized "It was one of those moments your perception changes … a watershed for me" (Carroll & Hatekenaka, 2001, p. 74). It was not sufficient to say that the culture had to change and assume it had: Steps needed to be taken that showed the commitment to the new way of working. In response to this event, senior management created an Executive Review Board to examine any disciplinary actions, but its scope broadened to become a forum for discussion and learning about difficult management issues.

Over time, through continual communication, creation of new forums for discussion, encouragement of contributions from all levels of the workforce, training of all managers, and willingness to admit error and vulnerability (a major culture change because managers were evaluated on having no problems), participation and initiative spread through the organization. Of course, some managers could not change and had to be moved aside, but this was done fairly and openly. As employees came to believe that management was doing what was best for the plant and for their jobs, they began to contribute their ideas and efforts to the common problem of rebuilding Millstone. This process took 2 years but resulted in sufficient change for the NRC to agree that a safety-conscious work environment had been established and demonstrated to its satisfaction and to authorize restart of one of the nuclear power units. (A year later, a second unit was restarted.)

Although these actions and changes targeted behaviors and espoused values, without conscious planning they engaged deeper cultural values and assumptions and helped tilt them toward the desired ones. Millstone's traditional cultural strengths of **excellence, professional integrity**, and **safety**, which had been in a kind of culture war with cost issues and the assumption that good managers had no problems, became linked in support of new values of **openness** and **mutual respect** through rewarding people that came forward with safety concerns and by having managers actively demonstrate the new values. For example, the meaning of **excellence** had been based on assumptions such as "**excellent managers have no problems"** and **"excellent engineers know everything they need to know**." Through training, role models, and experience with many months of change at Millstone, the meaning of **excellence** shifted to encompass new behaviors connecting to assumptions such as "**excellent managers want to hear about problems and surprises to prevent more serious problems"** and **"excellent engineers use their knowledge to constantly learn and improve." Professional integrity** shifted from "**professionals have deep knowledge in their field of training**" to include "**professionals speak up to keep everyone informed"** and **"professionals listen to and learn from other professionals to enhance safety**." Over time, these new interpretations proved more successful than the previous ones at getting work done, achieving performance and safety goals, and reducing conflict, and they were incorporated into the organization's culture.

CULTURE AND CHANGE IN THE HEALTH CARE SYSTEM

The culture of health care is distinctive; some aspects are readily apparent to outsiders. Popular TV shows portray hospitals as fast-paced, intense, high-stakes, and very personal settings. Doctors, nurses, technicians, administrators, pharmacists, scientist–researchers, students, patients, families, lawyers, emergency workers, and maintenance workers collide under near-battlefield conditions. We see the long hours, the high stress, the demanding training of students, the status hierarchies, the interdependence of professional specialties, the decisions made under uncertainty, and the challenges of balancing excellent care, discovery of new knowledge, and efficient use of time and money. This has been portrayed as high drama in *ER* and as low comedy in *Scrubs*.

The Institute of Medicine (Corrigan et al., 2001) report titled *Crossing the Quality Chasm* gave a good summary in chapter 3 of some of the cultural "rules" or patterns of behavior that characterize health care today, along with suggestions for tomorrow's desired rules. The current rules are stated as a mixture of behaviors (e.g., **care is based primarily on visits; professionals control care; the system reacts to needs**) and espoused values (e.g., **secrecy is necessary; decision making is based on training and experience**). They focus on the current importance of autonomy, professionalism, individuality, and secrecy. The suggestions for desired rules focus on the need for openness, patient-centered care, teamwork, and systems thinking.

However, what is missing is the third level in the analysis of culture: the underlying assumptions connecting rules to each other and to desired outcomes. For example, summaries of behavior and values stated as "**professionals control care; decision making is based on training and experience, and preference is given to professional roles over the system**" are likely linked by an underlying assumption that "**knowledge about best care resides in individual professionals who therefore have to be responsible for their patients**."

New rules such as "**the patient is the source of control; knowledge is shared and information flows freely; safety is a system property; cooperation among clinicians is a priority**" make little sense in the context of current assumptions. New shared assumptions are needed, such as **"knowledge is shared and information flows freely"** because **"different caregivers and the patient and family must all contribute to successful care, which is customized according to patient needs and values**."

To change the rules, we must therefore search within the current culture for elements that can be retained, strengthened, reframed, and linked to new desired behaviors and values. New behaviors have to be tried out and shown to be effective, so that new assumptions will emerge to embed these lessons in the culture. We need to understand how these new rules will interact with the existing underlying assumptions and work to address or change previously existing assumptions that might diminish the effectiveness of the new practices. We illustrate our approach with the cultural element of **autonomy** and then give several examples of culture change from health care organizations.

AUTONOMY AS AN EXAMPLE OF A CULTURAL ELEMENT IN HEALTH CARE

As mentioned previously, autonomy is a central aspect of the medical professional subculture, but it is also part of other health care professions. Nurses, for example, are expected to solve problems without asking their supervisors for help (Tucker & Edmondson, 2003).

Autonomy, by itself, can be a cultural strength. For decades, if not centuries, the well-being of patients has depended on the skills of the individual practitioner. Training, licensing rules, folklore, stories about heroes, honored role models, and personal experience reinforce the assumption that individual skill is the single most important determinant of health care outcomes. The traditional way to avoid medical error is to hire highly qualified doctors and nurses who will avoid mistakes (Tucker & Edmondson, 2003). Health care professionals take initiative, make difficult decisions rapidly under great stress, take responsibility for their own learning, and feel personally engaged in their work

But autonomy can also be a weakness. For example, in thinking "**this is not a team endeavor,**" doctors assume they have to be "ironmen" who can do everything themselves, learn everything themselves, and work long hours without sleep (Kellogg, 2003). If no one, including colleagues, feels comfortable telling a physician how to practice medicine (Tucker & Edmondson, 2003), it is challenging to standardize practices that vary by region, hospital, and physician, upwards of 85% of which have not been tested empirically (Millenson, 1997). If it is true that "among all of the skills for improvement, the most crucial one may be the skill to cooperate across traditional boundaries" (Berwick & Nolan, 1998, p. 291), then how do we change individualistic behavior into being "part of the team as opposed to being the sole decision-maker" (Parker & Wertheimer, 1997, p. 451)? Teams cannot just be imposed on autonomous individuals.

Equally troubling is the assumption that **asking for help is a fault**, which limits learning opportunities and encourages palliative fixes to problems. For example, when nurses do ask for help, they ask people they are close to rather than those best equipped to solve the problem (Tucker & Edmondson, 2003). **Autonomy** easily becomes attached to **secrecy** when people assume that **speaking up will bring sanctions and outside interference** in their practice (Berwick, 1998; Edmondson, 2003; Edmondson, Roberto, & Tucker, 2001).

Rather than try to reduce or eliminate autonomy in health care, we believe it is necessary to understand the assumptions underlying autonomous behavior so that we can leverage strengths and reduce weaknesses. Indeed, the assumptions around autonomy have created a strong sense of personal responsibility and willingness to take initiative that may often keep a dysfunctional system from collapsing completely and will be critical to overcome fear of change.

How can we reinterpret autonomy, building on understandings from various medical and nursing specialties, to encourage speaking up, teamwork, and shared purpose? The current understanding of autonomy seems closely linked to status and power: It is a zero-sum game about control over professional turf where any show of weakness is a defeat. If we disengage autonomy from status and power, perhaps we can tilt toward other cultural elements such as excellence, caring, and professionalism. **Autonomy** could then mean that **everyone brings their unique experience to the problem at hand and the group then identifies the necessary knowledge, wherever it is.** This changes the assumption that **asking for help is a fault** to **asking for help is what a responsible and caring professional does.**

Such shifts in assumptions are not accomplished by exhortations and logical arguments. People have to try out new ways of working and begin to see their effectiveness and frame the reasons for success in terms of these new meanings. Leaders play an important role in providing role models and interpretations that help move this process forward and that offer support to those taking risks. For example, stories about operating room teams in which nurses and technicians feel free to point out issues to the surgeon and respected surgeons welcome their contributions and learn from collaboration help to shift assumptions. As success stories are shared in a supportive context, and more people try out new behaviors and find them effective, there is a gradual shift in "the way we do things around here," a reduced threat for trying new behaviors, and new identities and assumptions. Over time it becomes inconceivable for a surgeon *not* to ask for input from everyone in the operating room.

CULTURE CHANGE IN HEALTH CARE

An interesting example of culture change in health care that started from senior leadership was the shift toward a focus on patient safety at Minneapolis Children's Hospitals and Clinics championed by Julie Morath, who became chief operating officer in 1999 (Edmondson, Roberto, et al., 2001). She framed the following vision for patient safety: "The culture of health care must be one of everyone working together to understand safety, identify risks, and report them without fear of blame. We must look at ways of changing the whole system when we manage to zero defects" (Edmondson, Roberto, et al., p. 5). While vision and guiding ideas can create a sense of shared purpose, shape thinking, and orient the organization toward particular values and criteria for success, these have to be connected to people's current understandings and exemplified by concrete "new ways of working" (Schein, 1999). For example, Morath changed the language from threatening terms such as *errors* and *investigations* to *accidents* and *analysis*. These words help separate reporting of events from blame, but they also invoke the language of scientific analysis and thereby draw on the cultural respect for science and engage participants' scientific curiosity and interest in innovation. Taking advantage of these cultural assumptions helps to tilt the culture toward a safety orientation.

The change team made sure that executive leaders also provided a psychologically safe environment for the changes and mentored line leaders and informal network leaders who supported the detailed work of change. Leaders acted as role models and teachers to help demonstrate and interpret new behaviors. Brock Nelson, chief executive officer of Children's Hospitals and Clinics, openly described his "personal epiphany" in being able to enact a new policy of disclosing more information and personally apologizing to a family who had lost a teenage child who had initially been misdiagnosed. This helped to balance the medical culture of infallibility, where acknowledging any failure "can meet with resistance among physicians and managers for whom success is the only acceptable result" (Berwick, 1998, p. 654). It also maintained consistency with the value of autonomy, by allowing health professionals to deal directly with patients and their families.

Senior executives can articulate and champion new values, but that will have little impact without opportunities for people to work together on common problems. Morath created structures such as the Patient Safety Steering Committee to provide resources and venues for cross-disciplinary participation, allowing people to work together to learn how to solve common problems in terms of the new values. At Children's Hospitals, even front-line employees found ways take action by drawing on their personal commitment, informal networks, and modest support from management. For example, a clinical nurse specialist and a pharmacist in hematology/oncology, with support from the pharmacy manager, started a safety action team of cross-functional front-line service workers to meet

TABLE 45–1. Health Care Cultural Resources

Desired Cultural Elements	*Existing Cultural Strengths*
Informed: Each health care contributor has current knowledge of safety factors.	*Scientific Inquiry*: Health care practice is more effective when based on evidence
Reporting: Everyone is able to speak up about problems and surprises.	*Self-Criticism*: We learn faster by attending to every detail of our performance.
Just: Everyone feels fairly treated by everyone else (e.g., not unfairly blamed).	*Responsibility*: The individual is responsible for all aspects of patient care.
Learning: There is more to know, so we are always learning, even from failure.	*Training*: We learn by repeated practice and feedback from experienced mentors.
Teamwork: Health care demands teamwork from complementary professionals.	*Caring*: Health care is a helping profession that focuses on people's well-being.
Quality: We measure our individual and collective performance to improve.	*Excellence*: We want to be the best at what we do and be recognized for it.
	Dedication: Excellence demands long hours and selfless focus on the work.

monthly to discuss medication safety issues. Senior leadership's consistent and frequent message that patient safety was a priority allowed these employees to act autonomously but also to use management support to overcome numerous barriers such as status differences, heavy workloads, and general resistance to change. The concept spread to other departments and then became an organizational initiative for every clinical unit manager.

A very different example of culture change illustrates the variety of subcultures in health care and the initiation of change within subcultures. In a study of implementation of new technology for minimally invasive cardiac surgery at 16 hospitals, it became clear that the key to successful implementation of the technology was not high-level management support but rather the way the surgical team leaders fostered an atmosphere of learning, including acknowledgment of doubt, encouragement of communication, and real-time team learning (Pisano, Bohmer, & Edmondson, 2001). Researchers observed how some teams were able to create a dialogue in the operating room that truly took advantage of the possibilities of the new technology and turned it into a success. In other teams, team leaders assumed it was their expertise alone that mattered rather than engaging the whole team in new work practices, and then failed to make the most of the new technology or abandoned it. Yet given the typically high autonomy in surgical teams, we might guess that success in some teams would only spread to other teams if the reasons for better performance were shared and understood and the surgical team leaders had support to try new behaviors.

In Table 45–1 we offer some suggestions for cultural elements that could be strengthened in health care, drawing on Reason's (1997) ideas about safety culture and the Institute of Medicine (Corrigan et al., 2001; Kohn et al., 2000) reports. The right-hand column lists some of the existing cultural strengths that could be used as leverage for change. For example, in trying to create a reporting culture in which everyone is comfortable raising concerns, even in the face of status hierarchies (as in Edmondson's example of operating room teams or the Millstone nuclear power station), we could draw on the current cultural elements of scientific empiricism (facts are important), responsibility, caring (about patient outcomes), self-criticism, and excellence to facilitate the change.

However, such an argument in the abstract is not of much use unless it is attached to a real change effort, such as developing a near-miss reporting system and experiencing the challenges and benefits it would provide (including the benefits of new working relationships among different groups involved in the new system; cf. Carroll, 1998). Until the new ideas prove to be consistently successful, they will not become embedded in the culture. Beer, Eisenstat, and Spector (1990) reported that change programs that focus on attitudes often fail. But energy can be directed toward change by drawing on existing attitudes and values, as the leadership of Dana-Farber Cancer Institute did when they announced, following Betsy Lehman's tragic death from a cancer treatment overdose, that they would become a national leader in safety just as they had become a national leader in research, drawing on their cultural values of excellence and scientific inquiry (Bohmer, 2003). Of course, these are only the beginnings of ideas and examples for an approach to culture change.

Each hospital or other health care organization has to find its own way to move forward. Each organization starts out with different strengths and opportunities, as well as challenges. As Schein (1992, 1999) pointed out, it is difficult and often counterproductive to work on culture change by directly trying to design the new culture in the abstract. Instead, it is possible to create opportunities for people to work together on real problems in ways that draw on cultural strengths and allow new ways of working to emerge, be experimented with, and become part of the new culture. Although attempting a cultural analysis disconnected from specific change issues may not be helpful, it may be useful to treat culture change examples from other organizations as "cases" that can prompt an open exchange of viewpoints across disciplines and units regarding the relevance of those cases for their problems and their organization (E. Schein, personal communication, April 18, 2004). Cultural survey instruments such as the ones discussed in this volume (Madsen, Andersen, & Itoh, this volume) can offer another way to identify issues and strengths and spark dialogue among diverse groups in interpreting the meaning of the survey results (Carroll, 1998). It takes creativity to turn that understanding into an action plan, and it takes real discipline to enact new behaviors and consistently communicate what is happening in new terms. But as shown by the Children's Hospitals and Millstone Station examples, senior managers can build on existing cultural strengths to start the process, although culture change takes considerable time and persistence.

SUGGESTIONS AND CONCLUSIONS

We have argued in this chapter that change will not last unless it is rooted in culture, that is, until it becomes part of "the way we do things around here." Opposing the existing culture may not be the best approach. Instead, we suggest ways to identify some strengths of the culture that we want to keep and build on those strengths to tilt the culture toward the desired state. This requires broad participation focused around change efforts for real problems that people care about. New behaviors need to be experimented with and shown to be successful so that new cultural assumptions will emerge. This process takes time, and it is important that leadership act as role models, practicing the new rules and rewarding those who use them.

The role of leadership is crucial. Senior leaders can make a difference by providing resources and protecting those taking well-intentioned risks. Leaders also may have to mandate new goals and values and, sometimes, even coerce behavior, but linking these to current strengths will reduce resistance. Leaders have to be willing to wait and be persistent while changes slowly take root. And, critically, leaders need to walk the talk, admit mistakes and doubts, and make clear that all are learning.

We sketch only the beginnings of an analysis of culture in health care, partly because every health care organization is different and therefore has to work with its own cultural elements in the context of practical change efforts. But there are themes that are likely to emerge repeatedly, such as the reinterpretation of professional autonomy and individuality to support teamwork, reporting of problems, and learning. Autonomy and individual responsibility are tremendous strengths that should not be dismissed or replaced but rather appreciated for what they offer. If we dig deeper into the culture, we can address the assumptions that give meaning to autonomy and individuality and connect these concepts to actions and values. Leaders can then begin to reframe new actions in terms that link to existing concepts but gradually tilt the balance of cultural elements to shift actions, values, and underlying assumptions.

References

Adler, P., Riley, P., Kwon, S.-W., Lee, B., & Satrasala, R. (2003). Performance improvement capability: Keys to accelerating performance improvement in hospitals. *California Management Review, 45,* 12–33.

Barley, S. R. (1986). Technology as an occasion for structuring: Evidence from observations of CT scanners and the social order of radiology departments. *Administrative Science Quarterly, 31,* 78–108.

Beckhard, R., & Pritchard, W. (1992). *Changing the essence: The art of creating and leading fundamental change in organizations.* San Francisco, CA: Jossey-Bass.

Beer, M., Eisenstat, R. A., & Spector, B. (1990). Why change programs don't produce change. *Harvard Business Review, 68,* 158.

Beer, M., & Nohria, N. (2000a). Cracking the code of change. *Harvard Business Review, 78,* 133–141.

Beer, M., & Nohria, N. (2000b). Resolving the tension between Theories E and O of change. In M. Beer & N. Nohria (Eds.), *Breaking the code of change* (pp. 1–33). Boston: Harvard Business School Press.

Berwick, D. M. (1998). Developing and testing changes in delivery of care. *Annals of Internal Medicine, 128,* 651–656.

Berwick, D. M., & Nolan, T. W. (1998). Physicians as leaders in improving health care: A new series in *Annals of Internal Medicine. Annals of Internal Medicine, 128,* 289–292.

Bohmer, R. (2003). *The Dana Farber Cancer Institute.* Boston: Harvard Business School.

Carroll, J. S. (1998). Safety culture as an ongoing process: Culture surveys as opportunities for enquiry and change. *Work and Stress, 12,* 272–284.

Carroll, J. S., & Hatekenaka, S. (2001). Driving organizational change in the midst of crisis. *MIT Sloan Management Review, 42,* 70–79.

Carroll, J. S., Rudolph, J. W., & Hatekenaka, S. (2002). Learning from experience in high-hazard organizations. *Research in Organizational Behavior, 24,* 87–137.

Cooperrider, D. L., Sorensen, P. F., Whitney, D., & Yaeger, T. F. (2000). Appreciative inquiry: Rethinking human organization toward a positive theory of change. Champaign, IL: Stipes Publishing.

Corrigan, J. M., Donaldson, M. S., Kohn, L. T., Maguire, S. K., & Pike, K. C. (Eds.). (2001). *Crossing the quality chasm: A new health system for the 21st century.* Washington DC: National Academy Press.

Dyer, W. G. J. (1986). *Culture change in family firms.* San Francisco, CA: Jossey-Bass.

Edmondson, A. C. (2003). Speaking up in the operating room: How team leaders promote learning in interdisciplinary action teams. *Journal of Management Studies, 40,* 1419–1452.

Edmondson, A. C. (2004). Learning from failure in health care: Frequent opportunities, pervasive barriers. *Quality and Safety in Health Care. 13*(6), 3–9.

Edmondson, A. C., Bohmer, R. M., & Pisano, G. (2001). Disrupted routines: Team learning and new technology implementation in hospitals. *Administrative Science Quarterly, 46,* 685–716.

Edmondson, A. C., Roberto, M., & Tucker, A. L. (2001). *Children's Hospitals and Clinics.* Boston: Harvard Business School Press.

Helmreich, R. L., & Merritt, A. C. (1998). *Culture at work in aviation and medicine: National, organizational and professional influences.* Brookfield, VT: Ashgate.

Kanter, R. M., Stein, B., & Jick, T. (1992). *The challenge of organizational change: How companies experience it and leaders guide it.* New York: Free Press.

Keating, E., Oliva, R., Repenning, N., Rockart, S., & Sterman, J. (1999). Overcoming the improvement paradox. *European Management Journal, 17,* 120–134.

Kellogg, K. (2003). *On the cutting edge: Changes in institutional logics and identity displays among surgical residents.* Unpublished manuscript, MIT Sloan School of Management, Cambridge, MA.

Kohn, L. T., Corrigan, J. M., & Donaldson, M. S. (Eds.). (2000). *To err is human: Building a safer health care system.* Washington DC: National Academy Press.

Kotter, J. P. (1996). *Leading change.* Boston: Harvard Business School Press.

Martin, J. (1992). *Cultures in organizations: Three perspectives.* New York: Oxford University Press.

Millenson, M. L. (1997). *Demanding medical excellence: Doctors and accountability in the information age.* Chicago: University of Chicago Press.

Millman, M. (1977). *The unkindest cut: Life in the backrooms of medicine.* New York: Morrow Quill.

Parker, D. L., & Wertheimer, D. S. (1997). Incorporating internal medicine residents into an interdisciplinary geriatric assessment team. *Academic Medicine, 72,* 451.

Pisano, G. P., Bohmer, R. M. J., & Edmondson, A. C. (2001). Organizational differences in rates of learning: Evidence from the adoption of minimally invasive cardiac surgery. *Management Science, 47,* pp. 752–768.

Reason, J. T. (1997). *Managing the risks of organizational accidents.* Brookfield, VT: Ashgate.

Schein, E. H. (1992). *Organizational culture and leadership* (2nd ed.). San Francisco, CA: Jossey-Bass.

Schein, E. H. (1999). *The corporate culture survival guide: Sense and nonsense about culture change* (1st ed.). San Francisco, CA: Jossey-Bass.

Schein, E. H. (2002). Models and tools for stability and change in human systems. *Reflections, 4,* 34–46.

Tucker, A. L., & Edmondson, A. (2003). Why hospitals don't learn from failures: Organizational and psychological dynamics that inhibit system change. *California Management Review, 45,* 55–72.

Weeks, J. R. (2004). *Unpopular culture: The ritual of complaining in a British bank.* Chicago: University of Chicago Press.

Section

IX

SPECIFIC APPLICATIONS

46

HUMAN FACTORS AND ERGONOMICS IN INTENSIVE CARE—A PROCESS-ORIENTED-APPROACH

Claus Backhaus and Wolfgang Friesdorf
Technical University of Berlin, Germany

INTENSIVE CARE UNIT (ICU) AS A WORK SYSTEM

Ergonomic work design is very important for the effectiveness of intensive medical work systems. Optimization of the work environment, work processes, and medical technologies help to decrease staff work load and create work that is good for the well-being and health of the staff (Friesdorf, Göbel & Buss, 2004). Ergonomics examines the interactions of human beings and other system elements and aims to increase and optimize personal well-being and the safety and efficiency of the system (Ergonomics Association Executive Council, 2002).

Human Factors System Approach

Friesdorf (1990) described the ICU as a point of convergence of different elements within the system—physicians and nurses (representatives of health care providers), patients, and machines—within the limits of a work system (see Figure 46–1). It is the task of the system to improve and stabilize the patient's state of health. Therefore, work processes take place within the work system in which information, energy, or material are transferred. All of these elements represent the patient's treatment process (Bridger & Poluta, 1998). The interactions within the work system depend on factors in the physical and social environment.

This system approach allows one to fix arbitrary limits to the system being examined. The work system model can be adapted to the particular issue at stake (Luczak, 1997). Similarly, work processes within the work system can be examined at different levels of abstraction and details. The aim of ergonomic optimization is both the improvement of interaction between single elements in the system (microergonomics) and the improvement of organizational and structural requirements in completing the work task (macroergonomics; Carayon, 2003; Gosbee, 1997; Hendrik, 1997; Staggers, 2003).

ICU as a Complex System

The ICU work environment has characteristics of a complex sociotechnical system (Friesdorf & Göbel, 2003; Institut für Technik-Folgenabschätzung, 2002):

- A. high quantity of information and data is processed (e.g., information about the physiological state and monitoring data).
- Single elements of the work system are linked up and interact with each other. The patient influences the diagnosis and the therapy as an active participant.

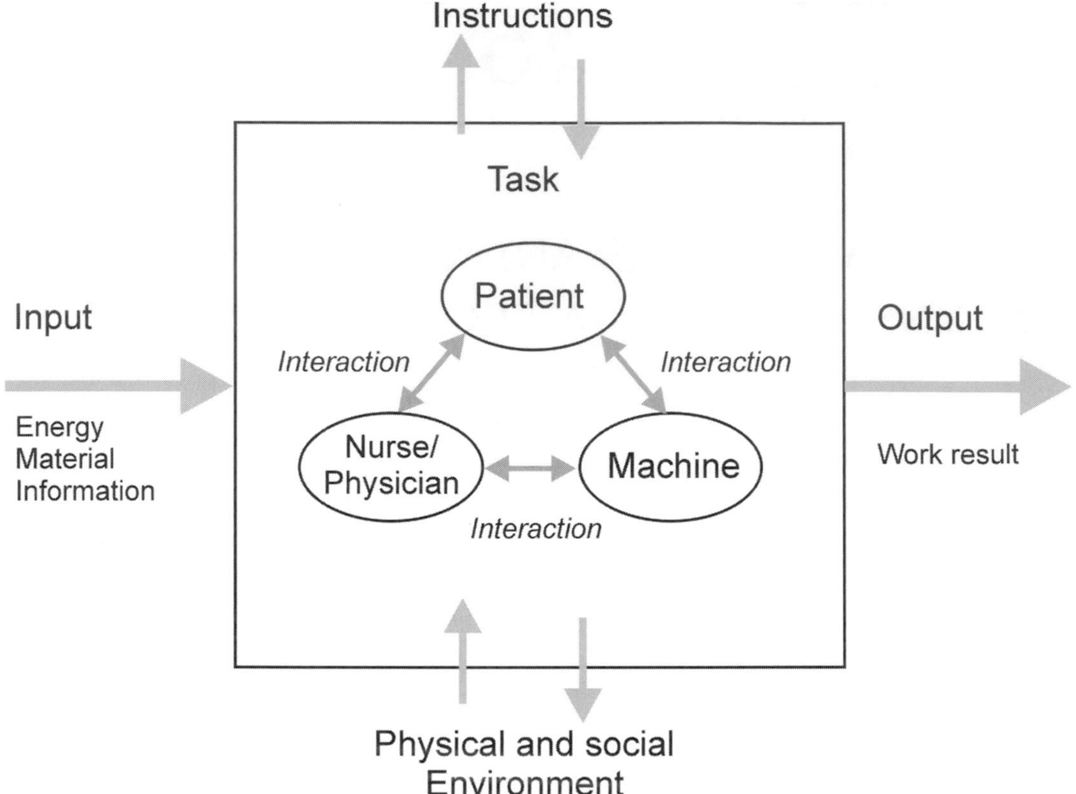

Figure 46–1. Intensive medical work system as a model.

- The work system is not clearly predictable in its conduct. Many of the patient's physiological processes cannot be measured directly by sensors and can be described only through models.
- The state of the system changes dynamically (i.e., a change in state takes place within the system even without outside influences, e.g., changes in pharmacokinetics and metabolism of the patient).
- The task changes within the work system (e.g., at first the patient state must be stabilized before specific diagnostic and therapeutic interventions can be started).

Through a small extension of the limits of the ergonomic system approach, the high complexity of ICU work environments becomes apparent (see Figure 46–2). The number of possible interactions increases much faster than the number of system elements being examined. This becomes obvious through a simplified description of two patients, two physicians or nurses, and only two technical devices per patient (Göbel, Backhaus, & Friesdorf, 2002).

The integration and dynamics of the ICU work system make an ergonomic optimization of single-system elements more difficult (e.g., medical devices are often optimized for single use only and not for interacting in a complex work system).

The Patient Therapy Process in ICU

A stabilization and improvement of the state of health of the critically ill patient is reached by various measures of diagnosis, therapy, and monitoring happening during the patient therapy process. The complex convergence of single medical interventions and the individuality of the human organism continuously demand an individual adaptation of the intensive medical treatment to the patient's state of health. Using a simplified model, the patient's therapy process in the ICU can be thought of

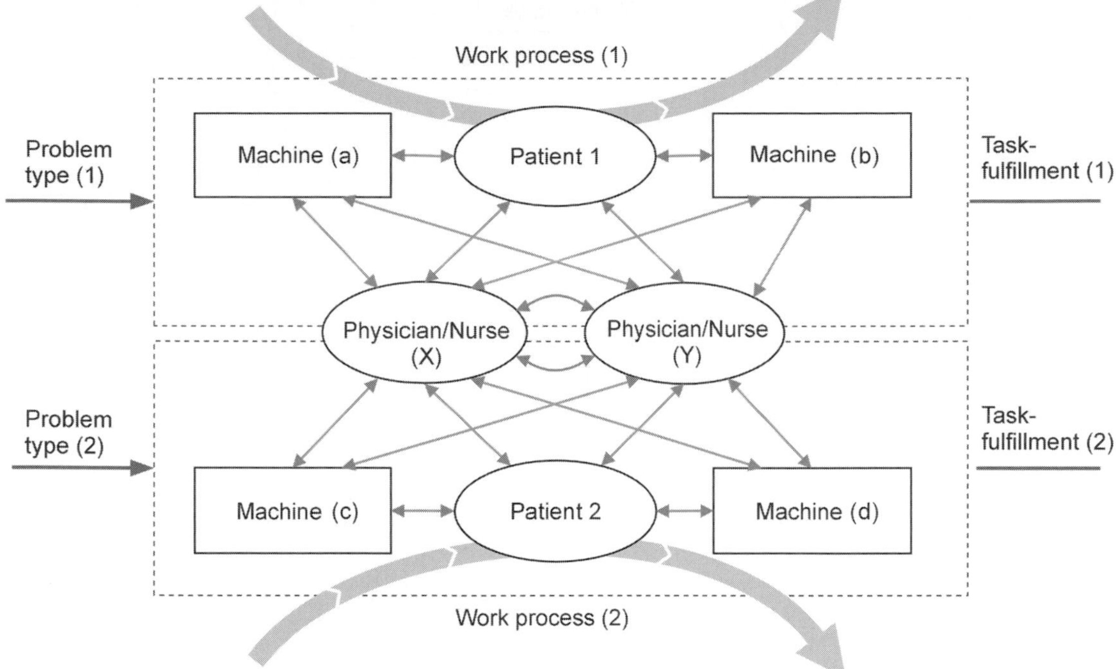

Figure 46–2. Complexity of intensive medical work system.

as a convergence of different cognitive and performance-oriented action cycles at the operative, strategic, and normative levels (see Figure 46–3; Backhaus, 2004). At the strategic level of treatment, the planning of the treatment happens in the form of an iterative procedure between the diagnosis or therapeutic interventions and the reaction by the patient. The normative cycle of performance serves primarily to generate knowledge (e.g., evidence-based medicine). Moreover, it serves to regulate and standardize medical performance (e.g., state of the art in medicine). An ergonomic optimization of the normative cycle of performance is only possible via macroergonomics. To optimize the operative cycle of performance, various microergonomic methods are available (Association for the Advancement of Medical Instrumentation, 1999; Dumas, 1996; U.S. Food and Drug Administration, 1999). For the ergonomic optimization of the strategic cycle performance, it is necessary to consider both medical normative knowledge as well as knowledge and experience from the operative cycle of performance. The optimization of the strategic cycle performance is oriented toward the planning and regulation of the patient's treatment and thus toward clinical work processes. It corresponds to the process ergonomics of the ICU work system (Backhaus & Friesdorf, 2004).

ERGONOMIC ASPECTS

Workplace Design in ICU

Within the context of workplace design creation, it is important to consider good accessibility to the patient and material in the workplace. For the design of physical working conditions, specific knowledge of anthropometrics is necessary. As a part of the discipline of ergonomics, anthropometry deals with anatomic body and functional measurements of people (Botha & Bridger, 1998; Jürgens, Aune, & Pieper, 1989; Schmidtke, 1993). Due to the fact that work in an ICU takes place in an upright position, the staff's height and the resulting tangible areas and functions are of great importance (see Figure 46–4).

This data is available in numerous technical norms and standardizations (Deutsches Institut für

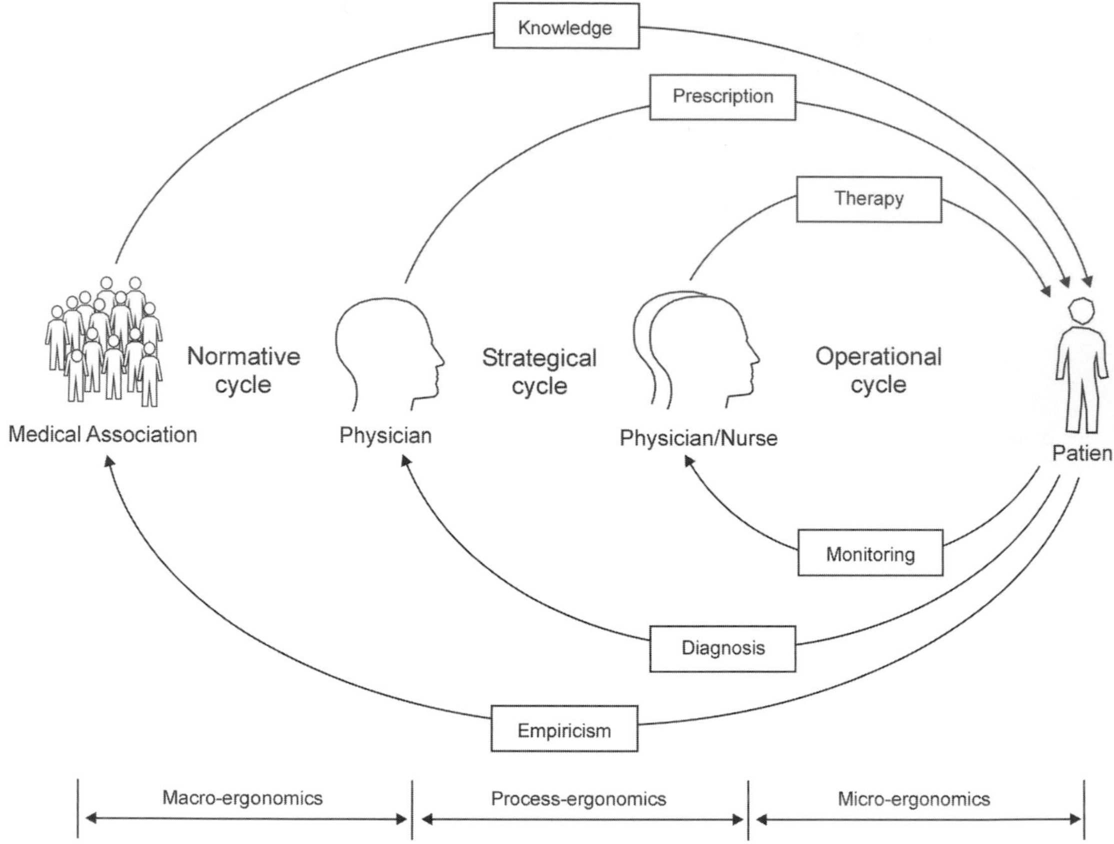

Figure 46–3. Model of patient therapy process in ICU.

Normung, 1986; International Standard Organization, 1996b, 2000, 2002; Jürgens, 1997). An example of the interpretation of an integrated intensive care workplace for the 5th percentile woman and the 50th percentile man is given in Figure 46–4 (Friesdorf & Hecker, 1991).

Established conventions have to be considered for the arrangement of technical devices. Especially in situations involving critical time management, it is important that the arrangements of work materials are compatible with the learned, established conventions of the user because they support a fast and safe discovery of information and technical devices (Friesdorf, Groß-Alltag, Konichezky, Schwilk, & Fett, 1994). These conventions are presented in Figure 46–5.

In addition, technical devices used in ICUs need to be compatible in their usability. In the arrangement of a workplace by single ergonomically designed technical devices whose usability is poorly coordinated, the danger of misuse increases, especially in situations of a critical time management. In these situations, the user tends to prefer well-known and more frequently usability rules instead of following less practiced and less internalized rules (Hoyos, 1974).

The functional grouping of technical devices increases the clarity of the workplace and in the past led to the development of integral concepts of operation (e.g., fluid management).

Context and Environment Factors

The efficiency of ICUs is influenced to a large degree by context and environmental factors. Because the ICU patient's organism is vitally threatened, it is hypersensitive to all kinds of influences surrounding it. Light, climate, and noise, as well as psychological pressure caused by staff and social

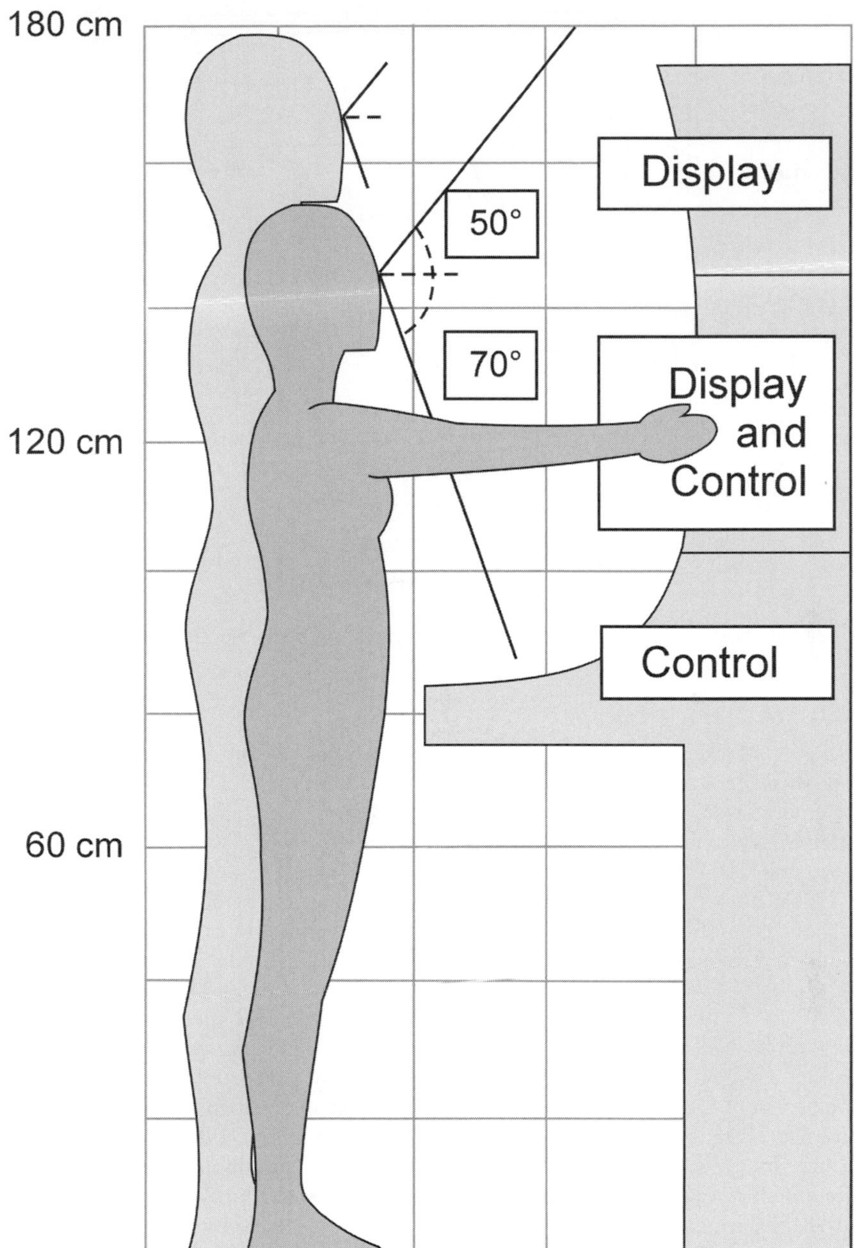

Figure 46–4. Anthropometrical design of an integrated intensive medical workplace.

contact with the patient, have a direct impact on the success of treatment (Donchin & Seagull, 2002). To design ergonomic intensive medical workplaces that also fit patients' needs, one has to consider the respective various guidelines for environmental factors.

This is particularly true regarding exposure to noise in ICUs. The average noise exposure at the patient's bed has increased because of the increased mechanization of intensive medical workplaces and the respective increase of acoustic alarm signals (McKinnon, 1984; Xiao, Mackenzie, Seagull, & Jaberi,

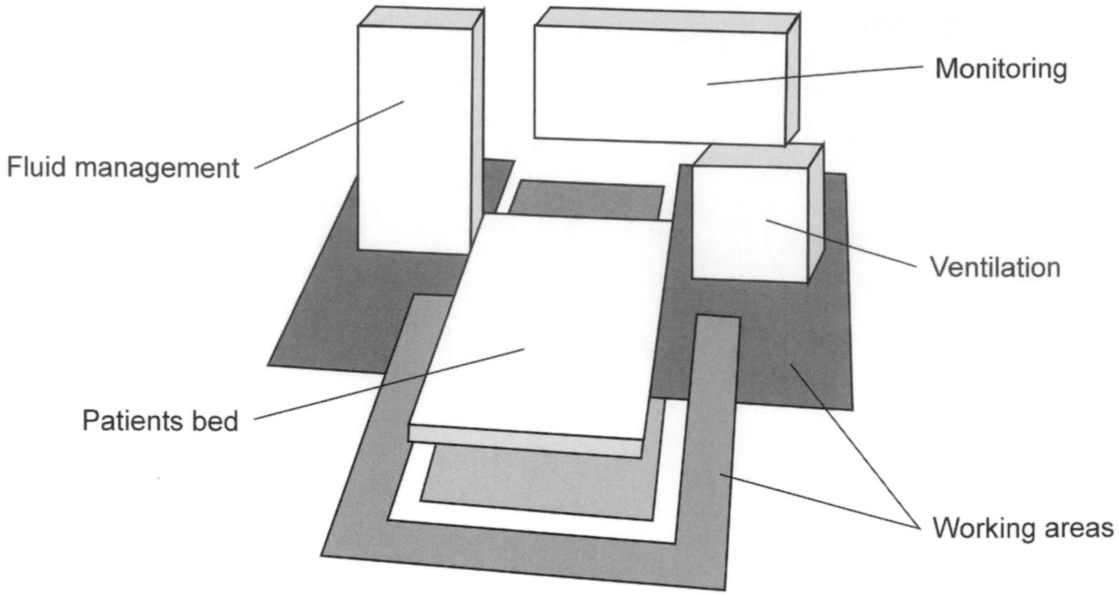

Figure 46–5. Established conventions for the arrangement of medical technology in an ICU.

2000; Xiao, Seagull, Mackenzie, Wickens, & Via, 2000; see also Buss & Friesdorf, this volume). Guidelines for environmental factors of work systems are provided in numerous standards (Deutsches Institut für Normung, 2004; International Standard Organization, 1996a, 2004) A detailed description of typical context and environment factors in health care can be found in Alvarado (this volume).

Medical Device Design in ICU

The ergonomic design of intensive medical devices with regard to real work processes involves the most efficient way of preventing incidents during the patient's treatment (Dhillon, 1986; Gaba, Fish, & Howard, 1998; Scharmer & Siegel, 1997) The support of the therapy process through the functionality of a technical device (process support) and the availability of process support for the user describing the usability of a technical device is of particular importance. Both factors define the usefulness of a medical technical device and have to reach a balance (see Figure 46–6).

The usability of an intensive medical technical device becomes better the safer and easier the dialogue or the interaction between user and technical device is. For a given task and a unified group of users, it is a measure for a fast (efficient), simple (effective), and satisfying use of a technical device (Bevan, Kirakowski, & Maissel, 1991; Dumas & Redish, 1993; Ravden & Johnson, 1989; Staggers, 2003).

To assess usability, various methods of usability engineering are provided (Mayhew, 1999; Nielsen, 1992; Whiteside, Bennett, & Holtzblatt, 1988; Wiklund, 1993a). For the evaluation of the usability of intensive medical technical devices, specific methods have proven their worth: user questioning, usability inspection methods, and usability tests (Backhaus, Papanikolaou, Kuhnigk & Friesdorf, 2001; Wiklund, 1993b, 1995).

User questioning gives information on what clinical staff think about a product. User questioning is conducted with questionnaires or interviews (Stanton & Young, 1998; Stoessel, 2002). For the evaluation of usability, there are several standardized questionnaires (Brooke, 1996; Jordan, 1998; Kirakowski, 1996). Another approach to evaluate the usability of intensive care devices are inspection methods. They discover potential user deficits in a product very early through the employment of suitable ergonomic or usability specialists. Thus the effort put into specific user questioning or user testing can be decreased (Virzi, 1997). The most important usability inspection methods for medical

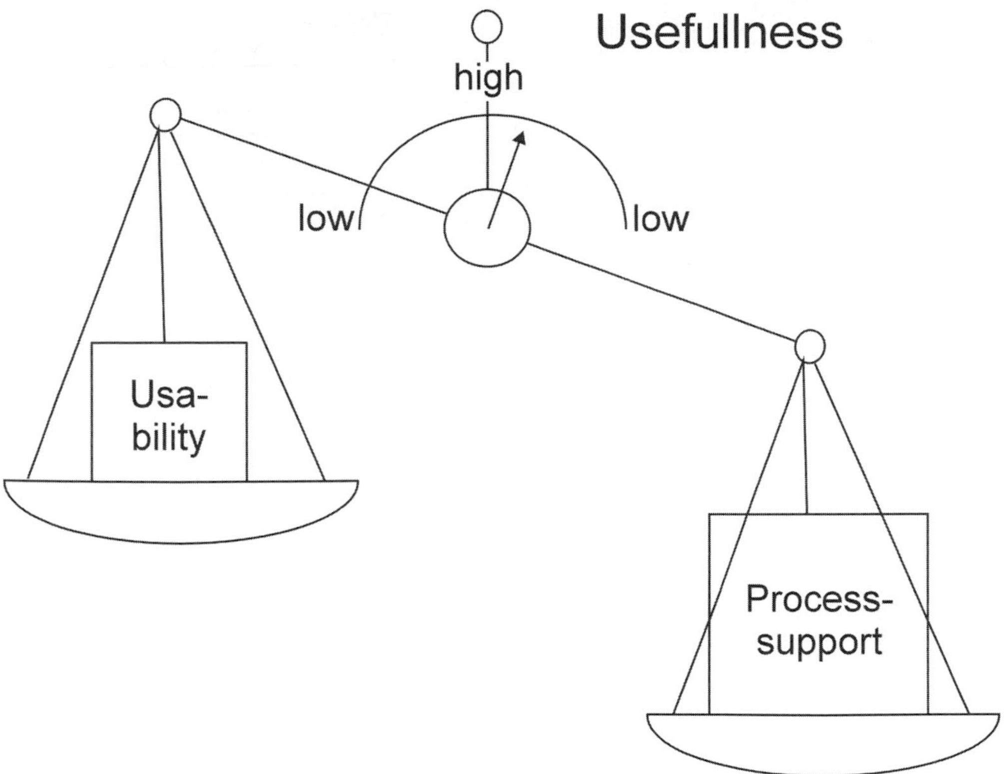

FIGURE 46–6. Usefulness represented as a balance between usability and process support for intensive care devices.

device design are heuristic evaluation, described by Nielsen and Molich (1990), and the cognitive walk-through, described by C. Lewis, Polson, Wharton, and Rieman (1990) and Wharton, Bradford, Jeffries, and Franzke (1992). Usability inspection methods can be used in a specifically effective way in the evaluation of intensive medical device design (Jeffries, Miller, Wharton, & Uyeda, 1991; Nielsen, 1989, 1994).

The most reliable method of assessing the usability of intensive care device is a usability test (Dumas, 1996; Nielsen & Phillips, 1993; Whiteside et al., 1988; Wiklund, 1993b). The use of a medical device is subdivided into separate performance categories that have to be processed by those experimenting. During the experiment, the processing of individual performances is observed and evaluated. Here, the appearance of wrong and insecure use or a long processing time is of particular interest as it reveals performances with user deficiencies. Moreover, they show which parts of the user's work area have to be revised (Rosenbaum, 1989; Sawyer, 2002; Spool, Snyder, & Robinson, 1996; Stanton & Baber, 1994; Wiklund, 1993a).

Another frequently used variation of a usability test for the evaluation of intensive care devices is the think-aloud technique. To get more detailed information on the user's conduct, the individuals is asked to think aloud while using the device. Thus, the experiment leader has the chance to get an impression of the interaction strategy of the user (Denning, Hoiem, Simpson, & Sullivan, 1990; Ericson & Simon, 1984; Nielsen, 1993; Nisbett & Wilson, 1977; Wilson & Corlett, 1995).

ERGONOMICS IN INTENSIVE CARE—A PROCESS-ORIENTED DESIGN APPROACH

The ergonomic design of intensive medical work groups is done through the process-oriented medical technology in clinical systems (PROMEDIKS) methodology (Backhaus, Marsolek, & Friesdorf, 2003; Backhaus, Semar, & Friesdorf, 2004). The PROMEDIKS methodology consists of four components:

1. process analysis and visualization.
2. assessment of process support.
3. assessment of usability.
4. evaluation of usefulness.

Process Analysis and Process Design

The aim of the process analysis is to define and delimit the technical work processes, analyze the clinical contents of the processes, and assess the process structure as well as the typical strong and weak points. In discussions with clinical users, the work processes that have to be analyzed are defined. Moreover, their strong points, ending points, and most important interfaces are described.

Through an open, nonparticipatory observation, the processed operations, performances, and decisions of the work processes are assessed. Here, events occurring sequentially are shown along a horizontal time axis chronologically from left to right. Operations occurring at the same time or alternative performances or processes as well as branches of work processes are presented along a vertical axis. Besides the observed operations, the persons performing, the work environment, the presence of the patient, the documents or systems of documentation used, specific ethical aspects, and observed strong and weak points are assessed and documented. The assessed dates are registered in the process flow diagrams (Marsolek & Friesdorf, this volume).

Depending on the question, it is possible to summarize single operations in logical process modules. This raises the degree of abstraction of the analysis and reduces the complexity of the visualized work processes. The documentation of the observed processes constitutes the base for their participative analysis. Therefore, the assessed process flows are discussed with the staff in separate interviews or group discussions. Thus, typical strong and weak points of the process flows can be completed or corrected. The procedure is repeated iteratively until the work process being analyzed is shown in an adequate precision and is confirmed in its validity by the staff.

TABLE 46–1. Evaluation Grades for Process Support

Process Support	*Characteristics*	*Value*
Complete	Complete functional support by the employed technology	2
Limited	Partial or only optionally available support by employed technology	1
None	No support	0

Process Support

Process support is assessed on the basis of the created process flows diagrammed. In a further analysis, each process module is examined according to the question about whether useful support through the employment of medical technology is possible. Here, one has to find out if single operations and processes can be improved in their efficiency or effective realization or might even need to be replaced completely. The result of this procedure is the potential process support, or the process support that could be possible in the observed treatment process.

In a second analysis step, the current given process support of the examined medical technology is assessed. Here, in an analogous way, which of the potentially supported process modules are supported by the examined medical technical device and to what extent this support takes place is assessed. For the evaluation of the process support, three evaluation grades are used (see Table 46–1).

The documentation of the analysis results from the grading of the process modules that potentially support and the process modules that are actually supported in the process flow diagrams.

Usability

For the usability evaluation, only process modules that allow a potential process support are considered. For the efficiency of the work process, process modules with standard employments are of great importance. To examine the safety and reliability of the use of technology, process modules that include a higher risk for a misuse have to be considered. This risk results from the combination of the following criteria:

- aggravating conditions of use for the employment of medical technology causing

TABLE 46–2. Suitability of Chosen Methods and Accessible Usability Attributes

	Suitability			
	Risk of Use	*Product Development*	*Processing Investment*	*Usability Attributes*
User questioning	Low	Early	Middle	User's acceptance, user requirements
Experts questioning	Middle	Middle	Little	Potential use deficits, ergonomic design deficits
User tests	High	Late	High	Misuse and weak points, processing time

an increased probability of occurrence of misuse (e.g., time-critical employment of medical technology in emergency situations, a high interaction density, and the like).

- specific endangerment connected to the employment of medical technology leading to an extensive damage of misuse (e.g., direct effect of an operation on the patient's state of health, user, or others).

The identified process modules are marked in the process flow diagrams and are summarized as clinical use cases. Here, it is important that all influencing factors of the work system that could have a relevant effect on the interaction of the user with the medical technology must be considered. On the basis of the assessed use cases, one continues with the analysis and evaluation of the usability.

The choice of a usability method is dependent on the question or given topic being examined. The expected risk in use of the technical device, the given capacity of the procedure, and the state of product development or the then resulting availability of models, mock-ups, or prototypes are important influences on the choice of the method. Table 46–2 shows the suitability of chosen usability methods.

To classify the examined usability attributes, the determination of a suitable range of values follows. Its aim is the transparent and unified visualization of the usability evaluation. Because of the fact that the extension of the results strongly depends on the examined question and the applied method, a determination of a generally valid quantitative range of values is not possible. Alternatively, a qualitative range of values that allows a classification of the assessed usability core sizes in a three-phase "ordinal scale" is recommended (Backhaus, Friesdorf & Zschernack, 2001; Deutsches Institut für Normung, 1995). Here, the assessed test or questioning results are classified according to one of three grades. The three grades of evaluation and their method-specific characteristics are presented in Table 46–3.

The advantage of the three-scale evaluation scheme is the easy classification of the assessed core sizes and clear presentation. Existing evaluation systems can be transferred very easily, if needed. The required number of experimental participants for the chosen method depends on how the examined random sample can be differentiated from a basic total population by specific characteristics. For a group that can be isolated easily, a lower number of experimental participants are necessary with the same quality of results. Recommendations for the size of the random sample for usability tests are given in Table 46–4 (Nielsen, 1989; Nielsen & Landauer, 1993; J. R. Lewis, 1994; Virzi, 1992). If specific requirements are required to achieve quality results, the size of the random sample should be raised appropriately (Table 46–4).

Usefulness

The evaluation of the usefulness is achieved by bringing together the results of the process support and the usability. Synthesis can be achieved qualitatively or quantitatively.

In a qualitative synthesis of the results, a documentation and discussion take place in regard to their relevance to the process. This is subdivided into three categories (see Table 46–5). Not completed or incomplete process modules (cf. Table 46–1) that consist of a high process relevance represent specifically serious deficiencies in the process support and should be stressed specifically.

Analogously, the evaluation of the middle usability deficits follows. This results from the assessed evaluation grade (see Table 46–3) and the effect of the operation on the whole process (performance relevance), which is assessed analogously

TABLE 46–3. Three-Step Evaluation Scheme for Usability

Color	Description	User Questioning	Specific Characteristics Expert Questioning	User Test	Value
Red	Large employment deficiencies measurements necessary	Urgent user requirement or unacceptable product characteristic	Large ergonomic lack of design	Great misuse; operation blockage; cannot go on working; long processing time	3
Yellow	Potential employment deficiencies; measurements useful	Recommended user requirement or disturbing product characteristic	Bad ergonomic design; uncomfortable	Light misuse; testing behavior; able to go on working	2
Green	No deficiency; no measurements necessary	User request; acceptable or positive product characteristic	Ergonomically well or sufficient design	No employment problems; safe and fast proceeding of operations	1

TABLE 46–4. Recommended Size of Random Sample for the Employment of Chosen Usability Methods

Distinctive Characteristics of the Random Sample Group	User Questioning		Experts Questioning		User Test	
	Single Interview	Focus Group	Heuristic Evaluation	Cognitive Walkthrough	Observation	Think Aloud
Good (for instance medical/nursing staff of a special discipline)	7–15	3–5	1–3	1–3	3–5	3–5
Middle (for instance medical/nursing staff)	15–30	5–7	3–5	3–5	5–10	5–7
Modest (for instance general staff in the health care system)	>30	7–9	5–7	5–7	>10	>7

with the process relevance (cf. Table 46–5). Deficits leading to a safety risk for the patient (e.g., for the patient's state of health) should be specifically considered in the total evaluation and should be presented adequately. In the context of a comparative evaluation, a quantitative synthesis of the results is recommended. Here, the single results are quantified on the basis of a referential size representing an ideal objective. For the process support, the referential size is calculated by the potentially possible process support (all process modules are supported completely) and the process relevance of the process modules considered. Based on this, a numerical value can be calculated resulting from the sum of all single products of the process relevance and the maximum process support. If one refers the assessed process support of the examined medical technical device to the utmost achievable (potential) process support, one achieves—multiplied by 100—the quantified process support

TABLE 46–5. Grades of Evaluation According to Process Relevance

Process Relevance	*Characteristic*	*Value*
High	Function or partial process is needed in every work process; lack of support causes more work effort for the user	3
Middle	Function or partial process is needed rarely to frequently; missing support distinctively leads to more work effort for the user	2
Modest	Process is needed rarely; missing support leads to less comfort only	1

percentage. The correlation described is also presented in Equation 1.

Equation 1: Quantification of process support

$$QPU = \frac{\Sigma(PR_i \times PU_{ai})}{\Sigma(PR_i \times PU_{max})} \times 100\%$$

PUai = process support of the observed process module

PRi = process relevance of the observed process module

PUmax = maximum process support

QPU = quantified process support (%)

The results of the usability are quantified analogously. The referential value is created on the basis of an optimal usability of the technical device in the examined use case. The usability referential value results from the sum of all single products of the performance relevance (cf. Table 46–5) and the maximum evaluation of performance (see also Table 46–3). The usability value results from the sum of the products of the evaluated performance and the relevance of the performance. To convert the evaluation of the performance into a positive evaluation range, the assessed point values are subtracted by 4. The relation of both point values (see Table 46–5) multiplied by 100 gives the quantified usability of the examined medical technical device as a percentage value. The correlation described is presented in Equation 2.

Equation 2: Quantification of usability

$$QU = \frac{\Sigma(TR_i \times (4 - UB_{ai}))}{\Sigma(TR_i \times UB_{max})} \times 100\%$$

UBai = usability evaluation of the examined performance

TRi = performance relevance

UBmax = maximum usability evaluation (value = 3)

QU = quantified usability (%)

The synthesis of the single values results from the calculation of the usefulness value of the examined product. To prefer medical products with balanced total evaluations, the calculation is done with the help of the hyperbole method (Pahl & Beitz, 1993). The correlation described is presented in Equation 3.

Equation 3: Quantification of usefulness

$$GT = \sqrt{QPU \times QU}$$

GT = usefulness value (%)

QU = quantified usability (%)

QPU = quantified process support (%)

The usability value can be represented by a numerical value in a coordinate system (see Figure 46–7). The quantitative presentation of the examined value is completed by a description of the most important strong and weak points of the examined product.

EXAMPLE OF AN EVALUATION OF A PATIENT DATA MANAGEMENT SYSTEM FOR ANAESTHESIA AND INTENSIVE CARE

Situation

The documentation of physiological patient data accompanying the anaesthesia and anaesthesio

TABLE 46–6. Process Support of the Examined Anesthesiological Data Management Systems and Process Relevance of the Process Modules

	Number of Identified Process Modules		
Proces Relevance	*Potential Process Support*	*Process Support System 1*	*Process Support System 2*
High	7	7	7
Middle	14	1	11
Low	11	0	5
Total	32	8	23

TABLE 46–7. Results of the Evaluated Performances of the User Test

	Performances			
	System 1		*System 2*	
Evaluated Performances	*Phase 1*	*Phase 2*	*Phase 1*	*Phase 2*
Bad	1	0	1	0
Critical	14	5	13	2
Good	18	28	21	33
Total	33		35	

logical operations on the patient belong to the routine of the anaesthetist during an operation. Documentation systems (paper supported) that have been used are continuously replaced by electronic anaesthesiological data management systems. Besides the running and obtaining costs, the acquisition of such a system demands usefulness in its later use as an important criterion of selection. Within the context of the investment decision of an anesthesiological unit, the usability of two anesthesiological data management systems should be assessed and evaluated (Backhaus, Marsolek, & Friesdorf, 2003; Marsolek, Backhaus, & Friesdorf, 2003).

Method

The evaluation was done with the help of PROMEDIKS methodology. The necessary process analysis was processed at the University Hospital in Fukuoka (Japan). Two preanaesthesia phases and four anaesthesia induction phases were observed actively as representative process phases and represented in the process flow diagrams. These were analyzed and actively verified with the staff of the unit for anaesthesiology ($n = 3$).

To examine the usability, the anaesthesiological data management systems were installed in a simulated operating room environment and evaluated by anaesthesiologists with the aid of the think-aloud technique. As an application scenario, an induction phase was simulated on a male patient with a slight initial illness (American Society of Anesthesiologists risk group II). As an application group, anaesthesiologists without existing knowledge of the use of anaesthesiological documentation systems were chosen ($n = 20$). The assignment of test participants to the systems was uniformly randomized. The user test was divided into two phases: Phase 1 involved experiment processing without an introduction, and Phase 2 involved experiment processing with a standardized introduction (the test participants received a standardized introduction to the use of the examined system after the first experiment processing).

Both experiments were videotaped. The completed performances were evaluated in a three-scale evaluation process. The assessed single results of the partial performances were summarized in the form of an average value. In a pretest, the feasibility and quality of the given experimental design were examined.

Results

The process analysis produced 32 process modules for the *preanaesthesia phases* and 23 process modules for the *anaesthesia induction phases.*

	System 1		System 2	
QPU = quantified process support	44%		88%	
QU = Quantified Usability	80%	92%	77%	93%
GT = usability value	59,3%	63,6%	82,3%	90,5%
	Phase 1	Phase 2	Phase 1	Phase 2

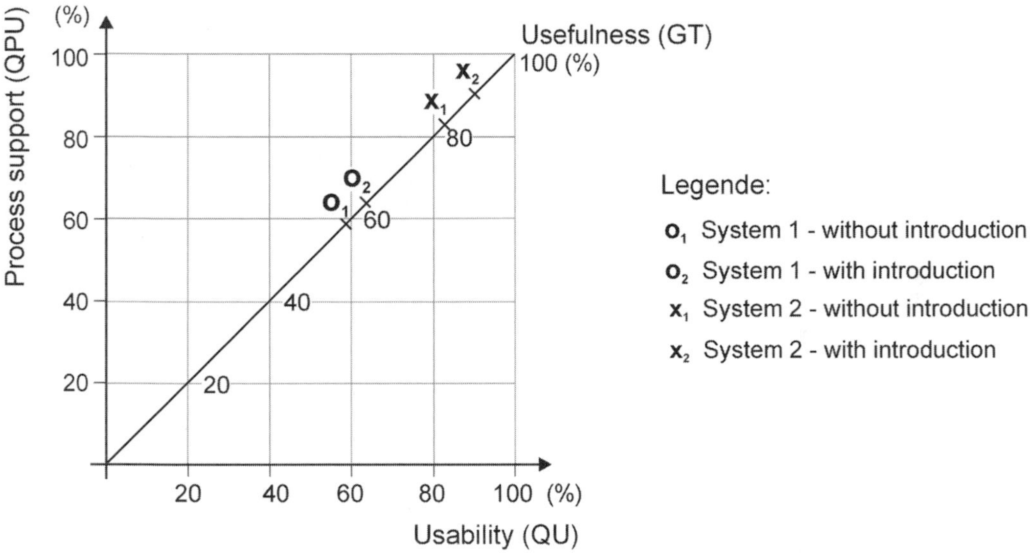

Figure 46–7. Results of the quantified process support and usability as well as presentation of usefulness values in usefulness diagram.

The analysis result of the process support shows that, in total, there were 32 of the 55 assessed process modules that could be supported through the employment of an anaesthesiological data management system. The process support of the anaesthesiological documentation system (System 1 and System 2); the process relevance of the supported process modules are presented in Table 46–6.

In the process analysis, eight usability relevant process modules were assessed and summarized in the application scenario "anaesthesia introduction." To process the experiment, 33 partial performances were differentiated in System 1. In System 2 there were 35 partial performances. The results of the user tests are presented in Table 46–7.

Discussion

The results of the process support and the usability were quantified for discussion. The results of the quantification and presentation of the measured usefulness values in the usefulness diagram are represented in Figure 46–7.

Looking at the usefulness diagram (Figure 46–7), it becomes apparent that System 2 achieves much better usefulness values. Looking at individual results, one can see that the difference is caused primarily by the fact that System 1 supports the work systems insufficiently. The anaesthesia data management System 1 does not support the process units at all. The assessed deficiencies in the process support for System 2 are of less importance, because the nonsupported process modules contain only small process relevance. On the basis of the assessed deficiencies, one can define future requirements for the further support of the medical work processes. The connection of the system to already existing clinical data management systems and the extra support of secondary performances (e.g., the ordering of medicine or medical technology) especially are of special interest and are very desirable. Moreover, through the supply of internal checklists (technical device checks or workplace checks, medication, and so on), an important aid to inexperienced anaesthetists can be guaranteed. Although System 2 contains a larger process support and a therefore larger resulting degree of functions, the differences in the usability of the examined data management systems in experimental Phase 2 (standardized introduction) are practically meaningless.

One can conclude that the evaluated System 2 represents a more developed product. The process support is good; the deficiencies in its use concentrate on only small, clearly definable components. These should be explained to the user through use of an introduction. Thus System 2 has a much better usefulness and should be chosen in the case of acquisition.

CONCLUSION

ICUs are complex sociotechnical work systems. They are characterized by a high data information throughput, a high and interdependency of different system components, intransparent system structures and self-dynamic system changes caused by the work object *patient.* In addition, the work task always has to follow the changing requirements of the work object patient very closely. These characteristics of the work system ICU call for an integrated and system-oriented ergonomic optimization based on the patient treatment processes. To measure and improve the ergonomic quality of ICU systems and devices, usefulness is an important factor that depends not only on a system's or device's ability to support the patient treatment process (its process support), but also on the availability of this support for the user (its usability). The described PROMEDIKS approach allows an iterative analysis, evaluation, and improvement of the usefulness of ICU systems and devices based on the patient treatment processes, because the improvement of a system's or device's usefulness is one of the best opportunities to enhance safety and reliability in the ICU work system.

References

Association for the Advancement of Medical Instrumentation. (1999). Human factors design process for medical devices: Part 1. *Human Factors Engineering Guidelines and Preferred Practice for the Design of Medical Devices.* Arlington, VA: AAMI—HE

Backhaus, C. (2004). *Entwicklung einer methodik zur analyse und bewertung der gebrauchstauglichkeit von medizintechnik.* [Development of an ergonomic approach to analyze and evaluate the usefulness of medical devices] Dissertation der Fakultät V der TU, Berlin. Retrieved December 12, 2005 from the World Wide Web: http://edocs.tu-berlin.de/diss/2004/backhaus_claus.htm

Backhaus, C., & Friesdorf, W. (2004). Prozessergonomie zur gestaltung von medizintechnik. [Process ergonomics for medical device design] *Dokumentation des 50. Kongress der Gesellschaft für Arbeitswissenschaft* (p. 662). Zürich: GfA Press.

Backhaus, C., Friesdorf, W., & Zschernack, S. (2001). Bedienbarkeit von infusionstherapiegeräten. [Usability of infusion pump systems] *Medizinprodukte Journal, 8,* 95–99.

Backhaus, C., Marsolek, I., & Friesdorf, W. (2003). Fitness for use– sichere und effiziente medizintechnik. [fitness for use - safe and efficent medical devices] *Dokumentation des 5. Arbeitswissenschaftlichen Symposium der TU Berlin* (pp. 102–112). Berlin: TU Berlin.

Backhaus, C., Marsolek, I., Köth, H., & Friesdorf W. (2003). Evaluating the usability of data management systems for the OR. *Journal of Clinical Monitoring and Computing, 17,* 467–468.

Backhaus, C., Papanikolaou, M., Kuhnigk, S., & Friesdorf, W. (2001). Usability engineering—Eine methodenübersicht zur anwendergerechten gestaltung von medizinprodukten. [Usability Engineering – methods for user friendly device design] *mt-Medizintechnik, 121,* 133–138.

Backhaus, C., Semar, B., & Friesdorf, W. (2004). PROMEDIKS—Prozessorientierte medizintechnik in klinischen systemen. [Process oriented medical devices in clinical systems] *Dokumentation des 5. Würzburger Medizintechnik Kongress* (pp. 131–133). Wetzlar: EURITIM-Verlag.

Bevan, N., Kirakowski, J., & Maissel, J. (1991). What is usability? *Proceedings of the Fourth International Conference on HCI, Stuttgart.* Retrieved December 5, 2005, from the World Wide Web: http://www.usability.serco.com/papers/whatis92.pdf

Botha, W. E., & Bridger, R. S. (1998), Anthropometric variability, equipment usability and musculoskeletal pain in a group of nurses in the Western Cape. *Applied Ergonomics, 6,* 481–490.

Bridger, R. S., & Poluta, M. A. (1998). Ergonomics: Introducing human factors into the clinical setting. *Journal of Clinical Engineering, 5/6,* 180–188.

Brooke, J. (1996). SUS—Quick and dirty usability scale. In P. W. Jordan, B. Thomas, B. A. Weerdmeester, & I. L. McClelland (Eds.), *Usability evaluation in industry* (pp. 189–194). London: Taylor & Francis.

Carayon, P. (2003). Macroergonomics in quality of care and patient safety. In H. Luczak & K. J. Zink (Eds.), *Human factors in organizational design and management* (Vol. 2, pp. 21–24). Santa Monica: IEA Press.

Denning, S., Hoiem, D., Simpson, M., & Sullivan, K. (1990). The value of thinking-aloud protocols in industry: A case study at Microsoft Corporation. In *Proceedings of the 34th annual meeting of the Human Factors Society* (pp. 1285–1283). Orlando, FL: HFS Press.

Deutsches Institut für Normung. (1986). Körpermaße des Menschen—Werte. [Anthropometric data] *DIN 33402–2.* Berlin: Beuth.

Deutsches Institut für Normung. (1995). Ergonomische Gestaltungsgrundsätze—Begriffe und allgemeine Leitsätze. [Basic standards for ergonomic design - terms and guidelines] *DIN 614–1.* Berlin: Beuth.

Deutsches Institut für Normung. (2004): Beleuchtung mit künstlichem Licht—Beleuchtung im Gesundheitswesen. [Illumination with artificial light] *DIN EN 5035–3.* Berlin: Beuth.

Dhillon B. S. (1986): *Human reliability with human factors.* New York: Pergamon.

Donchin, Y., & Seagull, F. J. (2002). The hostile environment of the intensive care unit. *Current Opinion in Critical Care, 8,* 316–320.

Dumas, J., & Redish, J. C. (1993). *A practical guide to usability testing.* Norwood, NJ: Ablex.

Dumas, J. S. (1996). The process of human factors engineering: Usability testing. In EDITOR (Ed.), *Human factors in medical devices: Design, regulation, and patient safety* (pp. 41Œ44). Arlington, VA: Association for the Advancement of Medical Instrumentation.

Ergonomics Association Executive Council. (2002). About IEA. Retrieved December 8, 2005, from the World Wide Web: http://www.iea.cc/about/html

Ericson, K. A., & Simon, H. A. (1984). *Protocol analysis: Verbal reports as data.* Cambridge, MA: MIT Press.

Friesdorf, W. (1990). Patient-Arzt-Maschine-System (PAMS). [Patient-Physician-Machine System] In W. Friesdorf, B. Schwilk, & J. Hähnel (Eds.), *Ergonomie in der Intensivmedizin* (pp. 39–42). Melsungen: Bibliomed.

Friesdorf, W., & Göbel, M. (2003). Safety and reliability of clinical work processes. *Proceedings of the 50th Annual GfA Conference* (pp. 669–672). Linz: GFA Press.

Friesdorf, W., Göbel, M., & Buss, B. (2004). Gestaltung hochtechnischer Arbeitsplätze im Gesundheitswesen. In B. Zimolong & U. Konradt [Design of complex workplaces in health care] (Eds.), *Enzyklopädie des Psychologie, Ingenieurpsychologie* (pp. 115–41). Göttingen: Hogrefe-Verlag.

Friesdorf, W., Groß-Alltag, F., Konichezky, S., Schwilk, B., & Fett, P. (1994). Lessons learned while building an integrated ICU workstation. *International Journal of Clinical Monitoring and Computing, 11,* 89–97.

Friesdorf, W., & Hecker, E. (1991). Ergonomie. [Human Factors] In H. Hutten (Ed.), *Biomedizinische Technik—Medizinische Sondergebiete* (pp. 276–279). Berlin: Springer.

Gaba, D. M., Fish, K. J., & Howard, S. K. (1998). *Zwischenfälle in der aästhesie—pävention und management.* Jena: Gustav-Fischer.

Göbel, M., Backhaus, C., & Friesdorf, W. (2002). Ergonomische Aspekte in der Intensivpflege. [Ergonomic aspects in intensive care] In K. D. Neander, G. Meyer, & H. Friesacher (Eds.), *Handbuch der Intensivpflege* (pp. 47–53). Landsberg/Lech: Ecomed.

Gosbee, J. W. (1997). The discovery phase of medical device design: A blend of intuition, creativity and science. *Medical Device and Diagnostic Industry, 19,* 79–82.

Hendrik, H. W. (1997). Organisational design and macroergonomics. In G. Salvendy (Ed.), *Handbook of human factors and ergonomics* (pp. 594–636). New York: Wiley.

Hoyos, C. (1974). Kompatibilität. [Compatibility] In H. Schmidtke (Ed.), *Ergonomie* (pp. 93–99). München: Hanser Verlag.

Institut für Technikfolgen-Abschätzung. (2002). *Evidenzbasierte Bedarfsplanung für Intensivbetten—Ein Assessment, Teil 1: Stand des Wissens.* [Evidence based planning in intensive care] Retrieved August 20, 2004, from the World Wide Web: http://www. oeaw.ac.at/ita/ebene5/d2–2b23.pdf

International Standard Organization. (1996a). Akustik—Angabe und Nachprüfung von Geräuschemissionswerten von Maschinen und Geräten. [Basic human body measurements for technological design] *ISO 4871.* Berlin: Beuth.

International Standard Organization. (1996b): Wesentliche Maße des menschlichen Körpers für die technische Gestaltung. [Acoustics - Declaration and verification of noise emission values of machinery and equipment] *ISO 7250.* Berlin: Beuth.

International Standard Organization. (2000). Ergonomic design for the safety of machinery: Part 3: Anthropometric data. *ISO 15534–3.* Berlin: Beuth.

International Standard Organization. (2002). Safety of machinery: Anthropometric requirements for the design of workstations at machinery. *ISO 14738.* Berlin: Beuth.

International Standard Organization. (2004). Ergonomie der thermischen Umgebung—Strategie zur Risikobeurteilung zur Abwendung von Stress oder Unbehagen unter thermischen Arbeitsbedingungen. [Ergonomics of the thermal environment - Risk assessment strategy for the prevention of stress or discomfort in thermal working conditions] *ISO 15265.* Berlin: Beuth.

Jeffries, R. J., Miller, J. R., Wharton, C., & Uyeda, K. M. (1991). User interface evaluation in the real world: A comparison of four techniques. In *Proceedings of the Conference on Human Factors in Computing Systems '91 Conference* (pp. 119–124). New York: Association for Computing Machinery.

Jordan, P. W. (1998). Human factors for pleasure in product use. *Applied Ergonomics, 29,* 25–33.

Jürgens, H. W. (1997). Körpermaße. [Body metrics] In Bundesamt für Wehrtechnik und Beschaffung. Schmidtke, H. (Ed.), *Handbuch der Ergonomie* (pp. 14–79). München: Hanser.

Jürgens, H. W., Aune, I. A., & Pieper, U. (1989) Internationaler anthropometrischer Datenatlas. [International anthropometric datas] *Schriftenreihe der Bundesanstalt für Arbeitsschutz Fb 587.*Dortmund: Bundesanstalt für Arbeitsschutz.

Kirakowski, J. (1996). The Software Usability Measurement Inventory: Background and usage. In P. W. Jordan, B. Thomas, B. A. Weerdmeester, & I. L. McClelland (Eds.), *Usability evaluation in industry* (pp. 169–177). London: Taylor & Francis.

Lewis, C., Polson, P., Wharton, C., & Rieman, J. (1990). Testing a walkthrough methodology for theory-based design of walk-up-and-use interfaces. In *Proceedings of the ACM CHI '90 Conference* (pp. 235–241). New York: Association for Computing Machinery.

Lewis, J. R. (1994). Sample sizes for usability studies: Additional considerations. *Human Factors, 36,* 368–378.

Luczak, H. (1997). Arbeitswissenschaft als Disziplin. [Ergonomic as a discipline] In H. Luczak, & W. Volpert (Eds.), *Handbuch Arbeitswissenschaft* (pp. 13–17). Stuttgart: Schäfer-Poeschel.

Marsolek, I., Backhaus, C., & Friesdorf, W. (2003). Clinical data management systems: Defining the process flow requirements. *Journal of Clinical Monitoring and Computing, 17,* 484.

Mayhew, D. J. (1999). *The usability engineering lifecycle.* San Francisco, CA: Morgan Kaufmann.

McKinnon, S. (1984). Maximizing your ICU patient's sensory and perceptual environment. *Canadian Nurse, 79,* 41–45.

Nielsen, J. (1992). The usability engineering lifecycle. *IEEE Computer, 25*(3), 12–22.

Nielsen, J. (1993): *Usability engineering.* London: Academic.

Nielsen, J. (1994). Heuristic evaluation. In J. Nielsen & R. L. Mack (Eds.), *Usability inspection methods* (pp. 25–61). New York: Wiley.

Nielsen, J., & Landauer, K. L. (1993). A mathematical model of the finding of usability problems. In *Proceedings of the INTERCHI '93 Conference* (pp. 214–221). New York: Association for Computing Machinery.

Nielsen, J., & Molich, R. (1990). Heuristic evaluation of user interfaces. In *Proceedings of the ACM CHI '90 Conference* (pp. 249–256). New York: Association for Computing Machinery.

Nielsen, J., & Phillips, V. (1993). Estimating the relative usability of two interfaces: Heuristic, formal, and empirical methods compared. In *Proceedings of the INTERCH '93 Conference* (pp. 214–221). New York: Association for Computing Machinery.

Nielsen, J. (1989). Usability testing at a discount. In G. Salvendy & M. S. Smith (Eds.), *Designing and using human–computer interfaces and knowledge based systems* (pp. 394–401). Amsterdam, the Netherlands: Elsevier.

Nisbett, R. E., & Wilson, T. D. (1977). Telling more than we can know: Verbal reports on mental processes. *Psychological Review, 84,* 231–259.

Pahl, G., & Beitz, W. (1993): *Konstruktionslehre. [Engineering Design]* Berlin: Springer.

Ravden, S., & Johnson, G. (1989). *Evaluating usability of human–computer interfaces: A practical method.* New York: Wiley.

Rosenbaum, S. (1989). Usability evaluations versus usability testing: When and why? *IEEE Transactions on Professional Communication, 32,* 210–216.

Sawyer, D. (2002). *Do it by design: An introduction to human factors in medical design.* Retrieved December 5, 2005, from the World Wide Web: http://www.fda.gov/cdrh/humfac/ doit.html

Scharmer, E. G., & Siegel, E. (1997). Fehlermöglichkeiten im Umgang mit Narkosegeräten und deren Vermeidung. [Mistakes and their prevention while using anesthesia machines] *Der Anästhesist, 10,* 880–888.

Schmidtke, H. (1993). Arbeitsplatzgestaltung. [Workplace design] In H. Schmidtke (Ed.), *Ergonomie* (pp. 502–520). München: Hanser Verlag.

Spool, J. M., Snyder, C., & Robinson, M. (1996). Smarter usability testing: Practical techniques for developing products. In *Proceedings of the ACM CHI '96 Conference* (pp. 365–366). New York: Association for Computing Machinery.

Staggers, N. (2003). Human factors: Imperative concepts for information systems in critical care. *AACN Clinical Issues, 14,* 310–319.

Stanton, N. A., & Baber, C. (1994). A pragmatic approach to the design of an evaluation of user interfaces. In *Tutorial notes for the Ergonomics Society annual conference* (pp. 19–22). Warwick: University of Warwick.

Stanton, N. A., & Young, M. (1998). Is utility in the mind of the beholder? A study of ergonomics methods. *Applied Ergonomics, 29,* 41–54.

Stoessel, S. (2002). Methoden des testing im usability engineering. [Test methods for Usability Engineering] In M. Beier & V. Gizycki (Eds.), *Usability: Nutzerfreundliches web-design* (pp. 79–96). Berlin: Springer.

U.S. Food and Drug Administration. (1999). *Device use safety: Incorporating human factors in risk management.* Rockville, MD: Author.

Virzi, R. A. (1992). Refining the test phase of usability evaluation: How many subjects is enough? *Human Factors, 34,* 457–468.

Virzi, R. A. (1997). Usability inspection methods. In M. G. Helander, T. K. Landauer, & P. V. Prabhu (Eds.), *Handbook of human–computer interaction* (pp. 705–715). Amsterdam, the Netherlands: Elsevier.

Wharton, C., Bradford, J., Jeffries, R., & Franzke, M. (1992). Applying cognitive walkthroughs to more complex user interfaces: Experiences, issues and recommendations. *Proceedings of the ACM CHI '92 Conference* (pp. 381–388). New York: Association for Computing Machinery.

Whiteside, J., Bennett, J., & Holtzblatt, K. (1988). Usability engineering: Our experience and evolution. In M. Helander (Ed.), *Handbook of human–computer interaction* (pp. 791–817). Amsterdam, the Netherlands: North-Holland.

Wiklund, M. E. (1995). *Medical device and equipment design: Usability engineering and ergonomics.* Buffalo Grove, IL: Interpharm.

Wiklund, M. E. (1993a). How to implement usability engineering. *Medical Device and Diagnostic Industry, 15,* 68–73.

Wiklund, M. E. (1993b). Usability tests of medical products as a prelude to the clinical trial. *Medical Device and Diagnostic Industry, 3,* 68–73.

Wilson, J., & Corlett, N (1995). *Evaluation of human work.* London: Taylor & Francis

Xiao, Y., Mackenzie, C. F., Seagull, J., & Jaberi, M. (2000). Managing the monitors: An analysis of alarm silencing activities during an anaesthetic procedure. In *Proceedings of the IEA 2000/HFES 2000 Congress* (pp. 250–253). San Diego, CA: Human Factors and Ergonomics Society.

Xioa, Y., Seagull, F. J., Mackenzie, C., Wickens, C., & Via, D. (2001). *Auditory warning signals in critical care settings* (Final Report prepared for National Patient Safety Foundation). Unpublished manuscript.

47

HUMAN FACTORS AND ERGONOMICS IN THE EMERGENCY DEPARTMENT

Robert L Wears and Shawna J Perry
University of Florida, Jacksonville

HEALTH CARE WORK AS SYSTEMS

It has become popular among those interested in safety and quality to take a "systems view" of health care processes, but the views of the systems that follow are wishful and simplistic, based on rationalized, engineered systems that are quite different from the actual world of health care work "in the wild" (Hutchins, 1996).

One or more physiological subsystems are embedded in every health care work system—in fact, these subsystems are the focus of the larger work system. Although this in itself is not unusual (e.g., chemical plants have a chemical process at the heart of their system), there are several important ways in which it differs from simpler systems: degree of knowledge, degraded functional states, homeostatic activity, equifinality, remoteness, unstoppability, and mortality.

Degree of Knowledge

Compared to what is known about the fundamental processes in a nuclear power plant, a jet engine, or a refinery, we know very little about the physiological processes of the human body in health and disease. Engineered systems have well-documented designs and are deterministic in at least a large part of their parameter space. In particular, after an abnormality, engineered systems can be deeply investigated (e.g., completely disassembled), if necessary, to provide a detailed understanding of the anomaly. Although they may occasionally manifest emergent or unexpected properties, they are fundamentally knowable from scientific and engineering first principles. Physiologic systems, on the other hand, are fundamentally unknown from first principles. After an abnormality, detailed investigation of the physiology is not often possible; even postmortem examinations sometimes fail to explain the clinical course. Thus reasoning about what is occurring in such systems, or predicting what state they will be in next, is substantively more difficult because of this fundamental lack of knowledge.

Degraded Functional State

In contrast to the processes on which the systems approach was originally based, the physiologic processes in the health care setting are always functioning abnormally (i.e., in a degraded state). Even the "worried well" have, at the very least, malfunctioning sensors.

Homeostatic Activity

Physiological systems differ from physical or chemical processes in that they actively try to counteract perceived malfunctions. Sometimes these homeostatic processes are beneficial, and sometimes they produce further problems. The point is that in health care, there are always two actors involved: the patient's own physiology striving to return to a normal state and the care team's interventions aimed at returning to a normal state. These processes interact in unpredictable ways. Thus, physiologic systems are adaptive in ways that engineered systems are not. Although this is largely a benefit, it is also an additional complicating factor.

Equifinality

A given physiologic status can be arrived at in a variety of different ways, and the trajectory of previous events and interventions can be important. Although equifinality is not unique to physiologic processes, it is a fundamental characteristic, in that it means that knowledge of the current status of the process is not always sufficient for effective or efficient intervention or control.

Remoteness

Fundamental physiologic processes are not observed directly but are made manifest by a relatively small number of proxy variables (symptoms, physical signs, and so on), so that uncertainty about the process itself is always high.

Unstoppability

Unlike most other operations that can (albeit at great expense) often be completely stopped, disassembled, repaired, reassembled, and restarted, core physiologic processes can never be stopped. The system must be repaired while it is still running.

Mortality

Whereas every airplane crash is unambiguously a failure, all human beings will eventually die, and most of those deaths will occur in proximity to health care (Gaba, 2000a, 2000b). The ability to identify a relatively small set of unambiguous events that substantially encompass the universe of possible failures has been a critical factor in advancing safety in most other fields (Schulman, 2002), but it is not possible in health care.

EMERGENCY DEPARTMENT (ED) WORK AS A SYSTEM

Applying human factors and ergonomics principles to the ED requires that we first understand the ED as a complex sociotechnical system. At the simplest and most abstract level, the ED is a basic transformation system that accepts inputs (patients), applies transformation processes to them (care), and delivers outputs (patients, hopefully in improved states). However, there are several complexities that make ED work particularly difficult to understand, manage, or improve, particularly when contrasted with other types of production or service systems. In particular the ED system is open, unbounded, and complexly interactive.

Open System

Although many health care activities can be viewed as closed, or nearly closed, systems, EDs are almost completely open systems. Their spatial boundaries may be well defined, but their temporal, administrative, and procedural boundaries are vague to nonexistent. EDs must be open to inputs from the outside world, without limit and without prior arrangement. The inputs are generally unrestricted, in the sense that they may span the entire spectrum of possible diseases and injuries. In addition, EDs are open to inputs from within the hospital; for example, when outpatients "crash" in radiology, or other similar areas, they are typically brought to the ED for stabilization until a suitable inpatient bed can be found. The hospital influences the EDs processes in many other ways besides adding to its inputs. When inpatient beds or nurses are in short supply, a standard output channel is blocked, and patients who normally would be transferred out of the ED to a ward instead remain in the ED and receive their inpatient care there. Similarly, delays in laboratory, radiology, or consulting services impact the ED's internal processes directly.

The openness of the ED system means that work in EDs is almost exclusively event driven. Long-term planning is probabilistic, short-term planning is immediate and highly contextual, and there is very little in between.

Unbounded System

In addition to being open to myriad outside influences, the ED is unbounded in several dimensions. It is temporally unbounded in that it must maintain 24/7/365 operations at all times; EDs do not shut down temporarily, even for major refitting or renovations. In contrast to other units, which have physical limits on the number of patients that can be

TABLE 47–1. Examples of Human Factors Issues in the ED

Issue	*Example/Impact*
Event driven work process: Work space and resources already in hour basis	Minor bus accident results in need to treat large number of patients who arrive simultaneously, quickly outstripping the staff, supplies, and equipment caring for patients difficult to predict on hour-to the ED
Patient type is highly variable causing difficulty identifying workload or services needed	A chief complaint of abdominal pain will require differing degrees of evaluation depending on attributes such as age, past medical history, severity of illness, cultural/ language barriers, and so on
Outcomes coupled in varying degrees to other microsystems within the hospital	Through-put of all patients is slowed (halted for some) with failure of a piece of laboratory equipment resulting in extended length of stay as blood tests are sent to another hospital to be performed
Work of emergency care requires many individual that are highly variable in number	Each patient is seen by a clinician who will generate a series of "orders" or actions tasksnecessary to care for the patient. Each order results in a series of tasks to be performed by the staff (typically six to seven per order). Number of orders ranges 1 to 10 per patient.
Complex sociotechnical workspace involving a number of specialities and skill levels working as a team to provide care	Emergency care in the ED may be provided by physicians, nurses, physician assistants, paramedics, nurse practitioners, or pediatricians and requires at times the involvement of the same skill types from other specialties (i.e. surgeons, dialysis nurses); standardization of work processes and cultural expectations is difficult
Lack of standardization of work tools	Tracking of patient through-put can be performed by a wide range of cognitive artifacts, including handwritten notes, status boards, software programs, human memory
Transitions or turnover of care are necessary	Shiftwork is required due to need for 24 hour operations resulting in unstructured transfer of information, authority, and responsibility for patient care
ED workspace rarely designed for the work performed there	Many EDs reside in recycled clinical treatment areas designed for the work of other specialties (e.g., outpatient clinic, x-ray suites, inpatient wards)

Note. ED = Emergency Department.

accepted, the ED is unbounded in the sense that there is no natural limit to the number of patients for which it can become responsible (Derlet, Richards, & Kravitz, 2001; Richardson, Asplin, & Lowe, 2002; Schull, Szalai, Schwartz, & Redelmeier, 2001). In fact, when other units are full, the excess is held in the ED, leading one department chairman to remark, somewhat puckishly, that "the ED is the only infinitely expansible part of the hospital" (James Adams, personal communication, December 12, 2002). Finally, it is unbounded in the types of patients and conditions it handles, with the sole exception being major surgical procedures.

Complexly Interactive System

An important property of the ED as a system is that its elements, particularly its patients, interact and affect each other's processes in complicated and unpredictable ways. This may not be immediately apparent, because patients generally do not directly interact with one another, but it is clearly the case that one patient's presence can affect the care given another in ways more complicated than simple queuing effects (Boutros & Redelmeier, 2000; Magid, Asplin, & Wears, 2004; Schull, Vermeulen, Slaughter, Morrison, & Daly, in press).

THE ED AS A SOCIOTECHNICAL SYSTEM

EDs as organizations are simultaneously social (i.e., consisting of people, norms, values, culture, climate) and technical (consisting of tools, facilities, equipment, technology, procedures). The social and technical components are deeply interrelated and interdependent; every change in one area affects the other, perhaps after a variable time delay. This reciprocal determinism of the social and technical aspects of the ED is influenced further by external forces (economics, regulation, litigation) and, to a limited extent, also influences them. Thus there are three aspects of the sociotechnical system of the ED: social, technical, and external. Table 47–1 provides some examples of how these aspects of the ED affect its work.

As is typical in health care, these components have tended to be managed separately in EDs, with predictable consequences due to suboptimization and the unexpected consequences of interactions with ignored areas. In what follows, we highlight some of the more salient human factors issues in ED care, following a framework proposed by Vincent (Vincent et al., 2000; Vincent, Taylor-Adams, & Stanhope, 1998) that bears a striking resemblance to the macroergonomic framework offered by Carayon (2003). Although we use a decomposition method to simplify the issues, the reader should keep in mind how the social, technical, and external aspects of the ED affect and are affected by these issues. In addition, to date little work has been done to improve these aspects of emergency care; thus we raise more problems than solutions.

HUMAN FACTORS ISSUES IN EDs

Patient

Patient factors affect safety in all areas of health care. Language barriers, patients' physiologic status, and their general physical condition (e.g., obesity), affect the difficulty and riskiness of the work clinicians do. An additional patient domain issue that is highly specific to emergency care is that patients and providers are generally complete strangers when the encounter begins; there is no common history or shared ground, so issues of lack of confidence or trust complicate the relationship and may affect the accuracy of historical information given to the provider or compliance with follow-up.

Provider

Selection and training. Emergency care providers are generally rigorously selected, highly trained, and strongly motivated, so further effort on selection and training seems unlikely to yield major improvements, for they have largely been maximized already. However, there are several provider areas in which emergency care falls short of what has been done in other environments.

First, although the initial training and assessment of emergency care professionals is extensive and rigorous, their continuing development has been largely ignored. Once certified, there are no routine mechanisms by which the provider's physical or mental fitness for the job is reassessed. Subsequent training is haphazard at best and is not provided for in the routine workplace; providers are expected to accomplish this on their own time and at their own expense. Given the rapid pace of change in health care, this seems to leave much to chance.

Board certification for emergency physicians has been available since 1980, but because the specialty is so new, it has only become a de facto requirement in a few metropolitan areas. There are roughly 15,000 full-time emergency physicians in the United States, of whom only about half are board certified (Moorhead et al, 2002). Thus in a practical sense, board certification cannot be a requirement for employment because there are not enough board-certified physicians to fill the available positions.

Emergency medicine was one of the first specialties to require recertification of its diplomates. Initially this was required at 10-year intervals; in 2000, this was changed to a continuous certification model wherein certain activities are required yearly to qualify for recertification in the tenth year.

Certification is also available for emergency nurses but seems to be less commonly pursued. Prehospital personnel (emergency medical technicians) and technicians have their own certification levels and requirements.

Fatigue and wellness. Gaba (1998) has pointed out that in every high hazard industry except medicine, the default assumption has been that long work hours and circadian disruption lead to health, performance, and safety problems and that the burden of proof is placed on organizations to demonstrate that their staffing and scheduling patterns are safe; however in health care, the burden of proof is reversed—any staffing or scheduling practice is assumed to be safe unless convincing evidence to the contrary is presented. In general, there has been a great reluctance in health care to accept data on fatigue, shift work, and circadian disruption from other settings, despite the absence of any compelling reason to believe that doctors and nurses are immune to these issues. Despite this thinking, the general association of fatigue to decreased performance seems well established (Frank, 2002; Jha, Duncan, & Bates, 2001). It has often been asserted that the performance level of someone who has been up all night approximates that of someone who is legally intoxicated (Bonnet, 2000).

Emergency medicine was the first specialty to begin to limit physician work hours in shifts. Originally, EDs were staffed using the traditional medical "on call" model in which physicians conducted their normal work during business hours but were occasionally on call for emergency cases, particularly at night or on weekends. As the demand for emergency care grew, ED work for physicians assumed a shift-work model similar to that for nurses. This model is now dominant, and the general expectation in most EDs is that physicians and nurses will be awake and at work at all times during the night (in other words, there are no "call rooms" available for rest in slack times) and that they will go home when their shift ends. However, there are no formal limits on continuous hours of service, or total hours per week, or total number of hours off between shifts, with the sole exception of residents in training, who are subject to an 80 hour per week rule. However, the effects of this limitation are currently unknown, because unexpected effects due to decreased continuity of care and shifting workload to other, unregulated workers have been noted (Laine, Goldman, Soukup, & Hayes, 1993; Petersen, Brennan, O'Neil, Cook, & Lee, 1994).

The adoption of shift work in EDs limits the effects of fatigue when shifts are limited to 8 to 12 hours or less (which is almost universally current practice; Thomas, Schwartz, & Whitehead, 1994). However, the effects of circadian disruption due to shift work still must be dealt with (Frank, 2002). Research in other domains (and some studies in health care settings) has shown rather consistently that shift work is associated with several negative effects (Costa, 1996), principally in the areas of sleep, performance, psychological and social health, and physical health.

Night workers get roughly 25% to 30% less sleep than day or evening workers, and their sleep is of poorer quality (Smith-Coggins, Rosekind, Hurd, & Buccino, 1994; Tepas & Carvalhais, 1990). Independent of this sleep deprivation, circadian dysynchrony reduces cognitive and work performance, particularly when the work does not bring its own intrinsic arousal stimulus. Thus, whereas ED workers on night shifts may be able to rise to the occasion for acute, exciting, and interesting cases, their performance is likely to suffer when the work is routine or repetitive (Folkard & Tucker, 2003; Howard et al, 2003; Monk, 1990; Monk & Carrier, 1997; Paley & Tepas, 1994; Smith-Coggins et al, 1994). This performance decrement is dramatically underscored by the number of dramatic accidents occurring on night shifts, including the Challenger explosion (Vaughan, 1996), the Bhopal chemical plant disaster (Perrow, 1994), the Exxon Valdez grounding, the Three Mile Island accident (Perrow, 1984), and the Chernobyl meltdown.

Shift work is associated with mood disturbances, irritability, and (if sustained for long periods of time), increased rates of substance abuse, divorce, and suicide (Folkard & Monk, 1985; Tepas et al., 2004). In addition, shift workers tend be more socially isolated (Monk & Folkard, 1992). A particular concern is that shift work is often cited as the primary reason for leaving emergency medicine (Hall, Wakeman, Levy, & Khoury, 1992).

Finally, long-term shift-work is associated with poorer physical health. Shift workers have higher rates of accidents, ulcers, hypertension, and coronary disease; in addition, conditions such as diabetes and epilepsy are exacerbated by shift work (Frank, 2002). Night-shift workers have a substantially greater risk of road traffic accidents on their way home from work than the general population (Monk, Folkard, & Wedderburn, 1996), a fact that medicine seems only recently to be learning (Barger et al., 2005).

Several factors are known to decrease workers' ability to tolerate circadian disruption (Tepas & Monk, 1987). Tolerance goes down with age, especially over 50 (Tepas, Duchon, & Gersten, 1993). Moonlighting (working a second job) is another risk factor, probably related to increased fatigue; and "morning larks" do not tolerate shift work as well as "night owls." However, individual variation in tolerance is high, and the deleterious effects of shift work in emergency care may be at least partially mitigated by self-selection; individuals who do not tolerate shift work well may discover this in their training and subsequently choose other fields (Steele, Ma, Watson, & Thomas, 2000; Steele, McNamara, Smith-Coggins, & Watson, 1997).

Professional societies have begun to address these issues, albeit in a hesitant manner (American College of Emergency Physicians, 1995). As the physician and nursing workforce in EDs age, the effects of shift work and fatigue can be expected to become more concerning. General recommendations have been made to minimize consecutive nights, use clockwise (forward) shift rotations, allow 24 to 48 hr off duty after a night shift, and avoid use of sedatives or stimulants, with the possible exception of caffeine (Frank, 2002). External regulators have not addressed this area at all.

Task and Tools

Emergency care suffers from the same sorts of difficulties with medical devices as do other areas of care. Devices used quite commonly in emergency care (such as monitor defibrillators) have seldom been evaluated for usability, but what few data are available indicate persistent problems in design (Fairbanks, Caplan, Shah, Marks, & Bishop, 2004). Similarly, formal procedures have generally undergone no empirical evaluation and are thus used as loose guides rather than precise specifications.

Emergency care has produced several unique devices to help workers cope with the demands of emergency care. The Broselow tape was developed as an aide to nonpediatric emergency caregivers who must care for critically ill or injured children (Luten, Broselow, Wears, & Blackwelder, 1998; Luten et al, 1992). Although it has not been formally assessed from a human factors point of view, there is some evidence suggesting its use can be associated with improved performance (Shah, Frush, Luo, & Wears, 2003). ED status boards are quite interesting developments. EDs commonly use status boards as tools for managing their clinical work. These tools are not specifically mandated by any external authority but instead were developed spontaneously and locally as the practice of emergency medicine grew more complex over the years. Thus, they can be considered cognitive artifacts produced by ED workers themselves to help them do their job better, with little or no outside influence. Status boards frequently take the form of large, manually updated whiteboards on which patient locations are represented as rows in a grid and columns are used to provide a variety of relevant data. They are typically located in a centrally accessible work area, such as the nursing station, where they can be used by all ED staff (i.e., not just physicians or nurses). Status boards began as simple tracking devices, for example displaying patient name and location, but because they afforded further annotation, they have evolved to include a great deal of additional information. This additional data might include some of any of the following:

- Patient demographic information, such as chief complaint, arrival time, or length of stay.
- Staff information, such as responsible ED physician, nurse, or technician, private physician.
- Process information, such as pending procedures, laboratory or imaging studies, completed laboratory or radiology work, whether seen by ED physician yet, partial plans (e.g., nebs × 3 or 4 hr obs).
- Patient status information, such as pending consultations, admission–discharge status, bed status.
- Additional patient information, such pregnancy, neutropenia, isolation requirements, allergies, or name conflict alerts, and so on.

Status boards are important tools for providing safe care in the ED. They do this by supporting shared memory, latent processes, collaboration, shared cognition, communication, and coordination. For example, status boards are particularly useful in reducing interruptions (Chisholm, Collison, Nelson, & Cordell, 2000; Chisholm, Dornfeld, Nelson, & Cordell, 2001; Chisholm, Pencek, Cordell, & Nelson, 1998; Coiera, Jayasuriya, Hardy, Bannan, & Thorpe, 2002) by supporting asynchronous communication among ED staff in a manner that reduces the memory burden on both sender and receiver (Parker & Coiera, 2000).

Because the manual format is flexible, the status board has often been extended to include other types of information. For example, patients not physically in the ED are often represented on the status board; transfers expected from other hospitals, referrals from physicians' offices, or ED patients who are away from the department for prolonged procedures, such as arteriography or dialysis, may still be represented on the board but not associated with any specific patient location. Similarly, information affecting the unit as a whole but not associated with any specific patient is also commonly displayed. This might include, for example, the status of critical care beds in the hospital, the current on-call consultants and their contact information, ED beds that are closed for some reason, and so on.

Because status boards were developed and have evolved locally, they are exquisitely situated in their own local environments and have not been standardized across different EDs. In addition, they are densely encoded, making use of color, cryptic abbreviations, and symbolic markings that are known to experienced staff. Newcomers assimilate this encoding during on-the-job training, but this skill is seldom remarked and never assessed, so there is a potential for hidden misunderstandings of the more idiosyncratic elements.

As designed artifacts produced by the workers themselves and unmodified by the demands of external forces, status boards offer a view into the implicit theories of the workers (Bisantz & Ockerman, 2003; Carroll & Campbell, 1989; Woods, 1998).

A task issue that seems fairly specific to EDs is the difficulty in maintaining sufficient experience with the most difficult and dangerous procedures that might be expected. For example, the most dangerous portions of anesthesia are induction and recovery, but anesthesiologists execute these procedures several times per day and thus are extremely facile. In contrast, the most dangerous procedures in emergency care (rapid sequence induction, cricothyrotomy, thrombolysis) are required on the order of weekly to yearly. It is unlikely that even the best-trained practitioners will be at their peak skill level when called on to perform these procedures. Simulation and similar venues may offer some utility in remedying this problem, but their role remains to be established.

Another important task issue is the ubiquity of information gaps in emergency care (Cook, Render, & Woods, 2000; Stiell, Forster, Stiell, & van Walraven, 2003). Hospital records, especially discharge summaries, details of past medical history, and other important information is often difficult to access in an expedient fashion. Even referral notes sent in by family doctors with the patient may not reach the emergency physician. Emergency physicians must then act on the basis of incomplete or erroneous information and become used to doing so, so that information is sometimes not even sought when it might have been retrieved.

Work Team

Teamwork. Patient care in the ED is not a solitary activity. Caregivers from multiple professions work together in a pattern of distributed, collaborative activity. This creates additional tasks of communication and coordination among the workers in addition to the specific clinical work to do. Physicians, nurses, and technicians are the three professionals involved in virtually every ED setting. In specific instances, they may be supplemented by a wide variety of additional professionals: pharmacists, respiratory therapists, clerks, radiology technicians, and so on. It is one of the strange anomalies of health care that all these professionals are trained and evaluated in their separate silos but are expected to work smoothly together. Consequently, in most situations, they work as a group, not as a team.

Some initial studies of teamwork (or lack of it) in the ED have suggested that teamwork failures could play a substantial role in the genesis of adverse events. A chart review study of malpractice claims identified an average of almost nine teamwork failures per case (Risser et al, 1999), suggesting that improved teamwork might avoid or mitigate some adverse events. This is supported indirectly by survey work indicating poor teamwork attitudes in health care (Sexton, Thomas, & Helmreich, 2000).

The widespread acceptance of teamwork training in aviation has led to attempts to introduce similar training in emergency care (Simon et al, 1998). The MedTeams project (Simon, Morey, Locke, Risser, & Langford, 1997) was a multi-institutional effort to adapt the principles of aviation teamwork training to the ED setting, develop and pilot test a training program, and finally assess the effects of teamwork training on performance. They reported that teamwork behaviors improved and observed errors decreased after training, although perceived workload did not change (Morey, Simon, Jay, & Rice, 2003; Morey et al., 2002). However, teamwork training has not been generally adopted in EDs, although considerable interest in it remains (Firth-Cozens, 2001; Firth-Cozens & Moss, 1998; Jay et al, 2000; Shapiro & Morchi, 2002; Shapiro et al, 2004; Small et al, 1999).

Shift changes have long been thought to be particularly hazardous times in emergency care (Salluzo, Mayer, Strauss, & Kidd, 1997), but surprisingly little has been written about them. However, there is at least some evidence that shift changes may actually contribute to the recovery from incipient failures (Wears, Perry, Shapiro, Beach, & Behara, 2003). Patterson, Roth, Woods, Chow, and Gomes (2004) noted 21 strategies used in high hazard industries to improve the effectiveness and efficiency of shift changes, and observational studies in five North American EDs noted that 8 of these strategies were used frequently, 4 sporadically, and 9 not at all (Behara et al., in press). Thus, there is the potential to improve the effectiveness of shift change turnovers by adopting some of these unused strategies, such as limiting interruptions during turnover or requiring the oncoming party to verbalize the "picture" of the patient they have been given.

Shift changes in the ED can be viewed as a specific example of a large problem in both EDs and other areas of care: the successful transfer of information, authority, and responsibility from one set of caregivers to another. Table 47–2 illustrates some different types of transitions occurring in

Table 47–2. The Types and Characteristics of Transitions of ED Care

From	To	Setting	Typical # of Patients	Time Investment
ED physician	ED physician	Shift change	Many	High
ED nurse	ED nurse	Shift change	Many	High
Paramedic	ED staff	Ambulance arrival	One	Low
ED physician	Consulting specialist	Referral or admission	One	Moderate
ED staff	Inpatient staff	Admission	One	Moderate
ED physician or nurse	Patient and/or family	Disposition	One	Variable

Note. ED = Emergency Department

emergency care. However, transitions in general and shift changes in particular appear to be extremely complex processes that have been studied very little.

Communication and coordination beyond the ED. Emergency care is by definition episodic care; the continuing care of patients must be passed on to either inpatient services or outpatient facilities. This transition across organizational boundaries would seem fraught, because the parties to the exchange do not share a great deal of common ground (Coiera, 2000). Like shift change, these interactions have barely been studied. Matthews, Harvey, Schuster, and Durso (2002) examined transitions from the ED to the internal medicine service and found that they contained little reference to formal guidelines or standards for care and that their perceived quality depended heavily on the confidence one participant had in the other. Other important areas, such as the ambulance to ED transition, or the ED to home care transition, to our knowledge have not been studied.

Communication and coordination in the reverse direction have likewise been little investigated, although ED effectiveness depends heavily on the level of responsiveness of consulting services, laboratory, and radiology.

Culture. Health professionals have been noted to have social and cultural characteristics that sometimes work against the goal of safe care. Although these studies have not focused on the ED, emergency practitioners would seem to share many of these attributes, because they are to some extent reinforced by the ED environment.

One particularly problematic attribute is the presence of what has been called "patching" behavior and the value attributed to it (Tucker & Edmondson, 2002, 2003). When health care workers encounter problems in their daily routine (which is, on average, about once per hour), they typically use patches to quickly deal with it and get on with their job. For example, a missing device might be "borrowed" from another unit, rather than reordered from the supply room. This solves the immediate problem but moves it to another unit; and the underlying cause remains untouched. The problem is seldom brought to anyone else's attention so it can be dealt with in a more permanent and proactive way. The ED, with its time pressure and unpredictability, encourages this sort of "muddling through" behavior.

Thus the unit culture tolerates—in fact expects—that things will not work as designed, and workers take some pride in the clever ways in which they can devise "work-arounds" to meet their goals. Although this ability is one of the reasons that the ED can function at all, it is counterproductive in the long run. In addition, it means that attempts to engineer forcing functions into devices so they cannot be used incorrectly (e.g., so that a tube intended to supply compressed air to a blood pressure cuff cannot be mistakenly connected to an intravenous line [Institute for Safe Medication Practices, 2003]) may be defeated by workers who expect things not to work together and see part of their job as finding ways for them to do so.

Work Environment

Most EDs are ergonomic nightmares (Wears & Perry, 2002). Few, if any, were designed using human factors principles, and in many the original design has undergone repeated renovation and modification over the years. In addition, overcrowding in EDs has reached epidemic proportions in the past 5 years, due to reductions in hospital capacity, economic pressures, and shortages of nurses to staff inpatient beds (Derlet et al, 2001; Richardson et al, 2002; Schull et al, 2001).

Noise is also a problem in many EDs (Barlas, Sama, Ward, & Lesser, 2001; Brown, Gough, Bryan-Berg, & Hunt, 1997; Topf, 2000), although

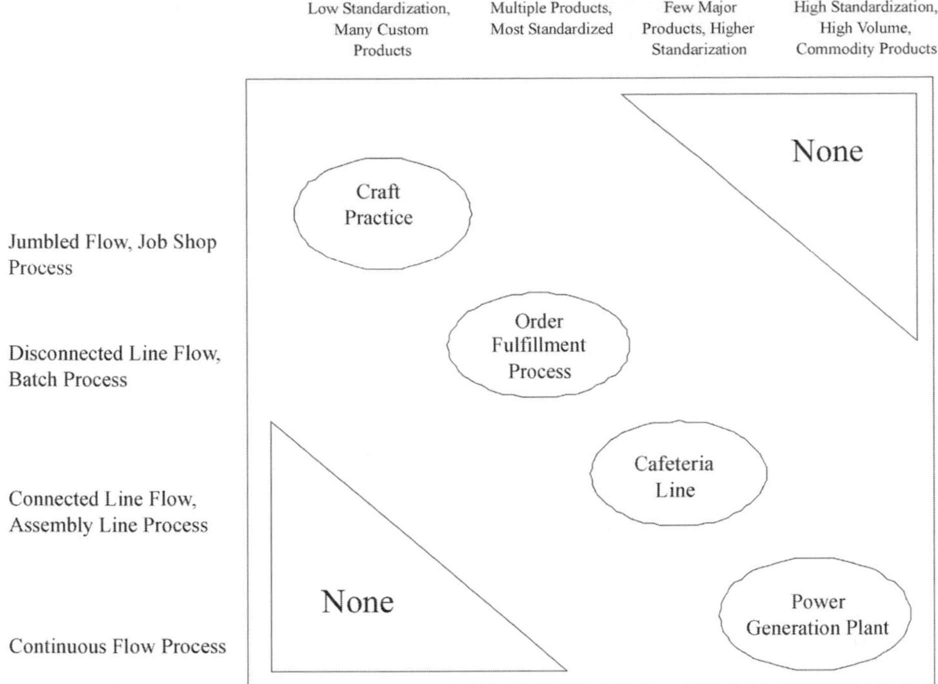

Figure 47–1. Process–product matrix for a system. The horizontal axis represents the type of products produced, from highly customized, "one off" products on the left, to highly standardized, fungible, "commodity" products on the right. The vertical axis represents the type of production processes best suited to those products, from highly variable "job shop" processes at the top, to continuous flow processes at the bottom. The triangles represent points where the mismatch between product and process is so great that no systems operate in these areas. The ovals show prototypical examples of good matches between product and process.

some institutions have begun to introduce noise abatement measures in an attempt to improve patient satisfaction. In the prehospital setting, noise is an occupational hazard, as emergency medical technicians demonstrate greater than expected hearing loss (Johnson, Hammond, & Sherman, 1980; Pepe, Jerger, Miller, & Jerger, 1985).

Patient privacy is also an important issue in many EDs (Barlas et al, 2001). There is a clear goal conflict here between the desire to support patients' privacy and the need to be able to view patients when not physically at their bedside.

Organization and Management

Effective care in EDs is much more an issue of organization and management than in other areas of care in which individual knowledge and skill play more prominent roles. Many clinical problems in emergency care are time-sensitive, but demand for care is highly variable, fluctuating, and only probabilistically predictable. Thus EDs must be organized, resourced, and staffed to handle variability in not only the quantity but also the types of problems encountered. This requires long-term planning and effort at a higher level than that of the individual practitioner. It is common for ED managers to use analogies to assembly-line systems (e.g., the Toyota manufacturing system [Womack, Jones, & Roos, 1990]) as models for the organization and management of EDs. However, an analysis of the "product–process" matrix (Hayes & Wheelwright, 1979) shown in Figure 47–1 suggests that this model is not likely to fit emergency care well, which seems to have much more in common with a job shop model. Two implications follow from this assessment: first, that EDs will necessarily require organizational slack (excess capacity) to function adequately, and second, that assembly line models of organization and management will inevitably be mismatched to the realities of ED work.

Because of the uncertainties of their inputs, EDs typically use global outcomes to assess performance, for example, average waiting time, proportion of patients leaving before completing treatment, or the proportion of unscheduled returns to the ED. However, there are hidden difficulties in this practice. Elements of one patient's care can affect many others (Boutros & Redelmeier, 2000) in unforeseen and unpredictable ways, so focusing improvement efforts at the individual patient level can lead to suboptimization and be counterproductive, making global outcome measures more attractive. However, a focus on global optimization is naïve and raises important ethical issues. A dependence on global outcome measures as been criticized as "agricultural" (Asch & Hershey, 1995; Rothwell, 2005). An agricultural intervention is presumably concerned only with optimizing the total yield, not with the individual well-being of each individual plant or animal. Unfortunately, not only is the balance of global optimization and individual suboptimization not known for EDs, the issue is largely unrecognized and undiscussed in any substantive terms.

Institutional Context

There are several unique aspects to the institutional milieu in which EDs operate, in addition to the general problems of economic pressures, hospital closings and mergers, litigation, and so on that affect health care generally. EDs are not allowed to make decisions to accept or not accept patients. Federal legislation (the COBRA/EMTALA laws; Cross, 1992; Derlet & Nishio, 1990; Frew, Roush, & LaGreca, 1988; Wanerman, 2002) mandates that any person presenting to an ED be evaluated at least to the extent necessary to determine his or her stability and must be stabilized before being transferred or referred elsewhere. It also mandates that a hospital or ED receiving a request to accept a transfer may not decline the transfer if they provide the sort of service the patient requires. For example, a head-injured patient may be transferred from one hospital to another, once stabilized, if the first hospital does not offer the services the patient requires (i.e., neurosurgery). If the second hospital does offer neurosurgery, it may not refuse to accept the patient in transfer.

The constraints of COBRA/EMTALA are not widely known in other parts of the health care organization; this leads to the potential for conflict when EDs accept patients whom hospitals or specialists might prefer be sent away.

SUMMARY

EDs are dynamic, uncertain, distributed, collaborative settings for complex technical work. A slight paraphrase from Rochlin (1997) seems an apt description of the problems they face:

For a caregiver in an emergency, deprived of experience, unsure of context, and pressed into action only when something has already gone wrong, with an overabundance of information and no mechanism for interpreting it, avoiding a mistake may be as much a matter of good luck as good training. (p. 128)

In such a complex, reciprocally interactive setting, supporting the human workers who must detect, make sense of, and repair the gaps in the system to ensure its safe and effective function would seem to be a priority. Such support, however, can be developed only with careful, meticulous research that avoids a narrow, keyhole view of the ED (Cook & Woods, 2003; Nemeth, Cook, & Woods, 2004).

Macroergonomic Balance

Given multiple opportunities for improving work in the ED, how might managers and human factors experts best proceed? The problem seems too large and too multifaceted to be attacked on all fronts at once, even if the resources to do so were readily available. One promising approach is to attempt to "balance" domains that are either primarily negative or relatively intractable to improvement (given realistic amounts of time and effort) with others in which meaningful improvements are realistically possible (Smith & Carayon, 2001; Smith & Sainfort, 1989). Of the seven domains discussed in this chapter, five are potentially amenable to improvement (we take patient factors as a given, and although the institutional context is in principle changeable, it is beyond the scope of ED and hospital managers). Thus balancing the positives and negatives among the domains of the individual workers, the tasks and tools, the team, the work environment, and the organization should be the strategic focus of improvement efforts.

References

American College of Emergency Physicians. (1995). Emergency physician shift work. *Annals of Emergency Medicine, 25,* 864.

Asch, D. A., & Hershey, J. C. (1995). Why some health policies don't make sense at the bedside. *Annals of Internal Medicine, 122,* 846–850.

Barger, L. K., Cade, B. E., Ayas, N. T., Cronin, J. W., Rosner, B., Speizer, F. E., et al. (2005). Extended work shifts and the risk of motor vehicle crashes among interns. *New England Journal of Medicine, 352,* 125–134.

Barlas, D., Sama, A. E., Ward, M. F., & Lesser, M. L. (2001). Comparison of the auditory and visual privacy of emergency department treatment areas with curtains versus those with solid walls. *Annals of Emergency Medicine, 38,* 135–139.

Behara, R., Wears, R. L., Perry, S. J., Eisenberg, E., Murphy, A. G., Vanderhoef, M., et al. (2005). Conceptual framework for the safety of handovers. In K. Henriksen (Ed.), *Advances in patient safety.* Rockville, MD: Agency for Healthcare Research and Quality, Department of Defense. 309–321.

Bisantz, A. M., & Ockerman, J. J. (2003). Lessons from a focus on artifacts and implicit theories: Case studies in analysis and design. In E. Hollnagel (Ed.), *Handbook of cognitive task design.* Mahwah, NJ: Lawrence Erlbaum Associates, Inc.

Bonnet, M. H. (2000). Sleep deprivation. In M. Kryger, T. Roth, & W. C. Dement (Eds.), *Principles and practice of sleep medicine* . Philadelphia, PA: Saunders.

Boutros, F., & Redelmeier, D. A. (2000). Effects of trauma cases on the care of patients who have chest pain in an emergency department. *The Journal of Trauma, 48,* 649–653.

Brown, L. H., Gough, J. E., Bryan-Berg, D. M., & Hunt, R. C. (1997). Assessment of breath sounds during ambulance transport. *Annals of Emergency Medicine, 29,* 228–231.

Carayon, P. (2003). Macroergonomics in quality of care and patient safety. In H. Luczak & K. J. Zink (Eds.), *Human factors in organizational design and management* (pp. 21–34). Santa Monica, CA: IEA Press.

Carroll, J. M., & Campbell, R. L. (1989). Artifacts as psychological theories. *Behavior and Information Technology, 8,* 247–256.

Chisholm, C. D., Collison, E. K., Nelson, D. R., & Cordell, W. H. (2000). Emergency department work place interruptions: Are emergency physicians multi-tasking or interrupt driven? *Academic Emergency Medicine, 7,* 1239–1243.

Chisholm, C. D., Dornfeld, A., Nelson, D., & Cordell, W. (2001). Work interrupted: A comparison of workplace interruptions in emergency departments and primary care offices. *Annals of Emergency Medicine, 38,* 146–151.

Chisholm, C. D., Pencek, A. M., Cordell, W. H., & Nelson, D. R. (1998). Interruptions and task performance in emergency departments compared with primary care offices. *Academic Emergency Medicine, 5,* 470.

Coiera, E. W. (2000). When conversation is better than computation. *Journal of the American Medical Informatics Association, 7,* 277–286.

Coiera, E. W., Jayasuriya, R. A., Hardy, J., Bannan, A., & Thorpe, M. E. (2002). Communication loads on clinical staff in the emergency department. *Medical Journal of Australia, 176,* 415–418.

Cook, R. I., Render, M., & Woods, D. D. (2000). Gaps in the continuity of care and progress on patient safety. *British Medical Journal, 320,* 791–794.

Cook, R. I., & Woods, D. D. (2003, October). *The messy details: Insights from technical work studies in health care.* Paper presented at the Proceedings of the Human Factors and Ergonomics Society 47th annual meeting, Denver, CO.

Costa, G. (1996). The impact of shift and night work on health. *Applied Ergonomics, 27,* 9.

Cross, L. A. (1992). Pressure on the emergency department: The expanding right to medical care. *Annals of Emergency Medicine, 21,* 1266–1272.

Derlet, R. W., & Nishio, D. A. (1990). Refusing care to patients who present to an emergency department. *Annals of Emergency Medicine, 19,* 262–267.

Derlet, R., Richards, J., & Kravitz, R. (2001). Frequent overcrowding in U.S. emergency departments. *Academic Emergency Medicine, 8,* 151–155.

Fairbanks, R. J., Caplan, S., Shah, M. N., Marks, A., & Bishop, P. (2004, October). *Defibrillator usability study among paramedics.* Paper presented at the Proceedings of the Human Factors and Ergonomics Society 48th annual meeting, New Orleans, LA.

Firth-Cozens, J. (2001). Multidisciplinary teamwork: The good, bad, and everything in between. *Quality & Safety in Health Care, 10,* 65–66.

Firth-Cozens, J., & Moss, F. (1998). Hours, sleep, teamwork, and stress. Sleep and teamwork matter as much as hours in reducing doctors' stress. *British Medical Journal, 317,* 1335–1336.

Folkard, S., & Monk, T. H. (Eds.). (1985). *Hours of work: Temporal factors in work scheduling.* Chichester, England: Wiley.

Folkard, S., & Tucker, P. (2003). Shift work, safety and productivity. *Occupational Medicine, 53,* 95–101.

Frank, J. R. (2002). Shiftwork and emergency medicine practice. *Canadian Journal of Emergency Medicine, 4,* 421–428.

Frew, S. A., Roush, W. R., & LaGreca, K. (1988). COBRA: Implications for emergency medicine. *Annals of Emergency Medicine, 17,* 835–837.

Gaba, D. M. (1998, November). *Physician work hours: The "sore thumb" of organizational safety in tertiary health care.* Paper presented at the Proceedings of the Second Annenberg Conference on Enhancing Patient Safety and Reducing Errors in Health Care, Rancho Mirage, CA.

Gaba, D. M. (2000a). *Re: Aviation fatality data.* Retrieved January 21, 2000, from the World Wide Web: http://www.bmj.com/cgi/eletters/319/7203/136#EL5

Gaba, D. M. (2000b). Structural and organizational issues in patient safety: A comparison of health care to other high-hazard industries. *California Management Review, 43,* 83–102.

Hall, K. N., Wakeman, M. A., Levy, R. C., & Khoury, J. (1992). Factors associated with career longevity in residency-trained emergency physicians. *Annals of Emergency Medicine, 21,* 291–297.

Hayes, R. H., & Wheelwright, S. C. (1979). Linking manufacturing process and product life cycle. *Harvard Business Review, 57,* 2–9.

Howard, S. K., Gaba, D. M., Smith, B. E., Weinger, M. B., Herndon, C., Keshavacharya, S., et al.. (2003). Simulation study of rested versus sleep-deprived anesthesiologists. *Anesthesiology, 98,* 1345–1355.

Hutchins, E. (1996). *Cognition in the wild.* Cambridge, MA: MIT Press.

Institute for Safe Medication Practices. (2003). Blood pressure monitor tubing may connect to IV ports. *ISMP Medication Safety Alert, 8,* 1–2.

Jay, G., Small, S., Langford, V., Jagminas, L., Shapiro, M., Suner, S., et al. (2000). Teamwork training supported by high fidelity simulation training improves emergency department team behaviors [Abstract]. *Academic Emergency Medicine, 7,* 519.

Jha, A. K., Duncan, B. W., & Bates, D. W. (2001). *Fatigue, sleepiness, and medical errors: No making health care safer: A critical analysis of patient safety practices* (Evidence Report, Technology Assessment No. 43, AHRQ Publication 01–E058). Rockville, MD: Agency for Healthcare Research and Quality.

Johnson, D. W., Hammond, R. J., & Sherman, R. E. (1980). Hearing in an ambulance paramedic population. *Annals of Emergency Medicine, 9,* 557–561.

Laine, C., Goldman, L., Soukup, J. R., & Hayes, J. G. (1993). The impact of a regulation restricting medical house staff working hours on the quality of patient care. *Journal of the American Medical Association, 269,* 374–378.

Luten, R. C., Broselow, J., Wears, R. L., & Blackwelder, B. (1998, November). *Broselow–Luten system: Color-coded zones for pediatric emergencies and implications for universal application.* Paper presented at the Proceedings of the Second Annenberg Conference on Enhancing Patient Safety and Reducing Error in Health Care, Rancho Mirage, CA.

Luten, R. C., Wears, R., Broselow, J., Zaritsky, A., Barnett, T. M., Lee, T., et al. (1992). Length-based endotracheal tube and emergency equipment in pediatrics. *Annals of Emergency Medicine, 21,* 900–904. (Published erratum appears in *Annals of Emergency Medicine, 22* [1993], 155)

Magid, D., Asplin, B. R., & Wears, R. L. (2004). The quality gap: Searching for the consequences of emergency department crowding. *Annals of Emergency Medicine, 44,* 586–588.

Matthews, A. L., Harvey, C. M., Schuster, R. J., & Durso, F. T. (2002, October). *Emergency physician to admitting physician handovers: An exploratory study.* Paper presented at the Proceedings of the Human Factors and Ergonomics Society 46th annual meeting, Baltimore, MD.

Monk, T. H. (1990). Shiftworker performance. *Occupational Medicine, 5,* 183–198.

Monk, T. H., & Carrier, J. (1997). Speed of mental processing in the middle of the night. *Sleep, 20,* 399–401.

Monk, T. H., & Folkard, S. (1992). *Making shiftwork tolerable.* London: Taylor & Francis.

Monk, T. H., Folkard, S., & Wedderburn, A. I. (1996). Maintaining safety and high performance on shiftwork. *Applied Ergonomics, 27,* 17.

Moorhead, J., Gallery, M., Hirshkorn, C., Barnaby, D., Barsan, W., Conrad, L., et al. (2002). A study of the workforce in emergency medicine: 1999. *Annals of Emergency Medicine, 40,* 3.

Morey, J. C., Simon, R., Jay, G. D., & Rice, M. M. (2003, November). *A transition from aviation crew resource management to hospital emergency departments: The MedTeams story.* Paper presented at the Proceedings of the 12th International Symposium on Aviation Psychology, Columbus, OH.

Morey, J. C., Simon, R., Jay, G. D., Wears, R. L., Salisbury, M. L., & Berns, S. D. (2002). Error reduction and performance improvement in the emergency department through formal teamwork training: Evaluation results of the MedTeams project. *Health Services Research, 37,* 1553–1581.

Nemeth, C. P., Cook, R. I., & Woods, D. D. (2004). The messy details: Insights from the study of technical work in health care. *IEEE Transactions on Systems, Man, and Cybernetics: Part A, 34,* 689–692.

Paley, M. J., & Tepas, D. I. (1994). Fatigue and the shiftworker: Firefighters working on a rotating shift schedule. *Human Factors, 36,* 269–284.

Parker, J., & Coiera, E. (2000). Improving clinical communication: A view from psychology. *Journal of the American Medical Informatics Association, 7,* 453–461.

Patterson, E. S., Roth, E. M., Woods, D. D., Chow, R., & Gomes, J. O. (2004). Handoff strategies in settings with high consequences for failure: Lessons for health care operations. *International Journal for Quality Health Care, 16,* 125–132.

Pepe, P. E., Jerger, J., Miller, R. H., & Jerger, S. (1985). Accelerated hearing loss in urban emergency medical services firefighters. *Annals of Emergency Medicine, 14,* 438–442.

Perrow, C. (1984). *Normal accidents: Living with high-risk technologies.* New York: Basic Books.

Perrow, C. (1994). The limits of safety: The enhancement of a theory of accidents. *Journal of Contingencies and Crisis Management, 2,* 212–220.

Petersen, L. A., Brennan, T. A., O'Neil, A. C., Cook, E. F., & Lee, T. H. (1994). Does housestaff discontinuity of care increase the risk for preventable adverse events? *Annals of Internal Medicine, 121,* 866–872.

Richardson, L. D., Asplin, B. R., & Lowe, R. A. (2002). Emergency department crowding as a health policy issue: Past development, future directions. *Annals of Emergency Medicine, 40,* 388–393.

Risser, D. T., Rice, M. M., Salisbury, M. L., Simon, R., Jay, G. D., & Berns, S. D. (1999). The potential for improved teamwork to reduce medical errors in the emergency department. *Annals of Emergency Medicine, 34,* 373–383.

Rochlin, G. I. (1997). *Trapped in the Net: The unanticipated consequences of computerization.* Princeton, NJ: Princeton University Press.

Rothwell, P. M. (2005). Treating individuals: 2. Subgroup analysis in randomised controlled trials: Importance, indications, and interpretation. *Lancet, 365,* 176–186.

Salluzo, R. F., Mayer, T. A., Strauss, R. W., & Kidd, P. (1997). *Emergency department management: Principles and applications.* St Louis, MO: Mosby.

Schull, M. J., Szalai, J. P., Schwartz, B., & Redelmeier, D. A. (2001). Emergency department overcrowding following systematic hospital restructuring: Trends at twenty hospitals over ten years. *Academic Emergency Medicine, 8,* 1037–1043.

Schull, M. J., Morrison, L. J., Vermeulen, M., Redelmeier, D. A. (2003). Emergency department overcrowding and ambulance transport delays for patients with chest pain. *Canadian Medical Association, 168*(3), 277–283.

Schulman, P. R. (2002). Medical errors: How reliable is reliability theory? In M. M. Rosenthal & K. M. Sutcliffe (Eds.), *Medical error: What do we know? What do we do?* (pp. 200–216). San Francisco, CA: Jossey-Bass.

Sexton, J. B., Thomas, E. J., & Helmreich, R. L. (2000). Error, stress, and teamwork in medicine and aviation: Cross sectional surveys. *British Medical Journal, 320,* 745–749.

Shah, A., Frush, K., Luo, X., & Wears, R. L. (2003). Effect of an intervention standardization system on pediatric dosing and equipment size determination. *Archives of Pediatrics and Adolescent Medicine, 157,* 229–236.

Shapiro, M. J., Morey, J. C., Small, S. D., Langford, V., Kaylor, C. J., Jagminas, L., et al.. (2004). Simulation based teamwork training for emergency department staff: Does it improve clinical team performance when added to an existing didactic teamwork curriculum? *Quality and Safety in Healthcare, 13,* 417–421.

Shapiro, M., & Morchi, R. (2002). High-fidelity medical simulation and teamwork training to enhance medical student performance in cardiac resuscitation. *Academic Emergency Medicine, 9,* 1055–1056.

Simon, R., Morey, J. C., Rice, M., Rogers, L., Jay, G. D., Salisbury, M., et al. (1998, October). *Reducing errors in emergency medicine through team performance: The MedTeams project.* Paper presented at the Proceedings of the Second Annenberg Conference on Enhancing Patient Safety and Reducing Errors in Health Care, Rancho Mirage, CA.

Simon, R., Morey, J., Locke, A., Risser, D. T., & Langford, V. (1997). *Full-scale development of the Emergency Team Coordination Course and evaluation measures.* Andover, MA: Dynamics Research Corporations.

Small, S. D., Wuerz, R. C., Simon, R., Shapiro, N., Conn, A., & Setnik, G. (1999). Demonstration of high-fidelity simulation team training for emergency medicine. *Academic Emergency Medicine, 6,* 312–323.

Smith-Coggins, R., Rosekind, M. R., Hurd, S., & Buccino, K. R. (1994). Relationship of day versus night sleep to physician performance and mood. *Annals of Emergency Medicine, 24,* 928–934.

Smith, M. J., & Carayon, P. (2001). Balance theory of job design. In W. Karwowski (Ed.), *International encyclopedia of ergonomics and human factors* (pp. 1181–1184). London: Taylor & Francis.

Smith, M. J., & Sainfort, P. C. (1989). A balance theory of job design for stress reduction. *International Journal of Industrial Ergonomics, 4,* 67–79.

Steele, M. T., Ma, O. J., Watson, W. A., & Thomas, H. A., Jr. (2000). Emergency medicine residents' shiftwork tolerance and preference. *Academic Emergency Medicine, 7,* 670–673.

Steele, M. T., McNamara, R. M., Smith-Coggins, R., & Watson, W. A. (1997). Morningness–eveningness preferences of emergency medicine residents are skewed toward eveningness. *Academic Emergency Medicine, 4,* 699–705.

Stiell, A., Forster, A. J., Stiell, I. G., & van Walraven, C. (2003). Prevalence of information gaps in the emergency department and the effect on patient outcomes. *Canadian Medical Association Journal, 169,* 1023–1028.

Tepas, D. I., Barnes-Farrell, J. L., Bobko, N., Fischer, F. M., Iskra-Golec, I., & Kaliterna, L. (2004). The impact of night work on subjective reports of well-being: An exploratory study of health care workers from five nations. *Rev Saude Publica, 38*(Suppl.), 26–31.

Tepas, D. I., & Carvalhais, A. B. (1990). Sleep patterns of shiftworkers. *Occupational Medicine, 5,* 199–208.

Tepas, D. I., Duchon, J. C., & Gersten, A. H. (1993). Shiftwork and the older worker. *Experimental Aging Research, 19,* 295–320.

Tepas, D. I., & Monk, T. H. (1987). Work schedules. In G. Salvendy (Ed.), *Handbook of human factors* (pp. 819–843). New York: Wiley.

Thomas, H., Jr., Schwartz, E., & Whitehead, D. C. (1994). Eight-versus 12-hour shifts: Implications for emergency physicians. *Annals of Emergency Medicine, 23,* 1096–1100.

Topf, M. (2000). Hospital noise pollution: An environmental stress model to guide research and clinical interventions. *Journal of Advanced Nursing, 31,* 520–528.

Tucker, A. L., & Edmondson, A. C. (2002). When problem solving prevents organizational learning. *Journal of Organizational Change Management, 15,* 122–137.

Tucker, A. L., & Edmondson, A. C. (2003). Why hospitals don't learn from failures: Organizational and psychological dynamics that inhibit system change. *California Management Review, 45,* 55–72.

Vaughan, D. (1996). *The Challenger launch decision: Risky technology, culture and deviance at NASA.* Chicago: University of Chicago Press.

Vincent, C., Taylor-Adams, S., Chapman, E. J., Hewett, D., Prior, S., Strange, P., et al. (2000). How to investigate and analyse clinical incidents: Clinical risk unit and association of litigation and risk management protocol. *British Medical Journal, 320,* 777–781.

Vincent, C., Taylor-Adams, S., & Stanhope, N. (1998). Framework for analysing risk and safety in clinical medicine. *British Medical Journal, 316,* 1154–1157.

Wanerman, R. (2002). The EMTALA paradox. Emergency Medical Treatment and Labor Act. *Annals of Emergency Medicine, 40,* 464–469.

Wears, R. L., & Perry, S. J. (2002). Human factors and ergonomics in the emergency department. *Annals of Emergency Medicine, 40,* 206–212.

Wears, R. L., Perry, S. J., Shapiro, M., Beach, C., & Behara, R. (2003, October). *Shift changes among emergency physicians: Best of times, worst of times.* Paper presented at the Proceedings of the Human Factors and Ergonomics Society 47th annual meeting, Denver, CO.

Womack, J. P., Jones, D. T., & Roos, D. (1990). *The machine that changed the world.* New York: HarperCollins.

Woods, D. D. (1998). Designs are hypotheses about how artifacts shape cognition and collaboration. *Ergonomics, 41,* 168–173.

48

HUMAN FACTORS AND ERGONOMICS IN PEDIATRICS

Matthew Scanlon
Medical College of Wisconsin

The U.S. Census Bureau estimates that 25% of the U.S. resident population consists of children less than 18 years of age (Yax, 2004). Unfortunately, a corresponding quarter of U.S. resources devoted to research do not go to furthering an understanding of this segment of the population. A much-used cliché in pediatrics is that children are not simply little adults. Children have many distinctive characteristics that require specific consideration separate from adult patients. Just as an appreciation and understanding of the unique characteristics of children is critical to providing safe medical care, so is the same understanding necessary when considering application of human factors and ergonomic principles to this population.

Children differ from adults in a multitude of ways. These differences extend to their dependency on adults, their cognitive and maturational development, and their physiologic differences. Pediatric patient safety differs from that of patient safety in adults as a result of the aforementioned differences from adults. Again, any attempt to apply concepts of human factors to the issues of pediatric patient safety requires an understanding of how children are different.

Unfortunately, little is written on the application of human factors as it relates to children. A review of leading textbooks of human factors finds no references to either pediatrics or children in their respective indexes (Salvendy, 1997; Wickens, Gordons,& Liu, 1998). Although children are mentioned in passing in each text, neither address the needs of children in any formal fashion. This chapter identifies the unique characteristics of both children and pediatric patient safety and addresses the limited literature on pediatric human factors.

HOW CHILDREN ARE DIFFERENT

The study of pediatrics is based on the central concept that from birth through early adulthood, children are dynamic. The extent of change that children experience includes size, gross and fine motor development, cognitive development, speech and psychological development, as well changes in physiology. As a result of the dynamic nature of children, the concept of "normal" is a moving target. Appropriate weight, heart rate, or speech is all age-dependent.

Beyond developing an understanding of the dynamic nature of healthy children, the study of pediatrics further explores the range of illness in children. The impact of a given illness on a child may vary dramatically depending on where they are in their growth and development. Consequently, the leading causes of illness and mortality in children also vary by age group. Unfortunately, and of potential interest to a human factors audience, accidental or unintentional injury is the leading cause of death in children ages 1 to 4 years, 5 to 14 years, and 14 years and older (Kochanek & Smith, 2004).

In addition to the constantly changing nature of children, another commonality in pediatrics is the dependency of children on adults for care. This point is important because until adolescents (and arguably beyond), children require the provision of nutrition and shelter from others. Access to medical care is dependent on care providers seeking care for the children in their care. This care is independent of any attempt to nurture or socialize a child. From a pediatric perspective, a patient safety perspective, and a human factors perspective, the implications of this dependency cannot be overemphasized. Children are unable to effectively advocate for themselves, make choices about their diet or lifestyle, or select products or environments that may decrease their risk of harm. For example, motor vehicle accidents are the leading cause of death by unintentional injury in children ages 1 to 14 years (Kochanek & Smith, 2004). These are not children who are electing to drive recklessly or unrestrained. Similarly, the national trend toward pediatric obesity largely reflects care decisions of adults. Obesity prevalence has increased from 15% to 22.5% in 6- to 11-year-olds (Troiano et al., 1995). Although there are genetic influences on the susceptibility to obesity, a significant contribution to the development of obesity is the consumption of calories in excess of energy expenditure. Adult caretakers dictate food purchasing and portion decisions, as well as behaviors such as exercise or television viewing. In neither the case of motor vehicle accidents nor the case of childhood obesity are children in a position to make decisions that could directly improve their health and reduce risk. The importance of a child's dependency on adults is critical when considering the challenges and solutions of patient safety, particularly from a human factors perspective.

NORMAL PEDIATRIC GROWTH AND DEVELOPMENT

Changes in Weight and Length

The dramatic changes witnessed from before birth through adolescence fall into the category of growth and development. A discussion of abnormal development is beyond the scope of this work. Additionally, the biological and psychosocial determinants of normal growth and development are the subjects of other chapters and textbooks. To this end, interested readers are directed to recommended readings listed at the end of this chapter.

From the time of delivery, dramatic growth and physiologic change occur in children. Parallel to the physical changes are psychosocial changes that allow for interaction with their environment and others. The rate of growth and development is not steady-state. Whereas newborns gain an average of 30 grams of weight a day and 3.5 centimeters of length a month, this growth begins to slow around 3 months of age (Needlman, 2004). A graphic illustration of the changes in rates of growth can be seen using the invaluable tool of growth charts used to document growth of children (Figures 48–1 and 48–2). The growth charts illustrate the ranges of growth that can be seen by weight and length or height. By following serial measurements over time, health care providers can screen for poor growth or abnormal growth patterns that may indicate the development of acute but subtle illness, chronic illnesses, poor nutrition, or psychosocial deprivation.

Change in Head Size

Beyond weight and length, a child's head circumference normally demonstrates significant growth during the first 3 years of life (Figure 48–3). This growth in head circumference normally reflects normal brain growth. Not surprisingly, failure of normal brain development can lead to abnormally small head size, or microcephaly. Derangements of brain development or the elaborate system of fluid that bathes the brain can also lead to inappropriately large head size (macrocephaly). Under normal circumstances, an infant or child's head is disproportionately large relative to the rest of his or her body when compared with adults. For example, an infant's head may account for 15% of its total body mass. In contrast, an adult's head typically accounts for 3% of his or her total body mass (Kauffman, 1997). This disproportionate mass, coupled with relatively weak neck musculature, places infants and young children at greater risk for traumatic acceleration and deceleration injuries to the brain.

Other Physiological Changes During Development

Just as a child's growth is reflected by externally visible changes, similar changes happen at a

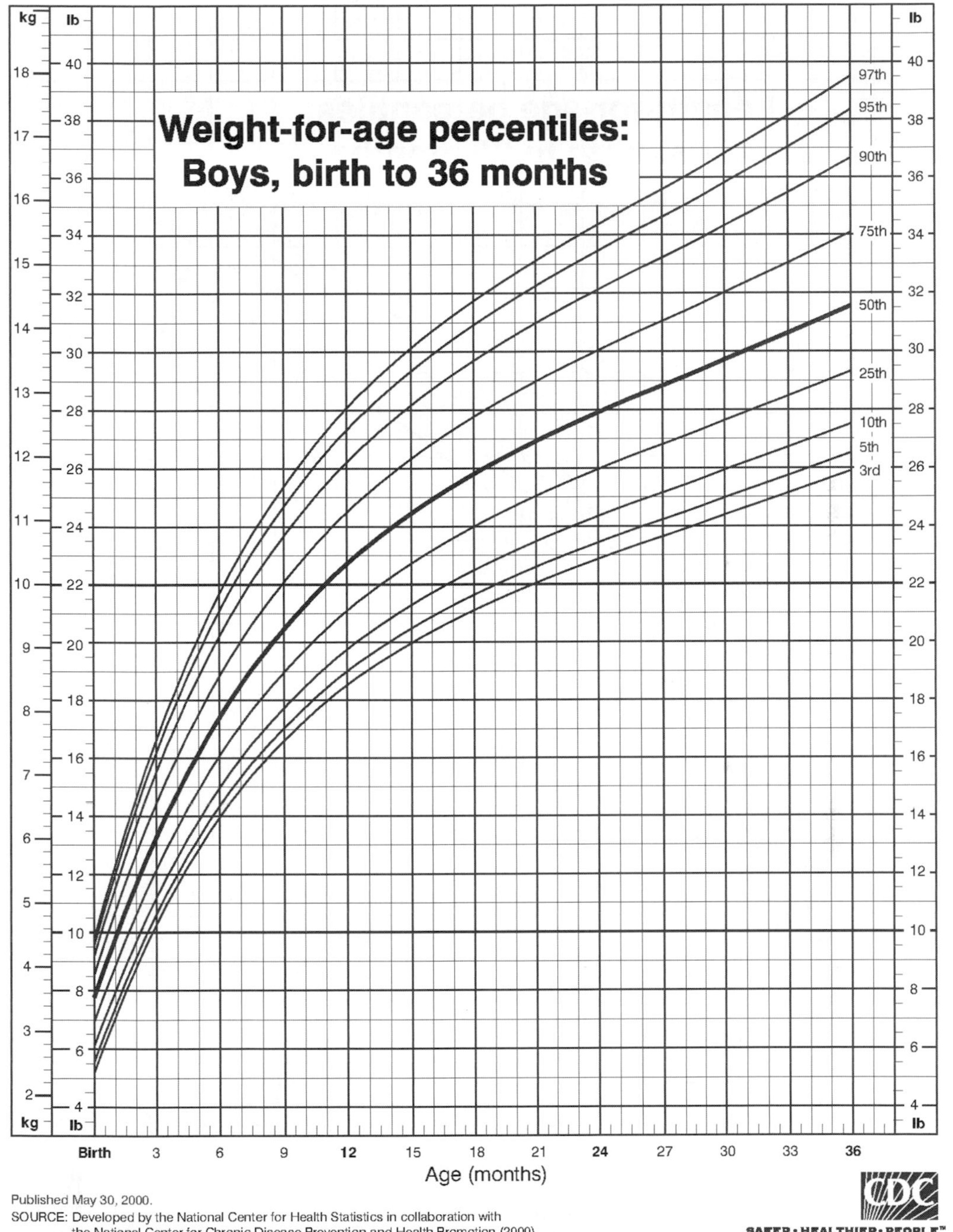

Figure 48–1. Change in weight with age in boys during the first 3 years of life demonstrating varying rate of increase. Source: National Center for Health Statistics. (2000). CDC growth charts: United States. Retrieved March 15, 2004, from the World Wide Web: http://www.cdc.gov/growthcharts/

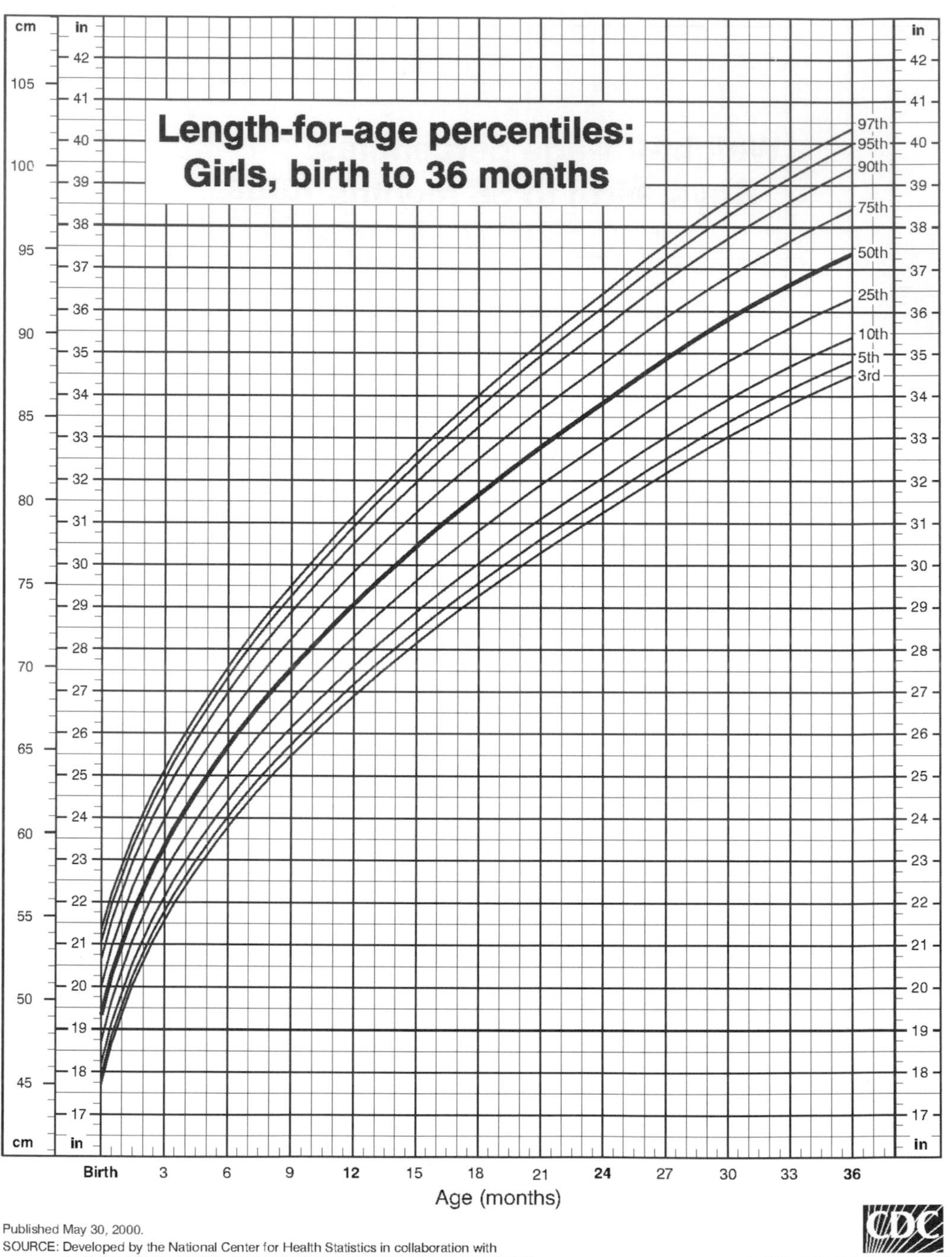

Figure 48–2. Change in length with age in girls during the first 3 years of life demonstrating varying rate of increase. Source: National Center for Health Statistics. (2000). CDC growth charts: United States. Retrieved March 15, 2004, from the World Wide Web: http://www.cdc.gov/growthcharts/

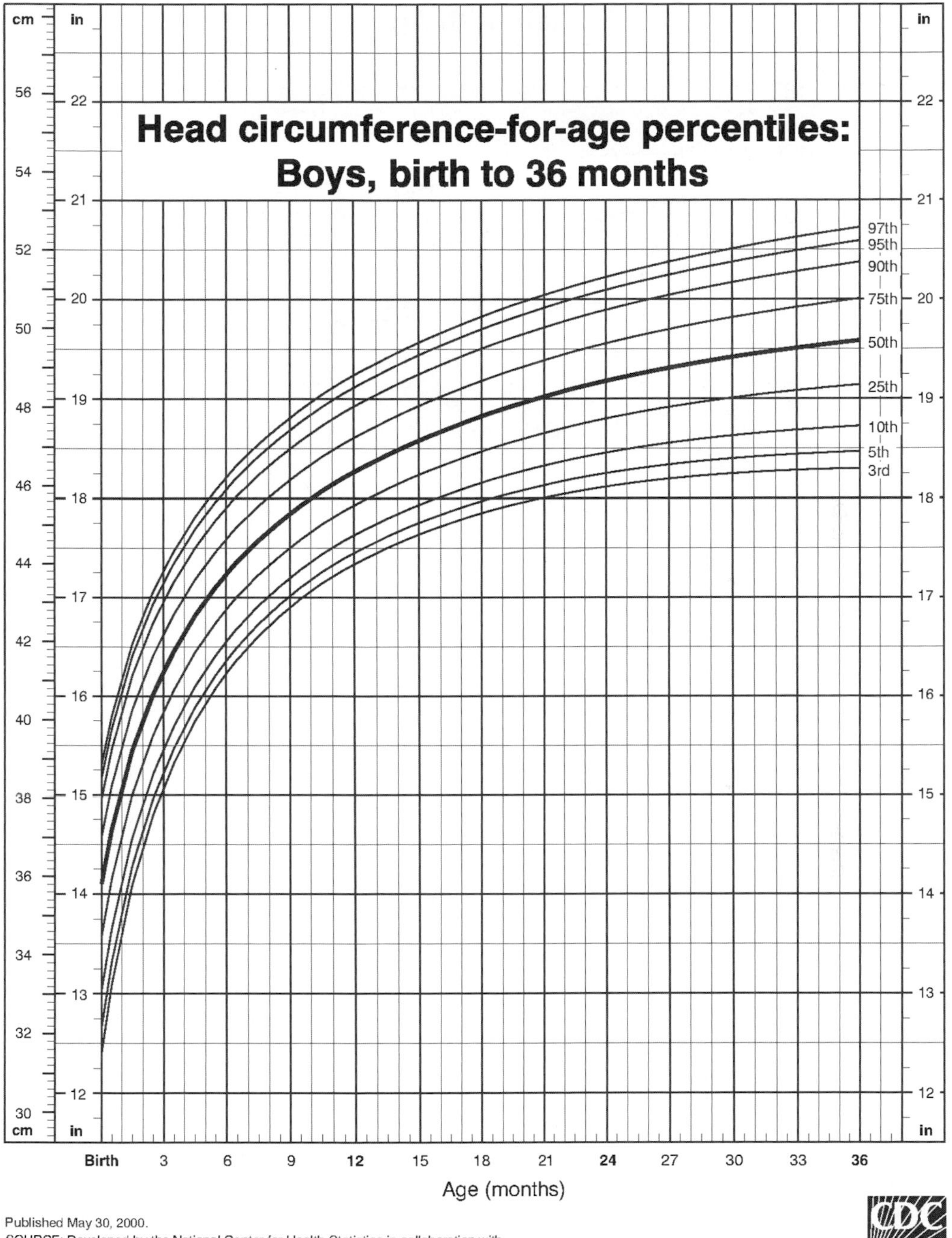

Figure 48–3. Varying rate of normal head growth in boys during the first 3 years of life. Source: National Center for Health Statistics. (2000). CDC growth charts: United States. Retrieved March 15, 2004, from the World Wide Web: http://www.cdc.gov/growthcharts/

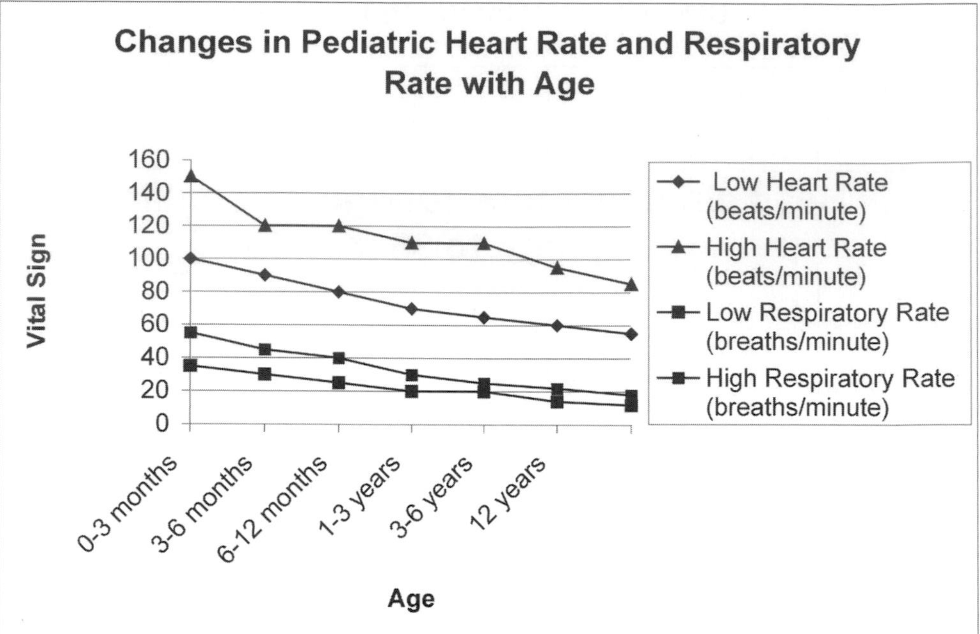

Figure 48–4. Normal decline in mean heart and respiratory rates over time in children.

physiologic level. Maturation of an individual can be detected in bone growth and ossification, as well as dental development. More dramatic, particularly to health care providers, are the dramatic changes in vital signs that normally occur in children. Vital signs are measurements of physiologic parameters that are widely used to assess the health or lack of health of a patient (adult or pediatric). Specifically, body temperature, blood pressure, and heart and respiratory rates are assessed in both health screening and ongoing patient care.

The use of vital signs is hardly unique to pediatrics; adult care providers are acutely aware of the importance of these data elements. However, unlike adult values, the pediatric values change dramatically with age changes (Figure 48–4). Additionally, within a given age-based "normal" range, there can be dramatic variations based on variables such as sleep, level of pain or agitation, and activity. Pathologic conditions may also dramatically influence a pediatric patient's vital signs. Of note, the typical causes for an abnormality of a vital sign in a child may be quite different than that of an adult. For example, abnormally low heart rates (bradycardia) are most commonly the result of intrinsic heart disease in adult populations. However, the leading cause in children is hypoxia or inadequate oxygen resulting from some respiratory compromise. Only through understanding the normal age-based variation in vital signs can a health care provider begin to assess the true status of a pediatric patient.

A child's metabolism undergoes significant changes during development. Metabolic changes, combined with changes in body size and composition that alter distribution of substances, and changes in organ development and function that may affect absorption and elimination all influence a child's response to medications and toxins (Reed, 2004). A consequence of these changes is that medication dosing in children is either by weight or body surface area. The need for weight-based calculations to accurately dose medications carries significant implications for both patient safety and the design of technological solutions.

Another important consideration is the nature of skin in children. The skin is a person's largest organ, providing numerous functions, including fluid homeostasis, temperature regulation, and protection from infection. Like the other parameters identified in this chapter, the skin of children differs from adults and changes with age. First, the relative increase in body surface area to body mass in children when compared to adults (3:1) results in the potential for much greater fluid loss. As a result, fluid management is often more complex in pediatric patients with skin injuries. A second important difference between adults and children is the composition of skin. Children less than 2 years of age have less subcutaneous fat and thinner skin layers, which contribute to both heat and fluid loss and vulnerability to thermal injury (O'Connor & Besner, 2003). Brief immersion (5 sec) in hot water that is tolerable to adults (140° F) will result in a full thickness burn to a child that would otherwise require a 2-min immersion at 126° F (Lucchesi, 2004). As a result of these differences, a design with heat exposure tolerance thresholds that would be viewed as acceptable in an adult setting might lead to significant injury in children.

Additional Motor Considerations in Children

Dramatic gross motor changes occur in children during the first year of life. These include the ability to roll over at approximately 4 to 6 months of age, the ability to crawl at 8 to 9 months of age, and the ability to walk at around 1 year of age. The constantly and often surprising changes in mobility have implications for both accidental injury and exposure to things previously believed to be inaccessible to children. Beyond a year of life, a child's ability to ambulate improves along a learning curve. It is not unusual for a child to run at 16 months of age; however, this running is often associated with many falls. Fine motor changes parallel the gross motor changes with children moving from crude grasping at 3 months, through reaching for objects and transferring objects between the hands at 4 and 6 months, to a finger–thumb "pincer" grasp between 8 and 9 months.

Children have additional challenges relating to imbalance. Imbalance is influenced by poor gross motor coordination and high centers of gravity because of relatively large head to body mass ratios and high upper segment to lower segment ratios at young ages. The upper to lower segment ratio looks at the proportion of one's body above the pubic symphysis to that below this landmark. The ratio decreases with time from 1.7 as a newborn to 1.3 at 3 years of age until reaching a ratio of 1 at approximately 7 years of age (Needlman, 2004). The higher center of gravity in younger children is another contributing factor to unintentional injuries, which may be aggravated by the design of a child's environment.

Cognitive and Speech Development in Children

In the setting of a stimulating environment, children flourish with rapid cognitive development. Infants begin with habituation to environmental stimulation and quickly learn to identify both the faces and voices of caretakers. By 4 months of age, children discover the world beyond their care providers. This discovery includes the phenomenon of discovering one's self, as well as an appreciation of cause and effect relations. Another component of world discovery is the placement of any object encountered by an infant in his or her mouth. The oral exploration marks only the start of intense exploration of the surrounding environment that continues for years. By 18 months, the ability to realize an object still exists though out of sight, or object constancy, is permanent. Cause—effect relations are better understood at this age.

Between 2 and 5 years, children experience magical thinking with egocentrism. The latter is an inability to take on another's perspective. The former leads to flawed views of causality, often as a result of the observation of coincidence. By 5 years of age, children transition through this prelogical phase to cognition that is heavily rule-based and concrete. Concrete thinking continues up until adolescence, when a child may begin developing formal logical thinking. However, the development of formal thinking, which includes logical evaluation, consideration of abstract processes, and the application of principles, is not a forgone conclusion. Some adolescents never make this transition and remain concrete thinkers throughout life.

Of note, the majority of these cognitive milestones are often associated with emotional turmoil and behavioral changes. For example, sleep disruption is common with the development of new skills in the preschool age groups.

Normal speech development occurs in relation to cognitive development. Sound recognition begins within the first month of life, with generation of cooing noises occurring by 4 months of age. By 1 year of age, children usually have several words, including, typically, "mama" or "dada." Vocabulary continues to grow with the capability to speak up to 200 words and multiword phrases between 2 and 3 years of age. At 2 years old, usually 50% of a child's words are intelligible to family, increasing to 75% at three years old. By 4 years most speech is intelligible to strangers. From this point, the complexity of language skills continues to advance (Dixon, 2000).

Speech development has several implications. First, the relative unintelligibility of children at young ages limit their ability to seek assistance or articulate concerns, discomforts, or fears. Second, by 4 and 5 years of age, children are able to communicate with family members; however, the ability to generate phrases and sentences that can appear relatively well thought out is misrepresentative of their ability to reason. The incongruity between "adult-like speech" and "adult thought processes" can create significant tension between children and adults.

Obesity in Children

The problem of obesity in children has been viewed as an epidemic. Children are now experiencing diseases such as hypercholesterolemia, high blood pressure, and type 2 diabetes mellitus, which had previously been considered only adult diseases (Donahue, 2004). A frequent consequence of these illnesses in adults is the development of coronary artery disease. Other significant morbidities of obesity include respiratory diseases, cancer, and orthopedic and joint injury, as well as poor self-esteem. Beyond the patient-level morbidity and mortality, there are significant economic costs associated with obesity. Cost evaluation in 1995 placed the combined direct and indirect cost of obesity at $99 billion in the United States (Wolf & Colditz, 1998). This figured has been increased to $117 billion based on year 2000 estimates (U.S Department of Health and Human Services, 2001).

Obesity is a multifactorial problem. The genetic components are not entirely understood. Another critical determinant in the development of obesity is a mismatch between caloric intake and energy expenditure. Other factors found to be positively predictive of pediatric obesity include parental obesity and time spent viewing television (Crespo et al., 2001).

The long-term impact of this epidemic is not yet understood. However, as the problem of obesity is leading to major changes in the physical characteristic and health of children, any evaluation of patient safety or human factors must consider this dilemma.

PEDIATRIC PATIENT SAFETY

Pediatric patient safety shares many commonalities with patient safety efforts focused on adult health care. There remains a dearth of fundamental "basic science" research, particularly related to pediatrics (Perrin & Bloom, 2004). Issues of culture change, communication, leadership, reporting, error identification and reduction, and human factors are important to both adult and pediatric safety work. However, many of the same unique characteristics of children, which led to a separate medical specialty, support the need for separate considerations in patient safety. This section explores what is known about pediatric patient safety with special attention to the unique characteristics of children that might have more importance.

Epidemiology of Pediatric Health Care Error and Injury

There have been few systematic investigations of the scope of pediatrics errors and injuries. A study of the utilization of medical care by children referenced finding 0.8% of pediatric discharges associated with a complication (McCormick, 2000). One study using administrative data to assess the risk to hospitalized children found 1.81 to 2.96 medical errors per 100 discharges (Slonim et al., 2003). This work used ICD–9 coding to identify medical errors. The authors identified many of the limitations of using such administrative data, including underreporting, identifying cause and effect related to variables such as length of stay, and a lack of clinical and physiological data. These findings are comparable to other published work, which estimated the rate of adverse events in children less than 5 years of age at 2.7 per 100 discharges (Brennan et al., 1991).

A second study applied the Agency for Healthcare Research and Quality's Patient Safety Indicators to administrative pediatric hospitalization data (Miller, Elixhauser, & Zhan, 2003). Miller et al. found rates of patient safety events identified using the Agency for Healthcare Research and Quality tool comparable to that of adults. Aside from the previously discussed issues of using administrative datasets, the work is limited by the facts that this tool was not designed for pediatrics and that subsequent changes in it may make these findings invalid (Scanlon, in press).

Even less is known about pediatric medical errors in an ambulatory setting. A review of current knowledge confirms the lack of research (Miller et al., 2004). Miller et al. identified a several medical errors that are common in pediatric ambulatory settings. These errors tend to fall into one of three categories of medical errors: underuse, misuse, and overuse of medical care. One specific error identified was the lack of immunization of children. U.S. statistics demonstrate that immunization rates in children in 2002 were only 78% (National Immunization Program, 2003). The result is preventable illness that can be life threatening and add to the burden of morbidity and cost. Another medical error identified in ambulatory settings was misinterpretation of pediatric echocardiograms by adult providers (Stanger, Silverman, & Foster, 1999; Ward & Purdie, 2001). These could be viewed as misuse errors that occur because the appropriate individuals are not performing and interpreting studies. Another source of ambulatory errors important to the pediatric population is the over prescription of antibiotics for viral illnesses. Antibiotics are only effective in treating infections caused by bacteria. Research has extrapolated that 6.5 million prescriptions are written annually for children who were diagnosed with a viral upper respiratory infection, or the common cold (Nyquist et al., 1998). This figure accounts for 12% of prescriptions written each year for children. These medications are unnecessary, can lead to adverse drug reactions, and, independent of side effects or errors in the prescribing, dispensing, and administration process, likely add dramatically to the cost of health care. What may aggravate this incredible overuse of medications is the fact that large numbers of medications are administered to children in school or child care settings where there is often no providers skilled in delivering medications.

Medication Errors in Pediatric Patients

Of all the possible medical errors that occur in children, the most is known about medication errors. A study of medication errors in children found that the rate of preventable adverse drug events was similar to adult studies (Kaushal, Bates, Landrigan, McKenna, & Clapp 2001). However, the rate of potential adverse drug events (errors that occurred yet did not reach the patient) was three times higher. These potential adverse drug events were most common in neonatal intensive care units, a finding that is consistent with previous work demonstrating a higher number of medication errors in neonatal and pediatric intensive care units (Folli et al, 1987; Raju et al., 1989).

The provision of medications depends on three main steps: medication prescription, medication dispensing, and medication administration. Each of these steps is associated with errors; in pediatrics each step has additional risks unique to delivering medications for children.

Correct prescribing is the first step to safe medication delivery. A major factor in pediatric prescription errors is the fact that medication dosing is weight-based. Prescribing medication requires the correct identification of the patient's weight. An incorrect weight can lead to miscalculation of medication dosing. Similarly, confusing the units of weight, pounds, and kilograms can lead to doubling over- or underdosing. The identification of accurate weight takes on added importance as patients often experience changes in weight over time. A single weight used as if it is a constant value can lead to dosing errors. Additional factors that have an adverse impact on pediatric medication prescriptions include decimal place errors and computational errors. Because of the weight-based calculation, which may involve small weights, decimals take on added importance. Misplacing a decimal by one place in either direction can lead to a tenfold under- or overdose. Research reveals that pediatric resident physicians are prone to making computational errors when prescribing medications (Potts, 1996; Rowe, Koren, & Koren, 1998). As a result, recent efforts have targeted the element of calculation in prescribing infusions of medications (see the Strategies for Improving Pediatric Patient Safety section). The net effect is that pediatric patients are at an increased risk for prescription errors. The work by Kaushal, Bates, et al. (2001) found nearly 80% of potential adverse drug events occurred at the ordering phase.

In addition to inpatient concerns, ambulatory prescription is also a problem. Beyond the previously discussed dilemma of overprescription of antibiotics, a study of 1,532 pediatric emergency department records found that 10.1% of the charts reflected a prescribing error (Kozer et al., 2002). In 1998, there were 124.3 million pediatric office visits, with 52.9% resulting in prescriptions. If the pediatric emergency department numbers apply, there is likely a significant medication prescription error rate with associated potential injury rate in the pediatric outpatient setting.

Medication dispensing creates another set of challenges. Unit doses can rarely be used for medications because of the weight-based nature of dosing. Consequently, pharmacists and nurses are required to draw up correct amounts from standard vials or containers of medication. Drawing up non-standard doses manually introduces a new step during which errors can occur. Potential safeguards built into manufacturer's packaging (bar coding or otherwise) are lost when a medication is drawn up into a separate container. Another dispensing hazard occurs when the medication is administered in both an enteral and a parenteral route. Repackaging or drawing up medications may result in using the wrong syringe and thus lead to parenteral administration of an enteral medication. Finally, repackaging in the dispensing phase creates a potential for mislabeling a medication, resulting in the wrong patient receiving the medication.

Administering medications also represents a source of errors in pediatrics. The potential for mixing up patients or routes of administration can occur. Additionally, pumps designed primarily for adults must be used in pediatrics. Many of these devices are not designed for the volumes or doses that are used in pediatrics, and they create hazards through features that seem inconsequential in the relatively standardized adult patient population for which they are designed. The majority of pediatric patients are unable to recognize and intervene if they are about to receive a wrong medication or dose. Although adults may be able to prevent an error by asking "Why am I getting a red pill? Mine are normally blue," most pediatric patients are unable cognitively and verbally to protect themselves.

Regardless of where a failure occurs in the medication process, children remain at greater risk. The weight-based nature of dosing, combined with decimal errors, increases the likelihood of errors that are of a factor of 10 or more from the intended dose, resulting in overdoses than an adult might see. Children are considered to have less physiologic reserve than most adults; a given medication error in a child may yield worse consequences than a proportionately equivalent error in adults. This lower reserve reflects both less physiologic tolerance to stress and a limited ability to "buffer" the effect.

The challenges of delivering medications to children are many and, unfortunately, not readily aided by technological solutions available. Few, if any, computerized physician order entry (CPOE) systems provide "off the shelf" rules for pediatric dosing, much less safety checks. Similarly, bar-coding technology has added challenges of creating bar codes that fit a wide range of patient sizes that may be encountered in pediatrics. For instance, many wristbands that are marked with bar codes may be unreadable to bar code scanning devices because of dressings or curvature of bands because of small extremities (Piehl, 2003. Personal Communication). Also, repackaging medications for pediatric use often circumvents safeguards offered by pharmaceutical company bar coding. The need to repackage doses and thus relabel with bar codes introduces yet another source of potential errors. These multiple challenges, and their limited technological solutions, create an environment of increased risk to pediatric patients.

Strategies for Improving Pediatric Patient Safety

The unique challenges of pediatric patient safety have not been lost on all in the patient safety field. Various authors have identified specific strategies that might lead to improved safety. One strategy identified five specific topics that required concentrated attention (Perrin & Bloom, 2004). These topics included (a) understanding and improving communication as it relates to safety, (b) understanding the value of technology with attention to implementation, (c) advancing pediatric patient safety as a topic of importance, (d) acquiring additional knowledge beyond basic epidemiology of safety issues, and (e) further defining the true priorities in pediatric patient safety.

Another group has advocated for three main steps focusing on pediatric residency training programs: (a) education related to patient safety, (b) increased reporting of errors, and (c) emphasis on fixing systems rather than people (Napper, Battles, & Fargason, 2003). Napper et al. admitted their goals may appear overly simplistic. However, an emphasis on these goals would be a major change in the current thinking of many training programs.

A third set of strategies has been described for preventing medical errors in pediatrics (Fernandez & Gillis-Ring 2003). These strategies include systems analysis, critical incident root cause analysis, the use of clinical pharmacists and computer-assisted decision making, changes in the production and distribution of pharmaceuticals, and education for families and providers. Much of this work further centers on reporting, analysis, and categorization of errors as a basis for designing solutions.

The value of these identified strategies is unknown. However, specific consideration of the role of human factors and ergonomics is necessary to better frame any solutions to the challenges of pediatric patient safety. What little is known related to human factors and pediatrics is considered in the next section.

HUMAN FACTORS, ERGONOMICS, AND PEDIATRICS

To continue a theme underlying this chapter, disappointingly little has been written about human factors and ergonomics as it relates to children. With some imagination, work on design for people with functional limitations might be applied to the pediatric population (Vanderheiden, 1997). Although Vanderheiden did not specifically identify children in his work, he addressed the need for attention to physical, cognitive, and language impairments in design. This work focused on limitations as a result of a cause (older age, disease, injury, or genetic abnormalities) rather than limitations as a preexisting though transient state. In light of the physical, cognitive, and language developmental limitations that children may face at different ages, some of Vanderheiden's concepts bear consideration. Of note, the solution he advocated for is a concept of universal design, or the design of products that "can be effectively and efficiently used by people with. ... no limitations as well as those operating with functional limitations" (p. 2014).

The Case for Applying Human Factors Concepts to Children

Before extrapolating the work of Vanderheiden (1997) and others to a pediatric population, it is worth considering whether such effort is necessary. One of the most vivid illustrations of a lack of attention to human factors when designing for children is in the toy industry. The U.S. Consumer Product Safety Commission ([CPSC], 2004) is a federal body charged with protecting the public from unreasonable risks of death and injury because of consumer products. The CPSC Web site holds a wealth of data illustrating the hazards of poor design for children.

One category tracked is deaths from toys. Although the death of a child from any cause is tragic, a death because of a toy seems even more egregious. In the CPSC report for 2001, 25 deaths were reported in children less than 15 years of age (McDonald, 2002a). Of these, 9 deaths were linked to choking or asphyxiation from a toy. Recognizing that young children have a developmental drive to explore their world by placing new objects in their mouth, while also having narrow airways prone to obstruction, the CPSC has created a 313-page document to guide manufacturers in preventing these deaths (Smith, 2002). Other work by the CPSC details over 69,500 emergency room visits in children less than 5 years of age because of injuries from nursery products (McDonald, 2002b).

The medical literature also provides additional evidence for addressing the human factors needs of children. A study of playground injuries found that nearly 50% could have been prevented through changes to the equipment and playground structure (Petridou, 2002). The incidence of playground injuries among boys and girls was 7 per 1,000 and 4 per 1,000, respectively, with a total of 777 injuries identified. The number of preventable injuries would be increased through the use of protective equipment and shoes in this setting.

Aside from the known morbidity and mortality of children from play-related activity, an increasing use of computers by younger children creates another setting in which design of interfaces and recognition of ergonomic implications may be more important.

An Approach to Applying Pediatric Considerations to Human Factors and Ergonomics

Wickens and colleagues (1998) identified the critical areas in which a child-focused consideration could drive research in human factors, ergonomics, and children. In addition to identifying the cognition, variation in human size and shape, the authors also touched on the issues of biomechanics and work (or play) physiology (Wickens et al., 1998). Applying the differences described at the beginning of this chapter, Wickens et al.'s outline serves as a template for identifying central pediatric issues that require either application of existing knowledge or research to acquire the needed knowledge. This, in turn, can provide a basis for extending Vanderheiden's (1997) universal design concept in the opposite direction of the aged to the young. Table 48–1 displays the critical design concepts identified by Vanderheiden. For each of these

TABLE 48–1. Human Factors and Pediatrics: Possible Design Implications

Human Factors Consideration	*Pediatric Specific Issues*	*Design Implications for Children*[a]
Sensory changes	Development of sight and hearing in young children	Use existing knowledge of imperfect senses
Variation in size and shape	Small size can circumvent safety features, limit use of interface	Consider smaller extremes of size for design
Variation in biomechanics	Limited strength, leverage can limit use	Consider consequences of users with developmentally impaired strength
Work (play) physiology	Developmentally varied vital signs and energy stores limit endurance	Consider broader range of users and endurance
Cognitive considerations	Magical vs. Concrete vs. Abstract thinking, innate curiosity	Consider consequences of misuse due to cognitive factors
Language considerations	Varied ability to read, comprehend, and express self	Consider limited abilities in design of interface, directions and safety labeling

[a]Design considerations predicated on desired use of device/process by children. Alternatively, consider safety features to prohibit use/misuse by children.

concepts, the pediatric specific issues are listed, along with possible design considerations to optimize use by and with children.

HUMAN FACTORS, ERGONOMICS, AND PEDIATRIC PATIENT SAFETY

Some of the existing strategies for improving patient safety were identified previously (see the previous section on Strategies for Improving Pediatric Patient Safety). Although implied at times, none of the strategies expressly identified a need for applied human factors concepts to efforts to improve pediatric patient safety. The need for such consideration is based on several observations.

In 1984, a published report described a case series of what initially was viewed as neonatal sepsis (Solomon et al., 1984). On further investigation, it was found that the illnesses were the result of a repeated medication error in a neonatal hospital unit. Because of nearly identical labels, racemic epinephrine was incorrectly administered rather than vitamin E. Labeling confusion is not a problem restricted to pediatrics. However, the issue is how medications are administered to pediatric patients, with small doses often being administered to multiple patients from a single container, rather than unit doses, which are more widely used in adults. Furthermore, the case is important because pediatric differences in metabolism and the relative dose-to-patient size were potential factors in the poisoning.

Two articles from the Great Ormand Street Hospital for Children NHS Trust in London, England, addressed specific safety implications in the performance of surgery for neonatal congenital heart defects. The first study described findings of 243 surgeries over an 18-month time period (de Leval et al., 2000). Central to the work is the importance of both major and minor human failures independent of patient risk factors. Major failures had great potential for patient death yet could be compensated for with prevention of poor outcome. Minor human failures, however, were often unrecognized, and thus no compensation occurred. The authors found a link in failures to negative outcomes, though the significance of any given minor failure varied by specific patient. Specific factors that were noted as critical for compensating for major failures included the diagnostic skill and problem-solving skills of a surgeon and his or her ability to communicate with the operative team.

The second work focused on the importance of reporting and near-miss identification in the performance of neonatal cardiac surgery (Carthey, de Leval, & Reason, 2001). Again, the work identified strategies for identifying human factors in a pediatric operative setting with the intent of preventing or compensating for human errors. Neither work is limited to pediatrics in terms of its findings. To that end, the studies could have been performed in adult settings with no attention to pediatric implications. However, in the complex environment of the repair of congenital heart defects, the studies lend credence for the need to apply human factors in pediatrics.

Another publication examined poor outcomes of pediatric procedural sedation with attention to human factors (Cote et al., 2000). Using critical incident analysis, Cote et al. found several important factors that could have led to improved outcomes. Identified variables associated with poor outcome (death or permanent neurologic injury) included sedation in nonhospital settings, inadequate and inconsistent monitoring of physiologic status (vital signs and oxygenation), and inadequate resuscitation. Other factors included inadequate staff to monitor the child postsedation, inadequate equipment, and medication errors. The significance of this work to a consideration of human factors in pediatric patient safety is the number of variables that could have been corrected with correct design. Monitoring procedures, equipment, and staff can be standardized, but a presedation recognition of the age-based variation in children is essential. Similarly, as children vary in size and weight, so do resuscitation equipment and medication doses. Inattention to the need for a range of equipment and insufficient familiarity and training with the equipment creates unnecessary hazards.

A final example of the need for human factor considerations in ensuring pediatric patient safety is the U.S. Food and Drug Administration alert describing preventable risk to children from excess radiation exposure during computed tomography (CT) scans (Fiegal, 2001). Citing literature that described an increased lifetime risk of cancer from the radiation exposure during CT scans (Brenner et al., 2001), the Food and Drug Administration identified how a "one-size-fits-all" approach to imaging creates risk. The unnecessary radiation exposure could be addressed through readily available interventions, including reduction of tube current, adjusting scanning axes, and more aggressive screening for the appropriate use of CT imaging. Additionally, knowledge of human factors could be used to develop charts or reference tools of ideal pediatric CT settings for a range of weights, body diameters, and anatomic regions. None of this is possible without first recognizing the needs specific to children and then applying concepts of human factors to this problem.

HUMAN FACTORS, TECHNOLOGY, AND PEDIATRIC PATIENT SAFETY

There is little evidence to suggest that principles of human factors and user-centered design have been effectively applied to the information technology solutions offered to address medical errors. Some of the identified challenges to implementation of CPOE include the need for weight-based dosing and age-dependent laboratory normal values (Kaushal, Barker & Bates, 2001). A systematic review of information technology and its potential for pediatric patient safety identified significant potential despite little specific pediatric evidence for numerous technologies (Johnson & Davison, 2004). Included in Johnson and Davison's work was a proposed research agenda for pediatric applications of safety information technology. However, nowhere in the review (including in the research agenda) was mention given to human factors consideration.

The need for consideration of human factors in the design and implementation is highlighted by case studies reported from a pediatric center with a 3-year history of CPOE (Scanlon, in press). Two cases of medical errors that resulted from a lack of human factors consideration in the design and implementation are described. Specific human factor principles that were not addressed in the design of the commercially available CPOE system include an overreliance on memory for correct performance, inattention to facilitating discrimination of similar data elements, failure to recognize the implications of mental models, inadequate decision support, and insufficient feedback to users of the system.

The lack of attention to human factors is further illustrated by the flowchart describing the first time dose of a single intravenous medication using a CPOE system (Figure 48–5.) This high-level flow chart outlines the steps required for using a commercially available CPOE system in a pediatric center beginning with a provider conceiving of prescribing a medication up to a nurse administering the medication. The actual administration process is not included in the flowchart because of the extensive complexity and variation in this portion of the medication process. The flowchart does not describe the individual steps within the CPOE system such as logging into the system and navigating the screens necessary to perform the flowcharted steps. The sheer complexity of this process, coupled with the case studies of error involving the same system (Scanlon, in press), suggest that proposed benefits of CPOE may not be readily realized without careful application of human factor principles.

The observation that a neglect of human factors may lead to imperfect technology that in turn may create rather than reduce risk of patient harm should not be surprising. The significance of human factors to medical device and technology has previously been described (Murff, 2001). One important potential effect of introducing technology—patient safety or other—to a health care environment is the

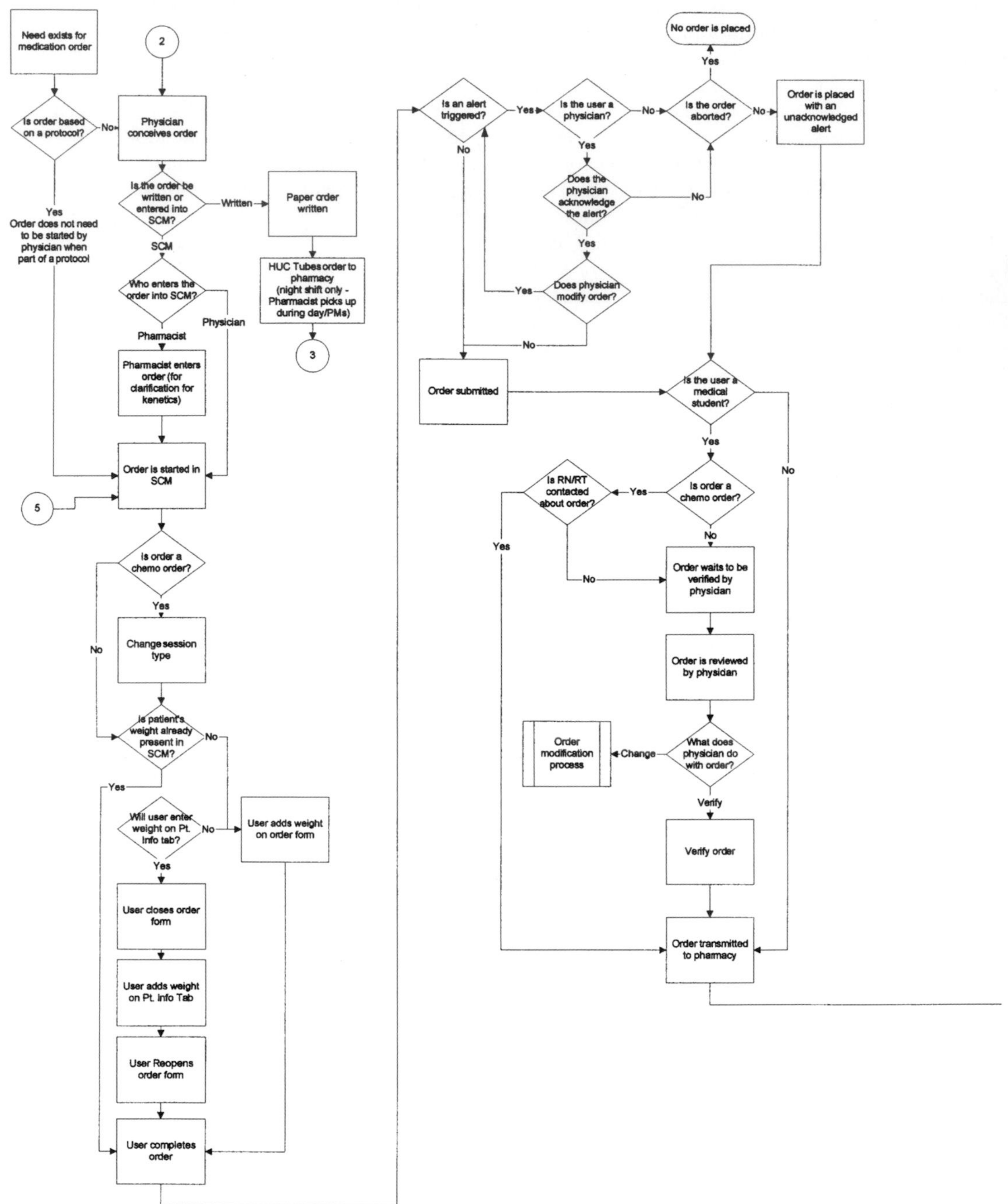
First Time IV Ordering/Administration Process
Need exists for medication order
Is order based on a protocol?
No
Yes
Order does not need to be started by physician when part of a protocol
2
Physician conceives order
Is the order be written or entered into SCM?
Written
Paper order written
HUC Tubes order to pharmacy (night shift only - Pharmacist picks up during day/PMs)
3
SCM
Who enters the order into SCM?
Physician
Pharmacist
Pharmacist enters order (for clarification for kenetics)
Order is started in SCM
5
Is order a chemo order?
Yes
No
Change session type
Is patient's weight already present in SCM?
No
Yes
Will user enter weight on Pt. Info tab?
No
User adds weight on order form
Yes
User closes order form
User adds weight on Pt. Info Tab
User Reopens order form
User completes order
Is an alert triggered?
Yes
No
Is the user a physician?
No
Yes
Does the physician acknowledge the alert?
No
Yes
Does physician modify order?
Yes
No
Is the order aborted?
Yes
No order is placed
No
Order is placed with an unacknowledged alert
Order submitted
Is the user a medical student?
Yes
No
Is order a chemo order?
Yes
Is RN/RT contacted about order?
Yes
No
No
Order waits to be verified by physician
Order is reviewed by physician
What does physician do with order?
Change
Order modification process
Verify
Verify order
Order transmitted to pharmacy

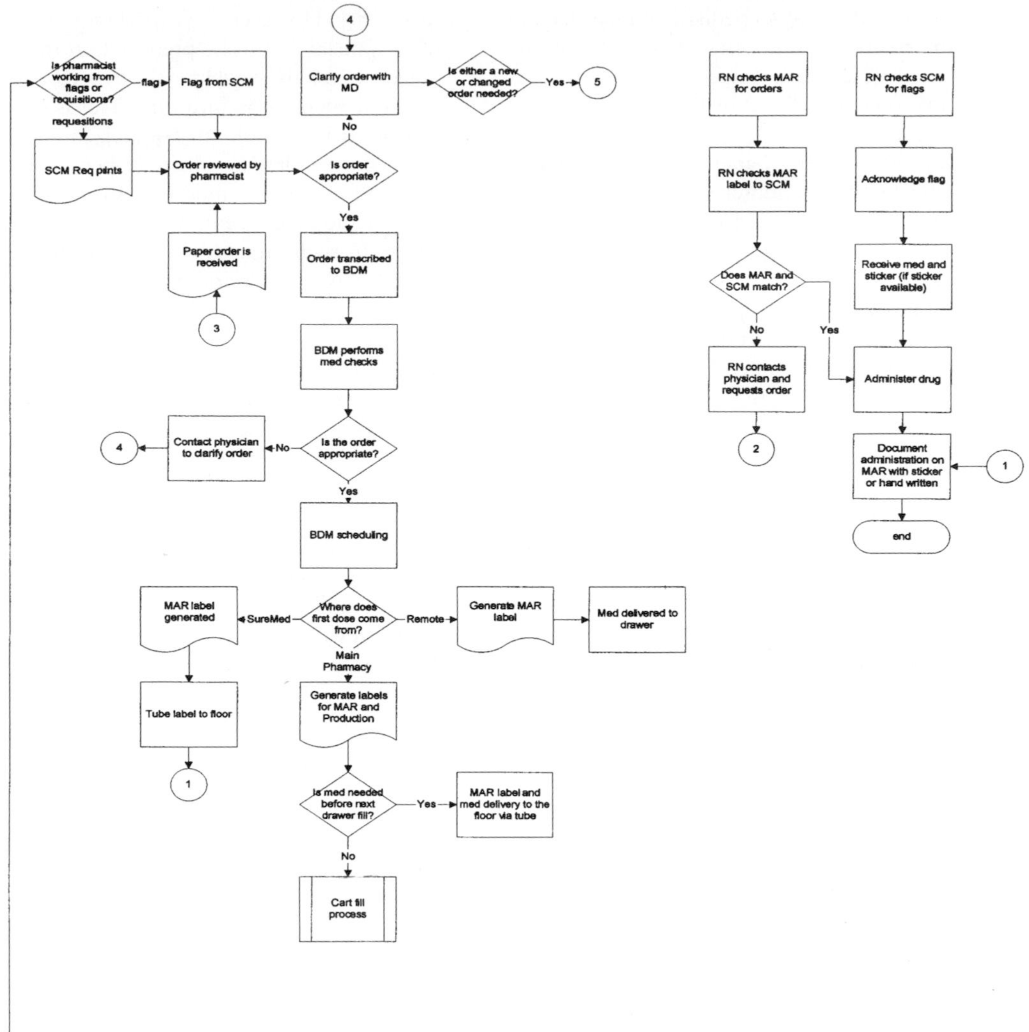

Figure 48–5. High-level flowchart of ordering, dispensing, and administration of the initial, single dose of an intravenous medication using CPOE.

change to the way work is performed. Consistent with human factor concepts, technology should facilitate the performance of work. Instead, technology may introduce new cognitive requirements, disrupt workflow, and ultimately distract from activities with a resultant decrease in safety.

Strategies for Improving Pediatric Patient Safety

It is tempting to lay charges of "child neglect" on both the medical, patient safety, and human factors communities when considering the relative lack of resources dedicated to pediatrics relative to population distribution. Beyond any brief satisfaction that complaining might bring, the larger issue of how human factors relate to pediatric patient safety remains largely unanswered.

In fact, it is just as easy to suggest the pediatric patient safety community is not attending to the importance of human factors. The Institute of Medicine's landmark work *To Err is Human* (Kohn, Corrigan, & Donaldson, 2000) made a point of identifying human factors as a critical aspect of improving patient safety. Unfortunately, the pediatric patient safety strategies identified to date (see earlier discussion) have failed to emphasize the need for more attention to pediatric human factors and its intersection with patient safety. All of the identified strategies have value and can be necessary but insufficient without further appreciation of human factors.

Instead, the pediatric patient safety community is following its adult counterparts in focusing on errors and technology without understanding the importance of human factors and design. The patient safety community has largely ignored the intersection of safe practice, economic factors, and workload (Cook 2003; Rasmussen, 1997). Without an understanding of the interrelation between how work is done and the presence or absence of patient safety, a meaningful application of human factors principles in pediatric patient safety is unlikely to be realized.

Central to this thought is that how work is designed is critical to making health care safe. For instance, designing a CPOE system to meet the needs of the end user, as well as the environment in which it will be used, could eliminate many of the problems described in the previous section. The converse implication holds true. Lack of understanding of how work is performed may lead to patient safety solutions—technological or other—that actually introduce risk.

An example of this latter scenario is represented by current recommendations by the Joint Commission on the Accreditation of Healthcare Organizations (2004) regarding the preparation of infusions of medications. The Joint Commission on the Accreditation of Healthcare Organization, in response to recommendations from the Institute for Safe Medication Practice (Grissinger, 2003), has ruled that preparation of infusions using a process called the "Rule of 6" (RO6) is unsafe and that its use should be stopped. The recommendation to stop this process is based on the observation that, in many care settings, individual nurses use the RO6 to prepare medications, often at a patient's bedside. The variation introduced in the preparation by multiple practitioners, coupled with the risk of mathematical errors, creates a situation of increased risk. Instead of the RO6, organizations are guided to use standard concentrations. The rationale is that by standardizing this step (medication preparation and dispensing), errors will be reduced at both the preparation and administration phases.

In this light, the Joint Commission on the Accreditation of Healthcare Organizations goal of eliminating the RO6 seems prudent and designed to reduce risk. However, the expectation to eliminate to RO6 appears to also be premised on the belief that organizations use equally risky processes for preparation of medication infusions. This is not necessarily true. It is possible that organizations have taken steps to standardize their respective medication preparation process independent of whether they utilize standard concentrations or the RO6 in preparation. In such organizations, changing a given step in the name of "standardization" for safety may actually introduce variation and risk to this process.

This phenomenon was identified using failure modes and effects analysis to explore the consequences of changing an existing medication infusion using a standardized RO6 process to one using "standard" concentrations (Scanlon, 2004). The failure modes and effects analysis identified five new steps that would be introduced through the process change, each associated with significant potential risk. These findings are not surprising in the context of a complicated process (Figure 48–5) that had been previously standardized where possible to reduce risk. Rather than reducing risk, the proposed regulatory changes may amplify risk by disregarding the need to standardize entire processes rather than one or two steps in a process. Only through developing a thorough understanding of how work is done and then designing a safer process will the unintended consequence of risk be avoided.

CONCLUSIONS

The care of children is complex and challenging due, in no small part, to the variation inherent in the patients because of age-based changes in physical and intellectual development. Consequently, improving the safety of children has added complexity when compared to adult care. The numerous unique characteristics in children limit their ability to protect themselves and increase their vulnerability to medical errors and harm. Health care systems and processes, when designed, rarely consider the needs of pediatric patients but instead focus on adult care.

Existing efforts to improve pediatric patient safety have focused on improving communication, education, and reporting; developing a better understand of the epidemiology of errors; and introducing technological solutions. All are important components that may yet ignore the importance of human factors in pediatric patient safety. This oversight is largely understandable; the dearth of literature on human factors likely reflects a more global lack of appreciation for much needed knowledge. Increased recognition of the need to understand work in both pediatric and adult health care, and then design safety solutions that either are consistent with or improve the way work is performed, is critical to lasting improvement.

The pediatric, patient safety, and human factor communities have a clear challenge ahead: Develop a recognition and understanding of each other's disciplines and then apply it in a multidisciplinary manner. Only then can the work of truly improving pediatric patient safety take place.

References

Brennan, T. A., Leape, L. L., Laird, N. M., Hebert L., Localio A.R., Lawthers A.G., et al. (1991). Incidence of adverse events and negligence in hospitalized patients. *New England Journal of Medicine, 324,* 370–376.

Brenner, D. J., Elliston, C. D., Hall, E. J., Berdon, W. E., et al. (2001). Estimated risks of radiation-induced fatal cancer from pediatric CT. *American Journal of Roentgeneology, 176,* 289–296.

Carthey, J., de Leval, M. R., & Reason, J. T. (2001). The human factor in cardiac surgery: Errors and near misses in high technology medical domain. *Annals of Thoracic Surgery, 72,* 300–305.

Cook, R. I. (2003). *Lessons from the war on cancer: The need for basic research on safety.* Testimony submitted to the Agency for Healthcare Research and Quality Second National Summit on Patient Safety Research (pp. 1–6). Retrieved December 5, 2003, from the World Wide Web: http://www.ctlab.org

Cote, C. J., Notterman, D. A., Weinberg, J. A., & McCloskey, C., et al. (2000). Adverse sedation in pediatrics: A critical incident analysis of contributing factors. *Pediatrics, 105,* 805–814.

Crespo, C. J., Smit, E., Troiano, R. P., Bartlett, S. J., Macera, C. A., & Andersen, R. E. (2001). Television watching, energy intake, and obesity in US children: Results from the Third National Health and Nutrition Examination Survey, 1988–1994. *Archives of Pediatric & Adolescent Medicine, 155,* 360–65.

de Leval, M. R., Carthey, J., Wright, D. J., Farewell, V. T., & Reason, J. T. (2000). Human factors and cardiac surgery: A multicenter study. *Journal of Thoracic and Cardiovascular Surgery, 4,* 661–672.

Dixon, S. D. (2000). Two years: Language emerges. In S. D. Dixon & M. T. Stein (Eds.), *Encounters with children: Pediatric behavior and development* (4th ed., pp. 382–409). St. Louis, MO: Mosby.

Donahue, P. A. (2004). Obesity. In R. E. Behrman, R. M. Kliegman, & H. B. Jenson (Eds.), *Nelson textbook of pediatrics* (17th ed., pp. 173–177). Philadelphia: Saunders.

Fernandez, C. V., & Gillis-Ring, J. (2003). Strategies for the prevention of medical error in pediatrics. *Journal of Pediatrics, 143,* 155–162.

Fiegal, D. W. (2001). *FDA public health notification: Reducing radiation risk from computed tomography for pediatric and small adult patients.* Washington, DC: U.S. Food and Drug Administration, Center for Devices and Radiological Health.

Folli, H. L., Poole, R. L., Benitz, W. E., & Russo, J. C. (1987). Medication error prevention by clinical pharmacists in two children's hospitals. *Pediatrics, 79,* 718–722.

Grissinger, M. (2003). Medication errors: "Rule of 6" not optimal for patient safety. *P & T, 28,* 234. Retrieved March 21, 2004, from the World Wide Web: http://www.jcaho.org/accredited+organizations/patient+safety/04+npsg/04_faqs.htm

Johnson, K. B., & Davison, C. L. (2004). Information technology: Its importance to child safety. *Ambulatory Pediatrics, 4,* 64–72.

Joint Commission on the Accreditation of Healthcare Organizations. (2006). *National Patient Safety Goals: 2006 Implementation Expectations.* Retrieved June 8, 2006, from http://www.jointcommission.org/NR/rdonlyres/DDE15942-8A19-4674-9F3B-C6AE2477072A/0/06_NPSG_IE.pdf

Kauffman, B. A. (1997). Head injury and intracranial pressure. In K. T. Oldham, P. M. Colombani, & R. P. Foglia (Eds.), *Surgery of infants and children: Scientific principles and practice* (pp. 417–427). Philadelphia:. Lippincott Raven.

Kaushal, R., Barker, K. N., & Bates, D. W. (2001). How can information technology improve patient safety and reduce medication errors in children's health care? *Archives of Pediatric & Adolescent Medicine, 155,* 1002–1007.

Kaushal, R., Bates, D. W., Landrigan, C., McKenna, K. J., Clapp, M. D., (2001). Medication errors and adverse drug events in pediatric inpatients. *Journal of the American Medical Association, 285,* 2114–2120.

Kochanek, K. A., & Smith, B. L. (2004). Deaths: Preliminary data for 2002. In *National Vital Statistic Report* (Vol. 52, pp. 1–48) Hyattsville, MD: National Center for Health Statistics. Retrieved March 15, 2004, from the World Wide Web: http://www.cdc.gov/nchs/data/nvsr/nvsr52_13.pdf

Kohn, L. T., Corrigan, J., & Donaldson, M. S. (Eds.), (2000). *To err is human: Building a safer health system.* Washington, DC: National Academy Press.

Kozer E., Scolnik, D., Macpherson, A., Keays, T., Shi, K., Luk, T., et al. (2002). Variables associated with medication errors in pediatric emergency medicine. *Pediatrics, 110,* 737–742.

Lucchesi, M. (2004). Burns, thermal. Retrieved March 19, 2004, from the World Wide Web: http://www.emedicine.com/ped/topic301.htm

McCormick, M. C., Kass, B., Elixhauser, A., Thompson, J., & Simpson, L. (2000). Annual report on access to and utilization of health care for children and youth in the United States–1999. *Pediatrics, 105,* 219–230.

McDonald, J. (2002a). *Nursery product-related injuries and deaths to children under age five.* Washington, DC: U.S. Consumer Product Safety Commission. Retrieved April 2, 2004, from the World Wide Web: http://www.cpsc.gov/library/nursry02.pdf

McDonald, J. (2002b). *Toy-related deaths and injuries, calendar year 2001.* Washington, DC: U.S. Consumer Product Safety Commission. Retrieved April 2, 2004, from the World Wide Web: http://www.cpsc.gov/library/toydth01.pdf

Miller, M. R., Elixhauser, A., &Zhan, C. (2003). Patient safety events during pediatric hospitalizations. *Pediatrics, 111,* 1358–1366.

Miller, M. R., Pronovost, P. J., & Burstin, H. P. (2004). Pediatric patient safety in the ambulatory setting. *Ambulatory Pediatrics, 4,* 47–54.

Murff, H. J., Gosbee, J. W., & Bates, D. J. (2001). Human Factors and Medical Devices, in *Making Health Care Safer: A Critical Analysis of Patient Safety Practices.* (Evidence Report/ Technology Assessment: Number 43. AHRQ Publication No. 01-E058). Rockville, MD: Agency for Healthcare Research and Quality. Retrieved from http://www.ahrq.gov/clinic/ptsafety/ July, 2001.

Napper, C., Battles, J. B., & Fargason, C. (2003). Pediatrics and patient safety. *Journal of Pediatrics, 142,* 359–360.

National Center for Health Statistics. (2000). *CDC growth charts: United States.* Retrieved March 15, 2004, from the World Wide Web: http://www.cdc.gov/growthcharts/

National Immunization Program. (2003). *Immunization coverage in the U.S.: Results from National Immunization Survey.* Retrieved March 15, 2004, from the World Wide Web: http://www.cdc.gov/nip/coverage/default.htm#NIS

Needlman, R. D. (2004). Growth and development. In R. E. Behrman, R. M. Kliegman, & H. B. Jenson (Eds.), *Nelson textbook of pediatrics* (17th ed., pp. 23–66). Philadelphia: Saunders.

Nyquist, A. C., Gonzales, R., Steiner, J. F., & Sande, M.A. (1998). Antibiotic prescribing for children with colds, upper respiratory tract infections, and bronchitis. *Journal of the American Medical Association, 279,* 875–877.

O'Connor, A., & Besner, G. E. (2003). *Burns: Surgical perspective.* Retrieved March 7, 2004, from the World Wide Web: http://www.emedicine.com/ped/topic2929.htm

Perrin, J., & Bloom, S. R. (2004). Promoting safety in child and adolescent health care: Conference overview. *Ambulatory Pediatrics, 4,* 43–46.

Petridou, E., Sibert, J., Dedoukou, X., Skalkidis, I., & Trichopoulos, D. (2002). Injuries in public and private playgrounds: the relative contribution of structural, equipment and human factors. *Acta Paeditrica, 9,* 691–697.

Potts, M. J., & Phelan, K. W. (1996). Deficiencies in calculation and applied mathematics skills in pediatrics among primary care interns. *Archives of Pediatric & Adolescent Medicine, 150,* 748–752.

Raju, T. N. K., Kecskes, S., Thornton, J. P., Perry, M., & Feldman, S. (1989). Medication errors in neonatal and paediatric intensive-care units. *The Lancet, 2,* 374–376.

Rasmussen, J. (1997). Risk management in a dynamic society: A modeling problem. *Safety Science, 27,* 183–213.

Rowe, C., Koren, T., & Koren, G. (1998). Errors by paediatric residents in calculating drug doses. *Archives of Disease in Childhood, 79,* 56–58.

Salvendy, G.(Ed.). (1997). *Handbook of human factors and ergonomics* (2nd ed.). New York: Wiley.

Scanlon, M. C. (2004, May). Use of failure modes and effects analysis to evaluate the introduction of variation to a medication infusion preparation process through regulatory "standardization." Poster presented at the Sixth Annual National Patient Safety Congress, Boston.

Scanlon, M. C., Miller, M., Harris, J. M., Schulz, K., & Sedman, A. (in press). Targeted chart review of pediatric patient safety events identified by the AHRQ PSI methodology. *Journal of Patient Safety.*

Scanlon, M. C. (2004). Computer physician order entry and the real world: We're just human. *Joint Commission Journal of Quality and Safety, 30(6)* 342–346.

Slonim, A. D., LaFleur, B. J., Ahmed, W., Joseph, J. G. (2003). Hospital-reported medical errors in children. *Pediatrics, 111,* 617–621.

Smith, T. P. (Ed.). (2002). Age determination guidelines: Relating children's ages to toy characteristics and play behavior. Washington, DC: United States Consumer Product Safety Commission. Retrieved April 1, 2004, from the World Wide Web: http://www.cpsc.gov/businfo/adg.pdf

Solomon, S. L., Wallace, E. M., Ford-Jones, E. L., Baker, W. M., Martone, W. J., Kopin, I. J., et al. (1984). Medication errors with inhalant epinephrine mimicking an epidemic of neonatal sepsis. *New England Journal of Medicine, 310,* 166–170.

Stanger P., Silverman,, N. H., & Foster, E. (1999). Diagnostic accuracy of pediatric echocardiograms performed in adult laboratories. *American Journal of Cardiology, 83,* 908–914.

Troiano, R. P., Flegal, K. M., Kuczmarski. R. J., Campbell, S. M., & Johnson, C. L. (1995). Overweight prevalence and trends for children and adolescents. *Archives of Pediatric & Adolescent Medicine, 149,* 1085–1091.

U.S. Consumer Product Safety Commission (2004), About CPSC. Retrieved March 15, 2004 from http://www.cpsc.gov / about/about.html

U.S. Department of Health and Human Services. (2001) *The Surgeon General's call to action to prevent and decrease overweight and obesity.* Rockville, MD: U.S. Department of Health and Human Services, Public Health Service, Office of the Surgeon General. Retrieved March 1, 2004, from http://www.surgeongeneral.gov/library.

Vanderheiden, G. C. (1997). Design for people with functional limitations resulting from disability, aging or circumstance In G. Salvendy (Ed.). (1997). *Handbook of human factors and ergonomics* (2nd ed., pp. 2010–2052). New York: Wiley.

Ward, C. J., & Purdie, J. (2001). Diagnostic accuracy of paediatric echocardiograms interpreted by individuals other than paediatric cardiologists. *Journal of Paeditric Child Health, 37,* 331–336.

Wickens, C. D., Gordons, S. E., & Liu, Y. (1998). *An introduction to human factors engineering.* New York: Addison Wesley Longman.

Wolf, A. M., & Colditz, G. A. (1998). Current estimates of the economic cost of obesity in the United States. *Obesity Research, 6,* 97–106.

Yax, L. K. (2004). Annual estimates of the resident population by selected age groups for the United States and States: July1, 2003 and April 1, 2000. U.S. Census Bureau population estimates. Retrieved March 15, 2004, from the World Wide Web: http://eire.census.gov/popest/data/states/tables/ST-EST2003-01res.xls

49

HUMAN FACTORS IN HOME CARE

Teresa Zayas-Cabán and Patricia Flatley Brennan
University of Wisconsin, Madison

HUMAN FACTORS IN HOME CARE

The household is fast becoming the central site for health and health care. Indeed, contemporary health care relies more often on what happens in the home environment for achieving the goals of the U.S. government's "Healthy People 2010" (Department of Health and Human Services (2000) than on anything that occurs in hospitals, clinics, or long-term care facilities. From maintaining basic nutrition and exercise routines to managing complex disease processes, laypeople use the home and its resources to carry out the practices, activities, and tasks needed to promote health and recover from disease. Additionally, clinicians and health professionals rely on a self-motivated, well-resourced patient to do what is needed to insure maximum benefits from modern medicine.

The current migration of health care to the home is haphazard. Judicious application of human factors engineering principles may make this migration more successful, insuring optimal well-being for all individuals (i.e., patients, families and friends, and providers). Human factors and ergonomics has much to offer to further the understanding of health and health care in the home and to insure the design and engineering of the physical environments, information systems, and goods and services that could be used to foster health and health care goals.

BACKGROUND: HEALTH AND HEALTH CARE IN THE HOME

Health is a state of optimal well-being, encompassing physical, psychological, spiritual, and social dimensions. All people engage in behaviors and habits that are health promoting (e.g., exercise, meditation) as well as some that are health-destroying (e.g. smoking, excess alcohol consumption).Individuals in their optimal states of well-being function to the extent possible as enabled by their innate resources coupled with the support offered by family caregivers, pharmaceuticals, and medical assistive devices such as walkers, canes, or prosthetics.

Health care is the totality of goods and services provided by professionals in the health industry in such a way as to promote wellness, mitigate harm, or foster recovery from disease and injury. Health care is provided by trained professionals, such as nurses, physicians, social workers, pharmacists, and so on, who bring their talents of assessment and therapeutics to bear on the problems of individuals. Laypeople—"patients"—participate as partners in care and have rights to adequate information, shared roles in decision making, and critical responsibilities to carry out health and healing practices in collaboration with professionals. Much of health care occurs in institutions, such as clinics and hospitals; however, since the mid 1980s, the episodes of institution-based care have become shorter and shorter, leaving much of the health care process to occur in the home (Aliotta & Andre, 1997; C. E. Smith, Mayer, Parkhurst, Perkins, & Pingleton, 1991; Youngblut, Brennan, & Swegart, 1994). Health care in the home represents a purposeful application of the talents of home residents and resources to accomplish health care goals.

It is useful to characterize explicitly the work involved, the actors involved, and the home setting to understand where and how human factors engineering methods can insure optimum use of the environment, the home dwellers, and the associated devices to achieve health care goals.

What Kind of Health Work Is Done at Home?

Home care activities can be grouped into three major areas: maintenance of physical and social well-being, managing health information, and carrying out therapeutics as needed for the care and treatment of illnesses and injuries.

Contemporary health care encompasses a range of activities from self-help to self-care to disease management. *Self-help* activities include the attitudes and practices individuals initiate and follow in attempts to promote well-being. These include following a balanced diet, maintaining a balance of rest and activity, and engaging in the mental stimulation necessary to promote cognitive function. Self-help activities generally fall out of the purview of health professionals, but their importance in insuring health and avoiding disease is well documented.

Individuals engage in *self-care* activities as a part of a complete health care regime. Self-care activities include that which an individual does in concert with professional advice and intervention with the goal of promoting health. Self-care includes primary and secondary prevention activities recognized and endorsed by health professionals (U.S. Preventative Services Task Force, 1996). Primary prevention addresses the avoidance of disease processes and includes such things as obtaining vaccines and prophylaxis on a timely cycle. Secondary prevention activities target screening for early detection of disease. Although the boundary between self-help and self-care is quite indistinct, the fundamental difference arises from the critical role played by professionals, who not only possess the essential knowledge base regarding self-care activities but also control access to many of the services needed for it.

What brings self-help and self-care under the rubric of home care is the fact that the tasks and activities encompassed under these efforts take place in the home and these practices, although essential to health and well-being, fall largely under the volition and control of the individual. As these activities represent a type of nonpaid work, the concepts and methods from human factors engineering and ergonomics may be evoked in a manner to insure that this work occurs in an optimal fashion.

What most individuals think of when they hear the phrase "home care" would fall under *disease management*, defined by the U.S. Preventative Services Task Force (1996) as "tertiary prevention," the treatment of disease and mitigation of their consequences. Disease management activities are largely directed by professionals, and although they rely on the volition and participation of the layperson, these activities are directed toward the handling or recovery from a specific disease process. Changing bandages, using sensors and monitors to record physiological status, and adhering to drug management routines all are types of disease management activities. Some individuals are able to accomplish disease management on their own; others rely on the informal assistance of family members and friends.

Who Is Involved in Health and Health Care in the Home?

There are many people involved in health care in the home. First and foremost are individuals, sick or well, who engage in health behaviors. That is, home care participants include those traditionally thought of as ill, needing assistance for physical or mental disabilities. It is useful, however, to consider as participants in health care in the home all persons who take deliberate action in support of achieving or protecting health. So it is important to consider traditional health behaviors, such as dressing changes and exercises, as well as more common household tasks, such as grocery shopping, that take on new meaning when they are conducted in a purposeful manner to address a health need.

The physical and mental capabilities of these individuals vary widely, due in part to the developmental stage of the person (i.e. child, adult, elder) as well as to their health status and changes occurring following illness or injury. For example, when conducting a study to understand which demographic, health, and disabilities of Americans over 70 years of age were associated with living in a home with certain modifications, Tabbarah,

Silverstein, and Seeman (2000) identified subgroups with varying functionality and opportunities within the aging population, which could lead to reframing how this particular user group is characterized.

Family members or friends sometimes provide personal care assistance. These informal caregivers aid others by cooking meals, assisting in personal hygiene, and aiding in mobility. Although rarely trained for the many tasks of caregiving, these individuals develop robust strategies to manage everything from communicating with physicians and pharmacies, managing assistive equipment in the home such as respirators or infusion pumps, and providing the social and emotional support needed by the ill person. Informal caregivers are at risk for fatigue, musculoskeletal injuries from improper positioning and body mechanics during personal care activities, and a sense of isolation that arises from the difficulties of maintaining social contacts while managing the burdens of care (Canam, 1993; Dokken & Sydnor-Greenberg, 1998; C. E. Smith et al., 1991; Youngblut et al., 1994).

Professional caregivers, nurses, physical therapists, and nursing assistants make episodic visits to some people at home. Patients who are unable to leave their homes may have visiting nursing care, in which professionals come to the home to minister to the patient's needs. The work of these health professions and the human factor engineering strategies to support them, although similar to those in institutional care settings, have some special characteristics due to the physical environment of the home. The human factors considerations of the work of professionals in the home is well addressed in many publications (see, for example, Brulin et al., 1998; Brulin, Winkvist, & Langendoen, 2000; Wunderlich & Kohler, 2001) and draws well from other chapters in this volume. Thus, this chapter focuses more directly on the needs, experiences, and human factors considerations of the home dwellers.

Home and Households

Hindus (1999) emphasized the importance of including the home in engineering research. She commented that the home is too economically important to ignore and provides a rich research field that could improve daily living for many users. She explained, though, that homes are fundamentally different from workplaces. Houses are not designed for technology, and home technologies must be designed so as to not present a hazard for babies, pets, and elders. She suggests that consumers' "motivations, concerns, resources, and decisions" can differ from workers'. Decision making and value-setting, she commented, are different in households than those in corporate organizations.

Home care occurs in residential living spaces that differ from formal health care facilities, such as hospitals and nursing homes, in several important ways. First, contemporary culture ascribes meaning to the home as an intimate space of residence where access and egress are controlled by the individuals who live there. Homes may include apartment-type dwellings as well as freestanding units. Homes generally provide places for shelter and leisure and are increasingly becoming places of work. Homes are, for all people, sick or well, sites where health care practices occur. For some people managing chronic or complex illnesses, homes become a place in which durable medical goods, supplies, medications, and assistive devices are stored and must be integrated within a specific environment.

Venkatesh (1996) and Venkatesh and Mazumdar (1999) identified three environments of interest in the home: (a) the social environment, in which family behavior occurs, configured in terms of family members who adopt and use the technology and made up of the social structure of the household and the activities performed within it; (b) the physical environment; and (c) the technical environment, the space in which household technologies are used, which consists of the configuration of household technologies, involves family attitudes and levels of satisfaction with technologies, and represents the nature of technological environment within the household. Characterizing these three environments provides a reasonable starting place for industrial engineers to insure optimal use of the environment for the health care goals of the individual or family.

The *social* environment of the home is described in social network terms. Households may consist of individuals or families—individuals who share legal or biological relationships—or unrelated individuals with varying degrees of social ties. Additional dimensions of the social structure of households include their permeability—the number and variety of persons who interact with household members—and their network adequacy—the ability of the social contacts to help household members meet instrumental and emotional needs.

The *physical* environment of the home is characterized by a circumscribed physical space, which is devoted to the use of an individual or set of household dwellers. Like other environments more familiar to industrial engineers, such as factories or offices, the physical environment of the home can be described by space, objects held within the environment, temperature, lighting, and odor. Understanding the physical environment of the home requires awareness of the community context in which the house is located. Aspects such as density; proximity to resources and heath services; and accessibility of infrastructure such as sanitation, communication, and transportation also describe the physical environment of the household.

The *technical* environment of the home is important insofar as it provides the infrastructure on which technical solutions can be built. The technical environment of the home includes the mechanical, electrical, and communications framework, which governs the ability of technical solutions to "fit into" a household.

WHY HUMAN FACTORS?

This volume addresses human factors and ergonomics in health care and patient safety, including issues of errors, safety, devices, workflow, and organization. In this chapter, we consider the full range of application of human factors to the full range of health care activities that occur in the home, exploring the progress from "Do solutions useful in other environments work in the home?" to the creation of solutions that might work best in the home. There are few human factors studies in the home, yet we present examples that illustrate how human factors can be used to assess and design the home environment. This chapter should help the reader to design de novo approaches or technologies to create novel solutions for the home as well as to identify for the home solutions that also might be extensible to other environments.

It is important to not just take methodologies that are helpful in understanding a situation and apply them to the home; it is essential to make the bridge between these understandings and some design direction. The purpose needs to be more than simply understanding what people actually do and why they do it; human factors engineers also need to think about how it might be done better and look for an opportunity for design. This means not simply understanding how a task done in a lab or office transfers to a home environment, but how the physical, social, and technical environment of the home shapes the need for the task and the ways it can be done.

Goals of Human Factors and Ergonomics in Home Care

Human factors and ergonomics is a branch of industrial engineering that identifies how to best bring together people, environments, technologies, and organizations to ensure optimal outcome of work. Although historically focused on industrial systems, human factors and ergonomics have much to offer home care. Specifically, human factors and ergonomic frameworks and methods can be applied to design optimally functioning home care systems, to reengineer the home from a place of social intercourse to one more purposefully directed toward accomplishing home care goals. Additionally, many of the issues of concern to human factors and ergonomics, including safety, quality, and efficiency, are also relevant to the home care environment.

Computer technologies are increasingly applied to make work in the home more feasible, efficient, and effective. Information systems link individuals in the home to health professionals and to relevant and timely health care information and provide them with electronic tools that help them manage the complexities of health care. All frameworks of human factors, from macroergonomics to human–computer interaction (HCI), can help guide the design, choice, placement, and deployment of computer systems in support of home health care.

Frameworks

Human factors and ergonomic approaches draw from underlying theories or models—how physical work occurs and how participants in a social environment interact. Therefore, it is useful to use models to understand and optimize health care in the home. Social science theories are critical to understanding the dimensions of home care amenable to human factors and ergonomics. Several models provide useful insights, including organizational behavior, social–cognitive theories, and human performance. However, they are limited by their scope—too abstract for the home situation (i.e., organizational theories) or too detailed for the many complex tasks that make up health care in the home (i.e., cognitive theories).

Macroergonomics

Macroergonomics is an approach to industrial engineering that combines human factors and ergonomics into a unified framework (Hendrick & Kleiner, 2002). It guides the analysis, design, and evaluation of work systems but has many key concepts and dimensions that make it readily applicable to the home care situation. For more information, see the chapter on the historical perspective and overview of macroergonomics in this book.

For example, one key concept of macroergonomics is organization, the planned coordination (management) of two or more people (need allocation), functioning over time, through division of labor and authority (structure), to achieve common goal(s) (function). Applied to the home setting, the organizations are much less formally arranged than in industries, and the negotiation of the division of labor and authority arise more from family dynamics than from organizational hierarchies. The concepts of organizational design and structure, generally driven by the business purposes of the industry, must be modified and expanded in the home to be cognizant that the interpersonal structures exist for a variety of purposes (family dynamics, social exchange) and are rather immutable. Family decision-making strategies, degree of integration, and formalization of procedures may be in the household much less obvious although no less fixed than what one might discover in industry. Differentiation of roles and levels is much less in the home than in industry, so the human factors strategies employed must have the capacity to absorb this high level of complexity. Thus, for the purposes of home care, the extant structure of the household must serve as the basis for assessment and redesign but is rarely the target of reengineering activities.

Spatial dispersion of activities is another concept from macroergonomics that holds particular relevance in home health care. Household areas serve many purposes—for social interaction, leisure, exercise, nutritional support, storage of family goods, and as a safe haven for intimate activities. Households meet many physical and social needs. Two dimensions of spatial dispersion are relevant for human factors engineers: (a) health activities occur throughout the physical environment, with very few places restricted for a single purpose; and (b) the social and physical environments may not be completely aligned. The nonalignment of social and physical environments appears in two ways: multiple families housed within a single household and single families living in distinct households but uniting to manage health concerns at home.

Consistent with a macroergonomic framework (Hendrick & Kleiner, 2002), no prescriptions are advanced for the "proper" organization and structure of a household. The low levels of formalization and centralization likely facilitate the household dweller's accomplishment of goals and tasks that are unpredictable, lack structure, and may be best for engaging those most affected by the health concerns. Macroergonomic approaches lack sufficient direction to ensure the design of effective home care systems. For this reason, concepts from sociotechnical systems theories, which are consistent with and can be subsumed under macroergonomics, are described in the following section.

Sociotechnical Systems

Sociotechnical systems models may work best in understanding, modeling, and improving home care; they provide theories of person–environment fit and job and task design (Pasmore, Francis, Haldeman, & Shani, 1982).

Sociotechnical systems theory is derived from open systems theory and emphasizes the interrelatedness of the functioning of the social and technological subsystems of the organization and the relation of the organization as a whole to the environment in which it operates (Pasmore et al., 1982). The social subsystem is comprised not only of the people who "work" within the organization but also the relationships among them. The technical subsystem consists of the tools, techniques, procedures, skills, knowledge, and devices used by the members to accomplish the tasks of the organization. A sociotechnical systems perspective can then be used to observe the interaction between the people and the technology within the home. Using this approach might help develop models to better understand family to better drive research design and development of in-home environments for home care.

The similarity between work systems and family work systems suggests that a human factors perspective could be used to conduct in-home assessments related to home health care. Within sociotechnical systems models, the work systems model (M. Smith & Carayon-Sainfort, 1989) provides efficient guidance for design and evaluation.

Work systems model. One work system model, the balance model, provides awareness of not only what to study in the home but how to begin to make

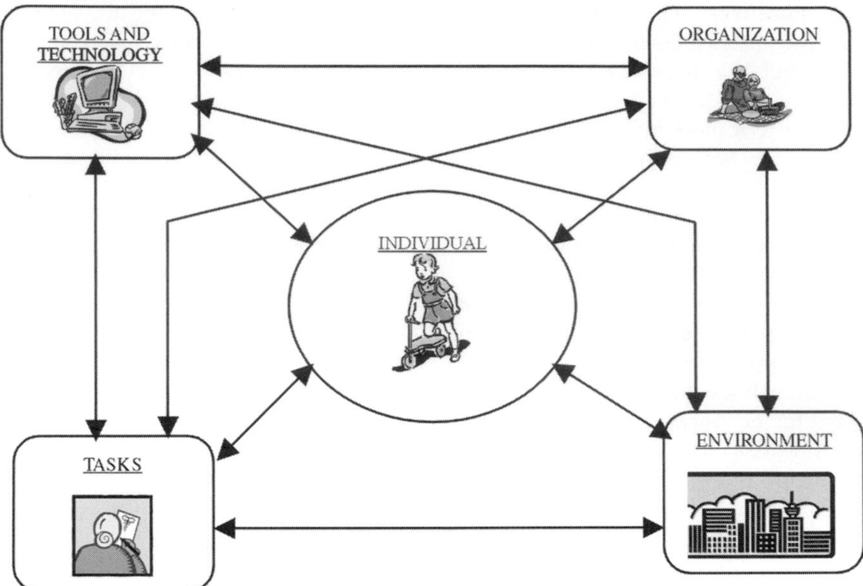

Figure 49–1. The balance model.

modifications. M. Smith and Carayon-Sainfort (1989) proposed this theory that conceptualizes job design based on the balance among job elements. The balance model is composed of five elements that offer greater detail about the environments than those proposed by Venkatesh and Mazumdar (1999). These elements are more specific and may help to explicitly address tasks that support the various jobs within the home. It also provides a holistic perspective used to analyze home jobs. At the center of the model is the individual with his or her physical characteristics, perceptions, personality, and behavior. The individual has a number of tools and technologies available to perform job tasks. These tasks are carried out in a "work" setting, which includes the physical and social environment and will affect the manner in which the tasks are carried out. Also, the organizational structure, which defines the nature and level of individual involvement, interaction, and control, will affect the way in which these tasks are performed.

The balance model's elements are all present in the home and family when carrying out any "job." For instance, human factors engineers can conceptualize the job to be studied as that of home health information management, which is comprised of a number of tasks. For example, one health information management task could be "writing down medical appointments." One or more family members or individuals (e.g., the mother) may carry out these tasks. The person who carries out these tasks and why they carry them out is a function of the family's organizational aspects (i.e., family rules that drive the way family activities are carried out) as well as the individual's own characteristics (i.e., his or her ability to carry out the task). In some households, for example, a mother who has good time management skills and is adept at record keeping might be responsible for all health coordination activities in the household and also happens to be very good at time management and record keeping. Thus, all aspects of the mother's role history and acceptance by the family may influence the way health information management occurs.

Furthermore, the way tasks are carried out are constrained by the home and community environment and will either be aided or constrained by tools or technologies available to the family. For example, because the mother wants to easily remember when appointments are scheduled, she writes them on a calendar kept in the kitchen cupboard where she sees it every morning and is also far from the reach from the small children.

Although the balance model was initially used to analyze the job of an individual, it is also considered a macroergonomics theory (Carayon & Smith, 2000). The balance model can be used to analyze the organization, which is what actually defines the smaller work systems. The model has been applied in the field of community ergonomics to provide a framework to guide stakeholders interested in working to improve the inner city to coordinate their efforts and develop cooperative approaches (M. J. Smith, Carayon, Smith, Cohen, & Upton, 1994).

Two other frameworks familiar to human factors engineers, physical ergonomics and HCI, also provide guidance for assessment and evaluation of the home as a work environment for health care.

Physical Ergonomics

Physical ergonomics uses a micro view of work, allowing intense attention to the psychomotor dimensions of specific activities and tasks common in health in the home. Relevant to this chapter are concepts arising from physical ergonomics, the design of equipment, biomechanics, and occupational health and safety.

There have been many studies that focus on physical ergonomics of the home and how these fit with its dwellers. Most of these address issues confronted by vulnerable populations, such as the elderly and disabled or homebound patients (D. B. D. Smith, 1990). Pinto et al. (2000) argued that the use of an ergonomics approach to home design may "develop an integrated strategy aimed at the well being and satisfaction of aging people" (p. 317). They described three areas of physiological decline in the elderly: muscular strength, posture, and movements. The authors suggested that the home environment presents several risk factors, which are compounded by the elderly's "psychophysical conditions." Some of the examples of practical recommendations for entrances and kitchens in home environments were to place furniture in the corners of a room or along the walls to avoid collisions and to place shelves and furniture so that they may be easily reached.

Voorbij and Steenbekkers (2002) examined the twisting force and hand configuration used to open a jar. The authors found that all participants used both hands while applying force: one on the lid and another to hold the jar. The results showed that with increased age, difficulty with completing the task increased. The authors found that the if torque required to open a jar was reduced to 2 Nm then only 2.4% of users 50 years of age or would have some difficulty opening a jar.

D. B. D. Smith (1990) provided an overview of research needs and design opportunities in human factors and aging. He used the areas of work, retirement, mobility, and the home environment to illustrate applications, all of which highlighted the themes of health, safety, technology, and the capabilities and limitations of age. D. B. D. Smith discussed several models proposed in this area that focused on a variety of issues, including identifying specific tasks and task components, adaptation, changing physical and mental capabilities, and accommodating to needs.

Other in-home ergonomics studies focus on design of equipment and how they interface with both patients and caregivers to support care and day-to-day household activities. Lathan, Kinsella, Rosen, Winters, and Trepagnier (1999), for example, described human factors issues in the use of telerehabilitation of persons with disabilities. The authors mentioned the need for human-centered design and pointed to specific design criteria to which the devices or systems must comply. The system operation must account for various impairments, including perceptual, cognitive, and motor (Jianwu, Jian, & Jing, 1996; Lathan et al., 1999). Lathan et al. commented that to build an effective device, the design process must take into account the goals, preferences, and abilities of their users.

There have also been studies that examine occupational health and safety issues surrounding home health work. These focus, though, on paid workers that come into the home to provide support to patients or caregivers (Delano & Hartman, 2001; Sitzman & Bloswick, 2002). Although information about paid workers is relevant to home care, it is beyond the scope of this chapter and is better addressed in the chapter on ergonomic programs and effective interventions in this volume.

Though these assessments are useful for focusing solely on the usability of technology and the physical aspects of the home environment, they focus only on one aspect that is relevant to home health activities: very explicit physical tasks. Examining the human factors and ergonomics aspects of home health activities requires a broader human factors approach that examines not only the physical aspect of the home but also the people who live in the home, their roles, the tasks they are trying to accomplish, and the resources they use to do so.

HCI *Approaches*

HCI is concerned with the design, evaluation, and implementation of interactive computing systems for humans (Hewett et al., 2004). As computers become more common in the home, they play an increasingly significant role in health and health care in the home. Laypeople use computers to gain access to health information, to communicate with clinicians and health care providers, and to manage the "business" of health care (claims processing, scheduling appointments, and so on).

Starren et al. (2002) reported on a system designed as a telemedicine intervention for diabetic patients. The application features use of an interactive video and computer monitoring system that helps people with diabetes monitor their blood sugar, communicate with their physicians, and review Web sites with nutritional advice (Starren et al., 2002). Additionally, computer chips are embedded in many assistive devices, such as home sensors and monitoring devices, providing an additional information management challenge in the health process.

The telemedicine application's usefulness was studied from an HCI perspective that focused on a usability evaluation of the system (Kaufman et al., 2003). The evaluation consisted of a cognitive walkthrough and field usability testing. The purpose of a cognitive walkthrough is to evaluate the cognitive processes of users performing a task. The method involved identifying sequences of actions and goals needed to accomplish a given task. The cognitive walkthrough was conducted by the experts and, in turn, was used to inform the tasks to be conducted during usability testing. Usability testing employs participants who are representative of a particular target population to evaluate the degree to which a product or system satisfies basic usability criteria. The researchers found some problems with the existing interface and system. For example, the screen display of the telemedicine unit included extensive amounts of text, some of which was in a small font and was difficult to read for some participants. The telemedicine evaluation also uncovered problems arising from a lack of familiarity with the device and the need for fine eye–hand coordination to perform some of the tasks. They also found significant differences between the underserved rural and urban populations; one of the main differences was level of education, which may have influenced both the acceptability of the technical intervention and the participant's familiarity with general computer skills. HCI models focus on the interaction between one or more humans and one or more computational machines and provide guidance for understanding the role these technologies play in home health care. However, because of the intense focus on a small component of tasks, HCI models may be too limited to fully understand the home "system." HCI models contribute best when the task is well understood and a computational machine solution is believed to be optimal. However, the broader macroergonomic models may provide frameworks that better encompass the physical and social aspects of technology use and the contexts in which the technology is used.

GATHERING INFORMATION: HOW TO APPLY HUMAN FACTORS ENGINEERING METHODS IN THE HOME

The Transition From the Workplace to the Home

Most application of human factors in organizational and job design has been carried out in traditional work settings, which, as discussed earlier, are different from homes in several key aspects. These differences have an effect on how human factors and ergonomics methods are applied in the home. The three key areas of modification of the analysis activities are: (a) involving household members in the design activities; (b) design under severe space, time, and financial limitations; and (c) the need to fit the design choices into the existing functional system of the household. These areas must be taken into consideration when conducting in-home human factors studies.

Assessment Techniques

Define the system and its objectives. Before any application of an assessment technique, it is essential to determine the system's objectives and define the system (Sanders & McCormick, 1993). Here, the balance model is used as a framework to give an overall picture of the home environment. For each of the elements, there are several aspects that human factors engineers assess to better understand the job being carried out and the potential for job redesign. These aspects are assessed using a number of human factors tools and methodologies. Figure 49–2 illustrates the criteria that are typically examined and the tools used to assess them. The tools or methods used by human factors engineers can be divided into three broad categories: (a) review of existing documentation; (b) direct observations,

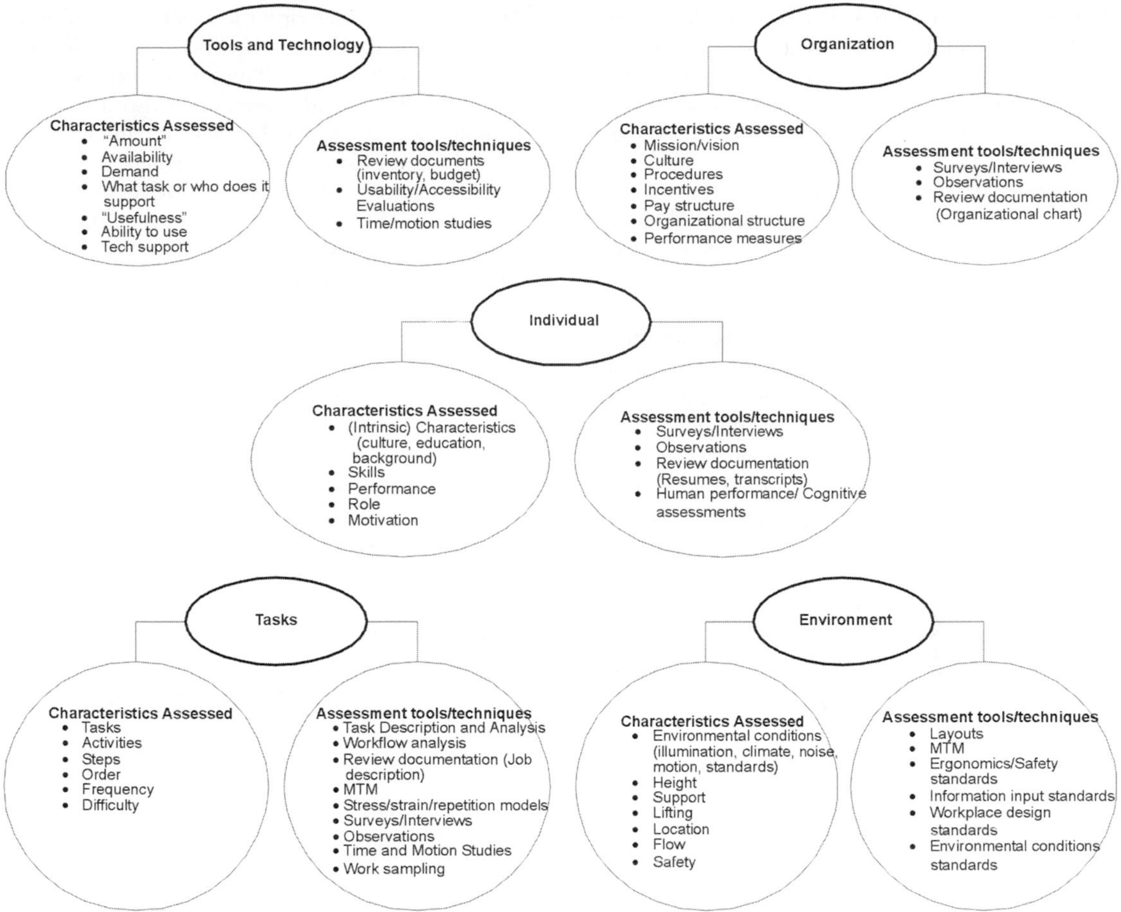

FIGURE 49–2 Organizes the major human factors concepts and their assessment methodologies around the balance model.

questionnaires, or interviews; and (c) use of specific techniques for certain aspects of work (i.e. case studies, time studies, work sampling, usability testing, and ergonomics assessment).

Analysis Techniques

Review of documentation. In traditional work settings, designers can also review supporting documentation, such as job applications, resumes, and letters of recommendation, to understand the worker's background, preparation, and skill set, and they can use performance measures to assess each individual's executions. Households generally have very few formal documentation systems. However, in terms of health concerns, there are many types of documentation that may exist. These include prescriptions, exercise routines, appointment reminders, contact information for clinicians, and health-related references. The documents tend to be stored in many locations around the house. The storage choices are generally driven by the anticipated future use of the information.

Reviewing documentation most likely occurs through interview and observations. It is critical that the engineer remain mindful that health information takes on very personal meaning for individuals, and any exploration of documents should be done in a manner that is sensitive to the private, personal nature of the information. The Advanced Technologies for Health @ Home project is the first study that has documented behaviors associated with information handling in the home. The overall goal of the project was to explore home health information management. Moen, Gregory, and Brennan (2004) presented study results that show a variety of information resources and remedies are in use to assist health management in these households. The resources they found are mostly paper-based and have a physical presence as visible tools to assist the household members. They also noted there is great variation in the types of devices and their locations.

Direct observations, questionnaires, and interviews. Human factors engineers have a wealth of techniques or methods they can use to assess individual characteristics. These commonly include the use of surveys, interviews, or observations. Relating specifically to health care in the home, then, human factors engineers will benefit most with the use of *observations, surveys, and interviews* to obtain information on the individuals that make up the home to assess aspects such as family composition, education, and roles and responsibilities.

The *individuals* are responsible for performing the tasks associated with their job. Traditionally job designers are interested in understanding the nature of these tasks, the steps needed to carry them out, the frequency of the tasks, and any difficulties that workers encounter when carrying out tasks. To fully understand the worker's responsibilities, job designers frequently review documentation about his or her position, such as a job description, to comprehend what the job entails. In the household, exploring the dimensions of these roles and the strategies used to carry out tasks can occur through interviews with household dwellers.

In Kiesler, Zdaniuk, Lundmark, and Kraut' (2000) study, families were given a computer and Internet access. The researchers later used a variety of data sources that included questionnaires, logs of requests for support, Internet use, and home interviews to examine the dynamics of technical support for computer use at home. The results showed that family members with high technical skill became the family technical gurus and were also able to influence the household's adoption of technology.

One of the most common techniques used in basic human factors design is *task* description and analysis. This process is used to list in sequence all the tasks that must be performed; each task is broken down into steps that it requires; and each step is then further analyzed. Workflow analysis is also often used to understand how processes within organizations are carried out and which individuals are involved. When analyzing the ergonomics of more repetitive, manual labor, designers use methods– time measurement and time and motion studies (Niebel & Freivalds, 1999). Work sampling is also a common technique used to assess the nature of the work and the tasks that compose it. Job designers and human factors engineers also make use of surveys, interviews, and observations to elicit from workers what their responsibilities are and how their daily tasks are carried out. These last tools lend themselves more to a home environment, where, for example, documents such as job descriptions do not exist.

Denham (2002) summarized three ethnographic studies that identify the ways family health is defined and practiced in three Appalachian communities. The study results showed that in the home, health management roles were assigned on the basis of skill level and knowledge of family members. The study aimed to understand the kinds of family routines that led to the production of family health. These included a number of routines or health behaviors such as dietary practices, sleep and rest patterns, avoidance behaviors, and so on. The routines, though, were not evaluated to determine whether they led to the production of health.

There have also been studies that examine the total workload of household members, which includes both paid and nonpaid work. In their study of the physical activities of female farmers in India, Nag and Chintharia (1985) examined how work changed during farming and nonfarming seasons. The study included 17 women who were interviewed and observed. The researchers found that household activities (i.e., gathering firewood, cooking, sweeping) required 4.7 hours in the farming season and 7.1 in the nonfarming season.

In another study that examined parents' total workload, Mardberg, Lundberg, and Frankenhaeuser (1991) aimed to develop and evaluate a questionnaire that assessed characteristics, perceived load, and positive and negative aspects from paid work and household work. The researchers mailed the questionnaire to 2,050 potential participants. The questionnaire covered the following areas: (a) general background, (b) job characteristics, (c) perception of job characteristics, (d) household, (e) child care, and (f) other duties. They used responses to conduct a factor analysis of the data and then developed what they found to be a reliable tool for measuring stress-related aspects of the total workload of white-collar workers.

Another study examined household work by administering a survey electronically. Marut and

Hedge (1999) wanted to explore the nature of household tasks and products. Participants were asked about general demographic information and the range of household tasks, including the equipment used for the tasks, time spent on each task, how tiring each task was, and sources of help for the tasks. The results showed that scrubbing, mopping, tidying, vacuuming, and doing laundry were perceived to be the five most tiring tasks. The authors suggested the results could serve to identify opportunities for the application of ergonomics design principles to the design of consumer products.

Workers must use *tools and technology* to complete their required tasks. Human factors engineers are interested in understanding the extent to which these tools support workers in completing their tasks. This includes the amount, availability, and demand for the resources and also the technology's usefulness, accessibility, and usability, as well as technology and repair support available to workers. Again, engineers could review documents on inventory and budgets to understand the availability of the technology. Usability and accessibility evaluations could help determine how good the human–machine fit is. Time and motion studies and work sampling can also be used to obtain information on technology use, frequency of use, and ease of use and its potential impact on performance. In the home, household members could be interviewed, observed, or both to gain an initial understanding of what technologies or tools are used to support health management.

Hindus, Mainwaring, Leduc, Hagstrom, and Bayley (2001) conducted studies to test the prototypes of the Casablanca study. The Study's goal was to develop communication technologies for the home. The researchers wanted to gain a better understanding of the usability, ease of use, and overall usefulness of the prototypes. Their pilot study entailed 16 in-depth interviews on where and how the technology was used. Two of these households were later revisited for longer observations. This study was then followed up with 35 additional in-home interviews and observations to understand technology use and attitudes toward technology. Both studies yielded important themes related to use home technology.

Norris, Lawrence, Hopkinson, and Wilson (1998) conducted a study to understand the causes of accidents with stepladders in Great Britain and identify potential improvements to their design. Their methods combined the use of surveys of accidents and the stepladder market. They also conducted observations of the use of the stepladders. Participants were asked to complete two domestic tasks. The results showed that few people followed the rules of safe stepladder use. The study results were then used in a nationwide publicity campaign in the United Kingdom to raise consumer awareness about safe stepladder use.

In another study conducted to understand family utilization of the personal computer and how computer use affects the relocation of other family applications and services, Frohlich, Dray, and Silverman (2001) conducted family in-home interviews. The authors explained that the interviews allowed them to explore all the family members' opinions on computer use and record the precise location of the computer and other technology inside the homes. This kind of study could then be complemented with more traditional human factors methods to assess either frequency of use or usability of the technology.

Jobs must be carried out in a physical *environment*, which may work to support or deter the worker's tasks. Ergonomists have mostly focused on assessing several environmental conditions such as illumination, climate, noise, and motion standards. Also when thinking of workstation design, human factors engineers focus on the task being carried out and the physical aspects of the workstation such as height, physical support, and lifting and how these may contribute to potential job-related stress or strain injuries. Safety also falls within the human factors environmental assessments focusing on safeguarding for potentially hazardous conditions. Finally, when thinking about the physical environment from a macroergonomics view, it is important to understand where tasks are carried out and overall flow of work within the organization. Human factors engineers have developed a number of standards with respect to motion and repetitive tasks, safety, information input, workplace design, and environmental conditions that are used to evaluate the current state of any physical environment within a workplace. Also, the use of layouts may aid in understanding overall flow of work within an organization.

In their study, Pinto et al. (2000) wanted to improve the relation between the aging user and the environment. When offering the practical recommendations, the illustrations used in the article showed the layout of the area of interest and illustrated the common habits or patterns of flow and use within the area. Their analysis not only recorded tasks for the task's sake but also aimed to understand the "artifact's role" in the task or job to be accomplished.

Similarly, the use of *layouts* in households can serve to aid in understanding of location of care activities. Because of the private nature of home spaces, layouts may seem to be a very invasive tool, and people may be threatened by the use of such a technique. This could, again, be complemented by the use of surveys and or interviews to gain an overall picture of health workflow within the home. Also, depending on the overall goal of the human factors assessment, other techniques related to workplace design could then be used.

Finally, the *organization* has a strong influence on overall functioning and how jobs are organized and carried out. Organizational characteristics that have been traditionally assessed in the field of human factors are the organization mission or vision, organizational culture, policies and procedures, incentives, pay structure, organizational structure, and performance measures. These characteristics can be assessed with the use of surveys or interviews, direct observations, and review of documentation relating to organizational structure, policies, and procedures.

As previously discussed, it is unlikely that there is documentation that describes the organizational structure of the home. Yet this is a very important aspect to be examined because it can influence all facets of home life. For example, Denham (2002) found that spiritual values and family traditions drive organizational aspects of the family concerning implementation strategies of health behaviors. Similar to the workplace, these are the ones that drive the way that roles are assigned and tasks are carried out. In terms of organizational elements, there is a formal task and work structure within the family. The mother assumes a prominent role in health issues, and there are member roles that determine who, why, where, how, and when to seek medical advice or services. These roles are tied to the family organizational structure as well as their norms of behavior. Furthermore, the roles are assigned on the basis of skill level and knowledge. The primary output in this case is the production of family health. This includes a number of routines or health behaviors such as dietary practices, sleep and rest patterns, avoidance behaviors, and so forth.

In a separate study, Honold (2000) defined the construct of culture within the HCI paradigm to develop a framework to elicit the influence of culture in product use. The researcher conducted a qualitative study of 35 households and used observations and interviews to identify eight factors that should be taken into consideration when defining product requirements in different cultures. Human factors engineers may be better served, then, by using interviews or surveys as well as observations to assess household culture and structure to understand how they affect home health care work.

Case Study

The following summarizes a case study conducted to gain a deeper understanding of the job of health information management and the tasks that support it to improve and maintain health. Data collection encompassed three stages: (a) in-home interview of the primary health information manager or person who makes most of the transactions related to health care within the household, (b) family interviews on home health information management tasks, and (c) observation of three of the tasks mentioned during the interviews. The following description includes a family profile, brief description of the distribution of health information throughout the home, and an example of a health information management task.

Family profile. The extended family of seven was composed of three major household subunits These subunits were (a) the primary health information manager and her spouse, (b) the health information manager's sibling, and (c) the health information manager's son along with his family. For this family, blood pressure was a major health concern, and although the view of the family health was very good, the primary health information manager considered herself to have fair health. The family used traditional as well as alternative medicine sources, which included using nopales, a cactus plant commonly known as the prickly pear. Although some of the family members had used a computer, the primary health information manager did not feel comfortable using e-mail, only used it sometimes, and felt uncomfortable accessing personal information over the Internet. For this household, Spanish was their first, and for some members their only, language, and this sometimes created problems during doctor visits. The primary health information manager mentioned that if she had an interpreter during doctor visits she would be able to understand all of the information. Also, the health information manager, who was a nurse, called her doctor and colleague and friend nurses in

TABLE 49–1 Locations and Devices by Health Information Type

Health Information Type	*Locations*	*Storage Devices*
Doctor's contact information	Kitchen	Post-it notes Refrigerator door
Prescription information/ medicines	Bedroom 1 Bedroom 2 Kitchen	Pill box Wardrobe Cookie jar Cupboard

Mexico when she had questions about a health concern. Even though each household subunit managed their own health information, other household members were aware of what information they kept and where they kept it. Furthermore, some health information management tasks, such as managing diets and grocery shopping, were done independently by each unit, whereas others, such as attending doctor appointments or measuring blood pressure levels, required support from other household members as well as other family members who did not currently live at home.

Distribution of health information. This family reported five different locations throughout the household in which health information was stored. They also made use of 12 different storage devices to store this information. Table 49–1 shows the number of locations and devices for two different types of health information.

Health information task: Checking blood pressure levels. Consuelo, her husband, and her brother, Roberto, all live in the same home. They all also happen to have high blood pressure. The three of them must check their blood pressure but they do this in different ways. At the beginning of the study, Consuelo would check her blood pressure in the morning before breakfast. She checked it monthly or every two months. She used her brother's blood pressure monitor and he assisted her in the task.

During the study period, Consuelo changed doctors and health insurance. This caused a change in her blood pressure medication, and she had to monitor her blood pressure more closely. The doctor asked her to check her pressure in the afternoons to see how it varied. She began checking her pressure every day around breakfast time in the kitchen. Consuelo then wrote down each reading on a piece of paper she stored in a kitchen cupboard. She then took the paper to her doctor whenever she had an appointments.

Roberto, on the other hand, kept a logbook of his readings in his bedroom, along with his monitor. He changed his logbook every seven months. He wrote down each reading along with the average. When he had high or low blood pressure readings, Roberto wrote down circumstances that might have caused it, such as tension or exercise. He also took medication every day to control his pressure.

Implications for design. The results described here have implications for human factors engineers to consider in the design of technology to support home health information management. For example, in this household, there were several subunits that managed their own health information as well as depended on other family members for support. Technology solutions, therefore, should take this into account and try to support their tasks and not create additional work. If not, the person might not be able to use it.

Although the family made use of technology and resources to manage its health information, these resources were generally paper based. This may have implications for design because, as illustrated in the results, these resources were distributed across the household and varied in levels of complexities. They were generally suited for the different kinds of information they stored and also took into consideration local family context, such as, for example, storing medications where children could not reach them. Careful consideration must be given then to the kinds of technology introduced into the home and what implications they have for ease of use.

The environment also played a crucial role in health information storage and use. Though limited by the household layout, there may be ownership associated with different spaces within the house. Storage in one location versus another may also be associated with anticipated level of use and privacy issues. For example, doctor appointments and calendars were kept in more visible places like kitchens or in portable devices like planners. This creates multiple levels of concern for designers in terms of distribution of use as well as access to information and privacy.

CONCLUSION

The home has become a relevant place for health care activities and is now an extension of the health care system. Many self-care practices such as diet, exercise, and complex disease management occur in the home. Many routine activities, such as shopping and family interaction, take on new meaning when there is a health care goal associated with them. In addition, as more and more patients are managing complex conditions at home, it is becoming the primary site of care for many persons. Yet, the complexities of health work at home are not well understood. Health care represents a purposeful application of the skills of home residents and resources to accomplish health care goals. Family or household members involved in health care activities at home must balance these activities with professional work, additional home work, and other household members. The family can be seen as a complex organization that works together to achieve family goals. This intersection of individuals, tasks or activities and environments pose an interesting challenge for human factors engineers to rethink the way they conceptualize work organizations and environments.

The use of a human factors perspective provides a framework that not only incorporates the analysis of the HCI but also encompasses the physical and social aspects of home health care work use and the contexts in which home health care activities occur. The goals of each specific human factors study in the home determines which assessment and analysis techniques will be more appropriate to achieve its objectives. For example, exploratory research conducted to better understand family roles in home health management might be better served by conducting in-depth, in-home observations. On the other hand, if the purpose of the study is to understand human factors issues in the use of a glucose meter, then designers could make use of a survey complemented with task observations to validate survey findings.

Home care is a central aspect of health care that can have implications for primary care as well as public health. Human factors has much to offer in designing for home environments and understanding how these environments fit within the larger health care system. The assessment techniques and examples presented in this chapter provide guidance for human factors engineers to begin to conceptualize the differences between home health care work and work done in traditional settings and how this may influence human factors assessment, analysis, and design.

References

Aliotta, S., & Andre, J. (1997). Case management and home health care: an integrated model. *Home Health Care Management Practice, 9,* 1–2.

Brulin, C., Gerdle, B., Granlund, B., Hoog, J., Knutson, A., & Sundelin, G. (1998). Physical and psychological work-related risk factors associated with musculoskeletal symptoms among home care personnel. *Scandinavian Journal of Caring Sciences, 12,* 104–110.

Brulin, C., Winkvist, A., & Langendoen, S. (2000). Stress from working conditions among home care personnel with musculoskeletal symptoms. *Journal of Advanced Nursing, 31,* 181–189.

Canam, C. (1993). Common adaptive tasks facing parents of children with chronic conditions. *Journal of Advanced Nursing, 18,* 46–53.

Carayon, P., & Smith, M. J. (2000). Work organization and ergonomics. *Applied Ergonomics, 31,* 649–662.

Delano, K. T., & Hartman, M. (2001, October). *Home health care: A needs assessment with design implications.* Paper presented at the Proceedings of the Human Factors and Ergonomics Society, CITY.

Denham, S. A. (2002). Family routines: A structural perspective for viewing family health. *Advanced Nursing Science, 24,* 60–64.

Department of Health and Human Services. (2000). *Healthy people 2010: Understanding and improving health* (2nd ed.). Washington, DC: Author.

Dokken, D. L., & Sydnor-Greenberg, N. (1998). Helping families mobilize their personal resources. *Pediatric Nursing, 24,* 66–69.

Frohlich, D. M., Dray, S., & Silverman, A. (2001). Breaking up is hard to do: Family perspectives on the future of the home PC. *International Journal of Human–Computer Studies, 54,* 701–724.

Hendrick, H. W., & Kleiner, B. M. (Eds.). (2002). *Macroergonomics: Theory, methods, and applications.* Mahwah, NJ: Lawrence Erlbaum Associates, Inc.

Hewett, T., Baecker, R., Card, S., Carey, T., Gasen, J., Mantei, M., et al. (2004). *Chapter 2: Human–computer interaction.* Retrieved

November 20, 2004, from the World Wide Web: http://sigchi.org/cdg/cdg2.html

Hindus, D. (1999, October). *The importance of homes in technology research.* Paper presented at the Second International Workshop, CoBuild 9, Pittsburgh.

Hindus, D., Mainwaring, S. D., Leduc, N., Hagstrom, A. E., & Bayley, O. (2001, April). *Casablanca: Designing social communication devices for the home.* Paper presented at the CHI 2001 Conference Proceedings. Seattle, Washington.

Honold, P. (2000). Culture and context: An empirical study for the development of a framework for the elicitation of cultural influence in product usage. *International Journal of Human–Computer Interaction, 12,* 327–345.

Kaufman, D. R., Patel, V. L., Hilliman, C., Morin, P. C., Pevzner, J., Weinstock, R. S., et al. (2003). Usability in the real world: Assessing medical information technologies in patients' homes. *Journal of Biomedical Informatics, 36,* 45–50.

Kiesler, S., Zdaniuk, B., Lundmark, V., & Kraut, R. (2000). Troubles with the Internet: The dynamics of help at home. *Human-Computer Interaction, 15,* 323–352.

Lathan, C. E., Kinsella, A., Rosen, M. J., Winters, J., & Trepagnier, C. (1999). Aspects of human factors engineering in home telemedicine and telerehabilitation systems. *Telemedicine Journal, 5,* 169–175.

Mardberg, B., Lundberg, U., & Frankenhaeuser, M. (1991). The total workload of parents employed in white-collar jobs: Construction of a questionnaire and a scoring system. *Scandinavian Journal of Psychology, 32,* 233–239.

Marut, M., & Hedge, A. (1999, September). *Ergonomic survey of household tasks and products.* Paper presented at the 43rd annual meeting of the Human Factors and Ergonomics Society, Houston, TX.

Moen, A., Gregory, J., & Brennan, P. F. (2004, September). *Cross-cultural factors necessary to enable design of flexible consumer health informatics systems (CHIS).* Paper presented at the IT in Health Care 2004 Conference, Portland, OR.

Nag, A., & Chintharia, S. (1985). Physical activities of women in farming and non-farming seasons. *Journal of Human Ergology, 14,* 65–70.

Niebel, B. W., & Freivalds, A. (1999). *Methods, standards, and work design* (10th ed.). Boston: McGraw-Hill.

Norris, B., Lawrence, K., Hopkinson, N., & Wilson, J. R. (1998). Using ergonomics investigations to improve stepladder safety. *International Journal for Consumer and Product Safety, 5,* 75–83.

Pasmore, W., Francis, C., Haldeman, J., & Shani, A. (1982). Sociotechnical systems: A North American reflection on empirical studies of the seventies. *Human Relations, 35,* 1179–1204.

Pinto, M. R., Medici, S. D., Sant, C. V., Bianchi, A., Zlotnicki, A., & Napoli, C. (2000). Ergonomics, gerontechnology, and design for the home environment. *Applied Ergonomics, 31,* 317–322.

Sanders, M. S., & McCormick, E. J. (1993). *Human factors in engineering and design* (7th ed.). New York: McGraw-Hill.

Sitzman, K., & Bloswick, D. (2002). Creative use of ergonomic principles in home care. *Home Healthcare Nurse, 20*(2), 98–103.

Smith, C. E., Mayer, L. S., Parkhurst, C., Perkins, S. B., & Pingleton, S. K. (1991). Adaptation in families with a member requiring mechanical ventilation at home. *Heart & Lung, 20,* 349–356.

Smith, D. B. D. (1990). Human factors and aging. An overview of research needs and application opportunities. *Human Factors, 32,* 509–526.

Smith, M., & Carayon-Sainfort, P. (1989). A balance theory of job design for stress reduction. *International Journal of Industrial Ergonomics, 4,* 67–69.

Smith, M. J., Carayon, P., Smith, J., Cohen, W., & Upton, J. (1994, October). *Community ergonomics: A theoretical model for rebuilding the inner city.* Paper presented at the Proceedings of the Human Factors and Ergonomics Society 38th annual meeting, Nashville, TN.

Starren, J., Hripcsak, G., Sengupta, S., Abbruscato, C. R., Knudson, P. E., Weinstock, R. S., et al. (2002). Columbia University Informatics for Diabetes Education and Telemedicine (IDEATel) Project: Technical implementation. *Journal of the American Medical Informatics Association, 9,* 25–26.

Tabbarah, M., Silverstein, M., & Seeman, T. (2000). A health and demographic profile of noninstitutionalized older Americans residing in environments with home modifications. *Journal of Aging and Health, 12,* 204–228.

U.S. Preventative Services Task Force. (1996). *U.S. Preventative Services Task Force guide to clinical preventive services* (2nd ed.). Washington, DC: Author.

Venkatesh, A. (1996). Computers and other interactive technologies for the home. *Communications of the ACM, 39*(12), 47–54.

Venkatesh, A., & Mazumdar, S. (1999, June). *New information technologies in the home: A study on uses, impacts, and design strategies.* Paper presented at Proceedings of the 30th annual conference of the Environmental Design Research Association, Orlando, FL.

Voorbij, A. I. M., & Steenbekkers, L. P. A. (2002). The twisting force of aged consumers when opening a jar. *Applied Ergonomics, 33,* 105–109.

Wunderlich, G. S., & Kohler, P. O. (Eds.). (2001). *Improving the quality of long-term care.* Washington, DC: National Academy Press.

Youngblut, J. M., Brennan, P. F., & Swegart, L. A. (1994). Families with medically fragile children: An exploratory study. *Pediatric Nursing, 20,* 463–468.

• 50 •

HUMAN FACTORS AND ERGONOMICS IN NURSING HOME CARE

David R. Zimmerman
Evgeniya (Jenya) Antonova
University of Wisconsin, Madison

In this chapter we address the application of human factors and ergonomic concepts to the provision of nursing home care in the United States. The evidence is clear that providing care to the impaired elderly in nursing homes presents some of the most difficult human factors challenges in the health care field. The work is physically demanding, emotionally draining, low-paying, and often carried out with inadequate staff. After providing an overview of nursing home care and describing the primary human resources required for it, we examine each of these human factor and ergonomic issues with a focus on what has been learned (and what has not) about how to identify and reduce human factor "stressors" in the provision of nursing home care.

Nursing home care in the United States is part of an eclectic, unintegrated system of long-term care, which the U.S. Congressional Budget Office (1999) defined as "the medical, social, personal care, and supportive services needed by those who have lost some capacity for self-care because of a chronic illness or condition". Nursing home care in other countries, such as Canada and most European nations, is typically a part of a better integrated set of long-term care, most of which is funded and provided directly by a government agency. Nevertheless, in most countries there are coordination issues between skilled nursing care and less intensive services on the continuum of long-term care. Structure of long-term care differs by country; see, for example, Harrington (2001), Kerrison and Pollock (2001), and Shapiro (2000).

Care services may be of short- or long-term nature and may be provided in a person's home, in the community, or in residential facilities. Home care services are the subject of another chapter in this volume and thus are not addressed here. Also, community-based programs are fundamentally different than residential settings along several important dimensions of service provision; we focus on residential settings, specifically nursing homes, which have the richest source of information on workforce and human factors issues among these types of settings.

Nursing homes, also called skilled nursing facilities or nursing facilities, offer the highest intensity level of long-term care. A skilled nursing facility is defined as "an institution ... which is primarily engaged in providing skilled nursing care and related services for residents who require medical or nursing care, or rehabilitation services for the rehabilitation of injured, disabled, or sick persons" (U.S. Social Security Administration, 2003). Nursing facilities offer care with lower skill requirements than skilled nursing facilities but still are at the higher end of the continuum of care intensity and skill level.

All forms of long-term care, including skilled nursing, are undergoing increasing pressure, primarily because of (a) major demographic trends and (b) dramatic improvement in medical care.

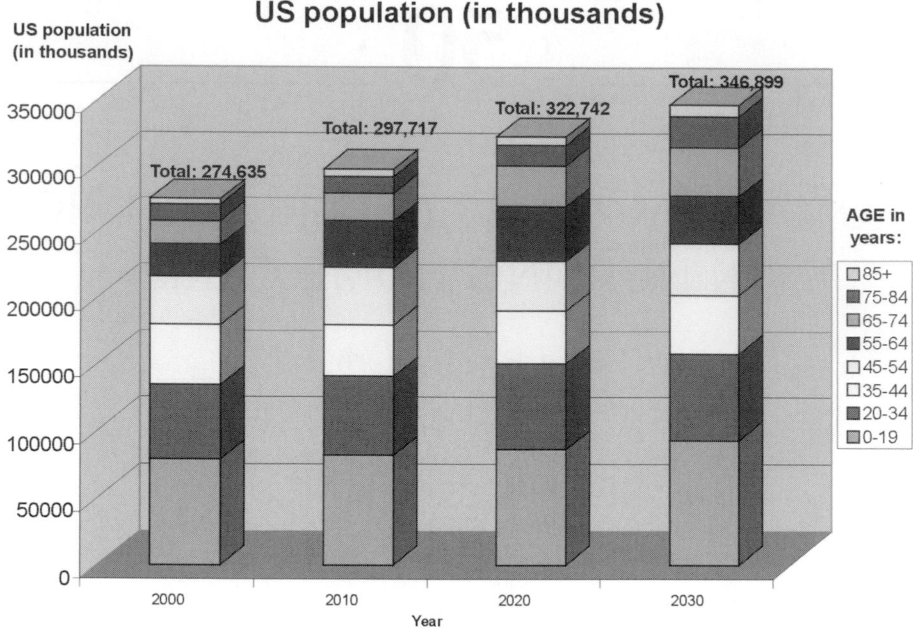

Figure 50–1. The U.S. population trend and prognosis.

OVERVIEW OF NURSING HOME CARE

The demand for long-term care, especially nursing home care, is expected to increase dramatically for the next 25 years because of (a) general population growth, (b) the aging of the baby-boom generation, and (c) improvements in medical technology that increase life expectancy but with a secondary impact of increasing the impairment level of those who live longer (White, Kim, & Deitz, 2001).

Demographic Trends and the Need for Long-Term Care

According to the Center for Medicare and Medicaid Services (CMS) and the U.S. Department of Labor of Bureau of Labor Statistics, as sited in White et al. (2001), the proportion of people older than 65 years in the U.S. population is expected to increase from 12.6% in 2000 to 20% in 2030, resulting in a population increase of more than 32 million persons. The number of people older than 85 years (those with the highest demand for the long-term care) is expected to more than double (see Figure 50–1).

The CMS is a federal agency within the U.S. Department of Health and Human Services. Formerly known as the Health Care Financing Administration, it is a federal agency responsible for administering both the Medicare and Medicaid programs.

Medicare is a federal program that pays for health services, but not all long-term care services, of people over 65 years old. Age is not the only eligibility criteria for the Medicare coverage; there are other criteria that apply to individuals and procedures. Medicaid is a program that is designed to provide health services coverage for people under the poverty level. Medicaid is operated by the state. The poverty level, as well as other rules related to coverage are set up by the state, although the program has some rules and regulations that are set at the federal level. More detailed information about CMS and the Medicare and Medicaid programs is available at the CMS Web site: http://www.cms.hhs.gov/.

Constructed based on the data from White, A., Kim, L., & Deitz, D. (2001). Chapter 8: Adequacy of nursing workforce to meet higher minimum nurse staffing requirements. In *Appropriateness of minimum nurse staffing ratios in nursing homes. Phase II: Final report.* Washington, DC: U.S. Department of

Health and Human Services. Retrieved August 25, 2004, from the World Wide Web: http://www.cms.hhs.gov/medicaid/reports/rp1201home.asp?

Although the population of those in greatest need of long-term care will grow dramatically, the availability of family caregivers is not expected to increase. Therefore, most of the burden will be put on the professional caregivers, which further explains the demand for long-term care generally and skilled nursing care (for example, nursing home care) in particular.

What Is Nursing Home Care?

Individuals enter a nursing home when they do not have a caregiver at home; they can no longer care for their own personal needs, such as eating, bathing, using the toilet, moving around, or taking medications (custodial care); or they cannot take care of themselves because of physical, emotional, or mental problems. Many persons entering a nursing home do so after being discharged from a hospital, either to receive short-term rehabilitation services or for chronic care.

Nursing homes provide private or shared rooms. Basic services provided in nursing homes are meals and snacks, usually planned with medical considerations addressed; housekeeping and linen service; personal care; some social activities; some transportation; professional service staff of other types (activity director, social worker); onsite medical staff and supervision; licensed nursing (at least one registered nurse [RN] available 8 hr per day); and nurse's aides (NAs) available 24 hr a day. In addition to the basic services, many nursing homes provide on-call physician services; physical, respiratory, occupational, rehabilitative, recreational, and speech therapy; medications; personal care items; laundry service; and hospice care.

Nursing home residents are generally frail and very old people, many of whom have mental impairments as well as physical disabilities. According to an American Health Care Association's ([AHCA], 2004c) report based on the Online Survey, Certification, and Reporting, average nursing home resident in the United States required assistance in almost four activities of daily living (ADL). On average, nearly half of nursing home residents suffer from dementia (AHCA, 2004c).

There are three levels of care provided by the skilled nursing facilities:

- Basic care—care required to maintain a resident's ADL, such as personal care, ambulation, supervision, and safety.
- Skilled care—care required for treatments and procedures provide by licensed nursing staff on a regular basis.
- Subacute care—(typically) more intensive care for someone who has had an acute condition due to an illness, injury, or surgery.

Some residents stay in a nursing home for only short period of time; others spend the rest of their lives there. A fundamental characteristic of nursing homes, one that has an impact on both care and human factors affecting that care, is that they are not only locations in which health care is provided; they also represent the actual home of the care recipient. It is where the care recipients live. This means that they require not only assistance with ADL but also fulfillment of their psychological, spiritual, and cultural needs, as well as their need for meaningful activities (Kayser-Jones, 2002). The care must be personalized, with more value placed on quality of life and quality of interactions between nursing staff and residents (Bowers, Esmond, & Jacobson, 2000). Increasing numbers of individuals are receiving their end-of-life care experience in nursing homes, which poses additional emotional burdens on the staff (Kayser-Jones, 2002). The severe impairment levels of residents leads to emotional issues for both themselves and the staff, physical demands on the staff, and in some cases very unpleasant environmental working conditions, with resident fecal and urine orders due to incontinence (Farmer, 1996).

The physical environment of a nursing home often is not well designed, leading to additional human factors problems. Many nursing homes are old, often having been converted from buildings used for other purposes and sometimes with minimum reconstruction. For example, doors are not wide enough to accommodate a wheelchair or the layout of a facility does not allow the placement of a nursing station in a convenient place.

Nursing Home Ownership, Reimbursement, and Regulation

According (OSCAR) to the Online Survey, Certification, and Reporting (OSCAR) data, among the more than 16,000 nursing facilities in the United

States, about 65.64% are for profit, 28.14% are government owned, and 6.22% of facilities have nonprofit status. About 14% of the facilities nationwide are operated by a chain.

Among total number of 1,771,544 nursing home beds, 126,400 (about 7%) are special care beds, including AIDS (1,706), Alzheimer's (93,763), hospice (3,287), ventilator (7,060), rehabilitation (15,743), and other special care beds (4,841; AHCA, 2004b).

In terms of the reimbursement structure, there are four main payment options: Medicare, Medicaid, private insurance carriers, and private funds. Of the total 1,443,073 nursing home residents, about one in eight (11.9%) are covered by Medicare, about two thirds are covered by Medicaid, and about 22% are covered private insurers or private funds (AHCA, 2004a). Thus, 8 out of 10 nursing home residents are cared for through government funding. A primary problem is that the nursing home reimbursement system is a complicated, eclectic, unintegrated mosaic of funding sources; providers can face a myriad of different private, federal, and state funding streams.

Nursing homes are heavily regulated, perhaps not surprising given the vulnerability of their populations. They are required to be certified (with stringent conditions) to be eligible for participation in the Medicare and Medicaid programs, and states require that they be formally licensed, sometimes with additional quality of care requirements. In addition, there are numerous U.S. Occupational Safety and Health Administration (OSHA) regulations that apply to nursing home care.

Nursing Home Structure and Staffing

Regulations have had important effects on both the structure of nursing homes and the level and mix of staffing in the facilities.

Organizational Structure

There is considerable variation in the organizational structure of nursing homes, due to the specifics of workflow organization, types of provided care, requirements of nursing home networks or chains, and other factors. In terms or organizational structure, an idea of how a nursing home is organized can be obtained from Figure 50–2.

Responsibility for managing the nursing home is the responsibility of the administrator (often called the executive director in today's parlance). The administrator oversees the day to day operations of the facility and is responsible for ultimate decision making regarding operations.

Nursing home departments are grouped according to the health care function its staff members perform. Nursing is the primary department, and most nursing homes have dietary, social services, activities, pharmacy, and therapy departments. Some nursing homes also offer more extensive rehabilitation or restorative nursing care. Other nursing home departments are organized on the basis of their operational function, such as maintenance and housekeeping.

Reprinted from a PowerPoint presentation "Improving Nursing Home Quality" by David R. Gifford, with permission of the author.

Nursing Home Staff

The employee group of a nursing home includes administrative and clerical staff; nursing staff (licensed and unlicensed); other care staff such as rehabilitation, dietary, social services, and activities; and workers in ancillary departments such as housekeeping and maintenance. Many facilities now employ marketing staff members and admission specialists as well.

The director of nursing (DON) is the primary care manager in the facility, responsible for the provision of nursing and other services at the nursing home, including planning, organizing, implementing, maintaining, and evaluating nursing services programs; outlining management responsibility for RNs, licensed practical nurses (LPNs), and NAs; and coordinating the care of nursing and related services, such as rehabilitation and resident activities. The DON typically reports directly to the administrator and often takes charge in his or her absence.

Other management staff includes the medical director, assistant DON, resident assessment coordinator, and a staff development (training) coordinator. Many of the responsibilities of these auxiliary staff relate to quality assurance and other activities to ensure compliance with federal and state regulations, coordination of reports to the Minimum Data Set, human resource management, and so on. The direct care staff members typically report to the DON or the administrator.

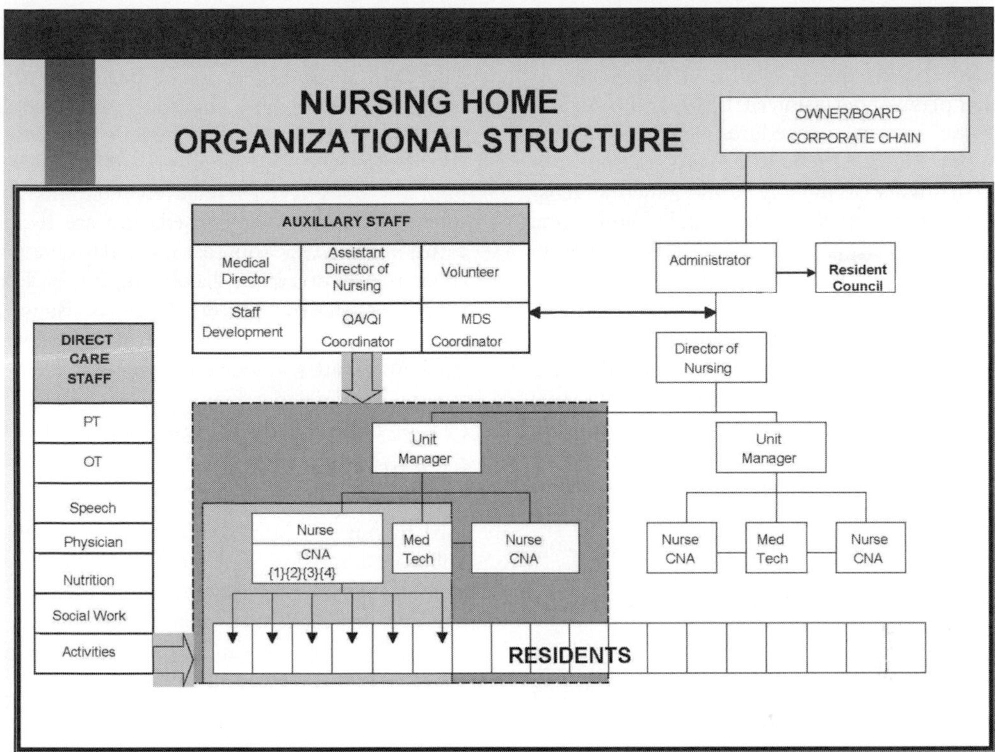

Figure 50–2. Nursing home organizational structure.

Federal law requires that all "nursing and nursing-related" tasks be performed by a licensed health professional or a certified nurse aide (CNA; Omnibus Budget Reconciliation Act of 1987 [OBRA], 1987). Therefore, most of the long-term direct care in nursing homes is provided by the nursing staff. Depending on the nature and complexity of the daily care, it is provided by either licensed or certified nursing staff. Licensed nursing staff includes RNs and LPNs. Certified nursing staff includes CNAs. Primary medical care is provided through periodic visits by physicians, physician assistants, or nurse practitioners. In addition to nursing and medical care, ancillary staff provide nutritional care, activities, speech and social services, nutrition, and occupational physical therapy.

Most nursing homes have a resident council, which under the Nursing Home Reform Act of 1987 (OBRA, 1987) is a right of residents and family members. Ideally the council is a self-organized and self-governing group of the long-term care residents who meet regularly to voice their needs and concerns and "to have input into the activities, policies, and issues affecting their lives in the facility" (Wunderlich & Kohler, 2000, p. 176).

Direct Nursing Staff

The primary driver of care and services is provided by or coordinated through the direct nursing staff at the facilities. In addition, most of the literature on human factors issues in nursing homes pertains to this group of care-providers. Therefore, the major thrust of our subsequent discussion in this chapter is focused on licensed and unlicensed direct nursing staff.

RNs. An RN is a person who practices professional nursing, including care for the ill, injured, or infirmed residents, or the maintenance of health or prevention of illness of other residents (Bureau of Labor Statistics [BLS], 2004). RNs are graduates of state-accredited nursing programs and are licensed by a state's Board of Nursing.

RNs are the primary care managers in nursing homes. Their responsibilities typically include assessment of residents' health, development of treatment plans, supervision of LPNs and NAs, and performing invasive procedures, such as starting intravenous fluids. Often, RNs work in specialty-care departments of nursing homes, such as rehabilitation units for residents with strokes and severe injuries, but they also spend much of their time on administrative and supervisory tasks (BLS, 2004).

LPNs. A LPN (sometimes called a licensed vocational nurse) provides care of convalescent, subacutely or chronically ill persons. Practical nursing does not require any substantial nursing skill, knowledge, or training; or the application of nursing principles based on biological, physical, or social sciences; or an understanding of cause and effect in the process of care" (Board of Nursing, 2003). LPNs are graduates of a state-accredited practical nursing program, with a license granted by a state's Board of Nursing.

LPNs in nursing homes provide basic bedside care to residents, for example taking vital signs such as temperature, blood pressure, pulse, and respiration; collecting samples for testing; performing routine laboratory tests; feeding patients; and recording food and fluid intake and output. They also prepare and give injections and enemas, monitor catheters, apply dressings, treat bedsores, give alcohol rubs and massages, and perform other procedures. LPNs also can be responsible for monitoring residents and reporting a need for medical intervention to supervising doctors or RNs. In addition to providing routine beside care, LPNs in nursing care facilities help evaluate residents' needs, develop care plans, and supervise the care provided by NAs (BLS, 2004), and they sometimes supervise CNAs.

CNAs. CNAs or aides perform routine but critically important tasks in the provision of care to residents. Their responsibilities include answering patients' call lights; helping patients with eating, dressing, bathing, getting in and out of bed, and walking; delivering messages; serving meals; and a variety of other functions. They also can provide basic bedside care and skin care to residents and monitor vital signs. CNAs are the principal caregivers for and have the most direct contact with residents. Because some residents may stay in a nursing care facility for months or even years the, CNAs develop strong ongoing relationships with them and become emotionally involved in the care.

Estimates show that close to 90% or direct resident care in nursing homes is provided by NAs. The "bed and body" work of nursing assistants is hard, stressful, highly injurious, often unpleasant work for relatively low pay, benefits, or recognition (Bowers Esmond, & Jacobson, 2003). Besides low pay, the aides receive relatively little training, are often inadequately supervised, and are required to care for more residents that they can serve properly (Bowers, Esmond, & Jacobson, 2003; Takeuchi, Burke, & McGeary, 1986; Riggs & Rantz, 2001; Schnelle, Simmons, Cretin, Feuerberg, Hu, & Keeler, 2001; Waxman, Carner, & Berkenstock, 1984).

Characteristics of Nursing Care Jobs in Nursing Homes

Jobs in nursing homes, across both licensed and unlicensed nursing staff and in ancillary services such as dietary and therapy departments, present uncommon challenges due to the characteristics and level of impairment of the resident population.

Working Conditions

Nursing home residents need round-the-clock care, requiring staff to work evenings, nights, weekends, and holidays. Most full-time aides and nurses work are on 40-hr schedules, but an increasing number of them work regular overtime—in many cases double shifts—and many hold more than one job. CNAs often have to skip breaks because of their physical inability to finish all their work on their shift and feeling guilty about leaving residents without care (Bowers et al., 2003).

Nurses and especially CNAs spend many hours standing and walking and often face heavy workloads. Moving patients in and out of beds and helping them to stand or walk puts them at a high risk of back injuries. Staff members also may face hazards from some infections and diseases, such as hepatitis.

CNAs face unpleasant duties, such as emptying bedpans and changing soiled bed linens every day. Often the residents may be disoriented, distressed, irritable, or uncooperative, which makes work of nursing staff, especially aides, more difficult. Despite the fact that their jobs are physically and emotionally very demanding, many nurses and aides find satisfaction from assisting elderly people and the close relationship that they establish with the residents (Bowers et al., 2003).

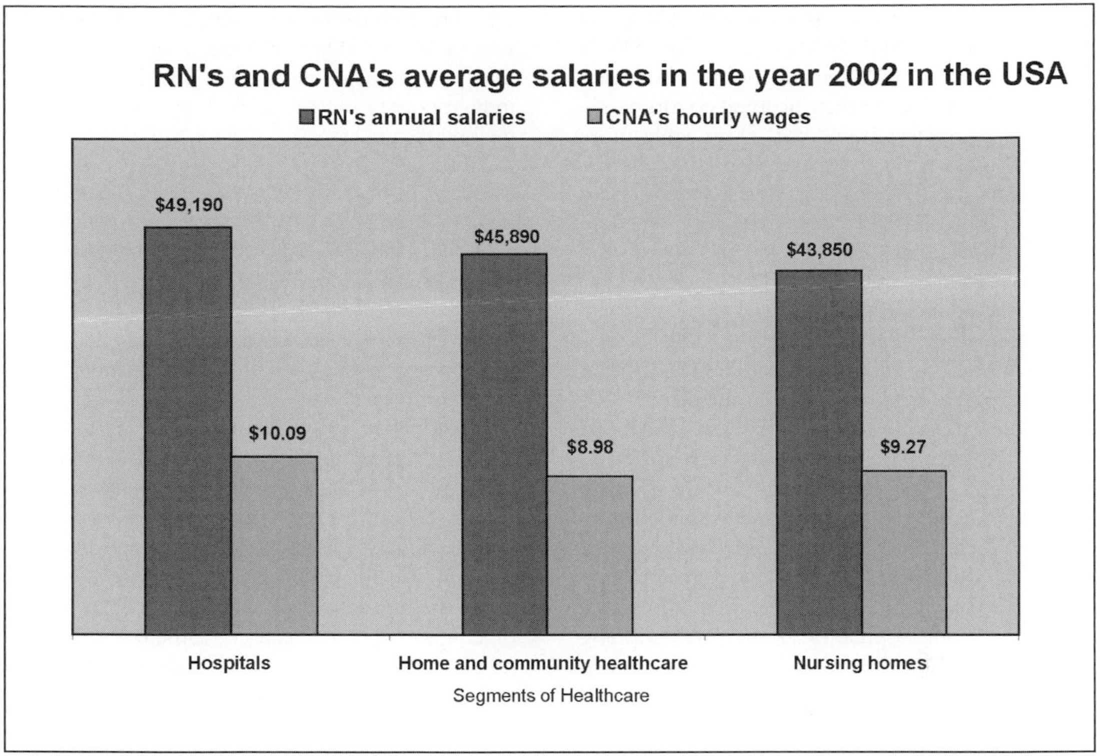

Figure 50–3. Average salaries of RNs and CNAs in the United States in the 2002.

Job Design Factors

The perception in the past was that long-term care jobs require mostly low-tech, unskilled labor with little specialized training and few qualifications in order to address the long-term nursing needs of residents (Davis, 1991; Kane et al., 1983). This is certainly not as true today, as residents are admitted to nursing homes with more acute illnesses, require more complex care, and have difficult emotional and cognitive problems. Their responses to medical and nursing treatment can vary, making it difficult to predict sudden drastic changes in residents' functional status that require immediate care, such as hospitalization. Although some nursing tasks, such as those involved in pressure ulcer prevention, are more standardized and uniform, most of the tasks are less stable. The nursing function is characterized by a high level of uncertainty, instability, and variability in the process of care (Alexander & Randolph, 1985; Perrow, 1967). The uncertainty of resident status variability of how to respond to problems makes nursing job tasks more complex (Banaszak-Holl & Hines, 1996).

Wages

Unfortunately, wages and economic benefits of nursing home workers have not kept pace with the increasing resident impairment levels and need for more complex care.

The wages of nursing home personnel are among the lowest in the nursing profession, at all levels. In 2002 the average annual salary of a nursing home RN was $43,850, which was close to the lowest 25th percentile of all RNs (see Figure 50–3). An average RN in a nursing home earned annually 10% less than the median of the RN workforce and 31% less than the top quartile. The situation for LPNs is very similar.

Constructed based on the data from the Bureau of Labor Statistics, U.S. Department of Labor, (2004). *Occupational outlook handbook, 2004–05 edition.* Washington, DC: Author.

The situation is even worse for CNAs working in nursing homes. Their average wages in nursing facilities were below the lowest 25th percentile of all CNAs in the country in 2002 (BLS, 2004). The average wage of a CNA in a nursing home was

$9.27 per hour, which was 9% lower than wage of CNAs in hospitals and 23% less than top 25% of CNAs. It is clear that nursing home CNAs are underpaid compared to their hospital counterparts, even though nursing home care is arguably more physically and emotionally stressful, especially for CNAs. The U.S. Health Resources and Services Administration (HRSA) reported that the U.S. General Accounting office (GAO) had found that large proportion of nurses reported "decreased job satisfaction, ... increased pressure to accomplish work, increased required overtime, and stress-related illness" (Scanlon, 2001, as cited in HRSA, 2001, p. 25). Although the aides do the most heavy, dirty, and unpleasant work (Bowers et al., 2003) and constantly face patients' death and suffering (Kayser-Jones, 2002), oftentimes unable to help them, they are paid very little and provided very few or no benefits (Stone, 2001). As Stone noted: "ironically, while these [frontline] workers are delivering essential care to some of the most vulnerable segments of our population, their peers flipping burgers at McDonalds make more, have much more financial security, and are treated with much more respect" (Stone, 2001, p. 49).

Job dissatisfaction is a primary reason for nurse retention problems. Nursing staff simply leaves nursing homes to find better working conditions. They either move to a hospital setting (AHCA, 2001) or simply leave the profession (Bowers et al., 2000).

UNDERSTAFFING AND TURNOVER: A PRIMARY STRESSOR

The already challenging nature of nursing home work is exacerbated by widespread understaffing and turnover, putting even more stress on workers in the long-term care field. There is mixed but powerful evidence that these staffing challenges in turn have had deleterious effects on the quality of nursing home care.

Impact of Staffing on Quality of Care

Concern over the quality of care in U.S. nursing homes has continued unabated for more than 25 years, among consumers and their advocates, policymakers, researchers, the general public, and even health care providers themselves (Institute of Medicine, 1996; Louwe & Kramer 2001; National Institute of Nursing Research, 1994; Wunderlich & Kohler, 2000). Although these groups have different perspectives, they agree that the long-term care industry needs better solutions to improve resident quality of care.

Multiple stakeholders also agree that inadequate staffing and high turnover have contributed to poor nursing home care (Bliesmer, Smayling, Kane & Shannon, 1992; Bowers et al., 2000; CMS, 2001; Harrington & Swan, 2003; Harrington, Zimmerman, Karon, Robinson, & Beutel, 2000; Kayser-Jones, 2002; Munroe, 1990; Wunderlich & Kohler, 2000). There is a rich set of studies on the relation between staffing and quality of care, work that is summarized in Table 50–1. This voluminous set of studies has produced mixed results concerning the strength of the relation between staffing and quality. Nevertheless, overall there is reasonably strong evidence that staffing is related to residents' reduced quality of life, as well as the presence of poor outcomes and so-called inappropriate practices, such as excessive use of antipsychotic medications, catheters, and feeding tubes; general drug overuse; poor restorative care practices; and poor care planning.

Residents' disruptive behavior such as aggression and resisting care is often a reflection of poor care practices or the lack of supervision of residents (Louwe & Kramer, 2001; Taylor, Ray, & Meador, 1990). The instances of residents' disruptive behavior were affected by availability of RN staff—most experienced nursing professionals, as well as other staff categories and overall staffing levels in the facility (Anderson, Hsieh, & Su, 1998).

Licensed staff level, especially RNs, was found to be important to avoid the negative residents' outcomes (Anderson et al., 1998; Harrington et al., 2000; Ramsay, Sainfort, & Zimmerman, 1995). However, aides staffing levels were found to be important for the outcomes more associated with the chronic as opposed to postacute residents: functional improvement, incidents of pressure ulcers, resisting care manifestations, skin trauma, and weight loss (Kramer & Fish, 2001). Congestive heart failure in residents was found to be associated with NA staffing levels only (Kramer & Fish, 2001).

The number of cited deficiencies an part of the nursing home survey (inspection process) is another measure of quality of care in nursing homes. Overall deficiencies, especially quality of care deficiencies, were associated with short nursing staffing (Harrington et al., 2000). Resident rights deficiencies were affected by low RN staffing (Harrington et al., 2000).

TABLE 50–1. Association Between Nursing Home Staffing and Process and Outcomes of the Long-Term Care

Process Variables	*Structure Staffing Variables*	*Studies*
Use of physical restraints	RN proportion	Anderson et al., 1998
		Zinn, 1993a
	Licensed	Ramsay, Sainfort, & Zimmerman, 1995
	CNA	Phillips, Hawes & Fries, 1993
		Ramsay et al., 1995
	All nursing	Bowers, 2001
		Castle, 2002
		Phillips et al., 1996
Use of wrist mitten	RN	Anderson, 1998
Use of psychotropic medications	All Nursing staff	Bowers et al., 2001
General drug use	Licensed, NA	Ramsay et al., 1995
Poor restorative care practices	Licensed, NA	Ramsay et al., 1995
Poor care planning	Licensed, NA	Ramsay et al., 1995
Use of catheters	RN	Zinn, 1993a
Outcome variables		
Probability of discharge home	RN	Braun, 1991a
	All nursing, RN, LPN, NA	Linn, 1977 a
	Licensed	Bliesmer, Smayling, Kane & Shannon, 1998
Mortality	RN	Zinn, 1993 a
	All nursing, RN, LPN, NA	Linn, 1977 a
	Licensed	Bliesmer et al., 1998
Hospitalization rates	RN	Braun, 1991 a
	RN turnover rates	Zimmerman, Grube Baldini, Hebel, Sloane & Magazine 2002
Infection rates	RN turnover raters	Zimmerman, et al. 2002
	NA	Zimmerman et al., 2002
Disruptive behavior of residents	All nursing	Rudman, 1993 a
	RN	Kolanowski 1994 a
Verbal aggression	RN	Anderson et al., 1998
Physical aggression	RN	Anderson et al., 1998
Resisting care	NA, licensed, RN	Kramer & Fish, 2001
Other disruptive behavior	RN	Anderson et al., 1998
	All Nursing staff	Berlowitz et al., 1999
		Bowers, 2001
Resident outcomes		
Contractures	RN	Anderson et al., 1998
Dehydration	RN	Anderson et al., 1998
		Kayser-Jones, 2002
Fracture in past 3 months	RN	Anderson et al., 1998
Pressure ulcers	All Nursing staff	Berlowitz et al., 1999
	RN	Graber, 1993
		Kramer & Fish, 2001
	NA	Kramer & Fish, 2001
Poor skin integrity	Licensed, NA	Ramsay et al., 1995
Skin trauma	NA, licensed	Kramer & Fish, 2001
Congestive heart falure	NA	Kramer & Fish, 2001
Electrolyte imbalance	NA, licensed, RN	Kramer & Fish, 2001
Respiratory infections	Licensed	Kramer & Fish, 2001
Sepsis	NA, licensed, RN	Kramer & Fish, 2001
Urinary tract infections	NA, licensed, RN	Kramer & Fish, 2001
Functional improvement	All nursing	Rudman, 1993[a]
		Spector 1991[a]
		Linn, 1977[a]
	NA	Linn, 1977*
		Kramer & Fish, 2001 Rohrer 1993*
	LPN	Linn, 1977*
		Rohrer 1993*
	RN	Linn, 1977*

TABLE 50–1. (Continued)

Process Variables	Structure Staffing Variables	Studies
		Kramer & Fish, 2001
		Rohrer 1993*
	Licensed	Rohrer 1993*
		Kramer & Fish, 2001
Incidence of decline in ADLs	Licensed	Bliesmer et al., 1998
Constrained mobility and function	Licensed, NA	Ramsay et al., 1995
Urinary incontinence	All nursing	Mueller & Cain, 2002
Malnutrition among residents	All nursing, professional nursing	Kayser-Jones, 2002
Weight loss among residents	All nursing	Kayser-Jones, 2002
	NA, licensed	Kramer & Fish, 2001
Other suboptimal outcomes	Licensed, NA	Ramsay et al., 1995
Survey deficiencies		
Quality of care deficiencies	RN, NA	Harrington et al., 2000
	RN	Graber, 1993*
		Munroe 1990
	All nursing	Johnson-Pawlson, 1996*
Physical health - poor care	RN, LPN, NA	Cherry, 1991*
Resident right deficiencies	RN	Johnson-Pawlson, 1996*
Total deficiencies	RN, NA	Harrington et al., 2000
	All Nursing	Johnson-Pawlson, 1996*
Other deficiencies	RN, LPN, NA, other staff	Harrington et al., 2000
Long-term care pressure ulcers cost	RN, NA	Hendrix et al., 2000

[a]as cited from Anderson (1998).

Licensed staff level, especially RNs, was found to be important in order to avoid the negative residents' outcomes (Anderson, 1998; Harrington et al., 2000; Ramsay et al., 1995). However, aides staffing levels were found to be important.

Note. RN = registered nurse; CNA = certified nursing assistant; NA = nursing assistant; LPN = licensed practical nurse

It is difficult to establish clear patterns with respect to the impact of staffing on quality of care, as summarized in Table 50–1. Researchers and clinicians have attempted to identify areas of care that are more sensitive to staffing, but there is not widespread agreement on this topic. Nevertheless, there is ample evidence to indicate that both the quality of care—in terms of clinical care and functional status—and the quality of resident life in nursing homes is impacted by both the number of staff and the training, experience, competence, and management of that staff.

Staffing Quality of Care Studies

Ramsay, Sainfort, and Zimmerman (1995) applied a structure–process–outcome model (Donabedian, 1980) and found that staffing levels appeared to affect care outcomes both directly and indirectly, through the process of care. Examples of the significant direct effects of elements of structure, process, and outcome are given in Figure 50–4.

Adopted with slight changes from Ramsay, J.D., Sainfort, F., & Zimmerman, D. (1995). An empirical test of the structure, process, and outcome quality paradigm using resident-based, nursing facility assessment data. *American Journal of Medical Quality, 10,* 63–75.

Louwe and Kramer (2001) used a qualitative approach to understand the ways in which attributes of staffing influence the quality of nursing home care. The variation in staffing ratios across institutions for licensed nursing staff was very large for both CNAs, which ranged from 1:3 to 1:27, and for licensed staff, which ranged from 1:6 to 1:49. The study also found that the quality of resident care in nursing facilities was affected by nursing staff levels. The staff in the understaffed facilities was simply physically unable to provide needed care and adequate monitoring to all residents. Quality of resident care, however, suffered in some acceptably staffed facilities but not in all. Management practices were related to the allocation of nursing staff and short staffing as well, which in turn affected the quality of resident care. One of the strongest factors that influenced the quality of resident care was the level of expertise, skills, and knowledge of the individual direct care nurses.

Minimum Staffing Levels

There is a federal regulation (OBRA, 1987) that each nursing home must employ a DON and have

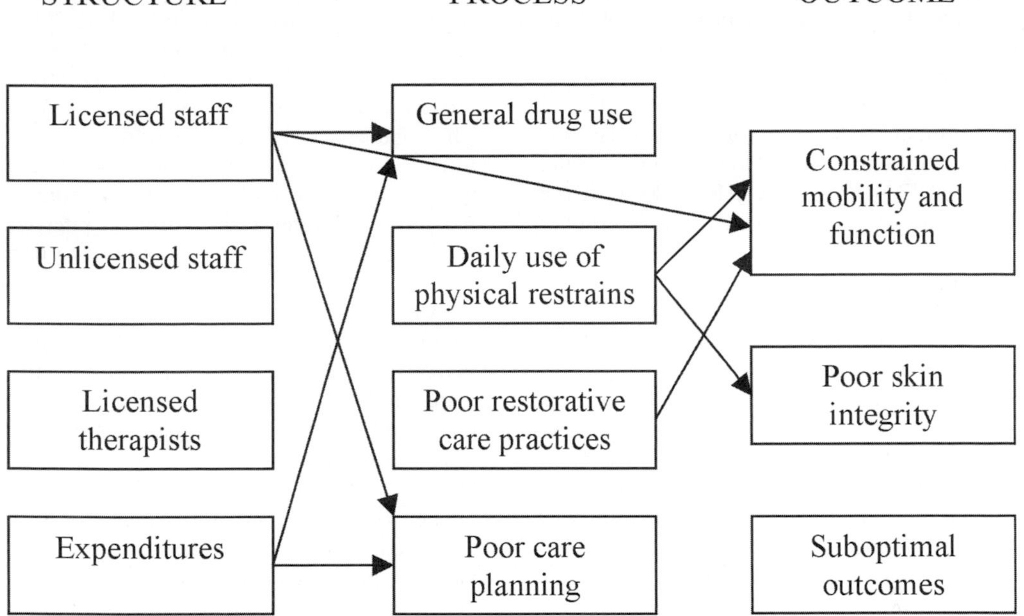

Figure 50–4. Significant direct effects of the structure, process, and outcome elements of nursing home care.

an RN on duty for 8 hr every day and an RN or LPN on duty around the clock. Beyond that, the federal government does not have any nationwide minimum staffing requirement. Thc requirement is simply that "sufficient nursing staff to provide nursing and related services to attain or maintain the highest practicable level of physical, mental, and psychosocial well-being of each resident" (OBRA, 1987, p. 185). The right to establish minimum staffing requirements is delegated to the states. However, two recent studies commissioned by CMS (2000, 2001) found that neither average staffing levels nor minimal state staffing requirements reflect acceptable staffing levels necessary to provide adequate quality of care. Moreover, even in the presence of minimum staffing requirements, there is no universal generally accepted staffing level.

Kramer and Fish (2001), based on their research, established staffing ratio thresholds below which quality of care could be compromised. Shnelle at al. (2001) utilized a modeling approach to estimate NA staffing levels necessary to provide a level of care consistent with Federal quality standards. Both researchers found that most nursing homes do not meet these thresholds. Shnelle et al. reported that "staffing levels that are similar to those reported in many of the nation's NFs (ratio of 8:1 day shift; 10:1 evening shift; 20:1 night shift) results in very long waits for services, and no assistance during meals for many residents, even when staff works hard" (p. 28). Chronic understaffing of the long-term care across the country was also reported by another study (White, Kim, & Deitz, 2001).

Understaffing, Nursing Shortage, and Turnover

There are several factors that contribute to understaffing in nursing homes. A major one is the acute shortage of nurses in the United States generally. This shortage has hit the nursing home industry particularly hard. Current vacancy rates in nursing homes reported by AHCA in 2002 were high for all staffing categories. On average, almost one in eight RN positions were vacant, whereas the LPN vacancy rate was one in seven and that of CNAs' was about two in twenty-five (ACHA, 2002). Industry surveys also reveal that it is becoming much more difficult to recruit staff in nursing homes (ACHA, 2002).

Turnover has been identified as one of the main contributors to the nursing home understaffing problem (Harrington & Swan, 2003). In the 1980s, studies in different regions of the United States reported annual turnover rates ranging from 40% to 400%. More recently reported turnover rates for nursing homes are around 70% to 100% annually. Half of all NAs leave their jobs within first 6 months, and many leave within weeks or even days after being hired (Banaszak-Holl & Hines, 1996; Riggs & Rantz, 2001). A recent AHCA (2004) survey revealed the following nationwide average turnover rates: staff RNs - 48.9%, ranging from 27.8% (District of Columbia) to 97.03% (Mississippi), LPNs - 48.9%, ranging from 21.2% (Vermont) to 74.3% (Arizona) and CNAs - 71.1% ranging from 20.7% (Hawaii) to 123.5 (Oklahoma).

Turnover also has been identified as a positive predictor of high workload and stress and low job satisfaction among nursing personnel (Bowers et al., 2003). Other research has shown that the turnover has a negative impact on the organization, which includes replacement costs (including training), lost productivity, compromised quality of care, and lowered staff morale (Banaszak-Holl & Hines, 1996; Cohen-Mansfield, 1997).

Clearly the problem of understaffing, job stress, and turnover is a circular one. Understaffing is reported to be the biggest problem related to the on-the-job stress in long-term care. People leave their jobs either for other facilities with lower workload or for jobs in other sectors (Jerrard, 2004).

Job stress, in turn, has been associated with higher turnover (Anderson, Issel, & McDaniel, 1997; Bloom, Alexander, & Nichols, 1992; Gaddy & Bechtel, 1995). Constant replacement of workers leads to a situation in which care is provided by inexperienced, incompetent, dissatisfied, and stressed-out workers. Therefore the understaffing, stress, and turnover create a closed loop (see Figure 50–5).

Factors Affecting Turnover

A rich source of insight into human factors issues affecting nursing home staff is the literature on factors affecting staff turnover, because turnover is often a result of factors that cause dissatisfaction in nursing home staff members. Banaszak-Holl and Hines (1996) examined job design and organizational factors that were associated with the turnover of NAs. The primary job design factors were the extent of aide training, involvement in determining and managing residents' care, and stability in aide's job assignments. Nursing homes in which NAs were asked to give advice or simply to discuss residents' care plans with aides had 30% lower turnover and the facilities in which "aides were involved in care plan meetings showed 50% higher retention" (Banaszak-Holl & Hines, 1996, p. 515). The organizational factors included proprietary status, facility size, payer mix, and resident case mix within the home and whether the home was a member of a larger chain. Turnover rates were 1.7 times higher in for-profit as opposed to nonprofit homes.

Another study (Anderson et al., 1997) divided factors associated with turnover into three categories: those that are within (a) minimal, (b) moderate, and (c) substantial facility control. The moderately controllable organizational variables were represented by profit margin (lower vs. higher), facility size, case mix, and occupancy rate. Higher profit margins were believed to permit improved work quality of the nursing staff; smaller facilities permitted closer working relations and greater attachment to the organization; more complicated case mix reflected residents whose conditions were unstable, which creates higher stress; and higher occupancy created higher demand on nursing staff "who must consistently work at higher capacity, potentially creating severe stress and strain" (Anderson et al, 1997, p. 72).

Important substantially controllable factors were skill mix, workload, administrative resources, and availability of RNs to residents. Skill mix, measured by ratio of RNs to other nursing staff, reflected the amount of professional supervisory attention available, which reduced uncertainty among CNAs. Higher workloads led to more physical demands and less ability of the staff to pay adequate attention to resident needs, which not only caused greater anxiety among staff but also led to a higher chance of medical (including medication) errors, an overlooked need for clinical intervention, and generally lower perceived job effectiveness. Availability of administrative personnel support functions and resources were supposed to increase attention to worker concerns and find better systems for doing work. The availability of RNs to residents, which represented clinical resources, resulted in more RN support available to CNAs.

The authors assumed that different categories of nursing staff can have different training and skills, professional and career orientations, social and economic situations, and alternative job opportunities and therefore might respond differently to

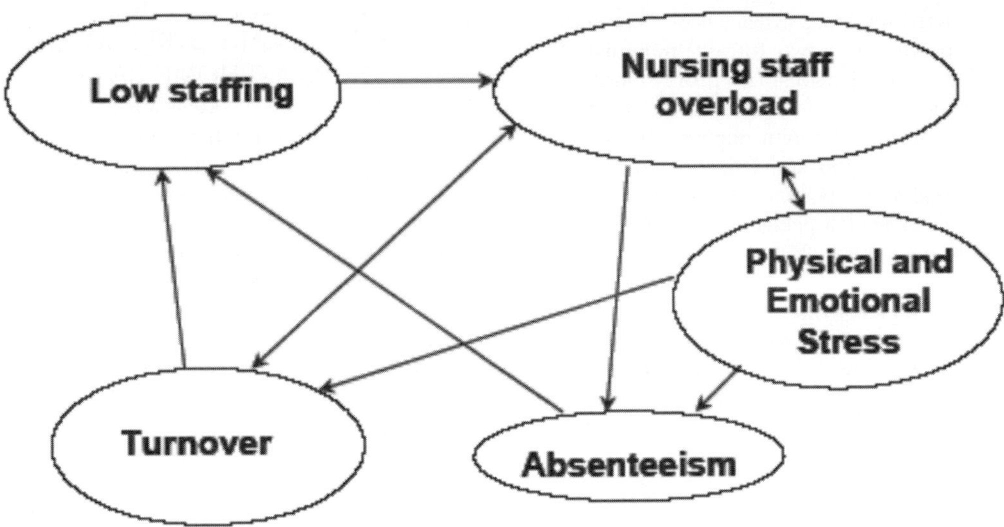

Figure 50–5. Interrelation of understaffing, stress, and turnover.

organizational characteristics. Thus, RNs, LPNs, and NAs were studied as separate caregiving groups.

Harrington and Swan (2003) studied facility's characteristics that affect nursing home staffing decisions and turnover. The facility characteristics were combined into two main groups: exogenous factors, reflecting extrinsic causes that are outside the facility and beyond its control; and endogenous factors, reflecting "organizational characteristics that are critical mediators of organizational decisions" (Harrington & Swan, 2003, p. 385; see also Banaszak-Holl, Zinn, & Mor, 1996; Brannon, Zinn, Mor, & Davis, 2002; Zinn, Mor, Castle, Intrator, & Brannon, 1999) and that are related to the nursing home itself and are within its control. The endogenous variables considered in the study were: total RN nursing hours per resident day, total nursing turnover rate, and resident ADL dependency. The authors found that lower turnover rates were more likely to be in not-for-profit facilities, facilities with lower proportion of Medicaid residents, rural facilities, smaller facilities, and facilities that had lower occupancy rates.

Brannon et al. (2002) explored factors that differentiated facilities with very high and very low turnover rates using a three-category taxonomy (Cotton & Tuttle, 1986): job factors, organizational factors, and marked environment factors. Among the job factors, leadership (having supervisors trained in management) was associated with a lower likelihood of being in the low turnover group of facilities. The authors reasoned that trained management was more likely to set expectations and provide feedback, which urged poor-performance workers to quit.

Among organizational factors, RN turnover had a significant positive linear relation with NA turnover. Administrator span of control (defined as number of departments reported to an administrator) and unionized environment were associated with being in the lower turnover group. Facilities that were a training site, had for-profit ownership status, or were part of a multifacility firm were more likely to be in the high-turnover rate group.

Bowers et al. (2003) used a qualitative approach to study factors affecting aides' decisions to leave their jobs. They found that the most important contributors to this decision were aide dissatisfaction; practices related to staff absenteeism, training, and orientation; facility's staffing policies; and low compensation. The findings on dissatisfaction were particularly interesting. One of the most common feelings shared by the aides was dissatisfaction, specifically the feeling of being unappreciated and unvalued. The feeling of being unappreciated was caused by the discrepancy between organization's rhetoric and its policies and practices, i.e., the perception that what was said by managers and supervisors contradicted what was actually done. The messages were perceived in two forms: "minimizing" and "leveling" at the personal and professional levels. Minimizing reflected assumptions about CNAs as a group. Professional minimizing was the lack of respect of an aide's job and failure to recognize their knowledge and experience. Personal minimizing reflected a "general disparagement of aide's character—the belief that they lack integrity, intelligence, and commitment (Bowers et al., 2003, p. 39).

Leveling resulted from applying the general assumptions of minimizing to individual aides. The professional leveling was defined as management's failure to distinguish between individual aides based on their occupational skills. The personal leveling occurred when the management and supervisors failed to distinguish CNAs in terms of their personal features: honesty, intelligence, and the level of commitment. Although facility managers and supervisors made claims that they respected CNAs and recognized their efforts, their actual actions, based on facility's policies and procedures and instances of personal interaction, often indicated the opposite.

Nursing home policies for determining staffing levels and reward systems for CNAs were found to lack the respect for CNAs and the work they do. This was perceived by the aides through facilities staff policies: rotating work assignments, use of pool staff, and constant recruiting of new personnel without an attempt to find "the right kind of person." It was not the low salary but the lack of a differential approach to rewarding employees that contributed to the aides' feeling unvalued. Both existing monetary and nonmonetary reward systems neither recognized nor accounted for aides' individual performance and level of experience. Administrators often perceived compensation issue as "not enough," whereas aides' viewed it more like "not fair" and "no recognition."

The study also found that supervisory relationships were central to the problem of turnover. Interactions of aides with their supervisors often made them feel personally and professionally dismissed. Aides reported that they felt invisible to the RNs, not respected by them, and treated as stupid.

OTHER PRIMARY STRESSORS IN NURSING HOME CARE

The discussion thus far has provided ample evidence to support the contention that caring for the impaired elderly in nursing homes is a very stressful job. A very simple taxonomy to categorize the sources of stress is to divide them into (a) physical demands and stress sources and (b) emotional sources of stress. We discuss each of these types in the following, with an example of how the sources of stress have been addressed in research and practical guidelines.

Physical Demands and Stress: The Safety Factor

A comprehensive review of nursing home quality has noted the importance of safety as a primary consideration—to both residents and caregivers (Wunderlich & Kohler, 2000). Safety must be built into the nursing home culture. The lack of safety approach leads to both resident and staff injury and stress (Kohn, Corringan, & Sonaldson, 1999). The culture of safety is associated with an increase in patient safety and employee satisfaction (Cooper, 2000). Implementation of a supportive culture within the nursing homes will result in increased staff satisfaction and increased staff commitment, and consequently will improve staff retention and resident outcomes (Riggs & Rantz, 2001). For additional information on physical ergonomics in health care, see the chapters on physical ergonomics in health care and evidence-based strategies for safe patient handling and movement in this volume.

OSHA Regulations

The OSHA[2] has clearly recognized the importance of maintaining a safe nursing home workplace, and it has been a leader in both research and the development of guidelines on this critical topic. Nursing homes are among the top three U.S. private industries in terms of the number and the incidence rates of nonfatal occupational injuries and illnesses, with

[2]The OSHA is located within the U.S. Department of Labor. OSHA's mission is to assure the safety and health of America's workers by setting and enforcing standards; providing training, outreach, and education; establishing partnerships; and encouraging continual improvement in workplace safety and health (see the OSHA Web site: http://www.osha.gov).

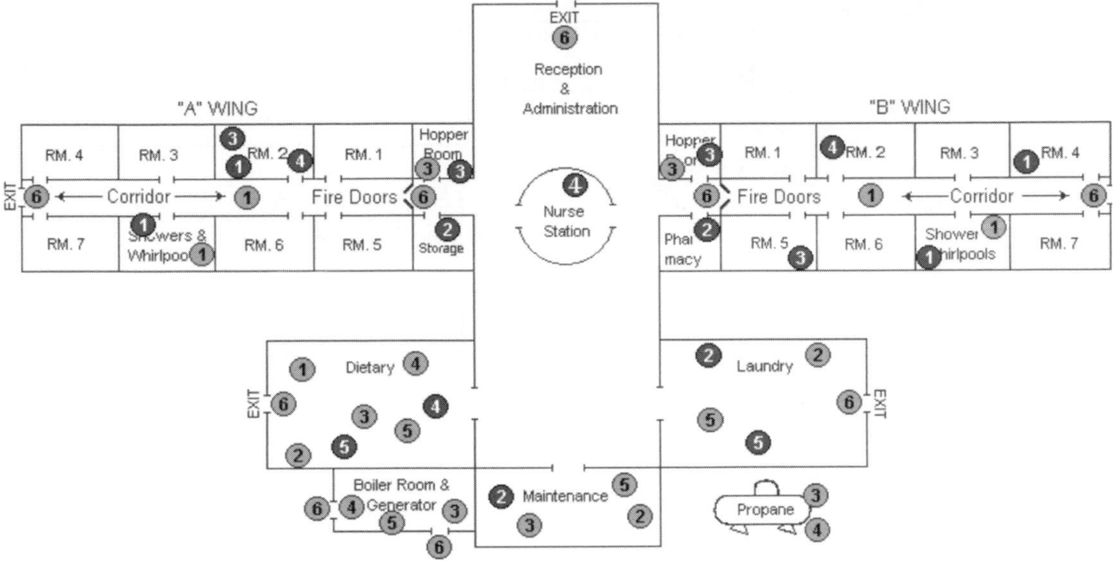

Figure 50–6. Anatomy of a nursing home with potential hazards.

more than 200,000 injuries and illnesses in nursing homes in 2000 and an incident rate of almost one in seven during that same year (BLS, 2004). A recent BLS study noted that NAs, orderlies, attendants, and RNs have higher rates of work-related muscular-skeletal disorders, resulting in more time away from work than any other occupation (Fragala, 2001).

Caring for the impaired elderly is physically demanding. Nursing home residents rely on their caregivers' assistance in everyday activities and, in some cases, are entirely reliant on caregivers for mobility, which increases the amount of manual lifting of nursing home residents and is associated with an increased risk of occupational injury, especially back injuries. The complexity of a nursing home layout and range of provided care and services also creates numerous potential health and safety hazards. The OSHA (2003) has identified more than ten types of potential hazards that exist in a nursing home facility, with every area of a nursing home associated with at least one hazard (as portrayed in Figure 50–6). Fore example, a lifting hazard is a potential health hazard that caregivers face. The caregivers face the need for lifting on an everyday, routine basis. They might need to lift a resident who fell or help a resident with some ADL that he or she cannot do otherwise because of poor health (moving from a bed to a stretcher, bathing, changing posture from lying to sitting, and so on). Besides the residents, the caregivers often have to lift and carry things and belongings, some equipment or heavy tools, or even laundry bags. Very often workers do this lifting without mechanical or human assistance. The lack of available mechanical assistance could result from limited funding or inappropriate facility layout. The lack of human help might result from understaffing, when basically there is no one available to help, or inefficient staff allocations, such as the absence of team work assignments.

To educate nursing home caregivers on identifying the hazards and possible solutions for preventing injuries resulting from those hazards, the OSHA has developed a manual that provides guidance information for developing a comprehensive safety and health program. The manual is called "Nursing Home E-Tool" and includes recommendations for good industry practice. The e-tool addresses identification of hazards and prevention of accidents in the following areas: blood borne pathogens, ergonomics, dietary, laundry, maintenance, nurses' station, pharmacy, tuberculosis, housekeeping, whirlpool/shower, and workplace violence.

Reprinted from the OSHA Web site: http://www.osha.gov/SLTC/nursinghome/nursinghome_map_gif.html.

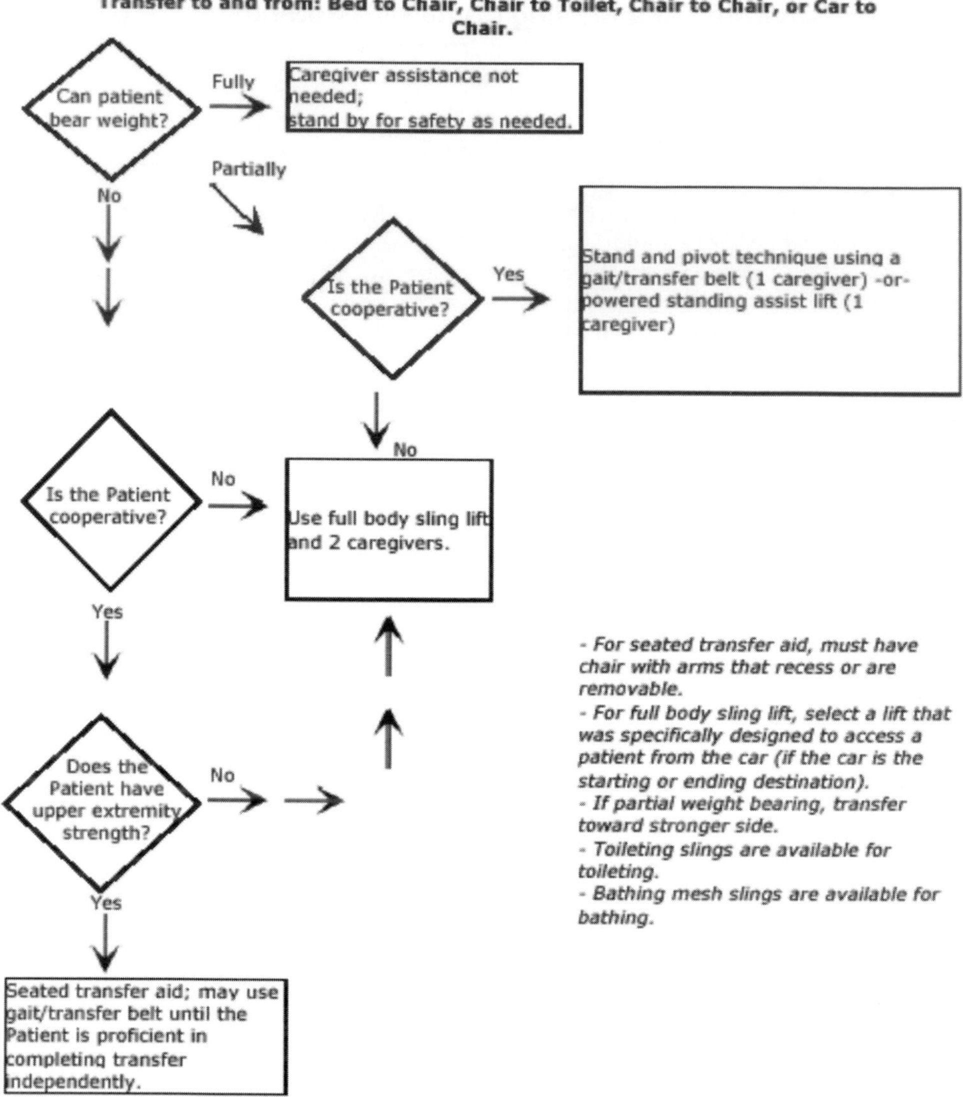

Figure 50–7. A flow chart of patient transfer.

To specifically address ergonomics issues in nursing homes, OSHA (2003) released detailed ergonomics guidelines for the nursing home industry, with the intent of reducing caregiver injuries. For example, the guidelines for MSDs (musculoskeletal disorders) include low back pain, sciatica, rotator cuff injuries, epicondylitis, carpal tunnel syndrome, and others (OSHA, 2003).

Ergonomic risk is also associated with the tasks that are done with repetitive motions—often in awkward postures—using a great deal of force and lifting heavy objects. The risk is greatly increased in the presence of overexertion, multiple lifts, lifting with no help, lifting uncooperative and confused residents, increased resident weight, unrealistic expectation about employee physical capabilities, increased distance to be moved, and ineffective employee training in body mechanics and proper lifting techniques.

An example of an OSHA flowchart, in the area of repositioning residents from sitting to standing is provided in Figure 50–7. The guidelines also address lateral transfer to and from bed, stretcher, and trolley; transfer to and from chair and stretcher;

reposition in bed (side-to-side and up); reposition in chair (wheelchair and dependency chair); transfer up from the floor.

Reprinted from the OSHA Web site: http://www.osha.gov/ergonomics/guidelines/nursinghome/final_nh_guidelines.html

OSHA regulations are also an important part of the nursing home industry. The regulations emphasize the necessity of addressing health and safety hazards in nursing homes in management leadership and employee participation, workplace analysis, hazard prevention and control medical management, and training. The guidelines are not mandatory and their implementation is voluntary. The AHCA and the American Association of Homes and Services for the Aging strongly support the guidelines, noting that they demonstrate an understanding of the complexities involved in applying ergonomics to everyday nursing tasks (Croasmun, 2003). However, the OSHA guidelines also raise cost questions, which have not gone unnoticed by the industry, regarding the unwillingness of Medicaid and Medicare to increase reimbursement to cover the substantial expense of nursing home care. For example, the AHCA estimated the total cost for all LTC facilities that adopt the OSHA recommendations at $1.2 billion for the first year (Schwartz, 2000).

In general, it is safe to say that government policy and educational programs, as well as regulations, have placed primary emphasis on how to mitigate the physical demands and reduce the level of physical injuries to nursing home staff. Other stressors, such as emotional factors, have not garnered as much attention, yet they can be just as or even more debilitating in terms of making the nursing home work environment unfavorable.

Emotional Stressors

Caring for the impaired elderly is an extremely emotionally draining and sometimes depressing occupation. Direct caregivers must function in an environment in which they observe constant and sometimes inevitable physical decline, cognitive impairment, and depression and grieving in the residents they serve. Yet most of these caregivers are toiling in this low-paid and demanding profession because they have a strong emotional attachment to the elderly in general and their residents in particular. The bonds they form with their residents are difficult to exaggerate. All this occurs in a work structure characterized by a high level of uncertainty, instability, and variability in the process of care. Bowers and colleagues summarized the problem well:

> The frontline caregiving staffs are devoted professionals, who struggle to provide high quality of care and life to the residents. They suffer from the fact that they are unable to give enough attention to the residents and communicate with them due to the lack of time and work overload. Many nurses and aides quit their careers in the long-term care or leave for other facilities with lower workload because of the feeling of inability to satisfy residents' needs.

The Example of Stressful Resident Behavior

An example of the emotional stress faced by caregivers is the difficulty of treating residents with behavioral symptoms, such as such as aggression, wandering, yelling and screaming, resisting care, and so forth. The suggested strategies to reduce stress in addressing this problem have general applicability to most other sources of stress as well. AHCA (2004) has estimated that almost one third of nursing home residents manifest disruptive behavior symptoms. The use of the term *manifest* is important, because in almost all cases the behavioral symptoms are, in fact, manifestations of other, sometimes undetected, issues with which the resident is dealing. These issues may be clinical or nonclinical in nature and source, and they may be exacerbated by the demented state of the resident. They can be temporary, as in delirium, or signs of a more permanent impairment.

There are many causes of behavioral manifestations, including deep personal loss of a relationship, loss of control over personal care and environment, loss of independence, and a feeling of hopelessness, helplessness, and frustration (Taylor et al., 1990). Also, often Alzheimer's residents do not like changes. If nursing staff fails to maintain stability of resident assignments because of rotating assignments or high turnover rates, residents may become agitated and start resisting care. Aggressive behavior also can occur as the result of a misunderstanding of staff's intentions, feeling of powerless, and other causes, or because the residents feel their privacy is being invaded.

Residents with behavioral symptoms put an extra burden on their caregivers and can cause stress (Banaszak-Holl, et al., 1996; Rodney, 2000). Not only the aggression itself, but "perceiving the possibility of aggressive behavior by the resident as threatening was found related to a high stress level in nurses" (Rodney, 2000, p. 172).

Reducing the Stress Caused by Resident Behavior

Efforts to reduce the stress caused by resident behavior symptoms have focused on two strategies: (a) reducing the behavioral symptoms themselves and (b) helping staff members cope with the behavioral symptoms. Preventive strategies are preferred, of course, because they also are beneficial for the residents, improving their quality of life.

Possible system solutions to help prevent and manage behavioral symptoms focus on both the process of care and the structure of the nursing home environment. For an example of the best practice recommendations see, for example, Taylor, et al. (1990) and American Geriatric Society and American Association for Geriatric Psychiatry (2003). The recommendations, however, focus mostly on the clinical aspects of resident care. Thus, we summarize both our own general recommendations and the recommendations from the previously mentioned studies, with an emphasis on organizational process of care.

Several other researchers have focused on the prevention of disruptive behavior in residents (Taylor et al., 1990). Reducing the stress caused during caregiving itself will decrease disruptive behavior as well. Suggested strategies include making sure that demands of care are matched with residents' abilities to eliminate confusion; watching for signs of resident anxiety; and reducing distractions such as noise, light, and other interference from other people. Taylor et al. noted that good communication with the resident is especially important and recommended that three communication factors play an important role: approach and verbal and nonverbal messages.

Disruptive behavior symptoms cannot be eliminated completely from the nursing home environment; many of them are often unavoidable because of their nature. Therefore, the process of care and the organizational structure of a nursing home should be able to protect the staff from the emotional and physical stress caused by the behavior. Researchers have studied a variety of strategies for dealing with the stress caused by disruptive behaviors:

- Environmental and physical layout solutions, such as wider doors to accommodate wheelchairs and sufficient lighting.
- Visual and audio aide sound detector devices that turn on soft, calm music and shaped and colored informational signs on a resident's bed to provide cues to staff.
- Training on dealing with disruptive behavior, focusing on verbal and nonverbal aspects of communication with residents with decreased cognitive skills, how to react, how to stay calm and not take residents' actions and words personally, and frequent reinforcement to aides that their work is valuable.
- Establishing more consistent care through a comprehensive care plan for each resident (Berg & Hallberg, 1994).
- More involvement of CNAs in care plan development to ensure that they perceive that their input in the process of care is appreciated and that their experience and expertise is valued (Bowers et al., 2003).
- Job design that allows for involving aides in decision making, which has the potential to enhance motivation and job satisfaction among NAs, reduce turnover, and improve staff retention (Riggs & Rantz, 2001).

Structural changes can also be important in improving stress. The following strategies have been suggested by researchers:

- Encourage teamwork and emphasize multidisciplinary team input, which has been shown to reduce caregivers' physical and emotional stress and burnout, increase their job satisfaction, and improve resident outcomes (Park et al., 2004).
- Empower NAs in their work by involving them in decision making, providing some flexibility in their work, and permitting them to help each other out (Banaszak-Holl & Hines, 1996; Riggs & Rantz, 2001).
- Provide professional and supervisory support to the staff from mental health specialists, physicians, gerontologist, and other health care professionals. Nursing staff, which cares for residents with severe dementia, have less stress if they receive systematic clinical supervision along with implementing careful resident care planning (Berg & Hallberg, 1994).
- Maintain stability of resident assignments to help residents and staff get to know each other better and learn to read signs

of anxiety more effectively. Frequent rotations convey a message that the aides are undistinguishable from one another in terms of their professional skills (Bowers et al., 2003).

- Provide decent working conditions to the staff and try to eliminate overwhelming assignments.

Changing the Culture

Increasingly, researchers and practitioners alike are beginning to realize the magnitude of the cultural change needed to reduce the stress of nursing home staff. Researchers have found that the workload itself is not the driving factor that makes the nursing staff leave, because caring for residents is one of the sources of job satisfaction for them (Bowers et al., 2003). Nurses even admit that they prefer to work understaffed rather that surrounded with noncommitted colleagues. It is mainly lack of respect and support embedded in the nursing home's culture that accounts for aides' job stress and job dissatisfaction. The direct caregivers should be constantly reminded that their work is appreciated and their experience and expertise valued.

Riggs and Rantz (2001) have developed a model of staff support to improve retention in long-term care. They stress four job characteristics that reflect a supportive environment:

(1) task identity, the completion of a piece of work in its entirety; (2) task significance, the impact of the job on others; (3) skill variety, performance of a variety of activities using different skills and talents; and (4) autonomy, freedom, independence, and discretion allowed the employee in decision making. (p. 47)

The need of a supportive organizational climate is a prevailing theme in long-term care. Support needs to be provided for all categories of nursing staff and should address informational, technological, and social needs of staff. Many aides and nurses are single mothers who have many personal problems in addition to high job stress. Support and help in dealing with personal issues will improve their life satisfaction, reduce absenteeism, and improve the quality of their work (Riggs & Rantz, 2001).

CONCLUSION

It has been long-recognized that providing care to the impaired elderly in nursing homes is one of the most difficult occupations in all of health care: It is physically demanding, with an extremely high injury rate; it is emotionally draining, with workers surrounded by grief, depression, and physical and cognitive deterioration; it is among the lowest paid occupations in health care; and dedicated staff members often toil under conditions of inadequate help, task uncertainty, and a lack of support and respect. Yet many in the field have an almost messianic passion for the work, which in itself leads to sometimes profound angst because of a perception of failed mission. We have tried to identify the primary sources of stress for these committed staff members and to chronicle and summarize the abundant but complex knowledge on what seems to work best in alleviating stress. Certainly staffing considerations play a fundamental role here, and to some extent alleviating understaffing and its consequences are a requisite for meaningful progress in addressing other human factors issues. Beyond that, there are clearly defined physical and safety factors that need to be addressed and less clearly defined but no less critical organizational, structural, "involvement," and support and respect factors that need to be further studied and translated into practice to reduce the inherent stress faced by nursing home staff.

References

Alexander, J. W., & Randolph, W. A. (1985). The fit between technology and structure as a predictor of performance in nursing subunits. *Academy of Management Journal, 28,* 844–859.

American Geriatric Society, & American Association for Geriatric Psychiatry. (2003). Consensus statement on improving the quality of mental health care in U.S. nursing homes: Management of depression and behavioral symptoms associated with dementia. *Journal of the American Geriatrics Society, 51,* 1287–1298.

American Health Care Association. (2004a). *Nursing facility patients by payer—Percentage of patients, CMS OSCAR data current surveys.* Retrieved August 18, 2004, from the World Wide Web: http://www.ahca.org/research/oscar/rpt_special_care_beds_200406.pdf

American Health Care Association. (2004b). *Nursing facility special care beds.* Retrieved August 18, 2004, from the World Wide Web: http://www.ahca.org/research/oscar/rpt_payer_200406.pdf

American Health Care Association. (2004c). *Nursing facility total, average and median number of patients per facility and ADL dependence.* Retrieved August 17, 2004, from the World Wide Web: http://www.ahca.org/research/index.html

American Health Care Association (AHCA) (2002). *Results of 2002 AHCA survey of nursing staff vacancy and turnover in nursing homes.* Retrieved on August 18, 2004, from http://www.ahca.org/research/rpt_vts2002_final.pdf

Anderson, R. A., Hsieh, P. C., & Su, H. F. (1998). Resource allocation and resident outcomes in nursing homes: Comparisons between the best and worst. *Research in Nursing and Health; 21,* 297–313.

Anderson, R. A., Issel, L. M., & McDaniel, R. R. (1997). Nursing staff turnover in nursing homes: A new look. *Public Administration Quarterly, 21,* 69–96.

Banaszak-Holl, J., & Hines, M. A. (1996). Factors associated with nursing home turnover. *The Gerontologist, 36,* 512–517.

Banaszak-Holl, J., Zinn, J. S., & Mor, V. (1996). The impact of market and organizational characteristics on nursing care facility innovation: A resource dependency perspective. *Health Services Research, 31,* 97–118.

Berg, A., & Hallberg, I. R. (1994). Nurses' creativity, tedium and burnout during 1 year of clinical supervision and implementation of individually planned nursing care: Comparisons between a ward for severely demented patients and a similar control ward. *Journal of Advanced Nursing, 20,* 742.

Berlowitz, D. R., Anderson, J. J., Brandeis, G. H., Lehner, L. A., Brand, H. K., Ash, A. S., & Moskowitz, M. A. (1999). Pressure ulcer development in the VA: characteristics of nursing homes providing best care. *American Journal of Medical Quality, 14,* 39–44.

Bliesmer, M. M., Smayling, M., Kane, R., & Shannon, I. (1992). The relationship between organization and quality. *The Gerontologist, 32,* 360–366.

Bliesmer, M. M., Smayling, M., Kane, R. L., & Shannon, I. (1998). The relationship between nursing staffing levels and nursing home outcomes. *Journal of Aging & Health, 10,* 351–371.

Bloom, J. R., Alexander, J. A., & Nichols, B. A.: (1992). The effect of social organization of work on the voluntary turnover rate of hospital nurses in the United States. *Social Science and Medicine, 12,* 1413–1224.

Board of Nursing, Department of Regulations and Licensing, State of Wisconsin. (2003). *Wisconsin statures and administrative code relating to the practice of nursing.* Retrieved July 16, 2004, from the World Wide Web: http://Drl.Wi.Gov/Boards/Nur/Code/Codebook.Pdf

Bowers, B. J., Esmond, S., & Jacobson, N. (2000). The relationship between staffing and quality in long-term care facilities: Exploring the views of nurse aides. *Journal of Nursing Care Quality, 14,* 55–64.

Bowers, B. J., Lauring, C., Jacobson, N. (2001). How nurses manage time and work in long-term care. *Journal of Advanced Nursing. 33*(4), 484-491.

Bowers, B. J., Esmond, S., & Jacobson, N. (2003). Turnover reinterpreted: CNAs talk about why they leave. *Journal of Gerontological Nursing, 29,* 36–39.

Brannon, D., Zinn, J. S., Mor, V., & Davis, J. (2002). An exploration of job, organizational, and environmental factors associated with high and low nursing assistant turnover. *The Gerontologist, 41,* 159–168.

Bureau of Labor Statistics, U.S. Department of Labor. (2004). *Occupational outlook handbook, 2004–05 edition.* Retrieved July 14, 2004, from the World Wide Web: http://www.bls.gov/oco/ocos083.htm

Castle, N. G. (2002, October). Characteristics of nursing homes that are restraint free. *Gerontologist, 38,* 181–188.

Center for Medicare and Medicaid Services. (2000). *Report to Congress: Appropriateness of minimum nurse staffing ratios in nursing homes.* Washington, DC: U.S. Department of Health and Human Services. Retrieved August 25, 2004, from the World Wide Web: http://www.cms.hhs.gov/medicaid/reports/rp700hmp.asp

Center for Medicare and Medicaid Services. (2001). *Appropriateness of minimum nurse staffing ratios in nursing homes. Phase II: Final report.* Washington, DC: U.S. Department of Health and Human Services. Retrieved August 25, 2004, from the World Wide Web: http://www.cms.hhs.gov/medicaid/reports/rp1201home.asp?

Cohen-Mansfield, J. (1997). Turnover among nursing home staff: A review. *Nursing Management, 28,* 59–62.

Cooper, M. D. (2000). Towards a model of safety culture. *Safety Science, 36,* 111–136.

Cotton, J., & Tuttle, J. (1986). Employee turnover: A meta-analysis and review with implication for research. *Academy of Management Review, 11,* 55–70.

Croasmun, J. (2003). *OSHA releases final nursing home guidelines.* Retrieved June 20, 2004, from the World Wide Web: http://www.ergoweb.com/news/detail.cfm?id'700

Davis, M. A. (1991). On nursing home quality: A review and analysis. *Medical Care Review, 48,* 129–166.

Donabedian, A. (1980). *Explorations in quality assessment and monitoring: The definitions of quality and approaches to its assessment* (Vol. 1). Ann Arbor, MI, Health Administration Press.

Farmer, B. C. (1996). *A nursing home and its organizational structure: An ethnography.* Westport, CT: Auburn House.

Fragala, G. (2001). After the ergonomics rule, what next? *Nursing Homes, 50,* 22–25.

Gaddy, T., & Bechtel, A. (1995). Nonlicensed employee turnover in a long-term care facility. *Health Care Supervisor, 13,* 54–60.

Graber, D. R. (1993). The influence of nursing home characteristics and task environment on complaints and survey performance. *Dissertation Abstracts International* (Publication no. AAT 9324038)

Harrington, C. (2001). Residential nursing facilities in the United States. *British Medical Journal, 323,* 507–510.

Harrington, C., & Swan, J. H. (2003). Nursing home staffing, turnover, and case mix. *Medical Care Research and Review, 60,* 366–392.

Harrington, C., Zimmerman, D., Karon, S. L., Robinson, J., & Beutel, P. (2000). Nursing home staffing and its relationship to deficiencies. *Journal of Gerontology: Social Sciences, 55,* S278–S287. Health Care Financing Administration (HSFA), (1994). Medicare and Medicaid Programs: Survey, Certification and Enforcement of Skilled Nursing Facilities and Nursing Facilities, Final Rule. *Federal Register, 59*(217). (42 CFR, parts 401and 498.) Accessed June 8, 2006 from http://www.medicare.gov/NHCompare/Static/Related/DataCollection.asp?dest=NAV%7CHome%7CDataDetails%7CDataCollection.

Health Resources and Services Administration (HRSA), Division of Nursing. (2001, November). *Nurse education and practice. First report to the Secretary of Health and Human Services and the Congress.* U.S. Department of Health and Human Services, Health Resources and Services Administration, Bureau of Health Professions, Division of Nursing, Rockville, Maryland.

Institute of Medicine (IOM) (1986). Takeuchi, J., Burke, R., & McGeary, M. (Eds.) *Improving the quality of care in nursing homes.* Washington, DC: National Academy Press.

Jerrard, J. (2004). Stressed staff: The negative effects of working in LTC—plus ideas to help staff manage stress. *Caring for the Ages, 4,* 14–15. Retrieved August 5, 2004, from the World Wide Web: http://www.amda.com/caring/october2003/staff.htm

Kohn, L. T., Corringan, J. M., & Sonaldson, M. S. (Eds.). (1999). *To err is human: Building a safer health system.* Washington, DC: National Academy Press.

Kane, R. L., Bell, R., Riegler, S., Wilson, A., & Keeler, E. (1983). Predicting the outcomes of nursing home patients. *The Gerontologist, 23,* 200–206.

Kayser-Jones, J. (2002). The experience of dying: An ethnographic nursing home study. *The Gerontologist, 42,* 11–19.

Kerrison, S. H., & Pollock, A. M. (2001). Caring for older people in the private sector in England. *British Medical Journal, 323,* 566–569.

Kramer, A. M., & Fish, R. (2001). Chapter 2: The relationship between nurse staffing levels and the quality of nursing home care. In Center for Medicare and Medicaid Services, *Appropriateness of minimum nurse staffing ratios in nursing homes. Phase II: Final report.* Washington, DC: U.S. Department of Health and Human Services. Retrieved August 25, 2004, from the World Wide Web: http://www.cms.hhs.gov/medicaid/reports/rp1201home.asp?

Louwe, H., & Kramer, A. (2001). Chapter 6: Case studies of nursing facility staffing issues and quality of care. In Center for Medicare and Medicaid Services, *Appropriateness of minimum nurse staffing ratios in nursing homes. Phase II: Final report.* Washington, DC: U.S. Department of Health and Human Services. Retrieved August 25, 2004, from the World Wide Web: http://www.cms.hhs.gov/medicaid/reports/rp1201home.asp?

Mueller, C., & Cain, H. (2002). Comprehensive management of urinary incontinence through quality improvement efforts. *Geriatric Nursing, 23,* 82–87.

Munroe, D. J. (1990). The influence of registered nursing staffing on the quality of nursing home care. *Research in Nursing and Health, 13,* 263–270.

National Institute of Nursing Research. (1994). Quality in nursing home care. In *Long-term care for older persons: Clinical problems and issues.* Washington, DC: National Institute of Health. Retrieved June 15, 2004, from the World Wide Web: http://ninr.nih.gov/ninr/research/vol3/TOC.htm

Occupational Safety and Health Administration. (2003). *Safety and health topics: Ergonomics: Guidelines for nursing homes.* Washington, DC: U.S. Department of Labor. Retrieved August, 17, 2004, from the World Wide Web: http://www.osha.gov/ergonomics/guidelines/nursinghome/index.html

Occupational Safety and Health Administration. (YEAR). *Nursing home e-tool.* Washington, DC: U.S. Department of Labor. Retrieved MONTH DAY, YEAR, from the World Wide Web: http://www.osha.gov/SLTC/etools/nursinghome/index.html

Omnibus Budget Reconciliation Act of 1987 (OBRA). Public Law 100-203. Subtitle C: Nursing Home Reform. Signed by President Regan, Washington, DC, December 22, 1987.

Park, M., Delaney, C., Maas, M., & Reed, D. (2004). Nursing and health care management and policy. Using a nursing minimum data set with older patients with dementia in an acute care setting. *Journal of Advanced Nursing, 47,* 329.

Perrow, C. (1967). A framework for comparative analysis of organizations. *American Sociological Review, 32,* 194–208.

Phillips, C. D., Hawes, C., & Fries, B. E. (1993). Reducing the use of physical restraints in the home: Will it increase costs? *American Journal of Public Health, 83,* 342–348.

Phillips, C. D., Hawes, C., Mor, V., Fries, B. E., Morris, J. N., & Nennstiel, M. E. (1996). Facility and area variation affecting the use of physical restraints in nursing homes. *Medical Care, 34,* 1149–1162.

Ramsay, J. D., Sainfort, F., & Zimmerman, D. (1995). An empirical test of the structure, process, and outcome quality paradigm using resident-based, nursing facility assessment data. *American Journal of Medical Quality, 10,* 63–75.

Riggs, C. J., & Rantz, M. J. (2001). A model of staff support to improve retention in long-term care. *Nursing Administration Quarterly, 25,* 43–55.

Rodney, V. (2000). Nurse stress associated with aggression in people with dementia: Its relationship to hardiness, cognitive appraisal and coping. *Journal of Advanced Nursing, 31,* 172–180.

Scanlon, W. J. (2001, May 17). *Nursing workforce: Recruitment and retention of nurses and nurse aides is a growing concern: Testimony before the committee on health, education, labor and pensions, U.S. Senate.* U.S. General Accounting Office, GAO-01-750T.

Schnelle, J. F., Simmons, S. F., Cretin, S., Feuerberg, M., Hu, Y.-F., & Keeler, E. (2001). Chapter 3: Minimum nurse aide staffing required to implement best practice care in nursing facilities. In Center for Medicare and Medicaid Services, *Appropriateness of minimum nurse staffing ratios in nursing homes. Phase II: Final report.* Washington, DC: U.S. Department of Health and Human Services. Retrieved August 25, 2004, from the World Wide Web: http://www.cms.hhs.gov/medicaid/reports/rp1201home.asp?

Schwartz, R. M. (2000). Ergonomics nurses vs. nursing homes. *Nursing Homes, 49,* 9.

Shapiro, E. (2000). Community and long-term facility care in Canada. *Journal of Health and Human Services Administration, 22,* 436–451.

Stone, R. I . (2001). Research on frontline workers in long-term care. *Generations, 25,* 49–57.

Takeuchi, J., Burke, R., & McGeary, M. (Eds.). (1986). *Improving the quality of care in nursing homes.* Washington, DC: National Academy Press.

Taylor, J. A., Ray, W. A., & Meador, K. G. (1990). *Managing behavioral symptoms in nursing home residents: A manual for nursing home staff.* Nashville, TN: Vanderbilt University School of Medicine, Department of Preventive Medicine.

U.S. Congressional Budget Office. (1999). *CBO memorandum: Projections of expenditures for long-term care services for the elderly.* Retrieved August 18, 2004, from the World Wide Web: http://www.cbo.gov/showdoc.cfm?index'1123&sequence'0

U.S. General Accountability Office. (2000). *Report to congressional committees: Skilled nursing facilities. Available data show average nursing staff time changed little after Medicare payment increase.* Washington, DC: U.S. Government Printing Office. Retrieved June 16, 2004, from the World Wide Web: http://www.gao.gov/new.items/d03176.pdf

U.S. Social Security Administration. (2003). Requirements for, and assuring quality of care in, skilled nursing facilities. In *The Social Security Act as amended with related enhancements* (Vol. 1, Sec. 1819) [42 U.S.C. 1395i–3]. Retrieved July 15, 2004, from the World Wide Web: http://www.ssa.gov/OP_Home/ssact/title18/1819.htm#fn069

Waxman, H. W., Carner, E. A., & Berkenstock, G. (1984). Job turnover and job satisfaction among nursing home aides. *The Gerontologist, 24,* 503–509.

White, A., Kim, L., & Deitz, D. (2001). Chapter 8: Adequacy of nursing workforce to meet higher minimum nurse staffing requirements. In Center for Medicare and Medicaid Services, *Appropriateness of minimum nurse staffing ratios in nursing homes.*

Phase II: Final report. Washington, DC: U.S. Department of Health and Human Services. Retrieved August 25, 2004, from the World Wide Web: http://www.cms.hhs.gov/medicaid/reports/rp1201home.asp?

Wunderlich, G. S., & Kohler, P. O. (Eds.). (2000). *Improving the quality of long-term care.* Washington, DC: National Academy Press.

Zimmerman, S., Gruber-Baldini, A. L., Hebel, J. R., Sloane, P. D., & Magaziner, J. (2002). Nursing home facility risk factors for infection and hospitalization: Importance of registered nurse turnover, administration, and social factors. *Journal of the American Geriatrics Society, 50,* 1987–1995.

Zinn, J. S., Mor, C., Castle, N., Intrator O., & Brannon D. (1999). Organizational and environmental factors associated with nursing home participations in managed care. *Health Services Research, 33,* 1753–1767.

• 51 •

HUMAN FACTORS AND ERGONOMICS IN PRIMARY CARE

John W. Beasley, Kamisha Hamilton Escoto, and Ben-Tzion Karsh
University of Wisconsin, Madison

DEFINITION OF PRIMARY CARE

Though primary care is one of the most complex components of the health care system, little research has explored the human factors issues that are distinctive and relevant to primary care for the design of systems to ensure the safety of patients. Though primary care differs from referral care in a number of ways, the existing research has shown that primary care suffers from many of the same types of medical errors, especially medication errors, as hospital-based care. Following a discussion of what distinguishes primary care from other types of care, the evidence of patient safety problems in primary care is reviewed. Next, 12 vignettes illustrating typical issues in primary care are provided along with discussion of the vignettes. Finally, the vignettes are analyzed to uncover the important human factors and ergonomics (HFE) issues in primary care. We hope that readers will be able to use the information from this chapter to identify HFE problems in their own practice so that they can be eliminated or reduced through good design.

The Institute of Medicine (Donaldson, Yordy, & Lohr, 1996) provided a formal definition of primary care: "Primary Care is the provision of integrated accessible health care services by clinicians who are accountable for addressing a large majority of personal health care needs, developing a sustained partnership with patients and practicing in the context of family and the community" (p. 1). Captured in the definition are the four components of primary care that make up its essential nature: first-contact care ("accessible"), longitudinal care ("sustained partnership"), comprehensive care ("majority of personal health care needs"), and coordinated care ("context of family and the community") (Beasley et al., 1983; Donaldson et al., 1996; Starfield, 1993). First-contact care is not entirely unique to defining primary care as not all first-contact events represent primary care—for example, a cholesterol screening at a health fair or asking a friend with medical expertise for health advice are not examples of primary-care encounters (Donaldson et al., 1996; Kovner & Jonas, 2002). However, primary care is the entryway into, as well as the patient's home in, the health care system. Longitudinal care that is continuous over time defines the care of the patient, rather than care limited to a specific disease process (e.g., cancer) or disease episode (e.g., appendicitis; Starfield, 1993). Care continues through different stages of a patient's life and in various settings, ranging from hospital nurseries to clinics to nursing homes. The focus of care is on the individual, regardless of the type of care needed, and is provided by a single individual or team of health professionals who must also act as advocates for their patients (Donaldson et al., 1996).

Comprehensive care is the integrated care that is provided for most of the common problems in the population (Starfield, 1993). Primary-care clinicians must be able to utilize other health professionals and resources when this would be helpful for evaluation and treatment. The coordination function is extremely important and consists of integrating the care that takes place through referrals, with different procedures, or through various therapies (Starfield, 1993). The necessity for coordination places additional demands on the primary-care clinician. For example, extra attention is required to ensure medical records are inclusive of pertinent information generated from other levels of care (Starfield, 1998). Well-coordinated care is critical if care is to be achieved in a cost-effective and safe manner (De Maeseneer, De Prins, Gosset, & Heyerick, 2003).

Put simply, primary care is the system of care to which patients can bring all their health problems and expect to receive care or referral, be given guidance through the health care system, develop a continuous relationship with clinicians, and obtain valuable information on disease prevention and health promotion. But these very aspects of primary care make providing services challenging. The need for clinicians and support staff to cope with a wide range of problems can lead to more chances for diagnostic and therapeutic errors. On the other hand, the flexibility of these systems allows for comprehensive management, the integration of preventive care with acute care, and "one-stop shopping" for patients who can get most of their needs cared for by one clinician in one location.

Primary care is provided by several types of health care professionals, including physicians, nurse practitioners, midwives, and physician assistants (Kovner & Jonas, 2002). The physician caregivers are family and general practitioners (Starfield, 1994) and also include physicians in internal medicine, general pediatrics, general geriatrics, and obstetrics and gynecology (to the extent this latter group of physicians provides patients with primary care). Primary-care clinicians practice in both outpatient and inpatient settings, though most patient encounters occur in outpatient settings. It is worth noting that the quality and costs of health care in the United States and elsewhere correlate most directly with the number of primary-care physicians rather than the number of specialists (Baicker & Chandra, 2004; Macinko, Starfield, & Shi, 2003; Starfield, 2001). In this chapter, we consider outpatient primary care only and do not address human factors and ergonomics (HFE) in all ambulatory care, as that would include ambulatory surgery and other forms of specialty ambulatory care.

WHAT DISTINGUISHES PRIMARY CARE FROM OTHER FIELDS OF MEDICINE?

Primary care is dissimilar from the other components in the health care system, such as secondary and tertiary care. One difference is that services utilized are frequently initiated by patients, whereas this is less common in secondary and tertiary care (Starfield, 1998). Also, primary care relies on both the biomedical and social sciences, is information intensive, requires different strategies for decision making from that in referral specialties, thrives on a sustained personal relationship between the patient and clinician, and aims to integrate mental health, physical health, and social issues (Donaldson et al., 1996) to a greater extent than secondary or tertiary care.

McWhinney (1996) noted that the criticality of the clinician–patient relationship is distinctive in primary care. This relationship provides for a focus on the individual rather than the illness, with the clinician in most cases knowing the individual before knowing what his or her illness may be (McWhinney, 1996). Building this relationship is vital to providing safe and effective care, and research shows that patients value the consideration of both their life and illness in services received by their physicians (Tarrant, Windridge, Baker, & Freeman, 2003).

Because primary care includes first-contact care, the symptoms presenting in primary care may have a low predictive value for any specific disease. Therefore the decision-making strategies of primary-care clinicians will be, appropriately, different from those of referral physicians. Those problems referred to specialist practice may be better defined as they have already been through a "filter" (Rosser, 1996; Starfield, 1993, 1998). Primary-care physicians deal with multiple problems and complaints, know the patient's character and pattern of health complaints, and can tolerate uncertainty, whereas referral physicians tend to focus on selected problems and more commonly have brief relationships with patients (Sox, 1996). Primary-care physicians have the benefit of making diagnoses through the combination of knowledge of the individual and his or her value system, as well as scientific diagnostic criteria (Rosser, 1996). These contexts make a difference in decisions made during diagnoses, interpretation of test results, and treatment—all relevant to safe and effective patient care. Thus, outpatient or primary care may be more subject to errors of omissions, whereas specialty care may be more prone to errors of commission (Starfield, 1998; Tierney, 2003).

PATIENT SAFETY IN PRIMARY CARE

A compelling case has been made for the existence of medical errors and patient safety problems in inpatient settings (Brennan et al., 1991; Kohn, Corrigan, & Donaldson, 2000; Leape et al., 1991), but much less is known about patient safety problems in outpatient primary-care settings. Researchers have suggested that the reasons for the lack of data on outpatient safety problems stem from outpatients administering their own medications, the infrequency of communication between patients and physicians, and inadequacy of outpatient care documentation (Gandhi et al., 2000, 2003). What is known, though, suggests that medical errors and preventable adverse events do occur in outpatient primary care settings and affect children, adults, and the elderly (Gandhi et al., 2003; Goulding, 2004; Kaushal, Barker, & Bates, 2001). The prevalence of preventable incidents or adverse events in primary care may be quite high, and recent evidence suggests that over half of them may be preventable (Bhasale, Miller, Reid, & Britt, 1998; Fischer, Fetters, Munro, & Goldman, 1997; Sandars & Esmail, 2003). Patient safety problems such adverse drug events (ADEs) are likely to increase, at least among the elderly, given the ever-increasing number of available and prescribed drugs (Tierney, 2003).

Most of the research to date on primary-care patient safety has focused on the adult population and has often focused on the elderly. The focus on the elderly is not surprising because they consume about one third of all medications in the United States and are more susceptible to ADEs (Huang et al., 2002). Results from recent studies suggest that polypharmacy occurs in nearly 20% of patients ages 65 and older, at least one inappropriate prescription is given in 7% to 10% of clinic visits, and the likelihood of having an inappropriate prescription increases as the number of prescriptions per visit increases (Goulding, 2004; Huang et al., 2002). Research on elderly outpatients has also shown that of all ADEs, 38% are life-threatening or fatal and 42% of the life-threatening or fatal ADEs are preventable (Gurwitz et al., 2003).

Elderly adults are not the only outpatients at risk for medical errors. Ghandi and colleagues (2000) recently found that 25% of surveyed adult outpatients over the age of 18 had an ADE. Of those, nearly 40% were considered preventable or at least ameliorable (Gandhi et al., 2003). Just as in the studies of the elderly, the number of medications that a patient took was associated with the risk of having an event related to a medication. In a similar study of 20- to 75-year-olds, 18% of patients reported a drug complication, though chart review of those same patients suggested that only 3% had drug complications (Gandhi et al., 2000).

The causes and categories of errors in primary care are similar to those in inpatient care. In Gandhi et al.'s study, ameliorable ADEs were found to be due to poor communication, such as patients not informing physicians of symptoms or physicians not responding to reported symptoms. Preventable ADEs in that study were associated with prescribing errors (Gandhi et al., 2003). Dovey and colleagues (2002) found a much larger range of causes, including administrative failures, investigation failures, treatment delivery lapses, miscommunication, payment system problems, errors in the execution of a clinical task, wrong treatment decision, and wrong diagnosis. Recent literature reviews have found basically the same error categories and causes (Elder & Dovey, 2002; Sandars & Esmail, 2003), though recently the problem of missing clinical information has been highlighted as well (Smith et al., 2005).

The consequences of error in primary care are typically less severe than in inpatient care. They have been shown to include delayed care (21%), financial or time cost to patients (9%), patient dissatisfaction (12%), patient became ill (7%), and patient was admitted to hospital (3%) (Dovey et al., 2002). In another study, almost 50% of the preventable adverse events resulted in emotional or financial harm to the patient (Fischer et al., 1997). These results suggest that errors in primary care have consequences to patients that can be severe.

It is quite clear from the data reviewed as well as from the funding committed to medical errors in outpatient settings by the U.S. Agency for Healthcare Research and Quality that serious patient safety problems in outpatient care exist. To demonstrate the nature of the patient safety problems found in primary care, the next section provides 12 brief vignettes describing typical primary-care situations. Following the section with the vignettes is a final section in which HFE hazards and design flaws implicit in the vignettes are identified and discussed.

WHAT HAPPENS IN PRIMARY CARE?

The following section describes some of the issues that must be taken into account as efforts to improve patient safety in primary care are undertaken. The vignettes that follow are based on the first author's 30+ years of experience as a family physician and reflect the substance of actual typical encounters.

1. Patients Require the Integration of Care for Multiple Problems

> Mr. J. is 84 years old and has anemia, diabetes, chronic renal failure, hypertension, atherosclerotic heart disease, arthritis, and fatigue. He takes eight different medications. He is in the clinic for a 15-min follow-up visit.

There is a common, although incorrect, notion that most primary-care patients come for care of a single, well-defined, acute medical problem. In fact, patients in primary care have, on average, slightly more than three discrete problems addressed at each visit (Beasley et al., 2004).

The task of the caregiver is to set management priorities and integrate care for existing conditions, to address new problems as they arise, and to integrate preventive care. Adding to the complexity is the fact that between 7% and 20% of visits include discussion and medical decisions about someone other than the identified patient—the "secondary" patient (Beasley et al., 2004; Orzano et al., 2001; Zyzanski, Stange, Langa, & Flocke, 1998). The decision making becomes complicated—almost Byzantine—and there are both positive and negative outcomes related to integrating the care for multiple problems.

The issue of the complexity engendered by patients having multiple problems is compounded by the issue of multiple agendas for the prioritization of care—agendas that may compete with each other. These agendas may arise from the caregiver (what he or she thinks is best), the patient (what's bothering him or her the most), the managed care organization (the expense of treatments), guidelines (often promulgated as the "ideal" for care for a single disease), or the threat of malpractice allegations. Although a setting of comprehensive care offers the potential to integrate multiple agendas, the attempt to do so may lead to some of them not being fully addressed.

2. Diversity of Clinical Needs Make Standardization of Care Processes Difficult

> Mrs. K. has diabetes and is in for her annual exam. She will have a pap smear and a flexible sigmoidoscopy. She has had a cough for 3 months. She has problems with insomnia. She also has a small skin lesion on her face that needs to be removed.

As seen in this example, the issues the clinician must address are diverse and range from medical to surgical to psychosocial. This requires that more types of in-clinic equipment be available for diagnostic testing and therapeutic interventions. A range of different types of laboratory and other testing procedures is required as is a large set of consultative and support resources. This diversity of problems demands a broad range of knowledge and skills among both clinical and support personnel who need to have knowledge of these diverse medical problems and the relevant diagnostic and therapeutic strategies. This diversity also requires flexibility in scheduling and makes the establishment and use of clinical protocols and routines difficult.

3. Time Limitations Are Important

> As the doctor gets up to leave the room at the close of a routine 15-min visit to monitor a 75-year-old patient's hypertension, insomnia, and arthritis, his wife mentions: "By the way, when he gets up in the morning, he is very confused and sees things that aren't there."

Given the medical complexity—and the need to negotiate priorities—time management becomes problematic. As in this vignette, this is not uncommonly accentuated by what is termed the "hand on the doorknob" phenomenon, whereby a major problem suddenly appears at the end of the allocated time. The need to accomplish medical tasks, listen to the patient, and complete associated administrative tasks (e.g., record keeping, billing, telephone calls) all require significant time. Although primary-care physicians feel as if there is less time for patient visits in recent years, this is probably not really the case (Mechanic, McAlpine, & Rosenthal, 2001). Rather, the complexity of medical management, the need to negotiate with patients to achieve shared decision making (Sheridan, Harris, & Woolf, 2004), and the administrative demands have all increased. Obviously, in many ways the increase in expectations for what will happen during a primary-care visit is good, as care is more comprehensive and patients have more input into their care. Nonetheless, the phenomenon of increasing demands has increased the complexity and the number of different tasks required of the clinician and the system. Despite this, however, the differences in care between high volume practices and lower volume practices, although real, are not great (Zyzanski et al., 1998).

In one sense, there is more, rather than less, time for patients in primary care because, in contrast to many other specialties such as surgery or psychiatry, the work with patients and their problems occurs in continuity over a period of time that is often measured in years. The continuity of care is not limited either by short periods or by the presence or absence of a disease, as would be, for example, the case for surgery for appendicitis. This continuity over time is used to identify and work to resolve problems and establish an effective relationship with the patient, which Balint (1957) termed the "mutual investment company of physician and patient." The continuity afforded by delivering care over long periods of time allows for a better chance to develop working relationships, to "fine-tune" care, and to avoid or recognize errors.

4. Multiple Places, Systems, and People Are Involved in the Care of Patients

> Mr. J. has vascular disease. He was living at his home until he came to the clinic with an acute loss of circulation to his leg and was transferred to the hospital for an amputation. He was then admitted to a nursing home and subsequently back to his own home where he receives support consisting of family and home nursing care. His insurance has just changed and he needs to switch to a new physician and new home health caregivers as well as use different brands of medication that are on the new formulary. If he is rehospitalized, it will be at a different hospital.

Primary-care clinicians interact with a wide variety of organizations and systems. Problems in these interactions account for about 24% of errors reported in family practices (Dovey et al., 2002). These organizations include multiple institutions such as hospitals and nursing homes, a variety of support agencies such as laboratories and x-ray units, home health agencies and family and community resources, as well as multiple payors for care. Most of the time this wealth of resources not only facilitates care but is actually absolutely necessary for the optimal care of patients. However, the involvement of different institutions and systems greatly increases the need for clinicians and their support staff to maintain effective coordination and communication. This in turn requires them to be knowledgeable about these institutions and systems. All of this takes significant time and attention on the part of clinicians and their staff.

Simultaneously, there is the need to assure that the care is coordinated within the clinical care team itself (Grumbach & Bodenheimer, 2004) and between organizations as the patient moves between institutions and systems (Starfield, 2003). In an ideal world, both the patient and his or her medical information would move smoothly and safely from one venue of care to the next. This is rarely the case as the multiplicity of organizations and payors has added greatly to the complexity of the care process. In part due to the sheer volume of data, important information (e.g., x-ray results, treatment plans) may be lost.

The situation may also be complicated by differing requirements by various payors. For example, insurance company A may have one brand of a type of drug available in their formulary and company B may well have another. Policies of payors directed at reducing costs by demanding that different types of care (e.g., mental health) utilize specific physicians or groups rather than the physicians or groups who provide the primary care may have the effect of reducing the overall quality of care (Ray, Daugherty, & Meador, 2003).

5. Communication With Patients Occurs in Many Ways

> Mr. J. leaves an e-mail for his doctor regarding his son's visit to an urgent care clinic for his asthma. The doctor gets it at home that evening. The doctor responds with a call from her cell phone while on her way to the hospital the next morning and then telephones in a prescription to the pharmacy. Later, when Mr. J. is in the clinic for his own health care, much of the time is taken up by further discussion of his son's problems.

It is typical for primary-care clinicians to have multiple modes of contact with patients in addition to the direct "in the office" contact. This vignette illustrates the multiple methods of communication to which could be added communication through other members of the primary-care clinical team (e.g., a medical assistant or pharmacist) or consultants. As in this example, it is not uncommon for the medical record to be unavailable and for no record of the discussions to be created, as was the case with the phone call and the clinic visit for

Mr. J. The communication between primary health care professionals and patients is also about many different types of topics. These relate to clinical status, response to treatments, or questions about administrative issues. Different response times are required, ranging from immediate, such as in the case of acute chest pain, to communication that is not at all urgent, such as conveying normal laboratory results.

6. Information—It's Feast and Famine

> Mrs. T. is a long-standing patient whose chart is in two 2-in thick volumes and contains physicians' notes, laboratory and x-ray reports, consultation reports, and hospital summaries in multiple formats scattered in various places in the chart. She was recently discharged from a regional hospital, but the clinical summary, laboratory reports, and medication changes from that hospitalization are not available when she comes to the office for follow-up. She is upset and crying and her husband who is also in the room is frustrated.

Errors related to information handling accounted for 24% of errors reported in one survey of family practice errors (Dovey et al., 2002), and a more recent study found that primary-care clinicians reported missing clinical information in nearly 14% of visits (including laboratory results, ((6.1%), letters/dictation (5.4%), radiology results (3.8%), history and physical examination (3.7%), and medications (3.2%; Smith et al., 2005)). As the vignette illustrates, there are many circumstances in which important medical record information is not available as well as many opportunities for information to be lost. However, it is at least as common for the clinician to be confronted with too much data that often is not well organized. Not only is the caregiver getting data from a chart (electronic or paper), but he or she will also need to recall what he or she was told by one of the medical assistants before the visit started, listen to the patient, pick up nonverbal cues, and perhaps listen to a relative who is also in the room as well.

The dilemma of having both excess data, which may or may not be integrated into information, and at the same time having potentially important data missing is a combination that impairs medical decision making. The sheer mass of data presented to clinicians, especially when multiple medical problems are involved, makes retrieval difficult and time pressures greater.

7. Context and Family Are Important

> Mr. S. is a dairy farmer who is an alcoholic. He is hospitalized for withdrawal but wants to leave to "take care of my farm." His neighbors agree to keep his herd milked and the barn cleaned out if he undergoes long-term treatment. His wife threatens to leave him unless he stays. He decides to stay for treatment.

Primary care takes place in the context of the family and the larger community, and both can either help or hinder care. The involvement of the patient's family and community is especially important when the patient is a child or very elderly, and where issues of infirmity or cognitive impairment may play a role. This involvement may take the form of transporting the patient to the office, providing history to the clinician, helping the patient understand treatment, assisting the patient in reporting problems, or monitoring medications or the response to treatment. A lack of family support may lead to more tests and hospitalizations (Pantell et al., 2004). At the same time, family members may inhibit communication, as might be the case when a parent accompanies an adolescent, thus making the discussion of issues such as sexual behavior more difficult.

Although dysfunctional families can render the best therapeutic efforts ineffective as well as create new problems such as abuse and depression, it is vastly more common for families to watch out for patients, to report additional problems to the physician, and to provide needed support. As with families, the patient's peers or community groups (e.g., a church group or Alcoholics Anonymous) can often provide support. However, it is also possible for peers to hinder care, as may be the case with a patient whose close friends are also alcoholic or a relative or friend who discourages the implementation of a care plan. Although the development of formal "family profiles" has been touted as a way for clinicians to recognize and act on these contextual issues, in reality much of the clinician's knowledge of contextual issues is developed over time through the care of the patient and his or her family (Beasley & Longenecker, 1983).

8. First-Contact Care Is Different

> Sammy N. is a 2-month-old child whose mother brings him in because he has a fever.

Many acutely presenting problems in primary care are either minor or self-limited, such as common viral infections or minor injuries that require minimal diagnostic or therapeutic intervention. However, there is a continuing need to be alert for rare but serious diseases such as meningitis, which are very uncommon. Even when a patient has a rare but serious condition, he or she tends to present to the primary-care setting in the early stages when the typical signs and symptoms are not fully developed. For example, early meningitis can present without the typical physical finding of a stiff neck.

When the probability of a serious acute disease is low, as is often the case in primary care, diagnostic testing is less useful. This is because all tests have some false-positive results. In a population of patients with a low prevalence of a particular disease, positive tests have a lower predictive value for that disease and may lead to excessive additional testing and treatment. For this reason, published guidelines for care of children, as in the previous vignette, are often not followed in primary-care practices. There may be no discernable detriments to patient welfare despite the decision to deviate from the guidelines (Pantell et al., 2004).

9. Relationships and Continuity Are Important

> Ms. T. is a poorly controlled diabetic who is schizophrenic, lives alone, and does not like to leave her apartment. She does not trust her physician and is worried that her medications are "poisoning" her and so is not reliable in taking them.

For patients in intensive care units, adherence to a prescribed medication is not an issue. They get their medications with little or no input on their part. However, for a patient who is in a primary-care setting, he or she is in control and a variety of factors may either facilitate or impede the care process. The development of a therapeutic and trusting relationship may well take precedence over the need to address other problems such as diabetes, and this makes prioritization of care important. In this vignette, the first priority has to be establishing some sort of working relationship with the patient. This prioritization according to the needs of the individual is an integral part of concurrent management of multiple problems and is often termed "treating the whole patient." It takes time for the caregiver to understand where the patient is "coming from" and what his or her priorities are for diagnosis and treatment; these are critical steps if there is to be a successful therapeutic alliance. As noted in a later vignette, the increasingly multicultural nature of our society, with or without language barriers, makes this an increasingly challenging task.

Continuity of care by a single clinician is generally seen as critical to the care process, but the literature is somewhat conflicting (De Maeseneer et al., 2003; Gill, Mainous, Diamond, & Lenhard, 2003; Parchman & Burge, 2004; Parkerton, Smith, & Straley, 2004). On balance it appears that the development of therapeutic relationships is important to at least some patients, and indeed it may even have intrinsic value. Balint (1957) has suggested that for many conditions "the doctor is the drug." There can also, however, be risks to a close patient–physician relationship. For example, the clinician may be reluctant to risk disruption of a warm relationship by pushing for needed procedures such as a diagnostic work-up for blood in the urine or addressing problems such as alcoholism. The closeness of the relationship can also engender problems if maintenance of that relationship becomes too important. For example, a physician may be reluctant to discuss a patient's problem with alcohol for fear of alienating the patient.

10. There Are Limitations on the Use of Therapeutics, Procedures, and Technology

> Mrs. B. is 78 years old and in generally good health. However, she has chronic atrial fibrillation and should, according to guidelines, be on coumadin to reduce her risk of a stroke. However, she doesn't want to take that "rat poison" or come in for the laboratory tests that would be required as she feels "just fine."

The omission of recommended care can also be considered an error. The ability of primary-care clinicians, on whom most of the responsibility for providing care rests, to meet the guidelines for recommended care is not good (McGlynn et al., 2003). There are multiple reasons for this. Some relate simply to a failure to organize a practice to deliver the needed care. Others relate to patients themselves and their multiple problems, which can result in competing demands. In some cases there are limitations posed by the patient's physiology, such as drug allergies or intolerance to certain categories of medications such as medications to treat coronary artery disease. The need to be reasonable, not only in terms of costs but also in terms of what patients can tolerate, places limits on what can be accomplished both for prevention and therapy and in some cases makes following guidelines impractical (Yarnall, Pollak, Ostbye, Krause, & Michener, 2003).

The limitations on what can be accomplished may also relate, as in this vignette, to patient preferences, which may be quite reasonable. (One patient once told the first author "If you give me one more pill I won't even have to eat breakfast!") At other times, less rational patient preferences can contribute to substandard care. For example, some technologies may not be acceptable to patients even when of proven benefit. Screening to detect colon cancer is a prime example. On the other hand, some screening procedures, such as testing for prostate cancer, which is subject is great controversy about risk versus benefit, are nonetheless more widely used (Sirovich, Schwartz, & Woloshin, 2003). This may be due primarily to patient request. Enthusiasm over new technologies, even when not proven to be helpful, may lead to excessive interventions, and one of the functions of primary-care clinicians is to help prevent the promiscuous use of unneeded and even harmful interventions (Franks, Clancy, & Nutting, 1992; Mold & Stein, 1986).

11. There Is a Great Variability of Patients and Their Needs—One Size Does Not Fit All

> During a day's clinic session, Dr. K. sees four patients with coronary artery disease. The first has good insurance, is highly motivated, is exercising regularly and is following his diet and taking his medications. His care will meet the guidelines for the care of coronary artery disease. The second patient is depressed and living alone. She cannot sleep. She continues to smoke. She has severe arthritis and cannot exercise. She is on Medicare and would like to use medications but cannot afford them. The third also has lung cancer. The lung cancer will limit her life to the point that treatment of the heart disease is no longer important. The final patient is a 63-year-old immigrant from the African country of Chad who speaks no English; her husband who accompanies her has limited English skills. She is seen on an emergent basis because of an episode of chest pain and is admitted immediately to the hospital.

Even when the underlying medical problem is the same, the approach to the patient cannot be easily standardized. In this vignette, the management of the coronary artery disease will be entirely different for each patient. In the second patient, the depression is probably the most critical issue and will appropriately consume most of the caregiver's attention. In the third patient, the focus will be on the cancer and end-of-life issues, and the coronary artery disease will not enter into the picture. For the fourth patient, there may not only be a potential language barrier but also subtle (and perhaps unperceived) differences in concepts of disease, the nature of care, and the expectations of the caregiver. Problems can arise from the lack of trust than can occur if the caregiver is of a different culture—or is perceived to be biased in his or her treatment of patients. There is clear evidence of major and persisting health care and health outcome disparities related to racial, ethnic, and social variables in the United States (Smedley, Stith, & Nelson, 2003). The variety of needed responses and interventions, even when a single major disease is present, makes standardization of protocols and procedures seem impractical.

12. Slack (Loose Coupling) in the System Can Allow for Error Recovery

> Dr. B. is not available Monday when Mr. M. is seen by one of his partners who orders a laboratory test. She instructs the patient to return Tuesday morning so a fasting sample can be taken. Mr. M. makes an appointment with the receptionist to have this done. The receptionist notifies the medical records staff to have his chart available Tuesday morning. When she draws the blood, the laboratory technician notes that the patient is anemic and on reviewing the chart notes that some

tests Dr. B usually gets for anemia were not requested. She asks the medical assistant to have Dr. B contact her and when he does she asks him if he wants these additional tests. He agrees this is a good idea and the laboratory technician calls the hospital laboratory and arranges for the tests to be done on the existing sample, thus saving the patient an additional blood draw while assuring that the right tests are done.

Although health care delivery is complex, and therefore arguably prone to failure, systems elements (e.g., clinicians, clinic staff, lab technicians, and so on) often do work together to prevent or recover from failures. In this case, proper tests were not ordered, possibly because the patient was handed off to another physician. However, although not ordering the right tests was an error of omission, the laboratory technician caught the error and set in motion a process for correcting the problem.

TABLE 51–1. Human Factors Issues and Related Vignettes.

Human Factors Issue	*Vignette Examples*
System design	2, 4, 6, 12
Technology design and selection	2, 10
Communication	5, 12
Memory, information processing demands	3, 6, 8, 9, 11
Work pressure and work load	1, 3, 5, 6
Complexity and coupling	7, 11, 12

HFE PROBLEMS IN PRIMARY CARE

The 12 primary-care vignettes highlight factors faced by primary-care professionals. In this section, we analyze the vignettes to identify the HFE issues implicit in each (see Table 51–1). Dozens of HFE issues were identified, including those related to memory, information processing, standardization, simplification, forcing functions, work pressure and workload, organizational design, uncertainty, information access, technology acceptance, and usability. These topics are among the many that can influence human performance and safety (Sanders & McCormick, 1993; Wickens, Lee, Liu, & Becker, 2004). Of the many HFE principles present in the vignettes, we focus on some of the more important problems. These are system design, technology selection and design, memory and information-processing demands, work pressure and workload, and complexity and coupling.

System Design

The vignettes described illustrate the fact that the primary-care system is indeed a complex sociotechnical system (Hendrick & Kleiner, 2001; Pasmore, 1988). There is a social system, made up of physicians, nurses, administrators, nursing assistants, patients, patient family members, and others, and there is a technical system, comprised of equipment, tools, technologies, and processes that are used to diagnose, treat, and monitor. Both the social and the technical system elements interact in complex manners and exist within an external environment. This external environment can refer to a patient care room, an entire clinic, a health care system, or regulatory bodies. To deal with the complexities described in all of the vignettes (but see Vignettes 2, 4, 6, or 12 for excellent illustrations), proper system design requires that the social and technical system elements be designed to complement each other and meet the demands of the external environment (Pasmore, 1988).

One specific system design problem illustrated by the vignettes is that these systems are not "designed for errors." A process that is designed for errors is one that facilitates error or hazard identification and thus prevention. In the absence of good system design, multiple latent conditions can lead to organizational failures (Reason, 1997). Generally, latent conditions include unworkable procedures, less than adequate equipment, gaps in communication, and so on or those elements that are not necessarily visible and facilitate the occurrence of errors (Reason, 1997). Consider Vignette 4 discussing the extensive interactions in which primary-care clinicians must engage to coordinate the care of their patients. Latent conditions may exist within the logistics required for Mr. J. to move between facilities as well as the coordination required in the change of insurance, medications, and a new physician. These may occur in the process of ensuring proper supports are available at home, at the interface between the care systems, and in the course of developing relationships with new caregivers. In Vignette 12 the design of the system did allow for an error of omission to be caught; in that case having a technician review test requests was a way to design for errors, albeit a poor way that relies on human behavior (i.e., reliable and accurate human checking).

Designing for errors is accomplished by improving information access and organization throughout the system so that people in the system know the status of the system at any given time. This might include, for example, knowing the status of patients (e.g., who they are being seen by, what facility they are currently at, their tests, their diagnoses, and their compliance levels). Having this feedback is critical for a system to function optimally, because feedback allows the people in the system to respond and adapt to changes.

The typical state of diagnostic testing processes offers a useful example of the need to design for errors because of how often tests are lost or missed. In many primary-care clinics, after a diagnostic test is ordered, physicians have to remember to look up the results. This system often leads to missing test results. To redesign this system, a simple paper or electronic system could be designed to track the status of diagnostic tests through the preanalytic, analytic, and postanalytic phases and alert physicians when tests results are available, even if they are not stat orders (Hallock, Alper, & Karsh, 2003). With an electronic system, outstanding or unreviewed tests can be presented on a home page after physicians log in.

Other system design principles from the sociotechnical system literature can help provide guidance to improve system design within the primary-care organization and between primary-care organizations and external organizations. These principles include ensuring that the design fits the needs of all stakeholders, including clinicians, patients, and managers; is simple and makes problems visible; and ensures that the various system components are compatible with one another (Clegg, 2000).

Technology Selection and Design

Although patient's physiology or preference may make the use of a technology unacceptable, there are additional criteria governing physician acceptance and use of technology (consider Vignette 10, for example). The selection of technology, especially new medical equipment and devices, should also include an evaluation of its use by health care professionals and in some cases by patients. Technologies used by health care professionals must support their cognitive and physical performance. Research has shown that designs must be perceived by end users as both usable (i.e., cognitively and physically easy to use, conforms with end user mental models) and useful (i.e., will help the end users do their jobs better or faster) if they are to be accepted, adopted, or used (Davis, 1989; Venkatesh, Morris, Davis, & Davis, 2003). Human factors research contributes here by providing criteria for addressing technology usability as well as identifying how well the new system will integrate into existing systems, including size, space, compatibility, and task difficulty (Welch, 1998).

Another human factors concept of technology design is known as functional allocation. Functional allocation refers to the allocation of tasks to machines or people. Tasks that can be improved with automation should be given to machines, and those that make the most of human strengths should be given to people. Proper functional allocation assigns boring, repetitive, computationally intensive, or unsafe tasks to machines and judgment and caring tasks to people. The goal of proper functional allocation is to allow the individuals in the system to utilize technology as tools for supporting their performance. The lack of heavy emphasis on machines or technologies in the vignettes could be viewed as a problem with functional allocation. The primary-care physician has complex judgments and decisions to make, trust to establish, family members to involve, and care to provide. These are all tasks that have a strong social component and are not easily automated—nor should they be. However, technologies could be used to improve decision making, facilitate communication, and improve access to information to improve the quality of care.

Communication

The patient–physician relationship has been continuously transforming due to the advent of various communication technologies, presenting both opportunities and challenges (Mandl, Kohane, & Brandt, 1998). Communication with patients, as demonstrated in all of the vignettes, deals with many different types of issues. These relate to clinical status, response to treatments, or questions about administrative matters. Although the potential to provide care for patients is amplified due to the opportunity to use electronic communication, the physician–patient relationship, the cornerstone of primary-care practice, could be negatively impacted.

E-mail can positively affect the relationship by, for example, keeping the lines of communication open and increasing access to care, as well as enabling patient education (Mandl et al., 1998; Patt, Houston, Jenckes, Sands, & Ford, 2003). There is also the possibility, however, of lost trust, which could harm the relationship (Patt et al., 2003). It has been found that one of the greatest barriers to patient use of e-mail is that they would rather speak with a real person. Other barriers include patients preferring to communicate with their physician by phone, concern that the message would get lost, or the time it might take to receive a response (Moyer, Stern, Dobias, Cox, & Katz, 2002).

In primary-care practice, different response times are required, ranging from immediate, such as in the case of acute chest pain, to a longer time, such as when conveying normal laboratory results. Formal policies typically do not exist that govern how e-mail should be used with patients, thus issues such as appropriate response time and content remain obscure (Patt et al., 2003) and may create opportunities for errors. From the physician's perspective, e-mail can be burdensome. Physicians have been found to describe e-mail as overused by patients, time consuming, and redundant (Patt et al., 2003). Moyer et al. (2002) also found that physicians feared being overwhelmed by e-mail if they were to give out their e-mail address. The key, again, is in the design; communications systems must be designed to meet the needs of the clinicians and patients and integrate with other systems that facilitate health care delivery.

Another communication issue, for example in Vignette 5, is information access. In this vignette, patient–physician interaction occurs in three locations. Patient care decisions require information, and clinician access to relevant and complete patient documentation is essential to avoid the occurrence of errors (Spath, 2000). On one hand, a multimodal communication system decreases the response time of the system, increases the access to care, and opens channels of communication that would not otherwise be available if communication occurred only face to face in the office setting with the caregiver. Nevertheless, flexibility and variety in the modes of communications can create problems in retrieving information and in the recording of data, leading to information being lost or not acted on. With the physician receiving e-mail messages at home, addressing needs from a cell phone while commuting, and then providing care in person, mechanisms should be in place to ensure that appropriate information is available and new information properly documented into the patient's record.

Vignette 12 also demonstrates that there are complex communication networks that do not directly involve the patient but rather are among clinic staff or among staff from more than one health care organization. The design of the communication may facilitate or inhibit access to individuals in the system. One could imagine a system designed such that call backs and confirmations would not be necessary, but until that time, it is important that each individual in the system know with whom he or she should communicate, how to access them, and at what times.

Memory, Information Processing Demands

The demands in primary care call for intensive use of cognitive resources. The nature of primary-care practice as well as the overall diversity of the patient population, both cultural and in needs and abilities, have been established in the vignettes (e.g., 3, 6, 8, 9, 11). The clinic visit, consisting of patients with multiple problems, requires the physician to employ a combination of long-term memory, working memory, and knowledge-based problem solving. A physician may have to keep patient information in working memory while the patient is explaining the purpose of his or her visit. Simultaneously, the physician may have to pay attention to a 2-in thick medical chart while trying to locate information relevant to the patient's complaint. If the physician cannot find the information he or she is looking for, then long-term memory will be accessed to try to recall the pertinent information. Such a situation is both typical and hampers good decision making because of how attentional and memory resources are divided. Human factors design recommends simplification, standardization, constraints, and forcing functions as a few strategies to alleviate the demands on memory and attention (Wickens et al., 2004).

Process simplification permits users to concentrate on the decision-critical aspects of their jobs instead of expending time and energy navigating the process at hand. Within primary care, there are many complexities that work against simplification, including the multiple places, people, and systems all interacting in complex ways to influence the safety of care. Similarly, knowing that patient charts are likely to be incomplete or unorganized promotes the need to rely on memory. However, there are areas identified in the vignettes in which simplification and reduced reliance on memory could be of use, if only in certain situations. For example, well-designed information technologies could help to simplify information management, such as the integration of diverse patient information into a single electronic chart.

Although standardization frequently applies to equipment and displays in equipment design, standardization can also pertain to procedures such as the way patient information is stored. Standardization of processes reduces reliance on memory and can help to simplify processes. The vignettes illuminated several ways that standardization could help primary-care professionals focus on patient care. For example, standardized protocols (designed to be sufficiently flexible for different patient needs) that might assist patients in identifying all the problems they are concerned about at the beginning of the visit could help physicians better plan diagnosis and treatment.

It is important to also understand what standardization is not. Standardization is not telling clinicians how to do their jobs, should not remove decision making about patient care from clinicians, and is not meant to take control from clinicians. In fact, a well-designed standardized process must recognize and accommodate the large range of patients and patient care situations for which the standardized process must be used. That necessarily means that there cannot be one best way of providing care because of the diversity of patients and patients' problems.

Constraints and forcing functions should also be used to reduce reliance on memory. These help to ensure that certain types of errors cannot be made because the user is forced not to make the error or is constrained from making the error. An example of a constraint is when after hitting "delete" on a computer one is prompted with "Are you sure you want to delete this file?" An example of a forcing function would be designing the coupling of an oxygen line to be incompatible with the coupling for an air line. Because of the relatively low reliance on technology in direct patient care and the variability of patients and procedures, there are few forcing functions in primary care. What forcing functions do exist tend to be for the protection of clinical staff and include such things as positioning the control panel for the x-ray machine so that the technician must be behind a shield to operate the x-ray or interlocks on laboratory centrifuges so that they cannot be opened while still spinning. In the former case the users are asked to confirm their decision, whereas in the latter the user is prevented from making the error through good design. Both techniques reduce reliance on memory and vigilance and simplify processes because they protect against the wrong decision. For example, a well-designed decision support system could help to either prevent or warn a physician from choosing the wrong antibiotic based on the bacteriology, choosing the wrong antibiotic based on the presence of an allergy, or choosing the wrong dose and form.

Work Pressure and Workload

Several of the early vignettes, such as 1, 3, 5, and 6 highlighted the problem of time pressure and workload. The context in these vignettes ranged from a multiplicity of health problems to be addressed with strict time restrictions to the necessity for physician proficiency to manage complex patient circumstances. Both work pressure and workload can impact human performance and safety via their impact on psychological stress (Murphy, DuBois, & Hurrell, 1986). Work pressure and workload can lead to short-term effects such as anxiety, fatigue, and reduced motivation, which in turn can lead to impaired reaction time, accuracy, and decision making. Such impaired human performance can further lead to unsafe behaviors, which can result in accidents and injuries. In the aforementioned vignettes, the time pressure could have resulted in missed diagnoses, poor treatment decisions, missed follow-up opportunities, or incorrect prescriptions. The same type of human factors design principles that can help to reduce reliance on memory, such as standardization, memory aids, and better information access, can also help to alleviate workload and work pressure precisely because they reduce the need for the health care professional to spend time recalling, locating, or processing information. There are also organizational and job-design changes that can help reduce workload and work pressure. Depending on the needs of the particular primary-care system, these could include team-based care approaches that better distribute patient care tasks or more appropriate scheduling systems. For additional information, see the chapters on job stress and burnout in health care in this volume.

Complexity and Coupling

Among the various elements around which an organization or system can vary, two that have important implications for safety are complexity and coupling (Perrow, 1984). The extent to which a system is complex or linear refers to whether the processes flow in a linear, anticipated fashion or whether there are unplanned or unanticipated interactions between system elements that can have unanticipated consequences. Highly coupled systems are those in which actions or processes follow rapidly from one another in a manner that permits little intervention by humans. Loosely coupled systems are those that have processes that develop more slowly and thus permit intervention at various stages because problems are more visible and there is time to intervene before a problem gets out of hand. It has been argued that systems that are both complex and tightly coupled are at a high risk for accidents (Perrow, 1984), though this theory, known as normal accident theory, has many documented limitations (Hopkins, 1999). The vignettes of primary-care encounters demonstrate that primary care is exceedingly complex (see, e.g., Vignettes 7 and 11) but also loosely coupled. The loose coupling stems from the fact that there is considerable lag time between many decisions and subsequent actions. This loose coupling can be exploited to the benefit of the system by designing in hazard detection mechanisms (Vignette 12). Many already exist. For example, because many steps occur between a primary-care physician ordering a medication and a patient taking it, there is time for erroneous prescriptions to be caught by the physician, pharmacist, patient, or patient family member.

CONCLUSION

The complexity of primary care results in many factors combining that can affect physician performance. These include knowledge factors related to solving problems, factors governing control of attention, and trade-off decisions that must be made when conflicting goals are present (Bogner, 1994). They also include patient factors, payor factors, clinic design, workflow, and technological factors. The different factors must integrate at both the individual and clinical team level for successful system operation, and they exist in the context of situational demands and resources and constraints that are posed by the organization (Bogner, 1994). In this chapter we have shown that primary care has its own unique patient safety challenges that can be addressed through good HFE design. The 12 vignettes illustrated some, but certainly not all, of the variety and complexity of primary-care encounters. From those vignettes, HFE issues were identified that played a role in either promoting situations that were hazardous to patients or helped to reduce the likelihood of hazards and errors. Readers will hopefully be able to use the information from this chapter to look at their systems of primary care through different lenses—lenses that make visible the system design factors that inhibit or facilitate good clinician performance.

References

Baicker, K., & Chandra, A. (2004). Medicare spending, The physician workforce, and beneficiaries' quality of care. *Health Affairs,* —Web exclusive. http://content.healthaffairs.org/cgi/content/full/hl thaff.w4.184vl/DC1. Retrieved July 27, 2004.

Balint, M. (1957). *The doctor, his patient and the illness.* New York: International Universities Press.

Beasley, J. W., Hankey, T. H., Erickson, R., Stange, K., Mundt, M., Elliott, M., et al. (2004). How many problems do family physicians manage at each encounter? A WReN study. *Annals of Family Medicine, 2,* 405–410.

Beasley, J. W., Hansen, M. F., Ganiere, D. S., Currie, B. F., Westgard, D. E., Connerly, P. W., et al. (1983). Ten central elements of family practice. *Journal of Family Practice, 16,* 551–555.

Beasley, J. W., & Longenecker, R. (1983). The patient/family profile. In R. B. Taylor (Ed.), *Fundamentals of family medicine* (pp. 335–340). New York: Springer-Verlag.

Bhasale, A. L., Miller, G. C., Reid, S. E., & Britt, H. C. (1998). Analysing potential harm in Australian general practice: An incident-monitoring study. *Medical Journal of Australia, 169,* 73–76.

Bogner, M. (1994). *Human error in medicine.* Hillsdale, NJ: Lawrence Erlbaum Associates, Inc.

Brennan, T. A., Leape, L. L., Laird, N. M., Hebert, L., Localio, A. R., Lawthers, A. G., et al. (1991). Incidence of adverse events and negligence in hospitalized patients—Results of the Harvard Medical-Practice Study: I. *New England Journal of Medicine, 324,* 370–376.

Clegg, C. (2000). Sociotechnical principles for system design. *Applied Ergonomics, 31,* 463–477.

Davis, F. D. (1989). Perceived usefulness, perceived ease of use, and user acceptance of information technology. *MIS Quarterly, 13,* 319–340.

De Maeseneer, J., De Prins, L., Gosset, C., & Heyerick, J. (2003). Provider continuity in family medicine: Does it make a difference for total health care costs? *Annals of Family Medicine, 1,* 144–148.

Donaldson, M., Yordy, K., & Lohr, K. (1996). *Primary care America's health in a new era.* Washington, DC: National Academy Press.

Dovey, S. M., Meyers, D. S., Phillips, R. L., Jr., Green, L. A., Fryer, G. E., Galliher, J. M., et al. (2002). A preliminary taxonomy of medical errors in family practice. *Quality & Safety in Health Care, 11,* 233–238.

Elder, N. C., & Dovey, S. M. (2002). Classification of medical errors and preventable adverse events in primary care: A synthesis of the literature *Journal of Family Practice, 51,* 927–932. (Erratum appears in *Journal of Family Practice, 51* [2002], 1079)

Fischer, G., Fetters, M. D., Munro, A. P., & Goldman, E. B. (1997). Adverse events in primary care identified from a risk-management database. *Journal of Family Practice, 45,* 40–46.

Franks, P., Clancy, C. M., & Nutting, P. A. (1992). Gatekeeping revisited—Protecting patients from overtreatment. *New England Journal of Medicine, 327,* 424–429.

Gandhi, T. K., Burstin, H. R., Cook, E. F., Puopolo, A. L., Haas, J. S., Brennan, T. A., et al. (2000). Drug complications in outpatients. *Journal of General Internal Medicine, 15,* 149–154.

Gandhi, T. K., Weingart, S. N., Borus, J., Seger, A. C., Peterson, J., Burdick, E., et al. (2003). Adverse drug events in ambulatory care. *New England Journal of Medicine, 348,* 1556–1564.

Gill, J. M., Mainous, A. G., III, Diamond, J. J., & Lenhard, M. J. (2003). Impact of provider continuity on quality of care for persons with diabetes mellitus. *Annals of Family Medicine, 1,* 162–170.

Goulding, M. R. (2004). Inappropriate medication prescribing for elderly ambulatory care patients. *Archives of Internal Medicine, 164,* 305–312.

Grumbach, K., & Bodenheimer, T. (2004). Can health care teams improve primary care practice? *Journal of the American Medical Association, 291,* 1246–1251.

Gurwitz, J. H., Field, T. S., Harrold, L. R., Rothschild, J., Debellis, K., Seger, A. C., et al. (2003). Incidence and preventability of adverse drug events among older persons in the ambulatory setting. *Journal of the American Medical Association, 289,* 1107–1116.

Hallock, M., Alper, S., & Karsh, B. (2003). Process improvement in an outpatient clinic: Application of sociotechnical systems analysis. In *Proceedings of the Human Factors and Ergonomics Society 47th annual meeting* (pp. 1406–1410).

Hendrick, H., & Kleiner, B. (2001). *Macroergonomics: An introduction to work system design.* Santa Monica, CA: Human Factors and Ergonomics Society.

Hopkins, A. (1999). The limits of normal accident theory. *Safety Science, 32,* 2–3, 93–102.

Huang, B., Bachmann, K. A., He, X., Chen, R., McAllister, J. S., & Wang, T. (2002). Inappropriate prescriptions for the aging population of the United States: An analysis of the National Ambulatory Medical Care Survey, 1997. *Pharmacoepidemiology & Drug Safety, 11,* 127–134.

Kaushal, R., Barker, K. N., & Bates, D. W. (2001). How can information technology improve patient safety and reduce medication errors in children's health care? *Archives of Pediatrics & Adolescent Medicine, 155,* 1002–1007.

Kohn, L. T., Corrigan, J. M., & Donaldson, M. S. (Eds.). (2000). *To err is human: Building a safer health system.* Washington DC: National Academy Press.

Kovner, A., & Jonas, S. (2002). *Health care delivery in the United States.* New York: Springer.

Leape, L. L., Brennan, T. A., Laird, N., Lawthers, A. G., Localio, A. R., Barnes, B. A., et al. (1991). The nature of adverse events in hospitalized patients—Results of the Harvard Medical-Practice Study: I. *New England Journal of Medicine, 324,* 377–384.

Macinko, J., Starfield, B., & Shi, L. (2003). The contribution of primary care systems to health outcomes within organization for Economic Cooperation and Development (OECD) countries 1970–1998. *Health Services Research, 38,* 831–865.

Mandl, K., Kohane, I., & Brandt, A. (1998). Electronic patient–physician communication: problems and promise. *Annals of Internal Medicine, 129,* 495–500.

McGlynn, E. A., Asch, S. M., Adams, J., Keesey, J., Hicks, J., DeCristofaro, A., et al. (2003). The quality of health care delivered to adults in the United States. *New England Journal of Medicine, 348,* 2635–2645.

McWhinney, I. (1996). The importance of being different. *British Journal of General Practice, 46,* 433–436.

Mechanic, D., McAlpine, D. D., & Rosenthal, M. (2001). Are patients' office visits with physicians getting shorter? *New England Journal of Medicine, 344,* 198–204.

Mold, J. W., & Stein, H. F. (1986). The cascade effect in the clinical care of patients. *New England Journal of Medicine, 314,* 512–514.

Moyer, C., Stern, D., Dobias, K., Cox, D., & Katz, S. (2002). Bridging the electronic divide: Patient and provider perspectives on e-mail communication in primary care. *American Journal of Managed Care, 8,* 427–433.

Murphy, L. R., DuBois, D., & Hurrell, J. J. (1986). Accident reduction through stress management. *Journal of Business and Psychology, 1,* 5–18.

Orzano, A. J., Gregory, P. M., Nutting, P. A., Werner, J. J., Flocke, S. A., & Stange, K. C. (2001). Care of the secondary patient in family practice. A report from the Ambulatory Sentinel Practice Network. *Journal of Family Practice, 50,* 113–116.

Pantell, R. H., Newman, T. B., Bernzweig, J., Bergman, D. A., Takayama, J. I., Segal, M., et al. (2004). Management and outcomes of care of fever in early infancy. *Journal of the American Medical Association, 291,* 1203–1212.

Parchman, M. L., & Burge, S. K. (2004). The patient–physician relationship, primary care attributes and preventive services. *Family Medicine, 36,* 22–27.

Parkerton, P. H., Smith, D. G., & Straley, H. L. (2004). Primary care practice coordination versus physician continuity. *Family Medicine, 36,* 15–21.

Pasmore, W. (1988). *Designing effective organizations: The sociotechnical systems perspective.* New York: Wiley.

Patt, M., Houston, T., Jenckes, M., Sands, D., & Ford, D. (2003). Doctors who are using e-mail with their patients: A qualitative exploration. *Journal of Medical Internet Research, 5,* e9.

Perrow, C. (1984). *Normal accidents: Living with high risk technologies.* New York: Basic Books.

Ray, W. A., Daugherty, J. R., & Meador, K. G. (2003). Effect of a mental health "carve-out" program on the continuity of antipsychotic therapy. *New England Journal of Medicine, 348,* 1885–1894.

Reason, J. (1997). *Managing the risks of organizational accidents.* Aldershot, England: Ashgate.

Rosser, W. (1996). Approach to diagnosis by primary care clinicians and specialists: Is there a difference? *The Journal of Family Practice, 42,* 139–144.

Sandars, J., & Esmail, A. (2003). The frequency and nature of medical error in primary care: Understanding the diversity across studies. *Family Practice, 20,* 231–236.

Sanders, M. S., & McCormick, E. J. (1993). *Human factors in engineering and design* (7th ed.). New York: McGraw-Hill.

Sheridan, S. L., Harris, R. P., & Woolf, S. H. (2004). Shared decision making about screening and chemoprevention. A suggested approach from the U.S. Preventive Services Task Force. *American Journal of Preventive Medicine, 26,* 56–66.

Sirovich, B. E., Schwartz, L. M., & Woloshin, S. (2003). Screening men for prostate and colorectal cancer in the United States: Does practice reflect the evidence? *Journal of the American Medical Association, 289,* 1414–1420.

Smedley, B. D., Stith, A. Y., & Nelson, A. R. (Eds.). (2003). *Unequal treatment: Confronting racial and ethnic disparities in health care.* Washington DC: National Academy Press.

Smith, P. C., Rodrig, A. G., Bublitz, C., Parnes, B., Dickinson, L. M., Van Vorst, R., et al. (2005). Missing clinical information during primary care visits. *Journal of the American Medical Association, 293,* 565–571.

Sox, H. (1996). Decision making: A comparison of referral practice and primary care. *Journal of Family Practice, 42,* 155–160.

Spath, P. (2000). *Error reduction in health care: A systems approach to improving patient safety.* San Francisco, CA: Jossey-Bass.

Starfield, B. (1993). Primary care. *Journal of Ambulatory Care Management, 16,* 27–37.

Starfield, B. (1994). Is primary care essential? *The Lancet, 344,* 1129–1133.

Starfield, B. (1998). *Primary care: Balancing health needs, services, and technology.* New York: Oxford University Press.

Starfield, B. (2001). New paradigms for quality in primary care. *British Journal of General Practice, 51,* 303–309.

Starfield, B. (2003). Primary and specialty care interfaces: The imperative of disease continuity. *British Journal of General Practice, 53,* 723–729.

Tarrant, C., Windridge, M., Baker, R., & Freeman, G. (2003). Qualitative study of the meaning of personal care in general practice. *British Medical Journal, 326,* 1310–1318.

Tierney, W. M. (2003). Adverse outpatient drug events—A problem and an opportunity. *New England Journal of Medicine, 348,* 1587–1589.

Venkatesh, V., Morris, M. G., Davis, G. B., & Davis, F. D. (2003). User acceptance of information technology: Toward a unified view. *MIS Quarterly, 27,* 425–478.

Welch, D. (1998). Human factors in the health care facility. *Biomedical Instrumentation & Technology, 32,* 311–316.

Wickens, C., Lee, J., Liu, Y., & Becker, S. (2004). *An introduction to human factors engineering* (2nd ed.). Upper Saddle River, NJ: Pearson Prentice Hall.

Yarnall, K. S., Pollak, K. I., Ostbye, T., Krause, K. M., & Michener, J. L. (2003). Primary care: Is there enough time for prevention? *American Journal of Public Health, 93,* 635–641.

Zyzanski, S. J., Stange, K. C., Langa, D., & Flocke, S. A. (1998). Trade-offs in high-volume primary care practice. *Journal of Family Practice, 46,* 397–402.

AUTHOR INDEX

B

E

F

G

H

I

L

M

N

Q

R

T

U

X

Y

Z

SUBJECT INDEX

A

C

F

G

I

J

K

L

M

N

O

P

Q

R

S

T

transactional leadership, 32

transformational leadership, 32–33

Y